Johns Hopkins Studies in the Mathematical Sciences
in association with the Department of Mathematical Sciences,
The Johns Hopkins University

Matrix Computations

Matrix Computations

Fourth Edition

Gene H. Golub
Department of Computer Science
Stanford University

Charles F. Van Loan
Department of Computer Science
Cornell University

The Johns Hopkins University Press
Baltimore

First edition 1983
Second edition 1989
Third edition 1996
Fourth edition 2013

The Johns Hopkins University Press
2715 North Charles Street
Baltimore, Maryland 21218-4363
www.press.jhu.edu

Library of Congress Control Number: 2012943449

ISBN 13: 978-1-4214-0794-4 (hc)
ISBN 10: 1-4214-0794-9 (hc)
ISBN 13: 978-1-4214-0859-0 (eb)
ISBN 10: 1-4214-0859-7 (eb)

A catalog record for this book is available from the British Library.

MATLAB® is a registered trademark of The Mathworks Inc.

Special discounts are available for bulk purchases of this book. For more information, please contact Special Sales at 410-516-6936 or specialsales@press.jhu.edu.

The Johns Hopkins University Press uses environmentally friendly book materials, including recycled text paper that is composed of at least 30 percent post-consumer waste, whenever possible.

To

ALSTON S. HOUSEHOLDER

AND

JAMES H. WILKINSON

Contents

Preface   xi
Global References   xiii
Other Books   xv
Useful URLs   xix
Common Notation   xxi

1   Matrix Multiplication   1

1.1   Basic Algorithms and Notation   2
1.2   Structure and Efficiency   14
1.3   Block Matrices and Algorithms   22
1.4   Fast Matrix-Vector Products   33
1.5   Vectorization and Locality   43
1.6   Parallel Matrix Multiplication   49

2   Matrix Analysis   63

2.1   Basic Ideas from Linear Algebra   64
2.2   Vector Norms   68
2.3   Matrix Norms   71
2.4   The Singular Value Decomposition   76
2.5   Subspace Metrics   81
2.6   The Sensitivity of Square Systems   87
2.7   Finite Precision Matrix Computations   93

3   General Linear Systems   105

3.1   Triangular Systems   106
3.2   The LU Factorization   111
3.3   Roundoff Error in Gaussian Elimination   122
3.4   Pivoting   125
3.5   Improving and Estimating Accuracy   137
3.6   Parallel LU   144

4   Special Linear Systems   153

4.1   Diagonal Dominance and Symmetry   154
4.2   Positive Definite Systems   159

4.3 Banded Systems 176
4.4 Symmetric Indefinite Systems 186
4.5 Block Tridiagonal Systems 196
4.6 Vandermonde Systems 203
4.7 Classical Methods for Toeplitz Systems 208
4.8 Circulant and Discrete Poisson Systems 219

5 Orthogonalization and Least Squares 233

5.1 Householder and Givens Transformations 234
5.2 The QR Factorization 246
5.3 The Full-Rank Least Squares Problem 260
5.4 Other Orthogonal Factorizations 274
5.5 The Rank-Deficient Least Squares Problem 288
5.6 Square and Underdetermined Systems 298

6 Modified Least Squares Problems and Methods 303

6.1 Weighting and Regularization 304
6.2 Constrained Least Squares 313
6.3 Total Least Squares 320
6.4 Subspace Computations with the SVD 327
6.5 Updating Matrix Factorizations 334

7 Unsymmetric Eigenvalue Problems 347

7.1 Properties and Decompositions 348
7.2 Perturbation Theory 357
7.3 Power Iterations 365
7.4 The Hessenberg and Real Schur Forms 376
7.5 The Practical QR Algorithm 385
7.6 Invariant Subspace Computations 394
7.7 The Generalized Eigenvalue Problem 405
7.8 Hamiltonian and Product Eigenvalue Problems 420
7.9 Pseudospectra 426

8 Symmetric Eigenvalue Problems 439

8.1 Properties and Decompositions 440
8.2 Power Iterations 450
8.3 The Symmetric QR Algorithm 458
8.4 More Methods for Tridiagonal Problems 467
8.5 Jacobi Methods 476
8.6 Computing the SVD 486
8.7 Generalized Eigenvalue Problems with Symmetry 497

9 Functions of Matrices 513

9.1 Eigenvalue Methods 514
9.2 Approximation Methods 522
9.3 The Matrix Exponential 530
9.4 The Sign, Square Root, and Log of a Matrix 536

10 Large Sparse Eigenvalue Problems 545

10.1 The Symmetric Lanczos Process 546
10.2 Lanczos, Quadrature, and Approximation 556
10.3 Practical Lanczos Procedures 562
10.4 Large Sparse SVD Frameworks 571
10.5 Krylov Methods for Unsymmetric Problems 579
10.6 Jacobi-Davidson and Related Methods 589

11 Large Sparse Linear System Problems 597

11.1 Direct Methods 598
11.2 The Classical Iterations 611
11.3 The Conjugate Gradient Method 625
11.4 Other Krylov Methods 639
11.5 Preconditioning 650
11.6 The Multigrid Framework 670

12 Special Topics 681

12.1 Linear Systems with Displacement Structure 681
12.2 Structured-Rank Problems 691
12.3 Kronecker Product Computations 707
12.4 Tensor Unfoldings and Contractions 719
12.5 Tensor Decompositions and Iterations 731

Index 747

Preface

My thirty-year book collaboration with Gene Golub began in 1977 at a matrix computation workshop held at Johns Hopkins University. His interest in my work at the start of my academic career prompted the writing of GVL1. Sadly, Gene died on November 16, 2007. At the time we had only just begun to talk about GVL4. While writing these pages, I was reminded every day of his far-reaching impact and professional generosity. This edition is a way to thank Gene for our collaboration and the friendly research community that his unique personality helped create.

It has been sixteen years since the publication of the third edition—a power-of-two reminder that what we need to know about matrix computations is growing exponentially! Naturally, it is impossible to provide in-depth coverage of all the great new advances and research trends. However, with the relatively recent publication of so many excellent textbooks and specialized volumes, we are able to complement our brief treatments with useful pointers to the literature. That said, here are the new features of GVL4:

Content

The book is about twenty-five percent longer. There are new sections on fast transforms (§1.4), parallel LU (§3.6), fast methods for circulant systems and discrete Poisson systems (§4.8), Hamiltonian and product eigenvalue problems (§7.8), pseudospectra (§7.9), the matrix sign, square root, and logarithm functions (§9.4), Lanczos and quadrature (§10.2), large-scale SVD (§10.4), Jacobi-Davidson (§10.6), sparse direct methods (§11.1), multigrid (§11.6), low displacement rank systems (§12.1), structured-rank systems (§12.2), Kronecker product problems (§12.3), tensor contractions (§12.4), and tensor decompositions (§12.5).

New topics at the subsection level include recursive block LU (§3.2.11), rook pivoting (§3.4.7), tournament pivoting (§3.6.3), diagonal dominance (§4.1.1), recursive block structures (§4.2.10), band matrix inverse properties (§4.3.8), divide-and-conquer strategies for block tridiagonal systems (§4.5.4), the cross product and various point/plane least squares problems (§5.3.9), the polynomial eigenvalue problem (§7.7.9), and the structured quadratic eigenvalue problem (§8.7.9).

Substantial upgrades include our treatment of floating-point arithmetic (§2.7), LU roundoff error analysis (§3.3.1), LS sensitivity analysis (§5.3.6), the generalized singular value decomposition (§6.1.6 and §8.7.4), and the CS decomposition (§8.7.6).

References

The annotated bibliographies at the end of each section remain. Because of space limitations, the master bibliography that was included in previous editions is now available through the book website. References that are historically important have been retained because old ideas have a way of resurrecting themselves. Plus, we must never forget the 1950's and 1960's! As mentioned above, we have the luxury of being able to draw upon an expanding library of books on matrix computations. A mnemonic-based citation system has been incorporated that supports these connections to the literature.

Examples

Non-illuminating, small-n numerical examples have been removed from the text. In their place is a modest suite of MATLAB demo scripts that can be run to provide insight into critical theorems and algorithms. We believe that this is a much more effective way to build intuition. The scripts are available through the book website.

Algorithmic Detail

It is important to have an algorithmic sense and an appreciation for high-performance matrix computations. After all, it is the clever exploitation of advanced architectures that account for much of the field's soaring success. However, the algorithms that we "formally" present in the book must never be considered as even prototype implementations. Clarity and communication of the big picture are what determine the level of detail in our presentations. Even though specific strategies for specific machines are beyond the scope of the text, we hope that our style promotes an ability to reason about memory traffic overheads and the importance of data locality.

Acknowledgements

I would like to thank everybody who has passed along typographical errors and suggestions over the years. Special kudos to the Cornell students in CS 4220, CS 6210, and CS 6220, where I used preliminary versions of GVL4. Harry Terkelson earned big bucks through through my ill-conceived $5-per-typo program!

A number of colleagues and students provided feedback and encouragement during the writing process. Others provided inspiration through their research and books. Thank you all: Diego Accame, David Bindel, Åke Björck, Laura Bolzano, Jim Demmel, Jack Dongarra, Mark Embree, John Gilbert, David Gleich, Joseph Grcar, Anne Greenbaum, Nick Higham, Ilse Ipsen, Bo Kågström, Vel Kahan, Tammy Kolda, Amy Langville, Julian Langou, Lek-Heng Lim, Nicola Mastronardi, Steve McCormick, Mike McCourt, Volker Mehrmann, Cleve Moler, Dianne O'Leary, Michael Overton, Chris Paige, Beresford Parlett, Stefan Ragnarsson, Lothar Reichel, Yousef Saad, Mike Saunders, Rob Schreiber, Danny Sorensen, Pete Stewart, Gil Strang, Francoise Tisseur, Nick Trefethen, Raf Vandebril, and Jianlin Xia.

Chris Paige and Mike Saunders were especially helpful with the editing of Chapters 10 and 11.

Vincent Burke, Jennifer Mallet, and Juliana McCarthy at Johns Hopkins University Press provided excellent support during the production process. Jennifer Slater did a terrific job of copy-editing. Of course, I alone am responsible for all mistakes and oversights.

Finally, this book would have been impossible to produce without my great family and my 4AM writing companion: Henry the Cat!

Charles F. Van Loan
Ithaca, New York
July, 2012

Global References

A number of books provide broad coverage of the field and are cited multiple times. We identify these global references using mnemonics. Bibliographic details are given in the Other Books section that follows.

AEP	**Wilkinson:**	*Algebraic Eigenvalue Problem*
ANLA	**Demmel:**	*Applied Numerical Linear Algebra*
ASNA	**Higham:**	*Accuracy and Stability of Numerical Algorithms*, second edition
EOM	**Chatelin:**	*Eigenvalues of Matrices*
FFT	**Van Loan:**	*Computational Frameworks for the Fast Fourier Transform*
FOM	**Higham:**	*Functions of Matrices*
FMC	**Watkins:**	*Fundamentals of Matrix Computations*
IMC	**Stewart:**	*Introduction to Matrix Computations*
IMK	**van der Vorst:**	*Iterative Krylov Methods for Large Linear Systems*
IMSL	**Greenbaum:**	*Iterative Methods for Solving Linear Systems*
ISM	**Axelsson:**	*Iterative Solution Methods*
IMSLE	**Saad:**	*Iterative Methods for Sparse Linear Systems*, second edition
LCG	**Meurant:**	*The Lanczos and Conjugate Gradient Algorithms* ...
MA	**Horn and Johnson:**	*Matrix Analysis*
MABD	**Stewart:**	*Matrix Algorithms: Basic Decompositions*
MAE	**Stewart:**	*Matrix Algorithms Volume II: Eigensystems*
MEP	**Watkins:**	*The Matrix Eigenvalue Problem: GR and Krylov Subspace Methods*
MPT	**Stewart and Sun:**	*Matrix Perturbation Theory*
NLA	**Trefethen and Bau:**	*Numerical Linear Algebra*
NMA	**Ipsen:**	*Numerical Matrix Analysis: Linear Systems and Least Squares*
NMLE	**Saad:**	*Numerical Methods for Large Eigenvalue Problems*, revised edition
NMLS	**Björck:**	*Numerical Methods for Least Squares Problems*
NMSE	**Kressner:**	*Numerical Methods for General and Structured Eigenvalue Problems*
SAP	**Trefethen and Embree:**	*Spectra and Pseudospectra*
SEP	**Parlett:**	*The Symmetric Eigenvalue Problem*
SLAS	**Forsythe and Moler:**	*Computer Solution of Linear Algebraic Systems*
SLS	**Lawson and Hanson:**	*Solving Least Squares Problems*
TMA	**Horn and Johnson:**	*Topics in Matrix Analysis*

LAPACK *LAPACK Users' Guide,* third edition
E. Anderson, Z. Bai, C. Bischof, S. Blackford, J. Demmel, J. Dongarra, J. Du Croz, A. Greenbaum, S. Hammarling, A. McKenney, and D. Sorensen.

scaLAPACK *ScaLAPACK Users' Guide*
L.S. Blackford, J. Choi, A. Cleary, E. D'Azevedo, J. Demmel, I. Dhillon, J. Dongarra, S. Hammarling, G. Henry, A. Petitet, K. Stanley, D. Walker, and R. C. Whaley.

LIN_TEMPLATES *Templates for the Solution of Linear Systems* ...
R. Barrett, M.W. Berry, T.F. Chan, J. Demmel, J. Donato, J. Dongarra, V. Eijkhout, R. Pozo, C. Romine, and H. van der Vorst.

EIG_TEMPLATES *Templates for the Solution of Algebraic Eigenvalue Problems* ...
Z. Bai, J. Demmel, J. Dongarra, A. Ruhe, and H. van der Vorst.

Other Books

The following volumes are a subset of a larger, ever-expanding library of textbooks and monographs that are concerned with matrix computations and supporting areas. The list of references below captures the evolution of the field and its breadth. Works that are more specialized are cited in the annotated bibliographies that appear at the end of each section in the chapters.

Early Landmarks

V.N. Faddeeva (1959). *Computational Methods of Linear Algebra,* Dover, New York.

E. Bodewig (1959). *Matrix Calculus,* North-Holland, Amsterdam.

J.H. Wilkinson (1963). *Rounding Errors in Algebraic Processes,* Prentice-Hall, Englewood Cliffs, NJ.

A.S. Householder (1964). *Theory of Matrices in Numerical Analysis,* Blaisdell, New York. Reprinted in 1974 by Dover, New York.

L. Fox (1964). *An Introduction to Numerical Linear Algebra,* Oxford University Press, Oxford.

J.H. Wilkinson (1965). *The Algebraic Eigenvalue Problem,* Clarendon Press, Oxford.

General Textbooks on Matrix Computations

G.W. Stewart (1973). *Introduction to Matrix Computations,* Academic Press, New York.

R.J. Goult, R.F. Hoskins, J.A. Milner, and M.J. Pratt (1974). *Computational Methods in Linear Algebra,* John Wiley and Sons, New York.

W.W. Hager (1988). *Applied Numerical Linear Algebra,* Prentice-Hall, Englewood Cliffs, NJ.

P.G. Ciarlet (1989). *Introduction to Numerical Linear Algebra and Optimisation,* Cambridge University Press, Cambridge.

P.E. Gill, W. Murray, and M.H. Wright (1991). *Numerical Linear Algebra and Optimization, Vol. 1,* Addison-Wesley, Reading, MA.

A. Jennings and J.J. McKeowen (1992). *Matrix Computation,* second edition, John Wiley and Sons, New York.

L.N. Trefethen and D. Bau III (1997). *Numerical Linear Algebra,* SIAM Publications, Philadelphia, PA.

J.W. Demmel (1997). *Applied Numerical Linear Algebra,* SIAM Publications, Philadelphia, PA.

A.J. Laub (2005). *Matrix Analysis for Scientists and Engineers,* SIAM Publications, Philadelphia, PA.

B.N. Datta (2010). *Numerical Linear Algebra and Applications,* second edition, SIAM Publications, Philadelphia, PA.

D.S. Watkins (2010). *Fundamentals of Matrix Computations,* John Wiley and Sons, New York.

A.J. Laub (2012). *Computational Matrix Analysis,* SIAM Publications, Philadelphia, PA.

Linear Equation and Least Squares Problems

G.E. Forsythe and C.B. Moler (1967). *Computer Solution of Linear Algebraic Systems,* Prentice-Hall, Englewood Cliffs, NJ.

A. George and J.W-H. Liu (1981). *Computer Solution of Large Sparse Positive Definite Systems.* Prentice-Hall, Englewood Cliffs, NJ.

I.S. Duff, A.M. Erisman, and J.K. Reid (1986). *Direct Methods for Sparse Matrices,* Oxford University Press, New York.

R.W. Farebrother (1987). *Linear Least Squares Computations,* Marcel Dekker, New York.

C.L. Lawson and R.J. Hanson (1995). *Solving Least Squares Problems,* SIAM Publications, Philadelphia, PA.

Å. Björck (1996). *Numerical Methods for Least Squares Problems,* SIAM Publications, Philadelphia, PA.

G.W. Stewart (1998). *Matrix Algorithms: Basic Decompositions,* SIAM Publications, Philadelphia, PA.

N.J. Higham (2002). *Accuracy and Stability of Numerical Algorithms,* second edition, SIAM Publications, Philadelphia, PA.

T.A. Davis (2006). *Direct Methods for Sparse Linear Systems,* SIAM Publications, Philadelphia, PA.

I.C.F. Ipsen (2009). *Numerical Matrix Analysis: Linear Systems and Least Squares,* SIAM Publications, Philadelphia, PA.

Eigenvalue Problems

A.R. Gourlay and G.A. Watson (1973). *Computational Methods for Matrix Eigenproblems,* John Wiley & Sons, New York.

F. Chatelin (1993). *Eigenvalues of Matrices,* John Wiley & Sons, New York.

B.N. Parlett (1998). *The Symmetric Eigenvalue Problem,* SIAM Publications, Philadelphia, PA.

G.W. Stewart (2001). *Matrix Algorithms Volume II: Eigensystems,* SIAM Publications, Philadelphia, PA.

L. Komzsik (2003). *The Lanczos Method: Evolution and Application,* SIAM Publications, Philadelphia, PA.

D. Kressner (2005). *Numerical Methods for General and Structured Eigenvalue Problems,* Springer, Berlin.

D.S. Watkins (2007). *The Matrix Eigenvalue Problem: GR and Krylov Subspace Methods,* SIAM Publications, Philadelphia, PA.

Y. Saad (2011). *Numerical Methods for Large Eigenvalue Problems,* revised edition, SIAM Publications, Philadelphia, PA.

Iterative Methods

R.S. Varga (1962). *Matrix Iterative Analysis,* Prentice-Hall, Englewood Cliffs, NJ.

D.M. Young (1971). *Iterative Solution of Large Linear Systems,* Academic Press, New York.

L.A. Hageman and D.M. Young (1981). *Applied Iterative Methods,* Academic Press, New York.

J. Cullum and R.A. Willoughby (1985). *Lanczos Algorithms for Large Symmetric Eigenvalue Computations, Vol. I Theory,* Birkhaüser, Boston.

J. Cullum and R.A. Willoughby (1985). *Lanczos Algorithms for Large Symmetric Eigenvalue Computations, Vol. II Programs,* Birkhaüser, Boston.

W. Hackbusch (1994). *Iterative Solution of Large Sparse Systems of Equations,* Springer-Verlag, New York.

O. Axelsson (1994). *Iterative Solution Methods,* Cambridge University Press.

A. Greenbaum (1997). *Iterative Methods for Solving Linear Systems,* SIAM Publications, Philadelphia, PA.

Y. Saad (2003). *Iterative Methods for Sparse Linear Systems,* second edition, SIAM Publications, Philadelphia, PA.

H. van der Vorst (2003). *Iterative Krylov Methods for Large Linear Systems,* Cambridge University Press, Cambridge, UK.

G. Meurant (2006). *The Lanczos and Conjugate Gradient Algorithms: From Theory to Finite Precision Computations*, SIAM Publications, Philadelphia, PA.

Special Topics/Threads

L.N. Trefethen and M. Embree (2005). *Spectra and Pseudospectra—The Behavior of Nonnormal Matrices and Operators*, Princeton University Press, Princeton and Oxford.

R. Vandebril, M. Van Barel, and N. Mastronardi (2007). *Matrix Computations and Semiseparable Matrices I: Linear Systems*, Johns Hopkins University Press, Baltimore, MD.

R. Vandebril, M. Van Barel, and N. Mastronardi (2008). *Matrix Computations and Semiseparable Matrices II: Eigenvalue and Singular Value Methods*, Johns Hopkins University Press, Baltimore, MD.

N.J. Higham (2008) *Functions of Matrices*, SIAM Publications, Philadelphia, PA.

Collected Works

R.H. Chan, C. Greif, and D.P. O'Leary, eds. (2007). *Milestones in Matrix Computation: Selected Works of G.H. Golub, with Commentaries*, Oxford University Press, Oxford.

M.E. Kilmer and D.P. O'Leary, eds. (2010). *Selected Works of G.W. Stewart*, Birkhauser, Boston, MA.

Implementation

B.T. Smith, J.M. Boyle, Y. Ikebe, V.C. Klema, and C.B. Moler (1970). *Matrix Eigensystem Routines: EISPACK Guide,* second edition, Lecture Notes in Computer Science, Vol. 6, Springer-Verlag, New York.

J.H. Wilkinson and C. Reinsch, eds. (1971). *Handbook for Automatic Computation, Vol. 2, Linear Algebra*, Springer-Verlag, New York.

B.S. Garbow, J.M. Boyle, J.J. Dongarra, and C.B. Moler (1972). *Matrix Eigensystem Routines: EISPACK Guide Extension,* Lecture Notes in Computer Science, Vol. 51, Springer-Verlag, New York.

J.J Dongarra, J.R. Bunch, C.B. Moler, and G.W. Stewart (1979). *LINPACK Users' Guide*, SIAM Publications, Philadelphia, PA.

K. Gallivan, M. Heath, E. Ng, B. Peyton, R. Plemmons, J. Ortega, C. Romine, A. Sameh, and R. Voigt (1990). *Parallel Algorithms for Matrix Computations*, SIAM Publications, Philadelphia, PA.

R. Barrett, M.W. Berry, T.F. Chan, J. Demmel, J. Donato, J. Dongarra, V. Eijkhout, R. Pozo, C. Romine, and H. van der Vorst (1993). *Templates for the Solution of Linear Systems: Building Blocks for Iterative Methods*, SIAM Publications, Philadelphia, PA.

L.S. Blackford, J. Choi, A. Cleary, E. D'Azevedo, J. Demmel, I. Dhillon, J. Dongarra, S. Hammarling, G. Henry, A. Petitet, K. Stanley, D. Walker, and R.C. Whaley (1997). *ScaLAPACK Users' Guide*, SIAM Publications, Philadelphia, PA.

J.J. Dongarra, I.S. Duff, D.C. Sorensen, and H.A. van der Vorst (1998). *Numerical Linear Algebra on High-Performance Computers*, SIAM Publications, Philadelphia, PA.

E. Anderson, Z. Bai, C. Bischof, S. Blackford, J. Demmel, J. Dongarra, J. Du Croz, A. Greenbaum, S. Hammarling, A. McKenney, and D. Sorensen (1999). *LAPACK Users' Guide,* third edition, SIAM Publications, Philadelphia, PA.

Z. Bai, J. Demmel, J. Dongarra, A. Ruhe, and H. van der Vorst (2000). *Templates for the Solution of Algebraic Eigenvalue Problems: A Practical Guide*, SIAM Publications, Philadelphia, PA.

V.A. Barker, L.S. Blackford, J. Dongarra, J. Du Croz, S. Hammarling, M. Marinova, J. Wasniewski, and P. Yalamov (2001). *LAPACK95 Users' Guide*, SIAM Publications, Philadelphia.

MATLAB

D.J. Higham and N.J. Higham (2005). *MATLAB Guide,* second edition, SIAM Publications, Philadelphia, PA.

R. Pratap (2006). *Getting Started with Matlab 7*, Oxford University Press, New York.

C.F. Van Loan and D. Fan (2009). *Insight Through Computing: A Matlab Introduction to Computational Science and Engineering*, SIAM Publications, Philadelphia, PA.

Matrix Algebra and Analysis

R. Horn and C. Johnson (1985). *Matrix Analysis*, Cambridge University Press, New York.

G.W. Stewart and J. Sun (1990). *Matrix Perturbation Theory*, Academic Press, San Diego.

R. Horn and C. Johnson (1991). *Topics in Matrix Analysis*, Cambridge University Press, New York.

D.S. Bernstein (2005). *Matrix Mathematics, Theory, Facts, and Formulas with Application to Linear Systems Theory*, Princeton University Press, Princeton, NJ.

L. Hogben (2006). *Handbook of Linear Algebra*, Chapman and Hall, Boca Raton, FL.

Scientific Computing/Numerical Analysis

G.W. Stewart (1996). *Afternotes on Numerical Analysis*, SIAM Publications, Philadelphia, PA.

C.F. Van Loan (1997). *Introduction to Scientific Computing: A Matrix-Vector Approach Using Matlab*, Prentice Hall, Upper Saddle River, NJ.

G.W. Stewart (1998). *Afternotes on Numerical Analysis: Afternotes Goes to Graduate School,* SIAM Publications, Philadelphia, PA.

M.T. Heath (2002). *Scientific Computing: An Introductory Survey,* second edition), McGraw-Hill, New York.

C.B. Moler (2008) *Numerical Computing with MATLAB,* revised reprint, SIAM Publications, Philadelphia, PA.

G. Dahlquist and Å. Björck (2008). *Numerical Methods in Scientific Computing,* Vol. 1, SIAM Publications, Philadelphia, PA.

U. Ascher and C. Greif (2011). *A First Course in Numerical Methods*, SIAM Publications, Philadelphia, PA.

Useful URLs

GVL4

MATLAB demo scripts and functions, master bibliography, list of errata.

http://www.cornell.edu/cv/GVL4

Netlib

Huge repository of numerical software including LAPACK.

http://www.netlib.org/index.html

Matrix Market

Test examples for matrix algorithms.

http://math.nist.gov/MatrixMarket/

Matlab Central

Matlab functions, demos, classes, toolboxes, videos.

http://www.mathworks.com/matlabcentral/

University of Florida Sparse Matrix Collections

Thousands of sparse matrix examples in several formats.

http://www.cise.ufl.edu/research/sparse/matrices/

Pseudospectra Gateway

Grapical tools for pseudospectra.

http://www.cs.ox.ac.uk/projects/pseudospectra/

ARPACK

Software for large sparse eigenvalue problems

http://www.caam.rice.edu/software/ARPACK/

Innovative Computing Laboratory

State-of-the-art high performance matrix computations.

http://icl.cs.utk.edu/

Common Notation

$\mathbb{R}, \mathbb{R}^n, \mathbb{R}^{m \times n}$	set of real numbers, vectors, and matrices (p. 2)
$\mathbb{C}, \mathbb{C}^n, \mathbb{C}^{m \times n}$	set of complex numbers, vectors, and matrices (p. 13)
$a_{ij}, A(i,j), [A]_{ij}$	(i,j) entry of a matrix (p. 2)
$\mathbf{u}$	unit roundoff (p. 96)
$\mathsf{fl}(\cdot)$	floating point operator (p. 96)
$\| x \|_p$	p-norm of a vector (p. 68)
$\| A \|_p, \| A \|_F$	p-norm and Frobenius norm of a matrix (p. 71)
$\mathsf{length}(x)$	dimension of a vector (p. 236)
$\kappa_p(A)$	p-norm condition (p. 87)
$|A|$	absolute value of a matrix (p. 91)
A^T, A^H	transpose and conjugate transpose (p. 2, 13)
$\mathsf{house}(x)$	Householder vector (p. 236)
$\mathsf{givens}(a,b)$	cosine-sine pair (p. 240)
x_{LS}	minimum-norm least squares solution (p. 260)
$\mathsf{ran}(A)$	range of a matrix (p. 64)
$\mathsf{null}(A)$	nullspace of a matrix (p. 64)
$\mathsf{span}\{v_1, \ldots, v_n\}$	span defined by vectors (p. 64)
$\mathsf{dim}(S)$	dimension of a subspace (p. 64)
$\mathsf{rank}(A)$	rank of a matrix (p. 65)
$\mathsf{det}(A)$	determinant of a matrix (p. 66)
$\mathsf{tr}(A)$	trace of a matrix (p. 327)
$\mathsf{vec}(A)$	vectorization of a matrix (p. 28)
$\mathsf{reshape}(A, p, q)$	reshaping a matrix (p. 28)
$\mathsf{Re}(A), \mathsf{Im}(A)$	real and imaginary parts of a matrix (p. 13)
$\mathsf{diag}(d_1, \ldots, d_n)$	diagonal matrix (p. 18)
I_n	n-by-n identity matrix (p. 19)
e_i	ith column of the identity matrix (p. 19)
$\mathcal{E}_n, \mathcal{D}_n, \mathcal{P}_{p,q}$	exchange, downshift, and perfect shuffle permutations (p. 20)
$\sigma_i(A)$	ith largest singular value (p. 77)
$\sigma_{\max}(A), \sigma_{\min}(A)$	largest and smallest singular value (p. 77)
$\mathsf{dist}(S_1, S_2)$	distance between two subspaces (p. 82)
$\mathsf{sep}(A_1, A_2)$	separation between two matrices (p. 360)
$\lambda(A)$	set of eigenvalues (p. 66)
$\lambda_i(A)$	ith largest eigenvalue of a symmetric matrix (p. 66)
$\lambda_{\max}(A), \lambda_{\min}(A)$	largest and smallest eigenvalue of a symmetric matrix (p. 66)
$\rho(A)$	spectral radius (p. 349)
$\mathcal{K}(A, q, j)$	Krylov subspace (p. 548)

Matrix Computations

Chapter 1

Matrix Multiplication

1.1   Basic Algorithms and Notation

1.2   Structure and Efficiency

1.3   Block Matrices and Algorithms

1.4   Fast Matrix-Vector Products

1.5   Vectorization and Locality

1.6   Parallel Matrix Multiplication

The study of matrix computations properly begins with the study of various matrix multiplication problems. Although simple mathematically, these calculations are sufficiently rich to develop a wide range of essential algorithmic skills.

In §1.1 we examine several formulations of the matrix multiplication update problem $C = C + AB$. Partitioned matrices are introduced and used to identify linear algebraic "levels" of computation.

If a matrix has special properties, then various economies are generally possible. For example, a symmetric matrix can be stored in half the space of a general matrix. A matrix-vector product may require much less time to execute if the matrix has many zero entries. These matters are considered in §1.2.

A block matrix is a matrix whose entries are themselves matrices. The "language" of block matrices is developed in §1.3. It supports the easy derivation of matrix factorizations by enabling us to spot patterns in a computation that are obscured at the scalar level. Algorithms phrased at the block level are typically rich in matrix-matrix multiplication, the operation of choice in many high-performance computing environments. Sometimes the block structure of a matrix is recursive, meaning that the block entries have an exploitable resemblance to the overall matrix. This type of connection is the foundation for "fast" matrix-vector product algorithms such as various fast Fourier transforms, trigonometric transforms, and wavelet transforms. These calculations are among the most important in all of scientific computing and are discussed in §1.4. They provide an excellent opportunity to develop a facility with block matrices and recursion.

1

The last two sections set the stage for effective, "large-n" matrix computations. In this context, data locality affects efficiency more than the volume of actual arithmetic. Having an ability to reason about memory hierarchies and multiprocessor computation is essential. Our goal in §1.5 and §1.6 is to build an appreciation for the attendant issues without getting into system-dependent details.

Reading Notes

The sections within this chapter depend upon each other as follows:

$$\S1.1 \quad \rightarrow \quad \S1.2 \quad \rightarrow \quad \S1.3 \quad \rightarrow \quad \S1.4$$
$$\downarrow$$
$$\S1.5 \quad \rightarrow \quad \S1.6$$

Before proceeding to later chapters, §1.1, §1.2, and §1.3 are essential. The fast transform ideas in §1.4 are utilized in §4.8 and parts of Chapters 11 and 12. The reading of §1.5 and §1.6 can be deferred until high-performance linear equation solving or eigenvalue computation becomes a topic of concern.

1.1   Basic Algorithms and Notation

Matrix computations are built upon a hierarchy of linear algebraic operations. Dot products involve the scalar operations of addition and multiplication. Matrix-vector multiplication is made up of dot products. Matrix-matrix multiplication amounts to a collection of matrix-vector products. All of these operations can be described in algorithmic form or in the language of linear algebra. One of our goals is to show how these two styles of expression complement each other. Along the way we pick up notation and acquaint the reader with the kind of thinking that underpins the matrix computation area. The discussion revolves around the matrix multiplication problem, a computation that can be organized in several ways.

1.1.1   Matrix Notation

Let $\mathbb{R}$ designate the set of real numbers. We denote the vector space of all m-by-n real matrices by $\mathbb{R}^{m \times n}$:

$$A \in \mathbb{R}^{m \times n} \quad \Longleftrightarrow \quad A = (a_{ij}) = \begin{bmatrix} a_{11} & \cdots & a_{1n} \\ \vdots & & \vdots \\ a_{m1} & \cdots & a_{mn} \end{bmatrix}, \quad a_{ij} \in \mathbb{R}$$

If a capital letter is used to denote a matrix (e.g., A, B, Δ), then the corresponding lower case letter with subscript ij refers to the (i, j) entry (e.g., a_{ij}, b_{ij}, δ_{ij}). Sometimes we designate the elements of a matrix with the notation $[\, A\,]_{ij}$ or $A(i, j)$.

1.1.2   Matrix Operations

Basic matrix operations include *transposition* ($\mathbb{R}^{m \times n} \rightarrow \mathbb{R}^{n \times m}$),

$$C = A^T \qquad \Longrightarrow \qquad c_{ij} = a_{ji},$$

addition ($\mathbb{R}^{m \times n} \times \mathbb{R}^{m \times n} \to \mathbb{R}^{m \times n}$),

$$C = A + B \quad \Longrightarrow \quad c_{ij} = a_{ij} + b_{ij},$$

scalar-matrix multiplication ($\mathbb{R} \times \mathbb{R}^{m \times n} \to \mathbb{R}^{m \times n}$),

$$C = \alpha A \quad \Longrightarrow \quad c_{ij} = \alpha a_{ij},$$

and *matrix-matrix multiplication* ($\mathbb{R}^{m \times p} \times \mathbb{R}^{p \times n} \to \mathbb{R}^{m \times n}$),

$$C = AB \quad \Longrightarrow \quad c_{ij} = \sum_{k=1}^{p} a_{ik} b_{kj}.$$

Pointwise matrix operations are occasionally useful, especially pointwise multiplication ($\mathbb{R}^{m \times n} \times \mathbb{R}^{m \times n} \to \mathbb{R}^{m \times n}$),

$$C = A .* B \quad \Longrightarrow \quad c_{ij} = a_{ij} b_{ij}$$

and pointwise division ($\mathbb{R}^{m \times n} \times \mathbb{R}^{m \times n} \to \mathbb{R}^{m \times n}$),

$$C = A ./ B \quad \Longrightarrow \quad c_{ij} = a_{ij} / b_{ij}.$$

Of course, for pointwise division to make sense, the "denominator matrix" must have nonzero entries.

1.1.3 Vector Notation

Let $\mathbb{R}^n$ denote the vector space of real n-vectors:

$$x \in \mathbb{R}^n \quad \Longleftrightarrow \quad x = \begin{bmatrix} x_1 \\ \vdots \\ x_n \end{bmatrix} \quad x_i \in \mathbb{R}.$$

We refer to x_i as the ith component of x. Depending upon context, the alternative notations $[x]_i$ and $x(i)$ are sometimes used.

Notice that we are identifying $\mathbb{R}^n$ with $\mathbb{R}^{n \times 1}$ and so the members of $\mathbb{R}^n$ are *column* vectors. On the other hand, the elements of $\mathbb{R}^{1 \times n}$ are *row* vectors:

$$x \in \mathbb{R}^{1 \times n} \quad \Longleftrightarrow \quad x = [x_1, \ldots, x_n].$$

If x is a column vector, then $y = x^T$ is a row vector.

1.1.4 Vector Operations

Assume that $a \in \mathbb{R}$, $x \in \mathbb{R}^n$, and $y \in \mathbb{R}^n$. Basic vector operations include *scalar-vector multiplication*,

$$z = ax \quad \Longrightarrow \quad z_i = a x_i,$$

vector addition,

$$z = x + y \quad \Longrightarrow \quad z_i = x_i + y_i,$$

and the *inner product* (or *dot product*),

$$c = x^T y \qquad \Longrightarrow \qquad c = \sum_{i=1}^{n} x_i y_i.$$

A particularly important operation, which we write in update form, is the *saxpy*:

$$y = ax + y \qquad \Longrightarrow \qquad y_i = ax_i + y_i$$

Here, the symbol "=" is used to denote assignment, not mathematical equality. The vector y is being updated. The name "saxpy" is used in LAPACK, a software package that implements many of the algorithms in this book. "Saxpy" is a mnemonic for "scalar a x plus y." See LAPACK.

Pointwise vector operations are also useful, including *vector multiplication*,

$$z = x.*y \qquad \Longrightarrow \qquad z_i = x_i y_i,$$

and *vector division*,

$$z = x./y \qquad \Longrightarrow \qquad z_i = x_i / y_i.$$

1.1.5 The Computation of Dot Products and Saxpys

Algorithms in the text are expressed using a stylized version of the MATLAB language. Here is our first example:

Algorithm 1.1.1 (Dot Product) If $x, y \in \mathbb{R}^n$, then this algorithm computes their dot product $c = x^T y$.

$$
\begin{aligned}
&c = 0 \\
&\textbf{for } i = 1{:}n \\
&\qquad c = c + x(i)y(i) \\
&\textbf{end}
\end{aligned}
$$

It is clear from the summation that the dot product of two n-vectors involves n multiplications and n additions. The dot product operation is an "$O(n)$" operation, meaning that the amount of work scales linearly with the dimension. The saxpy computation is also $O(n)$:

Algorithm 1.1.2 (Saxpy) If $x, y \in \mathbb{R}^n$ and $a \in \mathbb{R}$, then this algorithm overwrites y with $y + ax$.

$$
\begin{aligned}
&\textbf{for } i = 1{:}n \\
&\qquad y(i) = y(i) + ax(i) \\
&\textbf{end}
\end{aligned}
$$

We stress that the algorithms in this book are encapsulations of important computational ideas and are not to be regarded as "production codes."

1.1.6 Matrix-Vector Multiplication and the Gaxpy

Suppose $A \in \mathbb{R}^{m \times n}$ and that we wish to compute the update

$$y = y + Ax$$

where $x \in \mathbb{R}^n$ and $y \in \mathbb{R}^m$ are given. This generalized saxpy operation is referred to as a *gaxpy*. A standard way that this computation proceeds is to update the components one-at-a-time:

$$y_i = y_i + \sum_{j=1}^{n} a_{ij} x_j, \qquad i = 1{:}m.$$

This gives the following algorithm:

Algorithm 1.1.3 (Row-Oriented Gaxpy) If $A \in \mathbb{R}^{m \times n}$, $x \in \mathbb{R}^n$, and $y \in \mathbb{R}^m$, then this algorithm overwrites y with $Ax + y$.

> **for** $i = 1{:}m$
>   **for** $j = 1{:}n$
> $y(i) = y(i) + A(i,j)x(j)$
>   **end**
> **end**

Note that this involves $O(mn)$ work. If each dimension of A is doubled, then the amount of arithmetic increases by a factor of 4.

An alternative algorithm results if we regard Ax as a linear combination of A's columns, e.g.,

$$\begin{bmatrix} 1 & 2 \\ 3 & 4 \\ 5 & 6 \end{bmatrix} \begin{bmatrix} 7 \\ 8 \end{bmatrix} = \begin{bmatrix} 1 \cdot 7 + 2 \cdot 8 \\ 3 \cdot 7 + 4 \cdot 8 \\ 5 \cdot 7 + 6 \cdot 8 \end{bmatrix} = 7 \begin{bmatrix} 1 \\ 3 \\ 5 \end{bmatrix} + 8 \begin{bmatrix} 2 \\ 4 \\ 6 \end{bmatrix} = \begin{bmatrix} 23 \\ 53 \\ 83 \end{bmatrix}.$$

Algorithm 1.1.4 (Column-Oriented Gaxpy) If $A \in \mathbb{R}^{m \times n}$, $x \in \mathbb{R}^n$, and $y \in \mathbb{R}^m$, then this algorithm overwrites y with $Ax + y$.

> **for** $j = 1{:}n$
>   **for** $i = 1{:}m$
> $y(i) = y(i) + A(i,j) \cdot x(j)$
>   **end**
> **end**

Note that the inner loop in either gaxpy algorithm carries out a saxpy operation. The column version is derived by rethinking what matrix-vector multiplication "means" at the vector level, but it could also have been obtained simply by interchanging the order of the loops in the row version.

1.1.7 Partitioning a Matrix into Rows and Columns

Algorithms 1.1.3 and 1.1.4 access the data in A by row and by column, respectively. To highlight these orientations more clearly, we introduce the idea of a *partitioned matrix*.

From one point of view, a matrix is a stack of row vectors:

$$A \in \mathbb{R}^{m \times n} \qquad \Longleftrightarrow \qquad A = \begin{bmatrix} r_1^T \\ \vdots \\ r_m^T \end{bmatrix}, \quad r_k \in \mathbb{R}^n. \qquad (1.1.1)$$

This is called a *row partition* of A. Thus, if we row partition

$$\begin{bmatrix} 1 & 2 \\ 3 & 4 \\ 5 & 6 \end{bmatrix},$$

then we are choosing to think of A as a collection of rows with

$$r_1^T = [\, 1 \ \ 2 \,], \qquad r_2^T = [\, 3 \ \ 4 \,], \qquad r_3^T = [\, 5 \ \ 6 \,].$$

With the row partitioning (1.1.1), Algorithm 1.1.3 can be expressed as follows:

> **for** $i = 1{:}m$
> $\qquad y_i = y_i + r_i^T x$
> **end**

Alternatively, a matrix is a collection of column vectors:

$$A \in \mathbb{R}^{m \times n} \qquad \Longleftrightarrow \qquad A = [\, c_1 \,|\, \cdots \,|\, c_n \,], \quad c_k \in \mathbb{R}^m. \qquad (1.1.2)$$

We refer to this as a *column partition* of A. In the 3-by-2 example above, we thus would set c_1 and c_2 to be the first and second columns of A, respectively:

$$c_1 = \begin{bmatrix} 1 \\ 3 \\ 5 \end{bmatrix}, \qquad c_2 = \begin{bmatrix} 2 \\ 4 \\ 6 \end{bmatrix}.$$

With (1.1.2) we see that Algorithm 1.1.4 is a saxpy procedure that accesses A by columns:

> **for** $j = 1{:}n$
> $\qquad y = y + x_j c_j$
> **end**

In this formulation, we appreciate y as a running vector sum that undergoes repeated saxpy updates.

1.1.8 The Colon Notation

A handy way to specify a column or row of a matrix is with the "colon" notation. If $A \in \mathbb{R}^{m \times n}$, then $A(k, :)$ designates the kth row, i.e.,

$$A(k, :) = [a_{k1}, \ldots, a_{kn}].$$

The kth column is specified by

$$A(:,k) = \begin{bmatrix} a_{1k} \\ \vdots \\ a_{mk} \end{bmatrix}.$$

With these conventions we can rewrite Algorithms 1.1.3 and 1.1.4 as

> **for** $i = 1{:}m$
> $\qquad y(i) = y(i) + A(i,:) \cdot x$
> **end**

and

> **for** $j = 1{:}n$
> $\qquad y = y + x(j) \cdot A(:,j)$
> **end**

respectively. By using the colon notation, we are able to suppress inner loop details and encourage vector-level thinking.

1.1.9 The Outer Product Update

As a preliminary application of the colon notation, we use it to understand the *outer product update*

$$A = A + xy^T, \qquad A \in \mathbb{R}^{m \times n},\ x \in \mathbb{R}^m,\ y \in \mathbb{R}^n.$$

The outer product operation xy^T "looks funny" but is perfectly legal, e.g.,

$$\begin{bmatrix} 1 \\ 2 \\ 3 \end{bmatrix} \begin{bmatrix} 4 & 5 \end{bmatrix} = \begin{bmatrix} 4 & 5 \\ 8 & 10 \\ 12 & 15 \end{bmatrix}.$$

This is because xy^T is the product of two "skinny" matrices and the number of columns in the left matrix x equals the number of rows in the right matrix y^T. The entries in the outer product update are prescribed by

> **for** $i = 1{:}m$
> $\qquad$ **for** $j = 1{:}n$
> $\qquad\qquad a_{ij} = a_{ij} + x_i y_j$
> $\qquad$ **end**
> **end**

This involves $O(mn)$ arithmetic operations. The mission of the j loop is to add a multiple of y^T to the ith row of A, i.e.,

> **for** $i = 1{:}m$
> $\qquad A(i,:) = A(i,:) + x(i) \cdot y^T$
> **end**

On the other hand, if we make the i-loop the inner loop, then its task is to add a multiple of x to the jth column of A:

> **for** $j = 1{:}n$
> $A(:,j) = A(:,j) + y(j){\cdot}x$
> **end**

Note that both implementations amount to a set of saxpy computations.

1.1.10   Matrix-Matrix Multiplication

Consider the 2-by-2 matrix-matrix multiplication problem. In the dot product formulation, each entry is computed as a dot product:

$$\begin{bmatrix} 1 & 2 \\ 3 & 4 \end{bmatrix}\begin{bmatrix} 5 & 6 \\ 7 & 8 \end{bmatrix} = \begin{bmatrix} 1\cdot 5 + 2\cdot 7 & 1\cdot 6 + 2\cdot 8 \\ 3\cdot 5 + 4\cdot 7 & 3\cdot 6 + 4\cdot 8 \end{bmatrix}.$$

In the saxpy version, each column in the product is regarded as a linear combination of left-matrix columns:

$$\begin{bmatrix} 1 & 2 \\ 3 & 4 \end{bmatrix}\begin{bmatrix} 5 & 6 \\ 7 & 8 \end{bmatrix} = \left[5\begin{bmatrix} 1 \\ 3 \end{bmatrix} + 7\begin{bmatrix} 2 \\ 4 \end{bmatrix}, \ 6\begin{bmatrix} 1 \\ 3 \end{bmatrix} + 8\begin{bmatrix} 2 \\ 4 \end{bmatrix} \right].$$

Finally, in the outer product version, the result is regarded as the sum of outer products:

$$\begin{bmatrix} 1 & 2 \\ 3 & 4 \end{bmatrix}\begin{bmatrix} 5 & 6 \\ 7 & 8 \end{bmatrix} = \begin{bmatrix} 1 \\ 3 \end{bmatrix}\begin{bmatrix} 5 & 6 \end{bmatrix} + \begin{bmatrix} 2 \\ 4 \end{bmatrix}\begin{bmatrix} 7 & 8 \end{bmatrix}.$$

Although equivalent mathematically, it turns out that these versions of matrix multiplication can have very different levels of performance because of their memory traffic properties. This matter is pursued in §1.5. For now, it is worth detailing the various approaches to matrix multiplication because it gives us a chance to review notation and to practice thinking at different linear algebraic levels. To fix the discussion, we focus on the matrix-matrix update computation:

$$C = C + AB, \qquad C \in \mathbb{R}^{m\times n},\ A \in \mathbb{R}^{m\times r},\ B \in \mathbb{R}^{r\times n}.$$

The update $C = C + AB$ is considered instead of just $C = AB$ because it is the more typical situation in practice.

1.1.11   Scalar-Level Specifications

The starting point is the familiar triply nested loop algorithm:

Algorithm 1.1.5 (ijk **Matrix Multiplication**) If $A \in \mathbb{R}^{m\times r}$, $B \in \mathbb{R}^{r\times n}$, and $C \in \mathbb{R}^{m\times n}$ are given, then this algorithm overwrites C with $C + AB$.

> **for** $i = 1{:}m$
>   **for** $j = 1{:}n$
>   **for** $k = 1{:}r$
> $C(i,j) = C(i,j) + A(i,k){\cdot}B(k,j)$
>   **end**
>   **end**
> **end**

This computation involves $O(mnr)$ arithmetic. If the dimensions are doubled, then work increases by a factor of 8.

Each loop index in Algorithm 1.1.5 has a particular role. (The subscript i names the row, j names the column, and k handles the dot product.) Nevertheless, the ordering of the loops is arbitrary. Here is the (mathematically equivalent) jki variant:

for $j = 1{:}n$
    for $k = 1{:}r$
        for $i = 1{:}m$
$$C(i,j) = C(i,j) + A(i,k)B(k,j)$$
        end
    end
end

Altogether, there are six ($= 3!$) possibilities:

$$ijk, \quad jik, \quad ikj, \quad jki, \quad kij, \quad kji.$$

Each features an inner loop operation (dot product or saxpy) and each has its own pattern of data flow. For example, in the ijk variant, the inner loop oversees a dot product that requires access to a row of A and a column of B. The jki variant involves a saxpy that requires access to a column of C and a column of A. These attributes are summarized in Table 1.1.1 together with an interpretation of what is going on when

Loop Order	Inner Loop	Inner Two Loops	Inner Loop Data Access
ijk	dot	vector × matrix	A by row, B by column
jik	dot	matrix × vector	A by row, B by column
ikj	saxpy	row gaxpy	B by row, C by row
jki	saxpy	column gaxpy	A by column, C by column
kij	saxpy	row outer product	B by row, C by row
kji	saxpy	column outer product	A by column, C by column

Table 1.1.1. *Matrix multiplication: loop orderings and properties*

the middle and inner loops are considered together. Each variant involves the same amount of arithmetic, but accesses the A, B, and C data differently. The ramifications of this are discussed in §1.5.

1.1.12 A Dot Product Formulation

The usual matrix multiplication procedure regards $A{\cdot}B$ as an array of dot products to be computed one at a time in left-to-right, top-to-bottom order. This is the idea behind Algorithm 1.1.5 which we rewrite using the colon notation to highlight the mission of the innermost loop:

Algorithm 1.1.6 (Dot Product Matrix Multiplication) If $A \in \mathbb{R}^{m \times r}$, $B \in \mathbb{R}^{r \times n}$, and $C \in \mathbb{R}^{m \times n}$ are given, then this algorithm overwrites C with $C + AB$.

> **for** $i = 1{:}m$
>> **for** $j = 1{:}n$
>>> $C(i, j) = C(i, j) + A(i, :) \cdot B(:, j)$
>>
>> **end**
>
> **end**

In the language of partitioned matrices, if

$$A = \begin{bmatrix} a_1^T \\ \vdots \\ a_m^T \end{bmatrix} \qquad \text{and} \qquad B = [\, b_1 \,|\, \cdots \,|\, b_n \,],$$

then Algorithm 1.1.6 has this interpretation:

> **for** $i = 1{:}m$
>> **for** $j = 1{:}n$
>>> $c_{ij} = c_{ij} + a_i^T b_j$
>>
>> **end**
>
> **end**

Note that the purpose of the j-loop is to compute the ith row of the update. To emphasize this we could write

> **for** $i = 1{:}m$
>> $c_i^T = c_i^T + a_i^T B$
>
> **end**

where

$$C = \begin{bmatrix} c_1^T \\ \vdots \\ c_m^T \end{bmatrix}$$

is a row partitioning of C. To say the same thing with the colon notation we write

> **for** $i = 1{:}m$
>> $C(i, :) = C(i, :) + A(i, :) \cdot B$
>
> **end**

Either way we see that the inner two loops of the ijk variant define a transposed gaxpy operation.

1.1.13   A Saxpy Formulation

Suppose A and C are column-partitioned as follows:

$$A = [\, a_1 \,|\, \cdots \,|\, a_r \,], \qquad C = [\, c_1 \,|\, \cdots \,|\, c_n \,].$$

By comparing jth columns in $C = C + AB$ we see that

$$c_j = c_j + \sum_{k=1}^{r} a_k b_{kj}, \qquad j = 1{:}n.$$

These vector sums can be put together with a sequence of saxpy updates.

Algorithm 1.1.7 (Saxpy Matrix Multiplication) If the matrices $A \in \mathbb{R}^{m \times r}$, $B \in \mathbb{R}^{r \times n}$, and $C \in \mathbb{R}^{m \times n}$ are given, then this algorithm overwrites C with $C + AB$.

> **for** $j = 1{:}n$
> > **for** $k = 1{:}r$
> > > $C(:,j) = C(:,j) + A(:,k) \cdot B(k,j)$
> >
> > **end**
>
> **end**

Note that the k-loop oversees a gaxpy operation:

> **for** $j = 1{:}n$
> > $C(:,j) = C(:,j) + AB(:,j)$
>
> **end**

1.1.14 An Outer Product Formulation

Consider the kij variant of Algorithm 1.1.5:

> **for** $k = 1{:}r$
> > **for** $j = 1{:}n$
> > > **for** $i = 1{:}m$
> > > > $C(i,j) = C(i,j) + A(i,k) \cdot B(k,j)$
> > >
> > > **end**
> >
> > **end**
>
> **end**

The inner two loops oversee the outer product update

$$C = C + a_k b_k^T$$

where

$$A = [\, a_1 \mid \cdots \mid a_r \,] \quad \text{and} \quad B = \begin{bmatrix} b_1^T \\ \vdots \\ b_r^T \end{bmatrix} \tag{1.1.3}$$

with $a_k \in \mathbb{R}^m$ and $b_k \in \mathbb{R}^n$. This renders the following implementation:

Algorithm 1.1.8 (Outer Product Matrix Multiplication) If the matrices $A \in \mathbb{R}^{m \times r}$, $B \in \mathbb{R}^{r \times n}$, and $C \in \mathbb{R}^{m \times n}$ are given, then this algorithm overwrites C with $C + AB$.

> **for** $k = 1{:}r$
> > $C = C + A(:,k) \cdot B(k,:)$
>
> **end**

Matrix-matrix multiplication is a sum of outer products.

1.1.15   Flops

One way to quantify the volume of work associated with a computation is to count flops. A *flop* is a floating point add, subtract, multiply, or divide. The number of flops in a given matrix computation is usually obtained by summing the amount of arithmetic associated with the most deeply nested statements. For matrix-matrix multiplication, e.g., Algorithm 1.1.5, this is the 2-flop statement

$$C(i,j) = C(i,j) + A(i,k) \cdot B(k,j).$$

If $A \in \mathbb{R}^{m \times r}$, $B \in \mathbb{R}^{r \times n}$, and $C \in \mathbb{R}^{m \times n}$, then this statement is executed mnr times. Table 1.1.2 summarizes the number of flops that are required for the common operations detailed above.

Operation	Dimension	Flops
$\alpha = x^T y$	$x, y \in \mathbb{R}^n$	$2n$
$y = y + ax$	$a \in \mathbb{R}$, $x, y \in \mathbb{R}^n$	$2n$
$y = y + Ax$	$A \in \mathbb{R}^{m \times n}$, $x \in \mathbb{R}^n$, $y \in \mathbb{R}^m$	$2mn$
$A = A + yx^T$	$A \in \mathbb{R}^{m \times n}$, $x \in \mathbb{R}^n$, $y \in \mathbb{R}^m$	$2mn$
$C = C + AB$	$A \in \mathbb{R}^{m \times r}$, $B \in \mathbb{R}^{r \times n}$, $C \in \mathbb{R}^{m \times n}$	$2mnr$

Table 1.1.2. *Important flop counts*

1.1.16   Big-Oh Notation/Perspective

In certain settings it is handy to use the "Big-Oh" notation when an order-of-magnitude assessment of work suffices. (We did this in §1.1.1.) Dot products are $O(n)$, matrix-vector products are $O(n^2)$, and matrix-matrix products are $O(n^3)$. Thus, to make efficient an algorithm that involves a mix of these operations, the focus should typically be on the highest order operations that are involved as they tend to dominate the overall computation.

1.1.17   The Notion of "Level" and the BLAS

The dot product and saxpy operations are examples of *level-1* operations. Level-1 operations involve an amount of data and an amount of arithmetic that are linear in the dimension of the operation. An m-by-n outer product update or a gaxpy operation involves a quadratic amount of data ($O(mn)$) and a quadratic amount of work ($O(mn)$). These are *level-2* operations. The matrix multiplication update $C = C + AB$ is a *level-3* operation. Level-3 operations are quadratic in data and cubic in work.

Important level-1, level-2, and level-3 operations are encapsulated in the "BLAS," an acronym that stands for Basic Linear Algebra Subprograms. See LAPACK. The design of matrix algorithms that are rich in level-3 BLAS operations is a major preoccupation of the field for reasons that have to do with data reuse (§1.5).

1.1.18 Verifying a Matrix Equation

In striving to understand matrix multiplication via outer products, we essentially established the matrix equation

$$AB = \sum_{k=1}^{r} a_k b_k^T, \tag{1.1.4}$$

where the a_k and b_k are defined by the partitionings in (1.1.3).

Numerous matrix equations are developed in subsequent chapters. Sometimes they are established algorithmically as above and other times they are proved at the ij-component level, e.g.,

$$\left[\sum_{k=1}^{r} a_k b_k^T \right]_{ij} = \sum_{k=1}^{r} \left[a_k b_k^T \right]_{ij} = \sum_{k=1}^{r} a_{ik} b_{kj} = [AB]_{ij}.$$

Scalar-level verifications such as this usually provide little insight. However, they are sometimes the only way to proceed.

1.1.19 Complex Matrices

On occasion we shall be concerned with computations that involve complex matrices. The vector space of m-by-n complex matrices is designated by $\mathbb{C}^{m \times n}$. The scaling, addition, and multiplication of complex matrices correspond exactly to the real case. However, transposition becomes *conjugate transposition*:

$$C = A^H \implies c_{ij} = \bar{a}_{ji}.$$

The vector space of complex n-vectors is designated by $\mathbb{C}^n$. The dot product of complex n-vectors x and y is prescribed by

$$s = x^H y = \sum_{i=1}^{n} \bar{x}_i y_i.$$

If $A = B + iC \in \mathbb{C}^{m \times n}$, then we designate the real and imaginary parts of A by $\mathsf{Re}(A) = B$ and $\mathsf{Im}(A) = C$, respectively. The conjugate of A is the matrix $\bar{A} = (\bar{a}_{ij})$.

Problems

P1.1.1 Suppose $A \in \mathbb{R}^{n \times n}$ and $x \in \mathbb{R}^r$ are given. Give an algorithm for computing the first column of $M = (A - x_1 I) \cdots (A - x_r I)$.

P1.1.2 In a conventional 2-by-2 matrix multiplication $C = AB$, there are eight multiplications: $a_{11}b_{11}$, $a_{11}b_{12}$, $a_{21}b_{11}$, $a_{21}b_{12}$, $a_{12}b_{21}$, $a_{12}b_{22}$, $a_{22}b_{21}$, and $a_{22}b_{22}$. Make a table that indicates the order that these multiplications are performed for the ijk, jik, kij, ikj, jki, and kji matrix multiplication algorithms.

P1.1.3 Give an $O(n^2)$ algorithm for computing $C = (xy^T)^k$ where x and y are n-vectors.

P1.1.4 Suppose $D = ABC$ where $A \in \mathbb{R}^{m \times n}$, $B \in \mathbb{R}^{n \times p}$, and $C \in \mathbb{R}^{p \times q}$. Compare the flop count of an algorithm that computes D via the formula $D = (AB)C$ versus the flop count for an algorithm that computes D using $D = A(BC)$. Under what conditions is the former procedure more flop-efficient than the latter?

P1.1.5 Suppose we have real n-by-n matrices C, D, E, and F. Show how to compute real n-by-n matrices A and B with just three real n-by-n matrix multiplications so that

$$A + iB = (C + iD)(E + iF).$$

Hint: Compute $W = (C + D)(E - F)$.

P1.1.6 Suppose $W \in \mathbb{R}^{n \times n}$ is defined by

$$w_{ij} = \sum_{p=1}^{n} \sum_{q=1}^{n} x_{ip} y_{pq} z_{qj}$$

where $X, Y, Z \in \mathbb{R}^{n \times n}$. If we use this formula for each w_{ij} then it would require $O(n^4)$ operations to set up W. On the other hand,

$$w_{ij} = \sum_{p=1}^{n} x_{ip} \left(\sum_{q=1}^{n} y_{pq} z_{qj} \right) = \sum_{p=1}^{n} x_{ip} u_{pj}$$

where $U = YZ$. Thus, $W = XU = XYZ$ and only $O(n^3)$ operations are required.

Use this methodology to develop an $O(n^3)$ procedure for computing the n-by-n matrix A defined by

$$a_{ij} = \sum_{k_1=1}^{n} \sum_{k_2=1}^{n} \sum_{k_3=1}^{n} E(k_1, i) F(k_1, i) G(k_2, k_1) H(k_2, k_3) F(k_2, k_3) G(k_3, j)$$

where $E, F, G, H \in \mathbb{R}^{n \times n}$. Hint. Transposes and pointwise products are involved.

Notes and References for §1.1

For an appreciation of the BLAS and their foundational role, see:

C.L. Lawson, R.J. Hanson, D.R. Kincaid, and F.T. Krogh (1979). "Basic Linear Algebra Subprograms for FORTRAN Usage," *ACM Trans. Math. Softw. 5*, 308–323.

J.J. Dongarra, J. Du Croz, S. Hammarling, and R.J. Hanson (1988). "An Extended Set of Fortran Basic Linear Algebra Subprograms," *ACM Trans. Math. Softw. 14*, 1–17.

J.J. Dongarra, J. Du Croz, I.S. Duff, and S.J. Hammarling (1990). "A Set of Level 3 Basic Linear Algebra Subprograms," *ACM Trans. Math. Softw. 16*, 1–17.

B. Kågström, P. Ling, and C. Van Loan (1991). "High-Performance Level-3 BLAS: Sample Routines for Double Precision Real Data," in *High Performance Computing II*, M. Durand and F. El Dabaghi (eds.), North-Holland, Amsterdam, 269–281.

L.S. Blackford, J. Demmel, J. Dongarra, I. Duff, S. Hammarling, G. Henry, M. Heroux, L. Kaufman, A. Lumsdaine, A. Petitet, R. Pozo, K. Remington, and R.C. Whaley (2002). "An Updated Set of Basic Linear Algebra Subprograms (BLAS)", *ACM Trans. Math. Softw. 28*, 135–151.

The order in which the operations in the matrix product $A_1 \cdots A_r$ are carried out affects the flop count if the matrices vary in dimension. (See P1.1.4.) Optimization in this regard requires dynamic programming, see:

T.H. Corman, C.E. Leiserson, R.L. Rivest, and C. Stein (2001). *Introduction to Algorithms*, MIT Press and McGraw-Hill, 331–339.

1.2 Structure and Efficiency

The efficiency of a given matrix algorithm depends upon several factors. Most obvious and what we treat in this section is the amount of required arithmetic and storage. How to reason about these important attributes is nicely illustrated by considering examples that involve triangular matrices, diagonal matrices, banded matrices, symmetric matrices, and permutation matrices. These are among the most important types of structured matrices that arise in practice, and various economies can be realized if they are involved in a calculation.

1.2.1 Band Matrices

A matrix is *sparse* if a large fraction of its entries are zero. An important special case is the *band matrix*. We say that $A \in \mathbb{R}^{m \times n}$ has *lower bandwidth* p if $a_{ij} = 0$ whenever $i > j + p$ and *upper bandwidth* q if $j > i + q$ implies $a_{ij} = 0$. Here is an example of an 8-by-5 matrix that has lower bandwidth 1 and upper bandwidth 2:

$$
A = \begin{bmatrix}
\times & \times & \times & 0 & 0 \\
\times & \times & \times & \times & 0 \\
0 & \times & \times & \times & \times \\
0 & 0 & \times & \times & \times \\
0 & 0 & 0 & \times & \times \\
0 & 0 & 0 & 0 & \times \\
0 & 0 & 0 & 0 & 0 \\
0 & 0 & 0 & 0 & 0
\end{bmatrix}.
$$

The $\times$'s designate arbitrary nonzero entries. This notation is handy to indicate the structure of a matrix and we use it extensively. Band structures that occur frequently are tabulated in Table 1.2.1.

Type of Matrix	Lower Bandwidth	Upper Bandwidth
Diagonal	0	0
Upper triangular	0	$n - 1$
Lower triangular	$m - 1$	0
Tridiagonal	1	1
Upper bidiagonal	0	1
Lower bidiagonal	1	0
Upper Hessenberg	1	$n - 1$
Lower Hessenberg	$m - 1$	1

Table 1.2.1. *Band terminology for m-by-n matrices*

1.2.2 Triangular Matrix Multiplication

To introduce band matrix "thinking" we look at the matrix multiplication update problem $C = C + AB$ where A, B, and C are each n-by-n and upper triangular. The 3-by-3 case is illuminating:

$$
AB = \begin{bmatrix}
a_{11}b_{11} & a_{11}b_{12} + a_{12}b_{22} & a_{11}b_{13} + a_{12}b_{23} + a_{13}b_{33} \\
0 & a_{22}b_{22} & a_{22}b_{23} + a_{23}b_{33} \\
0 & 0 & a_{33}b_{33}
\end{bmatrix}.
$$

It suggests that the product is upper triangular and that its upper triangular entries are the result of abbreviated inner products. Indeed, since $a_{ik}b_{kj} = 0$ whenever $k < i$ or $j < k$, we see that the update has the form

$$
c_{ij} = c_{ij} + \sum_{k=i}^{j} a_{ik}b_{kj}
$$

for all i and j that satisfy $i \leq j$. This yields the following algorithm:

Algorithm 1.2.1 (Triangular Matrix Multiplication) Given upper triangular matrices $A, B, C \in \mathbb{R}^{n \times n}$, this algorithm overwrites C with $C + AB$.

> **for** $i = 1{:}n$
> > **for** $j = i{:}n$
> > > **for** $k = i{:}j$
> > > > $C(i, j) = C(i, j) + A(i, k) \cdot B(k, j)$
> > >
> > > **end**
> >
> > **end**
>
> **end**

1.2.3   The Colon Notation—Again

The dot product that the k-loop performs in Algorithm 1.2.1 can be succinctly stated if we extend the colon notation introduced in §1.1.8. If $A \in \mathbb{R}^{m \times n}$ and the integers p, q, and r satisfy $1 \leq p \leq q \leq n$ and $1 \leq r \leq m$, then

$$A(r, p{:}q) = [\, a_{rp} \,|\, \cdots \,|\, a_{rq} \,] \in \mathbb{R}^{1 \times (q-p+1)} \, .$$

Likewise, if $1 \leq p \leq q \leq m$ and $1 \leq c \leq n$, then

$$A(p{:}q, c) = \begin{bmatrix} a_{pc} \\ \vdots \\ a_{qc} \end{bmatrix} \in \mathbb{R}^{q-p+1}.$$

With this notation we can rewrite Algorithm 1.2.1 as

> **for** $i = 1{:}n$
> > **for** $j = i{:}n$
> > > $C(i, j) = C(i, j) + A(i, i{:}j) \cdot B(i{:}j, j)$
> >
> > **end**
>
> **end**

This highlights the abbreviated inner products that are computed by the innermost loop.

1.2.4   Assessing Work

Obviously, upper triangular matrix multiplication involves less arithmetic than full matrix multiplication. Looking at Algorithm 1.2.1, we see that c_{ij} requires $2(j - i + 1)$ flops if $(i \leq j)$. Using the approximations

$$\sum_{p=1}^{q} p = \frac{q(q+1)}{2} \approx \frac{q^2}{2}$$

and

$$\sum_{p=1}^{q} p^2 = \frac{q^3}{3} + \frac{q^2}{2} + \frac{q}{6} \approx \frac{q^3}{3},$$

we find that triangular matrix multiplication requires one-sixth the number of flops as full matrix multiplication:

$$\sum_{i=1}^{n}\sum_{j=i}^{n} 2(j-i+1) = \sum_{i=1}^{n}\sum_{j=1}^{n-i+1} 2j \approx \sum_{i=1}^{n} \frac{2(n-i+1)^2}{2} = \sum_{i=1}^{n} i^2 \approx \frac{n^3}{3}.$$

We throw away the low-order terms since their inclusion does not contribute to what the flop count "says." For example, an exact flop count of Algorithm 1.2.1 reveals that precisely $n^3/3 + n^2 + 2n/3$ flops are involved. For large n (the typical situation of interest) we see that the exact flop count offers no insight beyond the simple $n^3/3$ accounting.

Flop counting is a necessarily crude approach to the measurement of program efficiency since it ignores subscripting, memory traffic, and other overheads associated with program execution. We must not infer too much from a comparison of flop counts. We cannot conclude, for example, that triangular matrix multiplication is six times faster than full matrix multiplication. Flop counting captures just one dimension of what makes an algorithm efficient in practice. The equally relevant issues of vectorization and data locality are taken up in §1.5.

1.2.5 Band Storage

Suppose $A \in \mathbb{R}^{n \times n}$ has lower bandwidth p and upper bandwidth q and assume that p and q are much smaller than n. Such a matrix can be stored in a $(p+q+1)$-by-n array $A.band$ with the convention that

$$a_{ij} = A.band(i-j+q+1, j) \tag{1.2.1}$$

for all (i, j) that fall inside the band, e.g.,

$$
\begin{bmatrix}
a_{11} & a_{12} & a_{13} & 0 & 0 & 0 \\
a_{21} & a_{22} & a_{23} & a_{24} & 0 & 0 \\
0 & a_{32} & a_{33} & a_{34} & a_{35} & 0 \\
0 & 0 & a_{43} & a_{44} & a_{45} & a_{46} \\
0 & 0 & 0 & a_{54} & a_{55} & a_{56} \\
0 & 0 & 0 & 0 & a_{65} & a_{66}
\end{bmatrix}
\Rightarrow
\begin{bmatrix}
* & * & a_{13} & a_{24} & a_{35} & a_{46} \\
* & a_{12} & a_{23} & a_{34} & a_{45} & a_{56} \\
a_{11} & a_{22} & a_{33} & a_{44} & a_{55} & a_{66} \\
a_{21} & a_{32} & a_{43} & a_{54} & a_{65} & *
\end{bmatrix}.
$$

Here, the "$*$" entries are unused. With this data structure, our column-oriented gaxpy algorithm (Algorithm 1.1.4) transforms to the following:

Algorithm 1.2.2 (Band Storage Gaxpy) Suppose $A \in \mathbb{R}^{n \times n}$ has lower bandwidth p and upper bandwidth q and is stored in the $A.band$ format (1.2.1). If $x, y \in \mathbb{R}^n$, then this algorithm overwrites y with $y + Ax$.

> **for** $j = 1{:}n$
> $\quad \alpha_1 = \max(1, j-q),\ \alpha_2 = \min(n, j+p)$
> $\quad \beta_1 = \max(1, q+2-j),\ \beta_2 = \beta_1 + \alpha_2 - \alpha_1$
> $\quad y(\alpha_1{:}\alpha_2) = y(\alpha_1{:}\alpha_2) + A.band(\beta_1{:}\beta_2, j)x(j)$
> **end**

Notice that by storing A column by column in $A.band$, we obtain a column-oriented saxpy procedure. Indeed, Algorithm 1.2.2 is derived from Algorithm 1.1.4 by recognizing that each saxpy involves a vector with a small number of nonzeros. Integer arithmetic is used to identify the location of these nonzeros. As a result of this careful zero/nonzero analysis, the algorithm involves just $2n(p+q+1)$ flops with the assumption that p and q are much smaller than n.

1.2.6   Working with Diagonal Matrices

Matrices with upper and lower bandwidth zero are *diagonal*. If $D \in \mathbb{R}^{m \times n}$ is diagonal, then we use the notation

$$D = \mathrm{diag}(d_1, \ldots, d_q), \quad q = \min\{m, n\} \quad \Longleftrightarrow \quad d_i = d_{ii}.$$

Shortcut notations when the dimension is clear include $\mathrm{diag}(d)$ and $\mathrm{diag}(d_i)$. Note that if $D = \mathrm{diag}(d) \in \mathbb{R}^{n \times n}$ and $x \in \mathbb{R}^n$, then $Dx = d.*x$. If $A \in \mathbb{R}^{m \times n}$, then premultiplication by $D = \mathrm{diag}(d_1, \ldots, d_m) \in \mathbb{R}^{m \times m}$ scales rows,

$$B = DA \quad \Longleftrightarrow \quad B(i, :) = d_i \cdot A(i, :), \; i = 1{:}m$$

while post-multiplication by $D = \mathrm{diag}(d_1, \ldots, d_n) \in \mathbb{R}^{n \times n}$ scales columns,

$$B = AD \quad \Longleftrightarrow \quad B(:, j) = d_j \cdot A(:, j), \; j = 1{:}n.$$

Both of these special matrix-matrix multiplications require mn flops.

1.2.7   Symmetry

A matrix $A \in \mathbb{R}^{n \times n}$ is *symmetric* if $A^T = A$ and *skew-symmetric* if $A^T = -A$. Likewise, a matrix $A \in \mathbb{C}^{n \times n}$ is *Hermitian* if $A^H = A$ and *skew-Hermitian* if $A^H = -A$. Here are some examples:

Symmetric:
$\begin{bmatrix} 1 & 2 & 3 \\ 2 & 4 & 5 \\ 3 & 5 & 6 \end{bmatrix}$,
    Hermitian:
$\begin{bmatrix} 1 & 2-3i & 4-5i \\ 2+3i & 6 & 7-8i \\ 4+5i & 7+8i & 9 \end{bmatrix}$,

Skew-Symmetric:
$\begin{bmatrix} 0 & -2 & 3 \\ 2 & 0 & -5 \\ -3 & 5 & 0 \end{bmatrix}$,
    Skew-Hermitian:
$\begin{bmatrix} i & -2+3i & -4+5i \\ 2+3i & 6i & -7+8i \\ 4+5i & 7+8i & 9i \end{bmatrix}$.

For such matrices, storage requirements can be halved by simply storing the lower triangle of elements, e.g.,

$$A = \begin{bmatrix} 1 & 2 & 3 \\ 2 & 4 & 5 \\ 3 & 5 & 6 \end{bmatrix} \quad \Leftrightarrow \quad A.vec = \begin{bmatrix} 1 & 2 & 3 & 4 & 5 & 6 \end{bmatrix}.$$

For general n, we set

$$A.vec((n - j/2)(j - 1) + i) = a_{ij} \quad 1 \le j \le i \le n. \tag{1.2.2}$$

Here is a column-oriented gaxpy with the matrix A represented in $A.vec$.

Algorithm 1.2.3 (Symmetric Storage Gaxpy) Suppose $A \in \mathbb{R}^{n \times n}$ is symmetric and stored in the $A.vec$ style (1.2.2). If $x, y \in \mathbb{R}^n$, then this algorithm overwrites y with $y + Ax$.

> **for** $j = 1{:}n$
>   **for** $i = 1{:}j-1$
> $y(i) = y(i) + A.vec((i-1)n - i(i-1)/2 + j)x(j)$
>   **end**
>   **for** $i = j{:}n$
> $y(i) = y(i) + A.vec((j-1)n - j(j-1)/2 + i)x(j)$
>   **end**
> **end**

This algorithm requires the same $2n^2$ flops that an ordinary gaxpy requires.

1.2.8 Permutation Matrices and the Identity

We denote the n-by-n *identity matrix* by I_n, e.g.,

$$
I_4 = \begin{bmatrix} 1 & 0 & 0 & 0 \\ 0 & 1 & 0 & 0 \\ 0 & 0 & 1 & 0 \\ 0 & 0 & 0 & 1 \end{bmatrix}.
$$

We use the notation e_i to designate the ith column of I_n. If the rows of I_n are reordered, then the resulting matrix is said to be a *permutation matrix*, e.g.,

$$
P = \begin{bmatrix} 0 & 1 & 0 & 0 \\ 0 & 0 & 0 & 1 \\ 0 & 0 & 1 & 0 \\ 1 & 0 & 0 & 0 \end{bmatrix}. \tag{1.2.3}
$$

The representation of an n-by-n permutation matrix requires just an n-vector of integers whose components specify where the 1's occur. For example, if $v \in \mathbb{R}^n$ has the property that v_i specifies the column where the "1" occurs in row i, then $y = Px$ implies that $y_i = x_{v_i}$, $i = 1{:}n$. In the example above, the underlying v-vector is $v = [\, 2\, 4\, 3\, 1\,]$.

1.2.9 Specifying Integer Vectors and Submatrices

For permutation matrix work and block matrix manipulation (§1.3) it is convenient to have a method for specifying structured integer vectors of subscripts. The MATLAB colon notation is again the proper vehicle and a few examples suffice to show how it works. If $n = 8$, then

$$
\begin{aligned}
v = 1{:}2{:}n &\quad\Longrightarrow\quad v = [\,1\,3\,5\,7\,], \\
v = n{:}{-}1{:}1 &\quad\Longrightarrow\quad v = [\,8\,7\,6\,5\,4\,3\,2\,1\,], \\
v = [\,(1{:}2{:}n)\ (2{:}2{:}n)\,] &\quad\Longrightarrow\quad v = [\,1\,3\,5\,7\,2\,4\,6\,8\,].
\end{aligned}
$$

Suppose $A \in \mathbb{R}^{m \times n}$ and that $v \in \mathbb{R}^r$ and $w \in \mathbb{R}^s$ are integer vectors with the property that $1 \leq v_i \leq m$ and $1 \leq w_i \leq n$. If $B = A(v, w)$, then $B \in \mathbb{R}^{r \times s}$ is the matrix defined by $b_{ij} = a_{v_i, w_j}$ for $i = 1{:}r$ and $j = 1{:}s$. Thus, if $A \in \mathbb{R}^{8 \times 8}$, then

$$
A(1{:}2{:}8, 2{:}2{:}8) = \begin{bmatrix} a_{12} & a_{14} & a_{16} & a_{18} \\ a_{32} & a_{34} & a_{36} & a_{38} \\ a_{52} & a_{54} & a_{56} & a_{58} \\ a_{72} & a_{74} & a_{76} & a_{78} \end{bmatrix}.
$$

1.2.10   Working with Permutation Matrices

Using the colon notation, the 4-by-4 permutation matrix in (1.2.3) is defined by $P = I_4(v, :)$ where $v = [\,2\ 4\ 3\ 1\,]$. In general, if $v \in \mathbb{R}^n$ is a permutation of the vector $1{:}n = [1, 2, \ldots, n]$ and $P = I_n(v, :)$, then

$$
y = Px \quad \Longrightarrow \quad y = x(v) \quad \Longrightarrow \quad y_i = x_{v_i}, \ i = 1{:}n
$$

$$
y = P^T x \quad \Longrightarrow \quad y(v) = x \quad \Longrightarrow \quad y_{v_i} = x_i, \ i = 1{:}n
$$

The second result follows from the fact that v_i is the *row* index of the "1" in column i of P^T. Note that $P^T(Px) = x$. The inverse of a permutation matrix is its transpose.

The action of a permutation matrix on a given matrix $A \in \mathbb{R}^{m \times n}$ is easily described. If $P = I_m(v, :)$ and $Q = I_n(w, :)$, then $PAQ^T = A(v, w)$. It also follows that $I_n(v, :) \cdot I_n(w, :) = I_n(w(v), :)$. Although permutation operations involve no flops, they move data and contribute to execution time, an issue that is discussed in §1.5.

1.2.11   Three Famous Permutation Matrices

The *exchange permutation* $\mathcal{E}_n$ turns vectors upside down, e.g.,

$$
y = \mathcal{E}_4 x = \begin{bmatrix} 0 & 0 & 0 & 1 \\ 0 & 0 & 1 & 0 \\ 0 & 1 & 0 & 0 \\ 1 & 0 & 0 & 0 \end{bmatrix} \begin{bmatrix} x_1 \\ x_2 \\ x_3 \\ x_4 \end{bmatrix} = \begin{bmatrix} x_4 \\ x_3 \\ x_2 \\ x_1 \end{bmatrix}.
$$

In general, if $v = n{:}-1{:}1$, then the n-by-n exchange permutation is given by $\mathcal{E}_n = I_n(v, :)$. No change results if a vector is turned upside down twice and thus, $\mathcal{E}_n^T \mathcal{E}_n = \mathcal{E}_n^2 = I_n$.

The *downshift permutation* $\mathcal{D}_n$ pushes the components of a vector down one notch with wraparound, e.g.,

$$
y = \mathcal{D}_4 x = \begin{bmatrix} 0 & 0 & 0 & 1 \\ 1 & 0 & 0 & 0 \\ 0 & 1 & 0 & 0 \\ 0 & 0 & 1 & 0 \end{bmatrix} \begin{bmatrix} x_1 \\ x_2 \\ x_3 \\ x_4 \end{bmatrix} = \begin{bmatrix} x_4 \\ x_1 \\ x_2 \\ x_3 \end{bmatrix}.
$$

In general, if $v = [\,(2{:}n)\ 1\,]$, then the n-by-n downshift permutation is given by $\mathcal{D}_n = I_n(v, :)$. Note that $\mathcal{D}_n^T$ can be regarded as an *upshift* permutation.

The *mod-p perfect shuffle permutation* $\mathcal{P}_{p,r}$ treats the components of the input vector $x \in \mathbb{R}^n$, $n = pr$, as cards in a deck. The deck is cut into p equal "piles" and reassembled by taking one card from each pile in turn. Thus, if $p = 3$ and $r = 4$, then the piles are $x(1{:}4)$, $x(5{:}8)$, and $x(9{:}12)$ and

$$y = \mathcal{P}_{3,4}x = I_{pr}([\,1\ 5\ 9\ 2\ 6\ 10\ 3\ 7\ 11\ 4\ 8\ 12\,],:)x = \begin{bmatrix} x(1{:}4{:}12) \\ x(2{:}4{:}12) \\ x(3{:}4{:}12) \\ x(4{:}4{:}12) \end{bmatrix}.$$

In general, if $n = pr$, then

$$\mathcal{P}_{p,r} = I_n([\,(1{:}r{:}n)\ (2{:}r{:}n)\ \cdots\ (r{:}r{:}n)],:)$$

and it can be shown that

$$\mathcal{P}_{p,r}^T = I_n([\,(1{:}p{:}n)\ (2{:}p{:}n)\ \cdots\ (p{:}p{:}n)\,],:). \tag{1.2.4}$$

Continuing with the card deck metaphor, $\mathcal{P}_{p,r}^T$ reassembles the card deck by placing all the x_i having $i \bmod p = 1$ first, followed by all the x_i having $i \bmod p = 2$ second, and so on.

Problems

P1.2.1 Give an algorithm that overwrites A with A^2 where $A \in \mathbb{R}^{n \times n}$. How much extra storage is required? Repeat for the case when A is upper triangular.

P1.2.2 Specify an algorithm that computes the first column of the matrix $M = (A - \lambda_1 I) \cdots (A - \lambda_r I)$ where $A \in \mathbb{R}^{n \times n}$ is upper Hessenberg and $\lambda_1, \ldots, \lambda_r$ are given scalars. How many flops are required assuming that $r \ll n$?

P1.2.3 Give a column saxpy algorithm for the n-by-n matrix multiplication problem $C = C + AB$ where A is upper triangular and B is lower triangular.

P1.2.4 Extend Algorithm 1.2.2 so that it can handle rectangular band matrices. Be sure to describe the underlying data structure.

P1.2.5 If $A = B + iC$ is Hermitian with $B \in \mathbb{R}^{n \times n}$, then it is easy to show that $B^T = B$ and $C^T = -C$. Suppose we represent A in an array $A.herm$ with the property that $A.herm(i,j)$ houses b_{ij} if $i \geq j$ and c_{ij} if $j > i$. Using this data structure, write a matrix-vector multiply function that computes $\mathsf{Re}(z)$ and $\mathsf{Im}(z)$ from $\mathsf{Re}(x)$ and $\mathsf{Im}(x)$ so that $z = Ax$.

P1.2.6 Suppose $X \in \mathbb{R}^{n \times p}$ and $A \in \mathbb{R}^{n \times n}$ are given and that A is symmetric. Give an algorithm for computing $B = X^T A X$ assuming that both A and B are to be stored using the symmetric storage scheme presented in §1.2.7.

P1.2.7 Suppose $a \in \mathbb{R}^n$ is given and that $A \in \mathbb{R}^{n \times n}$ has the property that $a_{ij} = a_{|i-j|+1}$. Give an algorithm that overwrites y with $y + Ax$ where $x, y \in \mathbb{R}^n$ are given.

P1.2.8 Suppose $a \in \mathbb{R}^n$ is given and that $A \in \mathbb{R}^{n \times n}$ has the property that $a_{ij} = a_{((i+j-1) \bmod n)+1}$. Give an algorithm that overwrites y with $y + Ax$ where $x, y \in \mathbb{R}^n$ are given.

P1.2.9 Develop a compact storage scheme for symmetric band matrices and write the corresponding gaxpy algorithm.

P1.2.10 Suppose $A \in \mathbb{R}^{n \times n}$, $u \in \mathbb{R}^n$, and $v \in \mathbb{R}^n$ are given and that $k \leq n$ is an integer. Show how to compute $X \in \mathbb{R}^{n \times k}$ and $Y \in \mathbb{R}^{n \times k}$ so that $(A + uv^T)^k = A^k + XY^T$. How many flops are required?

P1.2.11 Suppose $x \in \mathbb{R}^n$. Write a single-loop algorithm that computes $y = \mathcal{D}_n^k x$ where k is a positive integer and $\mathcal{D}_n$ is defined in §1.2.11.

P1.2.12 (a) Verify (1.2.4). (b) Show that $\mathcal{P}_{p,r}^T = \mathcal{P}_{r,p}$.

P1.2.13 The number of n-by-n permutation matrices is $n!$. How many of these are symmetric?

Notes and References for §1.2

See LAPACK for a discussion about appropriate data structures when symmetry and/or bandedness is present in addition to

F.G. Gustavson (2008). "The Relevance of New Data Structure Approaches for Dense Linear Algebra in the New Multi-Core/Many-Core Environments," in *Proceedings of the 7th international Conference on Parallel Processing and Applied Mathematics*, Springer-Verlag, Berlin, 618–621.

The exchange, downshift, and perfect shuffle permutations are discussed in Van Loan (FFT).

1.3   Block Matrices and Algorithms

A block matrix is a matrix whose entries are themselves matrices. It is a point of view. For example, an 8-by-15 matrix of scalars can be regarded as a 2-by-3 block matrix with 4-by-5 entries. Algorithms that manipulate matrices at the block level are often more efficient because they are richer in level-3 operations. The derivation of many important algorithms is often simplified by using block matrix notation.

1.3.1   Block Matrix Terminology

Column and row partitionings (§1.1.7) are special cases of matrix blocking. In general, we can partition both the rows and columns of an m-by-n matrix A to obtain

$$
A = \left[\begin{array}{ccc} A_{11} & \dots & A_{1r} \\ \vdots & & \vdots \\ A_{q1} & \cdots & A_{qr} \end{array} \right] \begin{array}{c} m_1 \\ \\ m_q \end{array}
$$
$$
\quad n_1 \qquad n_r
$$

where $m_1 + \cdots + m_q = m$, $n_1 + \cdots + n_r = n$, and $A_{\alpha\beta}$ designates the (α, β) block (submatrix). With this notation, block $A_{\alpha\beta}$ has dimension m_α-by-n_β and we say that $A = (A_{\alpha\beta})$ is a q-by-r block matrix.

Terms that we use to describe well-known band structures for matrices with scalar entries have natural block analogs. Thus,

$$
\mathrm{diag}(A_{11}, A_{22}, A_{33}) = \left[\begin{array}{ccc} A_{11} & 0 & 0 \\ 0 & A_{22} & 0 \\ 0 & 0 & A_{33} \end{array} \right]
$$

is *block diagonal* while the matrices

$$
L = \left[\begin{array}{ccc} L_{11} & 0 & 0 \\ L_{21} & L_{22} & 0 \\ L_{31} & L_{32} & L_{33} \end{array} \right], \quad U = \left[\begin{array}{ccc} U_{11} & U_{12} & U_{13} \\ 0 & U_{22} & U_{23} \\ 0 & 0 & U_{33} \end{array} \right], \quad T = \left[\begin{array}{ccc} T_{11} & T_{12} & 0 \\ T_{21} & T_{22} & T_{23} \\ 0 & T_{32} & T_{33} \end{array} \right],
$$

are, respectively, *block lower triangular, block upper triangular,* and *block tridiagonal*. The blocks *do not* have to be square in order to use this *block sparse* terminology.

1.3.2 Block Matrix Operations

Block matrices can be scaled and transposed:

$$
\mu \begin{bmatrix} A_{11} & A_{12} \\ A_{21} & A_{22} \\ A_{31} & A_{32} \end{bmatrix} = \begin{bmatrix} \mu A_{11} & \mu A_{12} \\ \mu A_{21} & \mu A_{22} \\ \mu A_{31} & \mu A_{32} \end{bmatrix},
$$

$$
\begin{bmatrix} A_{11} & A_{12} \\ A_{21} & A_{22} \\ A_{31} & A_{32} \end{bmatrix}^T = \begin{bmatrix} A_{11}^T & A_{21}^T & A_{31}^T \\ A_{12}^T & A_{22}^T & A_{32}^T \end{bmatrix}.
$$

Note that the transpose of the original (i, j) block becomes the (j, i) block of the result. Identically blocked matrices can be added by summing the corresponding blocks:

$$
\begin{bmatrix} A_{11} & A_{12} \\ A_{21} & A_{22} \\ A_{31} & A_{32} \end{bmatrix} + \begin{bmatrix} B_{11} & B_{12} \\ B_{21} & B_{22} \\ B_{31} & B_{32} \end{bmatrix} = \begin{bmatrix} A_{11} + B_{11} & A_{12} + B_{12} \\ A_{21} + B_{21} & A_{22} + B_{22} \\ A_{31} + B_{31} & A_{32} + B_{32} \end{bmatrix}.
$$

Block matrix multiplication requires more stipulations about dimension. For example, if

$$
\begin{bmatrix} A_{11} & A_{12} \\ A_{21} & A_{22} \\ A_{31} & A_{32} \end{bmatrix} \begin{bmatrix} B_{11} & B_{12} \\ B_{21} & B_{22} \end{bmatrix} = \begin{bmatrix} A_{11}B_{11} + A_{12}B_{21} & A_{11}B_{12} + A_{12}B_{22} \\ A_{21}B_{11} + A_{22}B_{21} & A_{21}B_{12} + A_{22}B_{22} \\ A_{31}B_{11} + A_{32}B_{21} & A_{31}B_{12} + A_{32}B_{22} \end{bmatrix}
$$

is to make sense, then the column dimensions of A_{11}, A_{21}, and A_{31} must each be equal to the row dimension of both B_{11} and B_{12}. Likewise, the column dimensions of A_{12}, A_{22}, and A_{32} must each be equal to the row dimensions of both B_{21} and B_{22}.

Whenever a block matrix addition or multiplication is indicated, it is assumed that the row and column dimensions of the blocks satisfy all the necessary constraints. In that case we say that the operands are *partitioned conformably* as in the following theorem.

Theorem 1.3.1. *If*

$$
A = \begin{bmatrix} A_{11} & \cdots & A_{1s} \\ \vdots & & \vdots \\ A_{q1} & \cdots & A_{qs} \end{bmatrix} \begin{matrix} m_1 \\ \\ m_q \end{matrix}, \qquad B = \begin{bmatrix} B_{11} & \cdots & B_{1r} \\ \vdots & & \vdots \\ B_{s1} & \cdots & B_{sr} \end{bmatrix} \begin{matrix} p_1 \\ \\ p_s \end{matrix},
$$

$$
 \; p_1 \qquad p_s \qquad\qquad\qquad\qquad n_1 \qquad n_r
$$

and we partition the product $C = AB$ as follows,

$$
C = \begin{bmatrix} C_{11} & \cdots & C_{1r} \\ \vdots & & \vdots \\ C_{q1} & \cdots & C_{qr} \end{bmatrix} \begin{matrix} m_1 \\ \\ m_q \end{matrix},
$$

$$
 \; n_1 \qquad n_r
$$

then for $\alpha = 1{:}q$ and $\beta = 1{:}r$ we have $C_{\alpha\beta} = \displaystyle\sum_{\gamma=1}^{s} A_{\alpha\gamma} B_{\gamma\beta}$.

Proof. The proof is a tedious exercise in subscripting. Suppose $1 \leq \alpha \leq q$ and $1 \leq \beta \leq r$. Set $M = m_1 + \cdots + m_{\alpha-1}$ and $N = n_1 + \cdots n_{\beta-1}$. It follows that if $1 \leq i \leq m_\alpha$ and $1 \leq j \leq n_\beta$ then

$$
[C_{\alpha\beta}]_{ij} = \sum_{k=1}^{p_1+\cdots p_s} a_{M+i,k} b_{k,N+j} = \sum_{\gamma=1}^{s} \sum_{k=p_1+\cdots+p_{\gamma-1}+1}^{p_1+\cdots+p_\gamma} a_{M+i,k} b_{k,N+j}
$$

$$
= \sum_{\gamma=1}^{s} \sum_{k=1}^{p_\gamma} [A_{\alpha\gamma}]_{ik} [B_{\gamma\beta}]_{kj} = \sum_{\gamma=1}^{s} [A_{\alpha\gamma} B_{\gamma\beta}]_{ij} = \left[\sum_{\gamma=1}^{s} A_{\alpha\gamma} B_{\gamma\beta} \right]_{ij}.
$$

Thus, $C_{\alpha\beta} = A_{\alpha,1} B_{1,\beta} + \cdots + A_{\alpha,s} B_{s,\beta}$. $\square$

If you pay attention to dimension and remember that matrices do not commute, i.e., $A_{11} B_{11} + A_{12} B_{21} \neq B_{11} A_{11} + B_{21} A_{12}$, then block matrix manipulation is just ordinary matrix manipulation with the a_{ij}'s and b_{ij}'s written as A_{ij}'s and B_{ij}'s!

1.3.3 Submatrices

Suppose $A \in \mathbb{R}^{m \times n}$. If $\alpha = [\alpha_1, \ldots, \alpha_s]$ and $\beta = [\beta_1, \ldots, \beta_t]$ are integer vectors with distinct components that satisfy $1 \leq \alpha_i \leq m$, and $1 \leq \beta_i \leq n$, then

$$
A(\alpha, \beta) = \begin{bmatrix} a_{\alpha_1,\beta_1} & \cdots & a_{\alpha_1,\beta_t} \\ \vdots & \ddots & \vdots \\ a_{\alpha_s,\beta_1} & \cdots & a_{\alpha_s,\beta_t} \end{bmatrix}
$$

is an s-by-t *submatrix* of A. For example, if $A \in \mathbb{R}^{8 \times 6}$, $\alpha = [2\,4\,6\,8]$, and $\beta = [4\,5\,6]$, then

$$
A(\alpha, \beta) = \begin{bmatrix} a_{24} & a_{25} & a_{26} \\ a_{44} & a_{45} & a_{46} \\ a_{64} & a_{65} & a_{66} \\ a_{84} & a_{85} & a_{86} \end{bmatrix}.
$$

If $\alpha = \beta$, then $A(\alpha, \beta)$ is a *principal submatrix*. If $\alpha = \beta = 1{:}k$ and $1 \leq k \leq \min\{m, n\}$, then $A(\alpha, \beta)$ is a *leading principal submatrix*.

If $A \in \mathbb{R}^{m \times n}$ and

$$
A = \begin{bmatrix} A_{11} & \ldots & A_{1s} \\ \vdots & & \vdots \\ A_{q1} & \cdots & A_{qs} \end{bmatrix} \begin{matrix} m_1 \\ \\ m_q \end{matrix} \quad ,
$$
$$
\quad\quad n_1 \quad\quad n_r
$$

then the colon notation can be used to specify the individual blocks. In particular,

$$
A_{ij} = A(\tau + 1{:}\tau + m_i, \mu + 1{:}\mu + n_j)
$$

where $\tau = m_1 + \cdots + m_{i-1}$ and $\mu = n_1 + \cdots + n_{j-1}$. Block matrix notation is valuable for the way in which it hides subscript range expressions.

1.3.4   The Blocked Gaxpy

As an exercise in block matrix manipulation and submatrix designation, we consider two block versions of the gaxpy operation $y = y + Ax$ where $A \in \mathbb{R}^{m \times n}$, $x \in \mathbb{R}^n$, and $y \in \mathbb{R}^m$. If

$$
A = \begin{bmatrix} A_1 \\ \vdots \\ A_q \end{bmatrix} \begin{matrix} m_1 \\ \\ m_q \end{matrix}
\qquad \text{and} \qquad
y = \begin{bmatrix} y_1 \\ \vdots \\ y_q \end{bmatrix} \begin{matrix} m_1 \\ \\ m_q \end{matrix},
$$

then

$$
\begin{bmatrix} y_1 \\ \vdots \\ y_q \end{bmatrix} = \begin{bmatrix} y_1 \\ \vdots \\ y_q \end{bmatrix} + \begin{bmatrix} A_1 \\ \vdots \\ A_q \end{bmatrix} x,
$$

and we obtain

> $\alpha = 0$
> **for** $i = 1{:}q$
>> $idx = \alpha + 1 : \alpha + m_i$
>> $y(idx) = y(idx) + A(idx,:) \cdot x$
>> $\alpha = \alpha + m_i$
> **end**

The assignment to $y(idx)$ corresponds to $y_i = y_i + A_i x$. This row-blocked version of the gaxpy computation breaks the given gaxpy into q "shorter" gaxpys. We refer to A_i as the ith block row of A.

Likewise, with the partitionings

$$
A = \begin{bmatrix} A_1 \mid \cdots \mid A_r \end{bmatrix}
\qquad \text{and} \qquad
x = \begin{bmatrix} x_1 \\ \vdots \\ x_r \end{bmatrix} \begin{matrix} n_1 \\ \\ n_r \end{matrix},
$$

we see that

$$
y = y + \begin{bmatrix} A_1 \mid \cdots \mid A_r \end{bmatrix} \begin{bmatrix} x_1 \\ \vdots \\ x_r \end{bmatrix} = y + \sum_{j=1}^{r} A_j x_j
$$

and we obtain

> $\beta = 0$
> **for** $j = 1{:}r$
>> $jdx = \beta + 1 : \beta + n_j$
>> $y = y + A(:,jdx) \cdot x(jdx)$
>> $\beta = \beta + n_j$
> **end**

The assignment to y corresponds to $y = y + A_j x_j$. This column-blocked version of the gaxpy computation breaks the given gaxpy into r "thinner" gaxpys. We refer to A_j as the jth block column of A.

1.3.5   Block Matrix Multiplication

Just as ordinary, scalar-level matrix multiplication can be arranged in several possible ways, so can the multiplication of block matrices. To illustrate this with a minimum of subscript clutter, we consider the update

$$C = C + AB$$

where we regard $A = (A_{\alpha\beta})$, $B = (B_{\alpha\beta})$, and $C = (C_{\alpha\beta})$ as N-by-N block matrices with ℓ-by-ℓ blocks. From Theorem 1.3.1 we have

$$C_{\alpha\beta} \; = \; C_{\alpha\beta} \; + \; \sum_{\gamma=1}^{N} A_{\alpha\gamma} B_{\gamma\beta}, \qquad \alpha = 1{:}N, \quad \beta = 1{:}N.$$

If we organize a matrix multiplication procedure around this summation, then we obtain a block analog of Algorithm 1.1.5:

> **for** $\alpha = 1{:}N$
> > $i = (\alpha - 1)\ell + 1{:}\alpha\ell$
> > **for** $\beta = 1{:}N$
> > > $j = (\beta - 1)\ell + 1{:}\beta\ell$
> > > **for** $\gamma = 1{:}N$
> > > > $k = (\gamma - 1)\ell + 1{:}\gamma\ell$
> > > > $C(i,j) = C(i,j) \; + \; A(i,k)\cdot B(k,j)$
> > > **end**
> > **end**
> **end**

Note that, if $\ell = 1$, then $\alpha \equiv i$, $\beta \equiv j$, and $\gamma \equiv k$ and we revert to Algorithm 1.1.5.

Analogously to what we did in §1.1, we can obtain different variants of this procedure by playing with loop orders and blocking strategies. For example, corresponding to

$$\begin{bmatrix} C_{11} & \cdots & C_{1N} \\ \vdots & \ddots & \vdots \\ C_{N1} & \cdots & C_{NN} \end{bmatrix} + \begin{bmatrix} A_1 \\ \vdots \\ A_N \end{bmatrix} \begin{bmatrix} B_1 & \cdots & B_N \end{bmatrix}$$

where $A_i \in \mathbb{R}^{\ell \times n}$ and $B_j \in \mathbb{R}^{n \times \ell}$, we obtain the following block outer product computation:

> **for** $i = 1{:}N$
> > **for** $j = 1{:}N$
> > > $C_{ij} = C_{ij} + A_i B_j$
> > **end**
> **end**

1.3.6 The Kronecker Product

It is sometimes the case that the entries in a block matrix A are all scalar multiples of the same matrix. This means that A is a *Kronecker product*. Formally, if $B \in \mathbb{R}^{m_1 \times n_1}$ and $C \in \mathbb{R}^{m_2 \times n_2}$, then their Kronecker product $B \otimes C$ is an m_1-by-n_1 block matrix whose (i,j) block is the m_2-by-n_2 matrix $b_{ij}C$. Thus, if

$$
A = \begin{bmatrix} b_{11} & b_{12} \\ b_{21} & b_{22} \\ b_{31} & b_{32} \end{bmatrix} \otimes \begin{bmatrix} c_{11} & c_{12} & c_{13} \\ c_{21} & c_{22} & c_{23} \\ c_{31} & c_{32} & c_{33} \end{bmatrix}
$$

then

$$
A = \left[\begin{array}{ccc|ccc}
b_{11}c_{11} & b_{11}c_{12} & b_{11}c_{13} & b_{12}c_{11} & b_{12}c_{12} & b_{12}c_{13} \\
b_{11}c_{21} & b_{11}c_{22} & b_{11}c_{23} & b_{12}c_{21} & b_{12}c_{22} & b_{12}c_{23} \\
b_{11}c_{31} & b_{11}c_{32} & b_{11}c_{33} & b_{12}c_{31} & b_{12}c_{32} & b_{12}c_{33} \\
\hline
b_{21}c_{11} & b_{21}c_{12} & b_{21}c_{13} & b_{22}c_{11} & b_{22}c_{12} & b_{22}c_{13} \\
b_{21}c_{21} & b_{21}c_{22} & b_{21}c_{23} & b_{22}c_{21} & b_{22}c_{22} & b_{22}c_{23} \\
b_{21}c_{31} & b_{21}c_{32} & b_{21}c_{33} & b_{22}c_{31} & b_{22}c_{32} & b_{22}c_{33} \\
\hline
b_{31}c_{11} & b_{31}c_{12} & b_{31}c_{13} & b_{32}c_{11} & b_{32}c_{12} & b_{32}c_{13} \\
b_{31}c_{21} & b_{31}c_{22} & b_{31}c_{23} & b_{32}c_{21} & b_{32}c_{22} & b_{32}c_{23} \\
b_{31}c_{31} & b_{31}c_{32} & b_{31}c_{33} & b_{32}c_{31} & b_{32}c_{32} & b_{32}c_{33}
\end{array} \right] .
$$

This type of highly structured blocking occurs in many applications and results in dramatic economies when fully exploited.

Note that if B has a band structure, then $B \otimes C$ "inherits" that structure at the block level. For example, if

$$
B \text{ is } \left\{ \begin{array}{l} \text{diagonal} \\ \text{tridiagonal} \\ \text{lower triangular} \\ \text{upper triangular} \end{array} \right\} \text{ then } B \otimes C \text{ is } \left\{ \begin{array}{l} \text{block diagonal} \\ \text{block tridiagonal} \\ \text{block lower triangular} \\ \text{block upper triangular} \end{array} \right\} .
$$

Important Kronecker product properties include:

$$
(B \otimes C)^T = B^T \otimes C^T, \tag{1.3.1}
$$

$$
(B \otimes C)(D \otimes F) = BD \otimes CF, \tag{1.3.2}
$$

$$
(B \otimes C)^{-1} = B^{-1} \otimes C^{-1}, \tag{1.3.3}
$$

$$
B \otimes (C \otimes D) = (B \otimes C) \otimes D. \tag{1.3.4}
$$

Of course, the products BD and CF must be defined for (1.3.2) to make sense. Likewise, the matrices B and C must be nonsingular in (1.3.3).

In general, $B \otimes C \neq C \otimes B$. However, there is a connection between these two matrices via the perfect shuffle permutation that is defined in §1.2.11. If $B \in \mathbb{R}^{m_1 \times n_1}$ and $C \in \mathbb{R}^{m_2 \times n_2}$, then

$$
P(B \otimes C)Q^T = C \otimes B \tag{1.3.5}
$$

where $P = \mathcal{P}_{m_1, m_2}$ and $Q = \mathcal{P}_{n_1, n_2}$.

1.3.7 Reshaping Kronecker Product Expressions

A matrix-vector product in which the matrix is a Kronecker product is "secretly" a matrix-matrix-matrix product. For example, if $B \in \mathbb{R}^{3 \times 2}$, $C \in \mathbb{R}^{m \times n}$, and $x_1, x_2 \in \mathbb{R}^n$, then

$$\left[\begin{array}{c} y_1 \\ y_2 \\ y_3 \end{array} \right] = (B \otimes C) \left[\begin{array}{c} x_1 \\ x_2 \end{array} \right] = \left[\begin{array}{cc} b_{11}C & b_{12}C \\ b_{21}C & b_{22}C \\ b_{31}C & b_{32}C \end{array} \right] \left[\begin{array}{c} x_1 \\ x_2 \end{array} \right]$$

$$= \left[\begin{array}{c} b_{11}Cx_1 + b_{12}Cx_2 \\ b_{21}Cx_1 + b_{22}Cx_2 \\ b_{31}Cx_1 + b_{32}Cx_2 \end{array} \right]$$

where $y_1, y_2, y_3 \in \mathbb{R}^m$. On the other hand, if we define the matrices

$$X = [\, x_1 \ x_2 \,] \quad \text{and} \quad Y = [\, y_1 \ y_2 \ y_3 \,],$$

then $Y = CXB^T$.

To be precise about this reshaping, we introduce the vec operation. If $X \in \mathbb{R}^{m \times n}$, then $\mathsf{vec}(X)$ is an nm-by-1 vector obtained by "stacking" X's columns:

$$\mathsf{vec}(X) = \left[\begin{array}{c} X(:,1) \\ \vdots \\ X(:,n) \end{array} \right].$$

If $B \in \mathbb{R}^{m_1 \times n_1}$, $C \in \mathbb{R}^{m_2 \times n_2}$, and $X \in \mathbb{R}^{n_1 \times m_2}$, then

$$Y = CXB^T \ \Leftrightarrow \ \mathsf{vec}(Y) = (B \otimes C)\mathsf{vec}(X). \tag{1.3.6}$$

Note that if $B, C, X \in \mathbb{R}^{n \times n}$, then $Y = CXB^T$ costs $O(n^3)$ to evaluate while the disregard of Kronecker structure in $y = (B \otimes C)x$ leads to an $O(n^4)$ calculation. This is why reshaping is central for effective Kronecker product computation. The reshape operator is handy in this regard. If $A \in \mathbb{R}^{m \times n}$ and $m_1 n_1 = mn$, then

$$B = \mathsf{reshape}(A, m_1, n_1)$$

is the m_1-by-n_1 matrix defined by $\mathsf{vec}(B) = \mathsf{vec}(A)$. Thus, if $A \in \mathbb{R}^{3 \times 4}$, then

$$\mathsf{reshape}(A, 2, 6) = \left[\begin{array}{cccccc} a_{11} & a_{31} & a_{22} & a_{13} & a_{33} & a_{24} \\ a_{21} & a_{12} & a_{32} & a_{23} & a_{14} & a_{34} \end{array} \right].$$

1.3.8 Multiple Kronecker Products

Note that $A = B \otimes C \otimes D$ can be regarded as a block matrix whose entries are block matrices. In particular, $b_{ij}c_{k\ell}D$ is the (k, ℓ) block of A's (i, j) block.

As an example of a multiple Kronecker product computation, let us consider the calculation of $y = (B \otimes C \otimes D)x$ where $B, C, D \in \mathbb{R}^{n \times n}$ and $x \in \mathbb{R}^N$ with $N = n^3$. Using (1.3.6) it follows that

$$\mathsf{reshape}(y, n^2, n) = (C \otimes D) \cdot \mathsf{reshape}(x, n^2, n) \cdot B^T.$$

Thus, if

$$F = \mathsf{reshape}(x, n^2, n) \cdot B^T,$$

then $G = (C \otimes D)F \in \mathbb{R}^{n^2 \times n}$ can computed column-by-column using (1.3.6):

$$G(:, k) = \mathsf{reshape}(D \cdot \mathsf{reshape}(F(:, k), n, n) \cdot C^T, n^2, 1) \qquad k = 1{:}n.$$

It follows that $y = \mathsf{reshape}(G, N, 1)$. A careful accounting reveals that $6n^4$ flops are required. Ordinarily, a matrix-vector product of this dimension would require $2n^6$ flops.

The Kronecker product has a prominent role to play in tensor computations and in §13.1 we detail more of its properties.

1.3.9   A Note on Complex Matrix Multiplication

Consider the complex matrix multiplication update

$$C_1 + iC_2 = (C_1 + iC_2) + (A_1 + iA_2)(B_1 + iB_2)$$

where all the matrices are real and $i^2 = -1$. Comparing the real and imaginary parts we conclude that

$$\begin{bmatrix} C_1 \\ C_2 \end{bmatrix} = \begin{bmatrix} C_1 \\ C_2 \end{bmatrix} + \begin{bmatrix} A_1 & -A_2 \\ A_2 & A_1 \end{bmatrix} \begin{bmatrix} B_1 \\ B_2 \end{bmatrix}.$$

Thus, complex matrix multiplication corresponds to a structured real matrix multiplication that has expanded dimension.

1.3.10   Hamiltonian and Symplectic Matrices

While on the topic of 2-by-2 block matrices, we identify two classes of structured matrices that arise at various points later on in the text. A matrix $M \in \mathbb{R}^{2n \times 2n}$ is a *Hamiltonian matrix* if it has the form

$$M = \begin{bmatrix} A & G \\ F & -A^T \end{bmatrix}$$

where $A, F, G \in \mathbb{R}^{n \times n}$ and F and G are symmetric. Hamiltonian matrices arise in optimal control and other application areas. An equivalent definition can be given in terms of the permutation matrix

$$J = \begin{bmatrix} 0 & I_n \\ -I_n & 0 \end{bmatrix}.$$

In particular, if

$$JMJ^T = -M^T,$$

then M is Hamiltonian. A related class of matrices are the symplectic matrices. A matrix $S \in \mathbb{R}^{2n \times 2n}$ is *symplectic* if

$$S^T J S = J.$$

If

$$S = \begin{bmatrix} S_{11} & S_{12} \\ S_{21} & S_{22} \end{bmatrix}$$

where the blocks are n-by-n, then it follows that both $S_{11}^T S_{21}$ and $S_{22}^T S_{12}$ are symmetric and $S_{11}^T S_{22} = I_n + S_{21}^T S_{12}$.

1.3.11 Strassen Matrix Multiplication

We conclude this section with a completely different approach to the matrix-matrix multiplication problem. The starting point in the discussion is the 2-by-2 block matrix product

$$\begin{bmatrix} C_{11} & C_{12} \\ C_{21} & C_{22} \end{bmatrix} = \begin{bmatrix} A_{11} & A_{12} \\ A_{21} & A_{22} \end{bmatrix} \begin{bmatrix} B_{11} & B_{12} \\ B_{21} & B_{22} \end{bmatrix}$$

where each block is square. In the ordinary algorithm, $C_{ij} = A_{i1}B_{1j} + A_{i2}B_{2j}$. There are 8 multiplies and 4 adds. Strassen (1969) has shown how to compute C with just 7 multiplies and 18 adds:

$$
\begin{aligned}
P_1 &= (A_{11} + A_{22})(B_{11} + B_{22}), \\
P_2 &= (A_{21} + A_{22})B_{11}, \\
P_3 &= A_{11}(B_{12} - B_{22}), \\
P_4 &= A_{22}(B_{21} - B_{11}), \\
P_5 &= (A_{11} + A_{12})B_{22}, \\
P_6 &= (A_{21} - A_{11})(B_{11} + B_{12}), \\
P_7 &= (A_{12} - A_{22})(B_{21} + B_{22}), \\
C_{11} &= P_1 + P_4 - P_5 + P_7, \\
C_{12} &= P_3 + P_5, \\
C_{21} &= P_2 + P_4, \\
C_{22} &= P_1 + P_3 - P_2 + P_6.
\end{aligned}
$$

These equations are easily confirmed by substitution. Suppose $n = 2m$ so that the blocks are m-by-m. Counting adds and multiplies in the computation $C = AB$, we find that conventional matrix multiplication involves $(2m)^3$ multiplies and $(2m)^3 - (2m)^2$ adds. In contrast, if Strassen's algorithm is applied *with conventional multiplication at the block level*, then $7m^3$ multiplies and $7m^3 + 11m^2$ adds are required. If $m \gg 1$, then the Strassen method involves about $7/8$ the arithmetic of the fully conventional algorithm.

Now recognize that we can recur on the Strassen idea. In particular, we can apply the Strassen algorithm to each of the half-sized block multiplications associated with the P_i. Thus, if the original A and B are n-by-n and $n = 2^q$, then we can repeatedly apply the Strassen multiplication algorithm. At the bottom "level," the blocks are 1-by-1.

Of course, there is no need to recur down to the $n = 1$ level. When the block size gets sufficiently small, ($n \le n_{\min}$), it may be sensible to use conventional matrix multiplication when finding the P_i. Here is the overall procedure:

Algorithm 1.3.1 (Strassen Matrix Multiplication) Suppose $n = 2^q$ and that $A \in \mathbb{R}^{n \times n}$ and $B \in \mathbb{R}^{n \times n}$. If $n_{\min} = 2^d$ with $d \le q$, then this algorithm computes $C = AB$ by applying Strassen procedure recursively.

> **function** $C = \mathsf{strass}(A, B, n, n_{\min})$
>> **if** $n \le n_{\min}$
>>> $C = AB$ $\qquad$ (conventionally computed)
>> **else**
>>> $m = n/2; \; u = 1{:}m; \; v = m + 1{:}n$
>>>
>>> $P_1 = \mathsf{strass}(A(u,u) + A(v,v), B(u,u) + B(v,v), m, n_{\min})$
>>>
>>> $P_2 = \mathsf{strass}(A(v,u) + A(v,v), B(u,u), m, n_{\min})$
>>>
>>> $P_3 = \mathsf{strass}(A(u,u), B(u,v) - B(v,v), m, n_{\min})$
>>>
>>> $P_4 = \mathsf{strass}(A(v,v), B(v,u) - B(u,u), m, n_{\min})$
>>>
>>> $P_5 = \mathsf{strass}(A(u,u) + A(u,v), B(v,v), m, n_{\min})$
>>>
>>> $P_6 = \mathsf{strass}(A(v,u) - A(u,u), B(u,u) + B(u,v), m, n_{\min})$
>>>
>>> $P_7 = \mathsf{strass}(A(u,v) - A(v,v), B(v,u) + B(v,v), m, n_{\min})$
>>>
>>> $C(u,u) = P_1 + P_4 - P_5 + P_7$
>>>
>>> $C(u,v) = P_3 + P_5$
>>>
>>> $C(v,u) = P_2 + P_4$
>>>
>>> $C(v,v) = P_1 + P_3 - P_2 + P_6$
>> **end**

Unlike any of our previous algorithms, strass is recursive. Divide and conquer algorithms are often best described in this fashion. We have presented strass in the style of a MATLAB function so that the recursive calls can be stated with precision.

The amount of arithmetic associated with strass is a complicated function of n and $n_{\min}$. If $n_{\min} \gg 1$, then it suffices to count multiplications as the number of additions is roughly the same. If we just count the multiplications, then it suffices to examine the deepest level of the recursion as that is where all the multiplications occur. In strass there are $q - d$ subdivisions and thus 7^{q-d} conventional matrix-matrix multiplications to perform. These multiplications have size $n_{\min}$ and thus strass involves about $s = (2^d)^3 7^{q-d}$ multiplications compared to $c = (2^q)^3$, the number of multiplications in the conventional approach. Notice that

$$\frac{s}{c} = \left(\frac{2^d}{2^q}\right)^3 7^{q-d} = \left(\frac{7}{8}\right)^{q-d}.$$

If $d = 0$, i.e., we recur on down to the 1-by-1 level, then

$$s = (7/8)^q c = 7^q = n^{\log_2 7} \approx n^{2.807}.$$

Thus, asymptotically, the number of multiplications in Strassen's method is $O(n^{2.807})$. However, the number of additions (relative to the number of multiplications) becomes significant as $n_{\min}$ gets small.

Problems

P1.3.1 Rigorously prove the following block matrix equation:

$$
\begin{bmatrix} A_{11} & \cdots & A_{1r} \\ \vdots & \ddots & \vdots \\ A_{q1} & \cdots & A_{qr} \end{bmatrix}^T = \begin{bmatrix} A_{11}^T & \cdots & A_{q1}^T \\ \vdots & \ddots & \vdots \\ A_{1r}^T & \cdots & A_{qr}^T \end{bmatrix}.
$$

P1.3.2 Suppose $M \in \mathbb{R}^{n \times n}$ is Hamiltonian. How many flops are required to compute $N = M^2$?

P1.3.3 What can you say about the 2-by-2 block structure of a matrix $A \in \mathbb{R}^{2n \times 2n}$ that satisfies $\mathcal{E}_{2n} A \mathcal{E}_{2n} = A^T$ where $\mathcal{E}_{2n}$ is the exchange permutation defined in §1.2.11. Explain why A is symmetric about the "antidiagonal" that extends from the $(2n, 1)$ entry to the $(1, 2n)$ entry.

P1.3.4 Suppose

$$
A = \begin{bmatrix} 0 & B \\ B^T & 0 \end{bmatrix}
$$

where $B \in \mathbb{R}^{n \times n}$ is upper bidiagonal. Describe the structure of $T = PAP^T$ where $P = \mathcal{P}_{2,n}$ is the perfect shuffle permutation defined in §1.2.11.

P1.3.5 Show that if B *and* C are each permutation matrices, then $B \otimes C$ is also a permutation matrix.

P1.3.6 Verify Equation (1.3.5).

P1.3.7 Verify that if $x \in \mathbb{R}^m$ and $y \in \mathbb{R}^n$, then $y \otimes x = \mathbf{vec}(xy^T)$.

P1.3.8 Show that if $B \in \mathbb{R}^{p \times p}$, $C \in \mathbb{R}^{q \times q}$, and

$$
x = \begin{bmatrix} x_1 \\ \vdots \\ x_p \end{bmatrix} \qquad x_i \in \mathbb{R}^q,
$$

then

$$
x^T (B \otimes C) x = \sum_{i=1}^{p} \sum_{j=1}^{p} b_{ij} \left(x_i^T C x_j \right).
$$

P1.3.9 Suppose $A^{(k)} \in \mathbb{R}^{n_k \times n_k}$ for $k = 1{:}r$ and that $x \in \mathbb{R}^n$ where $n = n_1 \cdots n_r$. Give an efficient algorithm for computing $y = \left(A^{(r)} \otimes \cdots \otimes A^{(2)} \otimes A^{(1)} \right) x$.

P1.3.10 Suppose n is even and define the following function from $\mathbb{R}^n$ to $\mathbb{R}$:

$$
f(x) = x(1{:}2{:}n)^T x(2{:}2{:}n) = \sum_{i=1}^{n/2} x_{2i-1} x_{2i}.
$$

(a) Show that if $x, y \in \mathbb{R}^n$ then

$$
x^T y = \sum_{i=1}^{n/2} (x_{2i-1} + y_{2i})(x_{2i} + y_{2i-1}) - f(x) - f(y).
$$

(b) Now consider the n-by-n matrix multiplication $C = AB$. Give an algorithm for computing this product that requires $n^3/2$ multiplies once f is applied to the rows of A and the columns of B. See Winograd (1968) for details.

P1.3.12 Adapt strass so that it can handle square matrix multiplication of any order. Hint: If the "current" A has odd dimension, append a zero row and column.

P1.3.13 Adapt strass so that it can handle nonsquare products, e.g., $C = AB$ where $A \in \mathbb{R}^{m \times r}$ and $B \in \mathbb{R}^{r \times n}$. Is it better to augment A and B with zeros so that they become square and equal in size or to "tile" A and B with square submatrices?

P1.3.14 Let W_n be the number of flops that strass requires to compute an n-by-n product where n is a power of 2. Note that $W_2 = 25$ and that for $n \geq 4$

$$
W_n = 7W_{n/2} + 18(n/2)^2
$$

Show that for every $\epsilon > 0$ there is a constant c_ϵ so $W_n \leq c_\epsilon n^{\omega+\epsilon}$ where $\omega = \log_2 7$ and n is any power of two.

P1.3.15 Suppose $B \in \mathbb{R}^{m_1 \times n_1}$, $C \in \mathbb{R}^{m_2 \times n_2}$, and $D \in \mathbb{R}^{m_3 \times n_3}$. Show how to compute the vector $y = (B \otimes C \otimes D)x$ where $x \in \mathbb{R}^n$ and $n = n_1 n_2 n_3$ is given. Is the order of operations important from the flop point of view?

Notes and References for §1.3

Useful references for the Kronecker product include Horn and Johnson (TMA, Chap. 4), Van Loan (FFT), and:

C.F. Van Loan (2000). "The Ubiquitous Kronecker Product," *J. Comput. Appl. Math., 123*, 85–100.

For quite some time fast methods for matrix multiplication have attracted a lot of attention within computer science, see:

S. Winograd (1968). "A New Algorithm for Inner Product," *IEEE Trans. Comput. C-17*, 693–694.
V. Strassen (1969). "Gaussian Elimination is not Optimal," *Numer. Math. 13*, 354–356.
V. Pan (1984). "How Can We Speed Up Matrix Multiplication?," *SIAM Review 26*, 393–416.
I. Kaporin (1999). "A Practical Algorithm for Faster Matrix Multiplication," *Num. Lin. Alg. 6*, 687–700.
H. Cohn, R. Kleinberg, B. Szegedy, and C. Umans (2005). "Group-theoretic Algorithms for Matrix Multiplication," *Proceeedings of the 2005 Conference on the Foundations of Computer Science (FOCS)*, 379–388.
J. Demmel, I. Dumitriu, O. Holtz, and R. Kleinberg (2007). "Fast Matrix Multiplication is Stable," *Numer. Math. 106*, 199–224.
P. D'Alberto and A. Nicolau (2009). "Adaptive Winograd's Matrix Multiplication," *ACM Trans. Math. Softw. 36*, Article 3.

At first glance, many of these methods do not appear to have practical value. However, this has proven not to be the case, see:

D. Bailey (1988). "Extra High Speed Matrix Multiplication on the Cray-2," *SIAM J. Sci. Stat. Comput. 9*, 603–607.
N.J. Higham (1990). "Exploiting Fast Matrix Multiplication within the Level 3 BLAS," *ACM Trans. Math. Softw. 16*, 352–368.
C.C. Douglas, M. Heroux, G. Slishman, and R.M. Smith (1994). "GEMMW: A Portable Level 3 BLAS Winograd Variant of Strassen's Matrix-Matrix Multiply Algorithm," *J. Comput. Phys. 110*, 1–10.

Strassen's algorithm marked the beginning of a search for the fastest possible matrix multiplication algorithm from the complexity point of view. The *exponent of matrix multiplication* is the smallest number ω such that, for all $\epsilon > 0$, $O(n^{\omega+\epsilon})$ work suffices. The best known value of ω has decreased over the years and is currently around 2.4. It is interesting to speculate on the existence of an $O(n^{2+\epsilon})$ procedure.

1.4   Fast Matrix-Vector Products

In this section we refine our ability to think at the block level by examining some matrix-vector products $y = Ax$ in which the n-by-n matrix A is so highly structured that the computation can be carried out with many fewer than the usual $O(n^2)$ flops. These results are used in §4.8.

1.4.1   The Fast Fourier Transform

The *discrete Fourier transform* (DFT) of a vector $x \in \mathbb{C}^n$ is a matrix-vector product

$$y = F_n x$$

where the *DFT matrix* $F_n = (f_{kj}) \in \mathbb{C}^{n \times n}$ is defined by

$$f_{kj} = \omega_n^{(k-1)(j-1)} \tag{1.4.1}$$

with

$$\omega_n = \exp(-2\pi i/n) = \cos(2\pi/n) - i \cdot \sin(2\pi/n). \tag{1.4.2}$$

Here is an example:

$$F_4 = \begin{bmatrix} 1 & 1 & 1 & 1 \\ 1 & \omega_4 & \omega_4^2 & \omega_4^3 \\ 1 & \omega_4^2 & \omega_4^4 & \omega_4^6 \\ 1 & \omega_4^3 & \omega_4^6 & \omega_4^9 \end{bmatrix} = \begin{bmatrix} 1 & 1 & 1 & 1 \\ 1 & -i & -1 & i \\ 1 & -1 & 1 & -1 \\ 1 & i & -1 & -i \end{bmatrix}.$$

The DFT is ubiquitous throughout computational science and engineering and one reason has to do with the following property:

> If n is highly composite, then it is possible to carry out the DFT in many fewer than the $O(n^2)$ flops required by conventional matrix-vector multiplication.

To illustrate this we set $n = 2^t$ and proceed to develop the *radix-2 fast Fourier transform*.

The starting point is to examine the block structure of an even-order DFT matrix after its columns are reordered so that the odd-indexed columns come first. Consider the case

$$F_8 = \begin{bmatrix} 1 & 1 & 1 & 1 & 1 & 1 & 1 & 1 \\ 1 & \omega & \omega^2 & \omega^3 & \omega^4 & \omega^5 & \omega^6 & \omega^7 \\ 1 & \omega^2 & \omega^4 & \omega^6 & 1 & \omega^2 & \omega^4 & \omega^6 \\ 1 & \omega^3 & \omega^6 & \omega & \omega^4 & \omega^7 & \omega^2 & \omega^5 \\ 1 & \omega^4 & 1 & \omega^4 & 1 & \omega^4 & 1 & \omega^4 \\ 1 & \omega^5 & \omega^2 & \omega^7 & \omega^4 & \omega & \omega^6 & \omega^3 \\ 1 & \omega^6 & \omega^4 & \omega^2 & 1 & \omega^6 & \omega^4 & \omega^2 \\ 1 & \omega^7 & \omega^6 & \omega^5 & \omega^4 & \omega^3 & \omega^2 & \omega \end{bmatrix} \qquad (\omega = \omega_8).$$

(Note that ω_8 is a root of unity so that high powers simplify, e.g., $[F_8]_{4,7} = \omega^{3 \cdot 6} = \omega^{18} = \omega^2$.) If $cols = [1\ 3\ 5\ 7\ 2\ 4\ 6\ 8]$, then

$$F_8(:, cols) = \left[\begin{array}{cccc|cccc} 1 & 1 & 1 & 1 & 1 & 1 & 1 & 1 \\ 1 & \omega^2 & \omega^4 & \omega^6 & \omega & \omega^3 & \omega^5 & \omega^7 \\ 1 & \omega^4 & 1 & \omega^4 & \omega^2 & \omega^6 & \omega^2 & \omega^6 \\ 1 & \omega^6 & \omega^4 & \omega^2 & \omega^3 & \omega & \omega^7 & \omega^5 \\ \hline 1 & 1 & 1 & 1 & -1 & -1 & -1 & -1 \\ 1 & \omega^2 & \omega^4 & \omega^6 & -\omega & -\omega^3 & -\omega^5 & -\omega^7 \\ 1 & \omega^4 & 1 & \omega^4 & -\omega^2 & -\omega^6 & -\omega^2 & -\omega^6 \\ 1 & \omega^6 & \omega^4 & \omega^2 & -\omega^3 & -\omega & -\omega^7 & -\omega^5 \end{array} \right].$$

The lines through the matrix are there to help us think of $F_8(:, cols)$ as a 2-by-2 matrix with 4-by-4 blocks. Noting that $\omega^2 = \omega_8^2 = \omega_4$, we see that

$$F_8(:, cols) = \left[\begin{array}{c|c} F_4 & \Omega_4 F_4 \\ \hline F_4 & -\Omega_4 F_4 \end{array} \right].$$

where $\Omega_4 = \text{diag}(1, \omega_8, \omega_8^2, \omega_8^3)$. It follows that if $x \in \mathbb{R}^8$, then

$$F_8 x = F_8(:, cols) \cdot x(cols) = \left[\begin{array}{c|c} F_4 & \Omega_4 F_4 \\ \hline F_4 & -\Omega_4 F_4 \end{array} \right] \left[\begin{array}{c} x(1{:}2{:}8) \\ x(2{:}2{:}8) \end{array} \right] = \left[\begin{array}{c|c} I_4 & \Omega_4 \\ \hline I_4 & -\Omega_4 \end{array} \right] \left[\begin{array}{c} F_4 x(1{:}2{:}8) \\ F_4 x(2{:}2{:}8) \end{array} \right].$$

Thus, by simple scalings we can obtain the 8-point DFT $y = F_8 x$ from the 4-point DFTs $y_T = F_4 \cdot x(1{:}2{:}8)$ and $y_B = F_4 \cdot x(2{:}2{:}8)$. In particular,

$$y(1{:}4) = y_T + d .* y_B,$$
$$y(5{:}8) = y_T - d .* y_B$$

where

$$d = \left[\begin{array}{c} 1 \\ \omega \\ \omega^2 \\ \omega^3 \end{array} \right].$$

More generally, if $n = 2m$, then $y = F_n x$ is given by

$$y(1{:}m) = y_T + d .* y_B,$$
$$y(m+1{:}n) = y_T - d .* y_B$$

where $d = \left[\begin{array}{cccc} 1, & \omega_n, & \ldots, & \omega_n^{m-1} \end{array} \right]^T$ and

$$y_T = F_m x(1{:}2{:}n),$$
$$y_B = F_m x(2{:}2{:}n).$$

For $n = 2^t$, we can recur on this process until $n = 1$, noting that $F_1 x = x$.

Algorithm 1.4.1 If $x \in \mathbb{C}^n$ and $n = 2^t$, then this algorithm computes the discrete Fourier transform $y = F_n x$.

> **function** $y = \text{fft}(x, n)$
>   **if** $n = 1$
> $y = x$
>   **else**
> $m = n/2$
> $y_T = \text{fft}(x(1{:}2{:}n), m)$
> $y_B = \text{fft}(x(2{:}2{:}n), m)$
> $\omega = \exp(-2\pi i/n)$
> $d = \left[\begin{array}{cccc} 1, & \omega, & \cdots, & \omega^{m-1} \end{array} \right]^T$
> $z = d .* y_B$
> $y = \left[\begin{array}{c} y_T + z \\ y_T - z \end{array} \right]$
>   **end**

The flop analysis of fft requires an assessment of complex arithmetic and the solution of an interesting recursion. We first observe that the multiplication of two complex numbers involves six (real) flops while the addition of two complex numbers involves two flops. Let f_n be the number of flops that fft needs to produce the DFT of $x \in \mathbb{C}^n$. Scrutiny of the method reveals that

$$
\left.\begin{array}{c} y_T \\ y_B \\ d \\ z \\ y \end{array}\right\} \text{ requires } \left\{\begin{array}{c} f_m \text{ flops} \\ f_m \text{ flops} \\ 6m \text{ flops} \\ 6m \text{ flops} \\ 2n \text{ flops} \end{array}\right.
$$

where $n = 2m$. Thus,

$$
f_n = 2f_m + 8n \qquad (f_1 = 0).
$$

Conjecturing that $f_n = c \cdot n \log_2(n)$ for some constant c, it follows that

$$
f_n = c \cdot n \log_2(n) = 2c \cdot m \log_2(m) + 8n = c \cdot n(\log_2(n) - 1) + 8n,
$$

from which we conclude that $c = 8$. Thus, fft requires $8n \log_2(n)$ flops. Appreciate the speedup over conventional matrix-vector multiplication. If $n = 2^{20}$, it is a factor of about 10,000. We mention that the fft flop count can be reduced to $5n \log_2(n)$ by precomputing $\omega_n, \ldots, \omega_n^{n/2-1}$. See P1.4.1.

1.4.2   Fast Sine and Cosine Transformations

In the *discrete sine transform* (DST) problem, we are given real values $x_1, \ldots, x_{m-1}$ and compute

$$
y_k = \sum_{j=1}^{m-1} \sin\left(\frac{kj\pi}{m}\right) x_j \tag{1.4.3}
$$

for $k = 1{:}m - 1$. In the *discrete cosine transform* (DCT) problem, we are given real values $x_0, x_1, \ldots, x_m$ and compute

$$
y_k = \frac{x_0}{2} + \sum_{j=1}^{m-1} \cos\left(\frac{kj\pi}{m}\right) x_j + \frac{(-1)^k x_m}{2} \tag{1.4.4}
$$

for $k = 0{:}m$. Note that the sine and cosine evaluations "show up" in the DFT matrix. Indeed, for $k = 0{:}2m - 1$ and $j = 0{:}2m - 1$ we have

$$
[F_{2m}]_{k+1,j+1} = \omega_{2m}^{kj} = \cos\left(\frac{kj\pi}{m}\right) - i \sin\left(\frac{kj\pi}{m}\right). \tag{1.4.5}
$$

This suggests (correctly) that there is an exploitable connection between each of these trigonometric transforms and the DFT. The key observation is to block properly the real and imaginary parts of F_{2m}. To that end, define the matrices $S_r \in \mathbb{R}^{r \times r}$ and $C_r \in \mathbb{R}^{r \times r}$ by

$$
\begin{aligned}
[S_r]_{kj} &= \sin\left(\frac{kj\pi}{r+1}\right), \\
&\qquad\qquad\qquad\qquad k = 1{:}r, \ j = 1{:}r. \\
[C_r]_{kj} &= \cos\left(\frac{kj\pi}{r+1}\right),
\end{aligned} \tag{1.4.6}
$$

Recalling from §1.2.11 the definition of the exchange permutation $\mathcal{E}_n$, we have

Theorem 1.4.1. *Let m be a positive integer and define the vectors e and v by*

$$e^T = (\underbrace{1, 1, \ldots, 1}_{m-1}), \qquad v^T = (\underbrace{-1, 1, \ldots, (-1)^{m-1}}_{m-1}).$$

If $E = \mathcal{E}_{m-1}$, $C = C_{m-1}$, and $S = S_{m-1}$, then

$$F_{2m} = \begin{bmatrix} 1 & e^T & 1 & e^T \\ e & C - iS & v & (C + iS)E \\ 1 & v^T & (-1)^m & v^T E \\ e & E(C + iS) & Ev & E(C - iS)E \end{bmatrix}. \tag{1.4.7}$$

Proof. It is clear from (1.4.5) that $F_{2m}(:, 1)$, $F_{2m}(1, :)$, $F_{2m}(:, m+1)$, and $F_{2m}(m+1, :)$ are correctly specified. It remains for us to show that equation (1.4.7) holds in blocks positions (2,2), (2,4), (4,2), and (4,4). The (2,2) verification is straightforward:

$$[F_{2m}(2{:}m, 2{:}m)]_{kj} = \cos\left(\frac{kj\pi}{m}\right) - i\sin\left(\frac{kj\pi}{m}\right)$$
$$= [C - iS]_{kj}.$$

A little trigonometry is required to verify correctness in the (2,4) position:

$$[F_{2m}(2{:}m, m+2{:}2m)]_{kj} = \cos\left(\frac{k(m+j)\pi}{m}\right) - i\sin\left(\frac{k(m+j)\pi}{m}\right)$$

$$= \cos\left(\frac{kj\pi}{m} + k\pi\right) - i\sin\left(\frac{kj\pi}{m} + k\pi\right)$$

$$= \cos\left(-\frac{kj\pi}{m} + k\pi\right) + i\sin\left(-\frac{kj\pi}{m} + k\pi\right)$$

$$= \cos\left(\frac{(k(m-j)\pi}{m}\right) + i\sin\left(\frac{k(m-j)\pi}{m}\right)$$

$$= [(C + iS)E]_{kj}.$$

We used the fact that post-multiplying a matrix by the permutation $E = \mathcal{E}_{m-1}$ has the effect of reversing the order of its columns. The recipes for $F_{2m}(m+2{:}2m, 2{:}m)$ and $F_{2m}(m+2{:}2m, m+2{:}2m)$ are derived similarly. $\square$

Using the notation of the theorem, we see that the sine transform (1.4.3) is a matrix-vector product

$$y(1{:}m{-}1) = \text{DST}(m{-}1) \cdot x(1{:}m{-}1)$$

where
$$\text{DST}(m-1) = S_{m-1}. \tag{1.4.8}$$

If $x = x(1:m-1)$ and

$$x_{\sin} = \begin{bmatrix} 0 \\ x \\ 0 \\ -Ex \end{bmatrix} \in \mathbb{R}^{2m}, \tag{1.4.9}$$

then since $e^T E = e$ and $E^2 = E$ we have

$$\frac{i}{2}F_{2m}x_{\sin} = \frac{i}{2}\begin{bmatrix} 1 & e^T & 1 & e^T \\ e & C-iS & v & (C+iS)E \\ 1 & v^T & (-1)^m & v^T E \\ e & E(C+iS) & Ev & E(C-iS)E \end{bmatrix}\begin{bmatrix} 0 \\ x \\ 0 \\ -Ex \end{bmatrix}$$

$$= \frac{i}{2}\begin{bmatrix} e^T x - e^T E x \\ -2iSx \\ v^T x - v^T E^2 x \\ i(ESx + ESE^2 x) \end{bmatrix} = \begin{bmatrix} 0 \\ Sx \\ 0 \\ -ESx \end{bmatrix}.$$

Thus, the DST of $x(1:m-1)$ is a scaled subvector of $F_{2m}x_{\sin}$.

Algorithm 1.4.2 The following algorithm assigns the DST of $x_1,\ldots,x_{m-1}$ to y.

Set up the vector $x_{\sin}$ defined by (1.4.9).

Use fft (e.g., Algorithm 1.4.1) to compute $\tilde{y} = F_{2m}x_{\sin}$

$y = i \cdot \tilde{y}(2:m)/2$

This computation involves $O(m\log_2(m))$ flops. We mention that the vector $x_{\sin}$ is real and highly structured, something that would be exploited in a truly efficient implementation.

Now let us consider the discrete cosine transform defined by (1.4.4). Using the notation from Theorem 1.4.1, the DCT is a matrix-vector product

$$y(0:m) = \text{DCT}(m+1) \cdot x(0:m)$$

where

$$\text{DCT}(m+1) = \begin{bmatrix} 1/2 & e^T & 1/2 \\ e/2 & C_{m-1} & v/2 \\ 1/2 & v^T & (-1)^m/2 \end{bmatrix} \tag{1.4.10}$$

If $\tilde{x} = x(1:m-1)$ and

$$x_{\cos} = \begin{bmatrix} x_0 \\ \tilde{x} \\ x_m \\ E\tilde{x} \end{bmatrix} \in \mathbb{R}^{2m}, \tag{1.4.11}$$

then

$$
\frac{1}{2}F_{2m}x_{\cos} = \frac{1}{2}
\begin{bmatrix}
1 & e^T & 1 & e^T \\
e & C - iS & v & (C + iS)E \\
1 & v^T & (-1)^m & v^T E \\
e & E(C + iS) & Ev & E(C - iS)E
\end{bmatrix}
\begin{bmatrix}
x_0 \\
\tilde{x} \\
x_m \\
E\tilde{x}
\end{bmatrix}
$$

$$
=
\begin{bmatrix}
(x_0/2) & + & e^T\tilde{x} & + & (x_m/2) \\
(x_0/2)e & + & C\tilde{x} & + & (x_m/2)v \\
(x_0/2) & + & v^T\tilde{x} & + & (-1)^m(x_m/2) \\
(x_0/2)e & + & EC\tilde{x} & + & (x_m/2)Ev
\end{bmatrix}.
$$

Notice that the top three components of this block vector define the DCT of $x(0{:}m)$. Thus, the DCT is a scaled subvector of $F_{2m}x_{\cos}$.

Algorithm 1.4.3 The following algorithm assigns to $y \in \mathbb{R}^{m+1}$ the DCT of $x_0, \ldots, x_m$.

Set up the vector $x_{\cos} \in \mathbb{R}^{2m}$ defined by (1.4.11).

Use fft (e.g., Algorithm 1.4.1) to compute $\tilde{y} = F_{2m}x_{\cos}$

$y = \tilde{y}(1{:}m+1)/2$

This algorithm requires $O(m \log m)$ flops, but as with Algorithm 1.4.2, it can be more efficiently implemented by exploiting symmetries in the vector x_{cos}.

We mention that there are important variants of the DST and the DCT that can be computed fast:

$$
\text{DST-II:} \quad y_k = \sum_{j=1}^{m} \sin\left(\frac{k(2j-1)\pi}{2m}\right) x_j, \qquad k = 1{:}m,
$$

$$
\text{DST-III:} \quad y_k = \sum_{j=1}^{m} \sin\left(\frac{(2k-1)j\pi}{2m}\right) x_j, \qquad k = 1{:}m,
$$

$$
\text{DST-IV:} \quad y_k = \sum_{j=1}^{m} \sin\left(\frac{(2k-1)(2j-1)\pi}{2m}\right) x_j, \qquad k = 1{:}m,
$$

$$
\text{DCT-II:} \quad y_k = \sum_{j=0}^{m-1} \cos\left(\frac{k(2j-1)\pi}{2m}\right) x_j, \qquad k = 0{:}m-1,
$$

(1.4.12)

$$
\text{DCT-III:} \quad y_k = \frac{x_0}{2} = \sum_{j=1}^{m-1} \cos\left(\frac{(2k-1)j\pi}{2m}\right) x_j, \qquad k = 0{:}m-1,
$$

$$
\text{DCT-IV:} \quad y_k = \sum_{j=0}^{m-1} \cos\left(\frac{(2k-1)(2j-1)\pi}{2m}\right) x_j, \qquad k = 0{:}m-1.
$$

For example, if $\tilde{y} \in \mathbb{R}^{2m-1}$ is the DST of $\tilde{x} = [\, x_1, 0, x_2, 0, \ldots, 0, x_{m-1}, x_m \,]^T$, then $\tilde{y}(1{:}m)$ is the DST-II of $x \in \mathbb{R}^m$. See Van Loan (FFT) for further details.

1.4.3   The Haar Wavelet Transform

If $n = 2^t$, then the *Haar wavelet transform* $y = W_n x$ is a matrix-vector product in which the transform matrix $W_n \in \mathbb{R}^{n \times n}$ is defined recursively:

$$
W_n = \begin{cases} \left[\; W_m \otimes \begin{pmatrix} 1 \\ 1 \end{pmatrix} \; \middle| \; I_m \otimes \begin{pmatrix} 1 \\ -1 \end{pmatrix} \; \right] & \text{if } n = 2m, \\[2ex] [\,1\,] & \text{if } n = 1. \end{cases}
$$

Here are some examples:

$$
W_2 = \left[\begin{array}{c|c} 1 & 1 \\ \hline 1 & -1 \end{array} \right],
$$

$$
W_4 = \left[\begin{array}{cc|cc} 1 & 1 & 1 & 0 \\ 1 & 1 & -1 & 0 \\ \hline 1 & -1 & 0 & 1 \\ 1 & -1 & 0 & -1 \end{array} \right],
$$

$$
W_8 = \left[\begin{array}{cccc|cccc} 1 & 1 & 1 & 0 & 1 & 0 & 0 & 0 \\ 1 & 1 & 1 & 0 & -1 & 0 & 0 & 0 \\ 1 & 1 & -1 & 0 & 0 & 1 & 0 & 0 \\ 1 & 1 & -1 & 0 & 0 & -1 & 0 & 0 \\ \hline 1 & -1 & 0 & 1 & 0 & 0 & 1 & 0 \\ 1 & -1 & 0 & 1 & 0 & 0 & -1 & 0 \\ 1 & -1 & 0 & -1 & 0 & 0 & 0 & 1 \\ 1 & -1 & 0 & -1 & 0 & 0 & 0 & -1 \end{array} \right].
$$

An interesting block pattern emerges if we reorder the rows of W_n so that the odd-indexed rows come first:

$$
\mathcal{P}_{2,m}^T W_n = \begin{bmatrix} W_m & I_m \\ W_m & -I_m \end{bmatrix} = (W_2 \otimes I_m) \begin{bmatrix} W_m & 0 \\ 0 & I_m \end{bmatrix}. \tag{1.4.13}
$$

Thus, if $x \in \mathbb{R}^n$, $x_T = x(1{:}m)$, and $x_B = x(m+1{:}n)$, then

$$
y = W_n x = \mathcal{P}_{2,m} \begin{bmatrix} I_m & I_m \\ I_m & -I_m \end{bmatrix} \begin{bmatrix} W_m & 0 \\ 0 & I_m \end{bmatrix} \begin{bmatrix} x_T \\ x_B \end{bmatrix}
$$

$$
= \mathcal{P}_{2,m} \begin{bmatrix} W_m x_T + x_B \\ W_m x_T - x_B \end{bmatrix}.
$$

In other words,

$$
y(1{:}2{:}n) = W_m x_T + x_B, \qquad y(2{:}2{:}n) = W_m x_T - x_B.
$$

This points the way to a fast recursive procedure for computing $y = W_n x$.

Algorithm 1.4.4 (Haar Wavelet Transform) If $x \in \mathbb{R}^n$ and $n = 2^t$, then this algorithm computes the Haar transform $y = W_n x$.

> **function** $y = \text{fht}(x, n)$
> > **if** $n = 1$
> > > $y = x$
> >
> > **else**
> > > $m = n/2$
> > > $z = \text{fht}(x(1{:}m), m)$
> > > $y(1{:}2{:}m) = z + x(m + 1{:}n)$
> > > $y(2{:}2{:}m) = z - x(m + 1{:}n)$
> >
> > **end**

It can be shown that this algorithm requires $2n$ flops.

Problems

P1.4.1 Suppose $w = \left[1, \, \omega_n, \, \omega_n^2, \, \ldots, \, \omega_n^{n/2-1}\right]$ where $n = 2^t$. Using the colon notation, express

$$\left[1, \, \omega_r, \, \omega_r^2, \, \ldots, \, \omega_r^{r/2-1}\right]$$

as a subvector of w where $r = 2^q$, $q = 1{:}t$. Rewrite Algorithm 1.4.1 with the assumption that w is precomputed. Show that this maneuver reduces the flop count to $5n \log_2 n$.

P1.4.2 Suppose $n = 3m$ and examine

$$G = \left[\, F_n(:, 1{:}3{:}n - 1) \mid F_n(:, 2{:}3{:}n - 1) \mid F_n(:, 3{:}3{:}n - 1) \, \right]$$

as a 3-by-3 block matrix, looking for scaled copies of F_m. Based on what you find, develop a recursive radix-3 FFT analogous to the radix-2 implementation in the text.

P1.4.3 If $n = 2^t$, then it can be shown that $F_n = (A_t \Gamma_t) \cdots (A_1 \Gamma_1)$ where for $q = 1{:}t$

$$L_q = 2^q, \quad r_q = n/L_q,$$

$$A_q = I_{r_q} \otimes \begin{bmatrix} I_{L_{q-1}} & \Omega_q \\ I_{L_{q-1}} & -\Omega_q \end{bmatrix},$$

$$\Gamma_q = \mathcal{P}_{2, r_q} \otimes I_{L_{q-1}},$$

$$\Omega_q = \text{diag}(1, \omega_{L_q}, \ldots, \omega_{L_q}^{L_{q-1}-1}).$$

Note that with this factorization, the DFT $y = F_n x$ can be computed as follows:

> $y = x$
> **for** $q = 1{:}t$
> > $y = A_q(\Gamma_q y)$
>
> **end**

Fill in the details associated with the y updates and show that a careful implementation requires $5n \log_2(n)$ flops.

P1.4.4 What fraction of the components of W_n are zero?

P1.4.5 Using (1.4.13), verify by induction that if $n = 2^t$, then the Haar tranform matrix W_n has the factorization $W_n = H_t \cdots H_1$ where

$$H_q = \begin{bmatrix} \mathcal{P}_{2, L_*} & 0 \\ 0 & I_{n-L} \end{bmatrix} \begin{bmatrix} W_2 \otimes I_{L_*} & 0 \\ 0 & I_{n-L} \end{bmatrix} \qquad L = 2^q, \, L_* = L/2.$$

Thus, the computation of $y = W_n x$ may proceed as follows:

$$y = x$$
$$\textbf{for } q = 1{:}t$$
$$\qquad y = H_q y$$
$$\textbf{end}$$

Fill in the details associated with the update $y = H_q y$ and confirm that $W_n x$ costs $2n$ flops.

P1.4.6   Using (1.4.13), develop an $O(n)$ procedure for solving $W_n y = x$ where $x \in \mathbb{R}^n$ is given and $n = 2^t$.

Notes and References for §1.4

In Van Loan (FFT) the FFT family of algorithms is described in the language of matrix-factorizations. A discussion of various fast trigonometric transforms is also included. See also:

W.L. Briggs and V.E. Henson (1995). *The DFT: An Owners' Manual for the Discrete Fourier Transform*, SIAM Publications, Philadelphia, PA.

The design of a high-performance FFT is a nontrivial task. An important development in this regard is a software tool known as "the fastest Fourier transform in the west":

M. Frigo and S.G. Johnson (2005). "The Design and Implementation of FFTW3", *Proceedings of the IEEE, 93*, 216–231.

It automates the search for the "right" FFT given the underlying computer architecture. FFT references that feature interesting factorization and approximation ideas include:

A. Edelman, P. McCorquodale, and S. Toledo (1998). "The Future Fast Fourier Transform?," *SIAM J. Sci. Comput. 20*, 1094–1114.
A. Dutt and and V. Rokhlin (1993). "Fast Fourier Transforms for Nonequally Spaced Data," *SIAM J. Sci. Comput. 14*, 1368–1393.
A. F. Ware (1998). "Fast Approximate Fourier Transforms for Irregularly Spaced Data," *SIAM Review 40*, 838 –856.
N. Nguyen and Q.H. Liu (1999). "The Regular Fourier Matrices and Nonuniform Fast Fourier Transforms," *SIAM J. Sci. Comput. 21*, 283–293.
A. Nieslony and G. Steidl (2003). "Approximate Factorizations of Fourier Matrices with Nonequispaced Knots," *Lin. Alg. Applic. 366*, 337–351.
L. Greengard and J.–Y. Lee (2004). "Accelerating the Nonuniform Fast Fourier Transform," *SIAM Review 46*, 443–454.
K. Ahlander and H. Munthe-Kaas (2005). "Applications of the Generalized Fourier Transform in Numerical Linear Algebra," *BIT 45*, 819–850.

The fast multipole method and the fast Gauss transform represent another type of fast transform that is based on a combination of clever blocking and approximation.

L. Greengard and V. Rokhlin (1987). "A Fast Algorithm for Particle Simulation," *J. Comput. Phys. 73*, 325–348.
X. Sun and N.P. Pitsianis (2001). "A Matrix Version of the Fast Multipole Method," *SIAM Review 43*, 289–300.
L. Greengard and J. Strain (1991). "The Fast Gauss Transform," *SIAM J. Sci. Stat. Comput. 12*, 79–94.
M. Spivak, S.K. Veerapaneni, and L. Greengard (2010). "The Fast Generalized Gauss Transform," *SIAM J. Sci. Comput. 32*, 3092–3107.
X. Sun and Y. Bao (2003). "A Kronecker Product Representation of the Fast Gauss Transform," *SIAM J. Matrix Anal. Applic. 24*, 768–786.

The Haar transform is a simple example of a wavelet transform. The wavelet idea has had a profound impact throughout computational science and engineering. In many applications, wavelet basis functions work better than the sines and cosines that underly the DFT. Excellent monographs on this subject include

I Daubechies (1992). *Ten Lectures on Wavelets*, SIAM Publications, Philadelphia, PA.
G. Strang (1993). "Wavelet Transforms Versus Fourier Transforms," *Bull. AMS 28*, 288–305.
G. Strang and T. Nguyan (1996). *Wavelets and Filter Banks*, Wellesley-Cambridge Press.

1.5 Vectorization and Locality

When it comes to designing a high-performance matrix computation, it is not enough simply to minimize flops. Attention must be paid to how the arithmetic units interact with the underlying memory system. Data structures are an important part of the picture because not all matrix layouts are "architecture friendly." Our aim is to build a practical appreciation for these issues by presenting various simplified models of execution. These models are *qualitative* and are just informative pointers to complex implementation issues.

1.5.1 Vector Processing

An individual floating point operation typically requires several cycles to complete. A 3-cycle addition is depicted in Figure 1.5.1. The input scalars x and y proceed along

Figure 1.5.1. *A 3-Cycle adder*

a computational "assembly line," spending one cycle at each of three work "stations." The sum z emerges after three cycles. Note that, during the execution of a single, "free standing" addition, only one of the three stations would be active at any particular instant.

Vector processors exploit the fact that a vector operation is a very regular sequence of scalar operations. The key idea is *pipelining*, which we illustrate using the vector addition computation $z = x + y$. With pipelining, the x and y vectors are streamed through the addition unit. Once the pipeline is filled and steady state reached, a z-vector component is produced every cycle, as shown in Figure 1.5.2. In

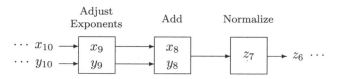

Figure 1.5.2. *Pipelined addition*

this case, we would anticipate vector processing to proceed at about three times the rate of scalar processing.

A vector processor comes with a repertoire of *vector instructions*, such as vector add, vector multiply, vector scale, dot product, and saxpy. These operations take place in *vector registers* with input and output handled by *vector load* and *vector store* instructions. An important attribute of a vector processor is the length v_L of the vector registers that carry out the vector operations. A length-n vector operation must be broken down into subvector operations of length v_L or less. Here is how such a partitioning might be managed for a vector addition $z = x + y$ where x and y are n-vectors:

$first = 1$

while $first \leq n$

$\qquad last = \min\{n, first + v_L - 1\}$

$\qquad$ Vector load: $r_1 \leftarrow x(first{:}last)$

$\qquad$ Vector load: $r_2 \leftarrow y(first{:}last)$ (1.5.1)

$\qquad$ Vector add: $r_1 = r_1 + r_2$

$\qquad$ Vector store: $z(first{:}last) \leftarrow r_1$

$\qquad first = last + 1$

end

The vector addition is a register-register operation while the "flopless" movement of data to and from the vector registers is identified with the left arrow "$\leftarrow$". Let us model the number of cycles required to carry out the various steps in (1.5.1). For clarity, assume that n is very large and an integral multiple of v_L, thereby making it safe to ignore the final cleanup pass through the loop.

Regarding the vectorized addition $r_1 = r_1 + r_2$, assume it takes τ_{add} cycles to fill the pipeline and that once this happens, a component of z is produced each cycle. It follows that

$$N_{\text{arith}} = \left(\frac{n}{v_L}\right)(\tau_{\text{add}} + v_L) = \left(\frac{\tau_{\text{add}}}{v_L} + 1\right) n$$

accounts for the total number cycles that (1.5.1) requires for arithmetic.

For the vector loads and stores, assume that $\tau_{\text{data}} + v_L$ cycles are required to transport a length-v_L vector from memory to a register or from a register to memory, where τ_{data} is the number of cycles required to fill the data pipeline. With these assumptions we see that

$$N_{\text{data}} = 3\left(\frac{n}{v_L}\right)(\tau_{\text{data}} + v_L) = 3\left(\frac{\tau_{\text{add}}}{v_L} + 1\right) n$$

specifies the number of cycles that are required by (1.5.1) to get data to and from the registers.

The arithmetic-to-data-motion ratio

$$N_{\text{arith}}/N_{\text{data}} = \frac{\tau_{\text{add}} + v_L}{3(\tau_{\text{data}} + v_L)}$$

and the total cycles sum

$$N_{\text{arith}} + N_{\text{data}} = \left(\frac{\tau_{\text{arith}} + 3\tau_{\text{data}}}{v_L} + 4\right) n$$

are illuminating statistics, but they are not necessarily good predictors of performance. In practice, vector loads, stores, and arithmetic are "overlapped" through the chaining together of various pipelines, a feature that is not captured by our model. Nevertheless, our simple analysis is a preliminary reminder that data motion is an important factor when reasoning about performance.

1.5.2   Gaxpy versus Outer Product

Two algorithms that involve the same number of flops can have substantially different data motion properties. Consider the n-by-n gaxpy

$$y = y + Ax$$

and the n-by-n outer product update

$$A = A + yx^T.$$

Both of these level-2 operations involve $2n^2$ flops. However, if we assume (for clarity) that $n = v_L$, then we see that the gaxpy computation

$$r_x \leftarrow x$$
$$r_y \leftarrow y$$
for $j = 1{:}n$
$$\qquad r_a \leftarrow A(:,j)$$
$$\qquad r_y \;=\; r_y + r_a r_x(j)$$
end
$$y \leftarrow r_y$$

requires $(3 + n)$ load/store operations while for the outer product update

$$r_x \leftarrow x$$
$$r_y \leftarrow y$$
for $j = 1{:}n$
$$\qquad r_a \leftarrow A(:,j)$$
$$\qquad r_a \;=\; r_a + r_y r_x(j)$$
$$\qquad A(:,j) \leftarrow r_a$$
end

the corresponding count is $(2 + 2n)$. Thus, the data motion overhead for the outer product update is worse by a factor of 2, a reality that could be a factor in the design of a high-performance matrix computation.

1.5.3   The Relevance of Stride

The time it takes to load a vector into a vector register may depend greatly on how the vector is laid out in memory, a detail that we did not consider in §1.5.1. Two concepts help frame the issue. A vector is said to have *unit stride* if its components are contiguous in memory. A matrix is said to be stored in *column-major order* if its columns have unit stride.

Let us consider the matrix multiplication update calculation

$$C = C + AB$$

where it is assumed that the matrices $C \in \mathbb{R}^{m \times n}$, $A \in \mathbb{R}^{m \times r}$, and $B \in \mathbb{R}^{r \times n}$ are stored in column-major order. Suppose the loading of a unit-stride vector proceeds much more quickly than the loading of a non-unit-stride vector. If so, then the implementation

for $j = 1{:}n$
    for $k = 1{:}r$
        $C(:,j) = C(:,j) + A(:,k) \cdot B(k,j)$
    end
end

which accesses C, A, and B by column would be preferred to

for $i = 1{:}m$
    for $j = 1{:}n$
        $C(i,j) = C(i,j) + A(i,:) \cdot B(:,j)$
    end
end

which accesses C and A by row. While this example points to the possible importance of stride, it is important to keep in mind that the penalty for non-unit-stride access varies from system to system and may depend upon the value of the stride itself.

1.5.4   Blocking for Data Reuse

Matrices reside in memory but *memory has levels*. A typical arrangement is depicted in Figure 1.5.3. The *cache* is a relatively small high-speed memory unit that sits

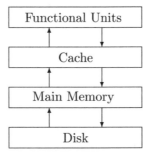

Figure 1.5.3. *A memory hierarchy*

just below the functional units where the arithmetic is carried out. During a matrix computation, matrix elements move up and down the memory hierarchy. The *cache*, which is a small high-speed memory situated in between the functional units and main memory, plays a particularly critical role. The overall design of the hierarchy varies from system to system. However, two maxims always apply:

- Each level in the hierarchy has a limited capacity and for economic reasons this capacity usually becomes smaller as we ascend the hierarchy.

- There is a cost, sometimes relatively great, associated with the moving of data between two levels in the hierarchy.

The efficient implementation of a matrix algorithm requires an ability to reason about the flow of data between the various levels of storage.

To develop an appreciation for cache utilization we again consider the update $C = C + AB$ where each matrix is n-by-n and blocked as follows:

$$C = \begin{bmatrix} C_{11} & \cdots & C_{1r} \\ \vdots & \ddots & \vdots \\ C_{qr} & \cdots & C_{qr} \end{bmatrix} \quad A = \begin{bmatrix} A_{11} & \cdots & A_{1p} \\ \vdots & \ddots & \vdots \\ A_{qr} & \cdots & A_{qp} \end{bmatrix} \quad B = \begin{bmatrix} B_{11} & \cdots & B_{1r} \\ \vdots & \ddots & \vdots \\ B_{pr} & \cdots & B_{pr} \end{bmatrix}.$$

Assume that these three matrices reside in main memory and that we plan to update C block by block:

$$C_{ij} = C_{ij} + \sum_{k=1}^{p} A_{ik}B_{kj}.$$

The data in the blocks must be brought up to the functional units via the cache which we assume is large enough to hold a C-block, an A-block, and a B-block. This enables us to structure the computation as follows:

> **for** $i = 1{:}q$
>   **for** $j = 1{:}r$
>   Load C_{ij} from main memory into cache
>   **for** $k = 1{:}p$
>   Load A_{ik} from main memory into cache
>   Load B_{kj} from main memory into cache   (1.5.4)
> $C_{ij} = C_{ij} + A_{ik}B_{kj}$
>   **end**
>   Store C_{ij} in main mcmory.
>   **end**
> **end**

The question before us is how to choose the blocking parameters q, r, and p so as to minimize memory traffic to and from the cache. Assume that the cache can hold M floating point numbers and that $M \ll 3n^2$, thereby forcing us to block the computation. We assume that

$$\left. \begin{array}{c} C_{ij} \\ A_{ik} \\ B_{kj} \end{array} \right\} \text{ is roughly } \left\{ \begin{array}{l} (n/q)\text{-by-}(n/r) \\ (n/q)\text{-by-}(n/p) \\ (n/p)\text{-by-}(n/r) \end{array} \right. .$$

We say "roughly" because if q, r, or p does not divide n, then the blocks are not quite uniformly sized, e.g.,

$$A = \begin{bmatrix} \times & \times & \times & \times & \times & \times & \times & \times & \times & \times \\ \times & \times & \times & \times & \times & \times & \times & \times & \times & \times \\ \times & \times & \times & \times & \times & \times & \times & \times & \times & \times \\ \times & \times & \times & \times & \times & \times & \times & \times & \times & \times \\ \times & \times & \times & \times & \times & \times & \times & \times & \times & \times \\ \times & \times & \times & \times & \times & \times & \times & \times & \times & \times \\ \times & \times & \times & \times & \times & \times & \times & \times & \times & \times \\ \times & \times & \times & \times & \times & \times & \times & \times & \times & \times \\ \times & \times & \times & \times & \times & \times & \times & \times & \times & \times \\ \times & \times & \times & \times & \times & \times & \times & \times & \times & \times \end{bmatrix}, \quad \begin{array}{l} n = 10, \\[4pt] q = 3, \\[4pt] p = 4. \end{array}$$

However, nothing is lost in glossing over this detail since our aim is simply to develop an intuition about cache utilization for large-n problems. Thus, we are led to impose the following constraint on the blocking parameters:

$$\left(\frac{n}{q}\right)\left(\frac{n}{r}\right) + \left(\frac{n}{q}\right)\left(\frac{n}{p}\right) + \left(\frac{n}{p}\right)\left(\frac{n}{r}\right) \leq M. \qquad (1.5.5)$$

Proceeding with the optimization, it is reasonable to *maximize* the amount of arithmetic associated with the update $C_{ij} = C_{ij} + A_{ik}B_{kj}$. After all, we have moved matrix data from main memory to cache and should make the most of the investment. This leads to the problem of maximizing $2n^3/(qrp)$ subject to the constraint (1.5.5). A straightforward Lagrange multiplier argument leads us to conclude that

$$q_{\text{opt}} = p_{\text{opt}} = r_{\text{opt}} \approx \sqrt{\frac{n^2}{3M}}. \qquad (1.5.6)$$

That is, each block of C, A, and B should be approximately square and occupy about one-third of the cache.

Because blocking affects the amount of memory traffic in a matrix computation, it is of paramount importance when designing a high-performance implementation. In practice, things are never as simple as in our model example. The optimal choice of q_{opt}, r_{opt}, and p_{opt} will also depend upon transfer rates between memory levels and upon all the other architecture factors mentioned earlier in this section. Data structures are also important; storing a matrix by block rather than in column-major order could enhance performance.

Problems

P1.5.1 Suppose $A \in \mathbb{R}^{n \times n}$ is tridiagonal and that the elements along its subdiagonal, diagonal, and superdiagonal are stored in vectors $e(1{:}n-1)$, $d(1{:}n)$, and $f(2{:}n)$. Give a vectorized implementation of the n-by-n gaxpy $y = y + Ax$. Hint: Make use of the vector multiplication operation.

P1.5.2 Give an algorithm for computing $C = C + A^T BA$ where A and B are n-by-n and B is symmetric. Innermost loops should oversee unit-stride vector operations.

P1.5.3 Suppose $A \in \mathbb{R}^{m \times n}$ is stored in column-major order and that $m = m_1 M$ and $n = n_1 N$. Regard A as an M-by-N block matrix with m_1-by-n_1 blocks. Give an algorithm for storing A in a vector $A.block(1{:}mn)$ with the property that each block A_{ij} is stored contiguously in column-major order.

Notes and References for §1.5

References that address vector computation include:

J.J. Dongarra, F.G. Gustavson, and A. Karp (1984). "Implementing Linear Algebra Algorithms for Dense Matrices on a Vector Pipeline Machine," *SIAM Review 26*, 91–112.
B.L. Buzbee (1986) "A Strategy for Vectorization," *Parallel Comput. 3*, 187–192.
K. Gallivan, W. Jalby, U. Meier, and A.H. Sameh (1988). "Impact of Hierarchical Memory Systems on Linear Algebra Algorithm Design," *Int. J. Supercomput. Applic. 2*, 12–48.
J.J. Dongarra and D. Walker (1995). "Software Libraries for Linear Algebra Computations on High Performance Computers," *SIAM Review 37*, 151–180.

One way to realize high performance in a matrix computation is to design algorithms that are rich in matrix multiplication and then implement those algorithms using an optimized level-3 BLAS library. For details on this philosophy and its effectiveness, see:

B. Kågström, P. Ling, and C. Van Loan (1998). "GEMM-based Level-3 BLAS: High-Performance Model Implementations and Performance Evaluation Benchmark," *ACM Trans. Math. Softw. 24*, 268–302.

M.J. Dayde and I.S. Duff (1999), "The RISC BLAS: A Blocked Implementation of Level 3 BLAS for RISC Processors," *ACM Trans. Math. Softw. 25*, 316–340.

E. Elmroth, F. Gustavson, I. Jonsson, and B. Kågström (2004). "Recursive Blocked Algorithms and Hybrid Data Structures for Dense Matrix Library Software," *SIAM Review 46*, 3–45.

K. Goto and R. Van De Geign (2008). "Anatomy of High-Performance Matrix Multiplication," *ACM Trans. Math. Softw. 34*, 12:1–12:25.

Advanced data structures that support high performance matrix computations are discussed in:

F.G. Gustavson (1997). "Recursion Leads to Automatic Variable Blocking for Dense Linear Algebra Algorithms," *IBM J. Res. Dev. 41*, 737–755.

V. Valsalam and A. Skjellum (2002). "A Framework for High-Performance Matrix Multiplication Based on Hierarchical Abstractions, Algorithms, and Optimized Low-Level Kernels," *Concurrency Comput. Pract. Exper. 14*, 805–839.

S.R. Chatterjee, P. Patnala, and M. Thottethodi (2002). "Recursive Array Layouts and Fast Matrix Multiplication," *IEEE Trans. Parallel. Distrib. Syst. 13*, 1105–1123.

F.G. Gustavson (2003). "High-Performance Linear Algebra Algorithms Using Generalized Data Structures for Matrices," *IBM J. Res. Dev. 47*, 31–54.

N. Park, B. Hong, and V.K. Prasanna (2003). "Tiling, Block Data Layout, and Memory Hierarchy Performance," *IEEE Trans. Parallel Distrib. Systems, 14*, 640–654.

J.A. Gunnels, F.G. Gustavson, G.M. Henry, and R.A. van de Geijn (2005). "A Family of High-Performance Matrix Multiplication Algorithms," *PARA 2004, LNCS 3732*, 256–265.

P. D'Alberto and A. Nicolau (2009). "Adaptive Winograd's Matrix Multiplications," *ACM Trans. Math. Softw. 36*, 3:1–3:23.

A great deal of effort has gone into the design of software tools that automatically block a matrix computation for high performance, e.g.,

S. Carr and R.B. Lehoucq (1997) "Compiler Blockability of Dense Matrix Factorizations," *ACM Trans. Math. Softw. 23*, 336–361.

J.A. Gunnels, F. G. Gustavson, G.M. Henry, and R. A. van de Geijn (2001). "FLAME: Formal Linear Algebra Methods Environment," *ACM Trans. Math. Softw. 27*, 422–455.

P. Bientinesi, J.A. Gunnels, M.E. Myers, E. Quintana-Orti, and R.A. van de Geijn (2005). "The Science of Deriving Dense Linear Algebra Algorithms," *ACM Trans. Math. Softw. 31*, 1–26.

J. Demmel, J. Dongarra, V. Eijkhout, E. Fuentes, A. Petitet, R. Vuduc, R.C. Whaley, and K. Yelick (2005). "Self-Adapting Linear Algebra Algorithms and Software,", *Proc. IEEE 93*, 293–312.

K. Yotov, X.Li, G. Ren, M. Garzaran, D. Padua, K. Pingali, and P. Stodghill (2005). "Is Search Really Necessary to Generate High-Performance BLAS?," *Proc. IEEE 93*, 358–386.

For a rigorous treatment of communication lower bounds in matrix computations, see:

G. Ballard, J. Demmel, O. Holtz, and O. Schwartz (2011). "Minimizing Communication in Numerical Linear Algebra," *SIAM J. Matrix Anal. Applic. 32*, 866–901.

1.6 Parallel Matrix Multiplication

The impact of matrix computation research in many application areas depends upon the development of parallel algorithms that *scale*. Algorithms that scale have the property that they remain effective as problem size grows and the number of involved processors increases. Although powerful new programming languages and related system tools continue to simplify the process of implementing a parallel matrix computation, being able to "think parallel" is still important. This requires having an intuition about load balancing, communication overhead, and processor synchronization.

1.6.1   A Model Computation

To illustrate the major ideas associated with parallel matrix computations, we consider the following *model computation*:

> Given $C \in \mathbb{R}^{m \times n}$, $A \in \mathbb{R}^{m \times r}$, and $B \in \mathbb{R}^{r \times n}$, effectively compute the matrix multiplication update $C = C + AB$ assuming the availability of p processors. Each processor has its own *local memory* and executes its own *local program*.

The matrix multiplication update problem is a good choice because it is an inherently parallel computation and because it is at the heart of many important algorithms that we develop in later chapters.

The design of a parallel procedure begins with the breaking up of the given problem into smaller parts that exhibit a measure of independence. In our problem we assume the blocking

$$
C = \begin{bmatrix} C_{11} & \cdots & C_{1N} \\ \vdots & \ddots & \vdots \\ C_{M1} & \cdots & C_{MN} \end{bmatrix}, \; A = \begin{bmatrix} A_{11} & \cdots & A_{1R} \\ \vdots & \ddots & \vdots \\ A_{M1} & \cdots & A_{MR} \end{bmatrix}, \; B = \begin{bmatrix} B_{11} & \cdots & B_{1N} \\ \vdots & \ddots & \vdots \\ B_{R1} & \cdots & B_{RN} \end{bmatrix}, \quad (1.6.1)
$$

$$
m = m_1 M, \qquad r = r_1 R, \qquad n = n_1 N
$$

with $C_{ij} \in \mathbb{R}^{m_1 \times n_1}$, $A_{ij} \in \mathbb{R}^{m_1 \times r_1}$, and $B_{ij} \in \mathbb{R}^{r_1 \times n_1}$. It follows that the $C + AB$ update partitions nicely into MN smaller *tasks*:

$$
\text{Task}(i,j): \quad C_{ij} = C_{ij} + \sum_{k=1}^{R} A_{ik} B_{kj}. \quad (1.6.2)
$$

Note that the block-block products $A_{ik} B_{kj}$ are all the same size.

Because the tasks are naturally double-indexed, we double index the available processors as well. Assume that $p = p_{\text{row}} p_{\text{col}}$ and designate the (i,j)th processor by $\text{Proc}(i,j)$ for $i = 1{:}p_{\text{row}}$ and $j = 1{:}p_{\text{col}}$. The double indexing of the processors is just a notation and is *not* a statement about their physical connectivity.

1.6.2   Load Balancing

An effective parallel program equitably partitions the work among the participating processors. Two subdivision strategies for the model computation come to mind. The *2-dimensional block distribution* assigns contiguous block updates to each processor. See Figure 1.6.1. Alternatively, we can have $\text{Proc}(\mu, \tau)$ oversee the update of C_{ij} for $i = \mu{:}p_{\text{row}}{:}M$ and $j = \tau{:}p_{\text{col}}{:}N$. This is called the *2-dimensional block-cyclic distribution*. See Figure 1.6.2. For the displayed example, both strategies assign twelve C_{ij} updates to each processor and each update involves R block-block multiplications, i.e., $12(2m_1 n_1 r_1)$ flops. Thus, from the flop point of view, both strategies are *load balanced*, by which we mean that the amount of arithmetic computation assigned to each processor is roughly the same.

Proc(1,1)	Proc(1,2)	Proc(1,3)
$\left\{\begin{array}{ccc} C_{11} & C_{12} & C_{13} \\ C_{21} & C_{22} & C_{23} \\ C_{31} & C_{32} & C_{33} \\ C_{41} & C_{42} & C_{43} \end{array}\right\}$	$\left\{\begin{array}{ccc} C_{14} & C_{15} & C_{16} \\ C_{24} & C_{25} & C_{26} \\ C_{34} & C_{35} & C_{36} \\ C_{44} & C_{45} & C_{46} \end{array}\right\}$	$\left\{\begin{array}{ccc} C_{17} & C_{18} & C_{19} \\ C_{27} & C_{28} & C_{29} \\ C_{37} & C_{38} & C_{39} \\ C_{47} & C_{48} & C_{49} \end{array}\right\}$
Proc(2,1)	Proc(2,2)	Proc(2,3)
$\left\{\begin{array}{ccc} C_{51} & C_{52} & C_{53} \\ C_{61} & C_{62} & C_{63} \\ C_{71} & C_{72} & C_{73} \\ C_{81} & C_{82} & C_{83} \end{array}\right\}$	$\left\{\begin{array}{ccc} C_{54} & C_{55} & C_{56} \\ C_{64} & C_{65} & C_{66} \\ C_{74} & C_{75} & C_{76} \\ C_{84} & C_{85} & C_{86} \end{array}\right\}$	$\left\{\begin{array}{ccc} C_{57} & C_{58} & C_{59} \\ C_{67} & C_{68} & C_{69} \\ C_{77} & C_{78} & C_{79} \\ C_{87} & C_{88} & C_{89} \end{array}\right\}$

Figure 1.6.1. *The block distribution of tasks*
($M = 8$, $p_{\text{row}} = 2$, $N = 9$, and $p_{\text{col}} = 3$).

Proc(1,1)	Proc(1,2)	Proc(1,3)
$\left\{\begin{array}{ccc} C_{11} & C_{14} & C_{17} \\ C_{31} & C_{34} & C_{37} \\ C_{51} & C_{54} & C_{57} \\ C_{71} & C_{74} & C_{77} \end{array}\right\}$	$\left\{\begin{array}{ccc} C_{12} & C_{15} & C_{18} \\ C_{32} & C_{35} & C_{38} \\ C_{52} & C_{55} & C_{58} \\ C_{72} & C_{75} & C_{78} \end{array}\right\}$	$\left\{\begin{array}{ccc} C_{13} & C_{16} & C_{19} \\ C_{33} & C_{36} & C_{39} \\ C_{53} & C_{56} & C_{59} \\ C_{73} & C_{76} & C_{79} \end{array}\right\}$
Proc(2,1)	Proc(2,2)	Proc(2,3)
$\left\{\begin{array}{ccc} C_{21} & C_{24} & C_{27} \\ C_{41} & C_{44} & C_{47} \\ C_{61} & C_{64} & C_{67} \\ C_{81} & C_{84} & C_{87} \end{array}\right\}$	$\left\{\begin{array}{ccc} C_{22} & C_{25} & C_{28} \\ C_{42} & C_{45} & C_{48} \\ C_{62} & C_{65} & C_{68} \\ C_{82} & C_{85} & C_{88} \end{array}\right\}$	$\left\{\begin{array}{ccc} C_{23} & C_{26} & C_{29} \\ C_{43} & C_{46} & C_{49} \\ C_{63} & C_{66} & C_{69} \\ C_{83} & C_{86} & C_{89} \end{array}\right\}$

Figure 1.6.2. *The block-cyclic distribution of tasks*
($M = 8$, $p_{\text{row}} = 2$, $N = 9$, and $p_{\text{col}} = 3$).

If M is not a multiple of p_{row} or if N is not a multiple of p_{col}, then the distribution of work among processors is no longer balanced. Indeed, if

$$M = \alpha_1 p_{\text{row}} + \beta_1, \qquad 0 \le \beta_1 < p_{\text{row}},$$
$$N = \alpha_2 p_{\text{col}} + \beta_2, \qquad 0 \le \beta_2 < p_{\text{col}},$$

then the number of block-block multiplications per processor can range from $\alpha_1 \alpha_2 R$ to $(\alpha_1 + 1)(\alpha_2 + 1)R$. However, this variation is insignificant in a large-scale computation with $M \gg p_{\text{row}}$ and $N \gg p_{\text{col}}$:

$$\frac{(\alpha_1 + 1)(\alpha_2 + 1)R}{(\alpha_1 \alpha_2)R} = 1 + O\left(\frac{p_{\text{row}}}{M} + \frac{p_{\text{col}}}{N}\right).$$

We conclude that both the block distribution and the block-cyclic distribution strategies are load balanced for the general $C + AB$ update.

This is *not* the case for certain block-sparse situations that arise in practice. If A is block lower triangular and B is block upper triangular, then the amount of work associated with Task(i, j) depends upon i and j. Indeed from (1.6.2) we have

$$C_{ij} = C_{ij} + \sum_{k=1}^{\min\{i,j,R\}} A_{ik} B_{kj}.$$

A very uneven allocation of work for the block distribution can result because the number of flops associated with Task(i, j) increases with i and j. The tasks assigned to Proc$(p_{\text{row}}, p_{\text{col}})$ involve the most work while the tasks assigned to Proc(1,1) involve the least. To illustrate the ratio of workloads, set $M = N = R = \tilde{M}$ and assume that $p_{\text{row}} = p_{\text{col}} = \tilde{p}$ divides $\tilde{M}$. It can be shown that

$$\frac{\text{Flops assigned to Proc}(\tilde{p}, \tilde{p})}{\text{Flops assigned to Proc}(1, 1)} = O(\tilde{p}) \qquad (1.6.3)$$

if we assume $\tilde{M}/\tilde{p} \gg 1$. Thus, load balancing does not depend on problem size and gets worse as the number of processors increase.

This is not the case for the block-cyclic distribution. Again, Proc(1,1) and Proc$(\tilde{p}, \tilde{p})$ are the least busy and most busy processors. However, now it can be verified that

$$\frac{\text{Flops assigned to Proc}(\tilde{p}, \tilde{p})}{\text{Flops assigned to Proc}(1, 1)} = 1 + O\left(\frac{\tilde{p}}{\tilde{M}}\right), \qquad (1.6.4)$$

showing that the allocation of work becomes increasingly balanced as the problem size grows.

Another situation where the block-cyclic distribution of tasks is preferred is the case when the first q block rows of A are zero and the first q block columns of B are zero. This situation arises in several important matrix factorization schemes. Note from Figure 1.6.1 that if q is large enough, then some processors have absolutely nothing to do if tasks are assigned according to the block distribution. On the other hand, the block-cyclic distribution is load balanced, providing further justification for this method of task distribution.

1.6.3 Data Motion Overheads

So far the discussion has focused on load balancing from the flop point of view. We now turn our attention to the costs associated with data motion and processor coordination. How does a processor get hold of the data it needs for an assigned task? How does a processor know enough to wait if the data it needs is the output of a computation being performed by another processor? What are the overheads associated with data transfer and synchronization and how do they compare to the costs of the actual arithmetic?

The importance of data locality is discussed in §1.5. However, in a parallel computing environment, the data that a processor needs can be "far away," and if that is the case too often, then it is possible to lose the multiprocessor advantage. Regarding synchronization, time spent waiting for another processor to finish a calculation is time lost. Thus, the design of an effective parallel computation involves paying attention to the number of synchronization points and their impact. Altogether, this makes it difficult to model performance, especially since an individual processor can typically compute and communicate at the same time. Nevertheless, we forge ahead with our analysis of the model computation to dramatize the cost of data motion relative to flops. For the remainder of this section we assume:

(a) The block-cyclic distribution of tasks is used to ensure that arithmetic is load balanced.

(b) Individual processors can perform the computation $C_{ij} = C_{ij} + A_{ik}B_{kj}$ at a rate of F flops per second. Typically, a processor will have its own local memory hierarchy and vector processing capability, so F is an attempt to capture in a single number all the performance issues that we discussed in §1.5.

(c) The time required to move η floating point numbers into or out of a processor is $\alpha + \beta\eta$. In this model, the parameters α and β respectively capture the *latency* and *bandwidth* attributes associated with data transfer.

With these simplifications we can roughly assess the effectiveness of assigning p processors to the update computation $C = C + AB$.

Let $T_{\mathrm{arith}}(p)$ be the time that each processor must spend doing arithmetic as it carries out its share of the computation. It follows from assumptions (a) and (b) that

$$T_{\mathrm{arith}}(p) \approx \frac{2mnr}{pF}. \tag{1.6.5}$$

Similarly, let $T_{\mathrm{data}}(p)$ be the time that each processor must spend acquiring the data it needs to perform its tasks. Ordinarily, this quantity would vary significantly from processor to processor. However, the implementation strategies outlined below have the property that the communication overheads are roughly the same for each processor. It follows that if $T_{\mathrm{arith}}(p) + T_{\mathrm{data}}(p)$ approximates the total execution time for the p-processor solution, then the quotient

$$S(p) = \frac{T_{\mathrm{arith}}(1)}{T_{\mathrm{arith}}(p) + T_{\mathrm{data}}(p)} = \frac{p}{1 + \dfrac{T_{\mathrm{data}}(p)}{T_{\mathrm{arith}}(p)}} \tag{1.6.6}$$

is a reasonable measure of *speedup*. Ideally, the assignment of p processors to the $C = C + AB$ update would reduce the single-processor execution time by a factor of p. However, from (1.6.6) we see that $S(p) < p$ with the compute-to-communicate ratio $T_{\text{data}}(p)/T_{\text{arith}}(p)$ explaining the degradation. To acquire an intuition about this all-important quotient, we need to examine more carefully the data transfer properties associated with each task.

1.6.4   Who Needs What

If a processor carries out Task(i, j), then at some time during the calculation, blocks $C_{ij}, A_{i1}, \ldots, A_{iR}, B_{1j}, \ldots, B_{Rj}$ must find their way into its local memory. Given assumptions (a) and (c), Table 1.6.1 summarizes the associated data transfer overheads for an individual processor:

Required Blocks			Data Transfer Time per Block
C_{ij}	$i = \mu{:}p_{\text{row}}{:}M$	$j = \tau{:}p_{\text{col}}{:}N$	$\alpha + \beta m_1 n_1$
A_{ij}	$i = \mu{:}p_{\text{row}}{:}M$	$j = 1{:}R$	$\alpha + \beta m_1 r_1$
B_{ij}	$i = 1{:}R$	$j = \tau{:}p_{\text{col}}{:}N$	$\alpha + \beta r_1 n_1$

TABLE 1.6.1. *Communication overheads for* $\text{Proc}(\mu, \tau)$

It follows that if

$$\gamma_C = \text{total number of required } C\text{-block transfers,} \tag{1.6.7}$$

$$\gamma_A = \text{total number of required } A\text{-block transfers,} \tag{1.6.8}$$

$$\gamma_B = \text{total number of required } B\text{-block transfers,} \tag{1.6.9}$$

then

$$T_{\text{data}}(p) \approx \gamma_C(\alpha + \beta m_1 n_1) + \gamma_A(\alpha + \beta m_1 r_1) + \gamma_B(\alpha + \beta r_1 n_1),$$

and so from from (1.6.5) we have

$$\frac{T_{\text{data}}(p)}{T_{\text{arith}}(p)} \approx \frac{Fp}{2}\left(\alpha\frac{\gamma_C + \gamma_A + \gamma_B}{mnr} + \beta\left(\frac{\gamma_C}{MNr} + \frac{\gamma_A}{MnR} + \frac{\gamma_B}{mNR}\right)\right). \tag{1.6.10}$$

To proceed further with our analysis, we need to estimate the γ-factors (1.6.7)–(1.6.9), and that requires assumptions about how the underlying architecture stores and accesses the matrices A, B, and C.

1.6.5   The Shared-Memory Paradigm

In a *shared-memory system* each processor has access to a common, global memory. See Figure 1.6.3. During program execution, data flows to and from the global memory and this represents a significant overhead that we proceed to assess. Assume that the matrices C, A, and B are in global memory at the start and that $\text{Proc}(\mu, \tau)$ executes the following:

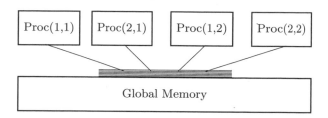

Figure 1.6.3. *A four-processor shared-memory system*

$$
\begin{aligned}
&\textbf{for } i = \mu{:}p_{\text{row}}{:}M \\
&\qquad \textbf{for } j = \tau{:}p_{\text{col}}{:}N \\
&\qquad\qquad C^{(\text{loc})} \leftarrow C_{ij} \\
&\qquad\qquad \textbf{for } k = 1{:}R \\
&\qquad\qquad\qquad A^{(\text{loc})} \leftarrow A_{ik} \\
&\qquad\qquad\qquad B^{(\text{loc})} \leftarrow B_{kj} \qquad\qquad\text{(Method 1)} \\
&\qquad\qquad\qquad C^{(\text{loc})} = C^{(\text{loc})} + A^{(\text{loc})} B^{(\text{loc})} \\
&\qquad\qquad \textbf{end} \\
&\qquad\qquad C_{ij} \leftarrow C^{(\text{loc})} \\
&\qquad \textbf{end} \\
&\textbf{end}
\end{aligned}
$$

As a reminder of the interactions between global and local memory, we use the "$\leftarrow$" notation to indicate data transfers between these memory levels and the "loc" superscript to designate matrices in local memory. The block transfer statistics (1.6.7)-(1.6.9) for Method 1 are given by

$$
\begin{aligned}
\gamma_C &\approx 2(MN/p), \\
\gamma_A &\approx R(MN/p), \\
\gamma_B &\approx R(MN/p),
\end{aligned}
$$

and so from (1.6.10) we obtain

$$
\frac{T_{\text{data}}(p)}{T_{\text{arith}}(p)} \approx \frac{F}{2}\left(\alpha\frac{2 + 2R}{m_1 n_1 r} \;+\; \beta\left(\frac{2}{r} + \frac{1}{n_1} + \frac{1}{m_1}\right)\right). \tag{1.6.11}
$$

By substituting this result into (1.6.6) we conclude that (a) speed-up degrades as the flop rate F increases and (b) speedup improves if the communication parameters α and β decrease or the block dimensions m_1, n_1, and r_1 increase. Note that the communicate-to-compute ratio (1.6.11) for Method 1 does not depend upon the number of processors.

Method 1 has the property that it is only necessary to store one C-block, one A-block, and one B-block in local memory at any particular instant, i.e., $C^{(\mathrm{loc})}$, $A^{(\mathrm{loc})}$, and $B^{(\mathrm{loc})}$. Typically, a processor's local memory is much smaller than global memory, so this particular solution approach is attractive for problems that are very large relative to local memory capacity. However, there is a hidden cost associated with this economy because in Method 1, each A-block is loaded N/p_{col} times and each B-block is loaded M/p_{row} times. This redundancy can be eliminated if each processor's local memory is large enough to house simultaneously all the C-blocks, A-blocks, and B-blocks that are required by its assigned tasks. Should this be the case, then the following method involves much less data transfer:

> **for** $k = 1{:}R$
> $\qquad A_{ik}^{(\mathrm{loc})} \leftarrow A_{ik} \qquad (i = \mu{:}p_{\mathrm{row}}{:}M)$
> $\qquad B_{kj}^{(\mathrm{loc})} \leftarrow B_{kj} \qquad (j = \tau{:}p_{\mathrm{col}}{:}N)$
> **end**
>
> **for** $i = \mu{:}p_{\mathrm{row}}{:}M$
> $\qquad$ **for** $j = \tau{:}p_{\mathrm{col}}{:}N$
> $\qquad\qquad C^{(\mathrm{loc})} \leftarrow C_{ij}$
> $\qquad\qquad$ **for** $k = 1{:}R$ $\qquad\qquad\qquad\qquad\qquad\qquad\qquad$ (Method 2)
> $\qquad\qquad\qquad C^{(\mathrm{loc})} = C^{(\mathrm{loc})} + A_{ik}^{(\mathrm{loc})} B_{kj}^{(\mathrm{loc})}$
> $\qquad\qquad$ **end**
> $\qquad\qquad C_{ij} \leftarrow C^{(\mathrm{loc})}$
> $\qquad$ **end**
> **end**

The block transfer statistics γ_C', γ_A', and γ_B', for Method 2 are more favorable than for Method 1. It can be shown that

$$\gamma_C' = \gamma_C, \qquad \gamma_A' = \gamma_A f_{\mathrm{col}}, \qquad \gamma_B' = \gamma_B f_{\mathrm{row}}, \qquad (1.6.12)$$

where the quotients $f_{\mathrm{col}} = p_{\mathrm{col}}/N$ and $f_{\mathrm{row}} = p_{\mathrm{row}}/M$ are typically much less than unity. As a result, the communicate-to-compute ratio for Method 2 is given by

$$\frac{T_{\mathrm{data}}(p)}{T_{\mathrm{arith}}(p)} \approx \frac{F}{2}\left(\alpha \frac{2 + R\,(f_{\mathrm{col}} + f_{\mathrm{row}})}{m_1 n_1 r} + \beta\left(\frac{2}{r} + \frac{1}{n_1}f_{\mathrm{col}} + \frac{1}{m_1}f_{\mathrm{row}}\right)\right), \qquad (1.6.13)$$

which is an improvement over (1.6.11). Methods 1 and 2 showcase the trade-off that frequently exists between local memory capacity and the overheads that are associated with data transfer.

1.6.6   Barrier Synchronization

The discussion in the previous section assumes that C, A, and B are available in global memory at the start. If we extend the model computation so that it includes the multiprocessor initialization of these three matrices, then an interesting issue arises. How does a processor "know" when the initialization is complete and it is therefore safe to begin its share of the $C = C + AB$ update?

Answering this question is an occasion to introduce a very simple synchronization construct known as the *barrier*. Suppose the C-matrix is initialized in global memory by assigning to each processor some fraction of the task. For example, $\mathrm{Proc}(\mu, \tau)$ could do this:

> **for** $i = \mu{:}p_{\mathrm{row}}{:}M$
>> **for** $j = \tau{:}p_{\mathrm{col}}{:}N$
>>> Compute the (i, j) block of C and store in $C^{(\mathrm{loc})}$.
>>> $C_{ij} \leftarrow C^{(\mathrm{loc})}$
>> **end**
> **end**

Similar approaches can be taken for the setting up of $A = (A_{ij})$ and $B = (B_{ij})$. Even if this partitioning of the initialization is load balanced, it cannot be assumed that each processor completes its share of the work at exactly the same time. This is where the barrier synchronization is handy. Assume that $\mathrm{Proc}(\mu, \tau)$ executes the following:

$$
\begin{array}{llll}
\text{Initialize } C_{ij}, & i = \mu : p_{\mathrm{row}} : M, & j = \tau : p_{\mathrm{col}} : N \\
\text{Initialize } A_{ij}, & i = \mu : p_{\mathrm{row}} : M, & j = \tau : p_{\mathrm{col}} : R \\
\text{Initialize } B_{ij}, & i = \mu : p_{\mathrm{row}} : R, & j = \tau : p_{\mathrm{col}} : N & (1.6.14) \\
\text{barrier} \\
\text{Update } C_{ij}, & i = \mu : p_{\mathrm{row}} : M, & j = \tau : p_{\mathrm{col}} : N
\end{array}
$$

To understand the **barrier** command, regard a processor as being either "blocked" or "free." Assume in (1.6.14) that all processors are free at the start. When it executes the **barrier** command, a processor becomes blocked and suspends execution. After the last processor is blocked, all the processors return to the free state and resume execution. In (1.6.14), the **barrier** does not allow the C_{ij} updating via Methods 1 or 2 to begin until all three matrices are fully initialized in global memory.

1.6.7 The Distributed-Memory Paradigm

In a *distributed-memory system* there is no global memory. The data is collectively housed in the local memories of the individual processors which are connected to form a network. There are many possible network topologies. An example is displayed in Figure 1.6.4. The cost associated with sending a message from one processor to another is likely to depend upon how "close" they are in the network. For example, with the torus in Figure 1.6.4, a message from $\mathrm{Proc}(1,1)$ to $\mathrm{Proc}(1,4)$ involves just one "hop" while a message from $\mathrm{Proc}(1,1)$ to $\mathrm{Proc}(3,3)$ would involve four.

Regardless, the message-passing costs in a distributed memory system have a serious impact upon performance just as the interactions with global memory affect performance in a shared memory system. Our goal is to approximate these costs as they might arise in the model computation. For simplicity, we make no assumptions about the underlying network topology.

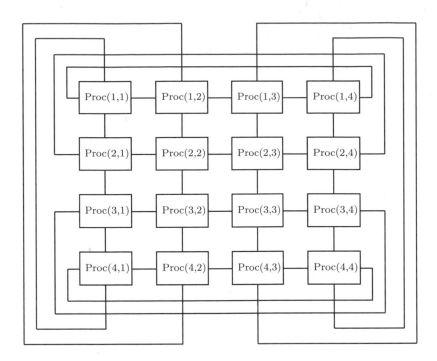

Figure 1.6.4. *A 2-Dimensional Torus*

Let us first assume that $M = N = R = p_{\text{row}} = p_{\text{col}} = 2$ and that the C, A, and B matrices are distributed as follows:

Proc(1,1)
C_{11}, A_{11}, B_{11}

Proc(1,2)
C_{12}, A_{12}, B_{12}

Proc(2,1)
C_{21}, A_{21}, B_{21}

Proc(2,2)
C_{22}, A_{22}, B_{22}

Assume that Proc(i, j) oversees the update of C_{ij} and notice that the required data for this computation is not entirely local. For example, Proc(1,1) needs to receive a copy of A_{12} from Proc(1,2) and a copy of B_{21} from Proc(2,1) before it can complete the update $C_{11} = C_{11} + A_{11}B_{11} + A_{12}B_{21}$. Likewise, it must send a copy of A_{11} to Proc(1,2) and a copy of B_{11} to Proc(2,1) so that they can carry out their respective updates. Thus, the local programs executing on each processor involve a mix of computational steps and message-passing steps:

Proc(1,1)
Send a copy of A_{11} to Proc(1,2)
Receive a copy of A_{12} from Proc(1,2)
Send a copy of B_{11} to Proc(2,1)
Receive a copy of B_{21} from Proc(2,1)
$C_{11} = C_{11} + A_{11}B_{11} + A_{12}B_{21}$

Proc(1,2)
Send a copy of A_{12} to Proc(1,1)
Receive a copy of A_{11} from Proc(1,1)
Send a copy of B_{12} to Proc(2,2)
Receive a copy of B_{22} from Proc(2,2)
$C_{12} = C_{12} + A_{11}B_{12} + A_{12}B_{22}$

Proc(2,1)
Send a copy of A_{21} to Proc(2,2)
Receive a copy of A_{22} from Proc(2,2)
Send a copy of B_{21} to Proc(1,1)
Receive a copy of B_{11} from Proc(1,1)
$C_{21} = C_{21} + A_{21}B_{11} + A_{22}B_{21}$

Proc(2,2)
Send a copy of A_{22} to Proc(2,1)
Receive a copy of A_{21} from Proc(2,1)
Send a copy of B_{22} to Proc(1,2)
Receive a copy of B_{12} from Proc(1,2)
$C_{22} = C_{22} + A_{21}B_{12} + A_{22}B_{22}$

This informal specification of the local programs does a good job delineating the duties of each processor, but it hides several important issues that have to do with the timeline of execution. (a) Messages do not necessarily arrive at their destination in the order that they were sent. How will a receiving processor know if it is an A-block or a B-block? (b) Receive-a-message commands can block a processor from proceeding with the rest of its calculations. As a result, it is possible for a processor to wait forever for a message that its neighbor never got around to sending. (c) Overlapping computation with communication is critical for performance. For example, after A_{11} arrives at Proc(1,2), the "half" update $C_{12} = C_{12} + A_{11}B_{12}$ can be carried out while the wait for B_{22} continues.

As can be seen, distributed-memory matrix computations are quite involved and require powerful systems to manage the packaging, tagging, routing, and reception of messages. The discussion of such systems is outside the scope of this book. Nevertheless, it is instructive to go beyond the above 2-by-2 example and briefly anticipate the data transfer overheads for the general model computation. Assume that $\text{Proc}(\mu, \tau)$ houses these matrices:

$$C_{ij}, \quad i = \mu : p_{\text{row}} : M, \quad j = \tau : p_{\text{col}} : N,$$
$$A_{ij}, \quad i = \mu : p_{\text{row}} : M, \quad j = \tau : p_{\text{col}} : R,$$
$$B_{ij}, \quad i = \mu : p_{\text{row}} : R, \quad j = \tau : p_{\text{col}} : N.$$

From Table 1.6.1 we conclude that if $\text{Proc}(\mu, \tau)$ is to update C_{ij} for $i = \mu : p_{\text{row}} : M$ and $j = \tau : p_{\text{col}} : N$, then it must

(a) For $i = \mu : p_{\text{row}} : M$ and $j = \tau : p_{\text{col}} : R$, send a copy of A_{ij} to

$$\text{Proc}(\mu, 1), \dots, \text{Proc}(\mu, \tau - 1), \text{Proc}(\mu, \tau + 1), \dots, \text{Proc}(\mu, p_{\text{col}}).$$

Data transfer time $\approx (p_{\text{col}} - 1)(M/p_{\text{row}})(R/p_{\text{col}}) (\alpha + \beta m_1 r_1)$

(b) For $i = \mu : p_{\text{row}} : R$ and $j = \tau : p_{\text{col}} : N$, send a copy of B_{ij} to

$$\text{Proc}(1, \tau), \dots, \text{Proc}(\mu - 1, \tau), \text{Proc}(\mu + 1, \tau), \dots, \text{Proc}(p_{\text{row}}, \tau).$$

Data transfer time $\approx (p_{\text{row}} - 1)(R/p_{\text{row}})(N/p_{\text{col}}) (\alpha + \beta r_1 n_1)$

(c) Receive copies of the A-blocks that are sent by processors

$$\text{Proc}(\mu, 1), \ldots, \ \text{Proc}(\mu, \tau - 1), \ \text{Proc}(\mu, \tau + 1), \ldots, \ \text{Proc}(\mu, p_{\text{col}}).$$

Data transfer time $\approx (p_{\text{col}} - 1)(M/p_{\text{row}})(R/p_{\text{col}}) \, (\alpha + \beta m_1 r_1)$

(d) Receive copies of the B-blocks that are sent by processors

$$\text{Proc}(1, \tau), \ldots, \ \text{Proc}(\mu - 1, \tau), \ \text{Proc}(\mu + 1, \tau), \ldots, \ \text{Proc}(p_{\text{row}}, \tau).$$

Data transfer time $\approx (p_{\text{row}} - 1)(R/p_{\text{row}})(N/p_{\text{col}}) \, (\alpha + \beta r_1 n_1)$

Let T_{data} be the summation of these data transfer overheads and recall that $T_{\text{arith}} = (2mnr)/(Fp)$ since arithmetic is evenly distributed around the processor network. It follows that

$$\frac{T_{\text{data}}(p)}{T_{\text{arith}}(p)} \approx F\left(\alpha \left(\frac{p_{\text{col}}}{m_1 r_1 n} + \frac{p_{\text{row}}}{m r_1 n_1}\right) + \beta \left(\frac{p_{\text{col}}}{n} + \frac{p_{\text{row}}}{m}\right)\right). \tag{1.6.15}$$

Thus, as problem size grows, this ratio tends to zero and speedup approaches p according to (1.6.6).

1.6.8 Cannon's Algorithm

We close with a brief description of the Cannon (1969) matrix multiplication scheme. The method is an excellent way to showcase the toroidal network displayed in Figure 1.6.4 together with the idea of "nearest-neighbor" thinking which is quite important in distributed matrix computations. For clarity, let us assume that $A = (A_{ij})$, $B = (B_{ij})$, and $C = (C_{ij})$ are 4-by-4 block matrices with n_1-by-n_1 blocks. Define the matrices

$$A^{(1)} = \begin{bmatrix} A_{11} & A_{12} & A_{13} & A_{14} \\ A_{22} & A_{23} & A_{24} & A_{21} \\ A_{33} & A_{34} & A_{31} & A_{32} \\ A_{44} & A_{41} & A_{42} & A_{43} \end{bmatrix}, \quad B^{(1)} = \begin{bmatrix} B_{11} & B_{22} & B_{33} & B_{44} \\ B_{21} & B_{32} & B_{43} & B_{14} \\ B_{31} & B_{42} & B_{13} & B_{24} \\ B_{41} & B_{12} & B_{23} & B_{34} \end{bmatrix},$$

$$A^{(2)} = \begin{bmatrix} A_{14} & A_{11} & A_{12} & A_{13} \\ A_{21} & A_{22} & A_{23} & A_{24} \\ A_{32} & A_{33} & A_{34} & A_{31} \\ A_{43} & A_{44} & A_{41} & A_{42} \end{bmatrix}, \quad B^{(2)} = \begin{bmatrix} B_{41} & B_{12} & B_{23} & B_{34} \\ B_{11} & B_{22} & B_{33} & B_{44} \\ B_{21} & B_{32} & B_{43} & B_{14} \\ B_{31} & B_{42} & B_{13} & B_{24} \end{bmatrix},$$

$$A^{(3)} = \begin{bmatrix} A_{13} & A_{14} & A_{11} & A_{12} \\ A_{24} & A_{21} & A_{22} & A_{23} \\ A_{31} & A_{32} & A_{33} & A_{34} \\ A_{42} & A_{43} & A_{44} & A_{41} \end{bmatrix}, \quad B^{(3)} = \begin{bmatrix} B_{31} & B_{42} & B_{13} & B_{24} \\ B_{41} & B_{12} & B_{23} & B_{34} \\ B_{11} & B_{22} & B_{33} & B_{44} \\ B_{21} & B_{32} & B_{43} & B_{14} \end{bmatrix},$$

$$A^{(4)} = \begin{bmatrix} A_{12} & A_{13} & A_{14} & A_{11} \\ A_{23} & A_{24} & A_{21} & A_{22} \\ A_{34} & A_{31} & A_{32} & A_{33} \\ A_{41} & A_{42} & A_{43} & A_{44} \end{bmatrix}, \quad B^{(4)} = \begin{bmatrix} B_{21} & B_{32} & B_{43} & B_{14} \\ B_{31} & B_{42} & B_{13} & B_{24} \\ B_{41} & B_{12} & B_{23} & B_{34} \\ B_{11} & B_{22} & B_{33} & B_{44} \end{bmatrix},$$

and note that

$$C_{ij} = A_{ij}^{(1)} B_{ij}^{(1)} + A_{ij}^{(2)} B_{ij}^{(2)} + A_{ij}^{(3)} B_{ij}^{(3)} + A_{ij}^{(4)} B_{ij}^{(4)}. \tag{1.6.16}$$

Refer to Figure 1.6.4 and assume that $\text{Proc}(i,j)$ is in charge of computing C_{ij} and that at the start it houses both $A_{ij}^{(1)}$ and $B_{ij}^{(1)}$. The message passing required to support the updates

$$C_{ij} = C_{ij} + A_{ij}^{(1)} B_{ij}^{(1)}, \qquad (1.6.17)$$

$$C_{ij} = C_{ij} + A_{ij}^{(2)} B_{ij}^{(2)}, \qquad (1.6.18)$$

$$C_{ij} = C_{ij} + A_{ij}^{(3)} B_{ij}^{(3)}, \qquad (1.6.19)$$

$$C_{ij} = C_{ij} + A_{ij}^{(4)} B_{ij}^{(4)}, \qquad (1.6.20)$$

involves communication with $\text{Proc}(i,j)$'s four neighbors in the toroidal network. To see this, define the block downshift permutation

$$P = \begin{bmatrix} 0 & 0 & 0 & I_{n_1} \\ I_{n_1} & 0 & 0 & 0 \\ 0 & I_{n_1} & 0 & 0 \\ 0 & 0 & I_{n_1} & 0 \end{bmatrix}$$

and observe that $A^{(k+1)} = A^{(k)} P^T$ and $B^{(k+1)} = P B^{(k)}$. That is, the transition from $A^{(k)}$ to $A^{(k+1)}$ involves shifting A-blocks to the right one column (with wraparound) while the transition from $B^{(k)}$ to $B^{(k+1)}$ involves shifting the B-blocks down one row (with wraparound). After each update (1.6.17)–(1.6.20), the housed A-block is passed to $\text{Proc}(i,j)$'s "east" neighbor and the next A-block is received from its "west" neighbor. Likewise, the housed B-block is sent to its "south" neighbor and the next B-block is received from its "north" neighbor.

Of course, the Cannon algorithm can be implemented on any processor network. But we see from the above that it is particularly well suited when there are toroidal connections for then communication is always between adjacent processors.

Problems

P1.6.1 Justify Equations (1.6.3) and (1.6.4).

P1.6.2 Contrast the two task distribution strategies in §1.6.2 for the case when the first q block rows of A are zero and the first q block columns of B are zero.

P1.6.3 Verify Equations (1.6.13) and (1.6.15).

P1.6.4 Develop a shared memory method for overwriting A with A^2 where it is assumed that $A \in \mathbb{R}^{n \times n}$ resides in global memory at the start.

P1.6.5 Develop a shared memory method for computing $B = A^T A$ where it is assumed that $A \in \mathbb{R}^{m \times n}$ resides in global memory at the start and that B is stored in global memory at the end.

P1.6.6 Prove (1.6.16) for general N. Use the block downshift matrix to define $A^{(i)}$ and $B^{(i)}$.

Notes and References for §1.6

To learn more about the practical implementation of parallel matrix multiplication, see `scaLAPACK` as well as:

L. Cannon (1969). "A Cellular Computer to Implement the Kalman Filter Algorithm," PhD Thesis, Montana State University, Bozeman, MT.

K. Gallivan, W. Jalby, and U. Meier (1987). "The Use of BLAS3 in Linear Algebra on a Parallel Processor with a Hierarchical Memory," *SIAM J. Sci. Stat. Comput. 8*, 1079–1084.

P. Bjørstad, F. Manne, T.Sørevik, and M. Vajteršic (1992). "Efficient Matrix Multiplication on SIMD Computers," *SIAM J. Matrix Anal. Appl. 13*, 386–401.

S.L. Johnsson (1993). "Minimizing the Communication Time for Matrix Multiplication on Multiprocessors," *Parallel Comput. 19*, 1235–1257.

K. Mathur and S.L. Johnsson (1994). "Multiplication of Matrices of Arbitrary Shape on a Data Parallel Computer," *Parallel Comput. 20*, 919–952.

J. Choi, D.W. Walker, and J. Dongarra (1994) "Pumma: Parallel Universal Matrix Multiplication Algorithms on Distributed Memory Concurrent Computers," *Concurrency: Pract. Exper. 6*, 543–570.

R.C. Agarwal, F.G. Gustavson, and M. Zubair (1994). "A High-Performance Matrix-Multiplication Algorithm on a Distributed-Memory Parallel Computer, Using Overlapped Communication," *IBM J. Res. Devel. 38*, 673–681.

D. Irony, S. Toledo, and A. Tiskin (2004). "Communication Lower Bounds for Distributed Memory Matrix Multiplication," *J. Parallel Distrib. Comput. 64*, 1017–1026.

Lower bounds for communication overheads are important as they establish a target for implementers, see:

G. Ballard, J. Demmel, O. Holtz, and O. Schwartz (2011). "Minimizing Communication in Numerical Linear Algebra," *SIAM. J. Matrix Anal. Applic. 32*, 866–901.

Matrix transpose in a distributed memory environment is surprisingly complex. The study of this central, no-flop calculation is a reminder of just how important it is to control the costs of data motion. See

S.L. Johnsson and C.T. Ho (1988). "Matrix Transposition on Boolean N-cube Configured Ensemble Architectures," *SIAM J. Matrix Anal. Applic. 9*, 419–454.

J. Choi, J.J. Dongarra, and D.W. Walker (1995). "Parallel Matrix Transpose Algorithms on Distributed Memory Concurrent Computers," *Parallel Comput. 21*, 1387–1406.

The parallel matrix computation literature is a vast, moving target. Ideas come and go with shifts in architectures. Nevertheless, it is useful to offer a small set of references that collectively trace the early development of the field:

D. Heller (1978). "A Survey of Parallel Algorithms in Numerical Linear Algebra," *SIAM Review 20*, 740–777.

J.M. Ortega and R.G. Voigt (1985). "Solution of Partial Differential Equations on Vector and Parallel Computers," *SIAM Review 27*, 149–240.

D.P. O'Leary and G.W. Stewart (1985). "Data Flow Algorithms for Parallel Matrix Computations," *Commun. ACM 28*, 841–853.

J.J. Dongarra and D.C. Sorensen (1986). "Linear Algebra on High Performance Computers," *Appl. Math. Comput. 20*, 57–88.

M.T. Heath, ed. (1987). *Hypercube Multiprocessors*, SIAM Publications, Philadelphia, PA.

Y. Saad and M.H. Schultz (1989). "Data Communication in Parallel Architectures," *J. Dist. Parallel Comput. 11*, 131–150.

J.J. Dongarra, I. Duff, D. Sorensen, and H. van der Vorst (1990). *Solving Linear Systems on Vector and Shared Memory Computers*, SIAM Publications, Philadelphia, PA.

K.A. Gallivan, R.J. Plemmons, and A.H. Sameh (1990). "Parallel Algorithms for Dense Linear Algebra Computations," *SIAM Review 32*, 54–135.

J.W. Demmel, M.T. Heath, and H.A. van der Vorst (1993). "Parallel Numerical Linear Algebra," in *Acta Numerica 1993*, Cambridge University Press.

A. Edelman (1993). "Large Dense Numerical Linear Algebra in 1993: The Parallel Computing Influence," *Int. J. Supercomput. Applic. 7*, 113–128.

Chapter 2

Matrix Analysis

2.1   Basic Ideas from Linear Algebra

2.2   Vector Norms

2.3   Matrix Norms

2.4   The Singular Value Decomposition

2.5   Subspace Metrics

2.6   The Sensitivity of Square Systems

2.7   Finite Precision Matrix Computations

The analysis and derivation of algorithms in the matrix computation area requires a facility with linear algebra. Some of the basics are reviewed in §2.1. Norms are particularly important, and we step through the vector and matrix cases in §2.2 and §2.3. The ubiquitous singular value decomposition is introduced in §2.4 and then used in the next section to define the CS decomposition and its ramifications for the measurement of subspace separation. In §2.6 we examine how the solution to a linear system $Ax = b$ changes if A and b are perturbed. It is the ideal setting for introducing the concepts of problem sensitivity, backward error analysis, and condition number. These ideas are central throughout the text. To complete the chapter we develop a model of finite-precision floating point arithmetic based on the IEEE standard. Several canonical examples of roundoff error analysis are offered.

Reading Notes

Familiarity with matrix manipulation consistent with §1.1–§1.3 is essential. The sections within this chapter depend upon each other as follows:

$$
\begin{array}{ccccccc}
 & & & & & \S2.5 & \\
 & & & & & \uparrow & \\
\S2.1 & \rightarrow & \S2.2 & \rightarrow & \S2.3 & \rightarrow & \S2.4 \\
 & & & & & \downarrow & \\
 & & & & & \S2.6 & \rightarrow \quad \S2.7
\end{array}
$$

Complementary references include Forsythe and Moler (SLAS), Stewart (IMC), Horn and Johnson (MA), Stewart (MABD), Ipsen (NMA), and Watkins (FMC). Fundamentals of matrix analysis that are specific to least squares problems and eigenvalue problems appear in later chapters.

2.1   Basic Ideas from Linear Algebra

This section is a quick review of linear algebra. Readers who wish a more detailed coverage should consult the references at the end of the section.

2.1.1   Independence, Subspace, Basis, and Dimension

A set of vectors $\{a_1, \ldots, a_n\}$ in $\mathbb{R}^m$ is *linearly independent* if $\sum_{j=1}^{n} \alpha_j a_j = 0$ implies $\alpha(1{:}n) = 0$. Otherwise, a nontrivial combination of the a_i is zero and $\{a_1, \ldots, a_n\}$ is said to be *linearly dependent*.

A *subspace* of $\mathbb{R}^m$ is a subset that is also a vector space. Given a collection of vectors $a_1, \ldots, a_n \in \mathbb{R}^m$, the set of all linear combinations of these vectors is a subspace referred to as the *span* of $\{a_1, \ldots, a_n\}$:

$$\mathsf{span}\{a_1, \ldots, a_n\} = \Big\{ \sum_{j=1}^{n} \beta_j a_j : \beta_j \in \mathbb{R} \Big\}.$$

If $\{a_1, \ldots, a_n\}$ is independent and $b \in \mathsf{span}\{a_1, \ldots, a_n\}$, then b is a unique linear combination of the a_j.

If $S_1, \ldots, S_k$ are subspaces of $\mathbb{R}^m$, then their sum is the subspace defined by $S = \{\, a_1 + a_2 + \cdots + a_k : a_i \in S_i,\ i = 1{:}k\,\}$. S is said to be a *direct sum* if each $v \in S$ has a unique representation $v = a_1 + \cdots + a_k$ with $a_i \in S_i$. In this case we write $S = S_1 \oplus \cdots \oplus S_k$. The intersection of the S_i is also a subspace, $S = S_1 \cap S_2 \cap \cdots \cap S_k$.

The subset $\{a_{i_1}, \ldots, a_{i_k}\}$ is a *maximal linearly independent subset* of $\{a_1, \ldots, a_n\}$ if it is linearly independent and is not properly contained in any linearly independent subset of $\{a_1, \ldots, a_n\}$. If $\{a_{i_1}, \ldots, a_{i_k}\}$ is maximal, then $\mathsf{span}\{a_1, \ldots, a_n\} = \mathsf{span}\{a_{i_1}, \ldots, a_{i_k}\}$ and $\{a_{i_1}, \ldots, a_{i_k}\}$ is a *basis* for $\mathsf{span}\{a_1, \ldots, a_n\}$. If $S \subseteq \mathbb{R}^m$ is a subspace, then it is possible to find independent basic vectors $a_1, \ldots, a_k \in S$ such that $S = \mathsf{span}\{a_1, \ldots, a_k\}$. All bases for a subspace S have the same number of elements. This number is the *dimension* and is denoted by $\mathsf{dim}(S)$.

2.1.2   Range, Null Space, and Rank

There are two important subspaces associated with an m-by-n matrix A. The *range* of A is defined by

$$\mathsf{ran}(A) = \{y \in \mathbb{R}^m : y = Ax \text{ for some } x \in \mathbb{R}^n\}$$

and the *nullspace* of A is defined by

$$\mathsf{null}(A) = \{x \in \mathbb{R}^n : Ax = 0\}.$$

If $A = [\, a_1 \mid \cdots \mid a_n \,]$ is a column partitioning, then

$$\mathsf{ran}(A) = \mathsf{span}\{a_1, \ldots, a_n\}.$$

The *rank* of a matrix A is defined by

$$\text{rank}(A) = \dim(\text{ran}(A)).$$

If $A \in \mathbb{R}^{m \times n}$, then

$$\dim(\text{null}(A)) + \text{rank}(A) = n.$$

We say that $A \in \mathbb{R}^{m \times n}$ is *rank deficient* if $\text{rank}(A) < \min\{m, n\}$. The rank of a matrix is the maximal number of linearly independent columns (or rows).

2.1.3 Matrix Inverse

If A and X are in $\mathbb{R}^{n \times n}$ and satisfy $AX = I$, then X is the *inverse* of A and is denoted by A^{-1}. If A^{-1} exists, then A is said to be *nonsingular*. Otherwise, we say A is *singular*. The inverse of a product is the reverse product of the inverses:

$$(AB)^{-1} = B^{-1} A^{-1}. \tag{2.1.1}$$

Likewise, the transpose of the inverse is the inverse of the transpose:

$$(A^{-1})^T = (A^T)^{-1} \equiv A^{-T}. \tag{2.1.2}$$

2.1.4 The Sherman-Morrison-Woodbury Formula

The identity

$$B^{-1} = A^{-1} - B^{-1}(B - A)A^{-1} \tag{2.1.3}$$

shows how the inverse changes if the matrix changes. The *Sherman-Morrison-Woodbury formula* gives a convenient expression for the inverse of the matrix $(A + UV^T)$ where $A \in \mathbb{R}^{n \times n}$ and U and V are n-by-k:

$$(A + UV^T)^{-1} = A^{-1} - A^{-1}U(I + V^T A^{-1} U)^{-1} V^T A^{-1}. \tag{2.1.4}$$

A rank-k correction to a matrix results in a rank-k correction of the inverse. In (2.1.4) we assume that both A and $(I + V^T A^{-1} U)$ are nonsingular.

The $k = 1$ case is particularly useful. If $A \in \mathbb{R}^{n \times n}$ is nonsingular, $u, v \in \mathbb{R}^n$, and $\alpha = 1 + v^T A^{-1} u \neq 0$, then

$$(A + uv^T)^{-1} = A^{-1} - \frac{1}{\alpha} A^{-1} uv^T A^{-1}. \tag{2.1.5}$$

This is referred to as the *Sherman-Morrison formula*.

2.1.5 Orthogonality

A set of vectors $\{x_1, \ldots, x_p\}$ in $\mathbb{R}^m$ is *orthogonal* if $x_i^T x_j = 0$ whenever $i \neq j$ and *orthonormal* if $x_i^T x_j = \delta_{ij}$. Intuitively, orthogonal vectors are maximally independent for they point in totally different directions.

A collection of subspaces $S_1, \ldots, S_p$ in $\mathbb{R}^m$ is *mutually orthogonal* if $x^T y = 0$ whenever $x \in S_i$ and $y \in S_j$ for $i \neq j$. The *orthogonal complement* of a subspace $S \subseteq \mathbb{R}^m$ is defined by

$$S^\perp = \{y \in \mathbb{R}^m : y^T x = 0 \text{ for all } x \in S\}.$$

It is not hard to show that $\mathsf{ran}(A)^{\perp} = \mathsf{null}(A^T)$. The vectors $v_1, \ldots, v_k$ form an *orthonormal* basis for a subspace $S \subseteq \mathbb{R}^m$ if they are orthonormal and span S.

A matrix $Q \in \mathbb{R}^{m \times m}$ is said to be *orthogonal* if $Q^T Q = I$. If $Q = [\, q_1 \,|\, \cdots \,|\, q_m \,]$ is orthogonal, then the q_i form an orthonormal basis for $\mathbb{R}^m$. It is always possible to extend such a basis to a full orthonormal basis $\{v_1, \ldots, v_m\}$ for $\mathbb{R}^m$:

Theorem 2.1.1. *If $V_1 \in \mathbb{R}^{n \times r}$ has orthonormal columns, then there exists $V_2 \in \mathbb{R}^{n \times (n-r)}$ such that*

$$V = [\, V_1 \,|\, V_2 \,]$$

is orthogonal. Note that $\mathsf{ran}(V_1)^{\perp} = \mathsf{ran}(V_2)$.

Proof. This is a standard result from introductory linear algebra. It is also a corollary of the QR factorization that we present in §5.2.　□

2.1.6　The Determinant

If $A = (a) \in \mathbb{R}^{1 \times 1}$, then its *determinant* is given by $\mathsf{det}(A) = a$. The determinant of $A \in \mathbb{R}^{n \times n}$ is defined in terms of order-$(n-1)$ determinants:

$$\mathsf{det}(A) = \sum_{j=1}^{n} (-1)^{j+1} a_{1j} \mathsf{det}(A_{1j}).$$

Here, A_{1j} is an $(n-1)$-by-$(n-1)$ matrix obtained by deleting the first row and jth column of A. Well-known properties of the determinant include $\mathsf{det}(AB) = \mathsf{det}(A)\mathsf{det}(B)$, $\mathsf{det}(A^T) = \mathsf{det}(A)$, and $\mathsf{det}(cA) = c^n \mathsf{det}(A)$ where $A, B \in \mathbb{R}^{n \times n}$ and $c \in \mathbb{R}$. In addition, $\mathsf{det}(A) \neq 0$ if and only if A is nonsingular.

2.1.7　Eigenvalues and Eigenvectors

Until we get to the main eigenvalue part of the book (Chapters 7 and 8), we need a handful of basic properties so that we can fully appreciate the singular value decomposition (§2.4), positive definiteness (§4.2), and various fast linear equation solvers (§4.8).

The *eigenvalues* of $A \in \mathbb{C}^{n \times n}$ are the zeros of the *characteristic polynomial*

$$p(x) = \mathsf{det}(A - xI).$$

Thus, every n-by-n matrix has n eigenvalues. We denote the set of A's eigenvalues by

$$\lambda(A) = \{\, x : \mathsf{det}(A - xI) = 0 \,\}.$$

If the eigenvalues of A are real, then we index them from largest to smallest as follows:

$$\lambda_n(A) \leq \cdots \leq \lambda_2(A) \leq \lambda_1(A).$$

In this case, we sometimes use the notation $\lambda_{\max}(A)$ and $\lambda_{\min}(A)$ to denote $\lambda_1(A)$ and $\lambda_n(A)$ respectively.

If $X \in \mathbb{C}^{n \times n}$ is nonsingular and $B = X^{-1}AX$, then A and B are *similar*. If two matrices are similar, then they have exactly the same eigenvalues.

If $\lambda \in \lambda(A)$, then there exists a nonzero vector x so that $Ax = \lambda x$. Such a vector is said to be an *eigenvector* for A associated with λ. If $A \in \mathbb{C}^{n \times n}$ has n independent eigenvectors $x_1, \ldots, x_n$ and $Ax_i = \lambda_i x_i$ for $i = 1{:}n$, then A is *diagonalizable*. The terminology is appropriate for if

$$X = [\, x_1 \,|\, \cdots \,|\, x_n \,],$$

then

$$X^{-1}AX = \operatorname{diag}(\lambda_1, \ldots, \lambda_n).$$

Not all matrices are diagonalizable. However, if $A \in \mathbb{R}^{n \times n}$ is symmetric, then there exists an orthogonal Q so that

$$Q^T A Q = \operatorname{diag}(\lambda_1, \ldots, \lambda_n). \tag{2.1.6}$$

This is called the *Schur decomposition*. The largest and smallest eigenvalues of a symmetric matrix satisfy

$$\lambda_{\max}(A) = \max_{x \neq 0} \frac{x^T A x}{x^T x} \tag{2.1.7}$$

and

$$\lambda_{\min}(A) = \min_{x \neq 0} \frac{x^T A x}{x^T x}. \tag{2.1.8}$$

2.1.8 Differentiation

Suppose α is a scalar and that $A(\alpha)$ is an m-by-n matrix with entries $a_{ij}(\alpha)$. If $a_{ij}(\alpha)$ is a differentiable function of α for all i and j, then by $\dot{A}(\alpha)$ we mean the matrix

$$\dot{A}(\alpha) = \frac{d}{d\alpha} A(\alpha) = \left(\frac{d}{d\alpha} a_{ij}(\alpha) \right) = (\dot{a}_{ij}(\alpha)).$$

Differentiation is a useful tool that can sometimes provide insight into the sensitivity of a matrix problem.

Problems

P2.1.1 Show that if $A \in \mathbb{R}^{m \times n}$ has rank p, then there exists an $X \in \mathbb{R}^{m \times p}$ and a $Y \in \mathbb{R}^{n \times p}$ such that $A = XY^T$, where $\operatorname{rank}(X) = \operatorname{rank}(Y) = p$.

P2.1.2 Suppose $A(\alpha) \in \mathbb{R}^{m \times r}$ and $B(\alpha) \in \mathbb{R}^{r \times n}$ are matrices whose entries are differentiable functions of the scalar α. (a) Show

$$\frac{d}{d\alpha}[A(\alpha)B(\alpha)] = \left[\frac{d}{d\alpha} A(\alpha) \right] B(\alpha) + A(\alpha) \left[\frac{d}{d\alpha} B(\alpha) \right].$$

(b) Assuming $A(\alpha)$ is always nonsingular, show

$$\frac{d}{d\alpha} \left[A(\alpha)^{-1} \right] = -A(\alpha)^{-1} \left[\frac{d}{d\alpha} A(\alpha) \right] A(\alpha)^{-1}.$$

P2.1.3 Suppose $A \in \mathbb{R}^{n \times n}$, $b \in \mathbb{R}^n$ and that $\phi(x) = \frac{1}{2} x^T A x - x^T b$. Show that the gradient of ϕ is given by $\nabla\phi(x) = \frac{1}{2}(A^T + A)x - b$.

P2.1.4 Assume that both A and $A + uv^T$ are nonsingular where $A \in \mathbb{R}^{n \times n}$ and $u, v \in \mathbb{R}^n$. Show that if x solves $(A + uv^T)x = b$, then it also solves a perturbed right-hand-side problem of the form $Ax = b + \alpha u$. Give an expression for α in terms of A, u, and v.

P2.1.5 Show that a triangular orthogonal matrix is diagonal.

P2.1.6 Suppose $A \in \mathbb{R}^{n \times n}$ is symmetric and nonsingular and define

$$\tilde{A} = A + \alpha(uu^T + vv^T) + \beta(uv^T + vu^T)$$

where $u, v \in \mathbb{R}^n$ and $\alpha, \beta \in \mathbb{R}$. Assuming that $\tilde{A}$ is nonsingular, use the Sherman-Morrison-Woodbury formula to develop a formula for $\tilde{A}^{-1}$.

P2.1.7 Develop a symmetric version of the Sherman-Morrison-Woodbury formula that characterizes the inverse of $A + USU^T$ where $A \in \mathbb{R}^{n \times n}$ and $S \in \mathbb{R}^{k \times k}$ are symmetric and $U \in \mathbb{R}^{n \times k}$.

P2.1.8 Suppose $Q \in \mathbb{R}^{n \times n}$ is orthogonal and $z \in \mathbb{R}^n$. Give an efficient algorithm for setting up an m-by-m matrix $A = (a_{ij})$ defined by $a_{ij} = v^T (Q^i)^T (Q^j) v$.-

P2.1.9 Show that if S is real and $S^T = -S$, then $I - S$ is nonsingular and the matrix $(I - S)^{-1}(I + S)$ is orthogonal. This is known as the *Cayley transform* of S.

P2.1.10 Refer to §1.3.10. (a) Show that if $S \in \mathbb{R}^{2n \times 2n}$ is symplectic, then S^{-1} exists and is also symplectic. (b) Show that if $M \in \mathbb{R}^{2n \times 2n}$ is Hamiltonian and $S \in \mathbb{R}^{2n \times 2n}$ is symplectic, then the matrix $M_1 = S^{-1}MS$ is Hamiltonian.

P2.1.11 Use (2.1.6) to prove (2.1.7) and (2.1.8).

Notes and References for §2.1

In addition to Horn and Johnson (MA) and Horn and Johnson (TMA), the following introductory applied linear algebra texts are highly recommended:

R. Bellman (1997). *Introduction to Matrix Analysis,* Second Edition, SIAM Publications, Philadelphia, PA.

C. Meyer (2000). *Matrix Analysis and Applied Linear Algebra,* SIAM Publications, Philadelphia, PA.

D. Lay (2005). *Linear Algebra and Its Applications,* Third Edition, Addison-Wesley, Reading, MA.

S.J. Leon (2007). *Linear Algebra with Applications,* Seventh Edition, Prentice-Hall, Englewood Cliffs, NJ.

G. Strang (2009). *Introduction to Linear Algebra,* Fourth Edition, SIAM Publications, Philadelphia, PA.

2.2   Vector Norms

A norm on a vector space plays the same role as absolute value: it furnishes a distance measure. More precisely, $\mathbb{R}^n$ together with a norm on $\mathbb{R}^n$ defines a metric space rendering the familiar notions of neighborhood, open sets, convergence, and continuity.

2.2.1   Definitions

A *vector norm* on $\mathbb{R}^n$ is a function $f : \mathbb{R}^n \to \mathbb{R}$ that satisfies the following properties:

$$
\begin{aligned}
f(x) &\geq 0, & x \in \mathbb{R}^n, \quad (f(x) = 0, \text{ iff } x = 0), \\
f(x + y) &\leq f(x) + f(y), & x, y \in \mathbb{R}^n, \\
f(\alpha x) &= |\alpha| f(x), & \alpha \in \mathbb{R}, x \in \mathbb{R}^n.
\end{aligned}
$$

We denote such a function with a double bar notation: $f(x) = \| x \|$. Subscripts on the double bar are used to distinguish between various norms. A useful class of vector norms are the *p-norms* defined by

$$\| x \|_p = (|x_1|^p + \cdots + |x_n|^p)^{\frac{1}{p}}, \qquad p \geq 1 .\qquad (2.2.1)$$

The $1-$, $2-$, and $\infty-$ norms are the most important:

$$
\begin{aligned}
\| x \|_1 &= |x_1| + \cdots + |x_n|, \\
\| x \|_2 &= \left(|x_1|^2 + \cdots + |x_n|^2\right)^{\frac{1}{2}} = \left(x^T x\right)^{\frac{1}{2}}, \\
\| x \|_\infty &= \max_{1 \leq i \leq n} |x_i|.
\end{aligned}
$$

A *unit vector* with respect to the norm $\| \cdot \|$ is a vector x that satisfies $\| x \| = 1$.

2.2.2 Some Vector Norm Properties

A classic result concerning p-norms is the *Hölder inequality*:

$$|x^T y| \leq \| x \|_p \| y \|_q \qquad \frac{1}{p} + \frac{1}{q} = 1. \qquad (2.2.2)$$

A very important special case of this is the *Cauchy-Schwarz inequality*:

$$|x^T y| \leq \| x \|_2 \| y \|_2. \qquad (2.2.3)$$

All norms on $\mathbb{R}^n$ are *equivalent* , i.e., if $\| \cdot \|_\alpha$ and $\| \cdot \|_\beta$ are norms on $\mathbb{R}^n$, then there exist positive constants c_1 and c_2 such that

$$c_1 \| x \|_\alpha \leq \| x \|_\beta \leq c_2 \| x \|_\alpha \qquad (2.2.4)$$

for all $x \in \mathbb{R}^n$. For example, if $x \in \mathbb{R}^n$, then

$$\| x \|_2 \leq \| x \|_1 \leq \sqrt{n} \, \| x \|_2, \qquad (2.2.5)$$

$$\| x \|_\infty \leq \| x \|_2 \leq \sqrt{n} \, \| x \|_\infty, \qquad (2.2.6)$$

$$\| x \|_\infty \leq \| x \|_1 \leq n \, \| x \|_\infty. \qquad (2.2.7)$$

Finally, we mention that the 2-norm is preserved under orthogonal transformation. Indeed, if $Q \in \mathbb{R}^{n \times n}$ is orthogonal and $x \in \mathbb{R}^n$, then

$$\| Qx \|_2^2 = (Qx)^T (Qx) = (x^T Q^T)(Qx) = x^T (Q^T Q)x = x^T x = \| x \|_2^2.$$

2.2.3 Absolute and Relative Errors

Suppose $\hat{x} \in \mathbb{R}^n$ is an approximation to $x \in \mathbb{R}^n$. For a given vector norm $\| \cdot \|$ we say that

$$\epsilon_{\mathrm{abs}} = \| \hat{x} - x \|$$

is the *absolute error* in $\hat{x}$. If $x \neq 0$, then

$$\epsilon_{\mathrm{rel}} = \frac{\| \hat{x} - x \|}{\| x \|}$$

prescribes the *relative error* in $\hat{x}$. Relative error in the ∞-norm can be translated into a statement about the number of correct significant digits in $\hat{x}$. In particular, if

$$\frac{\|\hat{x} - x\|_\infty}{\|x\|_\infty} \approx 10^{-p},$$

then the largest component of $\hat{x}$ has approximately p correct significant digits. For example, if $x = [\,1.234\ \ .05674\,]^T$ and $\hat{x} = [\,1.235\ \ .05128\,]^T$, then $\|\hat{x} - x\|_\infty/\|x\|_\infty \approx .0043 \approx 10^{-3}$. Note than $\hat{x}_1$ has about three significant digits that are correct while only one significant digit in $\hat{x}_2$ is correct.

2.2.4   Convergence

We say that a sequence $\{x^{(k)}\}$ of n-vectors *converges* to x if

$$\lim_{k \to \infty} \|x^{(k)} - x\| = 0.$$

Because of (2.2.4), convergence in *any* particular norm implies convergence in *all* norms.

Problems

P2.2.1   Show that if $x \in \mathbb{R}^n$, then $\lim_{p \to \infty} \|x\|_p = \|x\|_\infty$.

P2.2.2   By considering the inequality $0 \le (ax + by)^T(ax + by)$ for suitable scalars a and b, prove (2.2.3).

P2.2.3   Verify that $\|\cdot\|_1$, $\|\cdot\|_2$, and $\|\cdot\|_\infty$ are vector norms.

P2.2.4   Verify (2.2.5)-(2.2.7). When is equality achieved in each result?

P2.2.5   Show that in $\mathbb{R}^n$, $x^{(i)} \to x$ if and only if $x_k^{(i)} \to x_k$ for $k = 1{:}n$.

P2.2.6   Show that for any vector norm on $\mathbb{R}^n$ that $|\,\|x\| - \|y\|\,| \le \|x - y\|$.

P2.2.7   Let $\|\cdot\|$ be a vector norm on $\mathbb{R}^m$ and assume $A \in \mathbb{R}^{m \times n}$. Show that if $\mathrm{rank}(A) = n$, then $\|x\|_A = \|Ax\|$ is a vector norm on $\mathbb{R}^n$.

P2.2.8   Let x and y be in $\mathbb{R}^n$ and define $\psi{:}\mathbb{R} \to \mathbb{R}$ by $\psi(\alpha) = \|x - \alpha y\|_2$. Show that ψ is minimized if $\alpha = x^T y/y^T y$.

P2.2.9   Prove or disprove:

$$v \in \mathbb{R}^n \ \Rightarrow\ \|v\|_1 \|v\|_\infty \le \frac{1 + \sqrt{n}}{2} \|v\|_2^2.$$

P2.2.10   If $x \in \mathbb{R}^3$ and $y \in \mathbb{R}^3$ then it can be shown that $|x^T y| = \|x\|_2 \|y\|_2 |\cos(\theta)|$ where θ is the angle between x and y. An analogous result exists for the *cross product* defined by

$$x \times y \ = \ \begin{bmatrix} x_2 y_3 - x_3 y_2 \\ x_3 y_1 - x_1 y_3 \\ x_1 y_2 - x_2 y_1 \end{bmatrix}.$$

In particular, $\|x \times y\|_2 = \|x\|_2 \|y\|_2 |\sin(\theta)|$. Prove this.

P2.2.11   Suppose $x \in \mathbb{R}^n$ and $y \in \mathbb{R}^m$. Show that

$$\|x \otimes y\|_p = \|x\|_p \|y\|_p$$

for $p = 1, 2$, and ∞.

Notes and References for §2.2

Although a vector norm is "just" a generalization of the absolute value concept, there are some noteworthy subtleties:

J.D. Pryce (1984). "A New Measure of Relative Error for Vectors," *SIAM J. Numer. Anal.* **21**, 202–221.

2.3   Matrix Norms

The analysis of matrix algorithms requires use of matrix norms. For example, the quality of a linear system solution may be poor if the matrix of coefficients is "nearly singular." To quantify the notion of near-singularity, we need a measure of distance on the space of matrices. Matrix norms can be used to provide that measure.

2.3.1   Definitions

Since $\mathbb{R}^{m \times n}$ is isomorphic to $\mathbb{R}^{mn}$, the definition of a matrix norm should be equivalent to the definition of a vector norm. In particular, $f : \mathbb{R}^{m \times n} \to \mathbb{R}$ is a matrix norm if the following three properties hold:

$$
\begin{aligned}
f(A) &\geq 0, & A &\in \mathbb{R}^{m \times n}, & (f(A) = 0 \text{ iff } A = 0) \\
f(A + B) &\leq f(A) + f(B), & A, B &\in \mathbb{R}^{m \times n}, \\
f(\alpha A) &= |\alpha| f(A), & \alpha &\in \mathbb{R}, A \in \mathbb{R}^{m \times n}.
\end{aligned}
$$

As with vector norms, we use a double bar notation with subscripts to designate matrix norms, i.e., $\| A \| = f(A)$.

The most frequently used matrix norms in numerical linear algebra are the Frobenius norm

$$
\| A \|_F = \sqrt{\sum_{i=1}^{m} \sum_{j=1}^{n} |a_{ij}|^2} \tag{2.3.1}
$$

and the p-norms

$$
\| A \|_p = \sup_{x \neq 0} \frac{\| Ax \|_p}{\| x \|_p}. \tag{2.3.2}
$$

Note that the matrix p-norms are defined in terms of the vector p-norms discussed in the previous section. The verification that (2.3.1) and (2.3.2) are matrix norms is left as an exercise. It is clear that $\| A \|_p$ is the p-norm of the largest vector obtained by applying A to a unit p-norm vector:

$$
\| A \|_p = \sup_{x \neq 0} \left\| A \left(\frac{x}{\| x \|_p} \right) \right\|_p = \max_{\| x \|_p = 1} \| Ax \|_p.
$$

It is important to understand that (2.3.2) defines a family of norms—the 2-norm on $\mathbb{R}^{3 \times 2}$ is a different function from the 2-norm on $\mathbb{R}^{5 \times 6}$. Thus, the easily verified inequality

$$
\| AB \|_p \leq \| A \|_p \| B \|_p, \qquad A \in \mathbb{R}^{m \times n}, B \in \mathbb{R}^{n \times q} \tag{2.3.3}
$$

is really an observation about the relationship between three different norms. Formally, we say that norms f_1, f_2, and f_3 on $\mathbb{R}^{m \times q}$, $\mathbb{R}^{m \times n}$, and $\mathbb{R}^{n \times q}$ are *mutually consistent* if for all matrices $A \in \mathbb{R}^{m \times n}$ and $B \in \mathbb{R}^{n \times q}$ we have $f_1(AB) \leq f_2(A) f_3(B)$, or, in subscript-free norm notation:

$$
\| AB \| \leq \| A \| \| B \|. \tag{2.3.4}
$$

Not all matrix norms satisfy this property. For example, if $\| A \|_\Delta = \max |a_{ij}|$ and

$$A = B = \begin{bmatrix} 1 & 1 \\ 1 & 1 \end{bmatrix},$$

then $\| AB \|_\Delta > \| A \|_\Delta \| B \|_\Delta$. For the most part, we work with norms that satisfy (2.3.4).

The p-norms have the important property that for every $A \in \mathbb{R}^{m \times n}$ and $x \in \mathbb{R}^n$ we have

$$\| Ax \|_p \leq \| A \|_p \| x \|_p.$$

More generally, for any vector norm $\| \cdot \|_\alpha$ on $\mathbb{R}^n$ and $\| \cdot \|_\beta$ on $\mathbb{R}^m$ we have $\| Ax \|_\beta \leq \| A \|_{\alpha,\beta} \| x \|_\alpha$ where $\| A \|_{\alpha,\beta}$ is a matrix norm defined by

$$\| A \|_{\alpha,\beta} = \sup_{x \neq 0} \frac{\| Ax \|_\beta}{\| x \|_\alpha}. \tag{2.3.5}$$

We say that $\| \cdot \|_{\alpha,\beta}$ is *subordinate* to the vector norms $\| \cdot \|_\alpha$ and $\| \cdot \|_\beta$. Since the set $\{x \in \mathbb{R}^n : \| x \|_\alpha = 1\}$ is compact and $\| \cdot \|_\beta$ is continuous, it follows that

$$\| A \|_{\alpha,\beta} = \max_{\|x\|_\alpha = 1} \| Ax \|_\beta = \| Ax_* \|_\beta \tag{2.3.6}$$

for some $x_* \in \mathbb{R}^n$ having unit α-norm.

2.3.2 Some Matrix Norm Properties

The Frobenius and p-norms (especially $p = 1, 2, \infty$) satisfy certain inequalities that are frequently used in the analysis of a matrix computation. If $A \in \mathbb{R}^{m \times n}$ we have

$$\| A \|_2 \leq \| A \|_F \leq \sqrt{\min\{m,n\}} \| A \|_2, \tag{2.3.7}$$

$$\max_{i,j} |a_{ij}| \leq \| A \|_2 \leq \sqrt{mn} \max_{i,j} |a_{ij}|, \tag{2.3.8}$$

$$\| A \|_1 = \max_{1 \leq j \leq n} \sum_{i=1}^m |a_{ij}|, \tag{2.3.9}$$

$$\| A \|_\infty = \max_{1 \leq i \leq m} \sum_{j=1}^n |a_{ij}|, \tag{2.3.10}$$

$$\frac{1}{\sqrt{n}} \| A \|_\infty \leq \| A \|_2 \leq \sqrt{m} \| A \|_\infty, \tag{2.3.11}$$

$$\frac{1}{\sqrt{m}} \| A \|_1 \leq \| A \|_2 \leq \sqrt{n} \| A \|_1. \tag{2.3.12}$$

If $A \in \mathbb{R}^{m \times n}$, $1 \leq i_1 \leq i_2 \leq m$, and $1 \leq j_1 \leq j_2 \leq n$, then

$$\| A(i_1{:}i_2, j_1{:}j_2) \|_p \leq \| A \|_p. \tag{2.3.13}$$

The proofs of these relationships are left as exercises. We mention that a sequence $\{A^{(k)}\} \in \mathbb{R}^{m \times n}$ *converges* if there exists a matrix $A \in \mathbb{R}^{m \times n}$ such that

$$\lim_{k \to \infty} \| A^{(k)} - A \| = 0.$$

The choice of norm is immaterial since all norms on $\mathbb{R}^{m \times n}$ are equivalent.

2.3.3 The Matrix 2-Norm

A nice feature of the matrix 1-norm and the matrix ∞-norm is that they are easy, $O(n^2)$ computations. (See (2.3.9) and (2.3.10).) The calculation of the 2-norm is considerably more complicated.

Theorem 2.3.1. *If $A \in \mathbb{R}^{m \times n}$, then there exists a unit 2-norm n-vector z such that $A^T A z = \mu^2 z$ where $\mu = \| A \|_2$.*

Proof. Suppose $z \in \mathbb{R}^n$ is a unit vector such that $\| Az \|_2 = \| A \|_2$. Since z maximizes the function

$$g(x) = \frac{1}{2} \frac{\| Ax \|_2^2}{\| x \|_2^2} = \frac{1}{2} \frac{x^T A^T A x}{x^T x}$$

it follows that it satisfies $\nabla g(z) = 0$ where ∇g is the gradient of g. A tedious differentiation shows that for $i = 1{:}n$

$$\frac{\partial g(z)}{\partial z_i} = \left[(z^T z) \sum_{j=1}^{n} (A^T A)_{ij} z_j - (z^T A^T A z) z_i \right] \Big/ (z^T z)^2 .$$

In vector notation this says that $A^T A z = (z^T A^T A z) z$. The theorem follows by setting $\mu = \| Az \|_2$. $\square$

The theorem implies that $\| A \|_2^2$ is a zero of $p(\lambda) = \det(A^T A - \lambda I)$. In particular,

$$\| A \|_2 = \sqrt{\lambda_{\max}(A^T A)}$$

We have much more to say about eigenvalues in Chapters 7 and 8. For now, we merely observe that 2-norm computation is iterative and a more involved calculation than those of the matrix 1-norm or ∞-norm. Fortunately, if the object is to obtain an order-of-magnitude estimate of $\| A \|_2$, then (2.3.7), (2.3.8), (2.3.11), or (2.3.12) can be used.

As another example of norm analysis, here is a handy result for 2-norm estimation.

Corollary 2.3.2. *If $A \in \mathbb{R}^{m \times n}$, then $\| A \|_2 \leq \sqrt{\| A \|_1 \| A \|_\infty}$.*

Proof. If $z \neq 0$ is such that $A^T A z = \mu^2 z$ with $\mu = \| A \|_2$, then $\mu^2 \| z \|_1 = \| A^T A z \|_1 \leq \| A^T \|_1 \| A \|_1 \| z \|_1 = \| A \|_\infty \| A \|_1 \| z \|_1$. $\square$

2.3.4   Perturbations and the Inverse

We frequently use norms to quantify the effect of perturbations or to prove that a sequence of matrices converges to a specified limit. As an illustration of these norm applications, let us quantify the change in A^{-1} as a function of change in A.

Lemma 2.3.3. *If $F \in \mathbb{R}^{n \times n}$ and $\| F \|_p < 1$, then $I - F$ is nonsingular and*

$$(I - F)^{-1} = \sum_{k=0}^{\infty} F^k$$

with

$$\| (I - F)^{-1} \|_p \leq \frac{1}{1 - \| F \|_p}.$$

Proof. Suppose $I - F$ is singular. It follows that $(I - F)x = 0$ for some nonzero x. But then $\| x \|_p = \| Fx \|_p$ implies $\| F \|_p \geq 1$, a contradiction. Thus, $I - F$ is nonsingular. To obtain an expression for its inverse consider the identity

$$\left(\sum_{k=0}^{N} F^k \right) (I - F) = I - F^{N+1}.$$

Since $\| F \|_p < 1$ it follows that $\lim_{k \to \infty} F^k = 0$ because $\| F^k \|_p \leq \| F \|_p^k$. Thus,

$$\left(\lim_{N \to \infty} \sum_{k=0}^{N} F^k \right) (I - F) = I.$$

It follows that $(I - F)^{-1} = \lim_{N \to \infty} \sum_{k=0}^{N} F^k$. From this it is easy to show that

$$\| (I - F)^{-1} \|_p \leq \sum_{k=0}^{\infty} \| F \|_p^k = \frac{1}{1 - \| F \|_p}$$

completing the proof of the theorem. $\square$

Note that $\| (I - F)^{-1} - I \|_p \leq \| F \|_p / (1 - \| F \|_p)$ is a consequence of the lemma. Thus, if $\epsilon \ll 1$, then $O(\epsilon)$ perturbations to the identity matrix induce $O(\epsilon)$ perturbations in the inverse. In general, we have

Theorem 2.3.4. *If A is nonsingular and $r \equiv \| A^{-1}E \|_p < 1$, then $A + E$ is nonsingular and*

$$\| (A + E)^{-1} - A^{-1} \|_p \leq \frac{\| E \|_p \| A^{-1} \|_p^2}{1 - r}.$$

Proof. Note that $A + E = (I + F)A$ where $F = -EA^{-1}$. Since $\| F \|_p = r < 1$, it follows from Lemma 2.3.3 that $I + F$ is nonsingular and $\| (I + F)^{-1} \|_p \leq 1/(1 - r)$.

Thus, $(A + E)^{-1} = A^{-1}(I + F)^{-1}$ is nonsingular and

$$(A + E)^{-1} - A^{-1} = A^{-1}(A - (A + E))(A + E)^{-1} = -A^{-1}EA^{-1}(I + F)^{-1}.$$

The theorem follows by taking norms. $\square$

2.3.5 Orthogonal Invariance

If $A \in \mathbb{R}^{m \times n}$ and the matrices $Q \in \mathbb{R}^{m \times m}$ and $Z \in \mathbb{R}^{n \times n}$ are orthogonal, then

$$\| QAZ \|_F = \| A \|_F \qquad (2.3.14)$$

and

$$\| QAZ \|_2 = \| A \|_2 . \qquad (2.3.15)$$

These properties readily follow from the orthogonal invariance of the vector 2-norm. For example,

$$\| QA \|_F^2 = \sum_{j=1}^n \| QA(:,j) \|_2^2 = \sum_{j=1}^n \| A(:,j) \|_2^2 = \| A \|_F^2$$

and so $\| Q(AZ) \|_F^2 = \| (AZ) \|_F^2 = \| Z^T A^T \|_F^2 = \| A^T \|_F^2 = \| A \|_F^2.$

Problems

P2.3.1 Show $\| AB \|_p \leq \| A \|_p \| B \|_p$ where $1 \leq p \leq \infty$.

P2.3.2 Let B be any submatrix of A. Show that $\| B \|_p \leq \| A \|_p$.

P2.3.3 Show that if $D = \text{diag}(\mu_1, \ldots, \mu_k) \in \mathbb{R}^{m \times n}$ with $k = \min\{m, n\}$, then $\| D \|_p = \max |\mu_i|$.

P2.3.4 Verify (2.3.7) and (2.3.8).

P2.3.5 Verify (2.3.9) and (2.3.10).

P2.3.6 Verify (2.3.11) and (2.3.12).

P2.3.7 Show that if $0 \neq s \in \mathbb{R}^n$ and $E \in \mathbb{R}^{n \times n}$, then

$$\left\| E \left(I - \frac{ss^T}{s^T s} \right) \right\|_F^2 = \| E \|_F^2 - \frac{\| Es \|_2^2}{s^T s} .$$

P2.3.8 Suppose $u \in \mathbb{R}^m$ and $v \in \mathbb{R}^n$. Show that if $E = uv^T$, then $\| E \|_F = \| E \|_2 = \| u \|_2 \| v \|_2$ and $\| E \|_\infty \leq \| u \|_\infty \| v \|_1$.

P2.3.9 Suppose $A \in \mathbb{R}^{m \times n}$, $y \in \mathbb{R}^m$, and $0 \neq s \in \mathbb{R}^n$. Show that $E = (y - As)s^T/s^T s$ has the smallest 2-norm of all m-by-n matrices E that satisfy $(A + E)s = y$.

P2.3.10 Verify that there exists a scalar $c > 0$ such that

$$\| A \|_{\Delta, c} = \max_{i, j} c|a_{ij}|$$

satisfies the submultiplicative property (2.3.4) for matrix norms on $\mathbb{R}^{n \times n}$. What is the smallest value for such a constant? Referring to this value as c_*, exhibit nonzero matrices B and C with the property that $\| BC \|_{\Delta, c_*} = \| B \|_{\Delta, c_*} \| C \|_{\Delta, c_*}$.

P2.3.11 Show that if A and B are matrices, then $\| A \otimes B \|_F = \| A \|_F \| B \|_F$.

Notes and References for §2.3

For further discussion of matrix norms, see Stewart (IMC) as well as:

F.L. Bauer and C.T. Fike (1960). "Norms and Exclusion Theorems," *Numer. Math. 2*, 137–144.
L. Mirsky (1960). "Symmetric Gauge Functions and Unitarily Invariant Norms," *Quart. J. Math. 11*, 50–59.
A.S. Householder (1964). *The Theory of Matrices in Numerical Analysis*, Dover Publications, New York.
N.J. Higham (1992). "Estimating the Matrix p-Norm," *Numer. Math. 62*, 539–556.

2.4   The Singular Value Decomposition

It is fitting that the first matrix decomposition that we present in the book is the singular value decomposition (SVD). The practical and theoretical importance of the SVD is hard to overestimate. It has a prominent role to play in data analysis and in the characterization of the many matrix "nearness problems."

2.4.1   Derivation

The SVD is an orthogonal matrix reduction and so the 2-norm and Frobenius norm figure heavily in this section. Indeed, we can prove the existence of the decomposition using some elementary facts about the 2-norm developed in the previous two sections.

Theorem 2.4.1 (Singular Value Decomposition). *If A is a real m-by-n matrix, then there exist orthogonal matrices*

$$U = [\, u_1 \mid \cdots \mid u_m \,] \in \mathbb{R}^{m \times m} \quad and \quad V = [\, v_1 \mid \cdots \mid v_n \,] \in \mathbb{R}^{n \times n}$$

such that

$$U^T A V = \Sigma = \mathrm{diag}(\sigma_1, \ldots, \sigma_p) \in \mathbb{R}^{m \times n}, \qquad p = \min\{m, n\},$$

where $\sigma_1 \geq \sigma_2 \geq \ldots \geq \sigma_p \geq 0$.

Proof. Let $x \in \mathbb{R}^n$ and $y \in \mathbb{R}^m$ be unit 2-norm vectors that satisfy $Ax = \sigma y$ with $\sigma = \| A \|_2$. From Theorem 2.1.1 there exist $V_2 \in \mathbb{R}^{n \times (n-1)}$ and $U_2 \in \mathbb{R}^{m \times (m-1)}$ so $V = [\, x \mid V_2 \,] \in \mathbb{R}^{n \times n}$ and $U = [\, y \mid U_2 \,] \in \mathbb{R}^{m \times m}$ are orthogonal. It is not hard to show that

$$U^T A V = \begin{bmatrix} \sigma & w^T \\ 0 & B \end{bmatrix} \equiv A_1$$

where $w \in \mathbb{R}^{n-1}$ and $B \in \mathbb{R}^{(m-1) \times (n-1)}$. Since

$$\left\| A_1 \left(\begin{bmatrix} \sigma \\ w \end{bmatrix} \right) \right\|_2^2 \geq (\sigma^2 + w^T w)^2$$

we have $\| A_1 \|_2^2 \geq (\sigma^2 + w^T w)$. But $\sigma^2 = \| A \|_2^2 = \| A_1 \|_2^2$, and so we must have $w = 0$. An obvious induction argument completes the proof of the theorem.   □

The σ_i are the *singular values* of A, the u_i are the *left singular vectors* of A, and the v_i are *right singular vectors* of A. Separate visualizations of the SVD are required depending upon whether A has more rows or columns. Here are the 3-by-2 and 2-by-3 examples:

$$
\begin{bmatrix} u_{11} & u_{12} & u_{13} \\ u_{21} & u_{22} & u_{23} \\ u_{31} & u_{32} & u_{33} \end{bmatrix}^T
\begin{bmatrix} a_{11} & a_{12} \\ a_{21} & a_{22} \\ a_{31} & a_{32} \end{bmatrix}
\begin{bmatrix} v_{11} & v_{12} \\ v_{21} & v_{22} \end{bmatrix}
= \begin{bmatrix} \sigma_1 & 0 \\ 0 & \sigma_2 \\ 0 & 0 \end{bmatrix},
$$

$$
\begin{bmatrix} u_{11} & u_{12} \\ u_{21} & u_{22} \end{bmatrix}^T
\begin{bmatrix} a_{11} & a_{12} & a_{13} \\ a_{21} & a_{22} & a_{23} \end{bmatrix}
\begin{bmatrix} v_{11} & v_{12} & v_{13} \\ v_{21} & v_{22} & v_{23} \\ v_{31} & v_{32} & v_{33} \end{bmatrix}
= \begin{bmatrix} \sigma_1 & 0 & 0 \\ 0 & \sigma_2 & 0 \end{bmatrix}.
$$

In later chapters, the notation $\sigma_i(A)$ is used to designate the ith largest singular value of a matrix A. The largest and smallest singular values are important and for them we also have a special notation:

$$\sigma_{\max}(A) = \text{the largest singular value of matrix } A,$$
$$\sigma_{\min}(A) = \text{the smallest singular value of matrix } A.$$

2.4.2 Properties

We establish a number of important corollaries to the SVD that are used throughout the book.

Corollary 2.4.2. *If $U^T A V = \Sigma$ is the SVD of $A \in \mathbb{R}^{m \times n}$ and $m \geq n$, then for $i = 1{:}n$ $Av_i = \sigma_i u_i$ and $A^T u_i = \sigma_i v_i$.*

Proof. Compare columns in $AV = U\Sigma$ and $A^T U = V\Sigma^T$. $\square$

There is a nice geometry behind this result. The singular values of a matrix A are the lengths of the semiaxes of the hyperellipsoid E defined by $E = \{\, Ax : \| x \|_2 = 1 \,\}$. The semiaxis directions are defined by the u_i and their lengths are the singular values.

It follows immediately from the corollary that

$$A^T A v_i = \sigma_i^2 v_i, \tag{2.4.1}$$

$$A A^T u_i = \sigma_i^2 u_i \tag{2.4.2}$$

for $i = 1{:}n$. This shows that there is an intimate connection between the SVD of A and the eigensystems of the symmetric matrices $A^T A$ and $A A^T$. See §8.6 and §10.4.

The 2-norm and the Frobenius norm have simple SVD characterizations.

Corollary 2.4.3. *If $A \in \mathbb{R}^{m \times n}$, then*

$$\| A \|_2 = \sigma_1, \qquad \| A \|_F = \sqrt{\sigma_1^2 + \cdots + \sigma_p^2},$$

where $p = \min\{m, n\}$.

Proof. These results follow immediately from the fact that $\| U^T A V \| = \| \Sigma \|$ for both the 2-norm and the Frobenius norm. $\square$

We show in §8.6 that if A is perturbed by a matrix E, then no singular value can move by more than $\| E \|_2$. The following corollary identifies two useful instances of this result.

Corollary 2.4.4. *If $A \in \mathbb{R}^{m \times n}$ and $E \in \mathbb{R}^{m \times n}$, then*

$$\sigma_{\max}(A + E) \leq \sigma_{\max}(A) + \| E \|_2,$$
$$\sigma_{\min}(A + E) \geq \sigma_{\min}(A) - \| E \|_2.$$

Proof. Using Corollary 2.4.2 it is easy to show that

$$\sigma_{\min}(A) \cdot \| x \|_2 \leq \| Ax \|_2 \leq \sigma_{\max}(A) \cdot \| x \|_2.$$

The required inequalities follow from this result.   □

If a column is added to a matrix, then the largest singular value increases and the smallest singular value decreases.

Corollary 2.4.5. *If $A \in \mathbb{R}^{m \times n}$, $m > n$, and $z \in \mathbb{R}^m$, then*

$$\sigma_{\max}\left([\, A \,|\, z \,] \right) \geq \sigma_{\max}(A),$$
$$\sigma_{\min}\left([\, A \,|\, z \,] \right) \leq \sigma_{\min}(A).$$

Proof. Suppose $A = U\Sigma V^T$ is the SVD of A and let $x = V(:, 1)$ and $\tilde{A} = [\, A \,|\, z \,]$. Using Corollary 2.4.4, we have

$$\sigma_{\max}(A) = \| Ax \|_2 = \left\| \tilde{A} \begin{bmatrix} x \\ 0 \end{bmatrix} \right\|_2 \leq \sigma_{\max}(\tilde{A}).$$

The proof that $\sigma_{\min}(A) \geq \sigma_{\min}(\tilde{A})$ is similar.   □

The SVD neatly characterizes the rank of a matrix and orthonormal bases for both its nullspace and its range.

Corollary 2.4.6. *If A has r positive singular values, then $\mathsf{rank}(A) = r$ and*

$$\mathsf{null}(A) = \mathsf{span}\{v_{r+1}, \ldots, v_n\},$$
$$\mathsf{ran}(A) = \mathsf{span}\{u_1, \ldots, u_r\}.$$

Proof. The rank of a diagonal matrix equals the number of nonzero diagonal entries. Thus, $\mathsf{rank}(A) = \mathsf{rank}(\Sigma) = r$. The assertions about the nullspace and range follow from Corollary 2.4.2.   □

If A has rank r, then it can be written as the sum of r rank-1 matrices. The SVD gives us a particularly nice choice for this expansion.

Corollary 2.4.7. *If $A \in \mathbb{R}^{m \times n}$ and* $\mathsf{rank}(A) = r$, *then*

$$A = \sum_{i=1}^{r} \sigma_i u_i v_i^T.$$

Proof. This is an exercise in partitioned matrix multiplication:

$$(U\Sigma)V^T = \left(\left[\begin{array}{c|c|c|c|c|c} \sigma_1 u_1 & \sigma_2 u_2 & \cdots & \sigma_r u_r & 0 & \cdots & 0 \end{array} \right] \right) \begin{bmatrix} v_1^T \\ \vdots \\ v_n^T \end{bmatrix} = \sum_{i=1}^{r} \sigma_i u_i v_i^T. \qquad \square$$

The intelligent handling of rank degeneracy is an important topic that we discuss in Chapter 5. The SVD has a critical role to play because it can be used to identify nearby matrices of lesser rank.

Theorem 2.4.8 (The Eckhart-Young Theorem). *If $k < r = \mathrm{rank}(A)$ and*

$$A_k = \sum_{i=1}^{k} \sigma_i u_i v_i^T, \qquad (2.4.3)$$

then

$$\min_{\mathsf{rank}(B)=k} \| A - B \|_2 = \| A - A_k \|_2 = \sigma_{k+1}. \qquad (2.4.4)$$

Proof. Since $U^T A_k V = \mathrm{diag}(\sigma_1, \ldots, \sigma_k, 0, \ldots, 0)$ it follows that A_k is rank k. Moreover, $U^T(A - A_k)V = \mathrm{diag}(0, \ldots, 0, \sigma_{k+1}, \ldots, \sigma_p)$ and so $\| A - A_k \|_2 = \sigma_{k+1}$.

Now suppose $\mathsf{rank}(B) = k$ for some $B \in \mathbb{R}^{m \times n}$. It follows that we can find orthonormal vectors $x_1, \ldots, x_{n-k}$ so $\mathsf{null}(B) = \mathsf{span}\{x_1, \ldots, x_{n-k}\}$. A dimension argument shows that

$$\mathsf{span}\{x_1, \ldots, x_{n-k}\} \cap \mathsf{span}\{v_1, \ldots, v_{k+1}\} \neq \{0\}.$$

Let z be a unit 2-norm vector in this intersection. Since $Bz = 0$ and

$$Az = \sum_{i=1}^{k+1} \sigma_i(v_i^T z) u_i,$$

we have

$$\| A - B \|_2^2 \geq \| (A - B)z \|_2^2 = \| Az \|_2^2 = \sum_{i=1}^{k+1} \sigma_i^2 (v_i^T z)^2 \geq \sigma_{k+1}^2,$$

completing the proof of the theorem. $\square$

Note that this theorem says that the smallest singular value of A is the 2-norm distance of A to the set of all rank-deficient matrices. We also mention that the matrix A_k defined in (2.4.3) is the closest rank-k matrix to A in the Frobenius norm.

2.4.3   The Thin SVD

If $A = U\Sigma V^T \in \mathbb{R}^{m \times n}$ is the SVD of A and $m \geq n$, then

$$A = U_1 \Sigma_1 V^T$$

where

$$U_1 \; = \; U(:, 1{:}n) \; = \; [\, u_1 \mid \cdots \mid u_n \,] \in \mathbb{R}^{m \times n}$$

and

$$\Sigma_1 \; = \; \Sigma(1{:}n, 1{:}n) \; = \; \operatorname{diag}(\sigma_1, \ldots, \sigma_n) \in \mathbb{R}^{n \times n}.$$

We refer to this abbreviated version of the SVD as the *thin SVD*.

2.4.4   Unitary Matrices and the Complex SVD

Over the complex field the unitary matrices correspond to the orthogonal matrices. In particular, $Q \in \mathbb{C}^{n \times n}$ is *unitary* if $Q^H Q = Q Q^H = I_n$. Unitary transformations preserve both the 2-norm and the Frobenius norm. The SVD of a complex matrix involves unitary matrices. If $A \in \mathbb{C}^{m \times n}$, then there exist unitary matrices $U \in \mathbb{C}^{m \times m}$ and $V \in \mathbb{C}^{n \times n}$ such that

$$U^H A V = \operatorname{diag}(\sigma_1, \ldots, \sigma_p) \; \in \mathbb{R}^{m \times n} \qquad p = \min\{m, n\}$$

where $\sigma_1 \geq \sigma_2 \geq \ldots \geq \sigma_p \geq 0$. All of the real SVD properties given above have obvious complex analogs.

Problems

P2.4.1 Show that if $Q = Q_1 + iQ_2$ is unitary with $Q_1, Q_2 \in \mathbb{R}^{n \times n}$, then the 2n-by-2n real matrix

$$Z = \left[\begin{array}{cc} Q_1 & -Q_2 \\ Q_2 & Q_1 \end{array} \right]$$

is orthogonal.

P2.4.2 Prove that if $A \in \mathbb{R}^{m \times n}$, then

$$\sigma_{\max}(A) \; = \; \max_{\substack{y \,\in\, \mathbb{R}^m \\ x \,\in\, \mathbb{R}^n}} \; \frac{y^T A x}{\| x \|_2 \| y \|_2} \, .$$

P2.4.3 For the 2-by-2 matrix $A = \left[\begin{array}{cc} w & x \\ y & z \end{array} \right]$, derive expressions for $\sigma_{\max}(A)$ and $\sigma_{\min}(A)$ that are functions of w, x, y, and z.

P2.4.4 Show that any matrix in $\mathbb{R}^{m \times n}$ is the limit of a sequence of full rank matrices.

P2.4.5 Show that if $A \in \mathbb{R}^{m \times n}$ has rank n, then $\| A(A^T A)^{-1} A^T \|_2 = 1$.

P2.4.6 What is the nearest rank-1 matrix to

$$A = \left[\begin{array}{cc} 1 & M \\ 0 & 1 \end{array} \right]$$

in the Frobenius norm?

P2.4.7 Show that if $A \in \mathbb{R}^{m \times n}$, then $\| A \|_F \leq \sqrt{\operatorname{rank}(A)} \, \| A \|_2$, thereby sharpening (2.3.7).

P2.4.8 Suppose $A \in \mathbb{R}^{n \times n}$. Give an SVD solution to the following problem:

$$\min_{\det(B) = |\det(A)|} \| A - B \|_F.$$

P2.4.9 Show that if a nonzero row is added to a matrix, then both the largest and smallest singular values increase.

P2.4.10 Show that if θ_u and θ_v are real numbers and

$$A = \left[\begin{array}{cc} \cos(\theta_u) & \sin(\theta_u) \\ \cos(\theta_v) & \sin(\theta_v) \end{array} \right],$$

then $U^T A V = \Sigma$ where

$$U = \left[\begin{array}{cc} \cos(\pi/4) & -\sin(\pi/4) \\ \sin(\pi/4) & \cos(\pi/4) \end{array} \right], \; V = \left[\begin{array}{cc} \cos(a) & -\sin(a) \\ \sin(a) & \cos(a) \end{array} \right],$$

and $\Sigma = \text{diag}(\sqrt{2}\cos(b), \sqrt{2}\sin(b))$ with $a = (\theta_v + \theta_u)/2$ and $b = (\theta_v - \theta_u)/2$.

Notes and References for §2.4

Forsythe and Moler (SLAS) offer a good account of the SVD's role in the analysis of the $Ax = b$ problem. Their proof of the decomposition is more traditional than ours in that it makes use of the eigenvalue theory for symmetric matrices. Historical SVD references include:

E. Beltrami (1873). "Sulle Funzioni Bilineari," *Gionale di Mathematiche 11*, 98–106.

C. Eckart and G. Young (1939). "A Principal Axis Transformation for Non-Hermitian Matrices," *Bull. AMS 45*, 118–21.

G.W. Stewart (1993). "On the Early History of the Singular Value Decomposition," *SIAM Review 35*, 551–566.

One of the most significant developments in scientific computation has been the increased use of the SVD in application areas that require the intelligent handling of matrix rank. This work started with:

C. Eckart, and G. Young (1936). "The Approximation of One Matrix by Another of Lower Rank," *Psychometrika 1*, 211–218.

For generalizations of the SVD to infinite dimensional Hilbert space, see:

I.C. Gohberg and M.G. Krein (1969). *Introduction to the Theory of Linear Non-Self Adjoint Operators*, Amer. Math. Soc., Providence, RI.

F. Smithies (1970). *Integral Equations*, Cambridge University Press, Cambridge.

Reducing the rank of a matrix as in Corollary 2.4.6 when the perturbing matrix is constrained is discussed in:

J.W. Demmel (1987). "The Smallest Perturbation of a Submatrix which Lowers the Rank and Constrained Total Least Squares Problems, *SIAM J. Numer. Anal. 24*, 199–206.

G.H. Golub, A. Hoffman, and G.W. Stewart (1988). "A Generalization of the Eckart-Young-Mirsky Approximation Theorem." *Lin. Alg. Applic. 88/89*, 317–328.

G.A. Watson (1988). "The Smallest Perturbation of a Submatrix which Lowers the Rank of the Matrix," *IMA J. Numer. Anal. 8*, 295–304.

2.5 Subspace Metrics

If the object of a computation is to compute a matrix or a vector, then norms are useful for assessing the accuracy of the answer or for measuring progress during an iteration. If the object of a computation is to compute a subspace, then to make similar comments we need to be able to quantify the distance between two subspaces. Orthogonal projections are critical in this regard. After the elementary concepts are established we discuss the CS decomposition. This is an SVD-like decomposition that is handy when we have to compare a pair of subspaces.

2.5.1   Orthogonal Projections

Let $S \subseteq \mathbb{R}^n$ be a subspace. $P \in \mathbb{R}^{n \times n}$ is the *orthogonal projection* onto S if $\mathsf{ran}(P) = S$, $P^2 = P$, and $P^T = P$. From this definition it is easy to show that if $x \in \mathbb{R}^n$, then $Px \in S$ and $(I - P)x \in S^\perp$.

If P_1 and P_2 are each orthogonal projections, then for any $z \in \mathbb{R}^n$ we have

$$\| (P_1 - P_2)z \|_2^2 = (P_1 z)^T (I - P_2)z + (P_2 z)^T (I - P_1)z.$$

If $\mathsf{ran}(P_1) = \mathsf{ran}(P_2) = S$, then the right-hand side of this expression is zero, showing that the orthogonal projection for a subspace is unique. If the columns of $V = [\, v_1 \,|\, \cdots \,|\, v_k \,]$ are an orthonormal basis for a subspace S, then it is easy to show that $P = VV^T$ is the unique orthogonal projection onto S. Note that if $v \in \mathbb{R}^n$, then $P = vv^T/v^T v$ is the orthogonal projection onto $S = \mathsf{span}\{v\}$.

2.5.2   SVD-Related Projections

There are several important orthogonal projections associated with the singular value decomposition. Suppose $A = U\Sigma V^T \in \mathbb{R}^{m \times n}$ is the SVD of A and that $r = \mathsf{rank}(A)$. If we have the U and V partitionings

$$U = [\, U_r \,|\, \tilde{U}_r \,] \atop {\quad r \quad m-r} \qquad , \qquad V = [\, V_r \,|\, \tilde{V}_r \,] \atop {\quad r \quad n-r} \quad ,$$

then

$$
\begin{aligned}
V_r V_r^T &= \quad \text{projection on to } \mathsf{null}(A)^\perp = \mathsf{ran}(A^T), \\
\tilde{V}_r \tilde{V}_r^T &= \quad \text{projection on to } \mathsf{null}(A), \\
U_r U_r^T &= \quad \text{projection on to } \mathsf{ran}(A), \\
\tilde{U}_r \tilde{U}_r^T &= \quad \text{projection on to } \mathsf{ran}(A)^\perp = \mathsf{null}(A^T).
\end{aligned}
$$

2.5.3   Distance Between Subspaces

The one-to-one correspondence between subspaces and orthogonal projections enables us to devise a notion of distance between subspaces. Suppose S_1 and S_2 are subspaces of $\mathbb{R}^n$ and that $\dim(S_1) = \dim(S_2)$. We define the *distance* between these two spaces by

$$\mathsf{dist}(S_1, S_2) = \| P_1 - P_2 \|_2 \tag{2.5.1}$$

where P_i is the orthogonal projection onto S_i. The distance between a pair of subspaces can be characterized in terms of the blocks of a certain orthogonal matrix.

Theorem 2.5.1. *Suppose*

$$W = [\, W_1 \,|\, W_2 \,] \atop {\quad k \quad n-k} \qquad , \qquad Z = [\, Z_1 \,|\, Z_2 \,] \atop {\quad k \quad n-k} \quad ,$$

are n-by-n orthogonal matrices. If $S_1 = \mathsf{ran}(W_1)$ and $S_2 = \mathsf{ran}(Z_1)$, then

$$\mathsf{dist}(S_1, S_2) = \| W_1^T Z_2 \|_2 = \| Z_1^T W_2 \|_2.$$

Proof. We first observe that

$$\text{dist}(S_1, S_2) = \| W_1 W_1^T - Z_1 Z_1^T \|_2$$

$$= \| W^T (W_1 W_1^T - Z_1 Z_1^T) Z \|_2$$

$$= \left\| \begin{bmatrix} 0 & W_1^T Z_2 \\ -W_2^T Z_1 & 0 \end{bmatrix} \right\|_2 .$$

Note that the matrices $W_2^T Z_1$ and $W_1^T Z_2$ are submatrices of the orthogonal matrix

$$Q = \begin{bmatrix} Q_{11} & Q_{12} \\ Q_{21} & Q_{22} \end{bmatrix} \equiv \begin{bmatrix} W_1^T Z_1 & W_1^T Z_2 \\ W_2^T Z_1 & W_2^T Z_2 \end{bmatrix} = W^T Z. \qquad (2.5.2)$$

Our goal is to show that $\| Q_{21} \|_2 = \| Q_{12} \|_2$. Since Q is orthogonal it follows from

$$Q \begin{bmatrix} x \\ 0 \end{bmatrix} = \begin{bmatrix} Q_{11} x \\ Q_{21} x \end{bmatrix}$$

that $1 = \| Q_{11} x \|_2^2 + \| Q_{21} x \|_2^2$ for all unit 2-norm $x \in \mathbb{R}^k$. Thus,

$$\| Q_{21} \|_2^2 = \max_{\|x\|_2 = 1} \| Q_{21} x \|_2^2 = 1 - \min_{\|x\|_2 = 1} \| Q_{11} x \|_2^2 = 1 - \sigma_{\min}(Q_{11})^2.$$

Analogously, by working with Q^T (which is also orthogonal) it is possible to show that

$$\| Q_{12}^T \|_2^2 = 1 - \sigma_{\min}(Q_{11}^T)^2,$$

and therefore

$$\| Q_{12} \|_2^2 = 1 - \sigma_{\min}(Q_{11})^2.$$

Thus, $\| Q_{21} \|_2 = \| Q_{12} \|_2$. $\square$

Note that if S_1 and S_2 are subspaces in $\mathbb{R}^n$ with the same dimension, then

$$0 \leq \text{dist}(S_1, S_2) \leq 1.$$

It is easy to show that

$$\text{dist}(S_1, S_2) = 0 \Rightarrow S_1 = S_2,$$
$$\text{dist}(S_1, S_2) = 1 \Rightarrow S_1 \cap S_2^{\perp} \neq \{0\}.$$

A more refined analysis of the blocks of the matrix Q in (2.5.2) sheds light on the difference between a pair of subspaces. A special, SVD-like decomposition for orthogonal matrices is required.

2.5.4   The CS Decomposition

The blocks of an orthogonal matrix partitioned into 2-by-2 form have highly related SVDs. This is the gist of the *CS decomposition*. We prove a very useful special case first.

Theorem 2.5.2 (The CS Decomposition (Thin Version)). *Consider the matrix*

$$Q = \begin{bmatrix} Q_1 \\ Q_2 \end{bmatrix}, \qquad Q_1 \in \mathbb{R}^{m_1 \times n_1}, \ Q_2 \in \mathbb{R}^{m_2 \times n_1},$$

where $m_1 \geq n_1$ and $m_2 \geq n_1$. If the columns of Q are orthonormal, then there exist orthogonal matrices $U_1 \in \mathbb{R}^{m_1 \times m_1}$, $U_2 \in \mathbb{R}^{m_2 \times m_2}$, and $V_1 \in \mathbb{R}^{n_1 \times n_1}$ such that

$$\begin{bmatrix} U_1 & 0 \\ 0 & U_2 \end{bmatrix}^T \begin{bmatrix} Q_1 \\ Q_2 \end{bmatrix} V_1 = \begin{bmatrix} C \\ S \end{bmatrix}$$

where

$$C_0 = \mathrm{diag}(\cos(\theta_1), \ldots, \cos(\theta_{n_1})) \in \mathbb{R}^{m_1 \times n_1},$$

$$S_0 = \mathrm{diag}(\sin(\theta_1), \ldots, \sin(\theta_{n_1})) \in \mathbb{R}^{m_2 \times n_1},$$

and

$$0 \leq \theta_1 \leq \theta_2 \leq \cdots \leq \theta_{n_1} \leq \frac{\pi}{2}.$$

Proof. Since $\| Q_1 \|_2 \leq \| Q \|_2 = 1$, the singular values of Q_1 are all in the interval $[0, 1]$. Let

$$U_1^T Q_1 V_1 = C_0 = \mathrm{diag}(c_1, \ldots, c_{n_1}) = \begin{bmatrix} I_t & 0 \\ 0 & \Sigma \end{bmatrix} \begin{matrix} t \\ m_1 - t \end{matrix}$$
$$ \begin{matrix} t & n_1 - t \end{matrix}$$

be the SVD of Q_1 where we assume

$$1 = c_1 = \cdots = c_t > c_{t+1} \geq \cdots \geq c_{n_1} \geq 0.$$

To complete the proof of the theorem we must construct the orthogonal matrix U_2. If

$$Q_2 V_1 = \underset{t \quad\quad n_1 - t}{[\, W_1 \mid W_2 \,]},$$

then

$$\begin{bmatrix} U_1 & 0 \\ 0 & I_{m_2} \end{bmatrix}^T \begin{bmatrix} Q_1 \\ Q_2 \end{bmatrix} V_1 = \begin{bmatrix} I_t & 0 \\ 0 & \Sigma \\ W_1 & W_2 \end{bmatrix}.$$

Since the columns of this matrix have unit 2-norm, $W_1 = 0$. The columns of W_2 are nonzero and mutually orthogonal because

$$W_2^T W_2 = I_{n_1 - t} - \Sigma^T \Sigma \equiv \mathrm{diag}(1 - c_{t+1}^2, \ldots, 1 - c_{n_1}^2)$$

is nonsingular. If $s_k = \sqrt{1 - c_k^2}$ for $k = 1{:}n_1$, then the columns of

$$Z = W_2 \operatorname{diag}(1/s_{t+1}, \ldots, 1/s_n)$$

are orthonormal. By Theorem 2.1.1 there exists an orthogonal matrix $U_2 \in \mathbb{R}^{m_2 \times m_2}$ with $U_2(:, t+1{:}n_1) = Z$. It is easy to verify that

$$U_2^T Q_2 V_1 = \operatorname{diag}(s_1, \ldots, s_{n_1}) \equiv S_0.$$

Since $c_k^2 + s_k^2 = 1$ for $k = 1{:}n_1$, it follows that these quantities are the required cosines and sines. $\square$

By using the same techniques it is possible to prove the following, more general version of the decomposition:

Theorem 2.5.3 (CS Decomposition). *Suppose*

$$Q = = \begin{array}{c} \left[\begin{array}{cc} Q_{11} & Q_{12} \\ Q_{21} & Q_{22} \end{array} \right] \begin{array}{c} m_1 \\ m_2 \end{array} \\ \end{array}$$
$$n_1 \quad\; n_2$$

is a square orthogonal matrix and that $m_1 \geq n_1$ and $m_1 \geq m_2$. Define the nonnegative integers p and q by $p = \max\{0, n_1 - m_2\}$ and $q = \max\{0, m_2 - n_1\}$. There exist orthogonal $U_1 \in \mathbb{R}^{m_1 \times m_1}$, $U_2 \in \mathbb{R}^{m_2 \times m_2}$, $V_1 \in \mathbb{R}^{n_1 \times n_1}$, and $V_2 \in \mathbb{R}^{n_2 \times n_2}$ such that if

$$U = \left[\begin{array}{c|c} U_1 & 0 \\ \hline 0 & U_2 \end{array} \right] \quad \text{and} \quad V = \left[\begin{array}{c|c} V_1 & 0 \\ \hline 0 & V_2 \end{array} \right],$$

then

$$U^T Q V = \left[\begin{array}{cc|ccc} I & 0 & 0 & 0 & 0 \\ 0 & C & S & 0 & 0 \\ 0 & 0 & 0 & 0 & I \\ \hline 0 & S & -C & 0 & 0 \\ 0 & 0 & 0 & I & 0 \end{array} \right] \begin{array}{c} p \\ n_1 - p \\ m_1 - n_1 \\ n_1 - p \\ q \end{array}$$
$$p \quad n_1 - p \quad n_1 - p \quad q \quad m_1 - n_1$$

where

$$C = \operatorname{diag}(\cos(\theta_{p+1}), \ldots, \cos(\theta_{n_1})) = \operatorname{diag}(c_{p+1}, \ldots, c_{n_1}),$$

$$S = \operatorname{diag}(\sin(\theta_{p+1}), \ldots, \sin(\theta_{n_1})) = \operatorname{diag}(s_{p+1}, \ldots, s_{n_1}),$$

and $0 \leq \theta_{p+1} \leq \cdots \leq \theta_{n_1} \leq \pi/2$.

Proof. See Paige and Saunders (1981) for details. $\square$

We made the assumptions $m_1 \geq n_1$ and $m_1 \geq m_2$ for clarity. Through permutation and transposition, any 2-by-2 block orthogonal matrix can be put into the form required by the theorem. Note that the blocks in the transformed Q, i.e., the $U_i^T Q_{ij} V_j$, are diagonal-like but not necessarily diagonal. Indeed, as we have presented it, the CS decomposition gives us four unnormalized SVDs. If Q_{21} has more rows than columns, then $p = 0$ and the reduction looks like this (for example):

$$
U^T Q V = \left[\begin{array}{ccc|cccc}
c_1 & 0 & s_1 & 0 & 0 & 0 & 0 \\
0 & c_2 & 0 & s_2 & 0 & 0 & 0 \\
0 & 0 & 0 & 0 & 0 & 1 & 0 \\
0 & 0 & 0 & 0 & 0 & 0 & 1 \\
\hline
s_1 & 0 & -c_1 & 0 & 0 & 0 & 0 \\
0 & s_2 & 0 & -c_2 & 0 & 0 & 0 \\
0 & 0 & 0 & 0 & 1 & 0 & 0
\end{array}\right].
$$

On the other hand, if Q_{21} has more columns than rows, then $q = 0$ and the decomposition has the form

$$
U^T Q V = \left[\begin{array}{ccc|cc}
1 & 0 & 0 & 0 & 0 \\
0 & c_2 & 0 & s_2 & 0 \\
0 & 0 & c_3 & 0 & s_3 \\
\hline
0 & s_2 & 0 & -c_2 & 0 \\
0 & 0 & s_3 & 0 & -c_3
\end{array}\right].
$$

Regardless of the partitioning, the essential message of the CS decomposition is that the SVDs of the Q-blocks are highly related.

Problems

P2.5.1 Show that if P is an orthogonal projection, then $Q = I - 2P$ is orthogonal.

P2.5.2 What are the singular values of an orthogonal projection?

P2.5.3 Suppose $S_1 = \text{span}\{x\}$ and $S_2 = \text{span}\{y\}$, where x and y are unit 2-norm vectors in $\mathbb{R}^2$. Working only with the definition of $\text{dist}(\cdot, \cdot)$, show that $\text{dist}(S_1, S_2) = \sqrt{1 - (x^T y)^2}$, verifying that the distance between S_1 and S_2 equals the sine of the angle between x and y.

P2.5.4 Refer to §1.3.10. Show that if $Q \in \mathbb{R}^{2n \times 2n}$ is orthogonal and symplectic, then Q has the form

$$
Q = \left[\begin{array}{cc}
Q_1 & Q_2 \\
-Q_2 & Q_1
\end{array}\right], \qquad Q_1, Q_2 \in \mathbb{R}^{n \times n}.
$$

P2.5.5 Suppose $P \in \mathbb{R}^{n \times n}$ and $P^2 = P$. Show that $\| P \|_2 > 1$ if $\text{null}(P)$ is *not* a subspace of $\text{ran}(A)^{\perp}$. Such a matrix is called an *oblique projector*. See Stewart (2011).

Notes and References for §2.5

The computation of the CS decomposition is discussed in §8.7.6. For a discussion of its analytical properties, see:

C. Davis and W. Kahan (1970). "The Rotation of Eigenvectors by a Perturbation III," *SIAM J. Numer. Anal. 7*, 1–46.

G.W. Stewart (1977). "On the Perturbation of Pseudo-Inverses, Projections and Linear Least Squares Problems," *SIAM Review 19*, 634–662.

C.C. Paige and M. Saunders (1981). "Toward a Generalized Singular Value Decomposition," *SIAM J. Numer. Anal. 18*, 398–405.

C.C. Paige and M. Wei (1994). "History and Generality of the CS Decomposition," *Lin. Alg. Applic. 208/209*, 303–326.

A detailed numerical discussion of oblique projectors (P2.5.5) is given in:

G.W. Stewart (2011). "On the Numerical Analysis of Oblique Projectors," *SIAM J. Matrix Anal. Applic. 32*, 309–348.

2.6 The Sensitivity of Square Systems

We use tools developed in previous sections to analyze the linear system problem $Ax = b$ where $A \in \mathbb{R}^{n \times n}$ is nonsingular and $b \in \mathbb{R}^n$. Our aim is to examine how perturbations in A and b affect the solution x. Higham (ASNA) offers a more detailed treatment.

2.6.1 An SVD Analysis

If

$$A = \sum_{i=1}^{n} \sigma_i u_i v_i^T = U \Sigma V^T$$

is the SVD of A, then

$$x = A^{-1}b = (U\Sigma V^T)^{-1}b = \sum_{i=1}^{n} \frac{u_i^T b}{\sigma_i} v_i. \tag{2.6.1}$$

This expansion shows that small changes in A or b can induce relatively large changes in x if σ_n is small.

It should come as no surprise that the magnitude of σ_n should have a bearing on the sensitivity of the $Ax = b$ problem. Recall from Theorem 2.4.8 that σ_n is the 2-norm distance from A to the set of singular matrices. As the matrix of coefficients approaches this set, it is intuitively clear that the solution x should be increasingly sensitive to perturbations.

2.6.2 Condition

A precise measure of linear system sensitivity can be obtained by considering the parameterized system

$$(A + \epsilon F)x(\epsilon) = b + \epsilon f, \qquad x(0) = x,$$

where $F \in \mathbb{R}^{n \times n}$ and $f \in \mathbb{R}^n$. If A is nonsingular, then it is clear that $x(\epsilon)$ is differentiable in a neighborhood of zero. Moreover, $\dot{x}(0) = A^{-1}(f - Fx)$ and so the Taylor series expansion for $x(\epsilon)$ has the form

$$x(\epsilon) = x + \epsilon \dot{x}(0) + O(\epsilon^2).$$

Using any vector norm and consistent matrix norm we obtain

$$\frac{\| x(\epsilon) - x \|}{\| x \|} \leq |\epsilon| \, \| A^{-1} \| \left\{ \frac{\| f \|}{\| x \|} + \| F \| \right\} + O(\epsilon^2). \tag{2.6.2}$$

For square matrices A define the *condition number* $\kappa(A)$ by

$$\kappa(A) = \| A \| \, \| A^{-1} \| \tag{2.6.3}$$

with the convention that $\kappa(A) = \infty$ for singular A. From $\| b \| \leq \| A \| \, \| x \|$ and (2.6.2) it follows that

$$\frac{\| x(\epsilon) - x \|}{\| x \|} \leq \kappa(A)(\rho_A + \rho_b) + O(\epsilon^2) \tag{2.6.4}$$

where

$$\rho_A = |\epsilon| \frac{\|F\|}{\|A\|} \quad \text{and} \quad \rho_b = |\epsilon| \frac{\|f\|}{\|b\|}$$

represent the relative errors in A and b, respectively. Thus, the relative error in x can be $\kappa(A)$ times the relative error in A and b. In this sense, the condition number $\kappa(A)$ quantifies the sensitivity of the $Ax = b$ problem.

Note that $\kappa(\cdot)$ depends on the underlying norm and subscripts are used accordingly, e.g.,

$$\kappa_2(A) = \|A\|_2 \|A^{-1}\|_2 = \frac{\sigma_{\max}(A)}{\sigma_{\min}(A)}. \tag{2.6.5}$$

Thus, the 2-norm condition of a matrix A measures the elongation of the hyperellipsoid $\{Ax : \|x\|_2 = 1\}$.

We mention two other characterizations of the condition number. For p-norm condition numbers, we have

$$\frac{1}{\kappa_p(A)} = \min_{A+\Delta A \text{ singular}} \frac{\|\Delta A\|_p}{\|A\|_p}. \tag{2.6.6}$$

This result may be found in Kahan (1966) and shows that $\kappa_p(A)$ measures the relative p-norm distance from A to the set of singular matrices.

For any norm, we also have

$$\kappa(A) = \lim_{\epsilon \to 0} \sup_{\|\Delta A\| \leq \epsilon \|A\|} \frac{\|(A + \Delta A)^{-1} - A^{-1}\|}{\epsilon} \frac{1}{\|A^{-1}\|}. \tag{2.6.7}$$

This imposing result merely says that the condition number is a normalized *Fréchet derivative* of the map $A \to A^{-1}$. Further details may be found in Rice (1966). Recall that we were initially led to $\kappa(A)$ through differentiation.

If $\kappa(A)$ is large, then A is said to be an *ill-conditioned* matrix. Note that this is a norm-dependent property.[1] However, any two condition numbers $\kappa_\alpha(\cdot)$ and $\kappa_\beta(\cdot)$ on $\mathbb{R}^{n \times n}$ are equivalent in that constants c_1 and c_2 can be found for which

$$c_1 \kappa_\alpha(A) \leq \kappa_\beta(A) \leq c_2 \kappa_\alpha(A), \qquad A \in \mathbb{R}^{n \times n}.$$

For example, on $\mathbb{R}^{n \times n}$ we have

$$\frac{1}{n} \kappa_2(A) \leq \kappa_1(A) \leq n\kappa_2(A),$$

$$\frac{1}{n} \kappa_\infty(A) \leq \kappa_2(A) \leq n\kappa_\infty(A), \tag{2.6.8}$$

$$\frac{1}{n^2} \kappa_1(A) \leq \kappa_\infty(A) \leq n^2 \kappa_1(A).$$

Thus, if a matrix is ill-conditioned in the α-norm, it is ill-conditioned in the β-norm modulo the constants c_1 and c_2 above.

For any of the p-norms, we have $\kappa_p(A) \geq 1$. Matrices with small condition numbers are said to be *well-conditioned*. In the 2-norm, orthogonal matrices are perfectly conditioned because if Q is orthogonal, then $\kappa_2(Q) = \|Q\|_2 \|Q^T\|_2 = 1$.

---

[1]It also depends upon the definition of "large." The matter is pursued in §3.5

2.6.3 Determinants and Nearness to Singularity

It is natural to consider how well determinant size measures ill-conditioning. If $\det(A) = 0$ is equivalent to singularity, is $\det(A) \approx 0$ equivalent to near singularity? Unfortunately, there is little correlation between $\det(A)$ and the condition of $Ax = b$. For example, the matrix B_n defined by

$$B_n = \begin{bmatrix} 1 & -1 & \cdots & -1 \\ 0 & 1 & \cdots & -1 \\ \vdots & \vdots & \ddots & \vdots \\ 0 & 0 & \cdots & 1 \end{bmatrix} \in \mathbb{R}^{n \times n} \tag{2.6.9}$$

has unit determinant, but $\kappa_\infty(B_n) = n \cdot 2^{n-1}$. On the other hand, a very well-conditioned matrix can have a very small determinant. For example,

$$D_n = \text{diag}(10^{-1}, \ldots, 10^{-1}) \in \mathbb{R}^{n \times n}$$

satisfies $\kappa_p(D_n) = 1$ although $\det(D_n) = 10^{-n}$.

2.6.4 A Rigorous Norm Bound

Recall that the derivation of (2.6.4) was valuable because it highlighted the connection between $\kappa(A)$ and the rate of change of $x(\epsilon)$ at $\epsilon = 0$. However, it is a little unsatisfying because it is contingent on ϵ being "small enough" and because it sheds no light on the size of the $O(\epsilon^2)$ term. In this and the next subsection we develop some additional $Ax = b$ perturbation theorems that are completely rigorous.

We first establish a lemma that indicates in terms of $\kappa(A)$ when we can expect a perturbed system to be nonsingular.

Lemma 2.6.1. *Suppose*

$$Ax = b, \qquad\qquad A \in \mathbb{R}^{n \times n}, \ 0 \neq b \in \mathbb{R}^n,$$

$$(A + \Delta A)y = b + \Delta b, \qquad \Delta A \in \mathbb{R}^{n \times n}, \ \Delta b \in \mathbb{R}^n,$$

with $\| \Delta A \| \leq \epsilon \| A \|$ *and* $\| \Delta b \| \leq \epsilon \| b \|$. *If* $\epsilon \kappa(A) = r < 1$, *then* $A + \Delta A$ *is nonsingular and*

$$\frac{\| y \|}{\| x \|} \leq \frac{1+r}{1-r}.$$

Proof. Since $\| A^{-1} \Delta A \| \leq \epsilon \| A^{-1} \| \| A \| = r < 1$ it follows from Theorem 2.3.4 that $(A + \Delta A)$ is nonsingular. Using Lemma 2.3.3 and the equality

$$(I + A^{-1}\Delta A)y = x + A^{-1}\Delta b$$

we find

$$\| y \| \leq \| (I + A^{-1}\Delta A)^{-1} \| \left(\| x \| + \epsilon \| A^{-1} \| \| b \| \right)$$

$$\leq \frac{1}{1-r} \left(\| x \| + \epsilon \| A^{-1} \| \| b \| \right) = \frac{1}{1-r} \left(\| x \| + r \frac{\| b \|}{\| A \|} \right).$$

Since $\| b \| = \| Ax \| \le \| A \| \, \| x \|$ it follows that

$$\| y \| \le \frac{1}{1-r} \left(\| x \| + r \| x \| \right)$$

and this establishes the required inequality.   □

We are now set to establish a rigorous $Ax = b$ perturbation bound.

Theorem 2.6.2. *If the conditions of Lemma 2.6.1 hold, then*

$$\frac{\| y - x \|}{\| x \|} \le \frac{2\epsilon}{1-r} \kappa(A). \tag{2.6.10}$$

Proof. Since

$$y - x = A^{-1}\Delta b - A^{-1}\Delta A y \tag{2.6.11}$$

we have

$$\| y - x \| \le \epsilon \| A^{-1} \| \, \| b \| + \epsilon \| A^{-1} \| \, \| A \| \, \| y \|.$$

Thus,

$$\frac{\| y - x \|}{\| x \|} \le \epsilon \kappa(A) \frac{\| b \|}{\| A \| \, \| x \|} + \epsilon \kappa(A) \frac{\| y \|}{\| x \|} \le \epsilon \left(1 + \frac{1+r}{1-r} \right) \kappa(A),$$

from which the theorem readily follows.   □

A small example helps put this result in perspective. The $Ax = b$ problem

$$\begin{bmatrix} 1 & 0 \\ 0 & 10^{-6} \end{bmatrix} \begin{bmatrix} x_1 \\ x_2 \end{bmatrix} = \begin{bmatrix} 1 \\ 10^{-6} \end{bmatrix}$$

has solution $x = [\, 1 \, , \, 1 \,]^T$ and condition $\kappa_\infty(A) = 10^6$. If $\Delta b = [\, 10^{-6} \, , \, 0 \,]^T$, $\Delta A = 0$, and $(A + \Delta A)y = b + \Delta b$, then $y = [\, 1 + 10^{-6} \, , \, 1 \,]^T$ and the inequality (2.6.10) says

$$10^{-6} = \frac{\| x - y \|_\infty}{\| x \|_\infty} \ll \frac{\| \Delta b \|_\infty}{\| b \|_\infty} \kappa_\infty(A) = 10^{-6} 10^6 = 1.$$

Thus, the upper bound in (2.6.10) can be a gross overestimate of the error induced by the perturbation.

On the other hand, if $\Delta b = (\, 0 \, , \, 10^{-6} \,)^T$, $\Delta A = 0$, and $(A + \Delta A)y = b + \Delta b$, then this inequality says that

$$\frac{10^0}{10^0} \le 2 \times 10^{-6} 10^6.$$

Thus, there are perturbations for which the bound in (2.6.10) is essentially attained.

2.6.5   More Refined Bounds

An interesting refinement of Theorem 2.6.2 results if we extend the notion of absolute value to matrices:

$$F = (f_{ij}) \in \mathbb{R}^{m \times n} \quad \Rightarrow \quad |F| = (|f_{ij}|) \in \mathbb{R}^{m \times n}.$$

This notation together with a matrix-level version of "$\leq$" makes it easy to specify componentwise error bounds. If $F, G \in \mathbb{R}^{m \times n}$, then

$$|F| \leq |G| \quad \Leftrightarrow \quad |f_{ij}| \leq |g_{ij}|$$

for all i and j. Also note that if $F \in \mathbb{R}^{m \times q}$ and $G \in \mathbb{R}^{q \times n}$, then $|FG| \leq |F| \cdot |G|$. With these definitions and facts we obtain the following refinement of Theorem 2.6.2.

Theorem 2.6.3. *Suppose*

$$Ax = b, \qquad A \in \mathbb{R}^{n \times n}, \ 0 \neq b \in \mathbb{R}^n,$$

$$(A + \Delta A)y = b + \Delta b \quad \Delta A \in \mathbb{R}^{n \times n}, \ \Delta b \in \mathbb{R}^n,$$

and that $|\Delta A| \leq \epsilon |A|$ and $|\Delta b| \leq \epsilon |b|$. If $\delta \kappa_\infty(A) = r < 1$, then $(A + \Delta A)$ is nonsingular and

$$\frac{\| y - x \|_\infty}{\| x \|_\infty} \leq \frac{2\epsilon}{1 - r} \cdot \| \, |A^{-1}| \, |A| \, \|_\infty. \tag{2.6.12}$$

Proof. Since $\| \Delta A \|_\infty \leq \epsilon \| A \|_\infty$ and $\| \Delta b \|_\infty \leq \epsilon \| b \|_\infty$ the conditions of Lemma 2.6.1 are satisfied in the infinity norm. This implies that $A + \Delta A$ is nonsingular and

$$\frac{\| y \|_\infty}{\| x \|_\infty} \leq \frac{1 + r}{1 - r}.$$

Now using (2.6.11) we find

$$|y - x| \leq |A^{-1}| \, |\Delta b| + |A^{-1}| \, |\Delta A| \, |y|$$

$$\leq \epsilon |A^{-1}| \, |b| + \epsilon |A^{-1}| \, |A| \, |y| \leq \epsilon |A^{-1}| \, |A| \, (|x| + |y|).$$

If we take norms, then

$$\| y - x \|_\infty \leq \epsilon \| \, |A^{-1}| \, |A| \, \|_\infty \left(\| x \|_\infty + \frac{1 + r}{1 - r} \| x \|_\infty \right).$$

The theorem follows upon division by $\| x \|_\infty$. $\square$

The quantity $\| \, |A^{-1}| \, |A| \, \|_\infty$ is known as the *Skeel condition number* and there are examples where it is considerably less than $\kappa_\infty(A)$. In these situations, (2.6.12) is more informative than (2.6.10).

Norm bounds are frequently good enough when assessing error, but sometimes it is desirable to examine error at the component level. Oettli and Prager (1964) have an interesting result that indicates if an approximate solution $\hat{x} \in \mathbb{R}^n$ to the n-by-n

system $Ax = b$ satisfies a perturbed system with prescribed structure. Consider the problem of finding $\Delta A \in \mathbb{R}^{n \times n}$, $\Delta b \in \mathbb{R}^n$, and $\omega \geq 0$ such that

$$(A + \Delta A)\hat{x} = b + \Delta b \qquad |\Delta A| \leq \omega |E|, \ |\Delta b| \leq \omega |f|. \qquad (2.6.13)$$

where $E \in \mathbb{R}^{n \times n}$ and $f \in \mathbb{R}^n$ are given. With proper choice of E and f, the perturbed system can take on certain qualities. For example, if $E = A$ and $f = b$ and ω is small, then $\hat{x}$ satisfies a nearby system in the componentwise sense. The authors show that for a given A, b, $\hat{x}$, E, and f the smallest ω possible in (2.6.13) is given by

$$\omega_{\min} = \max_{1 \leq i \leq n} \frac{|A\hat{x} - b|_i}{(|E| \cdot |\hat{x}| + |f|)_i}.$$

If $A\hat{x} = b$, then $\omega_{\min} = 0$. On the other hand, if $\omega_{\min} = \infty$, then $\hat{x}$ does not satisfy any system of the prescribed perturbation structure.

Problems

P2.6.1 Show that if $\| I \| \geq 1$, then $\kappa(A) \geq 1$.

P2.6.2 Show that for a given norm, $\kappa(AB) \leq \kappa(A)\kappa(B)$ and that $\kappa(\alpha A) = \kappa(A)$ for all nonzero α.

P2.6.3 Relate the 2-norm condition of $X \in \mathbb{R}^{m \times n}$ $(m \geq n)$ to the 2-norm condition of the matrices

$$B = \left[\begin{array}{cc} I_m & X \\ 0 & I_n \end{array} \right] \qquad \text{and} \qquad C = \left[\begin{array}{c} X \\ I_n \end{array} \right].$$

P2.6.4 Suppose $A \in \mathbb{R}^{n \times n}$ is nonsingular. Assume for a particular i and j that there is no way to make A singular by changing the value of a_{ij}. What can you conclude about A^{-1}? Hint: Use the Sherman-Morrison formula.

P2.6.5 Suppose $A \in \mathbb{R}^{n \times n}$ is nonsingular, $b \in \mathbb{R}^n$, $Ax = b$, and $C = A^{-1}$. Use the Sherman-Morrison formula to show that

$$\frac{\partial x_k}{\partial a_{ij}} = -x_j c_{ki}.$$

Notes and References for §2.6

The condition concept is thoroughly investigated in:

J. Rice (1966). "A Theory of Condition," *SIAM J. Numer. Anal. 3*, 287–310.
W. Kahan (1966). "Numerical Linear Algebra," *Canadian Math. Bull. 9*, 757–801.

References for componentwise perturbation theory include:

W. Oettli and W. Prager (1964). "Compatibility of Approximate Solutions of Linear Equations with Given Error Bounds for Coefficients and Right Hand Sides," *Numer. Math. 6*, 405–409.
J.E. Cope and B.W. Rust (1979). "Bounds on Solutions of Systems with Accurate Data," *SIAM J. Numer. Anal. 16*, 950–63.
R.D. Skeel (1979). "Scaling for Numerical Stability in Gaussian Elimination," *J. ACM 26*, 494–526.
J.W. Demmel (1992). "The Componentwise Distance to the Nearest Singular Matrix," *SIAM J. Matrix Anal. Applic. 13*, 10–19.
D.J. Higham and N.J. Higham (1992). "Componentwise Perturbation Theory for Linear Systems with Multiple Right-Hand Sides," *Lin. Alg. Applic. 174*, 111–129.
N.J. Higham (1994). "A Survey of Componentwise Perturbation Theory in Numerical Linear Algebra," in *Mathematics of Computation 1943–1993: A Half Century of Computational Mathematics*, W. Gautschi (ed.), Volume 48 of Proceedings of Symposia in Applied Mathematics, American Mathematical Society, Providence, RI.

S. Chandrasekaren and I.C.F. Ipsen (1995). "On the Sensitivity of Solution Components in Linear Systems of Equations," *SIAM J. Matrix Anal. Applic. 16*, 93–112.

S.M. Rump (1999). "Ill-Conditioned Matrices Are Componentwise Near to Singularity," *SIAM Review 41*, 102–112.

The reciprocal of the condition number measures how near a given $Ax = b$ problem is to singularity. The importance of knowing how near is a given problem to a difficult or insoluble problem has come to be appreciated in many computational settings, see:

A. Laub(1985). "Numerical Linear Algebra Aspects of Control Design Computations," *IEEE Trans. Autom. Control. AC-30*, 97–108.

J.W. Demmel (1987). "On the Distance to the Nearest Ill-Posed Problem," *Numer. Math. 51*, 251–289.

N.J. Higham (1989). "Matrix Nearness Problems and Applications," in *Applications of Matrix Theory*, M.J.C. Gover and S. Barnett (eds.), Oxford University Press, Oxford, UK, 1–27.

Much has been written about problem sensitivity from the statistical point of view, see:

J.W. Demmel (1988). "The Probability that a Numerical Analysis Problem is Difficult," *Math. Comput. 50*, 449–480.

G.W. Stewart (1990). "Stochastic Perturbation Theory," *SIAM Review 32*, 579–610.

C. S. Kenney, A.J. Laub, and M.S. Reese (1998). "Statistical Condition Estimation for Linear Systems," *SIAM J. Sci. Comput. 19*, 566–583.

The problem of minimizing $\kappa_2(A + UV^T)$ where UV^T is a low-rank matrix is discussed in:

C. Greif and J.M. Varah (2006). "Minimizing the Condition Number for Small Rank Modifications," *SIAM J. Matrix Anal. Applic. 29*, 82–97.

2.7 Finite Precision Matrix Computations

Rounding errors are part of what makes the field of matrix computations so challenging. In this section we describe a model of floating point arithmetic and then use it to develop error bounds for floating point dot products, saxpys, matrix-vector products, and matrix-matrix products.

2.7.1 A 3-digit Calculator

Suppose we have a base-10 calculator that represents nonzero numbers in the following style:

$$x = \pm d_0.d_1 d_2 \times 10^e \qquad \text{where} \qquad \begin{cases} 1 \leq d_0 \leq 9, \\ 0 \leq d_1 \leq 9, \\ 0 \leq d_2 \leq 9, \\ -9 \leq e \leq 9. \end{cases}$$

Let us call these numbers *floating point* numbers. After playing around a bit we make a number of observations:

- The precision of the calculator has to do with the "length" of the *significand* $d_0.d_1 d_2$. For example, the number π would be represented as 3.14×10^0, which has a relative error approximately equal to 10^{-3}.

- There is not enough "room" to store exactly the results from most arithmetic operations between floating point numbers. Sums and products like

$$(1.23 \times 10^6) + (4.56 \times 10^4) = 1275600,$$
$$(1.23 \times 10^1) * (4.56 \times 10^2) = 5608.8$$

involve more than three significant digits. Results must be rounded in order to "fit" the 3-digit format, e.g., round(1275600) $= 1.28 \times 10^6$, round(5608.8) $= 5.61 \times 10^3$.

- If zero is to be a floating point number (and it must be), then we need a special convention for its representation, e.g., 0.00×10^0.

- In contrast to the real numbers, there is a smallest positive floating point number ($N_{\min} = 1.00 \times 10^{-9}$) and there is a largest positive floating point number ($N_{\max} = 9.99 \times 10^9$).

- Some operations yield answers whose exponents exceed the 1-digit allocation, e.g., $(1.23 \times 10^4) * (4.56 \times 10^7)$ and $(1.23 \times 10^{-2})/(4.56 \times 10^8)$.

- The set of floating point numbers is finite. For the toy calculator there are $2 \times 9 \times 10 \times 10 \times 19 + 1 = 34201$ floating point numbers.

- The spacing between the floating point numbers varies. Between 1.00×10^e and $1.00 \times 10^{e+1}$ the spacing is 10^{e-2}.

The careful design and analysis of a floating point computation requires an understanding of these inexactitudes and limitations. How are results rounded? How accurate is floating point arithmetic? What can we say about a sequence of floating point operations?

2.7.2 IEEE Floating Point Arithmetic

To build a solid, practical understanding of finite precision computation, we set aside our toy, motivational base-10 calculator and consider the key ideas behind the widely accepted IEEE floating point standard. The IEEE standard includes a 32-bit single format and a 64-bit double format. We will illustrate concepts using the latter as an example because typical accuracy requirements make it the format of choice.

The importance of having a standard for floating point arithmetic that is upheld by hardware manufacturers cannot be overstated. After all, floating point arithmetic is the foundation upon which all of scientific computing rests. The IEEE standard promotes software reliability and enables numerical analysts to make rigorous statements about computed results. Our discussion is based on the excellent book by Overton (2001).

The 64-bit double format allocates a single bit for the sign of the floating point number, 52 bits for the mantissa , and eleven bits for the exponent:

$$x : \boxed{\pm \mid a_1 a_2 \ldots a_{11} \mid b_1 b_2 \ldots b_{52}}. \qquad (2.7.1)$$

The "formula" for the value of this representation depends upon the exponent bits:

If $a_1 \ldots a_{11}$ is neither all 0's nor all 1's, then x is a *normalized* floating point number with value

$$x = \pm(1.b_1 b_2 \ldots b_{52})_2 \times 2^{(a_1 a_2 \ldots a_{11})_2 - 1023}. \qquad (2.7.2)$$

The "1023 *bias*" in the exponent supports the graceful inclusion of various "unnormalized" floating numbers which we describe shortly. Several important quantities capture the finiteness of the representation. The *machine epsilon* is the gap between 1 and the next largest floating point number. Its value is $2^{-52} \approx 10^{-16}$ for the double format. Among the positive normalized floating point numbers, $N_{\min} = 2^{-1022} \approx 10^{-308}$ is the smallest and $N_{\max} = (2 - 2^{-52})2^{1023} \approx 10^{308}$ is the largest. A real number x is within the *normalized range* if $N_{\min} \le |x| \le N_{\max}$.

If $a_1 \ldots a_{11}$ is all 0's, then the value of the representation (2.7.1) is

$$x = \pm(0.b_1 b_2 \ldots b_{52})_2 \times 2^{(a_1 a_2 \ldots a_{11})_2 - 1022} \qquad (2.7.3)$$

This includes 0 and the *subnormal* floating point numbers. This feature creates a uniform spacing of the floating point numbers between $-N_{\min}$ and $+N_{\min}$.

If $a_1 \ldots a_{11}$ is all 1's, then the encoding (2.7.1) represents `inf` for $+\infty$, `-inf` for $-\infty$, or `NaN` for "not-a-number." The determining factor is the value of the b_i. (If the b_i are not all zero, then the value of x is `NaN`.) Quotients like $1/0$, $-1/0$, and $0/0$ produce these special floating point numbers instead of prompting program termination.

There are four rounding modes: *round down* (toward $-\infty$), *round up* (toward $+\infty$), *round-toward-zero*, and *round-toward-nearest*. We focus on round-toward-nearest since it is the mode almost always used in practice.

If a real number x is outside the range of the normalized floating point numbers then

$$\text{round}(x) = \begin{cases} -\infty & \text{if } x < -N_{\max}, \\ +\infty & \text{if } x > N_{\max}. \end{cases}$$

Otherwise, the rounding process depends upon its floating point "neighbors":

x_- is the nearest floating point number to x that is $\le x$,

x_+ is the nearest floating point number to x that is $\ge x$.

Define $d_- = x - x_-$ and $d_+ = x_+ - x$ and let "lsb" stand for "least significant bit." If $N_{\min} \le |x| \le N_{\max}$, then

$$\text{round}(x) = \begin{cases} x_- & \text{if } d_- < d_+ \text{ or } d_- = d_+ \text{ and } \text{lsb}(x_-) = 0, \\ x_+ & \text{if } d_+ < d_- \text{ or } d_+ = d_- \text{ and } \text{lsb}(x_+) = 0. \end{cases}$$

The tie-breaking criteria is well-defined because x_- and x_+ are adjacent floating point numbers and so must differ in their least significant bit.

Regarding the accuracy of the round-to-nearest strategy, suppose x is a real number that satisfies $N_{\min} \le |x| \le N_{\max}$. Thus,

$$|\text{round}(x) - x| \le \frac{2^{-52}}{2} 2^e \le \frac{2^{-52}}{2} |x|$$

which says that relative error is bounded by half of the machine epsilon:

$$\frac{|\text{round}(x) - x|}{|x|} \le 2^{-53}.$$

The IEEE standard stipulates that each arithmetic operation be *correctly rounded*, meaning that the computed result is the rounded version of the exact result. The implementation of correct rounding is far from trivial and requires registers that are equipped with several extra bits of precision.

We mention that the IEEE standard also requires correct rounding in the square root operation, the remainder operation, and various format conversion operations.

2.7.3   The "fl" Notation

With intuition gleaned from the toy calculator example and an understanding of IEEE arithmetic, we are ready to move on to the roundoff analysis of some basic algebraic calculations. The challenge when presenting the effects of finite precision arithmetic in this section and throughout the book is to communicate essential behavior without excessive detail. To that end we use the notation $\text{fl}(\cdot)$ to identify a floating point storage and/or computation. Unless exceptions are a critical part of the picture, we freely invoke the fl notation without mentioning "$-\infty$," "∞," "NaN," etc.

If $x \in \mathbb{R}$, then $\text{fl}(x)$ is its floating point representation and we assume that

$$\text{fl}(x) = x(1 + \delta), \qquad |\delta| \leq \mathbf{u}, \tag{2.7.4}$$

where $\mathbf{u}$ is the *unit roundoff* defined by

$$\mathbf{u} = \frac{1}{2} \times (\text{gap between 1 and next largest floating point number}). \tag{2.7.5}$$

The unit roundoff for IEEE single format is about 10^{-7} and for double format it is about 10^{-16}.

If x and y are floating point numbers and "op" is any of the four arithmetic operations, then $\text{fl}(x \text{ op } y)$ is the floating point result from the floating point op. Following Trefethen and Bau (NLA), the *fundamental axiom of floating point arithmetic* is that

$$\text{fl}(x \text{ op } y) = (x \text{ op } y)(1 + \delta), \qquad |\delta| \leq \mathbf{u}, \tag{2.7.6}$$

where x and y are floating point numbers and the "op" inside the fl operation means "floating point operation." This shows that there is small relative error associated with individual arithmetic operations:

$$\frac{|\text{fl}(x \text{ op } y) - (x \text{ op } y)|}{|x \text{ op } y|} \leq \mathbf{u}, \qquad x \text{ op } y \neq 0.$$

Again, unless it is particularly relevant to the discussion, it will be our habit not to bring up the possibilities of an exception arising during the floating point operation.

2.7.4   Become a Floating Point Thinker

It is a good idea to have a healthy respect for the subleties of floating point calculation. So before we proceed with our first serious roundoff error analysis we offer three maxims to keep in mind when designing a practical matrix computation. Each reinforces the distinction between computer arithmetic and exact arithmetic.

Maxim 1. *Order is Important.*

Floating point arithmetic is not associative. For example, suppose

$$x = 1.24 \times 10^0, \qquad y = -1.23 \times 10^0, \qquad z = 1.00 \times 10^{-3}.$$

Using toy calculator arithmetic we have

$$\text{fl}(\text{fl}(x + y) + z)) = 1.10 \times 10^{-2}$$

while
$$\text{fl}(x + \text{fl}(y + z)) = 1.00 \times 10^{-2}.$$

A consequence of this is that mathematically equivalent algorithms may produce different results in floating point.

Maxim 2. *Larger May Mean Smaller.*

Suppose we want to compute the derivative of $f(x) = \sin(x)$ using a divided difference. Calculus tells us that $d = (\sin(x+h) - \sin(x))/h$ satisfies $|d - \cos(x)| = O(h)$ which argues for making h as small as possible. On the other hand, any roundoff error sustained in the sine evaluations is magnified by $1/h$. By setting $h = \sqrt{\mathbf{u}}$, the sum of the calculus error and roundoff error is approximately minimized. In other words, a value of h much greater than $\mathbf{u}$ renders a much smaller overall error. See Overton(2001, pp. 70–72).

Maxim 3. *A Math Book Is Not Enough.*

The explicit coding of a textbook formula is not always the best way to design an effective computation. As an example, we consider the quadratic equation $x^2 - 2px - q = 0$ where both p and q are positive. Here are two methods for computing the smaller (necessarily real) root:

$$\text{Method 1:} \quad r_{\min} = p - \sqrt{p^2 + q},$$

$$\text{Method 2:} \quad r_{\min} = \frac{q}{p + \sqrt{p^2 + q}}.$$

The first method is based on the familiar quadratic formula while the second uses the fact that $-q$ is the product of $r_{\min}$ and the larger root. Using IEEE double format arithmetic with input $p = 12345678$ and $q = 1$ we obtain these results:

$$\text{Method 1:} \quad r_{\min} = -4.097819328308106 \times 10^{-8},$$

$$\text{Method 2:} \quad r_{\min} = -4.050000332100021 \times 10^{-8} \quad \text{(correct)}.$$

Method 1 produces an answer that has almost no correct significant digits. It attempts to compute a small number by subtracting a pair of nearly equal large numbers. Almost all correct significant digits in the input data are lost during the subtraction, a phenomenon known as *catastrophic cancellation*. In contrast, Method 2 produces an answer that is correct to full machine precision. It computes a small number as a division of one number by a much larger number. See Forsythe (1970).

Keeping these maxims in mind does not guarantee the production of accurate, reliable software, but it helps.

2.7.5 Application: Storing a Real Matrix

Suppose $A \in \mathbb{R}^{m \times n}$ and that we wish to quantify the errors associated with its floating point representation. Denoting the stored version of A by $\text{fl}(A)$, we see that

$$[\text{fl}(A)]_{ij} = \text{fl}(a_{ij}) = a_{ij}(1 + \epsilon_{ij}), \qquad |\epsilon_{ij}| \leq \mathbf{u}, \tag{2.7.7}$$

for all i and j, i.e.,

$$|\mathsf{fl}(A) - A| \leq \mathbf{u}|A|\,.$$

A relation such as this can be easily turned into a norm inequality, e.g.,

$$\|\,\mathsf{fl}(A) - A\,\|_1 \leq \mathbf{u}\|\,A\,\|_1.$$

However, when quantifying the rounding errors in a matrix manipulation, the absolute value notation is sometimes more informative because it provides a comment on each entry.

2.7.6   Roundoff in Dot Products

We begin our study of finite precision matrix computations by considering the rounding errors that result in the standard dot product algorithm:

$$
\begin{aligned}
&s = 0 \\
&\mathbf{for}\ k = 1{:}n \\
&\qquad s = s + x_k y_k \\
&\mathbf{end}
\end{aligned}
\tag{2.7.8}
$$

Here, x and y are n-by-1 floating point vectors.

In trying to quantify the rounding errors in this algorithm, we are immediately confronted with a notational problem: the distinction between computed and exact quantities. If the underlying computations are clear, we shall use the $\mathsf{fl}(\cdot)$ operator to signify computed quantities. Thus, $\mathsf{fl}(x^T y)$ denotes the computed output of (2.7.8). Let us bound $|\mathsf{fl}(x^T y) - x^T y|$. If

$$s_p = \mathsf{fl}\left(\sum_{k=1}^{p} x_k y_k\right),$$

then $s_1 = x_1 y_1 (1 + \delta_1)$ with $|\delta_1| \leq \mathbf{u}$ and for $p = 2{:}n$

$$
\begin{aligned}
s_p &= \mathsf{fl}(s_{p-1} + \mathsf{fl}(x_p y_p)) \\
&= (s_{p-1} + x_p y_p (1 + \delta_p))(1 + \epsilon_p) \qquad |\delta_p|, |\epsilon_p| \leq \mathbf{u}.
\end{aligned}
\tag{2.7.9}
$$

A little algebra shows that

$$\mathsf{fl}(x^T y) = s_n = \sum_{k=1}^{n} x_k y_k (1 + \gamma_k)$$

where

$$(1 + \gamma_k) = (1 + \delta_k) \prod_{j=k}^{n} (1 + \epsilon_j)$$

with the convention that $\epsilon_1 = 0$. Thus,

$$|\mathsf{fl}(x^T y) - x^T y| \leq \sum_{k=1}^{n} |x_k y_k||\gamma_k|.
\tag{2.7.10}$$

To proceed further, we must bound the quantities $|\gamma_k|$ in terms of $\mathbf{u}$. The following result is useful for this purpose.

Lemma 2.7.1. *If* $(1 + \alpha) = \prod_{k=1}^{n}(1 + \alpha_k)$ *where* $|\alpha_k| \leq \mathbf{u}$ *and* $n\mathbf{u} \leq .01$, *then* $|\alpha| \leq 1.01n\mathbf{u}$.

Proof. See Higham (ASNA, p. 75). $\square$

Application of this result to (2.7.10) under the "reasonable" assumption $n\mathbf{u} \leq .01$ gives

$$|\mathsf{fl}(x^T y) - x^T y| \leq 1.01n\mathbf{u}|x|^T|y|. \qquad (2.7.11)$$

Notice that if $|x^T y| \ll |x|^T|y|$, then the relative error in $\mathsf{fl}(x^T y)$ may not be small.

2.7.7 Alternative Ways to Quantify Roundoff Error

An easier but less rigorous way of bounding α in Lemma 2.7.1 is to say $|\alpha| \leq n\mathbf{u}+O(\mathbf{u}^2)$. With this convention we have

$$|\mathsf{fl}(x^T y) - x^T y| \leq n\mathbf{u}|x|^T|y| + O(\mathbf{u}^2). \qquad (2.7.12)$$

Other ways of expressing the same result include

$$|\mathsf{fl}(x^T y) - x^T y| \leq \phi(n)\mathbf{u}|x|^T|y| \qquad (2.7.13)$$

and

$$|\mathsf{fl}(x^T y) - x^T y| \leq cn\,\mathbf{u}|x|^T|y|, \qquad (2.7.14)$$

where $\phi(n)$ is a "modest" function of n and c is a constant of order unity.

We shall not express a preference for any of the error bounding styles shown in (2.7.11)–(2.7.14). This spares us the necessity of translating the roundoff results that appear in the literature into a fixed format. Moreover, paying overly close attention to the details of an error bound is inconsistent with the "philosophy" of roundoff analysis. As Wilkinson (1971, p. 567) says,

> There is still a tendency to attach too much importance to the precise error bounds obtained by an a priori error analysis. In my opinion, the bound itself is usually the least important part of it. The main object of such an analysis is to expose the potential instabilities, if any, of an algorithm so that hopefully from the insight thus obtained one might be led to improved algorithms. Usually the bound itself is weaker than it might have been because of the necessity of restricting the mass of detail to a reasonable level and because of the limitations imposed by expressing the errors in terms of matrix norms. A priori bounds are not, in general, quantities that should be used in practice. Practical error bounds should usually be determined by some form of a posteriori error analysis, since this takes full advantage of the statistical distribution of rounding errors and of any special features, such as sparseness, in the matrix.

It is important to keep these perspectives in mind.

2.7.8   Roundoff in Other Basic Matrix Computations

It is easy to show that if A and B are floating point matrices and α is a floating point number, then

$$\text{fl}(\alpha A) = \alpha A + E, \qquad |E| \leq \mathbf{u}|\alpha A|, \tag{2.7.15}$$

and

$$\text{fl}(A + B) = (A + B) + E, \qquad |E| \leq \mathbf{u}|A + B|. \tag{2.7.16}$$

As a consequence of these two results, it is easy to verify that computed saxpy's and outer product updates satisfy

$$\text{fl}(y + \alpha x) = y + \alpha x + z, \qquad |z| \leq \mathbf{u}\left(|y| + 2|\alpha x|\right) + O(\mathbf{u}^2), \tag{2.7.17}$$

$$\text{fl}(C + uv^T) = C + uv^T + E, \qquad |E| \leq \mathbf{u}\left(|C| + 2|uv^T|\right) + O(\mathbf{u}^2). \tag{2.7.18}$$

Using (2.7.11) it is easy to show that a dot-product-based multiplication of two floating point matrices A and B satisfies

$$\text{fl}(AB) = AB + E, \qquad |E| \leq n\mathbf{u}|A||B| + O(\mathbf{u}^2). \tag{2.7.19}$$

The same result applies if a gaxpy or outer product based procedure is used. Notice that matrix multiplication does not necessarily give small relative error since $|AB|$ may be much smaller than $|A||B|$, e.g.,

$$\begin{bmatrix} 1 & 1 \\ 0 & 0 \end{bmatrix} \begin{bmatrix} 1 & 0 \\ -.99 & 0 \end{bmatrix} = \begin{bmatrix} .01 & 0 \\ 0 & 0 \end{bmatrix}.$$

It is easy to obtain norm bounds from the roundoff results developed thus far. If we look at the 1-norm error in floating point matrix multiplication, then it is easy to show from (2.7.19) that

$$\| \text{fl}(AB) - AB \|_1 \leq n\mathbf{u}\| A \|_1 \| B \|_1 + O(\mathbf{u}^2). \tag{2.7.20}$$

2.7.9   Forward and Backward Error Analyses

Each roundoff bound given above is the consequence of a *forward error analysis*. An alternative style of characterizing the roundoff errors in an algorithm is accomplished through a technique known as *backward error analysis*. Here, the rounding errors are related to the input data rather than the answer. By way of illustration, consider the $n = 2$ version of triangular matrix multiplication. It can be shown that:

$$\text{fl}(AB) = \begin{bmatrix} a_{11}b_{11}(1 + \epsilon_1) & (a_{11}b_{12}(1 + \epsilon_2) + a_{12}b_{22}(1 + \epsilon_3))(1 + \epsilon_4) \\ 0 & a_{22}b_{22}(1 + \epsilon_5) \end{bmatrix}$$

where $|\epsilon_i| \leq \mathbf{u}$, for $i = 1{:}5$. However, if we define

$$\hat{A} = \begin{bmatrix} a_{11} & a_{12}(1 + \epsilon_3)(1 + \epsilon_4) \\ 0 & a_{22}(1 + \epsilon_5) \end{bmatrix}$$

and

$$\hat{B} = \begin{bmatrix} b_{11}(1 + \epsilon_1) & b_{12}(1 + \epsilon_2)(1 + \epsilon_4) \\ & \\ 0 & b_{22} \end{bmatrix},$$

then it is easily verified that $\mathrm{fl}(AB) = \hat{A}\hat{B}$. Moreover,

$$\hat{A} = A + E, \qquad |E| \leq 2\mathbf{u}|A| + O(\mathbf{u}^2),$$

$$\hat{B} = B + F, \qquad |F| \leq 2\mathbf{u}|B| + O(\mathbf{u}^2).$$

which shows that the computed product is the exact product of slightly perturbed A and B.

2.7.10   Error in Strassen Multiplication

In §1.3.11 we outlined a recursive matrix multiplication procedure due to Strassen. It is instructive to compare the effect of roundoff in this method with the effect of roundoff in any of the conventional matrix multiplication methods of §1.1.

It can be shown that the Strassen approach (Algorithm 1.3.1) produces a $\hat{C} = \mathrm{fl}(AB)$ that satisfies an inequality of the form (2.7.20). This is perfectly satisfactory in many applications. However, the $\hat{C}$ that Strassen's method produces does *not* always satisfy an inequality of the form (2.7.19). To see this, suppose that

$$A = B = \begin{bmatrix} .99 & .0010 \\ .0010 & .99 \end{bmatrix}$$

and that we execute Algorithm 1.3.1 using 2-digit floating point arithmetic. Among other things, the following quantities are computed:

$$\hat{P}_3 = \mathrm{fl}(.99(.001 - .99)) = -.98,$$

$$\hat{P}_5 = \mathrm{fl}((.99 + .001).99) = .98,$$

$$\hat{c}_{12} = \mathrm{fl}(\hat{P}_3 + \hat{P}_5) = 0.0.$$

In exact arithmetic $c_{12} = 2(.001)(.99) = .00198$ and thus Algorithm 1.3.1 produces a $\hat{c}_{12}$ with no correct significant digits. The Strassen approach gets into trouble in this example because small off-diagonal entries are combined with large diagonal entries. Note that in conventional matrix multiplication the sums $b_{12} + b_{22}$ and $a_{11} + a_{12}$ do not arise. For that reason, the contribution of the small off-diagonal elements is not lost in this example. Indeed, for the above A and B a conventional matrix multiplication gives $\hat{c}_{12} = .0020$.

Failure to produce a componentwise accurate $\hat{C}$ can be a serious shortcoming in some applications. For example, in Markov processes the a_{ij}, b_{ij}, and c_{ij} are transition probabilities and are therefore nonnegative. It may be critical to compute c_{ij} accurately if it reflects a particularly important probability in the modeled phenomenon. Note that if $A \geq 0$ and $B \geq 0$, then conventional matrix multiplication produces a product $\hat{C}$ that has small componentwise relative error:

$$|\hat{C} - C| \leq n\mathbf{u}|A|\,|B| + O(\mathbf{u}^2) = n\mathbf{u}|C| + O(\mathbf{u}^2)\,.$$

This follows from (2.7.19). Because we cannot say the same for the Strassen approach, we conclude that Algorithm 1.3.1 is not attractive for *certain* nonnegative matrix multiplication problems *if* relatively accurate $\hat{c}_{ij}$ are required.

Extrapolating from this discussion we reach two fairly obvious but important conclusions:

- Different methods for computing the same quantity can produce substantially different results.

- Whether or not an algorithm produces satisfactory results depends upon the type of problem solved and the goals of the user.

These observations are clarified in subsequent chapters and are intimately related to the concepts of algorithm stability and problem condition. See §3.4.10.

2.7.11   Analysis of an Ideal Equation Solver

A nice way to conclude this chapter and to anticipate the next is to analyze the quality of a "make-believe" $Ax = b$ solution process in which all floating point operations are performed exactly *except* the storage of the matrix A and the right-hand-side b. It follows that the computed solution $\hat{x}$ satisfies

$$(A + E)\hat{x} = (b + e), \qquad \| E \|_\infty \leq \mathbf{u} \| A \|_\infty, \quad \| e \|_\infty \leq \mathbf{u} \| b \|_\infty . \qquad (2.7.21)$$

where

$$\mathsf{fl}(b) = b + e, \qquad \mathsf{fl}(A) = A + E.$$

If $\mathbf{u} \kappa_\infty(A) \leq \frac{1}{2}$ (say), then by Theorem 2.6.2 it can be shown that

$$\frac{\| x - \hat{x} \|_\infty}{\| x \|_\infty} \leq 4\mathbf{u} \kappa_\infty(A) . \qquad (2.7.22)$$

The bounds (2.7.21) and (2.7.22) are "best possible" norm bounds. No general ∞-norm error analysis of a linear equation solver that requires the storage of A and b can render sharper bounds. As a consequence, we cannot justifiably criticize an algorithm for returning an inaccurate $\hat{x}$ if A is ill-conditioned relative to the unit roundoff, e.g., $\mathbf{u} \kappa_\infty(A) \approx 1$. On the other hand, we have every "right" to pursue the development of a linear equation solver that renders the exact solution to a nearby problem in the style of (2.7.21).

Problems

P2.7.1 Show that if (2.7.8) is applied with $y = x$, then $\mathsf{fl}(x^T x) = x^T x(1 + \alpha)$ where $|\alpha| \leq n\mathbf{u} + O(\mathbf{u}^2)$.

P2.7.2   Prove (2.7.4) assuming that $\mathsf{fl}(x)$ is the nearest floating point number to $x \in \mathbb{R}$.

P2.7.3   Show that if $E \in \mathbb{R}^{m \times n}$ with $m \geq n$, then $\| |E| \|_2 \leq \sqrt{n} \| E \|_2$. This result is useful when deriving norm bounds from absolute value bounds.

P2.7.4   Assume the existence of a square root function satisfying $\mathsf{fl}(\sqrt{x}) = \sqrt{x}(1 + \epsilon)$ with $|\epsilon| \leq \mathbf{u}$. Give an algorithm for computing $\| x \|_2$ and bound the rounding errors.

P2.7.5   Suppose A and B are n-by-n upper triangular floating point matrices. If $\hat{C} = \mathsf{fl}(AB)$ is computed using one of the conventional §1.1 algorithms, does it follow that $\hat{C} = \hat{A}\hat{B}$ where $\hat{A}$ and $\hat{B}$ are close to A and B?

P2.7.6 Suppose A and B are n-by-n floating point matrices and that $\| \, |A^{-1}| \, |A| \, \|_\infty = \tau$. Show that if $\hat{C} = \mathrm{fl}(AB)$ is obtained using any of the §1.1 algorithms, then there exists a $\hat{B}$ so that $\hat{C} = A\hat{B}$ and $\| \, \hat{B} - B \, \|_\infty \leq n\mathbf{u}\tau\| \, B \, \|_\infty + O(\mathbf{u}^2)$.

P2.7.7 Prove (2.7.19).

P2.7.8 For the IEEE double format, what is the largest power of 10 that can be represented exactly? What is the largest integer that can be represented exactly?

P2.7.9 For $k = 1{:}62$, what is the largest power of 10 that can be stored exactly if k bits are are allocated for the mantissa and $63 - k$ are allocated for the exponent?

P2.7.10 Consider the quadratic equation

$$q(\lambda) \;=\; \det\left(\begin{bmatrix} w - \lambda & x \\ x & z - \lambda \end{bmatrix} \right).$$

This quadratic has two real roots r_1 and r_2. Assume that $|r_1 - z| \leq |r_2 - z|$. Give an algorithm that computes r_1 to full machine precision.

Notes and References for §2.7

For an excellent, comprehensive treatment of IEEE arithmetic and its implications, see:

M.L. Overton (2001). *Numerical Computing with IEEE Arithmetic*, SIAM Publications, Philadelphia, PA.

The following basic references are notable for the floating point insights that they offer: Wilkinson (AEP), Stewart (IMC), Higham (ASNA), and Demmel (ANLA). For high-level perspectives we recommend:

J.H. Wilkinson (1963). *Rounding Errors in Algebraic Processes*, Prentice-Hall, Englewood Cliffs, NJ.
G.E. Forsythe (1970). "Pitfalls in Computation or Why a Math Book is Not Enough," *Amer. Math. Monthly 77*, 931–956.
J.H. Wilkinson (1971). "Modern Error Analysis," *SIAM Review 13*, 548–68.
U.W. Kulisch and W.L. Miranker (1986). "The Arithmetic of the Digital Computer," *SIAM Review 28*, 1–40.
F. Chaitin-Chatelin and V. Frayseé (1996). *Lectures on Finite Precision Computations*, SIAM Publications, Philadelphia, PA.

The design of production software for matrix computations requires a detailed understanding of finite precision arithmetic, see:

J.W. Demmel (1984). "Underflow and the Reliability of Numerical Software," *SIAM J. Sci. Stat. Comput. 5*, 887–919.
W.J. Cody (1988). "ALGORITHM 665 MACHAR: A Subroutine to Dynamically Determine Machine Parameters," *ACM Trans. Math. Softw. 14*, 303–311.
D. Goldberg (1991). "What Every Computer Scientist Should Know About Floating Point Arithmetic," *ACM Surveys 23*, 5–48.

Other developments in error analysis involve interval analysis, the building of statistical models of roundoff error, and the automating of the analysis itself:

J. Larson and A. Sameh (1978). "Efficient Calculation of the Effects of Roundoff Errors," *ACM Trans. Math. Softw. 4*, 228–36.
W. Miller and D. Spooner (1978). "Software for Roundoff Analysis, II," *ACM Trans. Math. Softw. 4*, 369–90.
R.E. Moore (1979). *Methods and Applications of Interval Analysis*, SIAM Publications, Philadelphia, PA.
J.M. Yohe (1979). "Software for Interval Arithmetic: A Reasonable Portable Package," *ACM Trans. Math. Softw. 5*, 50–63.

The accuracy of floating point summation is detailed in:

S.M. Rump, T. Ogita, and S. Oishi (2008). "Accurate Floating-Point Summation Part I: Faithful Rounding," *SIAM J. Sci. Comput. 31*, 189–224.

S.M. Rump, T. Ogita, and S. Oishi (2008). "Accurate Floating-Point Summation Part II: Sign, K-fold Faithful and Rounding to Nearest," *SIAM J. Sci. Comput. 31*, 1269–1302.

For an analysis of the Strassen algorithm and other "fast" linear algebra procedures, see:

R.P. Brent (1970). "Error Analysis of Algorithms for Matrix Multiplication and Triangular Decomposition Using Winograd's Identity," *Numer. Math. 16,* 145–156.

W. Miller (1975). "Computational Complexity and Numerical Stability," *SIAM J. Comput. 4,* 97–107.

N.J. Higham (1992). "Stability of a Method for Multiplying Complex Matrices with Three Real Matrix Multiplications," *SIAM J. Matrix Anal. Applic. 13,* 681–687.

J.W. Demmel and N.J. Higham (1992). "Stability of Block Algorithms with Fast Level-3 BLAS," *ACM Trans. Math. Softw. 18,* 274–291.

B. Dumitrescu (1998). "Improving and Estimating the Accuracy of Strassen's Algorithm," *Numer. Math. 79,* 485–499.

The issue of extended precision has received considerable attention. For example, a superaccurate dot product results if the summation can be accumulated in a register that is "twice as wide" as the floating representation of vector components. The overhead may be tolerable in a given algorithm if extended precision is needed in only a few critical steps. For insights into this topic, see:

R.P. Brent (1978). "A Fortran Multiple Precision Arithmetic Package," *ACM Trans. Math. Softw. 4,* 57–70.

R.P. Brent (1978). "Algorithm 524 MP, a Fortran Multiple Precision Arithmetic Package," *ACM Trans. Math. Softw. 4,* 71–81.

D.H. Bailey (1993). "Algorithm 719: Multiprecision Translation and Execution of FORTRAN Programs," *ACM Trans. Math. Softw. 19,* 288–319.

X.S. Li, J.W. Demmel, D.H. Bailey, G. Henry, Y. Hida, J. Iskandar, W. Kahan, S.Y. Kang, A. Kapur, M.C. Martin, B.J. Thompson, T. Tung, and D.J. Yoo (2002). "Design, Implementation and Testing of Extended and Mixed Precision BLAS," *ACM Trans. Math. Softw. 28,* 152–205.

J.W. Demmel and Y. Hida (2004). "Accurate and Efficient Floating Point Summation," *SIAM J. Sci. Comput. 25,* 1214–1248.

M. Baboulin, A. Buttari, J. Dongarra, J. Kurzak, J. Langou, J. Langou, P. Luszczek, and S. Tomov (2009). "Accelerating Scientific Computations with Mixed Precision Algorithms," *Comput. Phys. Commun. 180,* 2526–2533.

Chapter 3

General Linear Systems

3.1   **Triangular Systems**

3.2   **The LU Factorization**

3.3   **Roundoff Error in Gaussian Elimination**

3.4   **Pivoting**

3.5   **Improving and Estimating Accuracy**

3.6   **Parallel LU**

The problem of solving a linear system $Ax = b$ is central to scientific computation. In this chapter we focus on the method of Gaussian elimination, the algorithm of choice if A is square, dense, and unstructured. Other methods are applicable if A does not fall into this category, see Chapter 4, Chapter 11, §12.1, and §12.2. Solution procedures for triangular systems are discussed first. These are followed by a derivation of Gaussian elimination that makes use of Gauss transformations. The process of eliminating unknowns from equations is described in terms of the factorization $A = LU$ where L is lower triangular and U is upper triangular. Unfortunately, the derived method behaves poorly on a nontrivial class of problems. An error analysis pinpoints the difficulty and sets the stage for a discussion of pivoting, a permutation strategy that keeps the numbers "nice" during the elimination. Practical issues associated with scaling, iterative improvement, and condition estimation are covered. A framework for computing the LU factorization in parallel is developed in the final section.

Reading Notes

Familiarity with Chapter 1, §§2.1–2.5, and §2.7 is assumed. The sections within this chapter depend upon each other as follows:

$$
\begin{array}{ccccccc}
 & & & & & & \S 3.5 \\
 & & & & & & \uparrow \\
\S 3.1 & \rightarrow & \S 3.2 & \rightarrow & \S 3.3 & \rightarrow & \S 3.4 \\
 & & & & & & \downarrow \\
 & & & & & & \S 3.6
\end{array}
$$

Useful global references include Forsythe and Moler (SLAS), Stewart(MABD), Higham (ASNA), Watkins (FMC), Trefethen and Bau (NLA), Demmel (ANLA), and Ipsen (NMA).

3.1 Triangular Systems

Traditional factorization methods for linear systems involve the conversion of the given square system to a triangular system that has the same solution. This section is about the solution of triangular systems.

3.1.1 Forward Substitution

Consider the following 2-by-2 lower triangular system:

$$\left[\begin{array}{cc} \ell_{11} & 0 \\ \ell_{21} & \ell_{22} \end{array}\right]\left[\begin{array}{c} x_1 \\ x_2 \end{array}\right] = \left[\begin{array}{c} b_1 \\ b_2 \end{array}\right].$$

If $\ell_{11}\ell_{22} \neq 0$, then the unknowns can be determined sequentially:

$$x_1 = b_1/\ell_{11},$$
$$x_2 = (b_2 - \ell_{21}x_1)/\ell_{22}.$$

This is the 2-by-2 version of an algorithm known as *forward substitution*. The general procedure is obtained by solving the ith equation in $Lx = b$ for x_i:

$$x_i = \left(b_i - \sum_{j=1}^{i-1}\ell_{ij}x_j\right)\Bigg/ \ell_{ii}.$$

If this is evaluated for $i = 1{:}n$, then a complete specification of x is obtained. Note that at the ith stage the dot product of $L(i, 1{:}i - 1)$ and $x(1{:}i - 1)$ is required. Since b_i is involved only in the formula for x_i, the former may be overwritten by the latter.

Algorithm 3.1.1 (Row-Oriented Forward Substitution) If $L \in \mathbb{R}^{n\times n}$ is lower triangular and $b \in \mathbb{R}^n$, then this algorithm overwrites b with the solution to $Lx = b$. L is assumed to be nonsingular.

$b(1) = b(1)/L(1, 1)$
for $i = 2{:}n$
    $b(i) = (b(i) - L(i, 1{:}i - 1){\cdot}b(1{:}i - 1))/L(i, i)$
end

This algorithm requires n^2 flops. Note that L is accessed by row. The computed solution $\hat{x}$ can be shown to satisfy

$$(L + F)\hat{x} = b \qquad |F| \leq n\mathbf{u}|L| + O(\mathbf{u}^2). \tag{3.1.1}$$

For a proof, see Higham (ASNA, pp. 141-142). It says that the computed solution exactly satisfies a slightly perturbed system. Moreover, each entry in the perturbing matrix F is small relative to the corresponding element of L.

3.1.2   Back Substitution

The analogous algorithm for an upper triangular system $Ux = b$ is called *back substitution*. The recipe for x_i is prescribed by

$$x_i = \left(b_i - \sum_{j=i+1}^{n} u_{ij} x_j \right) \bigg/ u_{ii}$$

and once again b_i can be overwritten by x_i.

Algorithm 3.1.2 (Row-Oriented Back Substitution) If $U \in \mathbb{R}^{n \times n}$ is upper triangular and $b \in \mathbb{R}^n$, then the following algorithm overwrites b with the solution to $Ux = b$. U is assumed to be nonsingular.

$b(n) = b(n)/U(n,n)$
for $i = n-1 : -1 : 1$
$\quad b(i) = (b(i) - U(i, i+1{:}n){\cdot}b(i+1{:}n))/U(i,i)$
end

This algorithm requires n^2 flops and accesses U by row. The computed solution $\hat{x}$ obtained by the algorithm can be shown to satisfy

$$(U + F)\hat{x} = b, \qquad |F| \leq n\mathbf{u}|U| + O(\mathbf{u}^2). \qquad (3.1.2)$$

3.1.3   Column-Oriented Versions

Column-oriented versions of the above procedures can be obtained by reversing loop orders. To understand what this means from the algebraic point of view, consider forward substitution. Once x_1 is resolved, it can be removed from equations 2 through n leaving us with the reduced system

$$L(2{:}n, 2{:}n)x(2{:}n) = b(2{:}n) - x(1){\cdot}L(2{:}n, 1).$$

We next compute x_2 and remove it from equations 3 through n, etc. Thus, if this approach is applied to

$$\begin{bmatrix} 2 & 0 & 0 \\ 1 & 5 & 0 \\ 7 & 9 & 8 \end{bmatrix} \begin{bmatrix} x_1 \\ x_2 \\ x_3 \end{bmatrix} = \begin{bmatrix} 6 \\ 2 \\ 5 \end{bmatrix},$$

we find $x_1 = 3$ and then deal with the 2-by-2 system

$$\begin{bmatrix} 5 & 0 \\ 9 & 8 \end{bmatrix} \begin{bmatrix} x_2 \\ x_3 \end{bmatrix} = \begin{bmatrix} 2 \\ 5 \end{bmatrix} - 3 \begin{bmatrix} 1 \\ 7 \end{bmatrix} = \begin{bmatrix} -1 \\ -16 \end{bmatrix}.$$

Here is the complete procedure with overwriting.

Algorithm 3.1.3 (Column-Oriented Forward Substitution) If the matrix $L \in \mathbb{R}^{n \times n}$ is lower triangular and $b \in \mathbb{R}^n$, then this algorithm overwrites b with the solution to $Lx = b$. L is assumed to be nonsingular.

for $j = 1{:}n - 1$
$$b(j) = b(j)/L(j, j)$$
$$b(j + 1{:}n) = b(j + 1{:}n) - b(j) \cdot L(j + 1{:}n, j)$$
end
$$b(n) = b(n)/L(n, n)$$

It is also possible to obtain a column-oriented saxpy procedure for back substitution.

Algorithm 3.1.4 (Column-Oriented Back Substitution) If $U \in \mathbb{R}^{n \times n}$ is upper triangular and $b \in \mathbb{R}^n$, then this algorithm overwrites b with the solution to $Ux = b$. U is assumed to be nonsingular.

for $j = n{:} -1{:}2$
$$b(j) = b(j)/U(j, j)$$
$$b(1{:}j - 1) = b(1{:}j - 1) - b(j) \cdot U(1{:}j - 1, j)$$
end
$$b(1) = b(1)/U(1, 1)$$

Note that the dominant operation in both Algorithms 3.1.3 and 3.1.4 is the saxpy operation. The roundoff behavior of these implementations is essentially the same as for the dot product versions.

3.1.4   Multiple Right-Hand Sides

Consider the problem of computing a solution $X \in \mathbb{R}^{n \times q}$ to $LX = B$ where $L \in \mathbb{R}^{n \times n}$ is lower triangular and $B \in \mathbb{R}^{n \times q}$. This is the *multiple-right-hand-side* problem and it amounts to solving q separate triangular systems, i.e., $LX(:, j) = B(:, j)$, $j = 1{:}q$. Interestingly, the computation can be blocked in such a way that the resulting algorithm is rich in matrix multiplication, assuming that q and n are large enough. This turns out to be important in subsequent sections where various block factorization schemes are discussed.

It is sufficient to consider just the lower triangular case as the derivation of block back substitution is entirely analogous. We start by partitioning the equation $LX = B$ as follows:

$$
\begin{bmatrix}
L_{11} & 0 & \cdots & 0 \\
L_{21} & L_{22} & \cdots & 0 \\
\vdots & \vdots & \ddots & \vdots \\
L_{N1} & L_{N2} & \cdots & L_{NN}
\end{bmatrix}
\begin{bmatrix}
X_1 \\
X_2 \\
\vdots \\
X_N
\end{bmatrix}
=
\begin{bmatrix}
B_1 \\
B_2 \\
\vdots \\
B_N
\end{bmatrix}.
\tag{3.1.3}
$$

Assume that the diagonal blocks are square. Paralleling the development of Algorithm 3.1.3, we solve the system $L_{11}X_1 = B_1$ for X_1 and then remove X_1 from block equations 2 through N:

$$
\begin{bmatrix}
L_{22} & 0 & \cdots & 0 \\
L_{32} & L_{33} & \cdots & 0 \\
\vdots & \vdots & \ddots & \vdots \\
L_{N2} & L_{N3} & \cdots & L_{NN}
\end{bmatrix}
\begin{bmatrix}
X_2 \\
X_3 \\
\vdots \\
X_N
\end{bmatrix}
=
\begin{bmatrix}
B_2 \\
B_3 \\
\vdots \\
B_N
\end{bmatrix}
-
\begin{bmatrix}
L_{21} \\
L_{31} \\
\vdots \\
L_{N1}
\end{bmatrix}
X_1.
$$

Continuing in this way we obtain the following block forward elimination scheme:

    for $j = 1{:}N$
        Solve $L_{jj}X_j = B_j$
        for $i = j + 1{:}N$ (3.1.4)
            $B_i = B_i - L_{ij}X_j$
        end
    end

Notice that the i-loop oversees a single block saxpy update of the form

$$
\begin{bmatrix} B_{j+1} \\ \vdots \\ B_N \end{bmatrix} = \begin{bmatrix} B_{j+1} \\ \vdots \\ B_N \end{bmatrix} - \begin{bmatrix} L_{j+1,j} \\ \vdots \\ L_{N,j} \end{bmatrix} X_j .
$$

To realize level-3 performance, the submatrices in (3.1.3) must be sufficiently large in dimension.

3.1.5   The Level-3 Fraction

It is handy to adopt a measure that quantifies the amount of matrix multiplication in a given algorithm. To this end we define the *level-3 fraction* of an algorithm to be the fraction of flops that occur in the context of matrix multiplication. We call such flops *level-3 flops*.

    Let us determine the level-3 fraction for (3.1.4) with the simplifying assumption that $n = rN$. (The same conclusions hold with the unequal blocking described above.) Because there are N applications of r-by-r forward elimination (the level-2 portion of the computation) and n^2 flops overall, the level-3 fraction is approximately given by

$$
1 - \frac{Nr^2}{n^2} = 1 - \frac{1}{N}.
$$

Thus, for large N almost all flops are level-3 flops. It makes sense to choose N as large as possible subject to the constraint that the underlying architecture can achieve a high level of performance when processing block saxpys that have width $r = n/N$ or greater.

3.1.6   Nonsquare Triangular System Solving

The problem of solving nonsquare, m-by-n triangular systems deserves some attention. Consider the lower triangular case when $m \geq n$, i.e.,

$$
\begin{bmatrix} L_{11} \\ L_{21} \end{bmatrix} x = \begin{bmatrix} b_1 \\ b_2 \end{bmatrix} \qquad \begin{array}{ll} L_{11} \in \mathbb{R}^{n \times n}, & b_1 \in \mathbb{R}^n, \\ L_{21} \in \mathbb{R}^{(m-n) \times n}, & b_2 \in \mathbb{R}^{m-n}. \end{array}
$$

Assume that L_{11} is lower triangular and nonsingular. If we apply forward elimination to $L_{11}x = b_1$, then x solves the system provided $L_{21}(L_{11}^{-1}b_1) = b_2$. Otherwise, there is no solution to the overall system. In such a case least squares minimization may be appropriate. See Chapter 5.

    Now consider the lower triangular system $Lx = b$ when the number of columns n exceeds the number of rows m. We can apply forward substitution to the square system $L(1{:}m, 1{:}m)x(1{:}m, 1{:}m) = b$ and prescribe an arbitrary value for $x(m+1{:}n)$. See §5.6 for additional comments on systems that have more unknowns than equations. The handling of nonsquare upper triangular systems is similar. Details are left to the reader.

3.1.7   The Algebra of Triangular Matrices

A *unit* triangular matrix is a triangular matrix with 1's on the diagonal. Many of the triangular matrix computations that follow have this added bit of structure. It clearly poses no difficulty in the above procedures.

For future reference we list a few properties about products and inverses of triangular and unit triangular matrices.

- The inverse of an upper (lower) triangular matrix is upper (lower) triangular.

- The product of two upper (lower) triangular matrices is upper (lower) triangular.

- The inverse of a unit upper (lower) triangular matrix is unit upper (lower) triangular.

- The product of two unit upper (lower) triangular matrices is unit upper (lower) triangular.

Problems

P3.1.1   Give an algorithm for computing a nonzero $z \in \mathbb{R}^n$ such that $Uz = 0$ where $U \in \mathbb{R}^{n \times n}$ is upper triangular with $u_{nn} = 0$ and $u_{11} \cdots u_{n-1,n-1} \neq 0$.

P3.1.2   Suppose $L = I_n - N$ is unit lower triangular where $N \in \mathbb{R}^{n \times n}$. Show that

$$L^{-1} = I_n + N + N^2 + \cdots + N^{n-1}.$$

What is the value of $\| L^{-1} \|_F$ if $N_{ij} = 1$ for all $i > j$?

P3.1.3   Write a detailed version of (3.1.4). Do not assume that N divides n.

P3.1.4   Prove all the facts about triangular matrices that are listed in §3.1.7.

P3.1.5   Suppose $S, T \in \mathbb{R}^{n \times n}$ are upper triangular and that $(ST - \lambda I)x = b$ is a nonsingular system. Give an $O(n^2)$ algorithm for computing x. Note that the explicit formation of $ST - \lambda I$ requires $O(n^3)$ flops. Hint: Suppose

$$S_+ = \left[\begin{array}{cc} \sigma & u^T \\ 0 & S_c \end{array} \right], \quad T_+ = \left[\begin{array}{cc} \tau & v^T \\ 0 & T_c \end{array} \right], \quad b_+ = \left[\begin{array}{c} \beta \\ b_c \end{array} \right],$$

where $S_+ = S(k-1{:}n, k-1{:}n)$, $T_+ = T(k-1{:}n, k-1{:}n)$, $b_+ = b(k-1{:}n)$, and $\sigma, \tau, \beta \in \mathbb{R}$. Show that if we have a vector x_c such that

$$(S_c T_c - \lambda I)x_c = b_c$$

and $w_c = T_c x_c$ is available, then

$$x_+ = \left[\begin{array}{c} \gamma \\ x_c \end{array} \right], \qquad \gamma = \frac{\beta - \sigma v^T x_c - u^T w_c}{\sigma \tau - \lambda}$$

solves $(S_+ T_+ - \lambda I)x_+ = b_+$. Observe that x_+ and $w_+ = T_+ x_+$ each require $O(n-k)$ flops.

P3.1.6   Suppose the matrices $R_1, \ldots, R_p \in \mathbb{R}^{n \times n}$ are all upper triangular. Give an $O(pn^2)$ algorithm for solving the system $(R_1 \cdots R_p - \lambda I)x = b$ assuming that the matrix of coefficients is nonsingular. Hint. Generalize the solution to the previous problem.

P3.1.7   Suppose $L, K \in \mathbb{R}^{n \times n}$ are lower triangular and $B \in \mathbb{R}^{n \times n}$. Give an algorithm for computing $X \in \mathbb{R}^{n \times n}$ so that $LXK = B$.

Notes and References for §3.1

The accuracy of a computed solution to a triangular system is often surprisingly good, see:

N.J. Higham (1989). "The Accuracy of Solutions to Triangular Systems," *SIAM J. Numer. Anal. 26*, 1252–1265.

Solving systems of the form $(T_p \cdots T_1 - \lambda I)x = b$ where each T_i is triangular is considered in:

C.D. Martin and C.F. Van Loan (2002). "Product Triangular Systems with Shift," *SIAM J. Matrix Anal. Applic. 24*, 292–301.

The trick to obtaining an $O(pn^2)$ procedure that does not involve any matrix-matrix multiplications is to look carefully at the back-substitution recursions. See P3.1.6.

A survey of parallel triangular system solving techniques and their stabilty is given in:

N.J. Higham (1995). "Stability of Parallel Triangular System Solvers," *SIAM J. Sci. Comput. 16*, 400–413.

3.2   The LU Factorization

Triangular system solving is an easy $O(n^2)$ computation. The idea behind Gaussian elimination is to convert a given system $Ax = b$ to an equivalent triangular system. The conversion is achieved by taking appropriate linear combinations of the equations. For example, in the system

$$3x_1 + 5x_2 = 9,$$
$$6x_1 + 7x_2 = 4,$$

if we multiply the first equation by 2 and subtract it from the second we obtain

$$3x_1 + 5x_2 = \quad 9,$$
$$-3x_2 = -14.$$

This is $n = 2$ Gaussian elimination. Our objective in this section is to describe the procedure in the language of matrix factorizations. This means showing that the algorithm computes a unit lower triangular matrix L and an upper triangular matrix U so that $A = LU$, e.g.,

$$\begin{bmatrix} 3 & 5 \\ 6 & 7 \end{bmatrix} = \begin{bmatrix} 1 & 0 \\ 2 & 1 \end{bmatrix} \begin{bmatrix} 3 & 5 \\ 0 & -3 \end{bmatrix}.$$

The solution to the original $Ax = b$ problem is then found by a two-step triangular solve process:

$$Ly = b, \quad Ux = y \quad \Longrightarrow \quad Ax = LUx = Ly = b. \tag{3.2.1}$$

The LU factorization is a "high-level" algebraic description of Gaussian elimination. Linear equation solving is not about the matrix vector product $A^{-1}b$ but about computing LU and using it effectively; see §3.4.9. Expressing the outcome of a matrix algorithm in the "language" of matrix factorizations is a productive exercise, one that is repeated many times throughout this book. It facilitates generalization and highlights connections between algorithms that can appear very different at the scalar level.

3.2.1 Gauss Transformations

To obtain a factorization description of Gaussian elimination as it is traditionally presented, we need a matrix description of the zeroing process. At the $n = 2$ level, if $v_1 \neq 0$ and $\tau = v_2/v_1$, then

$$
\begin{bmatrix} 1 & 0 \\ -\tau & 1 \end{bmatrix} \begin{bmatrix} v_1 \\ v_2 \end{bmatrix} = \begin{bmatrix} v_1 \\ 0 \end{bmatrix}.
$$

More generally, suppose $v \in \mathbb{R}^n$ with $v_k \neq 0$. If

$$
\tau^T = [\, \underbrace{0, \ldots, 0}_{k}, \tau_{k+1}, \ldots, \tau_n \,], \qquad \tau_i = \frac{v_i}{v_k}, \quad i = k+1{:}n,
$$

and we define

$$
M_k = I_n - \tau e_k^T, \tag{3.2.2}
$$

then

$$
M_k v =
\begin{bmatrix}
1 & \cdots & 0 & 0 & \cdots & 0 \\
\vdots & \ddots & \vdots & \vdots & & \vdots \\
0 & & 1 & 0 & & 0 \\
0 & & -\tau_{k+1} & 1 & & 0 \\
\vdots & \vdots & \vdots & \vdots & \ddots & \vdots \\
0 & \cdots & -\tau_n & 0 & \cdots & 1
\end{bmatrix}
\begin{bmatrix}
v_1 \\ \vdots \\ v_k \\ v_{k+1} \\ \vdots \\ v_n
\end{bmatrix}
=
\begin{bmatrix}
v_1 \\ \vdots \\ v_k \\ 0 \\ \vdots \\ 0
\end{bmatrix}.
$$

A matrix of the form $M_k = I_n - \tau e_k^T \in \mathbb{R}^{n \times n}$ is a *Gauss transformation* if the first k components of $\tau \in \mathbb{R}^n$ are zero. Such a matrix is unit lower triangular. The components of $\tau(k+1{:}n)$ are called *multipliers*. The vector τ is called the *Gauss vector*.

3.2.2 Applying Gauss Transformations

Multiplication by a Gauss transformation is particularly simple. If $C \in \mathbb{R}^{n \times r}$ and $M_k = I_n - \tau e_k^T$ is a Gauss transformation, then

$$
M_k C = (I_n - \tau e_k^T)C = C - \tau (e_k^T C) = C - \tau C(k,:)
$$

is an outer product update. Since $\tau(1{:}k) = 0$ only $C(k+1{:}n, :)$ is affected and the update $C = M_k C$ can be computed row by row as follows:

> **for** $i = k+1{:}n$
> $\quad C(i,:) = C(i,:) - \tau_i \cdot C(k,:)$
> **end**

This computation requires $2(n-k)r$ flops. Here is an example:

$$
C = \begin{bmatrix} 1 & 4 & 7 \\ 2 & 5 & 8 \\ 3 & 6 & 10 \end{bmatrix}, \quad \tau = \begin{bmatrix} 0 \\ 1 \\ -1 \end{bmatrix} \quad \Longrightarrow \quad (I - \tau e_1^T)C = \begin{bmatrix} 1 & 4 & 7 \\ 1 & 1 & 1 \\ 4 & 10 & 17 \end{bmatrix}.
$$

3.2.3 Roundoff Properties of Gauss Transformations

If $\hat{\tau}$ is the computed version of an exact Gauss vector τ, then it is easy to verify that

$$\hat{\tau} = \tau + e, \qquad |e| \leq \mathbf{u}|\tau|.$$

If $\hat{\tau}$ is used in a Gauss transform update and $\mathrm{fl}((I_n - \hat{\tau}e_k^T)C)$ denotes the computed result, then

$$\mathrm{fl}\left((I_n - \hat{\tau}e_k^T)C\right) = (I - \tau e_k^T)C + E,$$

where

$$|E| \leq 3\mathbf{u}\left(|C| + |\tau||C(k,:)|\right) + O(\mathbf{u}^2).$$

Clearly, if τ has large components, then the errors in the update may be large in comparison to $|C|$. For this reason, care must be exercised when Gauss transformations are employed, a matter that is pursued in §3.4.

3.2.4 Upper Triangularizing

Assume that $A \in \mathbb{R}^{n \times n}$. Gauss transformations $M_1, \ldots, M_{n-1}$ can usually be found such that $M_{n-1} \cdots M_2 M_1 A = U$ is upper triangular. To see this we first look at the $n = 3$ case. Suppose

$$A = \begin{bmatrix} 1 & 4 & 7 \\ 2 & 5 & 8 \\ 3 & 6 & 10 \end{bmatrix}$$

and note that

$$M_1 = \begin{bmatrix} 1 & 0 & 0 \\ -2 & 1 & 0 \\ -3 & 0 & 1 \end{bmatrix} \quad \Rightarrow \quad M_1 A = \begin{bmatrix} 1 & 4 & 7 \\ 0 & -3 & -6 \\ 0 & -6 & -11 \end{bmatrix}.$$

Likewise, in the second step we have

$$M_2 = \begin{bmatrix} 1 & 0 & 0 \\ 0 & 1 & 0 \\ 0 & -2 & 1 \end{bmatrix} \quad \Rightarrow \quad M_2(M_1 A) = \begin{bmatrix} 1 & 4 & 7 \\ 0 & -3 & -6 \\ 0 & 0 & 1 \end{bmatrix}.$$

Extrapolating from this example to the general n case we conclude two things.

- At the start of the kth step we have a matrix $A^{(k-1)} = M_{k-1} \cdots M_1 A$ that is upper triangular in columns 1 through $k - 1$.

- The multipliers in the kth Gauss transform M_k are based on $A^{(k-1)}(k+1:n, k)$ and $a_{kk}^{(k-1)}$ must be nonzero in order to proceed.

Noting that complete upper triangularization is achieved after $n - 1$ steps, we obtain the following rough draft of the overall process:

$A^{(1)} = A$

for $k = 1:n - 1$

    For $i = k + 1:n$, determine the multipliers $\tau_i^{(k)} = a_{ik}^{(k)}/a_{kk}^{(k)}$.   (3.2.3)

    Apply $M_k = I - \tau^{(k)}e_k^T$ to obtain $A^{(k+1)} = M_k A^{(k)}$.

end

For this process to be well-defined, the matrix entries $a_{11}^{(1)}, a_{22}^{(2)}, \ldots, a_{n-1,n-1}^{(n-1)}$ must be nonzero. These quantities are called *pivots*.

3.2.5   Existence

If no zero pivots are encountered in (3.2.3), then Gauss transformations $M_1, \ldots, M_{n-1}$ are generated such that $M_{n-1} \cdots M_1 A = U$ is upper triangular. It is easy to check that if $M_k = I_n - \tau^{(k)} e_k^T$, then its inverse is prescribed by $M_k^{-1} = I_n + \tau^{(k)} e_k^T$ and so

$$A = LU \tag{3.2.4}$$

where

$$L = M_1^{-1} \cdots M_{n-1}^{-1}. \tag{3.2.5}$$

It is clear that L is a unit lower triangular matrix because each M_k^{-1} is unit lower triangular. The factorization (3.2.4) is called the *LU factorization*.

The LU factorization may not exist. For example, it is impossible to find l_{ij} and u_{ij} so

$$\begin{bmatrix} 1 & 2 & 3 \\ 2 & 4 & 7 \\ 3 & 5 & 3 \end{bmatrix} = \begin{bmatrix} 1 & 0 & 0 \\ \ell_{21} & 1 & 0 \\ \ell_{31} & \ell_{32} & 1 \end{bmatrix} \begin{bmatrix} u_{11} & u_{12} & u_{13} \\ 0 & u_{22} & u_{23} \\ 0 & 0 & u_{33} \end{bmatrix}.$$

To see this, equate entries and observe that we must have $u_{11} = 1$, $u_{12} = 2$, $\ell_{21} = 2$, $u_{22} = 0$, and $\ell_{31} = 3$. But then the (3,2) entry gives us the contradictory equation $5 = \ell_{31} u_{12} + \ell_{32} u_{22} = 6$. For this example, the pivot $a_{22}^{(1)} = a_{22} - (a_{21}/a_{11}) a_{12}$ is zero.

It turns out that the kth pivot in (3.2.3) is zero if $A(1{:}k, 1{:}k)$ is singular. A submatrix of the form $A(1{:}k, 1{:}k)$ is called a *leading principal submatrix*.

Theorem 3.2.1. (LU Factorization). *If $A \in \mathbb{R}^{n \times n}$ and $\det(A(1{:}k, 1{:}k)) \neq 0$ for $k = 1{:}n-1$, then there exists a unit lower triangular $L \in \mathbb{R}^{n \times n}$ and an upper triangular $U \in \mathbb{R}^{n \times n}$ such that $A = LU$. If this is the case and A is nonsingular, then the factorization is unique and $\det(A) = u_{11} \cdots u_{nn}$.*

Proof. Suppose $k - 1$ steps in (3.2.3) have been executed. At the beginning of step k the matrix A has been overwritten by $M_{k-1} \cdots M_1 A = A^{(k-1)}$. Since Gauss transformations are unit lower triangular, it follows by looking at the leading k-by-k portion of this equation that

$$\det(A(1{:}k, 1{:}k)) = a_{11}^{(k-1)} \cdots a_{kk}^{(k-1)}. \tag{3.2.6}$$

Thus, if $A(1{:}k, 1{:}k)$ is nonsingular, then the kth pivot $a_{kk}^{(k-1)}$ is nonzero.

As for uniqueness, if $A = L_1 U_1$ and $A = L_2 U_2$ are two LU factorizations of a nonsingular A, then $L_2^{-1} L_1 = U_2 U_1^{-1}$. Since $L_2^{-1} L_1$ is unit lower triangular and $U_2 U_1^{-1}$ is upper triangular, it follows that both of these matrices must equal the identity. Hence, $L_1 = L_2$ and $U_1 = U_2$. Finally, if $A = LU$, then

$$\det(A) = \det(LU) = \det(L)\det(U) = \det(U).$$

It follows that $\det(A) = u_{11} \cdots u_{nn}$. $\square$

3.2.6 L Is the Matrix of Multipliers

It turns out that the construction of L is not nearly so complicated as Equation (3.2.5) suggests. Indeed,

$$
\begin{aligned}
L &= M_1^{-1} \cdots M_{n-1}^{-1} \\
&= \left(I_n - \tau^{(1)} e_1^T \right)^{-1} \cdots \left(I_n - \tau^{(n-1)} e_{n-1}^T \right)^{-1} \\
&= \left(I_n + \tau^{(1)} e_1^T \right) \cdots \left(I_n + \tau^{(n-1)} e_{n-1}^T \right) \\
&= I_n + \sum_{k=1}^{n-1} \tau^{(k)} e_k^T
\end{aligned}
$$

showing that

$$
L(k+1{:}n, k) = \tau^{(k)}(k+1{:}n) \qquad k = 1{:}n-1. \tag{3.2.7}
$$

In other words, the kth column of L is defined by the multipliers that arise in the k-th step of (3.2.3). Consider the example in §3.2.4:

$$
\tau^{(1)} = \begin{bmatrix} 0 \\ 2 \\ 3 \end{bmatrix}, \quad \tau^{(2)} = \begin{bmatrix} 0 \\ 0 \\ 2 \end{bmatrix} \quad \Rightarrow \quad \begin{bmatrix} 1 & 4 & 7 \\ 2 & 5 & 8 \\ 3 & 6 & 10 \end{bmatrix} = \begin{bmatrix} 1 & 0 & 0 \\ 2 & 1 & 0 \\ 3 & 2 & 1 \end{bmatrix} \begin{bmatrix} 1 & 4 & 7 \\ 0 & -3 & -6 \\ 0 & 0 & 1 \end{bmatrix}.
$$

3.2.7 The Outer Product Point of View

Since the application of a Gauss transformation to a matrix involves an outer product, we can regard (3.2.3) as a sequence of outer product updates. Indeed, if

$$
A = \begin{bmatrix} \alpha & w^T \\ v & B \end{bmatrix} \begin{matrix} 1 \\ n-1 \end{matrix}
$$

$$ \begin{matrix} 1 & n-1 \end{matrix}$$

then the first step in Gaussian elimination results in the decomposition

$$
\begin{bmatrix} \alpha & w^T \\ z & B \end{bmatrix} = \begin{bmatrix} 1 & 0 \\ z/\alpha & I_{n-1} \end{bmatrix} \begin{bmatrix} 1 & 0 \\ 0 & B - zw^T/\alpha \end{bmatrix} \begin{bmatrix} \alpha & w^T \\ 0 & I_{n-1} \end{bmatrix}.
$$

Steps 2 through $n-1$ compute the LU factorization

$$
B - zw^T/\alpha = L_1 U_1
$$

for then

$$
A = \begin{bmatrix} 1 & 0 \\ z/\alpha & I_{n-1} \end{bmatrix} \begin{bmatrix} 1 & 0 \\ 0 & L_1 U_1 \end{bmatrix} \begin{bmatrix} \alpha & w^T \\ 0 & I_{n-1} \end{bmatrix} = \begin{bmatrix} 1 & 0 \\ z/\alpha & L_1 \end{bmatrix} \begin{bmatrix} \alpha & w^T \\ 0 & U_1 \end{bmatrix} \equiv LU.
$$

3.2.8   Practical Implementation

Let us consider the efficient implementation of (3.2.3). First, because zeros have already been introduced in columns 1 through $k-1$, the Gauss transformation update need only be applied to columns k through n. Of course, we need not even apply the kth Gauss transform to $A(:, k)$ since we know the result. So the efficient thing to do is simply to update $A(k+1{:}n, k+1{:}n)$. Also, the observation (3.2.7) suggests that we can overwrite $A(k+1{:}n, k)$ with $L(k+1{:}n, k)$ since the latter houses the multipliers that are used to zero the former. Overall we obtain:

Algorithm 3.2.1 (Outer Product LU)   Suppose $A \in \mathbb{R}^{n \times n}$ has the property that $A(1{:}k, 1{:}k)$ is nonsingular for $k = 1{:}n-1$. This algorithm computes the factorization $A = LU$ where L is unit lower triangular and U is upper triangular. For $i = 1{:}n-1$, $A(i, i{:}n)$ is overwritten by $U(i, i{:}n)$ while $A(i+1{:}n, i)$ is overwritten by $L(i+1{:}n, i)$.

> **for** $k = 1{:}n-1$
>
> $\quad \rho = k + 1{:}n$
>
> $\quad A(\rho, k) = A(\rho, k)/A(k, k)$
>
> $\quad A(\rho, \rho) = A(\rho, \rho) - A(\rho, k) \cdot A(k, \rho)$
>
> **end**

This algorithm involves $2n^3/3$ flops and it is one of several formulations of *Gaussian elimination*. Note that the k-th step involves an $(n-k)$-by-$(n-k)$ outer product.

3.2.9   Other Versions

Similar to matrix-matrix multiplication, Gaussian elimination is a triple-loop procedure that can be arranged in several ways. Algorithm 3.2.1 corresponds to the "kij" version of Gaussian elimination if we compute the outer product update row by row:

> **for** $k = 1{:}n-1$
>
> $\quad A(k+1{:}n, k) = A(k+1{:}n, k)/A(k, k)$
>
> $\quad$ **for** $i = k+1{:}n$
>
> $\quad\quad$ **for** $j = k+1{:}n$
>
> $\quad\quad\quad A(i, j) = A(i, j) - A(i, k) \cdot A(k, j)$
>
> $\quad\quad$ **end**
>
> $\quad$ **end**
>
> **end**

There are five other versions: kji, ikj, ijk, jik, and jki. The last of these results in an implementation that features a sequence of gaxpys and forward eliminations which we now derive at the vector level.

The plan is to compute the jth columns of L and U in step j. If $j = 1$, then by comparing the first columns in $A = LU$ we conclude that

$$L(2{:}n, j) = A(2{:}n, 1)/A(1, 1)$$

and $U(1, 1) = A(1, 1)$. Now assume that $L(:, 1{:}j-1)$ and $U(1{:}j-1, 1{:}j-1)$ are known. To get the jth columns of L and U we equate the jth columns in the equation $A = LU$

and infer from the vector equation $A(:,j) = LU(:,j)$ that

$$A(1{:}j-1,j) \;=\; L(1{:}j-1,1{:}j-1){\cdot}U(1{:}j-1,j)$$

and

$$A(j{:}n,j) \;=\; \sum_{k=1}^{j} L(j{:}n,k){\cdot}U(k,j).$$

The first equation is a lower triangular linear system that can be solved for the vector $U(1{:}j-1,j)$. Once this is accomplished, the second equation can be rearranged to produce recipes for $U(j,j)$ and $L(j+1{:}n,j)$. Indeed, if we set

$$v(j{:}n) = A(j{:}n,j) - \sum_{k=1}^{j-1} L(j{:}n,k)U(k,j)$$

$$= A(j{:}n,j) - L(j{:}n,1{:}j-1){\cdot}U(1{:}j-1,j),$$

then $L(j+1{:}n,j) = v(j+1{:}n)/v(j)$ and $U(j,j) = v(j)$. Thus, $L(j+1{:}n,j)$ is a scaled gaxpy and we obtain the following alternative to Algorithm 3.2.1:

Algorithm 3.2.2 (Gaxpy LU) Suppose $A \in \mathbb{R}^{n \times n}$ has the property that $A(1{:}k,1{:}k)$ is nonsingular for $k = 1{:}n-1$. This algorithm computes the factorization $A = LU$ where L is unit lower triangular and U is upper triangular.

> Initialize L to the identity and U to the zero matrix.
> **for** $j = 1{:}n$
> > **if** $j = 1$
> > > $v = A(:,1)$
> > **else**
> > > $\tilde{a} = A(:,j)$
> > > Solve $L(1{:}j-1,1{:}j-1){\cdot}z = \tilde{a}(1{:}j-1)$ for $z \in \mathbb{R}^{j-1}$.
> > > $U(1{:}j-1,j) = z$
> > > $v(j{:}n) = \tilde{a}(j{:}n) - L(j{:}n,1{:}j-1){\cdot}z$
> > **end**
> > $U(j,j) = v(j)$
> > $L(j+1{:}n,j) = v(j+1{:}n)/v(j)$
> **end**

(We chose to have separate arrays for L and U for clarity; it is not necessary in practice.) Algorithm 3.2.2 requires $2n^3/3$ flops, the same volume of floating point work required by Algorithm 3.2.1. However, from §1.5.2 there is less memory traffic associated with a gaxpy than with an outer product, so the two implementations could perform differently in practice. Note that in Algorithm 3.2.2, the original $A(:,j)$ is untouched until step j.

The terms *right-looking* and *left-looking* are sometimes applied to Algorithms 3.2.1 and 3.2.2. In the outer-product implementation, after $L(k{:}n,k)$ is determined, the columns to the right of $A(:,k)$ are updated so it is a right-looking procedure. In contrast, subcolumns to the left of $A(:,k)$ are accessed in gaxpy LU before $L(k+1{:}n,k)$ is produced so that implementation left-looking.

3.2.10   The LU Factorization of a Rectangular Matrix

The LU factorization of a rectangular matrix $A \in \mathbb{R}^{n \times r}$ can also be performed. The $n > r$ case is illustrated by

$$\left[\begin{array}{cc} 1 & 2 \\ 3 & 4 \\ 5 & 6 \end{array}\right] = \left[\begin{array}{cc} 1 & 0 \\ 3 & 1 \\ 5 & 2 \end{array}\right] \left[\begin{array}{cc} 1 & 2 \\ 0 & -2 \end{array}\right]$$

while

$$\left[\begin{array}{ccc} 1 & 2 & 3 \\ 4 & 5 & 6 \end{array}\right] = \left[\begin{array}{cc} 1 & 0 \\ 4 & 1 \end{array}\right] \left[\begin{array}{ccc} 1 & 2 & 3 \\ 0 & -3 & -6 \end{array}\right]$$

depicts the $n < r$ situation. The LU factorization of $A \in \mathbb{R}^{n \times r}$ is guaranteed to exist if $A(1{:}k, 1{:}k)$ is nonsingular for $k = 1{:}\min\{n, r\}$.

The square LU factorization algorithms above needs only minor alterations to handle the rectangular case. For example, if $n > r$, then Algorithm 3.2.1 modifies to the following:

> **for** $k = 1{:}r$
>> $\rho = k + 1{:}n$
>> $A(\rho, k) = A(\rho, k)/A(k, k)$
>> **if** $k < r$ $\qquad\qquad\qquad\qquad\qquad\qquad\qquad\qquad\qquad\qquad$ (3.2.8)
>>> $\mu = k + 1{:}r$
>>> $A(\rho, \mu) = A(\rho, \mu) - A(\rho, k) \cdot A(k, \mu)$
>> **end**
> **end**

This calculation requires $nr^2 - r^3/3$ flops. Upon completion, A is overwritten by the strictly lower triangular portion of $L \in \mathbb{R}^{n \times r}$ and the upper triangular portion of $U \in \mathbb{R}^{r \times r}$.

3.2.11   Block LU

It is possible to organize Gaussian elimination so that matrix multiplication becomes the dominant operation. Partition $A \in \mathbb{R}^{n \times n}$ as follows:

$$A = \left[\begin{array}{cc} A_{11} & A_{12} \\ A_{21} & A_{22} \end{array}\right] \begin{array}{c} r \\ n-r \end{array}$$
$$\quad\ \ r \quad\ \ n-r$$

where r is a blocking parameter. Suppose we compute the LU factorization

$$\left[\begin{array}{c} A_{11} \\ A_{21} \end{array}\right] = \left[\begin{array}{c} L_{11} \\ L_{21} \end{array}\right] U_{11}.$$

Here, $L_{11} \in \mathbb{R}^{r \times r}$ is unit lower triangular and $U_{11} \in \mathbb{R}^{r \times r}$ is upper triangular and assumed to be nonsingular. If we solve $L_{11} U_{12} = A_{12}$ for $U_{12} \in \mathbb{R}^{r \times n-r}$, then

$$\left[\begin{array}{cc} A_{11} & A_{12} \\ A_{21} & A_{22} \end{array}\right] = \left[\begin{array}{cc} L_{11} & 0 \\ L_{21} & I_{n-r} \end{array}\right] \left[\begin{array}{cc} I_r & 0 \\ 0 & \tilde{A} \end{array}\right] \left[\begin{array}{cc} U_{11} & U_{12} \\ 0 & I_{n-r} \end{array}\right],$$

where

$$\tilde{A} = A_{22} - L_{21}U_{12} = A_{22} - A_{21}A_{11}^{-1}A_{12} \qquad (3.2.9)$$

is the *Schur complement* of A_{11} in A. Note that if

$$\tilde{A} = L_{22}U_{22}$$

is the LU factorization of $\tilde{A}$, then

$$A = \begin{bmatrix} L_{11} & 0 \\ L_{21} & L_{22} \end{bmatrix} \begin{bmatrix} U_{11} & U_{12} \\ 0 & U_{22} \end{bmatrix}$$

is the LU factorization of A. This lays the groundwork for a recursive implementation.

Algorithm 3.2.3 (Recursive Block LU) Suppose $A \in \mathbb{R}^{n \times n}$ has an LU factorization and r is a positive integer. The following algorithm computes unit lower triangular $L \in \mathbb{R}^{n \times n}$ and upper triangular $U \in \mathbb{R}^{n \times n}$ so $A = LU$.

> **function** $[L, U] = \mathsf{BlockLU}(A, n, r)$
> > **if** $n \leq r$
> > > Compute the LU factorization $A = LU$ using (say) Algorithm 3.2.1.
> >
> > **else**
> > > Use (3.2.8) to compute the LU factorization $A(:, 1{:}r) = \begin{bmatrix} L_{11} \\ L_{21} \end{bmatrix} U_{11}$.
> > >
> > > Solve $L_{11}U_{12} = A(1{:}r, r+1{:}n)$ for U_{12}.
> > >
> > > $\tilde{A} = A(r+1{:}n, r+1{:}n) - L_{21}U_{12}$
> > >
> > > $[L_{22}, U_{22}] = \mathsf{BlockLU}(\tilde{A}, n-r, r)$
> > >
> > > $L = \begin{bmatrix} L_{11} & 0 \\ L_{21} & L_{22} \end{bmatrix}, \quad U = \begin{bmatrix} U_{11} & U_{12} \\ 0 & U_{22} \end{bmatrix}$
> >
> > **end**
>
> **end**

The following table explains where the flops come from:

Activity	Flops
L_{11}, L_{21}, U_{11}	$nr^2 - r^3/3$
U_{12}	$(n-r)r^2$
$\tilde{A}$	$2(n-r)^2$

If $n \gg r$, then there are a total of about $2n^3/3$ flops, the same volume of atithmetic as Algorithms 3.2.1 and 3.2.2. The vast majority of these flops are the level-3 flops associated with the production of $\tilde{A}$.

The actual level-3 fraction, a concept developed in §3.1.5, is more easily derived from a nonrecursive implementation. Assume for clarity that $n = Nr$ where N is a positive integer and that we want to compute

$$\begin{bmatrix} A_{11} & \cdots & A_{1N} \\ \vdots & \ddots & \vdots \\ A_{N1} & \cdots & A_{NN} \end{bmatrix} = \begin{bmatrix} L_{11} & \cdots & 0 \\ \vdots & \ddots & \vdots \\ L_{N1} & \cdots & L_{NN} \end{bmatrix} \begin{bmatrix} U_{11} & \cdots & U_{1N} \\ \vdots & \ddots & \vdots \\ 0 & \cdots & U_{NN} \end{bmatrix} \qquad (3.2.10)$$

where all blocks are r-by-r. Analogously to Algorithm 3.2.3 we have the following.

Algorithm 3.2.4 (Nonrecursive Block LU) Suppose $A \in \mathbb{R}^{n \times n}$ has an LU factorization and r is a positive integer. The following algorithm computes unit lower triangular $L \in \mathbb{R}^{n \times n}$ and upper triangular $U \in \mathbb{R}^{n \times n}$ so $A = LU$.

> **for** $k = 1{:}N$
>> Rectangular Gaussian elimination:
>> $$\begin{bmatrix} A_{kk} \\ \vdots \\ A_{Nk} \end{bmatrix} = \begin{bmatrix} L_{kk} \\ \vdots \\ L_{Nk} \end{bmatrix} U_{kk}$$
>> Multiple right hand side solve:
>> $$L_{kk} \begin{bmatrix} U_{k,k+1} \mid \cdots \mid U_{kN} \end{bmatrix} = \begin{bmatrix} A_{k,k+1} \mid \cdots \mid A_{kN} \end{bmatrix}$$
>> Level-3 updates:
>> $$A_{ij} = A_{ij} - L_{ik}U_{kj}, \quad i = k+1{:}N, \, j = k+1{:}N$$
>
> **end**

Here is the flop situation during the kth pass through the loop:

Activity	Flops
Gaussian elimination	$(N - k + 1)r^3 - r^3/3$
Multiple RHS solve	$(N - k)r^3$
Level-3 updates	$2(N - k)^2 r^2$

Summing these quantities for $k = 1{:}N$ we find that the level-3 fraction is approximately

$$\frac{2n^3/3}{2n^3/3 + n^2 r} = 1 - \frac{3}{2N}.$$

Thus, for large N almost all arithmetic takes place in the context of matrix multiplication. This ensures a favorable amount of data reuse as discussed in §1.5.4.

Problems

P3.2.1 Verify Equation (3.2.6).

P3.2.2 Suppose the entries of $A(\epsilon) \in \mathbb{R}^{n \times n}$ are continuously differentiable functions of the scalar ϵ. Assume that $A \equiv A(0)$ and all its principal submatrices are nonsingular. Show that for sufficiently small ϵ, the matrix $A(\epsilon)$ has an LU factorization $A(\epsilon) = L(\epsilon)U(\epsilon)$ and that $L(\epsilon)$ and $U(\epsilon)$ are both continuously differentiable.

P3.2.3 Suppose we partition $A \in \mathbb{R}^{n \times n}$

$$A = \begin{bmatrix} A_{11} & A_{12} \\ A_{21} & A_{22} \end{bmatrix}$$

where A_{11} is r-by-r and nonsingular. Let S be the Schur complement of A_{11} in A as defined in (3.2.9). Show that after r steps of Algorithm 3.2.1, $A(r + 1{:}n, r + 1{:}n)$ houses S. How could S be obtained after r steps of Algorithm 3.2.2?

P3.2.4 Suppose $A \in \mathbb{R}^{n \times n}$ has an LU factorization. Show how $Ax = b$ can be solved without storing the multipliers by computing the LU factorization of the n-by-$(n+1)$ matrix $[A\ b]$.

P3.2.5 Describe a variant of Gaussian elimination that introduces zeros into the columns of A in the order, $n: -1{:}2$ and which produces the factorization $A = UL$ where U is unit upper triangular and L is lower triangular.

P3.2.6 Matrices in $\mathbb{R}^{n \times n}$ of the form $N(y,k) = I - y e_k^T$ where $y \in \mathbb{R}^n$ are called *Gauss-Jordan transformations*. (a) Give a formula for $N(y,k)^{-1}$ assuming it exists. (b) Given $x \in \mathbb{R}^n$, under what conditions can y be found so $N(y,k)x = e_k$? (c) Give an algorithm using Gauss-Jordan transformations that overwrites A with A^{-1}. What conditions on A ensure the success of your algorithm?

P3.2.7 Extend Algorithm 3.2.2 so that it can also handle the case when A has more rows than columns.

P3.2.8 Show how A can be overwritten with L and U in Algorithm 3.2.2. Give a 3-loop specification so that unit stride access prevails.

P3.2.9 Develop a version of Gaussian elimination in which the innermost of the three loops oversees a dot product.

Notes and References for §3.2

The method of Gaussian elimination has a long and interesting history, see:

J.F. Grcar (2011). "How Ordinary Elimination Became Gaussian Elimination," *Historica Mathematica, 38*, 163–218.

J.F. Grcar (2011). "Mathematicians of Gaussian Elimination," *Notices of the AMS 58*, 782–792.

Schur complements (3.2.9) arise in many applications. For a survey of both practical and theoretical interest, see:

R.W. Cottle (1974). "Manifestations of the Schur Complement," *Lin. Alg. Applic. 8*, 189–211.

Schur complements are known as "Gauss transforms" in some application areas. The use of Gauss-Jordan transformations (P3.2.6) is detailed in Fox (1964). See also:

T. Dekker and W. Hoffman (1989). "Rehabilitation of the Gauss-Jordan Algorithm," *Numer. Math. 54*, 591–599.

As we mentioned, inner product versions of Gaussian elimination have been known and used for some time. The names of Crout and Doolittle are associated with these techniques, see:

G.E. Forsythe (1960). "Crout with Pivoting," *Commun. ACM 3*, 507–508.

W.M. McKeeman (1962). "Crout with Equilibration and Iteration," *Commun. ACM. 5*, 553–555.

Loop orderings and block issues in LU computations are discussed in:

J.J. Dongarra, F.G. Gustavson, and A. Karp (1984). "Implementing Linear Algebra Algorithms for Dense Matrices on a Vector Pipeline Machine," *SIAM Review 26*, 91–112.

J.M. Ortega (1988). "The ijk Forms of Factorization Methods I: Vector Computers," *Parallel Comput. 7*, 135–147.

D.H. Bailey, K.Lee, and H.D. Simon (1991). "Using Strassen's Algorithm to Accelerate the Solution of Linear Systems," *J. Supercomput. 4*, 357–371.

J.W. Demmel, N.J. Higham, and R.S. Schreiber (1995). "Stability of Block LU Factorization," *Numer. Lin. Alg. Applic. 2*, 173–190.

Suppose $A = LU$ and $A + \Delta A = (L + \Delta L)(U + \Delta U)$ are LU factorizations. Bounds on the perturbations ΔL and ΔU in terms of ΔA are given in:

G.W. Stewart (1997). "On the Perturbation of LU and Cholesky Factors," *IMA J. Numer. Anal. 17*, 1–6.

X.-W. Chang and C.C. Paige (1998). "On the Sensitivity of the LU factorization," *BIT 38*, 486–501.

In certain limited domains, it is possible to solve linear systems exactly using rational arithmetic. For a snapshot of the challenges, see:

P. Alfeld and D.J. Eyre (1991). "The Exact Analysis of Sparse Rectangular Linear Systems," *ACM Trans. Math. Softw. 17*, 502–518.
P. Alfeld (2000). "Bivariate Spline Spaces and Minimal Determining Sets," *J. Comput. Appl. Math. 119*, 13–27.

3.3 Roundoff Error in Gaussian Elimination

We now assess the effect of rounding errors when the algorithms in the previous two sections are used to solve the linear system $Ax = b$. A much more detailed treatment of roundoff error in Gaussian elimination is given in Higham (ASNA).

3.3.1 Errors in the LU Factorization

Let us see how the error bounds for Gaussian elimination compare with the ideal bounds derived in §2.7.11. We work with the infinity norm for convenience and focus our attention on Algorithm 3.2.1, the outer product version. The error bounds that we derive also apply to the gaxpy formulation (Algorithm 3.2.2). Our first task is to quantify the roundoff errors associated with the computed triangular factors.

Theorem 3.3.1. *Assume that A is an n-by-n matrix of floating point numbers. If no zero pivots are encountered during the execution of Algorithm 3.2.1, then the computed triangular matrices $\hat{L}$ and $\hat{U}$ satisfy*

$$\hat{L}\hat{U} = A + H, \tag{3.3.1}$$

$$|H| \leq 2(n-1)\mathbf{u}\left(|A| + |\hat{L}||\hat{U}|\right) + O(\mathbf{u}^2). \tag{3.3.2}$$

Proof. The proof is by induction on n. The theorem obviously holds for $n = 1$. Assume that $n \geq 2$ and that the theorem holds for all $(n-1)$-by-$(n-1)$ floating point matrices. If A is partitioned as follows

$$A = \begin{bmatrix} \alpha & w^T \\ v & B \end{bmatrix} \begin{matrix} 1 \\ n{-}1 \end{matrix}$$
$$\quad\; 1 \quad\; n{-}1$$

then the first step in Algorithm 3.2.1 is to compute

$$\hat{z} = \text{fl}(v/\alpha), \qquad \hat{C} = \text{fl}(\hat{z}w^T), \qquad \hat{A}_1 = \text{fl}(B - \hat{C}),$$

from which we conclude that

$$\hat{z} = v/\alpha + f, \tag{3.3.3}$$

$$|f| \leq \mathbf{u}|v/\alpha|, \tag{3.3.4}$$

$$\hat{C} = \hat{z}w^T + F_1, \tag{3.3.5}$$

$$|F_1| \leq \mathbf{u}|\hat{z}||w^T|, \tag{3.3.6}$$

$$\hat{A}_1 = B - (\hat{z}w^T + F_1) + F_2, \tag{3.3.7}$$

$$|F_2| \leq \mathbf{u}\left(|B| + |\hat{z}||w^T|\right) + O(\mathbf{u}^2), \tag{3.3.8}$$

$$|\hat{A}_1| \leq |B| + |\hat{z}||w^T| + O(\mathbf{u}). \tag{3.3.9}$$

The algorithm proceeds to compute the LU factorization of $\hat{A}_1$. By induction, the computed factors $\hat{L}_1$ and $\hat{U}_1$ satisfy

$$\hat{L}_1\hat{U}_1 = \hat{A}_1 + H_1 \tag{3.3.10}$$

where

$$|H_1| \leq 2(n-2)\mathbf{u}\left(|\hat{A}_1| + |\hat{L}_1||\hat{U}_1|\right) + O(\mathbf{u}^2). \tag{3.3.11}$$

If

$$\hat{L} = \begin{bmatrix} 1 & 0 \\ \hat{z} & \hat{L}_1 \end{bmatrix}, \qquad \hat{U} = \begin{bmatrix} \alpha & w^T \\ 0 & \hat{U}_1 \end{bmatrix},$$

then it is easy to verify that

$$\hat{L}\hat{U} = A + H$$

where

$$H = \begin{bmatrix} 0 & 0 \\ \alpha f & H_1 - F_1 + F_2 \end{bmatrix}. \tag{3.3.12}$$

To prove the theorem we must verify (3.3.2), i.e.,

$$|H| \leq 2(n-1)\mathbf{u} \begin{bmatrix} 2|\alpha| & 2|w^T| \\ |v| + |\alpha||f| & |B| + |\hat{L}_1||\hat{U}_1| + |\hat{z}||w^T| \end{bmatrix} + O(\mathbf{u}^2).$$

Considering (3.3.12), this is obviously the case if

$$|H_1| + |F_1| + |F_2| \leq 2(n-1)\mathbf{u}\left(|B| + |\hat{z}||w^T| + |\hat{L}_1||\hat{U}_1|\right) + O(\mathbf{u}^2). \tag{3.3.13}$$

Using (3.3.9) and (3.3.11) we have

$$|H_1| \leq 2(n-2)\mathbf{u}\left(|B| + |\hat{z}||w^T| + |\hat{L}_1||\hat{U}_1|\right) + O(\mathbf{u}^2),$$

while (3.3.6) and (3.3.8) imply

$$|F_1| + |F_2| \leq \mathbf{u}(|B| + 2|\hat{z}||w|) + O(\mathbf{u}^2).$$

These last two results establish (3.3.13) and therefore the theorem. $\square$

We mention that if A is m-by-n, then the theorem applies with n replaced by the smaller of n and m in Equation 3.3.2.

3.3.2   Triangular Solving with Inexact Triangles

We next examine the effect of roundoff error when $\hat{L}$ and $\hat{U}$ are used by the triangular system solvers of §3.1.

Theorem 3.3.2. *Let $\hat{L}$ and $\hat{U}$ be the computed LU factors obtained by Algorithm 3.2.1 when it is applied to an n-by-n floating point matrix A. If the methods of §3.1 are used to produce the computed solution $\hat{y}$ to $\hat{L}y = b$ and the computed solution $\hat{x}$ to $\hat{U}x = \hat{y}$, then $(A + E)\hat{x} = b$ with*

$$|E| \le n\mathbf{u}\left(2|A| + 4|\hat{L}||\hat{U}|\right) \; + \; O(\mathbf{u}^2). \tag{3.3.14}$$

Proof. From (3.1.1) and (3.1.2) we have

$$(\hat{L} + F)\hat{y} \; = \; b, \qquad |F| \; \le \; n\mathbf{u}|\hat{L}| + O(\mathbf{u}^2),$$
$$(\hat{U} + G)\hat{x} \; = \; \hat{y}, \qquad |G| \; \le \; n\mathbf{u}|\hat{U}| + O(\mathbf{u}^2),$$

and thus

$$(\hat{L} + F)(\hat{U} + G)\hat{x} \; = \; (\hat{L}\hat{U} + F\hat{U} + \hat{L}G + FG)\hat{x} = b.$$

If follows from Theorem 3.3.1 that $\hat{L}\hat{U} = A + H$ with

$$|H| \; \le \; 2(n - 1)\mathbf{u}(|A| + |\hat{L}||\hat{U}|) + O(\mathbf{u}^2),$$

and so by defining

$$E \; = \; H + F\hat{U} + \hat{L}G + FG$$

we find $(A + E)\hat{x} = b$. Moreover,

$$|E| \; \le \; |H| + |F|\,|\hat{U}| \; + \; |\hat{L}|\,|G| \; + \; O(\mathbf{u}^2)$$
$$\le \; 2n\mathbf{u}\left(|A| + |\hat{L}||\hat{U}|\right) + 2n\mathbf{u}\left(|\hat{L}||\hat{U}|\right) + O(\mathbf{u}^2),$$

completing the proof of the theorem.   □

If it were not for the possibility of a large $|\hat{L}||\hat{U}|$ term, (3.3.14) would compare favorably with the ideal bound (2.7.21). (The factor n is of no consequence, cf. the Wilkinson quotation in §2.7.7.) Such a possibility exists, for there is nothing in Gaussian elimination to rule out the appearance of small pivots. If a small pivot is encountered, then we can expect large numbers to be present in $\hat{L}$ and $\hat{U}$.

We stress that small pivots are not necessarily due to ill-conditioning as the example

$$A = \begin{bmatrix} \epsilon & 1 \\ 1 & 0 \end{bmatrix} = \begin{bmatrix} 1 & 0 \\ 1/\epsilon & 1 \end{bmatrix} \begin{bmatrix} \epsilon & 1 \\ 0 & -1/\epsilon \end{bmatrix}$$

shows. Thus, Gaussian elimination can give arbitrarily poor results, even for well-conditioned problems. The method is unstable. For example, suppose 3-digit floating point arithmetic is used to solve

$$\begin{bmatrix} .001 & 1.00 \\ 1.00 & 2.00 \end{bmatrix} \begin{bmatrix} x_1 \\ x_2 \end{bmatrix} = \begin{bmatrix} 1.00 \\ 3.00 \end{bmatrix}.$$

(See §2.7.1.) Applying Gaussian elimination we get

$$\hat{L} = \begin{bmatrix} 1 & 0 \\ 1000 & 1 \end{bmatrix}, \qquad \hat{U} = \begin{bmatrix} .001 & 1 \\ 0 & -1000 \end{bmatrix},$$

and a calculation shows that

$$\hat{L}\hat{U} = \begin{bmatrix} .001 & 1 \\ 1 & 2 \end{bmatrix} + \begin{bmatrix} 0 & 0 \\ 0 & -2 \end{bmatrix} \equiv A + H.$$

If we go on to solve the problem using the triangular system solvers of §3.1, then using the same precision arithmetic we obtain a computed solution $\hat{x} = [0 \ , \ 1]^T$. This is in contrast to the exact solution $x = [1.002\dots, .998\dots]^T$.

Problems

P3.3.1 Show that if we drop the assumption that A is a floating point matrix in Theorem 3.3.1, then Equation 3.3.2 holds with the coefficient "2" replaced by "3."

P3.3.2 Suppose A is an n-by-n matrix and that $\hat{L}$ and $\hat{U}$ are produced by Algorithm 3.2.1. (a) How many flops are required to compute $\| \, |\hat{L}| \, |\hat{U}| \, \|_\infty$? (b) Show $\mathsf{fl}(|\hat{L}||\hat{U}|) \le (1 + 2n\mathbf{u})|\hat{L}||\hat{U}| + O(\mathbf{u}^2)$.

Notes and References for §3.3

The original roundoff analysis of Gaussian elimination appears in:

J.H. Wilkinson (1961). "Error Analysis of Direct Methods of Matrix Inversion," *J. ACM 8,* 281–330.

Various improvements and insights regarding the bounds and have been made over the years, see:

B.A. Chartres and J.C. Geuder (1967). "Computable Error Bounds for Direct Solution of Linear Equations," *J. ACM 14,* 63–71.
J.K. Reid (1971). "A Note on the Stability of Gaussian Elimination," *J. Inst. Math. Applic. 8,* 374–75.
C.C. Paige (1973). "An Error Analysis of a Method for Solving Matrix Equations," *Math. Comput. 27,* 355–59.
H.H. Robertson (1977). "The Accuracy of Error Estimates for Systems of Linear Algebraic Equations," *J. Inst. Math. Applic. 20,* 409–14.
J.J. Du Croz and N.J. Higham (1992). "Stability of Methods for Matrix Inversion," *IMA J. Numer. Anal. 12,* 1–19.
J.M. Banoczi, N.C. Chiu, G.E. Cho, and I.C.F. Ipsen (1998). "The Lack of Influence of the Right–Hand Side on the Accuracy of Linear System Solution," *SIAM J. Sci. Comput. 20,* 203–227.
P. Amodio and F. Mazzia (1999). "A New Approach to Backward Error Analysis of LU Factorization *BIT 39,* 385–402.

An interesting account of von Neuman's contributions to the numerical analysis of Gaussian elimination is detailed in:

J.F. Grcar (2011). "John von Neuman's Analysis of Gaussian Elimination and the Origins of Modern Numerical Analysis," *SIAM Review 53,* 607–682.

3.4 Pivoting

The analysis in the previous section shows that we must take steps to ensure that no large entries appear in the computed triangular factors $\hat{L}$ and $\hat{U}$. The example

$$A = \begin{bmatrix} .0001 & 1 \\ 1 & 1 \end{bmatrix} = \begin{bmatrix} 1 & 0 \\ 10000 & 1 \end{bmatrix} \begin{bmatrix} .0001 & 1 \\ 0 & -9999 \end{bmatrix} = LU$$

correctly identifies the source of the difficulty: relatively small pivots. A way out of this difficulty is to interchange rows. For example, if P is the permutation

$$P = \begin{bmatrix} 0 & 1 \\ 1 & 0 \end{bmatrix}$$

then

$$PA = \begin{bmatrix} 1 & 1 \\ .0001 & 1 \end{bmatrix} = \begin{bmatrix} 1 & 0 \\ .0001 & 1 \end{bmatrix} \begin{bmatrix} 1 & 1 \\ 0 & .9999 \end{bmatrix} = LU.$$

Observe that the triangular factors have modestly sized entries.

In this section we show how to determine a permuted version of A that has a reasonably stable LU factorization. There are several ways to do this and they each corresponds to a different pivoting strategy. Partial pivoting, complete pivoting, and rook pivoting are considered. The efficient implementation of these strategies and their properties are discussed. We begin with a few comments about permutation matrices that can be used to swap rows or columns.

3.4.1   Interchange Permutations

The stabilizations of Gaussian elimination that are developed in this section involve data movements such as the interchange of two matrix rows. In keeping with our desire to describe all computations in "matrix terms," we use permutation matrices to describe this process. (Now is a good time to review §1.2.8–§1.2.11.) *Interchange permutations* are particularly important. These are permutations obtained by swapping two rows in the identity, e.g.,

$$\Pi = \begin{bmatrix} 0 & 0 & 0 & 1 \\ 0 & 1 & 0 & 0 \\ 0 & 0 & 1 & 0 \\ 1 & 0 & 0 & 0 \end{bmatrix}.$$

Interchange permutations can be used to describe row and column swapping. If $A \in \mathbb{R}^{4 \times 4}$, then $\Pi \cdot A$ is A with rows 1 and 4 interchanged while $A \cdot \Pi$ is A with columns 1 and 4 swapped.

If $P = \Pi_m \cdots \Pi_1$ and each Π_k is the identity with rows k and $piv(k)$ interchanged, then $piv(1{:}m)$ encodes P. Indeed, $x \in \mathbb{R}^n$ can be overwritten by Px as follows:

> **for** $k = 1{:}m$
> $x(k) \leftrightarrow x(piv(k))$
> **end**

Here, the "$\leftrightarrow$" notation means "swap contents." Since each Π_k is symmetric, we have $P^T = \Pi_1 \cdots \Pi_m$. Thus, the *piv* representation can also be used to overwrite x with $P^T x$:

> **for** $k = m{:} - 1{:}1$
> $x(k) \leftrightarrow x(piv(k))$
> **end**

We remind the reader that although no floating point arithmetic is involved in a permutation operation, permutations move data and have a nontrivial effect upon performance.

3.4.2 Partial Pivoting

Interchange permutations can be used in LU computations to guarantee that no multiplier is greater than 1 in absolute value. Suppose

$$A = \begin{bmatrix} 3 & 17 & 10 \\ 2 & 4 & -2 \\ 6 & 18 & -12 \end{bmatrix}.$$

To get the smallest possible multipliers in the first Gauss transformation, we need a_{11} to be the largest entry in the first column. Thus, if Π_1 is the interchange permutation

$$\Pi_1 = \begin{bmatrix} 0 & 0 & 1 \\ 0 & 1 & 0 \\ 1 & 0 & 0 \end{bmatrix}$$

then

$$\Pi_1 A = \begin{bmatrix} 6 & 18 & -12 \\ 2 & 4 & -2 \\ 3 & 17 & 10 \end{bmatrix}.$$

It follows that

$$M_1 = \begin{bmatrix} 1 & 0 & 0 \\ -1/3 & 1 & 0 \\ -1/2 & 0 & 1 \end{bmatrix} \implies M_1 \Pi_1 A = \begin{bmatrix} 6 & 18 & -12 \\ 0 & -2 & 2 \\ 0 & 8 & 16 \end{bmatrix}.$$

To obtain the smallest possible multiplier in M_2, we need to swap rows 2 and 3. Thus, if

$$\Pi_2 = \begin{bmatrix} 1 & 0 & 0 \\ 0 & 0 & 1 \\ 0 & 1 & 0 \end{bmatrix} \quad \text{and} \quad M_2 = \begin{bmatrix} 1 & 0 & 0 \\ 0 & 1 & 0 \\ 0 & 1/4 & 1 \end{bmatrix},$$

then

$$M_2 \Pi_2 M_1 \Pi_1 A = \begin{bmatrix} 6 & 18 & -12 \\ 0 & 8 & 16 \\ 0 & 0 & 6 \end{bmatrix}.$$

For general n we have

> **for** $k = 1{:}n - 1$
>> Find an interchange permutation $\Pi_k \in \mathbb{R}^{n \times n}$ that swaps
>> $A(k, k)$ with the largest element in $|A(k{:}n, k)|$.
>> $A = \Pi_k A$ (3.4.1)
>> Determine the Gauss transformation $M_k = I_n - \tau^{(k)} e_k^T$ such that if
>> v is the kth column of $M_k A$, then $v(k + 1{:}n) = 0$.
>> $A = M_k A$
> **end**

This particular row interchange strategy is called *partial pivoting* and upon completion, we have

$$M_{n-1} \Pi_{n-1} \cdots M_1 \Pi_1 A = U \tag{3.4.2}$$

where U is upper triangular. As a consequence of the partial pivoting, no multiplier is larger than one in absolute value.

3.4.3   Where is L?

It turns out that (3.4.1) computes the factorization

$$PA = LU \tag{3.4.3}$$

where $P = \Pi_{n-1} \cdots \Pi_1$, U is upper triangular, and L is unit lower triangular with $|\ell_{ij}| \leq 1$. We show that $L(k+1:n, k)$ is a permuted version of M_k's multipliers. From (3.4.2) it can be shown that

$$\tilde{M}_{n-1} \cdots \tilde{M}_1 P A \;=\; U \tag{3.4.4}$$

where

$$\tilde{M}_k \;=\; (\Pi_{n-1} \cdots \Pi_{k+1}) M_k (\Pi_{k+1} \cdots \Pi_{n-1}) \tag{3.4.5}$$

for $k = 1:n-1$. For example, in the $n = 4$ case we have

$$\tilde{M}_3 \tilde{M}_2 \tilde{M}_1 P A \;=\; M_3 \cdot (\Pi_3 M_2 \Pi_3) \cdot (\Pi_3 \Pi_2 M_1 \Pi_2 \Pi_3) \cdot (\Pi_3 \Pi_2 \Pi_1) A$$

since the Π_i are symmetric. Moreover,

$$\tilde{M}_k \;=\; (\Pi_{n-1} \cdots \Pi_{k+1}) \cdot (I_n - \tau^{(k)} e_k^T) \cdot (\Pi_{k+1} \cdots \Pi_{n-1}) \;=\; I_n - \tilde{\tau}^{(k)} e_k^T$$

with $\tilde{\tau}^{(k)} = \Pi_{n-1} \cdots \Pi_{k+1} \tau^{(k)}$. This shows that $\tilde{M}_k$ is a Gauss transformation. The transformation from $\tau^{(k)}$ to $\tilde{\tau}^{(k)}$ is easy to implement in practice.

Algorithm 3.4.1 (Outer Product LU with Partial Pivoting) This algorithm computes the factorization $PA = LU$ where P is a permutation matrix encoded by $piv(1:n-1)$, L is unit lower triangular with $|\ell_{ij}| \leq 1$, and U is upper triangular. For $i = 1:n$, $A(i, i:n)$ is overwritten by $U(i, i:n)$ and $A(i+1:n, i)$ is overwritten by $L(i+1:n, i)$. The permutation P is given by $P = \Pi_{n-1} \cdots \Pi_1$ where Π_k is an interchange permutation obtained by swapping rows k and $piv(k)$ of I_n.

> **for** $k = 1:n-1$
>> Determine μ with $k \leq \mu \leq n$ so $|A(\mu, k)| = \| A(k:n, k) \|_\infty$
>> $piv(k) = \mu$
>> $A(k, :) \leftrightarrow A(\mu, :)$
>> **if** $A(k, k) \neq 0$
>>> $\rho = k+1:n$
>>> $A(\rho, k) = A(\rho, k)/A(k, k)$
>>> $A(\rho, \rho) = A(\rho, \rho) - A(\rho, k) A(k, \rho)$
>> **end**
> **end**

The *floating point* overhead associated with partial pivoting is minimal from the standpoint of arithmetic as there are only $O(n^2)$ comparisons associated with the search for the pivots. The overall algorithm involves $2n^3/3$ flops.

If Algorithm 3.4.1 is applied to

$$
A = \begin{bmatrix} 3 & 17 & 10 \\ 2 & 4 & -2 \\ 6 & 18 & -12 \end{bmatrix},
$$

then upon completion

$$
A = \begin{bmatrix} 6 & 18 & -12 \\ 1/2 & 8 & 16 \\ 1/3 & -1/4 & 6 \end{bmatrix}
$$

and $piv = [3, 3]$. These two quantities encode all the information associated with the reduction:

$$
\begin{bmatrix} 1 & 0 & 0 \\ 0 & 0 & 1 \\ 0 & 1 & 0 \end{bmatrix} \begin{bmatrix} 0 & 0 & 1 \\ 0 & 1 & 0 \\ 1 & 0 & 0 \end{bmatrix} A = \begin{bmatrix} 1 & 0 & 0 \\ 1/2 & 1 & 0 \\ 1/3 & -1/4 & 1 \end{bmatrix} \begin{bmatrix} 6 & 18 & -12 \\ 0 & 8 & 16 \\ 0 & 0 & 6 \end{bmatrix}.
$$

To compute the solution to $Ax = b$ after invoking Algorithm 3.4.1, we solve $Ly = Pb$ for y and $Ux = y$ for x. Note that b can be overwritten by Pb as follows

for $k = 1{:}n - 1$
$\qquad b(k) \leftrightarrow b(piv(k))$
end

We mention that if Algorithm 3.4.1 is applied to the problem,

$$
\begin{bmatrix} .001 & 1.00 \\ 1.00 & 2.00 \end{bmatrix} \begin{bmatrix} x_1 \\ x_2 \end{bmatrix} = \begin{bmatrix} 1.00 \\ 3.00 \end{bmatrix},
$$

using 3-digit floating point arithmetic, then

$$
P = \begin{bmatrix} 0 & 1 \\ 1 & 0 \end{bmatrix}, \quad \hat{L} = \begin{bmatrix} 1.00 & 0 \\ .001 & 1.00 \end{bmatrix}, \quad \hat{U} = \begin{bmatrix} 1.00 & 2.00 \\ 0 & 1.00 \end{bmatrix},
$$

and $\hat{x} = [1.00, .996]^T$. Recall from §3.3.2 that if Gaussian elimination without pivoting is applied to this problem, then the computed solution has $O(1)$ error.

We mention that Algorithm 3.4.1 *always* runs to completion. If $A(k{:}n, k) = 0$ in step k, then $M_k = I_n$.

3.4.4   The Gaxpy Version

In §3.2 we developed outer product and gaxpy schemes for computing the LU factorization. Having just incorporated pivoting in the outer product version, it is equally straight forward to do the same with the gaxpy approach. Referring to Algorithm 3.2.2, we simply search the vector $|v(j{:}n)|$ in that algorithm for its maximal element and proceed accordingly.

Algorithm 3.4.2 (Gaxpy LU with Partial Pivoting) This algorithm computes the factorization $PA = LU$ where P is a permutation matrix encoded by $piv(1:n-1)$, L is unit lower triangular with $|\ell_{ij}| \leq 1$, and U is upper triangular. For $i = 1:n$, $A(i, i:n)$ is overwritten by $U(i, i:n)$ and $A(i+1:n, i)$ is overwritten by $L(i+1:n, i)$. The permutation P is given by $P = \Pi_{n-1} \cdots \Pi_1$ where Π_k is an interchange permutation obtained by swapping rows k and $piv(k)$ of I_n.

> Initialize L to the identity and U to the zero matrix.
> **for** $j = 1:n$
> > **if** $j = 1$
> > > $v = A(:, 1)$
> > **else**
> > > $\tilde{a} = \Pi_{j-1} \cdots \Pi_1 A(:, j)$
> > > Solve $L(1:j-1, 1:j-1)z = \tilde{a}(1:j-1)$ for $z \in \mathbb{R}^{j-1}$
> > > $U(1:j-1, j) = z,\ v(j:n) = \tilde{a}(j:n) - L(j:n, 1:j-1) \cdot z$
> > **end**
> > Determine μ with $j \leq \mu \leq n$ so $|v(\mu)| = \| v(j:n) \|_\infty$ and set $piv(j) = \mu$
> > $v(j) \leftrightarrow v(\mu),\ L(j, 1:j-1) \leftrightarrow L(\mu, 1:j-1),\ U(j, j) = v(j)$
> > **if** $v(j) \neq 0$
> > > $L(j+1:n, j) = v(j+1:n)/v(j)$
> > **end**
> **end**

As with Algorithm 3.4.1, this procedure requires $2n^3/3$ flops and $O(n^2)$ comparisons.

3.4.5   Error Analysis and the Growth Factor

We now examine the stability that is obtained with partial pivoting. This requires an accounting of the rounding errors that are sustained during elimination and during the triangular system solving. Bearing in mind that there are no rounding errors associated with permutation, it is not hard to show using Theorem 3.3.2 that the computed solution $\hat{x}$ satisfies $(A + E)\hat{x} = b$ where

$$|E| \leq n\mathbf{u} \left(2|A| + 4\hat{P}^T|\hat{L}||\hat{U}| \right) + O(\mathbf{u}^2). \tag{3.4.6}$$

Here we are assuming that $\hat{P}$, $\hat{L}$, and $\hat{U}$ are the computed analogs of P, L, and U as produced by the above algorithms. Pivoting implies that the elements of $\hat{L}$ are bounded by one. Thus $\| \hat{L} \|_\infty \leq n$ and we obtain the bound

$$\| E \|_\infty \leq n\mathbf{u} \left(2\| A \|_\infty + 4n\| \hat{U} \|_\infty \right) + O(\mathbf{u}^2). \tag{3.4.7}$$

The problem now is to bound $\| \hat{U} \|_\infty$. Define the *growth factor* ρ by

$$\rho = \max_{i,j,k} \frac{|\hat{a}_{ij}^{(k)}|}{\| A \|_\infty} \tag{3.4.8}$$

where $\hat{A}^{(k)}$ is the computed version of the matrix $A^{(k)} = M_k \Pi_k \cdots M_1 \Pi_1 A$. It follows that

$$\| E \|_\infty \leq 6n^3 \rho \| A \|_\infty \mathbf{u} + O(\mathbf{u}^2). \tag{3.4.9}$$

Whether or not this compares favorably with the ideal bound (2.7.20) hinges upon the size of the growth factor of ρ. (The factor n^3 is not an operating factor in practice and may be ignored in this discussion.)

The growth factor measures how large the A-entries become during the process of elimination. Whether or not we regard Gaussian elimination with partial pivoting is safe to use depends upon what we can say about this quantity. From an average-case point of view, experiments by Trefethen and Schreiber (1990) suggest that ρ is usually in the vicinity of $n^{2/3}$. However, from the worst-case point of view, ρ can be as large as 2^{n-1}. In particular, if $A \in \mathbb{R}^{n \times n}$ is defined by

$$a_{ij} = \begin{cases} 1 & \text{if } i = j \text{ or } j = n, \\ -1 & \text{if } i > j, \\ 0 & \text{otherwise}, \end{cases}$$

then there is no swapping of rows during Gaussian elimination with partial pivoting. We emerge with $A = LU$ and it can be shown that $u_{nn} = 2^{n-1}$. For example,

$$\begin{bmatrix} 1 & 0 & 0 & 1 \\ -1 & 1 & 0 & 1 \\ -1 & -1 & 1 & 1 \\ -1 & -1 & -1 & 1 \end{bmatrix} = \begin{bmatrix} 1 & 0 & 0 & 0 \\ -1 & 1 & 0 & 0 \\ -1 & -1 & 1 & 0 \\ -1 & -1 & -1 & 1 \end{bmatrix} \begin{bmatrix} 1 & 0 & 0 & 1 \\ 0 & 1 & 0 & 2 \\ 0 & 0 & 1 & 4 \\ 0 & 0 & 0 & 8 \end{bmatrix}.$$

Understanding the behavior of ρ requires an intuition about what makes the U-factor large. Since $PA = LU$ implies $U = L^{-1}PA$ it would appear that the size of L^{-1} is relevant. However, Stewart (1997) discusses why one can expect the L-factor to be well conditioned.

Although there is still more to understand about ρ, the consensus is that serious element growth in Gaussian elimination with partial pivoting is *extremely* rare. *The method can be used with confidence.*

3.4.6 Complete Pivoting

Another pivot strategy called *complete pivoting* has the property that the associated growth factor bound is considerably smaller than 2^{n-1}. Recall that in partial pivoting, the kth pivot is determined by scanning the current subcolumn $A(k{:}n, k)$. In complete pivoting, the largest entry in the current submatrix $A(k{:}n, k{:}n)$ is permuted into the (k, k) position. Thus, we compute the upper triangularization

$$M_{n-1} \Pi_{n-1} \cdots M_1 \Pi_1 A \Gamma_1 \cdots \Gamma_{n-1} = U.$$

In step k we are confronted with the matrix

$$A^{(k-1)} = M_{k-1} \Pi_{k-1} \cdots M_1 \Pi_1 A \Gamma_1 \cdots \Gamma_{k-1}$$

and determine interchange permutations Π_k and Γ_k such that

$$\left| \left(\Pi_k A^{(k-1)} \Gamma_k \right)_{kk} \right| = \max_{k \leq i,j \leq n} \left| \left(\Pi_k A^{(k-1)} \Gamma_k \right)_{ij} \right|.$$

Algorithm 3.4.3 (Outer Product LU with Complete Pivoting) This algorithm computes the factorization $PAQ^T = LU$ where P is a permutation matrix encoded by $piv(1{:}n-1)$, Q is a permutation matrix encoded by $colpiv(1{:}n-1)$, L is unit lower triangular with $|\ell_{ij}| \leq 1$, and U is upper triangular. For $i = 1{:}n$, $A(i, i{:}n)$ is overwritten by $U(i, i{:}n)$ and $A(i+1{:}n, i)$ is overwritten by $L(i+1{:}n, i)$. The permutation P is given by $P = \Pi_{n-1} \cdots \Pi_1$ where Π_k is an interchange permutation obtained by swapping rows k and $rowpiv(k)$ of I_n. The permutation Q is given by $Q = \Gamma_{n-1} \cdots \Gamma_1$ where Γ_k is an interchange permutation obtained by swapping rows k and $colpiv(k)$ of I_n.

> **for** $k = 1{:}n-1$
>> Determine μ with $k \leq \mu \leq n$ and λ with $k \leq \lambda \leq n$ so
>>> $|A(\mu, \lambda)| = \max\{ |A(i,j)| : i = k{:}n,\ j = k{:}n \}$
>>
>> $rowpiv(k) = \mu$
>>
>> $A(k, 1{:}n) \leftrightarrow A(\mu, 1{:}n)$
>>
>> $colpiv(k) = \lambda$
>>
>> $A(1{:}n, k) \leftrightarrow A(1{:}n, \lambda)$
>>
>> **if** $A(k,k) \neq 0$
>>> $\rho = k+1{:}n$
>>>
>>> $A(\rho, k) = A(\rho, k)/A(k, k)$
>>>
>>> $A(\rho, \rho) = A(\rho, \rho) - A(\rho, k)A(k, \rho)$
>>
>> **end**
>
> **end**

This algorithm requires $2n^3/3$ flops and $O(n^3)$ comparisons. Unlike partial pivoting, complete pivoting involves a significant floating point arithmetic overhead because of the two-dimensional search at each stage.

With the factorization $PAQ^T = LU$ in hand the solution to $Ax = b$ proceeds as follows:

Step 1. Solve $Lz = Pb$ for z.

Step 2. Solve $Uy = z$ for y.

Step 3. Set $x = Q^T y$.

The *rowpiv* and *colpiv* representations can be used to form Pb and Qy, respectively.

Wilkinson (1961) has shown that in exact arithmetic the elements of the matrix $A^{(k)} = M_k \Pi_k \cdots M_1 \Pi_1 A \Gamma_1 \cdots \Gamma_k$ satisfy

$$|a_{ij}^{(k)}| \leq k^{1/2}(2 \cdot 3^{1/2} \cdots k^{1/(k-1)})^{1/2}\max|a_{ij}|. \tag{3.4.10}$$

The upper bound is a rather slow-growing function of k. This fact coupled with vast empirical evidence suggesting that ρ is always modestly sized (e.g., $\rho = 10$) permit us to conclude that *Gaussian elimination with complete pivoting is stable*. The method solves a nearby linear system $(A + E)\hat{x} = b$ in the sense of (2.7.21). However, in general there is little reason to choose complete pivoting over partial pivoting. A possible exception is when A is rank deficient. In principal, complete pivoting can be used to reveal the rank of a matrix. Suppose $\text{rank}(A) = r < n$. It follows that at the beginning of step

$r + 1$, $A(r+1{:}n, r+1{:}n) = 0$. This implies that $\Pi_k = \Gamma_k = M_k = I$ for $k = r + 1{:}n$ and so the algorithm can be terminated after step r with the following factorization in hand:

$$PAQ^T = LU = \begin{bmatrix} L_{11} & 0 \\ L_{21} & I_{n-r} \end{bmatrix} \begin{bmatrix} U_{11} & U_{12} \\ 0 & 0 \end{bmatrix}.$$

Here, L_{11} and U_{11} are r-by-r and L_{21} and U_{12}^T are $(n - r)$-by-r. Thus, Gaussian elimination with complete pivoting can in principle be used to determine the rank of a matrix. Nevertheless, roundoff errors make the probability of encountering an exactly zero pivot remote. In practice one would have to "declare" A to have rank k if the pivot element in step $k+1$ was sufficiently small. The numerical rank determination problem is discussed in detail in §5.5.

3.4.7 Rook Pivoting

A third type of LU stablization strategy called *rook pivoting* provides an interesting alternative to partial pivoting and complete pivoting. As with complete pivoting, it computes the factorization $PAQ = LU$. However, instead of choosing as pivot the largest value in $|A(k{:}n, k{:}n)|$, it searches for an element of that submatrix that is maximal in both its row *and* column. Thus, if

$$A(k{:}n, k{:}n) = \begin{bmatrix} 24 & 36 & 13 & 61 \\ 42 & 67 & 72 & 50 \\ 38 & 11 & 36 & 43 \\ 52 & 37 & 48 & 16 \end{bmatrix},$$

then "72" would be identified by complete pivoting while "52," "72," or "61" would be acceptable with the rook pivoting strategy. To implement rook pivoting, the scan-and-swap portion of Algorithm 3.4.3 is changed to

$\mu = k$, $\lambda = k$, $\tau = |a_{\mu\lambda}|$, $s = 0$
while $\tau < \| (A(k{:}n, \lambda) \|_\infty \ \vee \ \tau < \| (A(\mu, k{:}n) \|_\infty$
    if $\mathrm{mod}(s, 2) = 0$
        Update μ so that $|a_{\mu\lambda}| = \| (A(k{:}n, \lambda) \|_\infty$ with $k \le \mu \le n$.
    else
        Update λ so that $|a_{\mu\lambda}| = \| (A(\mu, k{:}n) \|_\infty$ with $k \le \lambda \le n$.
    end
    $s = s + 1$
end
$rowpiv(k) = \mu$, $A(k,:) \leftrightarrow A(\mu,:)$ $colpiv(k) = \lambda$, $A(:,k) \leftrightarrow A(:,\lambda)$

The search for a larger $|a_{\mu\lambda}|$ involves alternate scans of $A(k{:}n, \lambda)$ and $A(\mu, k{:}n)$. The value of τ is monotone increasing and that ensures termination of the **while**-loop. In theory, the exit value of s could be $O((n - k)^2)$, but in practice its value is $O(1)$. See Chang (2002). The bottom line is that rook pivoting represents the same $O(n^2)$ overhead as partial pivoting, but that it induces the same level of reliability as complete pivoting.

3.4.8   A Note on Underdetermined Systems

If $A \in \mathbb{R}^{m \times n}$ with $m < n$, rank$(A) = m$, and $b \in \mathbb{R}^m$, then the linear system $Ax = b$ is said to be *underdetermined*. Note that in this case there are an infinite number of solutions. With either complete or rook pivoting, it is possible to compute an LU factorization of the form

$$PAQ^T = L[\,U_1\,|\,U_2\,] \tag{3.4.11}$$

where P and Q are permutations, $L \in \mathbb{R}^{m \times m}$ is unit lower triangular, and $U_1 \in \mathbb{R}^{m \times m}$ is nonsingular and upper triangular. Note that

$$Ax = b \;\;\Leftrightarrow\;\; (PAQ^T)(Qx) = (Pb) \;\;\Leftrightarrow\;\; L[\,U_1\,|\,U_2\,]\begin{bmatrix} z_1 \\ z_2 \end{bmatrix} = L(U_1 z_1 + U_2 z_2) = c$$

where $c = Pb$ and

$$\begin{bmatrix} z_1 \\ z_2 \end{bmatrix} = Qx.$$

This suggests the following solution procedure:

Step 1. Solve $Ly = Pb$ for $y \in \mathbb{R}^m$.

Step 2. Choose $z_2 \in \mathbb{R}^{n-m}$ and solve $U_1 z_1 = y - U_2 z_2$ for z_1.

Step 3. Set

$$x = Q^T \begin{bmatrix} z_1 \\ z_2 \end{bmatrix}.$$

Setting $z_2 = 0$ is a natural choice. We have more to say about underdetermined systems in §5.6.2.

3.4.9   The LU Mentality

We offer three examples that illustrate how to think in terms of the LU factorization when confronted with a linear equation situation.

   Example 1. Suppose A is nonsingular and n-by-n and that B is n-by-p. Consider the problem of finding X (n-by-p) so $AX = B$. This is the *multiple right hand side problem.* If $X = [\,x_1\,|\cdots|\,x_p\,]$ and $B = [\,b_1\,|\cdots|\,b_p\,]$ are column partitions, then

> Compute $PA = LU$
> **for** $k = 1{:}p$
>> Solve $Ly = Pb_k$ and then $Ux_k = y$.   (3.4.12)
> **end**

If $B = I_n$, then we emerge with an approximation to A^{-1} .

   Example 2. Suppose we want to overwrite b with the solution to $A^k x = b$ where $A \in \mathbb{R}^{n \times n}$, $b \in \mathbb{R}^n$, and k is a positive integer. One approach is to compute $C = A^k$ and then solve $Cx = b$. However, the matrix multiplications can be avoided altogether:

Compute $PA = LU$.

for $j = 1{:}k$

Overwrite b with the solution to $Ly = Pb$.   (3.4.13)

Overwrite b with the solution to $Ux = b$.

end

As in Example 1, the idea is to get the LU factorization "outside the loop."

Example 3. Suppose we are given $A \in \mathbb{R}^{n \times n}$, $d \in \mathbb{R}^n$, and $c \in \mathbb{R}^n$ and that we want to compute $s = c^T A^{-1} d$. One approach is to compute $X = A^{-1}$ as discussed in (i) and then compute $s = c^T X d$. However, it is more economical to proceed as follows:

Compute $PA = LU$.

Solve $Ly = Pd$ and then $Ux = y$.

$s = c^T x$

An "A^{-1}" in a formula almost always means "solve a linear system" and almost never means "compute A^{-1}."

3.4.10   A Model Problem for Numerical Analysis

We are now in possession of a very important and well-understood algorithm (Gaussian elimination) for a very important and well-understood problem (linear equations). Let us take advantage of our position and formulate more abstractly what we mean by "problem sensitivity" and "algorithm stability." Our discussion follows Higham (ASNA, §1.5–1.6), Stewart (MA, §4.3), and Trefethen and Bau (NLA, Lectures 12, 14, 15, and 22).

A *problem* is a function $f{:}D \to S$ from "data/input space" D to "solution/output space" S. A *problem instance* is f together with a particular $d \in D$. We assume D and S are normed vector spaces. For linear systems, D is the set of matrix-vector pairs (A, b) where $A \in \mathbb{R}^{n \times n}$ is nonsingular and $b \in \mathbb{R}^n$. The function f maps (A, b) to $A^{-1}b$, an element of S. For a particular A and b, $Ax = b$ is a problem instance.

A *perturbation theory* for the problem f sheds light on the difference between $f(d)$ and $f(d + \Delta d)$ where $d \in D$ and $d + \Delta d \in D$. For linear systems, we discussed in §2.6 the difference between the solution to $Ax = b$ and the solution to $(A + \Delta A)(x + \Delta x) = (b + \Delta b)$. We bounded $\| \Delta x \|/\| x \|$ in terms of $\| \Delta A \|/\| A \|$ and $\| \Delta b \|/\| b \|$.

The *conditioning* of a problem refers to the behavior of f under perturbation at d. A *condition number* of a problem quantifies the rate of change of the solution with respect to the input data. If small changes in d induce relatively large changes in $f(d)$, then that problem instance is *ill-conditioned*. If small changes in d do not induce relatively large changes in $f(d)$, then that problem instance is *well-conditioned*. Definitions for "small" and "large" are required. For linear systems we showed in §2.6 that the magnitude of the condition number $\kappa(A) = \| A \|\| A^{-1} \|$ determines whether an $Ax = b$ problem is ill-conditioned or well-conditioned. One might say that a linear equation problem is well-conditioned if $\kappa(A) \approx O(1)$ and ill-conditioned if $\kappa(A) \approx O(1/\mathbf{u})$.

An *algorithm* for computing $f(d)$ produces an approximation $\tilde{f}(d)$. Depending on the situation, it may be necessary to identify a particular software implementation of the underlying method. The $\tilde{f}$ function for Gaussian elimination with partial pivoting, Gaussian elimination with rook pivoting, and Gaussian elimination with complete pivoting are all different.

An algorithm for computing $f(d)$ is *stable* if for some small Δd, the computed solution $\tilde{f}(d)$ is close to $f(d+\Delta d)$. A stable algorithm *nearly* solves a nearby problem. An algorithm for computing $f(d)$ is *backward stable* if for some small Δd, the computed solution $\tilde{f}(d)$ satisfies $\tilde{f}(d) = f(d+\Delta d)$. A backward stable algorithm *exactly* solves a nearby problem. Applied to a given linear system $Ax = b$, Gaussian elimination with complete pivoting is backward stable because the computed solution $\tilde{x}$ satisfies

$$(A + \Delta)\tilde{x} = b$$

and $\|\Delta\|/\|A\| \approx O(\mathbf{u})$. On the other hand, if b is specified by a matrix-vector product $b = Mv$, then

$$(A + \Delta)\tilde{x} = Mv + \delta$$

where $\|\Delta\|/\|A\| \approx O(\mathbf{u})$ and $\delta/(\|M\|\|v\|) \approx O(\mathbf{u})$. Here, the underlying f is defined by $f:(A, M, v) \to A^{-1}(Mv)$. In this case the algorithm is stable but not backward stable.

Problems

P3.4.1 Let $A = LU$ be the LU factorization of n-by-n A with $|\ell_{ij}| \leq 1$. Let a_i^T and u_i^T denote the ith rows of A and U, respectively. Verify the equation

$$u_i^T = a_i^T - \sum_{j=1}^{i-1} \ell_{ij} u_j^T$$

and use it to show that $\|U\|_\infty \leq 2^{n-1}\|A\|_\infty$. (Hint: Take norms and use induction.)

P3.4.2 Show that if $PAQ = LU$ is obtained via Gaussian elimination with complete pivoting, then no element of $U(i, i{:}n)$ is larger in absolute value than $|u_{ii}|$. Is this true with rook pivoting?

P3.4.3 Suppose $A \in \mathbb{R}^{n \times n}$ has an LU factorization and that L and U are known. Give an algorithm which can compute the (i, j) entry of A^{-1} in approximately $(n - j)^2 + (n - i)^2$ flops.

P3.4.4 Suppose $\hat{X}$ is the computed inverse obtained via (3.4.12). Give an upper bound for $\|A\hat{X} - I\|_F$.

P3.4.5 Extend Algorithm 3.4.3 so that it can produce the factorization (3.4.11). How many flops are required?

Notes and References for §3.4

Papers concerned with element growth and pivoting include:

C.W. Cryer (1968). "Pivot Size in Gaussian Elimination," *Numer. Math. 12*, 335–345.

J.K. Reid (1971). "A Note on the Stability of Gaussian Elimination," *J.Inst. Math. Applic. 8*, 374–375.

P.A. Businger (1971). "Monitoring the Numerical Stability of Gaussian Elimination," *Numer. Math. 16*, 360–361.

A.M. Cohen (1974). "A Note on Pivot Size in Gaussian Elimination," *Lin. Alg. Applic. 8*, 361–68.

A.M. Erisman and J.K. Reid (1974). "Monitoring the Stability of the Triangular Factorization of a Sparse Matrix," *Numer. Math. 22*, 183–186.

J. Day and B. Peterson (1988). "Growth in Gaussian Elimination," *Amer. Math. Monthly 95*, 489–513.

N.J. Higham and D.J. Higham (1989). "Large Growth Factors in Gaussian Elimination with Pivoting," *SIAM J. Matrix Anal. Applic. 10*, 155–164.

L.N. Trefethen and R.S. Schreiber (1990). "Average-Case Stability of Gaussian Elimination," *SIAM J. Matrix Anal. Applic. 11*, 335–360.

N. Gould (1991). "On Growth in Gaussian Elimination with Complete Pivoting," *SIAM J. Matrix Anal. Applic. 12*, 354–361.

A. Edelman (1992). "The Complete Pivoting Conjecture for Gaussian Elimination is False," *Mathematica J. 2*, 58–61.

S.J. Wright (1993). "A Collection of Problems for Which Gaussian Elimination with Partial Pivoting is Unstable," *SIAM J. Sci. Stat. Comput. 14*, 231–238.

L.V. Foster (1994). "Gaussian Elimination with Partial Pivoting Can Fail in Practice," *SIAM J. Matrix Anal. Applic. 15*, 1354–1362.

A. Edelman and W. Mascarenhas (1995). "On the Complete Pivoting Conjecture for a Hadamard Matrix of Order 12," *Lin. Multilin. Alg. 38*, 181–185.

J.M. Pena (1996). "Pivoting Strategies Leading to Small Bounds of the Errors for Certain Linear Systems," *IMA J. Numer. Anal. 16*, 141–153.

J.L. Barlow and H. Zha (1998). "Growth in Gaussian Elimination, Orthogonal Matrices, and the 2-Norm," *SIAM J. Matrix Anal. Applic. 19*, 807–815.

P. Favati, M. Leoncini, and A. Martinez (2000). "On the Robustness of Gaussian Elimination with Partial Pivoting," *BIT 40*, 62–73.

As we mentioned, the size of L^{-1} is relevant to the growth factor. Thus, it is important to have an understanding of triangular matrix condition, see:

D. Viswanath and L.N. Trefethen (1998). "Condition Numbers of Random Triangular Matrices," *SIAM J. Matrix Anal. Applic. 19*, 564–581.

The connection between small pivots and near singularity is reviewed in:

T.F. Chan (1985). "On the Existence and Computation of LU Factorizations with Small Pivots," *Math. Comput. 42*, 535–548.

A pivot strategy that we did not discuss is *pairwise pivoting*. In this approach, 2-by-2 Gauss transformations are used to zero the lower triangular portion of A. The technique is appealing in certain multiprocessor environments because only adjacent rows are combined in each step, see:

D. Sorensen (1985). "Analysis of Pairwise Pivoting in Gaussian Elimination," *IEEE Trans. Comput. C-34*, 274–278.

A related type of pivoting called *tournament pivoting* that is of interest in distributed memory computing is outlined in §3.6.3. For a discussion of rook pivoting and its properties, see:

L.V. Foster (1997). "The Growth Factor and Efficiency of Gaussian Elimination with Rook Pivoting," *J. Comput. Appl. Math., 86*, 177–194.

G. Poole and L. Neal (2000). "The Rook's Pivoting Strategy," *J. Comput. Appl. Math. 123*, 353–369.

X-W Chang (2002) "Some Features of Gaussian Elimination with Rook Pivoting," *BIT 42*, 66–83.

3.5 Improving and Estimating Accuracy

Suppose we apply Gaussian elimination with partial pivoting to the n-by-n system $Ax = b$ and that IEEE double precision arithmetic is used. Equation (3.4.9) essentially says that if the growth factor is modest then the computed solution $\hat{x}$ satisfies

$$(A + E)\hat{x} = b, \qquad \| E \|_\infty \approx \mathbf{u}\| A \|_\infty. \tag{3.5.1}$$

In this section we explore the practical ramifications of this result. We begin by stressing the distinction that should be made between residual size and accuracy. This is followed by a discussion of scaling, iterative improvement, and condition estimation. See Higham (ASNA) for a more detailed treatment of these topics.

We make two notational remarks at the outset. The infinity norm is used throughout since it is very handy in roundoff error analysis and in practical error estimation. Second, whenever we refer to "Gaussian elimination" in this section we really mean Gaussian elimination with some stabilizing pivot strategy such as partial pivoting.

3.5.1   Residual Size versus Accuracy

The *residual* of a computed solution $\hat{x}$ to the linear system $Ax = b$ is the vector $b - A\hat{x}$. A small residual means that $A\hat{x}$ effectively "predicts" the right hand side b. From Equation 3.5.1 we have $\| b - A\hat{x} \|_\infty \approx \mathbf{u}\| A \|_\infty \| \hat{x} \|_\infty$ and so we obtain

Heuristic I. *Gaussian elimination produces a solution $\hat{x}$ with a relatively small residual.*

Small residuals do not imply high accuracy. Combining Theorem 2.6.2 and (3.5.1), we see that

$$\frac{\| \hat{x} - x \|_\infty}{\| x \|_\infty} \approx \mathbf{u}\kappa_\infty(A) . \tag{3.5.2}$$

This justifies a second guiding principle.

Heuristic II. *If the unit roundoff and condition satisfy $\mathbf{u} \approx 10^{-d}$ and $\kappa_\infty(A) \approx 10^q$, then Gaussian elimination produces a solution $\hat{x}$ that has about $d - q$ correct decimal digits.*

If $\mathbf{u}\,\kappa_\infty(A)$ is large, then we say that A is ill-conditioned with respect to the machine precision.

As an illustration of the Heuristics I and II, consider the system

$$\begin{bmatrix} .986 & .579 \\ .409 & .237 \end{bmatrix} \begin{bmatrix} x_1 \\ x_2 \end{bmatrix} = \begin{bmatrix} .235 \\ .107 \end{bmatrix}$$

in which $\kappa_\infty(A) \approx 700$ and $x = [\,2, \, -3\,]^T$. Here is what we find for various machine precisions:

$\mathbf{u}$	$\hat{x}_1$	$\hat{x}_2$	$\dfrac{\| \hat{x} - x \|_\infty}{\| x \|_\infty}$	$\dfrac{\| b - A\hat{x} \|_\infty}{\| A \|_\infty \| \hat{x} \|_\infty}$
10^{-3}	2.11	-3.17	$5 \cdot 10^{-2}$	$2.0 \cdot 10^{-3}$
10^{-4}	1.986	-2.975	$8 \cdot 10^{-3}$	$1.5 \cdot 10^{-4}$
10^{-5}	2.0019	-3.0032	$1 \cdot 10^{-3}$	$2.1 \cdot 10^{-6}$
10^{-6}	2.00025	-3.00094	$3 \cdot 10^{-4}$	$4.2 \cdot 10^{-7}$

Whether or not to be content with the computed solution $\hat{x}$ depends on the requirements of the underlying source problem. In many applications accuracy is not important but small residuals are. In such a situation, the $\hat{x}$ produced by Gaussian elimination is probably adequate. On the other hand, if the number of correct digits in $\hat{x}$ is an issue, then the situation is more complicated and the discussion in the remainder of this section is relevant.

3.5.2   Scaling

Let β be the machine base (typically $\beta = 2$) and define the diagonal matrices $D_1 = \mathrm{diag}(\beta^{r_1}, \ldots, \beta^{r_n})$ and $D_2 = \mathrm{diag}(\beta^{c_1}, \ldots, \beta^{c_n})$. The solution to the n-by-n linear system $Ax = b$ can be found by solving the *scaled system* $(D_1^{-1} A D_2)y = D_1^{-1}b$ using

Gaussian elimination and then setting $x = D_2 y$. The scalings of A, b, and y require only $O(n^2)$ flops and may be accomplished without roundoff. Note that D_1 scales equations and D_2 scales unknowns.

It follows from Heuristic II that if $\hat{x}$ and $\hat{y}$ are the computed versions of x and y, then

$$\frac{\| D_2^{-1}(\hat{x} - x) \|_\infty}{\| D_2^{-1} x \|_\infty} = \frac{\| \hat{y} - y \|_\infty}{\| y \|_\infty} \approx \mathbf{u}\kappa_\infty(D_1^{-1}AD_2). \tag{3.5.3}$$

Thus, if $\kappa_\infty(D_1^{-1}AD_2)$ can be made considerably smaller than $\kappa_\infty(A)$, then we might expect a correspondingly more accurate $\hat{x}$, provided errors are measured in the "D_2" norm defined by $\| z \|_{D_2} = \| D_2^{-1} z \|_\infty$. This is the objective of scaling. Note that it encompasses two issues: the condition of the scaled problem and the appropriateness of appraising error in the D_2-norm.

An interesting but very difficult mathematical problem concerns the exact minimization of $\kappa_p(D_1^{-1}AD_2)$ for general diagonal D_i and various p. Such results as there are in this direction are not very practical. This is hardly discouraging, however, when we recall that (3.5.3) is a heuristic result, it makes little sense to minimize exactly a heuristic bound. What we seek is a fast, approximate method for improving the quality of the computed solution $\hat{x}$.

One technique of this variety is *simple row scaling*. In this scheme D_2 is the identity and D_1 is chosen so that each row in $D_1^{-1}A$ has approximately the same ∞-norm. Row scaling reduces the likelihood of adding a very small number to a very large number during elimination—an event that can greatly diminish accuracy.

Slightly more complicated than simple row scaling is *row-column equilibration*. Here, the object is to choose D_1 and D_2 so that the ∞-norm of each row and column of $D_1^{-1}AD_2$ belongs to the interval $[1/\beta, 1]$ where β is the base of the floating point system. For work along these lines, see McKeeman (1962).

It cannot be stressed too much that simple row scaling and row-column equilibration do not "solve" the scaling problem. Indeed, either technique can render a worse $\hat{x}$ than if no scaling whatever is used. The ramifications of this point are thoroughly discussed in Forsythe and Moler (SLE, Chap. 11). The basic recommendation is that the scaling of equations and unknowns must proceed on a problem-by-problem basis. General scaling strategies are unreliable. It is best to scale (if at all) on the basis of what the source problem proclaims about the significance of each a_{ij}. Measurement units and data error may have to be considered.

3.5.3 Iterative Improvement

Suppose $Ax = b$ has been solved via the partial pivoting factorization $PA = LU$ and that we wish to improve the accuracy of the computed solution $\hat{x}$. If we execute

$$r = b - A\hat{x}$$

Solve $Ly = Pr$.

Solve $Uz = y$.

$$x_{\text{new}} = \hat{x} + z$$

$$\tag{3.5.4}$$

then in exact arithmetic $Ax_{\text{new}} = A\hat{x} + Az = (b - r) + r = b$. Unfortunately, the naive floating point execution of these formulae renders an x_{new} that is no more accurate than $\hat{x}$. This is to be expected since $\hat{r} = \text{fl}(b - A\hat{x})$ has few, if any, correct significant digits. (Recall Heuristic I.) Consequently, $\hat{z} = \text{fl}(A^{-1}r) \approx A^{-1} \cdot$ noise $\approx$ noise is a very poor correction *from the standpoint of improving the accuracy of* $\hat{x}$. However, Skeel (1980) has an error analysis that indicates when (3.5.4) gives an improved x_{new} *from the standpoint of backward error*. In particular, if the quantity

$$\tau = \left(\| \, |A| \, |A^{-1}| \, \|_\infty \right) \left(\max_i (|A||x|)_i \, / \, \min_i (|A||x|)_i \right)$$

is not too big, then (3.5.4) produces an x_{new} such that $(A + E)x_{new} = b$ for very small E. Of course, if Gaussian elimination with partial pivoting is used, then the computed $\hat{x}$ already solves a nearby system. However, this may not be the case for certain pivot strategies used to preserve sparsity. In this situation, the *fixed precision iterative improvement* step (3.5.4) can be worthwhile and cheap. See Arioli, Demmel, and Duff (1988).

In general, for (3.5.4) to produce a more accurate x, it is necessary to compute the residual $b - A\hat{x}$ with extended precision floating point arithmetic. Typically, this means that if t-digit arithmetic is used to compute $PA = LU$, x, y, and z, then $2t$-digit arithmetic is used to form $b - A\hat{x}$. The process can be iterated. In particular, once we have computed $PA = LU$ and initialize $x = 0$, we repeat the following:

$$r = b - Ax \text{ (higher precision)}$$
$$\text{Solve } Ly = Pr \text{ for } y \text{ and } Uz = y \text{ for } z. \qquad (3.5.5)$$
$$x = x + z$$

We refer to this process as *mixed-precision iterative improvement*. The original A must be used in the high-precision computation of r. The basic result concerning the performance of (3.5.5) is summarized in the following heuristic:

Heuristic III. *If the machine precision* $\mathbf{u}$ *and condition satisfy* $\mathbf{u} = 10^{-d}$ *and* $\kappa_\infty(A) \approx 10^q$, *then after* k *executions of (3.5.5),* x *has approximately* $\min\{d, k(d-q)\}$ *correct digits if the residual computation is performed with precision* $\mathbf{u}^2$.

Roughly speaking, if $\mathbf{u}\,\kappa_\infty(A) \leq 1$, then iterative improvement can ultimately produce a solution that is correct to full (single) precision. Note that the process is relatively cheap. Each improvement costs $O(n^2)$, to be compared with the original $O(n^3)$ investment in the factorization $PA = LU$. Of course, no improvement may result if A is badly conditioned with respect to the machine precision.

3.5.4   Condition Estimation

Suppose that we have solved $Ax = b$ via $PA = LU$ and that we now wish to ascertain the number of correct digits in the computed solution $\hat{x}$. It follows from Heuristic II that in order to do this we need an estimate of the condition $\kappa_\infty(A) = \| A \|_\infty \| A^{-1} \|_\infty$. Computing $\| A \|_\infty$ poses no problem as we merely use the $O(n^2)$ formula (2.3.10). The challenge is with respect to the factor $\| A^{-1} \|_\infty$. Conceivably, we could estimate this quantity by $\| \hat{X} \|_\infty$, where $\hat{X} = [\, \hat{x}_1 \, | \cdots | \, \hat{x}_n \,]$ and $\hat{x}_i$ is the computed solution to $Ax_i = e_i$. (See §3.4.9.) The trouble with this approach is its expense: $\hat{\kappa}_\infty = \| A \|_\infty \| \hat{X} \|_\infty$ costs about three times as much as $\hat{x}$.

The central problem of *condition estimation* is how to estimate reliably the condition number in $O(n^2)$ flops assuming the availability of $PA = LU$ or one of the factorizations that are presented in subsequent chapters. An approach described in Forsythe and Moler (SLE, p. 51) is based on iterative improvement and the heuristic

$$\mathbf{u}\kappa_\infty(A) \approx \| z \|_\infty / \| x \|_\infty$$

where z is the first correction of x in (3.5.5).

Cline, Moler, Stewart, and Wilkinson (1979) propose an approach to the condition estimation problem thatis based on the implication

$$Ay = d \implies \| A^{-1} \|_\infty \geq \| y \|_\infty / \| d \|_\infty.$$

The idea behind their estimator is to choose d so that the solution y is large in norm and then set

$$\hat{\kappa}_\infty = \| A \|_\infty \| y \|_\infty / \| d \|_\infty.$$

The success of this method hinges on how close the ratio $\| y \|_\infty / \| d \|_\infty$ is to its maximum value $\| A^{-1} \|_\infty$.

Consider the case when $A = T$ is upper triangular. The relation between d and y is completely specified by the following column version of back substitution:

$p(1{:}n) = 0$

for $k = n{:} - 1{:}1$

  Choose $d(k)$.

  $y(k) = (d(k) - p(k))/T(k,k)$ (3.5.6)

  $p(1{:}k - 1) = p(1{:}k - 1) + y(k)T(1{:}k - 1, k)$

end

Normally, we use this algorithm to solve a *given* triangular system $Ty = d$. However, in the condition estimation setting we are free to pick the right-hand side d subject to the "constraint" that y is large relative to d.

One way to encourage growth in y is to choose $d(k)$ from the set $\{-1, +1\}$ so as to maximize $y(k)$. If $p(k) \geq 0$, then set $d(k) = -1$. If $p(k) < 0$, then set $d(k) = +1$. In other words, (3.5.6) is invoked with $d(k) = -\text{sign}(p(k))$. Overall, the vector d has the form $d(1{:}n) = [\pm 1, \ldots, \pm 1]^T$. Since this is a unit vector, we obtain the estimate $\hat{\kappa}_\infty = \| T \|_\infty \| y \|_\infty$.

A more reliable estimator results if $d(k) \in \{-1, +1\}$ is chosen so as to encourage growth both in $y(k)$ and the running sum update $p(1{:}k - 1, k) + T(1{:}k - 1, k)y(k)$. In particular, at step k we compute

$$y(k)^+ = (1 - p(k))/T(k,k),$$
$$s(k)^+ = |y(k)^+| + \| p(1{:}k - 1) + T(1{:}k - 1, k)y(k)^+ \|_1,$$
$$y(k)^- = (-1 - p(k))/T(k,k),$$
$$s(k)^- = |y(k)^-| + \| p(1{:}k - 1) + T(1{:}k - 1, k)y(k)^- \|_1,$$

and set

$$
y(k) = \begin{cases} y(k)^+ & \text{if } s(k)^+ \geq s(k)^-, \\[2mm] y(k)^- & \text{if } s(k)^+ < s(k)^-. \end{cases}
$$

This gives the following procedure.

Algorithm 3.5.1 (Condition Estimator) Let $T \in \mathbb{R}^{n \times n}$ be a nonsingular upper triangular matrix. This algorithm computes unit ∞-norm y and a scalar κ so $\| Ty \|_\infty \approx 1/\| T^{-1} \|_\infty$ and $\kappa \approx \kappa_\infty(T)$

> $p(1{:}n) = 0$
> **for** $k = n{:} -1{:}1$
> > $y(k)^+ = (1 - p(k))/T(k,k)$
> > $y(k)^- = (-1 - p(k))/T(k,k)$
> > $p(k)^+ = p(1{:}k-1) + T(1{:}k-1,k)y(k)^+$
> > $p(k)^- = p(1{:}k-1) + T(1{:}k-1,k)y(k)^-$
> > **if** $|y(k)^+| + \| p(k)^+ \|_1 \geq |y(k)^-| + \| p(k)^- \|_1$
> > > $y(k) = y(k)^+$
> > > $p(1{:}k-1) = p(k)^+$
> > **else**
> > > $y(k) = y(k)^-$
> > > $p(1{:}k-1) = p(k)^-$
> > **end**
> **end**
> $\kappa = \| y \|_\infty \| T \|_\infty$
> $y = y/\| y \|_\infty$

The algorithm involves several times the work of ordinary back substitution.

We are now in a position to describe a procedure for estimating the condition of a square nonsingular matrix A whose $PA = LU$ factorization is available:

> *Step 1.* Apply the lower triangular version of Algorithm 3.5.1 to U^T and obtain a large-norm solution to $U^T y = d$.
>
> *Step 2.* Solve the triangular systems $L^T r = y$, $Lw = Pr$, and $Uz = w$.
>
> *Step 3.* Set $\hat{\kappa}_\infty = \| A \|_\infty \| z \|_\infty / \| r \|_\infty$.

Note that $\| z \|_\infty \leq \| A^{-1} \|_\infty \| r \|_\infty$. The method is based on several heuristics. First, if A is ill-conditioned and $PA = LU$, then it is usually the case that U is correspondingly ill-conditioned. The lower triangle L tends to be fairly well-conditioned. Thus, it is more profitable to apply the condition estimator to U than to L. The vector r, because it solves $A^T P^T r = d$, tends to be rich in the direction of the left singular vector associated with $\sigma_{\min}(A)$. Right-hand sides with this property render large solutions to the problem $Az = r$.

In practice, it is found that the condition estimation technique that we have outlined produces adequate order-of-magnitude estimates of the true condition number.

Problems

P3.5.1 Show by example that there may be more than one way to equilibrate a matrix.

P3.5.2 Suppose $P(A + E) = \hat{L}\hat{U}$, where P is a permutation, $\hat{L}$ is lower triangular with $|\hat{\ell}_{ij}| \leq 1$, and $\hat{U}$ is upper triangular. Show that $\hat{\kappa}_{\infty}(A) \geq \| A \|_{\infty}/(\| E \|_{\infty} + \mu)$ where $\mu = \min |\hat{u}_{ii}|$. Conclude that if a small pivot is encountered when Gaussian elimination with pivoting is applied to A, then A is ill-conditioned. The converse is not true. (Hint: Let A be the matrix B_n defined in (2.6.9)).

P3.5.3 (Kahan (1966)) The system $Ax = b$ where

$$
A = \begin{bmatrix} 2 & -1 & 1 \\ -1 & 10^{-10} & 10^{-10} \\ 1 & 10^{-10} & 10^{-10} \end{bmatrix}, \qquad b = \begin{bmatrix} 2(1 + 10^{-10}) \\ -10^{-10} \\ 10^{-10} \end{bmatrix}
$$

has solution $x = [10^{-10} \ -1 \ 1]^T$. (a) Show that if $(A + E)y = b$ and $|E| \leq 10^{-8}|A|$, then $|x - y| \leq 10^{-7}|x|$. That is, small relative changes in A's entries do not induce large changes in x even though $\kappa_{\infty}(A) = 10^{10}$. (b) Define $D = \text{diag}(10^{-5}, 10^5, 10^5)$. Show that $\kappa_{\infty}(DAD) \leq 5$. (c) Explain what is going on using Theorem 2.6.3.

P3.5.4 Consider the matrix:

$$
T = \begin{bmatrix} 1 & 0 & M & -M \\ 0 & 1 & -M & M \\ 0 & 0 & 1 & 0 \\ 0 & 0 & 0 & 1 \end{bmatrix} \qquad M \in \mathbb{R}.
$$

What estimate of $\kappa_{\infty}(T)$ is produced when (3.5.6) is applied with $d(k) = -\text{sign}(p(k))$? What estimate does Algorithm 3.5.1 produce? What is the true $\kappa_{\infty}(T)$?

P3.5.5 What does Algorithm 3.5.1 produce when applied to the matrix B_n given in (2.6.9)?

Notes and References for §3.5

The following papers are concerned with the scaling of $Ax = b$ problems:

F.L. Bauer (1963). "Optimally Scaled Matrices," *Numer. Math. 5*, 73–87.
P.A. Businger (1968). "Matrices Which Can Be Optimally Scaled," *Numer. Math. 12*, 346–48.
A. van der Sluis (1969). "Condition Numbers and Equilibration Matrices," *Numer. Math. 14*, 14–23.
A. van der Sluis (1970). "Condition, Equilibration, and Pivoting in Linear Algebraic Systems," *Numer. Math. 15*, 74–86.
C. McCarthy and G. Strang (1973). "Optimal Conditioning of Matrices," *SIAM J. Numer. Anal. 10*, 370–388.
T. Fenner and G. Loizou (1974). "Some New Bounds on the Condition Numbers of Optimally Scaled Matrices," *J. ACM 21*, 514–524.
G.H. Golub and J.M. Varah (1974). "On a Characterization of the Best L_2-Scaling of a Matrix," *SIAM J. Numer. Anal. 11*, 472–479.
R. Skeel (1979). "Scaling for Numerical Stability in Gaussian Elimination," *J. ACM 26*, 494–526.
R. Skeel (1981). "Effect of Equilibration on Residual Size for Partial Pivoting," *SIAM J. Numer. Anal. 18*, 449–55.
V. Balakrishnan and S. Boyd (1995). "Existence and Uniqueness of Optimal Matrix Scalings," *SIAM J. Matrix Anal. Applic. 16*, 29–39.

Part of the difficulty in scaling concerns the selection of a norm in which to measure errors. An interesting discussion of this frequently overlooked point appears in:

W. Kahan (1966). "Numerical Linear Algebra," *Canadian Math. Bull. 9*, 757–801.

For a rigorous analysis of iterative improvement and related matters, see:

C.B. Moler (1967). "Iterative Refinement in Floating Point," *J. ACM 14*, 316-371.
M. Jankowski and M. Wozniakowski (1977). "Iterative Refinement Implies Numerical Stability," *BIT 17*, 303–311.
R.D. Skeel (1980). "Iterative Refinement Implies Numerical Stability for Gaussian Elimination," *Math. Comput. 35*, 817–832.
N.J. Higham (1997). "Iterative Refinement for Linear Systems and LAPACK," *IMA J. Numer. Anal. 17*, 495–509.

A. Dax (2003). "A Modified Iterative Refinement Scheme," *SIAM J. Sci. Comput. 25*, 1199–1213.

J. Demmel, Y. Hida, W. Kahan, X.S. Li, S. Mukherjee, and E.J. Riedy (2006). "Error Bounds from Extra-Precise Iterative Refinement," *ACM Trans. Math. Softw. 32*, 325–351.

The condition estimator that we described is given in:

A.K. Cline, C.B. Moler, G.W. Stewart, and J.H. Wilkinson (1979). "An Estimate for the Condition Number of a Matrix," *SIAM J. Numer. Anal. 16*, 368-75.

Other references concerned with the condition estimation problem include:

C.G. Broyden (1973). "Some Condition Number Bounds for the Gaussian Elimination Process," *J. Inst. Math. Applic. 12*, 273–286.

F. Lemeire (1973). "Bounds for Condition Numbers of Triangular Value of a Matrix," *Lin. Alg. Applic. 11*, 1–2.

D.P. O'Leary (1980). "Estimating Matrix Condition Numbers," *SIAM J. Sci. Stat. Comput. 1*, 205–209.

A.K. Cline, A.R. Conn, and C. Van Loan (1982). "Generalizing the LINPACK Condition Estimator," in *Numerical Analysis* , J.P. Hennart (ed.), Lecture Notes in Mathematics No. 909, Springer-Verlag, New York.

A.K. Cline and R.K. Rew (1983). "A Set of Counter examples to Three Condition Number Estimators," *SIAM J. Sci. Stat. Comput. 4*, 602–611.

W. Hager (1984). "Condition Estimates," *SIAM J. Sci. Stat. Comput. 5*, 311–316.

N.J. Higham (1987). "A Survey of Condition Number Estimation for Triangular Matrices," *SIAM Review 29*, 575–596.

N.J. Higham (1988). "FORTRAN Codes for Estimating the One-Norm of a Real or Complex Matrix with Applications to Condition Estimation (Algorithm 674)," *ACM Trans. Math. Softw. 14*, 381–396.

C.H. Bischof (1990). "Incremental Condition Estimation," *SIAM J. Matrix Anal. Applic. 11*, 312–322.

G. Auchmuty (1991). "A Posteriori Error Estimates for Linear Equations," *Numer. Math. 61*, 1–6.

N.J. Higham (1993). "Optimization by Direct Search in Matrix Computations," *SIAM J. Matrix Anal. Applic. 14*, 317–333.

D.J. Higham (1995). "Condition Numbers and Their Condition Numbers," *Lin. Alg. Applic. 214*, 193–213.

G.W. Stewart (1997). "The Triangular Matrices of Gaussian Elimination and Related Decompositions," *IMA J. Numer. Anal. 17*, 7–16.

3.6   Parallel LU

In §3.2.11 we show how to organize a block version of Gaussian elimination (without pivoting) so that the overwhelming majority of flops occur in the context of matrix multiplication. It is possible to incorporate partial pivoting and maintain the same level-3 fraction. After stepping through the derivation we proceed to show how the process can be effectively parallelized using the block-cyclic distribution ideas that were presented in §1.6.

3.6.1   Block LU with Pivoting

Throughout this section assume $A \in \mathbb{R}^{n \times n}$ and for clarity that $n = rN$:

$$A = \begin{bmatrix} A_{11} & \cdots & A_{1N} \\ \vdots & \ddots & \vdots \\ A_{N1} & \cdots & A_{NN} \end{bmatrix} \qquad A_{i,j} \in \mathbb{R}^{r \times r}. \qquad (3.6.1)$$

We revisit Algorithm 3.2.4 (nonrecursive block LU) and show how to incorporate partial pivoting.

The first step starts by applying scalar Gaussian elimination with partial pivoting to the first block column. Using an obvious rectangular matrix version of Algorithm 3.4.1 we obtain the following factorization:

$$
P_1 \begin{bmatrix} A_{11} \\ A_{21} \\ \vdots \\ A_{N1} \end{bmatrix} = \begin{bmatrix} L_{11} \\ L_{21} \\ \vdots \\ L_{N1} \end{bmatrix} U_{11}.
\tag{3.6.2}
$$

In this equation, $P_1 \in \mathbb{R}^{n \times n}$ is a permutation, $L_{11} \in \mathbb{R}^{r \times r}$ is unit lower triangular, and $U_{11} \in \mathbb{R}^{r \times r}$ is upper triangular.

The next task is to compute the first block row of U. To do this we set

$$
P_1 A = \begin{bmatrix} \tilde{A}_{11} & \cdots & \tilde{A}_{1N} \\ \vdots & \ddots & \vdots \\ \tilde{A}_{N1} & \cdots & \tilde{A}_{NN} \end{bmatrix}, \qquad \tilde{A}_{i,j} \in \mathbb{R}^{r \times r},
\tag{3.6.3}
$$

and solve the lower triangular multiple-right-hand-side problem

$$
L_{11} \begin{bmatrix} U_{12} \mid \cdots \mid U_{1N} \end{bmatrix} = \begin{bmatrix} \tilde{A}_{12} \mid \cdots \mid \tilde{A}_{1N} \end{bmatrix}
\tag{3.6.4}
$$

for $U_{12}, \ldots, U_{1N} \in \mathbb{R}^{r \times r}$. At this stage it is easy to show that we have the partial factorization

$$
P_1 A = \left[\begin{array}{c|cccc} L_{11} & 0 & \cdots & 0 \\ \hline L_{21} & I_r & \cdots & 0 \\ \vdots & \vdots & \ddots & \vdots \\ L_{N1} & 0 & \cdots & I_r \end{array} \right] \cdot \left[\begin{array}{c|c} I_r & 0 \\ \hline 0 & A^{(\mathrm{new})} \end{array} \right] \left[\begin{array}{c|cccc} U_{11} & U_{12} & \cdots & U_{1N} \\ \hline 0 & I_r & \cdots & 0 \\ \vdots & \vdots & \ddots & \vdots \\ 0 & 0 & \cdots & I_r \end{array} \right]
$$

where

$$
A^{(\mathrm{new})} = \begin{bmatrix} \tilde{A}_{22} & \cdots & \tilde{A}_{2N} \\ \vdots & \ddots & \vdots \\ \tilde{A}_{N2} & \cdots & \tilde{A}_{NN} \end{bmatrix} - \begin{bmatrix} L_{21} \\ \vdots \\ L_{N1} \end{bmatrix} \begin{bmatrix} U_{12} \mid \cdots \mid U_{1N} \end{bmatrix}.
\tag{3.6.5}
$$

Note that the computation of $A^{(\mathrm{new})}$ is a level-3 operation as it involves one matrix multiplication per A-block.

The remaining task is to compute the pivoted LU factorization of $A^{(\mathrm{new})}$. Indeed, if

$$
P^{(\mathrm{new})} A^{(\mathrm{new})} = L^{(\mathrm{new})} U^{(\mathrm{new})}
$$

and

$$
P^{(\mathrm{new})} \begin{bmatrix} L_{21} \\ \vdots \\ L_{N1} \end{bmatrix} = \begin{bmatrix} \tilde{L}_{21} \\ \vdots \\ \tilde{L}_{N1} \end{bmatrix},
$$

then

$$
PA = \begin{bmatrix} \begin{array}{c|ccc} L_{11} & 0 & \cdots & 0 \\ \hline \tilde{L}_{21} & & & \\ \vdots & & L^{(\text{new})} & \\ \tilde{L}_{N1} & & & \end{array} \end{bmatrix} \begin{bmatrix} \begin{array}{c|ccc} U_{11} & U_{12} & \cdots & U_{1N} \\ \hline 0 & & & \\ \vdots & & U^{(\text{new})} & \\ 0 & & & \end{array} \end{bmatrix}
$$

is the pivoted block LU factorization of A with

$$
P = \begin{bmatrix} I_r & 0 \\ 0 & P^{(\text{new})} \end{bmatrix} P_1 .
$$

In general, the processing of each block column in A is a four-part calculation:

Part A. Apply rectangular Gaussian Elimination with partial pivoting to a block column of A. This produces a permutation, a block column of L, and a diagonal block of U. See (3.6.2).

Part B. Apply the Part A permutation to the "rest of A." See (3.6.3).

Part C. Complete the computation of U's next block row by solving a lower triangular multiple right-hand-side problem. See (3.6.4).

Part D. Using the freshly computed L-blocks and U-blocks, update the "rest of A." See (3.6.5).

The precise formulation of the method with overwriting is similar to Algorithm 3.2.4 and is left as an exercise.

3.6.2   Parallelizing the Pivoted Block LU Algorithm

Recall the discussion of the block-cyclic distribution in §1.6.2 where the parallel computation of the matrix multiplication update $C = C + AB$ was outlined. To provide insight into how the pivoted block LU algorithm can be parallelized, we examine a representative step in a small example that also makes use of the block-cyclic distribution.

Assume that $N = 8$ in (3.6.1) and that we have a p_{row}-by-p_{col} processor network with $p_{\text{row}} = 2$ and $p_{\text{col}} = 2$. At the start, the blocks of $A = (A_{ij})$ are cyclically distributed as shown in Figure 3.6.1. Assume that we have carried out two steps of block LU and that the computed L_{ij} and U_{ij} have overwritten the corresponding A-blocks. Figure 3.6.2 displays the situation at the start of the third step. Blocks that are to participate in the Part A factorization

$$
P_3 \begin{bmatrix} A_{33} \\ \vdots \\ A_{83} \end{bmatrix} = \begin{bmatrix} L_{33} \\ \vdots \\ L_{83} \end{bmatrix} U_{33}
$$

are highlighted. Typically, p_{row} processors are involved and since the blocks are each r-by-r, there are r steps as shown in (3.6.6).

Proc(0,0)	Proc(0,1)	Proc(0,0)	Proc(0,1)	Proc(0,0)	Proc(0,1)	Proc(0,0)	Proc(0,1)
A_{11}	A_{12}	A_{13}	A_{14}	A_{15}	A_{16}	A_{17}	A_{18}
Proc(1,0)	Proc(1,1)	Proc(1,0)	Proc(1,1)	Proc(1,0)	Proc(1,1)	Proc(1,0)	Proc(1,1)
A_{21}	A_{22}	A_{23}	A_{24}	A_{25}	A_{26}	A_{27}	A_{28}
Proc(0,0)	Proc(0,1)	Proc(0,0)	Proc(0,1)	Proc(0,0)	Proc(0,1)	Proc(0,0)	Proc(0,1)
A_{31}	A_{32}	A_{33}	A_{34}	A_{35}	A_{36}	A_{37}	A_{38}
Proc(1,0)	Proc(1,1)	Proc(1,0)	Proc(1,1)	Proc(1,0)	Proc(1,1)	Proc(1,0)	Proc(1,1)
A_{41}	A_{42}	A_{43}	A_{44}	A_{45}	A_{46}	A_{47}	A_{48}
Proc(0,0)	Proc(0,1)	Proc(0,0)	Proc(0,1)	Proc(0,0)	Proc(0,1)	Proc(0,0)	Proc(0,1)
A_{51}	A_{52}	A_{53}	A_{54}	A_{55}	A_{56}	A_{57}	A_{58}
Proc(1,0)	Proc(1,1)	Proc(1,0)	Proc(1,1)	Proc(1,0)	Proc(1,1)	Proc(1,0)	Proc(1,1)
A_{61}	A_{62}	A_{63}	A_{64}	A_{65}	A_{66}	A_{67}	A_{68}
Proc(0,0)	Proc(0,1)	Proc(0,0)	Proc(0,1)	Proc(0,0)	Proc(0,1)	Proc(0,0)	Proc(0,1)
A_{71}	A_{72}	A_{73}	A_{74}	A_{75}	A_{76}	A_{77}	A_{78}
Proc(1,0)	Proc(1,1)	Proc(1,0)	Proc(1,1)	Proc(1,0)	Proc(1,1)	Proc(1,0)	Proc(1,1)
A_{81}	A_{82}	A_{83}	A_{84}	A_{85}	A_{86}	A_{87}	A_{88}

Figure 3.6.1.

Part A:

Proc(0,0)	Proc(0,1)	Proc(0,0)	Proc(0,1)	Proc(0,0)	Proc(0,1)	Proc(0,0)	Proc(0,1)
U_{11} L_{11}	U_{12}	U_{13}	U_{14}	U_{15}	U_{16}	U_{17}	U_{18}
Proc(1,0)	Proc(1,1)	Proc(1,0)	Proc(1,1)	Proc(1,0)	Proc(1,1)	Proc(1,0)	Proc(1,1)
L_{21}	U_{22} L_{22}	U_{23}	U_{24}	U_{25}	U_{26}	U_{27}	U_{28}
Proc(0,0)	Proc(0,1)	Proc(0,0)	Proc(0,1)	Proc(0,0)	Proc(0,1)	Proc(0,0)	Proc(0,1)
L_{31}	L_{32}	A_{33}	A_{34}	A_{35}	A_{36}	A_{37}	A_{38}
Proc(1,0)	Proc(1,1)	Proc(1,0)	Proc(1,1)	Proc(1,0)	Proc(1,1)	Proc(1,0)	Proc(1,1)
L_{41}	L_{42}	A_{43}	A_{44}	A_{45}	A_{46}	A_{47}	A_{48}
Proc(0,0)	Proc(0,1)	Proc(0,0)	Proc(0,1)	Proc(0,0)	Proc(0,1)	Proc(0,0)	Proc(0,1)
L_{51}	L_{52}	A_{53}	A_{54}	A_{55}	A_{56}	A_{57}	A_{58}
Proc(1,0)	Proc(1,1)	Proc(1,0)	Proc(1,1)	Proc(1,0)	Proc(1,1)	Proc(1,0)	Proc(1,1)
L_{61}	L_{62}	A_{63}	A_{64}	A_{65}	A_{66}	A_{67}	A_{68}
Proc(0,0)	Proc(0,1)	Proc(0,0)	Proc(0,1)	Proc(0,0)	Proc(0,1)	Proc(0,0)	Proc(0,1)
L_{71}	L_{72}	A_{73}	A_{74}	A_{75}	A_{76}	A_{77}	A_{78}
Proc(1,0)	Proc(1,1)	Proc(1,0)	Proc(1,1)	Proc(1,0)	Proc(1,1)	Proc(1,0)	Proc(1,1)
L_{81}	L_{82}	A_{83}	A_{84}	A_{85}	A_{86}	A_{87}	A_{88}

Figure 3.6.2.

for $j = 1{:}r$

     Columns $A_{kk}(:,j), \ldots, A_{N,k}(:,j)$ are assembled in
        the processor housing A_{kk}, the "pivot processor"

     The pivot processor determines the required row interchange and
        the Gauss transform vector

     The swapping of the two A-rows may require the involvement of
        two processors in the network

     The appropriate part of the Gauss vector together with   (3.6.6)
        $A_{kk}(j, j{:}r)$ is sent by the pivot processor to the
        processors that house $A_{k+1,k}, \ldots, A_{N,k}$

     The processors that house $A_{kk}, \ldots, A_{N,k}$ carry out their
        share of the update, a local computation

end

Upon completion, the parallel execution of Parts B and C follow. In the Part B computation, those blocks that may be involved in the row swapping have been highlighted. See Figure 3.6.3. This overhead generally engages the entire processor network, although communication is local to each processor column.

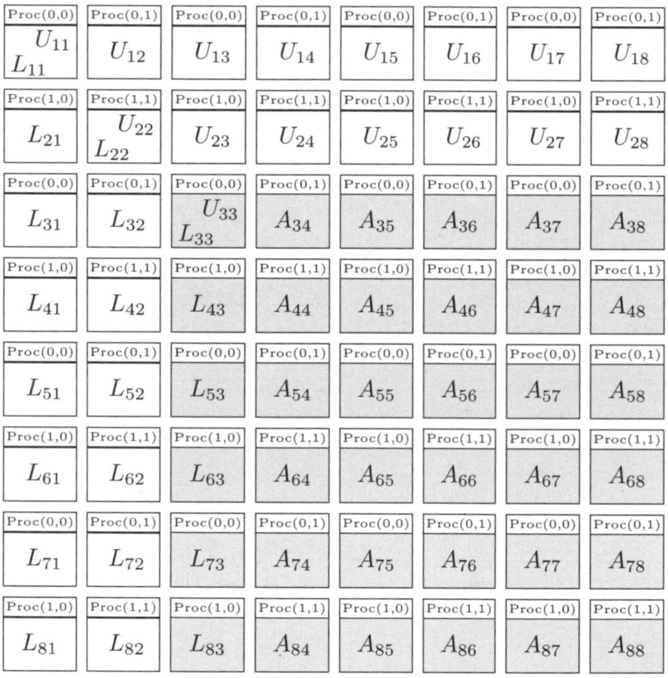

Figure 3.6.3.

Note that Part C involves just a single processor row while the "big" level-three update that follows typically involves the entire processor network. See Figures 3.6.4 and 3.6.5.

Part C:

Proc(0,0) U_{11} L_{11}	Proc(0,1) U_{12}	Proc(0,0) U_{13}	Proc(0,1) U_{14}	Proc(0,0) U_{15}	Proc(0,1) U_{16}	Proc(0,0) U_{17}	Proc(0,1) U_{18}
Proc(1,0) L_{21}	Proc(1,1) U_{22} L_{22}	Proc(1,0) U_{23}	Proc(1,1) U_{24}	Proc(1,0) U_{25}	Proc(1,1) U_{26}	Proc(1,0) U_{27}	Proc(1,1) U_{28}
Proc(0,0) L_{31}	Proc(0,1) L_{32}	Proc(0,0) U_{33} L_{33}	Proc(0,1) A_{34}	Proc(0,0) A_{35}	Proc(0,1) A_{36}	Proc(0,0) A_{37}	Proc(0,1) A_{38}
Proc(1,0) L_{41}	Proc(1,1) L_{42}	Proc(1,0) L_{43}	Proc(1,1) A_{44}	Proc(1,0) A_{45}	Proc(1,1) A_{46}	Proc(1,0) A_{47}	Proc(1,1) A_{48}
Proc(0,0) L_{51}	Proc(0,1) L_{52}	Proc(0,0) L_{53}	Proc(0,1) A_{54}	Proc(0,0) A_{55}	Proc(0,1) A_{56}	Proc(0,0) A_{57}	Proc(0,1) A_{58}
Proc(1,0) L_{61}	Proc(1,1) L_{62}	Proc(1,0) L_{63}	Proc(1,1) A_{64}	Proc(1,0) A_{65}	Proc(1,1) A_{66}	Proc(1,0) A_{67}	Proc(1,1) A_{68}
Proc(0,0) L_{71}	Proc(0,1) L_{72}	Proc(0,0) L_{73}	Proc(0,1) A_{74}	Proc(0,0) A_{75}	Proc(0,1) A_{76}	Proc(0,0) A_{77}	Proc(0,1) A_{78}
Proc(1,0) L_{81}	Proc(1,1) L_{82}	Proc(1,0) L_{83}	Proc(1,1) A_{84}	Proc(1,0) A_{85}	Proc(1,1) A_{86}	Proc(1,0) A_{87}	Proc(1,1) A_{88}

Figure 3.6.4.

Part D:

Proc(0,0) U_{11} L_{11}	Proc(0,1) U_{12}	Proc(0,0) U_{13}	Proc(0,1) U_{14}	Proc(0,0) U_{15}	Proc(0,1) U_{16}	Proc(0,0) U_{17}	Proc(0,1) U_{18}
Proc(1,0) L_{21}	Proc(1,1) U_{22} L_{22}	Proc(1,0) U_{23}	Proc(1,1) U_{24}	Proc(1,0) U_{25}	Proc(1,1) U_{26}	Proc(1,0) U_{27}	Proc(1,1) U_{28}
Proc(0,0) L_{31}	Proc(0,1) L_{32}	Proc(0,0) U_{33} L_{33}	Proc(0,1) A_{34}	Proc(0,0) A_{35}	Proc(0,1) A_{36}	Proc(0,0) A_{37}	Proc(0,1) A_{38}
Proc(1,0) L_{41}	Proc(1,1) L_{42}	Proc(1,0) L_{43}	Proc(1,1) A_{44}	Proc(1,0) A_{45}	Proc(1,1) A_{46}	Proc(1,0) A_{47}	Proc(1,1) A_{48}
Proc(0,0) L_{51}	Proc(0,1) L_{52}	Proc(0,0) L_{53}	Proc(0,1) A_{54}	Proc(0,0) A_{55}	Proc(0,1) A_{56}	Proc(0,0) A_{57}	Proc(0,1) A_{58}
Proc(1,0) L_{61}	Proc(1,1) L_{62}	Proc(1,0) L_{63}	Proc(1,1) A_{64}	Proc(1,0) A_{65}	Proc(1,1) A_{66}	Proc(1,0) A_{67}	Proc(1,1) A_{68}
Proc(0,0) L_{71}	Proc(0,1) L_{72}	Proc(0,0) L_{73}	Proc(0,1) A_{74}	Proc(0,0) A_{75}	Proc(0,1) A_{76}	Proc(0,0) A_{77}	Proc(0,1) A_{78}
Proc(1,0) L_{81}	Proc(1,1) L_{82}	Proc(1,0) L_{83}	Proc(1,1) A_{84}	Proc(1,0) A_{85}	Proc(1,1) A_{86}	Proc(1,0) A_{87}	Proc(1,1) A_{88}

Figure 3.6.5.

The communication overhead associated with Part D is masked by the matrix multiplications that are performed on each processor.

This completes the $k = 3$ step of parallel block LU with partial pivoting. The process can obviously be repeated on the trailing 5-by-5 block matrix. The virtues of the block-cyclic distribution are revealed through the schematics. In particular, the dominating level-3 step (Part D) is load balanced for all but the last few values of k. Subsets of the processor grid are used for the "smaller," level-2 portions of the computation.

We shall not attempt to predict the fraction of time that is devoted to these computations or the propagation of the interchange permutations. Enlightenment in this direction requires benchmarking.

3.6.3   Tournament Pivoting

The decomposition via partial pivoting in Step A requires a lot of communication. An alternative that addresses this issue involves a strategy called *tournament pivoting*. Here is the main idea. Suppose we want to compute $PW = LU$ where the blocks of

$$
W = \begin{bmatrix} W_1 \\ W_2 \\ W_3 \\ W_4 \end{bmatrix} \in \mathbb{R}^{n \times r}
$$

are distributed around some network of processors. Assume that each W_i has many more rows than columns. The goal is to choose r rows from W that can serve as pivot rows. If we compute the "local" factorizations

$$
P_1 W_1 = L_1 U_1, \qquad P_2 W_2 = L_2 U_2, \qquad P_3 W_3 = L_3 U_3, \qquad P_4 W_4 = L_4 U_4,
$$

via Gaussian elimination with partial pivoting, then the top r rows of the matrices $P_1 W_1$, $P_2 W_2$, $P_3 W_3$, are $P_4 W_4$ are pivot row candidates. Call these square matrices W_1', W_2', W_3', and W_4' and note that we have reduced the number of possible pivot rows from n to $4r$.

Next we compute the factorizations

$$
P_{12} W_{12}' = P_{12} \begin{bmatrix} W_1' \\ W_2' \end{bmatrix} = L_{12} U_{12},
$$

$$
P_{34} W_{34}' = P_{34} \begin{bmatrix} W_3' \\ W_4' \end{bmatrix} = L_{34} U_{34},
$$

and recognize that the top r rows of $P_{12} W_{12}'$ and the top r rows of $P_{34} W_{34}'$ are even better pivot row candidates. Assemble these $2r$ rows into a matrix W_{1234} and compute

$$
P_{1234} W_{1234} = L_{1234} U_{1234}.
$$

The top r rows of $P_{1234} W_{1234}$ are then the chosen pivot rows for the LU reduction of W.

Of course, there are communication overheads associated with each round of the "tournament," but the volume of interprocessor data transfers is much reduced. See Demmel, Grigori, and Xiang (2010).

Problems

P3.6.1 In §3.6.1 we outlined a single step of block LU with partial pivoting. Specify a complete version of the algorithm.

P3.6.2 Regarding parallel block LU with partial pivoting, why is it better to "collect" all the permutations in Part A before applying them across the remaining block columns? In other words, why not propagate the Part A permutations as they are produced instead of having Part B, a separate permutation application step?

P3.6.3 Review the discussion about parallel shared memory computing in §1.6.5 and §1.6.6. Develop a shared memory version of Algorithm 3.2.1. Designate one processor for computation of the multipliers and a load-balanced scheme for the rank-1 update in which all the processors participate. A **barrier** is necessary because the rank-1 update cannot proceed until the multipliers are available. What if partial pivoting is incorporated?

Notes and References for §3.6

See the `scaLAPACK` manual for a discussion of parallel Gaussian elimination as well as:

J. Ortega (1988). *Introduction to Parallel and Vector Solution of Linear Systems*, Plenum Press, New York.

K. Gallivan, W. Jalby, U. Meier, and A.H. Sameh (1988). "Impact of Hierarchical Memory Systems on Linear Algebra Algorithm Design," *Int. J. Supercomput. Applic. 2*, 12–48.

J. Dongarra, I. Duff, D. Sorensen, and H. van der Vorst (1990). *Solving Linear Systems on Vector and Shared Memory Computers*, SIAM Publications, Philadelphia, PA.

Y. Robert (1990). *The Impact of Vector and Parallel Architectures on the Gaussian Elimination Algorithm*, Halsted Press, New York.

J. Choi, J.J. Dongarra, L.S. Osttrouchov, A.P. Petitet, D.W. Walker, and R.C. Whaley (1996). "Design and Implementation of the ScaLAPACK LU, QR, and Cholesky Factorization Routines," *Scientific Programming, 5*, 173–184.

X.S. Li (2005). "An Overview of SuperLU: Algorithms, Implementation, and User Interface," *ACM Trans. Math. Softw. 31*, 302–325.

S. Tomov, J. Dongarra, and M. Baboulin (2010). "Towards Dense Linear Algebra for Hybrid GPU Accelerated Manycore Systems," *Parallel Comput. 36*, 232–240.

The tournament pivoting strategy is a central feature of the optimized LU implementation discussed in:

J. Demmel, L. Grigori, and H. Xiang (2011). "CALU: A Communication Optimal LU Factorization Algorithm," *SIAM J. Matrix Anal. Applic. 32*, 1317-1350.

E. Solomonik and J. Demmel (2011). "Communication-Optimal Parallel 2.5D Matrix Multiplication and LU Factorization Algorithms," *Euro-Par 2011 Parallel Processing Lecture Notes in Computer Science, 2011, Volume 6853/2011*, 90–109.

Chapter 4

Special Linear Systems

4.1 Diagonal Dominance and Symmetry

4.2 Positive Definite Systems

4.3 Banded Systems

4.4 Symmetric Indefinite Systems

4.5 Block Tridiagonal Systems

4.6 Vandermonde Systems

4.7 Classical Methods for Toeplitz Systems

4.8 Circulant and Discrete Poisson Systems

It is a basic tenet of numerical analysis that solution procedures should exploit structure whenever it is present. In numerical linear algebra, this translates into an expectation that algorithms for general linear systems can be streamlined in the presence of such properties as symmetry, definiteness, and bandedness. Two themes prevail:

- There are important classes of matrices for which it is safe not to pivot when computing the LU or a related factorization.

- There are important classes of matrices with highly structured LU factorizations that can be computed quickly, sometimes, *very quickly*.

Challenges arise when a fast, but unstable, LU factorization is available.

Symmetry and diagonal dominance are prime examples of exploitable matrix structure and we use these properties to introduce some key ideas in §4.1. In §4.2 we examine the case when A is both symmetric and positive definite, deriving the stable Cholesky factorization. Unsymmetric positive definite systems are also investigated. In §4.3, banded versions of the LU and Cholesky factorizations are discussed and this is followed in §4.4 with a treatment of the symmetric indefinite problem. Block matrix ideas and sparse matrix ideas come together when the matrix of coefficients is block tridiagonal. This important class of systems receives a special treatment in §4.5.

Classical methods for Vandermonde and Toeplitz systems are considered in §4.6 and §4.7. In §4.8 we connect the fast transform discussion in §1.4 to the problem of solving circulant systems and systems that arise when the Poisson problem is discretized using finite differences.

Before we get started, we clarify some terminology associated with structured problems that pertains to this chapter and beyond. Banded matrices and block-banded matrices are examples of *sparse matrices*, meaning that the vast majority of their entries are zero. Linear equation methods that are appropriate when the zero-nonzero pattern is more arbitrary are discussed in Chapter 11. Toeplitz, Vandermonde, and circulant matrices are *data sparse*. A matrix $A \in \mathbb{R}^{m \times n}$ is data sparse if it can be parameterized with many fewer than $O(mn)$ numbers. Cauchy-like systems and semiseparable systems are considered in §12.1 and §12.2.

Reading Notes

Knowledge of Chapters 1, 2, and 3 is assumed. Within this chapter there are the following dependencies:

$$
\begin{array}{ccccccc}
\S4.1 & \rightarrow & \S4.2 & \rightarrow & \S4.3 & \rightarrow & \S4.4 \\
\downarrow & & & & \downarrow & & \\
\S4.6 & & & & \S4.5 & \rightarrow & \S4.7 & \rightarrow & \S4.8
\end{array}
$$

Global references include Stewart(MABD), Higham (ASNA), Watkins (FMC), Trefethen and Bau (NLA), Demmel (ANLA), and Ipsen (NMA).

4.1   Diagonal Dominance and Symmetry

Pivoting is a serious concern in the context of high-performance computing because the cost of moving data around rivals the cost of computation. Equally important, pivoting can destroy exploitable structure. For example, if A is symmetric, then it involves half the data of a general A. Our intuition (correctly) tells us that we should be able to solve a symmetric $Ax = b$ problem with half the arithmetic. However, in the context of Gaussian elimination with pivoting, symmetry can be destroyed at the very start of the reduction, e.g.,

$$
\begin{bmatrix} 0 & 0 & 1 \\ 0 & 1 & 0 \\ 1 & 0 & 0 \end{bmatrix}
\begin{bmatrix} a & b & c \\ b & d & e \\ c & e & f \end{bmatrix}
=
\begin{bmatrix} c & e & f \\ b & d & e \\ a & b & c \end{bmatrix}.
$$

Taking advantage of symmetry and other patterns and identifying situations where pivoting is unnecessary are typical activities in the realm of structured $Ax = b$ solving. The goal is to expose computational shortcuts and to justify their use through analysis.

4.1.1   Diagonal Dominance and the LU Factorization

If A's diagonal entries are large compared to its off-diagonal entries, then we anticipate that it is safe to compute $A = LU$ without pivoting. Consider the $n = 2$ case:

$$
\begin{bmatrix} a & b \\ c & d \end{bmatrix}
=
\begin{bmatrix} 1 & 0 \\ c/a & 1 \end{bmatrix}
\begin{bmatrix} a & b \\ 0 & d - (c/a)b \end{bmatrix}.
$$

If a and d "dominate" b and c in magnitude, then the elements of L and U will be nicely bounded. To quantify this we make a definition. We say that $A \in \mathbb{R}^{n \times n}$ is *row diagonally dominant* if

$$|a_{ii}| \geq \sum_{\substack{j=1 \\ j \neq i}}^{n} |a_{ij}|, \qquad i = 1{:}n . \tag{4.1.1}$$

Similarly, *column diagonal dominance* means that $|a_{jj}|$ is larger than the sum of all off-diagonal element magnitudes in the same column. If these inequalities are strict, then A is *strictly (row/column) diagonally dominant*. A diagonally dominant matrix can be singular, e.g., the 2-by-2 matrix of 1's. However, if a nonsingular matrix is diagonally dominant, then it has a "safe" LU factorization.

Theorem 4.1.1. *If A is nonsingular and column diagonally dominant, then it has an LU factorization and the entries in $L = (\ell_{ij})$ satisfy $|l_{ij}| \leq 1$.*

Proof. We proceed by induction. The theorem is obviously true if $n = 1$. Assume that it is true for $(n-1)$-by-$(n-1)$ nonsingular matrices that are column diagonally dominant. Partition $A \in \mathbb{R}^{n \times n}$ as follows:

$$A = \begin{bmatrix} \alpha & w^T \\ v & C \end{bmatrix}, \qquad \alpha \in \mathbb{R},\ v, w \in \mathbb{R}^{n-1},\ C \in \mathbb{R}^{(n-1) \times (n-1)}.$$

If $\alpha = 0$, then $v = 0$ and A is singular. Thus, $\alpha \neq 0$ and we have the factorization

$$\begin{bmatrix} \alpha & w^T \\ v & C \end{bmatrix} = \begin{bmatrix} 1 & 0 \\ v/\alpha & I \end{bmatrix} \begin{bmatrix} 1 & 0 \\ 0 & B \end{bmatrix} \begin{bmatrix} \alpha & w^T \\ 0 & I \end{bmatrix}, \tag{4.1.2}$$

where

$$B = C - \frac{1}{\alpha} v w^T.$$

Since $\det(A) = \alpha \cdot \det(B)$, it follows that B is nonsingular. It is also column diagonally dominant because

$$\sum_{\substack{i=1 \\ i \neq j}}^{n-1} |b_{ij}| = \sum_{\substack{i=1 \\ i \neq j}}^{n-1} |c_{ij} - v_i w_j / \alpha| \leq \sum_{\substack{i=1 \\ i \neq j}}^{n-1} |c_{ij}| + \frac{|w_j|}{|\alpha|} \sum_{\substack{i=1 \\ i \neq j}}^{n-1} |v_i|$$

$$< (|c_{jj}| - |w_j|) + \frac{|w_j|}{|\alpha|}(|\alpha| - |v_j|) \leq \left| c_{jj} - \frac{w_j v_j}{\alpha} \right| = |b_{jj}|.$$

By induction, B has an LU factorization $L_1 U_1$ and so from (4.1.2) we have

$$A = \begin{bmatrix} 1 & 0 \\ v/\alpha & L_1 \end{bmatrix} \begin{bmatrix} \alpha & w^T \\ 0 & U_1 \end{bmatrix} \equiv LU.$$

The entries in $|v/\alpha|$ are bounded by 1 because A is column diagonally dominant. By induction, the same can be said about the entries in $|L_1|$. Thus, the entries in $|L|$ are all bounded by 1 completing the proof. $\square$

The theorem shows that Gaussian elimination without pivoting is a stable solution procedure for a column diagonally dominant matrix. If the diagonal elements strictly dominate the off-diagonal elements, then we can actually bound $\| A^{-1} \|$.

Theorem 4.1.2. *If $A \in \mathbb{R}^{n \times n}$ and*

$$\delta = \min_{1 \le j \le n} \left(|a_{jj}| - \sum_{\substack{i=1 \\ i \ne j}}^{n} |a_{ij}| \right) > 0 \qquad (4.1.3)$$

then

$$\| A^{-1} \|_1 \le 1/\delta.$$

Proof. Define $D = \mathrm{diag}(a_{11}, \ldots, a_{nn})$ and $E = A - D$. If e is the column n-vector of 1's, then

$$e^T |E| \le e^T |D| - \delta e^T.$$

If $x \in \mathbb{R}^n$, then $Dx = Ax - Ex$ and

$$|D|\,|x| \le |Ax| + |E|\,|x|.$$

Thus,

$$e^T |D|\,|x| \le e^T |Ax| + e^T |E|\,|x| \le \| Ax \|_1 + \left(e^T |D| - \delta e^T \right) |x|$$

and so $\delta\| x \|_1 = \delta e^T |x| \le \| Ax \|_1$. The bound on $\| A^{-1} \|_1$ follows from the fact that for any $y \in \mathbb{R}^n$,

$$\delta\| A^{-1}y \|_1 \le \| A(A^{-1}y) \|_1 = \| y \|_1. \qquad \square$$

The "dominance" factor δ defined in (4.1.3) is important because it has a bearing on the condition of the linear system. Moreover, if it is too small, then diagonal dominance may be lost during the elimination process because of roundoff. That is, the computed version of the B matrix in (4.1.2) may not be column diagonally dominant.

4.1.2   Symmetry and the LDL^T Factorization

If A is symmetric and has an LU factorization $A = LU$, then L and U have a connection. For example, if $n = 2$ we have

$$\begin{bmatrix} a & c \\ c & d \end{bmatrix} = \begin{bmatrix} 1 & 0 \\ c/a & 1 \end{bmatrix} \cdot \begin{bmatrix} a & c \\ 0 & d - (c/a)c \end{bmatrix}$$

$$= \begin{bmatrix} 1 & 0 \\ c/a & 1 \end{bmatrix} \cdot \left(\begin{bmatrix} a & 0 \\ 0 & d - (c/a)c \end{bmatrix} \begin{bmatrix} 1 & c/a \\ 0 & 1 \end{bmatrix} \right).$$

It appears that U is a row scaling of L^T. Here is a result that makes this precise.

Theorem 4.1.3. (LDL^T Factorization) *If $A \in \mathbb{R}^{n \times n}$ is symmetric and the principal submatrix $A(1{:}k, 1{:}k)$ is nonsingular for $k = 1{:}n - 1$, then there exists a unit lower triangular matrix L and a diagonal matrix*

$$D = \text{diag}(d_1, \ldots, d_n)$$

such that $A = LDL^T$. The factorization is unique.

Proof. By Theorem 3.2.1 we know that A has an LU factorization $A = LU$. Since the matrix

$$L^{-1}AL^{-T} = UL^{-T}$$

is both symmetric and upper triangular, it must be diagonal. The theorem follows by setting $D = UL^{-T}$ and the uniqueness of the LU factorization. □

Note that once we have the LDL^T factorization, then solving $Ax = b$ is a 3-step process:

$$Lz = b, \qquad Dy = z, \qquad L^T x = y.$$

This works because $Ax = L(D(L^T x)) = L(Dy) = Lz = b$.

Because there is only one triangular matrix to compute, it is not surprising that the factorization $A = LDL^T$ requires half as many flops to compute as $A = LU$. To see this we derive a Gaxpy-rich procedure that, for $j = 1{:}n$, computes $L(j+1{:}n, j)$ and d_j in step j. Note that

$$A(j{:}n, j) = L(j{:}n, 1{:}j) \cdot v(1{:}j)$$

where

$$v(1{:}j) = \begin{bmatrix} d_1 \ell_{j1} \\ d_2 \ell_{j2} \\ \vdots \\ d_{j-1} \ell_{j,j-1} \\ d_j \end{bmatrix}.$$

From this we conclude that

$$d_j = a_{jj} - \sum_{k=1}^{j-1} d_k \ell_{jk}^2.$$

With d_j available, we can rearrange the equation

$$A(j+1{:}n, j) = L(j+1{:}n, 1{:}j) \cdot v(1{:}j)$$

$$= L(j+1{:}n, 1{:}j-1) \cdot v(1{:}j-1) + d_j \cdot L(j+1{:}n, j)$$

to get a recipe for $L(j+1{:}n, j)$:

$$L(j+1{:}n, j) = \frac{1}{d_j} \left(A(j+1{:}n, j) - L(j+1{:}n, 1{:}j-1) \cdot v(1{:}j-1) \right).$$

Properly sequenced, we obtain the following overall procedure:

for $j = 1{:}n$
    for $i = 1{:}j - 1$
        $v(i) = L(j, i) \cdot d(i)$
    end
    $d(j) = A(j, j) - L(j, 1{:}j - 1) \cdot v(1{:}j - 1)$
    $L(j + 1{:}n, j) = (A(j + 1{:}n, j) - L(j + 1{:}n, 1{:}j - 1) \cdot v(1{:}j - 1))/d(j)$
end

With overwriting we obtain the following procedure.

Algorithm 4.1.1 (LDL^{T}) If $A \in \mathbb{R}^{n \times n}$ is symmetric and has an LU factorization, then this algorithm computes a unit lower triangular matrix L and a diagonal matrix $D = \mathrm{diag}(d_1, \ldots, d_n)$ so $A = LDL^T$. The entry a_{ij} is overwritten with ℓ_{ij} if $i > j$ and with d_i if $i = j$.

for $j = 1{:}n$
    for $i = 1{:}j - 1$
        $v(i) = A(j, i)A(i, i)$
    end
    $A(j, j) = A(j, j) - A(j, 1{:}j - 1) \cdot v(1{:}j - 1)$
    $A(j + 1{:}n, j) = (A(j + 1{:}n, j) - A(j + 1{:}n, 1{:}j - 1) \cdot v(1{:}j - 1))/A(j, j)$
end

This algorithm requires $n^3/3$ flops, about half the number of flops involved in Gaussian elimination.

The computed solution $\hat{x}$ to $Ax = b$ obtained via Algorithm 4.1.1 and the usual triangular system solvers of §3.1 can be shown to satisfy a perturbed system $(A+E)\hat{x} = b$, where

$$|E| \leq n\mathbf{u}\left(2|A| + 4|\hat{L}||\hat{D}||\hat{L}^T|\right) + O(\mathbf{u}^2) \qquad (4.1.4)$$

and $\hat{L}$ and $\hat{D}$ are the computed versions of L and D, respectively.

As in the case of the LU factorization considered in the previous chapter, the upper bound in (4.1.4) is without limit unless A has some special property that guarantees stability. In the next section, we show that if A is symmetric and positive definite, then Algorithm 4.1.1 not only runs to completion, but is extremely stable. If A is symmetric but not positive definite, then, as we discuss in §4.4, it is necessary to consider alternatives to the LDL^T factorization.

Problems

P4.1.1 Show that if all the inequalities in (4.1.1) are strict inequalities, then A is nonsingular.

P4.1.2 State and prove a result similar to Theorem 4.1.2 that applies to a row diagonally dominant matrix. In particular, show that $\| A^{-1} \|_{\infty} \leq 1/\delta$ where δ measures the strength of the row diagonal dominance as defined in Equation 4.1.3.

P4.1.3 Suppose A is column diagonally dominant, symmetric, and nonsingular and that $A = LDL^T$.

What can you say about the size of entries in L and D? Give the smallest upper bound you can for $\| L \|_1$.

Notes and References for §4.1

The unsymmetric analog of Algorithm 4.1.2 is related to the methods of Crout and Doolittle. See Stewart (IMC, pp. 131–149) and also:

G.E. Forsythe (1960). "Crout with Pivoting," *Commun. ACM 3*, 507–508.

W.M. McKeeman (1962). "Crout with Equilibration and Iteration," *Commun. ACM 5*, 553–555.

H.J. Bowdler, R.S. Martin, G. Peters, and J.H. Wilkinson (1966), "Solution of Real and Complex Systems of Linear Equations," *Numer. Math. 8*, 217–234.

Just as algorithms can be tailored to exploit structure, so can error analysis and perturbation theory:

C. de Boor and A. Pinkus (1977). "A Backward Error Analysis for Totally Positive Linear Systems," *Numer. Math. 27*, 485–490.

J.R. Bunch, J.W. Demmel, and C.F. Van Loan (1989). "The Strong Stability of Algorithms for Solving Symmetric Linear Systems," *SIAM J. Matrix Anal. Applic. 10*, 494–499.

A. Barrlund (1991). "Perturbation Bounds for the LDL^T and LU Decompositions," *BIT 31*, 358–363.

D.J. Higham and N.J. Higham (1992). "Backward Error and Condition of Structured Linear Systems," *SIAM J. Matrix Anal. Applic. 13*, 162–175.

J.M. Peña (2004). "LDU Decompositions with L and U Well Conditioned," *ETNA 18*, 198–208.

J-G. Sun (2004). "A Note on Backward Errors for Structured Linear Systems," *Numer. Lin. Alg. 12*, 585–603.

R. Cantó, P. Koev, B. Ricarte, and M. Urbano (2008). "LDU Factorization of Nonsingular Totally Positive Matrices," *SIAM J. Matrix Anal. Applic. 30*, 777–782.

Numerical issues that associated with the factorization of a diagonaly dominant matrix are discussed in:

J.M. Peña (1998). "Pivoting Strategies Leading to Diagonal Dominance by Rows," *Numer. Math. 81*, 293–304.

M. Mendoza, M. Raydan, and P. Tarazaga (1999). "Computing the Nearest Diagonally Dominant Matrix," *Numer. Lin. Alg. 5*, 461–474.

A. George and K.D. Ikramov (2005). "Gaussian Elimination Is Stable for the Inverse of a Diagonally Dominant Matrix," *Math. Comput. 73*, 653–657.

J.M. Peña (2007). "Strict Diagonal Dominance and Optimal Bounds for the Skeel Condition Number," *SIAM J. Numer. Anal. 45*, 1107–1108.

F. Dopico and P. Koev (2011). "Perturbation Theory for the LDU Factorization and Accurate Computations for Diagonally Dominant Matrices," *Numer. Math. 119*, 337–371.

4.2 Positive Definite Systems

A matrix $A \in \mathbb{R}^{n \times n}$ is *positive definite* if $x^T A x > 0$ for all nonzero $x \in \mathbb{R}^n$, *positive semidefinite* if $x^T A x \geq 0$ for all $x \in \mathbb{R}^n$, and *indefinite* if we can find $x, y \in \mathbb{R}^n$ so $(x^T A x)(y^T A y) < 0$. Symmetric positive definite systems constitute one of the most important classes of special $Ax = b$ problems. Consider the 2-by-2 symmetric case. If

$$A = \begin{bmatrix} \alpha & \beta \\ \beta & \gamma \end{bmatrix}$$

is positive definite then

$$
\begin{array}{rcllcl}
x & = & [\,1,\,0\,]^T & \Rightarrow & x^T A x & = & \alpha > 0, \\
x & = & [\,0,1\,]^T & \Rightarrow & x^T A x & = & \gamma > 0, \\
x & = & [\,1,\,1\,]^T & \Rightarrow & x^T A x & = & \alpha + 2\beta + \gamma > 0, \\
x & = & [\,1,\,-1\,]^T & \Rightarrow & x^T A x & = & \alpha - 2\beta + \gamma > 0.
\end{array}
$$

The last two equations imply $|\beta| \leq (\alpha+\gamma)/2$. From these results we see that the largest entry in A is on the diagonal and that it is positive. This turns out to be true in general. (See Theorem 4.2.8 below.) A symmetric positive definite matrix has a diagonal that is sufficiently "weighty" to preclude the need for pivoting. A special factorization called the Cholesky factorization is available for such matrices. It exploits both symmetry and definiteness and its implementation is the main focus of this section. However, before those details are pursued we discuss unsymmetric positive definite matrices. This class of matrices is important in its own right and and presents interesting pivot-related issues.

4.2.1 Positive Definiteness

Suppose $A \in \mathbb{R}^{n \times n}$ is positive definite. It is obvious that a positive definite matrix is nonsingular for otherwise we could find a nonzero x so $x^T A x = 0$. However, much more is implied by the positivity of the *quadratic form* $x^T A x$ as the following results show.

Theorem 4.2.1. *If $A \in \mathbb{R}^{n \times n}$ is positive definite and $X \in \mathbb{R}^{n \times k}$ has rank k, then $B = X^T A X \in \mathbb{R}^{k \times k}$ is also positive definite.*

Proof. If $z \in \mathbb{R}^k$ satisfies $0 \geq z^T B z = (Xz)^T A(Xz)$, then $Xz = 0$. But since X has full column rank, this implies that $z = 0$. $\square$

Corollary 4.2.2. *If A is positive definite, then all its principal submatrices are positive definite. In particular, all the diagonal entries are positive.*

Proof. If v is an integer length-k vector with $1 \leq v_1 < \cdots < v_k \leq n$, then $X = I_n(:, v)$ is a rank-k matrix made up of columns $v_1, \ldots, v_k$ of the identity. It follows from Theorem 4.2.1 that $A(v, v) = X^T A X$ is positive definite. $\square$

Theorem 4.2.3. *The matrix $A \in \mathbb{R}^{n \times n}$ is positive definite if and only if the symmetric matrix*

$$T = \frac{A + A^T}{2}$$

has positive eigenvalues.

Proof. Note that $x^T A x = x^T T x$. If $Tx = \lambda x$ then $x^T A x = \lambda \cdot x^T x$. Thus, if A is positive definite then λ is positive. Conversely, suppose T has positive eigenvalues and $Q^T T Q = \text{diag}(\lambda_i)$ is its Schur decomposition. (See §2.1.7.) It follows that if $x \in \mathbb{R}^n$ and $y = Q^T x$, then

$$x^T A x = x^T T x = y^T (Q^T T Q) y = \sum_{k=1}^{n} \lambda_k y_k^2 > 0,$$

completing the proof of the theorem. $\square$

Corollary 4.2.4. *If A is positive definite, then it has an LU factorization and the diagonal entries of U are positive.*

Proof. From Corollary 4.2.2, it follows that the submatrices $A(1{:}k, 1{:}k)$ are nonsingular for $k = 1{:}n$ and so from Theorem 3.2.1 the factorization $A = LU$ exists. If we apply Theorem 4.2.1 with $X = (L^{-1})^T = L^{-T}$, then $B = X^T A X = L^{-1}(LU)L^{-1} = UL^{-T}$ is positive definite and therefore has positive diagonal entries. The corollary follows because L^{-T} is unit upper triangular and this implies $b_{ii} = u_{ii}$, $i = 1{:}n$. $\quad\square$

The mere existence of an LU factorization does not mean that its computation is advisable because the resulting factors may have unacceptably large elements. For example, if $\epsilon > 0$, then the matrix

$$A = \begin{bmatrix} \epsilon & m \\ -m & \epsilon \end{bmatrix} = \begin{bmatrix} 1 & 0 \\ -m/\epsilon & 1 \end{bmatrix} \begin{bmatrix} \epsilon & m \\ 0 & 1 + m^2/\epsilon \end{bmatrix}$$

is positive definite. However, if $m/\epsilon \gg 1$, then it appears that some kind of pivoting is in order. This prompts us to pose an interesting question. Are there conditions that guarantee when it is safe to compute the LU-without-pivoting factorization of a positive definite matrix?

4.2.2 Unsymmetric Positive Definite Systems

The positive definiteness of a general matrix A is inherited from its *symmetric part*:

$$T = \frac{A + A^T}{2}.$$

Note that for any square matrix we have $A = T + S$ where

$$S = \frac{A - A^T}{2}$$

is the *skew-symmetric* part of A. Recall that a matrix S is *skew symmetric* if $S^T = -S$. If S is skew-symmetric, then $x^T S x = 0$ for all $x \in \mathbb{R}^n$ and $s_{ii} = 0$, $i = 1{:}n$. It follows that A is positive definite if and only if its symmetric part is positive definite.

The derivation and analysis of methods for positive definite systems require an understanding about how the symmetric and skew-symmetric parts interact during the LU process.

Theorem 4.2.5. *Suppose*

$$A = \begin{bmatrix} \alpha & v^T \\ v & B \end{bmatrix} + \begin{bmatrix} 0 & -w^T \\ w & C \end{bmatrix}$$

is positive definite and that $B \in \mathbb{R}^{(n-1)\times(n-1)}$ is symmetric and $C \in \mathbb{R}^{(n-1)\times(n-1)}$ is skew-symmetric. Then it follows that

$$A = \begin{bmatrix} 1 & 0 \\ (v + w)/\alpha & I \end{bmatrix} \begin{bmatrix} \alpha & (v - w)^T \\ 0 & B_1 + C_1 \end{bmatrix} \tag{4.2.1}$$

where

$$B_1 = B - \frac{1}{\alpha} \left(vv^T - ww^T \right) \tag{4.2.2}$$

is symmetric positive definite and

$$C_1 = C - \frac{1}{\alpha} \left(wv^T - vw^T \right) \tag{4.2.3}$$

is skew-symmetric.

Proof. Since $\alpha \neq 0$ it follows that (4.2.1) holds. It is obvious from their definitions that B_1 is symmetric and that C_1 is skew-symmetric. Thus, all we have to show is that B_1 is positive definite i.e.,

$$0 < z^T B_1 z = z^T Bz - \frac{1}{\alpha} \left(v^T z \right)^2 + \frac{1}{\alpha} \left(w^T z \right)^2 \tag{4.2.4}$$

for all nonzero $z \in \mathbb{R}^{n-1}$. For any $\mu \in \mathbb{R}$ and $0 \neq z \in \mathbb{R}^{n-1}$ we have

$$0 < \begin{bmatrix} \mu \\ z \end{bmatrix}^T A \begin{bmatrix} \mu \\ z \end{bmatrix} = \begin{bmatrix} \mu \\ z \end{bmatrix}^T \begin{bmatrix} \alpha & v^T \\ v & B \end{bmatrix} \begin{bmatrix} \mu \\ z \end{bmatrix}$$

$$= \mu^2 \alpha + 2\mu v^T z + z^T Bz.$$

If $\mu = -(v^T z)/\alpha$, then

$$0 < z^T Bz - \frac{1}{\alpha} \left(v^T z \right)^2,$$

which establishes the inequality (4.2.4).   □

From (4.2.1) we see that if $B_1 + C_1 = L_1 U_1$ is the LU factorization, then $A = LU$ where

$$L = \begin{bmatrix} 1 & 0 \\ (v + w)/\alpha & L_1 \end{bmatrix} \begin{bmatrix} \alpha & (v - w)^T \\ 0 & U_1 \end{bmatrix}.$$

Thus, the theorem shows that triangular factors in $A = LU$ are nicely bounded if S is not too big compared to T^{-1}. Here is a result that makes this precise:

Theorem 4.2.6. *Let $A \in \mathbb{R}^{n \times n}$ be positive definite and set $T = (A + A^T)/2$ and $S = (A - A^T)/2$. If $A = LU$ is the LU factorization, then*

$$\| \, |L||U| \, \|_F \leq n \left(\| \, T \, \|_2 + \| \, ST^{-1}S \, \|_2 \right). \tag{4.2.5}$$

Proof. See Golub and Van Loan (1979).   □

The theorem suggests when it is safe not to pivot. Assume that the computed factors $\hat{L}$ and $\hat{U}$ satisfy

$$\| \, |\hat{L}||\hat{U}| \, \|_F \leq c \| \, |L||U| \, \|_F, \tag{4.2.6}$$

where c is a constant of modest size. It follows from (4.2.1) and the analysis in §3.3 that if these factors are used to compute a solution to $Ax = b$, then the computed solution $\hat{x}$ satisfies $(A + E)\hat{x} = b$ with

$$\| E \|_F \leq \mathbf{u} \left(2n\| A \|_F + 4cn^2 \left(\| T \|_2 + \| ST^{-1}S \|_2 \right) \right) + O(\mathbf{u}^2). \tag{4.2.7}$$

It is easy to show that $\| T \|_2 \leq \| A \|_2$, and so it follows that if

$$\Omega = \frac{\| ST^{-1}S \|_2}{\| A \|_2} \tag{4.2.8}$$

is not too large, then it is safe not to pivot. In other words, the norm of the skew-symmetric part S has to be modest relative to the condition of the symmetric part T. Sometimes it is possible to estimate Ω in an application. This is trivially the case when A is symmetric for then $\Omega = 0$.

4.2.3   Symmetric Positive Definite Systems

If we apply the above results to a symmetric positive definite matrix we know that the factorization $A = LU$ exists and is stable to compute. The computation of the factorization $A = LDL^T$ via Algorithm 4.1.2 is also stable and exploits symmetry. However, for symmetric positive definite systems it is often handier to work with a variation of LDL^T.

Theorem 4.2.7 (Cholesky Factorization). *If $A \in \mathbb{R}^{n \times n}$ is symmetric positive definite, then there exists a unique lower triangular $G \in \mathbb{R}^{n \times n}$ with positive diagonal entries such that $A = GG^T$.*

Proof. From Theorem 4.1.3, there exists a unit lower triangular L and a diagonal

$$D = \text{diag}(d_1, \ldots, d_n)$$

such that $A = LDL^T$. Theorem 4.2.1 tells us that $L^{-1}AL^{-T} = D$ is positive definite. Thus, the d_k are positive and the matrix $G = L \, \text{diag}(\sqrt{d_1}, \ldots, \sqrt{d_n})$ is real and lower triangular with positive diagonal entries. It also satisfies $A = GG^T$. Uniqueness follows from the uniqueness of the LDL^T factorization. $\square$

The factorization $A = GG^T$ is known as the *Cholesky factorization* and G is the *Cholesky factor*. Note that if we compute the Cholesky factorization and solve the triangular systems $Gy = b$ and $G^Tx = y$, then $b = Gy = G(G^Tx) = (GG^T)x = Ax$.

4.2.4   The Cholesky Factor is not a Square Root

A matrix $X \in \mathbb{R}^{n \times n}$ that satisfies $A = X^2$ is a *square root of A*. Note that if A symmetric, positive definite, and not diagonal, then its Cholesky factor is *not* a square root. However, if $A = GG^T$ and $X = U\Sigma U^T$ where $G = U\Sigma V^T$ is the SVD, then

$$X^2 = (U\Sigma U^T)(U\Sigma U^T) = U\Sigma^2 U^T = (U\Sigma V^T)(U\Sigma V^T)^T = GG^T = A.$$

Thus, a symmetric positive definite matrix A has a symmetric positive definite square root denoted by $A^{1/2}$. We have more to say about matrix square roots in §9.4.2.

4.2.5   A Gaxpy-Rich Cholesky Factorization

Our proof of the Cholesky factorization in Theorem 4.2.7 is constructive. However, we can develop a more effective procedure by comparing columns in $A = GG^T$. If $A \in \mathbb{R}^{n \times n}$ and $1 \le j \le n$, then

$$A(:,j) = \sum_{k=1}^{j} G(j,k) \cdot G(:,k).$$

This says that

$$G(j,j)G(:,j) \;=\; A(:,j) - \sum_{k=1}^{j-1} G(j,k) \cdot G(:,k) \;\equiv\; v. \qquad (4.2.9)$$

If the first $j-1$ columns of G are known, then v is computable. It follows by equating components in (4.2.9) that

$$G(j{:}n,j) = v(j{:}n)/\sqrt{v(j)}$$

and so we obtain

> **for** $j = 1{:}n$
> $\quad v(j{:}n) = A(j{:}n, j)$
> $\quad$ **for** $k = 1{:}j-1$
> $\quad\quad v(j{:}n) = v(j{:}n) - G(j,k) \cdot G(j{:}n, k)$
> $\quad$ **end**
> $\quad G(j{:}n, j) = v(j{:}n)/\sqrt{v(j)}$
> **end**

It is possible to arrange the computations so that G overwrites the lower triangle of A.

Algorithm 4.2.1 (Gaxpy Cholesky) Given a symmetric positive definite $A \in \mathbb{R}^{n \times n}$, the following algorithm computes a lower triangular G such that $A = GG^T$. For all $i \ge j$, $G(i,j)$ overwrites $A(i,j)$.

> **for** $j = 1{:}n$
> $\quad$ **if** $j > 1$
> $\quad\quad A(j{:}n,j) = A(j{:}n,j) - A(j{:}n, 1{:}j-1) \cdot A(j, 1{:}j-1)^T$
> $\quad$ **end**
> $\quad A(j{:}n,j) = A(j{:}n,j)/\sqrt{A(j,j)}$
> **end**

This algorithm requires $n^3/3$ flops.

4.2.6   Stability of the Cholesky Process

In exact arithmetic, we know that a symmetric positive definite matrix has a Cholesky factorization. Conversely, if the Cholesky process runs to completion with strictly positive square roots, then A is positive definite. Thus, to find out if a matrix A is positive definite, we merely try to compute its Cholesky factorization using any of the methods given above.

The situation in the context of roundoff error is more interesting. The numerical stability of the Cholesky algorithm roughly follows from the inequality

$$g_{ij}^2 \leq \sum_{k=1}^{i} g_{ik}^2 = a_{ii}.$$

This shows that the entries in the Cholesky triangle are nicely bounded. The same conclusion can be reached from the equation $\| G \|_2^2 = \| A \|_2$.

The roundoff errors associated with the Cholesky factorization have been extensively studied in a classical paper by Wilkinson (1968). Using the results in this paper, it can be shown that if $\hat{x}$ is the computed solution to $Ax = b$, obtained via the Cholesky process, then $\hat{x}$ solves the perturbed system

$$(A + E)\hat{x} = b \qquad \| E \|_2 \leq c_n \mathbf{u} \| A \|_2,$$

where c_n is a small constant that depends upon n. Moreover, Wilkinson shows that if $q_n \mathbf{u} \kappa_2(A) \leq 1$ where q_n is another small constant, then the Cholesky process runs to completion, i.e., no square roots of negative numbers arise.

It is important to remember that symmetric positive definite linear systems can be ill-conditioned. Indeed, the eigenvalues and singular values of a symmetric positive definite matrix are the same. This follows from (2.4.1) and Theorem 4.2.3. Thus,

$$\kappa_2(A) = \frac{\lambda_{\max}(A)}{\lambda_{\min}(A)}.$$

The eigenvalue $\lambda_{\min}(A)$ is the "distance to trouble" in the Cholesky setting. This prompts us to consider a permutation strategy that steers us away from using small diagonal elements that jeopardize the factorization process.

4.2.7 The LDL^T Factorization with Symmetric Pivoting

With an eye towards handling ill-conditioned symmetric positive definite systems, we return to the LDL^T factorization and develop an outer product implementation with pivoting. We first observe that if A is symmetric and P_1 is a permutation, then $P_1 A$ is *not* symmetric. On the other hand, $P_1 A P_1^T$ is symmetric suggesting that we consider the following factorization:

$$P_1 A P_1^T = \begin{bmatrix} \alpha & v^T \\ v & B \end{bmatrix} = \begin{bmatrix} 1 & 0 \\ v/\alpha & I_{n-1} \end{bmatrix} \begin{bmatrix} \alpha & 0 \\ 0 & \tilde{A} \end{bmatrix} \begin{bmatrix} 1 & 0 \\ v/\alpha & I_{n-1} \end{bmatrix}^T$$

where

$$\tilde{A} = B - \frac{1}{\alpha} v v^T.$$

Note that with this kind of *symmetric pivoting*, the new (1,1) entry α is some diagonal entry a_{ii}. Our plan is to choose P_1 so that α is the largest of A's diagonal entries. If we apply the same strategy recursively to $\tilde{A}$ and compute

$$\tilde{P} \tilde{A} \tilde{P}^T = \tilde{L} \tilde{D} \tilde{L}^T,$$

then we emerge with the factorization

$$PAP^T = LDL^T \tag{4.2.10}$$

where

$$P = \begin{bmatrix} 1 & 0 \\ 0 & \tilde{P} \end{bmatrix} P_1, \qquad L = \begin{bmatrix} 1 & 0 \\ v/\alpha & \tilde{L} \end{bmatrix}, \qquad D = \begin{bmatrix} \alpha & 0 \\ 0 & \tilde{D} \end{bmatrix}.$$

By virtue of this pivot strategy,

$$d_1 \geq d_2 \geq \cdots \geq d_n > 0.$$

Here is a nonrecursive implementation of the overall algorithm:

Algorithm 4.2.2 (Outer Product LDL^T with Pivoting) Given a symmetric positive semidefinite $A \in \mathbb{R}^{n \times n}$, the following algorithm computes a permutation P, a unit lower triangular L, and a diagonal matrix $D = \text{diag}(d_1, \ldots, d_n)$ so $PAP^T = LDL^T$ with $d_1 \geq d_2 \geq \cdots \geq d_n > 0$. The matrix element a_{ij} is overwritten by d_i if $i = j$ and by ℓ_{ij} if $i > j$. $P = P_1 \cdots P_n$ where P_k is the identity with rows k and $piv(k)$ interchanged.

> **for** $k = 1{:}n$
> $\qquad piv(k) = j$ where $a_{jj} = \max\{a_{kk}, \ldots, a_{nn}\}$
> $\qquad A(k,:) \leftrightarrow A(j,:)$
> $\qquad A(:,k) \leftrightarrow A(:,j)$
> $\qquad \alpha = A(k,k)$
> $\qquad v = A(k+1{:}n, k)$
> $\qquad A(k+1{:}n, k) = v/\alpha$
> $\qquad A(k+1{:}n, k+1{:}n) = A(k+1{:}n, k+1{:}n) - vv^T/\alpha$
> **end**

If symmetry is exploited in the outer product update, then $n^3/3$ flops are required. To solve $Ax = b$ given $PAP^T = LDL^T$, we proceed as follows:

$$Lw = Pb, \qquad Dy = w, \qquad L^T z = y, \qquad x = P^T z.$$

We mention that Algorithm 4.2.2 can be implemented in a way that only references the lower trianglar part of A.

It is reasonable to ask why we even bother with the LDL^T factorization given that it appears to offer no real advantage over the Cholesky factorization. There are two reasons. First, it is more efficient in narrow band situations because it avoids square roots; see §4.3.6. Second, it is a graceful way to introduce factorizations of the form

$$PAP^T = \left(\begin{array}{c} \text{lower} \\ \text{triangular} \end{array} \right) \times \left(\begin{array}{c} \text{simple} \\ \text{matrix} \end{array} \right) \times \left(\begin{array}{c} \text{lower} \\ \text{triangular} \end{array} \right)^T,$$

where P is a permutation arising from a symmetry-exploiting pivot strategy. The symmetric indefinite factorizations that we develop in §4.4 fall under this heading as does the "rank revealing" factorization that we are about to discuss for semidefinite problems.

4.2.8   The Symmetric Semidefinite Case

A symmetric matrix $A \in \mathbb{R}^{n \times n}$ is *positive semidefinite* if

$$x^T A x \geq 0$$

for every $x \in \mathbb{R}^n$. It is easy to show that if $A \in \mathbb{R}^{n \times n}$ is symmetric and positive semidefinite, then its eigenvalues satisfy

$$0 = \lambda_n(A) = \cdots = \lambda_{r+1}(A) < \lambda_r(A) \leq \cdots \leq \lambda_1(A) \qquad (4.2.11)$$

where r is the rank of A. Our goal is to show that Algorithm 4.2.2 can be used to estimate r and produce a streamlined version of (4.2.10). But first we establish some useful properties.

Theorem 4.2.8. *If $A \in \mathbb{R}^{n \times n}$ is symmetric positive semidefinite, then*

$$|a_{ij}| \leq (a_{ii} + a_{jj})/2, \qquad (4.2.12)$$

$$|a_{ij}| \leq \sqrt{a_{ii} a_{jj}}, \qquad (i \neq j), \qquad (4.2.13)$$

$$\max |a_{ij}| = \max a_{ii}, \qquad (4.2.14)$$

$$a_{ii} = 0 \Rightarrow A(i,:) = 0, \ A(:,i) = 0. \qquad (4.2.15)$$

Proof. Let e_i denote the ith column of I_n. Since

$$x = e_i + e_j \ \Rightarrow \ 0 \leq x^T A x = a_{ii} + 2a_{ij} + a_{jj},$$

$$x = e_i - e_j \ \Rightarrow \ 0 \leq x^T A x = a_{ii} - 2a_{ij} + a_{jj},$$

it follows that

$$-2a_{ij} \leq a_{ii} + a_{jj},$$

$$2a_{ij} \leq a_{ii} + a_{jj}.$$

These two equations confirm (4.2.12), which in turn implies (4.2.14).
    To prove (4.2.13), set $x = \tau e_i + e_j$ where $\tau \in \mathbb{R}$. It follows that

$$0 < x^T A x = a_{ii}\tau^2 + 2a_{ij}\tau + a_{jj}$$

must hold for all τ. This is a quadratic equation in τ and for the inequality to hold, the discriminant $4a_{ij}^2 - 4a_{ii}a_{jj}$ must be negative, i.e., $|a_{ij}| \leq \sqrt{a_{ii}a_{jj}}$. The implication in (4.2.15) follows immediately from (4.2.13). $\square$

Let us examine what happens when Algorithm 4.2.2 is applied to a rank-r positive semidefinite matrix. If $k \leq r$, then after k steps we have the factorization

$$\tilde{P}A\tilde{P}^T - \begin{bmatrix} L_{11} & 0 \\ L_{21} & I_{n-k} \end{bmatrix} \begin{bmatrix} D_k & 0 \\ 0 & A_k \end{bmatrix} \begin{bmatrix} L_{11}^T & L_{21}^T \\ 0 & I_{n-k} \end{bmatrix} \qquad (4.2.16)$$

where $D_k = \text{diag}(d_1, \ldots, d_k) \in \mathbb{R}^{k \times k}$ and $d_1 \geq \cdots \geq d_k \geq 0$. By virtue of the pivot strategy, if $d_k = 0$, then A_k has a zero diagonal. Since A_k is positive semidefinite, it follows from (4.2.15) that $A_k = 0$. This contradicts the assumption that A has rank r *unless* $k = r$. Thus, if $k \leq r$, then $d_k > 0$. Moreover, we must have $A_r = 0$ since A has the same rank as $\text{diag}(D_r, A_r)$. It follows from (4.2.16) that

$$PAP^T = \begin{bmatrix} L_{11} \\ L_{21} \end{bmatrix} D_r \begin{bmatrix} L_{11}^T \mid L_{21}^T \end{bmatrix} \qquad (4.2.17)$$

where $D_r = \text{diag}(d_1, \ldots, d_r)$ has positive diagonal entries, $L_{11} \in \mathbb{R}^{r \times r}$ is unit lower triangular, and $L_{21} \in \mathbb{R}^{(n-r) \times r}$. If ℓ_j is the jth column of the L-matrix, then we can rewrite (4.2.17) as a sum of rank-1 matrices:

$$PAP^T = \sum_{j=1}^{r} d_j \ell_j \ell_j^T.$$

This can be regarded as a relatively cheap alternative to the SVD rank-1 expansion.

It is important to note that our entire semidefinite discussion has been an exact arithmetic discussion. In practice, a threshold tolerance for small diagonal entries has to be built into Algorithm 4.2.2. If the diagonal of the computed A_k in (4.2.16) is sufficiently small, then the loop can be terminated and $\tilde{r}$ can be regarded as the numerical rank of A. For more details, see Higham (1989).

4.2.9   Block Cholesky

Just as there are block methods for computing the LU factorization, so are there are block methods for computing the Cholesky factorization. Paralleling the derivation of the block LU algorithm in §3.2.11, we start by blocking $A = GG^T$ as follows

$$\begin{bmatrix} A_{11} & A_{21}^T \\ A_{21} & A_{22} \end{bmatrix} = \begin{bmatrix} G_{11} & 0 \\ G_{21} & G_{22} \end{bmatrix} \begin{bmatrix} G_{11} & 0 \\ G_{21} & G_{22} \end{bmatrix}^T. \qquad (4.2.18)$$

Here, $A_{11} \in \mathbb{R}^{r \times r}$, $A_{22} \in \mathbb{R}^{(n-r) \times (n-r)}$, r is a blocking parameter, and G is partitioned conformably. Comparing blocks in (4.2.18) we conclude that

$$A_{11} = G_{11} G_{11}^T,$$
$$A_{21} = G_{21} G_{11}^T,$$
$$A_{22} = G_{21} G_{21}^T + G_{22} G_{22}^T,$$

which suggests the following 3-step procedure:

Step 1: Compute the Cholesky factorization of A_{11} to get G_{11}.

Step 2: Solve a lower triangular multiple-right-hand-side system for G_{21}.

Step 3: Compute the Cholesky factor G_{22} of $A_{22} - G_{21} G_{21}^T = A_{22} - A_{21} A_{11}^{-1} A_{21}^T$.

In recursive form we obtain the following algorithm.

Algorithm 4.2.3 (Recursive Block Cholesky) Suppose $A \in \mathbb{R}^{n \times n}$ is symmetric positive definite and r is a positive integer. The following algorithm computes a lower triangular $G \in \mathbb{R}^{n \times n}$ so $A = GG^T$.

> **function** $G = \mathsf{BlockCholesky}(A, n, r)$
>> **if** $n \leq r$
>>> Compute the Cholesky factorization $A = GG^T$.
>>
>> **else**
>>> Compute the Cholesky factorization $A(1{:}r, 1{:}r) = G_{11}G_{11}^T$.
>>>
>>> Solve $G_{21}G_{11}^T = A(r+1{:}n, 1{:}r)$ for G_{21}.
>>>
>>> $\tilde{A} = A(r+1{:}n, r+1{:}n) - G_{21}G_{21}^T$
>>>
>>> $G_{22} = \mathsf{BlockCholesky}(\tilde{A}, n - r, r)$
>>>
>>> $G = \begin{bmatrix} G_{11} & 0 \\ G_{21} & G_{22} \end{bmatrix}$
>>
>> **end**
>
> **end**

If symmetry is exploited in the computation of $\tilde{A}$, then this algorithm requires $n^3/3$ flops. A careful accounting of flops reveals that the level-3 fraction is about $1 - 1/N^2$ where $N \approx n/r$. The "small" Cholesky computation for G_{11} and the "thin" solution process for G_{21} are dominated by the "large" level-3 update for $\tilde{A}$.

To develop a nonrecursive implementation, we assume for clarity that $n = Nr$ where N is a positive integer and consider the partitioning

$$\begin{bmatrix} A_{11} & \cdots & A_{1N} \\ \vdots & \ddots & \vdots \\ A_{N1} & \cdots & A_{NN} \end{bmatrix} = \begin{bmatrix} G_{11} & \cdots & 0 \\ \vdots & \ddots & \vdots \\ G_{N1} & \cdots & G_{NN} \end{bmatrix} \begin{bmatrix} G_{11} & \cdots & 0 \\ \vdots & \ddots & \vdots \\ G_{N1} & \cdots & G_{NN} \end{bmatrix}^T \quad (4.2.19)$$

where all blocks are r-by-r. By equating (i, j) blocks with $i \geq j$ it follows that

$$A_{ij} = \sum_{k=1}^{j} G_{ik}G_{jk}^T.$$

Define

$$S = A_{ij} - \sum_{k=1}^{j-1} G_{ik}G_{jk}^T = A_{ij} - [\, G_{i1} \mid \cdots \mid G_{i,j-1} \,] \begin{bmatrix} G_{j1}^T \\ \vdots \\ G_{j,j-1}^T \end{bmatrix}.$$

If $i = j$, then G_{jj} is the Cholesky factor of S. If $i > j$, then $G_{ij}G_{jj}^T = S$ and G_{ij} is the solution to a triangular multiple right hand side problem. Properly sequenced, these equations can be arranged to compute all the G-blocks.

Algorithm 4.2.4 (Nonrecursive Block Cholesky) Given a symmetric positive definite $A \in \mathbb{R}^{n \times n}$ with $n = Nr$ with blocking (4.2.19), the following algorithm computes a lower triangular $G \in \mathbb{R}^{n \times n}$ such that $A = GG^T$. The lower triangular part of A is overwritten by the lower triangular part of G.

> **for** $j = 1:N$
> > **for** $i = j:N$
> >
> > > Compute $S = A_{ij} - \sum_{k=1}^{j-1} G_{ik} G_{jk}^T.$
> > >
> > > **if** $i = j$
> > > > Compute Cholesky factorization $S = G_{jj} G_{jj}^T.$
> > >
> > > **else**
> > > > Solve $G_{ij} G_{jj}^T = S$ for G_{ij}.
> > >
> > > **end**
> > > $A_{ij} = G_{ij}.$
> >
> > **end**
>
> **end**

The overall process involves $n^3/3$ flops like the other Cholesky procedures that we have developed. The algorithm is rich in matrix multiplication with a level-3 fraction given by $1 - (1/N^2)$. The algorithm can be easily modified to handle the case when r does not divide n.

4.2.10 Recursive Blocking

It is instructive to look a little more deeply into the implementation of a block Cholesky factorization as it is an occasion to stress the importance of designing data structures that are tailored to the problem at hand. High-performance matrix computations are filled with tensions and tradeoffs. For example, a successful pivot strategy might balance concerns about stability and memory traffic. Another tension is between performance and memory constraints. As an example of this, we consider how to achieve level-3 performance in a Cholesky implementation given that the matrix is represented in *packed format*. This data structure houses the lower (or upper) triangular portion of a matrix $A \in \mathbb{R}^{n \times n}$ in a vector of length $N = n(n+1)/2$. The symvec arrangement stacks the lower triangular subcolumns, e.g.,

$$\text{symvec}(A) = [\, a_{11} \; a_{21} \; a_{31} \; a_{41} \; a_{22} \; a_{32} \; a_{42} \; a_{33} \; a_{43} \; a_{44} \,]^T. \tag{4.2.20}$$

This layout is not very friendly when it comes to block Cholesky calculations because the assembly of an A-block (say $A(i_1:i_2, j_1:j_2)$) involves irregular memory access patterns. To realize a high-performance matrix multiplication it is usually necessary to have the matrices laid out conventionally as full rectangular arrays that are contiguous in memory, e.g.,

$$\text{vec}(A) = [\, a_{11} \; a_{21} \; a_{31} \; a_{41} \; a_{12} \; a_{22} \; a_{32} \; a_{42} \; a_{13} \; a_{23} \; a_{33} \; a_{43} \; a_{14} \; a_{24} \; a_{34} \; a_{44} \,]^T. \tag{4.2.21}$$

(Recall that we introduced the vec operation in §1.3.7.) Thus, the challenge is to develop a high performance block algorithm that overwrites a symmetric positive definite A in packed format with its Cholesky factor G in packed format. Toward that end, we present the main ideas behind a *recursive* data structure that supports level-3 computation and is storage efficient. As memory hierarchies get deeper and more complex, recursive data structures are an interesting way to address the problem of blocking for performance.

The starting point is once again a 2-by-2 blocking of the equation $A = GG^T$:

$$
\begin{bmatrix} A_{11} & A_{12} \\ A_{21} & A_{22} \end{bmatrix} = \begin{bmatrix} G_{11} & 0 \\ G_{21} & G_{22} \end{bmatrix} \begin{bmatrix} G_{11} & 0 \\ G_{21} & G_{22} \end{bmatrix}^T .
$$

However, unlike in (4.2.18) where A_{11} has a chosen block size, we now assume that $A_{11} \in \mathbb{R}^{m \times m}$ where $m = \text{ceil}(n/2)$. In other words, the four blocks are roughly the same size. As before, we equate entries and identify the key subcomputations:

$G_{11}G_{11}^T = A_{11}$	half-sized Cholesky.
$G_{21}G_{11}^T = A_{21}$	multiple-right-hand-side triangular solve.
$\tilde{A}_{22} = A_{22} - G_{21}G_{21}^T$	symmetric matrix multiplication update.
$G_{22}G_{22}^T = \tilde{A}_{22}$	half-sized Cholesky.

Our goal is to develop a symmetry-exploiting, level-3-rich procedure that overwrites A with its Cholesky factor G. To do this we introduce the *mixed packed format*. An $n = 9$ example with $A_{11} \in \mathbb{R}^{5 \times 5}$ serves to distinguish this layout from the conventional packed format layout:

1				
2	10			
3	11	18		
4	12	19	25	
5	13	20	26	31
6	14	21	27	32	36
7	15	22	28	33	37	40
8	16	23	29	34	38	41	43
9	17	24	30	35	39	42	44	45

Packed format

1				
2	6			
3	7	10		
4	8	11	13	
5	9	12	14	15
16	20	24	28	32	36
17	21	25	29	33	37	40
18	22	26	30	34	38	41	43
19	23	27	31	35	39	42	44	45

Mixed packed format

Notice how the entries from A_{11} and A_{21} are shuffled with the conventional packed format layout. On the other hand, with the mixed packed format layout, the 15 entries that define A_{11} are followed by the 20 numbers that define A_{21} which in turn are followed by the 10 numbers that define A_{22}. The process can be repeated on A_{11} and

A_{22}:

1							
2	4						
3	5	6					
7	9	11	13				
8	10	12	14	15			
16	20	24	28	32	36		
17	21	25	29	33	37	38	
18	22	26	30	34	39	41	43
19	23	27	31	35	40	42	44	45

Thus, the key to this recursively defined data layout is the idea of representing square diagonal blocks in a *mixed packed format*. To be precise, recall the definition of vec and symvec in (4.2.20) and (4.2.21). If $C \in \mathbb{R}^{q \times q}$ is such a block, then

$$
\text{mixvec}(C) = \begin{bmatrix} \text{symvec}(C_{11}) \\ \text{vec}(C_{21}) \\ \text{symvec}(C_{22}) \end{bmatrix} \tag{4.2.22}
$$

where $m = \text{ceil}(q/2)$, $C_{11} = C(1{:}m, 1{:}m)$, $C_{22} = C(m+1{:}n, m+1{:}n)$, and $C_{21} = C(m+1{:}n, 1{:}m)$. Notice that since C_{21} is conventionally stored, it is ready to be engaged in a high-performance matrix multiplication.

We now outline a recursive, divide-and-conquer block Cholesky procedure that works with A in packed format. To achieve high performance the incoming A is converted to mixed format at each level of the recursion. Assuming the existence of a triangular system solve procedure TriSol (for the system $G_{21}G_{11}^T = A_{21}$) and a symmetric update procedure SymUpdate (for $A_{22} \leftarrow A_{22} - G_{21}G_{21}^T$) we have the following framework:

function $G = $ PackedBlockCholesky(A)

{A and G in packed format}

$n = \text{size}(A)$

if $n \le n_{\min}$

    G is obtained via any level-2, packed-format Cholesky method .

else

    Set $m = \text{ceil}(n/2)$ and overwrite A's packed-format representation
        with its mixed-format representation.

    $G_{11} = $ PackedBlockCholesky(A_{11})

    $G_{21} = $ TriSol(G_{11}, A_{21})

    $A_{22} = $ SymUpdate(A_{22}, G_{21})

    $G_{22} = $ PackedBlockCholesky(A_{22})

end

Here, $n_{\min}$ is a threshold dimension below which it is not possible to achieve level-3 performance. To take full advantage of the mixed format, the procedures TriSol and SymUpdate require a recursive design based on blockings that halve problem size. For example, TriSol should take the incoming packed format A_{11}, convert it to mixed format, and solve a 2-by-2 blocked system of the form

$$
\left[\begin{array}{c|c} X_1 & X_2 \end{array}\right] \left[\begin{array}{cc} L_{11} & 0 \\ L_{21} & L_{22} \end{array}\right]^T = \left[\begin{array}{c|c} B_1 & B_2 \end{array}\right].
$$

This sets up a recursive solution based on the half-sized problems

$$
X_1 L_{11}^T = B_1,
$$
$$
X_2 L_{22}^T = B_2 - X_1 L_{21}^T.
$$

Likewise, SymUpdate should take the incoming packed format A_{22}, convert it to mixed format, and block the required update as follows:

$$
\left[\begin{array}{cc} C_{11} & C_{21}^T \\ C_{21} & C_{22} \end{array}\right] = \left[\begin{array}{cc} C_{11} & C_{21}^T \\ C_{21} & C_{22} \end{array}\right] - \left[\begin{array}{c} Y_1 \\ Y_2 \end{array}\right] \left[\begin{array}{c} Y_1 \\ Y_2 \end{array}\right]^T.
$$

The evaluation is recursive and based on the half-sized updates

$$
C_{11} = C_{11} - Y_1 Y_1^T,
$$
$$
C_{21} = C_{21} - Y_2 Y_1^T,
$$
$$
C_{22} = C_{22} - Y_2 Y_2^T.
$$

Of course, if the incoming matrices are small enough relative to $n_{\min}$, then TriSol and SymUpdate carry out their tasks conventionally without any further subdivisions.

Overall, it can be shown that PackedBlockCholesky has a level-3 fraction approximately equal to $1 - O(n_{\min}/n)$.

Problems

P4.2.1 Suppose that $H = A + iB$ is Hermitian and positive definite with $A, B \in \mathbb{R}^{n \times n}$. This means that $x^H H x > 0$ whenever $x \neq 0$. (a) Show that

$$
C = \left[\begin{array}{cc} A & -B \\ B & A \end{array}\right]
$$

is symmetric and positive definite. (b) Formulate an algorithm for solving $(A + iB)(x + iy) = (b + ic)$, where b, c, x, and y are in $\mathbb{R}^n$. It should involve $8n^3/3$ flops. How much storage is required?

P4.2.2 Suppose $A \in \mathbb{R}^{n \times n}$ is symmetric and positive definite. Give an algorithm for computing an upper triangular matrix $R \in \mathbb{R}^{n \times n}$ such that $A = RR^T$.

P4.2.3 Let $A \in \mathbb{R}^{n \times n}$ be positive definite and set $T = (A + A^T)/2$ and $S = (A - A^T)/2$. (a) Show that $\| A^{-1} \|_2 \leq \| T^{-1} \|_2$ and $x^T A^{-1} x \leq x^T T^{-1} x$ for all $x \in \mathbb{R}^n$. (b) Show that if $A = LDM^T$, then $d_k \geq 1/\| T^{-1} \|_2$ for $k = 1{:}n$.

P4.2.4 Find a 2-by-2 real matrix A with the property that $x^T A x > 0$ for all real nonzero 2-vectors but which is not positive definite when regarded as a member of $\mathbb{C}^{2 \times 2}$.

P4.2.5 Suppose $A \in \mathbb{R}^{n \times n}$ has a positive diagonal. Show that if both A and A^T are strictly diagonally dominant, then A is positive definite.

P4.2.6 Show that the function $f(x) = \sqrt{x^T A x}/2$ is a vector norm on $\mathbb{R}^n$ if and only if A is positive definite.

P4.2.7 Modify Algorithm 4.2.1 so that if the square root of a negative number is encountered, then the algorithm finds a unit vector x so that $x^T A x < 0$ and terminates.

P4.2.8 Develop an outer product implementation of Algorithm 4.2.1 and a gaxpy implementation of Algorithm 4.2.2.

P4.2.9 Assume that $A \in \mathbb{C}^{n \times n}$ is Hermitian and positive definite. Show that if $a_{11} = \cdots = a_{nn} = 1$ and $|a_{ij}| < 1$ for all $i \neq j$, then $\mathrm{diag}(A^{-1}) \geq \mathrm{diag}((\mathrm{Re}(A))^{-1})$.

P4.2.10 Suppose $A = I + uu^T$ where $A \in \mathbb{R}^{n \times n}$ and $\| u \|_2 = 1$. Give explicit formulae for the diagonal and subdiagonal of A's Cholesky factor.

P4.2.11 Suppose $A \in \mathbb{R}^{n \times n}$ is symmetric positive definite and that its Cholesky factor is available. Let $e_k = I_n(:, k)$. For $1 \leq i < j \leq n$, let α_{ij} be the smallest real that makes $A + \alpha(e_i e_j^T + e_j e_i^T)$ singular. Likewise, let α_{ii} be the smallest real that makes $(A + \alpha e_i e_i^T)$ singular. Show how to compute these quantities using the Sherman-Morrison-Woodbury formula. How many flops are required to find all the α_{ij}?

P4.2.12 Show that if

$$M = \begin{bmatrix} A & B \\ B^T & C \end{bmatrix}$$

is symmetric positive definite and A and C are square, then

$$M^{-1} = \begin{bmatrix} A^{-1} + A^{-1} B S^{-1} B^T A^{-1} & -A^{-1} B S^{-1} \\ S^{-1} B^T A^{-1} & S^{-1} \end{bmatrix}, \qquad S = C - B^T A^{-1} B.$$

P4.2.13 Suppose $\sigma \in \mathbb{R}$ and $u \in \mathbb{R}^n$. Under what conditions can we find a matrix $X \in \mathbb{R}^{n \times n}$ so that $X(I + \sigma uu^T)X = I_n$? Give an efficient algorithm for computing X if it exists.

P4.2.14 Suppose $D = \mathrm{diag}(d_1, \ldots, d_n)$ with $d_i > 0$ for all i. Give an efficient algorithm for computing the largest entry in the matrix $(D + CC^T)^{-1}$ where $C \in \mathbb{R}^{n \times r}$. Hint: Use the Sherman-Morrison-Woodbury formula.

P4.2.15 Suppose $A(\lambda)$ has continuously differentiable entries and is always symmetric and positive definite. If $f(\lambda) = \log(\det(A(\lambda)))$, then how would you compute $f'(0)$?

P4.2.16 Suppose $A \in \mathbb{R}^{n \times n}$ is a rank-r symmetric positive semidefinite matrix. Assume that it costs one dollar to evaluate each a_{ij}. Show how to compute the factorization (4.2.17) spending only $O(nr)$ dollars on a_{ij} evaluation.

P4.2.17 The point of this problem is to show that from the complexity point of view, if you have a fast matrix multiplication algorithm, then you have an equally fast matrix inversion algorithm, and vice versa. (a) Suppose F_n is the number of flops required by some method to form the inverse of an n-by-n matrix. Assume that there exists a constant c_1 and a real number α such that $F_n \leq c_1 n^\alpha$ for all n. Show that there is a method that can compute the n-by-n matrix product AB with fewer than $c_2 n^\alpha$ flops where c_2 is a constant independent of n. Hint: Consider the inverse of

$$C = \begin{bmatrix} I_n & A & 0 \\ 0 & I_n & B \\ 0 & 0 & I_n \end{bmatrix}.$$

(b) Let G_n be the number of flops required by some method to form the n-by-n matrix product AB. Assume that there exists a constant c_1 and a real number α such that $G_n \leq c_1 n^\alpha$ for all n. Show that there is a method that can invert a nonsingular n-by-n matrix A with fewer than $c_2 n^\alpha$ flops where c_2 is a constant. Hint: First show that the result applies for triangular matrices by applying recursion to

$$\begin{bmatrix} G_{11} & 0 \\ G_{21} & G_{22} \end{bmatrix}^{-1} = \begin{bmatrix} G_{11}^{-1} & 0 \\ -G_{22}^{-1} G_{21} G_{11}^{-1} & G_{22}^{-1} \end{bmatrix}.$$

Then observe that for general A, $A^{-1} = A^T (AA^T)^{-1} = A^T G^{-T} G^{-1}$ where $AA^T = GG^T$ is the Cholesky factorization.

Notes and References for §4.2

For an in-depth theoretical treatment of positive definiteness, see:

R. Bhatia (2007). *Positive Definite Matrices*, Princeton University Press, Princeton, NJ.

The definiteness of the quadratic form $x^T A x$ can frequently be established by considering the mathematics of the underlying problem. For example, the discretization of certain partial differential operators gives rise to provably positive definite matrices. Aspects of the unsymmetric positive definite problem are discussed in:

A. Buckley (1974). "A Note on Matrices $A = I + H$, H Skew-Symmetric," *Z. Angew. Math. Mech.* *54*, 125–126.
A. Buckley (1977). "On the Solution of Certain Skew-Symmetric Linear Systems," *SIAM J. Numer. Anal. 14*, 566–570.
G.H. Golub and C. Van Loan (1979). "Unsymmetric Positive Definite Linear Systems," *Lin. Alg. Applic. 28*, 85–98.
R. Mathias (1992). "Matrices with Positive Definite Hermitian Part: Inequalities and Linear Systems," *SIAM J. Matrix Anal. Applic. 13*, 640–654.
K.D. Ikramov and A.B. Kucherov (2000). "Bounding the growth factor in Gaussian elimination for Buckley's class of complex symmetric matrices," *Numer. Lin. Alg. 7*, 269–274.

Complex symmetric matrices have the property that their real and imaginary parts are each symmetric. The following paper shows that if they are also positive definite, then the LDL^T factorization is safe to compute without pivoting:

S. Serbin (1980). "On Factoring a Class of Complex Symmetric Matrices Without Pivoting," *Math. Comput. 35*, 1231–1234.

Historically important Algol implementations of the Cholesky factorization include:

R.S. Martin, G. Peters, and J.H. Wilkinson (1965). "Symmetric Decomposition of a Positive Definite Matrix," *Numer. Math. 7*, 362–83.
R.S. Martin, G. Peters, and J.H. Wilkinson (1966). "Iterative Refinement of the Solution of a Positive Definite System of Equations," *Numer. Math. 8*, 203–16.
F.L. Bauer and C. Reinsch (1971). "Inversion of Positive Definite Matrices by the Gauss-Jordan Method," in *Handbook for Automatic Computation Vol. 2, Linear Algebra*, J.H. Wilkinson and C. Reinsch (eds.), Springer-Verlag, New York, 45–49.

For roundoff error analysis of Cholesky, see:

J.H. Wilkinson (1968). "A Priori Error Analysis of Algebraic Processes," *Proceedings of the International Congress on Mathematics*, Izdat. Mir, 1968, Moscow, 629–39.
J. Meinguet (1983). "Refined Error Analyses of Cholesky Factorization," *SIAM J. Numer. Anal. 20*, 1243–1250.
A. Kielbasinski (1987). "A Note on Rounding Error Analysis of Cholesky Factorization," *Lin. Alg. Applic. 88/89*, 487–494.
N.J. Higham (1990). "Analysis of the Cholesky Decomposition of a Semidefinite Matrix," in *Reliable Numerical Computation*, M.G. Cox and S.J. Hammarling (eds.), Oxford University Press, Oxford, U.K., 161–185.
J-Guang Sun (1992). "Rounding Error and Perturbation Bounds for the Cholesky and LDL^T Factorizations," *Lin. Alg. Applic. 173*, 77–97.

The floating point determination of positive definiteness is an interesting problem, see:

S.M. Rump (2006). "Verification of Positive Definiteness," *BIT 46*, 433–452.

The question of how the Cholesky triangle G changes when $A = GG^T$ is perturbed is analyzed in:

G.W. Stewart (1977). "Perturbation Bounds for the QR Factorization of a Matrix," *SIAM J. Num. Anal. 14*, 509–18.
Z. Dramač, M. Omladič, and K. Veselič (1994). "On the Perturbation of the Cholesky Factorization," *SIAM J. Matrix Anal. Applic. 15*, 1319–1332.
X-W. Chang, C.C. Paige, and G.W. Stewart (1996). "New Perturbation Analyses for the Cholesky Factorization," *IMA J. Numer. Anal. 16*, 457–484.

G.W. Stewart (1997) "On the Perturbation of LU and Cholesky Factors," *IMA J. Numer. Anal. 17*, 1–6.

Nearness/sensitivity issues associated with positive semidefiniteness are presented in:

N.J. Higham (1988). "Computing a Nearest Symmetric Positive Semidefinite Matrix," *Lin. Alg. Applic. 103*, 103–118.

The numerical issues associated with semi-definite rank determination are covered in:

P.C. Hansen and P.Y. Yalamov (2001). "Computing Symmetric Rank-Revealing Decompositions via Triangular Factorization," *SIAM J. Matrix Anal. Applic. 23*, 443–458.
M. Gu and L. Miranian (2004). "Strong Rank-Revealing Cholesky Factorization," *ETNA 17*, 76–92.

The issues that surround level-3 performance of packed-format Cholesky are discussed in:

F.G. Gustavson (1997). "Recursion Leads to Automatic Variable Blocking for Dense Linear-Algebra Algorithms," *IBM J. Res. Dev. 41*, 737–756.
F.G. Gustavson, A. Henriksson, I. Jonsson, B. Kågström, , and P. Ling (1998). "Recursive Blocked Data Formats and BLAS's for Dense Linear Algebra Algorithms," *Applied Parallel Computing Large Scale Scientific and Industrial Problems*, Lecture Notes in Computer Science, Springer-Verlag, 1541/1998, 195–206.
F.G. Gustavson and I. Jonsson (2000). "Minimal Storage High-Performance Cholesky Factorization via Blocking and Recursion," *IBM J. Res. Dev. 44*, 823–849.
B.S. Andersen, J. Wasniewski, and F.G. Gustavson (2001). "A Recursive Formulation of Cholesky Factorization of a Matrix in Packed Storage," *ACM Trans. Math. Softw. 27*, 214–244.
E. Elmroth, F. Gustavson, I. Jonsson, and B. Kågström, (2004). "Recursive Blocked Algorithms and Hybrid Data Structures for Dense Matrix Library Software," *SIAM Review 46*, 3–45.
F.G. Gustavson, J. Wasniewski, J.J. Dongarra, and J. Langou (2010). "Rectangular Full Packed Format for Cholesky's Algorithm: Factorization, Solution, and Inversion," *ACM Trans. Math. Softw. 37*, Article 19.

Other high-performance Cholesky implementations include:

F.G. Gustavson, L. Karlsson, and B. Kågström, (2009). "Distributed SBP Cholesky Factorization Algorithms with Near-Optimal Scheduling," *ACM Trans. Math. Softw. 36*, Article 11.
G. Ballard, J. Demmel, O. Holtz, and O. Schwartz (2010). "Communication-Optimal Parallel and Sequential Cholesky," *SIAM J. Sci. Comput. 32*, 3495–3523.
P. Bientinesi, B. Gunter, and R.A. van de Geijn (2008). "Families of Algorithms Related to the Inversion of a Symmetric Positive Definite Matrix," *ACM Trans. Math. Softw. 35*, Article 3.
M.D. Petković and P.S. Stanimirović (2009). "Generalized Matrix Inversion is not Harder than Matrix Multiplication," *J. Comput. Appl. Math. 230*, 270–282.

4.3 Banded Systems

In many applications that involve linear systems, the matrix of coefficients is *banded*. This is the case whenever the equations can be ordered so that each unknown x_i appears in only a few equations in a "neighborhood" of the ith equation. Recall from §1.2.1 that $A = (a_{ij})$ has *upper bandwidth* q if $a_{ij} = 0$ whenever $j > i + q$ and *lower bandwidth* p if $a_{ij} = 0$ whenever $i > j + p$. Substantial economies can be realized when solving banded systems because the triangular factors in LU, GG^T, and LDL^T are also banded.

4.3.1 Band LU Factorization

Our first result shows that if A is banded and $A = LU$, then L inherits the lower bandwidth of A and U inherits the upper bandwidth of A.

Theorem 4.3.1. *Suppose $A \in \mathbb{R}^{n \times n}$ has an LU factorization $A = LU$. If A has upper bandwidth q and lower bandwidth p, then U has upper bandwidth q and L has lower bandwidth p.*

Proof. The proof is by induction on n. Since

$$A = \begin{bmatrix} \alpha & w^T \\ v & B \end{bmatrix} = \begin{bmatrix} 1 & 0 \\ v/\alpha & I_{n-1} \end{bmatrix} \begin{bmatrix} 1 & 0 \\ 0 & B - vw^T/\alpha \end{bmatrix} \begin{bmatrix} \alpha & w^T \\ 0 & I_{n-1} \end{bmatrix}.$$

It is clear that $B - vw^T/\alpha$ has upper bandwidth q and lower bandwidth p because only the first q components of w and the first p components of v are nonzero. Let $L_1 U_1$ be the LU factorization of this matrix. Using the induction hypothesis and the sparsity of w and v, it follows that

$$L = \begin{bmatrix} 1 & 0 \\ v/\alpha & L_1 \end{bmatrix}, \qquad U = \begin{bmatrix} \alpha & w^T \\ 0 & U_1 \end{bmatrix}$$

have the desired bandwidth properties and satisfy $A = LU$. $\quad\square$

The specialization of Gaussian elimination to banded matrices having an LU factorization is straightforward.

Algorithm 4.3.1 (Band Gaussian Elimination) Given $A \in \mathbb{R}^{n \times n}$ with upper bandwidth q and lower bandwidth p, the following algorithm computes the factorization $A = LU$, assuming it exists. $A(i, j)$ is overwritten by $L(i, j)$ if $i > j$ and by $U(i, j)$ otherwise.

> **for** $k = 1{:}n - 1$
> > **for** $i = k + 1{:}\min\{k + p, n\}$
> > > $A(i, k) = A(i, k)/A(k, k)$
> >
> > **end**
> > **for** $j = k + 1{:}\min\{k + q, n\}$
> > > **for** $i = k + 1{:}\min\{k + p, n\}$
> > > > $A(i, j) = A(i, j) - A(i, k) \cdot A(k, j)$
> > >
> > > **end**
> >
> > **end**
>
> **end**

If $n \gg p$ and $n \gg q$, then this algorithm involves about $2npq$ flops. Effective implementations would involve band matrix data structures; see §1.2.5. A band version of Algorithm 4.1.1 (LDL^T) is similar and we leave the details to the reader.

4.3.2 Band Triangular System Solving

Banded triangular system solving is also fast. Here are the banded analogues of Algorithms 3.1.3 and 3.1.4:

Algorithm 4.3.2 (Band Forward Substitution) Let $L \in \mathbb{R}^{n \times n}$ be a unit lower triangular matrix with lower bandwidth p. Given $b \in \mathbb{R}^n$, the following algorithm overwrites b with the solution to $Lx = b$.

> **for** $j = 1{:}n$
> > **for** $i = j + 1{:}\min\{j + p, n\}$
> > > $b(i) = b(i) - L(i,j) \cdot b(j)$
> > **end**
> **end**

If $n \gg p$, then this algorithm requires about $2np$ flops.

Algorithm 4.3.3 (Band Back Substitution) Let $U \in \mathbb{R}^{n \times n}$ be a nonsingular upper triangular matrix with upper bandwidth q. Given $b \in \mathbb{R}^n$, the following algorithm overwrites b with the solution to $Ux = b$.

> **for** $j = n{:}-1{:}1$
> > $b(j) = b(j)/U(j,j)$
> > **for** $i = \max\{1, j - q\}{:}j - 1$
> > > $b(i) = b(i) - U(i,j) \cdot b(j)$
> > **end**
> **end**

If $n \gg q$, then this algorithm requires about $2nq$ flops.

4.3.3 Band Gaussian Elimination with Pivoting

Gaussian elimination with partial pivoting can also be specialized to exploit band structure in A. However, if $PA = LU$, then the band properties of L and U are not quite so simple. For example, if A is tridiagonal and the first two rows are interchanged at the very first step of the algorithm, then u_{13} is nonzero. Consequently, row interchanges expand bandwidth. Precisely how the band enlarges is the subject of the following theorem.

Theorem 4.3.2. *Suppose $A \in \mathbb{R}^{n \times n}$ is nonsingular and has upper and lower bandwidths q and p, respectively. If Gaussian elimination with partial pivoting is used to compute Gauss transformations*

$$M_j = I - \alpha^{(j)} e_j^T \qquad j = 1{:}n - 1$$

and permutations $P_1, \ldots, P_{n-1}$ such that $M_{n-1} P_{n-1} \cdots M_1 P_1 A = U$ is upper triangular, then U has upper bandwidth $p + q$ and $\alpha_i^{(j)} = 0$ whenever $i \le j$ or $i > j + p$.

Proof. Let $PA = LU$ be the factorization computed by Gaussian elimination with partial pivoting and recall that $P = P_{n-1} \cdots P_1$. Write $P^T = [\, e_{s_1} \,|\, \cdots \,|\, e_{s_n} \,]$, where $\{s_1, \ldots, s_n\}$ is a permutation of $\{1, 2, \ldots, n\}$. If $s_i > i + p$ then it follows that the leading i-by-i principal submatrix of PA is singular, since $[PA]_{ij} = a_{s_i,j}$ for $j = 1{:}s_i - p - 1$ and $s_i - p - 1 \ge i$. This implies that U and A are singular, a contradiction. Thus,

$s_i \leq i + p$ for $i = 1{:}n$ and therefore, PA has upper bandwidth $p + q$. It follows from Theorem 4.3.1 that U has upper bandwidth $p + q$. The assertion about the $\alpha^{(j)}$ can be verified by observing that M_j need only zero elements $(j + 1, j), \dots, (j + p, j)$ of the partially reduced matrix $P_j M_{j-1} P_{j-1} \cdots_1 P_1 A$. $\quad \square$

Thus, pivoting destroys band structure in the sense that U becomes "wider" than A's upper triangle, while nothing at all can be said about the bandwidth of L. However, since the jth column of L is a permutation of the jth Gauss vector α_j, it follows that L has at most $p + 1$ nonzero elements per column.

4.3.4 Hessenberg LU

As an example of an unsymmetric band matrix computation, we show how Gaussian elimination with partial pivoting can be applied to factor an upper Hessenberg matrix H. (Recall that if H is upper Hessenberg then $h_{ij} = 0$, $i > j + 1$.) After $k - 1$ steps of Gaussian elimination with partial pivoting we are left with an upper Hessenberg matrix of the form

$$
\begin{bmatrix}
\times & \times & \times & \times & \times \\
0 & \times & \times & \times & \times \\
0 & 0 & \times & \times & \times \\
0 & 0 & \times & \times & \times \\
0 & 0 & 0 & \times & \times
\end{bmatrix}, \qquad k = 3, n = 5.
$$

By virtue of the special structure of this matrix, we see that the next permutation, P_3, is either the identity or the identity with rows 3 and 4 interchanged. Moreover, the next Gauss transformation M_k has a single nonzero multiplier in the $(k + 1, k)$ position. This illustrates the kth step of the following algorithm.

Algorithm 4.3.4 (Hessenberg LU) Given an upper Hessenberg matrix $H \in \mathbb{R}^{n \times n}$, the following algorithm computes the upper triangular matrix $M_{n-1} P_{n-1} \cdots M_1 P_1 H = U$ where each P_k is a permutation and each M_k is a Gauss transformation whose entries are bounded by unity. $H(i, k)$ is overwritten with $U(i, k)$ if $i \leq k$ and by $-[M_k]_{k+1,k}$ if $i = k + 1$. An integer vector $piv(1{:}n - 1)$ encodes the permutations. If $P_k = I$, then $piv(k) = 0$. If P_k interchanges rows k and $k + 1$, then $piv(k) = 1$.

> **for** $k = 1{:}n - 1$
> > **if** $|H(k, k)| < |H(k + 1, k)|$
> > > $piv(k) = 1; \ H(k, k{:}n) \leftrightarrow H(k + 1, k{:}n)$
> >
> > **else**
> > > $piv(k) = 0$
> >
> > **end**
> > **if** $H(k, k) \neq 0$
> > > $\tau = H(k + 1, k)/H(k, k)$
> > >
> > > $H(k + 1, k + 1{:}n) = H(k + 1, k + 1{:}n) - \tau \cdot H(k, k + 1{:}n)$
> > >
> > > $H(k + 1, k) = \tau$
> >
> > **end**
>
> **end**

This algorithm requires n^2 flops.

4.3.5   Band Cholesky

The rest of this section is devoted to banded $Ax = b$ problems where the matrix A is also symmetric positive definite. The fact that pivoting is unnecessary for such matrices leads to some very compact, elegant algorithms. In particular, it follows from Theorem 4.3.1 that if $A = GG^T$ is the Cholesky factorization of A, then G has the same lower bandwidth as A. This leads to the following banded version of Algorithm 4.2.1.

Algorithm 4.3.5 (Band Cholesky)   Given a symmetric positive definite $A \in \mathbb{R}^{n \times n}$ with bandwidth p, the following algorithm computes a lower triangular matrix G with lower bandwidth p such that $A = GG^T$. For all $i \geq j$, $G(i,j)$ overwrites $A(i,j)$.

> **for** $j = 1{:}n$
>> **for** $k = \max(1, j - p){:}j - 1$
>>> $\lambda = \min(k + p, n)$
>>> $A(j{:}\lambda, j) = A(j{:}\lambda, j) - A(j, k){\cdot}A(j{:}\lambda, k)$
>> **end**
>> $\lambda = \min(j + p, n)$
>> $A(j{:}\lambda, j) = A(j{:}\lambda, j)/\sqrt{A(j, j)}$
> **end**

If $n \gg p$, then this algorithm requires about $n(p^2 + 3p)$ flops and n square roots. Of course, in a serious implementation an appropriate data structure for A should be used. For example, if we just store the nonzero lower triangular part, then a $(p + 1)$-by-n array would suffice.

   If our band Cholesky procedure is coupled with appropriate band triangular solve routines, then approximately $np^2 + 7np + 2n$ flops and n square roots are required to solve $Ax = b$. For small p it follows that the square roots represent a significant portion of the computation and it is preferable to use the LDL^T approach. Indeed, a careful flop count of the steps $A = LDL^T$, $Ly = b$, $Dz = y$, and $L^T x = z$ reveals that $np^2 + 8np + n$ flops and no square roots are needed.

4.3.6   Tridiagonal System Solving

As a sample narrow band LDL^T solution procedure, we look at the case of symmetric positive definite tridiagonal systems. Setting

$$
L = \begin{bmatrix}
1 & 0 & \cdots & 0 \\
\ell_1 & 1 & & \vdots \\
\vdots & \ddots & \ddots & 0 \\
0 & \cdots & \ell_{n-1} & 1
\end{bmatrix}
$$

and $D = \mathrm{diag}(d_1, \ldots, d_n)$, we deduce from the equation $A = LDL^T$ that

$$
\begin{aligned}
a_{11} &= d_1, \\
a_{k,k-1} &= \ell_{k-1} d_{k-1}, & k &= 2{:}n, \\
a_{kk} &= d_k + \ell_{k-1}^2 d_{k-1} = d_k + \ell_{k-1} a_{k,k-1}, & k &= 2{:}n.
\end{aligned}
$$

Thus, the d_i and ℓ_i can be resolved as follows:

$$d_1 = a_{11}$$
$$\textbf{for } k = 2{:}n$$
$$\quad \ell_{k-1} = a_{k,k-1}/d_{k-1}$$
$$\quad d_k = a_{kk} - \ell_{k-1}a_{k,k-1}$$
$$\textbf{end}$$

To obtain the solution to $Ax = b$ we solve $Ly = b$, $Dz = y$, and $L^Tx = z$. With overwriting we obtain

Algorithm 4.3.6 (Symmetric, Tridiagonal, Positive Definite System Solver) Given an n-by-n symmetric, tridiagonal, positive definite matrix A and $b \in \mathbb{R}^n$, the following algorithm overwrites b with the solution to $Ax = b$. It is assumed that the diagonal of A is stored in $\alpha(1{:}n)$ and the superdiagonal in $\beta(1{:}n-1)$.

$$\textbf{for } k = 2{:}n$$
$$\quad t = \beta(k-1), \ \beta(k-1) = t/\alpha(k-1), \ \alpha(k) = \alpha(k) - t{\cdot}\beta(k-1)$$
$$\textbf{end}$$
$$\textbf{for } k = 2{:}n$$
$$\quad b(k) = b(k) - \beta(k-1){\cdot}\beta(k-1)$$
$$\textbf{end}$$
$$b(n) = b(n)/\alpha(n)$$
$$\textbf{for } k = n-1{:}-1{:}1$$
$$\quad b(k) = b(k)/\alpha(k) - \beta(k){\cdot}b(k+1)$$
$$\textbf{end}$$

This algorithm requires $8n$ flops.

4.3.7   Vectorization Issues

The tridiagonal example brings up a sore point: narrow band problems and vectorization do not mix. However, it is sometimes the case that large, independent sets of such problems must be solved at the same time. Let us examine how such a computation could be arranged in light of the issues raised in §1.5. For simplicity, assume that we must solve the n-by-n unit lower bidiagonal systems

$$A^{(k)}x^{(k)} = b^{(k)}, \qquad k = 1{:}m,$$

and that $m \gg n$. Suppose we have arrays $E(1{:}n-1, 1{:}m)$ and $B(1{:}n, 1{:}m)$ with the property that $E(1{:}n-1, k)$ houses the subdiagonal of $A^{(k)}$ and $B(1{:}n, k)$ houses the kth right-hand side $b^{(k)}$. We can overwrite $b^{(k)}$ with the solution $x^{(k)}$ as follows:

$$\textbf{for } k = 1{:}m$$
$$\quad \textbf{for } i = 2{:}n$$
$$\quad\quad B(i,k) = B(i,k) - E(i-1,k){\cdot}B(i-1,k)$$
$$\quad \textbf{end}$$
$$\textbf{end}$$

This algorithm sequentially solves each bidiagonal system in turn. Note that the inner loop does not vectorize because of the dependence of $B(i, k)$ on $B(i-1, k)$. However, if we interchange the order of the two loops, then the calculation does vectorize:

> **for** $i = 2{:}n$
> $\qquad B(i, :) = B(i, :) - E(i-1, :) \,.* \, B(i-1, :)$
> **end**

A column-oriented version can be obtained simply by storing the matrix subdiagonals by row in E and the right-hand sides by row in B:

> **for** $i = 2{:}n$
> $\qquad B(:, i) = B(:, i) - E(:, i-1) \,.* \, B(:, i-1)$
> **end**

Upon completion, the transpose of solution $x^{(k)}$ is housed on $B(k, :)$.

4.3.8   The Inverse of a Band Matrix

In general, the inverse of a nonsingular band matrix A is full. However, the off-diagonal blocks of A^{-1} have low rank.

Theorem 4.3.3. *Suppose*

$$A = \begin{bmatrix} A_{11} & A_{12} \\ A_{21} & A_{22} \end{bmatrix}$$

is nonsingular and has lower bandwidth p and upper bandwidth q. Assume that the diagonal blocks are square. If

$$A^{-1} = X = \begin{bmatrix} X_{11} & X_{12} \\ X_{21} & X_{22} \end{bmatrix}$$

is partitioned conformably, then

$$\mathsf{rank}(X_{21}) \le p, \tag{4.3.1}$$

$$\mathsf{rank}(X_{12}) \le q. \tag{4.3.2}$$

Proof. Assume that A_{11} and A_{22} are nonsingular. From the equation $AX = I$ we conclude that

$$A_{21}X_{11} + A_{22}X_{21} = 0,$$
$$A_{11}X_{12} + A_{12}X_{22} = 0,$$

and so

$$\mathsf{rank}(X_{21}) = \mathsf{rank}(A_{22}^{-1}A_{21}X_{11}) \le \mathsf{rank}(A_{21})$$
$$\mathsf{rank}(X_{12}) = \mathsf{rank}(A_{11}^{-1}A_{12}X_{22}) \le \mathsf{rank}(A_{12}).$$

From the bandedness assumptions it follows that A_{21} has at most p nonzero rows and A_{12} has at most q nonzero rows. Thus, $\mathsf{rank}(A_{21}) \leq p$ and $\mathsf{rank}(A_{12}) \leq q$ which proves the theorem for the case when both A_{11} and A_{22} are nonsingular. A simple limit argument can be used to handle the situation when A_{11} and/or A_{22} are singular. See P4.3.11.   □

It can actually be shown that $\mathsf{rank}(A_{21}) = \mathsf{rank}(X_{21})$ and $\mathsf{rank}(A_{12}) = \mathsf{rank}(X_{12})$. See Strang and Nguyen (2004). As we will see in §11.5.9 and §12.2, the low-rank, off-diagonal structure identified by the theorem has important algorithmic ramifications.

4.3.9   Band Matrices with Banded Inverse

If $A \in \mathbb{R}^{n \times n}$ is a product

$$A = F_1 \cdots F_N \tag{4.3.3}$$

and each $F_i \in \mathbb{R}^{n \times n}$ is block diagonal with 1-by-1 and 2-by-2 diagonal blocks, then it follows that *both* A and

$$A^{-1} = F_N^{-1} \cdots F_1^{-1}$$

are banded, assuming that N is not too big. For example, if

$$
A =
\left[\begin{array}{c|cc|cc|cc|cc}
\times & 0 & 0 & 0 & 0 & 0 & 0 & 0 & 0 \\ \hline
0 & \times & \times & 0 & 0 & 0 & 0 & 0 & 0 \\
0 & \times & \times & 0 & 0 & 0 & 0 & 0 & 0 \\ \hline
0 & 0 & 0 & \times & \times & 0 & 0 & 0 & 0 \\
0 & 0 & 0 & \times & \times & 0 & 0 & 0 & 0 \\ \hline
0 & 0 & 0 & 0 & 0 & \times & \times & 0 & 0 \\
0 & 0 & 0 & 0 & 0 & \times & \times & 0 & 0 \\ \hline
0 & 0 & 0 & 0 & 0 & 0 & 0 & \times & \times \\
0 & 0 & 0 & 0 & 0 & 0 & 0 & \times & \times
\end{array}\right]
\left[\begin{array}{cc|cc|cc|cc|c}
\times & \times & 0 & 0 & 0 & 0 & 0 & 0 & 0 \\
\times & \times & 0 & 0 & 0 & 0 & 0 & 0 & 0 \\ \hline
0 & 0 & \times & \times & 0 & 0 & 0 & 0 & 0 \\
0 & 0 & \times & \times & 0 & 0 & 0 & 0 & 0 \\ \hline
0 & 0 & 0 & 0 & \times & \times & 0 & 0 & 0 \\
0 & 0 & 0 & 0 & \times & \times & 0 & 0 & 0 \\ \hline
0 & 0 & 0 & 0 & 0 & 0 & \times & \times & 0 \\
0 & 0 & 0 & 0 & 0 & 0 & \times & \times & 0 \\ \hline
0 & 0 & 0 & 0 & 0 & 0 & 0 & 0 & \times
\end{array}\right]
$$

then

$$
A =
\left[\begin{array}{ccccccccc}
\times & \times & 0 & 0 & 0 & 0 & 0 & 0 & 0 \\
\times & \times & \times & \times & 0 & 0 & 0 & 0 & 0 \\
\times & \times & \times & \times & 0 & 0 & 0 & 0 & 0 \\
0 & 0 & \times & \times & \times & \times & 0 & 0 & 0 \\
0 & 0 & \times & \times & \times & \times & 0 & 0 & 0 \\
0 & 0 & 0 & 0 & \times & \times & \times & \times & 0 \\
0 & 0 & 0 & 0 & \times & \times & \times & \times & 0 \\
0 & 0 & 0 & 0 & 0 & 0 & \times & \times & \times \\
0 & 0 & 0 & 0 & 0 & 0 & \times & \times & \times
\end{array}\right],
\quad
A^{-1} =
\left[\begin{array}{ccccccccc}
\times & \times & \times & 0 & 0 & 0 & 0 & 0 & 0 \\
\times & \times & \times & 0 & 0 & 0 & 0 & 0 & 0 \\
0 & \times & \times & \times & \times & 0 & 0 & 0 & 0 \\
0 & \times & \times & \times & \times & 0 & 0 & 0 & 0 \\
0 & 0 & 0 & \times & \times & \times & \times & 0 & 0 \\
0 & 0 & 0 & \times & \times & \times & \times & 0 & 0 \\
0 & 0 & 0 & 0 & 0 & \times & \times & \times & \times \\
0 & 0 & 0 & 0 & 0 & \times & \times & \times & \times \\
0 & 0 & 0 & 0 & 0 & 0 & 0 & \times & \times
\end{array}\right].
$$

Strang (2010a, 2010b) has pointed out a very important "reverse" fact. If A and A^{-1} are banded, then there is a factorization of the form (4.3.3) with relatively small N. Indeed, he shows that N is very small for certain types of matrices that arise in signal processing. An important consequence of this is that both the forward transform Ax and the inverse transform $A^{-1}x$ can be computed very fast.

Problems

P4.3.1 Develop a version of Algorithm 4.3.1 which assumes that the matrix A is stored in band format style. (See §1.2.5.)

P4.3.2 Show how the output of Algorithm 4.3.4 can be used to solve the upper Hessenberg system $Hx = b$.

P4.3.3 Show how Algorithm 4.3.4 could be used to solve a lower hessenberg system $Hx = b$.

P4.3.4 Give an algorithm for solving an unsymmetric tridiagonal system $Ax = b$ that uses Gaussian elimination with partial pivoting. It should require only four n-vectors of floating point storage for the factorization.

P4.3.5 (a) For $C \in \mathbb{R}^{n \times n}$ define the *profile indices* $m(C, i) = \min\{j : c_{ij} \neq 0\}$, where $i = 1{:}n$. Show that if $A = GG^T$ is the Cholesky factorization of A, then $m(A, i) = m(G, i)$ for $i = 1{:}n$. (We say that G has the same *profile* as A.) (b) Suppose $A \in \mathbb{R}^{n \times n}$ is symmetric positive definite with profile indices $m_i = m(A, i)$ where $i = 1{:}n$. Assume that A is stored in a one-dimensional array v as follows:

$$v = (a_{11}, a_{2,m_2}, \ldots, a_{22}, a_{3,m_3}, \ldots, a_{33}, \ldots, a_{n,m_n}, \ldots, a_{nn}).$$

Give an algorithm that overwrites v with the corresponding entries of the Cholesky factor G and then uses this factorization to solve $Ax = b$. How many flops are required? (c) For $C \in \mathbb{R}^{n \times n}$ define $p(C, i)$ $= \max\{j : c_{ij} \neq 0\}$. Suppose that $A \in \mathbb{R}^{n \times n}$ has an LU factorization $A = LU$ and that

$$m(A, 1) \quad \leq \quad m(A, 2) \quad \leq \quad \cdots \quad \leq \quad m(A, n),$$
$$p(A, 1) \quad \leq \quad p(A, 2) \quad \leq \quad \cdots \quad \leq \quad p(A, n).$$

Show that $m(A, i) = m(L, i)$ and $p(A, i) = p(U, i)$ for $i = 1{:}n$.

P4.3.6 Develop a gaxpy version of Algorithm 4.3.1.

P4.3.7 Develop a unit stride, vectorizable algorithm for solving the symmetric positive definite tridiagonal systems $A^{(k)} x^{(k)} = b^{(k)}$. Assume that the diagonals, superdiagonals, and right hand sides are stored by row in arrays D, E, and B and that $b^{(k)}$ is overwritten with $x^{(k)}$.

P4.3.8 Give an example of a 3-by-3 symmetric positive definite matrix whose tridiagonal part is not positive definite.

P4.3.9 Suppose a symmetric positive definite matrix $A \in \mathbb{R}^{n \times n}$ has the "arrow structure", e.g.,

$$A = \begin{bmatrix} \times & \times & \times & \times & \times \\ \times & \times & 0 & 0 & 0 \\ \times & 0 & \times & 0 & 0 \\ \times & 0 & 0 & \times & 0 \\ \times & 0 & 0 & 0 & \times \end{bmatrix}.$$

(a) Show how the linear system $Ax = b$ can be solved with $O(n)$ flops using the Sherman-Morrison-Woodbury formula. (b) Determine a permutation matrix P so that the Cholesky factorization

$$PAP^T = GG^T$$

can be computed with $O(n)$ flops.

P4.3.10 Suppose $A \in \mathbb{R}^{n \times n}$ is tridiagonal, positive definite, but not symmetric. Give an efficient algorithm for computing the largest entry of $|ST^{-1}S|$ where $S = (A - A^T)/2$ and $T = (A + A^T)/2$.

P4.3.11 Show that if $A \in \mathbb{R}^{n \times n}$ and $\epsilon > 0$, then there is a $B \in \mathbb{R}^{n \times n}$ such that $\| A - B \| \leq \epsilon$ and B has the property that all its principal submatrices are nonsingular. Use this result to formally complete the proof of Theorem 4.3.3.

P4.3.12 Give an upper bound on the bandwidth of the matrix A in (4.3.3).

P4.3.13 Show that A^T and A^{-1} have the same upper and lower bandwidths in (4.3.3).

P4.3.14 For the $A = F_1 F_2$ example in §4.3.9, show that $A(2{:}3, :)$, $A(4{:}5, :)$, $A(6{:}7, :), \ldots$ each consist of two singular 2-by-2 blocks.

Notes and References for §4.3

Representative papers on the topic of banded systems include:

R.S. Martin and J.H. Wilkinson (1965). "Symmetric Decomposition of Positive Definite Band Matrices," *Numer. Math. 7*, 355–61.

R. S. Martin and J.H. Wilkinson (1967). "Solution of Symmetric and Unsymmetric Band Equations and the Calculation of Eigenvalues of Band Matrices," *Numer. Math. 9*, 279–301.

E.L. Allgower (1973). "Exact Inverses of Certain Band Matrices," *Numer. Math. 21*, 279–284.

Z. Bohte (1975). "Bounds for Rounding Errors in the Gaussian Elimination for Band Systems," *J. Inst. Math. Applic. 16*, 133–142.

L. Kaufman (2007). "The Retraction Algorithm for Factoring Banded Symmetric Matrices," *Numer. Lin. Alg. Applic. 14*, 237–254.

C. Vomel and J. Slemons (2009). "Twisted Factorization of a Banded Matrix," *BIT 49*, 433–447.

Tridiagonal systems are particularly important, see:

C. Fischer and R.A. Usmani (1969). "Properties of Some Tridiagonal Matrices and Their Application to Boundary Value Problems," *SIAM J. Numer. Anal. 6*, 127–142.

D.J. Rose (1969). "An Algorithm for Solving a Special Class of Tridiagonal Systems of Linear Equations," *Commun. ACM 12*, 234–236.

M.A. Malcolm and J. Palmer (1974). "A Fast Method for Solving a Class of Tridiagonal Systems of Linear Equations," *Commun. ACM 17*, 14–17.

N.J. Higham (1986). "Efficient Algorithms for Computing the Condition Number of a Tridiagonal Matrix," *SIAM J. Sci. Stat. Comput. 7*, 150–165.

N.J. Higham (1990). "Bounding the Error in Gaussian Elimination for Tridiagonal Systems," *SIAM J. Matrix Anal. Applic. 11*, 521–530.

I.S. Dhillon (1998). "Reliable Computation of the Condition Number of a Tridiagonal Matrix in $O(n)$ Time," *SIAM J. Matrix Anal. Applic. 19*, 776–796.

I. Bar-On and M. Leoncini (2000). "Reliable Solution of Tridiagonal Systems of Linear Equations," *SIAM J. Numer. Anal. 38*, 1134–1153.

M.I. Bueno and F.M. Dopico (2004). "Stability and Sensitivity of Tridiagonal LU Factorization without Pivoting," *BIT 44*, 651–673.

J.R. Bunch and R.F. Marcia (2006). "A Simplified Pivoting Strategy for Symmetric Tridiagonal Matrices," *Numer. Lin. Alg. 13*, 865–867.

For a discussion of parallel methods for banded problems, see:

H.S. Stone (1975). "Parallel Tridiagonal Equation Solvers," *ACM Trans. Math. Softw. 1*, 289–307.

I. Bar-On, B. Codenotti and M. Leoncini (1997). "A Fast Parallel Cholesky Decomposition Algorithm for Tridiagonal Symmetric Matrices," *SIAM J. Matrix Anal. Applic. 18*, 403–418.

G.H. Golub, A.H. Sameh, and V. Sarin (2001). "A Parallel Balance Scheme for Banded Linear Systems," *Num. Lin. Alg. 8*, 297–316.

S. Rao and Sarita (2008). "Parallel Solution of Large Symmetric Tridiagonal Linear Systems," *Parallel Comput. 34*, 177–197.

Papers that are concerned with the structure of the inverse of a band matrix include:

E. Asplund (1959). "Inverses of Matrices $\{a_{ij}\}$ Which Satisfy $a_{ij} = 0$ for $j > i + p$," *Math. Scand. 7*, 57–60.

C.A. Micchelli (1992). "Banded Matrices with Banded Inverses," *J. Comput. Appl. Math. 41*, 281–300.

G. Strang and T. Nguyen (2004). "The Interplay of Ranks of Submatrices," *SIAM Review 46*, 637–648.

G. Strang (2010a). "Fast Transforms: Banded Matrices with Banded Inverses," *Proc. National Acad. Sciences 107*, 12413-12416.

G. Strang (2010b). "Banded Matrices with Banded Inverses and $A = LPU$," *Proceedings International Congress of Chinese Mathematicians*, Beijing.

A pivotal result in this arena is the *nullity theorem*, a more general version of Theorem 4.3.3, see:

R. Vandebril, M. Van Barel, and N. Mastronardi (2008). *Matrix Computations and Semiseparable Matrices, Volume I Linear Systems*, Johns Hopkins University Press, Baltimore, MD., 37–40.

4.4   Symmetric Indefinite Systems

Recall that a matrix whose quadratic form $x^T A x$ takes on both positive and negative values is *indefinite*. In this section we are concerned with symmetric indefinite linear systems. The LDL^T factorization is not always advisable as the following 2-by-2 example illustrates:

$$\begin{bmatrix} \epsilon & 1 \\ 1 & 0 \end{bmatrix} = \begin{bmatrix} 1 & 0 \\ 1/\epsilon & 1 \end{bmatrix} \begin{bmatrix} \epsilon & 0 \\ 0 & -1/\epsilon \end{bmatrix} \begin{bmatrix} 1 & 0 \\ 1/\epsilon & 1 \end{bmatrix}^T .$$

Of course, any of the pivot strategies in §3.4 could be invoked. However, they destroy symmetry and, with it, the chance for a "Cholesky speed" symmetric indefinite system solver. Symmetric pivoting, i.e., data reshufflings of the form $A \leftarrow PAP^T$, must be used as we discussed in §4.2.8. Unfortunately, symmetric pivoting does not always stabilize the LDL^T computation. If ϵ_1 *and* ϵ_2 are small, then regardless of P, the matrix

$$\tilde{A} = P \begin{bmatrix} \epsilon_1 & 1 \\ 1 & \epsilon_2 \end{bmatrix} P^T$$

has small diagonal entries and large numbers surface in the factorization. With symmetric pivoting, the pivots are always selected from the diagonal and trouble results if these numbers are small relative to what must be zeroed off the diagonal. Thus, LDL^T with symmetric pivoting cannot be recommended as a reliable approach to symmetric indefinite system solving. It seems that the challenge is to involve the off-diagonal entries in the pivoting process while at the same time maintaining symmetry.

In this section we discuss two ways to do this. The first method is due to Aasen (1971) and it computes the factorization

$$PAP^T = LTL^T, \tag{4.4.1}$$

where $L = (\ell_{ij})$ is unit lower triangular and T is tridiagonal. P is a permutation chosen such that $|\ell_{ij}| \le 1$. In contrast, the *diagonal pivoting method* due to Bunch and Parlett (1971) computes a permutation P such that

$$PAP^T = LDL^T, \tag{4.4.2}$$

where D is a direct sum of 1-by-1 and 2-by-2 pivot blocks. Again, P is chosen so that the entries in the unit lower triangular L satisfy $|\ell_{ij}| \le 1$. Both factorizations involve $n^3/3$ flops and once computed, can be used to solve $Ax = b$ with $O(n^2)$ work:

$$PAP^T = LTL^T, \ Lz = Pb, Tw = z, \ L^T y = w, \ x = P^T y \quad \Rightarrow \quad Ax = b,$$

$$PAP^T = LDL^T, \ Lz = Pb, \ Dw = z, \ L^T y = w, \ x = P^T y \quad \Rightarrow \quad Ax = b.$$

A few comments need to be made about the $Tw = z$ and $Dw = z$ systems that arise when these methods are invoked.

In Aasen's method, the symmetric indefinite tridiagonal system $Tw = z$ is solved in $O(n)$ time using band Gaussian elimination with pivoting. Note that there is no serious price to pay for the disregard of symmetry at this level since the overall process is $O(n^3)$.

In the diagonal pivoting approach, the $Dw = z$ system amounts to a set of 1-by-1 and 2-by-2 symmetric indefinite systems. The 2-by-2 problems can be handled via Gaussian elimination with pivoting. Again, there is no harm in disregarding symmetry during this $O(n)$ phase of the calculation. Thus, the central issue in this section is the efficient computation of the factorizations (4.4.1) and (4.4.2).

4.4.1   The Parlett-Reid Algorithm

Parlett and Reid (1970) show how to compute (4.4.1) using Gauss transforms. Their algorithm is sufficiently illustrated by displaying the $k = 2$ step for the case $n = 5$. At the beginning of this step the matrix A has been transformed to

$$
A^{(1)} \;=\; M_1 P_1 A P_1^T M_1^T \;=\; \begin{bmatrix} \alpha_1 & \beta_1 & 0 & 0 & 0 \\ \beta_1 & \alpha_2 & v_3 & v_4 & v_5 \\ 0 & v_3 & \times & \times & \times \\ 0 & v_4 & \times & \times & \times \\ 0 & v_5 & \times & \times & \times \end{bmatrix},
$$

where P_1 is a permutation chosen so that the entries in the Gauss transformation M_1 are bounded by unity in modulus. Scanning the vector $[\, v_3 \; v_4 \; v_5 \,]^T$ for its largest entry, we now determine a 3-by-3 permutation $\tilde{P}_2$ such that

$$
\tilde{P}_2 \begin{bmatrix} v_3 \\ v_4 \\ v_5 \end{bmatrix} = \begin{bmatrix} \tilde{v}_3 \\ \tilde{v}_4 \\ \tilde{v}_5 \end{bmatrix} \qquad \Rightarrow \qquad |\tilde{v}_3| = \max\{|\tilde{v}_3|, |\tilde{v}_4|, |\tilde{v}_5|\}.
$$

If this maximal element is zero, we set $M_2 = P_2 = I$ and proceed to the next step. Otherwise, we set $P_2 = \mathrm{diag}(I_2, \tilde{P}_2)$ and $M_2 = I - \alpha^{(2)} e_3^T$ with

$$
\alpha^{(2)} \;=\; \begin{bmatrix} 0 & 0 & 0 & \tilde{v}_4/\tilde{v}_3 & \tilde{v}_5/\tilde{v}_3 \end{bmatrix}^T.
$$

Observe that

$$
A^{(2)} \;=\; M_2 P_2 A^{(1)} P_2^T M_2^T \;=\; \begin{bmatrix} \alpha_1 & \beta_1 & 0 & 0 & 0 \\ \beta_1 & \alpha_2 & \tilde{v}_3 & 0 & 0 \\ 0 & \tilde{v}_3 & \times & \times & \times \\ 0 & 0 & \times & \times & \times \\ 0 & 0 & \times & \times & \times \end{bmatrix}.
$$

In general, the process continues for $n - 2$ steps leaving us with a tridiagonal matrix

$$
T \;=\; A^{(n-2)} \;=\; (M_{n-2}P_{n-2}\cdots M_1 P_1)A(M_{n-2}P_{n-2}\cdots M_1 P_1)^T.
$$

It can be shown that (4.4.1) holds with $P = P_{n-2}\cdots P_1$ and

$$
L = (M_{n-2}P_{n-2}\cdots M_1 P_1 P^T)^{-1}.
$$

Analysis of L reveals that its first column is e_1 and that its subdiagonal entries in column k with $k > 1$ are "made up" of the multipliers in M_{k-1}.

The efficient implementation of the Parlett-Reid method requires care when computing the update

$$A^{(k)} = M_k(P_k A^{(k-1)} P_k^T) M_k^T. \tag{4.4.3}$$

To see what is involved with a minimum of notation, suppose $B = B^T \in \mathbb{R}^{(n-k)\times(n-k)}$ has and that we wish to form

$$B_+ = (I - w e_1^T) B (I - w e_1^T)^T,$$

where $w \in \mathbb{R}^{n-k}$ and e_1 is the first column of I_{n-k}. Such a calculation is at the heart of (4.4.3). If we set

$$u = B e_1 - \frac{b_{11}}{2} w,$$

then $B_+ = B - w u^T - u w^T$ and its lower triangular portion can be formed in $2(n-k)^2$ flops. Summing this quantity as k ranges from 1 to $n-2$ indicates that the Parlett-Reid procedure requires $2n^3/3$ flops—twice the volume of work associated with Cholesky.

4.4.2   The Method of Aasen

An $n^3/3$ approach to computing (4.4.1) due to Aasen (1971) can be derived by reconsidering some of the computations in the Parlett-Reid approach. We examine the no-pivoting case first where the goal is to compute a unit lower triangular matrix L with $L(:,1) = e_1$ and a tridiagonal matrix

$$T = \begin{bmatrix} \alpha_1 & \beta_1 & & \cdots & & 0 \\ \beta_1 & \alpha_2 & \ddots & & & \vdots \\ & \ddots & \ddots & \ddots & & \\ & & \ddots & \ddots & \ddots & \\ \vdots & & & \ddots & \ddots & \beta_{n-1} \\ 0 & \cdots & & & \beta_{n-1} & \alpha_n \end{bmatrix}.$$

such that $A = LTL^T$. The Aasen method is structured as follows:

>**for** $j = 1{:}n$
>>$\{\alpha(1{:}j-1),\ \beta(1{:}j-1) \text{ and } L(:,1{:}j) \text{ are known}\}$
>>Compute α_j.
>>**if** $j \leq n-1$
>>>Compute β_j.
>>
>>**end**
>>**if** $j \leq n-2$
>>>Compute $L(j+2{:}n, j+1)$.
>>
>>**end**
>
>**end**

$\hspace{11cm}$ (4.4.4)

To develop recipes for α_j, β_j, and $L(j+2{:}n, j+1)$, we compare the jth columns in the equation $A = LH$ where $H = TL^T$. Noting that H is an upper Hessenberg matrix we obtain

$$A(:, j) = LH(:, j) = \sum_{k=1}^{j+1} L(:, k) \cdot h(k), \qquad (4.4.5)$$

where $h(1{:}j+1) = H(1{:}j+1, j)$ and we assume that $j \leq n - 1$. It follows that

$$h_{j+1} \cdot L(j+1{:}n, j+1) = v(j+1{:}n), \qquad (4.4.6)$$

where

$$v(j+1{:}n) = A(j+1{:}n, j) - L(j+1{:}n, 1{:}j) \cdot h(1{:}j). \qquad (4.4.7)$$

Since L is unit lower triangular and $L(:, 1{:}j)$ is known, this gives us a working recipe for $L(j+2{:}n, j+1)$ provided we know $h(1{:}j)$. Indeed, from (4.4.6) and (4.4.7) it is easy to show that

$$L(j+2{:}n, j+1) = v(j+2{:}n)/v(j+1). \qquad (4.4.8)$$

To compute $h(1{:}j)$ we turn to the equation $H = TL^T$ and examine its jth column. The case $j = 5$ amply displays what is going on:

$$
\begin{bmatrix} h_1 \\ h_2 \\ h_3 \\ h_4 \\ h_5 \\ h_6 \end{bmatrix}
=
\begin{bmatrix}
\alpha_1 & \beta_1 & 0 & 0 & 0 \\
\beta_1 & \alpha_2 & \beta_2 & 0 & 0 \\
0 & \beta_2 & \alpha_3 & \beta_3 & 0 \\
0 & 0 & \beta_3 & \alpha_4 & \beta_4 \\
0 & 0 & 0 & \beta_4 & \alpha_5 \\
0 & 0 & 0 & 0 & \beta_5
\end{bmatrix}
\begin{bmatrix} 0 \\ \ell_{52} \\ \ell_{53} \\ \ell_{54} \\ 1 \end{bmatrix}
=
\begin{bmatrix}
\beta_1 \ell_{52} \\
\alpha_2 \ell_{52} + \beta_2 \ell_{53} \\
\beta_2 \ell_{52} + \alpha_3 \ell_{53} + \beta_3 \ell_{54} \\
\beta_3 \ell_{53} + \alpha_4 \ell_{54} + \beta_4 \\
\beta_4 \ell_{54} + \alpha_5 \\
\beta_5
\end{bmatrix}
\qquad (4.4.9)
$$

At the start of step j, we know $\alpha(1{:}j-1)$, $\beta(1{:}j-1)$ and $L(:, 1{:}j)$. Thus, we can determine $h(1{:}j-1)$ as follows

$h_1 = \beta_1 \ell_{j2}$
for $k = 1{:}j-1$
$\qquad h_k = \beta_{k-1} \ell_{j,k-1} + \alpha_k \ell_{jk} + \beta_k \ell_{j,k+1} \qquad (4.4.10)$
end

Equation (4.4.5) gives us a formula for h_j:

$$h_j = A(j, j) - \sum_{k=1}^{j-1} L(j, k) h_k. \qquad (4.4.11)$$

From (4.4.9) we infer that

$$\alpha_j = h_j - \beta_{j-1} \ell_{j,j-1}, \qquad (4.4.12)$$

$$\beta_j = h_{j+1}. \qquad (4.4.13)$$

Combining these equations with (4.4.4), (4.4.7), (4.4.8), (4.4.10), and (4.4.11) we obtain the Aasen method without pivoting:

$$L = I_n$$
for $j = 1:n$
  if $j = 1$
    $\alpha_1 = a_{11}$
    $v(2{:}n) = A(2{:}n, 1)$
  else
    $h_1 = \beta_1 \cdot \ell_{j2}$
    for $k = 2:j - 1$
      $h_k = \beta_{k-1}\ell_{j,k-1} + \alpha_k \ell_{jk} + \beta_k \ell_{j,k+1}$
    end
    $h_j = a_{jj} - L(j, 1{:}j - 1) \cdot h(1{:}j - 1)$
    $\alpha_j = h_j - \beta_{j-1}\ell_{j,j-1}$ (4.4.14)
    $v(j + 1{:}n) = A(j + 1{:}n, j) - L(j + 1{:}n, 1{:}j) \cdot h(1{:}j)$
  end
  if $j <= n - 1$
    $\beta_j = v(j + 1)$
  end
  if $j <= n - 2$
    $L(j + 2{:}n, j + 1) = v(j + 2{:}n)/v(j + 1)$
  end
end

The dominant operation each pass through the j-loop is an $(n{-}j)$-by-j gaxpy operation. Accounting for the associated flops we see that the overall Aasen ccomputation involves $n^3/3$ flops, the same as for the Cholesky factorization.

As it now stands, the columns of L are scalings of the v-vectors in (4.4.14). If any of these scalings are large, i.e., if any $v(j + 1)$ is small, then we are in trouble. To circumvent this problem, it is only necessary to permute the largest component of $v(j+1{:}n)$ to the top position. Of course, this permutation must be suitably applied to the unreduced portion of A and the previously computed portion of L. With pivoting, Aasen's method is stable in the same sense that Gaussian elimination with partial pivoting is stable.

In a practical implementation of the Aasen algorithm, the lower triangular portion of A would be overwritten with L and T, e.g.,

$$A \leftarrow \begin{bmatrix} \alpha_1 & & & & \\ \beta_1 & \alpha_2 & & & \\ \ell_{32} & \beta_2 & \alpha_3 & & \\ \ell_{42} & \ell_{43} & \beta_3 & \alpha_4 & \\ \ell_{52} & \ell_{53} & \ell_{54} & \beta_4 & \alpha_5 \end{bmatrix}.$$

Notice that the columns of L are shifted left in this arrangement.

4.4.3 Diagonal Pivoting Methods

We next describe the computation of the block LDL^T factorization (4.4.2). We follow the discussion in Bunch and Parlett (1971). Suppose

$$P_1 A P_1^T = \begin{bmatrix} E & C^T \\ C & B \end{bmatrix} \begin{matrix} s \\ n-s \end{matrix}$$
$$\phantom{P_1 A P_1^T = \begin{bmatrix}} s \quad n-s$$

where P_1 is a permutation matrix and $s = 1$ or 2. If A is nonzero, then it is always possible to choose these quantities so that E is nonsingular, thereby enabling us to write

$$P_1 A P_1^T = \begin{bmatrix} I_s & 0 \\ CE^{-1} & I_{n-s} \end{bmatrix} \begin{bmatrix} E & 0 \\ 0 & B - CE^{-1}C^T \end{bmatrix} \begin{bmatrix} I_s & E^{-1}C^T \\ 0 & I_{n-s} \end{bmatrix}.$$

For the sake of stability, the s-by-s "pivot" E should be chosen so that the entries in

$$\tilde{A} = (\tilde{a}_{ij}) \equiv B - CE^{-1}C^T \tag{4.4.15}$$

are suitably bounded. To this end, let $\alpha \in (0, 1)$ be given and define the size measures

$$\mu_0 = \max_{i,j} |a_{ij}|,$$

$$\mu_1 = \max_i |a_{ii}|.$$

The Bunch-Parlett pivot strategy is as follows:

> **if** $\mu_1 \geq \alpha\mu_0$
>> $s = 1$
>> Choose P_1 so $|e_{11}| = \mu_1$.
> **else**
>> $s = 2$
>> Choose P_1 so $|e_{21}| = \mu_0$.
> **end**

It is easy to verify from (4.4.15) that if $s = 1$, then

$$|\tilde{a}_{ij}| \leq (1 + \alpha^{-1})\mu_0, \tag{4.4.16}$$

while $s = 2$ implies

$$|\tilde{a}_{ij}| \leq \frac{3 - \alpha}{1 - \alpha}\mu_0. \tag{4.4.17}$$

By equating $(1 + \alpha^{-1})^2$, the growth factor that is associated with two $s = 1$ steps, and $(3 - \alpha)/(1 - \alpha)$, the corresponding $s = 2$ factor, Bunch and Parlett conclude that $\alpha = (1 + \sqrt{17})/8$ is optimum from the standpoint of minimizing the bound on element growth.

The reductions outlined above can be repeated on the order-$(n - s)$ symmetric matrix $\tilde{A}$. A simple induction argument establishes that the factorization (4.4.2) exists and that $n^3/3$ flops are required if the work associated with pivot determination is ignored.

4.4.4   Stability and Efficiency

Diagonal pivoting with the above strategy is shown by Bunch (1971) to be as stable as Gaussian elimination with complete pivoting. Unfortunately, the overall process requires between $n^3/12$ and $n^3/6$ comparisons, since μ_0 involves a two-dimensional search at each stage of the reduction. The actual number of comparisons depends on the total number of 2-by-2 pivots but in general the Bunch-Parlett method for computing (4.4.2) is considerably slower than the technique of Aasen. See Barwell and George (1976).

This is not the case with the diagonal pivoting method of Bunch and Kaufman (1977). In their scheme, it is only necessary to scan two columns at each stage of the reduction. The strategy is fully illustrated by considering the very first step in the reduction:

$\alpha = (1 + \sqrt{17})/8$
$\lambda = |a_{r1}| = \max\{|a_{21}|, \dots, |a_{n1}|\}$
if $\lambda > 0$
    if $|a_{11}| \geq \alpha\lambda$
        Set $s = 1$ and $P_1 = I$.
    else
        $\sigma = |a_{pr}| = \max\{|a_{1r}|, \dots, |a_{r-1,r}|, |a_{r+1,r}|, \dots, |a_{nr}|\}$
        if $\sigma|a_{11}| \geq \alpha\lambda^2$
            Set $s = 1$ and $P_1 = I$
        elseif $|a_{rr}| \geq \alpha\sigma$
            Set $s = 1$ and choose P_1 so $(P_1^T A P_1)_{11} = a_{rr}$.
        else
            Set $s = 2$ and choose P_1 so $(P_1^T A P_1)_{21} = a_{rp}$.
        end
    end
end

Overall, the Bunch-Kaufman algorithm requires $n^3/3$ flops, $O(n^2)$ comparisons, and, like all the methods of this section, $n^2/2$ storage.

4.4.5   A Note on Equilibrium Systems

A very important class of symmetric indefinite matrices have the form

$$A = \begin{bmatrix} C & B \\ B^T & 0 \end{bmatrix} \begin{matrix} n \\ p \end{matrix} \qquad (4.4.18)$$
$$ \begin{matrix} n & p \end{matrix}$$

where C is symmetric positive definite and B has full column rank. These conditions ensure that A is nonsingular.

Of course, the methods of this section apply to A. However, they do not exploit its structure because the pivot strategies "wipe out" the zero (2,2) block. On the other hand, here is a tempting approach that does exploit A's block structure:

Step 1. Compute the Cholesky factorization $C = GG^T$.

Step 2. Solve $GK = B$ for $K \in \mathbb{R}^{n \times p}$.

Step 3. Compute the Cholesky factorization $HH^T = K^T K = B^T C^{-1} B$.

From this it follows that

$$A = \begin{bmatrix} G & 0 \\ K^T & H \end{bmatrix} \begin{bmatrix} G^T & K \\ 0 & -H^T \end{bmatrix}.$$

In principle, this triangular factorization can be used to solve the *equilibrium system*

$$\begin{bmatrix} C & B \\ B^T & 0 \end{bmatrix} \begin{bmatrix} x \\ y \end{bmatrix} = \begin{bmatrix} f \\ g \end{bmatrix}. \tag{4.4.19}$$

However, it is clear by considering steps (b) and (c) above that the accuracy of the computed solution depends upon $\kappa(C)$ and this quantity may be much greater than $\kappa(A)$. The situation has been carefully analyzed and various structure-exploiting algorithms have been proposed. A brief review of the literature is given at the end of the section.

It is interesting to consider a special case of (4.4.19) that clarifies what it means for an algorithm to be stable and illustrates how perturbation analysis can structure the search for better methods. In several important applications, $g = 0$, C is diagonal, and the solution subvector y is of primary importance. A manipulation shows that this vector is specified by

$$y = (B^T C^{-1} B)^{-1} B^T C^{-1} f. \tag{4.4.20}$$

Looking at this we are again led to believe that $\kappa(C)$ should have a bearing on the accuracy of the computed y. However, it can be shown that

$$\| (B^T C^{-1} B)^{-1} B^T C^{-1} \| \leq \psi_B \tag{4.4.21}$$

where the upper bound ψ_B is independent of C, a result that (correctly) suggests that y is *not* sensitive to perturbations in C. A stable method for computing this vector should respect this, meaning that the accuracy of the computed y should be independent of C. Vavasis (1994) has developed a method with this property. It involves the careful assembly of a matrix $V \in \mathbb{R}^{n \times (n-p)}$ whose columns are a basis for the nullspace of $B^T C^{-1}$. The n-by-n linear system

$$[\, B \mid V \,] \begin{bmatrix} y \\ q \end{bmatrix} = f$$

is then solved implying $f = By + Vq$. Thus, $B^T C^{-1} f = B^T C^{-1} By$ and (4.4.20) holds.

Problems

P4.4.1 Show that if all the 1-by-1 and 2-by-2 principal submatrices of an n-by-n symmetric matrix A are singular, then A is zero.

P4.4.2 Show that no 2-by-2 pivots can arise in the Bunch-Kaufman algorithm if A is positive definite.

P4.4.3 Arrange (4.4.14) so that only the lower triangular portion of A is referenced and so that $\alpha(j)$ overwrites $A(j,j)$ for $j = 1{:}n$, $\beta(j)$ overwrites $A(j+1,j)$ for $j = 1{:}n-1$, and $L(i,j)$ overwrites $A(i,j-1)$ for $j = 2{:}n-1$ and $i = j+1{:}n$.

P4.4.4 Suppose $A \in \mathbb{R}^{n \times n}$ is symmetric and strictly diagonally dominant. Give an algorithm that computes the factorization

$$\Pi A \Pi^T = \left[\begin{array}{cc} R & 0 \\ S & -M \end{array} \right] \left[\begin{array}{cc} R^T & S^T \\ 0 & M^T \end{array} \right]$$

where Π is a permuation and the diagonal blocks R and M are lower triangular.

P4.4.5 A symmetric matrix A is *quasidefinite* if it has the form

$$A = \left[\begin{array}{cc} A_{11} & A_{12} \\ A_{21} & -A_{22} \end{array} \right] \begin{array}{c} n \\ p \end{array}$$
$$\qquad\quad n \quad\ \ p$$

with A_{11} and A_{22} positive definite. (a) Show that such a matrix has an LDL^T factorization with the property that

$$D = \left[\begin{array}{cc} D_1 & 0 \\ 0 & -D_2 \end{array} \right]$$

where $D_1 \in \mathbb{R}^{n \times n}$ and $D_2 \in \mathbb{R}^{p \times p}$ have positive diagonal entries. (b) Show that if A is quasidefinite then all its principal submatrices are nonsingular. This means that PAP^T has an LDL^T factorization for any permutation matrix P.

P4.4.6 Prove (4.4.16) and (4.4.17).

P4.4.7 Show that $-(B^T C^{-1} B)^{-1}$ is the (2,2) block of A^{-1} where A is given by equation (4.4.18).

P4.4.8 The point of this problem is to consider a special case of (4.4.21). Define the matrix

$$M(\alpha) = (B^T C^{-1} B)^{-1} B^T C^{-1}$$

where $C = (I_n + \alpha e_k e_k^T)$, $\alpha > -1$, and $e_k = I_n(:,k)$. (Note that C is just the identity with α added to the (k,k) entry.) Assume that $B \in \mathbb{R}^{n \times p}$ has rank p and show that

$$M(\alpha) = (B^T B)^{-1} B^T \left(I_n - \frac{\alpha}{1 + \alpha w^T w} e_k w^T \right)$$

where

$$w = (I_n - B(B^T B)^{-1} B^T) e_k.$$

Show that if $\| w \|_2 = 0$ or $\| w \|_2 = 1$, then $\| M(\alpha) \|_2 = 1/\sigma_{\min}(B)$. Show that if $0 < \| w \|_2 < 1$, then

$$\| M(\alpha) \|_2 \le \max \left\{ \frac{1}{1 - \| w \|_2}, 1 + \frac{1}{\| w \|_2} \right\} \bigg/ \sigma_{\min}(B).$$

Thus, $\| M(\alpha) \|_2$ has an α-independent upper bound.

Notes and References for §4.4

The basic references for computing (4.4.1) are as follows:

J.O. Aasen (1971). "On the Reduction of a Symmetric Matrix to Tridiagonal Form," *BIT 11*, 233–242.

B.N. Parlett and J.K. Reid (1970). "On the Solution of a System of Linear Equations Whose Matrix Is Symmetric but not Definite," *BIT 10*, 386–397.

The diagonal pivoting literature includes:

J.R. Bunch and B.N. Parlett (1971). "Direct Methods for Solving Symmetric Indefinite Systems of Linear Equations," *SIAM J. Numer. Anal. 8*, 639–655.

J.R. Bunch (1971). "Analysis of the Diagonal Pivoting Method," *SIAM J. Numer. Anal. 8*, 656–680.

J.R. Bunch (1974). "Partial Pivoting Strategies for Symmetric Matrices," *SIAM J. Numer. Anal. 11*, 521–528.

J.R. Bunch, L. Kaufman, and B.N. Parlett (1976). "Decomposition of a Symmetric Matrix," *Numer. Math. 27,* 95–109.
J.R. Bunch and L. Kaufman (1977). "Some Stable Methods for Calculating Inertia and Solving Symmetric Linear Systems," *Math. Comput. 31,* 162–79.
M.T. Jones and M.L. Patrick (1993). "Bunch-Kaufman Factorization for Real Symmetric Indefinite Banded Matrices," *SIAM J. Matrix Anal. Applic. 14,* 553–559.

Because "future" columns must be scanned in the pivoting process, it is awkward (but possible) to obtain a gaxpy-rich diagonal pivoting algorithm. On the other hand, Aasen's method is naturally rich in gaxpys. Block versions of both procedures are possible. Various performance issues are discussed in:

V. Barwell and J.A. George (1976). "A Comparison of Algorithms for Solving Symmetric Indefinite Systems of Linear Equations," *ACM Trans. Math. Softw. 2,* 242–251.
M.T. Jones and M.L. Patrick (1994). "Factoring Symmetric Indefinite Matrices on High-Performance Architectures," *SIAM J. Matrix Anal. Applic. 15,* 273–283.

Another idea for a cheap pivoting strategy utilizes error bounds based on more liberal interchange criteria, an idea borrowed from some work done in the area of sparse elimination methods, see:

R. Fletcher (1976). "Factorizing Symmetric Indefinite Matrices," *Lin. Alg. Applic. 14,* 257–272.

Before using any symmetric $Ax = b$ solver, it may be advisable to equilibrate A. An $O(n^2)$ algorithm for accomplishing this task is given in:

J.R. Bunch (1971). "Equilibration of Symmetric Matrices in the Max-Norm," *J. ACM 18,* 566–572.
N.J. Higham (1997). "Stability of the Diagonal Pivoting Method with Partial Pivoting," *SIAM J. Matrix Anal. Applic. 18,* 52–65.

Procedures for skew-symmetric systems similar to the methods that we have presented in this section also exist:

J.R. Bunch (1982). "A Note on the Stable Decomposition of Skew Symmetric Matrices," *Math. Comput. 158,* 475–480.
J. Bunch (1982). "Stable Decomposition of Skew-Symmetric Matrices," *Math. Comput. 38,* 475–479.
P. Benner, R. Byers, H. Fassbender, V. Mehrmann, and D. Watkins (2000). "Cholesky-like Factorizations of Skew-Symmetric Matrices," *ETNA 11,* 85–93.

For a discussion of symmetric indefinite system solvers that are also banded or sparse, see:

C. Ashcraft, R.G. Grimes, and J.G. Lewis (1998). "Accurate Symmetric Indefinite Linear Equation Solvers," *SIAM J. Matrix Anal. Applic. 20,* 513–561.
S.H. Cheng and N.J. Higham (1998). "A Modified Cholesky Algorithm Based on a Symmetric Indefinite Factorization," *SIAM J. Matrix Anal. Applic. 19,* 1097–1110.
J. Zhao, W. Wang, and W. Ren (2004). "Stability of the Matrix Factorization for Solving Block Tridiagonal Symmetric Indefinite Linear Systems," *BIT 44,* 181–188.
H. Fang and D.P. O'Leary (2006). "Stable Factorizations of Symmetric Tridiagonal and Triadic Matrices," *SIAM J. Matrix Anal. Applic. 28,* 576–595.
D. Irony and S. Toledo (2006). "The Snap-Back Pivoting Method for Symmetric Banded Indefinite Matrices," *SIAM J. Matrix Anal. Applic. 28,* 398–424.

The equilibrium system literature is scattered among the several application areas where it has an important role to play. Nice overviews with pointers to this literature include:

G. Strang (1988). "A Framework for Equilibrium Equations," *SIAM Review 30,* 283–297.
S.A. Vavasis (1994). "Stable Numerical Algorithms for Equilibrium Systems," *SIAM J. Matrix Anal. Applic. 15,* 1108–1131.
P.E. Gill, M.A. Saunders, and J.R. Shinnerl (1996). "On the Stability of Cholesky Factorization for Symmetric Quasidefinite Systems," *SIAM J. Matrix Anal. Applic. 17,* 35–46.
G.H. Golub and C. Greif (2003). "On Solving Block-Structured Indefinite Linear Systems," *SIAM J. Sci. Comput. 24,* 2076–2092.

For a discussion of (4.4.21), see:

G.W. Stewart (1989). "On Scaled Projections and Pseudoinverses," *Lin. Alg. Applic. 112,* 189–193.

D.P. O'Leary (1990). "On Bounds for Scaled Projections and Pseudoinverses," *Lin. Alg. Applic. 132*, 115–117.

M.J. Todd (1990). "A Dantzig-Wolfe-like Variant of Karmarkar's Interior-Point Linear Programming Algorithm," *Oper. Res. 38*, 1006–1018.

An equilibrium system is a special case of a *saddle point system*. See §11.5.10.

4.5   Block Tridiagonal Systems

Block tridiagonal linear systems of the form

$$
\begin{bmatrix}
D_1 & F_1 & & \cdots & & 0 \\
E_1 & D_2 & \ddots & & & \vdots \\
& \ddots & \ddots & \ddots & & \\
\vdots & & \ddots & \ddots & F_{N-1} \\
0 & \cdots & & E_{N-1} & D_N
\end{bmatrix}
\begin{bmatrix}
x_1 \\ x_2 \\ \vdots \\ \vdots \\ x_N
\end{bmatrix}
=
\begin{bmatrix}
b_1 \\ b_2 \\ \vdots \\ \vdots \\ b_N
\end{bmatrix}. \tag{4.5.1}
$$

frequently arise in practice. We assume for clarity that all blocks are q-by-q. In this section we discuss both a block LU approach to this problem as well as a pair of divide-and-conquer schemes.

4.5.1   Block Tridiagonal LU Factorization

If

$$
A =
\begin{bmatrix}
D_1 & F_1 & & \cdots & & 0 \\
E_1 & D_2 & \ddots & & & \vdots \\
& & \ddots & \ddots & \ddots & \\
\vdots & & & \ddots & \ddots & F_{N-1} \\
0 & \cdots & & & E_{N-1} & D_N
\end{bmatrix} \tag{4.5.2}
$$

then by comparing blocks in

$$
A =
\begin{bmatrix}
I & & & \cdots & & 0 \\
L_1 & I & & & & \vdots \\
& \ddots & \ddots & & & \\
\vdots & & \ddots & & & \\
0 & \cdots & & L_{N-1} & I
\end{bmatrix}
\begin{bmatrix}
U_1 & F_1 & & \cdots & & 0 \\
0 & U_2 & \ddots & & & \vdots \\
& & \ddots & \ddots & \ddots & \\
\vdots & & & \ddots & \ddots & F_{N-1} \\
0 & \cdots & & & 0 & U_N
\end{bmatrix} \tag{4.5.3}
$$

we formally obtain the following algorithm for computing the L_i and U_i:

> $U_1 = D_1$
> **for** $i = 2{:}N$
>   Solve $L_{i-1}U_{i-1} = E_{i-1}$ for L_{i-1}.   (4.5.4)
> $U_i = D_i - L_{i-1}F_{i-1}$
> **end**

The procedure is defined as long as the U_i are nonsingular.

Having computed the factorization (4.5.3), the vector x in (4.5.1) can be obtained via block forward elimination and block back substitution:

$$
\begin{aligned}
&y_1 = b_1 \\
&\textbf{for } i = 2{:}N \\
&\qquad y_i = b_i - L_{i-1}y_{i-1} \\
&\textbf{end} \\
&\text{Solve } U_N x_N = y_N \text{ for } x_N \\
&\textbf{for } i = N - 1{:}-1{:}1 \\
&\qquad \text{Solve } U_i x_i = y_i - F_i x_{i+1} \text{ for } x_i \\
&\textbf{end}
\end{aligned}
\tag{4.5.5}
$$

To carry out both (4.5.4) and (4.5.5), each U_i must be factored since linear systems involving these submatrices are solved. This could be done using Gaussian elimination with pivoting. However, this does not guarantee the stability of the overall process.

4.5.2   Block Diagonal Dominance

In order to obtain satisfactory bounds on the L_i and U_i it is necessary to make additional assumptions about the underlying block matrix. For example, if we have

$$
\| \, D_i^{-1} \, \|_1 \, (\| \, F_{i-1} \, \|_1 + \| \, E_i \, \|_1) < 1, \qquad E_N \equiv F_0 \equiv 0,
\tag{4.5.6}
$$

for $i = 1{:}N$, then the factorization (4.5.3) exists and it is possible to show that the L_i and U_i satisfy the inequalities

$$
\| \, L_i \, \|_1 \le 1,
\tag{4.5.7}
$$

$$
\| \, U_i \, \|_1 \le \| \, A_n \, \|_1.
\tag{4.5.8}
$$

The conditions (4.5.6) define a type of *block diagonal dominance*.

4.5.3   Block-Cyclic Reduction

We next describe the method of *block-cyclic reduction* that can be used to solve some important special instances of the block tridiagonal system (4.5.1). For simplicity, we assume that A has the form

$$
A = \begin{bmatrix}
D & F & & \cdots & 0 \\
F & D & \ddots & & \vdots \\
& \ddots & \ddots & \ddots & \\
\vdots & & \ddots & \ddots & F \\
0 & \cdots & & F & D
\end{bmatrix} \in \mathbb{R}^{Nq \times Nq}
\tag{4.5.9}
$$

where F and D are q-by-q matrices that satisfy $DF = FD$. We also assume that $N = 2^k - 1$. These conditions hold in certain important applications such as the discretization of Poisson's equation on a rectangle. (See §4.8.4.)

The basic idea behind cyclic reduction is to halve repeatedly the dimension of the problem on hand repeatedly until we are left with a single q-by-q system for the unknown subvector $x_{2^{k-1}}$. This system is then solved by standard means. The previously eliminated x_i are found by a back-substitution process.

The general procedure is adequately illustrated by considering the case $N = 7$:

$$
\begin{aligned}
b_1 &= Dx_1 + Fx_2, \\
b_2 &= Fx_1 + Dx_2 + Fx_3, \\
b_3 &= \qquad\quad Fx_2 + Dx_3 + Fx_4, \\
b_4 &= \qquad\qquad\quad Fx_3 + Dx_4 + Fx_5, \\
b_5 &= \qquad\qquad\qquad\quad Fx_4 + Dx_5 + Fx_6, \\
b_6 &= \qquad\qquad\qquad\qquad\quad Fx_5 + Dx_6 + Fx_7, \\
b_7 &= \qquad\qquad\qquad\qquad\qquad\quad Fx_6 + Dx_7.
\end{aligned}
$$

For $i = 2$, 4, and 6 we multiply equations $i-1$, i, and $i+1$ by F, $-D$, and F, respectively, and add the resulting equations to obtain

$$
\begin{aligned}
(2F^2 - D^2)x_2 + \qquad\qquad F^2 x_4 \qquad\qquad\qquad\qquad &= F(b_1 + b_3) - Db_2, \\
F^2 x_2 + (2F^2 - D^2)x_4 + \qquad\qquad F^2 x_6 &= F(b_3 + b_5) - Db_4, \\
F^2 x_4 + (2F^2 - D^2)x_6 &= F(b_5 + b_7) - Db_6.
\end{aligned}
$$

Thus, with this tactic we have removed the odd-indexed x_i and are left with a reduced block tridiagonal system of the form

$$
\begin{aligned}
D^{(1)}x_2 + F^{(1)}x_4 \qquad\qquad\quad &= b_2^{(1)}, \\
F^{(1)}x_2 + D^{(1)}x_4 + F^{(1)}x_6 &= b_4^{(1)}, \\
F^{(1)}x_4 + D^{(1)}x_6 &= b_6^{(1)},
\end{aligned}
$$

where $D^{(1)} = 2F^2 - D^2$ and $F^{(1)} = F^2$ commute. Applying the same elimination strategy as above, we multiply these three equations respectively by $F^{(1)}$, $-D^{(1)}$, and $F^{(1)}$. When these transformed equations are added together, we obtain the single equation

$$
\left(2[F^{(1)}]^2 - D^{(1)^2} \right) x_4 = F^{(1)} \left(b_2^{(1)} + b_6^{(1)} \right) - D^{(1)} b_4^{(1)},
$$

which we write as

$$
D^{(2)} x_4 = b^{(2)}.
$$

This completes the cyclic reduction. We now solve this (small) q-by-q system for x_4. The vectors x_2 and x_6 are then found by solving the systems

$$
\begin{aligned}
D^{(1)} x_2 &= b_2^{(1)} - F^{(1)} x_4, \\
D^{(1)} x_6 &= b_6^{(1)} - F^{(1)} x_4.
\end{aligned}
$$

Finally, we use the first, third, fifth, and seventh equations in the original system to compute x_1, x_3, x_5, and x_7, respectively.

The amount of work required to perform these recursions for general N depends greatly upon the sparsity of the $D^{(p)}$ and $F^{(p)}$. In the worst case when these matrices are full, the overall flop count has order $\log(N)q^3$. Care must be exercised in order to ensure stability during the reduction. For further details, see Buneman (1969).

4.5.4 The SPIKE Framework

A bandwidth-p matrix $A \in \mathbb{R}^{Nq \times Nq}$ can also be regarded as a block tridiagonal matrix with banded diagonal blocks and low-rank off-diagonal blocks. Here is an example where $N = 4$, $q = 7$, and $p = 2$:

$$
A = \begin{bmatrix}
\times\times\times & & & \\
\times\times\times\times & & & \\
\times\times\times\times\times & & & \\
\times\times\times\times\times & & & \\
\times\times\times\times\times & & & \\
\times\times\times\times\,|\,\times & & & \\
\times\times\times\,|\,\times\times & & & \\
& \ddots & & \\
& & \ddots & \\
& & & \ddots
\end{bmatrix} \qquad (4.5.11)
$$

Note that the diagonal blocks have bandwidth p and the blocks along the subdiagonal and superdiagonal have rank p. The low rank of the off-diagonal blocks makes it possible to formulate a divide-and-conquer procedure known as the "SPIKE" algorithm. The method is of interest because it parallelizes nicely. Our brief discussion is based on Polizzi and Sameh (2007).

Assume for clarity that the diagonal blocks $D_1, \ldots, D_4$ are sufficiently well conditioned. If we premultiply the above matrix by the inverse of $\mathrm{diag}(D_1, D_2, D_3, D_4)$, then we obtain

$$
\tilde{A} = \begin{bmatrix}
I & S_1 & & \\
T_2 & I & S_2 & \\
& T_3 & I & S_3 \\
& & T_4 & I
\end{bmatrix} \qquad (4.5.12)
$$

With this maneuver, the original linear system

$$
\begin{bmatrix}
D_1 & F_1 & 0 & 0 \\
E_1 & D_2 & F_2 & 0 \\
0 & E_2 & D_3 & F_3 \\
0 & 0 & E_3 & D_4
\end{bmatrix}
\begin{bmatrix}
x_1 \\ x_2 \\ x_3 \\ x_4
\end{bmatrix}
=
\begin{bmatrix}
b_1 \\ b_2 \\ b_3 \\ b_4
\end{bmatrix},
\tag{4.5.13}
$$

which corresponds to (4.5.11), transforms to

$$
\begin{bmatrix}
I_7 & \tilde{F}_1 & 0 & 0 \\
\tilde{E}_1 & I_7 & \tilde{F}_2 & 0 \\
0 & \tilde{E}_2 & I_7 & \tilde{F}_3 \\
0 & 0 & \tilde{E}_3 & I_7
\end{bmatrix}
\begin{bmatrix}
x_1 \\ x_2 \\ x_3 \\ x_4
\end{bmatrix}
=
\begin{bmatrix}
\tilde{b}_1 \\ \tilde{b}_2 \\ \tilde{b}_3 \\ \tilde{b}_4
\end{bmatrix},
\tag{4.5.14}
$$

where $D_i \tilde{b}_i = b_i$, $D_i \tilde{F}_i = F_i$, and $D_{i+1} \tilde{E}_i = E_i$. Next, we refine the blocking (4.5.14) by turning each submatrix into a 3-by-3 block matrix and each subvector into a 3-by-1 block vector as follows:

$$
\left[
\begin{array}{ccc|ccc|ccc|ccc}
I_2 & 0 & 0 & K_1 & 0 & 0 & 0 & 0 & 0 & 0 & 0 & 0 \\
0 & I_3 & 0 & H_1 & 0 & 0 & 0 & 0 & 0 & 0 & 0 & 0 \\
0 & 0 & I_2 & G_1 & 0 & 0 & 0 & 0 & 0 & 0 & 0 & 0 \\
\hline
0 & 0 & R_1 & I_2 & 0 & 0 & K_2 & 0 & 0 & 0 & 0 & 0 \\
0 & 0 & S_1 & 0 & I_3 & 0 & H_2 & 0 & 0 & 0 & 0 & 0 \\
0 & 0 & T_1 & 0 & 0 & I_2 & G_2 & 0 & 0 & 0 & 0 & 0 \\
\hline
0 & 0 & 0 & 0 & 0 & R_2 & I_2 & 0 & 0 & K_3 & 0 & 0 \\
0 & 0 & 0 & 0 & 0 & S_2 & 0 & I_3 & 0 & H_3 & 0 & 0 \\
0 & 0 & 0 & 0 & 0 & T_2 & 0 & 0 & I_2 & G_3 & 0 & 0 \\
\hline
0 & 0 & 0 & 0 & 0 & 0 & 0 & 0 & R_3 & I_q & 0 & 0 \\
0 & 0 & 0 & 0 & 0 & 0 & 0 & 0 & S_3 & 0 & I_m & 0 \\
0 & 0 & 0 & 0 & 0 & 0 & 0 & 0 & T_3 & 0 & 0 & I_q
\end{array}
\right]
\begin{bmatrix}
w_1 \\ y_1 \\ z_1 \\ w_2 \\ y_2 \\ z_2 \\ w_3 \\ y_3 \\ z_3 \\ w_4 \\ y_4 \\ z_4
\end{bmatrix}
=
\begin{bmatrix}
c_1 \\ d_1 \\ f_1 \\ c_2 \\ d_2 \\ f_2 \\ c_3 \\ d_3 \\ f_3 \\ c_4 \\ d_4 \\ f_4
\end{bmatrix}.
\tag{4.5.15}
$$

The block rows and columns in this equation can be reordered to produce the following equivalent system:

$$
\left[
\begin{array}{cccccccc|cccc}
I_2 & 0 & K_1 & 0 & 0 & 0 & 0 & 0 & 0 & 0 & 0 & 0 \\
0 & I_2 & G_1 & 0 & 0 & 0 & 0 & 0 & 0 & 0 & 0 & 0 \\
0 & R_1 & I_2 & 0 & K_2 & 0 & 0 & 0 & 0 & 0 & 0 & 0 \\
0 & T_1 & 0 & I_2 & G_2 & 0 & 0 & 0 & 0 & 0 & 0 & 0 \\
0 & 0 & 0 & R_2 & I_2 & 0 & K_3 & 0 & 0 & 0 & 0 & 0 \\
0 & 0 & 0 & T_2 & 0 & I_2 & G_3 & 0 & 0 & 0 & 0 & 0 \\
0 & 0 & 0 & 0 & 0 & R_3 & I_2 & 0 & 0 & 0 & 0 & 0 \\
0 & 0 & 0 & 0 & 0 & T_3 & 0 & I_2 & 0 & 0 & 0 & 0 \\
\hline
0 & 0 & H_1 & 0 & 0 & 0 & 0 & 0 & I_3 & 0 & 0 & 0 \\
0 & S_1 & 0 & 0 & H_2 & 0 & 0 & 0 & 0 & I_3 & 0 & 0 \\
0 & 0 & 0 & S_2 & 0 & 0 & H_3 & 0 & 0 & 0 & I_3 & 0 \\
0 & 0 & 0 & 0 & 0 & S_3 & 0 & 0 & 0 & 0 & 0 & I_3
\end{array}
\right]
\begin{bmatrix}
w_1 \\ z_1 \\ w_2 \\ z_2 \\ w_3 \\ z_3 \\ w_4 \\ z_4 \\ y_1 \\ y_2 \\ y_3 \\ y_4
\end{bmatrix}
=
\begin{bmatrix}
c_1 \\ f_1 \\ c_2 \\ f_2 \\ c_3 \\ f_3 \\ c_4 \\ f_4 \\ d_1 \\ d_2 \\ d_3 \\ d_4
\end{bmatrix}.
\tag{4.5.16}
$$

If we assume that $N \gg 1$, then the (1,1) block is a relatively small banded matrix that define the z_i and w_i. Once these quantities are computed, then the remaining unknowns follow from a decoupled set of large matrix-vector multiplications, e.g., $y_1 = d_1 - H_1 w_2$, $y_2 = d_2 - S_1 z_1 - H_2 w_3$, $y_3 = d_3 - S_2 z_2 - H_3 w_4$, and $y_4 = d_4 - S_3 z_3$. Thus, in a four-processor execution of this method, there are (short) communications that involves the w_i and z_i and a lot of large, local gaxpy computations.

Problems

P4.5.1 (a) Show that a block diagonally dominant matrix is nonsingular. (b) Verify that (4.5.6) implies (4.5.7) and (4.5.8).

P4.5.2 Write a recursive function $x = \mathsf{CR}(D, F, N, b)$ that returns the solution to $Ax = b$ where A is specified by (4.5.9). Assume that $N = 2^k - 1$ for some positive integer k, $D, F \in \mathbb{R}^{q \times q}$, and $b \in \mathbb{R}^{Nq}$.

P4.5.3 How would you solve a system of the form

$$
\begin{bmatrix} D_1 & F_1 \\ E_1 & D_2 \end{bmatrix} \begin{bmatrix} x_1 \\ x_2 \end{bmatrix} = \begin{bmatrix} b_1 \\ b_2 \end{bmatrix}
$$

where D_1 and D_2 are diagonal and F_1 and E_1 are tridiagonal? Hint: Use the perfect shuffle permutation.

P4.5.4 In the simplified SPIKE framework that we presented in §4.5.4, we treat A as an N-by-N block matrix with q-by-q blocks. It is assumed that $A \in \mathbb{R}^{Nq \times Nq}$ has bandwidth p and that $p \ll q$. For this general case, describe the block sizes that result when the transition from (4.5.11) to (4.5.16) is carried out. Assuming that A's band is dense, what fraction of flops are gaxpy flops?

Notes and References for §4.5

The following papers provide insight into the various nuances of block matrix computations:

J.M. Varah (1972). "On the Solution of Block-Tridiagonal Systems Arising from Certain Finite-Difference Equations," *Math. Comput. 26*, 859–868.
R. Fourer (1984). "Staircase Matrices and Systems," *SIAM Review 26*, 1–71.
M.L. Merriam (1985). "On the Factorization of Block Tridiagonals With Storage Constraints," *SIAM J. Sci. Stat. Comput. 6*, 182-192.

The property of block diagonal dominance and its various implications is the central theme in:

D.G. Feingold and R.S. Varga (1962). "Block Diagonally Dominant Matrices and Generalizations of the Gershgorin Circle Theorem,"*Pacific J. Math. 12*, 1241–1250.
R.S. Varga (1976). "On Diagonal Dominance Arguments for Bounding $\| A^{-1} \|_\infty$," *Lin. Alg. Applic. 14*, 211–217.

Early methods that involve the idea of cyclic reduction are described in:

R.W. Hockney (1965). "A Fast Direct Solution of Poisson's Equation Using Fourier Analysis, "*J. ACM 12*, 95–113.
B.L. Buzbee, G.H. Golub, and C.W. Nielson (1970). "On Direct Methods for Solving Poisson's Equations," *SIAM J. Numer. Anal. 7*, 627–656.

The accumulation of the right-hand side must be done with great care, for otherwise there would be a significant loss of accuracy. A stable way of doing this is described in:

O. Buneman (1969). "A Compact Non-Iterative Poisson Solver," Report 294, Stanford University Institute for Plasma Research, Stanford, CA.

Other literature concerned with cyclic reduction includes:

F.W. Dorr (1970). "The Direct Solution of the Discrete Poisson Equation on a Rectangle," *SIAM Review 12*, 248–263.

B.L. Buzbee, F.W. Dorr, J.A. George, and G.H. Golub (1971). "The Direct Solution of the Discrete Poisson Equation on Irregular Regions," *SIAM J. Numer. Anal. 8*, 722–736.

F.W. Dorr (1973). "The Direct Solution of the Discrete Poisson Equation in $O(n^2)$ Operations," *SIAM Review 15*, 412–415.

P. Concus and G.H. Golub (1973). "Use of Fast Direct Methods for the Efficient Numerical Solution of Nonseparable Elliptic Equations," *SIAM J. Numer. Anal. 10*, 1103–1120.

B.L. Buzbee and F.W. Dorr (1974). "The Direct Solution of the Biharmonic Equation on Rectangular Regions and the Poisson Equation on Irregular Regions," *SIAM J. Numer. Anal. 11*, 753–763.

D. Heller (1976). "Some Aspects of the Cyclic Reduction Algorithm for Block Tridiagonal Linear Systems," *SIAM J. Numer. Anal. 13*, 484–496.

Various generalizations and extensions to cyclic reduction have been proposed:

P.N. Swarztrauber and R.A. Sweet (1973). "The Direct Solution of the Discrete Poisson Equation on a Disk," *SIAM J. Numer. Anal. 10*, 900–907.

R.A. Sweet (1974). "A Generalized Cyclic Reduction Algorithm," *SIAM J. Num. Anal. 11*, 506–20.

M.A. Diamond and D.L.V. Ferreira (1976). "On a Cyclic Reduction Method for the Solution of Poisson's Equation," *SIAM J. Numer. Anal. 13*, 54–70.

R.A. Sweet (1977). "A Cyclic Reduction Algorithm for Solving Block Tridiagonal Systems of Arbitrary Dimension," *SIAM J. Numer. Anal. 14*, 706–720.

P.N. Swarztrauber and R. Sweet (1989). "Vector and Parallel Methods for the Direct Solution of Poisson's Equation," *J. Comput. Appl. Math. 27*, 241–263.

S. Bondeli and W. Gander (1994). "Cyclic Reduction for Special Tridiagonal Systems," *SIAM J. Matrix Anal. Applic. 15*, 321–330.

A 2-by-2 block system with very thin (1,2) and (2,1) blocks is referred to as a *bordered linear system*. Special techniques for problems with this structure are discussed in:

W. Govaerts and J.D. Pryce (1990). "Block Elimination with One Iterative Refinement Solves Bordered Linear Systems Accurately," *BIT 30*, 490–507.

W. Govaerts (1991). "Stable Solvers and Block Elimination for Bordered Systems," *SIAM J. Matrix Anal. Applic. 12*, 469–483.

W. Govaerts and J.D. Pryce (1993). "Mixed Block Elimination for Linear Systems with Wider Borders," *IMA J. Numer. Anal. 13*, 161–180.

Systems that are block bidiagonal, block Hessenberg, and block triangular also occur, see:

G. Fairweather and I. Gladwell (2004). "Algorithms for Almost Block Diagonal Linear Systems," *SIAM Review 46*, 49–58.

U. von Matt and G. W. Stewart (1996). "Rounding Errors in Solving Block Hessenberg Systems," *Math. Comput. 65*, 115–135.

L. Gemignani and G. Lotti (2003). "Efficient and Stable Solution of M-Matrix Linear Systems of (Block) Hessenberg Form," *SIAM J. Matrix Anal. Applic. 24*, 852–876.

M. Hegland and M.R. Osborne (1998). "Wrap-Around Partitioning for Block Bidiagonal Linear Systems," *IMA J. Numer. Anal. 18*, 373–383.

T. Rossi and J. Toivanen (1999). "A Parallel Fast Direct Solver for Block Tridiagonal Systems with Separable Matrices of Arbitrary Dimension," *SIAM J. Sci. Comput. 20*, 1778–1793.

I.M. Spitkovsky and D. Yong (2000). "Almost Periodic Factorization of Certain Block Triangular Matrix Functions," *Math. Comput. 69*, 1053–1070.

The SPIKE framework supports many different options according to whether the band is sparse or dense. Also, steps have to be taken if the diagonal blocks are ill-conditioned, see:

E. Polizzi and A. Sameh (2007). "SPIKE: A Parallel Environment for Solving Banded Linear Systems," *Comput. Fluids 36*, 113–120.

C.C.K. Mikkelsen and M. Manguoglu (2008). "Analysis of the Truncated SPIKE Algorithm," *SIAM J. Matrix Anal. Applic. 30*, 1500–1519.

4.6 Vandermonde Systems

Suppose $x(0{:}n) \in \mathbb{R}^{n+1}$. A matrix $V \in \mathbb{R}^{(n+1)\times(n+1)}$ of the form

$$
V = V(x_0,\ldots,x_n) = \begin{bmatrix} 1 & 1 & \cdots & 1 \\ x_0 & x_1 & \cdots & x_n \\ \vdots & \vdots & & \vdots \\ x_0^n & x_1^n & \cdots & x_n^n \end{bmatrix}
$$

is said to be a *Vandermonde matrix*. Note that the discrete Fourier transform matrix (§1.4.1) is a very special complex Vandermonde matrix.

In this section, we show how the systems $V^T a = f = f(0{:}n)$ and $Vz = b = b(0{:}n)$ can be solved in $O(n^2)$ flops. *For convenience, vectors and matrices are subscripted from 0 in this section.*

4.6.1 Polynomial Interpolation: $V^T a = f$

Vandermonde systems arise in many approximation and interpolation problems. Indeed, the key to obtaining a fast Vandermonde solver is to recognize that solving $V^T a = f$ is equivalent to polynomial interpolation. This follows because if $V^T a = f$ and

$$
p(x) = \sum_{j=0}^{n} a_j x^j, \tag{4.6.1}
$$

then $p(x_i) = f_i$ for $i = 0{:}n$.

Recall that if the x_i are distinct then there is a unique polynomial of degree n that interpolates $(x_0, f_0),\ldots,(x_n, f_n)$. Consequently, V is nonsingular as long as the x_i are distinct. We assume this throughout the section.

The first step in computing the a_j of (4.6.1) is to calculate the Newton representation of the interpolating polynomial p:

$$
p(x) = \sum_{k=0}^{n} c_k \left(\prod_{i=0}^{k-1}(x - x_i) \right). \tag{4.6.2}
$$

The constants c_k are divided differences and may be determined as follows:

$$
\begin{aligned}
&c(0{:}n) = f(0{:}n) \\
&\textbf{for } k = 0{:}n{-}1 \\
&\qquad \textbf{for } i = n{:}{-}1{:}k{+}1 \\
&\qquad\qquad c_i = (c_i - c_{i-1})/(x_i - x_{i-k-1}) \\
&\qquad \textbf{end} \\
&\textbf{end}
\end{aligned} \tag{4.6.3}
$$

See Conte and deBoor (1980).

The next task is to generate the coefficients $a_0,\ldots,a_n$ in (4.6.1) from the Newton representation coefficients $c_0,\ldots,c_n$. Define the polynomials $p_n(x),\ldots,p_0(x)$ by the iteration

$$p_n(x) = c_n$$
$$\textbf{for } k = n - 1 : -1 : 0$$
$$\qquad p_k(x) = c_k + (x - x_k) \cdot p_{k+1}(x)$$
$$\textbf{end}$$

and observe that $p_0(x) = p(x)$. Writing

$$p_k(x) = a_k^{(k)} + a_{k+1}^{(k)} x + \cdots + a_n^{(k)} x^{n-k}$$

and equating like powers of x in the equation $p_k = c_k + (x - x_k) p_{k+1}$ gives the following recursion for the coefficients $a_i^{(k)}$:

$$a_n^{(n)} = c_n$$
$$\textbf{for } k = n-1 : -1 : 0$$
$$\qquad a_k^{(k)} = c_k - x_k a_{k+1}^{(k+1)}$$
$$\qquad \textbf{for } i = k+1 : n - 1$$
$$\qquad\qquad a_i^{(k)} = a_i^{(k+1)} - x_k a_{i+1}^{(k+1)}$$
$$\qquad \textbf{end}$$
$$\qquad a_n^{(k)} = a_n^{(k+1)}$$
$$\textbf{end}$$

Consequently, the coefficients $a_i = a_i^{(0)}$ can be calculated as follows:

$$a(0{:}n) = c(0{:}n)$$
$$\textbf{for } k = n-1 : -1 : 0$$
$$\qquad \textbf{for } i = k{:}n - 1 \qquad\qquad\qquad\qquad\qquad\qquad\qquad (4.6.4)$$
$$\qquad\qquad a_i = a_i - x_k a_{i+1}$$
$$\qquad \textbf{end}$$
$$\textbf{end}$$

Combining this iteration with (4.6.3) gives the following algorithm.

Algorithm 4.6.1 Given $x(0:n) \in \mathbb{R}^{n+1}$ with distinct entries and $f = f(0:n) \in \mathbb{R}^{n+1}$, the following algorithm overwrites f with the solution $a = a(0:n)$ to the Vandermonde system $V(x_0, \ldots, x_n)^T a = f$.

$$\textbf{for } k = 0{:}n - 1$$
$$\qquad \textbf{for } i = n : -1 : k+1$$
$$\qquad\qquad f(i) = (f(i) - f(i - 1))/(x(i) - x(i - k - 1))$$
$$\qquad \textbf{end}$$
$$\textbf{end}$$
$$\textbf{for } k = n - 1 : -1 : 0$$
$$\qquad \textbf{for } i = k : n - 1$$
$$\qquad\qquad f(i) = f(i) - f(i + 1) \cdot x(k)$$
$$\qquad \textbf{end}$$
$$\textbf{end}$$

This algorithm requires $5n^2/2$ flops.

4.6.2   The System $Vz = b$

Now consider the system $Vz = b$. To derive an efficient algorithm for this problem, we describe what Algorithm 4.6.1 does in matrix-vector language. Define the lower bidiagonal matrix $L_k(\alpha) \in \mathbb{R}^{(n+1)\times(n+1)}$ by

$$
L_k(\alpha) \;=\;
\left[
\begin{array}{c|ccccc}
I_k & & & 0 & & \\
\hline
 & 1 & 0 & \cdots & & 0 \\
 & -\alpha & 1 & & & \\
0 & 0 & \ddots & \ddots & & \\
 & \vdots & & \ddots & \ddots & \vdots \\
 & & & & \ddots & 1 \\
 & 0 & & \cdots & -\alpha & 1
\end{array}
\right]
$$

and the diagonal matrix D_k by

$$
D_k \;=\; \mathrm{diag}(\underbrace{1,\ldots,1}_{k+1}, x_{k+1} - x_0, \ldots, x_n - x_{n-k-1}).
$$

With these definitions it is easy to verify from (4.6.3) that, if $f = f(0:n)$ and $c = c(0:n)$ is the vector of divided differences, then

$$
c = U^T f
$$

where U is the upper triangular matrix defined by

$$
U^T = D_{n-1}^{-1} L_{n-1}(1) \cdots D_0^{-1} L_0(1).
$$

Similarly, from (4.6.4) we have

$$
a = L^T c,
$$

where L is the unit lower triangular matrix defined by

$$
L^T \;=\; L_0(x_0)^T \cdots L_{n-1}(x_{n-1})^T.
$$

It follows that $a = V^{-T} f$ is given by

$$
a \;=\; L^T U^T f.
$$

Thus,

$$
V^{-T} \;=\; L^T U^T
$$

which shows that Algorithm 4.6.1 solves $V^T a = f$ by tacitly computing the "UL factorization" of V^{-1}. Consequently, the solution to the system $Vz = b$ is given by

$$
z \;=\; V^{-1} b = U(Lb)
$$

$$
= \left(L_0(1)^T D_0^{-1} \cdots L_{n-1}(1)^T D_{n-1}^{-1}\right)\left(L_{n-1}(x_{n-1}) \cdots L_0(x_0) b\right).
$$

This observation gives rise to the following algorithm:

Algorithm 4.6.2 Given $x(0:n) \in \mathbb{R}^{n+1}$ with distinct entries and $b = b(0:n) \in \mathbb{R}^{n+1}$, the following algorithm overwrites b with the solution $z = z(0:n)$ to the Vandermonde system $V(x_0, \ldots, x_n)z = b$.

> **for** $k = 0:n-1$
>> **for** $i = n: -1:k+1$
>>> $b(i) = b(i) - x(k)b(i-1)$
>>
>> **end**
>
> **end**
> **for** $k = n-1: -1:0$
>> **for** $i = k+1:n$
>>> $b(i) = b(i)/(x(i) - x(i-k-1))$
>>
>> **end**
>> **for** $i = k:n-1$
>>> $b(i) = b(i) - b(i+1)$
>>
>> **end**
>
> **end**

This algorithm requires $5n^2/2$ flops.

Algorithms 4.6.1 and 4.6.2 are discussed and analyzed by Björck and Pereyra (1970). Their experience is that these algorithms frequently produce surprisingly accurate solutions, even if V is ill-conditioned.

We mention that related techniques have been developed and analyzed for *confluent Vandermonde systems*, e.g., systems of the form

$$
\begin{bmatrix} 1 & 1 & 0 & 1 \\ x_0 & x_1 & 1 & x_3 \\ x_0^2 & x_1^2 & 2x_1 & x_3^2 \\ x_0^3 & x_1^3 & 3x_1^2 & x_3^3 \end{bmatrix}^T \begin{bmatrix} a_0 \\ a_1 \\ a_2 \\ a_3 \end{bmatrix} = \begin{bmatrix} f_0 \\ f_1 \\ f_2 \\ f_3 \end{bmatrix}.
$$

See Higham (1990).

Problems

P4.6.1 Show that if $V = V(x_0, \ldots, x_n)$, then

$$
\det(V) = \prod_{n \geq i > j \geq 0} (x_i - x_j).
$$

P4.6.2 (Gautschi 1975) Verify the following inequality for the $n = 1$ case above:

$$
\| V^{-1} \|_\infty \leq \max_{0 \leq k \leq n} \prod_{\substack{i=0 \\ i \neq k}}^{n} \frac{1 + |x_i|}{|x_k - x_i|}.
$$

Equality results if the x_i are all on the same ray in the complex plane.

Notes and References for §4.6

Our discussion of Vandermonde linear systems is drawn from the following papers:

A. Björck and V. Pereyra (1970). "Solution of Vandermonde Systems of Equations," *Math. Comput.* *24*, 893–903.
A. Björck and T. Elfving (1973). "Algorithms for Confluent Vandermonde Systems," *Numer. Math.* *21*, 130–37.

The divided difference computations we discussed are detailed in:

S.D. Conte and C. de Boor (1980). *Elementary Numerical Analysis: An Algorithmic Approach,* Third Edition, McGraw-Hill, New York, Chapter 2.

Error analyses of Vandermonde system solvers include:

N.J. Higham (1987). "Error Analysis of the Björck-Pereyra Algorithms for Solving Vandermonde Systems," *Numer. Math. 50*, 613–632.
N.J. Higham (1988). "Fast Solution of Vandermonde-like Systems Involving Orthogonal Polynomials," *IMA J. Numer. Anal. 8*, 473–486.
N.J. Higham (1990). "Stability Analysis of Algorithms for Solving Confluent Vandermonde-like Systems," *SIAM J. Matrix Anal. Applic. 11*, 23–41.
S.G. Bartels and D.J. Higham (1992). "The Structured Sensitivity of Vandermonde-Like Systems," *Numer. Math. 62*, 17–34.
J.M. Varah (1993). "Errors and Perturbations in Vandermonde Systems," *IMA J. Numer. Anal. 13*, 1–12.

Interesting theoretical results concerning the condition of Vandermonde systems may be found in:

W. Gautschi (1975). "Norm Estimates for Inverses of Vandermonde Matrices," *Numer. Math. 23*, 337–347.
W. Gautschi (1975). "Optimally Conditioned Vandermonde Matrices," *Numer. Math. 24*, 1–12.
J-G. Sun (1998). "Bounds for the Structured Backward Errors of Vandermonde Systems," *SIAM J. Matrix Anal. Applic. 20*, 45–59.
B.K. Alpert (1996). "Condition Number of a Vandermonde Matrix," *SIAM Review 38*, 314–314.
B. Beckermann (2000). "The condition number of real Vandermonde, Krylov and positive definite Hankel matrices," *Numer. Math. 85*, 553–577.

The basic algorithms presented can be extended to cover confluent Vandermonde systems, block Vandermonde systems, and Vandermonde systems with other polynomial bases:

G. Galimberti and V. Pereyra (1970). "Numerical Differentiation and the Solution of Multidimensional Vandermonde Systems," *Math. Comput. 24*, 357–364.
G. Galimberti and V. Pereyra (1971). "Solving Confluent Vandermonde Systems of Hermitian Type," *Numer. Math. 18*, 44–60.
H. Van de Vel (1977). "Numerical Treatment of Generalized Vandermonde Systems of Equations," *Lin. Alg. Applic. 17*, 149–174.
G.H. Golub and W.P Tang (1981). "The Block Decomposition of a Vandermonde Matrix and Its Applications," *BIT 21*, 505–517.
D. Calvetti and L. Reichel (1992). "A Chebychev-Vandermonde Solver," *Lin. Alg. Applic. 172*, 219–229.
D. Calvetti and L. Reichel (1993). "Fast Inversion of Vandermonde-Like Matrices Involving Orthogonal Polynomials," *BIT 33*, 473–484.
H. Lu (1994). "Fast Solution of Confluent Vandermonde Linear Systems," *SIAM J. Matrix Anal. Applic. 15*, 1277–1289.
H. Lu (1996). "Solution of Vandermonde-like Systems and Confluent Vandermonde-like Systems," *SIAM J. Matrix Anal. Applic. 17*, 127–138.
M.-R. Skrzipek (2004). "Inversion of Vandermonde-Like Matrices," *BIT 44*, 291–306.
J.W. Demmel and P. Koev (2005). "The Accurate and Efficient Solution of a Totally Positive Generalized Vandermonde Linear System," *SIAM J. Matrix Anal. Applic. 27*, 142–152.

The displacement rank idea that we discuss in §12.1 can also be used to develop fast methods for Vandermonde systems.

4.7   Classical Methods for Toeplitz Systems

Matrices whose entries are constant along each diagonal arise in many applications and are called *Toeplitz matrices*. Formally, $T \in \mathbb{R}^{n \times n}$ is Toeplitz if there exist scalars $r_{-n+1}, \ldots, r_0, \ldots, r_{n-1}$ such that $a_{ij} = r_{j-i}$ for all i and j. Thus,

$$
T = \begin{bmatrix} r_0 & r_1 & r_2 & r_3 \\ r_{-1} & r_0 & r_1 & r_2 \\ r_{-2} & r_{-1} & r_0 & r_1 \\ r_{-3} & r_{-2} & r_{-1} & r_0 \end{bmatrix} = \begin{bmatrix} 3 & 1 & 7 & 6 \\ 4 & 3 & 1 & 7 \\ 0 & 4 & 3 & 1 \\ 9 & 0 & 4 & 3 \end{bmatrix}
$$

is Toeplitz. In this section we show that Toeplitz systems can be solved in $O(n^2)$ flops The discussion focuses on the important case when T is also symmetric and positive definite, but we also include a few comments about general Toeplitz systems. An alternative approach to Toeplitz system solving based on displacement rank is given in §12.1.

4.7.1   Persymmetry

The key fact that makes it possible to solve a Toeplitz system $Tx = b$ so fast has to do with the structure of T^{-1}. Toeplitz matrices belong to the larger class of *persymmetric matrices*. We say that $B \in \mathbb{R}^{n \times n}$ is persymmetric if

$$
\mathcal{E}_n B \mathcal{E}_n = B^T
$$

where $\mathcal{E}_n$ is the n-by-n exchange matrix defined in §1.2.11, e.g.,

$$
\mathcal{E}_4 = \begin{bmatrix} 0 & 0 & 0 & 1 \\ 0 & 0 & 1 & 0 \\ 0 & 1 & 0 & 0 \\ 1 & 0 & 0 & 0 \end{bmatrix}.
$$

If B is persymmetric, then $\mathcal{E}_n B$ is symmetric. This means that B is symmetric about its *antidiagonal*. Note that the inverse of a persymmetric matrix is also persymmetric:

$$
\mathcal{E}_n B^{-1} \mathcal{E}_n = (\mathcal{E}_n B \mathcal{E}_n)^{-1} = (B^T)^{-1} = (B^{-1})^T.
$$

Thus, the inverse of a nonsingular Toeplitz matrix is persymmetric.

4.7.2   Three Problems

Assume that we have scalars $r_1, \ldots, r_n$ such that for $k = 1{:}n$ the matrices

$$
T_k = \begin{bmatrix} 1 & r_1 & \cdots & r_{k-2} & r_{k-1} \\ r_1 & 1 & \ddots & & r_{k-2} \\ \vdots & \ddots & \ddots & \ddots & \vdots \\ r_{k-2} & & \ddots & \ddots & r_1 \\ r_{k-1} & r_{k-2} & \cdots & r_1 & 1 \end{bmatrix}
$$

are positive definite. (There is no loss of generality in normalizing the diagonal.) We
set out to describe three important algorithms:

- Durbin's algorithm for the *Yule-Walker problem* $T_n y = -[r_1, \ldots, r_n]^T$.

- Levinson's algorithm for the general right-hand-side problem $T_n x = b$.

- Trench's algorithm for computing $B = T_n^{-1}$.

4.7.3   Solving the Yule-Walker Equations

We begin by presenting Durbin's algorithm for the Yule-Walker equations which arise
in conjunction with certain linear prediction problems. Suppose for some k that sat-
isfies $1 \le k \le n - 1$ we have solved the kth order Yule-Walker system $T_k y = -r = -[r_1, \ldots, r_k]^T$. We now show how the $(k+1)$st order Yule-Walker system

$$\begin{bmatrix} T_k & \mathcal{E}_k r \\ r^T \mathcal{E}_k & 1 \end{bmatrix} \begin{bmatrix} z \\ \alpha \end{bmatrix} = -\begin{bmatrix} r \\ r_{k+1} \end{bmatrix}$$

can be solved in $O(k)$ flops. First observe that

$$z = T_k^{-1}(-r - \alpha \mathcal{E}_k r) = y - \alpha T_k^{-1} \mathcal{E}_k r$$

and

$$\alpha = -r_{k+1} - r^T \mathcal{E}_k z.$$

Since T_k^{-1} is persymmetric, $T_k^{-1}\mathcal{E}_k = \mathcal{E}_k T_k^{-1}$ and thus

$$z = y - \alpha \mathcal{E}_k T_k^{-1} r = y + \alpha \mathcal{E}_k y.$$

By substituting this into the above expression for α we find

$$\alpha = -r_{k+1} - r^T \mathcal{E}_k (y + \alpha \mathcal{E}_k y) = -(r_{k+1} + r^T \mathcal{E}_k y)/(1 + r^T y).$$

The denominator is positive because T_{k+1} is positive definite and because

$$\begin{bmatrix} I & \mathcal{E}_k y \\ 0 & 1 \end{bmatrix}^T \begin{bmatrix} T_k & \mathcal{E}_k r \\ r^T \mathcal{E}_k & 1 \end{bmatrix} \begin{bmatrix} I & \mathcal{E}_k y \\ 0 & 1 \end{bmatrix} = \begin{bmatrix} T_k & 0 \\ 0 & 1 + r^T y \end{bmatrix}.$$

We have illustrated the kth step of an algorithm proposed by Durbin (1960). It proceeds
by solving the Yule-Walker systems

$$T_k y^{(k)} = -r^{(k)} = -[r_1, \ldots, r_k]^T$$

for $k = 1{:}n$ as follows:

$$y^{(1)} = -r_1$$
$$\textbf{for } k = 1{:}n - 1$$
$$\beta_k = 1 + [r^{(k)}]^T y^{(k)}$$
$$\alpha_k = -(r_{k+1} + r^{(k)^T} \mathcal{E}_k y^{(k)})/\beta_k \qquad (4.7.1)$$
$$z^{(k)} = y^{(k)} + \alpha_k \mathcal{E}_k y^{(k)}$$
$$y^{(k+1)} = \left[\begin{array}{c} z^{(k)} \\ \alpha_k \end{array} \right]$$
$$\textbf{end}$$

As it stands, this algorithm would require $3n^2$ flops to generate $y = y^{(n)}$. It is possible, however, to reduce the amount of work even further by exploiting some of the above expressions:

$$\beta_k = 1 + [r^{(k)}]^T y^{(k)}$$
$$= 1 + \left[\begin{array}{c} r^{(k-1)} \\ r_k \end{array} \right]^T \left[\begin{array}{c} y^{(k-1)} + \alpha_{k-1}\mathcal{E}_{k-1} y^{(k-1)} \\ \alpha_{k-1} \end{array} \right]$$
$$= (1 + [r^{(k-1)}]^T y^{(k-1)}) + \alpha_{k-1}\left([r^{(k-1)}]^T \mathcal{E}_{k-1} y^{(k-1)} + r_k\right)$$
$$= \beta_{k-1} + \alpha_{k-1}(-\beta_{k-1}\alpha_{k-1})$$
$$= (1 - \alpha_{k-1}^2)\beta_{k-1}.$$

Using this recursion we obtain the following algorithm:

Algorithm 4.7.1 (Durbin) Given real numbers $r_0, r_1, \ldots, r_n$ with $r_0 = 1$ such that $T = (r_{|i-j|}) \in \mathbb{R}^{n \times n}$ is positive definite, the following algorithm computes $y \in \mathbb{R}^n$ such that $Ty = -[r_1, \ldots, r_n]^T$.

$$y(1) = -r(1); \ \beta = 1; \ \alpha = -r(1)$$
$$\textbf{for } k = 1{:}n - 1$$
$$\beta = (1 - \alpha^2)\beta$$
$$\alpha = -\left(r(k+1) + r(k{:}-1{:}1)^T y(1{:}k)\right)/\beta$$
$$z(1{:}k) = y(1{:}k) + \alpha y(k{:}-1{:}1)$$
$$y(1{:}k+1) = \left[\begin{array}{c} z(1{:}k) \\ \alpha \end{array} \right]$$
$$\textbf{end}$$

This algorithm requires $2n^2$ flops. We have included an auxiliary vector z for clarity, but it can be avoided.

4.7.4   The General Right-Hand-Side Problem

With a little extra work, it is possible to solve a symmetric positive definite Toeplitz system that has an arbitrary right-hand side. Suppose that we have solved the system

$$T_k x = b = [b_1, \ldots, b_k]^T \qquad (4.7.2)$$

for some k satisfying $1 \le k < n$ and that we now wish to solve

$$\begin{bmatrix} T_k & \mathcal{E}_k r \\ r^T \mathcal{E}_k & 1 \end{bmatrix} \begin{bmatrix} v \\ \mu \end{bmatrix} = \begin{bmatrix} b \\ b_{k+1} \end{bmatrix}. \tag{4.7.3}$$

Here, $r = [r_1, \dots, r_k]^T$ as above. Assume also that the solution to the order-k Yule-Walker system $T_k y = -r$ is also available. From $T_k v + \mu \mathcal{E}_k r = b$ it follows that

$$v = T_k^{-1}(b - \mu \mathcal{E}_k r) = x - \mu T_k^{-1} \mathcal{E}_k r = x + \mu \mathcal{E}_k y$$

and so

$$\mu = b_{k+1} - r^T \mathcal{E}_k v$$
$$= b_{k+1} - r^T \mathcal{E}_k x - \mu r^T y$$
$$= \left(b_{k+1} - r^T \mathcal{E}_k x \right) / \left(1 + r^T y \right).$$

Consequently, we can effect the transition from (4.7.2) to (4.7.3) in $O(k)$ flops.

Overall, we can efficiently solve the system $T_n x = b$ by solving the systems

$$T_k x^{(k)} = b^{(k)} = [b_1, \dots, b_k]^T$$

and

$$T_k y^{(k)} = -r^{(k)} = -[r_1, \dots, r_k]^T$$

"in parallel" for $k = 1{:}n$. This is the gist of the Levinson algorithm.

Algorithm 4.7.2 (Levinson) Given $b \in \mathbb{R}^n$ and real numbers $1 = r_0, r_1, \dots, r_n$ such that $T = (r_{|i-j|}) \in \mathbb{R}^{n \times n}$ is positive definite, the following algorithm computes $x \in \mathbb{R}^n$ such that $Tx = b$.

> $y(1) = -r(1)$; $x(1) = b(1)$; $\beta = 1$; $\alpha = -r(1)$
> **for** $k = 1{:}n-1$
> > $\beta = (1 - \alpha^2)\beta$
> > $\mu = \left(b(k+1) - r(1{:}k)^T x(k{:} -1{:}1) \right) / \beta$
> > $v(1{:}k) = x(1{:}k) + \mu \cdot y(k{:} -1{:}1)$
> > $x(1{:}k+1) = \begin{bmatrix} v(1{:}k) \\ \mu \end{bmatrix}$
> > **if** $k < n-1$
> > > $\alpha = -\left(r(k+1) + r(1{:}k)^T y(k{:} -1{:}1) \right) / \beta$
> > > $z(1{:}k) = y(1{:}k) + \alpha \cdot y(k{:} -1{:}1)$
> > > $y(1{:}k+1) = \begin{bmatrix} z(1{:}k) \\ \alpha \end{bmatrix}$
> > **end**
> **end**

This algorithm requires $4n^2$ flops. The vectors z and v are for clarity and can be avoided in a detailed implementation.

4.7.5   Computing the Inverse

One of the most surprising properties of a symmetric positive definite Toeplitz matrix T_n is that its complete inverse can be calculated in $O(n^2)$ flops. To derive the algorithm for doing this, partition T_n^{-1} as follows:

$$
T_n^{-1} = \left[\begin{array}{cc} A & Er \\ r^T E & 1 \end{array}\right]^{-1} = \left[\begin{array}{cc} B & v \\ v^T & \gamma \end{array}\right] \tag{4.7.4}
$$

where $A = T_{n-1}$, $E = \mathcal{E}_{n-1}$, and $r = [r_1, \ldots, r_{n-1}]^T$. From the equation

$$
\left[\begin{array}{cc} A & Er \\ r^T E & 1 \end{array}\right]\left[\begin{array}{c} v \\ \gamma \end{array}\right] = \left[\begin{array}{c} 0 \\ 1 \end{array}\right]
$$

it follows that $Av = -\gamma Er = -\gamma E(r_1, \ldots, r_{n-1})^T$ and $\gamma = 1 - r^T Ev$. If y solves the order-$(n-1)$ Yule-Walker system $Ay = -r$, then these expressions imply that

$$
\gamma = 1/(1 + r^T y),
$$

$$
v = \gamma Ey.
$$

Thus, the last row and column of T_n^{-1} are readily obtained.

It remains for us to develop working formulae for the entries of the submatrix B in (4.7.4). Since $AB + \mathcal{E}rv^T = I_{n-1}$, it follows that

$$
B = A^{-1} - (A^{-1}Er)v^T = A^{-1} + \frac{vv^T}{\gamma}.
$$

Now since $A = T_{n-1}$ is nonsingular and Toeplitz, its inverse is persymmetric. Thus,

$$
\begin{aligned}
b_{ij} &= (A^{-1})_{ij} + \frac{v_i v_j}{\gamma} \\
&= (A^{-1})_{n-j,n-i} + \frac{v_i v_j}{\gamma} \\
&= b_{n-j,n-i} - \frac{v_{n-j}v_{n-i}}{\gamma} + \frac{v_i v_j}{\gamma} \\
&= b_{n-j,n-i} + \frac{1}{\gamma}\left(v_i v_j - v_{n-j}v_{n-i}\right).
\end{aligned} \tag{4.7.5}
$$

This indicates that although B is not persymmetric, we can readily compute an element b_{ij} from its reflection across the northeast-southwest axis. Coupling this with the fact that A^{-1} is persymmetric enables us to determine B from its "edges" to its "interior."

Because the order of operations is rather cumbersome to describe, we preview the formal specification of the algorithm pictorially. To this end, assume that we know the last column and row of T_n^{-1}:

$$
T_n^{-1} = \left[\begin{array}{cccccc}
u & u & u & u & u & k \\
u & u & u & u & u & k \\
u & u & u & u & u & k \\
u & u & u & u & u & k \\
u & u & u & u & u & k \\
k & k & k & k & k & k
\end{array}\right].
$$

Here "u" and "k" denote the unknown and the known entries, respectively, and $n = 6$. Alternately exploiting the persymmetry of T_n^{-1} and the recursion (4.7.5), we can compute B, the leading $(n-1)$-by-$(n-1)$ block of T_n^{-1}, as follows:

$$
\overset{\text{persym}}{\longrightarrow}
\begin{bmatrix}
k & k & k & k & k & k \\
k & u & u & u & u & k \\
k & u & u & u & u & k \\
k & u & u & u & u & k \\
k & u & u & u & u & k \\
k & k & k & k & k & k
\end{bmatrix}
\overset{(4.7.5)}{\longrightarrow}
\begin{bmatrix}
k & k & k & k & k & k \\
k & u & u & u & k & k \\
k & u & u & u & k & k \\
k & u & u & u & k & k \\
k & k & k & k & k & k \\
k & k & k & k & k & k
\end{bmatrix}
\overset{\text{persym}}{\longrightarrow}
\begin{bmatrix}
k & k & k & k & k & k \\
k & k & k & k & k & k \\
k & k & u & u & k & k \\
k & k & u & u & k & k \\
k & k & k & k & k & k \\
k & k & k & k & k & k
\end{bmatrix}
$$

$$
\overset{(4.7.5)}{\longrightarrow}
\begin{bmatrix}
k & k & k & k & k & k \\
k & k & k & k & k & k \\
k & k & u & k & k & k \\
k & k & k & k & k & k \\
k & k & k & k & k & k \\
k & k & k & k & k & k
\end{bmatrix}
\overset{\text{persym}}{\longrightarrow}
\begin{bmatrix}
k & k & k & k & k & k \\
k & k & k & k & k & k \\
k & k & k & k & k & k \\
k & k & k & k & k & k \\
k & k & k & k & k & k \\
k & k & k & k & k & k
\end{bmatrix}.
$$

Of course, when computing a matrix that is both symmetric and persymmetric, such as T_n^{-1}, it is only necessary to compute the "upper wedge" of the matrix—e.g.,

$$
\begin{array}{cccccc}
\times & \times & \times & \times & \times & \times \\
 & \times & \times & \times & \times & \\
 & & \times & \times & &
\end{array}
\qquad (n = 6).
$$

With this last observation, we are ready to present the overall algorithm.

Algorithm 4.7.3 (Trench)   Given real numbers $1 = r_0, r_1, \ldots, r_n$ such that $T = (r_{|i-j|}) \in \mathbb{R}^{n \times n}$ is positive definite, the following algorithm computes $B = T_n^{-1}$. Only those b_{ij} for which $i \leq j$ and $i + j \leq n + 1$ are computed.

> Use Algorithm 4.7.1 to solve $T_{n-1} y = -(r_1, \ldots, r_{n-1})^T$.
> $\gamma = 1/(1 + r(1{:}n-1)^T y(1{:}n-1))$
> $v(1{:}n-1) = \gamma y(n-1{:}-1{:}1)$
> $B(1,1) = \gamma$
> $B(1, 2{:}n) = v(n-1{:}-1{:}1)^T$
> **for** $i = 2 : \mathsf{floor}((n-1)/2) + 1$
> > **for** $j = i{:}n-i+1$
> > > $B(i,j) = B(i-1, j-1) + (v(n+1-j)v(n+1-i) - v(i-1)v(j-1))/\gamma$
> > **end**
> **end**

This algorithm requires $13n^2/4$ flops.

4.7.6   Stability Issues

Error analyses for the above algorithms have been performed by Cybenko (1978), and we briefly report on some of his findings.

The key quantities turn out to be the α_k in (4.7.1). In exact arithmetic these scalars satisfy

$$|\alpha_k| < 1$$

and can be used to bound $\| T^{-1} \|_1$:

$$\max \left\{ \frac{1}{\prod\limits_{j=1}^{n-1}(1 - \alpha_j^2)} , \frac{1}{\prod\limits_{j=1}^{n-1}(1 - \alpha_j)} \right\} \leq \| T_n^{-1} \| \leq \prod\limits_{j=1}^{n-1} \frac{1 + |\alpha_j|}{1 - |\alpha_j|}. \tag{4.7.6}$$

Moreover, the solution to the Yule-Walker system $T_n y = -r(1{:}n)$ satisfies

$$\| y \|_1 = \left(\prod_{k=1}^{n}(1 + \alpha_k) \right) - 1 \tag{4.7.7}$$

provided all the α_k are nonnegative.

Now if $\hat{x}$ is the computed Durbin solution to the Yule-Walker equations, then the vector $r_D = T_n \hat{x} + r$ can be bounded as follows

$$\| r_D \| \approx \mathbf{u} \prod_{k=1}^{n}(1 + |\hat{\alpha}_k|),$$

where $\hat{\alpha}_k$ is the computed version of α_k. By way of comparison, since each $|r_i|$ is bounded by unity, it follows that $\| r_C \| \approx \mathbf{u}\| y \|_1$ where r_C is the residual associated with the computed solution obtained via the Cholesky factorization. Note that the two residuals are of comparable magnitude provided (4.7.7) holds. Experimental evidence suggests that this is the case even if some of the α_k are negative. Similar comments apply to the numerical behavior of the Levinson algorithm.

For the Trench method, the computed inverse $\hat{B}$ of T_n^{-1} can be shown to satisfy

$$\frac{\| T_n^{-1} - \hat{B} \|_1}{\| T_n^{-1} \|_1} \approx \mathbf{u} \prod_{k=1}^{n} \frac{1 + |\hat{\alpha}_k|}{1 - |\hat{\alpha}_k|} \, .$$

In light of (4.7.7) we see that the right-hand side is an approximate upper bound for $\mathbf{u}\| T_n^{-1} \|$ which is approximately the size of the relative error when T_n^{-1} is calculated using the Cholesky factorization.

4.7.7   A Toeplitz Eigenvalue Problem

Our discussion of the symmetric eigenvalue problem begins in Chapter 8. However, we are able to describe a solution procedure for an important Toeplitz eigenvalue problem that does not require the heavy machinery from that later chapter. Suppose

$$T = \begin{bmatrix} 1 & r^T \\ r & B \end{bmatrix}$$

is symmetric, positive definite, and Toeplitz with $r \in \mathbb{R}^{n-1}$. Cybenko and Van Loan (1986) show how to pair the Durbin algorithm with Newton's method to compute $\lambda_{\min}(T)$ assuming that

$$\lambda_{\min}(T) < \lambda_{\min}(B). \qquad (4.7.8)$$

This assumption is typically the case in practice. If

$$\begin{bmatrix} 1 & r^T \\ r & B \end{bmatrix} \begin{bmatrix} \alpha \\ y \end{bmatrix} = \lambda_{\min} \begin{bmatrix} \alpha \\ y \end{bmatrix},$$

then $y = -\alpha(B - \lambda_{\min}I)^{-1}r$, $\alpha \neq 0$, and

$$\alpha + r^T \left[-\alpha(B - \lambda_{\min}I)^{-1}r \right] = \lambda_{\min}\alpha.$$

Thus, $\lambda_{\min}$ is a zero of the rational function

$$f(\lambda) = 1 - \lambda - r^T(B - \lambda I)^{-1}r.$$

Note that if $\lambda < \lambda_{\min}(B)$, then

$$f'(\lambda) = -1 - \| (B - \lambda I)^{-1}r \|_2^2 \leq -1,$$
$$f''(\lambda) = -2r^T(B - \lambda I)^{-3}r \leq 0.$$

Using these facts it can be shown that if

$$\lambda_{\min}(T) \leq \lambda^{(0)} < \lambda_{\min}(B), \qquad (4.7.9)$$

then the Newton iteration

$$\lambda^{(k+1)} = \lambda^{(k)} - \frac{f(\lambda^{(k)})}{f'(\lambda^{(k)})} \qquad (4.7.10)$$

converges to $\lambda_{\min}(T)$ monotonically from the right. The iteration has the form

$$\lambda^{(k+1)} = \lambda^{(k)} + \frac{1 + r^T w - \lambda^{(k)}}{1 + w^T w},$$

where w solves the "shifted" Yule-Walker system

$$(B - \lambda^{(k)}I)w = -r.$$

Since $\lambda^{(k)} < \lambda_{\min}(B)$, this system is positive definite and the Durbin algorithm (Algorithm 4.7.1) can be applied to the normalized Toeplitz matrix $(B - \lambda^{(k)}I)/(1 - \lambda^{(k)})$.

The Durbin algorithm can also be used to determine a starting value $\lambda^{(0)}$ that satisfies (4.7.9). If that algorithm is applied to

$$T_\lambda = (T - \lambda I)/(1 - \lambda)$$

then it runs to completion if T_λ is positive definite. In this case, the β_k defined in (4.7.1) are all positive. On the other hand, if $k \leq n-1$, $\beta_k \leq 0$ and $\beta_1, \ldots, \beta_{k-1}$ are all positive, then it follows that $T_\lambda(1{:}k, 1{:}k)$ is positive definite but that $T_\lambda(1{:}k+1, k+1)$ is not. Let $m(\lambda)$ be the index of the first nonpositive β and observe that if $m(\lambda^{(0)}) = n-1$, then $B - \lambda^{(0)}I$ is positive definite and $T - \lambda^{(0)}I$ is not, thereby establishing (4.7.9). A bisection scheme can be formulated to compute $\lambda^{(0)}$ with this property:

$$L = 0$$
$$R = 1 - |r_1|$$
$$\mu = (L + R)/2$$
while $m(\mu) \neq n - 1$
$\quad$ **if** $m(\mu) < n - 1$
$\quad\quad R = \mu$
$\quad$ **else**   (4.7.11)
$\quad\quad L = \mu$
$\quad$ **end**
$\quad \mu = (L + R)/2$
end
$\lambda^{(0)} = \mu$

At all times during the iteration we have $m(L) \leq n - 1 \leq m(R)$. The initial value for R follows from the inequality

$$0 < \lambda_{\min}(T) < \lambda_{\min}(B) \leq \lambda_{\min} \left(\begin{bmatrix} 1 & r_1 \\ r_1 & 1 \end{bmatrix} \right) = 1 - |r_1|.$$

Note that the iterations in (4.7.10) and (4.7.11) involve at most $O(n^2)$ flops per pass. A heuristic argument that $O(\log n)$ iterations are required is given by Cybenko and Van Loan (1986).

4.7.8 Unsymmetric Toeplitz System Solving

We close with some remarks about unsymmetric Toeplitz system-solving. Suppose we are given scalars $r_1, \ldots, r_{n-1}$, $p_1, \ldots, p_{n-1}$, and $b_1, \ldots, b_n$ and that we want to solve a linear system $Tx = b$ of the form

$$\begin{bmatrix} 1 & r_1 & r_2 & r_3 & r_4 \\ p_1 & 1 & r_1 & r_2 & r_3 \\ p_2 & p_1 & 1 & r_1 & r_2 \\ p_3 & p_2 & p_1 & 1 & r_1 \\ p_4 & p_3 & p_2 & p_1 & 1 \end{bmatrix} \begin{bmatrix} x_1 \\ x_2 \\ x_3 \\ x_4 \\ x_5 \end{bmatrix} = \begin{bmatrix} b_1 \\ b_2 \\ b_3 \\ b_4 \\ b_5 \end{bmatrix} \qquad (n = 5).$$

Assume that $T_k = T(1{:}k, 1{:}k)$ is nonsingular for $k = 1{:}n$. It can shown that if we have the solutions to the k-by-k systems

$$\begin{aligned} T_k^T y &= -r = -[\, r_1 \; r_2 \; \cdots \; r_k \,]^T, \\ T_k w &= -p = -[\, p_1 \; p_2 \; \cdots \; p_k \,]^T, \\ T_k x &= b = [\, b_1 \; b_2 \; \cdots \; b_k \,]^T, \end{aligned} \qquad (4.7.12)$$

then we can obtain solutions to

$$
\begin{bmatrix} T_k & \mathcal{E}_k r \\ p^T \mathcal{E}_k & 1 \end{bmatrix}^T \begin{bmatrix} z \\ \alpha \end{bmatrix} = - \begin{bmatrix} r \\ r_{k+1} \end{bmatrix},
$$

$$
\begin{bmatrix} T_k & \mathcal{E}_k r \\ p^T \mathcal{E}_k & 1 \end{bmatrix} \begin{bmatrix} u \\ \nu \end{bmatrix} = - \begin{bmatrix} p \\ p_{k+1} \end{bmatrix}, \qquad (4.7.13)
$$

$$
\begin{bmatrix} T_k & \mathcal{E}_k r \\ p^T \mathcal{E}_k & 1 \end{bmatrix} \begin{bmatrix} v \\ \mu \end{bmatrix} = \begin{bmatrix} b \\ b_{k+1} \end{bmatrix}
$$

in $O(k)$ flops. The update formula derivations are very similar to the Levinson algorithm derivations in §4.7.3. Thus, if the process is repeated for $k = 1{:}n-1$, then we emerge with the solution to $Tx = T_n x = b$. Care must be exercised if a T_k matrix is singular or ill-conditioned. One strategy involves a *lookahead* idea. In this framework, one might transition from the T_k problem directly to the T_{k+2} problem if it is deemed that the T_{k+1} problem is dangerously ill-conditioned. See Chan and Hansen (1992). An alternative approach based on displacement rank is given in §12.1.

Problems

P4.7.1 For any $v \in \mathbb{R}^n$ define the vectors $v_+ = (v + \mathcal{E}_n v)/2$ and $v_- = (v - \mathcal{E}_n v)/2$. Suppose $A \in \mathbb{R}^{n \times n}$ is symmetric and persymmetric. Show that if $Ax = b$ then $Ax_+ = b_+$ and $Ax_- = b_-$.

P4.7.2 Let $U \in \mathbb{R}^{n \times n}$ be the unit upper triangular matrix with the property that $U(1{:}k-1, k) = \mathcal{E}_{k-1} y^{(k-1)}$ where $y^{(k)}$ is defined by (4.7.1). Show that $U^T T_n U = \mathrm{diag}(1, \beta_1, \ldots, \beta_{n-1})$.

P4.7.3 Suppose that $z \in \mathbb{R}^n$ and that $S \in \mathbb{R}^{n \times n}$ is orthogonal. Show that if $X = \left[z, \ Sz, \ \ldots, S^{n-1} z \right]$, then $X^T X$ is Toeplitz.

P4.7.4 Consider the LDL^T factorization of an n-by-n symmetric, tridiagonal, positive definite Toeplitz matrix. Show that d_n and $\ell_{n,n-1}$ converge as $n \to \infty$.

P4.7.5 Show that the product of two lower triangular Toeplitz matrices is Toeplitz.

P4.7.6 Give an algorithm for determining $\mu \in \mathbb{R}$ such that $T_n + \mu \left(e_n e_1^T + e_1 e_n^T \right)$ is singular. Assume $T_n = (r_{|i-j|})$ is positive definite, with $r_0 = 1$.

P4.7.7 Suppose $T \in \mathbb{R}^{n \times n}$ is symmetric, positive definite, and Toeplitz with unit diagonal. What is the smallest perturbation of the the ith diagonal that makes T semidefinite?

P4.7.8 Rewrite Algorithm 4.7.2 so that it does not require the vectors z and v.

P4.7.9 Give an algorithm for computing $\kappa_\infty(T_k)$ for $k = 1{:}n$.

P4.7.10 A p-by-p block matrix $A = (A_{ij})$ with m-by-m blocks is *block Toeplitz* if there exist $A_{-p+1}, \ldots, A_{-1}, A_0, A_1, \ldots, A_{p-1} \in \mathbb{R}^{m \times m}$ so that $A_{ij} = A_{i-j}$, e.g.,

$$
A = \begin{bmatrix} A_0 & A_1 & A_2 & A_3 \\ A_{-1} & A_0 & A_1 & A_2 \\ A_{-2} & A_{-1} & A_0 & A_1 \\ A_{-3} & A_{-2} & A_{-1} & A_0 \end{bmatrix}.
$$

(a) Show that there is a permutation Π such that

$$
\Pi^T A \Pi =: \begin{bmatrix} T_{11} & T_{12} & \cdots & T_{1m} \\ T_{21} & T_{22} & & \vdots \\ \vdots & & \ddots & \vdots \\ T_{m1} & \cdots & & T_{mm} \end{bmatrix}
$$

where each T_{ij} is p-by-p and Toeplitz. Each T_{ij} should be "made up" of (i,j) entries selected from the A_k matrices. (b) What can you say about the T_{ij} if $A_k = A_{-k}$, $k = 1{:}p - 1$?

P4.7.11 Show how to compute the solutions to the systems in (4.7.13) given that the solutions to the systems in (4.7.12) are available. Assume that all the matrices involved are nonsingular. Proceed to develop a fast unsymmetric Toeplitz solver for $Tx = b$ assuming that T's leading principal submatrices are all nonsingular.

P4.7.12 Consider the order-k Yule-Walker system $T_k y^{(k)} = -r^{(k)}$ that arises in (4.7.1). Show that if $y^{(k)} = [y_{k1}, \ldots, y_{kk}]^T$ for $k = 1{:}n - 1$ and

$$
L = \begin{bmatrix}
1 & 0 & 0 & 0 & \cdots & 0 \\
y_{11} & 1 & 0 & 0 & \cdots & 0 \\
y_{22} & y_{21} & 1 & 0 & \cdots & 0 \\
\vdots & \vdots & \vdots & \vdots & \ddots & \vdots \\
y_{n-1,n-1} & y_{n-1,n-2} & y_{n-1,n-3} & \cdots & y_{n-1,1} & 1
\end{bmatrix},
$$

then $L^T T_n L = \mathrm{diag}(1, \beta_1, \ldots, \beta_{n-1})$ where $\beta_k = 1 + r^{(k)^T} y^{(k)}$. Thus, the Durbin algorithm can be thought of as a fast method for computing and LDL^T factorization of T_n^{-1}.

P4.7.13 Show how the Trench algorithm can be used to obtain an initial bracketing interval for the bisection scheme (4.7.11).

Notes and References for §4.7

The original references for the three algorithms described in this section are as follows:

J. Durbin (1960). "The Fitting of Time Series Models," *Rev. Inst. Int. Stat. 28*, 233–243.

N. Levinson (1947). "The Weiner RMS Error Criterion in Filter Design and Prediction," *J. Math. Phys. 25*, 261–278.

W.F. Trench (1964). "An Algorithm for the Inversion of Finite Toeplitz Matrices," *J. SIAM 12*, 515–522.

As is true with the "fast algorithms" area in general, unstable Toeplitz techniques abound and caution must be exercised, see:

G. Cybenko (1978). "Error Analysis of Some Signal Processing Algorithms," PhD Thesis, Princeton University.

G. Cybenko (1980). "The Numerical Stability of the Levinson-Durbin Algorithm for Toeplitz Systems of Equations," *SIAM J. Sci. Stat. Comput. 1*, 303-319.

J.R. Bunch (1985). "Stability of Methods for Solving Toeplitz Systems of Equations," *SIAM J. Sci. Stat. Comput. 6*, 349–364.

E. Linzer (1992). "On the Stability of Solution Methods for Band Toeplitz Systems," *Lin. Alg. Applic. 170*, 1–32.

J.M. Varah (1994). "Backward Error Estimates for Toeplitz Systems," *SIAM J. Matrix Anal. Applic. 15*, 408–417.

A.W. Bojanczyk, R.P. Brent, F.R. de Hoog, and D.R. Sweet (1995). "On the Stability of the Bareiss and Related Toeplitz Factorization Algorithms," *SIAM J. Matrix Anal. Applic. 16*, 40–57.

M.T. Chu, R.E. Funderlic, and R.J. Plemmons (2003). "Structured Low Rank Approximation," *Lin. Alg. Applic. 366*, 157–172.

A. Bottcher and S. M. Grudsky (2004). "Structured Condition Numbers of Large Toeplitz Matrices are Rarely Better than Usual Condition Numbers," *Num. Lin. Alg. 12*, 95–102.

J.-G. Sun (2005). "A Note on Backwards Errors for Structured Linear Systems," *Numer. Lin. Alg. Applic. 12*, 585–603.

P. Favati, G. Lotti, and O. Menchi (2010). "Stability of the Levinson Algorithm for Toeplitz-Like Systems," *SIAM J. Matrix Anal. Applic. 31*, 2531–2552.

Papers concerned with the lookahead idea include:

T.F. Chan and P. Hansen (1992). "A Look-Ahead Levinson Algorithm for Indefinite Toeplitz Systems," *SIAM J. Matrix Anal. Applic. 13*, 490–506.

M. Gutknecht and M. Hochbruck (1995). "Lookahead Levinson and Schur Algorithms for Nonhermitian Toeplitz Systems," *Numer. Math. 70*, 181–227.

M. Van Barel and A. Bultheel (1997). "A Lookahead Algorithm for the Solution of Block Toeplitz Systems," *Lin. Alg. Applic. 266*, 291–335.

Various Toeplitz eigenvalue computations are presented in:

G. Cybenko and C. Van Loan (1986). "Computing the Minimum Eigenvalue of a Symmetric Positive Definite Toeplitz Matrix," *SIAM J. Sci. Stat. Comput. 7*, 123–131.
W.F. Trench (1989). "Numerical Solution of the Eigenvalue Problem for Hermitian Toeplitz Matrices," *SIAM J. Matrix Anal. Appl. 10*, 135–146.
H. Voss (1999). "Symmetric Schemes for Computing the Minimum Eigenvalue of a Symmetric Toeplitz Matrix," *Lin. Alg. Applic. 287*, 359–371.
A. Melman (2004). "Computation of the Smallest Even and Odd Eigenvalues of a Symmetric Positive-Definite Toeplitz Matrix," *SIAM J. Matrix Anal. Applic. 25*, 947–963.

4.8 Circulant and Discrete Poisson Systems

If $A \in \mathbb{C}^{n \times n}$ has a factorization of the form

$$V^{-1}AV = \Lambda = \mathrm{diag}(\lambda_1, \ldots, \lambda_n), \tag{4.8.1}$$

then the columns of V are eigenvectors and the λ_i are the corresponding eigenvalues[2]. In principle, such a decomposition can be used to solve a nonsingular $Au = b$ problem:

$$u = A^{-1}b = (V\Lambda V^{-1})^{-1}b = V(\Lambda^{-1}(V^{-1}b)). \tag{4.8.2}$$

However, if this solution framework is to rival the efficiency of Gaussian elimination or the Cholesky factorization, then V and Λ need to be very special. We say that A has a *fast eigenvalue decomposition* (4.8.1) if

(1) Matrix-vector products of the form $y = Vx$ require $O(n \log n)$ flops to evaluate.

(2) The eigenvalues $\lambda_1, \ldots, \lambda_n$ require $O(n \log n)$ flops to evaluate.

(3) Matrix-vector products of the form $\tilde{b} = V^{-1}b$ require $O(n \log n)$ flops to evaluate.

If these three properties hold, then it follows from (4.8.2) that $O(n \log n)$ flops are required to solve $Au = b$.

Circulant systems and related discrete Poisson systems lend themselves to this strategy and are the main concern of this section. In these applications, the V-matrices are associated with the discrete Fourier transform and various sine and cosine transforms. (Now is the time to review §1.4.1 and §1.4.2 and to recall that we have $n \log n$ methods for the DFT, DST, DST2, and DCT.) It turns out that fast methods exist for the inverse of these transforms and that is important because of (3). We will not be concerned with precise flop counts because in the fast transform "business", some n are friendlier than others from the efficiency point of view. While this issue may be important in practice, it is not something that we have to worry about in our brief, proof-of-concept introduction. Our discussion is modeled after §4.3–§4.5 in Van Loan (FFT) where the reader can find complete derivations and greater algorithmic detail. The interconnection between boundary conditions and fast transforms is a central theme and in that regard we also recommend Strang (1999).

---

[2]This section *does not* depend on Chapters 7 and 8 which deal with computing eigenvalues and eigenvectors. The eigensystems that arise in this section have closed-form expressions and thus the algorithms in those later chapters are not relevant to the discussion.

4.8.1   The Inverse of the DFT Matrix

Recall from §1.4.1 that the DFT matrix $F_n \in \mathbb{C}^{n \times n}$ is defined by

$$[F_n]_{kj} = \omega_n^{(k-1)(j-1)}, \qquad \omega_n = \cos\left(\frac{2\pi}{n}\right) - i \sin\left(\frac{2\pi}{n}\right).$$

It is easy to verify that

$$F_n^H = \bar{F}_n$$

and so for all p and q that satisfy $0 \le p < n$ and $0 \le q < n$ we have

$$F_n(:, p+1)^H F_n(:, q+1) = \sum_{k=0}^{n-1} \bar{\omega}_n^{kp} \omega_n^{kq} = \sum_{k=0}^{n-1} \omega_n^{k(q-p)}.$$

If $q = p$, then this sum equals n. Otherwise,

$$\sum_{k=0}^{n-1} \omega_n^{k(q-p)} = \frac{1 - \omega_n^{n(q-p)}}{1 - \omega_n^{q-p}} = \frac{1-1}{1 - \omega_n^{q-p}} = 0.$$

It follows that

$$n I_n = F_n^H F_n = \bar{F}_n F_n.$$

Thus, the DFT matrix is a scaled unitary matrix and

$$F_n^{-1} = \frac{1}{n} \bar{F}_n.$$

A fast Fourier transform procedure for $F_n x$ can be turned into a fast inverse Fourier transform procedure for $F_n^{-1} x$. Since

$$y = F_n^{-1} x = \frac{1}{n} \bar{F}_n x,$$

simply replace each reference to ω_n with a reference to $\bar{\omega}_n$ and scale. See Algorithm 1.4.1.

4.8.2   Circulant Systems

A *circulant matrix* is a Toeplitz matrix with "wraparound", e.g.,

$$C(z) = \begin{bmatrix} z_0 & z_4 & z_3 & z_2 & z_1 \\ z_1 & z_0 & z_4 & z_3 & z_2 \\ z_2 & z_1 & z_0 & z_4 & z_3 \\ z_3 & z_2 & z_1 & z_0 & z_4 \\ z_4 & z_3 & z_2 & z_1 & z_0 \end{bmatrix}.$$

We assume that the vector z is complex. Any circulant $C(z) \in \mathbb{C}^{n \times n}$ is a linear combination of $I_n, \mathcal{D}_n, \ldots, \mathcal{D}_n^{n-1}$ where $\mathcal{D}_n$ is the downshift permutation defined in §1.2.11. For example, if $n = 5$, then

$$\mathcal{D}_5 = \begin{bmatrix} 0 & 0 & 0 & 0 & 1 \\ 1 & 0 & 0 & 0 & 0 \\ 0 & 1 & 0 & 0 & 0 \\ 0 & 0 & 1 & 0 & 0 \\ 0 & 0 & 0 & 1 & 0 \end{bmatrix}$$

and

$$
\mathcal{D}_5^2 = \begin{bmatrix} 0 & 0 & 0 & 1 & 0 \\ 0 & 0 & 0 & 0 & 1 \\ 1 & 0 & 0 & 0 & 0 \\ 0 & 1 & 0 & 0 & 0 \\ 0 & 0 & 1 & 0 & 0 \end{bmatrix}, \quad
\mathcal{D}_5^3 = \begin{bmatrix} 0 & 0 & 1 & 0 & 0 \\ 0 & 0 & 0 & 1 & 0 \\ 0 & 0 & 0 & 0 & 1 \\ 1 & 0 & 0 & 0 & 0 \\ 0 & 1 & 0 & 0 & 0 \end{bmatrix}, \quad
\mathcal{D}_5^4 = \begin{bmatrix} 0 & 1 & 0 & 0 & 0 \\ 0 & 0 & 1 & 0 & 0 \\ 0 & 0 & 0 & 1 & 0 \\ 0 & 0 & 0 & 0 & 1 \\ 1 & 0 & 0 & 0 & 0 \end{bmatrix}.
$$

Thus, the 5-by-5 circulant matrix displayed above is given by

$$
C(z) = z_0 I + z_1 \mathcal{D}_n + z_2 \mathcal{D}_n^2 + z_3 \mathcal{D}_n^3 + z_4 \mathcal{D}_n^4.
$$

Note that $\mathcal{D}_5^5 = I_5$. More generally,

$$
z = \begin{bmatrix} z_0 \\ z_1 \\ \vdots \\ z_{n-1} \end{bmatrix} \quad \Rightarrow \quad C(z) = \sum_{k=0}^{n-1} z_k \mathcal{D}_n^k. \tag{4.8.3}
$$

Note that if $V^{-1} \mathcal{D}_n V = \Lambda$ is diagonal, then

$$
V^{-1} C(z) V = V^{-1} \left(\sum_{k=0}^{n-1} z_k \mathcal{D}_n^k \right) V = \sum_{k=0}^{n-1} z_k \left(V^{-1} \mathcal{D}_n V^{-1} \right)^k = \sum_{k=0}^{n-1} z_k \Lambda^k \tag{4.8.4}
$$

is diagonal. It turns out that the DFT matrix diagonalizes the downshift permutation.

Lemma 4.8.1. *If $V = F_n$, then $V^{-1} \mathcal{D}_n V = \Lambda = \mathrm{diag}(\lambda_1, \dots, \lambda_n)$ where*

$$
\lambda_{j+1} = \bar{\omega}_n^j = \cos\left(\frac{2j\pi}{n} \right) + i \sin\left(\frac{2j\pi}{n} \right)
$$

for $j = 0{:}n-1$.

Proof. For $j = 0{:}n-1$ we have

$$
\mathcal{D}_n F_n(:, j+1) = \mathcal{D}_n \begin{bmatrix} 1 \\ \omega_n^j \\ \omega_n^{2j} \\ \vdots \\ \omega_n^{(n-1)j} \end{bmatrix} = \begin{bmatrix} \omega_n^{(n-1)j} \\ 1 \\ \omega_n^j \\ \vdots \\ \omega_n^{(n-2)j} \end{bmatrix} = \bar{\omega}_n^j \begin{bmatrix} 1 \\ \omega_n^j \\ \omega_n^{2j} \\ \vdots \\ \omega_n^{(n-1)j} \end{bmatrix}.
$$

This vector is precisely $F_n \Lambda(:, j+1)$. Thus, $\mathcal{D}_n V = V\Lambda$, i.e., $V^{-1} \mathcal{D}_n V = \Lambda$. $\quad\square$

It follows from (4.8.4) that any circulant $C(z)$ is diagonalized by F_n and the eigenvalues of $C(z)$ can be computed fast.

Theorem 4.8.2. *Suppose* $z \in \mathbb{C}^n$ *and* $C(z)$ *are defined by (4.8.3). If* $V = F_n$ *and* $\lambda = \bar{F}_n z$, *then* $V^{-1} C(z) V = \mathrm{diag}(\lambda_1, \ldots, \lambda_n)$.

Proof. Define

$$
f = \begin{bmatrix} 1 \\ \bar{\omega}_n \\ \vdots \\ \bar{\omega}_n^{\,n-1} \end{bmatrix}
$$

and note that the columns of $\bar{F}_n$ are componentwise powers of this vector. In particular, $\bar{F}_n(:, k+1) = f.\hat{\ }k$ where $[f.\hat{\ }k]_j = f_j^k$. Since $\Lambda = \mathrm{diag}(f)$, it follows from Lemma 4.8.1 that

$$
V^{-1} C(z) V = \sum_{k=0}^{n-1} z_k \Lambda^k = \sum_{k=0}^{n-1} z_k \, \mathrm{diag}(f)^k = \sum_{k=0}^{n-1} z_k \, \mathrm{diag}(f.\hat{\ }k)
$$

$$
= \mathrm{diag}\left(\sum_{k=0}^{n-1} z_k \, f.\hat{\ }k \right) = \mathrm{diag}\left(\bar{F}_n z \right)
$$

completing the proof of the theorem $\square$

Thus, the eigenvalues of the circulant matrix $C(z)$ are the components of the vector $\bar{F}_n z$. Using this result we obtain the following algorithm.

Algorithm 4.8.1 If $z \in \mathbb{C}^n$, $y \in \mathbb{C}^n$, and $C(z)$ is nonsingular, then the following algorithm solves the linear system $C(z)x = y$.

> Use an FFT to compute $c = \bar{F}_n y$ and $d = \bar{F}_n z$.
> $w = c./d$
> Use an FFT to compute $u = F_n w$.
> $x = u/n$

This algorithm requires $O(n \log n)$ flops.

4.8.3   The Discretized Poisson Equation in One Dimension

We now turn our attention to a family of *real* matrices that have *real*, fast eigenvalue decompositions. The starting point in the discussion is the differential equation

$$
\frac{d^2 u}{dx^2} = -f(x) \qquad \alpha \le u(x) \le \beta, \tag{4.8.5}
$$

together with one of four possible specifications of $u(x)$ on the boundary.

Dirichlet-Dirichlet (DD):	$u(\alpha) = u_\alpha,$	$u(\beta) = u_\beta,$
Dirichlet-Neumann (DN):	$u(\alpha) = u_\alpha,$	$u'(\beta) = u'_\beta,$
Neumann-Neumann (NN):	$u'(\alpha) = u'_\alpha,$	$u'(\beta) = u'_\beta,$
Periodic (P):	$u(\alpha) = u(\beta).$	

By replacing the derivatives in (4.8.5) with divided differences, we obtain a system of linear equations. Indeed, if m is a positive integer and

$$h = \frac{\beta - \alpha}{m},$$

then for $i = 1{:}m - 1$ we have

$$\frac{\dfrac{u_{i+1} - u_i}{h} - \dfrac{u_i - u_{i-1}}{h}}{h} = \frac{u_{i-1} - 2u_i + u_{i+1}}{h^2} = -f_i \qquad (4.8.6)$$

where $f_i = f(\alpha + ih)$ and $u_i \approx u(\alpha + ih)$. To appreciate this discretization we display the linear equations that result when $m = 5$ for the various possible boundary conditions. The matrices $\mathcal{T}_n^{(DD)}$, $\mathcal{T}_n^{(DN)}$, $\mathcal{T}_n^{(NN)}$, and $\mathcal{T}_n^{(P)}$ are formally defined afterwards.

For the Dirichlet-Dirichlet problem, the system is 4-by-4 and tridiagonal:

$$\mathcal{T}_4^{(DD)} \cdot u(1{:}4) \equiv \begin{bmatrix} 2 & -1 & 0 & 0 \\ -1 & 2 & -1 & 0 \\ 0 & -1 & 2 & -1 \\ 0 & 0 & -1 & 2 \end{bmatrix} \begin{bmatrix} u_1 \\ u_2 \\ u_3 \\ u_4 \end{bmatrix} = \begin{bmatrix} h^2 f_1 + u_\alpha \\ h^2 f_2 \\ h^2 f_3 \\ h^2 f_4 + u_\beta \end{bmatrix}.$$

For the Dirichlet-Neumann problem the system is still tridiagonal, but u_5 joins $u_1, \ldots, u_4$ as an unknown:

$$\mathcal{T}_5^{(DN)} \cdot u(1{:}5) \equiv \begin{bmatrix} 2 & -1 & 0 & 0 & 0 \\ -1 & 2 & -1 & 0 & 0 \\ 0 & -1 & 2 & -1 & 0 \\ 0 & 0 & -1 & 2 & -1 \\ 0 & 0 & 0 & -2 & 2 \end{bmatrix} \begin{bmatrix} u_1 \\ u_2 \\ u_3 \\ u_4 \\ u_5 \end{bmatrix} = \begin{bmatrix} h^2 f_1 + u_\alpha \\ h^2 f_2 \\ h^2 f_3 \\ h^2 f_4 \\ 2h u'_\beta \end{bmatrix}.$$

The new equation on the bottom is derived from the approximation $u'(\beta) \approx (u_5 - u_4)/h$. (The scaling of this equation by 2 simplifies some of the derivations below.) For the Neumann-Neumann problem, u_5 *and* u_0 need to be determined:

$$\mathcal{T}_6^{(NN)} \cdot u(0{:}5) \equiv \begin{bmatrix} 2 & -2 & 0 & 0 & 0 & 0 \\ -1 & 2 & -1 & 0 & 0 & 0 \\ 0 & -1 & 2 & -1 & 0 & 0 \\ 0 & 0 & -1 & 2 & -1 & 0 \\ 0 & 0 & 0 & -1 & 2 & -1 \\ 0 & 0 & 0 & 0 & -2 & 2 \end{bmatrix} \begin{bmatrix} u_0 \\ u_1 \\ u_2 \\ u_3 \\ u_4 \\ u_5 \end{bmatrix} = \begin{bmatrix} -2h u'_\alpha \\ h^2 f_1 \\ h^2 f_2 \\ h^2 f_3 \\ h^2 f_3 \\ 2h u'_\beta \end{bmatrix}.$$

Finally, for the periodic problem we have

$$\mathcal{T}_5^{(P)} \cdot u(1{:}5) \equiv \begin{bmatrix} 2 & -1 & 0 & 0 & -1 \\ -1 & 2 & -1 & 0 & 0 \\ 0 & -1 & 2 & -1 & 0 \\ 0 & 0 & -1 & 2 & -1 \\ -1 & 0 & 0 & -1 & 2 \end{bmatrix} \begin{bmatrix} u_1 \\ u_2 \\ u_3 \\ u_4 \\ u_5 \end{bmatrix} = \begin{bmatrix} h^2 f_1 \\ h^2 f_2 \\ h^2 f_3 \\ h^2 f_4 \\ h^2 f_5 \end{bmatrix}.$$

The first and last equations use the conditions $u_0 = u_5$ and $u_1 = u_6$. These constraints follow from the assumption that u has period $\beta - \alpha$.

As we show below, the n-by-n matrix

$$
\mathcal{T}_n^{(DD)} = \begin{bmatrix} 2 & -1 & \cdots & 0 \\ -1 & 2 & \ddots & \vdots \\ \vdots & \ddots & \ddots & -1 \\ 0 & \cdots & -1 & 2 \end{bmatrix} \tag{4.8.7}
$$

and its low-rank adjustments

$$
\mathcal{T}_n^{(DN)} = \mathcal{T}_n^{(DD)} - e_n e_{n-1}^T, \tag{4.8.8}
$$

$$
\mathcal{T}_n^{(NN)} = \mathcal{T}_n^{(DD)} - e_n e_{n-1}^T - e_1 e_2^T, \tag{4.8.9}
$$

$$
\mathcal{T}_n^{(P)} = \mathcal{T}_n^{(DD)} - e_1 e_n^T - e_n e_1^T. \tag{4.8.10}
$$

have fast eigenvalue decompositions. However, the existence of $O(n \log n)$ methods for these systems is not very interesting because algorithms based on Gaussian elimination are faster: $O(n)$ versus $O(n \log n)$. Things get much more interesting when we discretize the 2-dimensional analogue of (4.8.5).

4.8.4   The Discretized Poisson Equation in Two Dimensions

To launch the 2D discussion, suppose $F(x, y)$ is defined on the rectangle

$$
R = \{(x, y) : \alpha_x \le x \le \beta_x, \ \alpha_y \le y \le \beta_y\}
$$

and that we wish to find a function u that satisfies

$$
\frac{\partial^2 u}{\partial x^2} + \frac{\partial^2 u}{\partial y^2} = -F(x, y) \tag{4.8.11}
$$

on R and has its value prescribed on the boundary of R. This is *Poisson's equation with Dirichlet boundary conditions.* Our plan is to approximate u at the grid points $(\alpha_x + ih_x, \alpha_y + jh_y)$ where $i = 1{:}m_1 - 1$, $j = 1{:}m_2 - 1$, and

$$
h_x = \frac{\beta_x - \alpha_x}{m_1} \qquad h_y = \frac{\beta_y - \alpha_y}{m_2}.
$$

Refer to Figure 4.8.1, which displays the case when $m_1 = 6$ and $m_2 = 5$. Notice that there are two kinds of grid points. The function u is known at the "$\bullet$" grid points on the boundary. The function u is to be determined at the "$\circ$" grid points in the interior. The interior grid points have been indexed in a top-to-bottom, left-to-right order. The idea is to have u_k approximate the value of $u(x, y)$ at grid point k.

As in the one-dimensional problem considered §4.8.3, we use divided differences to obtain a set of linear equations that define the unknowns. An interior grid point P has a north (N), east (E), south (S), and west (W) neighbor. Using this "compass point" notation we obtain the following approximation to (4.8.11) at P:

$$
\frac{\dfrac{u(E) - u(P)}{h_x} - \dfrac{u(P) - u(W)}{h_x}}{h_x} + \frac{\dfrac{u(N) - u(P)}{h_y} - \dfrac{u(P) - u(S)}{h_y}}{h_y} = -F(P)
$$

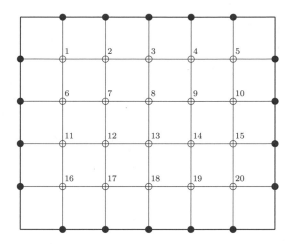

Figure 4.8.1. *A grid with $m_1 = 6$ and $m_2 = 5$.*

The x-partial and y-partial have been replaced by second-order divided differences. Assume for clarity that the horizontal and vertical grid spacings are equal, i.e., $h_x = h_y = h$. With this assumption, the linear equation at point P has the form

$$4u(P) - u(N) - u(E) - u(S) - u(W) = h^2 F(P).$$

In our example, there are 20 such equations. It should be noted that some of P's neighbors may be on the boundary, in which case the corresponding linear equation involves fewer than 5 unknowns. For example, if P is the third grid point then we see from Figure 4.8.1 that the north neighbor N is on the boundary. It follows that the associated linear equation has the form

$$4u(P) - u(E) - u(S) - u(W) = h^2 F(P) + u(N).$$

Reasoning like this, we conclude that the matrix of coefficients has the following block tridiagonal form

$$A = \begin{bmatrix} \mathcal{T}_5^{(DD)} & 0 & 0 & 0 \\ 0 & \mathcal{T}_5^{(DD)} & 0 & 0 \\ 0 & 0 & \mathcal{T}_5^{(DD)} & 0 \\ 0 & 0 & 0 & \mathcal{T}_5^{(DD)} \end{bmatrix} + \begin{bmatrix} 2I_5 & -I_5 & 0 & 0 \\ -I_5 & 2I_5 & -I_5 & 0 \\ 0 & -I_5 & 2I_5 & -I_5 \\ 0 & 0 & -I_5 & 2I_5 \end{bmatrix}.$$

i.e.,

$$A = I_4 \otimes \mathcal{T}_5^{(DD)} + \mathcal{T}_4^{(DD)} \otimes I_5.$$

Notice that the first matrix is associated with the x-partials while the second matrix is associated with the y-partials. The right-hand side in $Au - b$ is made up of F-evaluations and specified values of $u(x, y)$ on the boundary.

Extrapolating from our example, we conclude that the matrix of coefficients is an $(m_2 - 1)$-by-$(m_2 - 1)$ block tridiagonal matrix with $(m_1 - 1)$-by-$(m_1 - 1)$ blocks:

$$A = I_{m_2-1} \otimes \mathcal{T}_{m_1-1}^{(DD)} + \mathcal{T}_{m_2-1}^{(DD)} \otimes I_{m_1-1}.$$

Alternative specifications along the boundary lead to systems with similar structure, e.g.,

$$Au \equiv (I_{n_2} \otimes A_1 + A_2 \otimes I_{n_1}) u = b. \tag{4.8.12}$$

For example, if we impose Dirichet-Neumann, Neumann-Neumann, or periodic boundary conditions along the left and right edges of the rectangular domain R, then A_1 will equal $\mathcal{T}_{m_1}^{(DN)}$, $\mathcal{T}_{m_1+1}^{(NN)}$, or $\mathcal{T}_{m_1}^{(P)}$ accordingly. Likewise, if we impose Dirichet-Neumann, Neumann-Neumann, or periodic boundary conditions along the bottom and top edges of R, then A_2 will equal $\mathcal{T}_{m_2}^{(DN)}$, $\mathcal{T}_{m_2+1}^{(NN)}$, or $\mathcal{T}_{m_2}^{(P)}$. If the system (4.8.12) is nonsingular and A_1 and A_2 have fast eigenvalue decompositions, then it can be solved with just $O(N \log N)$ flops where $N = n_1 n_2$. To see why this is possible, assume that

$$V^{-1}A_1 V = D_1 = \text{diag}(\lambda_1, \ldots, \lambda_{n_1}), \tag{4.8.13}$$

$$W^{-1}A_2 W = D_2 = \text{diag}(\mu_1, \ldots, \mu_{n_2}) \tag{4.8.14}$$

are fast eigenvalue decompositions. Using facts about the Kronecker product that are set forth in §1.3.6–§1.3.8, we can reformulate (4.8.12) as a matrix equation

$$A_1 U + U A_2^T = B$$

where $U = \text{reshape}(u, n_1, n_2)$ and $B = \text{reshape}(b, n_1, n_2)$. Substituting the above eigenvalue decompositions into this equation we obtain

$$D_1 \tilde{U} + \tilde{U} D_2 = \tilde{B},$$

where $\tilde{U} = (\tilde{u}_{ij}) = V^{-1} U W^{-T}$ and $\tilde{B} = (\tilde{b}_{ij}) = V^{-1} B W^{-T}$. Note how easy it is to solve this transformed system because D_1 and D_2 are diagonal:

$$\tilde{u}_{ij} = \frac{\tilde{b}_{ij}}{\lambda_i + \mu_j} \qquad i = 1{:}n_1, \ j = 1{:}n_2.$$

For this to be well-defined, no eigenvalue of A_1 can be the negative of an eigenvalue of A_2. In our example, all the λ_i and μ_i are positive. Overall we obtain

Algorithm 4.8.2 (**Fast Poisson Solver Framework**) Assume that $A_1 \in \mathbb{R}^{n_1 \times n_1}$ and $A_2 \in \mathbb{R}^{n_2 \times n_2}$ have fast eigenvalue decompositions (4.8.13) and (4.8.14) and that the matrix $A = I_{n_2} \otimes A_1 + A_2 \otimes I_{n_1}$ is nonsingular. The following algorithm solves the linear system $Au = b$ where $b \in \mathbb{R}^{n_1 n_2}$.

$\quad \tilde{B} = (W^{-1}(V^{-1}B)^T)^T$ where $B = \text{reshape}(b, n_1, n_2)$
$\quad \textbf{for } i = 1{:}n_1$
$\quad\quad \textbf{for } j = 1{:}n_2$
$\quad\quad\quad \tilde{u}_{ij} = \tilde{b}_{ij}/(\lambda_i + \mu_j)$
$\quad\quad \textbf{end}$
$\quad \textbf{end}$
$\quad u = \text{reshape}(U, n_1 n_2, 1)$ where $U = (W(V\tilde{U})^T)^T$

The following table accounts for the work involved:

Operation	How Many?	Work
V^{-1} times n_1-vector	n_2	$O(n_2 \cdot n_1 \cdot \log n_1)$
W^{-1} times n_2-vector	n_1	$O(n_1 \cdot n_2 \cdot \log n_2)$
V times n_1-vector	n_2	$O(n_2 \cdot n_1 \cdot \log n_1)$
W times n_2-vector	n_1	$O(n_1 \cdot n_2 \cdot \log n_2)$

Adding up the operation counts, we see that $O(n_1 n_2 \log(n_1 n_2)) = O(N \log N)$ flops are required where $N = n_1 n_2$ is the size of the matrix A.

Below we show that the matrices $\mathcal{T}_n^{(DD)}$, $\mathcal{T}_n^{(DN)}$, $\mathcal{T}_n^{(NN)}$, and $\mathcal{T}_n^{(P)}$ have fast eigenvalue decompositions and this means that Algorithm 4.8.2 can be used to solve discrete Poisson systems. To appreciate the speedup over conventional methods, suppose $A_1 = \mathcal{T}_{n_1}^{(DD)}$ and $A_2 = \mathcal{T}_{n_2}^{(DD)}$. It can be shown that A is symmetric positive definite with bandwidth $n_1 + 1$. Solving $Au = b$ using Algorithm 4.3.5 (band Cholesky) would require $O(n_1^3 n_2) = O(N n_1^2)$ flops.

4.8.5 The Inverse of the DST and DCT Matrices

The eigenvector matrices for $\mathcal{T}_n^{(DD)}$, $\mathcal{T}_n^{(DN)}$, $\mathcal{T}_n^{(NN)}$, and $\mathcal{T}_n^{(P)}$ are associated with the fast trigonometric transforms presented in §1.4.2. It is incumbent upon us to show that the inverse of these transforms can also be computed fast. We do this for the discrete sine transform (DST) and the discrete cosine transform (DCT) and leave similar fast inverse verifications to the exercises at the end of the section.

By considering the blocks of the DFT matrix F_{2m}, we can determine the inverses of the transform matrices $\text{DST}(m-1)$ and $\text{DCT}(m+1)$. Recall from §1.4.2 that if $C_r \in \mathbb{R}^{r \times r}$ and $S_r \in \mathbb{R}^{r \times r}$ are defined by

$$[C_r]_{kj} = \cos\left(\frac{kj\pi}{r+1}\right), \qquad [S_r]_{kj} = \sin\left(\frac{kj\pi}{r+1}\right)$$

then

$$F_{2m} = \begin{bmatrix} 1 & e^T & 1 & e^T \\ e & C - iS & v & (C+iS)E \\ 1 & v^T & (-1)^m & v^T E \\ e & E(C+iS) & Ev & E(C-iS)E \end{bmatrix}$$

where $C = C_{m-1}$, $S = S_{m-1}$, $E = \mathcal{E}_{m-1}$, and

$$e^T = (\underbrace{1, 1, \ldots, 1}_{m-1}) \qquad v^T = (\underbrace{-1, 1, \ldots, (-1)^{m-1}}_{m-1}).$$

By comparing the (2,1), (2,2), (2,3), and (2,4) blocks in the equation $2mI = \bar{F}_{2m} F_{2m}$ we conclude that

$$0 = 2Ce + e + v,$$

$$2mI_{m-1} = 2C^2 + 2S^2 + ee^T + vv^T,$$

$$0 = 2Cv + e + (-1)^m v,$$

$$0 = 2C^2 - 2S^2 + ee^T + vv^T.$$

It follows that $2S^2 = mI_{m-1}$ and $2C^2 = mI_{m-1} - ee^T - vv^T$. Using these equations it is easy to verify that

$$S_{m-1}^{-1} = \frac{2}{m} S_{m-1}$$

and

$$\begin{bmatrix} 1/2 & e^T & 1/2 \\ e/2 & C_{m-1} & v/2 \\ 1/2 & v^T & (-1)^m/2 \end{bmatrix}^{-1} = \frac{2}{m} \begin{bmatrix} 1/2 & e^T & 1/2 \\ e/2 & C_{m-1} & v/2 \\ 1/2 & v^T & (-1)^m/2 \end{bmatrix}.$$

Thus, it follows from the definitions (1.4.8) and (1.4.10) that

$$V = \mathrm{DST}(m-1) \;\Rightarrow\; V^{-1} = \frac{2}{m}\,\mathrm{DST}(m-1),$$

$$V = \mathrm{DCT}(m+1) \;\Rightarrow\; V^{-1} = \frac{2}{m}\,\mathrm{DCT}(m+1).$$

In both cases, the inverse transform is a multiple of the "forward" transform and can be computed fast. See Algorithms 1.4.2 and 1.4.3.

4.8.6 Four Fast Eigenvalue Decompositions

The matrices $\mathcal{T}_n^{(DD)}$, $\mathcal{T}_n^{(DN)}$, $\mathcal{T}_n^{(NN)}$, and $\mathcal{T}_n^{(P)}$ do special things to vectors of sines and cosines.

Lemma 4.8.3. *Define the real n-vectors $s(\theta)$ and $c(\theta)$ by*

$$s(\theta) = \begin{bmatrix} s_1 \\ \vdots \\ s_n \end{bmatrix}, \qquad c(\theta) = \begin{bmatrix} c_0 \\ \vdots \\ c_{n-1} \end{bmatrix}, \tag{4.8.15}$$

where $s_k = \sin(k\theta)$ and $c_k = \cos(k\theta)$. If $e_k = I_n(:,k)$ and $\lambda = 4\sin^2(\theta/2)$, then

$$\mathcal{T}_n^{(DD)} \cdot s(\theta) = \lambda \cdot s(\theta) + s_{n+1} e_n, \tag{4.8.16}$$

$$\mathcal{T}_n^{(DD)} \cdot c(\theta) = \lambda \cdot c(\theta) + c_1 e_1 + c_n e_n, \tag{4.8.17}$$

$$\mathcal{T}_n^{(DN)} \cdot s(\theta) = \lambda \cdot s(\theta) + (s_{n+1} - s_{n-1}) e_n, \tag{4.8.18}$$

$$\mathcal{T}_n^{(NN)} \cdot c(\theta) = \lambda \cdot c(\theta) + (c_n - c_{n-2}) e_n, \tag{4.8.19}$$

$$\mathcal{T}_n^{(P)} \cdot s(\theta) = \lambda \cdot s(\theta) - s_n e_1 + (s_{n+1} - s_1) e_n, \tag{4.8.20}$$

$$\mathcal{T}_n^{(P)} \cdot c(\theta) = \lambda \cdot c(\theta) + (c_1 - c_{n-1}) e_1 + (c_n - 1) e_n. \tag{4.8.21}$$

Proof. The proof is mainly an exercise in using the trigonometric identities

$$s_{k-1} = c_1 s_k - s_1 c_k, \qquad c_{k-1} = c_1 c_k + s_1 s_k,$$
$$s_{k+1} = c_1 s_k + s_1 c_k, \qquad c_{k+1} = c_1 c_k - s_1 s_k.$$

For example, if $y = \mathcal{T}_n^{(DD)} s(\theta)$, then

$$y_k = \begin{cases} 2s_1 - s_2 = 2s_1(1 - c_1), & \text{if } k = 1, \\ -s_{k-1} + 2s_k - s_{k+1} = 2s_k(1 - c_1), & \text{if } 2 \leq k \leq n - 1, \\ -s_{n-1} + 2s_n = 2s_n(1 - c_1) + s_{n+1}, & \text{if } k = n. \end{cases}$$

Equation (4.8.16) follows since $(1 - c_1) = 1 - \cos(\theta) = 2\sin^2(\theta/2)$. The proof of (4.8.17) is similar while the remaining equations follow from Equations (4.8.8)–(4.8.10). □

Notice that (4.8.16)-(4.8.21) are eigenvector equations except for the "e_1" and "e_n" terms. By choosing the right value for θ, we can make these residuals disappear, thereby obtaining recipes for the eigensystems of $\mathcal{T}_n^{(DD)}$, $\mathcal{T}_n^{(DN)}$, $\mathcal{T}_n^{(NN)}$, and $\mathcal{T}_n^{(P)}$.

The Dirichlet-Dirichlet Matrix

If j is an integer and $\theta = j\pi/(n+1)$, then $s_{n+1} = \sin((n+1)\theta) = 0$. It follows from (4.8.16) that

$$\mathcal{T}_n^{(DD)} s(\theta_j) = 4\sin^2(\theta_j/2)s(\theta_j), \qquad \theta_j = \frac{j\pi}{n+1},$$

for $j = 1{:}n$. Thus, the columns of the matrix $V_n^{(DD)} \in \mathbb{R}^{n \times n}$ defined by

$$[V_n^{(DD)}]_{kj} = \sin\left(\frac{kj\pi}{n+1}\right)$$

are eigenvectors for $\mathcal{T}_n^{(DD)}$ and the corresponding eigenvalues are given by

$$\lambda_j = 4\sin^2\left(\frac{j\pi}{2(n+1)}\right),$$

for $j = 1{:}n$. Note that $V_n^{(DD)} = \text{DST}(n)$. It follows that $\mathcal{T}_n^{(DD)}$ has a fast eigenvalue decomposition.

The Dirichlet-Neumann Matrix

If j is an integer and $\theta = (2j-1)\pi/(2n)$, then $s_{n+1} - s_{n-1} = 2s_1 c_n = 0$. It follows from (4.8.18) that

$$\mathcal{T}_n^{(DN)} \cdot s(\theta_j) = 4\sin^2(\theta_j/2) \cdot s(\theta_j), \qquad \theta_j = \frac{(2j-1)\pi}{2n},$$

for $j = 1{:}n$. Thus, the columns of the matrix $V_n^{(DN)} \in \mathbb{R}^{n \times n}$ defined by

$$[V_n^{(DN)}]_{kj} = \sin\left(\frac{k(2j-1)\pi}{2n}\right)$$

are eigenvectors of the matrix $\mathcal{T}_n^{(DN)}$ and the corresponding eigenvalues are given by

$$\lambda_j = 4\sin^2\left(\frac{(2j-1)\pi}{4n}\right)$$

for $j = 1{:}n$. Comparing with (1.4.13) we see that that $V_n^{(DN)} = \text{DST2}(n)$. The inverse DST2 can be evaluated fast. See Van Loan (FFT, p. 242) for details, but also P4.8.11. It follows that $\mathcal{T}^{(DN)}$ has a fast eigenvalue decomposition.

The Neumann-Neumann Matrix

If j is an integer and $\theta = (j-1)\pi/(n-1)$, then $c_n - c_{n-2} = -2s_1 s_{n-1} = 0$. It follows from (4.8.19) that

$$\mathcal{T}_n^{(NN)} \cdot c(\theta_j) = 4\sin^2\left(\frac{\theta_j}{2}\right) \cdot c(\theta_j), \qquad \theta_j = \frac{(j-1)\pi}{n-1}.$$

Thus, the columns of the matrix $V_n^{(DN)} \in \mathbb{R}^{n\times n}$ defined by

$$[V_n^{(NN)}]_{kj} = \cos\left(\frac{(k-1)(j-1)\pi}{n-1}\right)$$

are eigenvectors of the matrix $\mathcal{T}_n^{(DN)}$ and the corresponding eigenvalues are given by

$$\lambda_j = 4\sin^2\left(\frac{(j-1)\pi}{2(n-1)}\right)$$

for $j = 1{:}n$. Comparing with (1.4.10) we see that

$$V_n^{(NN)} = \text{DCT}(n) \cdot \text{diag}(2, I_{n-2}, 2)$$

and therefore $\mathcal{T}^{(NN)}$ has a fast eigenvalue decomposition.

The Periodic Matrix

We can proceed to work out the eigenvalue decomposition for $\mathcal{T}_n^{(P)}$ as we did in the previous three cases, i.e., by zeroing the residuals in (4.8.20) and (4.8.21). However, $\mathcal{T}_n^{(P)}$ is a circulant matrix and so we know from Theorem 4.8.2 that

$$F_n^{-1}\mathcal{T}_n^{(P)} F_n = \text{diag}(\lambda_1, \ldots, \lambda_n)$$

where

$$\lambda = \bar{F}_n \begin{bmatrix} 2 \\ -1 \\ 0 \\ \vdots \\ -1 \end{bmatrix} = 2\bar{F}_n(:,1) - \bar{F}_n(:,2) - \bar{F}_n(:,n).$$

It can be shown that

$$\lambda_j = 4\sin^2\left(\frac{(j-1)\pi}{n}\right)$$

for $j = 1{:}n$. It follows that $\mathcal{T}_n^{(P)}$ has a fast eigenvalue decomposition. However, since this matrix is real it is preferable to have a real V-matrix. Using the facts that

$$\lambda_j = \lambda_{n+2-j} \tag{4.8.22}$$

and

$$\bar{F}_n(:,j) = F_n(:,(n+2-j)) \tag{4.8.23}$$

for $j = 2{:}n$, it can be shown that if $m = \mathsf{ceil}((n+1)/2)$ and

$$V_n^{(P)} = [\,\mathsf{Re}(F_n(:,1{:}m)) \mid \mathsf{Im}(F_n(:,m+1{:}n))\,] \tag{4.8.24}$$

then

$$\mathcal{T}_n^{(P)} V_n^{(P)}(:,j) = \lambda_j V_n^{(P)}(:,j) \tag{4.8.25}$$

for $j = 1{:}n$. Manipulations with this real matrix and its inverse can be carried out rapidly as discussed in Van Loan (FFT, Chap. 4).

4.8.7   A Note on Symmetry and Boundary Conditions

In our presentation, the matrices $\mathcal{T}_n^{(DN)}$ and $\mathcal{T}_n^{(NN)}$ are not symmetric. However, a simple diagonal similarity transformation changes this. For example, if $D = \mathrm{diag}(I_{n-1}, \sqrt{2})$, then $D^{-1}\mathcal{T}_n^{(DN)}D$ is symmetric. Working with symmetric second difference matrices has certain attractions, i.e., the automatic orthogonality of the eigenvector matrix. See Strang (1999).

Problems

P4.8.1 Suppose $z \in \mathbb{R}^n$ has the property that $z(2{:}n) = \mathcal{E}_{n-1}z(2{:}n)$. Show that $C(z)$ is symmetric and $\bar{F}_n z$ is real.

P4.8.2 As measured in the Frobenius norm, what is the nearest real circulant matrix to a given real Toeplitz matrix?

P4.8.3 Given $x, z \in \mathbb{C}^n$, show how to compute $y = C(z)\cdot x$ in $O(n \log n)$ flops. In this case, y is the *cyclic convolution* of x and z.

P4.8.4 Suppose $a = [\,a_{-n+1}, \ldots, a_{-1}, a_0, a_1, \ldots, a_{n-1}\,]$ and let $T = (t_{kj})$ be the n-by-n Toeplitz matrix defined by $t_{kj} = a_{k-j}$. Thus, if $a = [\,a_{-2}, a_{-1}, a_0, a_1, a_2\,]$, then

$$T = T(a) = \begin{bmatrix} a_0 & a_{-1} & a_{-2} \\ a_1 & a_0 & a_{-1} \\ a_2 & a_1 & a_0 \end{bmatrix}.$$

It is possible to "embed" T into a circulant, e.g.,

$$C = \begin{bmatrix} a_0 & a_{-1} & a_{-2} & 0 & 0 & 0 & a_2 & a_1 \\ a_1 & a_0 & a_{-1} & a_{-2} & 0 & 0 & 0 & a_2 \\ a_2 & a_1 & a_0 & a_{-1} & a_{-2} & 0 & 0 & 0 \\ 0 & a_2 & a_1 & a_0 & a_{-1} & a_{-2} & 0 & 0 \\ 0 & 0 & a_2 & a_1 & a_0 & a_{-1} & a_{-2} & 0 \\ 0 & 0 & 0 & a_2 & a_1 & a_0 & a_{-1} & a_{-2} \\ a_{-2} & 0 & 0 & 0 & a_2 & a_1 & a_0 & a_{-1} \\ a_{-1} & a_{-2} & 0 & 0 & 0 & a_2 & a_1 & a_0 \end{bmatrix}.$$

Given $a_{-n+1}, \ldots, a_{-1}, 1_0, a_1, \ldots, a_{n-1}$ and $m \geq 2n-1$, show how to construct a vector $v \in \mathbb{C}^m$ so that if $C = C(v)$, then $C(1{:}n, 1{:}n) = T$. Note that v is not unique if $m > 2n-1$.

P4.8.5 Complete the proof of Lemma 4.8.3.

P4.8.6 Show how to compute a Toeplitz-vector product $y = Tu$ in $n \log n$ time using the embedding idea outlined in the previous problem and the fact that circulant matrices have a fast eigenvalue decomposition.

P4.8.7 Give a complete specification of the vector b in (4.8.12) if $A_1 = \mathcal{T}_{n_1}^{(DD)}$, $A_2 = \mathcal{T}_{n_2}^{(DD)}$, and $u(x, y) = 0$ on the boundary of the rectangular domain R. In terms of the underlying grid, $n_1 = m_1 - 1$ and $n_2 = m_2 - 1$.

P4.8.8 Give a complete specification of the vector b in (4.8.12) if $A_1 = \mathcal{T}_{n_1}^{(DN)}$, $A_2 = \mathcal{T}_{n_2}^{(DN)}$, $u(x, y) = 0$ on the bottom and left edge of R, $u_x(x, y) = 0$ along the right edge of R, and $u_y(x, y) = 0$ along the top edge of R. In terms of the underlying grid, $n_1 = m_1$ and $n_2 = m_2$.

P4.8.9 Define a Neumann-Dirichlet matrix $\mathcal{T}_n^{(ND)}$ that would arise in conjunction with (4.8.5) if $u'(\alpha)$ and $u(\beta)$ were specified. Show that $\mathcal{T}_n^{(ND)}$ has a fast eigenvalue decomposition.

P4.8.10 . The matrices $\mathcal{T}_n^{(NN)}$ and $\mathcal{T}_n^{(P)}$ are singular. (a) Assuming that b is in the range of $A = I_{n_2} \otimes \mathcal{T}_{n_1}^{(P)} + \mathcal{T}_{n_2}^{(P)} \otimes I_{n_1}$, how would you solve the linear system $Au = b$ subject to the constraint that the mean of u's components is zero? Note that this constraint makes the system solvable. (b) Repeat part (a) replacing $\mathcal{T}_{n_1}^{(P)}$ with $\mathcal{T}_{n_1}^{(NN)}$ and $\mathcal{T}_{n_2}^{(P)}$ with $\mathcal{T}_{n_2}^{(NN)}$.

P4.8.11 Let V be the matrix that defines the DST2(n) transformation in (1.4.12). (a) Show that

$$V^T V = \frac{n}{2} I_n + \frac{1}{2} vv^T$$

where $v = [1, -1, 1, \ldots, (-1)^n]^T$. (b) Verify that

$$V^{-1} = \frac{2}{n} \left(I - \frac{1}{2n} vv^T \right) V^T.$$

(c) Show how to compute $V^{-1}x$ rapidly.

P4.8.12 Verify (4.8.22), (4.8.23), and (4.8.25).

P4.8.13 Show that if $V = V_{2m}^{(P)}$ defined in (4.8.24), then

$$V^T V = m \left(I_n + e_1 e_1^T + e_{m+1} e_{m+1}^T \right).$$

What can you say about $V^T V$ if $V = V_{2m-1}^{(P)}$?

Notes and References for §4.8

As we mentioned, this section is based on Van Loan (FFT). For more details about fast Poisson solvers, see:

R.W. Hockney (1965). "A Fast Direct Solution of Poisson's Equation Using Fourier Analysis," *J. Assoc. Comput. Mach.* *12*, 95–113.

B. Buzbee, G. Golub, and C. Nielson (1970). "On Direct Methods for Solving Poisson's Equation," *SIAM J. Numer. Anal.* *7*, 627–656.

F. Dorr (1970). "The Direct Solution of the Discrete Poisson Equation on a Rectangle," *SIAM Review* *12*, 248–263.

R. Sweet (1973). "Direct Methods for the Solution of Poisson's Equation on a Staggered Grid," *J. Comput. Phys.* *12*, 422–428.

P.N. Swarztrauber (1974). "A Direct Method for the Discrete Solution of Separable Elliptic Equations," *SIAM J. Numer. Anal.* *11*, 1136–1150.

P.N. Swarztrauber (1977). "The Methods of Cyclic Reduction, Fourier Analysis and Cyclic Reduction-Fourier Analysis for the Discrete Solution of Poisson's Equation on a Rectangle," *SIAM Review* *19*, 490–501.

There are actually eight variants of the discrete cosine transform each of which corresponds to the location of the Neumann conditions and how the divided difference approximations are set up. For a unified, matrix-based treatment, see:

G. Strang (1999). "The Discrete Cosine Transform," *SIAM Review 41*, 135–147.

Chapter 5

Orthogonalization and Least Squares

5.1 Householder and Givens Transformations

5.2 The QR Factorization

5.3 The Full-Rank Least Squares Problem

5.4 Other Orthogonal Factorizations

5.5 The Rank-Deficient Least Squares Problem

5.6 Square and Underdetermined Systems

This chapter is primarily concerned with the least squares solution of overdetermined systems of equations, i.e., the minimization of $\| Ax - b \|_2$ where $A \in \mathbb{R}^{m \times n}$, $b \in \mathbb{R}^m$, and $m \geq n$. The most reliable solution procedures for this problem involve the reduction of A to various canonical forms via orthogonal transformations. Householder reflections and Givens rotations are central to this process and we begin the chapter with a discussion of these important transformations. In §5.2 we show how to compute the factorization $A = QR$ where Q is orthogonal and R is upper triangular. This amounts to finding an orthonormal basis for the range of A. The QR factorization can be used to solve the full-rank least squares problem as we show in §5.3. The technique is compared with the method of normal equations after a perturbation theory is developed. In §5.4 and §5.5 we consider methods for handling the difficult situation when A is (nearly) rank deficient. QR with column pivoting and other rank-revealing procedures including the SVD are featured. Some remarks about underdetermined systems are offered in §5.6.

Reading Notes

Knowledge of chapters 1, 2, and 3 and §§4.1–§4.3 is assumed. Within this chapter there are the following dependencies:

$$\S5.1 \quad \rightarrow \quad \S5.2 \quad \rightarrow \quad \S5.3 \quad \rightarrow \quad \S5.4 \quad \rightarrow \quad \S5.5 \quad \rightarrow \quad \S5.6$$
$$\downarrow$$
$$\S5.4$$

For more comprehensive treatments of the least squares problem, see Björck (NMLS) and Lawson and Hansen (SLS). Other useful global references include Stewart (MABD), Higham (ASNA), Watkins (FMC), Trefethen and Bau (NLA), Demmel (ANLA), and Ipsen (NMA).

5.1   Householder and Givens Transformations

Recall that $Q \in \mathbb{R}^{m \times m}$ is *orthogonal* if

$$Q^T Q = Q Q^T = I_m.$$

Orthogonal matrices have an important role to play in least squares and eigenvalue computations. In this section we introduce Householder reflections and Givens rotations, the key players in this game.

5.1.1   A 2-by-2 Preview

It is instructive to examine the geometry associated with rotations and reflections at the $m = 2$ level. A 2-by-2 orthogonal matrix Q is a *rotation* if it has the form

$$Q \;=\; \left[\begin{array}{cc} \cos(\theta) & \sin(\theta) \\ -\sin(\theta) & \cos(\theta) \end{array} \right].$$

If $y = Q^T x$, then y is obtained by rotating x counterclockwise through an angle θ.

A 2-by-2 orthogonal matrix Q is a *reflection* if it has the form

$$Q \;=\; \left[\begin{array}{cc} \cos(\theta) & \sin(\theta) \\ \sin(\theta) & -\cos(\theta) \end{array} \right].$$

If $y = Q^T x = Q x$, then y is obtained by reflecting the vector x across the line defined by

$$S = \mathsf{span} \left\{ \left[\begin{array}{c} \cos(\theta/2) \\ \sin(\theta/2) \end{array} \right] \right\}.$$

Reflections and rotations are computationally attractive because they are easily constructed and because they can be used to introduce zeros in a vector by properly choosing the rotation angle or the reflection plane.

5.1.2   Householder Reflections

Let $v \in \mathbb{R}^m$ be nonzero. An m-by-m matrix P of the form

$$P = I - \beta v v^T, \qquad \beta = \frac{2}{v^T v} \tag{5.1.1}$$

is a *Householder reflection*. (Synonyms are Householder matrix and Householder transformation.) The vector v is the *Householder vector*. If a vector x is multiplied by P, then it is reflected in the hyperplane $\mathsf{span}\{v\}^{\perp}$. It is easy to verify that Householder matrices are symmetric and orthogonal.

Householder reflections are similar to Gauss transformations introduced in §3.2.1 in that they are rank-1 modifications of the identity and can be used to zero selected components of a vector. In particular, suppose we are given $0 \neq x \in \mathbb{R}^m$ and want

$$Px = \left(I - \frac{2vv^T}{v^T v} \right) x = x - \frac{2v^T x}{v^T v} v$$

to be a multiple of $e_1 = I_m(:, 1)$. From this we conclude that $v \in \text{span}\{x, e_1\}$. Setting

$$v = x + \alpha e_1$$

gives

$$v^T x = x^T x + \alpha x_1$$

and

$$v^T v = x^T x + 2\alpha x_1 + \alpha^2.$$

Thus,

$$Px = \left(1 - 2\frac{x^T x + \alpha x_1}{x^T x + 2\alpha x_1 + \alpha^2} \right) x - 2\alpha \frac{v^T x}{v^T v} e_1$$

$$= \left(\frac{\alpha^2 - \| x \|_2^2}{x^T x + 2\alpha x_1 + \alpha^2} \right) x - 2\alpha \frac{v^T x}{v^T v} e_1.$$

In order for the coefficient of x to be zero, we set $\alpha = \pm \| x \|_2$ for then

$$v = x \pm \| x \|_2 e_1 \Rightarrow Px = \left(I - 2\frac{vv^T}{v^T v} \right) x = \mp \| x \|_2 e_1. \qquad (5.1.2)$$

It is this simple determination of v that makes the Householder reflections so useful.

5.1.3   Computing the Householder Vector

There are a number of important practical details associated with the determination of a Householder matrix, i.e., the determination of a Householder vector. One concerns the choice of sign in the definition of v in (5.1.2). Setting

$$v_1 = x_1 - \| x \|_2$$

leads to the nice property that Px is a positive multiple of e_1. But this recipe is dangerous if x is close to a positive multiple of e_1 because severe cancellation would occur. However, the formula

$$v_1 = x_1 - \| x \|_2 = \frac{x_1^2 - \| x \|_2^2}{x_1 + \| x \|_2} = \frac{-(x_2^2 + \cdots + x_n^2)}{x_1 + \| x \|_2}$$

suggested by Parlett (1971) does not suffer from this defect in the $x_1 > 0$ case.

In practice, it is handy to normalize the Householder vector so that $v(1) = 1$. This permits the storage of $v(2:m)$ where the zeros have been introduced in x, i.e., $x(2:m)$. We refer to $v(2:m)$ as the *essential part* of the Householder vector. Recalling that $\beta = 2/v^T v$ and letting $\mathsf{length}(x)$ specify vector dimension, we may encapsulate the overall process as follows:

Algorithm 5.1.1 (Householder Vector) Given $x \in \mathbb{R}^m$, this function computes $v \in \mathbb{R}^m$ with $v(1) = 1$ and $\beta \in \mathbb{R}$ such that $P = I_m - \beta vv^T$ is orthogonal and $Px = \| x \|_2 e_1$.

> **function** $[v, \beta] = \mathsf{house}(x)$
>
> $$m = \mathsf{length}(x), \ \sigma = x(2{:}m)^T x(2{:}m), \ v = \left[\begin{array}{c} 1 \\ x(2{:}m) \end{array} \right]$$
>
> **if** $\sigma = 0$ and $x(1) >= 0$
>> $\beta = 0$
>
> **elseif** $\sigma = 0$ & $x(1) < 0$
>> $\beta = -2$
>
> **else**
>> $\mu = \sqrt{x(1)^2 + \sigma}$
>>
>> **if** $x(1) <= 0$
>>> $v(1) = x(1) - \mu$
>>
>> **else**
>>> $v(1) = -\sigma/(x(1) + \mu)$
>>
>> **end**
>>
>> $\beta = 2v(1)^2/(\sigma + v(1)^2)$
>>
>> $v = v/v(1)$
>
> **end**

Here, $\mathsf{length}(\cdot)$ returns the dimension of a vector. This algorithm involves about $3m$ flops. The computed Householder matrix that is orthogonal to machine precision, a concept discussed below.

5.1.4 Applying Householder Matrices

It is critical to exploit structure when applying $P = I - \beta vv^T$ to a matrix A. Premultiplication involves a matrix-vector product and a rank-1 update:

$$PA = \left(I - \beta vv^T\right) A = A - (\beta v)(v^T A).$$

The same is true for post-multiplication,

$$AP = A \left(I - \beta vv^T\right) = A - (Av)(\beta v)^T.$$

In either case, the update requires $4mn$ flops if $A \in \mathbb{R}^{m \times n}$. Failure to recognize this and to treat P as a general matrix increases work by an order of magnitude. *Householder updates never entail the explicit formation of the Householder matrix.*

In a typical situation, house is applied to a subcolumn or subrow of a matrix and $(I - \beta vv^T)$ is applied to a submatrix. For example, if $A \in \mathbb{R}^{m \times n}$, $1 \le j < n$, and $A(j{:}m, 1{:}j - 1)$ is zero, then the sequence

$$[v, \beta] = \mathsf{house}(A(j{:}m, j))$$
$$A(j{:}m, j{:}n) = A(j{:}m, j{:}n) - (\beta v)(v^T A(j{:}m, j{:}n))$$
$$A(j + 1{:}m, j) = v(2{:}m - j + 1)$$

applies $(I_{m-j+1} - \beta vv^T)$ to $A(j{:}m, 1{:}n)$ and stores the essential part of v where the "new" zeros are introduced.

5.1.5   Roundoff Properties

The roundoff properties associated with Householder matrices are very favorable. Wilkinson (AEP, pp. 152–162) shows that **house** produces a Householder vector $\hat{v}$ that is very close to the exact v. If $\hat{P} = I - 2\hat{v}\hat{v}^T/\hat{v}^T\hat{v}$ then

$$\| \hat{P} - P \|_2 = O(\mathbf{u}).$$

Moreover, the computed updates with $\hat{P}$ are close to the exact updates with P :

$$\mathrm{fl}(\hat{P}A) = P(A + E), \qquad \| E \|_2 = O(\mathbf{u}\| A \|_2),$$
$$\mathrm{fl}(A\hat{P}) = (A + E)P, \qquad \| E \|_2 = O(\mathbf{u}\| A \|_2).$$

For a more detailed analysis, see Higham(ASNA, pp. 357–361).

5.1.6   The Factored-Form Representation

Many Householder-based factorization algorithms that are presented in the following sections compute products of Householder matrices

$$Q = Q_1 Q_2 \cdots Q_n \qquad Q_j = I_m - \beta_j v^{(j)} [v^{(j)}]^T \qquad (5.1.3)$$

where $n \leq m$ and each $v^{(j)}$ has the form

$$v^{(j)} = [\underbrace{0,\ 0,\ldots 0,}_{j-1}\ 1\ v^{(j)}_{j+1},\ \ldots,\ v^{(j)}_m]^T .$$

It is usually not necessary to compute Q explicitly even if it is involved in subsequent calculations. For example, if $C \in \mathbb{R}^{m \times p}$ and we wish to compute $Q^T C$, then we merely execute the loop

> **for** $j = 1{:}n$
> $\qquad C = Q_j C$
> **end**

The storage of the Householder vectors $v^{(1)} \cdots v^{(n)}$ and the corresponding β_j amounts to a *factored-form representation* of Q.

To illustrate the economies of the factored-form representation, suppose we have an array A and that for $j = 1{:}n$, $A(j + 1{:}m, j)$ houses $v^{(j)}(j + 1{:}m)$, the essential part of the jth Householder vector. The overwriting of $C \in \mathbb{R}^{m \times p}$ with $Q^T C$ can then be implemented as follows:

> **for** $j = 1{:}n$
> $\qquad v(j{:}m) = \begin{bmatrix} 1 \\ A(j + 1{:}m, j) \end{bmatrix}$
> $\qquad \beta_j = 2/(1 + \| A(j + 1{:}m, j) \|_2^2)$ (5.1.4)
> $\qquad C(j{:}m, :) = C(j{:}m, :) - (\beta_j {\cdot} v(j{:}m)) \cdot (v(j{:}m)^T C(j{:}m, :))$
> **end**

This involves about $pn(2m - n)$ flops. If Q is explicitly represented as an m-by-m matrix, then $Q^T C$ would involve $2m^2 p$ flops. The advantage of the factored form representation is apparant if $n << m$.

Of course, in some applications, it is necessary to explicitly form Q (or parts of it). There are two possible algorithms for computing the matrix Q in (5.1.3):

Forward accumulation	*Backward accumulation*
$Q = I_m$	$Q = I_m$
for $j = 1:n$	**for** $j = n: -1:1$
$\quad Q = Q\,Q_j$	$\quad Q = Q_j Q$
end	**end**

Recall that the leading $(j - 1)$-by-$(j - 1)$ portion of Q_j is the identity. Thus, at the beginning of backward accumulation, Q is "mostly the identity" and it gradually becomes full as the iteration progresses. This pattern can be exploited to reduce the number of required flops. In contrast, Q is full in forward accumulation after the first step. For this reason, backward accumulation is cheaper and the strategy of choice. Here are the details with the proviso that we only need $Q(:, 1:k)$ where $1 \le k \le m$:

$$Q = I_m(:, 1:k)$$
$$\textbf{for } j = n: -1:1$$
$$v(j:m) = \begin{bmatrix} 1 \\ A(j+1:m, j) \end{bmatrix} \tag{5.1.5}$$
$$\beta_j = 2/(1 + \| A(j+1:m, j) \|_2^2)$$
$$Q(j:m, j:k) = Q(j:m, j:k) - (\beta_j v(j:m))(v(j:m)^T Q(j:m, j:k))$$
$$\textbf{end}$$

This involves about $4mnk - 2(m + k)n^2 + (4/3)n^3$ flops.

5.1.7   The WY Representation

Suppose $Q = Q_1 \cdots Q_r$ is a product of m-by-m Householder matrices. Since each Q_j is a rank-1 modification of the identity, it follows from the structure of the Householder vectors that Q is a rank-r modification of the identity and can be written in the form

$$Q = I_m - WY^T \tag{5.1.6}$$

where W and Y are m-by-r matrices. The key to computing the *WY representation* (5.1.6) is the following lemma.

Lemma 5.1.1. *Suppose* $Q = I_m - WY^T$ *is an m-by-m orthogonal matrix with* $W, Y \in \mathbb{R}^{m \times j}$. *If* $P = I_m - \beta vv^T$ *with* $v \in \mathbb{R}^m$ *and* $z = \beta Q v$, *then*

$$Q_+ = QP = I_m - W_+ Y_+^T$$

where $W_+ = [\, W \mid z \,]$ *and* $Y_+ = [\, Y \mid v \,]$ *are each m-by-$(j + 1)$.*

Proof. Since

$$QP = (I_m - WY^T)(I_m - \beta vv^T) = I_m - WY^T - \beta Qvv^T$$

it follows from the definition of z that

$$Q_+ = I_m - WY^T - zv^T = I_m - [W \mid z][Y \mid v]^T = I_m - W_+ Y_+^T. \qquad \square$$

By repeatedly applying the lemma, we can transition from a factored-form representation to a block representation.

Algorithm 5.1.2 Suppose $Q = Q_1 \cdots Q_r$ where the $Q_j = I_m - \beta_j v^{(j)} v^{(j)^T}$ are stored in factored form. This algorithm computes matrices $W, Y \in \mathbb{R}^{m \times r}$ such that $Q = I_m - WY^T$.

> $Y = v^{(1)}; \quad W = \beta_1 v^{(1)}$
> **for** $j = 2{:}r$
> > $z = \beta_j (I_m - WY^T) v^{(j)}$
> > $W = [W \mid z]$
> > $Y = [Y \mid v^{(j)}]$
> **end**

This algorithm involves about $2r^2 m - 2r^3/3$ flops if the zeros in the $v^{(j)}$ are exploited. Note that Y is merely the matrix of Householder vectors and is therefore unit lower triangular. Clearly, the central task in the generation of the WY representation (5.1.6) is the computation of the matrix W.

The block representation for products of Householder matrices is attractive in situations where Q must be applied to a matrix. Suppose $C \in \mathbb{R}^{m \times p}$. It follows that the operation

$$C = Q^T C = (I_m - WY^T)^T C = C - Y(W^T C)$$

is rich in level-3 operations. On the other hand, if Q is in factored form, then the formation of $Q^T C$ is just rich in the level-2 operations of matrix-vector multiplication and outer product updates. Of course, in this context, the distinction between level-2 and level-3 diminishes as C gets narrower.

We mention that the WY representation (5.1.6) is not a generalized Householder transformation from the geometric point of view. True block reflectors have the form

$$Q = I - 2VV^T$$

where $V \in \mathbb{R}^{n \times r}$ satisfies $V^T V = I_r$. See Schreiber and Parlett (1987).

5.1.8 Givens Rotations

Householder reflections are exceedingly useful for introducing zeros on a grand scale, e.g., the annihilation of all but the first component of a vector. However, in calculations where it is necessary to zero elements more selectively, *Givens rotations* are the transformation of choice. These are rank-2 corrections to the identity of the form

$$
G(i,k,\theta) \;=\; \begin{bmatrix}
1 & \cdots & 0 & \cdots & 0 & \cdots & 0 \\
\vdots & \ddots & \vdots & & \vdots & & \vdots \\
0 & \cdots & c & \cdots & s & \cdots & 0 \\
\vdots & & \vdots & \ddots & \vdots & & \vdots \\
0 & \cdots & -s & \cdots & c & \cdots & 0 \\
\vdots & & \vdots & & \vdots & \ddots & \vdots \\
0 & \cdots & 0 & \cdots & 0 & \cdots & 1
\end{bmatrix}
\begin{matrix} \\ \\ i \\ \\ k \\ \\ \\ \end{matrix}
\qquad (5.1.7)
$$

$$
\qquad\quad i \qquad\quad k
$$

where $c = \cos(\theta)$ and $s = \sin(\theta)$ for some θ. Givens rotations are clearly orthogonal.

Premultiplication by $G(i,k,\theta)^T$ amounts to a counterclockwise rotation of θ radians in the (i,k) coordinate plane. Indeed, if $x \in \mathbb{R}^m$ and

$$
y = G(i,k,\theta)^T x,
$$

then

$$
y_j \;=\; \begin{cases}
cx_i - sx_k, & j = i, \\
sx_i + cx_k, & j = k, \\
x_j, & j \neq i,k.
\end{cases}
$$

From these formulae it is clear that we can force y_k to be zero by setting

$$
c = \frac{x_i}{\sqrt{x_i^2 + x_k^2}}, \qquad s = \frac{-x_k}{\sqrt{x_i^2 + x_k^2}}. \qquad (5.1.8)
$$

Thus, it is a simple matter to zero a specified entry in a vector by using a Givens rotation. In practice, there are better ways to compute c and s than (5.1.8), e.g.,

Algorithm 5.1.3 Given scalars a and b, this function computes $c = \cos(\theta)$ and $s = \sin(\theta)$ so

$$
\begin{bmatrix} c & s \\ -s & c \end{bmatrix}^T \begin{bmatrix} a \\ b \end{bmatrix} = \begin{bmatrix} r \\ 0 \end{bmatrix}.
$$

> **function** $[c,s] = \text{givens}(a,b)$
>> **if** $b = 0$
>>> $c = 1;\; s = 0$
>> **else**
>>> **if** $|b| > |a|$
>>>> $\tau = -a/b;\; s = 1/\sqrt{1 + \tau^2};\; c = s\tau$
>>> **else**
>>>> $\tau = -b/a;\; c = 1/\sqrt{1 + \tau^2};\; s = c\tau$
>>> **end**
>> **end**

This algorithm requires 5 flops and a single square root. *Note that inverse trigonometric functions are not involved.*

5.1.9   Applying Givens Rotations

It is critical that the simple structure of a Givens rotation matrix be exploited when it is involved in a matrix multiplication. Suppose $A \in \mathbb{R}^{m \times n}$, $c = \cos(\theta)$, and $s = \sin(\theta)$. If $G(i, k, \theta) \in \mathbb{R}^{m \times m}$, then the update $A = G(i, k, \theta)^T A$ affects just two rows,

$$A([i, k], :) = \begin{bmatrix} c & s \\ -s & c \end{bmatrix}^T A([i, k], :),$$

and involves $6n$ flops:

> **for** $j = 1{:}n$
> $\qquad \tau_1 = A(i, j)$
> $\qquad \tau_2 = A(k, j)$
> $\qquad A(i, j) = c\tau_1 - s\tau_2$
> $\qquad A(k, j) = s\tau_1 + c\tau_2$
> **end**

Likewise, if $G(i, k, \theta) \in \mathbb{R}^{n \times n}$, then the update $A = AG(i, k, \theta)$ affects just two columns,

$$A(:, [i, k]) = A(:, [i, k]) \begin{bmatrix} c & s \\ -s & c \end{bmatrix},$$

and involves $6m$ flops:

> **for** $j = 1{:}m$
> $\qquad \tau_1 = A(j, i)$
> $\qquad \tau_2 = A(j, k)$
> $\qquad A(j, i) = c\tau_1 - s\tau_2$
> $\qquad A(j, k) = s\tau_1 + c\tau_2$
> **end**

5.1.10   Roundoff Properties

The numerical properties of Givens rotations are as favorable as those for Householder reflections. In particular, it can be shown that the computed $\hat{c}$ and $\hat{s}$ in **givens** satisfy

$$\begin{aligned}
\hat{c} &= c(1 + \epsilon_c), & \epsilon_c &= O(\mathbf{u}), \\
\hat{s} &= s(1 + \epsilon_s), & \epsilon_s &= O(\mathbf{u}).
\end{aligned}$$

If $\hat{c}$ and $\hat{s}$ are subsequently used in a Givens update, then the computed update is the exact update of a nearby matrix:

$$\begin{aligned}
\mathrm{fl}[\hat{G}(i, k, \theta)^T A] &= G(i, k, \theta)^T (A + E), & \| E \|_2 &\approx \mathbf{u} \| A \|_2, \\
\mathrm{fl}[A\hat{G}(i, k, \theta)] &= (A + E)G(i, k, \theta), & \| E \|_2 &\approx \mathbf{u} \| A \|_2.
\end{aligned}$$

Detailed error analysis of Givens rotations may be found in Wilkinson (AEP, pp. 131-39), Higham(ASNA, pp. 366–368), and Bindel, Demmel, Kahan, and Marques (2002).

5.1.11 Representing Products of Givens Rotations

Suppose $Q = G_1 \cdots G_t$ is a product of Givens rotations. As with Householder re-flections, it is sometimes more economical to keep Q in factored form rather than to compute explicitly the product of the rotations. Stewart (1976) has shown how to do this in a very compact way. The idea is to associate a single floating point number ρ with each rotation. Specifically, if

$$Z = \begin{bmatrix} c & s \\ -s & c \end{bmatrix}, \qquad c^2 + s^2 = 1,$$

then we define the scalar ρ by

> **if** $c = 0$
>> $\rho = 1$
>
> **elseif** $|s| < |c|$
>> $\rho = \text{sign}(c) \cdot s/2$ (5.1.9)
>
> **else**
>> $\rho = 2 \cdot \text{sign}(s)/c$
>
> **end**

Essentially, this amounts to storing $s/2$ if the sine is smaller and $2/c$ if the cosine is smaller. With this encoding, it is possible to reconstruct Z (or $-Z$) as follows:

> **if** $\rho = 1$
>> $c = 0;\ s = 1$
>
> **elseif** $|\rho| < 1$
>> $s = 2\rho;\ c = \sqrt{1 - s^2}$ (5.1.10)
>
> **else**
>> $c = 2/\rho;\ s = \sqrt{1 - c^2}$
>
> **end**

Note that the reconstruction of $-Z$ is not a problem, for if Z introduces a strategic zero then so does $-Z$. The reason for essentially storing the smaller of c and s is that the formula $\sqrt{1 - x^2}$ renders poor results if x is near unity. More details may be found in Stewart (1976). Of course, to "reconstruct" $G(i, k, \theta)$ we need i and k in addition to the associated ρ. This poses no difficulty if we agree to store ρ in the (i, k) entry of some array.

5.1.12 Error Propagation

An m-by-m floating point matrix $\hat{Q}$ is *orthogonal to working precision* if there exists an orthogonal $Q \in \mathbb{R}^{m \times m}$ such that

$$\| \hat{Q} - Q \| = O(\mathbf{u}).$$

A corollary of this is that

$$\| \hat{Q}^T \hat{Q} - I_m \| = O(\mathbf{u}).$$

The matrices defined by the floating point output of **house** and **givens** are orthogonal to working precision.

In many applications, sequences of Householders and/or Given transformations are generated and applied. In these settings, the rounding errors are nicely bounded. To be precise, suppose $A = A_0 \in \mathbb{R}^{m \times n}$ is given and that matrices $A_1, \ldots, A_p = B$ are generated via the formula

$$A_k = \mathrm{fl}(\hat{Q}_k A_{k-1} \hat{Z}_k), \qquad k = 1{:}p \,.$$

Assume that the above Householder and Givens algorithms are used for both the generation and application of the $\hat{Q}_k$ and $\hat{Z}_k$. Let Q_k and Z_k be the orthogonal matrices that would be produced in the absence of roundoff. It can be shown that

$$B \;=\; (Q_p \cdots Q_1)(A + E)(Z_1 \cdots Z_p), \tag{5.1.11}$$

where $\| \, E \, \|_2 \;\leq\; c \cdot \mathbf{u} \| \, A \, \|_2$ and c is a constant that depends mildly on n, m, and p. In other words, B is an exact orthogonal update of a matrix near to A. For a comprehensive error analysis of Householder and Givens computations, see Higham (ASNA, §19.3, §19.6).

5.1.13   The Complex Case

Most of the algorithms that we present in this book have complex versions that are fairly straightforward to derive from their real counterparts. (This is *not* to say that everything is easy and obvious at the implementation level.) As an illustration we briefly discuss complex Householder and complex Givens transformations.

Recall that if $A = (a_{ij}) \in \mathbb{C}^{m \times n}$, then $B = A^H \in \mathbb{C}^{n \times m}$ is its conjugate transpose. The 2-norm of a vector $x \in \mathbb{C}^n$ is defined by

$$\| \, x \, \|_2^2 \;=\; x^H x \;=\; |x_1|^2 + \cdots + |x_n|^2$$

and $Q \in \mathbb{C}^{n \times n}$ is *unitary* if $Q^H Q = I_n$. Unitary matrices preserve the 2-norm.

A complex Householder transformation is a unitary matrix of the form

$$P = I_m - \beta v v^H, \qquad 0 \neq v \in \mathbb{C}^m,$$

where $\beta = 2/v^H v$. Given a nonzero vector $x \in \mathbb{C}^m$, it is easy to determine v so that if $y = Px$, then $y(2{:}m) = 0$. Indeed, if

$$x_1 = r e^{i\theta}$$

where $r, \theta \in \mathbb{R}$ and

$$v = x \pm e^{i\theta} \| \, x \, \|_2 e_1, \qquad e_1 = I_m(:, 1),$$

then $Px = \mp e^{i\theta} \| \, x \, \|_2 e_1$. The sign can be determined to maximize $\| \, v \, \|_2$ for the sake of stability.

Regarding complex Givens rotations, it is easy to verify that a 2-by-2 matrix of the form

$$Q \;=\; \begin{bmatrix} \cos(\theta) & \sin(\theta) e^{i\phi} \\ -\sin(\theta) e^{-i\phi} & \cos(\theta) \end{bmatrix}$$

where $\theta, \phi \in \mathbb{R}$ is unitary. We show how to compute $c = \cos(\theta)$ and $s = \sin(\theta)e^{i\phi}$ so that

$$\begin{bmatrix} c & s \\ -\bar{s} & c \end{bmatrix}^H \begin{bmatrix} u \\ v \end{bmatrix} = \begin{bmatrix} r \\ 0 \end{bmatrix} \tag{5.1.12}$$

where $u = u_1 + iu_2$ and $v = v_1 + iv_2$ are given complex numbers. First, **givens** is applied to compute real cosine-sine pairs $\{c_\alpha, s_\alpha\}$, $\{c_\beta, s_\beta\}$, and $\{c_\theta, s_\theta\}$ so that

$$\begin{bmatrix} c_\alpha & s_\alpha \\ -s_\alpha & c_\alpha \end{bmatrix}^T \begin{bmatrix} u_1 \\ u_2 \end{bmatrix} = \begin{bmatrix} r_u \\ 0 \end{bmatrix},$$

$$\begin{bmatrix} c_\beta & s_\beta \\ -s_\beta & c_\beta \end{bmatrix}^T \begin{bmatrix} v_1 \\ v_2 \end{bmatrix} = \begin{bmatrix} r_v \\ 0 \end{bmatrix},$$

and

$$\begin{bmatrix} c_\theta & s_\theta \\ -s_\theta & c_\theta \end{bmatrix}^T \begin{bmatrix} r_u \\ r_v \end{bmatrix} = \begin{bmatrix} r \\ 0 \end{bmatrix}.$$

Note that $u = r_u e^{-i\alpha}$ and $v = r_v e^{-i\beta}$. If we set

$$e^{i\phi} = e^{i(\beta-\alpha)} = (c_\alpha c_\beta + s_\alpha s_\beta) + i(c_\alpha s_\beta - c_\beta s_\alpha),$$

$c = c_\theta$, and $s = s_\theta e^{i\phi}$, then

$$\bar{s}u + cv = s_\theta e^{-i\phi} r_u e^{-i\alpha} + c_\theta r_v e^{-i\beta} = e^{-i\beta}(s_\theta r_u + c_\theta r_v) = 0$$

which confirms (5.1.12).

Problems

P5.1.1 Let x and y be nonzero vectors in $\mathbb{R}^m$. Give an algorithm for determining a Householder matrix P such that Px is a multiple of y.

P5.1.2 Use Householder matrices to show that $\det(I + xy^T) = 1 + x^T y$ where x and y are given m-vectors.

P5.1.3 (a) Assume that $x, y \in \mathbb{R}^2$ have unit 2-norm. Give an algorithm that computes a Givens rotation Q so that $y = Q^T x$. Make effective use of **givens**. (b) Suppose x and y are unit vectors in $\mathbb{R}^m$. Give an algorithm using Givens transformations which computes an orthogonal Q such that $Q^T x = y$.

P5.1.4 By generalizing the ideas in §5.1.11, develop a compact representation scheme for complex givens rotations.

P5.1.5 Suppose that $Q = I - YTY^T$ is orthogonal where $Y \in \mathbb{R}^{m \times j}$ and $T \in \mathbb{R}^{j \times j}$ is upper triangular. Show that if $Q_+ = QP$ where $P = I - 2vv^T/v^T v$ is a Householder matrix, then Q_+ can be expressed in the form $Q_+ = I - Y_+ T_+ Y_+^T$ where $Y_+ \in \mathbb{R}^{m \times (j+1)}$ and $T_+ \in \mathbb{R}^{(j+1) \times (j+1)}$ is upper triangular. This is the main idea behind the *compact WY representation*. See Schreiber and Van Loan (1989).

P5.1.6 Suppose $Q_1 = I_m - Y_1 T_1 Y_1$ and $Q_2 = I_m - Y_2 T_2 Y_2^T$ are orthogonal where $Y_1 \in \mathbb{R}^{m \times r_1}$, $Y_2 \in \mathbb{R}^{m \times r_2}$, $T_1 \in \mathbb{R}^{r_1 \times r_1}$, and $T_2 \in \mathbb{R}^{r_2 \times r_2}$. Assume that T_1 and T_2 are upper triangular. Show how to compute $Y \in \mathbb{R}^{m \times r}$ and upper triangular $T \in \mathbb{R}^{r \times r}$ with $r = r_1 + r_2$ so that $Q_2 Q_1 = I_m - YTY^T$.

P5.1.7 Give a detailed implementation of Algorithm 5.1.2 with the assumption that $v^{(j)}(j + 1:m)$, the essential part of the jth Householder vector, is stored in $A(j + 1:m, j)$. Since Y is effectively represented in A, your procedure need only set up the W matrix.

P5.1.8 Show that if S is skew-symmetric ($S^T = -S$), then $Q = (I + S)(I - S)^{-1}$ is orthogonal. (The matrix Q is called the *Cayley transform* of S.) Construct a rank-2 S so that if x is a vector, then Qx is zero except in the first component.

P5.1.9 Suppose $P \in \mathbb{R}^{m \times m}$ satisfies $\| P^T P - I_m \|_2 = \epsilon < 1$. Show that all the singular values of P are in the interval $[1 - \epsilon, 1 + \epsilon]$ and that $\| P - UV^T \|_2 \leq \epsilon$ where $P = U \Sigma V^T$ is the SVD of P.

P5.1.10 Suppose $A \in \mathbb{R}^{2 \times 2}$. Under what conditions is the closest rotation to A closer than the closest reflection to A? Work with the Frobenius norm.

P5.1.11 How could Algorithm 5.1.3 be modified to ensure $r \geq 0$?

P5.1.12 (*Fast Givens Transformations*) Suppose

$$x = \begin{bmatrix} x_1 \\ x_2 \end{bmatrix} \quad \text{and} \quad D = \begin{bmatrix} d_1 & 0 \\ 0 & d_2 \end{bmatrix}$$

with d_1 and d_2 positive. Show how to compute

$$M_1 = \begin{bmatrix} \beta_1 & 1 \\ 1 & \alpha_1 \end{bmatrix}$$

so that if $y = M_1 x$ and $\tilde{D} = M_1^T D M_1$, then $y_2 = 0$ and $\tilde{D}$ is diagonal. Repeat with M_1 replaced by

$$M_2 = \begin{bmatrix} 1 & \alpha_2 \\ \beta_2 & 1 \end{bmatrix}.$$

(b) Show that either $\| M_1^T D M_1 \|_2 \leq 2 \| D \|_2$ or $\| M_2^T D M_2 \|_2 \leq 2 \| D \|_2$. (c) Suppose $x \in \mathbb{R}^m$ and that $D \in \mathbb{R}^{n \times n}$ is diagonal with positive diagonal entries. Given indices i and j with $1 \leq i < j \leq m$, show how to compute $M \in \mathbb{R}^{n \times n}$ so that if $y = Mx$ and $\tilde{D} = M^T D M$, then $y_j = 0$ and $\tilde{D}$ is diagonal with $\| \tilde{D} \|_2 \leq 2 \| D \|_2$. (d) From part (c) conclude that $Q = D^{1/2} M \tilde{D}^{-1/2}$ is orthogonal and that the update $y = Mx$ can be diagonally transformed to $(D^{1/2} y) = Q(D^{1/2} x)$.

Notes and References for §5.1

Householder matrices are named after A.S. Householder, who popularized their use in numerical analysis. However, the properties of these matrices have been known for quite some time, see:

H.W. Turnbull and A.C. Aitken (1961). *An Introduction to the Theory of Canonical Matrices*, Dover Publications, New York, 102–105.

Other references concerned with Householder transformations include:

A.R. Gourlay (1970). "Generalization of Elementary Hermitian Matrices," *Comput. J. 13*, 411–412.
B.N. Parlett (1971). "Analysis of Algorithms for Reflections in Bisectors," *SIAM Review 13*, 197–208.
N.K. Tsao (1975). "A Note on Implementing the Householder Transformations." *SIAM J. Numer. Anal. 12*, 53–58.
B. Danloy (1976). "On the Choice of Signs for Householder Matrices," *J. Comput. Appl. Math. 2*, 67–69.
J.J.M. Cuppen (1984). "On Updating Triangular Products of Householder Matrices," *Numer. Math. 45*, 403–410.
A.A. Dubrulle (2000). "Householder Transformations Revisited," *SIAM J. Matrix Anal. Applic. 22*, 33–40.
J.W. Demmel, M. Hoemmen, Y. Hida, and E.J. Riedy (2009). "Nonnegative Diagonals and High Performance On Low-Profile Matrices from Householder QR," *SIAM J. Sci. Comput. 31*, 2832–2841.

A detailed error analysis of Householder transformations is given in Lawson and Hanson (SLE, pp. 83–89). The basic references for block Householder representations and the associated computations include:

C.H. Bischof and C. Van Loan (1987). "The WY Representation for Products of Householder Matrices," *SIAM J. Sci. Stat. Comput. 8*, s2–s13.

B.N. Parlett and R. Schreiber (1988). "Block Reflectors: Theory and Computation," *SIAM J. Numer. Anal. 25*, 189–205.

R.S. Schreiber and C. Van Loan (1989). "A Storage-Efficient WY Representation for Products of Householder Transformations," *SIAM J. Sci. Stat. Comput. 10*, 52–57.

C. Puglisi (1992). "Modification of the Householder Method Based on the Compact WY Representation," *SIAM J. Sci. Stat. Comput. 13*, 723–726.

X. Sun and C.H. Bischof (1995). "A Basis-Kernel Representation of Orthogonal Matrices," *SIAM J. Matrix Anal. Applic. 16*, 1184–1196.

T. Joffrain, T.M. Low, E.S. Quintana-Orti, R. van de Geijn, and F.G. Van Zee (2006). "Accumulating Householder Transformations, Revisited," *ACM Trans. Math. Softw. 32*, 169–179.

M. Sadkane and A. Salam (2009). "A Note on Symplectic Block Reflectors," *ETNA 33*, 45–52.

Givens rotations are named after Wallace Givens. There are some subtleties associated with their computation and representation:

G.W. Stewart (1976). "The Economical Storage of Plane Rotations," *Numer. Math. 25*, 137–138.

D. Bindel, J. Demmel, W. Kahan, and O. Marques (2002). "On computing givens rotations reliably and efficiently," *ACM Trans. Math. Softw. 28*, 206–238.

It is possible to aggregate rotation transformations to achieve high performance, see:

B. Lang (1998). "Using Level 3 BLAS in Rotation–Based Algorithms," *SIAM J. Sci. Comput. 19*, 626–634.

Fast Givens transformations (see P5.1.11) are also referred to as *square-root-free Givens transformations*. (Recall that a square root must ordinarily be computed during the formation of Givens transformation.) There are several ways fast Givens calculations can be arranged, see:

M. Gentleman (1973). "Least Squares Computations by Givens Transformations without Square Roots," *J. Inst. Math. Appl. 12*, 329–336.

C.F. Van Loan (1973). "Generalized Singular Values With Algorithms and Applications," PhD Thesis, University of Michigan, Ann Arbor.

S. Hammarling (1974). "A Note on Modifications to the Givens Plane Rotation," *J. Inst. Math. Applic. 13*, 215–218.

J.H. Wilkinson (1977). "Some Recent Advances in Numerical Linear Algebra," in *The State of the Art in Numerical Analysis*, D.A.H. Jacobs (ed.), Academic Press, New York, 1–53.

A.A. Anda and H. Park (1994). "Fast Plane Rotations with Dynamic Scaling," *SIAM J. Matrix Anal. Applic. 15*, 162–174.

R.J. Hanson and T. Hopkins (2004). "Algorithm 830: Another Visit with Standard and Modified Givens Transformations and a Remark on Algorithm 539," *ACM Trans. Math. Softw. 20*, 86–94.

5.2   The QR Factorization

A rectangular matrix $A \in \mathbb{R}^{m \times n}$ can be factored into a product of an orthogonal matrix $Q \in \mathbb{R}^{m \times m}$ and an upper triangular matrix $R \in \mathbb{R}^{m \times n}$:

$$A = QR.$$

This factorization is referred to as the *QR factorization* and it has a central role to play in the linear least squares problem. In this section we give methods for computing QR based on Householder, block Householder, and Givens transformations. The QR factorization is related to the well-known Gram-Schmidt process.

5.2.1   Existence and Properties

We start with a constructive proof of the QR factorization.

Theorem 5.2.1 (QR Factorization). *If $A \in \mathbb{R}^{m \times n}$, then there exists an orthogonal $Q \in \mathbb{R}^{m \times m}$ and an upper triangular $R \in \mathbb{R}^{m \times n}$ so that $A = QR$.*

Proof. We use induction. Suppose $n = 1$ and that Q is a Householder matrix so that if $R = Q^T A$, then $R(2{:}m) = 0$. It follows that $A = QR$ is a QR factorization of A. For general n we partition A,

$$A = [\, A_1 \mid v \,],$$

where $v = A(:, n)$. By induction, there exists an orthogonal $Q_1 \in \mathbb{R}^{m \times m}$ so that $R_1 = Q_1^T A_1$ is upper triangular. Set $w = Q^T v$ and let $w(n{:}m) = Q_2 R_2$ be the QR factorization of $w(n{:}m)$. If

$$Q = Q_1 \begin{bmatrix} I_{n-1} & 0 \\ 0 & Q_2 \end{bmatrix},$$

then

$$A = Q \begin{bmatrix} R_1 & \dfrac{w(1{:}n-1)}{R_2} \end{bmatrix}$$

is a QR factorization of A. $\qquad\square$

The columns of Q have an important connection to the range of A and its orthogonal complement.

Theorem 5.2.2. *If $A = QR$ is a QR factorization of a full column rank $A \in \mathbb{R}^{m \times n}$ and*

$$A = [\, a_1 \mid \cdots \mid a_n \,],$$

$$Q = [\, q_1 \mid \cdots \mid q_m \,]$$

are column partitionings, then for $k = 1{:}n$

$$\mathsf{span}\{a_1, \ldots, a_k\} \ = \ \mathsf{span}\{q_1, \ldots, q_k\} \tag{5.2.1}$$

and $r_{kk} \neq 0$. Moreover, if $Q_1 = Q(1{:}m, 1{:}n)$, $Q_2 = Q(1{:}m, n+1{:}m)$, and $R_1 = R(1{:}n, 1{:}n)$, then

$$\begin{aligned} \mathsf{ran}(A) &= \mathsf{ran}(Q_1), \\ \mathsf{ran}(A)^{\perp} &= \mathsf{ran}(Q_2), \end{aligned}$$

and

$$A = Q_1 R_1. \tag{5.2.2}$$

Proof. Comparing the kth columns in $A = QR$ we conclude that

$$a_k \ = \ \sum_{i=1}^{k} r_{ik} q_i \ \in \ \mathsf{span}\{q_1, \ldots, q_k\}, \tag{5.2.3}$$

and so

$$\mathsf{span}\{a_1, \ldots, a_k\} \subseteq \mathsf{span}\{q_1, \ldots, q_k\}.$$

If $r_{kk} = 0$, then $a_1, \ldots, a_k$ are dependent. Thus, R cannot have a zero on its diagonal and so $\mathsf{span}\{a_1, \ldots, a_k\}$ has dimension k. Coupled with (5.2.3) this establishes (5.2.1). To prove (5.2.2) we note that

$$A = QR = \left[\begin{array}{c|c} Q_1 & Q_2 \end{array}\right] \left[\begin{array}{c} R_1 \\ 0 \end{array}\right] = Q_1 R_1. \quad \square$$

The matrices $Q_1 = Q(1{:}m, 1{:}n)$ and $Q_2 = Q(1{:}m, n{+}1{:}m)$ can be easily computed from a factored form representation of Q. We refer to (5.2.2) as the *thin QR factorization*. The next result addresses its uniqueness.

Theorem 5.2.3 (Thin QR Factorization). *Suppose $A \in \mathbb{R}^{m \times n}$ has full column rank. The thin QR factorization*

$$A = Q_1 R_1$$

is unique where $Q_1 \in \mathbb{R}^{m \times n}$ has orthonormal columns and R_1 is upper triangular with positive diagonal entries. Moreover, $R_1 = G^T$ where G is the lower triangular Cholesky factor of $A^T A$.

Proof. Since $A^T A = (Q_1 R_1)^T (Q_1 R_1) = R_1^T R_1$ we see that $G = R_1^T$ is the Cholesky factor of $A^T A$. This factor is unique by Theorem 4.2.7. Since $Q_1 = A R_1^{-1}$ it follows that Q_1 is also unique. $\square$

How are Q_1 and R_1 affected by perturbations in A? To answer this question we need to extend the notion of 2-norm condition to rectangular matrices. Recall from §2.6.2 that for square matrices, $\kappa_2(A)$ is the ratio of the largest to the smallest singular value. For rectangular matrices A with full column rank we continue with this definition:

$$\kappa_2(A) = \frac{\sigma_{\max}(A)}{\sigma_{\min}(A)}. \tag{5.2.4}$$

If the columns of A are nearly dependent, then this quotient is large. Stewart (1993) has shown that $O(\epsilon)$ relative error in A induces $O(\epsilon \cdot \kappa_2(A))$ error in Q_1 and R_1.

5.2.2   Householder QR

We begin with a QR factorization method that utilizes Householder transformations. The essence of the algorithm can be conveyed by a small example. Suppose $m = 6$, $n = 5$, and assume that Householder matrices H_1 and H_2 have been computed so that

$$H_2 H_1 A = \left[\begin{array}{ccccc} \times & \times & \times & \times & \times \\ 0 & \times & \times & \times & \times \\ 0 & 0 & \mathbf{\times} & \times & \times \\ 0 & 0 & \mathbf{\times} & \times & \times \\ 0 & 0 & \mathbf{\times} & \times & \times \\ 0 & 0 & \mathbf{\times} & \times & \times \end{array}\right].$$

Concentrating on the highlighted entries, we determine a Householder matrix $\tilde{H}_3 \in \mathbb{R}^{4 \times 4}$ such that

$$\tilde{H}_3 \left[\begin{array}{c} \mathbf{\times} \\ \mathbf{\times} \\ \mathbf{\times} \\ \mathbf{\times} \end{array}\right] = \left[\begin{array}{c} \times \\ 0 \\ 0 \\ 0 \end{array}\right].$$

If $H_3 = \text{diag}(I_2, \tilde{H}_3)$, then

$$H_3 H_2 H_1 A = \begin{bmatrix} \times & \times & \times & \times & \times \\ 0 & \times & \times & \times & \times \\ 0 & 0 & \times & \times & \times \\ 0 & 0 & 0 & \times & \times \\ 0 & 0 & 0 & \times & \times \\ 0 & 0 & 0 & \times & \times \end{bmatrix}.$$

After n such steps we obtain an upper triangular $H_n H_{n-1} \cdots H_1 A = R$ and so by setting $Q = H_1 \cdots H_n$ we obtain $A = QR$.

Algorithm 5.2.1 (Householder QR) Given $A \in \mathbb{R}^{m \times n}$ with $m \geq n$, the following algorithm finds Householder matrices $H_1, \ldots, H_n$ such that if $Q = H_1 \cdots H_n$, then $Q^T A = R$ is upper triangular. The upper triangular part of A is overwritten by the upper triangular part of R and components $j + 1{:}m$ of the jth Householder vector are stored in $A(j + 1{:}m, j), j < m$.

> **for** $j = 1{:}n$
> $\qquad [v, \beta] = \textsf{house}(A(j{:}m, j))$
> $\qquad A(j{:}m, j{:}n) = (I - \beta v v^T) A(j{:}m, j{:}n)$
> $\qquad$ **if** $j < m$
> $\qquad\qquad A(j + 1{:}m, j) = v(2{:}m - j + 1)$
> $\qquad$ **end**
> **end**

This algorithm requires $2n^2(m - n/3)$ flops.

To clarify how A is overwritten, if

$$v^{(j)} = [\ \underbrace{0, \ldots, 0}_{j-1}, 1, v_{j+1}^{(j)}, \ldots, v_m^{(j)}\ ]^T$$

is the jth Householder vector, then upon completion

$$A = \begin{bmatrix} r_{11} & r_{12} & r_{13} & r_{14} & r_{15} \\ v_2^{(1)} & r_{22} & r_{23} & r_{24} & r_{25} \\ v_3^{(1)} & v_3^{(2)} & r_{33} & r_{34} & r_{35} \\ v_4^{(1)} & v_4^{(2)} & v_4^{(3)} & r_{44} & r_{45} \\ v_5^{(1)} & v_5^{(2)} & v_5^{(3)} & v_5^{(4)} & r_{55} \\ v_6^{(1)} & v_6^{(2)} & v_6^{(3)} & v_6^{(4)} & v_6^{(5)} \end{bmatrix}.$$

If the matrix $Q = H_1 \cdots H_n$ is required, then it can be accumulated using (5.1.5). This accumulation requires $4(m^2 n - m n^2 + n^3/3)$ flops. Note that the β-values that arise in Algorithm 5.2.1 can be retrieved from the stored Householder vectors:

$$\beta_j = \frac{2}{1 + \| A(j + 1{:}m, j) \|^2}.$$

We mention that the computed upper triangular matrix $\hat{R}$ is the exact R for a nearby A in the sense that $Z^T(A + E) = \hat{R}$ where Z is some exact orthogonal matrix and $\| E \|_2 \approx \mathbf{u}\| A \|_2$.

5.2.3   Block Householder QR Factorization

Algorithm 5.2.1 is rich in the level-2 operations of matrix-vector multiplication and outer product updates. By reorganizing the computation and using the WY representation discussed in §5.1.7 we can obtain a level-3 procedure. The idea is to apply the underlying Householder transformations in clusters of size r. Suppose $n = 12$ and $r = 3$. The first step is to generate Householders H_1, H_2, and H_3 as in Algorithm 5.2.1. However, unlike Algorithm 5.2.1 where each H_i is applied across the entire remaining submatrix, we apply only H_1, H_2, and H_3 to $A(:, 1{:}3)$. After this is accomplished we generate the block representation $H_1 H_2 H_3 = I - W_1 Y_1^T$ and then perform the level-3 update

$$A(:, 4{:}12) = (I - W Y^T)A(:, 4{:}12).$$

Next, we generate H_4, H_5, and H_6 as in Algorithm 5.2.1. However, these transformations are not applied to $A(:, 7{:}12)$ until their block representation $H_4 H_5 H_6 = I - W_2 Y_2^T$ is found. This illustrates the general pattern.

Algorithm 5.2.2 (Block Householder QR) If $A \in \mathbb{R}^{m \times n}$ and r is a positive integer, then the following algorithm computes an orthogonal $Q \in \mathbb{R}^{m \times m}$ and an upper triangular $R \in \mathbb{R}^{m \times n}$ so that $A = QR$.

$Q = I_m$; $\lambda = 1$; $k = 0$

while $\lambda \leq n$

    $\tau \leftarrow \min(\lambda + r - 1, n)$; $k = k + 1$

    Use Algorithm 5.2.1, to upper triangularize $A(\lambda{:}m, \lambda{:}\tau)$,
        generating Householder matrices $H_\lambda, \ldots, H_\tau$.

    Use Algorithm 5.1.2 to get the block representation
        $I - W_k Y_k = H_\lambda \cdots H_\tau$.

    $A(\lambda{:}m, \tau + 1{:}n) = (I - W_k Y_k^T)^T A(\lambda{:}m, \tau + 1{:}n)$

    $Q(:, \lambda{:}m) = Q(:, \lambda{:}m)(I - W_k Y_k^T)$

    $\lambda = \tau + 1$

end

The zero-nonzero structure of the Householder vectors that define $H_\lambda, \ldots, H_\tau$ implies that the first $\lambda - 1$ rows of W_k and Y_k are zero. This fact would be exploited in a practical implementation.

The proper way to regard Algorithm 5.2.2 is through the partitioning

$$A = [A_1 | \cdots | A_N], \qquad N = \text{ceil}(n/r)$$

where block column A_k is processed during the kth step. In the kth step of the reduction, a block Householder is formed that zeros the subdiagonal portion of A_k. The remaining block columns are then updated.

The roundoff properties of Algorithm 5.2.2 are essentially the same as those for Algorithm 5.2.1. There is a slight increase in the number of flops required because of the W-matrix computations. However, as a result of the blocking, all but a small fraction of the flops occur in the context of matrix multiplication. In particular, the level-3 fraction of Algorithm 5.2.2 is approximately $1 - O(1/N)$. See Bischof and Van Loan (1987) for further details.

5.2.4 Block Recursive QR

A more flexible approach to blocking involves recursion. Suppose $A \in \mathbb{R}^{m \times n}$ and assume for clarity that A has full column rank. Partition the thin QR factorization of A as follows:

$$
\begin{bmatrix} A_1 \mid A_2 \end{bmatrix} = \begin{bmatrix} Q_1 \mid Q_2 \end{bmatrix} \begin{bmatrix} R_{11} & R_{12} \\ 0 & R_{22} \end{bmatrix}.
$$

where $n_1 = \text{floor}(n/2)$, $n_2 = n - n_1$, $A_1, Q_1 \in \mathbb{R}^{m \times n_1}$ and $A_2, Q_2 \in \mathbb{R}^{m \times n_2}$. From the equations $Q_1 R_{11} = A_1$, $R_{12} = Q_1^T A_2$, and $Q_2 R_{22} = A_2 - Q_1 R_{12}$ we obtain the following recursive procedure:

Algorithm 5.2.3 (Recursive Block QR) Suppose $A \in \mathbb{R}^{m \times n}$ has full column rank and n_b is a positive blocking parameter. The following algorithm computes $Q \in \mathbb{R}^{m \times n}$ with orthonormal columns and upper triangular $R \in \mathbb{R}^{n \times n}$ such that $A = QR$.

> **function** $[Q, R] = \text{BlockQR}(A, n, n_b)$
>> **if** $n \leq n_b$
>>> Use Algorithm 5.2.1 to compute the thin QR factorization $A = QR$.
>> **else**
>>> $n_1 = \text{floor}(n/2)$
>>>
>>> $[Q_1, R_{11}] = \text{BlockQR}(A(:, 1{:}n_1), n_1, n_b)$
>>>
>>> $R_{12} = Q_1^T A(:, n_1 + 1{:}n)$
>>>
>>> $A(:, n_1 + 1{:}n) = A(:, n_1 + 1{:}n) - Q_1 R_{12}$
>>>
>>> $[Q_2, R_{22}] = \text{BlockQR}(A(:, n_1 + 1{:}n), n - n_1, n_b)$
>>>
>>> $Q = [Q_1 \mid Q_2], \quad R = \begin{bmatrix} R_{11} & R_{12} \\ 0 & R_{22} \end{bmatrix}$
>> **end**
> **end**

This divide-and-conquer approach is rich in matrix-matrix multiplication and provides a framework for the effective parallel computation of the QR factorization. See Elmroth and Gustavson (2001). Key implementation ideas concern the representation of the Q-matrices and the incorporation of the §5.2.3 blocking strategies.

5.2.5   Givens QR Methods

Givens rotations can also be used to compute the QR factorization and the 4-by-3 case illustrates the general idea:

$$
\begin{bmatrix} \times & \times & \times \\ \times & \times & \times \\ \mathbf{x} & \times & \times \\ \mathbf{x} & \times & \times \end{bmatrix} \xrightarrow{(3,4)}
\begin{bmatrix} \times & \times & \times \\ \mathbf{x} & \times & \times \\ \mathbf{x} & \times & \times \\ 0 & \times & \times \end{bmatrix} \xrightarrow{(2,3)}
\begin{bmatrix} \mathbf{x} & \times & \times \\ \mathbf{x} & \times & \times \\ 0 & \times & \times \\ 0 & \times & \times \end{bmatrix} \xrightarrow{(1,2)}
$$

$$
\begin{bmatrix} \times & \times & \times \\ 0 & \times & \times \\ 0 & \mathbf{x} & \times \\ 0 & \mathbf{x} & \times \end{bmatrix} \xrightarrow{(3,4)}
\begin{bmatrix} \times & \times & \times \\ 0 & \mathbf{x} & \times \\ 0 & \mathbf{x} & \times \\ 0 & 0 & \times \end{bmatrix} \xrightarrow{(2,3)}
\begin{bmatrix} \times & \times & \times \\ 0 & \times & \times \\ 0 & 0 & \mathbf{x} \\ 0 & 0 & \mathbf{x} \end{bmatrix} \xrightarrow{(3,4)} R.
$$

We highlighted the 2-vectors that define the underlying Givens rotations. If G_j denotes the jth Givens rotation in the reduction, then $Q^T A = R$ is upper triangular, where $Q = G_1 \cdots G_t$ and t is the total number of rotations. For general m and n we have:

Algorithm 5.2.4 (Givens QR) Given $A \in \mathbb{R}^{m \times n}$ with $m \geq n$, the following algorithm overwrites A with $Q^T A = R$, where R is upper triangular and Q is orthogonal.

> **for** $j = 1{:}n$
> > **for** $i = m{:} -1{:}j+1$
> > > $[c, s] = \mathsf{givens}(A(i-1,j), A(i,j))$
> > >
> > > $A(i-1{:}i, j{:}n) = \begin{bmatrix} c & s \\ -s & c \end{bmatrix}^T A(i-1{:}i, j{:}n)$
> >
> > **end**
>
> **end**

This algorithm requires $3n^2(m - n/3)$ flops. Note that we could use the representation ideas from §5.1.11 to encode the Givens transformations that arise during the calculation. Entry $A(i, j)$ can be overwritten with the associated representation.

With the Givens approach to the QR factorization, there is flexibility in terms of the rows that are involved in each update and also the order in which the zeros are introduced. For example, we can replace the inner loop body in Algorithm 5.2.4 with

$$
[c, s] = \mathsf{givens}(A(j,j), A(i,j))
$$

$$
A([\, j \; i \,], j{:}n) = \begin{bmatrix} c & s \\ -s & c \end{bmatrix}^T A([\, j \; i \,], j{:}n)
$$

and still emerge with the QR factorization. It is also possible to introduce zeros by row. Whereas Algorithm 5.2.4 introduces zeros by column,

$$
\begin{bmatrix} \times & \times & \times \\ 3 & \times & \times \\ 2 & 5 & \times \\ 1 & 4 & 6 \end{bmatrix},
$$

the implementation

> **for** $i = 2{:}m$
>> **for** $j = 1{:}i-1$
>>> $[c, s] = \text{givens}(A(j,j), A(i,j))$
>>>
>>> $A([j\ i], j{:}n) = \begin{bmatrix} c & s \\ -s & c \end{bmatrix}^{T} A([j\ i], j{:}n)$
>>
>> **end**
>
> **end**

introduces zeros by row, e.g.,

$$\begin{bmatrix} \times & \times & \times \\ 1 & \times & \times \\ 2 & 3 & \times \\ 4 & 5 & 6 \end{bmatrix}.$$

5.2.6 Hessenberg QR via Givens

As an example of how Givens rotations can be used in a structured problem, we show how they can be employed to compute the QR factorization of an upper Hessenberg matrix. (Other structured QR factorizations are discussed in Chapter 6 and §11.1.8.) A small example illustrates the general idea. Suppose $n = 6$ and that after two steps we have computed

$$G(2,3,\theta_2)^{T} G(1,2,\theta_1)^{T} A = \begin{bmatrix} \times & \times & \times & \times & \times & \times \\ 0 & \times & \times & \times & \times & \times \\ 0 & 0 & \times & \times & \times & \times \\ 0 & 0 & \mathbf{x} & \times & \times & \times \\ 0 & 0 & 0 & \times & \times & \times \\ 0 & 0 & 0 & 0 & \times & \times \end{bmatrix}.$$

Next, we compute $G(3,4,\theta_3)$ to zero the current (4,3) entry, thereby obtaining

$$G(3,4,\theta_3)^{T} G(2,3,\theta_2)^{T} G(1,2,\theta_1)^{T} A = \begin{bmatrix} \times & \times & \times & \times & \times & \times \\ 0 & \times & \times & \times & \times & \times \\ 0 & 0 & \times & \times & \times & \times \\ 0 & 0 & 0 & \times & \times & \times \\ 0 & 0 & 0 & \times & \times & \times \\ 0 & 0 & 0 & 0 & \times & \times \end{bmatrix}.$$

Continuing in this way we obtain the following algorithm.

Algorithm 5.2.5 (Hessenberg QR) If $A \in \mathbb{R}^{n \times n}$ is upper Hessenberg, then the following algorithm overwrites A with $Q^{T} A = R$ where Q is orthogonal and R is upper triangular. $Q = G_1 \cdots G_{n-1}$ is a product of Givens rotations where G_j has the form $G_j = G(j, j+1, \theta_j)$.

for $j = 1{:}n - 1$

$\quad$ $[\, c, s \,] \; = \; \mathsf{givens}(A(j, j), A(j + 1, j))$

$$A(j{:}j + 1, j{:}n) \; = \; \left[\begin{array}{cc} c & s \\ -s & c \end{array} \right]^{T} A(j{:}j + 1, j{:}n)$$

end

This algorithm requires about $3n^2$ flops.

5.2.7   Classical Gram-Schmidt Algorithm

We now discuss two alternative methods that can be used to compute the thin QR factorization $A = Q_1 R_1$ directly. If $\mathsf{rank}(A) = n$, then equation (5.2.3) can be solved for q_k:

$$q_k \; = \; \left(a_k \; - \; \sum_{i=1}^{k-1} r_{ik} q_i \right) \Big/ r_{kk}.$$

Thus, we can think of q_k as a unit 2-norm vector in the direction of

$$z_k \; = \; a_k \; - \; \sum_{i=1}^{k-1} r_{ik} q_i$$

where to ensure $z_k \in \mathsf{span}\{q_1, \dots, q_{k-1}\}^{\perp}$ we choose

$$r_{ik} \; = \; q_i^T a_k, \qquad i = 1{:}k{-}1.$$

This leads to the *classical Gram-Schmidt* (CGS) algorithm for computing $A = Q_1 R_1$.

$\quad R(1, 1) \; = \; \|\, A(:, 1)\,\|_2$

$\quad Q(:, 1) \; = \; A(:, 1)/R(1, 1)$

$\quad$ **for** $k = 2{:}n$

$\quad\quad R(1{:}k - 1, k) \; = \; Q(1{:}m, 1{:}k - 1)^T A(1{:}m, k)$

$\quad\quad z \; = \; A(1{:}m, k) - Q(1{:}m, 1{:}k - 1) \cdot R(1{:}k - 1, k)$

$\quad\quad R(k, k) \; = \; \|\, z \,\|_2$

$\quad\quad Q(1{:}m, k) \; = \; z/R(k, k)$

$\quad$ **end**

In the kth step of CGS, the kth columns of both Q and R are generated.

5.2.8   Modified Gram-Schmidt Algorithm

Unfortunately, the CGS method has very poor numerical properties in that there is typically a severe loss of orthogonality among the computed q_i. Interestingly, a rearrangement of the calculation, known as *modified Gram-Schmidt* (MGS), leads to a more reliable procedure. In the kth step of MGS, the kth column of Q (denoted by q_k)

and the kth row of R (denoted by r_k^T) are determined. To derive the MGS method, define the matrix $A^{(k)} \in \mathbb{R}^{m \times (n-k+1)}$ by

$$[\, 0 \mid A^{(k)} \,] = A - \sum_{i=1}^{k-1} q_i r_i^T = \sum_{i=k}^{n} q_i r_i^T.$$

It follows that if

$$A^{(k)} = [\begin{matrix} z & \mid & B \end{matrix}]$$
$$\phantom{A^{(k)} = [\,} \underset{1}{} \underset{n-k}{}$$

then $r_{kk} = \| z \|_2$, $q_k = z/r_{kk}$, and $[r_{k,k+1}, \dots, r_{kn}] = q_k^T B$. We then compute the outer product $A^{(k+1)} = B - q_k [r_{k,k+1} \cdots r_{kn}]$ and proceed to the next step. This completely describes the kth step of MGS.

Algorithm 5.2.6 (Modified Gram-Schmidt) Given $A \in \mathbb{R}^{m \times n}$ with rank$(A) = n$, the following algorithm computes the thin QR factorization $A = Q_1 R_1$ where $Q_1 \in \mathbb{R}^{m \times n}$ has orthonormal columns and $R_1 \in \mathbb{R}^{n \times n}$ is upper triangular.

>**for** $k = 1{:}n$
>>$R(k,k) = \| A(1{:}m, k) \|_2$
>>
>>$Q(1{:}m, k) = A(1{:}m, k)/R(k,k)$
>>
>>**for** $j = k+1{:}n$
>>>$R(k,j) = Q(1{:}m, k)^T A(1{:}m, j)$
>>>
>>>$A(1{:}m, j) = A(1{:}m, j) - Q(1{:}m, k)R(k,j)$
>>
>>**end**
>
>**end**

This algorithm requires $2mn^2$ flops. It is not possible to overwrite A with both Q_1 and R_1. Typically, the MGS computation is arranged so that A is overwritten by Q_1 and the matrix R_1 is stored in a separate array.

5.2.9 Work and Accuracy

If one is interested in computing an orthonormal basis for ran(A), then the Householder approach requires $2mn^2 - 2n^3/3$ flops to get Q in factored form and another $2mn^2 - 2n^3/3$ flops to get the first n columns of Q. (This requires "paying attention" to just the first n columns of Q in (5.1.5).) Therefore, for the problem of finding an orthonormal basis for ran(A), MGS is about twice as efficient as Householder orthogonalization. However, Björck (1967) has shown that MGS produces a computed $\hat{Q}_1 = [\hat{q}_1 \mid \cdots \mid \hat{q}_n]$ that satisfies

$$\hat{Q}_1^T \hat{Q}_1 = I + E_{MGS}, \qquad \| E_{MGS} \|_2 \approx \mathbf{u}\, \kappa_2(A),$$

whereas the corresponding result for the Householder approach is of the form

$$\hat{Q}_1^T \hat{Q}_1 = I + E_H, \qquad \| E_H \|_2 \approx \mathbf{u}.$$

Thus, if orthonormality is critical, then MGS should be used to compute orthonormal bases only when the vectors to be orthogonalized are fairly independent.

We also mention that the computed triangular factor $\hat{R}$ produced by MGS satisfies $\| A - \hat{Q}\hat{R} \| \approx \mathbf{u}\| A \|$ and that there exists a Q with perfectly orthonormal columns such that $\| A - Q\hat{R} \| \approx \mathbf{u}\| A \|$. See Higham (ASNA, p. 379) and additional references given at the end of this section.

5.2.10   A Note on Complex Householder QR

Complex Householder transformations (§5.1.13) can be used to compute the QR factorization of a complex matrix $A \in \mathbb{C}^{m \times n}$. Analogous to Algorithm 5.2.1 we have

> **for** $j = 1{:}n$
> > Compute a Householder matrix Q_j so that $Q_j A$ is upper triangular
> > > through its first j columns.
> >
> > $A = Q_j A$
>
> **end**

Upon termination, A has been reduced to an upper triangular matrix $R \in \mathbb{C}^{m \times n}$ and we have $A = QR$ where $Q = Q_1 \cdots Q_n$ is unitary. The reduction requires about four times the number of flops as the real case.

Problems

P5.2.1 Adapt the Householder QR algorithm so that it can efficiently handle the case when $A \in \mathbb{R}^{m \times n}$ has lower bandwidth p and upper bandwidth q.

P5.2.2 Suppose $A \in \mathbb{R}^{n \times n}$ and let $\mathcal{E}$ be the exchange permutation $\mathcal{E}_n$ obtained by reversing the order of the rows in I_n. (a) Show that if $R \in \mathbb{R}^{n \times n}$ is upper triangular, then $L = \mathcal{E}R\mathcal{E}$ is lower triangular. (b) Show how to compute an orthogonal $Q \in \mathbb{R}^{n \times n}$ and a lower triangular $L \in \mathbb{R}^{n \times n}$ so that $A = QL$ assuming the availability of a procedure for computing the QR factorization.

P5.2.3 Adapt the Givens QR factorization algorithm so that the zeros are introduced by diagonal. That is, the entries are zeroed in the order $(m, 1)$, $(m - 1, 1)$, $(m, 2)$, $(m - 2, 1)$, $(m - 1, 2)$, $(m, 3)$, etc.

P5.2.4 Adapt the Givens QR factorization algorithm so that it efficiently handles the case when A is n-by-n and tridiagonal. Assume that the subdiagonal, diagonal, and superdiagonal of A are stored in $e(1{:}n-1)$, $a(1{:}n)$, $f(1{:}n-1)$, respectively. Design your algorithm so that these vectors are overwritten by the nonzero portion of T.

P5.2.5   Suppose $L \in \mathbb{R}^{m \times n}$ with $m \geq n$ is lower triangular. Show how Householder matrices $H_1, \ldots, H_n$ can be used to determine a lower triangular $L_1 \in \mathbb{R}^{n \times n}$ so that

$$H_n \cdots H_1 L = \begin{bmatrix} L_1 \\ 0 \end{bmatrix}.$$

Hint: The second step in the 6-by-3 case involves finding H_2 so that

$$H_2 \begin{bmatrix} \times & 0 & 0 \\ \times & \times & 0 \\ \times & \times & \times \\ \times & \times & 0 \\ \times & \times & 0 \\ \times & \times & 0 \end{bmatrix} = \begin{bmatrix} \times & 0 & 0 \\ \times & \times & 0 \\ \times & \times & \times \\ \times & 0 & 0 \\ \times & 0 & 0 \\ \times & 0 & 0 \end{bmatrix}$$

with the property that rows 1 and 3 are left alone.

P5.2.6 Suppose $A \in \mathbb{R}^{n \times n}$ and $D = \text{diag}(d_1, \ldots, d_n) \in \mathbb{R}^{n \times n}$. Show how to construct an orthogonal Q such that

$$Q^T A - DQ^T = R$$

is upper triangular. Do not worry about efficiency—this is just an exercise in QR manipulation.

P5.2.7 Show how to compute the QR factorization of the product

$$A = A_p \cdots A_2 A_1$$

without explicitly multiplying the matrices $A_1, \ldots, A_p$ together. Assume that each A_i is square. Hint: In the $p = 3$ case, write

$$Q_3^T A = Q_3^T A_3 Q_2 Q_2^T A_2 Q_1 Q_1^T A_1$$

and determine orthogonal Q_i so that $Q_i^T(A_i Q_{i-1})$ is upper triangular. ($Q_0 = I$.)

P5.2.8 MGS applied to $A \in \mathbb{R}^{m \times n}$ is numerically equivalent to the first step in Householder QR applied to

$$\tilde{A} = \begin{bmatrix} O_n \\ A \end{bmatrix}$$

where O_n is the n-by-n zero matrix. Verify that this statement is true after the first step of each method is completed.

P5.2.9 Reverse the loop orders in Algorithm 5.2.6 (MGS) so that R is computed column by column.

P5.2.10 How many flops are required by the complex QR factorization procedure outlined in §5.10?

P5.2.11 Develop a complex version of the Givens QR factorization in which the diagonal of R is nonnegative. See §5.1.13.

P5.2.12 Show that if $A \in \mathbb{R}^{n \times n}$ and $a_i = A(:, i)$, then

$$|\det(A)| \leq \| a_1 \|_2 \cdots \| a_n \|_2.$$

Hint: Use the QR factorization.

P5.2.13 Suppose $A \in \mathbb{R}^{m \times n}$ with $m \geq n$. Construct an orthogonal $Q \in \mathbb{R}^{(m+n) \times (m+n)}$ with the property that $Q(1{:}m, 1{:}n)$ is a scalar multiple of A. Hint. If $\alpha \in \mathbb{R}$ is chosen properly, then $I - \alpha^2 A^T A$ has a Cholesky factorization.

P5.2.14 Suppose $A \in \mathbb{R}^{m \times n}$. Analogous to Algorithm 5.2.4, show how fast Givens transformations (P5.1.12) can be used to compute $M \in \mathbb{R}^{m \times m}$ and a diagonal $D \in \mathbb{R}^{m \times m}$ with positive diagonal entries so that $M^T A = S$ is upper triangular and $M M^T = D$. Relate M and S to A's QR factors.

P5.2.15 (*Parallel Givens QR*) Suppose $A \in \mathbb{R}^{9 \times 3}$ and that we organize a Givens QR so that the subdiagonal entries are zeroed over the course of ten "time steps" as follows:

Step	Entries Zeroed		
$T = 1$	(9,1)		
$T = 2$	(8,1)		
$T = 3$	(7,1)	(9,2)	
$T = 4$	(6,1)	(8,2)	
$T = 5$	(5,1)	(7,2)	(9,3)
$T = 6$	(4,1)	(6,2)	(8,3)
$T = 7$	(3,1)	(5,2)	(7,3)
$T = 8$	(2,1)	(4,2)	(6,3)
$T = 9$		(3,2)	(5,3)
$T = 10$			(4,3)

Assume that a rotation in plane $(i - 1, i)$ is used to zero a matrix entry (i, j). It follows that the rotations associated with any given time step involve disjoint pairs of rows and may therefore be computed in parallel. For example, during time step $T = 6$, there is a (3,4), (5,6), and (7,8) rotation. Three separate processors could oversee the three updates. Extrapolate from this example to the m-by-n case and show how the QR factorization could be computed in $O(m + n)$ time steps. How many of those time steps would involve n "nonoverlapping" rotations?

Notes and References for §5.2

The idea of using Householder transformations to solve the least squares problem was proposed in:

A.S. Householder (1958). "Unitary Triangularization of a Nonsymmetric Matrix," *J. ACM 5*, 339–342.

The practical details were worked out in:

P. Businger and G.H. Golub (1965). "Linear Least Squares Solutions by Householder Transformations," *Numer. Math. 7*, 269–276.
G.H. Golub (1965). "Numerical Methods for Solving Linear Least Squares Problems," *Numer. Math. 7*, 206–216.

The basic references for Givens QR include:

W. Givens (1958). "Computation of Plane Unitary Rotations Transforming a General Matrix to Triangular Form," *SIAM J. Appl. Math. 6*, 26–50.
M. Gentleman (1973). "Error Analysis of QR Decompositions by Givens Transformations," *Lin. Alg. Applic. 10*, 189–197.

There are modifications for the QR factorization that make it more attractive when dealing with rank deficiency. See §5.4. Nevertheless, when combined with the condition estimation ideas in §3.5.4, the traditional QR factorization can be used to address rank deficiency issues:

L.V. Foster (1986). "Rank and Null Space Calculations Using Matrix Decomposition without Column Interchanges," *Lin. Alg. Applic. 74*, 47–71.

The behavior of the Q and R factors when A is perturbed is of interest. A main result is that the resulting changes in Q and R are bounded by the condition of A times the relative change in A, see:

G.W. Stewart (1977). "Perturbation Bounds for the QR Factorization of a Matrix," *SIAM J. Numer. Anal. 14*, 509–518.
H. Zha (1993). "A Componentwise Perturbation Analysis of the QR Decomposition," *SIAM J. Matrix Anal. Applic. 4*, 1124–1131.
G.W. Stewart (1993). "On the Perturbation of LU Cholesky, and QR Factorizations," *SIAM J. Matrix Anal. Applic. 14*, 1141–1145.
A. Barrlund (1994). "Perturbation Bounds for the Generalized QR Factorization," *Lin. Alg. Applic. 207*, 251–271.
J.-G. Sun (1995). "On Perturbation Bounds for the QR Factorization," *Lin. Alg. Applic. 215*, 95–112.
X.-W. Chang and C.C. Paige (2001). "Componentwise Perturbation Analyses for the QR factorization," *Numer. Math. 88*, 319–345.

Organization of the computation so that the entries in Q depend continuously on the entries in A is discussed in:

T.F. Coleman and D.C. Sorensen (1984). "A Note on the Computation of an Orthonormal Basis for the Null Space of a Matrix," *Mathematical Programming 29*, 234–242.

References for the Gram-Schmidt process and various ways to overcome its shortfalls include:

J.R. Rice (1966). "Experiments on Gram-Schmidt Orthogonalization," *Math. Comput. 20*, 325–328.
A. Björck (1967). "Solving Linear Least Squares Problems by Gram-Schmidt Orthogonalization," *BIT 7*, 1–21.
N.N. Abdelmalek (1971). "Roundoff Error Analysis for Gram-Schmidt Method and Solution of Linear Least Squares Problems," *BIT 11*, 345–368.
A. Ruhe (1983). "Numerical Aspects of Gram-Schmidt Orthogonalization of Vectors," *Lin. Alg. Applic. 52/53*, 591–601.
W. Jalby and B. Philippe (1991). "Stability Analysis and Improvement of the Block Gram-Schmidt Algorithm," *SIAM J. Sci. Stat. Comput. 12*, 1058–1073.
Å. Björck and C.C. Paige (1992). "Loss and Recapture of Orthogonality in the Modified Gram-Schmidt Algorithm," *SIAM J. Matrix Anal. Applic. 13*, 176–190.
A. Björck (1994). "Numerics of Gram-Schmidt Orthogonalization," *Lin. Alg. Applic. 197/198*, 297–316.
L. Giraud and J. Langou (2003). "A Robust Criterion for the Modified Gram-Schmidt Algorithm with Selective Reorthogonalization," *SIAM J. Sci. Comput. 25*, 417–441.
G.W. Stewart (2005). "Error Analysis of the Quasi-Gram–Schmidt Algorithm," *SIAM J. Matrix Anal. Applic. 27*, 493–506.

L. Giraud, J. Langou, M. Rozlonk, and J. van den Eshof (2005). "Rounding Error Analysis of the Classical Gram-Schmidt Orthogonalization Process," *Numer. Math. 101*, 87–100.

A. Smoktunowicz, J.L. Barlow and J. Langou (2006). "A Note on the Error Analysis of Classical Gram-Schmidt," *Numer. Math. 105*, 299–313.

Various high-performance issues pertaining to the QR factorization are discussed in:

B. Mattingly, C. Meyer, and J. Ortega (1989). "Orthogonal Reduction on Vector Computers," *SIAM J. Sci. Stat. Comput. 10*, 372–381.

P.A. Knight (1995). "Fast Rectangular Matrix Multiplication and the QR Decomposition," *Lin. Alg. Applic. 221*, 69–81.

J.J. Carrig, Jr. and G.L. Meyer (1997). "Efficient Householder QR Factorization for Superscalar Processors," *ACM Trans. Math. Softw. 23*, 362–378.

D. Vanderstraeten (2000). "An Accurate Parallel Block Gram-Schmidt Algorithm without Reorthogonalization," *Numer. Lin. Alg. 7*, 219–236.

E. Elmroth and F.G. Gustavson (2000). "Applying Recursion to Serial and Parallel QR Factorization Leads to Better Performance," *IBM J. Res. Dev. 44*, 605–624.

Many important high-performance implementation ideas apply equally to LU, Cholesky, and QR, see:

A. Buttari, J. Langou, J. Kurzak, and J. Dongarra (2009). "A Class of Parallel Tiled Linear Algebra Algorithms for Multicore Architectures," *Parallel Comput. 35*, 38–53.

J. Kurzak, H. Ltaief, and J. Dongarra (2010). "Scheduling Dense Linear Algebra Operations on Multicore Processors," *Concurrency Comput. Pract. Exper. 22*, 15–44.

J. Demmel, L. Grigori, M, Hoemmen, and J. Langou (2012). "Methods and Algorithms for Scientific Computing Communication-optimal Parallel and Sequential QR and LU Factorizations," *SIAM J. Sci. Comput. 34*, A206-A239.

Historical references concerned with parallel Givens QR include:

W.M. Gentleman and H.T. Kung (1981). "Matrix Triangularization by Systolic Arrays," *SPIE Proc. 298*, 19–26.

D.E. Heller and I.C.F. Ipsen (1983). "Systolic Networks for Orthogonal Decompositions," *SIAM J. Sci. Stat. Comput. 4*, 261–269.

M. Costnard, J.M. Muller, and Y. Robert (1986). "Parallel QR Decomposition of a Rectangular Matrix," *Numer. Math. 48*, 239–250.

L. Eldin and R. Schreiber (1986). "An Application of Systolic Arrays to Linear Discrete Ill-Posed Problems," *SIAM J. Sci. Stat. Comput. 7*, 892–903.

F.T. Luk (1986). "A Rotation Method for Computing the QR Factorization," *SIAM J. Sci. Stat. Comput. 7*, 452–459.

J.J. Modi and M.R.B. Clarke (1986). "An Alternative Givens Ordering," *Numer. Math. 43*, 83–90.

The QR factorization of a structured matrix is usually structured itself, see:

A.W. Bojanczyk, R.P. Brent, and F.R. de Hoog (1986). "QR Factorization of Toeplitz Matrices," *Numer. Math. 49*, 81–94.

S. Qiao (1986). "Hybrid Algorithm for Fast Toeplitz Orthogonalization," *Numer. Math. 53*, 351–366.

C.J. Demeure (1989). "Fast QR Factorization of Vandermonde Matrices," *Lin. Alg. Applic. 122/123/124*, 165–194.

L. Reichel (1991). "Fast QR Decomposition of Vandermonde-Like Matrices and Polynomial Least Squares Approximation," *SIAM J. Matrix Anal. Applic. 12*, 552–564.

D.R. Sweet (1991). "Fast Block Toeplitz Orthogonalization," *Numer. Math. 58*, 613–629.

Quantum computation has an interesting connection to complex Givens rotations and their application to vectors, see:

G. Cybenko (2001). "Reducing Quantum Computations to Elementary Unitary Transformations," *Comput. Sci. Eng. 3*, 27–32.

D.P. O'Leary and S.S. Bullock (2005). "QR Factorizations Using a Restricted Set of Rotations," *ETNA 21*, 20–27.

N.D. Mermin (2007). *Quantum Computer Science*, Cambridge University Press, New York.

5.3   The Full-Rank Least Squares Problem

Consider the problem of finding a vector $x \in \mathbb{R}^n$ such that $Ax = b$ where the *data matrix* $A \in \mathbb{R}^{m \times n}$ and the *observation vector* $b \in \mathbb{R}^m$ are given and $m \geq n$. When there are more equations than unknowns, we say that the system $Ax = b$ is *overdetermined.* Usually an overdetermined system has no exact solution since b must be an element of $\mathrm{ran}(A)$, a proper subspace of $\mathbb{R}^m$.

This suggests that we strive to minimize $\| Ax - b \|_p$ for some suitable choice of p. Different norms render different optimum solutions. For example, if $A = [\, 1,\, 1,\, 1\,]^T$ and $b = [\, b_1,\, b_2,\, b_3\,]^T$ with $b_1 \geq b_2 \geq b_3 \geq 0$, then it can be verified that

$$
\begin{aligned}
p &= 1 &\Rightarrow\quad x_{\mathrm{opt}} &= b_2, \\
p &= 2 &\Rightarrow\quad x_{\mathrm{opt}} &= (b_1 + b_2 + b_3)/3, \\
p &= \infty &\Rightarrow\quad x_{\mathrm{opt}} &= (b_1 + b_3)/2.
\end{aligned}
$$

Minimization in the 1-norm and infinity-norm is complicated by the fact that the function $f(x) = \| Ax - b \|_p$ is not differentiable for these values of p. However, there are several good techniques available for 1-norm and ∞-norm minimization. See Coleman and Li (1992), Li (1993), and Zhang (1993).

In contrast to general p-norm minimization, the *least squares* (LS) problem

$$
\min_{x \in \mathbb{R}^n} \| Ax - b \|_2 \tag{5.3.1}
$$

is more tractable for two reasons:

- $\phi(x) = \frac{1}{2}\| Ax - b \|_2^2$ is a differentiable function of x and so the minimizers of ϕ satisfy the gradient equation $\nabla\phi(x) = 0$. This turns out to be an easily constructed symmetric linear system which is positive definite if A has full column rank.

- The 2-norm is preserved under orthogonal transformation. This means that we can seek an orthogonal Q such that the equivalent problem of minimizing $\| (Q^T A)x - (Q^T b) \|_2$ is "easy" to solve.

In this section we pursue these two solution approaches for the case when A has full column rank. Methods based on normal equations and the QR factorization are detailed and compared.

5.3.1   Implications of Full Rank

Suppose $x \in \mathbb{R}^n$, $z \in \mathbb{R}^n$, $\alpha \in \mathbb{R}$, and consider the equality

$$
\| A(x + \alpha z) - b \|_2^2 \;=\; \| Ax - b \|_2^2 + 2\alpha z^T A^T (Ax - b) + \alpha^2 \| Az \|_2^2
$$

where $A \in \mathbb{R}^{m \times n}$ and $b \in \mathbb{R}^m$. If x solves the LS problem (5.3.1), then we must have $A^T(Ax - b) = 0$. Otherwise, if $z = -A^T(Ax - b)$ and we make α small enough, then we obtain the contradictory inequality $\| A(x + \alpha z) - b \|_2 < \| Ax - b \|_2$. We may also conclude that if x *and* $x + \alpha z$ are LS minimizers, then $z \in \mathrm{null}(A)$.

Thus, if A has full column rank, then there is a unique LS solution x_{LS} and it solves the symmetric positive definite linear system

$$
A^T A x_{\mathrm{LS}} = A^T b.
$$

These are called the *normal equations*. Note that if

$$\phi(x) = \frac{1}{2} \| Ax - b \|_2^2,$$

then

$$\nabla\phi(x) = A^T(Ax - b),$$

so solving the normal equations is tantamount to solving the gradient equation $\nabla\phi = 0$. We call

$$r_{\text{LS}} = b - Ax_{\text{LS}}$$

the *minimum residual* and we use the notation

$$\rho_{\text{LS}} = \| Ax_{\text{LS}} - b \|_2$$

to denote its size. Note that if ρ_{LS} is small, then we can do a good job "predicting" b by using the columns of A.

Thus far we have been assuming that $A \in \mathbb{R}^{m \times n}$ has full column rank, an assumption that is dropped in §5.5. However, even if $\mathsf{rank}(A) = n$, trouble can be expected if A is nearly rank deficient. The SVD can be used to substantiate this remark. If

$$A = U\Sigma V^T = \sum_{i=1}^{n} \sigma_i u_i v_i^T$$

is the SVD of a full rank matrix $A \in \mathbb{R}^{m \times n}$, then

$$\| Ax - b \|_2^2 = \| (U^T AV)(V^T x) - U^T b \|_2^2 = \sum_{i=1}^{n}(\sigma_i y_i - (u_i^T b))^2 + \sum_{i=n+1}^{m} (u_i^T b)^2$$

where $y = V^T x$. It follows that this summation is minimized by setting $y_i = u_i^T b/\sigma_i$, $i = 1{:}n$. Thus,

$$x_{\text{LS}} = \sum_{i=1}^{n} \frac{u_i^T b}{\sigma_i} v_i \tag{5.3.2}$$

and

$$\rho_{\text{LS}}^2 = \sum_{i=n+1}^{2} (u_i^T b)^2. \tag{5.3.3}$$

It is clear that the presence of small singular values means LS solution sensitivity. The effect of perturbations on the minimum sum of squares is less clear and requires further analysis which we offer below.

When assessing the quality of a computed LS solution $\hat{x}_{\text{LS}}$, there are two important issues to bear in mind:

- How close is $\hat{x}_{\text{LS}}$ to x_{LS}?

- How small is $\hat{r}_{\text{LS}} = b - A\hat{x}_{\text{LS}}$ compared to $r_{\text{LS}} = b - Ax_{\text{LS}}$?

The relative importance of these two criteria varies from application to application. In any case it is important to understand how x_{LS} and r_{LS} are affected by perturbations in A and b. Our intuition tells us that if the columns of A are nearly dependent, then these quantities may be quite sensitive. For example, suppose

$$
A = \begin{bmatrix} 1 & 0 \\ 0 & 10^{-6} \\ 0 & 0 \end{bmatrix}, \quad \delta A = \begin{bmatrix} 0 & 0 \\ 0 & 0 \\ 0 & 10^{-8} \end{bmatrix}, \quad b = \begin{bmatrix} 1 \\ 0 \\ 1 \end{bmatrix}, \quad \delta b = \begin{bmatrix} 0 \\ 0 \\ 0 \end{bmatrix},
$$

and that x_{LS} and $\hat{x}_{\mathrm{LS}}$ minimize $\| Ax - b \|_2$ and $\| (A + \delta A)x - (b + \delta b) \|_2$, respectively. If r_{LS} and $\hat{r}_{\mathrm{LS}}$ are the corresponding minimum residuals, then it can be shown that

$$
x_{\mathrm{LS}} = \begin{bmatrix} 1 \\ 0 \end{bmatrix}, \quad \hat{x}_{\mathrm{LS}} = \begin{bmatrix} 1 \\ .9999 \cdot 10^4 \end{bmatrix}, \quad r_{\mathrm{LS}} = \begin{bmatrix} 0 \\ 0 \\ 1 \end{bmatrix}, \quad \hat{r}_{\mathrm{LS}} = \begin{bmatrix} 0 \\ -.9999 \cdot 10^{-2} \\ .9999 \cdot 10^0 \end{bmatrix}.
$$

Recall that the 2-norm condition of a rectangular matrix is the ratio of its largest to smallest singular values. Since $\kappa_2(A) = 10^6$ we have

$$
\frac{\| \hat{x}_{\mathrm{LS}} - x_{\mathrm{LS}} \|_2}{\| x_{\mathrm{LS}} \|_2} \approx .9999 \cdot 10^4 \leq \kappa_2(A)^2 \frac{\| \delta A \|_2}{\| A \|_2} = 10^{12} \cdot 10^{-8}
$$

and

$$
\frac{\| \hat{r}_{\mathrm{LS}} - r_{\mathrm{LS}} \|_2}{\| b \|_2} \approx .7070 \cdot 10^{-2} \leq \kappa_2(A) \frac{\| \delta A \|_2}{\| A \|_2} = 10^6 \cdot 10^{-8}.
$$

The example suggests that the sensitivity of x_{LS} can depend upon $\kappa_2(A)^2$. Below we offer an LS perturbation theory that confirms the possibility.

5.3.2   The Method of Normal Equations

A widely-used method for solving the full-rank LS problem is the *method of normal equations*.

Algorithm 5.3.1 (Normal Equations) Given $A \in \mathbb{R}^{m \times n}$ with the property that $\mathrm{rank}(A) = n$ and $b \in \mathbb{R}^m$, this algorithm computes a vector x_{LS} that minimizes $\| Ax - b \|_2$.

   Compute the lower triangular portion of $C = A^T A$.

   Form the matrix-vector product $d = A^T b$.

   Compute the Cholesky factorization $C = GG^T$.

   Solve $Gy = d$ and $G^T x_{\mathrm{LS}} = y$.

This algorithm requires $(m + n/3)n^2$ flops. The normal equation approach is convenient because it relies on standard algorithms: Cholesky factorization, matrix-matrix multiplication, and matrix-vector multiplication. The compression of the m-by-n data matrix A into the (typically) much smaller n-by-n cross-product matrix C is attractive.

   Let us consider the accuracy of the computed normal equations solution $\hat{x}_{\mathrm{LS}}$. For clarity, assume that no roundoff errors occur during the formation of $C = A^T A$ and

$d = A^T b$. It follows from what we know about the roundoff properties of the Cholesky factorization (§4.2.6) that

$$(A^T A + E)\hat{x}_{\mathrm{LS}} = A^T b$$

where

$$\| E \|_2 \approx \mathbf{u} \| A^T \|_2 \| A \|_2 = \mathbf{u} \| A^T A \|_2.$$

Thus, we can expect

$$\frac{\| \hat{x}_{\mathrm{LS}} - x_{\mathrm{LS}} \|_2}{\| x_{\mathrm{LS}} \|_2} \approx \mathbf{u}\kappa_2(A^T A) = \mathbf{u}\kappa_2(A)^2. \tag{5.3.4}$$

In other words, the accuracy of the computed normal equations solution depends on the square of the condition. See Higham (ASNA, §20.4) for a detailed roundoff analysis of the normal equations approach.

It should be noted that the formation of $A^T A$ can result in a significant loss of information. If

$$A = \begin{bmatrix} 1 & 1 \\ \sqrt{\mathbf{u}} & 0 \\ 0 & \sqrt{\mathbf{u}} \end{bmatrix},$$

then $\kappa_2(A) \approx \sqrt{\mathbf{u}}$. However,

$$\mathsf{fl}(A^T A) = \begin{bmatrix} 1 & 1 \\ 1 & 1 \end{bmatrix}$$

is exactly singular. Thus, the method of normal equations can break down on matrices that are not particularly close to being numerically rank deficient.

5.3.3   LS Solution Via QR Factorization

Let $A \in \mathbb{R}^{m \times n}$ with $m \geq n$ and $b \in \mathbb{R}^m$ be given and suppose that an orthogonal matrix $Q \in \mathbb{R}^{m \times m}$ has been computed such that

$$Q^T A = R = \begin{bmatrix} R_1 \\ 0 \end{bmatrix} \begin{matrix} n \\ m-n \end{matrix} \tag{5.3.5}$$

is upper triangular. If

$$Q^T b = \begin{bmatrix} c \\ d \end{bmatrix} \begin{matrix} n \\ m-n \end{matrix}$$

then

$$\| Ax - b \|_2^2 = \| Q^T Ax - Q^T b \|_2^2 = \| R_1 x - c \|_2^2 + \| d \|_2^2$$

for any $x \in \mathbb{R}^n$. Since $\mathsf{rank}(A) = \mathsf{rank}(R_1) = n$, it follows that x_{LS} is defined by the upper triangular system

$$R_1 x_{\mathrm{LS}} = c.$$

Note that

$$\rho_{\mathrm{LS}} = \| d \|_2.$$

We conclude that the full-rank LS problem can be readily solved once we have computed the QR factorization of A. Details depend on the exact QR procedure. If Householder matrices are used and Q^T is applied in factored form to b, then we obtain

Algorithm 5.3.2 (Householder LS Solution) If $A \in \mathbb{R}^{m \times n}$ has full column rank and $b \in \mathbb{R}^m$, then the following algorithm computes a vector $x_{\mathrm{LS}} \in \mathbb{R}^n$ such that $\| Ax_{\mathrm{LS}} - b \|_2$ is minimum.

Use Algorithm 5.2.1 to overwrite A with its QR factorization.

for $j = 1{:}n$

$$v = \left[\begin{array}{c} 1 \\ A(j+1:m, j) \end{array} \right]$$

$$\beta = 2/v^T v$$

$$b(j{:}m) = b(j{:}m) - \beta(v^T b(j{:}m))v$$

end

Solve $R(1{:}n, 1{:}n){\cdot}x_{\mathrm{LS}} = b(1{:}n)$.

This method for solving the full-rank LS problem requires $2n^2(m - n/3)$ flops. The $O(mn)$ flops associated with the updating of b and the $O(n^2)$ flops associated with the back substitution are not significant compared to the work required to factor A.

It can be shown that the computed $\hat{x}_{\mathrm{LS}}$ solves

$$\min\| (A + \delta A)x - (b + \delta b) \|_2 \tag{5.3.6}$$

where

$$\| \delta A \|_F \leq (6m - 3n + 41)\, n\, \mathbf{u} \| A \|_F + O(\mathbf{u}^2) \tag{5.3.7}$$

and

$$\| \delta b \|_2 \leq (6m - 3n + 40)\, n\, \mathbf{u} \| b \|_2 + O(\mathbf{u}^2). \tag{5.3.8}$$

These inequalities are established in Lawson and Hanson (SLS, p. 90ff) and show that $\hat{x}_{\mathrm{LS}}$ satisfies a "nearby" LS problem. (We cannot address the relative error in $\hat{x}_{\mathrm{LS}}$ without an LS perturbation theory, to be discussed shortly.) We mention that similar results hold if Givens QR is used.

5.3.4   Breakdown in Near-Rank-Deficient Case

As with the method of normal equations, the Householder method for solving the LS problem breaks down in the back-substitution phase if $\mathsf{rank}(A) < n$. Numerically, trouble can be expected if $\kappa_2(A) = \kappa_2(R) \approx 1/\mathbf{u}$. This is in contrast to the normal equations approach, where completion of the Cholesky factorization becomes problematical once $\kappa_2(A)$ is in the neighborhood of $1/\sqrt{\mathbf{u}}$ as we showed above. Hence the claim in Lawson and Hanson (SLS, pp. 126–127) that for a fixed machine precision, a wider class of LS problems can be solved using Householder orthogonalization.

5.3.5   A Note on the MGS Approach

In principle, MGS computes the thin QR factorization $A = Q_1 R_1$. This is enough to solve the full-rank LS problem because it transforms the normal equation system

$(A^T A)x = A^T b$ to the upper triangular system $R_1 x = Q_1^T b$. But an analysis of this approach when $Q_1^T b$ is explicitly formed introduces a $\kappa_2(A)^2$ term. This is because the computed factor $\hat{Q}_1$ satisfies $\| \hat{Q}_1^T \hat{Q}_1 - I_n \|_2 \approx \mathbf{u}\kappa_2(A)$ as we mentioned in §5.2.9.

However, if MGS is applied to the augmented matrix

$$A_+ = [\, A \,|\, b \,] = [\, Q_1 \,|\, q_{n+1} \,] \begin{bmatrix} R_1 & z \\ 0 & \rho \end{bmatrix},$$

then $z = Q_1^T b$. Computing $Q_1^T b$ in this fashion and solving $R_1 x_{\rm LS} = z$ produces an LS solution $\hat{x}_{\rm LS}$ that is "just as good" as the Householder QR method. That is to say, a result of the form (5.3.6)–(5.3.8) applies. See Björck and Paige (1992).

It should be noted that the MGS method is slightly more expensive than Householder QR because it always manipulates m-vectors whereas the latter procedure deals with vectors that become shorter in length as the algorithm progresses.

5.3.6   The Sensitivity of the LS Problem

We now develop a perturbation theory for the full-rank LS problem that assists in the comparison of the normal equations and QR approaches. LS sensitivity analysis has a long and fascinating history. Grcar (2009, 2010) compares about a dozen different results that have appeared in the literature over the decades and the theorem below follows his analysis. It examines how the LS solution and its residual are affected by changes in A and b and thereby sheds light on the condition of the LS problem. Four facts about $A \in \mathbb{R}^{m \times n}$ are used in the proof, where it is assumed that $m > n$:

$$
\begin{aligned}
1 &= \| A(A^T A)^{-1} A^T \|_2, & \frac{1}{\sigma_n(A)} &= \| (A^T A)^{-1} A^T \|_2, \\
1 &= \| I - A(A^T A)^{-1} A^T \|_2, & \frac{1}{\sigma_n(A)^2} &= \| (A^T A)^{-1} \|_2.
\end{aligned}
\qquad (5.3.9)
$$

These equations are easily verified using the SVD.

Theorem 5.3.1. *Suppose that $x_{\rm LS}$, $r_{\rm LS}$, $\hat{x}_{\rm LS}$, and $\hat{r}_{\rm LS}$ satisfy*

$$\| Ax_{\rm LS} - b \|_2 = \min, \qquad r_{\rm LS} = b - Ax_{\rm LS},$$

$$\| (A + \delta A)\hat{x}_{\rm LS} - (b + \delta b) \|_2 = \min, \qquad \hat{r}_{\rm LS} = (b + \delta b) - (A + \delta A)\hat{x}_{\rm LS},$$

where A has rank n and $\| \delta A \|_2 < \sigma_n(A)$. Assume that b, $r_{\rm LS}$, and $x_{\rm LS}$ are not zero. Let $\theta_{\rm LS} \in (0, \pi/2)$ be defined by

$$\sin(\theta_{\rm LS}) = \frac{\| r_{\rm LS} \|_2}{\| b \|_2}.$$

If

$$\epsilon = \max \left\{ \frac{\| \delta A \|_2}{\| A \|_2}, \frac{\| \delta b \|_2}{\| b \|_2} \right\}$$

and

$$\nu_{\rm LS} = \frac{\| Ax_{\rm LS} \|_2}{\sigma_n(A)\| x_{\rm LS} \|_2}, \qquad (5.3.10)$$

then

$$\frac{\|\, \hat{x}_{\text{LS}} - x_{\text{LS}} \,\|_2}{\|\, x \,\|_2} \leq \epsilon \left\{ \frac{\nu_{\text{LS}}}{\cos(\theta_{\text{LS}})} + [\, 1 + \nu_{\text{LS}} \tan(\theta_{\text{LS}}) \,] \, \kappa_2(A) \right\} + O(\epsilon^2) \qquad (5.3.11)$$

and

$$\frac{\|\, \hat{r}_{\text{LS}} - r_{\text{LS}} \,\|_2}{\|\, r_{\text{LS}} \,\|_2} \leq \epsilon \left\{ \frac{1}{\sin(\theta_{\text{LS}})} + \left[\frac{1}{\nu_{\text{LS}} \tan(\theta_{\text{LS}})} + 1 \right] \kappa_2(A) \right\} + O(\epsilon^2). \qquad (5.3.12)$$

Proof. Let E and f be defined by $E = \delta A/\epsilon$ and $f = \delta b/\epsilon$. By Theorem 2.5.2 we have $\text{rank}(A + tE) = n$ for all $t \in [0, \epsilon]$. It follows that the solution $x(t)$ to

$$(A + tE)^T (A + tE) x(t) = (A + tE)^T (b + tf) \qquad (5.3.13)$$

is continuously differentiable for all $t \in [0, \epsilon]$. Since $x_{\text{LS}} = x(0)$ and $\hat{x}_{\text{LS}} = x(\epsilon)$, we have

$$\hat{x}_{\text{LS}} = x_{\text{LS}} + \epsilon \dot{x}(0) + O(\epsilon^2).$$

By taking norms and dividing by $\|\, x_{\text{LS}} \,\|_2$ we obtain

$$\frac{\|\, \hat{x}_{\text{LS}} - x_{\text{LS}} \,\|_2}{\|\, x_{\text{LS}} \,\|_2} = \epsilon \frac{\|\, \dot{x}(0) \,\|_2}{\|\, x_{\text{LS}} \,\|_2} + O(\epsilon^2). \qquad (5.3.14)$$

In order to bound $\|\, \dot{x}(0) \,\|_2$, we differentiate (5.3.13) and set $t = 0$ in the result. This gives

$$E^T A x_{\text{LS}} + A^T E x_{\text{LS}} + A^T A \dot{x}(0) = A^T f + E^T b,$$

i.e.,

$$\dot{x}(0) = (A^T A)^{-1} A^T (f - E x_{\text{LS}}) + (A^T A)^{-1} E^T r_{\text{LS}}. \qquad (5.3.15)$$

Using (5.3.9) and the inequalities $\|\, f \,\|_2 \leq \|\, b \,\|_2$ and $\|\, E \,\|_2 \leq \|\, A \,\|_2$, it follows that

$$\|\, \dot{x}(0) \,\| \leq \|\, (A^T A)^{-1} A^T f \,\|_2 + \|\, (A^T A)^{-1} A^T E x_{\text{LS}} \,\|_2 + \|\, (A^T A)^{-1} E^T r_{\text{LS}} \,\|_2$$

$$\leq \frac{\|\, b \,\|_2}{\sigma_n(A)} + \frac{\|\, A \,\|_2 \|\, x_{\text{LS}} \,\|_2}{\sigma_n(A)} + \frac{\|\, A \,\|_2 \|\, r_{\text{LS}} \,\|_2}{\sigma_n(A)^2}.$$

By substituting this into (5.3.14) we obtain

$$\frac{\|\, \hat{x}_{\text{LS}} - x_{\text{LS}} \,\|_2}{\|\, x_{\text{LS}} \,\|_2} \leq \epsilon \left(\frac{\|\, b \,\|_2}{\sigma_n(A) \|\, x_{\text{LS}} \,\|_2} + \frac{\|\, A \,\|_2}{\sigma_n(A)} + \frac{\|\, A \,\|_2 \|\, r_{\text{LS}} \,\|_2}{\sigma_n(A)^2 \|\, x_{\text{LS}} \,\|_2} \right) + O(\epsilon^2).$$

Inequality (5.3.11) follows from the definitions of $\kappa_2(A)$ and ν_{LS} and the identities

$$\cos(\theta_{\text{LS}}) = \frac{\|\, A x_{\text{LS}} \,\|_2}{\|\, b \,\|_2}, \qquad \tan(\theta_{\text{LS}}) = \frac{\|\, r_{\text{LS}} \,\|_2}{\|\, A x_{\text{LS}} \,\|_2}. \qquad (5.3.16)$$

The proof of the residual bound (5.3.12) is similar. Define the differentiable vector function $r(t)$ by

$$r(t) = (b + tf) - (A + tE) x(t)$$

and observe that $r_{\mathrm{LS}} = r(0)$ and $\hat{r}_{\mathrm{LS}} = r(\epsilon)$. Thus,

$$\frac{\| \hat{r}_{\mathrm{LS}} - r_{\mathrm{LS}} \|_2}{\| r_{\mathrm{LS}} \|_2} = \epsilon \frac{\| \dot{r}(0) \|_2}{\| r_{\mathrm{LS}} \|_2} + O(\epsilon^2). \tag{5.3.17}$$

From (5.3.15) we have

$$\dot{r}(0) = \left(I - A(A^T A)^{-1} A^T \right) (f - E x_{\mathrm{LS}}) - A(A^T A)^{-1} E^T r_{\mathrm{LS}}.$$

By taking norms, using (5.3.9) and the inequalities $\| f \|_2 \leq \| b \|_2$ and $\| E \|_2 \leq \| A \|_2$, we obtain

$$\| \dot{r}(0) \|_2 \leq \| b \|_2 + \| A \|_2 \| x_{\mathrm{LS}} \|_2 + \frac{\| A \|_2 \| r_{\mathrm{LS}} \|_2}{\sigma_n(A)}$$

and thus from (5.3.17) we have

$$\frac{\| \hat{r}_{\mathrm{LS}} - r_{\mathrm{LS}} \|_2}{\| r_{\mathrm{LS}} \|_2} \leq \frac{\| b \|_2}{\| r_{\mathrm{LS}} \|_2} + \frac{\| A \|_2 \| x_{\mathrm{LS}} \|_2}{\| r_{\mathrm{LS}} \|_2} + \frac{\| A \|_2}{\sigma_n(A)}.$$

The inequality (5.3.12) follows from the definitions of $\kappa_2(A)$ and ν_{LS} and the identities (5.3.16). $\quad\square$

It is instructive to identify conditions that turn the upper bound in (5.3.11) into a bound that involves $\kappa_2(A)^2$. The example in §5.3.1 suggests that this factor might figure in the definition of an LS condition number. However, the theorem shows that the situation is more subtle. Note that

$$\nu_{\mathrm{LS}} = \frac{\| A x_{\mathrm{LS}} \|_2}{\sigma_n(A) \| x_{\mathrm{LS}} \|_2} \leq \frac{\| A \|_2}{\sigma_n(A)} = \kappa_2(A).$$

The SVD expansion (5.3.2) suggests that if b has a modest component in the direction of the left singular vector u_n, then

$$\nu_{\mathrm{LS}} \approx \kappa_2(A).$$

If this is the case *and* θ_{LS} is sufficiently bounded away from $\pi/2$, then the inequality (5.3.11) essentially says that

$$\frac{\| \hat{x}_{\mathrm{LS}} - x_{\mathrm{LS}} \|_2}{\| x_{\mathrm{LS}} \|_2} \approx \epsilon \left(\kappa_2(A) + \frac{\rho_{\mathrm{LS}}}{\| b \|_2} \kappa_2(A)^2 \right). \tag{5.3.18}$$

Although this simple heuristic assessment of LS sensitivity is almost always applicable, it important to remember that the true condition of a particular LS problem depends on ν_{LS}, θ_{LS}, and $\kappa_2(A)$.

Regarding the perturbation of the residual, observe that the upper bound in the residual result (5.3.12) is less than the upper bound in the solution result (5.3.11) by a factor of $\nu_{\mathrm{LS}} \tan(\theta_{\mathrm{LS}})$. We also observe that if θ_{LS} is sufficiently bounded away from both 0 and $\pi/2$, then (5.3.12) essentially says that

$$\frac{\| \hat{r}_{\mathrm{LS}} - r_{\mathrm{LS}} \|_2}{\| r_{\mathrm{LS}} \|_2} \approx \epsilon \cdot \kappa_2(A). \tag{5.3.19}$$

For more insights into the subtleties behind Theorem 5.3.1., see Wedin (1973), Vandersluis (1975), Björck (NMLS, p. 30), Higham (ASNA, p. 382), and Grcar(2010).

5.3.7   Normal Equations Versus QR

It is instructive to compare the normal equation and QR approaches to the full-rank
LS problem in light of Theorem 5.3.1.

- The method of normal equations produces an $\hat{x}_{\mathrm{LS}}$ whose relative error depends
  on $\kappa_2(A)^2$, a factor that can be considerably larger than the condition number
  associated with a "small residual" LS problem.

- The QR approach (Householder, Givens, careful MGS) solves a nearby LS problem. Therefore, these methods produce a computed solution with relative error
  that is "predicted" by the condition of the underlying LS problem.

Thus, the QR approach is more appealing in situations where b is close to the span of
A's columns.

Finally, we mention two other factors that figure in the debate about QR versus
normal equations. First, the normal equations approach involves about half of the
arithmetic when $m \gg n$ and does not require as much storage, assuming that $Q(:, 1{:}n)$
is required. Second, QR approaches are applicable to a wider class of LS problems.
This is because the Cholesky solve in the method of normal equations is "in trouble"
if $\kappa_2(A) \approx 1/\sqrt{\mathbf{u}}$ while the R-solve step in a QR approach is in trouble only if $\kappa_2(A) \approx
1/\mathbf{u}$. Choosing the "right" algorithm requires having an appreciation for these tradeoffs.

5.3.8   Iterative Improvement

A technique for refining an approximate LS solution has been analyzed by Björck (1967,
1968). It is based on the idea that if

$$\begin{bmatrix} I_m & A \\ A^T & 0 \end{bmatrix} \begin{bmatrix} r \\ x \end{bmatrix} = \begin{bmatrix} b \\ 0 \end{bmatrix}, \qquad A \in \mathbb{R}^{m \times n},\ b \in \mathbb{R}^m, \tag{5.3.20}$$

then $\| b - Ax \|_2 = \min$. This follows because $r + Ax = b$ and $A^T r = 0$ imply $A^T Ax =
A^T b$. The above augmented system is nonsingular if $\mathsf{rank}(A) = n$, which we hereafter
assume. By casting the LS problem in the form of a square linear system, the iterative
improvement scheme §3.5.3 can be applied:

$r^{(0)} = 0,\ x^{(0)} = 0$

for $k = 0, 1, \ldots$

$$\begin{bmatrix} f^{(k)} \\ g^{(k)} \end{bmatrix} = \begin{bmatrix} b \\ 0 \end{bmatrix} - \begin{bmatrix} I & A \\ A^T & 0 \end{bmatrix} \begin{bmatrix} r^{(k)} \\ x^{(k)} \end{bmatrix}$$

$$\begin{bmatrix} I & A \\ A^T & 0 \end{bmatrix} \begin{bmatrix} p^{(k)} \\ z^{(k)} \end{bmatrix} = \begin{bmatrix} f^{(k)} \\ g^{(k)} \end{bmatrix}$$

$$\begin{bmatrix} r^{(k+1)} \\ x^{(k+1)} \end{bmatrix} = \begin{bmatrix} r^{(k)} \\ x^{(k)} \end{bmatrix} + \begin{bmatrix} p^{(k)} \\ z^{(k)} \end{bmatrix}$$

end

The residuals $f^{(k)}$ and $g^{(k)}$ must be computed in higher precision, and an original copy of A must be around for this purpose.

If the QR factorization of A is available, then the solution of the augmented system is readily obtained. In particular, if $A = QR$ and $R_1 = R(1{:}n, 1{:}n)$, then a system of the form

$$\begin{bmatrix} I & A \\ A^T & 0 \end{bmatrix} \begin{bmatrix} p \\ z \end{bmatrix} = \begin{bmatrix} f \\ g \end{bmatrix}$$

transforms to

$$\begin{bmatrix} I_n & 0 & R_1 \\ 0 & I_{m-n} & 0 \\ R_1^T & 0 & 0 \end{bmatrix} \begin{bmatrix} h \\ f_2 \\ z \end{bmatrix} = \begin{bmatrix} f_1 \\ f_2 \\ g \end{bmatrix}$$

where

$$Q^T f = \begin{bmatrix} f_1 \\ f_2 \end{bmatrix} \begin{matrix} n \\ m-n \end{matrix} \quad , \quad Q^T p = \begin{bmatrix} h \\ f_2 \end{bmatrix} \begin{matrix} n \\ m-n \end{matrix} \quad .$$

Thus, p and z can be determined by solving the triangular systems $R_1^T h = g$ and $R_1 z = f_1 - h$ and setting

$$p = Q \begin{bmatrix} h \\ f_2 \end{bmatrix} .$$

Assuming that Q is stored in factored form, each iteration requires $8mn - 2n^2$ flops.

The key to the iteration's success is that both the LS residual and solution are updated—not just the solution. Björck (1968) shows that if $\kappa_2(A) \approx \beta^q$ and t-digit, β-base arithmetic is used, then $x^{(k)}$ has approximately $k(t-q)$ correct base-β digits, provided the residuals are computed in double precision. Notice that it is $\kappa_2(A)$, not $\kappa_2(A)^2$, that appears in this heuristic.

5.3.9 Some Point/Line/Plane Nearness Problems in 3-Space

The fields of computer graphics and computer vision are replete with many interesting matrix problems. Below we pose three geometric "nearness" problems that involve points, lines, and planes in 3-space. Each is a highly structured least squares problem with a simple, closed-form solution. The underlying trigonometry leads rather naturally to the vector cross product, so we start with a quick review of this important operation.

The *cross product* of a vector $p \in \mathbb{R}^3$ with a vector $q \in \mathbb{R}^3$ is defined by

$$p \times q = \begin{bmatrix} p_2 q_3 - p_3 q_2 \\ p_3 q_1 - p_1 q_3 \\ p_1 q_2 - p_2 q_1 \end{bmatrix} .$$

This operation can be framed as a matrix-vector product. For any $v \in \mathbb{R}^3$, define the skew-symmetric matrix v^c by

$$v^c = \begin{bmatrix} 0 & -v_3 & v_2 \\ v_3 & 0 & -v_1 \\ -v_2 & v_1 & 0 \end{bmatrix} .$$

It follows that

$$p \times q \;=\; p^c \cdot q \;=\; -q^c \cdot p \;=\; -(q \times p).$$

Using the skew-symmetry of p^c and q^c, it is easy to show that

$$p \times q \;\in\; \mathsf{span}\{p,\, q\}^{\perp}. \tag{5.3.21}$$

Other properties include

$$(p \times q) \times r = (p^c \cdot q)^c r \;=\; \left(qp^T - pq^T\right) r \;=\; (p^T r){\cdot}q - (q^T r){\cdot}p, \tag{5.3.22}$$

$$(p \times q)^T (r \times s) = (p^c q)^T {\cdot} (r^c s) \;=\; \det([p\ q]^T [r\ s]), \tag{5.3.23}$$

$$p^c p^c = pp^T - \parallel p \parallel_2^2 {\cdot} I_3, \tag{5.3.24}$$

$$\parallel p^c q \parallel_2^2 = \parallel p \parallel_2^2 {\cdot} \parallel q \parallel_2^2 {\cdot} \left(1 - \left(\frac{p^T q}{\parallel p \parallel_2 {\cdot} \parallel q \parallel_2} \right)^2 \right). \tag{5.3.25}$$

We are now set to state the three problems and specify their *theoretical* solutions. For hints at how to establish the correctness of the solutions, see P5.3.13–P5.3.15.

Problem 1. Given a line L and a point y, find the point z^{opt} on L that is closest to y, i.e., solve

$$\min_{z \in L} \parallel z - y \parallel_2.$$

If L passes through distinct points p_1 and p_2, then it can be shown that

$$z^{\mathrm{opt}} \;=\; y + \frac{1}{v^T v}\, v^c v^c (y - p_1), \qquad v = p_2 - p_1. \tag{5.3.26}$$

Problem 2. Given lines L_1 and L_2, find the point z_1^{opt} on L_1 that is closest to L_2 and the point z_2^{opt} on L_2 that is closest to L_1, i.e., solve

$$\min_{z_1 \in L_1,\, z_2 \in L_2} \parallel z_1 - z_2 \parallel_2.$$

If L_1 passes through distinct points p_1 and p_2 and L_2 passes through distinct points q_1 and q_2, then it can be shown that

$$z_1^{\mathrm{opt}} = p_1 + \frac{1}{r^T r} \cdot vw^T \cdot r^c (q_1 - p_1), \tag{5.3.27}$$

$$z_2^{\mathrm{opt}} = q_1 + \frac{1}{r^T r} \cdot wv^T \cdot r^c (q_1 - p_1), \tag{5.3.28}$$

where $v = p_2 - p_1$, $w = q_2 - q_1$, and $r = v^c w$.

Problem 3. Given a plane P and a point y, find the point z^{opt} on P that is closest to y, i.e., solve

$$\min_{z \in P} \parallel z - y \parallel_2.$$

If P passes through three distinct points p_1, p_2, and p_3, then it can be shown that

$$z^{\text{opt}} = p_1 - \frac{1}{v^T v} \cdot v^c v^c (y - p_1) \tag{5.3.29}$$

where $v = (p_2 - p_1)^c (p_3 - p_1)$.

The nice closed-form solutions (5.3.26)–(5.3.29) are deceptively simple and great care must be exercised when computing with these formulae or their mathematical equivalents. See Kahan (2011).

Problems

P5.3.1 Assume $A^T A x = A^T b$, $(A^T A + F)\hat{x} = A^T b$, and $2\| F \|_2 \leq \sigma_n(A)^2$. Show that if $r = b - Ax$ and $\hat{r} = b - A\hat{x}$, then $\hat{r} - r = A(A^T A + F)^{-1} F x$ and

$$\| \hat{r} - r \|_2 \leq 2\,\kappa_2(A) \frac{\| F \|_2}{\| A \|_2} \| x \|_2.$$

P5.3.2 Assume that $A^T A x = A^T b$ and that $A^T A \hat{x} = A^T b + f$ where $\| f \|_2 \leq c\mathbf{u}\| A^T \|_2 \| b \|_2$ and A has full column rank. Show that

$$\frac{\| x - \hat{x} \|_2}{\| x \|_2} \leq c\mathbf{u}\kappa_2(A)^2 \frac{\| A^T \|_2 \| b \|_2}{\| A^T b \|}.$$

P5.3.3 Let $A \in \mathbb{R}^{m \times n}$ $(m \geq n)$, $w \in \mathbb{R}^n$, and define

$$B = \left[\begin{array}{c} A \\ w^T \end{array} \right].$$

Show that $\sigma_n(B) \geq \sigma_n(A)$ and $\sigma_1(B) \leq \sqrt{\| A \|_2^2 + \| w \|_2^2}$. Thus, the condition of a matrix may increase or decrease if a row is added.

P5.3.4 (Cline 1973) Suppose that $A \in \mathbb{R}^{m \times n}$ has rank n and that Gaussian elimination with partial pivoting is used to compute the factorization $PA = LU$, where $L \in \mathbb{R}^{m \times n}$ is unit lower triangular, $U \in \mathbb{R}^{n \times n}$ is upper triangular, and $P \in \mathbb{R}^{m \times m}$ is a permutation. Explain how the decomposition in P5.2.5 can be used to find a vector $z \in \mathbb{R}^n$ such that $\| Lz - Pb \|_2$ is minimized. Show that if $Ux = z$, then $\| Ax - b \|_2$ is minimum. Show that this method of solving the LS problem is more efficient than Householder QR from the flop point of view whenever $m \leq 5n/3$.

P5.3.5 The matrix $C = (A^T A)^{-1}$, where $\text{rank}(A) = n$, arises in many statistical applications. Assume that the factorization $A = QR$ is available. (a) Show $C = (R^T R)^{-1}$. (b) Give an algorithm for computing the diagonal of C that requires $n^3/3$ flops. (c) Show that

$$R = \left[\begin{array}{cc} \alpha & v^T \\ 0 & S \end{array} \right] \quad \Rightarrow \quad C = (R^T R)^{-1} = \left[\begin{array}{cc} (1 + v^T C_1 v)/\alpha^2 & -v^T C_1/\alpha \\ -C_1 v/\alpha & C_1 \end{array} \right]$$

where $C_1 = (S^T S)^{-1}$. (d) Using (c), give an algorithm that overwrites the upper triangular portion of R with the upper triangular portion of C. Your algorithm should require $2n^3/3$ flops.

P5.3.6 Suppose $A \in \mathbb{R}^{n \times n}$ is symmetric and that $r = b - Ax$ where $r, b, x \in \mathbb{R}^n$ and x is nonzero. Show how to compute a symmetric $E \in \mathbb{R}^{n \times n}$ with minimal Frobenius norm so that $(A + E)x = b$. Hint: Use the QR factorization of $[\, x \mid r \,]$ and note that $Ex = r \Rightarrow (Q^T E Q)(Q^T x) = Q^T r$.

P5.3.7 Points $P_1, \ldots, P_n$ on the x-axis have x-coordinates $x_1, \ldots, x_n$. We know that $x_1 = 0$ and wish to compute $x_2, \ldots, x_n$ given that we have estimates d_{ij} of the separations:

$$x_i - x_j \approx d_{ij}, \qquad 1 \leq i < j \leq n.$$

Using the method of normal equations, show how to minimize

$$\phi(x_1, \ldots, x_n) = \sum_{i=1}^{n-1} \sum_{j=i+1}^{n} (x_i - x_j - d_{ij})^2$$

subject to the constraint $x_1 = 0$.

P5.3.8 Suppose $A \in \mathbb{R}^{m \times n}$ has full rank and that $b \in \mathbb{R}^m$ and $c \in \mathbb{R}^n$ are given. Show how to compute $\alpha = c^T x_{\mathrm{LS}}$ without computing x_{LS} explicitly. Hint: Suppose Z is a Householder matrix such that $Z^T c$ is a multiple of $I_n(:,n)$. It follows that $\alpha = (Z^T c)^T y_{\mathrm{LS}}$ where y_{LS} minimizes $\| \tilde{A}y - b \|_2$ with $y = Z^T x$ and $\tilde{A} = AZ$.

P5.3.9 Suppose $A \in \mathbb{R}^{m \times n}$ and $b \in \mathbb{R}^m$ with $m \geq n$. How would you solve the full rank least squares problem given the availability of a matrix $M \in \mathbb{R}^{m \times m}$ such that $M^T A = S$ is upper triangular and $M^T M = D$ is diagonal?

P5.3.10 Let $A \in \mathbb{R}^{m \times n}$ have rank n and for $\alpha \geq 0$ define

$$M(\alpha) = \begin{bmatrix} \alpha I_m & A \\ A^T & 0 \end{bmatrix}.$$

Show that

$$\sigma_{m+n}(M(\alpha)) = \min \left\{ \alpha, \ -\frac{\alpha}{2} + \sqrt{\sigma_n(A)^2 + \left(\frac{\alpha}{2}\right)^2} \right\}$$

and determine the value of α that minimizes $\kappa_2(M(\alpha))$.

P5.3.11 Another iterative improvement method for LS problems is the following:

$$x^{(0)} = 0$$
$$\textbf{for } k = 0, 1, \dots$$
$$\qquad r^{(k)} = b - A x^{(k)} \quad \text{(double precision)}$$
$$\qquad \| A z^{(k)} - r^{(k)} \|_2 = \min$$
$$\qquad x^{(k+1)} = x^{(k)} + z^{(k)}$$
$$\textbf{end}$$

(a) Assuming that the QR factorization of A is available, how many flops per iteration are required?
(b) Show that the above iteration results by setting $g^{(k)} = 0$ in the iterative improvement scheme given in §5.3.8.

P5.3.12 Verify (5.3.21)–(5.3.25).

P5.3.13 Verify (5.3.26) noting that $L = \{ p_1 + \tau(p_2 - p_1) : \tau \in \mathbb{R} \}$.

P5.3.14 Verify (5.3.27) noting that the minimizer $\tau^{\mathrm{opt}} \in \mathbb{R}^2$ of $\| (p_1 - q_1) - [p_2 - p_1 \mid q_2 - q_1]\tau \|_2$ is relevant.

P5.3.15 Verify (5.3.29) noting that $P = \{ x : x^T((p_2 - p_1) \times (p_3 - p_1)) = 0 \}$.

Notes and References for §5.3

Some classical references for the least squares problem include:

F.L. Bauer (1965). "Elimination with Weighted Row Combinations for Solving Linear Equations and Least Squares Problems," *Numer. Math. 7*, 338–352.
G.H. Golub and J.H. Wilkinson (1966). "Note on the Iterative Refinement of Least Squares Solution," *Numer. Math. 9*, 139–148.
A. van der Sluis (1975). "Stability of the Solutions of Linear Least Squares Problem," *Numer. Math. 23*, 241–254.

The use of Gauss transformations to solve the LS problem has attracted some attention because they are cheaper to use than Householder or Givens matrices, see:

G. Peters and J.H. Wilkinson (1970). "The Least Squares Problem and Pseudo-Inverses," *Comput. J. 13*, 309–16.
A.K. Cline (1973). "An Elimination Method for the Solution of Linear Least Squares Problems," *SIAM J. Numer. Anal. 10*, 283–289.
R.J. Plemmons (1974). "Linear Least Squares by Elimination and MGS," *J. ACM 21*, 581–585.

The *seminormal equations* are given by $R^T Rx = A^T b$ where $A = QR$. It can be shown that by solving the seminormal equations an acceptable LS solution is obtained if one step of fixed precision iterative improvement is performed, see:

Å. Björck (1987). "Stability Analysis of the Method of Seminormal Equations," *Lin. Alg. Applic. 88/89*, 31–48.

Survey treatments of LS perturbation theory include Lawson and Hanson (SLS), Stewart and Sun (MPT), and Björck (NMLS). See also:

P.-A. Wedin (1973). "Perturbation Theory for Pseudoinverses," *BIT 13*, 217–232.

Å. Björck (1991). "Component-wise Perturbation Analysis and Error Bounds for Linear Least Squares Solutions," *BIT 31*, 238–244.

B. Waldén, R. Karlson, J. Sun (1995). "Optimal Backward Perturbation Bounds for the Linear Least Squares Problem," *Numerical Lin. Alg. Applic. 2*, 271–286.

J.-G. Sun (1996). "Optimal Backward Perturbation Bounds for the Linear Least-Squares Problem with Multiple Right-Hand Sides," *IMA J. Numer. Anal. 16*, 1–11.

J.-G. Sun (1997). "On Optimal Backward Perturbation Bounds for the Linear Least Squares Problem," *BIT 37*, 179–188.

R. Karlson and B. Waldén (1997). "Estimation of Optimal Backward Perturbation Bounds for the Linear Least Squares Problem," *BIT 37*, 862–869.

J.-G. Sun (1997). "On Optimal Backward Perturbation Bounds for the Linear Least Squares Problem," *BIT 37*, 179–188.

M. Gu (1998). "Backward Perturbation Bounds for Linear Least Squares Problems," *SIAM J. Matrix Anal. Applic. 20*, 363–372.

M. Arioli, M. Baboulin and S. Gratton (2007). "A Partial Condition Number for Linear Least Squares Problems," *SIAM J. Matrix Anal. Applic. 29*, 413–433.

M. Baboulin, J. Dongarra, S. Gratton, and J. Langou (2009). "Computing the Conditioning of the Components of a Linear Least-Squares Solution," *Num. Lin. Alg. Applic. 16*, 517–533.

M. Baboulin and S. Gratton (2009). "Using Dual Techniques to Derive Componentwise and Mixed Condition Numbers for a Linear Function of a Least Squares Solution," *BIT 49*, 3–19.

J. Grcar (2009). "Nuclear Norms of Rank-2 Matrices for Spectral Condition Numbers of Rank Least Squares Solutions," ArXiv:1003.2733v4.

J. Grcar (2010). "Spectral Condition Numbers of Orthogonal Projections and Full Rank Linear Least Squares Residuals," *SIAM J. Matrix Anal. Applic. 31*, 2934–2949.

Practical insights into the accuracy of a computed least squares solution can be obtained by applying the condition estimation ideas of §3.5. to the R matrix in $A = QR$ or the Cholesky factor of $A^T A$ should a normal equation approach be used. For a discussion of LS-specific condition estimation, see:

G.W. Stewart (1980). "The Efficient Generation of Random Orthogonal Matrices with an Application to Condition Estimators," *SIAM J. Numer. Anal. 17*, 403–9.

S. Gratton (1996). "On the Condition Number of Linear Least Squares Problems in a Weighted Frobenius Norm," *BIT 36*, 523–530.

C.S. Kenney, A.J. Laub, and M.S. Reese (1998). "Statistical Condition Estimation for Linear Least Squares," *SIAM J. Matrix Anal. Applic. 19*, 906–923.

Our restriction to least squares approximation is not a vote against minimization in other norms. There are occasions when it is advisable to minimize $\| Ax - b \|_p$ for $p = 1$ and ∞. Some algorithms for doing this are described in:

A.K. Cline (1976). "A Descent Method for the Uniform Solution to Overdetermined Systems of Equations," *SIAM J. Numer. Anal. 13*, 293–309.

R.H. Bartels, A.R. Conn, and C. Charalambous (1978). "On Cline's Direct Method for Solving Overdetermined Linear Systems in the L_∞ Sense," *SIAM J. Numer. Anal. 15*, 255–270.

T.F. Coleman and Y. Li (1992). "A Globally and Quadratically Convergent Affine Scaling Method for Linear L_1 Problems," *Mathematical Programming 56, Series A*, 189–222.

Y. Li (1993). "A Globally Convergent Method for L_p Problems," *SIAM J. Optim. 3*, 609–629.

Y. Zhang (1993). "A Primal-Dual Interior Point Approach for Computing the L_1 and L_∞ Solutions of Overdetermined Linear Systems," *J. Optim. Theory Applic. 77*, 323–341.

Iterative improvement in the least squares context is discussed in:

G.H. Golub and J.H. Wilkinson (1966). "Note on Iterative Refinement of Least Squares Solutions," *Numer. Math. 9*, 139–148.

Å. Björck and G.H. Golub (1967). "Iterative Refinement of Linear Least Squares Solutions by Householder Transformation," *BIT 7*, 322–337.

Å. Björck (1967). "Iterative Refinement of Linear Least Squares Solutions I," *BIT 7*, 257–278.

Å. Björck (1968). "Iterative Refinement of Linear Least Squares Solutions II," *BIT 8*, 8–30.

J. Gluchowska and A. Smoktunowicz (1999). "Solving the Linear Least Squares Problem with Very High Relative Acuracy," *Computing 45*, 345–354.

J. Demmel, Y. Hida, and E.J. Riedy (2009). "Extra-Precise Iterative Refinement for Overdetermined Least Squares Problems," *ACM Trans. Math. Softw. 35*, Article 28.

The following texts treat various geometric matrix problems that arise in computer graphics and vision:

A.S. Glassner (1989). *An Introduction to Ray Tracing*, Morgan Kaufmann, Burlington, MA.

R. Hartley and A. Zisserman (2004). *Multiple View Geometry in Computer Vision*, Second Edition, Cambridge University Press, New York.

M. Pharr and M. Humphreys (2010). *Physically Based Rendering, from Theory to Implementation*, Second Edition, Morgan Kaufmann, Burlington, MA.

For a numerical perspective, see:

W. Kahan (2008). "Computing Cross-Products and Rotations in 2- and 3-dimensional Euclidean Spaces," http://www.cs.berkeley.edu/ wkahan/MathH110/Cross.pdf.

5.4 Other Orthogonal Factorizations

Suppose $A \in \mathbb{R}^{m \times 4}$ has a thin QR factorization of the following form:

$$
A = [\, a_1, \, a_2, \, a_3, \, a_4 \,] = [\, q_1, \, q_2, \, q_3, \, q_4 \,]
\begin{bmatrix}
1 & 1 & 1 & 1 \\
0 & 0 & 1 & 1 \\
0 & 0 & 0 & 1 \\
0 & 0 & 0 & 1
\end{bmatrix} .
$$

Note that $\mathsf{ran}(A)$ has dimension 3 but does not equal $\mathsf{span}\{q_1, q_2, q_3\}$, $\mathsf{span}\{q_1, q_2, q_4\}$, $\mathsf{span}\{q_1, q_3, q_4\}$, or $\mathsf{span}\{q_2, q_3, q_4\}$ because a_4 does not belong to any of these subspaces. In this case, the QR factorization reveals neither the range nor the nullspace of A and the number of nonzeros on R's diagonal does *not* equal its rank. Moreover, the LS solution process based on the QR factorization (Algorithm 5.3.2) breaks down because the upper triangular portion of R is singular.

We start this section by introducing several decompositions that overcome these shortcomings. They all have the form $Q^T A Z = T$ where T is a structured block triangular matrix that sheds light on A's rank, range, and nullspace. We informally refer to matrix reductions of this form as *rank revealing*. See Chandrasekaren and Ipsen (1994) for a more precise formulation of the concept.

Our focus is on a modification of the QR factorization that involves column pivoting. The resulting R-matrix has a structure that supports rank estimation. To set the stage for updating methods, we briefly discus the ULV and UTV frameworks Updating is discussed in §6.5 and refers to the efficient recomputation of a factorization after the matrix undergoes a low-rank change.

All these methods can be regarded as inexpensive alternatives to the SVD, which represents the "gold standard" in the area of rank determination. Nothing "takes apart" a matrix so conclusively as the SVD and so we include an explanation of its airtight reliability. The computation of the full SVD, which we discuss in §8.6, begins with the reduction to bidiagonal form using Householder matrices. Because this decomposition is important in its own right, we provide some details at the end of this section.

5.4.1 Numerical Rank and the SVD

Suppose $A \in \mathbb{R}^{m \times n}$ has SVD $U^T A V = \Sigma = \text{diag}(\sigma_i)$. If $\text{rank}(A) = r < n$, then according to the exact arithmetic discussion of §2.4 the singular values $\sigma_{r+1}, \ldots, \sigma_n$ are zero and

$$A = \sum_{i=1}^{r} \sigma_k u_k v_k^T . \tag{5.4.1}$$

The exposure of rank degeneracy could not be more clear.

In Chapter 8 we describe the Golub-Kahan-Reinsch algorithm for computing the SVD. Properly implemented, it produces nearly orthogonal matrices $\widehat{U}$ and $\widehat{V}$ so that

$$\widehat{U}^T A \widehat{V} \approx \widehat{\Sigma} = \text{diag}(\widehat{\sigma}_1, \ldots, \widehat{\sigma}_n), \qquad \widehat{\sigma}_1 \geq \cdots \geq \widehat{\sigma}_n \geq 0.$$

(Other SVD procedures have this property as well.) Unfortunately, unless remarkable cancellation occurs, none of the computed singular values will be zero because of roundoff error. This forces an issue. On the one hand, we can adhere to the strict mathematical definition of rank, count the number of nonzero computed singular values, and conclude from

$$A \approx \sum_{i=1}^{n} \widehat{\sigma}_k \widehat{u}_k \widehat{v}_k^T \tag{5.4.2}$$

that A has full rank. However, working with every matrix as if it possessed full column rank is not particularly useful. It is more productive to liberalize the notion of rank by setting small computed singular values to zero in (5.4.2). This results in an approximation of the form

$$A \approx \sum_{i=1}^{\widehat{r}} \widehat{\sigma}_k \widehat{u}_k \widehat{v}_k^T , \qquad \widehat{r} \leq \widehat{n} \tag{5.4.3}$$

where we regard $\widehat{r}$ as the *numerical rank*. For this approach to make sense we need to guarantee that $|\widehat{\sigma}_i - \sigma_i|$ is small.

For a properly implemented Golub-Kahan-Reinsch SVD algorithm, it can be shown that

$$\widehat{U} = W + \Delta U, \ \ W^T W = I_m, \qquad \| \, \Delta U \, \|_2 \leq \epsilon,$$

$$\widehat{V} = Z + \Delta V, \ \ Z^T Z = I_n, \qquad \| \, \Delta V \, \|_2 \leq \epsilon, \tag{5.4.4}$$

$$\widehat{\Sigma} = W^T (A + \Delta A) Z, \qquad \| \, \Delta A \, \|_2 \leq \epsilon \| \, A \, \|_2,$$

where ϵ is a small multiple of $\mathbf{u}$, the machine precision. In other words, the SVD algorithm computes the singular values of a nearby matrix $A + \Delta A$.

Note that $\widehat{U}$ and $\widehat{V}$ are not necessarily close to their exact counterparts. However, we can show that $\widehat{\sigma}_k$ is close to σ_k as follows. Using Corollary 2.4.6 we have

$$\sigma_k \; = \; \min_{\mathrm{rank}(B)=k-1} \; \| \, A - B \, \|_2 \; = \; \min_{\mathrm{rank}(B)=k-1} \; \| \, (\widehat{\Sigma} - B) - E \, \|_2$$

where

$$E \; = \; W^T(\Delta A)Z$$

and

$$\| \, E \, \|_2 \; \leq \; \epsilon \| \, A \, \|_2 \; = \; \epsilon \sigma_1.$$

Since

$$\| \, \widehat{\Sigma} - B \, \| - \| \, E \, \| \; \leq \; \| \, \widehat{\Sigma} - B \, \| \; \leq \; \| \, \widehat{\Sigma} - B \, \| + \| \, E \, \|$$

and

$$\min_{\mathrm{rank}(B)=k-1} \; \| \, \widehat{\Sigma}_k - B \, \|_2 \; = \; \widehat{\sigma}_k,$$

it follows that

$$| \sigma_k - \widehat{\sigma}_k | \; \leq \; \epsilon \sigma_1$$

for $k = 1{:}n$. Thus, if A has rank r, then we can expect $n - r$ of the computed singular values to be small. Near rank deficiency in A cannot escape detection if the SVD of A is computed.

Of course, all this hinges on having a definition of "small." This amounts to choosing a tolerance $\delta > 0$ and declaring A to have numerical rank $\widehat{r}$ if the computed singular values satisfy

$$\widehat{\sigma}_1 \geq \cdots \geq \widehat{\sigma}_{\widehat{r}} > \delta \geq \widehat{\sigma}_{\widehat{r}+1} \geq \cdots \geq \widehat{\sigma}_n \, . \tag{5.4.5}$$

We refer to the integer $\widehat{r}$ as the δ-*rank of* A. The tolerance should be consistent with the machine precision, e.g., $\delta = \mathbf{u} \| \, A \, \|_\infty$. However, if the general level of relative error in the data is larger than $\mathbf{u}$, then δ should be correspondingly bigger, e.g., $\delta = 10^{-2} \| \, A \, \|_\infty$ if the entries in A are correct to two digits.

For a given δ it is important to stress that, although the SVD provides a great deal of rank-related insight, it does not change the fact that the determination of numerical rank is a sensitive computation. If the gap between $\widehat{\sigma}_{\widehat{r}}$ and $\widehat{\sigma}_{\widehat{r}+1}$ is small, then A is also close (in the δ sense) to a matrix with rank $\widehat{r} - 1$. Thus, the amount of confidence we have in the correctness of $\widehat{r}$ and in how we proceed to use the approximation (5.4.2) depends on the gap between $\widehat{\sigma}_{\widehat{r}}$ and $\widehat{\sigma}_{\widehat{r}+1}$.

5.4.2   QR with Column Pivoting

We now examine alternative rank-revealing strategies to the SVD starting with a modification of the Householder QR factorization procedure (Algorithm 5.2.1). In exact arithmetic, the modified algorithm computes the factorization

$$Q^T A \Pi \; = \; \begin{bmatrix} R_{11} & R_{12} \\ 0 & 0 \end{bmatrix} \begin{matrix} r \\ m-r \end{matrix} \tag{5.4.6}$$
$$\qquad\qquad\quad r \quad\;\; n-r$$

where $r = \mathsf{rank}(A)$, Q is orthogonal, R_{11} is upper triangular and nonsingular, and Π is a permutation. If we have the column partitionings $A\Pi = [\, a_{c_1} \,|\cdots|\, a_{c_n} \,]$ and $Q = [\, q_1 \,|\cdots|\, q_m \,]$, then for $k = 1{:}n$ we have

$$a_{c_k} = \sum_{i=1}^{\min\{r,k\}} r_{ik} q_i \in \mathsf{span}\{q_1, \ldots, q_r\}$$

implying

$$\mathsf{ran}(A) = \mathsf{span}\{q_1, \ldots, q_r\}.$$

To see how to compute such a factorization, assume for some k that we have computed Householder matrices $H_1, \ldots, H_{k-1}$ and permutations $\Pi_1, \ldots, \Pi_{k-1}$ such that

$$(H_{k-1} \cdots H_1) A (\Pi_1 \cdots \Pi_{k-1}) = R^{(k-1)} = \left[\begin{array}{cc} R_{11}^{(k-1)} & R_{12}^{(k-1)} \\ 0 & R_{22}^{(k-1)} \end{array} \right] \begin{array}{c} k-1 \\ m-k+1 \end{array} \qquad (5.4.7)$$
$$\phantom{(H_{k-1} \cdots H_1) A (\Pi_1 \cdots \Pi_{k-1}) = R^{(k-1)} = }\begin{array}{cc} k-1 & n-k+1 \end{array}$$

where $R_{11}^{(k-1)}$ is a nonsingular and upper triangular matrix. Now suppose that

$$R_{22}^{(k-1)} = [\, z_k^{(k-1)} \,|\cdots|\, z_n^{(k-1)} \,]$$

is a column partitioning and let $p \geq k$ be the smallest index such that

$$\| z_p^{(k-1)} \|_2 = \max\left\{ \| z_k^{(k-1)} \|_2, \ldots, \| z_n^{(k-1)} \|_2 \right\}. \qquad (5.4.8)$$

Note that if $\mathsf{rank}(A) = k-1$, then this maximum is zero and we are finished. Otherwise, let Π_k be the n-by-n identity with columns p and k interchanged and determine a Householder matrix H_k such that if

$$R^{(k)} = H_k R^{(k-1)} \Pi_k,$$

then $R^{(k)}(k+1{:}m, k) = 0$. In other words, Π_k moves the largest column in $R_{22}^{(k-1)}$ to the lead position and H_k zeroes all of its subdiagonal components.

The column norms do not have to be recomputed at each stage if we exploit the property

$$Q^T z = \left[\begin{array}{c} \alpha \\ w \end{array} \right] \begin{array}{c} 1 \\ s-1 \end{array} \qquad \Longrightarrow \qquad \| w \|_2^2 = \| z \|_2^2 - \alpha^2,$$

which holds for any orthogonal matrix $Q \in \mathbb{R}^{s \times s}$. This reduces the overhead associated with column pivoting from $O(mn^2)$ flops to $O(mn)$ flops because we can get the new column norms by updating the old column norms, e.g.,

$$\| z_j^{(k)} \|_2^2 = \| z_j^{(k-1)} \|_2^2 - r_{kj}^2 \qquad j = k+1{:}n.$$

Combining all of the above we obtain the following algorithm first presented by Businger and Golub (1965):

Algorithm 5.4.1 (Householder QR With Column Pivoting) Given $A \in \mathbb{R}^{m \times n}$ with $m \geq n$, the following algorithm computes $r = \mathsf{rank}(A)$ and the factorization (5.4.6) with $Q = H_1 \cdots H_r$ and $\Pi = \Pi_1 \cdots \Pi_r$. The upper triangular part of A is overwritten by the upper triangular part of R and components $j + 1{:}m$ of the jth Householder vector are stored in $A(j + 1{:}m, j)$. The permutation Π is encoded in an integer vector piv. In particular, Π_j is the identity with rows j and $piv(j)$ interchanged.

> **for** $j = 1{:}n$
> $\qquad c(j) = A(1{:}m, j)^T A(1{:}m, j)$
> **end**
> $r = 0$
> $\tau = \max\{c(1), \ldots, c(n)\}$
> **while** $\tau > 0$ and $r < n$
> $\qquad r = r + 1$
> $\qquad$ Find smallest k with $r \leq k \leq n$ so $c(k) = \tau$.
> $\qquad piv(r) = k$
> $\qquad A(1{:}m, r) \leftrightarrow A(1{:}m, k)$
> $\qquad c(r) \leftrightarrow c(k)$
> $\qquad [v, \beta] = \mathsf{house}(A(r{:}m, r))$
> $\qquad A(r{:}m, r{:}n) = (I_{m-r+1} - \beta v v^T) A(:r{:}m, r{:}n)$
> $\qquad A(r + 1{:}m, r) = v(2{:}m - r + 1)$
> $\qquad$ **for** $i = r + 1{:}n$
> $\qquad\qquad c(i) = c(i) - A(r, i)^2$
> $\qquad$ **end**
> $\qquad \tau = \max\{c(r + 1), \ldots, c(n)\}$
> **end**

This algorithm requires $4mnr - 2r^2(m + n) + 4r^3/3$ flops where $r = \mathsf{rank}(A)$.

5.4.3   Numerical Rank and $A\Pi = QR$

In principle, QR with column pivoting reveals rank. But how informative is the method in the context of floating point arithmetic? After k steps we have

$$\mathsf{fl}(H_k \cdots H_1 A \Pi_1 \cdots \Pi_k) = \widehat{R}^{(k)} = \begin{bmatrix} \widehat{R}_{11}^{(k)} & \widehat{R}_{12}^{(k)} \\ 0 & \widehat{R}_{22}^{(k)} \end{bmatrix} \begin{matrix} k \\ m-k \end{matrix} \qquad (5.4.9)$$
$$\qquad\qquad k \qquad n-k$$

If $\widehat{R}_{22}^{(k)}$ is suitably small in norm, then it is reasonable to terminate the reduction and declare A to have rank k. A typical termination criteria might be

$$\| \widehat{R}_{22}^{(k)} \|_2 \leq \epsilon_1 \| A \|_2$$

for some small machine-dependent parameter ϵ_1. In view of the roundoff properties associated with Householder matrix computation (cf. §5.1.12), we know that $\widehat{R}^{(k)}$ is the exact R-factor of a matrix $A + E_k$, where

$$\| E_k \|_2 \leq c_2 \| A \|_2, \qquad \epsilon_2 = O(\mathbf{u}).$$

Using Corollary 2.4.4 we have

$$\sigma_{k+1}(A + E_k) = \sigma_{k+1}(\widehat{R}^{(k)}) \leq \| \widehat{R}_{22}^{(k)} \|_2 .$$

Since $\sigma_{k+1}(A) \leq \sigma_{k+1}(A + E_k) + \| E_k \|_2$, it follows that

$$\sigma_{k+1}(A) \leq (\epsilon_1 + \epsilon_2) \| A \|_2.$$

In other words, a relative perturbation of $O(\epsilon_1 + \epsilon_2)$ in A can yield a rank-k matrix. With this termination criterion, we conclude that QR with column pivoting discovers rank deficiency *if* $\widehat{R}_{22}^{(k)}$ is small for some $k < n$. However, *it does not follow* that the matrix $\widehat{R}_{22}^{(k)}$ in (5.4.9) is small if rank$(A) = k$. There are examples of nearly rank deficient matrices whose R-factor look perfectly "normal." A famous example is the Kahan matrix

$$\mathsf{Kah}_n(s) \;=\; \mathrm{diag}(1, s, \ldots, s^{n-1}) \begin{bmatrix} 1 & -c & -c & \cdots & -c \\ 0 & 1 & -c & \cdots & -c \\ & & \ddots & & \vdots \\ \vdots & & & 1 & -c \\ 0 & & \cdots & & 1 \end{bmatrix}.$$

Here, $c^2 + s^2 = 1$ with $c, s > 0$. (See Lawson and Hanson (SLS, p. 31).) These matrices are unaltered by Algorithm 5.4.1 and thus $\| R_{22}^{(k)} \|_2 \geq s^{n-1}$ for $k = 1:n-1$. This inequality implies (for example) that the matrix $\mathsf{Kah}_{300}(.99)$ has no particularly small trailing principal submatrix since $s^{299} \approx .05$. However, a calculation shows that $\sigma_{300} = O(10^{-19})$.

Nevertheless, in practice, small trailing R-submatrices almost always emerge that correlate well with the underlying rank. In other words, it is almost always the case that $\widehat{R}_{22}^{(k)}$ is small if A has rank k.

5.4.4 Finding a Good Column Ordering

It is important to appreciate that Algorithm 5.4.1 is just one way to determine the column permutation Π. The following result sets the stage for a better way.

Theorem 5.4.1. *If $A \in \mathbb{R}^{m \times n}$ and $v \in \mathbb{R}^n$ is a unit 2-norm vector, then there exists a permutation Π so that the QR factorization*

$$A\Pi \;=\; QR$$

satisfies $|r_{nn}| \leq \sqrt{n}\sigma$ where $\sigma = \| Av \|_2$.

Proof. Suppose $\Pi \in \mathbb{R}^{n \times n}$ is a permutation such that if $w = \Pi^T v$, then

$$|w_n| = \max |v_i|.$$

Since w_n is the largest component of a unit 2-norm vector, $|w_n| \geq 1/\sqrt{n}$. If $A\Pi = QR$ is a QR factorization, then

$$\sigma = \| Av \|_2 = \| (Q^T A \Pi)(\Pi^T v) \|_2 = \| R(1{:}n, 1{:}n)w \|_2 \geq |r_{nn}w_n| \geq |r_{nn}|/\sqrt{n}. \quad \square$$

Note that if $v = v_n$ is the right singular vector corresponding to $\sigma_{\min}(A)$, then $|r_{nn}| \leq \sqrt{n}\sigma_n$. This suggests a framework whereby the column permutation matrix Π is based on an estimate of v_n:

> *Step 1.* Compute the QR factorization $A = Q_0 R_0$ and note that R_0 has the same right singular vectors as A.

> *Step 2.* Use condition estimation techniques to obtain a unit vector v with $\| R_0 v \|_2 \approx \sigma_n$.

> *Step 3.* Determine Π and the QR factorization $A\Pi = QR$.

See Chan (1987) for details about this approach to rank determination. The permutation Π can be generated as a sequence of swap permutations. This supports a very economical Givens rotation method for generating of Q and R from Q_0 and R_0.

5.4.5 More General Rank-Revealing Decompositions

Additional rank-revealing strategies emerge if we allow general orthogonal recombinations of the A's columns instead of just permutations. That is, we look for an orthogonal Z so that the QR factorization

$$AZ = QR$$

produces a rank-revealing R. To impart the spirit of this type of matrix reduction, we show how the rank-revealing properties of a given $AZ = QR$ factorization can be improved by replacing Z, Q, and R with

$$Z_{\text{new}} = ZZ_G, \qquad Q_{\text{new}} = QQ_G, \qquad R_{\text{new}} = Q_G^T R Z_G,$$

respectively, where Q_G and Z_G are products of Givens rotations and R_{new} is upper triangular. The rotations are generated by introducing zeros into a unit 2-norm n-vector v which we assume approximates the n-th right singular vector of AZ. In particular, if $Z_G^T v = e_n = I_n(:, n)$ and $\| Rv \|_2 \approx \sigma_n$, then

$$\| R_{\text{new}} e_n \|_2 = \| Q_G^T R Z_G e_n \|_2 = \| Q_G^T R v \|_2 = \| Rv \|_2 \approx \sigma_n$$

This says that the norm of the last column of R_{new} is approximately the smallest singular value of A, which is certainly one way to reveal the underlying matrix rank.

We use the case $n = 4$ to illustrate how the Givens rotations arise and why the overall process is economical. Because we are transforming v to e_n and not e_1, we need to "flip" the mission of the 2-by-2 rotations in the Z_G computations so that top components are zeroed, i.e.,

$$\begin{bmatrix} 0 \\ \times \end{bmatrix} = \begin{bmatrix} c & s \\ -s & c \end{bmatrix} \begin{bmatrix} \times \\ \times \end{bmatrix}.$$

This requires only a slight modification of Algorithm 5.1.3.

In the $n = 4$ case we start with

$$R = \begin{bmatrix} \times & \times & \times & \times \\ 0 & \times & \times & \times \\ 0 & 0 & \times & \times \\ 0 & 0 & 0 & \times \end{bmatrix} \qquad v = \begin{bmatrix} \times \\ \times \\ \times \\ \times \end{bmatrix}$$

and proceed to compute

$$Z_G = G_{12}G_{23}G_{34}$$

and

$$Q_G = H_{12}H_{23}H_{34}$$

as products of Givens rotations. The first step is to zero the top component of v with a "flipped" $(1,2)$ rotation and update R accordingly:

$$R \leftarrow RG_{12} = \begin{bmatrix} \times & \times & \times & \times \\ \times & \times & \times & \times \\ 0 & 0 & \times & \times \\ 0 & 0 & 0 & \times \end{bmatrix}, \qquad v \leftarrow G_{12}^T v = \begin{bmatrix} 0 \\ \times \\ \times \\ \times \end{bmatrix}.$$

To remove the unwanted subdiagonal in R, we apply a conventional (nonflipped) Givens rotation from the left to R (but not v):

$$R \leftarrow H_{12}^T R = \begin{bmatrix} \times & \times & \times & \times \\ 0 & \times & \times & \times \\ 0 & 0 & \times & \times \\ 0 & 0 & 0 & \times \end{bmatrix}, \qquad v = \begin{bmatrix} 0 \\ \times \\ \times \\ \times \end{bmatrix}.$$

The next step is analogous:

$$R \leftarrow RG_{23} = \begin{bmatrix} \times & \times & \times & \times \\ 0 & \times & \times & \times \\ 0 & \times & \times & \times \\ 0 & 0 & 0 & \times \end{bmatrix}, \qquad v \leftarrow G_{23}^T v = \begin{bmatrix} 0 \\ 0 \\ \times \\ \times \end{bmatrix}.$$

$$R \leftarrow H_{23}^T R = \begin{bmatrix} \times & \times & \times & \times \\ 0 & \times & \times & \times \\ 0 & 0 & \times & \times \\ 0 & 0 & 0 & \times \end{bmatrix}, \qquad v = \begin{bmatrix} 0 \\ 0 \\ \times \\ \times \end{bmatrix}.$$

And finally,

$$R \leftarrow RG_{34} = \begin{bmatrix} \times & \times & \times & \times \\ 0 & \times & \times & \times \\ 0 & 0 & \times & \times \\ 0 & 0 & \times & \times \end{bmatrix}, \qquad v = G_{34}^T v = \begin{bmatrix} 0 \\ 0 \\ 0 \\ \times \end{bmatrix},$$

$$R \;\leftarrow\; H_{34}^T R \;=\; \begin{bmatrix} \times & \times & \times & \times \\ 0 & \times & \times & \times \\ 0 & 0 & \times & \times \\ 0 & 0 & 0 & \times \end{bmatrix}, \qquad\qquad v \;=\; \begin{bmatrix} 0 \\ 0 \\ 0 \\ \times \end{bmatrix}.$$

The pattern is clear, for $i = 1{:}n - 1$, a $G_{i,i+1}$ is used to zero the current v_i and an $H_{i,i+1}$ is used to zero the current $r_{i+1,i}$. The overall transition from $\{Q,\, Z,\, R\}$ to $\{Q_{\text{new}},\, Z_{\text{new}},\, R_{\text{new}}\}$ involves $O(mn)$ flops. If the Givens rotations are kept in factored form, this flop count is reduced to $O(n^2)$. We mention that the ideas in this subsection can be iterated to develop matrix reductions that expose the structure of matrices whose rank is less than $n - 1$. "Zero-chasing" with Givens rotations is at the heart of many important matrix algorithms; see §6.3, §7.5, and §8.3.

5.4.6   The UTV Framework

As mentioned at the start of this section, we are interested in factorizations that are cheaper than the SVD but which provide the same high quality information about rank, range, and nullspace. Factorizations of this type are referred to as UTV *factorizations* where the "T" stands for triangular and the "U" and "V" remind us of the SVD and orthogonal U and V matrices of singular vectors.

The matrix T can be upper triangular (these are the URV factorizations) or lower triangular (these are the ULV factorizations). It turns out that in a particular application one may favor a URV approach over a ULV approach, see §6.3. Moreover, the two reductions have different approximation properties. For example, suppose $\sigma_k(A) > \sigma_{k+1}(A)$ and S is the subspace spanned by A's right singular vectors $v_{k+1}, \ldots, v_n$. Think of S as an approximate nullspace of A. Following Stewart (1993), if

$$U^T A V = R = \begin{bmatrix} R_{11} & R_{12} \\ 0 & R_{22} \end{bmatrix} \begin{matrix} k \\ m-k \end{matrix}$$
$$\begin{matrix} k & n-k \end{matrix}$$

and $V = [\, V_1 \mid V_2 \,]$ is partitioned conformably, then

$$\text{dist}(\text{ran}(V_2),\, S) \;\leq\; \frac{\|\, R_{12}\, \|_2}{(1 - \rho_R^2)\sigma_{\min}(R_{11})} \tag{5.4.10}$$

where

$$\rho_R \;=\; \frac{\|\, R_{22}\, \|_2}{\sigma_{\min}(R_{11})}$$

is assumed to be less than 1. On the other hand, in the ULV setting we have

$$U^T A V = L = \begin{bmatrix} L_{11} & 0 \\ L_{21} & L_{22} \end{bmatrix} \begin{matrix} k \\ m-k \end{matrix} \;\cdot$$
$$\begin{matrix} k & n-k \end{matrix}$$

If $V = [\, V_1 \mid V_2 \,]$ is partitioned conformably, then

$$\text{dist}(\text{ran}(V_2), S) \;\leq\; \rho_L \, \frac{\|\, L_{12} \,\|_2}{(1 - \rho_L^2)\sigma_{\min}(L_{11})} \tag{5.4.11}$$

where

$$\rho_L \;=\; \frac{\|\, L_{22} \,\|_2}{\sigma_{\min}(L_{11})}$$

is also assumed to be less than 1. However, in practice the ρ-factors in both (5.4.10) and (5.4.11) are often much less than 1. Observe that when this is the case, the upper bound in (5.4.11) is much smaller than the upper bound in (5.4.10).

5.4.7   Complete Orthogonal Decompositions

Related to the UTV framework is the idea of a *complete orthogonal factorization*. Here we compute orthogonal U and V such that

$$U^T A V \;=\; \begin{bmatrix} T_{11} & 0 \\ 0 & 0 \end{bmatrix} \begin{matrix} r \\ m-r \end{matrix} \tag{5.4.12}$$
$$\qquad\qquad\quad r \quad n-r$$

where $r = \text{rank}(A)$. The SVD is obviously an example of a decomposition that has this structure. However, a cheaper, two-step QR process is also possible. We first use Algorithm 5.4.1 to compute

$$U^T A \Pi \;=\; \begin{bmatrix} R_{11} & R_{12} \\ 0 & 0 \end{bmatrix} \begin{matrix} r \\ m-r \end{matrix}$$
$$\qquad\qquad\quad r \quad n-r$$

and then follow up with a second QR factorization

$$Q^T \begin{bmatrix} R_{11}^T \\ R_{12}^T \end{bmatrix} = \begin{bmatrix} S_1 \\ 0 \end{bmatrix}$$

via Algorithm 5.2.1. If we set $V = \Pi Q$, then (5.4.12) is realized with $T_{11} = S_1^T$. Note that two important subspaces are defined by selected columns of $U = [\, u_1 \mid \cdots \mid u_m \,]$ and $V = [\, v_1 \mid \cdots \mid v_n \,]$:

$$\text{ran}(A) = \text{span}\{u_1, \ldots, u_r\},$$
$$\text{null}(A) = \text{span}\{v_{r+1}, \ldots, v_n\}.$$

Of course, the computation of a complete orthogonal decomposition in practice would require the careful handling of numerical rank.

5.4.8   Bidiagonalization

There is one other two-sided orthogonal factorization that is important to discuss and that is the *bidiagonal factorization*. It is not a rank-revealing factorization per se, but it has a useful role to play because it rivals the SVD in terms of data compression.

Suppose $A \in \mathbb{R}^{m \times n}$ and $m \geq n$. The idea is to compute orthogonal U_B (m-by-m) and V_B (n-by-n) such that

$$
U_B^T A V_B \;=\; \left[\begin{array}{ccccc}
d_1 & f_1 & 0 & \cdots & 0 \\
0 & d_2 & f_2 & & 0 \\
\vdots & \ddots & \ddots & \ddots & \vdots \\
0 & \cdots & & d_{n-1} & f_{n-1} \\
0 & \cdots & & 0 & d_n \\
\hline
& & \multicolumn{3}{c}{\text{\Large 0}}
\end{array} \right] . \tag{5.4.13}
$$

$U_B = U_1 \cdots U_n$ and $V_B = V_1 \cdots V_{n-2}$ can each be determined as a product of Householder matrices, e.g.,

$$
\begin{bmatrix}
\times & \times & \times & \times \\
\times & \times & \times & \times \\
\times & \times & \times & \times \\
\times & \times & \times & \times \\
\times & \times & \times & \times
\end{bmatrix}
\xrightarrow{U_1}
\begin{bmatrix}
\times & \times & \times & \times \\
0 & \times & \times & \times \\
0 & \times & \times & \times \\
0 & \times & \times & \times \\
0 & \times & \times & \times
\end{bmatrix}
\xrightarrow{V_1}
$$

$$
\begin{bmatrix}
\times & \times & 0 & 0 \\
0 & \times & \times & \times \\
0 & \times & \times & \times \\
0 & \times & \times & \times \\
0 & \times & \times & \times
\end{bmatrix}
\xrightarrow{U_2}
\begin{bmatrix}
\times & \times & 0 & 0 \\
0 & \times & \times & \times \\
0 & 0 & \times & \times \\
0 & 0 & \times & \times \\
0 & 0 & \times & \times
\end{bmatrix}
\xrightarrow{V_2}
$$

$$
\begin{bmatrix}
\times & \times & 0 & 0 \\
0 & \times & \times & 0 \\
0 & 0 & \times & \times \\
0 & 0 & \times & \times \\
0 & 0 & \times & \times
\end{bmatrix}
\xrightarrow{U_3}
\begin{bmatrix}
\times & \times & 0 & 0 \\
0 & \times & \times & 0 \\
0 & 0 & \times & \times \\
0 & 0 & 0 & \times \\
0 & 0 & 0 & \times
\end{bmatrix}
\xrightarrow{U_4}
\begin{bmatrix}
\times & \times & 0 & 0 \\
0 & \times & \times & 0 \\
0 & 0 & \times & \times \\
0 & 0 & 0 & \times \\
0 & 0 & 0 & 0
\end{bmatrix} .
$$

In general, U_k introduces zeros into the kth column, while V_k zeros the appropriate entries in row k. Overall we have:

Algorithm 5.4.2 (Householder Bidiagonalization) Given $A \in \mathbb{R}^{m \times n}$ with $m \geq n$, the following algorithm overwrites A with $U_B^T A V_B = B$ where B is upper bidiagonal and $U_B = U_1 \cdots U_n$ and $V_B = V_1 \cdots V_{n-2}$. The essential part of U_j's Householder vector is stored in $A(j+1{:}m, j)$ and the essential part of V_j's Householder vector is stored in $A(j, j+2{:}n)$.

for $j = 1:n$

$\quad [v, \beta] = \mathsf{house}(A(j:m, j))$

$\quad A(j:m, j:n) = (I_{m\ j+1} - \beta vv^T)A(j:m, j:n)$

$\quad A(j+1:m, j) = v(2:m - j + 1)$

$\quad$ **if** $j \leq n - 2$

$\quad\quad [v, \beta] = \mathsf{house}(A(j, j+1:n)^T)$

$\quad\quad A(j:m, j+1:n) = A(j:m, j+1:n)(I_{n-j} - \beta vv^T)$

$\quad\quad A(j, j+2:n) = v(2:n - j)^T$

$\quad$ **end**

end

This algorithm requires $4mn^2 - 4n^3/3$ flops. Such a technique is used by Golub and Kahan (1965), where bidiagonalization is first described. If the matrices U_B and V_B are explicitly desired, then they can be accumulated in $4m^2n - 4n^3/3$ and $4n^3/3$ flops, respectively. The bidiagonalization of A is related to the tridiagonalization of $A^T A$. See §8.3.1.

5.4.9 R-Bidiagonalization

If $m \gg n$, then a faster method of bidiagonalization method results if we upper triangularize A first before applying Algorithm 5.4.2. In particular, suppose we compute an orthogonal $Q \in \mathbb{R}^{m \times m}$ such that

$$Q^T A = \begin{bmatrix} R_1 \\ 0 \end{bmatrix}$$

is upper triangular. We then bidiagonalize the square matrix R_1,

$$U_R^T R_1 V_B = B_1,$$

where U_R and V_B are orthogonal. If $U_B = Q \, \mathrm{diag}\,(U_R, I_{m-n})$, then

$$U^T AV = \begin{bmatrix} B_1 \\ 0 \end{bmatrix} \equiv B$$

is a bidiagonalization of A.

The idea of computing the bidiagonalization in this manner is mentioned by Lawson and Hanson (SLS, p. 119) and more fully analyzed by Chan (1982). We refer to this method as R-bidiagonalization and it requires $(2mn^2 + 2n^3)$ flops. This is less than the flop count for Algorithm 5.4.2 whenever $m \geq 5n/3$.

Problems

P5.4.1 Let $x, y \in \mathbb{R}^m$ and $Q \in \mathbb{R}^{m \times m}$ be given with Q orthogonal. Show that if

$$Q^T x = \begin{bmatrix} \alpha \\ u \end{bmatrix} \begin{matrix} 1 \\ m-1 \end{matrix} \quad , \qquad Q^T y = \begin{bmatrix} \beta \\ v \end{bmatrix} \begin{matrix} 1 \\ m-1 \end{matrix}$$

then $u^T v = x^T y - \alpha\beta$.

P5.4.2 Let $A = [\, a_1 \mid \cdots \mid a_n \,] \in \mathbb{R}^{m \times n}$ and $b \in \mathbb{R}^m$ be given. For any column subset $\{a_{c_1}, \ldots, a_{c_k}\}$ define

$$\text{res}\,([\, a_{c_1} \mid \cdots \mid a_{c_k} \,]) \;=\; \min_{x \, \in \, \mathbf{R}^k} \;\| \, [\, a_{c_1} \mid \cdots \mid a_{c_k} \,] x - b \,\|_2$$

Describe an alternative pivot selection procedure for Algorithm 5.4.1 such that if $QR = A\Pi = [\, a_{c_1} \mid \cdots \mid a_{c_n} \,]$ in the final factorization, then for $k = 1{:}n$:

$$\text{res}\,([\, a_{c_1} \mid \cdots \mid a_{c_k} \,]) \;=\; \min_{i \, \geq \, k} \;\text{res}\,\big([a_{c_1}, \ldots, a_{c_{k-1}}, a_{c_i}]\big)\,.$$

P5.4.3 Suppose $T \in \mathbb{R}^{n \times n}$ is upper triangular and $t_{kk} = \sigma_{min}(T)$. Show that $T(1{:}k-1, k) = 0$ and $T(k, k+1{:}n) = 0$.

P5.4.4   Suppose $A \in \mathbb{R}^{m \times n}$ with $m \geq n$. Give an algorithm that uses Householder matrices to compute an orthogonal $Q \in \mathbb{R}^{m \times m}$ so that if $Q^T A = L$, then $L(n+1{:}m, :) = 0$ and $L(1{:}n, 1{:}n)$ is lower triangular.

P5.4.5 Suppose $R \in \mathbb{R}^{n \times n}$ is upper triangular and $Y \in \mathbb{R}^{n \times j}$ has orthonormal columns and satisfies $\| \, RY \, \|_2 = \sigma$. Give an algorithm that computes orthogonal U and V, each products of Givens rotations, so that $U^T RV = R_{\text{new}}$ is upper triangular and $V^T Y = Y_{\text{new}}$ has the property that

$$Y_{\text{new}}(n - j + 1{:}n, :) = \text{diag}(\pm 1).$$

What can you say about $R_{\text{new}}(n - j + 1{:}n, n - j + 1{:}n)$?

P5.4.6   Give an algorithm for reducing a complex matrix A to *real* bidiagonal form using complex Householder transformations.

P5.4.7 Suppose $B \in \mathbb{R}^{n \times n}$ is upper bidiagonal with $b_{nn} = 0$. Show how to construct orthogonal U and V (product of Givens rotations) so that $U^T BV$ is upper bidiagonal with a zero nth column.

P5.4.8 Suppose $A \in \mathbb{R}^{m \times n}$ with $m < n$. Give an algorithm for computing the factorization

$$U^T AV = [\, B \mid O \,]$$

where B is an m-by-m upper bidiagonal matrix. (Hint: Obtain the form

$$\begin{bmatrix} \times & \times & 0 & 0 & 0 & 0 \\ 0 & \times & \times & 0 & 0 & 0 \\ 0 & 0 & \times & \times & 0 & 0 \\ 0 & 0 & 0 & \times & \times & 0 \end{bmatrix}.$$

using Householder matrices and then "chase" the $(m, m+1)$ entry up the $(m+1)$st column by applying Givens rotations from the right.)

P5.4.9 Show how to efficiently bidiagonalize an n-by-n upper triangular matrix using Givens rotations.

P5.4.10 Show how to upper bidiagonalize a tridiagonal matrix $T \in \mathbb{R}^{n \times n}$ using Givens rotations.

P5.4.11 Show that if $B \in \mathbb{R}^{n \times n}$ is an upper bidiagonal matrix having a repeated singular value, then B must have a zero on its diagonal or superdiagonal.

Notes and References for §5.4

QR with column pivoting was first discussed in:

P.A. Businger and G.H. Golub (1965). "Linear Least Squares Solutions by Householder Transformations," *Numer. Math. 7,* 269–276.

In matters that concern rank deficiency, it is helpful to obtain information about the smallest singular value of the upper triangular matrix R. This can be done using the techniques of §3.5.4 or those that are discussed in:

I. Karasalo (1974). "A Criterion for Truncation of the QR Decomposition Algorithm for the Singular Linear Least Squares Problem," *BIT 14,* 156–166.
N. Anderson and I. Karasalo (1975). "On Computing Bounds for the Least Singular Value of a Triangular Matrix," *BIT 15,* 1–4.

C.-T. Pan and P.T.P. Tang (1999). "Bounds on Singular Values Revealed by QR Factorizations," *BIT* 39, 740–756.

C.H. Bischof (1990). "Incremental Condition Estimation," *SIAM J. Matrix Anal. Applic., 11*, 312–322.

Revealing the rank of a matrix through a carefully implementated factorization has prompted a great deal of research, see:

T.F. Chan (1987). "Rank Revealing QR Factorizations," *Lin. Alg. Applic. 88/89*, 67–82.

T.F. Chan and P. Hansen (1992). "Some Applications of the Rank Revealing QR Factorization," *SIAM J. Sci. Stat. Comp. 13*, 727–741.

S. Chandrasekaren and I.C.F. Ipsen (1994). "On Rank-Revealing Factorizations," *SIAM J. Matrix Anal. Applic. 15*, 592–622.

M. Gu and S.C. Eisenstat (1996). "Efficient Algorithms for Computing a Strong Rank-Revealing QR Factorization," *SIAM J. Sci. Comput. 17*, 848–869.

G.W. Stewart (1999). "The QLP Approximation to the Singular Value Decomposition," *SIAM J. Sci. Comput. 20*, 1336–1348.

D.A. Huckaby and T.F. Chan (2005). "Stewart's Pivoted QLP Decomposition for Low-Rank Matrices," *Num. Lin. Alg. Applic. 12*, 153–159.

A. Dax (2008). "Orthogonalization via Deflation: A Minimum Norm Approach to Low-Rank Approximation of a Matrix," *SIAM J. Matrix Anal. Applic. 30*, 236–260.

Z. Drmač and Z. Bujanović (2008). "On the Failure of Rank-Revealing QR Factorization Software—A Case Study," *ACM Trans. Math. Softw. 35*, Article 12.

We have more to say about the UTV framework in §6.5 where updating is discussed. Basic references for what we cover in this section include:

G.W. Stewart (1993). "UTV Decompositions," in *Numerical Analysis 1993, Proceedings of the 15th Dundee Conference, June–July 1993*, Longman Scientic & Technical, Harlow, Essex, UK, 225–236.

P.A. Yoon and J.L. Barlow (1998) "An Efficient Rank Detection Procedure for Modifying the ULV Decomposition," *BIT 38*, 781–801.

J.L. Barlow, H. Erbay, and I. Slapnicar (2005). "An Alternative Algorithm for the Refinement of ULV Decompositions," *SIAM J. Matrix Anal. Applic. 27*, 198–211.

Column-pivoting makes it more difficult to achieve high performance when computing the QR factorization. However, it can be done:

C.H. Bischof and P.C. Hansen (1992). "A Block Algorithm for Computing Rank-Revealing QR Factorizations," *Numer. Algorithms 2*, 371-392.

C.H. Bischof and G. Quintana-Orti (1998). "Computing Rank-revealing QR factorizations of Dense Matrices," *ACM Trans. Math. Softw. 24*, 226–253.

C.H. Bischof and G. Quintana-Orti (1998). "Algorithm 782: Codes for Rank-Revealing QR factorizations of Dense Matrices," *ACM Trans. Math. Softw. 24*, 254–257.

G. Quintana-Orti, X. Sun, and C.H. Bischof (1998). "A BLAS–3 Version of the QR Factorization with Column Pivoting," *SIAM J. Sci. Comput. 19*, 1486–1494.

A carefully designed LU factorization can also be used to shed light on matrix rank:

T-M. Hwang, W-W. Lin, and E.K. Yang (1992). "Rank-Revealing LU Factorizations," *Lin. Alg. Applic. 175*, 115–141.

T.-M. Hwang, W.-W. Lin and D. Pierce (1997). "Improved Bound for Rank Revealing LU Factorizations," *Lin. Alg. Applic. 261*, 173–186.

L. Miranian and M. Gu (2003). "Strong Rank Revealing LU Factorizations," *Lin. Alg. Applic. 367*, 1–16.

Column pivoting can be incorporated into the modified Gram-Schmidt process, see:

A. Dax (2000). "A Modified Gram-Schmidt Algorithm with Iterative Orthogonalization and Column Pivoting," *Lin. Alg. Applic. 310*, 25–42.

M. Wei and Q. Liu (2003). "Roundoff Error Estimates of the Modified GramSchmidt Algorithm with Column Pivoting," *BIT 43*, 627–645.

Aspects of the complete orthogonal decomposition are discussed in:

R.J. Hanson and C.L. Lawson (1969). "Extensions and Applications of the Householder Algorithm for Solving Linear Least Square Problems," *Math. Comput. 23*, 787–812.

P.A. Wedin (1973). "On the Almost Rank-Deficient Case of the Least Squares Problem," *BIT 13*, 344–354.

G.H. Golub and V. Pereyra (1976). "Differentiation of Pseudo-Inverses, Separable Nonlinear Least Squares Problems and Other Tales," in *Generalized Inverses and Applications*, M.Z. Nashed (ed.), Academic Press, New York, 303–324.

The quality of the subspaces that are exposed through a complete orthogonal decomposition are analyzed in:

R.D. Fierro and J.R. Bunch (1995). "Bounding the Subspaces from Rank Revealing Two-Sided Orthogonal Decompositions," *SIAM J. Matrix Anal. Applic. 16*, 743–759.

R.D. Fierro (1996). "Perturbation Analysis for Two-Sided (or Complete) Orthogonal Decompositions," *SIAM J. Matrix Anal. Applic. 17*, 383–400.

The bidiagonalization is a particularly important decomposition because it typically precedes the computation of the SVD as we discuss in §8.6. Thus, there has been a strong research interest in its efficient and accurate computation:

B. Lang (1996). "Parallel Reduction of Banded Matrices to Bidiagonal Form," *Parallel Comput. 22*, 1–18.

J.L. Barlow (2002). "More Accurate Bidiagonal Reduction for Computing the Singular Value Decomposition," *SIAM J. Matrix Anal. Applic. 23*, 761–798.

J.L. Barlow, N. Bosner and Z. Drmač (2005). "A New Stable Bidiagonal Reduction Algorithm," *Lin. Alg. Applic. 397*, 35–84.

B.N. Parlett (2005). "A Bidiagonal Matrix Determines Its Hyperbolic SVD to Varied Relative Accuracy," *SIAM J. Matrix Anal. Applic. 26*, 1022–1057.

N. Bosner and J.L. Barlow (2007). "Block and Parallel Versions of One-Sided Bidiagonalization," *SIAM J. Matrix Anal. Applic. 29*, 927–953.

G.W. Howell, J.W. Demmel, C.T. Fulton, S. Hammarling, and K. Marmol (2008). "Cache Efficient Bidiagonalization Using BLAS 2.5 Operators," *ACM Trans. Math. Softw. 34*, Article 14.

H. Ltaief, J. Kurzak, and J. Dongarra (2010). "Parallel Two-Sided Matrix Reduction to Band Bidiagonal Form on Multicore Architectures," *IEEE Trans. Parallel Distrib. Syst. 21*, 417–423.

5.5   The Rank-Deficient Least Squares Problem

If A is rank deficient, then there are an infinite number of solutions to the LS problem. We must resort to techniques that incorporate numerical rank determination and identify a particular solution as "special." In this section we focus on using the SVD to compute the minimum norm solution and QR-with-column-pivoting to compute what is called the basic solution. Both of these approaches have their merits and we conclude with a subset selection procedure that combines their positive attributes.

5.5.1   The Minimum Norm Solution

Suppose $A \in \mathbb{R}^{m \times n}$ and $\mathsf{rank}(A) = r < n$. The rank-deficient LS problem has an infinite number of solutions, for if x is a minimizer and $z \in \mathsf{null}(A)$, then $x + z$ is also a minimizer. The set of all minimizers

$$\mathcal{X} = \{ x \in \mathbb{R}^n : \| Ax - b \|_2 = \min \}$$

is convex and so if $x_1, x_2 \in \mathcal{X}$ and $\lambda \in [0, 1]$, then

$$\| A(\lambda x_1 + (1 - \lambda)x_2) - b \|_2 \le \lambda \| Ax_1 - b \|_2 + (1 - \lambda) \| Ax_2 - b \|_2 = \min_{x \in \mathbf{R}^n} \| Ax - b \|_2.$$

Thus, $\lambda x_1 + (1 - \lambda)x_2 \in \mathcal{X}$. It follows that $\mathcal{X}$ has a unique element having minimum 2-norm and we denote this solution by x_{LS}. (Note that in the full-rank case, there is only one LS solution and so it must have minimal 2-norm. Thus, we are consistent with the notation in §5.3.)

Any complete orthogonal factorization (§5.4.7) can be used to compute x_{LS}. In particular, if Q and Z are orthogonal matrices such that

$$Q^T A Z = T = \begin{bmatrix} T_{11} & 0 \\ 0 & 0 \end{bmatrix} \begin{matrix} r \\ m-r \end{matrix} \quad , \quad r = \mathsf{rank}(A)$$
$$\begin{matrix} r & n-r \end{matrix}$$

then

$$\| Ax - b \|_2^2 = \| (Q^T AZ)Z^T x - Q^T b \|_2^2 = \| T_{11}w - c \|_2^2 + \| d \|_2^2$$

where

$$Z^T x = \begin{bmatrix} w \\ y \end{bmatrix} \begin{matrix} r \\ n-r \end{matrix} \quad , \qquad Q^T b = \begin{bmatrix} c \\ d \end{bmatrix} \begin{matrix} r \\ m-r \end{matrix} \quad .$$

Clearly, if x is to minimize the sum of squares, then we must have $w = T_{11}^{-1}c$. For x to have minimal 2-norm, y must be zero, and thus

$$x_{LS} = Z \begin{bmatrix} T_{11}^{-1}c \\ 0 \end{bmatrix}.$$

Of course, the SVD is a particularly revealing complete orthogonal decomposition. It provides a neat expression for x_{LS} and the norm of the minimum residual $\rho_{LS} = \| Ax_{LS} - b \|_2$.

Theorem 5.5.1. *Suppose $U^T AV = \Sigma$ is the SVD of $A \in \mathbb{R}^{m \times n}$ with $r = \mathsf{rank}(A)$. If $U = [\, u_1 \mid \cdots \mid u_m \,]$ and $V = [\, v_1 \mid \cdots \mid v_n \,]$ are column partitionings and $b \in \mathbb{R}^m$, then*

$$x_{LS} = \sum_{i=1}^{r} \frac{u_i^T b}{\sigma_i} v_i \qquad (5.5.1)$$

minimizes $\| Ax - b \|_2$ and has the smallest 2-norm of all minimizers. Moreover

$$\rho_{LS}^2 = \| Ax_{LS} - b \|_2^2 = \sum_{i=r+1}^{m} (u_i^T b)^2. \qquad (5.5.2)$$

Proof. For any $x \in \mathbb{R}^n$ we have

$$\| Ax - b \|_2^2 = \| (U^T AV)(V^T x) - U^T b \|_2^2 = \| \Sigma\alpha - U^T b \|_2^2$$
$$= \sum_{i=1}^{r}(\sigma_i \alpha_i - u_i^T b)^2 + \sum_{i=r+1}^{m} (u_i^T b)^2,$$

where $\alpha = V^T x$. Clearly, if x solves the LS problem, then $\alpha_i = (u_i^T b/\sigma_i)$ for $i = 1{:}r$. If we set $\alpha(r + 1{:}n) = 0$, then the resulting x has minimal 2-norm. $\square$

5.5.2   A Note on the Pseudoinverse

If we define the matrix $A^+ \in \mathbb{R}^{n \times m}$ by $A^+ = V\Sigma^+ U^T$ where

$$\Sigma^+ = \text{diag}\left(\frac{1}{\sigma_1}, \ldots, \frac{1}{\sigma_r}, 0, \ldots, 0\right) \in \mathbb{R}^{n \times m}, \qquad r = \text{rank}(A),$$

then $x_{LS} = A^+ b$ and $\rho_{LS} = \|(I - AA^+)b\|_2$. A^+ is referred to as the *pseudo-inverse* of A. It is the unique minimal Frobenius norm solution to the problem

$$\min_{X \in \mathbb{R}^{m \times n}} \| AX - I_m \|_F. \tag{5.5.3}$$

If $\text{rank}(A) = n$, then $A^+ = (A^T A)^{-1} A^T$, while if $m = n = \text{rank}(A)$, then $A^+ = A^{-1}$. Typically, A^+ is defined to be the unique matrix $X \in \mathbb{R}^{n \times m}$ that satisfies the four *Moore-Penrose conditions*:

$$
\begin{array}{llll}
\text{(i)} & AXA = A, & \text{(iii),} & (AX)^T = AX, \\
\text{(ii)} & XAX = X, & \text{(iv)} & (XA)^T = XA.
\end{array}
$$

These conditions amount to the requirement that AA^+ and A^+A be orthogonal projections onto $\text{ran}(A)$ and $\text{ran}(A^T)$, respectively. Indeed,

$$AA^+ = U_1 U_1^T$$

where $U_1 = U(1:m, 1:r)$ and

$$A^+A = V_1 V_1^T$$

where $V_1 = V(1:n, 1:r)$.

5.5.3   Some Sensitivity Issues

In §5.3 we examined the sensitivity of the full-rank LS problem. The behavior of x_{LS} in this situation is summarized in Theorem 5.3.1. If we drop the full-rank assumption, then x_{LS} is not even a continuous function of the data and small changes in A and b can induce arbitrarily large changes in $x_{LS} = A^+ b$. The easiest way to see this is to consider the behavior of the pseudoinverse. If A and δA are in $\mathbb{R}^{m \times n}$, then Wedin (1973) and Stewart (1975) show that

$$\| (A + \delta A)^+ - A^+ \|_F \leq 2\| \delta A \|_F \max \left\{ \| A^+ \|_2^2 , \| (A + \delta A)^+ \|_2^2 \right\}.$$

This inequality is a generalization of Theorem 2.3.4 in which perturbations in the matrix inverse are bounded. However, unlike the square nonsingular case, the upper bound does not necessarily tend to zero as δA tends to zero. If

$$
A = \begin{bmatrix} 1 & 0 \\ 0 & 0 \\ 0 & 0 \end{bmatrix} \qquad \text{and} \qquad \delta A = \begin{bmatrix} 0 & 0 \\ 0 & \epsilon \\ 0 & 0 \end{bmatrix}
$$

then

$$
A^+ = \begin{bmatrix} 1 & 0 & 0 \\ 0 & 0 & 0 \end{bmatrix} \qquad \text{and} \qquad (A + \delta A)^+ = \begin{bmatrix} 1 & 0 & 0 \\ 1 & 1/\epsilon & 0 \end{bmatrix},
$$

and

$$\| A^+ - (A + \delta A)^+ \|_2 = 1/\epsilon.$$

The numerical determination of an LS minimizer in the presence of such discontinuities is a major challenge.

5.5.4 The Truncated SVD Solution

Suppose $\widehat{U}$, $\widehat{\Sigma}$, and $\widehat{V}$ are the computed SVD factors of a matrix A and $\hat{r}$ is accepted as its δ-rank, i.e.,

$$\hat{\sigma}_n \leq \cdots \leq \hat{\sigma}_{\hat{r}} \leq \delta < \hat{\sigma}_{\hat{r}} \leq \cdots \leq \hat{\sigma}_1.$$

It follows that we can regard

$$x_{\hat{r}} \;=\; \sum_{i=1}^{\hat{r}} \frac{\hat{u}_i^T b}{\hat{\sigma}_i} \hat{v}_i$$

as an approximation to x_{LS}. Since $\| x_{\hat{r}} \|_2 \approx 1/\sigma_{\hat{r}} \leq 1/\delta$, then δ may also be chosen with the intention of producing an approximate LS solution with suitably small norm. In §6.2.1, we discuss more sophisticated methods for doing this.

If $\hat{\sigma}_{\hat{r}} \gg \delta$, then we have reason to be comfortable with $x_{\hat{r}}$ because A can then be unambiguously regarded as a rank($A_{\hat{r}}$) matrix (modulo δ).

On the other hand, $\{\hat{\sigma}_1, \ldots, \hat{\sigma}_n\}$ might not clearly split into subsets of small and large singular values, making the determination of $\hat{r}$ by this means somewhat arbitrary. This leads to more complicated methods for estimating rank, which we now discuss in the context of the LS problem. The issues are readily communicated by making two simplifying assumptions. Assume that $r = n$, and that $\Delta A = 0$ in (5.4.4), which implies that $W^T A Z = \widehat{\Sigma} = \Sigma$ is the SVD. Denote the ith columns of the matrices $\widehat{U}$, W, $\widehat{V}$, and Z by $\hat{u}_i$, w_i, $\hat{v}_i$, and z_i, respectively. Because

$$x_{LS} \;-\; x_{\hat{r}} = \sum_{i=1}^{n} \frac{w_i^T b}{\sigma_i} z_i \;-\; \sum_{i=1}^{\hat{r}} \frac{\hat{u}_i^T b}{\sigma_i} \hat{v}_i$$

$$= \sum_{i=1}^{\hat{r}} \frac{((w_i - \hat{u}_i)^T b) z_i \;+\; (\hat{u}_i^T b)(z_i - \hat{v}_i)}{\sigma_i} \;+\; \sum_{i=\hat{r}+1}^{n} \frac{w_i^T b}{\sigma_i} z_i$$

it follows from $\| w_i - \hat{u}_i \|_2 \leq \epsilon$, $\| \hat{u}_i \|_2 \leq 1 + \epsilon$, and $\| z_i - \hat{v}_i \|_2 \leq \epsilon$ that

$$\| x_{\hat{r}} - x_{LS} \|_2 \leq \frac{\hat{r}}{\sigma_{\hat{r}}} 2(1+\epsilon)\epsilon \| b \|_2 \;+\; \sqrt{\sum_{i=\hat{r}+1}^{n} \left(\frac{w_i^T b}{\sigma_i} \right)^2}.$$

The parameter $\hat{r}$ can be determined as that integer which minimizes the upper bound. Notice that the first term in the bound increases with $\hat{r}$, while the second decreases.

On occasions when minimizing the residual is more important than accuracy in the solution, we can determine $\hat{r}$ on the basis of how close we surmise $\| b - A x_{\hat{r}} \|_2$ is to the true minimum. Paralleling the above analysis, it can be shown that

$$\| b - A x_{\hat{r}} \|_2 \;\leq\; \| b - A x_{LS} \|_2 \;+\; (n - \hat{r}) \| b \|_2 + \epsilon \hat{r} \| b \|_2 \left(1 + (1+\epsilon) \frac{\hat{\sigma}_1}{\hat{\sigma}_{\hat{r}}} \right).$$

Again $\hat{r}$ could be chosen to minimize the upper bound. See Varah (1973) for practical details and also **LAPACK**.

5.5.5   Basic Solutions via QR with Column Pivoting

Suppose $A \in \mathbb{R}^{m \times n}$ has rank r. QR with column pivoting (Algorithm 5.4.1) produces the factorization $A\Pi = QR$ where

$$R = \begin{bmatrix} R_{11} & R_{12} \\ 0 & 0 \end{bmatrix} \begin{matrix} r \\ m-r \end{matrix} \quad .$$
$$\quad\quad\quad r \quad\quad n-r$$

Given this reduction, the LS problem can be readily solved. Indeed, for any $x \in \mathbb{R}^n$ we have

$$\| Ax - b \|_2^2 = \| (Q^T A\Pi)(\Pi^T x) - (Q^T b) \|_2^2 = \| R_{11}y - (c - R_{12}z) \|_2^2 + \| d \|_2^2,$$

where

$$\Pi^T x = \begin{bmatrix} y \\ z \end{bmatrix} \begin{matrix} r \\ n-r \end{matrix} \quad \text{and} \quad Q^T b = \begin{bmatrix} c \\ d \end{bmatrix} \begin{matrix} r \\ m-r \end{matrix} \quad .$$

Thus, if x is an LS minimizer, then we must have

$$x = \Pi \begin{bmatrix} R_{11}^{-1}(c - R_{12}z) \\ z \end{bmatrix} .$$

If z is set to zero in this expression, then we obtain the *basic solution*

$$x_B = \Pi \begin{bmatrix} R_{11}^{-1}c \\ 0 \end{bmatrix} .$$

Notice that x_B has at most r nonzero components and so Ax_B involves a subset of A's columns.

The basic solution is not the minimal 2-norm solution unless the submatrix R_{12} is zero since

$$\| x_{LS} \|_2 = \min_{z \in \mathbb{R}^{n-2}} \left\| x_B - \Pi \begin{bmatrix} R_{11}^{-1}R_{12} \\ -I_{n-r} \end{bmatrix} z \right\|_2 . \tag{5.5.4}$$

Indeed, this characterization of $\| x_{LS} \|_2$ can be used to show that

$$1 \le \frac{\| x_B \|_2}{\| x_{LS} \|_2} \le \sqrt{1 + \| R_{11}^{-1}R_{12} \|_2^2} . \tag{5.5.5}$$

See Golub and Pereyra (1976) for details.

5.5.6   Some Comparisons

As we mentioned, when solving the LS problem via the SVD, only Σ and V have to be computed assuming that the right hand side b is available. The table in Figure 5.5.1 compares the flop efficiency of this approach with the other algorithms that we have presented.

LS Algorithm	Flop Count
Normal equations	$mn^2 + n^3/3$
Householder QR	$n^3/3$
Modified Gram-Schmidt	$2mn^2$
Givens QR	$3mn^2 - n^3$
Householder Bidiagonalization	$4mn^2 - 2n^3$
R-Bidiagonalization	$2mn^2 + 2n^3$
SVD	$4mn^2 + 8n^3$
R-SVD	$2mn^2 + 11n^3$

Figure 5.5.1. *Flops associated with various least squares methods*

5.5.7 SVD-Based Subset Selection

Replacing A by $A_{\tilde{r}}$ in the LS problem amounts to filtering the small singular values and can make a great deal of sense in those situations where A is derived from noisy data. In other applications, however, rank deficiency implies redundancy among the factors that comprise the underlying model. In this case, the model-builder may not be interested in a predictor such as $A_{\tilde{r}}x_{\tilde{r}}$ that involves all n redundant factors. Instead, a predictor Ay may be sought where y has at most $\tilde{r}$ nonzero components. The position of the nonzero entries determines which columns of A, i.e., which factors in the model, are to be used in approximating the observation vector b. How to pick these columns is the problem of *subset selection*.

QR with column pivoting is one way to proceed. However, Golub, Klema, and Stewart (1976) have suggested a technique that heuristically identifies a more independent set of columns than are involved in the predictor Ax_B. The method involves both the SVD and QR with column pivoting:

Step 1. Compute the SVD $A = U\Sigma V^T$ and use it to determine a rank estimate $\tilde{r}$.

Step 2. Calculate a permutation matrix P such that the columns of the matrix $B_1 \in \mathbb{R}^{m \times \tilde{r}}$ in $AP = [\, B_1 \mid B_2 \,]$ are "sufficiently independent."

Step 3. Predict b with Ay where $y = P \begin{bmatrix} z \\ 0 \end{bmatrix}$ and $z \in \mathbb{R}^{\tilde{r}}$ minimizes $\| B_1 z - b \|_2$.

The second step is key. Because

$$\min_{z \in \mathbb{R}^{\tilde{r}}} \| B_1 z - b \|_2 \;=\; \| Ay - b \|_2 \;\geq\; \min_{x \in \mathbb{R}^n} \| Ax - b \|_2$$

it can be argued that the permutation P should be chosen to make the residual $r = (I - B_1 B_1^+)b$ as small as possible. Unfortunately, such a solution procedure can be unstable. For example, if

$$
A = \begin{bmatrix} 1 & 1 & 0 \\ 1 & 1+\epsilon & 1 \\ 0 & 0 & 1 \end{bmatrix}, \qquad b = \begin{bmatrix} 1 \\ -1 \\ 0 \end{bmatrix},
$$

$\tilde{r} = 2$, and $P = I$, then $\min \| B_1 z - b \|_2 = 0$, but $\| B_1^+ b \|_2 = O(1/\epsilon)$. On the other hand, any proper subset involving the third column of A is strongly independent but renders a much larger residual.

This example shows that there can be a trade-off between the independence of the chosen columns and the norm of the residual that they render. How to proceed in the face of this trade-off requires useful bounds on $\sigma_{\tilde{r}}(B_1)$, the smallest singular value of B_1.

Theorem 5.5.2. *Let the SVD of $A \in \mathbb{R}^{m \times n}$ be given by $U^T A V = \Sigma = \mathrm{diag}(\sigma_i)$ and define the matrix $B_1 \in \mathbb{R}^{m \times \tilde{r}}$, $\tilde{r} \leq \mathrm{rank}(A)$, by*

$$
AP = [\; B_1 \mid B_2 \;]
$$
$$
\underset{\tilde{r} \qquad\quad n-\tilde{r}}{}
$$

where $P \in \mathbb{R}^{n \times n}$ is a permutation. If

$$
P^T V = \begin{bmatrix} \tilde{V}_{11} & \tilde{V}_{12} \\ \tilde{V}_{21} & \tilde{V}_{22} \end{bmatrix} \begin{matrix} \tilde{r} \\ n-\tilde{r} \end{matrix}
$$
$$
\underset{\tilde{r} \qquad n-\tilde{r}}{}
$$
(5.5.6)

and $\tilde{V}_{11}$ is nonsingular, then

$$
\frac{\sigma_{\tilde{r}}(A)}{\| \tilde{V}_{11}^{-1} \|_2} \leq \sigma_{\tilde{r}}(B_1) \leq \sigma_{\tilde{r}}(A).
$$

Proof. The upper bound follows from Corollary 2.4.4. To establish the lower bound, partition the diagonal matrix of singular values as follows:

$$
\Sigma = \begin{bmatrix} \Sigma_1 & 0 \\ 0 & \Sigma_2 \end{bmatrix} \begin{matrix} \tilde{r} \\ m-\tilde{r} \end{matrix}.
$$
$$
\underset{\tilde{r} \quad n-\tilde{r}}{}
$$

If $w \in \mathbb{R}^{\tilde{r}}$ is a unit vector with the property that $\| B_1 w \|_2 = \sigma_{\tilde{r}}(B_1)$, then

$$
\sigma_{\tilde{r}}(B_1)^2 = \| B_1 w \|_2^2 = \left\| U \Sigma V^T P \begin{bmatrix} w \\ 0 \end{bmatrix} \right\|_2^2 = \| \Sigma_1 \tilde{V}_{11}^T w \|_2^2 + \| \Sigma_2 \tilde{V}_{12}^T w \|_2^2.
$$

The theorem now follows because $\| \Sigma_1 \tilde{V}_{11}^T w \|_2 \geq \sigma_{\tilde{r}}(A)/\| \tilde{V}_{11}^{-1} \|_2$. $\square$

This result suggests that in the interest of obtaining a sufficiently independent subset of columns, we choose the permutation P such that the resulting $\tilde{V}_{11}$ submatrix is as well-conditioned as possible. A heuristic solution to this problem can be obtained by computing the QR with column-pivoting factorization of the matrix $[\, V_{11}^T \ V_{21}^T \,]$, where

$$
V \;=\; \begin{bmatrix} V_{11} & V_{12} \\ V_{21} & V_{22} \end{bmatrix} \begin{matrix} \tilde{r} \\ n-\tilde{r} \end{matrix}
$$
$$
\begin{matrix} \tilde{r} & n-\tilde{r} \end{matrix}
$$

is a partitioning of the matrix V, A's matrix of right singular vectors. In particular, if we apply QR with column pivoting (Algorithm 5.4.1) to compute

$$
Q^T[\, V_{11}^T \ V_{21}^T \,]P \;=\; [\, R_{11} \mid R_{12}\,]
$$
$$
\begin{matrix} \tilde{r} & n-\tilde{r} \end{matrix}
$$

where Q is orthogonal, P is a permutation matrix, and R_{11} is upper triangular, then (5.5.6) implies

$$
\begin{bmatrix} \widetilde{V}_{11} \\ \widetilde{V}_{21} \end{bmatrix} \;=\; P^T \begin{bmatrix} V_{11} \\ V_{21} \end{bmatrix} \;=\; \begin{bmatrix} R_{11}^T Q^T \\ R_{12}^T Q^T \end{bmatrix}.
$$

Note that R_{11} is nonsingular and that $\| \widetilde{V}_{11}^{-1} \|_2 = \| R_{11}^{-1} \|_2$. Heuristically, column pivoting tends to produce a well-conditioned R_{11}, and so the overall process tends to produce a well-conditioned $\widetilde{V}_{11}$.

Algorithm 5.5.1   Given $A \in \mathbb{R}^{m \times n}$ and $b \in \mathbb{R}^m$ the following algorithm computes a permutation P, a rank estimate $\tilde{r}$, and a vector $z \in \mathbb{R}^{\tilde{r}}$ such that the first $\tilde{r}$ columns of $B = AP$ are independent and $\| B(:,1{:}\tilde{r})z - b \|_2$ is minimized.

    Compute the SVD $U^T A V = \operatorname{diag}(\sigma_1, \ldots, \sigma_n)$ and save V.

    Determine $\tilde{r} \le \operatorname{rank}(A)$.

    Apply QR with column pivoting: $Q^T V(:,1{:}\tilde{r})^T P = [\, R_{11} \mid R_{12}\,]$ and set

        $AP \;=\; [\, B_1 \mid B_2\,]$ with $B_1 \in \mathbb{R}^{m \times \tilde{r}}$ and $B_2 \in \mathbb{R}^{m \times (n-\tilde{r})}$.

    Determine $z \in \mathbb{R}^{\tilde{r}}$ such that $\| b - B_1 z \|_2 \;=\; \min$.

5.5.8   Column Independence Versus Residual Size

We return to the discussion of the trade-off between column independence and norm of the residual. In particular, to assess the above method of subset selection we need to examine the residual of the vector y that it produces

$$
r_y \;=\; b - Ay = b - B_1 z = (I - B_1 B_1^+)b.
$$

Here, $B_1 = B(:,1{:}\tilde{r})$ with $B = AP$. To this end, it is appropriate to compare r_y with

$$
r_{x_{\tilde{r}}} \;=\; b - Ax_{\tilde{r}}
$$

since we are regarding A as a rank-$\tilde{r}$ matrix and since $x_{\tilde{r}}$ solves the nearest rank-$\tilde{r}$ LS problem $\min \| A_{\tilde{r}}x - b \|_2$.

Theorem 5.5.3. *Assume that $U^T A V = \Sigma$ is the SVD of $A \in \mathbb{R}^{m \times n}$ and that r_y and $r_{x_{\tilde{r}}}$ are defined as above. If $\tilde{V}_{11}$ is the leading r-by-r principal submatrix of $P^T V$, then*

$$\| \, r_{x_{\tilde{r}}} - r_y \, \|_2 \; \leq \; \frac{\sigma_{\tilde{r}+1}(A)}{\sigma_{\tilde{r}}(A)} \| \, \tilde{V}_{11}^{-1} \, \|_2 \| \, b \, \|_2 .$$

Proof. Note that $r_{x_{\tilde{r}}} = (I - U_1 U_1^T)b$ and $r_y = (I - Q_1 Q_1^T)b$ where

$$U \; = \; [\; \underset{\tilde{r}}{U_1} \; | \; \underset{m - \tilde{r}}{U_2} \;]$$

is a partitioning of the matrix U and $Q_1 = B_1 (B_1^T B_1)^{-1/2}$. Using Theorem 2.6.1 we obtain

$$\| \, r_{x_{\tilde{r}}} - r_y \, \|_2 \; \leq \; \| \, U_1 U_1^T - Q_1 Q_1^T \, \|_2 \| \, b \, \|_2 \; = \; \| \, U_2^T Q_1 \, \|_2 \| \, b \, \|_2$$

while Theorem 5.5.2 permits us to conclude

$$\| \, U_2^T Q_1 \, \|_2 \; \leq \; \| \, U_2^T B_1 \, \|_2 \| \, (B_1^T B_1)^{-1/2} \, \|_2$$

$$\leq \; \sigma_{\tilde{r}+1}(A) \frac{1}{\sigma_{\tilde{r}}(B_1)} \; \leq \; \frac{\sigma_{\tilde{r}+1}(A)}{\sigma_{\tilde{r}}(A)} \| \, \tilde{V}_{11}^{-1} \, \|_2 ,$$

and this establishes the theorem.   □

Noting that

$$\| \, r_{x_{\tilde{r}}} - r_y \, \|_2 \; = \; \left\| B_1 y - \sum_{i=1}^{r} (u_i^T b) u_i \right\|_2$$

we see that Theorem 5.5.3 sheds light on how well $B_1 y$ can predict the "stable" component of b, i.e., $U_1^T b$. Any attempt to approximate $U_2^T b$ can lead to a large norm solution. Moreover, the theorem says that if $\sigma_{\tilde{r}+1}(A) \ll \sigma_{\tilde{r}}(A)$, then any reasonably independent subset of columns produces essentially the same-sized residual. On the other hand, if there is no well-defined gap in the singular values, then the determination of $\tilde{r}$ becomes difficult and the entire subset selection problem becomes more complicated.

Problems

P5.5.1 Show that if

$$A \; = \; \begin{bmatrix} T & S \\ 0 & 0 \end{bmatrix} \begin{matrix} r \\ m-r \end{matrix}$$
$$\quad \;\; r \quad n-r$$

where $r = \mathsf{rank}(A)$ and T is nonsingular, then

$$X \; = \; \begin{bmatrix} T^{-1} & 0 \\ 0 & 0 \end{bmatrix} \begin{matrix} r \\ n-r \end{matrix}$$
$$\quad \;\;\; r \quad m-r$$

satisfies $AXA = A$ and $(AX)^T = (AX)$. In this case, we say that X is a (1,3) *pseudoinverse* of A. Show that for general A, $x_B = Xb$ where X is a (1,3) pseudoinverse of A.

P5.5.2 Define $B(\lambda) \in \mathbb{R}^{n \times m}$ by

$$B(\lambda) \; = \; (A^T A + \lambda I)^{-1} A^T$$

where $\lambda > 0$. Show that

$$\| B(\lambda) - A^+ \|_2 \; = \; \frac{\lambda}{\sigma_r(A)[\sigma_r(A)^2 + \lambda]}, \qquad r = \text{rank}(A),$$

and therefore that $B(\lambda) \to A^+$ as $\lambda \to 0$.

P5.5.3 Consider the rank-deficient LS problem

$$\min_{y \in \mathbb{R}^r, \, z \in \mathbb{R}^{n-r}} \left\| \begin{bmatrix} R & S \\ 0 & 0 \end{bmatrix} \begin{bmatrix} y \\ z \end{bmatrix} - \begin{bmatrix} c \\ d \end{bmatrix} \right\|_2$$

where $R \in \mathbb{R}^{r \times r}$, $S \in \mathbb{R}^{r \times n-r}$, $y \in \mathbb{R}^r$, and $z \in \mathbb{R}^{n-r}$. Assume that R is upper triangular and nonsingular. Show how to obtain the minimum norm solution to this problem by computing an appropriate QR factorization without pivoting and then solving for the appropriate y and z.

P5.5.4 Show that if $A_k \to A$ and $A_k^+ \to A^+$, then there exists an integer k_0 such that $\text{rank}(A_k)$ is constant for all $k \geq k_0$.

P5.5.5 Show that if $A \in \mathbb{R}^{m \times n}$ has rank n, then so does $A + E$ if $\| E \|_2 \| A^+ \|_2 < 1$.

P5.5.6 Suppose $A \in \mathbb{R}^{m \times n}$ is rank deficient and $b \in \mathbb{R}^m$. Assume for $k = 0, 1, \ldots$ that $x^{(k+1)}$ minimizes

$$\phi_k(x) \; = \; \| Ax - b \|_2^2 + \lambda \| x - x^{(k)} \|_2^2$$

where $\lambda > 0$ and $x^{(0)} = 0$. Show that $x^{(k)} \to x_{LS}$.

P5.5.8 Suppose $A \in \mathbb{R}^{m \times n}$ and that $\| u^T A \|_2 = \sigma$ with $u^T u = 1$. Show that if $u^T(Ax - b) = 0$ for $x \in \mathbb{R}^n$ and $b \in \mathbb{R}^m$, then $\| x \|_2 \geq |u^T b|/\sigma$.

P5.5.9 In Equation (5.5.6) we know that the matrix $P^T V$ is orthogonal. Thus, $\| \tilde{V}_{11}^{-1} \|_2 = \| \tilde{V}_{22}^{-1} \|_2$ from the CS decomposition (Theorem 2.5.3). Show how to compute P by applying the QR with column-pivoting algorithm to $[\tilde{V}_{22}^T | \tilde{V}_{12}^T]$. (For $\tilde{r} > n/2$, this procedure would be more economical than the technique discussed in the text.) Incorporate this observation in Algorithm 5.5.1.

P5.5.10 Suppose $F \in \mathbb{R}^{m \times r}$ and $G \in \mathbb{R}^{n \times r}$ each have rank r. (a) Give an efficient algorithm for computing the minimum 2-norm minimizer of $\| FG^T x - b \|_2$ where $b \in \mathbb{R}^m$. (b) Show how to compute the vector x_B.

Notes and References for §5.5

For a comprehensive treatment of the pseudoinverse and its manipulation, see:

M.Z. Nashed (1976). *Generalized Inverses and Applications*, Academic Press, New York.
S.L. Campbell and C.D. Meyer (2009). *Generalized Inverses of Linear Transformations*, SIAM Publications, Philadelphia, PA.

For an analysis of how the pseudo-inverse is affected by perturbation, see:

P.A. Wedin (1973). "Perturbation Theory for Pseudo-Inverses," *BIT 13*, 217–232.
G.W. Stewart (1977). "On the Perturbation of Pseudo-Inverses, Projections, and Linear Least Squares," *SIAM Review 19*, 634–662.

Even for full rank problems, column pivoting seems to produce more accurate solutions. The error analysis in the following paper attempts to explain why:

L.S. Jennings and M.R. Osborne (1974). "A Direct Error Analysis for Least Squares," *Numer. Math. 22*, 322–332.

Various other aspects of the rank-deficient least squares problem are discussed in:

J.M. Varah (1973). "On the Numerical Solution of Ill-Conditioned Linear Systems with Applications to Ill-Posed Problems," *SIAM J. Numer. Anal. 10*, 257–67.
G.W. Stewart (1984). "Rank Degeneracy," *SIAM J. Sci. Stat. Comput. 5*, 403–413.
P.C. Hansen (1987). "The Truncated SVD as a Method for Regularization," *BIT 27*, 534–553.
G.W. Stewart (1987). "Collinearity and Least Squares Regression," *Stat. Sci. 2*, 68–100.

R.D. Fierro and P.C. Hansen (1995). "Accuracy of TSVD Solutions Computed from Rank-Revealing Decompositions," *Numer. Math. 70*, 453–472.

P.C. Hansen (1997). *Rank-Deficient and Discrete Ill-Posed Problems: Numerical Aspects of Linear Inversion*, SIAM Publications, Philadelphia, PA.

A. Dax and L. Elden (1998). "Approximating Minimum Norm Solutions of Rank-Deficient Least Squares Problems," *Numer. Lin. Alg. 5*, 79–99.

G. Quintana-Orti, E.S. Quintana-Orti, and A. Petitet (1998). "Efficient Solution of the Rank-Deficient Linear Least Squares Problem," *SIAM J. Sci. Comput. 20*, 1155–1163.

L.V. Foster (2003). "Solving Rank-Deficient and Ill-posed Problems Using UTV and QR Factorizations," *SIAM J. Matrix Anal. Applic. 25*, 582–600.

D.A. Huckaby and T.F. Chan (2004). "Stewart's Pivoted QLP Decomposition for Low-Rank Matrices," *Numer. Lin. Alg. 12*, 153–159.

L. Foster and R. Kommu (2006). "Algorithm 853: An Efficient Algorithm for Solving Rank-Deficient Least Squares Problems," *ACM Trans. Math. Softw. 32*, 157–165.

For a sampling of the subset selection literature, we refer the reader to:

H. Hotelling (1957). "The Relations of the Newer Multivariate Statistical Methods to Factor Analysis," *Brit. J. Stat. Psych. 10*, 69–79.

G.H. Golub, V. Klema and G.W. Stewart (1976). "Rank Degeneracy and Least Squares Problems," Technical Report TR-456, Department of Computer Science, University of Maryland, College Park, MD.

S. Van Huffel and J. Vandewalle (1987). "Subset Selection Using the Total Least Squares Approach in Collinearity Problems with Errors in the Variables," *Lin. Alg. Applic. 88/89*, 695–714.

M.R. Osborne, B. Presnell, and B.A. Turlach (2000). "A New Approach to Variable Selection in Least Squares Problems," *IMA J. Numer. Anal. 20*, 389–403.

5.6   Square and Underdetermined Systems

The orthogonalization methods developed in this chapter can be applied to square systems and also to systems in which there are fewer equations than unknowns. In this brief section we examine the various possibilities.

5.6.1   Square Systems

The least squares solvers based on the QR factorization and the SVD can also be used to solve square linear systems. Figure 5.6.1 compares the associated flop counts. It is

Method	Flops
Gaussian elimination	$2n^3/3$
Householder QR	$4n^3/3$
Modified Gram-Schmidt	$2n^3$
Singular value decomposition	$12n^3$

Figure 5.6.1. *Flops associated with various methods for square linear systems*

assumed that the right-hand side is available at the time of factorization. Although Gaussian elimination involves the least amount of arithmetic, there are three reasons why an orthogonalization method might be considered:

- The flop counts tend to exaggerate the Gaussian elimination advantage. When memory traffic and vectorization overheads are considered, the QR approach is comparable in efficiency.

- The orthogonalization methods have guaranteed stability; there is no "growth factor" to worry about as in Gaussian elimination.

- In cases of ill-conditioning, the orthogonal methods give an added measure of reliability. QR with condition estimation is very dependable and, of course, SVD is unsurpassed when it comes to producing a meaningful solution to a nearly singular system.

We are not expressing a strong preference for orthogonalization methods but merely suggesting viable alternatives to Gaussian elimination.

We also mention that the SVD entry in the above table assumes the availability of b at the time of decomposition. Otherwise, $20n^3$ flops are required because it then becomes necessary to accumulate the U matrix.

If the QR factorization is used to solve $Ax = b$, then we ordinarily have to carry out a back substitution: $Rx = Q^T b$. However, this can be avoided by "preprocessing" b. Suppose H is a Householder matrix such that $Hb = \beta e_n$ where e_n is the last column of I_n. If we compute the QR factorization of $(HA)^T$, then $A = H^T R^T Q^T$ and the system transforms to

$$R^T y = \beta e_n$$

where $y = Q^T x$. Since R^T is lower triangular, $y = (\beta/r_{nn})e_n$ and so

$$x = \frac{\beta}{r_{nn}} Q(:, n).$$

5.6.2   Underdetermined Systems

In §3.4.8 we discussed how Gaussian elimination with either complete pivoting or rook pivoting can be used to solve a full-rank, underdetermined linear system

$$Ax = b, \qquad A \in \mathbb{R}^{m \times n},\ b \in \mathbb{R}^m. \tag{5.6.1}$$

Various orthogonal factorizations can also be used to solve this problem. Notice that (5.6.1) either has no solution or has an infinity of solutions. In the second case, it is important to distinguish between algorithms that find the minimum 2-norm solution and those that do not. The first algorithm we present is in the latter category.

Assume that A has full row rank and that we apply QR with column pivoting to obtain

$$Q^T A \Pi = [\, R_1 \mid R_2 \,]$$

where $R_1 \in \mathbb{R}^{m \times m}$ is upper triangular and $R_2 \in \mathbb{R}^{m \times (n-m)}$. Thus, $Ax = b$ transforms to

$$(Q^T A \Pi)(\Pi^T x) = [\, R_1 \mid R_2 \,] \begin{bmatrix} z_1 \\ z_2 \end{bmatrix} = Q^T b$$

where

$$\Pi^T x\ =\ \begin{bmatrix} z_1 \\ z_2 \end{bmatrix}$$

with $z_1 \in \mathbb{R}^m$ and $z_2 \in \mathbb{R}^{(n-m)}$. By virtue of the column pivoting, R_1 is nonsingular because we are assuming that A has full row rank. One solution to the problem is therefore obtained by setting $z_1 = R_1^{-1}Q^Tb$ and $z_2 = 0$.

Algorithm 5.6.1   Given $A \in \mathbb{R}^{m \times n}$ with $\mathsf{rank}(A) = m$ and $b \in \mathbb{R}^m$, the following algorithm finds an $x \in \mathbb{R}^n$ such that $Ax = b$.

> Compute QR-with-column-pivoting factorization: $Q^TA\Pi = R$.
>
> Solve $R(1{:}m, 1{:}m)z_1 = Q^Tb$.
>
> Set $x = \Pi \begin{bmatrix} z_1 \\ 0 \end{bmatrix}$.

This algorithm requires $2m^2n - m^3/3$ flops. The minimum norm solution is not guaranteed. (A different Π could render a smaller z_1.) However, if we compute the QR factorization

$$A^T = QR = Q \begin{bmatrix} R_1 \\ 0 \end{bmatrix}$$

with $R_1 \in \mathbb{R}^{m \times m}$, then $Ax = b$ becomes

$$(QR)^Tx = \begin{bmatrix} R_1^T & | & 0 \end{bmatrix} \begin{bmatrix} z_1 \\ z_2 \end{bmatrix} = b,$$

where

$$Q^Tx = \begin{bmatrix} z_1 \\ z_2 \end{bmatrix}, \qquad z_1 \in \mathbb{R}^m, \; z_2 \in \mathbb{R}^{n-m}.$$

In this case the minimum norm solution *does* follow by setting $z_2 = 0$.

Algorithm 5.6.2   Given $A \in \mathbb{R}^{m \times n}$ with $\mathsf{rank}(A) = m$ and $b \in \mathbb{R}^m$, the following algorithm finds the minimum 2-norm solution to $Ax = b$.

> Compute the QR factorization $A^T = QR$.
>
> Solve $R(1{:}m, 1{:}m)^Tz = b$.
>
> Set $x = Q(:, 1{:}m)z$.

This algorithm requires at most $2m^2n - 2m^3/3$ flops.

   The SVD can also be used to compute the minimum norm solution of an underdetermined $Ax = b$ problem. If

$$A = \sum_{i=1}^{r} \sigma_i u_i v_i^T, \qquad r = \mathsf{rank}(A)$$

is the SVD of A, then

$$x = \sum_{i=1}^{r} \frac{u_i^T b}{\sigma_i} v_i.$$

As in the least squares problem, the SVD approach is desirable if A is nearly rank deficient.

5.6.3 Perturbed Underdetermined Systems

We conclude this section with a perturbation result for full-rank underdetermined systems.

Theorem 5.6.1. *Suppose* $\mathsf{rank}(A) = m \le n$ *and that* $A \in \mathbb{R}^{m \times n}$, $\delta A \in \mathbb{R}^{m \times n}$, $0 \ne b \in \mathbb{R}^m$, *and* $\delta b \in \mathbb{R}^m$ *satisfy*

$$\epsilon = \max\{\epsilon_A, \epsilon_b\} < \sigma_m(A),$$

where $\epsilon_A = \|\delta A\|_2 / \|A\|_2$ *and* $\epsilon_b = \|\delta b\|_2 / \|b\|_2$. *If* x *and* $\hat{x}$ *are minimum norm solutions that satisfy*

$$Ax = b, \qquad\qquad (A + \delta A)\hat{x} = b + \delta b,$$

then

$$\frac{\|\hat{x} - x\|_2}{\|x\|_2} \le \kappa_2(A)\,(\epsilon_A \min\{2, n - m + 1\} + \epsilon_b) + O(\epsilon^2).$$

Proof. Let E and f be defined by $\delta A / \epsilon$ and $\delta b / \epsilon$. Note that $\mathsf{rank}(A + tE) = m$ for all $0 < t < \epsilon$ and that

$$x(t) = (A + tE)^T \left((A + tE)(A + tE)^T\right)^{-1} (b + tf)$$

satisfies $(A + tE)x(t) = b + tf$. By differentiating this expression with respect to t and setting $t = 0$ in the result we obtain

$$\dot{x}(0) = \left(I - A^T(AA^T)^{-1}A\right) E^T (AA^T)^{-1}b + A^T(AA^T)^{-1}(f - Ex). \qquad (5.6.2)$$

Because

$$\|x\|_2 = \|A^T(AA^T)^{-1}b\|_2 \ge \sigma_m(A)\|(AA^T)^{-1}b\|_2,$$

$$\|I - A^T(AA^T)^{-1}A\|_2 = \min(1, n - m),$$

and

$$\frac{\|f\|_2}{\|x\|_2} \le \frac{\|f\|_2\|A\|_2}{\|b\|_2},$$

we have

$$\frac{\|\hat{x} - x\|_2}{\|x\|_2} = \frac{x(\epsilon) - x(0)}{\|x(0)\|_2} = \epsilon\frac{\|\dot{x}(0)\|_2}{\|x\|_2} + O(\epsilon^2)$$

$$\le \epsilon \min(1, n - m)\left\{\frac{\|E\|_2}{\|A\|_2} + \frac{\|f\|_2}{\|b\|_2} + \frac{\|E\|_2}{\|A\|_2}\right\} \kappa_2(A) + O(\epsilon^2),$$

from which the theorem follows. $\square$

Note that there is no $\kappa_2(A)^2$ factor as in the case of overdetermined systems.

Problems

P5.6.1 Derive equation (5.6.2).

P5.6.2 Find the minimal norm solution to the system $Ax = b$ where $A = [\,1\ 2\ 3\,]$ and $b = 1$.

P5.6.3 Show how triangular system solving can be avoided when using the QR factorization to solve an underdetermined system.

P5.6.4 Suppose $b, x \in \mathbb{R}^n$ are given and consider the following problems:

(a) Find an unsymmetric Toeplitz matrix T so $Tx = b$.

(b) Find a symmetric Toeplitz matrix T so $Tx = b$.

(c) Find a circulant matrix C so $Cx = b$.

Pose each problem in the form $Ap = b$ where A is a matrix made up of entries from x and p is the vector of sought-after parameters.

Notes and References for §5.6

For an analysis of linear equation solving via QR, see:

N.J. Higham (1991). "Iterative Refinement Enhances the Stability of QR Factorization Methods for Solving Linear Equations," *BIT 31*, 447–468.

Interesting aspects concerning singular systems are discussed in:

T.F. Chan (1984). "Deflated Decomposition Solutions of Nearly Singular Systems," *SIAM J. Numer. Anal. 21*, 738–754.

Papers concerned with underdetermined systems include:

R.E. Cline and R.J. Plemmons (1976). "L_2-Solutions to Underdetermined Linear Systems," *SIAM Review 18*, 92–106.

M.G. Cox (1981). "The Least Squares Solution of Overdetermined Linear Equations having Band or Augmented Band Structure," *IMA J. Numer. Anal. 1*, 3–22.

M. Arioli and A. Laratta (1985). "Error Analysis of an Algorithm for Solving an Underdetermined System," *Numer. Math. 46*, 255–268.

J.W. Demmel and N.J. Higham (1993). "Improved Error Bounds for Underdetermined System Solvers," *SIAM J. Matrix Anal. Applic. 14*, 1–14.

S. Jokar and M.E. Pfetsch (2008). "Exact and Approximate Sparse Solutions of Underdetermined Linear Equations," *SIAM J. Sci. Comput. 31*, 23–44.

The central matrix problem in the emerging field of compressed sensing is to solve an underdetermined system $Ax = b$ such that the 1-norm of x is minimized, see:

E. Candes, J. Romberg, and T. Tao (2006). "Robust Uncertainty Principles: Exact Signal Reconstruction from Highly Incomplete Frequency Information," *IEEE Trans. Information Theory 52*, 489–509.

D. Donoho (2006). "Compressed Sensing," *IEEE Trans. Information Theory 52*, 1289–1306.

This strategy tends to produce a highly sparse solution vector x.

Chapter 6

Modified Least Squares Problems and Methods

6.1 Weighting and Regularization

6.2 Constrained Least Squares

6.3 Total Least Squares

6.4 Subspace Computations with the SVD

6.5 Updating Matrix Factorizations

In this chapter we discuss an assortment of least square problems that can be solved using QR and SVD. We also introduce a generalization of the SVD that can be used to simultaneously diagonalize a pair of matrices, a maneuver that is useful in certain applications.

The first three sections deal with variations of the ordinary least squares problem that we treated in Chapter 5. The unconstrained minimization of $\| Ax - b \|_2$ does not always make a great deal of sense. How do we balance the importance of each equation in $Ax = b$? How might we control the size of x if A is ill-conditioned? How might we minimize $\| Ax - b \|_2$ over a proper subspace of $\mathbb{R}^n$? What if there are errors in the "data matrix" A in addition to the usual errors in the "vector of observations" b?

In §6.4 we consider a number of multidimensional subspace computations including the problem of determining the principal angles between a pair of given subspaces. The SVD plays a prominent role.

The final section is concerned with the updating of matrix factorizations. In many applications, one is confronted with a succession of least squares (or linear equation) problems where the matrix associated with the current step is highly related to the matrix associated with the previous step. This opens the door to updating strategies that can reduce factorization overheads by an order of magnitude.

Reading Notes

Knowledge of Chapter 5 is assumed. The sections in this chapter are independent of each other except that §6.1 should be read before §6.2. Excellent global references include Björck (NMLS) and Lawson and Hansen (SLS).

6.1   Weighting and Regularization

We consider two basic modifications to the linear least squares problem. The first concerns how much each equation "counts" in the $\| Ax - b \|_2$ minimization. Some equations may be more important than others and there are ways to produce approximate minimzers that reflect this. Another situation arises when A is ill-conditioned. Instead of minimizing $\| Ax - b \|_2$ with a possibly wild, large norm x-vector, we settle for a predictor Ax in which x is "nice" according to some regularizing metric.

6.1.1   Row Weighting

In ordinary least squares, the minimization of $\| Ax - b \|_2$ amounts to minimizing the sum of the squared discrepancies in each equation:

$$\| Ax - b \|^2 \; = \; \sum_{i=1}^{m} \left(a_i^T x - b_i\right)^2 .$$

We assume that $A \in \mathbb{R}^{m \times n}$, $b \in \mathbb{R}^m$, and $a_i^T = A(i,:)$. In the *weighted least squares problem* the discrepancies are scaled and we solve

$$\min_{x \in \mathbb{R}^n} \| D(Ax - b) \|^2 \; = \; \min_{x \in \mathbb{R}^n} \sum_{i=1}^{m} d_i^2 \left(a_i^T x - b_i\right)^2 \qquad (6.1.1)$$

where $D = \mathrm{diag}(d_1, \ldots, d_m)$ is nonsingular. Note that if x_D minimizes this summation, then it minimizes $\| \widetilde{A}x - \tilde{b} \|_2$ where $\widetilde{A} = DA$ and $\tilde{b} = Db$. Although there can be numerical issues associated with disparate weight values, it is generally possible to solve the weighted least squares problem by applying any Chapter 5 method to the "tilde problem." For example, if A has full column rank and we apply the method of normal equations, then we are led to the following positive definite system:

$$(A^T D^2 A)x_D \; = \; A^T D^2 b. \qquad (6.1.2)$$

Subtracting the unweighted system $A^T A x_{LS} = A^T b$ we see that

$$x_D - x_{LS} \; = \; (A^T D^2 A)^{-1} A^T (D^2 - I)(b - Ax_{LS}). \qquad (6.1.3)$$

Note that weighting has less effect if b is almost in the range of A.

At the component level, increasing d_k relative to the other weights stresses the importance of the kth equation and the resulting residual $r = b - Ax_D$ tends to be smaller in that component. To make this precise, define

$$D(\delta) \; = \; \mathrm{diag}(d_1, \ldots, d_{k-1}, \, d_k \sqrt{1 + \delta} \,, d_{k+1}, \ldots, d_m)$$

where $\delta > -1$. Assume that $x(\delta)$ minimizes $\| D(\delta)(Ax - b) \|_2$ and set

$$r_k(\delta) \; = \; e_k^T (b - Ax(\delta)) \; = \; b_k - a_k^T (A^T D(\delta)^2 A)^{-1} A^T D(\delta)^2 b$$

where $e_k = I_m(:, k)$. We show that the penalty for disagreement between $a_k^T x$ and b_k increases with δ. Since

$$\frac{d}{d\delta} \left[D(\delta)^2 \right] \; = \; d_k^2 e_k e_k^T$$

and

$$\frac{d}{d\delta}\left[(A^T D(\delta)^2 A)^{-1}\right] = -(A^T D(\delta)^2 A)^{-1}(A^T(d_k^2 e_k e_k^T)A)(A^T D(\delta)^2 A)^{-1},$$

it can be shown that

$$\frac{d}{d\delta}r_k(\delta) = -d_k^2\left(a_k^T(A^T D(\delta)^2 A)^{-1}a_k\right)r_k(\delta). \tag{6.1.4}$$

Assuming that A has full rank, the matrix $(A^T D(\delta)A)^{-1}$ is positive definite and so

$$\frac{d}{d\delta}[r_k(\delta)^2] = 2\,r_k(\delta)\cdot\frac{d}{d\delta}r_k(\delta) = -2d_k^2\left(a_k^T(A^T D(\delta)^2 A)^{-1}a_k\right)r_k(\delta)^2 < 0.$$

It follows that $|r_k(\delta)|$ is a monotone decreasing function of δ. Of course, the change in r_k when all the weights are varied at the same time is much more complicated.

Before we move on to a more general type of row weighting, we mention that (6.1.1) can be framed as a symmetric indefinite linear system. In particular, if

$$\begin{bmatrix} D^{-2} & A \\ A^T & 0 \end{bmatrix}\begin{bmatrix} r \\ x \end{bmatrix} = \begin{bmatrix} b \\ 0 \end{bmatrix}, \tag{6.1.5}$$

then x minimizes (6.1.1). Compare with (5.3.20).

6.1.2 Generalized Least Squares

In statistical data-fitting applications, the weights in (6.1.1) are often chosen to increase the relative importance of accurate measurements. For example, suppose the vector of observations b has the form $b_{\text{true}} + \Delta$ where Δ_i is normally distributed with mean zero and standard deviation σ_i. If the errors are uncorrelated, then it makes statistical sense to minimize (6.1.1) with $d_i = 1/\sigma_i$.

In more general estimation problems, the vector b is related to x through the equation

$$b = Ax + w \tag{6.1.6}$$

where the *noise vector* w has zero mean and a symmetric positive definite *covariance matrix* $\sigma^2 W$. Assume that W is known and that $W = BB^T$ for some $B \in \mathbb{R}^{m\times m}$. The matrix B might be given or it might be W's Cholesky triangle. In order that all the equations in (6.1.6) contribute equally to the determination of x, statisticians frequently solve the LS problem

$$\min_{x\in\mathbb{R}^n} \| B^{-1}(Ax - b) \|_2 . \tag{6.1.7}$$

An obvious computational approach to this problem is to form $\tilde{A} = B^{-1}A$ and $\tilde{b} = B^{-1}b$ and then apply any of our previous techniques to minimize $\| \tilde{A}x - \tilde{b} \|_2$. Unfortunately, if B is ill-conditioned, then x will be poorly determined by such a procedure.

A more stable way of solving (6.1.7) using orthogonal transformations has been suggested by Paige (1979a, 1979b). It is based on the idea that (6.1.7) is equivalent to the *generalized least squares problem*,

$$\min_{b=Ax+Bv} v^T v . \tag{6.1.8}$$

Notice that this problem is defined even if A and B are rank deficient. Although in the Paige technique can be applied when this is the case, we shall describe it under the assumption that both these matrices have full rank.

The first step is to compute the QR factorization of A:

$$Q^T A = \begin{bmatrix} R_1 \\ 0 \end{bmatrix}, \qquad Q = [\, Q_1 \mid Q_2 \,] \cdot \atop{}^{n \quad m-n}$$

Next, an orthogonal matrix $Z \in \mathbb{R}^{m \times m}$ is determined such that

$$(Q_2^T B)Z = \underset{n \quad m-n}{[\, 0 \mid \ S \,]}, \qquad Z = \underset{n \quad m-n}{[\, Z_1 \mid \ Z_2 \,]}$$

where S is upper triangular. With the use of these orthogonal matrices, the constraint in (6.1.8) transforms to

$$\begin{bmatrix} Q_1^T b \\ Q_2^T b \end{bmatrix} = \begin{bmatrix} R_1 \\ 0 \end{bmatrix} x + \begin{bmatrix} Q_1^T B Z_1 & Q_1^T B Z_2 \\ 0 & S \end{bmatrix} \begin{bmatrix} Z_1^T v \\ Z_2^T v \end{bmatrix} .$$

The bottom half of this equation determines v while the top half prescribes x:

$$Su = Q_2^T b, \qquad v = Z_2 u, \tag{6.1.9}$$

$$R_1 x = Q_1^T b - (Q_1^T B Z_1 Z_1^T + Q_1^T B Z_2 Z_2^T)v = Q_1^T b - Q_1^T B Z_2 u. \tag{6.1.10}$$

The attractiveness of this method is that all potential ill-conditioning is concentrated in the triangular systems (6.1.9) and (6.1.10). Moreover, Paige (1979b) shows that the above procedure is numerically stable, something that is not true of any method that explicitly forms $B^{-1}A$.

6.1.3 A Note on Column Weighting

Suppose $G \in \mathbb{R}^{n \times n}$ is nonsingular and define the G-norm $\| \cdot \|_G$ on $\mathbb{R}^n$ by

$$\| z \|_G = \| Gz \|_2 .$$

If $A \in \mathbb{R}^{m \times n}$, $b \in \mathbb{R}^m$, and we compute the minimum 2-norm solution y_{LS} to

$$\min_{x \in \mathbb{R}^n} \| (AG^{-1})y - b \|_2 ,$$

then $x_G = G^{-1}y_{LS}$ is a minimizer of $\| Ax - b \|_2$. If $\mathsf{rank}(A) < n$, then within the set of minimizers, x_G has the smallest G-norm.

The choice of G is important. Sometimes its selection can be based upon a priori knowledge of the uncertainties in A. On other occasions, it may be desirable to normalize the columns of A by setting

$$G = G_0 \equiv \mathrm{diag}(\| A(:,1) \|_2, \ldots, \| A(:,n) \|_2).$$

Van der Sluis (1969) has shown that with this choice, $\kappa_2(AG^{-1})$ is approximately minimized. Since the computed accuracy of y_{LS} depends on $\kappa_2(AG^{-1})$, a case can be made for setting $G = G_0$.

We remark that column weighting affects singular values. Consequently, a scheme for determining numerical rank may not return the same estimate when applied to A and AG^{-1}. See Stewart (1984).

6.1.4 Ridge Regression

In the *ridge regression problem* we are given $A \in \mathbb{R}^{m \times n}$ and $b \in \mathbb{R}^m$ and proceed to solve

$$\min_{x} \left\| \begin{bmatrix} A \\ \sqrt{\lambda}I \end{bmatrix} x - \begin{bmatrix} b \\ 0 \end{bmatrix} \right\|_2^2 = \min_{x} \| Ax - b \|_2^2 + \lambda \| x \|_2^2 . \tag{6.1.11}$$

where the value of the *ridge parameter* λ is chosen to "shape" the solution $x = x(\lambda)$ in some meaningful way. Notice that the normal equation system for this problem is given by

$$(A^T A + \lambda I)x = A^T b. \tag{6.1.12}$$

It follows that if

$$A = U\Sigma V^T = \sum_{i=1}^{r} \sigma_i u_i v_i^T \tag{6.1.13}$$

is the SVD of A, then (6.1.12) converts to

$$(\Sigma^T \Sigma + \lambda I_n)(V^T x) = \Sigma^T U^T b$$

and so

$$x(\lambda) = \sum_{i=1}^{r} \frac{\sigma_i \, u_i^T b}{\sigma_i^2 + \lambda} v_i. \tag{6.1.14}$$

By inspection, it is clear that

$$\lim_{\lambda \to 0} x(\lambda) = x_{LS}$$

and $\| x(\lambda) \|_2$ is a monotone decreasing function of λ. These two facts show how an ill-conditioned least squares solution can be *regularized* by judiciously choosing λ. The idea is to get sufficiently close to x_{LS} subject to the constraint that the norm of the ridge regression minimzer $x(\lambda)$ is sufficiently modest. Regularization in this context is all about the intelligent balancing of these two tensions.

The ridge parameter can also be chosen with an eye toward balancing the "impact" of each equation in the overdetermined system $Ax = b$. We describe a λ-selection procedure due to Golub, Heath, and Wahba (1979). Set

$$D_k = I - e_k e_k^T = \text{diag}(1, \ldots, 1, 0, 1, \ldots, 1) \in \mathbb{R}^{m \times m}$$

and let $x_k(\lambda)$ solve

$$\min_{x \in \mathbb{R}^n} \| D_k(Ax - b) \|_2^2 + \lambda \| x \|_2^2 . \tag{6.1.15}$$

Thus, $x_k(\lambda)$ is the solution to the ridge regression problem with the kth row of A and kth component of b deleted, i.e., the kth equation in the overdetermined system $Ax = b$ is deleted. Now consider choosing λ so as to minimize the *cross-validation weighted square error* $C(\lambda)$ defined by

$$C(\lambda) = \frac{1}{m} \sum_{k=1}^{m} w_k (a_k^T x_k(\lambda) - b_k)^2 .$$

Here, $w_1, \ldots, w_m$ are nonnegative weights and a_k^T is the kth row of A. Noting that

$$\| Ax_k(\lambda) - b \|_2^2 = \| D_k(Ax_k(\lambda) - b) \|_2^2 + (a_k^T x_k(\lambda) - b_k)^2,$$

we see that $\left(a_k^T x_k(\lambda) - b_k\right)^2$ is the increase in the sum of squares that results when the kth row is "reinstated." Minimizing $C(\lambda)$ is tantamount to choosing λ such that the final model is not overly dependent on any one experiment.

A more rigorous analysis can make this statement precise and also suggest a method for minimizing $C(\lambda)$. Assuming that $\lambda > 0$, an algebraic manipulation shows that

$$x_k(\lambda) = x(\lambda) + \frac{a_k^T x(\lambda) - b_k}{1 - z_k^T a_k} z_k \tag{6.1.16}$$

where $z_k = (A^T A + \lambda I)^{-1} a_k$ and $x(\lambda) = (A^T A + \lambda I)^{-1} A^T b$. Applying $-a_k^T$ to (6.1.16) and then adding b_k to each side of the resulting equation gives

$$r_k = b_k - a_k^T x_k(\lambda) = \frac{e_k^T (I - A(A^T A + \lambda I)^{-1} A^T) b}{e_k^T (I - A(A^T A + \lambda I)^{-1} A^T) e_k} . \tag{6.1.17}$$

Noting that the residual $r = [r_1, \ldots, r_m]^T = b - Ax(\lambda)$ is given by the formula

$$r = [I - A(A^T A + \lambda I)^{-1} A^T] b,$$

we see that

$$C(\lambda) = \frac{1}{m} \sum_{k=1}^{m} w_k \left(\frac{r_k}{\partial r_k / \partial b_k} \right)^2 . \tag{6.1.18}$$

The quotient $r_k/(\partial r_k/\partial b_k)$ may be regarded as an inverse measure of the "impact" of the kth observation b_k on the model. If $\partial r_k/\partial b_k$ is small, then this says that the error in the model's prediction of b_k is somewhat independent of b_k. The tendency for this to be true is lessened by basing the model on the λ^* that minimizes $C(\lambda)$.

The actual determination of λ^* is simplified by computing the SVD of A. Using the SVD (6.1.13) and Equations (6.1.17) and (6.1.18), it can be shown that

$$C(\lambda) = \frac{1}{m} \sum_{k=1}^{m} w_k \left[\frac{\tilde{b}_k - \sum_{j=1}^{r} u_{kj} \tilde{b}_j \left(\frac{\sigma_j^2}{\sigma_j^2 + \lambda} \right)}{1 - \sum_{j=1}^{r} u_{kj}^2 \left(\frac{\sigma_j^2}{\sigma_j^2 + \lambda} \right)} \right]^2 \tag{6.1.19}$$

where $\tilde{b} = U^T b$. The minimization of this expression is discussed in Golub, Heath, and Wahba (1979).

6.1.5 Tikhonov Regularization

In the *Tikhonov regularization problem*, we are given $A \in \mathbb{R}^{m \times n}$, $B \in \mathbb{R}^{n \times n}$, and $b \in \mathbb{R}^m$ and solve

$$\min_{x} \left\| \begin{bmatrix} A \\ \sqrt{\lambda}B \end{bmatrix} x - \begin{bmatrix} b \\ 0 \end{bmatrix} \right\|_2^2 = \min_{x} \| Ax - b \|_2^2 + \lambda \| Bx \|_2^2. \qquad (6.1.20)$$

The normal equations for this problem have the form

$$(A^T A + \lambda B^T B)x = A^T b. \qquad (6.1.21)$$

This system is nonsingular if $\mathsf{null}(A) \cap \mathsf{null}(B) = \{0\}$. The matrix B can be chosen in several ways. For example, in certain data-fitting applications second derivative smoothness can be promoted by setting $B = \mathcal{T}_{DD}$, the second difference matrix defined in Equation 4.8.7.

To analyze how A and B interact in the Tikhonov problem, it would be handy to transform (6.1.21) into an equivalent diagonal problem. For the ridge regression problem ($B = I_n$) the SVD accomplishes this task. For the Tikhonov problem, we need a generalization of the SVD that simultaneously diagonalizes both A and B.

6.1.6 The Generalized Singular Value Decomposition

The generalized singular value decomposition (GSVD) set forth in Van Loan (1974) provides a useful way to simplify certain two-matrix problems such as the Tychanov regularization problem.

Theorem 6.1.1 (Generalized Singular Value Decomposition). *Assume that* $A \in \mathbb{R}^{m_1 \times n_1}$ *and* $B \in \mathbb{R}^{m_2 \times n_1}$ *with* $m_1 \geq n_1$ *and*

$$r = \mathsf{rank}\left(\begin{bmatrix} A \\ B \end{bmatrix} \right).$$

There exist orthogonal $U_1 \in \mathbb{R}^{m_1 \times m_1}$ *and* $U_2 \in \mathbb{R}^{m_2 \times m_2}$ *and invertible* $X \in \mathbb{R}^{n_1 \times n_1}$ *such that*

$$U_1^T A X = D_A = \begin{bmatrix} I & 0 & 0 \\ 0 & \mathsf{diag}(\alpha_{p+1}, \ldots, \alpha_r) & 0 \\ 0 & 0 & 0 \end{bmatrix} \begin{matrix} p \\ r-p \\ m_1-r \end{matrix}, \qquad (6.1.22)$$
$$\begin{matrix} p & r-p & n_1-r \end{matrix}$$

$$U_2^T B X = D_B = \begin{bmatrix} 0 & 0 & 0 \\ 0 & \mathsf{diag}(\beta_{p+1}, \ldots, \beta_r) & 0 \\ 0 & 0 & 0 \end{bmatrix} \begin{matrix} p \\ r-p \\ m_2-r \end{matrix}, \qquad (6.1.23)$$
$$\begin{matrix} p & r-p & n_1-r \end{matrix}$$

where $p = \max\{r - m_2, 0\}$.

Proof. The proof makes use of the SVD and the CS decomposition (Theorem 2.5.3). Let

$$
\begin{bmatrix} A \\ B \end{bmatrix} = \begin{bmatrix} Q_{11} & Q_{12} \\ Q_{21} & Q_{22} \end{bmatrix} \begin{bmatrix} \Sigma_r & 0 \\ 0 & 0 \end{bmatrix} Z^T \tag{6.1.24}
$$

be the SVD where $\Sigma_r \in \mathbb{R}^{r \times r}$ is nonsingular, $Q_{11} \in \mathbb{R}^{m_1 \times r}$, and $Q_{21} \in \mathbb{R}^{m_2 \times r}$. Using the CS decomposition, there exist orthogonal matrices U_1 (m_1-by-m_1), U_2 (m_2-by-m_2), and V_1 (r-by-r) such that

$$
\begin{bmatrix} U_1 & 0 \\ 0 & U_2 \end{bmatrix}^T \begin{bmatrix} Q_{11} \\ Q_{21} \end{bmatrix} V_1 = \begin{bmatrix} D_A(:,1{:}r) \\ D_B(:,1{:}r) \end{bmatrix} \tag{6.1.25}
$$

where D_A and D_B have the forms specified by (6.1.21) and (6.1.22). It follows from (6.1.24) and (6.1.25) that

$$
\begin{bmatrix} U_1 & 0 \\ 0 & U_2 \end{bmatrix}^T \begin{bmatrix} A \\ B \end{bmatrix} Z = \begin{bmatrix} D_A(:,1{:}r) & U_1 Q_{12} \\ D_B(:,1{:}r) & U_2 Q_{22} \end{bmatrix} \begin{bmatrix} V_1^T \Sigma_r & 0 \\ 0 & 0 \end{bmatrix}
$$

$$
= \begin{bmatrix} D_A(:,1{:}r) & 0 \\ D_B(:,1{:}r) & 0 \end{bmatrix} \begin{bmatrix} V_1^T \Sigma_r & 0 \\ 0 & I_{n_1-r} \end{bmatrix}
$$

$$
= \begin{bmatrix} D_A \\ D_B \end{bmatrix} \begin{bmatrix} V_1^T \Sigma_r & 0 \\ 0 & I_{n_1-r} \end{bmatrix}.
$$

By setting

$$
X = Z \begin{bmatrix} V_1^T \Sigma_r & 0 \\ 0 & I_{n_1-r} \end{bmatrix}^{-1}
$$

the proof is complete.   □

Note that if $B = I_{n_1}$ and we set $X = U_2$, then we obtain the SVD of A. The GSVD is related to the generalized eigenvalue problem

$$
A^T A x = \mu^2 B^T B x
$$

which is considered in §8.7.4. As with the SVD, algorithmic issues cannot be addressed until we develop procedures for the symmetric eigenvalue problem in Chapter 8.

To illustrate the insight that can be provided by the GSVD, we return to the Tikhonov regularization problem (6.1.20). If B is square and nonsingular, then the GSVD defined by (6.1.22) and (6.1.23) transforms the system (6.1.21) to

$$
(D_A^T D_A + \lambda D_B^T D_B) y = D_A^T \tilde{b}
$$

where $x = Xy$, $\tilde{b} = U_1^T b$, and

$$
(D_A^T D_A + \lambda D_B^T D_B) = \operatorname{diag}(\alpha_1^2 + \lambda \beta_1^2, \dots, \alpha_n^2 + \lambda \beta_n^2).
$$

Thus, if

$$X = [\, x_1 \mid \cdots \mid x_n \,]$$

is a column partitioning, then

$$x(\lambda) = \sum_{k=1}^{n} \left(\frac{\alpha_k \tilde{b}_k}{\alpha_k^2 + \lambda \beta_k^2} \right) x_k \tag{6.1.26}$$

solves (6.1.20). The "calming influence" of the regularization is revealed through this representation. Use of λ to manage "trouble" in the direction of x_k depends on the values of α_k and β_k.

Problems

P6.1.1 Verify (6.1.4).

P6.1.2 What is the inverse of the matrix in (6.1.5)?

P6.1.3 Show how the SVD can be used to solve the generalized LS problem (6.1.8) if the matrices A and B are rank deficient.

P6.1.4 Suppose A is the m-by-1 matrix of 1's and let $b \in \mathbb{R}^m$. Show that the cross-validation technique with unit weights prescribes an optimal λ given by

$$\lambda = \left(\left(\frac{\tilde{b}}{s} \right)^2 - \frac{1}{m} \right)^{-1}$$

where $\tilde{b} = (b_1 + \cdots + b_m)/m$ and

$$s = \sum_{i=1}^{m} (b_i - \tilde{b})^2 / (m-1).$$

P6.1.5 Using the GSVD, give bounds for $\| x(\lambda) - x(0) \|$ and $\| Ax(\lambda) - b \|_2^2 - \| Ax(0) - b \|_2^2$ where $x(\lambda)$ is defined by (6.1.26).

Notes and References for §6.1

Row and column weighting in the LS problem is discussed in Lawson and Hanson (SLS, pp. 180-88). Other analyses include:

A. van der Sluis (1969). "Condition Numbers and Equilibration of Matrices," *Numer. Math. 14*, 14–23.

G.W. Stewart (1984). "On the Asymptotic Behavior of Scaled Singular Value and QR Decompositions," *Math. Comput. 43*, 483–490.

A. Forsgren (1996). "On Linear Least-Squares Problems with Diagonally Dominant Weight Matrices," *SIAM J. Matrix Anal. Applic. 17*, 763–788.

P.D. Hough and S.A. Vavasis (1997). "Complete Orthogonal Decomposition for Weighted Least Squares," *SIAM J. Matrix Anal. Applic. 18*, 551–555.

J.K. Reid (2000). "Implicit Scaling of Linear Least Squares Problems," *BIT 40*, 146–157.

For a discussion of cross-validation issues, see:

G.H. Golub, M. Heath, and G. Wahba (1979). "Generalized Cross-Validation as a Method for Choosing a Good Ridge Parameter," *Technometrics 21*, 215–23.

L. Eldén (1985). "A Note on the Computation of the Generalized Cross-Validation Function for Ill-Conditioned Least Squares Problems," *BIT 24*, 467–472.

Early references concerned with the generalized singular value decomposition include:

C.F. Van Loan (1976). "Generalizing the Singular Value Decomposition," *SIAM J. Numer. Anal. 13*, 76–83.

C.C. Paige and M.A. Saunders (1981). "Towards A Generalized Singular Value Decomposition," *SIAM J. Numer. Anal. 18*, 398–405.

The theoretical and computational aspects of the generalized least squares problem appear in:

C.C. Paige (1979). "Fast Numerically Stable Computations for Generalized Linear Least Squares Problems," *SIAM J. Numer. Anal. 16*, 165–171.
C.C. Paige (1979b). "Computer Solution and Perturbation Analysis of Generalized Least Squares Problems," *Math. Comput. 33*, 171–84.
S. Kourouklis and C.C. Paige (1981). "A Constrained Least Squares Approach to the General Gauss-Markov Linear Model," *J. Amer. Stat. Assoc. 76*, 620–625.
C.C. Paige (1985). "The General Limit Model and the Generalized Singular Value Decomposition," *Lin. Alg. Applic. 70*, 269–284.

Generalized factorizations have an important bearing on generalized least squares problems, see:

C.C. Paige (1990). "Some Aspects of Generalized QR Factorization," in *Reliable Numerical Computations*, M. Cox and S. Hammarling (eds.), Clarendon Press, Oxford.
E. Anderson, Z. Bai, and J. Dongarra (1992). "Generalized QR Factorization and Its Applications," *Lin. Alg. Applic. 162/163/164*, 243–271.

The development of regularization techniques has a long history, see:

L. Eldén (1977). "Algorithms for the Regularization of Ill-Conditioned Least Squares Problems," *BIT 17*, 134–45.
D.P. O'Leary and J.A. Simmons (1981). "A Bidiagonalization-Regularization Procedure for Large Scale Discretizations of Ill-Posed Problems," *SIAM J. Sci. Stat. Comput. 2*, 474–489.
L. Eldén (1984). "An Algorithm for the Regularization of Ill-Conditioned, Banded Least Squares Problems," *SIAM J. Sci. Stat. Comput. 5*, 237–254.
P.C. Hansen (1990). "Relations Between SVD and GSVD of Discrete Regularization Problems in Standard and General Form," *Lin.Alg. Applic. 141*, 165–176.
P.C. Hansen (1995). "Test Matrices for Regularization Methods," *SIAM J. Sci. Comput. 16*, 506–512.
A. Neumaier (1998). "Solving Ill–Conditioned and Singular Linear Systems: A Tutorial on Regularization," *SIAM Review 40*, 636–666.
P.C. Hansen (1998). *Rank-Deficient and Discrete Ill-Posed Problems: Numerical Aspects of Linear Inversion*, SIAM Publications, Philadelphia, PA.
M.E. Gulliksson and P.-A. Wedin (2000). "The Use and Properties of Tikhonov Filter Matrices," *SIAM J. Matrix Anal. Applic. 22*, 276–281.
M.E. Gulliksson, P.-A. Wedin, and Y. Wei (2000). "Perturbation Identities for Regularized Tikhonov Inverses and Weighted Pseudoinverses," *BIT 40*, 513–523.
T. Kitagawa, S. Nakata, and Y. Hosoda (2001). "Regularization Using QR Factorization and the Estimation of the Optimal Parameter," *BIT 41*, 1049–1058.
M.E. Kilmer and D.P. O'Leary. (2001). "Choosing Regularization Parameters in Iterative Methods for Ill-Posed Problems," *SIAM J. Matrix Anal. Applic. 22*, 1204–1221.
A. N. Malyshev (2003). "A Unified Theory of Conditioning for Linear Least Squares and Tikhonov Regularization Solutions," *SIAM J. Matrix Anal. Applic. 24*, 1186–1196.
M. Hanke (2006). "A Note on Tikhonov Regularization of Large Linear Problems," *BIT 43*, 449–451.
P.C. Hansen, J.G. Nagy, and D.P. OLeary (2006). *Deblurring Images: Matrices, Spectra, and Filtering*, SIAM Publications, Philadelphia, PA.
M.E. Kilmer, P.C. Hansen, and M.I. Español (2007). "A Projection-Based Approach to General-Form Tikhonov Regularization," *SIAM J. Sci. Comput. 29*, 315–330.
T. Elfving and I. Skoglund (2009). "A Direct Method for a Regularized Least-Squares Problem," *Num. Lin. Alg. Applic. 16*, 649–675.
I. Hnětynková and M. Plešinger (2009). "The Regularizing Effect of the Golub-Kahan Iterative Bidiagonalization and revealing the Noise level in Data," *BIT 49*, 669–696.
P.C. Hansen (2010). *Discrete Inverse Problems: Insight and Algorithms*, SIAM Publications, Philadelphia, PA.

6.2 Constrained Least Squares

In the least squares setting it is sometimes natural to minimize $\| Ax - b \|_2$ over a proper subset of $\mathbb{R}^n$. For example, we may wish to predict b as best we can with Ax subject to the constraint that x is a unit vector. Or perhaps the solution defines a fitting function $f(t)$ which is to have prescribed values at certain points. This can lead to an equality-constrained least squares problem. In this section we show how these problems can be solved using the QR factorization, the SVD, and the GSVD.

6.2.1 Least Squares Minimization Over a Sphere

Given $A \in \mathbb{R}^{m \times n}$, $b \in \mathbb{R}^m$, and a positive $\alpha \in \mathbb{R}$, we consider the problem

$$\min_{\|x\|_2 \leq \alpha} \| Ax - b \|_2. \tag{6.2.1}$$

This is an example of the *LSQI* (least squares with quadratic inequality constraint) problem. This problem arises in nonlinear optimization and other application areas. As we are soon to observe, the LSQI problem is related to the ridge regression problem discussed in §6.1.4.

Suppose

$$A = U \Sigma V^T = \sum_{i=1}^{r} \sigma_i u_i v_i^T \tag{6.2.2}$$

is the SVD of A which we assume to have rank r. If the unconstrained minimum norm solution

$$x_{LS} = \sum_{i=1}^{r} \frac{u_i^T b}{\sigma_i} v_i$$

satisfies $\| x_{LS} \|_2 \leq \alpha$, then it obviously solves (6.2.1). Otherwise,

$$\| x_{LS} \|_2^2 = \sum_{i=1}^{r} \left(\frac{u_i^T b}{\sigma_i} \right)^2 > \alpha^2, \tag{6.2.3}$$

and it follows that the solution to (6.2.1) is on the boundary of the constraint sphere. Thus, we can approach this constrained optimization problem using the method of Lagrange multipliers. Define the parameterized objective function ϕ by

$$\phi(x, \lambda) = \frac{1}{2} \| Ax - b \|_2^2 + \frac{\lambda}{2} \left(\| x \|_2^2 - \alpha^2 \right)$$

and equate its gradient to zero. This gives a shifted normal equation system:

$$(A^T A + \lambda I) \cdot x(\lambda) = A^T b.$$

The goal is to choose λ so that $\| x(\lambda) \|_2 = \alpha$. Using the SVD (6.2.2), this leads to the problem of finding a zero of the function

$$f(\lambda) = \| x(\lambda) \|_2^2 - \alpha^2 = \sum_{k=1}^{n} \left(\frac{\sigma_k u_k^T b}{\sigma_k^2 + \lambda} \right)^2 - \alpha^2.$$

This is an example of a *secular equation* problem. From (6.2.3), $f(0) > 0$. Since $f'(\lambda) < 0$ for $\lambda \geq 0$, it follows that f has a unique positive root λ_+. It can be shown that

$$\rho(\lambda) = \| Ax(\lambda) - b \|_2^2 = \| Ax_{LS} - b \|_2^2 + \sum_{i=1}^{r} \left(\frac{\lambda \, u_i^T b}{\sigma_i^2 + \lambda} \right)^2. \tag{6.2.4}$$

It follows that $x(\lambda_+)$ solves (6.2.1).

Algorithm 6.2.1 Given $A \in \mathbb{R}^{m \times n}$ with $m \geq n$, $b \in \mathbb{R}^m$, and $\alpha > 0$, the following algorithm computes a vector $x \in \mathbb{R}^n$ such that $\| Ax - b \|_2$ is minimum subject to the constraint that $\| x \|_2 \leq \alpha$.

> Compute the SVD $A = U\Sigma V^T$, save $V = [v_1 | \cdots | v_n]$, form $\tilde{b} = U^T b$,
> and determine $r = \mathsf{rank}(A)$.

if $\displaystyle\sum_{i=1}^{r} \left(\frac{\tilde{b}_i}{\sigma_i} \right)^2 > \alpha^2$

> Find $\lambda_+ > 0$ such that $\displaystyle\sum_{i=1}^{r} \left(\frac{\sigma_i \tilde{b}_i}{\sigma_i^2 + \lambda_+} \right)^2 = \alpha^2.$
>
> $x = \displaystyle\sum_{i=1}^{r} \left(\frac{\sigma_i \tilde{b}_i}{\sigma_i^2 + \lambda_+} \right) v_i$

else

> $x = \displaystyle\sum_{i=1}^{r} \left(\frac{\tilde{b}_i}{\sigma_i} \right) v_i$

end

The SVD is the dominant computation in this algorithm.

6.2.2   More General Quadratic Constraints

A more general version of (6.2.1) results if we minimize $\| Ax - b \|_2$ over an arbitrary hyperellipsoid:

$$\text{minimize } \| Ax - b \|_2 \qquad \text{subject to } \| Bx - d \|_2 \leq \alpha. \tag{6.2.5}$$

Here we are assuming that $A \in \mathbb{R}^{m_1 \times n_1}$, $b \in \mathbb{R}^{m_1}$, $B \in \mathbb{R}^{m_2 \times n_1}$, $d \in \mathbb{R}^{m_2}$, and $\alpha \geq 0$. Just as the SVD turns (6.2.1) into an equivalent diagonal problem, we can use the GSVD to transform (6.2.5) into a diagonal problem. In particular, if the GSVD of A and B is given by (6.1.22) and (6.2.23), then (6.2.5) is equivalent to

$$\text{minimize } \| D_A y - \tilde{b} \|_2 \qquad \text{subject to } \| D_B y - \tilde{d} \|_2 \leq \alpha \tag{6.2.6}$$

where

$$\tilde{b} = U_1^T b, \qquad \tilde{d} = U_2^T d, \qquad y = X^{-1} x.$$

The simple form of the objective function and the constraint equation facilitate the analysis. For example, if $\text{rank}(B) = m_2 < n_1$, then

$$\| D_A y - \tilde{b} \|_2^2 = \sum_{i=1}^{n_1} (\alpha_i y_i - \tilde{b}_i)^2 + \sum_{i=n_1+1}^{m_1} \tilde{b}_i^2 \qquad (6.2.7)$$

and

$$\| D_B y - \tilde{d} \|_2^2 = \sum_{i=1}^{m_2} (\beta_i y_i - \tilde{d}_i)^2 + \sum_{i=m_2+1}^{n_1} \tilde{d}_i^2 \leq \alpha^2. \qquad (6.2.8)$$

A Lagrange multiplier argument can be used to determine the solution to this transformed problem (if it exists).

6.2.3 Least Squares With Equality Constraints

We consider next the constrained least squares problem

$$\min_{Bx=d} \| Ax - b \|_2 \qquad (6.2.9)$$

where $A \in \mathbb{R}^{m_1 \times n_1}$ with $m_1 \geq n_1$, $B \in \mathbb{R}^{m_2 \times n_1}$ with $m_2 < n_1$, $b \in \mathbb{R}^{m_1}$, and $d \in \mathbb{R}^{m_2}$. We refer to this as the *LSE problem* (least squares with equality constraints). By setting $\alpha = 0$ in (6.2.5) we see that the LSE problem is a special case of the LSQI problem. However, it is simpler to approach the LSE problem directly rather than through Lagrange multipliers.

For clarity, we assume that both A and B have full rank. Let

$$Q^T B^T = \begin{bmatrix} R \\ 0 \end{bmatrix} \begin{matrix} n_1 \\ n_1 - m_2 \end{matrix}$$

be the QR factorization of B^T and set

$$AQ = [\, A_1 \,|\, A_2 \,] , \qquad Q^T x = \begin{bmatrix} y \\ z \end{bmatrix} \begin{matrix} m_2 \\ n_1 - m_2 \end{matrix}.$$
$$ \begin{matrix} m_2 \quad n_1 - m_2 \end{matrix}$$

It is clear that with these transformations (6.2.9) becomes

$$\min_{R^T y = d} \| A_1 y + A_2 z - b \|_2.$$

Thus, y is determined from the constraint equation $R^T y = d$ and the vector z is obtained by solving the unconstrained LS problem

$$\min_{z \in \mathbb{R}^{n_1 - m_2}} \| A_2 z - (b - A_1 y) \|_2.$$

Combining the above, we see that the following vector solves the LSE problem:

$$x = Q \begin{bmatrix} y \\ z \end{bmatrix}.$$

Algorithm 6.2.2 Suppose $A \in \mathbb{R}^{m_1 \times n_1}$, $B \in \mathbb{R}^{m_2 \times n_1}$, $b \in \mathbb{R}^{m_1}$, and $d \in \mathbb{R}^{m_2}$. If $\mathsf{rank}(A) = n_1$ and $\mathsf{rank}(B) = m_2 < n_1$, then the following algorithm minimizes $\| Ax - b \|_2$ subject to the constraint $Bx = d$.

Compute the QR factorization $B^T = QR$.

Solve $R(1{:}m_2, 1{:}m_2)^T \cdot y = d$ for y.

$A = AQ$

Find z so $\| A(:, m_2 + 1{:}n_1)z - (b - A(:, 1{:}m_2)\cdot y) \|_2$ is minimized.

$x = Q(:, 1{:}m_2)\cdot y + Q(:, m_2 + 1{:}n_1)\cdot z$.

Note that this approach to the LSE problem involves two QR factorizations and a matrix multiplication. If A and/or B are rank deficient, then it is possible to devise a similar solution procedure using the SVD instead of QR. Note that there may not be a solution if $\mathsf{rank}(B) < m_2$. Also, if $\mathsf{null}(A) \cap \mathsf{null}(B) \neq \{0\}$ and $d \in \mathsf{ran}(B)$, then the LSE solution is not unique.

6.2.4   LSE Solution Using the Augmented System

The LSE problem can also be approached through the method of Lagrange multipliers. Define the augmented objective function

$$f(x, \lambda) \; = \; \frac{1}{2} \| Ax - b \|_2^2 \, + \, \lambda^T (d - Bx), \qquad \lambda \in \mathbb{R}^{m_2},$$

and set to zero its gradient with respect to x:

$$A^T Ax - A^T b - B^T \lambda \; = \; 0.$$

Combining this with the equations $r = b - Ax$ and $Bx = d$ we obtain the symmetric indefinite linear system

$$\begin{bmatrix} 0 & A^T & B^T \\ A & I & 0 \\ B & 0 & 0 \end{bmatrix} \begin{bmatrix} x \\ r \\ \lambda \end{bmatrix} = \begin{bmatrix} 0 \\ b \\ d \end{bmatrix}. \tag{6.2.10}$$

This system is nonsingular if both A and B have full rank. The augmented system presents a solution framework for the sparse LSE problem.

6.2.5   LSE Solution Using the GSVD

Using the GSVD given by (6.1.22) and (6.1.23), we see that the LSE problem transforms to

$$\min_{D_B y = \tilde{d}} \| D_A y - \tilde{b} \|_2 \tag{6.2.11}$$

where $\tilde{b} = U_1^T b$, $\tilde{d} = U_2^T d$, and $y = X^{-1} x$. It follows that if $\mathsf{null}(A) \cap \mathsf{null}(B) = \{0\}$ and $X = [\, x_1 \,|\, \cdots \,|\, x_n \,]$, then

$$x \; = \; \sum_{i=1}^{m_2} \left(\frac{\tilde{d}_i}{\beta_i} \right) x_i \, + \, \sum_{i=m_2+1}^{n_1} \left(\frac{\tilde{b}_i}{\alpha_i} \right) x_i \tag{6.2.12}$$

solves the LSE problem.

6.2.6 LSE Solution Using Weights

An interesting way to obtain an approximate LSE solution is to solve the unconstrained LS problem

$$
\min_x \left\| \begin{bmatrix} A \\ \sqrt{\lambda}B \end{bmatrix} x - \begin{bmatrix} b \\ \sqrt{\lambda}d \end{bmatrix} \right\|_2 \tag{6.2.13}
$$

for large λ. (Compare with the Tychanov regularization problem (6.1.21).) Since

$$
\left\| \begin{bmatrix} A \\ \sqrt{\lambda}B \end{bmatrix} x - \begin{bmatrix} b \\ \sqrt{\lambda}d \end{bmatrix} \right\|_2^2 = \| Ax - b \|_2^2 + \lambda \| Bx - d \|^2,
$$

we see that there is a penalty for discrepancies among the constraint equations. To quantify this, assume that both A and B have full rank and substitute the GSVD defined by (6.1.22) and (6.1.23) into the normal equation system

$$
(A^T A + \lambda B^T B)x = A^T b + \lambda B^T d.
$$

This shows that the solution $x(\lambda)$ is given by $x(\lambda) = Xy(\lambda)$ where $y(\lambda)$ solves

$$
(D_A^T D_A + \lambda D_B^T D_B)y = D_A^T \tilde{b} + \lambda D_B^T \tilde{d}
$$

with $\tilde{b} = U_1^T b$ and $\tilde{d} = U_2^T d$. It follows that

$$
x(\lambda) = \sum_{i=1}^{m_2} \left(\frac{\alpha_i \tilde{b}_i + \lambda \beta_i \tilde{d}_i}{\alpha_i^2 + \lambda \beta_i^2} \right) x_i + \sum_{i=m_2+1}^{n_1} \left(\frac{\tilde{b}_i}{\alpha_i} \right) x_i
$$

and so from (6.2.13) we have

$$
x(\lambda) - x = \sum_{i=1}^{p} \frac{\alpha_i}{\beta_i} \left(\frac{\beta_i u_i^T b - \alpha_i v_i^T d}{\alpha_i^2 + \lambda^2 \beta_i^2} \right) x_i. \tag{6.2.14}
$$

This shows that $x(\lambda) \to x$ as $\lambda \to \infty$. The appeal of this approach to the LSE problem is that it can be implemented with unconstrained LS problem software. However, for large values of λ numerical problems can arise and it is necessary to take precautions. See Powell and Reid (1968) and Van Loan (1982).

Problems

P6.2.1 Is the solution to (6.2.1) always unique?

P6.2.2 Let $p_0(x), \ldots, p_n(x)$ be given polynomials and $(x_0, y_0), \ldots, (x_m, y_m)$ be a given set of coordinate pairs with $x_i \in [a, b]$. It is desired to find a polynomial $p(x) = \sum_{k=0}^{n} \alpha_k p_k(x)$ such that

$$
\phi(\alpha) = \sum_{i=0}^{m} (p(x_i) - y_i)^2
$$

is minimized subject to the constraint that

$$\int_a^b [p''(x)]^2 dx \approx h \sum_{i=0}^N \left(\frac{p(z_{i-1}) - 2p(z_i) + p(z_{i+1})}{h^2} \right)^2 \leq \alpha^2$$

where $z_i = a + ih$ and $b = a + Nh$. Show that this leads to an LSQI problem of the form (6.2.5) with $d = 0$.

P6.2.3 Suppose $Y = [\, y_1 \,|\, \cdots \,|\, y_k \,] \in \mathbb{R}^{m \times k}$ has the property that

$$Y^T Y = \text{diag}(d_1^2, \ldots, d_k^2), \qquad d_1 \geq d_2 \geq \cdots \geq d_k > 0.$$

Show that if $Y = QR$ is the QR factorization of Y, then R is diagonal with $|r_{ii}| = d_i$.

P6.2.4 (a) Show that if $(A^T A + \lambda I)x = A^T b$, $\lambda > 0$, and $\|\, x \,\|_2 = \alpha$, then $z = (Ax - b)/\lambda$ solves the *dual equations* $(AA^T + \lambda I)z = -b$ with $\|\, A^T z \,\|_2 = \alpha$. (b) Show that if $(AA^T + \lambda I)z = -b$, $\|\, A^T z \,\|_2 = \alpha$, then $x = -A^T z$ satisfies $(A^T A + \lambda I)x = A^T b$, $\|\, x \,\|_2 = \alpha$.

P6.2.5 Show how to compute y (if it exists) so that both (6.2.7) and (6.2.8) are satisfied.

P6.2.6 Develop an SVD version of Algorithm 6.2.2 that can handle the situation when A and/or B are rank deficient.

P6.2.7 Suppose

$$A = \left[\begin{array}{c} A_1 \\ A_2 \end{array} \right]$$

where $A_1 \in \mathbb{R}^{n \times n}$ is nonsingular and $A_2 \in \mathbb{R}^{(m-n) \times n}$. Show that

$$\sigma_{\min}(A) \geq \sqrt{1 + \sigma_{\min}(A_2 A_1^{-1})^2}\, \sigma_{\min}(A_1)\,.$$

P6.2.8 Suppose $p \geq m \geq n$ and that $A \in \mathbb{R}^{m \times n}$ and $B \in \mathbb{R}^{m \times p}$ Show how to compute orthogonal $Q \in \mathbb{R}^{m \times m}$ and orthogonal $V \in \mathbb{R}^{n \times n}$ so that

$$Q^T A = \left[\begin{array}{c} R \\ 0 \end{array} \right], \qquad Q^T BV = [\, 0 \,|\, S \,]$$

where $R \in \mathbb{R}^{n \times n}$ and $S \in \mathbb{R}^{m \times m}$ are upper triangular.

P6.2.9 Suppose $r \in \mathbb{R}^m$, $y \in \mathbb{R}^n$, and $\delta > 0$. Show how to solve the problem

$$\min_{E \in \mathbb{R}^{m \times n}\,,\, \|E\|_F \leq \delta} \|Ey - r\|_2$$

Repeat with "min" replaced by "max."

P6.2.10 Show how the constrained least squares problem

$$\min_{Bx = d} \|\, Ax - b \,\|_2 \qquad A \in \mathbb{R}^{m \times n}, \ B \in \mathbb{R}^{p \times n}, \ \text{rank}(B) = p$$

can be reduced to an unconstrained least square problem by performing p steps of Gaussian elimination on the matrix

$$\left[\begin{array}{c} B \\ A \end{array} \right] = \left[\begin{array}{cc} B_1 & B_2 \\ A_1 & A_2 \end{array} \right], \qquad B_1 \in \mathbb{R}^{p \times p}, \ \text{rank}(B_1) = p.$$

Explain. Hint: The Schur complement is of interest.

Notes and References for §6.2

The LSQI problem is discussed in:

G.E. Forsythe and G.H. Golub (1965). "On the Stationary Values of a Second-Degree Polynomial on the Unit Sphere," *SIAM J. App. Math. 14*, 1050–1068.

L. Eldén (1980). "Perturbation Theory for the Least Squares Problem with Linear Equality Constraints," *SIAM J. Numer. Anal. 17*, 338–350.

W. Gander (1981). "Least Squares with a Quadratic Constraint," *Numer. Math. 36*, 291–307.

L. Eldén (1983). "A Weighted Pseudoinverse, Generalized Singular Values, and Constrained Least Squares Problems," *BIT 22* , 487–502.

G.W. Stewart (1984). "On the Asymptotic Behavior of Scaled Singular Value and QR Decompositions," *Math. Comput. 43*, 483–490.

G.H. Golub and U. von Matt (1991). "Quadratically Constrained Least Squares and Quadratic Problems," *Numer. Math. 59*, 561–580.

T.F. Chan, J.A. Olkin, and D. Cooley (1992). "Solving Quadratically Constrained Least Squares Using Black Box Solvers," *BIT 32*, 481–495.

Secular equation root-finding comes up in many numerical linear algebra settings. For an algorithmic overview, see:

O.E. Livne and A. Brandt (2002). "N Roots of the Secular Equation in O(N) Operations," *SIAM J. Matrix Anal. Applic. 24*, 439–453.

For a discussion of the augmented systems approach to least squares problems, see:

Å. Björck (1992). "Pivoting and Stability in the Augmented System Method," *Proceedings of the 14th Dundee Conference*, D.F. Griffiths and G.A. Watson (eds.), Longman Scientific and Technical, Essex, U.K.

Å. Björck and C.C. Paige (1994). "Solution of Augmented Linear Systems Using Orthogonal Factorizations," *BIT 34*, 1–24.

References that are concerned with the method of weighting for the LSE problem include:

M.J.D. Powell and J.K. Reid (1968). "On Applying Householder's Method to Linear Least Squares Problems," *Proc. IFIP Congress*, pp. 122–26.

C. Van Loan (1985). "On the Method of Weighting for Equality Constrained Least Squares Problems," *SIAM J. Numer. Anal. 22*, 851–864.

J.L. Barlow and S.L. Handy (1988). "The Direct Solution of Weighted and Equality Constrained Least-Squares Problems," *SIAM J. Sci. Stat. Comput. 9*, 704–716.

J.L. Barlow, N.K. Nichols, and R.J. Plemmons (1988). "Iterative Methods for Equality Constrained Least Squares Problems," SIAM J. Sci. Stat. Comput. 9, 892–906.

J.L. Barlow (1988). "Error Analysis and Implementation Aspects of Deferred Correction for Equality Constrained Least-Squares Problems," *SIAM J. Numer. Anal. 25*, 1340–1358.

J.L. Barlow and U.B. Vemulapati (1992). "A Note on Deferred Correction for Equality Constrained Least Squares Problems," *SIAM J. Numer. Anal. 29*, 249–256.

M. Gulliksson and P.-Å. Wedin (1992). "Modifying the QR-Decomposition to Constrained and Weighted Linear Least Squares," *SIAM J. Matrix Anal. Applic. 13*, 1298–1313.

M. Gulliksson (1994). "Iterative Refinement for Constrained and Weighted Linear Least Squares," *BIT 34*, 239–253.

G. W. Stewart (1997). "On the Weighting Method for Least Squares Problems with Linear Equality Constraints," *BIT 37*, 961–967.

For the analysis of the LSE problem and related methods, see:

M. Wei (1992). "Perturbation Theory for the Rank-Deficient Equality Constrained Least Squares Problem," *SIAM J. Numer. Anal. 29*, 1462–1481.

M. Wei (1992). "Algebraic Properties of the Rank-Deficient Equality-Constrained and Weighted Least Squares Problems," *Lin. Alg. Applic. 161*, 27–44.

M. Gulliksson (1995). "Backward Error Analysis for the Constrained and Weighted Linear Least Squares Problem When Using the Weighted QR Factorization," *SIAM J. Matrix. Anal. Applic. 13*, 675–687.

M. Gulliksson (1995). "Backward Error Analysis for the Constrained and Weighted Linear Least Squares Problem When Using the Weighted *QR* Factorization," *SIAM J. Matrix Anal. Applic. 16*, 675–687.

J. Ding and W. Hang (1998). "New Perturbation Results for Equality-Constrained Least Squares Problems," *Lin. Alg. Applic. 272*, 181–192.

A.J. Cox and N.J. Higham (1999). "Accuracy and Stability of the Null Space Method for Solving the Equality Constrained Least Squares Problem," *BIT 39*, 34–50.

A.J. Cox and N.J. Higham (1999). "Row-Wise Backward Stable Elimination Methods for the Equality Constrained Least Squares Problem," *SIAM J. Matrix Anal. Applic. 21*, 313–326.

A.J. Cox and Nicholas J. Higham (1999). "Backward Error Bounds for Constrained Least Squares Problems," *BIT 39*, 210–227.

M. Gulliksson and P-A. Wedin (2000). "Perturbation Theory for Generalized and Constrained Linear Least Squares," *Num. Lin. Alg.* 7, 181–195.

M. Wei and A.R. De Pierro (2000). "Upper Perturbation Bounds of Weighted Projections, Weighted and Constrained Least Squares Problems," *SIAM J. Matrix Anal. Applic.* 21, 931–951.

E.Y. Bobrovnikova and S.A. Vavasis (2001). "Accurate Solution of Weighted Least Squares by Iterative Methods *SIAM. J. Matrix Anal. Applic.* 22, 1153–1174.

M. Gulliksson, X-Q.Jin, and Y-M. Wei (2002). "Perturbation Bounds for Constrained and Weighted Least Squares Problems," *Lin. Alg. Applic. 349,* 221–232.

6.3   Total Least Squares

The problem of minimizing $\| Ax - b \|_2$ where $A \in \mathbb{R}^{m \times n}$ and $b \in \mathbb{R}^m$ can be recast as follows:

$$\min_{b+r \,\in\, \mathsf{ran}(A)} \| r \|_2 \; . \tag{6.3.1}$$

In this problem, there is a tacit assumption that the errors are confined to the *vector of observations* b. If error is also present in the *data matrix* A, then it may be more natural to consider the problem

$$\min_{b+r \,\in\, \mathsf{ran}(A+E)} \| [\, E \mid r \,] \|_F \; . \tag{6.3.2}$$

This problem, discussed by Golub and Van Loan (1980), is referred to as the *total least squares* (TLS) problem. If a minimizing $[\, E_0 \mid r_0 \,]$ can be found for (6.3.2), then any x satisfying $(A + E_0)x = b + r_0$ is called a TLS solution. However, it should be realized that (6.3.2) may fail to have a solution altogether. For example, if

$$A = \begin{bmatrix} 1 & 0 \\ 0 & 0 \\ 0 & 0 \end{bmatrix}, \quad b = \begin{bmatrix} 1 \\ 1 \\ 1 \end{bmatrix}, \quad E_\epsilon = \begin{bmatrix} 0 & 0 \\ 0 & \epsilon \\ 0 & \epsilon \end{bmatrix},$$

then for all $\epsilon > 0$, $b \in \mathsf{ran}(A + E_\epsilon)$. However, there is no smallest value of $\| [\, E , r \,] \|_F$ for which $b + r \in \mathsf{ran}(A + E)$.

A generalization of (6.3.2) results if we allow multiple right-hand sides and use a weighted Frobenius norm. In particular, if $B \in \mathbb{R}^{m \times k}$ and the matrices

$$D = \mathrm{diag}(d_1, \ldots, d_m),$$

$$T = \mathrm{diag}(t_1, \ldots, t_{n+k})$$

are nonsingular, then we are led to an optimization problem of the form

$$\min_{B+R \,\in\, \mathsf{ran}(A+E)} \| D\,[\, E \mid R\,]\, T \|_F \tag{6.3.3}$$

where $E \in \mathbb{R}^{m \times n}$ and $R \in \mathbb{R}^{m \times k}$. If $[\, E_0 \mid R_0 \,]$ solves (6.3.3), then any $X \in \mathbb{R}^{n \times k}$ that satisfies

$$(A + E_0)X \; = \; (B + R_0)$$

is said to be a TLS solution to (6.3.3).

In this section we discuss some of the mathematical properties of the total least squares problem and show how it can be solved using the SVD. For a more detailed introduction, see Van Huffel and Vanderwalle (1991).

6.3.1 Mathematical Background

The following theorem gives conditions for the uniqueness and existence of a TLS solution to the multiple-right-hand-side problem.

Theorem 6.3.1. *Suppose $A \in \mathbb{R}^{m \times n}$ and $B \in \mathbb{R}^{m \times k}$ and that $D = \mathrm{diag}(d_1, \ldots, d_m)$ and $T = \mathrm{diag}(t_1, \ldots, t_{n+k})$ are nonsingular. Assume $m \geq n + k$ and let the SVD of*

$$C = D[\, A \mid B \,]T = [\, C_1 \mid C_2 \,]$$
$$ n \qquad k$$

be specified by $U^T C V = \mathrm{diag}(\sigma_1, \ldots, \sigma_{n+k}) = \Sigma$ where U, V, and Σ are partitioned as follows:

$$U = [\, U_1 \mid U_2 \,] \quad , \qquad V = \begin{bmatrix} V_{11} & V_{12} \\ V_{21} & V_{22} \end{bmatrix} \begin{matrix} n \\ k \end{matrix} \quad , \qquad \Sigma = \begin{bmatrix} \Sigma_1 & 0 \\ 0 & \Sigma_2 \end{bmatrix} \begin{matrix} n \\ k \end{matrix} \quad .$$
$$ n \quad k \phantom{\,] \quad , \qquad V = \begin{bmatrix} V_{11} \end{bmatrix}} n \quad k \phantom{\qquad \Sigma = \begin{bmatrix}} n \quad k$$

If $\sigma_n(C_1) > \sigma_{n+1}(C)$, then the matrix $[\, E_0 \mid R_0 \,]$ defined by

$$D[\, E_0 \mid R_0 \,]T = -U_2 \Sigma_2 [\, V_{12}^T \mid V_{22}^T \,] \tag{6.3.4}$$

solves (6.3.3). If $T_1 = \mathrm{diag}(t_1, \ldots, t_n)$ and $T_2 = \mathrm{diag}(t_{n+1}, \ldots, t_{n+k})$, then the matrix

$$X_{TLS} = -T_1 V_{12} V_{22}^{-1} T_2^{-1}$$

exists and is the unique TLS solution to $(A + E_0)X = B + R_0$.

Proof. We first establish two results that follow from the assumption $\sigma_n(C_1) > \sigma_{n+1}(C)$. From the equation $CV = U\Sigma$ we have

$$C_1 V_{12} + C_2 V_{22} = U_2 \Sigma_2.$$

We wish to show that V_{22} is nonsingular. Suppose $V_{22}x = 0$ for some unit 2-norm x. It follows from

$$V_{12}^T V_{12} + V_{22}^T V_{22} = I$$

that $\| V_{12}x \|_2 = 1$. But then

$$\sigma_{n+1}(C) \geq \| U_2 \Sigma_2 x \|_2 = \| C_1 V_{12}x \|_2 \geq \sigma_n(C_1),$$

a contradiction. Thus, the submatrix V_{22} is nonsingular. The second fact concerns the strict separation of $\sigma_n(C)$ and $\sigma_{n+1}(C)$. From Corollary 2.4.5, we have $\sigma_n(C) \geq \sigma_n(C_1)$ and so

$$\sigma_n(C) \geq \sigma_n(C_1) > \sigma_{n+1}(C).$$

We are now set to prove the theorem. If $\mathsf{ran}(B + R) \subset \mathsf{ran}(A + E)$, then there is an X (n-by-k) so $(A + E)X = B + R$, i.e.,

$$\{\, D[\, A \mid B \,]T + D[\, E \mid R \,]T \,\} T^{-1} \begin{bmatrix} X \\ -I_k \end{bmatrix} = 0. \tag{6.3.5}$$

Thus, the rank of the matrix in curly brackets is at most equal to n. By following the argument in the proof of Theorem 2.4.8, it can be shown that

$$\| D[E|R]T \|_F^2 \geq \sum_{i=n+1}^{n+k} \sigma_i(C)^2.$$

Moreover, the lower bound is realized by setting $[E|R] = [E_0|R_0]$. Using the inequality $\sigma_n(C) > \sigma_{n+1}(C)$, we may infer that $[E_0|R_0]$ is the unique minimizer.

To identify the TLS solution X_{TLS}, we observe that the nullspace of

$$\{D[A|B]T + D[E_0|R_0]T\} = U_1 \Sigma_1 [V_{11}^T | V_{21}^T]$$

is the range of $\begin{bmatrix} V_{12} \\ V_{22} \end{bmatrix}$. Thus, from (6.3.5)

$$T^{-1} \begin{bmatrix} X \\ -I_k \end{bmatrix} = \begin{bmatrix} V_{12} \\ V_{22} \end{bmatrix} S$$

for some k-by-k matrix S. From the equations $T_1^{-1}X = V_{12}S$ and $-T_2^{-1} = V_{22}S$ we see that $S = -V_{22}^{-1}T_2^{-1}$ and so

$$X = T_1 V_{12} S = -T_1 V_{12} V_{22}^{-1} T_2^{-1} = X_{TLS}. \quad \square$$

Note from the thin CS decomposition (Theorem 2.5.2) that

$$\| X \|_\tau^2 = \| V_{12} V_{22}^{-1} \|_2^2 = \frac{1 - \sigma_k(V_{22})^2}{\sigma_k(V_{22})^2}$$

where we define the "τ-norm" on $\mathbb{R}^{n \times k}$ by $\| Z \|_\tau = \| T_1^{-1} Z T_2 \|_2$.

If $\sigma_n(C_1) = \sigma_{n+1}(C)$, then the solution procedure implicit in the above proof is problematic. The TLS problem may have no solution or an infinite number of solutions. See §6.3.4 for suggestions as to how one might proceed.

6.3.2   Solving the Single Right Hand Side Case

We show how to maximize $\sigma_k(V_{22})$ in the important $k = 1$ case. Suppose the singular values of C satisfy $\sigma_{n-p} > \sigma_{n-p+1} = \cdots = \sigma_{n+1}$ and let $V = [v_1 | \cdots | v_{n+1}]$ be a column partitioning of V. If $\widetilde{Q}$ is a Householder matrix such that

$$V(:, n+1-p{:}n+1)\widetilde{Q} = \begin{bmatrix} W & z \\ 0 & \alpha \end{bmatrix} \begin{matrix} n \\ 1 \end{matrix},$$
$$\quad\quad\quad\quad\quad\quad\quad p \quad 1$$

then the last column of this matrix has the largest $(n+1)$st component of all the vectors in $\mathsf{span}\{v_{n+1-p}, \ldots, v_{n+1}\}$. If $\alpha = 0$, then the TLS problem has no solution. Otherwise

$$x_{TLS} = -T_1 z/(t_{n+1}\alpha).$$

Moreover,

$$
\begin{bmatrix} I_{n-1} & 0 \\ 0 & \tilde{Q} \end{bmatrix} U^T (D[\, A \mid b\,]T) V \begin{bmatrix} I_{n-p} & 0 \\ 0 & \tilde{Q} \end{bmatrix} = \Sigma
$$

and so

$$
D[\, E_0 \mid r_0\,]T = -D[\, A \mid b\,]T \begin{bmatrix} z \\ \alpha \end{bmatrix} [\, z^T \mid \alpha\,].
$$

Overall, we have the following algorithm:

Algorithm 6.3.1 Given $A \in \mathbb{R}^{m \times n}$ $(m > n)$, $b \in \mathbb{R}^m$, nonsingular $D = \text{diag}(d_1, \ldots, d_m)$, and nonsingular $T = \text{diag}(t_1, \ldots, t_{n+1})$, the following algorithm computes (if possible) a vector $x_{\text{TLS}} \in \mathbb{R}^n$ such that $(A + E_0)x_{\text{TLS}} = (b + r_0)$ and $\| D[\, E_0 \mid r_0\,]T \|_F$ is minimal.

Compute the SVD $U^T (D[\, A \mid b\,]T)V = \text{diag}(\sigma_1, \ldots, \sigma_{n+1})$ and save V.

Determine p such that $\sigma_1 \geq \cdots \geq \sigma_{n-p} > \sigma_{n-p+1} = \cdots = \sigma_{n+1}$.

Compute a Householder P such that if $\tilde{V} = VP$, then $\tilde{V}(n+1, n-p+1{:}n) = 0$.

if $\tilde{v}_{n+1,n+1} \neq 0$

    for $i = 1{:}n$

        $x_i = -t_i \tilde{v}_{i,n+1}/(t_{n+1}\tilde{v}_{n+1,n+1})$

    end

    $x_{\text{TLS}} = x$

end

This algorithm requires about $2mn^2 + 12n^3$ flops and most of these are associated with the SVD computation.

6.3.3 A Geometric Interpretation

It can be shown that the TLS solution x_{TLS} minimizes

$$
\psi(x) = \sum_{i=1}^{m} d_i^2 \left(\frac{|a_i^T x - b_i|^2}{x^T T_1^{-2} x + t_{n+1}^{-2}} \right) \tag{6.3.6}
$$

where a_i^T is the ith row of A and b_i is the ith component of b. A geometrical interpretation of the TLS problem is made possible by this observation. Indeed,

$$
\delta_i = \frac{|a_i^T x - b_i|^2}{x^T T_1^{-2} x + t_{n+1}^{-2}}
$$

is the square of the distance from

$$
\begin{bmatrix} a_i \\ b_i \end{bmatrix} \in \mathbb{R}^{n+1}
$$

to the nearest point in the subspace

$$
P_x = \left\{ \begin{bmatrix} a \\ b \end{bmatrix} : a \in \mathbb{R}^n, \ b \in \mathbb{R}, \ b = x^T a \right\}
$$

where the distance in $\mathbb{R}^{n+1}$ is measured by the norm $\| z \| = \| T z \|_2$. The TLS problem is essentially the problem of *orthogonal regression*, a topic with a long history. See Pearson (1901) and Madansky (1959).

6.3.4 Variations of the Basic TLS Problem

We briefly mention some modified TLS problems that address situations when additional constraints are imposed on the optimizing E and R and the associated TLS solution.

In the *restricted TLS problem*, we are given $A \in \mathbb{R}^{m \times n}$, $B \in \mathbb{R}^{m \times k}$, $P_1 \in \mathbb{R}^{m \times q}$, and $P_2 \in \mathbb{R}^{n+k \times r}$, and solve

$$\min_{B+R \subset \mathrm{ran}(A+E)} \| P_1^T [\, E \mid R \,] P_2 \|_F . \tag{6.3.7}$$

We assume that $q \le m$ and $r \le n + k$. An important application arises if some of the columns of A are error-free. For example, if the first s columns of A are error-free, then it makes sense to force the optimizing E to satisfy $E(:, 1{:}s) = 0$. This goal is achieved by setting $P_1 = I_m$ and $P_2 = I_{m+k}(:, s + 1{:}n + k)$ in the restricted TLS problem.

If a particular TLS problem has no solution, then it is referred to as a *nongeneric TLS problem*. By adding a constraint it is possible to produce a meaningful solution. For example, let $U^T [\, A \mid b \,] V = \Sigma$ be the SVD and let p be the largest index so $V(n + 1, p) \ne 0$. It can be shown that the problem

$$\min_{\substack{(A+E)x=b+r \\ [\, E \mid r \,] V(:,p+1:n+1)=0}} \| [\, E \mid r \,] \|_F \tag{6.3.8}$$

has a solution $[\, E_0 \mid r_0 \,]$ and the nongeneric TLS solution satisfies $(A + E_0)x + b + r_0$. See Van Huffel (1992).

In the *regularized TLS problem* additional constraints are imposed to ensure that the solution x is properly constrained/smoothed:

$$\min_{\substack{(A+E)x=b+r \\ \|Lx\|_2 \le \delta}} \| [\, E \mid r \,] \|_F . \tag{6.3.9}$$

The matrix $L \in \mathbb{R}^{n \times n}$ could be the identity or a discretized second-derivative operator. The regularized TLS problem leads to a Lagrange multiplier system of the form

$$(A^T A + \lambda_1 I + \lambda_2 L^T L)x = A^T b.$$

See Golub, Hansen, and O'Leary (1999) for more details. Another regularization approach involves setting the small singular values of $[A \mid b]$ to zero. This is the *truncated TLS problem* discussed in Fierro, Golub, Hansen, and O'Leary (1997).

Problems

P6.3.1 Consider the TLS problem (6.3.2) with nonsingular D and T. (a) Show that if $\mathrm{rank}(A) < n$, then (6.3.2) has a solution if and only if $b \in \mathrm{ran}(A)$. (b) Show that if $\mathrm{rank}(A) = n$, then (6.3.2) has no solution if $A^T D^2 b = 0$ and $|t_{n+1}|\,\|\,Db\,\|_2 \geq \sigma_n(DAT_1)$ where $T_1 = \mathrm{diag}(t_1, \ldots, t_n)$.

P6.3.2 Show that if $C = D[\,A\mid b\,]T = [\,A_1\mid d\,]$ and $\sigma_n(C) > \sigma_{n+1}(C)$, then x_{TLS} satisfies

$$(A_1^T A_1 - \sigma_{n+1}(C)^2 I)x_{\mathrm{TLS}} = A_1^T d.$$

Appreciate this as a "negatively shifted" system of normal equations.

P6.3.3 Show how to solve (6.3.2) with the added constraint that the first p columns of the minimizing E are zero. Hint: Compute the QR factorization of $A(:, 1{:}p)$.

P6.3.4 Show how to solve (6.3.3) given that D and T are general nonsingular matrices.

P6.3.5 Verify Equation (6.3.6).

P6.3.6 If $A \in \mathbb{R}^{m \times n}$ has full column rank and $B \in \mathbb{R}^{p \times n}$ has full row rank, show how to minimize

$$f(x) = \frac{\|\,Ax - b\,\|_2^2}{1 + x^T x}$$

subject to the constraint that $Bx = 0$.

P6.3.7 In the *data least squares problem*, we are given $A \in \mathbb{R}^{m \times n}$ and $b \in \mathbb{R}^m$ and minimize $\|\,E\,\|_F$ subject to the constraint that $b \in \mathrm{ran}(A + E)$. Show how to solve this problem. See Paige and Strakoš (2002b).

Notes and References for §6.3

Much of this section is based on:

G.H. Golub and C.F. Van Loan (1980). "An Analysis of the Total Least Squares Problem," *SIAM J. Numer. Anal. 17*, 883–93.

The idea of using the SVD to solve the TLS problem is set forth in:

G.H. Golub and C. Reinsch (1970). "Singular Value Decomposition and Least Squares Solutions," *Numer. Math. 14*, 403–420.
G.H. Golub (1973). "Some Modified Matrix Eigenvalue Problems," *SIAM Review 15*, 318–334.

The most comprehensive treatment of the TLS problem is:

S. Van Huffel and J. Vandewalle (1991). *The Total Least Squares Problem: Computational Aspects and Analysis*, SIAM Publications, Philadelphia, PA.

There are two excellent conference proceedings that cover just about everything you would like to know about TLS algorithms, generalizations, applications, and the associated statistical foundations:

S. Van Huffel (ed.) (1996). *Recent Advances in Total Least Squares Techniques and Errors in Variables Modeling*, SIAM Publications, Philadelphia, PA.
S. Van Huffel and P. Lemmerling (eds.) (2002) *Total Least Squares and Errors-in-Variables Modeling: Analysis, Algorithms, and Applications*, Kluwer Academic, Dordrecht, The Netherlands.

TLS is but one approach to the errors-in-variables problem, a subject that has a long and important history in statistics:

K. Pearson (1901). "On Lines and Planes of Closest Fit to Points in Space," *Phil. Mag. 2*, 559–72.
A. Wald (1940). "The Fitting of Straight Lines if Both Variables are Subject to Error," *Annals of Mathematical Statistics 11*, 284–300.
G.W. Stewart (2002). "Errors in Variables for Numerical Analysts," in *Recent Advances in Total Least Squares Techniques and Errors-in-Variables Modelling*, S. Van Huffel (ed.), SIAM Publications, Philadelphia PA, pp. 3–10,

In certain settings there are more economical ways to solve the TLS problem than the Golub-Kahan-Reinsch SVD algorithm:

S. Van Huffel and H. Zha (1993). "An Efficient Total Least Squares Algorithm Based On a Rank-Revealing Two-Sided Orthogonal Decomposition," *Numer. Alg. 4*, 101–133.
Å. Björck, P. Heggernes, and P. Matstoms (2000). "Methods for Large Scale Total Least Squares Problems," *SIAM J. Matrix Anal. Applic. 22*, 413–429.

H. Guo and R.A. Renaut (2005). "Parallel Variable Distribution for Total Least Squares," *Num. Lin. Alg. 12*, 859–876.

The condition of the TLS problem is analyzed in:

M. Baboulin and S. Gratton (2011). "A Contribution to the Conditioning of the Total Least-Squares Problem," *SIAM J. Matrix Anal. Applic. 32*, 685–699.

Efforts to connect the LS and TLS paradigms have lead to nice treatments that unify the presentation of both approaches:

B.D. Rao (1997). "Unified Treatment of LS, TLS, and Truncated SVD Methods Using a Weighted TLS Framework," in *Recent Advances in Total Least Squares Techniques and Errors-in-Variables Modelling*, S. Van Huffel (ed.), SIAM Publications, Philadelphia, PA., pp. 11–20.

C.C. Paige and Z. Strakoš (2002a). "Bounds for the Least Squares Distance Using Scaled Total Least Squares," *Numer. Math. 91*, 93–115.

C.C. Paige and Z. Strakoš (2002b). "Scaled Total Least Squares Fundamentals," *Numer. Math. 91*, 117–146.

X.-W. Chang, G.H. Golub, and C.C. Paige (2008). "Towards a Backward Perturbation Analysis for Data Least Squares Problems," *SIAM J. Matrix Anal. Applic. 30*, 1281–1301.

X.-W. Chang and D. Titley-Peloquin (2009). "Backward Perturbation Analysis for Scaled Total Least-Squares," *Num. Lin. Alg. Applic. 16*, 627–648.

For a discussion of the situation when there is no TLS solution or when there are multiple solutions, see:

S. Van Huffel and J. Vandewalle (1988). "Analysis and Solution of the Nongeneric Total Least Squares Problem," *SIAM J. Matrix Anal. Appl. 9*, 360–372.

S. Van Huffel (1992). "On the Significance of Nongeneric Total Least Squares Problems," *SIAM J. Matrix Anal. Appl. 13*, 20–35.

M. Wei (1992). "The Analysis for the Total Least Squares Problem with More than One Solution," *SIAM J. Matrix Anal. Appl. 13*, 746–763.

For a treatment of the multiple right hand side TLS problem, see:

I. Hnětynková, M. Plešinger, D.M. Sima, Z. Strakoš, and S. Van Huffel (2011). "The Total Least Squares Problem in AX ≈ B: A New Classification with the Relationship to the Classical Works," *SIAM J. Matrix Anal. Applic. 32*, 748–770.

If some of the columns of A are known exactly then it is sensible to force the TLS perturbation matrix E to be zero in the same columns. Aspects of this constrained TLS problem are discussed in:

J.W. Demmel (1987). "The Smallest Perturbation of a Submatrix which Lowers the Rank and Constrained Total Least Squares Problems," *SIAM J. Numer. Anal.* 24, 199–206.

S. Van Huffel and J. Vandewalle (1988). "The Partial Total Least Squares Algorithm," *J. Comput. App. Math. 21*, 333–342.

S. Van Huffel and J. Vandewalle (1989). "Analysis and Properties of the Generalized Total Least Squares Problem $AX \approx B$ When Some or All Columns in A are Subject to Error," *SIAM J. Matrix Anal. Applic. 10*, 294–315.

S. Van Huffel and H. Zha (1991). "The Restricted Total Least Squares Problem: Formulation, Algorithm, and Properties," *SIAM J. Matrix Anal. Applic. 12*, 292–309.

C.C. Paige and M. Wei (1993). "Analysis of the Generalized Total Least Squares Problem $AX = B$ when Some of the Columns are Free of Error," *Numer. Math. 65*, 177–202.

Another type of constraint that can be imposed in the TLS setting is to insist that the optimum perturbation of A have the same structure as A. For examples and related strategies, see:

J. Kamm and J.G. Nagy (1998). "A Total Least Squares Method for Toeplitz Systems of Equations," *BIT 38*, 560–582.

P. Lemmerling, S. Van Huffel, and B. De Moor (2002). "The Structured Total Least Squares Approach for Nonlinearly Structured Matrices," *Num. Lin. Alg. 9*, 321–332.

P. Lemmerling, N. Mastronardi, and S. Van Huffel (2003). "Efficient Implementation of a Structured Total Least Squares Based Speech Compression Method," *Lin. Alg. Applic. 366*, 295–315.

N. Mastronardi, P. Lemmerling, and S. Van Huffel (2004). "Fast Regularized Structured Total Least Squares Algorithm for Solving the Basic Deconvolution Problem," *Num. Lin. Alg. 12*, 201–209.

I. Markovsky, S. Van Huffel, and R. Pintelon (2005). "Block-Toeplitz/Hankel Structured Total Least Squares," *SIAM J. Matrix Anal. Applic. 26*, 1083–1099.

A. Beck and A. Ben-Tal (2005). "A Global Solution for the Structured Total Least Squares Problem with Block Circulant Matrices," *SIAM J. Matrix Anal. Applic. 27*, 238–255.

H. Fu, M.K. Ng, and J.L. Barlow (2006). "Structured Total Least Squares for Color Image Restoration," *SIAM J. Sci. Comput. 28*, 1100–1119.

As in the least squares problem, there are techniques that can be used to regularlize an otherwise "wild" TLS solution:

R.D. Fierro and J.R. Bunch (1994). "Collinearity and Total Least Squares," *SIAM J. Matrix Anal. Applic. 15*, 1167–1181.

R.D. Fierro, G.H. Golub, P.C. Hansen and D.P. O'Leary (1997). "Regularization by Truncated Total Least Squares," *SIAM J. Sci. Comput. 18*, 1223–1241.

G.H. Golub, P.C. Hansen, and D.P. O'Leary (1999). "Tikhonov Regularization and Total Least Squares," *SIAM J. Matrix Anal. Applic. 21*, 185–194.

R.A. Renaut and H. Guo (2004). "Efficient Algorithms for Solution of Regularized Total Least Squares," *SIAM J. Matrix Anal. Applic. 26*, 457–476.

D.M. Sima, S. Van Huffel, and G.H. Golub (2004). "Regularized Total Least Squares Based on Quadratic Eigenvalue Problem Solvers," *BIT 44*, 793–812.

N. Mastronardi, P. Lemmerling, and S. Van Huffel (2005). "Fast Regularized Structured Total Least Squares Algorithm for Solving the Basic Deconvolution Problem," *Num. Lin. Alg. Applic. 12*, 201–209.

S. Lu, S.V. Pereverzev, and U. Tautenhahn (2009). "Regularized Total Least Squares: Computational Aspects and Error Bounds," *SIAM J. Matrix Anal. Applic. 31*, 918–941.

Finally, we mention an interesting TLS problem where the solution is subject to a unitary constraint:

K.S. Arun (1992). "A Unitarily Constrained Total Least Squares Problem in Signal Processing," *SIAM J. Matrix Anal. Applic. 13*, 729–745.

6.4   Subspace Computations with the SVD

It is sometimes necessary to investigate the relationship between two given subspaces. How close are they? Do they intersect? Can one be "rotated" into the other? And so on. In this section we show how questions like these can be answered using the singular value decomposition.

6.4.1   Rotation of Subspaces

Suppose $A \in \mathbb{R}^{m \times p}$ is a data matrix obtained by performing a certain set of experiments. If the same set of experiments is performed again, then a different data matrix, $B \in \mathbb{R}^{m \times p}$, is obtained. In the *orthogonal Procrustes problem* the possibility that B can be rotated into A is explored by solving the following problem:

$$\text{minimize } \| A - BQ \|_F, \qquad \text{subject to } Q^T Q = I_p. \qquad (6.4.1)$$

We show that optimizing Q can be specified in terms of the SVD of $B^T A$. The *matrix trace* is critical to the derivation. The trace of a matrix is the sum of its diagonal entries:

$$\text{tr}(C) = \sum_{i=1}^{n} c_{ii}, \qquad C \in \mathbb{R}^{n \times n}.$$

It is easy to show that if C_1 and C_2 have the same row and column dimension, then

$$\text{tr}(C_1^T C_2) = \text{tr}(C_2^T C_1). \qquad (6.4.2)$$

Returning to the Procrustes problem (6.4.1), if $Q \in \mathbb{R}^{p \times p}$ is orthogonal, then

$$
\| A - BQ \|_F^2 = \sum_{k=1}^{p} \| A(:,k) - B \cdot Q(:,k) \|_2^2
$$

$$
= \sum_{k=1}^{p} \| A(:,k) \|_2^2 + \| BQ(:,k) \|_2^2 - 2Q(:,k)^T B^T A(:,k)
$$

$$
= \| A \|_F^2 + \| BQ \|_F^2 - 2 \sum_{k=1}^{p} \left[Q^T (B^T A) \right]_{kk}
$$

$$
= \| A \|_F^2 + \| B \|_F^2 - 2\mathsf{tr}(Q^T (B^T A)).
$$

Thus, (6.4.1) is equivalent to the problem

$$
\max_{Q^T Q = I_p} \mathsf{tr}(Q^T B^T A) .
$$

If $U^T (B^T A)V = \Sigma = \mathsf{diag}(\sigma_1, \ldots, \sigma_p)$ is the SVD of $B^T A$ and we define the orthogonal matrix Z by $Z = V^T Q^T U$, then by using (6.4.2) we have

$$
\mathsf{tr}(Q^T B^T A) = \mathsf{tr}(Q^T U \Sigma V^T) = \mathsf{tr}(Z \Sigma) = \sum_{i=1}^{p} z_{ii} \sigma_i \leq \sum_{i=1}^{p} \sigma_i .
$$

The upper bound is clearly attained by setting $Z = I_p$, i.e., $Q = UV^T$.

Algorithm 6.4.1 Given A and B in $\mathbb{R}^{m \times p}$, the following algorithm finds an orthogonal $Q \in \mathbb{R}^{p \times p}$ such that $\| A - BQ \|_F$ is minimum.

$C = B^T A$

Compute the SVD $U^T CV = \Sigma$ and save U and V.

$Q = UV^T$

We mention that if $B = I_p$, then the problem (6.4.1) is related to the *polar decomposition*. This decomposition states that any square matrix A has a factorization of the form $A = QP$ where Q is orthogonal and P is symmetric and positive semidefinite. Note that if $A = U \Sigma V^T$ is the SVD of A, then $A = (UV^T)(V \Sigma V^T)$ is its polar decomposition. For further discussion, see §9.4.3.

6.4.2   Intersection of Nullspaces

Let $A \in \mathbb{R}^{m \times n}$ and $B \in \mathbb{R}^{p \times n}$ be given, and consider the problem of finding an orthonormal basis for $\mathsf{null}(A) \cap \mathsf{null}(B)$. One approach is to compute the nullspace of the matrix

$$
C = \begin{bmatrix} A \\ B \end{bmatrix}
$$

since this is just what we want: $Cx = 0 \Leftrightarrow x \in \mathsf{null}(A) \cap \mathsf{null}(B)$. However, a more economical procedure results if we exploit the following theorem.

Theorem 6.4.1. *Suppose $A \in \mathbb{R}^{m \times n}$ and let $\{z_1, \ldots, z_t\}$ be an orthonormal basis for* null(A)*. Define $Z = [\, z_1 \mid \cdots \mid z_t \,]$ and let $\{w_1, \ldots, w_q\}$ be an orthonormal basis for* null(BZ) *where $B \in \mathbb{R}^{p \times n}$. If $W = [\, w_1 \mid \cdots \mid w_q \,]$, then the columns of ZW form an orthonormal basis for* null$(A) \cap$ null(B)*.*

Proof. Since $AZ = 0$ and $(BZ)W = 0$, we clearly have ran$(ZW) \subset$ null$(A) \cap$ null(B). Now suppose x is in both null(A) and null(B). It follows that $x = Za$ for some $0 \neq a \in \mathbb{R}^t$. But since $0 = Bx = BZa$, we must have $a = Wb$ for some $b \in \mathbb{R}^q$. Thus, $x = ZWb \in$ ran(ZW).   □

If the SVD is used to compute the orthonormal bases in this theorem, then we obtain the following procedure:

Algorithm 6.4.2 Given $A \in \mathbb{R}^{m \times n}$ and $B \in \mathbb{R}^{p \times n}$, the following algorithm computes and integer s and a matrix $Y = [\, y_1 \mid \cdots \mid y_s \,]$ having orthonormal columns which span null$(A) \cap$ null(B). If the intersection is trivial, then $s = 0$.

> Compute the SVD $U_A^T A V_A = \text{diag}(\sigma_i)$, save V_A, and set $r = $ rank(A).
> **if** $r < n$
>> $C = BV_A(:, r+1{:}n)$
>> Compute the SVD $U_C^T C V_C = \text{diag}(\gamma_i)$, save V_C, and set $q = $ rank(C).
>> **if** $q < n - r$
>>> $s = n - r - q$
>>> $Y = V_A(:, r+1{:}n)V_C(:, q+1{:}n-r)$
>> **else**
>>> $s = 0$
>> **end**
> **else**
>> $s = 0$
> **end**

The practical implementation of this algorithm requires an ability to reason about numerical rank. See §5.4.1.

6.4.3   Angles Between Subspaces

Let F and G be subspaces in $\mathbb{R}^m$ whose dimensions satisfy

$$p = \dim(F) \geq \dim(G) = q \geq 1.$$

The *principal angles* $\{\theta_i\}_{i=1}^q$ between these two subspaces and the associated *principal vectors* $\{f_i, g_i\}_{i=1}^q$ are defined recursively by

$$\cos(\theta_k) = f_k^T g_k = \max_{\substack{f \in F,\, \|f\|_2 = 1 \\ f^T[f_1, \ldots, f_{k-1}] = 0}} \max_{\substack{g \in G,\, \|g\|_2 = 1 \\ g^T[g_1, \ldots, g_{k-1}] = 0}} f^T g \, . \tag{6.4.3}$$

Note that the principal angles satisfy $0 \leq \theta_1 \leq \cdots \leq \theta_q \leq \pi/2$.. The problem of computing principal angles and vectors is oftentimes referred to as the *canonical correlation problem.*

Typically, the subspaces F and G are matrix ranges, e.g.,

$$F = \mathsf{ran}(A), \qquad A \in \mathbb{R}^{n \times p},$$
$$G = \mathsf{ran}(B), \qquad B \in \mathbb{R}^{n \times q}.$$

The principal vectors and angles can be computed using the QR factorization and the SVD. Let $A = Q_A R_A$ and $B = Q_B R_B$ be thin QR factorizations and assume that

$$Q_A^T Q_B = Y \Sigma Z^T = \sum_{i=1}^{q} \sigma_i y_i z_i^T$$

is the SVD of $Q_A^T Q_B \in \mathbb{R}^{p \times q}$. Since $\| Q_A^T Q_B \|_2 \leq 1$, all the singular values are between 0 and 1 and we may write $\sigma_i = \cos(\theta_i)$, $i = 1{:}q$. Let

$$Q_A Y = [\, f_1 \,|\, \cdots \,|\, f_p \,], \tag{6.4.4}$$
$$Q_B Z = [\, g_1 \,|\, \cdots \,|\, g_q \,] \tag{6.4.5}$$

be column partitionings of the matrices $Q_A Y \in \mathbb{R}^{n \times p}$ and $Q_B Z \in \mathbb{R}^{n \times q}$. These matrices have orthonormal columns. If $f \in F$ and $g \in G$ are unit vectors, then there exist unit vectors $u \in \mathbb{R}^p$ and $v \in \mathbb{R}^q$ so that $f = Q_A u$ and $g = Q_B v$. Thus,

$$f^T g = (Q_A u)^T (Q_B v) = u^T (Q_A^T Q_B) v = u^T (Y \Sigma Z^T) v$$
$$= (Y^T u)^T \Sigma (Z^T v) = \sum_{i=1}^{q} \sigma_i (y_i^T u)(z_i^T v). \tag{6.4.6}$$

This expression attains its maximal value of $\sigma_1 = \cos(\theta_1)$ by setting $u = y_1$ and $v = z_1$. It follows that $f = Q_A y_1 = f_1$ and $v = Q_B z_1 = g_1$.

Now assume that $k > 1$ and that the first $k - 1$ columns of the matrices in (6.4.4) and (6.4.5) are known, i.e., $f_1, \ldots, f_{k-1}$ and $g_1, \ldots, g_{k-1}$. Consider the problem of maximizing $f^T g$ given that $f = Q_A u$ and $g = Q_B v$ are unit vectors that satisfy

$$f^T [\, f_1 \,|\, \cdots \,|\, f_{k-1} \,] = 0,$$
$$g^T [\, g_1 \,|\, \cdots \,|\, g_{k-1} \,] = 0.$$

It follows from (6.4.6) that

$$f^T g = \sum_{i=k}^{q} \sigma_i (y_i^T u)(z_i^T v) \leq \sigma_k \sum_{i=k}^{q} |y_i^T u| \cdot |z_i^T v|.$$

This expression attains its maximal value of $\sigma_k = \cos(\theta_k)$ by setting $u = y_k$ and $v = z_k$. It follows from (6.4.4) and (6.4.5) that $f = Q_A y_k = f_k$ and $g = Q_B z_k = g_k$. Combining these observations we obtain

Algorithm 6.4.3 (Principal Angles and Vectors) Given $A \in \mathbb{R}^{m \times p}$ and $B \in \mathbb{R}^{m \times q}$ $(p \geq q)$ each with linearly independent columns, the following algorithm computes the cosines of the principal angles $\theta_1 \geq \cdots \geq \theta_q$ between $\mathsf{ran}(A)$ and $\mathsf{ran}(B)$. The vectors $f_1, \ldots, f_q$ and $g_1, \ldots, g_q$ are the associated principal vectors.

> Compute the thin QR factorizations $A = Q_A R_A$ and $B = Q_B R_B$.
>
> $C = Q_A^T Q_B$
>
> Compute the SVD $Y^T C Z = \mathsf{diag}(\cos(\theta_k))$.
>
> $Q_A Y(:, 1{:}q) = [\, f_1 \,|\, \cdots \,|\, f_q \,]$
>
> $Q_B Z(:, 1{:}q) = [\, g_1 \,|\, \cdots \,|\, g_q \,]$

The idea of using the SVD to compute the principal angles and vectors is due to Björck and Golub (1973). The problem of rank deficiency in A and B is also treated in this paper. Principal angles and vectors arise in many important statistical applications. The largest principal angle is related to the notion of distance between equidimensional subspaces that we discussed in §2.5.3. If $p = q$, then

$$\mathsf{dist}(F, G) \;=\; \sqrt{1 - \cos(\theta_p)^2} \;=\; \sin(\theta_p).$$

6.4.4   Intersection of Subspaces

In light of the following theorem, Algorithm 6.4.3 can also be used to compute an orthonormal basis for $\mathsf{ran}(A) \cap \mathsf{ran}(B)$ where $A \in \mathbb{R}^{m \times p}$ and $B \in \mathbb{R}^{m \times q}$

Theorem 6.4.2. Let $\{\cos(\theta_i)\}_{i=1}^{q}$ and $\{f_i, g_i\}_{i=1}^{q}$ be defined by Algorithm 6.4.3. If the index s is defined by $1 = \cos(\theta_1) = \cdots = \cos(\theta_s) > \cos(\theta_{s+1})$, then

$$\mathsf{ran}(A) \cap \mathsf{ran}(B) \;=\; \mathsf{span}\{f_1, \ldots, f_s\} \;=\; \mathsf{span}\{g_1, \ldots, g_s\}.$$

Proof. The proof follows from the observation that if $\cos(\theta_i) = 1$, then $f_i = g_i$. $\square$

The practical determination of the intersection dimension s requires a definition of what it means for a computed singular value to equal 1. For example, a computed singular value $\hat{\sigma}_i = \cos(\hat{\theta}_i)$ could be regarded as a unit singular value if $\hat{\sigma}_i \geq 1 - \delta$ for some intelligently chosen small parameter δ.

Problems

P6.4.1 Show that if A and B are m-by-p matrices, with $p \leq m$, then

$$\min_{Q^T Q = I_p} \| A - BQ \|_F^2 \;=\; \sum_{i=1}^{p} (\sigma_i(A)^2 - 2\sigma_i(B^T A) + \sigma_i(B)^2).$$

P6.4.2 Extend Algorithm 6.4.2 so that it computes an orthonormal basis for $\mathsf{null}(A_1) \cap \cdots \cap \mathsf{null}(A_s)$ where each matrix A_i has n columns.

P6.4.3 Extend Algorithm 6.4.3 so that it can handle the case when A and B are rank deficient.

P6.4.4 Verify Equation (6.4.2).

P6.4.5 Suppose $A, B \in \mathbb{R}^{m \times n}$ and that A has full column rank. Show how to compute a symmetric matrix $X \in \mathbb{R}^{n \times n}$ that minimizes $\| AX - B \|_F$. Hint: Compute the SVD of A.

P6.4.6 This problem is an exercise in F-norm optimization. (a) Show that if $C \in \mathbb{R}^{m \times n}$ and $e \in \mathbb{R}^m$ is a vector of ones, then $v = C^T e/m$ minimizes $\| C - ev^T \|_F$. (b) Suppose $A \in \mathbb{R}^{m \times n}$ and $B \in \mathbb{R}^{m \times n}$ and that we wish to solve

$$\min_{Q^T Q = I_n \, , \, v \in \mathbb{R}^n} \| A - (B + ev^T)Q \|_F$$

Show that $v_{\text{opt}} = (A-B)^T e/m$ and $Q_{\text{opt}} = U\Sigma V^T$ solve this problem where $B^T(I - ee^T/m)A = UV^T$ is the SVD.

P6.4.7 A 3-by-3 matrix H is *ROPR matrix* if $H = Q + xy^T$ where $Q \in \mathbb{R}^{3 \times 3}$ rotation and $x, y \in \mathbb{R}^3$. (A rotation matrix is an orthogonal matrix with unit determinant. "ROPR" stands for "rank-1 perturbation of a rotation.") ROPR matrices arise in computational photography and this problem highlights some of their properties. (a) If H is a ROPR matrix, then there exist rotations $U, V \in \mathbb{R}^{3 \times 3}$, such that $U^T HV = \text{diag}(\sigma_1, \sigma_2, \sigma_3)$ satisfies $\sigma_1 \geq \sigma_2 \geq |\sigma_3|$. (b) Show that if $Q \in \mathbb{R}^{3 \times 3}$ is a rotation, then there exist cosine-sine pairs $(c_i, s_i) = (\cos(\theta_i), \sin(\theta_i))$, $i = 1{:}3$ such that $Q = Q(\theta_1, \theta_2, \theta_3)$ where

$$Q(\theta_1, \theta_2, \theta_3) = \begin{bmatrix} 1 & 0 & 0 \\ 0 & c_1 & s_1 \\ 0 & -s_1 & c_1 \end{bmatrix} \begin{bmatrix} c_2 & s_2 & 0 \\ -s_2 & c_2 & 0 \\ 0 & 0 & 1 \end{bmatrix} \begin{bmatrix} 1 & 0 & 0 \\ 0 & c_3 & s_3 \\ 0 & -s_3 & c_3 \end{bmatrix}$$

$$= \begin{bmatrix} c_2 & s_2 c_3 & s_2 s_3 \\ -c_1 s_2 & c_1 c_2 c_3 - s_1 s_3 & c_1 c_2 s_3 + s_1 c_3 \\ s_1 s_2 & -s_1 c_2 c_3 - c_1 s_3 & -s_1 c_2 s_3 + c_1 c_3 \end{bmatrix} .$$

Hint: The Givens QR factorization involves three rotations. (c) Show that if

$$\begin{bmatrix} \sigma_1 & 0 & 0 \\ 0 & \sigma_2 & 0 \\ 0 & 0 & \sigma_3 \end{bmatrix} = Q(\theta_1, \theta_2, \theta_3) - xy^T, \qquad x, y \in \mathbb{R}^3$$

then xy^T must have the form

$$xy^T = \begin{bmatrix} s_2 \\ \mu c_1 \\ -\mu s_1 \end{bmatrix} \begin{bmatrix} -s_2/\mu \\ c_3 \\ s_3 \end{bmatrix}^T$$

for some $\mu \geq 0$ and

$$\begin{bmatrix} c_2 - \mu & 1 \\ 1 & c_2 - \mu \end{bmatrix} \begin{bmatrix} c_1 s_3 \\ s_1 c_3 \end{bmatrix} = \begin{bmatrix} 0 \\ 0 \end{bmatrix} .$$

(d) Show that the second singular value of a ROPR matrix is 1.

P6.4.8 Let $U_* \in \mathbb{R}^{n \times d}$ be a matrix with orthonormal columns whose span is a subspace S that we wish to estimate. Assume that $U_c \in \mathbb{R}^{n \times d}$ is a given matrix with orthonormal columns and regard $\text{ran}(U_c)$ as the "current" estimate of S. This problem examines what is required to get an improved estimate of S given the availability of a vector $v \in S$. (a) Define the vectors

$$w = U_c^T v, \qquad v_1 = U_c U_c^T v, \qquad v_2 = (I_n - U_c U_c^T)v,$$

and assume that each is nonzero. (a) Show that if

$$z_\theta = \left(\frac{\cos(\theta) - 1}{\| v_1 \| \| w \|} \right) v_1 + \left(\frac{\sin(\theta)}{\| v_2 \| \| w \|} \right) v_2$$

and

$$U_\theta = (I_n + z_\theta v^T)U_c,$$

then $U_\theta^T U_\theta = I_d$. Thus, $U_\theta U_\theta^T$ is an orthogonal projection. (b) Define the distance function

$$\text{dist}_F(\text{ran}(V), \text{ran}(W)) = \| VV^T - WW^T \|_F$$

where $V, W \in \mathbb{R}^{n \times d}$ have orthonormal columns and show

$$\text{dist}_F(\text{ran}(V), \text{ran}(W))^2 = 2(d - \| W^T V \|_F^2) = 2 \sum_{i=1}^{d} (1 - \sigma_i(W^T V)^2).$$

Note that $\text{dist}(\text{ran}(V), \text{ran}(W))^2 = 1 - \sigma_1(W^T V)^2$. (c) Show that

$$d_\theta^2 = d_c^2 - 2 \cdot \text{tr}(U_* U_*^T (U_\theta U_\theta^T - U_c U_c^T))$$

where $d_\theta = \text{dist}_F(\text{ran}(U_*), \text{ran}(U_\theta))$ and $d_c = \text{dist}_F(\text{ran}(U_*), \text{ran}(U_c))$. (d) Show that if

$$y_\theta = \cos(\theta) \frac{v_1}{\| v_1 \|} + \sin(\theta) \frac{v_2}{\| v_2 \|},$$

then

$$U_\theta U_\theta^T - U_c U_c^T = y_\theta y_\theta^T - \frac{v_1 v_1^T}{v_1^T v_1}$$

and

$$d_\theta^2 = d_c^2 + 2 \left(\frac{\| U_*^T v_1 \|_2^2}{\| v_1 \|_2^2} - \| U_*^T y_\theta \|_2^2 \right).$$

(e) Show that if θ minimizes this quantity, then

$$\sin(2\theta) \left(\frac{\| P_S v_2 \|^2}{\| v_2 \|_2^2} - \frac{\| P_S v_1 \|^2}{\| v_1 \|_2^2} \right) + \cos(2\theta) \frac{v_1^T P_S v_2}{\| v_1 \|_2 \| v_2 \|_2} = 0, \qquad P_S = U_* U_*^T.$$

Notes and References for §6.4

References for the Procrustes problem include:

B. Green (1952). "The Orthogonal Approximation of an Oblique Structure in Factor Analysis," *Psychometrika 17*, 429–40.

P. Schonemann (1966). "A Generalized Solution of the Orthogonal Procrustes Problem," *Psychometrika 31*, 1–10.

R.J. Hanson and M.J. Norris (1981). "Analysis of Measurements Based on the Singular Value Decomposition," *SIAM J. Sci. Stat. Comput. 2*, 363–374.

N.J. Higham (1988). "The Symmetric Procrustes Problem," *BIT 28*, 133–43.

H. Park (1991). "A Parallel Algorithm for the Unbalanced Orthogonal Procrustes Problem," *Parallel Comput. 17*, 913–923.

L.E. Andersson and T. Elfving (1997). "A Constrained Procrustes Problem," *SIAM J. Matrix Anal. Applic. 18*, 124–139.

L. Eldén and H. Park (1999). "A Procrustes Problem on the Stiefel Manifold," *Numer. Math. 82*, 599–619.

A.W. Bojanczyk and A. Lutoborski (1999). "The Procrustes Problem for Orthogonal Stiefel Matrices," *SIAM J. Sci. Comput. 21*, 1291–1304.

If $B = I$, then the Procrustes problem amounts to finding the closest orthogonal matrix. This computation is related to the polar decomposition problem that we consider in §9.4.3. Here are some basic references:

Å. Björck and C. Bowie (1971). "An Iterative Algorithm for Computing the Best Estimate of an Orthogonal Matrix," *SIAM J. Numer. Anal. 8*, 358–64.

N.J. Higham (1986). "Computing the Polar Decomposition with Applications," *SIAM J. Sci. Stat. Comput. 7*, 1160–1174.

Using the SVD to solve the angles-between-subspaces problem is discussed in:

Å. Björck and G.H. Golub (1973). "Numerical Methods for Computing Angles Between Linear Subspaces," *Math. Comput. 27*, 579–94.

L.M. Ewerbring and F.T. Luk (1989). "Canonical Correlations and Generalized SVD: Applications and New Algorithms," *J. Comput. Appl. Math. 27*, 37–52.

G.H. Golub and H. Zha (1994). "Perturbation Analysis of the Canonical Correlations of Matrix Pairs," *Lin. Alg. Applic. 210*, 3–28.

Z. Drmac (2000). "On Principal Angles between Subspaces of Euclidean Space," *SIAM J. Matrix Anal. Applic. 22*, 173–194.

A.V. Knyazev and M.E. Argentati (2002). "Principal Angles between Subspaces in an A–Based Scalar Product: Algorithms and Perturbation Estimates," *SIAM J. Sci. Comput. 23*, 2008–2040.

P. Strobach (2008). "Updating the Principal Angle Decomposition," *Numer. Math. 110*, 83–112.

In reduced-rank regression the object is to connect a matrix of signals to a matrix of noisey observations through a matrix that has specified low rank. An svd-based computational procedure that involves principal angles is discussed in:

L. Eldén and B. Savas (2005). "The Maximum Likelihood Estimate in Reduced-Rank Regression," *Num. Lin. Alg. Applic. 12*, 731–741,

The SVD has many roles to play in statistical computation, see:

S.J. Hammarling (1985). "The Singular Value Decomposition in Multivariate Statistics," *ACM SIGNUM Newsletter 20*, 2-25.

An algorithm for computing the rotation and rank-one matrix in P6.4.7 that define a given ROPR matrix is discussed in:

R. Schreiber, Z. Li, and H. Baker (2009). "Robust Software for Computing Camera Motion Parameters," *J. Math. Imaging Vision 33*, 1–9.

For a more details about the estimation problem associated with P6.4.8, see:

L. Balzano, R. Nowak, and B. Recht (2010). "Online Identification and Tracking of Subspaces from Highly Incomplete Information," *Proceedings of the Allerton Conference on Communication, Control, and Computing 2010*.

6.5   Updating Matrix Factorizations

In many applications it is necessary to refactor a given matrix $A \in \mathbb{R}^{m \times n}$ after it has undergone a small modification. For example, given that we have the QR factorization of a matrix A, we may require the QR factorization of the matrix $\widetilde{A}$ obtained from A by appending a row or column or deleting a row or column. In this section we show that in situations like these, it is much more efficient to "update" A's QR factorization than to generate the required QR factorization of $\widetilde{A}$ from scratch. Givens rotations have a prominent role to play. In addition to discussing various update-QR strategies, we show how to downdate a Cholesky factorization using hyperbolic rotations and how to update a rank-revealing ULV decomposition.

6.5.1   Rank-1 Changes

Suppose we have the QR factorization $QR = A \in \mathbb{R}^{n \times n}$ and that we need to compute the QR factorization $\widetilde{A} = A + uv^T = Q_1 R_1$ where $u, v \in \mathbb{R}^n$ are given. Observe that

$$\widetilde{A} \;=\; A + uv^T \;=\; Q(R + wv^T) \tag{6.5.1}$$

where $w = Q^T u$. Suppose rotations $J_{n-1}, \ldots, J_2, J_1$ are computed such that

$$J_1^T \cdots J_{n-1}^T w \;=\; \pm \| \, w \, \|_2 \, e_1.$$

where each J_k is a Givens rotation in planes k and $k + 1$. If these same rotations are applied to R, then

$$H = J_1^T \cdots J_{n-1}^T R \tag{6.5.2}$$

is upper Hessenberg. For example, in the $n = 4$ case we start with

$$
w \leftarrow \begin{bmatrix} \times \\ \times \\ \times \\ \times \end{bmatrix}, \qquad
R \leftarrow \begin{bmatrix} \times & \times & \times & \times \\ 0 & \times & \times & \times \\ 0 & 0 & \times & \times \\ 0 & 0 & 0 & \times \end{bmatrix},
$$

and then update as follows:

$$
w \leftarrow J_3^T w = \begin{bmatrix} \times \\ \times \\ \times \\ 0 \end{bmatrix}, \qquad
R \leftarrow J_3^T R = \begin{bmatrix} \times & \times & \times & \times \\ 0 & \times & \times & \times \\ 0 & 0 & \times & \times \\ 0 & 0 & \times & \times \end{bmatrix},
$$

$$
w \leftarrow J_2^T w = \begin{bmatrix} \times \\ \times \\ 0 \\ 0 \end{bmatrix}, \qquad
R \leftarrow J_2^T R = \begin{bmatrix} \times & \times & \times & \times \\ 0 & \times & \times & \times \\ 0 & \times & \times & \times \\ 0 & 0 & \times & \times \end{bmatrix},
$$

$$
w \leftarrow J_1^T w = \begin{bmatrix} \times \\ 0 \\ 0 \\ 0 \end{bmatrix}, \qquad
H \leftarrow J_1^T R = \begin{bmatrix} \times & \times & \times & \times \\ \times & \times & \times & \times \\ 0 & \times & \times & \times \\ 0 & 0 & \times & \times \end{bmatrix}.
$$

Consequently,

$$
(J_1^T \cdots J_{n-1}^T)(R + wv^T) = H \pm \| w \|_2 e_1 v^T = H_1 \tag{6.5.3}
$$

is also upper Hessenberg. Following Algorithm 5.2.4, we compute Givens rotations G_k, $k = 1{:}n-1$ such that $G_{n-1}^T \cdots G_1^T H_1 = R_1$ is upper triangular. Combining everything we obtain the QR factorization $\tilde{A} = A + uv^T = Q_1 R_1$ where

$$
Q_1 = Q J_{n-1} \cdots J_1 G_1 \cdots G_{n-1}.
$$

A careful assessment of the work reveals that about $26n^2$ flops are required.

The technique readily extends to the case when A is rectangular. It can also be generalized to compute the QR factorization of $A + UV^T$ where $U \in \mathbb{R}^{m \times p}$ and $V \in \mathbb{R}^{n \times p}$.

6.5.2   Appending or Deleting a Column

Assume that we have the QR factorization

$$
QR = A = [\, a_1 \,|\, \cdots \,|\, a_n \,], \qquad a_i \in \mathbb{R}^m, \tag{6.5.4}
$$

and for some k, $1 \le k \le n$, partition the upper triangular matrix $R \in \mathbb{R}^{m \times n}$ as follows:

$$
R = \left[\begin{array}{ccc} R_{11} & v & R_{13} \\ 0 & r_{kk} & w^T \\ 0 & 0 & R_{33} \end{array} \right] \begin{array}{c} k-1 \\ 1 \\ m-k \end{array}.
$$
$$
 \begin{array}{ccc} k-1 & 1 & n-k \end{array}
$$

Now suppose that we want to compute the QR factorization of

$$\widetilde{A} = [\, a_1 \mid \cdots \mid a_{k-1} \mid a_{k+1} \mid \cdots \mid a_n \,] \in \mathbb{R}^{m \times (n-1)} \, .$$

Note that $\widetilde{A}$ is just A with its kth column deleted and that

$$Q^T \widetilde{A} = \begin{bmatrix} R_{11} & R_{13} \\ 0 & w^T \\ 0 & R_{33} \end{bmatrix} = H$$

is upper Hessenberg, e.g.,

$$H = \begin{bmatrix} \times & \times & \times & \times & \times \\ 0 & \times & \times & \times & \times \\ 0 & 0 & \times & \times & \times \\ 0 & 0 & \times & \times & \times \\ 0 & 0 & 0 & \times & \times \\ 0 & 0 & 0 & 0 & \times \\ 0 & 0 & 0 & 0 & 0 \end{bmatrix}, \qquad m = 7,\ n = 6,\ k = 3.$$

Clearly, the unwanted subdiagonal elements $h_{k+1,k}, \ldots, h_{n,n-1}$ can be zeroed by a sequence of Givens rotations: $G_{n-1}^T \cdots G_k^T H = R_1$. Here, G_i is a rotation in planes i and $i + 1$ for $i = k{:}n - 1$. Thus, if $Q_1 = Q G_k \cdots G_{n-1}$ then $\widetilde{A} = Q_1 R_1$ is the QR factorization of $\widetilde{A}$.

The above update procedure can be executed in $O(n^2)$ flops and is very useful in certain least squares problems. For example, one may wish to examine the significance of the kth factor in the underlying model by deleting the kth column of the corresponding data matrix and solving the resulting LS problem.

Analogously, it is possible to update efficiently the QR factorization of a matrix after a column has been added. Assume that we have (6.5.4) but now want the QR factorization of

$$\widetilde{A} = [\, a_1 \mid \ldots \mid a_k \mid z \mid a_{k+1} \mid \ldots \mid a_n \,]$$

where $z \in \mathbb{R}^m$ is given. Note that if $w = Q^T z$ then

$$Q^T \widetilde{A} = [\, Q^T a_1 \mid \cdots \mid Q^T a_k \mid w \mid Q^T a_{k+1} \mid \cdots \mid Q^T a_n \,]$$

is upper triangular except for the presence of a "spike" in its $(k + 1)$st column, e.g.,

$$\widetilde{A} \leftarrow Q^T \widetilde{A} = \begin{bmatrix} \times & \times & \times & \times & \times & \times \\ 0 & \times & \times & \times & \times & \times \\ 0 & 0 & \times & \times & \times & \times \\ 0 & 0 & 0 & \times & \times & \times \\ 0 & 0 & 0 & \times & 0 & \times \\ 0 & 0 & 0 & \times & 0 & 0 \\ 0 & 0 & 0 & \times & 0 & 0 \end{bmatrix}, \qquad m = 7,\ n = 5,\ k = 3.$$

It is possible to determine a sequence of Givens rotations that restores the triangular form:

$$
\widetilde{A} \leftarrow J_6^T \widetilde{A} =
\begin{bmatrix}
\times & \times & \times & \times & \times & \times \\
0 & \times & \times & \times & \times & \times \\
0 & 0 & \times & \times & \times & \times \\
0 & 0 & 0 & \times & \times & \times \\
0 & 0 & 0 & \times & 0 & \times \\
0 & 0 & 0 & \times & 0 & 0 \\
0 & 0 & 0 & 0 & 0 & 0
\end{bmatrix},
\qquad
\widetilde{A} \leftarrow J_5^T \widetilde{A} =
\begin{bmatrix}
\times & \times & \times & \times & \times & \times \\
0 & \times & \times & \times & \times & \times \\
0 & 0 & \times & \times & \times & \times \\
0 & 0 & 0 & \times & \times & \times \\
0 & 0 & 0 & \times & 0 & \times \\
0 & 0 & 0 & 0 & 0 & \times \\
0 & 0 & 0 & 0 & 0 & 0
\end{bmatrix},
$$

$$
\widetilde{A} \leftarrow J_4^T \widetilde{A} =
\begin{bmatrix}
\times & \times & \times & \times & \times & \times \\
0 & \times & \times & \times & \times & \times \\
0 & 0 & \times & \times & \times & \times \\
0 & 0 & 0 & \times & \times & \times \\
0 & 0 & 0 & 0 & \times & \times \\
0 & 0 & 0 & 0 & 0 & \times \\
0 & 0 & 0 & 0 & 0 & 0
\end{bmatrix}.
$$

This update requires $O(mn)$ flops.

6.5.3   Appending or Deleting a Row

Suppose we have the QR factorization $QR = A \in \mathbb{R}^{m \times n}$ and now wish to obtain the QR factorization of

$$
\widetilde{A} = \begin{bmatrix} w^T \\ A \end{bmatrix}
$$

where $w \in \mathbb{R}^n$. Note that

$$
\mathrm{diag}(1, Q^T)\widetilde{A} = \begin{bmatrix} w^T \\ R \end{bmatrix} = H
$$

is upper Hessenberg. Thus, rotations $J_1, \ldots, J_n$ can be determined so $J_n^T \cdots J_1^T H = R_1$ is upper triangular. It follows that $\widetilde{A} = Q_1 R_1$ is the desired QR factorization, where $Q_1 = \mathrm{diag}(1, Q)J_1 \cdots J_n$. See Algorithm 5.2.5.

No essential complications result if the new row is added between rows k and $k + 1$ of A. Indeed, if

$$
\begin{bmatrix} A_1 \\ A_2 \end{bmatrix} = QR, \qquad A_1 \in \mathbb{R}^{k \times n}, \; A_2 \in \mathbb{R}^{(m-k) \times n},
$$

and

$$
P = \begin{bmatrix}
0 & 1 & 0 \\
I_k & 0 & 0 \\
0 & 0 & I_{m-k}
\end{bmatrix},
$$

then

$$
\mathrm{diag}(1, Q^T)P \begin{bmatrix} A_1 \\ w^T \\ A_2 \end{bmatrix} = \begin{bmatrix} w^T \\ R \end{bmatrix} = H
$$

is upper Hessenberg and we proceed as before.

Lastly, we consider how to update the QR factorization $QR = A \in \mathbb{R}^{m \times n}$ when the first row of A is deleted. In particular, we wish to compute the QR factorization of the submatrix A_1 in

$$A = \begin{bmatrix} z^T \\ A_1 \end{bmatrix} \begin{matrix} 1 \\ m-1 \end{matrix} .$$

(The procedure is similar when an arbitrary row is deleted.) Let q^T be the first row of Q and compute Givens rotations $G_1, \ldots, G_{m-1}$ such that $G_1^T \cdots G_{m-1}^T q = \alpha e_1$ where $\alpha = \pm 1$. Note that

$$H = G_1^T \cdots G_{m-1}^T R = \begin{bmatrix} v^T \\ R_1 \end{bmatrix} \begin{matrix} 1 \\ m-1 \end{matrix}$$

is upper Hessenberg and that

$$QG_{m-1} \cdots G_1 = \begin{bmatrix} \alpha & 0 \\ 0 & Q_1 \end{bmatrix}$$

where $Q_1 \in \mathbb{R}^{(m-1) \times (m-1)}$ is orthogonal. Thus,

$$A = \begin{bmatrix} z^T \\ A_1 \end{bmatrix} = (QG_{m-1} \cdots G_1)(G_1^T \cdots G_{m-1}^T R) = \begin{bmatrix} \alpha & 0 \\ 0 & Q_1 \end{bmatrix} \begin{bmatrix} v^T \\ R_1 \end{bmatrix}$$

from which we conclude that $A_1 = Q_1 R_1$ is the desired QR factorization.

6.5.4 Cholesky Updating and Downdating

Suppose we are given a symmtetric positive definite matrix $A \in \mathbb{R}^{n \times n}$ and its Cholesky factor G. In the *Cholesky updating problem*, the challenge is to compute the Cholesky factorization $\widetilde{A} = \widetilde{G}\widetilde{G}^T$ where

$$\widetilde{A} = A + zz^T, \qquad z \in \mathbb{R}^n. \tag{6.5.5}$$

Noting that

$$\widetilde{A} = \begin{bmatrix} G^T \\ z^T \end{bmatrix}^T \begin{bmatrix} G^T \\ z^T \end{bmatrix}, \tag{6.5.6}$$

we can solve this problem by computing a product of Givens rotations $Q = Q_1 \cdots Q_n$ so that

$$Q^T \begin{bmatrix} G^T \\ z^T \end{bmatrix} = \begin{bmatrix} R \\ 0 \end{bmatrix}, \qquad R \in \mathbb{R}^{n \times n} \tag{6.5.7}$$

is upper triangular. It follows that $\widetilde{A} = RR^T$ and so the updated Cholesky factor is given by $\widetilde{G} = R^T$. The zeroing sequence that produces R is straight forward, e.g.,

$$\begin{bmatrix} \times & \times & \times \\ 0 & \times & \times \\ 0 & 0 & \times \\ \times & \times & \times \end{bmatrix} \xrightarrow{Q_1} \begin{bmatrix} \times & \times & \times \\ 0 & \times & \times \\ 0 & 0 & \times \\ 0 & \times & \times \end{bmatrix} \xrightarrow{Q_2} \begin{bmatrix} \times & \times & \times \\ 0 & \times & \times \\ 0 & 0 & \times \\ 0 & 0 & \times \end{bmatrix} \xrightarrow{Q_3} \begin{bmatrix} \times & \times & \times \\ 0 & \times & \times \\ 0 & 0 & \times \\ 0 & 0 & 0 \end{bmatrix} .$$

The Q_k update involves only rows k and $n + 1$. The overall process is essentially the same as the strategy we outlined in the previous subsection for updating the QR factorization of a matrix when a row is appended.

The *Cholesky downdating problem* involves a different set of tools and a new set of numerical concerns. We are again given a Cholesky factorization $A = GG^T$ and a vector $z \in \mathbb{R}^n$. However, now the challenge is to compute the Cholesky factorization $\widetilde{A} = \widetilde{G}\widetilde{G}^T$ where

$$\widetilde{A} = A - zz^T \tag{6.5.8}$$

is presumed to be positive definite. By introducing the notion of a *hyperbolic rotation* we can develop a downdating framework that corresponds to the Givens-based updating framework. Define the matrix S as follows

$$S = \begin{bmatrix} I_n & 0 \\ 0 & -1 \end{bmatrix} \tag{6.5.9}$$

and note that

$$\widetilde{A} = GG^T - zz^T = \begin{bmatrix} G^T \\ z^T \end{bmatrix}^T S \begin{bmatrix} G^T \\ z^T \end{bmatrix}. \tag{6.5.10}$$

This corresponds to (6.5.6), but instead of computing the QR factorization (6.5.7), we seek a matrix $H \in \mathbb{R}^{(n+1)\times(n+1)}$ that satisfies two properties:

$$HSH^T = S, \tag{6.5.11}$$

$$H^T \begin{bmatrix} G^T \\ z^T \end{bmatrix} = \begin{bmatrix} R \\ 0 \end{bmatrix}, \qquad R \in \mathbb{R}^{n\times n} \text{ (upper triangular)}. \tag{6.5.12}$$

If this can be accomplished, then it follows from

$$\widetilde{A} = \left(H^T \begin{bmatrix} G^T \\ z^T \end{bmatrix} \right)^T \begin{bmatrix} I_n & 0 \\ 0 & -1 \end{bmatrix} \left(H^T \begin{bmatrix} G^T \\ z^T \end{bmatrix} \right) = R^T R$$

that the Cholesky factor of $\widetilde{A} = A - zz^T$ is given by $\widetilde{G} = R^T$. A matrix H that satisfies (6.5.11) is said to be *S-orthogonal*. Note that the product of S-orthogonal matrices is also S-orthogonal.

An important subset of the S-orthogonal matrices are the *hyperbolic rotations* and here is a 4-by-4 example:

$$H_2(\theta) = \begin{bmatrix} 1 & 0 & 0 & 0 \\ 0 & c & 0 & -s \\ 0 & 0 & 1 & 0 \\ 0 & -s & 0 & c \end{bmatrix}, \qquad c = \cosh(\theta), \, s = \sinh(\theta).$$

The S-orthogonality of this matrix follows from $\cosh(\theta)^2 - \sinh(\theta)^2 = 1$. In general, $H_k \in \mathbb{R}^{(n+1)\times(n+1)}$ is a hyperbolic rotation if it agrees with I_{n+1} except in four locations:

$$\begin{bmatrix} [H_k]_{k,k} & [H_k]_{k,n+1} \\ [H_k]_{n+1,k} & [H_k]_{n+1,n+1} \end{bmatrix} = \begin{bmatrix} \cosh(\theta) & -\sinh(\theta) \\ -\sinh(\theta) & \cosh(\theta) \end{bmatrix}.$$

Hyperbolic rotations look like Givens rotations and, not surprisingly, can be used to introduce zeros into a vector or matrix. However, upon consideration of the equation

$$
\begin{bmatrix} c & -s \\ -s & c \end{bmatrix} \begin{bmatrix} x_1 \\ x_2 \end{bmatrix} = \begin{bmatrix} r \\ 0 \end{bmatrix}, \qquad c^2 - s^2 = 1
$$

we see that the required cosh-sinh pair may not exist. Since we always have $|\cosh(\theta)| > |\sinh(\theta)|$, there is no real solution to $-sx_1 + cx_2 = 0$ if $|x_2| > |x_1|$. On the other hand, if $|x_1| > |x_2|$, then $\{c, s\} = \{\cosh(\theta), \sinh(\theta)\}$ can be computed as follows:

$$
\tau = \frac{x_2}{x_1}, \qquad c = \frac{1}{\sqrt{1 - \tau^2}}, \qquad s = c \cdot \tau. \tag{6.5.13}
$$

There are clearly numerical issues if $|x_1|$ is just slightly greater than $|x_2|$. However, it is possible to organize hyperbolic rotation computations successfully, see Alexander, Pan, and Plemmons (1988).

Putting these concerns aside, we show how the matrix H in (6.5.12) can be computed as a product of hyperbolic rotations $H = H_1 \cdots H_n$ just as the transforming Q in the updating problem is a product of Givens rotations. Consider the role of H_1 in the $n = 3$ case:

$$
\begin{bmatrix} c & 0 & 0 & -s \\ 0 & 1 & 0 & 0 \\ 0 & 0 & 1 & 0 \\ -s & 0 & 0 & c \end{bmatrix}^T \begin{bmatrix} g_{11} & g_{21} & g_{31} \\ 0 & g_{22} & g_{32} \\ 0 & 0 & g_{33} \\ z_1 & z_2 & z_3 \end{bmatrix} = \begin{bmatrix} \tilde{g}_{11} & \tilde{g}_{21} & \tilde{g}_{31} \\ 0 & g_{22} & g_{32} \\ 0 & 0 & g_{33} \\ 0 & z_2' & z_3' \end{bmatrix}.
$$

Since $\widetilde{A} = GG^T - zz^T$ is positive definite, $[\widetilde{A}]_{11} = g_{11}^2 - z_1^2 > 0$. It follows that $|g_{11}| > |z_1|$ which guarantees that the cosh-sinh computations (6.5.13) go through. For the overall process to be defined, we have to guarantee that hyperbolic rotations $H_2, \ldots, H_n$ can be found to zero out the bottom row in the matrix $[\, G^T \ z\,]^T$. The following theorem ensures that this is the case.

Theorem 6.5.1. *If*

$$
A = \begin{bmatrix} \alpha & v^T \\ v & B \end{bmatrix} = \begin{bmatrix} g_{11} & 0 \\ g_1 & G_1 \end{bmatrix} \begin{bmatrix} g_{11} & g_1^T \\ 0 & G_1^T \end{bmatrix}
$$

and

$$
\widetilde{A} = A - zz^T = A - \begin{bmatrix} \mu \\ w \end{bmatrix} \begin{bmatrix} \mu \\ w \end{bmatrix}^T
$$

are positive definite, then it is possible to determine $c = \cosh(\theta)$ and $s = \sinh(\theta)$ so

$$
\begin{bmatrix} c & 0 & -s \\ 0 & I_{n-1} & 0 \\ -s & 0 & c \end{bmatrix} \begin{bmatrix} g_{11} & g_1^T \\ 0 & G_1^T \\ \mu & w^T \end{bmatrix} = \begin{bmatrix} \tilde{g}_{11} & \tilde{g}_1^T \\ 0 & G_1^T \\ 0 & w_1^T \end{bmatrix}.
$$

Moreover, the matrix $\widetilde{A}_1 = G_1 G_1^T - w_1 w_1^T$ is positive definite.

Proof. The blocks in A's Cholesky factor are given by

$$g_{11} = \sqrt{\alpha}, \qquad g_1 = v/g_{11}, \qquad G_1 G_1^T = B - \frac{1}{\alpha} vv^T. \qquad (6.5.14)$$

Since $A - zz^T$ is positive definite, $a_{11} - z_1^2 = g_{11}^2 - \mu^2 > 0$ and so from (6.5.13) with $\tau = \mu/g_{11}$ we see that

$$c = \frac{\sqrt{\alpha}}{\sqrt{\alpha - \mu^2}}, \qquad s = \frac{\mu}{\sqrt{\alpha - \mu^2}}. \qquad (6.5.15)$$

Since $w_1 = -sg_1 + cw$ it follows from (6.5.14) and (6.5.15) that

$$\begin{aligned}
\tilde{A}_1 = G_1 G_1^T - w_1 w_1^T &= B - \frac{1}{\alpha} vv^T - (-sg_1 + cw)(-sg_1 + cw)^T \\
&= B - \frac{c^2}{\alpha} vv^T - c^2 ww^T + \frac{sc}{\sqrt{\alpha}}(vw^T + wv^T) \\
&= B - \frac{1}{\alpha - \mu^2} vv^T - \frac{\alpha}{\alpha - \mu^2} ww^T + \frac{\mu}{\alpha - \mu^2}(vw^T + wv^T).
\end{aligned}$$

It is easy to verify that this matrix is precisely the Schur complement of α in

$$\tilde{A} = A - zz^T = \begin{bmatrix} \alpha - \mu^2 & v^T - \mu w^T \\ v - \mu w & B - ww^T \end{bmatrix}$$

and is therefore positive definite. $\square$

The theorem provides the key step in an induction proof that the factorization (6.5.12) exists.

6.5.5 Updating a Rank-Revealing ULV Decomposition

We close with a discussion about updating a nullspace basis after one or more rows have been appended to the underlying matrix. We work with the ULV decomposition which is much more tractable than the SVD from the updating point of view. We pattern our remarks after Stewart(1993).

A *rank-revealing ULV decomposition* of a matrix $A \in \mathbb{R}^{m \times n}$ has the form

$$U^T A V = \begin{bmatrix} L \\ 0 \end{bmatrix} = \begin{bmatrix} L_{11} & 0 \\ L_{21} & L_{22} \\ 0 & 0 \end{bmatrix}, \qquad U^T U = I_m, \ V^T V = I_n \qquad (6.5.16)$$

where $L_{11} \in \mathbb{R}^{r \times r}$ and $L_{22} \in \mathbb{R}^{(n-r) \times (n-r)}$ are lower triangular and $\| L_{21} \|_2$ and $\| L_{22} \|_2$ are small compared to $\sigma_{\min}(L_{11})$. Such a decomposition can be obtained by applying QR with column pivoting

$$U^T A \Pi = \begin{bmatrix} R \\ 0 \end{bmatrix}, \qquad R \in \mathbb{R}^{n \times n}$$

followed by a QR factorization $V_1^T R^T = L^T$. In this case the matrix V in (6.5.16) is given by $V = \Pi V_1$. The parameter r is the estimated rank. Note that if

$$V = [\, V_1 \mid V_2 \,] \, , \qquad U = [\, U_1 \mid U_2 \,] \, ,$$
$$ {}_{r} \quad {}_{n-r} {}_{r} \quad {}_{m-r}$$

then the columns of V_2 define an approximate nullspace:

$$\| \, A V_2 \, \|_2 = \| \, U_2 L_{22} \, \|_2 = \| \, L_{22} \, \|_2.$$

Our goal is to produce cheaply a rank-revealing ULV decomposition for the row-appended matrix

$$\tilde{A} = \left[\begin{array}{c} A \\ z^T \end{array} \right],$$

In particular, we show how to revise L, V, and possibly r in $O(n^2)$ flops. Note that

$$\left[\begin{array}{cc} U & 0 \\ 0 & 1 \end{array} \right]^T \left[\begin{array}{c} A \\ z^T \end{array} \right] V = \left[\begin{array}{cc} L_{11} & 0 \\ L_{21} & L_{22} \\ 0 & 0 \\ w^T & y^T \end{array} \right].$$

We illustrate the key ideas through an example. Suppose $n = 7$ and $r = 4$. By permuting the rows so that the bottom row is just underneath L, we obtain

$$\left[\begin{array}{cc} L_{11} & 0 \\ L_{21} & L_{22} \\ w^T & y^T \end{array} \right] = \left[\begin{array}{cccc|ccc} \ell & 0 & 0 & 0 & 0 & 0 & 0 \\ \ell & \ell & 0 & 0 & 0 & 0 & 0 \\ \ell & \ell & \ell & 0 & 0 & 0 & 0 \\ \ell & \ell & \ell & \ell & 0 & 0 & 0 \\ \hline \epsilon & \epsilon & \epsilon & \epsilon & \epsilon & 0 & 0 \\ \epsilon & \epsilon & \epsilon & \epsilon & \epsilon & \epsilon & 0 \\ \epsilon & \epsilon & \epsilon & \epsilon & \epsilon & \epsilon & \epsilon \\ \hline w & w & w & w & y & y & y \end{array} \right].$$

The ϵ entries are small while the ℓ, w, and y entries are not. Next, a sequence of Givens rotations $G_7, \ldots, G_1$ are applied from the left to zero out the bottom row:

$$\left[\begin{array}{c} \tilde{L} \\ \hline 0 \end{array} \right] = \left[\begin{array}{ccccccc} \times & 0 & 0 & 0 & 0 & 0 & 0 \\ \times & \times & 0 & 0 & 0 & 0 & 0 \\ \times & \times & \times & 0 & 0 & 0 & 0 \\ \times & \times & \times & \times & 0 & 0 & 0 \\ \times & \times & \times & \times & \times & 0 & 0 \\ \times & \times & \times & \times & \times & \times & 0 \\ \times & \times & \times & \times & \times & \times & \times \\ \hline 0 & 0 & 0 & 0 & 0 & 0 & 0 \end{array} \right] = G_{17} \cdots G_{57} G_{67} \left[\begin{array}{cc} L_{11} & 0 \\ L_{21} & L_{22} \\ w^T & y^T \end{array} \right].$$

Because this zeroing process intermingles the (presumably large) entries of the bottom row with the entries from each of the other rows, the lower triangular form is typically *not* rank revealing. However, and this is key, we can restore the rank-revealing structure with a combination of condition estimation and Givens zero chasing.

Let us assume that with the added row, the new nullspace has dimension 2. With a reliable condition estimator we produce a unit 2-norm vector p such that

$$\| p^T \widetilde{L} \|_2 \approx \sigma_{\min}(\widetilde{L}).$$

(See §3.5.4). Rotations $\{U_{i,i+1}\}_{i=1}^{6}$ can be found such that

$$U_{67}^T U_{56}^T U_{45}^T U_{34}^T U_{23}^T U_{12}^T p = e_7 = I_7(:, 7).$$

Applying these rotations to $\widetilde{L}$ produces a lower Hessenberg matrix

$$H = U_{67}^T U_{56}^T U_{45}^T U_{34}^T U_{23}^T U_{12}^T \widetilde{L}.$$

Applying more rotations from the right restores H to a lower triangular form:

$$L_+ = H V_{12} V_{23} V_{34} V_{45} V_{56} V_{67}.$$

It follows that

$$e_7^T L_+ = \left(e_8^T H \right) V_{12} V_{23} V_{34} V_{45} V_{56} V_{67} = \left(p^T \widetilde{L} \right) V_{12} V_{23} V_{34} V_{45} V_{56} V_{67}$$

has approximate norm $\sigma_{\min}(\widetilde{L})$. Thus, we obtain a lower triangular matrix of the form

$$L_+ = \left[\begin{array}{cccccc|c} \times & 0 & 0 & 0 & 0 & 0 & 0 \\ \times & \times & 0 & 0 & 0 & 0 & 0 \\ \times & \times & \times & 0 & 0 & 0 & 0 \\ \times & \times & \times & \times & 0 & 0 & 0 \\ \times & \times & \times & \times & \times & 0 & 0 \\ \times & \times & \times & \times & \times & \times & 0 \\ \hline \epsilon & \epsilon & \epsilon & \epsilon & \epsilon & \epsilon & \epsilon \end{array} \right]$$

We can repeat the condition estimation and zero chasing on the leading 6-by-6 portion. Assuming that the nullspace of the augmented matrix has dimension two, this produces another row of small numbers:

$$\left[\begin{array}{ccccc|cc} \times & 0 & 0 & 0 & 0 & 0 & 0 \\ \times & \times & 0 & 0 & 0 & 0 & 0 \\ \times & \times & \times & 0 & 0 & 0 & 0 \\ \times & \times & \times & \times & 0 & 0 & 0 \\ \times & \times & \times & \times & \times & 0 & 0 \\ \hline \epsilon & \epsilon & \epsilon & \epsilon & \epsilon & \epsilon & 0 \\ \epsilon & \epsilon & \epsilon & \epsilon & \epsilon & \epsilon & \epsilon \end{array} \right].$$

This illustrates how we can restore any lower triangular matrix to rank-revealing form.

Problems

P6.5.1 Suppose we have the QR factorization for $A \in \mathbb{R}^{m \times n}$ and now wish to solve

$$\min_{x \in \mathbb{R}^n} \| (A + uv^T)x - b \|_2$$

where $u, b \in \mathbb{R}^m$ and $v \in \mathbb{R}^n$ are given. Give an algorithm for solving this problem that requires $O(mn)$ flops. Assume that Q must be updated.

P6.5.2 Suppose

$$A = \begin{bmatrix} c^T \\ B \end{bmatrix}, \qquad c \in \mathbb{R}^n, \ B \in \mathbb{R}^{(m-1) \times n}$$

has full column rank and $m > n$. Using the Sherman-Morrison-Woodbury formula show that

$$\frac{1}{\sigma_{\min}(B)} \leq \frac{1}{\sigma_{\min}(A)} + \frac{\| (A^T A)^{-1} c \|_2^2}{1 - c^T (A^T A)^{-1} c}.$$

P6.5.3 As a function of x_1 and x_2, what is the 2-norm of the hyperbolic rotation produced by (6.5.13)?

P6.5.4 Assume that

$$A = \begin{bmatrix} R & H \\ 0 & E \end{bmatrix}, \qquad \rho = \frac{\| E \|_2}{\sigma_{\min}(R)} < 1,$$

where R and E are square. Show that if

$$Q = \begin{bmatrix} Q_{11} & Q_{12} \\ Q_{21} & Q_{22} \end{bmatrix}$$

is orthogonal and

$$\begin{bmatrix} R & H \\ 0 & E \end{bmatrix} \begin{bmatrix} Q_{11} & Q_{12} \\ Q_{21} & Q_{22} \end{bmatrix} = \begin{bmatrix} R_1 & 0 \\ H_1 & E_1 \end{bmatrix},$$

then $\| H_1 \|_2 \leq \rho \| H \|_2$.

P6.5.5 Suppose $A \in \mathbb{R}^{m \times n}$ and $b \in \mathbb{R}^m$ with $m \geq n$. In the *indefinite least squares* (ILS) problem, the goal is to minimize

$$\phi(x) = (b - Ax)^T J(b - Ax),$$

where

$$S = \begin{bmatrix} I_p & 0 \\ 0 & -I_q \end{bmatrix}, \qquad p + q = m.$$

It is assumed that $p \geq 1$ and $q \geq 1$. (a) By taking the gradient of ϕ, show that the ILS problem has a unique solution if and only if $A^T S A$ is positive definite. (b) Assume that the ILS problem has a unique solution. Show how it can be found by computing the Cholesky factorization of $Q_1^T Q_1 - Q_2^T Q_2$ where

$$A = \begin{bmatrix} Q_1 \\ Q_2 \end{bmatrix}, \qquad Q_1 \in \mathbb{R}^{p \times n}, \ Q_2 \in \mathbb{R}^{q \times n}$$

is the thin QR factorization. (c) A matrix $Q \in \mathbb{R}^{m \times m}$ is *S-orthogonal* if $QSQ^T = S$ If

$$Q = \begin{bmatrix} Q_{11} & Q_{12} \\ Q_{21} & Q_{22} \end{bmatrix} \begin{matrix} p \\ q \end{matrix}$$
$$\quad\ p \quad\ q$$

is S-orthogonal, then by comparing blocks in the equation $Q^T S Q = S$ we have

$$Q_{11}^T Q_{11} = I_p + Q_{21}^T Q_{21}, \qquad Q_{11}^T Q_{12} = Q_{21}^T Q_{22}, \qquad Q_{22}^T Q_{22} = I_q + Q_{12}^T Q_{12}.$$

Thus, the singular values of Q_{11} and Q_{22} are never smaller than 1. Assume that $p \geq q$. By analogy with how the CS decomposition is established in §2.5.4, show that there exist orthogonal matrices U_1, U_2, V_1 and V_2 such that

$$\begin{bmatrix} U_1 & 0 \\ 0 & U_2 \end{bmatrix}^T Q \begin{bmatrix} V_1 & 0 \\ 0 & V_2 \end{bmatrix} = \left[\begin{array}{ccc|c} D & & 0 & (D^2 - I)^{1/2} \\ & 0 & I_{p-q} & 0 \\ \hline (D^2 - I_p)^{1/2} & 0 & & D \end{array} \right]$$

where $D = \text{diag}(d_1, \ldots, d_p)$ with $d_i \geq 1$, $i = 1{:}p$. This is the *hyperbolic CS decomposition* and details can be found in Stewart and Van Dooren (2006).

Notes and References for §6.5

The seminal matrix factorization update paper is:

P.E. Gill, G.H. Golub, W. Murray, and M.A. Saunders (1974). "Methods for Modifying Matrix Factorizations," *Math. Comput. 28,* 505–535.

Initial research into the factorization update problem was prompted by the development of quasi-Newton methods and the simplex method for linear programming. In these venues, a linear system must be solved in step k that is a low-rank perturbation of the linear system solved in step $k-1$, see:

R.H. Bartels (1971). "A Stabilization of the Simplex Method," *Numer. Math. 16,* 414–434.

P.E. Gill, W. Murray, and M.A. Saunders (1975). "Methods for Computing and Modifying the LDV Factors of a Matrix," *Math. Comput. 29,* 1051–1077.

D. Goldfarb (1976). "Factored Variable Metric Methods for Unconstrained Optimization," *Math. Comput. 30,* 796–811.

J.E. Dennis and R.B. Schnabel (1983). *Numerical Methods for Unconstrained Optimization and Nonlinear Equations,* Prentice-Hall, Englewood Cliffs, NJ.

W.W. Hager (1989). "Updating the Inverse of a Matrix," *SIAM Review 31,* 221–239.

S.K. Eldersveld and M.A. Saunders (1992). "A Block-LU Update for Large-Scale Linear Programming," *SIAM J. Matrix Anal. Applic. 13,* 191–201.

Updating issues in the least squares setting are discussed in:

J. Daniel, W.B. Gragg, L. Kaufman, and G.W. Stewart (1976). "Reorthogonaization and Stable Algorithms for Updating the Gram-Schmidt QR Factorization," *Math. Comput. 30,* 772–795.

S. Qiao (1988). "Recursive Least Squares Algorithm for Linear Prediction Problems," *SIAM J. Matrix Anal. Applic. 9,* 323–328.

Å. Björck, H. Park, and L. Eldén (1994). "Accurate Downdating of Least Squares Solutions," *SIAM J. Matrix Anal. Applic. 15,* 549–568.

S.J. Olszanskyj, J.M. Lebak, and A.W. Bojanczyk (1994). "Rank-k Modification Methods for Recursive Least Squares Problems," *Numer. Alg. 7,* 325–354.

L. Eldén and H. Park (1994). "Block Downdating of Least Squares Solutions," *SIAM J. Matrix Anal. Applic. 15,* 1018–1034.

Kalman filtering is a very important tool for estimating the state of a linear dynamic system in the presence of noise. An illuminating, stable implementation that involves updating the QR factorization of an evolving block banded matrix is given in:

C.C. Paige and M.A. Saunders (1977). "Least Squares Estimation of Discrete Linear Dynamic Systems Using Orthogonal Transformations," *SIAM J. Numer. Anal. 14,* 180–193.

The Cholesky downdating literature includes:

G.W. Stewart (1979). "The Effects of Rounding Error on an Algorithm for Downdating a Cholesky Factorization," *J. Inst. Math. Applic. 23,* 203–213.

A.W. Bojanczyk, R.P. Brent, P. Van Dooren, and F.R. de Hoog (1987). "A Note on Downdating the Cholesky Factorization," *SIAM J. Sci. Stat. Comput. 8,* 210–221.

C.-T. Pan (1993). "A Perturbation Analysis of the Problem of Downdating a Cholesky Factorization," *Lin. Alg. Applic. 183,* 103–115.

L. Eldén and H. Park (1994). "Perturbation Analysis for Block Downdating of a Cholesky Decomposition," *Numer. Math. 68,* 457–468.

M.R. Osborne and L. Sun (1999). "A New Approach to Symmetric Rank-One Updating," *IMA J. Numer. Anal. 19,* 497–507.

E.S. Quintana-Orti and R.A. Van Geijn (2008). "Updating an LU Factorization with Pivoting," *ACM Trans. Math. Softw. 35(2),* Article 11.

Hyperbolic tranformations have been successfully used in a number of settings:

G.H. Golub (1969). "Matrix Decompositions and Statistical Computation," in *Statistical Computation,* ed., R.C. Milton and J.A. Nelder, Academic Press, New York, pp. 365–397.

C.M. Rader and A.O. Steinhardt (1988). "Hyperbolic Householder Transforms," *SIAM J. Matrix Anal. Applic. 9,* 269–290.

S.T. Alexander, C.T. Pan, and R.J. Plemmons (1988). "Analysis of a Recursive Least Squares Hyperbolic Rotation Algorithm for Signal Processing," *Lin. Alg. and Its Applic. 98*, 3–40.

G. Cybenko and M. Berry (1990). "Hyperbolic Householder Algorithms for Factoring Structured Matrices," *SIAM J. Matrix Anal. Applic. 11*, 499–520.

A.W. Bojanczyk, R. Onn, and A.O. Steinhardt (1993). "Existence of the Hyperbolic Singular Value Decomposition," *Lin. Alg. Applic. 185*, 21–30.

S. Chandrasekaran, M. Gu, and A.H. Sayad (1998). "A Stable and Efficient Algorithm for the Indefinite Linear Least Squares Problem," *SIAM J. Matrix Anal. Applic. 20*, 354–362.

A.J. Bojanczyk, N.J. Higham, and H. Patel (2003a). "Solving the Indefinite Least Squares Problem by Hyperbolic QR Factorization," *SIAM J. Matrix Anal. Applic. 24*, 914–931.

A. Bojanczyk, N.J. Higham, and H. Patel (2003b). "The Equality Constrained Indefinite Least Squares Problem: Theory and Algorithms," *BIT 43*, 505–517.

M. Stewart and P. Van Dooren (2006). "On the Factorization of Hyperbolic and Unitary Transformations into Rotations," *SIAM J. Matrix Anal. Applic. 27*, 876–890.

N.J. Higham (2003). "J-Orthogonal Matrices: Properties and Generation," *SIAM Review 45*, 504–519.

High-performance issues associated with QR updating are discussed in:

B.C. Gunter and R.A. Van De Geijn (2005). "Parallel Out-of-Core Computation and Updating of the QR Factorization," *ACM Trans. Math. Softw. 31*, 60–78.

Updating and downdating the ULV and URV decompositions and related topics are covered in:

C.H. Bischof and G.M. Shroff (1992). "On Updating Signal Subspaces," *IEEE Trans. Signal Proc. 40*, 96–105.

G.W. Stewart (1992). "An Updating Algorithm for Subspace Tracking," *IEEE Trans. Signal Proc. 40*, 1535–1541.

G.W. Stewart (1993). "Updating a Rank-Revealing ULV Decomposition," *SIAM J. Matrix Anal. Applic. 14*, 494–499.

G.W. Stewart (1994). "Updating URV Decompositions in Parallel," *Parallel Comp. 20*, 151–172.

H. Park and L. Eldén (1995). "Downdating the Rank-Revealing URV Decomposition," *SIAM J. Matrix Anal. Applic. 16*, 138–155.

J.L. Barlow and H. Erbay (2009). "Modifiable Low-Rank Approximation of a Matrix," *Num. Lin. Alg. Applic. 16*, 833–860.

Other interesting update-related topics include the updating of condition estimates, see:

W.R. Ferng, G.H. Golub, and R.J. Plemmons (1991). "Adaptive Lanczos Methods for Recursive Condition Estimation," *Numerical Algorithms 1*, 1-20.

G. Shroff and C.H. Bischof (1992). "Adaptive Condition Estimation for Rank-One Updates of QR Factorizations," *SIAM J. Matrix Anal. Applic. 13*, 1264–1278.

D.J. Pierce and R.J. Plemmons (1992). "Fast Adaptive Condition Estimation," *SIAM J. Matrix Anal. Applic. 13*, 274–291.

and the updating of solutions to constrained least squares problems:

K. Schittkowski and J. Stoer (1979). "A Factorization Method for the Solution of Constrained Linear Least Squares Problems Allowing for Subsequent Data changes," *Numer. Math. 31*, 431–463.

Å. Björck (1984). "A General Updating Algorithm for Constrained Linear Least Squares Problems," *SIAM J. Sci. Stat. Comput. 5*, 394–402.

Finally, we mention the following paper concerned with SVD updating:

M. Moonen, P. Van Dooren, and J. Vandewalle (1992). "A Singular Value Decomposition Updating Algorithm," *SIAM J. Matrix Anal. Applic. 13*, 1015–1038.

Chapter 7

Unsymmetric Eigenvalue Problems

7.1   Properties and Decompositions

7.2   Perturbation Theory

7.3   Power Iterations

7.4   The Hessenberg and Real Schur Forms

7.5   The Practical QR Algorithm

7.6   Invariant Subspace Computations

7.7   The Generalized Eigenvalue Problem

7.8   Hamiltonian and Product Eigenvalue Problems

7.9   Pseudospectra

Having discussed linear equations and least squares, we now direct our attention to the third major problem area in matrix computations, the algebraic eigenvalue problem. The unsymmetric problem is considered in this chapter and the more agreeable symmetric case in the next.

Our first task is to present the decompositions of Schur and Jordan along with the basic properties of eigenvalues and invariant subspaces. The contrasting behavior of these two decompositions sets the stage for §7.2 in which we investigate how the eigenvalues and invariant subspaces of a matrix are affected by perturbation. Condition numbers are developed that permit estimation of the errors induced by roundoff.

The key algorithm of the chapter is the justly famous QR algorithm. This procedure is one of the most complex algorithms presented in the book and its development is spread over three sections. We derive the basic QR iteration in §7.3 as a natural generalization of the simple power method. The next two sections are devoted to making this basic iteration computationally feasible. This involves the introduction of the Hessenberg decomposition in §7.4 and the notion of origin shifts in §7.5.

The QR algorithm computes the real Schur form of a matrix, a canonical form that displays eigenvalues but not eigenvectors. Consequently, additional computations

347

usually must be performed if information regarding invariant subspaces is desired. In §7.6, which could be subtitled, "What to Do after the Real Schur Form is Calculated," we discuss various invariant subspace calculations that can be performed after the QR algorithm has done its job.

The next two sections are about Schur decomposition challenges. The generalized eigenvalue problem $Ax = \lambda Bx$ is the subject of §7.7. The challenge is to compute the Schur decomposition of $B^{-1}A$ without actually forming the indicated inverse or the product. The product eigenvalue problem is similar, only arbitrarily long sequences of products are considered. This is treated in §7.8 along with the Hamiltonian eigenproblem where the challenge is to compute a Schur form that has a special 2-by-2 block structure.

In the last section the important notion of pseudospectra is introduced. It is sometimes the case in unsymmetric matrix problems that traditional eigenvalue analysis fails to tell the "whole story" because the eigenvector basis is ill-conditioned. The pseudospectra framework effectively deals with this issue.

We mention that it is handy to work with complex matrices and vectors in the more theoretical passages that follow. Complex versions of the QR factorization, the singular value decomposition, and the CS decomposition surface in the discussion.

Reading Notes

Knowledge of Chapters 1–3 and §§5.1–§5.2 are assumed. Within this chapter there are the following dependencies:

$$§7.1 \;\rightarrow\; §7.2 \;\rightarrow\; §7.3 \;\rightarrow\; §7.4 \;\rightarrow\; §7.5 \;\rightarrow\; §7.6 \;\rightarrow\; §7.9$$
$$\downarrow \qquad \searrow$$
$$§7.7 \qquad §7.8$$

Excellent texts for the dense eigenproblem include Chatelin (EOM), Kressner (NMSE), Stewart (MAE), Stewart and Sun (MPA), Watkins (MEP), and Wilkinson (AEP).

7.1 Properties and Decompositions

In this section the background necessary to develop and analyze the eigenvalue algorithms that follow are surveyed. For further details, see Horn and Johnson (MA).

7.1.1 Eigenvalues and Invariant Subspaces

The *eigenvalues* of a matrix $A \in \mathbb{C}^{n \times n}$ are the n roots of its *characteristic polynomial* $p(z) = \det(zI - A)$. The set of these roots is called the *spectrum* of A and is denoted by

$$\lambda(A) \;=\; \{\, z : \det(zI - A) = 0 \,\}.$$

If $\lambda(A) = \{\lambda_1, \ldots, \lambda_n\}$, then

$$\det(A) \;=\; \lambda_1 \lambda_2 \cdots \lambda_n$$

and

$$\mathsf{tr}(A) \;=\; \lambda_1 + \cdots + \lambda_n$$

where the trace function, introduced in §6.4.1, is the sum of the diagonal entries, i.e.,

$$\text{tr}(A) = \sum_{i=1}^{n} a_{ii}.$$

These characterizations of the determinant and the trace follow by looking at the constant term and the coefficient of z^{n-1} in the characteristic polynomial.

Four other attributes associated with the spectrum of $A \in \mathbb{C}^{n \times n}$ include the

$$\text{Spectral Radius:} \quad \rho(A) = \max_{\lambda \in \lambda(A)} |\lambda|, \tag{7.1.1}$$

$$\text{Spectral Abscissa:} \quad \alpha(A) = \max_{\lambda \in \lambda(A)} \text{Re}(\lambda), \tag{7.1.2}$$

$$\text{Numerical Radius:} \quad r(A) = \max_{\lambda \in \lambda(A)} \{ |x^H A x| : \| x \|_2 = 1 \}, \tag{7.1.3}$$

$$\text{Numerical Range:} \quad W(A) = \{ x^H A x : \| x \|_2 = 1 \}. \tag{7.1.4}$$

The numerical range, which is sometimes referred to as the *field of values*, obviously includes $\lambda(A)$. It can be shown that $W(A)$ is convex.

If $\lambda \in \lambda(A)$, then the nonzero vectors $x \in \mathbb{C}^n$ that satisfy $Ax = \lambda x$ are *eigenvectors*. More precisely, x is a *right eigenvector* for λ if $Ax = \lambda x$ and a *left eigenvector* if $x^H A = \lambda x^H$. Unless otherwise stated, "eigenvector" means "right eigenvector."

An eigenvector defines a 1-dimensional subspace that is invariant with respect to premultiplication by A. A subspace $S \subseteq \mathbb{C}^n$ with the property that

$$x \in S \Longrightarrow Ax \in S$$

is said to be *invariant* (for A). Note that if

$$AX = XB, \qquad B \in \mathbb{C}^{k \times k}, \ X \in \mathbb{C}^{n \times k},$$

then $\text{ran}(X)$ is invariant and $By = \lambda y \Rightarrow A(Xy) = \lambda(Xy)$. Thus, if X has full column rank, then $AX = XB$ implies that $\lambda(B) \subseteq \lambda(A)$. If X is square and nonsingular, then A and $B = X^{-1}AX$ are *similar*, X is a *similarity transformation*, and $\lambda(A) = \lambda(B)$.

7.1.2 Decoupling

Many eigenvalue computations involve breaking the given problem down into a collection of smaller eigenproblems. The following result is the basis for these reductions.

Lemma 7.1.1. *If $T \in \mathbb{C}^{n \times n}$ is partitioned as follows,*

$$T = \begin{bmatrix} T_{11} & T_{12} \\ 0 & T_{22} \end{bmatrix} \begin{matrix} p \\ q \end{matrix}$$
$$\phantom{T = \begin{bmatrix} T_{11} \end{bmatrix}} \begin{matrix} p & q \end{matrix}$$

then $\lambda(T) = \lambda(T_{11}) \cup \lambda(T_{22})$.

Proof. Suppose

$$
Tx = \left[\begin{array}{cc} T_{11} & T_{12} \\ 0 & T_{22} \end{array} \right] \left[\begin{array}{c} x_1 \\ x_2 \end{array} \right] = \lambda \left[\begin{array}{c} x_1 \\ x_2 \end{array} \right]
$$

where $x_1 \in \mathbb{C}^p$ and $x_2 \in \mathbb{C}^q$. If $x_2 \neq 0$, then $T_{22}x_2 = \lambda x_2$ and so $\lambda \in \lambda(T_{22})$. If $x_2 = 0$, then $T_{11}x_1 = \lambda x_1$ and so $\lambda \in \lambda(T_{11})$. It follows that $\lambda(T) \subset \lambda(T_{11}) \cup \lambda(T_{22})$. But since both $\lambda(T)$ and $\lambda(T_{11}) \cup \lambda(T_{22})$ have the same cardinality, the two sets are equal. $\square$

7.1.3   Basic Unitary Decompositions

By using similarity transformations, it is possible to reduce a given matrix to any one of several canonical forms. The canonical forms differ in how they display the eigenvalues and in the kind of invariant subspace information that they provide. Because of their numerical stability we begin by discussing the reductions that can be achieved with unitary similarity.

Lemma 7.1.2. *If $A \in \mathbb{C}^{n \times n}$, $B \in \mathbb{C}^{p \times p}$, and $X \in \mathbb{C}^{n \times p}$ satisfy*

$$
AX = XB, \qquad \mathrm{rank}(X) = p, \tag{7.1.5}
$$

then there exists a unitary $Q \in \mathbb{C}^{n \times n}$ such that

$$
Q^H AQ = T = \left[\begin{array}{cc} T_{11} & T_{12} \\ 0 & T_{22} \end{array} \right] \begin{array}{c} p \\ n-p \end{array} \tag{7.1.6}
$$
$$
 \begin{array}{cc} p & n-p \end{array}
$$

and $\lambda(T_{11}) = \lambda(A) \cap \lambda(B)$.

Proof. Let

$$
X = Q \left[\begin{array}{c} R_1 \\ 0 \end{array} \right], \qquad Q \in \mathbb{C}^{n \times n}, \ R_1 \in \mathbb{C}^{p \times p}
$$

be a QR factorization of X. By substituting this into (7.1.5) and rearranging we have

$$
\left[\begin{array}{cc} T_{11} & T_{12} \\ T_{21} & T_{22} \end{array} \right] \left[\begin{array}{c} R_1 \\ 0 \end{array} \right] = \left[\begin{array}{c} R_1 \\ 0 \end{array} \right] B
$$

where

$$
Q^H AQ = \left[\begin{array}{cc} T_{11} & T_{12} \\ T_{21} & T_{22} \end{array} \right] \begin{array}{c} p \\ n-p \end{array} \ .
$$
$$
 \begin{array}{cc} p & n-p \end{array}
$$

By using the nonsingularity of R_1 and the equations $T_{21}R_1 = 0$ and $T_{11}R_1 = R_1 B$, we can conclude that $T_{21} = 0$ and $\lambda(T_{11}) = \lambda(B)$. The lemma follows because from Lemma 7.1.1 we have $\lambda(A) = \lambda(T) = \lambda(T_{11}) \cup \lambda(T_{22})$. $\square$

Lemma 7.1.2 says that a matrix can be reduced to block triangular form using unitary similarity transformations if we know one of its invariant subspaces. By induction we can readily establish the decomposition of Schur (1909).

Theorem 7.1.3 (Schur Decomposition). *If* $A \in \mathbb{C}^{n \times n}$, *then there exists a unitary* $Q \in \mathbb{C}^{n \times n}$ *such that*

$$Q^H A Q = T = D + N \tag{7.1.7}$$

where $D = \text{diag}(\lambda_1, \ldots, \lambda_n)$ *and* $N \in \mathbb{C}^{n \times n}$ *is strictly upper triangular. Furthermore,* Q *can be chosen so that the eigenvalues* λ_i *appear in any order along the diagonal.*

Proof. The theorem obviously holds if $n = 1$. Suppose it holds for all matrices of order $n - 1$ or less. If $Ax = \lambda x$ and $x \neq 0$, then by Lemma 7.1.2 (with $B = (\lambda)$) there exists a unitary U such that

$$U^H A U = \begin{bmatrix} \lambda & w^H \\ 0 & C \end{bmatrix} \begin{matrix} 1 \\ n-1 \end{matrix} .$$
$$\quad\quad\quad\quad\; 1 \quad n-1$$

By induction there is a unitary $\tilde{U}$ such that $\tilde{U}^H C \tilde{U}$ is upper triangular. Thus, if $Q = U \cdot \text{diag}(1, \tilde{U})$, then $Q^H A Q$ is upper triangular. $\quad\square$

If $Q = [\, q_1 \,|\, \cdots \,|\, q_n \,]$ is a column partitioning of the unitary matrix Q in (7.1.7), then the q_i are referred to as *Schur vectors*. By equating columns in the equations $AQ = QT$, we see that the Schur vectors satisfy

$$Aq_k = \lambda_k q_k + \sum_{i=1}^{k-1} n_{ik} q_i, \qquad k = 1{:}n. \tag{7.1.8}$$

From this we conclude that the subspaces

$$S_k = \text{span}\{q_1, \ldots, q_k\}, \qquad k = 1{:}n,$$

are invariant. Moreover, it is not hard to show that if $Q_k = [\, q_1 \,|\, \cdots \,|\, q_k \,]$, then $\lambda(Q_k^H A Q_k) = \{\lambda_1, \ldots, \lambda_k\}$. Since the eigenvalues in (7.1.7) can be arbitrarily ordered, it follows that there is at least one k-dimensional invariant subspace associated with each subset of k eigenvalues. Another conclusion to be drawn from (7.1.8) is that the Schur vector q_k is an eigenvector if and only if the kth column of N is zero. This turns out to be the case for $k = 1{:}n$ whenever $A^H A = A A^H$. Matrices that satisfy this property are called *normal*.

Corollary 7.1.4. $A \in \mathbb{C}^{n \times n}$ *is normal if and only if there exists a unitary* $Q \in \mathbb{C}^{n \times n}$ *such that* $Q^H A Q = \text{diag}(\lambda_1, \ldots, \lambda_n)$.

Proof. See P7.1.1. $\quad\square$

Note that if $Q^H A Q = T = \text{diag}(\lambda_i) + N$ is a Schur decomposition of a general n-by-n matrix A, then $\| N \|_F$ is independent of the choice of Q:

$$\| N \|_F^2 = \| A \|_F^2 - \sum_{i=1}^{n} |\lambda_i|^2 \equiv \Delta^2(A).$$

This quantity is referred to as A's *departure from normality*. Thus, to make T "more diagonal," it is necessary to rely on nonunitary similarity transformations.

7.1.4   Nonunitary Reductions

To see what is involved in nonunitary similarity reduction, we consider the block diagonalization of a 2-by-2 block triangular matrix.

Lemma 7.1.5. *Let $T \in \mathbb{C}^{n \times n}$ be partitioned as follows:*

$$T = \begin{bmatrix} T_{11} & T_{12} \\ 0 & T_{22} \end{bmatrix} \begin{matrix} p \\ q \end{matrix} \quad .$$
$$\quad\quad\quad p \quad\; q$$

Define the linear transformation $\phi : \mathbb{C}^{p \times q} \to \mathbb{C}^{p \times q}$ by

$$\phi(X) = T_{11}X - XT_{22}$$

where $X \in \mathbb{C}^{p \times q}$. Then ϕ is nonsingular if and only if $\lambda(T_{11}) \cap \lambda(T_{22}) = \emptyset$. If ϕ is nonsingular and Y is defined by

$$Y = \begin{bmatrix} I_p & Z \\ 0 & I_q \end{bmatrix}$$

where $\phi(Z) = -T_{12}$, then $Y^{-1}TY = \mathrm{diag}(T_{11}, T_{22})$.

Proof. Suppose $\phi(X) = 0$ for $X \neq 0$ and that

$$U^H X V = \begin{bmatrix} \Sigma_r & 0 \\ 0 & 0 \end{bmatrix} \begin{matrix} r \\ p-r \end{matrix}$$
$$\quad\quad\quad r \quad\; q-r$$

is the SVD of X with $\Sigma_r = \mathrm{diag}(\sigma_i)$, $r = \mathrm{rank}(X)$. Substituting this into the equation $T_{11}X = XT_{22}$ gives

$$\begin{bmatrix} A_{11} & A_{12} \\ A_{21} & A_{22} \end{bmatrix} \begin{bmatrix} \Sigma_r & 0 \\ 0 & 0 \end{bmatrix} = \begin{bmatrix} \Sigma_r & 0 \\ 0 & 0 \end{bmatrix} \begin{bmatrix} B_{11} & B_{12} \\ B_{21} & B_{22} \end{bmatrix}$$

where $U^H T_{11} U = (A_{ij})$ and $V^H T_{22} V = (B_{ij})$. By comparing blocks in this equation it is clear that $A_{21} = 0$, $B_{12} = 0$, and $\lambda(A_{11}) = \lambda(B_{11})$. Consequently, A_{11} and B_{11} have an eigenvalue in common and that eigenvalue is in $\lambda(T_{11}) \cap \lambda(T_{22})$. Thus, if ϕ is singular, then T_{11} and T_{22} have an eigenvalue in common. On the other hand, if $\lambda \in \lambda(T_{11}) \cap \lambda(T_{22})$, then we have eigenvector equations $T_{11}x = \lambda x$ and $y^H T_{22} = \lambda y^H$. A calculation shows that $\phi(xy^H) = 0$ confirming that ϕ is singular.

Finally, if ϕ is nonsingular, then $\phi(Z) = -T_{12}$ has a solution and

$$Y^{-1}TY = \begin{bmatrix} I_p & -Z \\ 0 & I_q \end{bmatrix} \begin{bmatrix} T_{11} & T_{12} \\ 0 & T_{22} \end{bmatrix} \begin{bmatrix} I_p & Z \\ 0 & I_q \end{bmatrix} = \begin{bmatrix} T_{11} & T_{11}Z - ZT_{22} + T_{12} \\ 0 & T_{22} \end{bmatrix}$$

has the required block diagonal form. $\square$

By repeatedly applying this lemma, we can establish the following more general result.

Theorem 7.1.6 (Block Diagonal Decomposition). *Suppose*

$$Q^H A Q = T = \begin{bmatrix} T_{11} & T_{12} & \cdots & T_{1q} \\ 0 & T_{22} & \cdots & T_{2q} \\ \vdots & \vdots & \ddots & \vdots \\ 0 & 0 & \cdots & T_{qq} \end{bmatrix} \tag{7.1.9}$$

is a Schur decomposition of $A \in \mathbb{C}^{n \times n}$ *and that the* T_{ii} *are square. If* $\lambda(T_{ii}) \cap \lambda(T_{jj}) = \emptyset$ *whenever* $i \neq j$, *then there exists a nonsingular matrix* $Y \in \mathbb{C}^{n \times n}$ *such that*

$$(QY)^{-1} A (QY) = \operatorname{diag}(T_{11}, \ldots, T_{qq}). \tag{7.1.10}$$

Proof. See P7.1.2. □

If each diagonal block T_{ii} is associated with a distinct eigenvalue, then we obtain

Corollary 7.1.7. *If* $A \in \mathbb{C}^{n \times n}$, *then there exists a nonsingular* X *such that*

$$X^{-1} A X = \operatorname{diag}(\lambda_1 I + N_1, \ldots, \lambda_q I + N_q) \qquad N_i \in \mathbb{C}^{n_i \times n_i} \tag{7.1.11}$$

where $\lambda_1, \ldots, \lambda_q$ *are distinct, the integers* $n_1, \ldots, n_q$ *satisfy* $n_1 + \cdots + n_q = n$, *and each* N_i *is strictly upper triangular.*

A number of important terms are connected with decomposition (7.1.11). The integer n_i is referred to as the *algebraic multiplicity* of λ_i. If $n_i = 1$, then λ_i is said to be *simple*. The *geometric multiplicity* of λ_i equals the dimensions of $\mathsf{null}(N_i)$, i.e., the number of linearly independent eigenvectors associated with λ_i. If the algebraic multiplicity of λ_i exceeds its geometric multiplicity, then λ_i is said to be a *defective eigenvalue*. A matrix with a defective eigenvalue is referred to as a *defective matrix*. Nondefective matrices are also said to be *diagonalizable*.

Corollary 7.1.8 (Diagonal Form). $A \in \mathbb{C}^{n \times n}$ *is nondefective if and only if there exists a nonsingular* $X \in \mathbb{C}^{n \times n}$ *such that*

$$X^{-1} A X = \operatorname{diag}(\lambda_1, \ldots, \lambda_n). \tag{7.1.12}$$

Proof. A is nondefective if and only if there exist independent vectors $x_1 \ldots x_n \in \mathbb{C}^n$ and scalars $\lambda_1, \ldots, \lambda_n$ such that $A x_i = \lambda_i x_i$ for $i = 1{:}n$. This is equivalent to the existence of a nonsingular $X = [\, x_1 \mid \cdots \mid x_n \,] \in \mathbb{C}^{n \times n}$ such that $AX = XD$ where $D = \operatorname{diag}(\lambda_1, \ldots, \lambda_n)$. □

Note that if y_i^H is the ith row of X^{-1}, then $y_i^H A = \lambda_i y_i^H$. Thus, the columns of X^{-H} are left eigenvectors and the columns of X are right eigenvectors.

If we partition the matrix X in (7.1.11),

$$X = \underset{n_1 \qquad\quad n_q}{[\, X_1 \mid \cdots \mid X_q \,]}$$

then $\mathbb{C}^n = \text{ran}(X_1) \oplus \ldots \oplus \text{ran}(X_q)$, a direct sum of invariant subspaces. If the bases for these subspaces are chosen in a special way, then it is possible to introduce even more zeroes into the upper triangular portion of $X^{-1}AX$.

Theorem 7.1.9 (Jordan Decomposition). *If $A \in \mathbb{C}^{n \times n}$, then there exists a non-singular $X \in \mathbb{C}^{n \times n}$ such that $X^{-1}AX = \text{diag}(J_1, \ldots, J_q)$ where*

$$
J_i = \begin{bmatrix} \lambda_i & 1 & & \cdots & 0 \\ 0 & \lambda_i & \ddots & & \vdots \\ & & \ddots & \ddots & \ddots \\ \vdots & & & \ddots & \ddots & 1 \\ 0 & \cdots & & & 0 & \lambda_i \end{bmatrix} \in \mathbb{C}^{n_i \times n_i}
$$

and $n_1 + \cdots + n_q = n$.

Proof. See Horn and Johnson (MA, p. 330) $\square$

The J_i are referred to as *Jordan blocks*. The number and dimensions of the Jordan blocks associated with each distinct eigenvalue are unique, although their ordering along the diagonal is not.

7.1.5   Some Comments on Nonunitary Similarity

The Jordan block structure of a defective matrix is difficult to determine numerically. The set of n-by-n diagonalizable matrices is dense in $\mathbb{C}^{n \times n}$, and thus, small changes in a defective matrix can radically alter its Jordan form. We have more to say about this in §7.6.5.

A related difficulty that arises in the eigenvalue problem is that a nearly defective matrix can have a poorly conditioned matrix of eigenvectors. For example, any matrix X that diagonalizes

$$
A = \begin{bmatrix} 1+\epsilon & 1 \\ 0 & 1-\epsilon \end{bmatrix}, \qquad 0 < \epsilon \ll 1, \tag{7.1.13}
$$

has a 2-norm condition of order $1/\epsilon$.

These observations serve to highlight the difficulties associated with ill-conditioned similarity transformations. Since

$$
\text{fl}(X^{-1}AX) = X^{-1}AX + E, \tag{7.1.14}
$$

where

$$
\| E \|_2 \approx \mathbf{u} \cdot \kappa_2(X) \| A \|_2, \tag{7.1.15}
$$

it is clear that large errors can be introduced into an eigenvalue calculation when we depart from unitary similarity.

7.1.6 Singular Values and Eigenvalues

Since the singular values of A and its Schur decomposition $Q^H A Q = \text{diag}(\lambda_i) + N$ are the same, it follows that

$$\sigma_{\min}(A) \leq \min_{1 \leq i \leq n} |\lambda_i| \leq \max_{1 \leq i \leq n} |\lambda_i| \leq \sigma_{\max}(A).$$

From what we know about the condition of triangular matrices, it may be the case that

$$\max_{1 \leq i,j \leq n} \frac{|\lambda_i|}{|\lambda_j|} \ll \kappa_2(A).$$

See §5.4.3. This is a reminder that for nonnormal matrices, eigenvalues do not have the "predictive power" of singular values when it comes to $Ax = b$ sensitivity matters. Eigenvalues of nonnormal matrices have other shortcomings, a topic that is the focus of §7.9.

Problems

P7.1.1 (a) Show that if $T \in \mathbb{C}^{n \times n}$ is upper triangular and normal, then T is diagonal. (b) Show that if A is normal and $Q^H A Q = T$ is a Schur decomposition, then T is diagonal. (c) Use (a) and (b) to complete the proof of Corollary 7.1.4.

P7.1.2 Prove Theorem 7.1.6 by using induction and Lemma 7.1.5.

P7.1.3 Suppose $A \in \mathbb{C}^{n \times n}$ has distinct eigenvalues. Show that if $Q^H A Q = T$ is its Schur decomposition and $AB = BA$, then $Q^H B Q$ is upper triangular.

P7.1.4 Show that if A and B^H are in $\mathbb{C}^{m \times n}$ with $m \geq n$, then

$$\lambda(AB) = \lambda(BA) \cup \{ \underbrace{0, \ldots, 0}_{m \ n} \}.$$

P7.1.5 Given $A \in \mathbb{C}^{n \times n}$, use the Schur decomposition to show that for every $\epsilon > 0$, there exists a diagonalizable matrix B such that $\| A - B \|_2 \leq \epsilon$. This shows that the set of diagonalizable matrices is dense in $\mathbb{C}^{n \times n}$ and that the Jordan decomposition is not a continuous matrix decomposition.

P7.1.6 Suppose $A_k \to A$ and that $Q_k^H A_k Q_k = T_k$ is a Schur decomposition of A_k. Show that $\{Q_k\}$ has a converging subsequence $\{Q_{k_i}\}$ with the property that

$$\lim_{i \to \infty} Q_{k_i} = Q$$

where $Q^H A Q = T$ is upper triangular. This shows that the eigenvalues of a matrix are continuous functions of its entries.

P7.1.7 Justify (7.1.14) and (7.1.15).

P7.1.8 Show how to compute the eigenvalues of

$$M = \begin{bmatrix} A & C \\ B & D \end{bmatrix} \begin{matrix} k \\ j \end{matrix}$$

where A, B, C, and D are given real diagonal matrices.

P7.1.9 Use the Jordan decomposition to show that if all the eigenvalues of a matrix A are strictly less than unity, then $\lim_{k \to \infty} A^k = 0$.

P7.1.10 The initial value problem

$$\begin{aligned} \dot{x}(t) &= y(t), & x(0) &= 1, \\ \dot{y}(t) &= -x(t), & y(0) &= 0, \end{aligned}$$

has solution $x(t) = \cos(t)$ and $y(t) = \sin(t)$. Let $h > 0$. Here are three reasonable iterations that can be used to compute approximations $x_k \approx x(kh)$ and $y_k \approx y(kh)$ assuming that $x_0 = 1$ and $y_k = 0$:

$$\text{Method 1:}\quad \begin{array}{rcl} x_{k+1} & = & x_k + hy_k, \\ y_{k+1} & = & y_k - hx_k, \end{array}$$

$$\text{Method 2:}\quad \begin{array}{rcl} x_{k+1} & = & x_k + hy_k, \\ y_{k+1} & = & y_k - hx_{k+1}, \end{array}$$

$$\text{Method 3:}\quad \begin{array}{rcl} x_{k+1} & = & x_k + hy_{k+1}, \\ y_{k+1} & = & y_k - hx_{k+1}. \end{array}$$

Express each method in the form

$$\left[\begin{array}{c} x_{k+1} \\ y_{k+1} \end{array} \right] = A_h \left[\begin{array}{c} x_k \\ y_k \end{array} \right]$$

where A_h is a 2-by-2 matrix. For each case, compute $\lambda(A_h)$ and use the previous problem to discuss $\lim x_k$ and $\lim y_k$ as $k \rightarrow \infty$.

P7.1.11 If $J \in \mathbb{R}^{d \times d}$ is a Jordan block, what is $\kappa_\infty(J)$?

P7.1.12 Suppose $A, B \in \mathbb{C}^{n \times n}$. Show that the $2n$-by-$2n$ matrices

$$M_1 = \left[\begin{array}{cc} AB & 0 \\ B & 0 \end{array} \right] \quad \text{and} \quad M_2 = \left[\begin{array}{cc} 0 & 0 \\ B & BA \end{array} \right]$$

are similar thereby showing that $\lambda(AB) = \lambda(BA)$.

P7.1.13 Suppose $A \in \mathbb{R}^{n \times n}$. We say that $B \in \mathbb{R}^{n \times n}$ is the *Drazin inverse* of A if (i) $AB = BA$, (ii) $BAB = B$, and (iii) the spectral radius of $A - ABA$ is zero. Give a formula for B in terms of the Jordan decomposition of A paying particular attention to the blocks associated with A's zero eigenvalues.

P7.1.14 Show that if $A \in \mathbb{R}^{n \times n}$, then $\rho(A) \geq (\sigma_1 \cdots \sigma_n)^{1/n}$ where $\sigma_1, \ldots, \sigma_n$ are the singular values of A.

P7.1.15 Consider the polynomial $q(x) = \det(I_n + xA)$ where $A \in \mathbb{R}^{n \times n}$. We wish to compute the coefficient of x^2. (a) Specify the coefficient in terms of the eigenvalues $\lambda_1, \ldots, \lambda_n$ of A. (b) Give a simple formula for the coefficient in terms of $\text{tr}(A)$ and $\text{tr}(A^2)$.

P7.1.16 Given $A \in \mathbb{R}^{2 \times 2}$, show that there exists a nonsingular $X \in \mathbb{R}^{2 \times 2}$ so $X^{-1}AX = A^T$. See Dubrulle and Parlett (2007).

Notes and References for §7.1

For additional discussion about the linear algebra behind the eigenvalue problem, see Horn and Johnson (MA) and:

L. Mirsky (1963). *An Introduction to Linear Algebra,* Oxford University Press, Oxford, U.K.

M. Marcus and H. Minc (1964). *A Survey of Matrix Theory and Matrix Inequalities,* Allyn and Bacon, Boston.

R. Bellman (1970). *Introduction to Matrix Analysis,* second edition, McGraw-Hill, New York.

I. Gohberg, P. Lancaster, and L. Rodman (2006). *Invariant Subspaces of Matrices with Applications,* SIAM Publications, Philadelphia, PA.

For a general discussion about the similarity connection between a matrix and its transpose, see:

A.A. Dubrulle and B.N. Parlett (2010). "Revelations of a Transposition Matrix," *J. Comp. and Appl. Math. 233,* 1217–1219.

The Schur decomposition originally appeared in:

I. Schur (1909). "On the Characteristic Roots of a Linear Substitution with an Application to the Theory of Integral Equations." *Math. Ann. 66,* 488-510 (German).

A proof very similar to ours is given in:

H.W. Turnbull and A.C. Aitken (1961). *An Introduction to the Theory of Canonical Forms,* Dover, New York, 105.

7.2   Perturbation Theory

The act of computing eigenvalues is the act of computing zeros of the characteristic polynomial. Galois theory tells us that such a process has to be iterative if $n > 4$ and so errors arise because of finite termination. In order to develop intelligent stopping criteria we need an informative perturbation theory that tells us how to think about approximate eigenvalues and invariant subspaces.

7.2.1   Eigenvalue Sensitivity

An important framework for eigenvalue computation is to produce a sequence of similarity transformations $\{X_k\}$ with the property that the matrices $X_k^{-1}AX_k$ are progressively "more diagonal." The question naturally arises, how well do the diagonal elements of a matrix approximate its eigenvalues?

Theorem 7.2.1 (Gershgorin Circle Theorem). *If $X^{-1}AX = D + F$ where $D = \mathrm{diag}(d_1, \ldots, d_n)$ and F has zero diagonal entries, then*

$$\lambda(A) \subseteq \bigcup_{i=1}^{n} D_i$$

where $D_i = \{z \in \mathbb{C} : |z - d_i| \leq \sum_{j=1}^{n} |f_{ij}|\}$.

Proof. Suppose $\lambda \in \lambda(A)$ and assume without loss of generality that $\lambda \neq d_i$ for $i = 1{:}n$. Since $(D - \lambda I) + F$ is singular, it follows from Lemma 2.3.3 that

$$1 \leq \| (D - \lambda I)^{-1} F \|_\infty = \sum_{j=1}^{n} \frac{|f_{kj}|}{|d_k - \lambda|}$$

for some k, $1 \leq k \leq n$. But this implies that $\lambda \in D_k$. $\square$

It can also be shown that if the Gershgorin disk D_i is isolated from the other disks, then it contains precisely one eigenvalue of A. See Wilkinson (AEP, pp. 71ff.).

For some methods it is possible to show that the computed eigenvalues are the exact eigenvalues of a matrix $A + E$ where E is small in norm. Consequently, we should understand how the eigenvalues of a matrix can be affected by small perturbations.

Theorem 7.2.2 (Bauer-Fike). *If μ is an eigenvalue of $A + E \in \mathbb{C}^{n \times n}$ and $X^{-1}AX = D = \mathrm{diag}(\lambda_1, \ldots, \lambda_n)$, then*

$$\min_{\lambda \in \lambda(A)} |\lambda - \mu| \leq \kappa_p(X)\| E \|_p$$

where $\| \cdot \|_p$ denotes any of the p-norms.

Proof. If $\mu \in \lambda(A)$, then the theorem is obviously true. Otherwise if the matrix $X^{-1}(A + E - \mu I)X$ is singular, then so is $I + (D - \mu I)^{-1}(X^{-1}EX)$. Thus, from

Lemma 2.3.3 we obtain

$$1 \;\leq\; \| \, (D - \mu I)^{-1}(X^{-1}EX) \, \|_p \;\leq\; \| \, (D - \mu I)^{-1} \, \|_p \| \, X \, \|_p \| \, E \, \|_p \| \, X^{-1} \, \|_p \, .$$

Since $(D - \mu I)^{-1}$ is diagonal and the p-norm of a diagonal matrix is the absolute value of the largest diagonal entry, it follows that

$$\| \, (D - \mu I)^{-1} \, \|_p \;=\; \max_{\lambda \in \lambda(A)} \frac{1}{|\lambda - \mu|},$$

completing the proof.   □

An analogous result can be obtained via the Schur decomposition:

Theorem 7.2.3. *Let $Q^H A Q = D + N$ be a Schur decomposition of $A \in \mathbb{C}^{n \times n}$ as in (7.1.7). If $\mu \in \lambda(A + E)$ and p is the smallest positive integer such that $|N|^p = 0$, then*

$$\min_{\lambda \in \lambda(A)} |\lambda - \mu| \;\leq\; \max\{\theta, \, \theta^{1/p}\}$$

where

$$\theta \;=\; \| \, E \, \|_2 \sum_{k=0}^{p-1} \| \, N \, \|_2^k \, .$$

Proof. Define

$$\delta \;=\; \min_{\lambda \in \lambda(A)} |\lambda - \mu| \;=\; \frac{1}{\| \, (\mu I - D)^{-1} \, \|_2} \, .$$

The theorem is clearly true if $\delta = 0$. If $\delta > 0$, then $I - (\mu I - A)^{-1}E$ is singular and by Lemma 2.3.3 we have

$$1 \leq \| \, (\mu I - A)^{-1}E \, \|_2 \;\leq\; \| \, (\mu I - A)^{-1} \, \|_2 \| \, E \, \|_2 \tag{7.2.1}$$

$$= \| \, ((\mu I - D) - N)^{-1} \, \|_2 \| \, E \, \|_2 \, .$$

Since $(\mu I - D)^{-1}$ is diagonal and $|N|^p = 0$, it follows that $((\mu I - D)^{-1}N)^p = 0$. Thus,

$$((\mu I - D) - N)^{-1} \;=\; \sum_{k=0}^{p-1} \left((\mu I - D)^{-1}N\right)^k (\mu I - D)^{-1}$$

and so

$$\| \, ((\mu I - D) - N)^{-1} \, \|_2 \;\leq\; \frac{1}{\delta} \sum_{k=0}^{p-1} \left(\frac{\| \, N \, \|_2}{\delta}\right)^k \, .$$

If $\delta > 1$, then

$$\| \, (\mu I - D) - N)^{-1} \, \|_2 \;\leq\; \frac{1}{\delta} \sum_{k=0}^{p-1} \| \, N \, \|_2^k$$

and so from (7.2.1), $\delta \le \theta$. If $\delta \le 1$, then

$$\| (\mu I - D) - N)^{-1} \|_2 \le \frac{1}{\delta^p} \sum_{k=0}^{p-1} \| N \|_2^k.$$

By using (7.2.1) again we have $\delta^p \le \theta$ and so $\delta \le \max\{\theta, \theta^{1/p}\}$. $\square$

Theorems 7.2.2 and 7.2.3 suggest that the eigenvalues of a nonnormal matrix may be sensitive to perturbations. In particular, if $\kappa_2(X)$ or $\| N \|_2^{p-1}$ is large, then small changes in A can induce large changes in the eigenvalues.

7.2.2 The Condition of a Simple Eigenvalue

Extreme eigenvalue sensitivity for a matrix A cannot occur if A is normal. On the other hand, nonnormality does not necessarily imply eigenvalue sensitivity. Indeed, a nonnormal matrix can have a mixture of well-conditioned and ill-conditioned eigenvalues. For this reason, it is beneficial to refine our perturbation theory so that it is applicable to individual eigenvalues and not the spectrum as a whole.

To this end, suppose that λ is a simple eigenvalue of $A \in \mathbb{C}^{n \times n}$ and that x and y satisfy $Ax = \lambda x$ and $y^H A = \lambda y^H$ with $\| x \|_2 = \| y \|_2 = 1$. If $Y^H A X = J$ is the Jordan decomposition with $Y^H = X^{-1}$, then y and x are nonzero multiples of $X(:,i)$ and $Y(:,i)$ for some i. It follows from $1 = Y(:,i)^H X(:,i)$ that $y^H x \ne 0$, a fact that we shall use shortly.

Using classical results from function theory, it can be shown that in a neighborhood of the origin there exist differentiable $x(\epsilon)$ and $\lambda(\epsilon)$ such that

$$(A + \epsilon F)x(\epsilon) = \lambda(\epsilon)x(\epsilon), \qquad \| F \|_2 = 1,$$

where $\lambda(0) = \lambda$ and $x(0) = x$. By differentiating this equation with respect to ϵ and setting $\epsilon = 0$ in the result, we obtain

$$A\dot{x}(0) + Fx = \dot{\lambda}(0)x + \lambda\dot{x}(0).$$

Applying y^H to both sides of this equation, dividing by $y^H x$, and taking absolute values gives

$$|\dot{\lambda}(0)| = \left| \frac{y^H F x}{y^H x} \right| \le \frac{1}{|y^H x|}.$$

The upper bound is attained if $F = yx^H$. For this reason we refer to the reciprocal of

$$s(\lambda) = |y^H x| \tag{7.2.2}$$

as the *condition of the eigenvalue* λ.

Roughly speaking, the above analysis shows that $O(\epsilon)$ perturbations in A can induce $\epsilon/s(\lambda)$ changes in an eigenvalue. Thus, if $s(\lambda)$ is small, then λ is appropriately regarded as ill-conditioned. Note that $s(\lambda)$ is the cosine of the angle between the left and right eigenvectors associated with λ and is unique only if λ is simple.

A small $s(\lambda)$ implies that A is near a matrix having a multiple eigenvalue. In particular, if λ is distinct and $s(\lambda) < 1$, then there exists an E such that λ is a repeated eigenvalue of $A + E$ and

$$\frac{\| E \|_2}{\| A \|_2} \leq \frac{s(\lambda)}{\sqrt{1 - s(\lambda)^2}} \, .$$

This result is proved by Wilkinson (1972).

7.2.3   Sensitivity of Repeated Eigenvalues

If λ is a repeated eigenvalue, then the eigenvalue sensitivity question is more complicated. For example, if

$$A \; = \; \left[\begin{array}{cc} 1 & a \\ 0 & 1 \end{array} \right] \qquad \text{and} \qquad F \; = \; \left[\begin{array}{cc} 0 & 0 \\ 1 & 0 \end{array} \right],$$

then $\lambda(A + \epsilon F) = \{1 \pm \sqrt{\epsilon a}\}$. Note that if $a \neq 0$, then it follows that the eigenvalues of $A + \epsilon F$ are not differentiable at zero; their rate of change at the origin is infinite. In general, if λ is a defective eigenvalue of A, then $O(\epsilon)$ perturbations in A can result in $O(\epsilon^{1/p})$ perturbations in λ if λ is associated with a p-dimensional Jordan block. See Wilkinson (AEP, pp. 77ff.) for a more detailed discussion.

7.2.4   Invariant Subspace Sensitivity

A collection of sensitive eigenvectors can define an insensitive invariant subspace provided the corresponding cluster of eigenvalues is isolated. To be precise, suppose

$$Q^H A Q \; = \; \left[\begin{array}{cc} T_{11} & T_{12} \\ 0 & T_{22} \end{array} \right] \begin{array}{l} r \\ n-r \end{array} \qquad\qquad (7.2.3)$$
$$ \begin{array}{cc} r & n-r \end{array}$$

is a Schur decomposition of A with

$$Q \; = \; [\, Q_1 \mid Q_2 \,] \quad .$$
$$ \begin{array}{cc} r & n-r \end{array} \qquad\qquad (7.2.4)$$

It is clear from our discussion of eigenvector perturbation that the sensitivity of the invariant subspace $\mathsf{ran}(Q_1)$ depends on the distance between $\lambda(T_{11})$ and $\lambda(T_{22})$. The proper measure of this distance turns out to be the smallest singular value of the linear transformation $X \to T_{11}X - XT_{22}$. (Recall that this transformation figures in Lemma 7.1.5.) In particular, if we define the *separation* between the matrices T_{11} and T_{22} by

$$\mathsf{sep}(T_{11}, T_{22}) \; = \; \min_{X \neq 0} \frac{\| T_{11}X - XT_{22} \|_F}{\| X \|_F}, \qquad\qquad (7.2.5)$$

then we have the following general result:

Theorem 7.2.4. *Suppose that (7.2.3) and (7.2.4) hold and that for any matrix $E \in \mathbb{C}^{n \times n}$ we partition $Q^H E Q$ as follows:*

$$Q^H E Q = \begin{bmatrix} E_{11} & E_{12} \\ E_{21} & E_{22} \end{bmatrix} \begin{matrix} r \\ n-r \end{matrix}$$
$$\phantom{Q^H E Q = \begin{bmatrix}} r \quad\; n-r$$

If $\mathsf{sep}(T_{11}, T_{22}) > 0$ *and*

$$\| E \|_F \left(1 + \frac{5\| T_{12} \|_F}{\mathsf{sep}(T_{11}, T_{22})} \right) \leq \frac{\mathsf{sep}(T_{11}, T_{22})}{5},$$

then there exists a $P \in \mathbb{C}^{(n-r) \times r}$ *with*

$$\| P \|_F \leq 4 \frac{\| E_{21} \|_F}{\mathsf{sep}(T_{11}, T_{22})}$$

such that the columns of $\widetilde{Q}_1 = (Q_1 + Q_2 P)(I + P^H P)^{-1/2}$ *are an orthonormal basis for a subspace invariant for* $A + E$.

Proof. This result is a slight recasting of Theorem 4.11 in Stewart (1973) which should be consulted for proof details. See also Stewart and Sun (MPA, p. 230). The matrix $(I + P^H P)^{-1/2}$ is the inverse of the square root of the symmetric positive definite matrix $I + P^H P$. See §4.2.4. □

Corollary 7.2.5. *If the assumptions in Theorem 7.2.4 hold, then*

$$\mathsf{dist}(\mathsf{ran}(Q_1), \mathsf{ran}(\widetilde{Q}_1)) \leq 4 \frac{\| E_{21} \|_F}{\mathsf{sep}(T_{11}, T_{22})}.$$

Proof. Using the SVD of P, it can be shown that

$$\| P(I + P^H P)^{-1/2} \|_2 \leq \| P \|_2 \leq \| P \|_F. \tag{7.2.6}$$

Since the required distance is the 2-norm of $Q_2^H \widetilde{Q}_1 = P(I + P^H P)^{-1/2}$, the proof is complete. □

Thus, the reciprocal of $\mathsf{sep}(T_{11}, T_{22})$ can be thought of as a condition number that measures the sensitivity of $\mathsf{ran}(Q_1)$ as an invariant subspace.

7.2.5 Eigenvector Sensitivity

If we set $r = 1$ in the preceding subsection, then the analysis addresses the issue of eigenvector sensitivity.

Corollary 7.2.6. *Suppose* $A, E \in \mathbb{C}^{n \times n}$ *and that* $Q = [\, q_1 \mid Q_2 \,] \in \mathbb{C}^{n \times n}$ *is unitary with* $q_1 \in \mathbb{C}^n$. *Assume*

$$Q^H A Q = \begin{bmatrix} \lambda & v^H \\ 0 & T_{22} \end{bmatrix} \begin{matrix} 1 \\ n-1 \end{matrix}, \qquad Q^H E Q = \begin{bmatrix} \epsilon & \gamma^H \\ \delta & E_{22} \end{bmatrix} \begin{matrix} 1 \\ n-1 \end{matrix}.$$
$$\phantom{Q^H A Q = \begin{bmatrix}} 1 \quad n-1 1 \quad n-1$$

(Thus, q_1 is an eigenvector.) If $\sigma = \sigma_{\min}(T_{22} - \lambda I) > 0$ and

$$\| E \|_F \left(1 + \frac{5\| v \|_2}{\sigma} \right) \leq \frac{\sigma}{5},$$

then there exists $p \in \mathbb{C}^{n-1}$ *with*

$$\| p \|_2 \leq 4\frac{\| \delta \|_2}{\sigma}$$

such that $\tilde{q}_1 = (q_1 + Q_2 p)/\sqrt{1 + p^H p}$ *is a unit 2-norm eigenvector for $A + E$. Moreover,*

$$\mathsf{dist}(\mathsf{span}\{q_1\}, \mathsf{span}\{\tilde{q}_1\}) \leq 4\frac{\| \delta \|_2}{\sigma}.$$

Proof. The result follows from Theorem 7.2.4, Corollary 7.2.5, and the observation that if $T_{11} = \lambda$, then $\mathsf{sep}(T_{11}, T_{22}) = \sigma_{\min}(T_{22} - \lambda I)$. $\square$

Note that $\sigma_{\min}(T_{22} - \lambda I)$ roughly measures the separation of λ from the eigenvalues of T_{22}. We have to say "roughly" because

$$\mathsf{sep}(\lambda, T_{22}) = \sigma_{\min}(T_{22} - \lambda I) \leq \min_{\mu \in \lambda(T_{22})} |\mu - \lambda|$$

and the upper bound can be a gross overestimate.

That the separation of the eigenvalues should have a bearing upon eigenvector sensitivity should come as no surprise. Indeed, if λ is a nondefective, repeated eigenvalue, then there are an infinite number of possible eigenvector bases for the associated invariant subspace. The preceding analysis merely indicates that this indeterminancy begins to be felt as the eigenvalues coalesce. In other words, the eigenvectors associated with nearby eigenvalues are "wobbly."

Problems

P7.2.1 Suppose $Q^H A Q = \mathrm{diag}(\lambda_1) + N$ is a Schur decomposition of $A \in \mathbb{C}^{n \times n}$ and define $\nu(A) = \| A^H A - A A^H \|_F$. The upper and lower bounds in

$$\frac{\nu(A)^2}{6\| A \|_F^2} \leq \| N \|_F^2 \leq \sqrt{\frac{n^3 - n}{12}}\nu(A)$$

are established by Henrici (1962) and Eberlein (1965), respectively. Verify these results for the case $n = 2$.

P7.2.2 Suppose $A \in \mathbb{C}^{n \times n}$ and $X^{-1} A X = \mathrm{diag}(\lambda_1, \ldots, \lambda_n)$ with distinct λ_i. Show that if the columns of X have unit 2-norm, then $\kappa_F(X)^2 = n(1/s(\lambda_1)^2 + \cdots + 1/s(\lambda_n)^2)$.

P7.2.3 Suppose $Q^H A Q = \mathrm{diag}(\lambda_i) + N$ is a Schur decomposition of A and that $X^{-1} A X = \mathrm{diag}(\lambda_i)$. Show $\kappa_2(X)^2 \geq 1 + (\| N \|_F/\| A \|_F)^2$. See Loizou (1969).

P7.2.4 If $X^{-1} A X = \mathrm{diag}(\lambda_i)$ and $|\lambda_1| \geq \cdots \geq |\lambda_n|$, then

$$\frac{\sigma_i(A)}{\kappa_2(X)} \leq |\lambda_i| \leq \kappa_2(X)\sigma_i(A).$$

Prove this result for the $n = 2$ case. See Ruhe (1975).

P7.2.5 Show that if $A = \begin{bmatrix} a & c \\ 0 & b \end{bmatrix}$ and $a \neq b$, then $s(a) = s(b) = (1 + |c/(a-b)|^2)^{-1/2}$.

P7.2.6 Suppose

$$A = \begin{bmatrix} \lambda & v^T \\ 0 & T_{22} \end{bmatrix}$$

and that $\lambda \notin \lambda(T_{22})$. Show that if $\sigma = \mathsf{sep}(\lambda, T_{22})$, then

$$s(\lambda) \;=\; \frac{1}{\sqrt{1 + \parallel (T_{22} - \lambda I)^{-1} v \parallel_2^2}} \;\leq\; \frac{\sigma}{\sqrt{\sigma^2 + \parallel v \parallel_2^2}} \; .$$

where $s(\lambda)$ is defined in (7.2.2).

P7.2.7 Show that the condition of a simple eigenvalue is preserved under unitary similarity transformations.

P7.2.8 With the same hypothesis as in the Bauer-Fike theorem (Theorem 7.2.2), show that

$$\min_{\lambda \in \lambda(A)} \; |\lambda - \mu| \;\leq\; \parallel |X^{-1}| \, |E| \, |X| \parallel_p.$$

P7.2.9 Verify (7.2.6).

P7.2.10 Show that if $B \in \mathbb{C}^{m \times m}$ and $C \in \mathbb{C}^{n \times n}$, then $\mathsf{sep}(B, C)$ is less than or equal to $|\lambda - \mu|$ for all $\lambda \in \lambda(B)$ and $\mu \in \lambda(C)$.

Notes and References for §7.2

Many of the results presented in this section may be found in Wilkinson (AEP), Stewart and Sun (MPA) as well as:

F.L. Bauer and C.T. Fike (1960). "Norms and Exclusion Theorems," *Numer. Math. 2*, 123–44.

A.S. Householder (1964). *The Theory of Matrices in Numerical Analysis*. Blaisdell, New York.

R. Bhatia (2007). *Perturbation Bounds for Matrix Eigenvalues*, SIAM Publications, Philadelphia, PA.

Early papers concerned with the effect of perturbations on the eigenvalues of a general matrix include:

A. Ruhe (1970). "Perturbation Bounds for Means of Eigenvalues and Invariant Subspaces," *BIT 10*, 343–54.

A. Ruhe (1970). "Properties of a Matrix with a Very Ill-Conditioned Eigenproblem," *Numer. Math. 15*, 57–60.

J.H. Wilkinson (1972). "Note on Matrices with a Very Ill-Conditioned Eigenproblem," *Numer. Math. 19*, 176–78.

W. Kahan, B.N. Parlett, and E. Jiang (1982). "Residual Bounds on Approximate Eigensystems of Nonnormal Matrices," *SIAM J. Numer. Anal. 19*, 470–484.

J.H. Wilkinson (1984). "On Neighboring Matrices with Quadratic Elementary Divisors," *Numer. Math. 44*, 1-21.

Wilkinson's work on nearest defective matrices is typical of a growing body of literature that is concerned with "nearness" problems, see:

A. Ruhe (1987). "Closest Normal Matrix Found!," *BIT 27*, 585-598.

J.W. Demmel (1987). "On the Distance to the Nearest Ill-Posed Problem," *Numer. Math. 51*, 251–289.

J.W. Demmel (1988). "The Probability that a Numerical Analysis Problem is Difficult," *Math. Comput. 50*, 449–480.

N.J. Higham (1989). "Matrix Nearness Problems and Applications," in *Applications of Matrix Theory*, M.J.C. Gover and S. Barnett (eds.), Oxford University Press, Oxford, 1–27.

A.N. Malyshev (1999). "A Formula for the 2-norm Distance from a Matrix to the Set of Matrices with Multiple Eigenvalues," *Numer. Math. 83*, 443–454.

J.-M. Gracia (2005). "Nearest Matrix with Two Prescribed Eigenvalues," *Lin. Alg. Applic. 401*, 277–294.

An important subset of this literature is concerned with nearness to the set of unstable matrices. A matrix is *unstable* if it has an eigenvalue with nonnegative real part. Controllability is a related notion, see:

C. Van Loan (1985). "How Near is a Stable Matrix to an Unstable Matrix?," *Contemp. Math. 47*, 465–477.

J.W. Demmel (1987). "A Counterexample for two Conjectures About Stability," *IEEE Trans. Autom. Contr. AC-32,* 340–342.

R. Byers (1988). "A Bisection Method for Measuring the distance of a Stable Matrix to the Unstable Matrices," *J. Sci. Stat. Comput. 9*, 875–881.

J.V. Burke and M.L. Overton (1992). "Stable Perturbations of Nonsymmetric Matrices," *Lin. Alg. Applic. 171*, 249–273.

C. He and G.A. Watson (1998). "An Algorithm for Computing the Distance to Instability," *SIAM J. Matrix Anal. Applic. 20*, 101–116.

M. Gu, E. Mengi, M.L. Overton, J. Xia, and J. Zhu (2006). "Fast Methods for Estimating the Distance to Uncontrollability," *SIAM J. Matrix Anal. Applic. 28*, 477–502.

Aspects of eigenvalue condition are discussed in:

C. Van Loan (1987). "On Estimating the Condition of Eigenvalues and Eigenvectors," *Lin. Alg. Applic.* 88/89, 715–732.

C.D. Meyer and G.W. Stewart (1988). "Derivatives and Perturbations of Eigenvectors," *SIAM J. Numer. Anal. 25*, 679–691.

G.W. Stewart and G. Zhang (1991). "Eigenvalues of Graded Matrices and the Condition Numbers of Multiple Eigenvalues," *Numer. Math. 58*, 703–712.

J.-G. Sun (1992). "On Condition Numbers of a Nondefective Multiple Eigenvalue," *Numer. Math. 61*, 265–276.

S.M. Rump (2001). "Computational Error Bounds for Multiple or Nearly Multiple Eigenvalues," *Lin. Alg. Applic. 324,* 209–226.

The relationship between the eigenvalue condition number, the departure from normality, and the condition of the eigenvector matrix is discussed in:

P. Henrici (1962). "Bounds for Iterates, Inverses, Spectral Variation and Fields of Values of Non-normal Matrices," *Numer. Math. 4*, 24–40.

P. Eberlein (1965). "On Measures of Non-Normality for Matrices," *AMS Monthly 72*, 995–996.

R.A. Smith (1967). "The Condition Numbers of the Matrix Eigenvalue Problem," *Numer. Math. 10* 232–240.

G. Loizou (1969). "Nonnormality and Jordan Condition Numbers of Matrices," *J. ACM 16*, 580–640.

A. van der Sluis (1975). "Perturbations of Eigenvalues of Non-normal Matrices," *Commun. ACM 18*, 30–36.

S.L. Lee (1995). "A Practical Upper Bound for Departure from Normality," *SIAM J. Matrix Anal. Applic. 16*, 462–468.

Gershgorin's theorem can be used to derive a comprehensive perturbation theory. The theorem itself can be generalized and extended in various ways, see:

R.S. Varga (1970). "Minimal Gershgorin Sets for Partitioned Matrices," *SIAM J. Numer. Anal. 7*, 493–507.

R.J. Johnston (1971). "Gershgorin Theorems for Partitioned Matrices," *Lin. Alg. Applic. 4*, 205–20.

R.S. Varga and A. Krautstengl (1999). "On Gergorin-type Problems and Ovals of Cassini," *ETNA 8*, 15–20.

R.S. Varga (2001). "Gergorin-type Eigenvalue Inclusion Theorems and Their Sharpness," *ETNA 12*, 113–133.

C. Beattie and I.C.F. Ipsen (2003). "Inclusion Regions for Matrix Eigenvalues," *Lin. Alg. Applic. 358*, 281–291.

In our discussion, the perturbations to the *A*-matrix are general. More can be said when the perturbations are structured, see:

G.W. Stewart (2001). "On the Eigensystems of Graded Matrices," *Numer. Math. 90*, 349–370.

J. Moro and F.M. Dopico (2003). "Low Rank Perturbation of Jordan Structure," *SIAM J. Matrix Anal. Applic. 25*, 495–506.

R. Byers and D. Kressner (2004). "On the Condition of a Complex Eigenvalue under Real Perturbations," *BIT 44*, 209–214.

R. Byers and D. Kressner (2006). "Structured Condition Numbers for Invariant Subspaces," *SIAM J. Matrix Anal. Applic. 28*, 326–347.

An absolute perturbation bound comments on the difference between an eigenvalue λ and its perturbation $\tilde{\lambda}$. A relative perturbation bound examines the quotient $|\lambda - \tilde{\lambda}|/|\lambda|$, something that can be very important when there is a concern about a small eigenvalue. For results in this direction consult:

R.-C. Li (1997). "Relative Perturbation Theory. III. More Bounds on Eigenvalue Variation," *Lin. Alg. Applic. 266*, 337–345.

S.C. Eisenstat and I.C.F. Ipsen (1998). "Three Absolute Perturbation Bounds for Matrix Eigenvalues Imply Relative Bounds," *SIAM J. Matrix Anal. Applic. 20*, 149–158.

S.C. Eisenstat and I.C.F. Ipsen (1998). "Relative Perturbation Results for Eigenvalues and Eigenvectors of Diagonalisable Matrices," *BIT 38*, 502–509.

I.C.F. Ipsen (1998). "Relative Perturbation Results for Matrix Eigenvalues and Singular Values," *Acta Numerica, 7*, 151–201.

I.C.F. Ipsen (2000). "Absolute and Relative Perturbation Bounds for Invariant Subspaces of Matrices," *Lin. Alg. Applic. 309*, 45–56.

I.C.F. Ipsen (2003). "A Note on Unifying Absolute and Relative Perturbation Bounds," *Lin. Alg. Applic. 358*, 239–253.

Y. Wei, X. Li, F. Bu, and F. Zhang (2006). "Relative Perturbation Bounds for the Eigenvalues of Diagonalizable and Singular Matrices–Application to Perturbation Theory for Simple Invariant Subspaces," *Lin. Alg. Applic. 419*, 765-771.

The eigenvectors and invariant subspaces of a matrix also "move" when there are perturbations. Tracking these changes is typically more challenging than tracking changes in the eigenvalues, see:

T. Kato (1966). *Perturbation Theory for Linear Operators,* Springer-Verlag, New York.

C. Davis and W.M. Kahan (1970). "The Rotation of Eigenvectors by a Perturbation, III," *SIAM J. Numer. Anal. 7*, 1–46.

G.W. Stewart (1971). "Error Bounds for Approximate Invariant Subspaces of Closed Linear Operators," *SIAM. J. Numer. Anal. 8*, 796–808.

G.W. Stewart (1973). "Error and Perturbation Bounds for Subspaces Associated with Certain Eigenvalue Problems," *SIAM Review 15*, 727–764.

J. Xie (1997). "A Note on the Davis-Kahan $sin(2\theta)$ Theorem," *Lin. Alg. Applic. 258*, 129–135.

S.M. Rump and J.-P.M. Zemke (2003). "On Eigenvector Bounds," *BIT 43*, 823–837.

Detailed analyses of the function $\mathsf{sep}(.,.)$ and the map $X \rightarrow AX + XA^T$ are given in:

J. Varah (1979). "On the Separation of Two Matrices," *SIAM J. Numer. Anal. 16*, 216–22.

R. Byers and S.G. Nash (1987). "On the Singular Vectors of the Lyapunov Operator," *SIAM J. Alg. Disc. Methods 8*, 59–66.

7.3 Power Iterations

Suppose that we are given $A \in \mathbb{C}^{n \times n}$ and a unitary $U_0 \in \mathbb{C}^{n \times n}$. Recall from §5.2.10 that the Householder QR factorization can be extended to complex matrices and consider the following iteration:

$$T_0 = U_0^H A U_0$$
$$\textbf{for } k = 1, 2, \ldots$$
$$\qquad T_{k-1} = U_k R_k \quad \text{(QR factorization)} \tag{7.3.1}$$
$$\qquad T_k = R_k U_k$$
$$\textbf{end}$$

Since $T_k = R_k U_k = U_k^H (U_k R_k) U_k = U_k^H T_{k-1} U_k$ it follows by induction that

$$T_k = (U_0 U_1 \cdots U_k)^H A (U_0 U_1 \cdots U_k). \tag{7.3.2}$$

Thus, each T_k is unitarily similar to A. Not so obvious, and what is a central theme of this section, is that the T_k almost always converge to upper triangular form, i.e., (7.3.2) almost always "converges" to a Schur decomposition of A.

Iteration (7.3.1) is called the *QR iteration,* and it forms the backbone of the most effective algorithm for computing a complete Schur decomposition of a dense general matrix. In order to motivate the method and to derive its convergence properties, two other eigenvalue iterations that are important in their own right are presented first: the power method and the method of orthogonal iteration.

7.3.1   The Power Method

Suppose $A \in \mathbb{C}^{n \times n}$ and $X^{-1}AX = \text{diag}(\lambda_1, \ldots, \lambda_n)$ with $X = [\, x_1 \mid \cdots \mid x_n \,]$. Assume that

$$|\lambda_1| > |\lambda_2| \geq \cdots \geq |\lambda_n|.$$

Given a unit 2-norm $q^{(0)} \in \mathbb{C}^n$, the *power method* produces a sequence of vectors $q^{(k)}$ as follows:

> **for** $k = 1, 2, \ldots$
> $\qquad z^{(k)} = A q^{(k-1)}$
> $\qquad q^{(k)} = z^{(k)} / \|\, z^{(k)} \,\|_2$ $\hfill$ (7.3.3)
> $\qquad \lambda^{(k)} = [q^{(k)}]^H A q^{(k)}$
> **end**

There is nothing special about using the 2-norm for normalization except that it imparts a greater unity on the overall discussion in this section.

Let us examine the convergence properties of the power iteration. If

$$q^{(0)} = a_1 x_1 + a_2 x_2 + \cdots + a_n x_n \tag{7.3.4}$$

and $a_1 \neq 0$, then

$$A^k q^{(0)} = a_1 \lambda_1^k \left(x_1 + \sum_{j=2}^{n} \frac{a_j}{a_1} \left(\frac{\lambda_j}{\lambda_1} \right)^k x_j \right).$$

Since $q^{(k)} \in \text{span}\{A^k q^{(0)}\}$ we conclude that

$$\text{dist}\left(\text{span}\{q^{(k)}\}, \text{span}\{x_1\} \right) = O\left(\left| \frac{\lambda_2}{\lambda_1} \right|^k \right).$$

It is also easy to verify that

$$|\, \lambda_1 - \lambda^{(k)} \,| = O\left(\left| \frac{\lambda_2}{\lambda_1} \right|^k \right). \tag{7.3.5}$$

Since λ_1 is larger than all the other eigenvalues in modulus, it is referred to as a *dominant eigenvalue.* Thus, the power method converges if λ_1 is dominant and if $q^{(0)}$ has a component in the direction of the corresponding *dominant eigenvector* x_1. The behavior of the iteration without these assumptions is discussed in Wilkinson (AEP, p. 570) and Parlett and Poole (1973).

In practice, the usefulness of the power method depends upon the ratio $|\lambda_2|/|\lambda_1|$, since it dictates the rate of convergence. The danger that $q^{(0)}$ is deficient in x_1 is less worrisome because rounding errors sustained during the iteration typically ensure that subsequent iterates have a component in this direction. Moreover, it is typically the case in applications that one has a reasonably good guess as to the direction of x_1. This guards against having a pathologically small coefficient a_1 in (7.3.4).

Note that the only thing required to implement the power method is a procedure for matrix-vector products. It is not necessary to store A in an n-by-n array. For this reason, the algorithm is of interest when the dominant eigenpair for a large sparse matrix is required. We have much more to say about large sparse eigenvalue problems in Chapter 10.

Estimates for the error $|\lambda^{(k)} - \lambda_1|$ can be obtained by applying the perturbation theory developed in §7.2.2. Define the vector

$$r^{(k)} = Aq^{(k)} - \lambda^{(k)}q^{(k)}$$

and observe that $(A + E^{(k)})q^{(k)} = \lambda^{(k)}q^{(k)}$ where $E^{(k)} = -r^{(k)}[q^{(k)}]^H$. Thus $\lambda^{(k)}$ is an eigenvalue of $A + E^{(k)}$ and

$$| \lambda^{(k)} - \lambda_1 | \approx \frac{\| E^{(k)} \|_2}{s(\lambda_1)} = \frac{\| r^{(k)} \|_2}{s(\lambda_1)}.$$

If we use the power method to generate approximate right *and* left dominant eigenvectors, then it is possible to obtain an estimate of $s(\lambda_1)$. In particular, if $w^{(k)}$ is a unit 2-norm vector in the direction of $(A^H)^k w^{(0)}$, then we can use the approximation $s(\lambda_1) \approx | w^{(k)^H} q^{(k)} |$.

7.3.2 Orthogonal Iteration

A straightforward generalization of the power method can be used to compute higher-dimensional invariant subspaces. Let r be a chosen integer satisfying $1 \leq r \leq n$. Given $A \in \mathbb{C}^{n \times n}$ and an n-by-r matrix Q_0 with orthonormal columns, the method of *orthogonal iteration* generates a sequence of matrices $\{Q_k\} \subseteq \mathbb{C}^{n \times r}$ and a sequence of eigenvalue estimates $\{\lambda_1^{(k)}, \ldots, \lambda_r^{(k)}\}$ as follows:

$$
\begin{aligned}
&\textbf{for } k = 1, 2, \ldots \\
&\quad Z_k = AQ_{k-1} \\
&\quad Q_k R_k = Z_k \qquad \text{(QR factorization)} \\
&\quad \lambda(Q_k^H A Q_k) = \{\lambda_1^{(k)}, \ldots, \lambda_r^{(k)}\} \\
&\textbf{end}
\end{aligned}
\qquad (7.3.6)
$$

Note that if $r = 1$, then this is just the power method (7.3.3). Moreover, the sequence $\{Q_k e_1\}$ is precisely the sequence of vectors produced by the power iteration with starting vector $q^{(0)} = Q_0 e_1$.

In order to analyze the behavior of this iteration, suppose that

$$Q^H A Q = T = \text{diag}(\lambda_i) + N, \qquad |\lambda_1| \geq |\lambda_2| \geq \cdots \geq |\lambda_n| \qquad (7.3.7)$$

is a Schur decomposition of $A \in \mathbb{C}^{n \times n}$. Assume that $1 \leq r < n$ and partition Q and T as follows:

$$Q = [\,Q_\alpha \mid Q_\beta\,]\underset{r \quad n-r}{} , \qquad T = \left[\begin{array}{cc} T_{11} & T_{12} \\ 0 & T_{22} \end{array}\right]\begin{array}{c} r \\ n-r \end{array} . \qquad (7.3.8)$$
$${r \qquad n-r}$$

If $|\lambda_r| > |\lambda_{r+1}|$, then the subspace $D_r(A) = \text{ran}(Q_\alpha)$ is referred to as a *dominant invariant subspace*. It is the unique invariant subspace associated with the eigenvalues $\lambda_1, \ldots, \lambda_r$. The following theorem shows that with reasonable assumptions, the subspaces $\text{ran}(Q_k)$ generated by (7.3.6) converge to $D_r(A)$ at a rate proportional to $|\lambda_{r+1}/\lambda_r|^k$.

Theorem 7.3.1. *Let the Schur decomposition of $A \in \mathbb{C}^{n \times n}$ be given by (7.3.7) and (7.3.8) with $n \geq 2$. Assume that $|\lambda_r| > |\lambda_{r+1}|$ and that $\mu \geq 0$ satisfies*

$$(1 + \mu)|\lambda_r| > \| N \|_F.$$

Suppose $Q_0 \in \mathbb{C}^{n \times r}$ has orthonormal columns and that d_k is defined by

$$d_k = \text{dist}(D_r(A), \text{ran}(Q_k)), \qquad k \geq 0.$$

If

$$d_0 < 1, \qquad (7.3.9)$$

then the matrices Q_k generated by (7.3.6) satisfy

$$d_k \leq (1+\mu)^{n-2} \cdot \left(1 + \frac{\| T_{12} \|_F}{\text{sep}(T_{11}, T_{22})}\right) \cdot \left[\frac{|\lambda_{r+1}| + \dfrac{\| N \|_F}{1+\mu}}{|\lambda_r| - \dfrac{\| N \|_F}{1+\mu}}\right]^k \cdot \frac{d_0}{\sqrt{1 - d_0^2}}. \qquad (7.3.10)$$

Proof. The proof is given in an appendix at the end of this section. ◻

The condition (7.3.9) ensures that the initial matrix Q_0 is not deficient in certain eigendirections. In particular, no vector in the span of Q_0's columns is orthogonal to $D_r(A^H)$. The theorem essentially says that if this condition holds and if μ is chosen large enough, then

$$\text{dist}(D_r(A), \text{ran}(Q_k)) \approx c \left|\frac{\lambda_{r+1}}{\lambda_r}\right|^k$$

where c depends on $\text{sep}(T_{11}, T_{22})$ and A's departure from normality.

It is possible to accelerate the convergence in orthogonal iteration using a technique described in Stewart (1976). In the accelerated scheme, the approximate eigenvalue $\lambda_i^{(k)}$ satisfies

$$|\lambda_i^{(k)} - \lambda_i| \approx \left|\frac{\lambda_{r+1}}{\lambda_i}\right|^k, \qquad i = 1{:}r.$$

(Without the acceleration, the right-hand side is $|\lambda_{i+1}/\lambda_i|^k$.) Stewart's algorithm involves computing the Schur decomposition of the matrices $Q_k^T A Q_k$ every so often. The method can be very useful in situations where A is large and sparse and a few of its largest eigenvalues are required.

7.3.3 The QR Iteration

We now derive the QR iteration (7.3.1) and examine its convergence. Suppose $r = n$ in (7.3.6) and the eigenvalues of A satisfy

$$|\lambda_1| > |\lambda_2| > \cdots > |\lambda_n|.$$

Partition the matrix Q in (7.3.7) and Q_k in (7.3.6) as follows:

$$Q = [\, q_1 \mid \cdots \mid q_n \,], \qquad Q_k = [\, q_1^{(k)} \mid \cdots \mid q_n^{(k)} \,].$$

If

$$\mathsf{dist}(D_i(A^H), \mathsf{span}\{q_1^{(0)}, \ldots, q_i^{(0)}\}) \; < \; 1, \qquad i = 1{:}n, \tag{7.3.11}$$

then it follows from Theorem 7.3.1 that

$$\mathsf{dist}(\, \mathsf{span}\{q_1^{(k)}, \ldots, q_i^{(k)}\}\,, \; \mathsf{span}\{q_1, \ldots, q_i\}\,) \; \to \; 0$$

for $i = 1{:}n$. This implies that the matrices T_k defined by

$$T_k \; = \; Q_k^H A Q_k$$

are converging to upper triangular form. Thus, it can be said that the method of orthogonal iteration computes a Schur decomposition provided the original iterate $Q_0 \in \mathbb{C}^{n \times n}$ is not deficient in the sense of (7.3.11).

The QR iteration arises naturally by considering how to compute the matrix T_k directly from its predecessor T_{k-1}. On the one hand, we have from (7.3.6) and the definition of T_{k-1} that

$$T_{k-1} = Q_{k-1}^H A Q_{k-1} = Q_{k-1}^H (A Q_{k-1}) = (Q_{k-1}^H Q_k) R_k.$$

On the other hand,

$$T_k = Q_k^H A Q_k = (Q_k^H A Q_{k-1})(Q_{k-1}^H Q_k) = R_k (Q_{k-1}^H Q_k).$$

Thus, T_k is determined by computing the QR factorization of T_{k-1} and then multiplying the factors together in reverse order, precisely what is done in (7.3.1).

Note that a single QR iteration is an $O(n^3)$ calculation. Moreover, since convergence is only linear (when it exists), it is clear that the method is a prohibitively expensive way to compute Schur decompositions. Fortunately these practical difficulties can be overcome as we show in §7.4 and §7.5.

7.3.4 LR Iterations

We conclude with some remarks about power iterations that rely on the LU factorization rather than the QR factorizaton. Let $G_0 \in \mathbb{C}^{n \times r}$ have rank r. Corresponding to (7.3.1) we have the following iteration:

$$
\begin{aligned}
&\textbf{for } k = 1, 2, \ldots \\
&\qquad Z_k = A G_{k-1} \\
&\qquad Z_k = G_k R_k \qquad \text{(LU factorization)} \\
&\textbf{end}
\end{aligned}
\tag{7.3.12}
$$

Suppose $r = n$ and that we define the matrices T_k by

$$T_k = G_k^{-1} A G_k. \tag{7.3.13}$$

It can be shown that if we set $L_0 = G_0$, then the T_k can be generated as follows:

$$T_0 = L_0^{-1} A L_0$$
for $k = 1, 2, \ldots$
$\qquad T_{k-1} = L_k R_k \qquad$ (LU factorization) $\hfill (7.3.14)$
$\qquad T_k = R_k L_k$
end

Iterations (7.3.12) and (7.3.14) are known as *treppeniteration* and the *LR iteration*, respectively. Under reasonable assumptions, the T_k converge to upper triangular form. To successfully implement either method, it is necessary to pivot. See Wilkinson (AEP, p. 602).

Appendix

In order to establish Theorem 7.3.1 we need the following lemma that bounds powers of a matrix and powers of its inverse.

Lemma 7.3.2. *Let $Q^H A Q = T = D + N$ be a Schur decomposition of $A \in \mathbb{C}^{n \times n}$ where D is diagonal and N strictly upper triangular. Let $\lambda_{\max}$ and $\lambda_{\min}$ denote the largest and smallest eigenvalues of A in absolute value. If $\mu \geq 0$, then for all $k \geq 0$ we have*

$$\| A^k \|_2 \leq (1 + \mu)^{n-1} \left(|\lambda_{\max}| + \frac{\| N \|_F}{1 + \mu} \right)^k. \tag{7.3.15}$$

If A is nonsingular and $\mu \geq 0$ satisfies $(1 + \mu)|\lambda_{\min}| > \| N \|_F$, then for all $k \geq 0$ we also have

$$\| A^{-k} \|_2 \leq (1 + \mu)^{n-1} \left(\frac{1}{|\lambda_{\min}| - \| N \|_F / (1 + \mu)} \right)^k. \tag{7.3.16}$$

Proof. For $\mu \geq 0$, define the diagonal matrix Δ by

$$\Delta = \text{diag}\, (1, (1 + \mu), (1 + \mu)^2, \ldots, (1 + \mu)^{n-1})$$

and note that $\kappa_2(\Delta) = (1 + \mu)^{n-1}$. Since N is strictly upper triangular, it is easy to verify that

$$\| \Delta N \Delta^{-1} \|_F \leq \frac{\| N \|_F}{1 + \mu}$$

and thus

$$\| A^k \|_2 = \| T^k \|_2 = \| \Delta^{-1} (D + \Delta N \Delta^{-1})^k \Delta \|_2$$
$$\leq \kappa_2(\Delta) \left(\| D \|_2 + \| \Delta N \Delta^{-1} \|_2 \right)^k \leq (1 + \mu)^{n-1} \left(|\lambda_{\max}| + \frac{\| N \|_F}{1 + \mu} \right)^k.$$

On the other hand, if A is nonsingular and $(1 + \mu)|\lambda_{\min}| > \| N \|_F$, then

$$\| \Delta D^{-1} N \Delta^{-1} \|_2 = \| D^{-1}(\Delta N \Delta^{-1}) \|_2 \leq \frac{1}{|\lambda_{\min}|} \| \Delta N \Delta^{-1} \|_F < 1.$$

Using Lemma 2.3.3 we obtain

$$\| A^{-k} \|_2 = \| T^{-k} \|_2 = \left\| \Delta^{-1}[(I + \Delta D^{-1} N \Delta^{-1})^{-1} D^{-1}]^k \Delta \right\|_2$$
$$\leq \kappa_2(\Delta) \left(\frac{\| D^{-1} \|_2}{1 - \| \Delta D^{-1} N \Delta^{-1} \|_2} \right)^k \leq (1 + \mu)^{n-1} \left(\frac{1}{|\mu| - \| N \|_F/(1 + \mu)} \right)^k$$

completing the proof of the lemma. $\square$

Proof of Theorem 7.3.1. By induction it is easy to show that the matrix Q_k in (7.3.6) satisfies

$$A^k Q_0 = Q_k(R_k \cdots R_1),$$

a QR factorization of $A^k Q_0$. By substituting the Schur decomposition (7.3.7)-(7.3.8) into this equation we obtain

$$T^k \begin{bmatrix} V_0 \\ W_0 \end{bmatrix} = \begin{bmatrix} V_k \\ W_k \end{bmatrix} (R_k \cdots R_1) \tag{7.3.17}$$

where

$$V_k = Q_\alpha^H Q_k, \qquad W_k = Q_\beta^H Q_k.$$

Our goal is to bound $\| W_k \|_2$ since by the definition of subspace distance given in §2.5.3 we have

$$\| W_k \|_2 = \mathsf{dist}(D_r(A), \mathsf{ran}(Q_k)). \tag{7.3.18}$$

Note from the thin CS decomposition (Theorem 2.5.2) that

$$1 = d_k^2 + \sigma_{\min}(V_k)^2. \tag{7.3.19}$$

Since T_{11} and T_{22} have no eigenvalues in common, Lemma 7.1.5 tells us that the Sylvester equation $T_{11} X - X T_{22} = -T_{12}$ has a solution $X \in \mathbb{C}^{r \times (n-r)}$ and that

$$\| X \|_F \leq \frac{\| T_{12} \|_F}{\mathsf{sep}(T_{11}, T_{22})}. \tag{7.3.20}$$

It follows that

$$\begin{bmatrix} I_r & X \\ 0 & I_{n-r} \end{bmatrix}^{-1} \begin{bmatrix} T_{11} & T_{12} \\ 0 & T_{22} \end{bmatrix} \begin{bmatrix} I_r & X \\ 0 & I_{n-r} \end{bmatrix} = \begin{bmatrix} T_{11} & 0 \\ 0 & T_{22} \end{bmatrix}.$$

By substituting this into (7.3.17) we obtain

$$\begin{bmatrix} T_{11}^k & 0 \\ 0 & T_{22}^k \end{bmatrix} \begin{bmatrix} V_0 - X W_0 \\ W_0 \end{bmatrix} = \begin{bmatrix} V_k - X W_k \\ W_k \end{bmatrix} (R_k \cdots R_1),$$

i.e.,

$$T_{11}^k(V_0 - XW_0) = (V_k - XW_k)(R_k \cdots R_1), \tag{7.3.21}$$

$$T_{22}^k W_0 = W_k(R_k \cdots R_1). \tag{7.3.22}$$

The matrix $I + XX^H$ is Hermitian positive definite and so it has a Cholesky factorization

$$I + XX^H = GG^H. \tag{7.3.23}$$

It is clear that

$$\sigma_{\min}(G) \geq 1. \tag{7.3.24}$$

If the matrix $Z \in \mathbb{C}^{n \times (n-r)}$ is defined by

$$Z = Q \begin{bmatrix} I_r \\ -X^H \end{bmatrix} G^{-H} = [\, Q_\alpha \, Q_\beta \,] \begin{bmatrix} I_r \\ -X^H \end{bmatrix} G^{-H} = (Q_\alpha - Q_\beta X^H)G^{-H},$$

then it follows from the equation $A^H Q = QT^H$ that

$$A^H(Q_\alpha - Q_\beta X^H) = (Q_\alpha - Q_\beta X^H)T_{11}^H. \tag{7.3.25}$$

Since $Z^H Z = I_r$ and $\mathsf{ran}(Z) = \mathsf{ran}(Q_\alpha - Q_\beta X^H)$, it follows that the columns of Z are an orthonormal basis for $D_r(A^H)$. Using the CS decomposition, (7.3.19), and the fact that $\mathsf{ran}(Q_\beta) = D_r(A^H)^\perp$, we have

$$\sigma_{\min}(Z^T Q_0)^2 = 1 - \mathsf{dist}(D_r(A^H), Q_0)^2 = 1 - \| Q_\beta^H Q_0 \|$$

$$= \sigma_{\min}(Q_\alpha^T Q_0)^2 = \sigma_{\min}(V_0)^2 = 1 - d_0^2 > 0.$$

This shows that

$$V_0 - XW_0 = [\, I_r \mid -X \,] \begin{bmatrix} Q_\alpha^H Q_0 \\ Q_\beta^H Q_0 \end{bmatrix} = (ZG^H)^H Q_0 = G(Z^H Q_0)$$

is nonsingular and together with (7.3.24) we obtain

$$\| (V_0 - XW_0)^{-1} \|_2 \leq \| G^{-1} \|_2 \| (Z^H Q_0)^{-1} \|_2 \leq \frac{1}{\sqrt{1 - d_0^2}}. \tag{7.3.26}$$

Manipulation of (7.3.19) and (7.3.20) yields

$$W_k = T_{22}^k W_0(R_k \cdots R_1)^{-1} = T_{22}^k W_0(V_0 - XW_0)^{-1}T_{11}^{-k}(V_k - XW_k).$$

The verification of (7.3.10) is completed by taking norms in this equation and using (7.3.18), (7.3.19), (7.3.20), (7.3.26), and the following facts:

$$\| T_{22}^k \|_2 \leq (1 + \mu)^{n-r-1}(|\lambda_{r+1}| + \| N \|_F/(1 + \mu))^k,$$

$$\| T_{11}^{-k} \|_2 \leq (1 + \mu)^{r-1}/(|\lambda_r| - \| N \|_F/(1 + \mu))^k,$$

$$\| V_k - XW_k \|_2 \leq \| V_k \|_2 + \| X \|_2 \| W_k \|_2 \leq 1 + \| T_{12} \|_F/\mathsf{sep}(T_{11}, T_{22}).$$

The bounds for $\| T_{22}^k \|_2$ and $\| T_{11}^{-k} \|_2$ follow from Lemma 7.3.2.

Problems

P7.3.1 Verify Equation (7.3.5).

P7.3.2 Suppose the eigenvalues of $A \in \mathbb{R}^{n \times n}$ satisfy $|\lambda_1| = |\lambda_2| > |\lambda_3| \geq \cdots \geq |\lambda_n|$ and that λ_1 and λ_2 are complex conjugates of one another. Let $S = \text{span}\{y, z\}$ where $y, z \in \mathbb{R}^n$ satisfy $A(y + iz) = \lambda_1(y + iz)$. Show how the power method with a real starting vector can be used to compute an approximate basis for S.

P7.3.3 Assume $A \in \mathbb{R}^{n \times n}$ has eigenvalues $\lambda_1, \ldots, \lambda_n$ that satisfy

$$\lambda = \lambda_1 = \lambda_2 = \lambda_3 = \lambda_4 > |\lambda_5| \geq \cdots \geq |\lambda_n|$$

where λ is positive. Assume that A has two Jordan blocks of the form.

$$\begin{bmatrix} \lambda & 1 \\ 0 & \lambda \end{bmatrix}.$$

Discuss the convergence properties of the power method when applied to this matrix and how the convergence might be accelerated.

P7.3.4 A matrix A is a *positive matrix* if $a_{ij} > 0$ for all i and j. A vector $v \in \mathbb{R}^n$ is a positive vector if $v_i > 0$ for all i. *Perron's theorem* states that if A is a positive square matrix, then it has a unique dominant eigenvalue equal to its spectral radius $\rho(A)$ and there is a positive vector x so that $Ax = \rho(A) \cdot x$. In this context, x is called the *Perron vector* and $\rho(A)$ is called the *Perron root*. Assume that $A \in \mathbb{R}^{n \times n}$ is positive and $q \in \mathbb{R}^n$ is positive with unit 2-norm. Consider the following implementation of the power method (7.3.3):

$z = Aq, \ \lambda = q^T z$
while $\| z - \lambda q \|_2 > \delta$
    $q = z, \ q = q/\| q \|_2, \ z = Aq, \ \lambda = q^T z$
end

(a) Adjust the termination criteria to guarantee (in principle) that the final λ and q satisfy $\tilde{A}q = \lambda q$, where $\| \tilde{A} - A \|_2 \leq \delta$ *and* $\tilde{A}$ is positive. (b) Applied to a positive matrix $A \in \mathbb{R}^{n \times n}$, the *Collatz-Wielandt formula* states that $\rho(A)$ is the maximum value of the function f defined by

$$f(x) = \min_{1 \leq i \leq n} \frac{y_i}{x_i}$$

where $x \in \mathbb{R}^n$ is positive and $y = Ax$. Does it follow that $f(Aq) \geq f(q)$? In other words, do the iterates $\{q^{(k)}\}$ in the power method have the property that $f(q^{(k)})$ increases monotonically to the Perron root, assuming that $q^{(0)}$ is positive?

P7.3.5 (Read the previous problem for background.) A matrix A is a *nonnegative matrix* if $a_{ij} \geq 0$ for all i and j. A matrix $A \in \mathbb{R}^{n \times n}$ is *reducible* if there is a permutation P so that $P^T A P$ is block triangular with two or more square diagonal blocks. A matrix that is not reducible is *irreducible*. The *Perron-Frobenius theorem* states that if A is a square, nonnegative, and irreducible, then $\rho(A)$, the Perron root, is an eigenvalue for A and there is a positive vector x, the Perron vector, so that $Ax = \rho(A) \cdot x$. Assume that $A_1, A_2, A_3 \in \mathbb{R}^{n \times n}$ are each positive and let the nonnegative matrix A be defined by

$$A = \begin{bmatrix} 0 & A_1 & 0 \\ 0 & 0 & A_2 \\ A_3 & 0 & 0 \end{bmatrix}.$$

(a) Show that A is irreducible. (b) Let $B = A_1 A_2 A_3$. Show how to compute the Perron root and vector for A from the Perron root and vector for B. (c) Show that A has other eigenvalues with absolute value equal to the Perron root. How could those eigenvalues and the associated eigenvectors be computed?

P7.3.6 (Read the previous two problems for background.) A nonnegative matrix $P \in \mathbb{R}^{n \times n}$ is *stochastic* if the entries in each column sum to 1. A vector $v \in \mathbb{R}^n$ is a *probability vector* if its entries are nonnegative and sum to 1. (a) Show that if $P \in \mathbb{R}^{n \times n}$ is stochastic and $v \in \mathbb{R}^n$ is a probability vector, then $w = Pv$ is also a probability vector. (b) The entries in a stochastic matrix $P \in \mathbb{R}^{n \times n}$ can be regarded as the *transition probabilities* associated with an n-state *Markov Chain*. Let v_j be the probability of being in state j at time $t = t_{\text{current}}$. In the Markov model, the probability of being in state i at time $t = t_{\text{next}}$ is given by

$$w_i = \sum_{j=1}^{n} p_{ij} v_j \qquad i = 1{:}n,$$

i.e., $w = Pv$. With the help of a biased coin, a surfer on the World Wide Web randomly jumps from page to page. Assume that the surfer is currently viewing web page j and that the coin comes up heads with probability α. Here is how the surfer determines the next page to visit:

Step 1. A coin is tossed.

Step 2. If it comes up heads and web page j has at least one outlink, then the next page to visit is randomly selected from the list of outlink pages.

Step 3. Otherwise, the next page to visit is randomly selected from the list of all possible pages.

Let $P \in \mathbb{R}^{n \times n}$ be the matrix of transition probabilities that define this random process. Specify P in terms of α, the vector of ones e, and the link matrix $H \in \mathbb{R}^{n \times n}$ defined by

$$h_{ij} = \begin{cases} 1 & \text{if there is a link on web page } j \text{ to web page } i \\ 0 & \text{otherwise} \end{cases}$$

Hints: The number of nonzero components in $H(:, j)$ is the number of outlinks on web page j. P is a convex combination of a very sparse sparse matrix and a very dense rank-1 matrix. (c) Detail how the power method can be used to determine a probability vector x so that $Px = x$. Strive to get as much computation "outside the loop" as possible. Note that in the limit we can expect to find the random surfer viewing web page i with probability x_i. Thus, a case can be made that more important pages are associated with the larger components of x. This is the basis of Google PageRank. If

$$x_{i_1} \geq x_{i_2} \geq \cdots \geq x_{i_n}$$

then web page i_k has page rank k.

P7.3.7 (a) Show that if $X \in \mathbb{C}^{n \times n}$ is nonsingular, then

$$\| A \|_X = \| X^{-1} A X \|_2$$

defines a matrix norm with the property that

$$\| AB \|_X \leq \| A \|_X \| B \|_X.$$

(b) Show that for any $\epsilon > 0$ there exists a nonsingular $X \in \mathbb{C}^{n \times n}$ such that

$$\| A \|_X = \| X^{-1} A X \|_2 \leq \rho(A) + \epsilon$$

where $\rho(A)$ is A's spectral radius. Conclude that there is a constant M such that

$$\| A^k \|_2 \leq M(\rho(A) + \epsilon)^k$$

for all non-negative integers k. (Hint: Set $X = Q \operatorname{diag}(1, a, \ldots, a^{n-1})$ where $Q^H A Q = D + N$ is A's Schur decomposition.)

P7.3.8 Verify that (7.3.14) calculates the matrices T_k defined by (7.3.13).

P7.3.9 Suppose $A \in \mathbb{C}^{n \times n}$ is nonsingular and that $Q_0 \in \mathbb{C}^{n \times p}$ has orthonormal columns. The following iteration is referred to as *inverse orthogonal iteration*.

> **for** $k = 1, 2, \ldots$
>   Solve $A Z_k = Q_{k-1}$ for $Z_k \in \mathbb{C}^{n \times p}$
> $Z_k = Q_k R_k$ (QR factorization)
> **end**

Explain why this iteration can usually be used to compute the p smallest eigenvalues of A in absolute value. Note that to implement this iteration it is necessary to be able to solve linear systems that involve A. If $p = 1$, the method is referred to as the *inverse power method*.

Notes and References for §7.3

For an excellent overview of the QR iteration and related procedures, see Watkins (MEP), Stewart (MAE), and Kressner (NMSE). A detailed, practical discussion of the power method is given in Wilkinson (AEP, Chap. 10). Methods are discussed for accelerating the basic iteration, for calculating nondominant eigenvalues, and for handling complex conjugate eigenvalue pairs. The connections among the various power iterations are discussed in:

B.N. Parlett and W.G. Poole (1973). "A Geometric Theory for the QR, LU, and Power Iterations," *SIAM J. Numer. Anal. 10*, 389–412.

The QR iteration was concurrently developed in:

J.G.F. Francis (1961). "The QR Transformation: A Unitary Analogue to the LR Transformation," *Comput. J. 4*, 265–71, 332–334.
V.N. Kublanovskaya (1961). "On Some Algorithms for the Solution of the Complete Eigenvalue Problem," *USSR Comput. Math. Phys. 3*, 637–657.

As can be deduced from the title of the first paper by Francis, the LR iteration predates the QR iteration. The former very fundamental algorithm was proposed by:

H. Rutishauser (1958). "Solution of Eigenvalue Problems with the LR Transformation," *Nat. Bur. Stand. Appl. Math. Ser. 49*, 47–81.

More recent, related work includes:

B.N. Parlett (1995). "The New qd Algorithms," *Acta Numerica 5*, 459–491.
C. Ferreira and B.N. Parlett (2009). "Convergence of the LR Algorithm for a One-Point Spectrum Tridiagonal Matrix," *Numer. Math. 113*, 417–431.

Numerous papers on the convergence and behavior of the QR iteration have appeared, see:

J.H. Wilkinson (1965). "Convergence of the LR, QR, and Related Algorithms," *Comput. J. 8*, 77–84.
B.N. Parlett (1965). "Convergence of the Q-R Algorithm," *Numer. Math. 7*, 187–93. (Correction in *Numer. Math. 10*, 163–164.)
B.N. Parlett (1966). "Singular and Invariant Matrices Under the QR Algorithm," *Math. Comput. 20*, 611–615.
B.N. Parlett (1968). "Global Convergence of the Basic QR Algorithm on Hessenberg Matrices," *Math. Comput. 22*, 803–817.
D.S. Watkins (1982). "Understanding the QR Algorithm," *SIAM Review 24*, 427–440.
T. Nanda (1985). "Differential Equations and the QR Algorithm," *SIAM J. Numer. Anal. 22*, 310–321.
D.S. Watkins (1993). "Some Perspectives on the Eigenvalue Problem," *SIAM Review 35*, 430–471.
D.S. Watkins (2008). "The QR Algorithm Revisited," *SIAM Review 50*, 133–145.
D.S. Watkins (2011). "Francis's Algorithm," *AMS Monthly 118*, 387–403.

A block analog of the QR iteration is discussed in:

M. Robbè and M. Sadkane (2005). "Convergence Analysis of the Block Householder Block Diagonalization Algorithm," *BIT 45*, 181–195.

The following references are concerned with various practical and theoretical aspects of simultaneous iteration:

H. Rutishauser (1970). "Simultaneous Iteration Method for Symmetric Matrices," *Numer. Math. 16*, 205–223.
M. Clint and A. Jennings (1971). "A Simultaneous Iteration Method for the Unsymmetric Eigenvalue Problem," *J. Inst. Math. Applic. 8*, 111-121.
G.W. Stewart (1976). "Simultaneous Iteration for Computing Invariant Subspaces of Non-Hermitian Matrices," *Numer. Math. 25*, 123–136.
A. Jennings (1977). *Matrix Computation for Engineers and Scientists*, John Wiley and Sons, New York.
Z. Bai and G.W. Stewart (1997). "Algorithm 776: SRRIT: a Fortran Subroutine to Calculate the Dominant Invariant Subspace of a Nonsymmetric Matrix," *ACM Trans. Math. Softw. 23*, 494–513.

Problems P7.3.4–P7.3.6 explore the relevance of the power method to the problem of computing the Perron root and vector of a nonnegative matrix. For further background and insight, see:

A. Berman and R.J. Plemmons (1994). *Nonnegative Matrices in the Mathematical Sciences*, SIAM Publications,Philadelphia, PA.
A.N. Langville and C.D. Meyer (2006). *Google's PageRank and Beyond*, Princeton University Press, Princeton and Oxford. .

The latter volume is outstanding in how it connects the tools of numerical linear algebra to the design and analysis of Web browsers. See also:

W.J. Stewart (1994). *Introduction to the Numerical Solution of Markov Chains*, Princeton University Press, Princeton, NJ.
M.W. Berry, Z. Drmač, and E.R. Jessup (1999). "Matrices, Vector Spaces, and Information Retrieval," *SIAM Review 41*, 335–362.
A.N. Langville and C.D. Meyer (2005). "A Survey of Eigenvector Methods for Web Information Retrieval," *SIAM Review 47*, 135–161.
A.N. Langville and C.D. Meyer (2006). "A Reordering for the PageRank Problem", *SIAM J. Sci. Comput. 27*, 2112–2120.
A.N. Langville and C.D. Meyer (2006). "Updating Markov Chains with an Eye on Google's PageRank," *SIAM J. Matrix Anal. Applic. 27*, 968–987.

7.4   The Hessenberg and Real Schur Forms

In this and the next section we show how to make the QR iteration (7.3.1) a fast, effective method for computing Schur decompositions. Because the majority of eigenvalue/invariant subspace problems involve real data, we concentrate on developing the real analogue of (7.3.1) which we write as follows:

$$
\begin{aligned}
&H_0 = U_0^T A U_0 \\
&\textbf{for } k = 1, 2, \ldots \\
&\quad H_{k-1} = U_k R_k \qquad \text{(QR factorization)} \\
&\quad H_k = R_k U_k \\
&\textbf{end}
\end{aligned}
\tag{7.4.1}
$$

Here, $A \in \mathbb{R}^{n \times n}$, each $U_k \in \mathbb{R}^{n \times n}$ is orthogonal, and each $R_k \in \mathbb{R}^{n \times n}$ is upper triangular. A difficulty associated with this real iteration is that the H_k can never converge to triangular form in the event that A has complex eigenvalues. For this reason, we must lower our expectations and be content with the calculation of an alternative decomposition known as the *real Schur decomposition*.

In order to compute the real Schur decomposition efficiently we must carefully choose the initial orthogonal similarity transformation U_0 in (7.4.1). In particular, if we choose U_0 so that H_0 is upper Hessenberg, then the amount of work per iteration is reduced from $O(n^3)$ to $O(n^2)$. The initial reduction to Hessenberg form (the U_0 computation) is a very important computation in its own right and can be realized by a sequence of Householder matrix operations.

7.4.1   The Real Schur Decomposition

A block upper triangular matrix with either 1-by-1 or 2-by-2 diagonal blocks is upper *quasi-triangular*. The real Schur decomposition amounts to a real reduction to upper quasi-triangular form.

Theorem 7.4.1 (Real Schur Decomposition). *If $A \in \mathbb{R}^{n \times n}$, then there exists an orthogonal $Q \in \mathbb{R}^{n \times n}$ such that*

$$Q^T A Q = \begin{bmatrix} R_{11} & R_{12} & \cdots & R_{1m} \\ 0 & R_{22} & \cdots & R_{2m} \\ \vdots & \vdots & \ddots & \vdots \\ 0 & 0 & \cdots & R_{mm} \end{bmatrix} \qquad (7.4.2)$$

where each R_{ii} is either a 1-by-1 matrix or a 2-by-2 matrix having complex conjugate eigenvalues.

Proof. The complex eigenvalues of A occur in conjugate pairs since the characteristic polynomial $\det(zI - A)$ has real coefficients. Let k be the number of complex conjugate pairs in $\lambda(A)$. We prove the theorem by induction on k. Observe first that Lemma 7.1.2 and Theorem 7.1.3 have obvious real analogs. Thus, the theorem holds if $k = 0$. Now suppose that $k \geq 1$. If $\lambda = \gamma + i\mu \in \lambda(A)$ and $\mu \neq 0$, then there exist vectors y and z in $\mathbb{R}^n (z \neq 0)$ such that $A(y + iz) = (\gamma + i\mu)(y + iz)$, i.e.,

$$A \begin{bmatrix} y & | & z \end{bmatrix} = \begin{bmatrix} y & | & z \end{bmatrix} \begin{bmatrix} \gamma & \mu \\ -\mu & \gamma \end{bmatrix}.$$

The assumption that $\mu \neq 0$ implies that y and z span a 2-dimensional, real invariant subspace for A. It then follows from Lemma 7.1.2 that an orthogonal $U \in \mathbb{R}^{n \times n}$ exists such that

$$U^T A U = \begin{bmatrix} T_{11} & T_{12} \\ 0 & T_{22} \end{bmatrix} \begin{matrix} 2 \\ n-2 \end{matrix}$$
$$\quad\;\; \begin{matrix} 2 & n-2 \end{matrix}$$

where $\lambda(T_{11}) = \{\lambda, \bar{\lambda}\}$. By induction, there exists an orthogonal $\tilde{U}$ so $\tilde{U}^T T_{22} \tilde{U}$ has the required structure. The theorem follows by setting $Q = U \cdot \mathrm{diag}(I_2, \tilde{U})$. $\square$

The theorem shows that any real matrix is orthogonally similar to an upper quasi-triangular matrix. It is clear that the real and imaginary parts of the complex eigenvalues can be easily obtained from the 2-by-2 diagonal blocks. Thus, it can be said that the real Schur decomposition is an eigenvalue-revealing decomposition.

7.4.2 A Hessenberg QR Step

We now turn our attention to the efficient execution of a single QR step in (7.4.1). In this regard, the most glaring shortcoming associated with (7.4.1) is that each step requires a full QR factorization costing $O(n^3)$ flops. Fortunately, the amount of work per iteration can be reduced by an order of magnitude if the orthogonal matrix U_0 is judiciously chosen. In particular, if $U_0^T A U_0 = H_0 = (h_{ij})$ is upper Hessenberg ($h_{ij} = 0$, $i > j + 1$), then each subsequent H_k requires only $O(n^2)$ flops to calculate. To see this we look at the computations $H = QR$ and $H_+ = RQ$ when H is upper Hessenberg. As described in §5.2.5, we can upper triangularize H with a sequence of $n - 1$ Givens rotations: $Q^T H \equiv G_{n-1}^T \cdots G_1^T H = R$. Here, $G_i = G(i, i+1, \theta_i)$. For the $n = 4$ case there are three Givens premultiplications:

$$
\begin{bmatrix} \times & \times & \times & \times \\ \times & \times & \times & \times \\ 0 & \times & \times & \times \\ 0 & 0 & \times & \times \end{bmatrix} \rightarrow \begin{bmatrix} \times & \times & \times & \times \\ 0 & \times & \times & \times \\ 0 & \times & \times & \times \\ 0 & 0 & \times & \times \end{bmatrix} \rightarrow \begin{bmatrix} \times & \times & \times & \times \\ 0 & \times & \times & \times \\ 0 & 0 & \times & \times \\ 0 & 0 & \times & \times \end{bmatrix} \rightarrow \begin{bmatrix} \times & \times & \times & \times \\ 0 & \times & \times & \times \\ 0 & 0 & \times & \times \\ 0 & 0 & 0 & \times \end{bmatrix}.
$$

See Algorithm 5.2.5. The computation $RQ = R(G_1 \cdots G_{n-1})$ is equally easy to implement. In the $n = 4$ case there are three Givens post-multiplications:

$$
\begin{bmatrix} \times & \times & \times & \times \\ 0 & \times & \times & \times \\ 0 & 0 & \times & \times \\ 0 & 0 & 0 & \times \end{bmatrix} \rightarrow \begin{bmatrix} \times & \times & \times & \times \\ \times & \times & \times & \times \\ 0 & 0 & \times & \times \\ 0 & 0 & 0 & \times \end{bmatrix} \rightarrow \begin{bmatrix} \times & \times & \times & \times \\ \times & \times & \times & \times \\ 0 & \times & \times & \times \\ 0 & 0 & 0 & \times \end{bmatrix} \rightarrow \begin{bmatrix} \times & \times & \times & \times \\ \times & \times & \times & \times \\ 0 & \times & \times & \times \\ 0 & 0 & \times & \times \end{bmatrix}.
$$

Overall we obtain the following algorithm:

Algorithm 7.4.1 If H is an n-by-n upper Hessenberg matrix, then this algorithm overwrites H with $H_+ = RQ$ where $H = QR$ is the QR factorization of H.

> **for** $k = 1{:}n - 1$
> $\quad [\, c_k \,,\, s_k \,] = \mathsf{givens}(H(k,k), H(k+1,k))$
> $\quad H(k{:}k+1, k{:}n) = \begin{bmatrix} c_k & s_k \\ -s_k & c_k \end{bmatrix}^T H(k{:}k+1, k{:}n)$
> **end**
> **for** $k = 1{:}n - 1$
> $\quad H(1{:}k+1, k{:}k+1) = H(1{:}k+1, k{:}k+1) \begin{bmatrix} c_k & s_k \\ -s_k & c_k \end{bmatrix}$
> **end**

Let $G_k = G(k, k+1, \theta_k)$ be the kth Givens rotation. It is easy to confirm that the matrix $Q = G_1 \cdots G_{n-1}$ is upper Hessenberg. Thus, $RQ = H_+$ is also upper Hessenberg. The algorithm requires about $6n^2$ flops, an order of magnitude more efficient than a full matrix QR step (7.3.1).

7.4.3   The Hessenberg Reduction

It remains for us to show how the *Hessenberg decomposition*

$$
U_0^T A U_0 = H, \qquad U_0^T U_0 = I \tag{7.4.3}
$$

can be computed. The transformation U_0 can be computed as a product of Householder matrices $P_1, \ldots, P_{n-2}$. The role of P_k is to zero the kth column below the subdiagonal. In the $n = 6$ case, we have

$$
\begin{bmatrix} \times & \times & \times & \times & \times & \times \\ \times & \times & \times & \times & \times & \times \\ \times & \times & \times & \times & \times & \times \\ \times & \times & \times & \times & \times & \times \\ \times & \times & \times & \times & \times & \times \\ \times & \times & \times & \times & \times & \times \end{bmatrix} \overset{P_1}{\rightarrow} \begin{bmatrix} \times & \times & \times & \times & \times & \times \\ \times & \times & \times & \times & \times & \times \\ 0 & \times & \times & \times & \times & \times \\ 0 & \times & \times & \times & \times & \times \\ 0 & \times & \times & \times & \times & \times \\ 0 & \times & \times & \times & \times & \times \end{bmatrix} \overset{P_2}{\rightarrow}
$$

$$
\begin{bmatrix}
\times & \times & \times & \times & \times & \times \\
\times & \times & \times & \times & \times & \times \\
0 & \times & \times & \times & \times & \times \\
0 & 0 & \times & \times & \times & \times \\
0 & 0 & \times & \times & \times & \times \\
0 & 0 & \times & \times & \times & \times
\end{bmatrix}
\xrightarrow{P_3}
\begin{bmatrix}
\times & \times & \times & \times & \times & \times \\
\times & \times & \times & \times & \times & \times \\
0 & \times & \times & \times & \times & \times \\
0 & 0 & \times & \times & \times & \times \\
0 & 0 & 0 & \times & \times & \times \\
0 & 0 & 0 & \times & \times & \times
\end{bmatrix}
\xrightarrow{P_4}
\begin{bmatrix}
\times & \times & \times & \times & \times & \times \\
\times & \times & \times & \times & \times & \times \\
0 & \times & \times & \times & \times & \times \\
0 & 0 & \times & \times & \times & \times \\
0 & 0 & 0 & \times & \times & \times \\
0 & 0 & 0 & 0 & \times & \times
\end{bmatrix}.
$$

In general, after $k-1$ steps we have computed $k-1$ Householder matrices $P_1, \ldots, P_{k-1}$ such that

$$
(P_1 \cdots P_{k-1})^T A (P_1 \cdots P_{k-1}) =
\begin{array}{c}
\begin{bmatrix}
B_{11} & B_{12} & B_{13} \\
B_{21} & B_{22} & B_{23} \\
0 & B_{32} & B_{33}
\end{bmatrix}
\begin{array}{l} k-1 \\ 1 \\ n-k \end{array} \\
\begin{array}{ccc} k-1 & 1 & n-k \end{array}
\end{array}
$$

is upper Hessenberg through its first $k-1$ columns. Suppose $\tilde{P}_k$ is an order-$(n-k)$ Householder matrix such that $\tilde{P}_k B_{32}$ is a multiple of $e_1^{(n-k)}$. If $P_k = \text{diag}(I_k, \tilde{P}_k)$, then

$$
(P_1 \cdots P_k)^T A (P_1 \cdots P_k) =
\begin{bmatrix}
B_{11} & B_{12} & B_{13}\tilde{P}_k \\
B_{21} & B_{22} & B_{23}\tilde{P}_k \\
0 & \tilde{P}_k B_{32} & \tilde{P}_k B_{33}\tilde{P}_k
\end{bmatrix}
$$

is upper Hessenberg through its first k columns. Repeating this for $k = 1{:}n-2$ we obtain

Algorithm 7.4.2 (Householder Reduction to Hessenberg Form) Given $A \in \mathbb{R}^{n \times n}$, the following algorithm overwrites A with $H = U_0^T A U_0$ where H is upper Hessenberg and U_0 is a product of Householder matrices.

> **for** $k = 1{:}n - 2$
> $\qquad [v, \beta] = \text{house}(A(k+1{:}n, k))$
> $\qquad A(k+1{:}n, k{:}n) = (I - \beta vv^T)A(k+1{:}n, k{:}n)$
> $\qquad A(1{:}n, k+1{:}n) = A(1{:}n, k+1{:}n)(I - \beta vv^T)$
> **end**

This algorithm requires $10n^3/3$ flops. If U_0 is explicitly formed, an additional $4n^3/3$ flops are required. The kth Householder matrix can be represented in $A(k+2{:}n, k)$. See Martin and Wilkinson (1968) for a detailed description.

The roundoff properties of this method for reducing A to Hessenberg form are very desirable. Wilkinson (AEP, p. 351) states that the computed Hessenberg matrix $\hat{H}$ satisfies

$$
\hat{H} = Q^T(A + E)Q,
$$

where Q is orthogonal and $\| E \|_F \le cn^2 \mathbf{u} \| A \|_F$ with c a small constant.

7.4.4 Level-3 Aspects

The Hessenberg reduction (Algorithm 7.4.2) is rich in level-2 operations: half gaxpys and half outer product updates. We briefly mention two ideas for introducing level-3 computations into the process.

The first involves a block reduction to block Hessenberg form and is quite straightforward. Suppose (for clarity) that $n = rN$ and write

$$
A = \left[\begin{array}{cc} A_{11} & A_{12} \\ A_{21} & A_{22} \end{array}\right] \begin{array}{c} r \\ n-r \end{array} \; .
$$
$$
\quad\;\; r \quad\; n-r
$$

Suppose that we have computed the QR factorization $A_{21} = \tilde{Q}_1 R_1$ and that $\tilde{Q}_1$ is in WY form. That is, we have $W_1, Y_1 \in \mathbb{R}^{(n-r)\times r}$ such that $\tilde{Q}_1 = I + W_1 Y_1^T$. (See §5.2.2 for details.) If $Q_1 = \text{diag}(I_r, \tilde{Q}_1)$ then

$$
Q_1^T A Q_1 = \left[\begin{array}{cc} A_{11} & A_{12}\tilde{Q}_1 \\ R_1 & \tilde{Q}_1^T A_{22}\tilde{Q}_1 \end{array}\right] \; .
$$

Notice that the updates of the (1,2) and (2,2) blocks are rich in level-3 operations given that $\tilde{Q}_1$ is in WY form. This fully illustrates the overall process as $Q_1^T A Q_1$ is block upper Hessenberg through its first block column. We next repeat the computations on the first r columns of $\tilde{Q}_1^T A_{22}\tilde{Q}_1$. After $N-1$ such steps we obtain

$$
H = U_0^T A U_0 = \left[\begin{array}{cccccc} H_{11} & H_{12} & \cdots & & \cdots & H_{1N} \\ H_{21} & H_{22} & \cdots & & \cdots & H_{2N} \\ 0 & \ddots & \ddots & & \cdots & \vdots \\ \vdots & \vdots & \ddots & \ddots & & \vdots \\ 0 & 0 & \cdots & H_{N,N-1} & H_{NN} \end{array}\right]
$$

where each H_{ij} is r-by-r and $U_0 = Q_1 \cdots Q_{N-2}$ with each Q_i in WY form. The overall algorithm has a level-3 fraction of the form $1 - O(1/N)$. Note that the subdiagonal blocks in H are upper triangular and so the matrix has lower bandwidth r. It is possible to reduce H to actual Hessenberg form by using Givens rotations to zero all but the first subdiagonal.

Dongarra, Hammarling, and Sorensen (1987) have shown how to proceed directly to Hessenberg form using a mixture of gaxpys and level-3 updates. Their idea involves minimal updating after each Householder transformation is generated. For example, suppose the first Householder P_1 has been computed. To generate P_2 we need just the second column of $P_1 A P_1$, not the full outer product update. To generate P_3 we need just the thirrd column of $P_2 P_1 A P_1 P_2$, etc. In this way, the Householder matrices can be determined using only gaxpy operations. No outer product updates are involved. Once a suitable number of Householder matrices are known they can be aggregated and applied in level-3 fashion.

For more about the challenges of organizing a high-performance Hessenberg reduction, see Karlsson (2011).

7.4.5   Important Hessenberg Matrix Properties

The Hessenberg decomposition is not unique. If Z is any n-by-n orthogonal matrix and we apply Algorithm 7.4.2 to $Z^T A Z$, then $Q^T A Q = H$ is upper Hessenberg where $Q = Z U_0$. However, $Q e_1 = Z(U_0 e_1) = Z e_1$ suggesting that H is unique once the first column of Q is specified. This is essentially the case provided H has no zero subdiagonal entries. Hessenberg matrices with this property are said to be *unreduced*. Here is important theorem that clarifies these issues.

Theorem 7.4.2 (Implicit Q Theorem). *Suppose* $Q = [\, q_1 \mid \cdots \mid q_n \,]$ *and* $V = [\, v_1 \mid \cdots \mid v_n \,]$ *are orthogonal matrices with the property that the matrices* $Q^T A Q = H$ *and* $V^T A V = G$ *are each upper Hessenberg where* $A \in \mathbb{R}^{n \times n}$. *Let* k *denote the smallest positive integer for which* $h_{k+1,k} = 0$, *with the convention that* $k = n$ *if* H *is unreduced. If* $q_1 = v_1$, *then* $q_i = \pm v_i$ *and* $|h_{i,i-1}| = |g_{i,i-1}|$ *for* $i = 2{:}k$. *Moreover, if* $k < n$, *then* $g_{k+1,k} = 0$.

Proof. Define the orthogonal matrix $W = [\, w_1 \mid \cdots \mid w_n \,] = V^T Q$ and observe that $GW = WH$. By comparing column $i - 1$ in this equation for $i = 2{:}k$ we see that

$$h_{i,i-1} w_i \;=\; G w_{i-1} \;-\; \sum_{j=1}^{i-1} h_{j,i-1} w_j.$$

Since $w_1 = e_1$, it follows that $[\, w_1 \mid \cdots \mid w_k \,]$ is upper triangular and so for $i = 2{:}k$ we have $w_i = \pm I_n(:,i) = \pm e_i$. Since $w_i = V^T q_i$ and $h_{i,i-1} = w_i^T G w_{i-1}$ it follows that $v_i = \pm q_i$ and

$$|h_{i,i-1}| = |q_i^T A q_{i-1}| = |v_i^T A v_{i-1}| = |g_{i,i-1}|$$

for $i = 2{:}k$. If $k < n$, then

$$g_{k+1,k} = e_{k+1}^T G e_k \;=\; \pm e_{k+1}^T G W e_k \;=\; \pm e_{k+1}^T W H e_k$$

$$= \pm e_{k+1}^T \sum_{i=1}^{k} h_{ik} W e_i \;=\; \pm \sum_{i=1}^{k} h_{ik} e_{k+1}^T e_i \;=\; 0,$$

completing the proof of the theorem. $\square$

The gist of the implicit Q theorem is that if $Q^T A Q = H$ and $Z^T A Z = G$ are each unreduced upper Hessenberg matrices and Q and Z have the same first column, then G and H are "essentially equal" in the sense that $G = D^{-1} H D$ where $D = \operatorname{diag}(\pm 1, \dots, \pm 1)$.

Our next theorem involves a new type of matrix called a *Krylov matrix*. If $A \in \mathbb{R}^{n \times n}$ and $v \in \mathbb{R}^n$, then the Krylov matrix $K(A, v, j) \in \mathbb{R}^{n \times j}$ is defined by

$$K(A, v, j) = [\, v \mid Av \mid \dots \mid A^{j-1} v \,].$$

It turns out that there is a connection between the Hessenberg reduction $Q^T A Q = H$ and the QR factorization of the Krylov matrix $K(A, Q(:,1), n)$.

Theorem 7.4.3. *Suppose* $Q \in \mathbb{R}^{n \times n}$ *is an orthogonal matrix and* $A \in \mathbb{R}^{n \times n}$. *Then* $Q^T A Q = H$ *is an unreduced upper Hessenberg matrix if and only if* $Q^T K(A, Q(:,1), n) = R$ *is nonsingular and upper triangular.*

Proof. Suppose $Q \in \mathbb{R}^{n \times n}$ is orthogonal and set $H = Q^T A Q$. Consider the identity

$$Q^T K(A, Q(:, 1), n) = \left[\, e_1 \mid H e_1 \mid \ldots \mid H^{n-1} e_1 \,\right] \equiv R.$$

If H is an unreduced upper Hessenberg matrix, then it is clear that R is upper triangular with $r_{ii} = h_{21} h_{32} \cdots h_{i,i-1}$ for $i = 2{:}n$. Since $r_{11} = 1$ it follows that R is nonsingular. To prove the converse, suppose R is upper triangular and nonsingular. Since $R(:, k+1) = H R(:, k)$ it follows that $H(:, k) \in \mathsf{span}\{\, e_1, \ldots, e_{k+1} \,\}$. This implies that H is upper Hessenberg. Since $r_{nn} = h_{21} h_{32} \cdots h_{n,n-1} \neq 0$ it follows that H is also unreduced.   □

Thus, there is more or less a correspondence between nonsingular Krylov matrices and orthogonal similarity reductions to unreduced Hessenberg form.

Our last result is about the geometric multiplicity of an eigenvalue of an unreduced upper Hessenberg matrix.

Theorem 7.4.4. *If λ is an eigenvalue of an unreduced upper Hessenberg matrix $H \in \mathbb{R}^{n \times n}$, then its geometric multiplicity is 1.*

Proof. For any $\lambda \in \mathbb{C}$ we have $\mathsf{rank}(A - \lambda I) \geq n - 1$ because the first $n - 1$ columns of $H - \lambda I$ are independent.   □

7.4.6 Companion Matrix Form

Just as the Schur decomposition has a nonunitary analogue in the Jordan decomposition, so does the Hessenberg decomposition have a nonunitary analog in the *companion matrix decomposition*. Let $x \in \mathbb{R}^n$ and suppose that the Krylov matrix $K = K(A, x, n)$ is nonsingular. If $c = c(0{:}n-1)$ solves the linear system $Kc = -A^n x$, then it follows that $AK = KC$ where C has the form

$$C = \begin{bmatrix} 0 & 0 & \cdots & 0 & -c_0 \\ 1 & 0 & \cdots & 0 & -c_1 \\ 0 & 1 & \cdots & 0 & -c_2 \\ \vdots & \vdots & \vdots & \vdots & \vdots \\ 0 & 0 & \cdots & 1 & -c_{n-1} \end{bmatrix}. \tag{7.4.4}$$

The matrix C is said to be a *companion matrix*. Since

$$\mathsf{det}(zI - C) = c_0 + c_1 z + \cdots + c_{n-1} z^{n-1} + z^n,$$

it follows that if K is nonsingular, then the decomposition $K^{-1} A K = C$ displays A's characteristic polynomial. This, coupled with the sparseness of C, leads to "companion matrix methods" in various application areas. These techniques typically involve:

   Step 1. Compute the Hessenberg decomposition $U_0^T A U_0 = H$.

   Step 2. Hope H is unreduced and set $Y = \left[\, e_1 \mid H e_1 \mid \ldots \mid H^{n-1} e_1 \,\right]$.

   Step 3. Solve $YC = HY$ for C.

Unfortunately, this calculation can be highly unstable. A is similar to an unreduced Hessenberg matrix only if each eigenvalue has unit geometric multiplicity. Matrices that have this property are called *nonderogatory*. It follows that the matrix Y above can be very poorly conditioned if A is close to a derogatory matrix.

A full discussion of the dangers associated with companion matrix computation can be found in Wilkinson (AEP, pp. 405ff.).

Problems

P7.4.1 Suppose $A \in \mathbb{R}^{n \times n}$ and $z \in \mathbb{R}^n$. Give a detailed algorithm for computing an orthogonal Q such that $Q^T A Q$ is upper Hessenberg and $Q^T z$ is a multiple of e_1. Hint: Reduce z first and then apply Algorithm 7.4.2.

P7.4.2 Develop a similarity reduction to Hessenberg form using Gauss transforms with pivoting. How many flops are required? See Businger (1969).

P7.4.3 In some situations, it is necessary to solve the linear system $(A + zI)x = b$ for many different values of $z \in \mathbb{R}$ and $b \in \mathbb{R}^n$. Show how this problem can be efficiently and stably solved using the Hessenberg decomposition.

P7.4.4 Suppose $H \in \mathbb{R}^{n \times n}$ is an unreduced upper Hessenberg matrix. Show that there exists a diagonal matrix D such that each subdiagonal element of $D^{-1}HD$ is equal to 1. What is $\kappa_2(D)$?

P7.4.5 Suppose $W, Y \in \mathbb{R}^{n \times n}$ and define the matrices C and B by

$$C = W + iY, \qquad B = \begin{bmatrix} W & -Y \\ Y & W \end{bmatrix}.$$

Show that if $\lambda \in \lambda(C)$ is real, then $\lambda \in \lambda(B)$. Relate the corresponding eigenvectors.

P7.4.6 Suppose

$$A = \begin{bmatrix} w & x \\ y & z \end{bmatrix}$$

is a real matrix having eigenvalues $\lambda \pm i\mu$, where μ is nonzero. Give an algorithm that stably determines $c = \cos(\theta)$ and $s = \sin(\theta)$ such that

$$\begin{bmatrix} c & s \\ -s & c \end{bmatrix}^T \begin{bmatrix} w & x \\ y & z \end{bmatrix} \begin{bmatrix} c & s \\ -s & c \end{bmatrix} = \begin{bmatrix} \lambda & \beta \\ \alpha & \lambda \end{bmatrix}$$

where $\alpha\beta = -\mu^2$.

P7.4.7 Suppose (λ, x) is a known eigenvalue-eigenvector pair for the upper Hessenberg matrix $H \in \mathbb{R}^{n \times n}$. Give an algorithm for computing an orthogonal matrix P such that

$$P^T H P = \begin{bmatrix} \lambda & w^T \\ 0 & H_1 \end{bmatrix}$$

where $H_1 \in \mathbb{R}^{(n-1) \times (n-1)}$ is upper Hessenberg. Compute P as a product of Givens rotations.

P7.4.8 Suppose $H \in \mathbb{R}^{n \times n}$ has lower bandwidth p. Show how to compute $Q \in \mathbb{R}^{n \times n}$, a product of Givens rotations, such that $Q^T H Q$ is upper Hessenberg. How many flops are required?

P7.4.9 Show that if C is a companion matrix with distinct eigenvalues $\lambda_1, \ldots, \lambda_n$, then $VCV^{-1} = \text{diag}(\lambda_1, \ldots, \lambda_n)$ where

$$V = \begin{bmatrix} 1 & \lambda_1 & \cdots & \lambda_1^{n-1} \\ 1 & \lambda_2 & \cdots & \lambda_2^{n-1} \\ \vdots & \vdots & \ddots & \vdots \\ 1 & \lambda_n & \cdots & \lambda_n^{n-1} \end{bmatrix}.$$

Notes and References for §7.4

The real Schur decomposition was originally presented in:

F.D. Murnaghan and A. Wintner (1931). "A Canonical Form for Real Matrices Under Orthogonal Transformations," *Proc. Nat. Acad. Sci. 17*, 417–420.

A thorough treatment of the reduction to Hessenberg form is given in Wilkinson (AEP, Chap. 6), and Algol procedures appear in:

R.S. Martin and J.H. Wilkinson (1968). "Similarity Reduction of a General Matrix to Hessenberg Form," *Numer. Math. 12*, 349–368.

Givens rotations can also be used to compute the Hessenberg decomposition, see:

W. Rath (1982). "Fast Givens Rotations for Orthogonal Similarity," *Numer. Math. 40*, 47–56.

The high-performance computation of the Hessenberg reduction is a major challenge because it is a two-sided factorization, see:

J.J. Dongarra, L. Kaufman, and S. Hammarling (1986). "Squeezing the Most Out of Eigenvalue Solvers on High Performance Computers," *Lin. Alg. Applic. 77*, 113–136.
J.J. Dongarra, S. Hammarling, and D.C. Sorensen (1989). "Block Reduction of Matrices to Condensed Forms for Eigenvalue Computations," *J. ACM 27*, 215–227.
M.W. Berry, J.J. Dongarra, and Y. Kim (1995). "A Parallel Algorithm for the Reduction of a Non-symmetric Matrix to Block Upper Hessenberg Form," *Parallel Comput. 21*, 1189–1211.
G. Quintana-Orti and R. Van De Geijn (2006). "Improving the Performance of Reduction to Hessenberg Form," *ACM Trans. Math. Softw. 32*, 180–194.
S. Tomov, R. Nath, and J. Dongarra (2010). "Accelerating the Reduction to Upper Hessenberg, Tridiagonal, and Bidiagonal Forms Through Hybrid GPU-Based Computing," *Parallel Comput. 36*, 645–654.
L. Karlsson (2011). "Scheduling of Parallel Matrix Computations and Data Layout Conversion for HPC and Multicore Architectures," PhD Thesis, University of Umeå.

Reaching the Hessenberg form via Gauss transforms is discussed in:

P. Businger (1969). "Reducing a Matrix to Hessenberg Form," *Math. Comput. 23*, 819–821.
G.W. Howell and N. Diaa (2005). "Algorithm 841: BHESS: Gaussian Reduction to a Similar Banded Hessenberg Form," *ACM Trans. Math. Softw. 31*, 166–185.

Some interesting mathematical properties of the Hessenberg form may be found in:

B.N. Parlett (1967). "Canonical Decomposition of Hessenberg Matrices," *Math. Comput. 21*, 223–227.

Although the Hessenberg decomposition is largely appreciated as a "front end" decomposition for the QR iteration, it is increasingly popular as a cheap alternative to the more expensive Schur decomposition in certain problems. For a sampling of applications where it has proven to be very useful, consult:

W. Enright (1979). "On the Efficient and Reliable Numerical Solution of Large Linear Systems of O.D.E.'s," *IEEE Trans. Autom. Contr. AC-24*, 905–908.
G.H. Golub, S. Nash and C. Van Loan (1979). "A Hessenberg-Schur Method for the Problem $AX + XB = C$," *IEEE Trans. Autom. Contr. AC-24*, 909–913.
A. Laub (1981). "Efficient Multivariable Frequency Response Computations," *IEEE Trans. Autom. Contr. AC-26*, 407–408.
C.C. Paige (1981). "Properties of Numerical Algorithms Related to Computing Controllability," *IEEE Trans. Auto. Contr. AC-26*, 130–138.
G. Miminis and C.C. Paige (1982). "An Algorithm for Pole Assignment of Time Invariant Linear Systems," *Int. J. Contr. 35*, 341–354.
C. Van Loan (1982). "Using the Hessenberg Decomposition in Control Theory," in *Algorithms and Theory in Filtering and Control*, D.C. Sorensen and R.J. Wets (eds.), Mathematical Programming Study No. 18, North Holland, Amsterdam, 102–111.
C.D. Martin and C.F. Van Loan (2006). "Solving Real Linear Systems with the Complex Schur Decomposition," *SIAM J. Matrix Anal. Applic. 29*, 177–183.

The advisability of posing polynomial root problems as companion matrix eigenvalue problem is discussed in:

A. Edelman and H. Murakami (1995). "Polynomial Roots from Companion Matrix Eigenvalues," *Math. Comput. 64*, 763–776.

7.5 The Practical QR Algorithm

We return to the Hessenberg QR iteration, which we write as follows:

$$H = U_0^T A U_0 \qquad \text{(Hessenberg reduction)}$$
$$\textbf{for } k = 1, 2, \ldots$$
$$\qquad H = UR \qquad \text{(QR factorization)} \qquad\qquad (7.5.1)$$
$$\qquad H = RU$$
$$\textbf{end}$$

Our aim in this section is to describe how the H's converge to upper quasi-triangular form and to show how the convergence rate can be accelerated by incorporating *shifts*.

7.5.1 Deflation

Without loss of generality we may assume that each Hessenberg matrix H in (7.5.1) is unreduced. If not, then at some stage we have

$$H = \begin{bmatrix} H_{11} & H_{12} \\ 0 & H_{22} \end{bmatrix} \begin{matrix} p \\ n-p \end{matrix}$$
$$\qquad\quad p \quad\; n-p$$

where $1 \leq p < n$ and the problem *decouples* into two smaller problems involving H_{11} and H_{22}. The term *deflation* is also used in this context, usually when $p = n - 1$ or $n - 2$.

In practice, decoupling occurs whenever a subdiagonal entry in H is suitably small. For example, if

$$|h_{p+1,p}| \leq c\mathbf{u}(|h_{pp}| + |h_{p+1,p+1}|) \qquad\qquad (7.5.2)$$

for a small constant c, then $h_{p+1,p}$ can justifiably be set to zero because rounding errors of order $\mathbf{u}\| H \|$ are typically present throughout the matrix anyway.

7.5.2 The Shifted QR Iteration

Let $\mu \in \mathbb{R}$ and consider the iteration:

$$H = U_0^T A U_0 \qquad \text{(Hessenberg reduction)}$$
$$\textbf{for } k = 1, 2, \ldots$$
$$\qquad \text{Determine a scalar } \mu.$$
$$\qquad H - \mu I = UR \qquad \text{(QR factorization)} \qquad\qquad (7.5.3)$$
$$\qquad H = RU + \mu I$$
$$\textbf{end}$$

The scalar μ is referred to as a *shift* . Each matrix H generated in (7.5.3) is similar to A, since

$$RU + \mu I \; = \; U^T (UR + \mu I)U \; = \; U^T HU.$$

If we order the eigenvalues λ_i of A so that

$$|\lambda_1 - \mu| \geq \cdots \geq |\lambda_n - \mu|,$$

and μ is fixed from iteration to iteration, then the theory of §7.3 says that the pth subdiagonal entry in H converges to zero with rate

$$\left| \frac{\lambda_{p+1} - \mu}{\lambda_p - \mu} \right|^k .$$

Of course, if $\lambda_p = \lambda_{p+1}$, then there is no convergence at all. But if, for example, μ is much closer to λ_n than to the other eigenvalues, then the zeroing of the $(n, n-1)$ entry is rapid. In the extreme case we have the following:

Theorem 7.5.1. *Let μ be an eigenvalue of an n-by-n unreduced Hessenberg matrix H. If*

$$\tilde{H} = RU + \mu I,$$

where $H - \mu I = UR$ is the QR factorization of $H - \mu I$, then $\tilde{h}_{n,n-1} = 0$ and $\tilde{h}_{nn} = \mu$.

Proof. Since H is an unreduced Hessenberg matrix the first $n - 1$ columns of $H - \mu I$ are independent, regardless of μ. Thus, if $UR = (H - \mu I)$ is the QR factorization then $r_{ii} \neq 0$ for $i = 1{:}n - 1$. But if $H - \mu I$ is singular, then $r_{11} \cdots r_{nn} = 0$. Thus, $r_{nn} = 0$ and $\tilde{H}(n, :) = [\, 0, \ldots, 0, \mu \,]$. □

The theorem says that if we shift by an exact eigenvalue, then in exact arithmetic deflation occurs in one step.

7.5.3   The Single-Shift Strategy

Now let us consider varying μ from iteration to iteration incorporating new information about $\lambda(A)$ as the subdiagonal entries converge to zero. A good heuristic is to regard h_{nn} as the best approximate eigenvalue along the diagonal. If we shift by this quantity during each iteration, we obtain the *single-shift QR iteration*:

> **for** $k = 1, 2, \ldots$
>
> $\quad \mu = H(n, n)$
>
> $\quad H - \mu I = UR \qquad$ (QR factorization) $\hspace{2cm}$ (7.5.4)
>
> $\quad H = RU + \mu I$
>
> **end**

If the $(n, n-1)$ entry converges to zero, it is likely to do so at a quadratic rate. To see this, we borrow an example from Stewart (IMC, p. 366). Suppose H is an unreduced upper Hessenberg matrix of the form

$$H = \begin{bmatrix} \times & \times & \times & \times & \times \\ \times & \times & \times & \times & \times \\ 0 & \times & \times & \times & \times \\ 0 & 0 & \times & \times & \times \\ 0 & 0 & 0 & \epsilon & h_{nn} \end{bmatrix}$$

and that we perform one step of the single-shift QR algorithm, i.e.,

$$UR = H - h_{nn}$$
$$\tilde{H} = RU + h_{nn}I.$$

After $n-2$ steps in the orthogonal reduction of $H - h_{nn}I$ to upper triangular form we obtain a matrix with the following structure:

$$H = \begin{bmatrix} \times & \times & \times & \times & \times \\ 0 & \times & \times & \times & \times \\ 0 & 0 & \times & \times & \times \\ 0 & 0 & 0 & a & b \\ 0 & 0 & 0 & \epsilon & 0 \end{bmatrix}.$$

It is not hard to show that

$$\tilde{h}_{n,n-1} = -\frac{\epsilon^2 b}{a^2 + \epsilon^2}.$$

If we assume that $\epsilon \ll a$, then it is clear that the new $(n, n-1)$ entry has order ϵ^2, precisely what we would expect of a quadratically converging algorithm.

7.5.4 The Double-Shift Strategy

Unfortunately, difficulties with (7.5.4) can be expected if at some stage the eigenvalues a_1 and a_2 of

$$G = \begin{bmatrix} h_{mm} & h_{mn} \\ h_{nm} & h_{nn} \end{bmatrix}, \qquad m = n-1, \tag{7.5.5}$$

are complex for then h_{nn} would tend to be a poor approximate eigenvalue.

A way around this difficulty is to perform two single-shift QR steps in succession using a_1 and a_2 as shifts:

$$H - a_1 I = U_1 R_1$$
$$H_1 = R_1 U_1 + a_1 I \tag{7.5.6}$$
$$H_1 - a_2 I = U_2 R_2$$
$$H_2 = R_2 U_2 + a_2 I$$

These equations can be manipulated to show that

$$(U_1 U_2)(R_2 R_1) = M \tag{7.5.7}$$

where M is defined by

$$M = (H - a_1 I)(H - a_2 I). \tag{7.5.8}$$

Note that M is a real matrix even if G's eigenvalues are complex since

$$M = H^2 - sH + tI$$

where

$$s = a_1 + a_2 = h_{mm} + h_{nn} = \text{tr}(G) \in \mathbb{R}$$

and
$$t = a_1 a_2 = h_{mm} h_{nn} - h_{mn} h_{nm} = \det(G) \in \mathbb{R}.$$

Thus, (7.5.7) is the QR factorization of a real matrix and we may choose U_1 and U_2 so that $Z = U_1 U_2$ is real orthogonal. It then follows that

$$H_2 = U_2^H H_1 U_2 = U_2^H (U_1^H H U_1) U_2 = (U_1 U_2)^H H (U_1 U_2) = Z^T H Z$$

is real.

Unfortunately, roundoff error almost always prevents an exact return to the real field. A real H_2 could be guaranteed if we

- explicitly form the real matrix $M = H^2 - sH + tI$,

- compute the real QR factorization $M = ZR$, and

- set $H_2 = Z^T H Z$.

But since the first of these steps requires $O(n^3)$ flops, this is not a practical course of action.

7.5.5   The Double-Implicit-Shift Strategy

Fortunately, it turns out that we can implement the double-shift step with $O(n^2)$ flops by appealing to the implicit Q theorem of §7.4.5. In particular we can effect the transition from H to H_2 in $O(n^2)$ flops if we

- compute Me_1, the first column of M;

- determine a Householder matrix P_0 such that $P_0(Me_1)$ is a multiple of e_1;

- compute Householder matrices $P_1, \ldots, P_{n-2}$ such that if

$$Z_1 = P_0 P_1 \cdots P_{n-2},$$

then $Z_1^T H Z_1$ is upper Hessenberg and the first columns of Z and Z_1 are the same.

Under these circumstances, the implicit Q theorem permits us to conclude that, if $Z^T H Z$ and $Z_1^T H Z_1$ are both unreduced upper Hessenberg matrices, then they are essentially equal. Note that if these Hessenberg matrices are not unreduced, then we can effect a decoupling and proceed with smaller unreduced subproblems.

Let us work out the details. Observe first that P_0 can be determined in $O(1)$ flops since $Me_1 = [x, \; y, \; z, \; 0, \ldots, 0]^T$ where

$$x = h_{11}^2 + h_{12} h_{21} - s h_{11} + t,$$
$$y = h_{21}(h_{11} + h_{22} - s),$$
$$z = h_{21} h_{32}.$$

Since a similarity transformation with P_0 only changes rows and columns 1, 2, and 3, we see that

$$P_0 H P_0 = \begin{bmatrix} \times & \times & \times & \times & \times & \times \\ \times & \times & \times & \times & \times & \times \\ \times & \times & \times & \times & \times & \times \\ \times & \times & \times & \times & \times & \times \\ 0 & 0 & 0 & \times & \times & \times \\ 0 & 0 & 0 & 0 & \times & \times \end{bmatrix} .$$

Now the mission of the Householder matrices $P_1, \ldots, P_{n-2}$ is to restore this matrix to upper Hessenberg form. The calculation proceeds as follows:

$$\begin{bmatrix} \times & \times & \times & \times & \times & \times \\ \times & \times & \times & \times & \times & \times \\ \times & \times & \times & \times & \times & \times \\ \times & \times & \times & \times & \times & \times \\ 0 & 0 & 0 & \times & \times & \times \\ 0 & 0 & 0 & 0 & \times & \times \end{bmatrix} \xrightarrow{P_1} \begin{bmatrix} \times & \times & \times & \times & \times & \times \\ \times & \times & \times & \times & \times & \times \\ 0 & \times & \times & \times & \times & \times \\ 0 & \times & \times & \times & \times & \times \\ 0 & \times & \times & \times & \times & \times \\ 0 & 0 & 0 & 0 & \times & \times \end{bmatrix} \xrightarrow{P_2}$$

$$\begin{bmatrix} \times & \times & \times & \times & \times & \times \\ \times & \times & \times & \times & \times & \times \\ 0 & \times & \times & \times & \times & \times \\ 0 & 0 & \times & \times & \times & \times \\ 0 & 0 & \times & \times & \times & \times \\ 0 & 0 & \times & \times & \times & \times \end{bmatrix} \xrightarrow{P_3} \begin{bmatrix} \times & \times & \times & \times & \times & \times \\ \times & \times & \times & \times & \times & \times \\ 0 & \times & \times & \times & \times & \times \\ 0 & 0 & \times & \times & \times & \times \\ 0 & 0 & 0 & \times & \times & \times \\ 0 & 0 & 0 & \times & \times & \times \end{bmatrix} \xrightarrow{P_4} \begin{bmatrix} \times & \times & \times & \times & \times & \times \\ \times & \times & \times & \times & \times & \times \\ 0 & \times & \times & \times & \times & \times \\ 0 & 0 & \times & \times & \times & \times \\ 0 & 0 & 0 & \times & \times & \times \\ 0 & 0 & 0 & 0 & \times & \times \end{bmatrix} .$$

Each P_k is the identity with a 3-by-3 or 2-by-2 Householder somewhere along its diagonal, e.g.,

$$P_1 = \begin{bmatrix} 1 & 0 & 0 & 0 & 0 & 0 \\ 0 & \times & \times & \times & 0 & 0 \\ 0 & \times & \times & \times & 0 & 0 \\ 0 & \times & \times & \times & 0 & 0 \\ 0 & 0 & 0 & 0 & 1 & 0 \\ 0 & 0 & 0 & 0 & 0 & 1 \end{bmatrix}, \quad P_2 = \begin{bmatrix} 1 & 0 & 0 & 0 & 0 & 0 \\ 0 & 1 & 0 & 0 & 0 & 0 \\ 0 & 0 & \times & \times & \times & 0 \\ 0 & 0 & \times & \times & \times & 0 \\ 0 & 0 & \times & \times & \times & 0 \\ 0 & 0 & 0 & 0 & 0 & 1 \end{bmatrix},$$

$$P_3 = \begin{bmatrix} 1 & 0 & 0 & 0 & 0 & 0 \\ 0 & 1 & 0 & 0 & 0 & 0 \\ 0 & 0 & 1 & 0 & 0 & 0 \\ 0 & 0 & 0 & \times & \times & \times \\ 0 & 0 & 0 & \times & \times & \times \\ 0 & 0 & 0 & \times & \times & \times \end{bmatrix}, \quad P_4 = \begin{bmatrix} 1 & 0 & 0 & 0 & 0 & 0 \\ 0 & 1 & 0 & 0 & 0 & 0 \\ 0 & 0 & 1 & 0 & 0 & 0 \\ 0 & 0 & 0 & \times & 0 & 0 \\ 0 & 0 & 0 & \times & \times & \times \\ 0 & 0 & 0 & 0 & \times & \times \end{bmatrix} .$$

The applicability of Theorem 7.4.3 (the implicit Q theorem) follows from the observation that $P_k e_1 = e_1$ for $k = 1{:}n - 2$ and that P_0 and Z have the same first column. Hence, $Z_1 e_1 = Z e_1$, and we can assert that Z_1 essentially equals Z provided that the upper Hessenberg matrices $Z^T H Z$ and $Z_1^T H Z_1$ are each unreduced.

The implicit determination of H_2 from H outlined above was first described by Francis (1961) and we refer to it as a *Francis QR step*. The complete Francis step is summarized as follows:

Algorithm 7.5.1 (Francis QR step)   Given the unreduced upper Hessenberg matrix $H \in \mathbb{R}^{n \times n}$ whose trailing 2-by-2 principal submatrix has eigenvalues a_1 and a_2, this algorithm overwrites H with $Z^T H Z$, where Z is a product of Householder matrices and $Z^T(H - a_1 I)(H - a_2 I)$ is upper triangular.

$m = n - 1$

{Compute first column of $(H - a_1 I)(H - a_2 I)$}

$s = H(m, m) + H(n, n)$

$t = H(m, m) \cdot H(n, n) - H(m, n) \cdot H(n, m)$

$x = H(1, 1) \cdot H(1, 1) + H(1, 2) \cdot H(2, 1) - s \cdot H(1, 1) + t$

$y = H(2, 1) \cdot (H(1, 1) + H(2, 2) - s)$

$z = H(2, 1) \cdot H(3, 2)$

for $k = 0{:}n - 3$

    $[v, \beta] = \mathsf{house}([x \ y \ z]^T)$

    $q = \max\{1, k\}.$

    $H(k + 1{:}k + 3, q{:}n) = (I - \beta v v^T) \cdot H(k + 1{:}k + 3, q{:}n)$

    $r = \min\{k + 4, n\}$

    $H(1{:}r, k + 1{:}k + 3) = H(1{:}r, k + 1{:}k + 3) \cdot (I - \beta v v^T)$
    $x = H(k + 2, k + 1)$

    $y = H(k + 3, k + 1)$

    if $k < n - 3$

        $z = H(k + 4, k + 1)$

    end

end

$[v, \beta] = \mathsf{house}([\, x \ y \,]^T)$

$H(n - 1{:}n, n - 2{:}n) = (I - \beta v v^T) \cdot H(n - 1{:}n, n - 2{:}n)$

$H(1{:}n, n - 1{:}n) = H(1{:}n, n - 1{:}n) \cdot (I - \beta v v^T)$

This algorithm requires $10n^2$ flops. If Z is accumulated into a given orthogonal matrix, an additional $10n^2$ flops are necessary.

7.5.6   The Overall Process

Reduction of A to Hessenberg form using Algorithm 7.4.2 and then iteration with Algorithm 7.5.1 to produce the real Schur form is the standard means by which the dense unsymmetric eigenproblem is solved. During the iteration it is necessary to monitor the subdiagonal elements in H in order to spot any possible decoupling. How this is done is illustrated in the following algorithm:

Algorithm 7.5.2 (QR Algorithm) Given $A \in \mathbb{R}^{n \times n}$ and a tolerance tol greater than the unit roundoff, this algorithm computes the real Schur canonical form $Q^T A Q = T$. If Q and T are desired, then T is stored in H. If only the eigenvalues are desired, then diagonal blocks in T are stored in the corresponding positions in H.

Use Algorithm 7.4.2 to compute the Hessenberg reduction

$H = U_0^T A U_0$ where $U_0 = P_1 \cdots P_{n-2}$.

If Q is desired form $Q = P_1 \cdots P_{n-2}$. (See §5.1.6.)

until $q = n$

Set to zero all subdiagonal elements that satisfy:

$|h_{i,i-1}| \leq$ tol$\cdot(|h_{ii}| + |h_{i-1,i-1}|)$.

Find the largest nonnegative q and the smallest non-negative p such that

$$H = \begin{bmatrix} H_{11} & H_{12} & H_{13} \\ 0 & H_{22} & H_{23} \\ 0 & 0 & H_{33} \end{bmatrix} \begin{matrix} p \\ n-p-q \\ q \end{matrix}$$
$$\quad\quad\quad\begin{matrix} p & n-p-q & q \end{matrix}$$

where H_{33} is upper quasi-triangular and H_{22} is unreduced.

if $q < n$

Perform a Francis QR step on H_{22}: $H_{22} = Z^T H_{22} Z$.

if Q is required
$Q = Q \cdot \text{diag}(I_p, Z, I_q)$
$H_{12} = H_{12} Z$
$H_{23} = Z^T H_{23}$
end

end

end

Upper triangularize all 2-by-2 diagonal blocks in H that have real eigenvalues and accumulate the transformations (if necessary).

This algorithm requires $25n^3$ flops if Q and T are computed. If only the eigenvalues are desired, then $10n^3$ flops are necessary. These flops counts are very approximate and are based on the empirical observation that on average only two Francis iterations are required before the lower 1-by-1 or 2-by-2 decouples.

The roundoff properties of the QR algorithm are what one would expect of any orthogonal matrix technique. The computed real Schur form $\hat{T}$ is orthogonally similar to a matrix near to A, i.e.,

$$Q^T (A + E) Q = \hat{T}$$

where $Q^T Q = I$ and $\| E \|_2 \approx \mathbf{u} \| A \|_2$. The computed $\hat{Q}$ is almost orthogonal in the sense that $\hat{Q}^T \hat{Q} = I + F$ where $\| F \|_2 \approx \mathbf{u}$.

The order of the eigenvalues along $\hat{T}$ is somewhat arbitrary. But as we discuss in §7.6, any ordering can be achieved by using a simple procedure for swapping two adjacent diagonal entries.

7.5.7   Balancing

Finally, we mention that if the elements of A have widely varying magnitudes, then A should be *balanced* before applying the QR algorithm. This is an $O(n^2)$ calculation in which a diagonal matrix D is computed so that if

$$
D^{-1}AD = [\, c_1 \mid \cdots \mid c_n \,] = \begin{bmatrix} r_1^T \\ \vdots \\ r_n^T \end{bmatrix}
$$

then $\| r_i \|_\infty \approx \| c_i \|_\infty$ for $i = 1{:}n$. The diagonal matrix D is chosen to have the form

$$
D = \mathrm{diag}(\beta^{i_1}, \ldots, \beta^{i_n})
$$

where β is the floating point base. Note that $D^{-1}AD$ can be calculated without roundoff. When A is balanced, the computed eigenvalues are usually more accurate although there are exceptions. See Parlett and Reinsch (1969) and Watkins(2006).

Problems

P7.5.1 Show that if $\bar{H} = Q^T HQ$ is obtained by performing a single-shift QR step with

$$
H = \begin{bmatrix} w & x \\ y & z \end{bmatrix},
$$

then $|\bar{h}_{21}| \le |y^2 x|/[(w - z)^2 + y^2]$.

P7.5.2 Given $A \in \mathbb{R}^{2\times 2}$, show how to compute a diagonal $D \in \mathbb{R}^{2\times 2}$ so that $\| D^{-1}AD \|_F$ is minimized.

P7.5.3 Explain how the single-shift QR step $H - \mu I = UR$, $\tilde{H} = RU + \mu I$ can be carried out implicitly. That is, show how the transition from H to $\tilde{H}$ can be carried out without subtracting the shift μ from the diagonal of H.

P7.5.4 Suppose H is upper Hessenberg and that we compute the factorization $PH = LU$ via Gaussian elimination with partial pivoting. (See Algorithm 4.3.4.) Show that $H_1 = U(P^T L)$ is upper Hessenberg and similar to H. (This is the basis of the *modified LR algorithm*.)

P7.5.5 Show that if $H = H_0$ is given and we generate the matrices H_k via $H_k - \mu_k I = U_k R_k$, $H_{k+1} = R_k U_k + \mu_k I$, then $(U_1 \cdots U_j)(R_j \cdots R_1) = (H - \mu_1 I) \cdots (H - \mu_j I)$.

Notes and References for §7.5

Historically important papers associated with the QR iteration include:

H. Rutishauser (1958). "Solution of Eigenvalue Problems with the LR Transformation," *Nat. Bur. Stand. App. Math. Ser. 49*, 47–81.

J.G.F. Francis (1961). "The QR Transformation: A Unitary Analogue to the LR Transformation, Parts I and II" *Comput. J. 4*, 265–72, 332–345.

V.N. Kublanovskaya (1961). "On Some Algorithms for the Solution of the Complete Eigenvalue Problem," *Vychisl. Mat. Mat. Fiz 1(4)*, 555–570.

R.S. Martin and J.H. Wilkinson (1968). "The Modified LR Algorithm for Complex Hessenberg Matrices," *Numer. Math. 12*, 369–376.

R.S. Martin, G. Peters, and J.H. Wilkinson (1970). "The QR Algorithm for Real Hessenberg Matrices," *Numer. Math. 14*, 219–231.

For a general insight, we recommend:

D.S. Watkins (1982). "Understanding the QR Algorithm," *SIAM Review 24*, 427–440.

D.S. Watkins (1993). "Some Perspectives on the Eigenvalue Problem," *SIAM Review 35*, 430–471.

D.S. Watkins (2008). "The QR Algorithm Revisited," *SIAM Review 50*, 133–145.
D.S. Watkins (2011). "Francis's Algorithm," *Amer. Math. Monthly 118*, 387–403.

Papers concerned with the convergence of the method, shifting, deflation, and related matters include:

P.A. Businger (1971). "Numerically Stable Deflation of Hessenberg and Symmetric Tridiagonal Matrices, *BIT 11*, 262–270.
D.S. Watkins and L. Elsner (1991). "Chasing Algorithms for the Eigenvalue Problem," *SIAM J. Matrix Anal. Applic. 12*, 374–384.
D.S. Watkins and L. Elsner (1991). "Convergence of Algorithms of Decomposition Type for the Eigenvalue Problem," *Lin. Alg. Applic. 143*, 19–47.
J. Erxiong (1992). "A Note on the Double-Shift QL Algorithm," *Lin. Alg. Applic. 171*, 121–132.
A.A. Dubrulle and G.H. Golub (1994). "A Multishift QR Iteration Without Computation of the Shifts," *Numer. Algorithms 7*, 173–181.
D.S. Watkins (1996). "Forward Stability and Transmission of Shifts in the QR Algorithm," *SIAM J. Matrix Anal. Applic. 16*, 469–487.
D.S. Watkins (1996). "The Transmission of Shifts and Shift Blurring in the QR algorithm," *Lin. Alg. Applic. 241–3*, 877–896.
D.S. Watkins (1998). "Bulge Exchanges in Algorithms of QR Type," *SIAM J. Matrix Anal. Applic. 19*, 1074–1096.
R. Vandebril (2011). "Chasing Bulges or Rotations? A Metamorphosis of the QR-Algorithm" *SIAM. J. Matrix Anal. Applic. 32*, 217–247.

Aspects of the balancing problem are discussed in:

E.E. Osborne (1960). "On Preconditioning of Matrices," *J. ACM 7*, 338–345.
B.N. Parlett and C. Reinsch (1969). "Balancing a Matrix for Calculation of Eigenvalues and Eigenvectors," *Numer. Math. 13*, 292–304.
D.S. Watkins (2006). "A Case Where Balancing is Harmful," *ETNA 23*, 1–4.

Versions of the algorithm that are suitable for companion matrices are discussed in:

D.A. Bini, F. Daddi, and L. Gemignani (2004). "On the Shifted QR iteration Applied to Companion Matrices," *ETNA 18*, 137–152.
M. Van Barel, R. Vandebril, P. Van Dooren, and K. Frederix (2010). "Implicit Double Shift QR-Algorithm for Companion Matrices," *Numer. Math. 116*, 177–212.

Papers that are concerned with the high-performance implementation of the QR iteration include:

Z. Bai and J.W. Demmel (1989). "On a Block Implementation of Hessenberg Multishift QR Iteration," *Int. J. High Speed Comput. 1*, 97–112.
R.A. Van De Geijn (1993). "Deferred Shifting Schemes for Parallel QR Methods," *SIAM J. Matrix Anal. Applic. 14*, 180–194.
D.S. Watkins (1994). "Shifting Strategies for the Parallel QR Algorithm," *SIAM J. Sci. Comput. 15*, 953–958.
G. Henry and R. van de Geijn (1996). "Parallelizing the QR Algorithm for the Unsymmetric Algebraic Eigenvalue Problem: Myths and Reality," *SIAM J. Sci. Comput. 17*, 870–883.
Z. Bai, J. Demmel, J. Dongarra, A. Petitet, H. Robinson, and K. Stanley (1997). "The Spectral Decomposition of Nonsymmetric Matrices on Distributed Memory Parallel Computers," *SIAM J. Sci. Comput. 18*, 1446–1461.
G. Henry, D.S. Watkins, and J. Dongarra (2002). "A Parallel Implementation of the Nonsymmetric QR Algorithm for Distributed Memory Architectures," *SIAM J. Sci. Comput. 24*, 284–311.
K. Braman, R. Byers, and R. Mathias (2002). "The Multishift QR Algorithm. Part I: Maintaining Well-Focused Shifts and Level 3 Performance," *SIAM J. Matrix Anal. Applic. 23*, 929–947.
K. Braman, R. Byers, and R. Mathias (2002). "The Multishift QR Algorithm. Part II: Aggressive Early Deflation," *SIAM J. Matrix Anal. Applic. 23*, 948–973.
M.R. Fahey (2003). "Algorithm 826: A Parallel Eigenvalue Routine for Complex Hessenberg Matrices," *ACM Trans. Math. Softw. 29*, 326–336.
D. Kressner (2005). "On the Use of Larger Bulges in the QR Algorithm," *ETNA 20*, 50–63.
D. Kressner (2008). "The Effect of Aggressive Early Deflation on the Convergence of the QR Algorithm," *SIAM J. Matrix Anal. Applic. 30*, 805–821.

7.6   Invariant Subspace Computations

Several important invariant subspace problems can be solved once the real Schur decomposition $Q^T A Q = T$ has been computed. In this section we discuss how to

- compute the eigenvectors associated with some subset of $\lambda(A)$,

- compute an orthonormal basis for a given invariant subspace,

- block-diagonalize A using well-conditioned similarity transformations,

- compute a basis of eigenvectors regardless of their condition, and

- compute an approximate Jordan canonical form of A.

Eigenvector/invariant subspace computation for sparse matrices is discussed in §7.3.1 and §7.3.2 as well as portions of Chapters 8 and 10.

7.6.1   Selected Eigenvectors via Inverse Iteration

Let $q^{(0)} \in \mathbb{R}^n$ be a given unit 2-norm vector and assume that $A - \mu I \in \mathbb{R}^{n \times n}$ is nonsingular. The following is referred to as *inverse iteration*:

> **for** $k = 1, 2, \ldots$
>
> > Solve $(A - \mu I) z^{(k)} = q^{(k-1)}$.
> > $q^{(k)} = z^{(k)} / \| z^{(k)} \|_2$ (7.6.1)
> > $\lambda^{(k)} = q^{(k)^T} A q^{(k)}$
>
> **end**

Inverse iteration is just the power method applied to $(A - \mu I)^{-1}$.

To analyze the behavior of (7.6.1), assume that A has a basis of eigenvectors $\{x_1, \ldots, x_n\}$ and that $A x_i = \lambda_i x_i$ for $i = 1{:}n$. If

$$q^{(0)} = \sum_{i=1}^{n} \beta_i x_i$$

then $q^{(k)}$ is a unit vector in the direction of

$$(A - \mu I)^{-k} q^{(0)} = \sum_{i=1}^{n} \frac{\beta_i}{(\lambda_i - \mu)^k} x_i.$$

Clearly, if μ is much closer to an eigenvalue λ_j than to the other eigenvalues, then $q^{(k)}$ is rich in the direction of x_j provided $\beta_j \neq 0$.

A sample stopping criterion for (7.6.1) might be to quit as soon as the residual

$$r^{(k)} = (A - \mu I) q^{(k)}$$

satisfies

$$\| r^{(k)} \|_\infty \leq c \mathbf{u} \| A \|_\infty$$ (7.6.2)

where c is a constant of order unity. Since

$$(A + E_k)q^{(k)} = \mu q^{(k)}$$

with $E_k = -r^{(k)}q^{(k)T}$, it follows that (7.6.2) forces μ and $q^{(k)}$ to be an exact eigenpair for a nearby matrix.

Inverse iteration can be used in conjunction with Hessenberg reduction and the QR algorithm as follows:

Step 1. Compute the Hessenberg decomposition $U_0^T A U_0 = H$.

Step 2. Apply the double-implicit-shift Francis iteration to H *without* accumulating transformations.

Step 3. For each computed eigenvalue λ whose corresponding eigenvector x is sought, apply (7.6.1) with $A = H$ and $\mu = \lambda$ to produce a vector z such that $Hz \approx \mu z$.

Step 4. Set $x = U_0 z$.

Inverse iteration with H is very economical because we do not have to accumulate transformations during the double Francis iteration. Moreover, we can factor matrices of the form $H - \lambda I$ in $O(n^2)$ flops, and (3) only one iteration is typically required to produce an adequate approximate eigenvector.

This last point is perhaps the most interesting aspect of inverse iteration and requires some justification since λ can be comparatively inaccurate if it is ill-conditioned. Assume for simplicity that λ is real and let

$$H - \lambda I = \sum_{i=1}^{n} \sigma_i u_i v_i^T = U\Sigma V^T$$

be the SVD of $H - \lambda I$. From what we said about the roundoff properties of the QR algorithm in §7.5.6, there exists a matrix $E \in \mathbb{R}^{n \times n}$ such that $H + E - \lambda I$ is singular and $\| E \|_2 \approx \mathbf{u}\| H \|_2$. It follows that $\sigma_n \approx \mathbf{u}\sigma_1$ and

$$\| (H - \hat{\lambda}I)v_n \|_2 \approx \mathbf{u}\sigma_1,$$

i.e., v_n is a good approximate eigenvector. Clearly if the starting vector $q^{(0)}$ has the expansion

$$q^{(0)} = \sum_{i=1}^{n} \gamma_i u_i$$

then

$$z^{(1)} = \sum_{i=1}^{n} \frac{\gamma_i}{\sigma_i} v_i$$

is "rich" in the direction v_n. Note that if $s(\lambda) \approx |u_n^T v_n|$ is small, then $z^{(1)}$ is rather deficient in the direction u_n. This explains (heuristically) why another step of inverse iteration is not likely to produce an improved eigenvector approximate, especially if λ is ill-conditioned. For more details, see Peters and Wilkinson (1979).

7.6.2   Ordering Eigenvalues in the Real Schur Form

Recall that the real Schur decomposition provides information about invariant subspaces. If

$$Q^T A Q = T = \begin{bmatrix} T_{11} & T_{12} \\ 0 & T_{22} \end{bmatrix} \begin{matrix} p \\ q \end{matrix}$$
$$\qquad\qquad\quad p \qquad q$$

and

$$\lambda(T_{11}) \cap \lambda(T_{22}) = \emptyset,$$

then the first p columns of Q span the unique invariant subspace associated with $\lambda(T_{11})$. (See §7.1.4.) Unfortunately, the Francis iteration supplies us with a real Schur decomposition $Q_F^T A Q_F = T_F$ in which the eigenvalues appear somewhat randomly along the diagonal of T_F. This poses a problem if we want an orthonormal basis for an invariant subspace whose associated eigenvalues are not at the top of T_F's diagonal. Clearly, we need a method for computing an orthogonal matrix Q_D such that $Q_D^T T_F Q_D$ is upper quasi-triangular with appropriate eigenvalue ordering.

A look at the 2-by-2 case suggests how this can be accomplished. Suppose

$$Q_F^T A Q_F = T_F = \begin{bmatrix} \lambda_1 & t_{12} \\ 0 & \lambda_2 \end{bmatrix}, \qquad \lambda_1 \neq \lambda_2$$

and that we wish to reverse the order of the eigenvalues. Note that

$$T_F x = \lambda_2 x$$

where

$$x = \begin{bmatrix} t_{12} \\ \lambda_2 - \lambda_1 \end{bmatrix}.$$

Let Q_D be a Givens rotation such that the second component of $Q_D^T x$ is zero. If

$$Q = Q_F Q_D,$$

then

$$(Q^T A Q) e_1 = Q_D^T T_F (Q_D e_1) = \lambda_2 Q_D^T (Q_D e_1) = \lambda_2 e_1.$$

The matrices A and $Q^T A Q$ have the same Frobenius norm and so it follows that the latter must have the following form:

$$Q^T A Q = \begin{bmatrix} \lambda_2 & \pm t_{12} \\ 0 & \lambda_1 \end{bmatrix}.$$

The swapping gets a little more complicated if T has 2-by-2 blocks along its diagonal. See Ruhe (1970) and Stewart (1976) for details.

By systematically interchanging adjacent pairs of eigenvalues (or 2-by-2 blocks), we can move any subset of $\lambda(A)$ to the top of T's diagonal. Here is the overall procedure for the case when there are no 2-by-2 bumps:

Algorithm 7.6.1 Given an orthogonal matrix $Q \in \mathbb{R}^{n \times n}$, an upper triangular matrix $T = Q^T A Q$, and a subset $\Delta = \{\lambda_1, \ldots, \lambda_p\}$ of $\lambda(A)$, the following algorithm computes an orthogonal matrix Q_D such that $Q_D^T T Q_D = S$ is upper triangular and $\{s_{11}, \ldots, s_{pp}\} = \Delta$. The matrices Q and T are overwritten by $Q Q_D$ and S, respectively.

> **while** $\{t_{11}, \ldots, t_{pp}\} \neq \Delta$
>> **for** $k = 1{:}n-1$
>>> **if** $t_{kk} \notin \Delta$ and $t_{k+1,k+1} \in \Delta$
>>>
>>> $$[\,c,\ s\,] = \mathsf{givens}(T(k,k+1), T(k+1,k+1) - T(k,k))$$
>>>
>>> $$T(k{:}k+1, k{:}n) = \begin{bmatrix} c & s \\ -s & c \end{bmatrix}^T T(k{:}k+1, k{:}n)$$
>>>
>>> $$T(1{:}k+1, k{:}k+1) = T(1{:}k+1, k{:}k+1) \begin{bmatrix} c & s \\ -s & c \end{bmatrix}$$
>>>
>>> $$Q(1{:}n, k{:}k+1) = Q(1{:}n, k{:}k+1) \begin{bmatrix} c & s \\ -s & c \end{bmatrix}$$
>>>
>>> **end**
>> **end**
> **end**

This algorithm requires $k(12n)$ flops, where k is the total number of required swaps. The integer k is never greater than $(n-p)p$.

Computation of invariant subspaces by manipulating the real Schur decomposition is extremely stable. If $\hat{Q} = [\,\hat{q}_1 \,|\, \cdots \,|\, \hat{q}_n\,]$ denotes the computed orthogonal matrix Q, then $\| \hat{Q}^T \hat{Q} - I \|_2 \approx \mathbf{u}$ and there exists a matrix E satisfying $\| E \|_2 \approx \mathbf{u} \| A \|_2$ such that $(A + E)\hat{q}_i \in \mathsf{span}\{\hat{q}_1, \ldots, \hat{q}_p\}$ for $i = 1{:}p$.

7.6.3 Block Diagonalization

Let

$$T = \begin{bmatrix} T_{11} & T_{12} & \cdots & T_{1q} \\ 0 & T_{22} & \cdots & T_{2q} \\ \vdots & \vdots & \ddots & \vdots \\ 0 & 0 & \cdots & T_{qq} \end{bmatrix} \begin{matrix} n_1 \\ n_2 \\ \\ n_q \end{matrix} \tag{7.6.3}$$

$$\qquad\quad n_1 \quad\ n_2 \qquad\ n_q$$

be a partitioning of some real Schur canonical form $Q^T A Q = T \in \mathbb{R}^{n \times n}$ such that $\lambda(T_{11}), \ldots, \lambda(T_{qq})$ are disjoint. By Theorem 7.1.6 there exists a matrix Y such that

$$Y^{-1} T Y = \mathrm{diag}(T_{11}, \ldots, T_{qq}).$$

A practical procedure for determining Y is now given together with an analysis of Y's sensitivity as a function of the above partitioning.

Partition $I_n = [E_1 | \cdots | E_q]$ conformably with T and define the matrix $Y_{ij} \in \mathbb{R}^{n \times n}$ as follows:

$$Y_{ij} = I_n + E_i Z_{ij} E_j^T, \qquad i < j, \ Z_{ij} \in \mathbb{R}^{n_i \times n_j}.$$

In other words, Y_{ij} looks just like the identity except that Z_{ij} occupies the (i, j) block position. It follows that if $Y_{ij}^{-1} T Y_{ij} = \bar{T} = (\bar{T}_{ij})$, then T and $\bar{T}$ are identical except that

$$
\begin{aligned}
\bar{T}_{ij} &= T_{ii} Z_{ij} - Z_{ij} T_{jj} + T_{ij}, \\
\bar{T}_{ik} &= T_{ik} - Z_{ij} T_{jk}, \qquad (k = j + 1{:}q), \\
\bar{T}_{kj} &= T_{ki} Z_{ij} + T_{kj}, \qquad (k = 1{:}i - 1) .
\end{aligned}
$$

Thus, T_{ij} can be zeroed provided we have an algorithm for solving the *Sylvester equation*

$$
FZ - ZG \ = \ C \tag{7.6.4}
$$

where $F \in \mathbb{R}^{p \times p}$ and $G \in \mathbb{R}^{r \times r}$ are given upper quasi-triangular matrices and $C \in \mathbb{R}^{p \times r}$.

Bartels and Stewart (1972) have devised a method for doing this. Let $C = [\, c_1 \,|\cdots|\, c_r \,]$ and $Z = [\, z_1 \,|\cdots|\, z_r \,]$ be column partitionings. If $g_{k+1,k} = 0$, then by comparing columns in (7.6.4) we find

$$
F z_k \ - \ \sum_{i=1}^{k} g_{ik} z_i \ = \ c_k.
$$

Thus, once we know $z_1, \ldots, z_{k-1}$, then we can solve the quasi-triangular system

$$
(F \ - \ g_{kk} I) \, z_k \ = \ c_k + \sum_{i=1}^{k-1} g_{ik} z_i
$$

for z_k. If $g_{k+1,k} \neq 0$, then z_k and z_{k+1} can be simultaneously found by solving the $2p$-by-$2p$ system

$$
\begin{bmatrix} F - g_{kk} I & -g_{mk} I \\ -g_{km} I & F - g_{mm} I \end{bmatrix} \begin{bmatrix} z_k \\ z_m \end{bmatrix} = \begin{bmatrix} c_k \\ c_m \end{bmatrix} + \sum_{i=1}^{k-1} \begin{bmatrix} g_{ik} z_i \\ g_{im} z_i \end{bmatrix} \tag{7.6.5}
$$

where $m = k + 1$. By reordering the equations according to the perfect shuffle permutation $(1, p + 1, 2, p + 2, \ldots, p, 2p)$, a banded system is obtained that can be solved in $O(p^2)$ flops. The details may be found in Bartels and Stewart (1972). Here is the overall process for the case when F and G are each triangular.

Algorithm 7.6.2 (Bartels-Stewart Algorithm) Given $C \in \mathbb{R}^{p \times r}$ and upper triangular matrices $F \in \mathbb{R}^{p \times p}$ and $G \in \mathbb{R}^{r \times r}$ that satisfy $\lambda(F) \cap \lambda(G) = \emptyset$, the following algorithm overwrites C with the solution to the equation $FZ - ZG = C$.

> **for** $k = 1{:}r$
> > $C(1{:}p, k) = C(1{:}p, k) + C(1{:}p, 1{:}k - 1) \cdot G(1{:}k - 1, k)$
> > Solve $(F - G(k, k) I) z = C(1{:}p, k)$ for z.
> > $C(1{:}p, k) = z$
>
> **end**

This algorithm requires $pr(p + r)$ flops. By zeroing the superdiagonal blocks in T in the appropriate order, the entire matrix can be reduced to block diagonal form.

Algorithm 7.6.3 Given an orthogonal matrix $Q \in \mathbb{R}^{n \times n}$, an upper quasi-triangular matrix $T = Q^T A Q$, and the partitioning (7.6.3), the following algorithm overwrites Q with QY where $Y^{-1}TY = \mathrm{diag}(T_{11}, \ldots, T_{qq})$.

> **for** $j = 2:q$
>> **for** $i = 1:j-1$
>>> Solve $T_{ii}Z - ZT_{jj} = -T_{ij}$ for Z using the Bartels-Stewart algorithm.
>>> **for** $k = j+1:q$
>>>> $T_{ik} = T_{ik} - ZT_{jk}$
>>> **end**
>>> **for** $k = 1:q$
>>>> $Q_{kj} = Q_{ki}Z + Q_{kj}$
>>> **end**
>> **end**
> **end**

The number of flops required by this algorithm is a complicated function of the block sizes in (7.6.3).

The choice of the real Schur form T and its partitioning in (7.6.3) determines the sensitivity of the Sylvester equations that must be solved in Algorithm 7.6.3. This in turn affects the condition of the matrix Y and the overall usefulness of the block diagonalization. The reason for these dependencies is that the relative error of the computed solution $\hat{Z}$ to

$$T_{ii}Z - ZT_{jj} = -T_{ij} \tag{7.6.6}$$

satisfies

$$\frac{\| \hat{Z} - Z \|_F}{\| Z \|_F} \approx \mathbf{u} \frac{\| T \|_F}{\mathsf{sep}(T_{ii}, T_{jj})}.$$

For details, see Golub, Nash, and Van Loan (1979). Since

$$\mathsf{sep}(T_{ii}, T_{jj}) = \min_{X \neq 0} \frac{\| T_{ii}X - XT_{jj} \|_F}{\| X \|_F} \leq \min_{\substack{\lambda \in \lambda(T_{ii}) \\ \mu \in \lambda(T_{jj})}} |\lambda - \mu|$$

there can be a substantial loss of accuracy whenever the subsets $\lambda(T_{ii})$ are insufficiently separated. Moreover, if Z satisfies (7.6.6) then

$$\| Z \|_F \leq \frac{\| T_{ij} \|_F}{\mathsf{sep}(T_{ii}, T_{jj})}.$$

Thus, large norm solutions can be expected if $\mathsf{sep}(T_{ii}, T_{jj})$ is small. This tends to make the matrix Y in Algorithm 7.6.3 ill-conditioned since it is the product of the matrices

$$Y_{ij} = \begin{bmatrix} I_{n_i} & Z \\ 0 & I_{n_j} \end{bmatrix}.$$

Note that $\kappa_F(Y_{ij}) = n_i^2 + n_j^2 + \| Z \|_F^2$.

Confronted with these difficulties, Bavely and Stewart (1979) develop an algorithm for block diagonalizing that dynamically determines the eigenvalue ordering and partitioning in (7.6.3) so that all the Z matrices in Algorithm 7.6.3 are bounded in norm by some user-supplied tolerance. Their research suggests that the condition of Y can be controlled by controlling the condition of the Y_{ij}.

7.6.4   Eigenvector Bases

If the blocks in the partitioning (7.6.3) are all 1-by-1, then Algorithm 7.6.3 produces a basis of eigenvectors. As with the method of inverse iteration, the computed eigenvalue-eigenvector pairs are exact for some "nearby" matrix. A widely followed rule of thumb for deciding upon a suitable eigenvector method is to use inverse iteration whenever fewer than 25% of the eigenvectors are desired.

We point out, however, that the real Schur form can be used to determine selected eigenvectors. Suppose

$$
Q^T A Q = \begin{array}{c} \\ \\ \\ \end{array}\left[\begin{array}{ccc} T_{11} & u & T_{13} \\ 0 & \lambda & v^T \\ 0 & 0 & T_{33} \end{array}\right] \begin{array}{c} k-1 \\ 1 \\ n-k \end{array}
$$
$$
\quad\quad\quad\quad k-1 \quad\; 1 \quad\; n-k
$$

is upper quasi-triangular and that $\lambda \notin \lambda(T_{11}) \cup \lambda(T_{33})$. It follows that if we solve the linear systems $(T_{11} - \lambda I)w = -u$ and $(T_{33} - \lambda I)^T z = -v$ then

$$
x = Q \left[\begin{array}{c} w \\ 1 \\ 0 \end{array}\right] \quad \text{and} \quad y = Q \left[\begin{array}{c} 0 \\ 1 \\ z \end{array}\right]
$$

are the associated right and left eigenvectors, respectively. Note that the condition of λ is prescribed by

$$
1/s(\lambda) = \sqrt{(1 + w^T w)(1 + z^T z)}.
$$

7.6.5   Ascertaining Jordan Block Structures

Suppose that we have computed the real Schur decomposition $A = QTQ^T$, identified clusters of "equal" eigenvalues, and calculated the corresponding block diagonalization $T = Y \cdot \text{diag}(T_{11}, \ldots, T_{qq})Y^{-1}$. As we have seen, this can be a formidable task. However, even greater numerical problems confront us if we attempt to ascertain the Jordan block structure of each T_{ii}. A brief examination of these difficulties will serve to highlight the limitations of the Jordan decomposition.

Assume for clarity that $\lambda(T_{ii})$ is real. The reduction of T_{ii} to Jordan form begins by replacing it with a matrix of the form $C = \lambda I + N$, where N is the strictly upper triangular portion of T_{ii} and where λ, say, is the mean of its eigenvalues.

Recall that the dimension of a Jordan block $J(\lambda)$ is the smallest nonnegative integer k for which $[J(\lambda) - \lambda I]^k = 0$. Thus, if $p_i = \text{dim}[\text{null}(N^i)]$, for $i = 0{:}n$, then $p_i - p_{i-1}$ equals the number of blocks in C's Jordan form that have dimension i or greater. A concrete example helps to make this assertion clear and to illustrate the role of the SVD in Jordan form computations.

Assume that C is 7-by-7. Suppose we compute the SVD $U_1^T N V_1 = \Sigma_1$ and "discover" that N has rank 3. If we order the singular values from small to large then it follows that the matrix $N_1 = V_1^T N V_1$ has the form

$$N_1 = \begin{bmatrix} 0 & K \\ 0 & L \end{bmatrix} \begin{matrix} 4 \\ 3 \end{matrix} \quad .$$
$$\quad\quad 4 \quad 3$$

At this point, we know that the geometric multiplicity of λ is 4—i.e, C's Jordan form has four blocks $(p_1 - p_0 = 4 - 0 = 4)$.

Now suppose $\tilde{U}_2^T L \tilde{V}_2 = \Sigma_2$ is the SVD of L and that we find that L has unit rank. If we again order the singular values from small to large, then $L_2 = \tilde{V}_2^T L \tilde{V}_2$ clearly has the following structure:

$$L_2 = \begin{bmatrix} 0 & 0 & a \\ 0 & 0 & b \\ 0 & 0 & c \end{bmatrix} .$$

However, $\lambda(L_2) = \lambda(L) = \{0, 0, 0\}$ and so $c = 0$. Thus, if

$$V_2 = \text{diag}(I_4, \tilde{V}_2)$$

then $N_2 = V_2^T N_1 V_2$ has the following form:

$$N_2 = \begin{bmatrix} 0 & 0 & 0 & 0 & \times & \times & \times \\ 0 & 0 & 0 & 0 & \times & \times & \times \\ 0 & 0 & 0 & 0 & \times & \times & \times \\ 0 & 0 & 0 & 0 & \times & \times & \times \\ 0 & 0 & 0 & 0 & 0 & 0 & a \\ 0 & 0 & 0 & 0 & 0 & 0 & b \\ 0 & 0 & 0 & 0 & 0 & 0 & 0 \end{bmatrix} .$$

Besides allowing us to introduce more zeros into the upper triangle, the SVD of L also enables us to deduce the dimension of the nullspace of N^2. Since

$$N_1^2 = \begin{bmatrix} 0 & KL \\ 0 & L^2 \end{bmatrix} = \begin{bmatrix} 0 & K \\ 0 & L \end{bmatrix} \begin{bmatrix} 0 & K \\ 0 & L \end{bmatrix}$$

and $\begin{bmatrix} K \\ L \end{bmatrix}$ has full column rank,

$$p_2 = \text{dim(null}(N^2)) = \text{dim(null}(N_1^2)) = 4 + \text{dim(null}(L)) = p_1 + 2.$$

Hence, we can conclude at this stage that the Jordan form of C has at least two blocks of dimension 2 or greater.

Finally, it is easy to see that $N_1^3 = 0$, from which we conclude that there is $p_3 - p_2 = 7 - 6 = 1$ block of dimension 3 or larger. If we define $V = V_1 V_2$ then it follows that the decomposition

$$
V^T C V = \left[\begin{array}{ccccccc}
\lambda & 0 & 0 & 0 & \times & \times & \times \\
0 & \lambda & 0 & 0 & \times & \times & \times \\
0 & 0 & \lambda & 0 & \times & \times & \times \\
0 & 0 & 0 & \lambda & \times & \times & \times \\
0 & 0 & 0 & 0 & \lambda & \times & a \\
0 & 0 & 0 & 0 & 0 & \lambda & 0 \\
0 & 0 & 0 & 0 & 0 & 0 & \lambda
\end{array}\right]
\begin{array}{l}
\left.\vphantom{\begin{array}{c}\\\\\\\\\end{array}}\right\} \text{four blocks of order 1 or larger} \\[1.2em]
\left.\vphantom{\begin{array}{c}\\\\\end{array}}\right\} \text{two blocks of order 2 or larger} \\[0.6em]
\left.\vphantom{\begin{array}{c}\\\end{array}}\right\} \text{one block of order 3 or larger}
\end{array}
$$

displays C's Jordan block structure: two blocks of order 1, one block of order 2, and one block of order 3.

To compute the Jordan decomposition it is necessary to resort to nonorthogonal transformations. We refer the reader to Golub and Wilkinson (1976), Kågström and Ruhe (1980a, 1980b), and Demmel (1983) for more details. The above calculations with the SVD amply illustrate that difficult rank decisions must be made at each stage and that the final computed block structure depends critically on those decisions.

Problems

P7.6.1   Give a complete algorithm for solving a real, n-by-n, upper quasi-triangular system $Tx = b$.

P7.6.2   Suppose $U^{-1}AU = \mathrm{diag}(\alpha_1, \ldots, \alpha_m)$ and $V^{-1}BV = \mathrm{diag}(\beta_1, \ldots, \beta_n)$. Show that if

$$\phi(X) = AX - XB,$$

then

$$\lambda(\phi) = \{\, \alpha_i - \beta_j : i = 1{:}m,\ j = 1{:}n \,\}.$$

What are the corresponding eigenvectors? How can these facts be used to solve $AX - XB = C$?

P7.6.3   Show that if $Z \in \mathbb{C}^{p \times q}$ and

$$
Y = \left[\begin{array}{cc}
I_p & Z \\
0 & I_q
\end{array}\right],
$$

then $\kappa_2(Y) = [2 + \sigma^2 + \sqrt{4\sigma^2 + \sigma^4}\,]/2$ where $\sigma = \|\, Z \,\|_2$.

P7.6.4   Derive the system (7.6.5).

P7.6.5   Assume that $T \in \mathbb{R}^{n \times n}$ is block upper triangular and partitioned as follows:

$$
T = \left[\begin{array}{ccc}
T_{11} & T_{12} & T_{13} \\
0 & T_{22} & T_{23} \\
0 & 0 & T_{33}
\end{array}\right],
\qquad
T \in \mathbb{R}^{n \times n}.
$$

Suppose that the diagonal block T_{22} is 2-by-2 with complex eigenvalues that are disjoint from $\lambda(T_{11})$ and $\lambda(T_{33})$. Give an algorithm for computing the 2-dimensional real invariant subspace associated with T_{22}'s eigenvalues.

P7.6.6   Suppose $H \in \mathbb{R}^{n \times n}$ is upper Hessenberg with a complex eigenvalue $\lambda + i \cdot \mu$. How could inverse iteration be used to compute $x, y \in \mathbb{R}^n$ so that $H(x + iy) = (\lambda + i\mu)(x + iy)$? Hint: Compare real and imaginary parts in this equation and obtain a $2n$-by-$2n$ real system.

Notes and References for §7.6

Much of the material discussed in this section may be found in the following survey paper:

G.H. Golub and J.H. Wilkinson (1976). "Ill-Conditioned Eigensystems and the Computation of the Jordan Canonical Form," *SIAM Review 18*, 578–619.

The problem of ordering the eigenvalues in the real Schur form is the subject of:

A. Ruhe (1970). "An Algorithm for Numerical Determination of the Structure of a General Matrix," *BIT 10*, 196–216.

G.W. Stewart (1976). "Algorithm 406: HQR3 and EXCHNG: Fortran Subroutines for Calculating and Ordering the Eigenvalues of a Real Upper Hessenberg Matrix," *ACM Trans. Math. Softw. 2*, 275–280.

J.J. Dongarra, S. Hammarling, and J.H. Wilkinson (1992). "Numerical Considerations in Computing Invariant Subspaces," *SIAM J. Matrix Anal. Applic. 13*, 145–161.

Z. Bai and J.W. Demmel (1993). "On Swapping Diagonal Blocks in Real Schur Form," *Lin. Alg. Applic. 186*, 73–95

Procedures for block diagonalization including the Jordan form are described in:

C. Bavely and G.W. Stewart (1979). "An Algorithm for Computing Reducing Subspaces by Block Diagonalization," *SIAM J. Numer. Anal. 16*, 359–367.

B. Kågström and A. Ruhe (1980a). "An Algorithm for Numerical Computation of the Jordan Normal Form of a Complex Matrix," *ACM Trans. Math. Softw. 6*, 398–419.

B. Kågström and A. Ruhe (1980b). "Algorithm 560 JNF: An Algorithm for Numerical Computation of the Jordan Normal Form of a Complex Matrix," *ACM Trans. Math. Softw. 6*, 437–443.

J.W. Demmel (1983). "A Numerical Analyst's Jordan Canonical Form," PhD Thesis, Berkeley.

N. Ghosh, W.W. Hager, and P. Sarmah (1997). "The Application of Eigenpair Stability to Block Diagonalization," *SIAM J. Numer. Anal. 34*, 1255–1268.

S. Serra-Capizzano, D. Bertaccini, and G.H. Golub (2005). "How to Deduce a Proper Eigenvalue Cluster from a Proper Singular Value Cluster in the Nonnormal Case," *SIAM J. Matrix Anal. Applic. 27*, 82–86.

Before we offer pointers to the literature associated with invariant subspace computation, we remind the reader that in §7.3 we discussed the power method for computing the dominant eigenpair and the method of orthogonal iteration that can be used to compute dominant invariant subspaces. Inverse iteration is a related idea and is the concern of the following papers:

J. Varah (1968). "The Calculation of the Eigenvectors of a General Complex Matrix by Inverse Iteration," *Math. Comput. 22*, 785–791.

J. Varah (1970). "Computing Invariant Subspaces of a General Matrix When the Eigensystem is Poorly Determined," *Math. Comput. 24*, 137–149.

G. Peters and J.H. Wilkinson (1979). "Inverse Iteration, Ill-Conditioned Equations, and Newton's Method," *SIAM Review 21*, 339–360.

I.C.F. Ipsen (1997). "Computing an Eigenvector with Inverse Iteration," *SIAM Review 39*, 254–291.

In certain applications it is necessary to track an invariant subspace as the matrix changes, see:

L. Dieci and M.J. Friedman (2001). "Continuation of Invariant Subspaces," *Num. Lin. Alg. 8*, 317–327.

D. Bindel, J.W. Demmel, and M. Friedman (2008). "Continuation of Invariant Subsapces in Large Bifurcation Problems," *SIAM J. Sci. Comput. 30*, 637–656.

Papers concerned with estimating the error in a computed eigenvalue and/or eigenvector include:

S.P. Chan and B.N. Parlett (1977). "Algorithm 517: A Program for Computing the Condition Numbers of Matrix Eigenvalues Without Computing Eigenvectors," *ACM Trans. Math. Softw. 3*, 186–203.

H.J. Symm and J.H. Wilkinson (1980). "Realistic Error Bounds for a Simple Eigenvalue and Its Associated Eigenvector," *Numer. Math. 35*, 113–126.

C. Van Loan (1987). "On Estimating the Condition of Eigenvalues and Eigenvectors," *Lin. Alg. Applic. 88/89*, 715–732.

Z. Bai, J. Demmel, and A. McKenney (1993). "On Computing Condition Numbers for the Nonsymmetric Eigenproblem," *ACM Trans. Math. Softw. 19*, 202–223.

Some ideas about improving computed eigenvalues, eigenvectors, and invariant subspaces may be found in:

J. Varah (1968). "Rigorous Machine Bounds for the Eigensystem of a General Complex Matrix," *Math. Comp. 22*, 793–801.

J.J. Dongarra, C.B. Moler, and J.H. Wilkinson (1983). "Improving the Accuracy of Computed Eigenvalues and Eigenvectors," *SIAM J. Numer. Anal. 20*, 23–46.

J.W. Demmel (1987). "Three Methods for Refining Estimates of Invariant Subspaces," *Comput. 38*, 43–57.

As we have seen, the sep(.,.) function is of great importance in the assessment of a computed invariant subspace. Aspects of this quantity and the associated Sylvester equation are discussed in:

J. Varah (1979). "On the Separation of Two Matrices," *SIAM J. Numer. Anal. 16*, 212–222.
R. Byers (1984). "A Linpack-Style Condition Estimator for the Equation $AX - XB^T = C$," *IEEE Trans. Autom. Contr. AC-29*, 926–928.
M. Gu and M.L. Overton (2006). "An Algorithm to Compute Sep$_\lambda$," *SIAM J. Matrix Anal. Applic. 28*, 348–359.
N.J. Higham (1993). "Perturbation Theory and Backward Error for AX - XB = C," *BIT 33*, 124–136.

Sylvester equations arise in many settings, and there are many solution frameworks, see:

R.H. Bartels and G.W. Stewart (1972). "Solution of the Equation $AX + XB = C$," *Commun. ACM 15*, 820–826.
G.H. Golub, S. Nash, and C. Van Loan (1979). "A Hessenberg-Schur Method for the Matrix Problem $AX + XB = C$," *IEEE Trans. Autom. Contr. AC-24*, 909–913.
K. Datta (1988). "The Matrix Equation $XA - BX = R$ and Its Applications," *Lin. Alg. Applic. 109*, 91–105.
B. Kågström and P. Poromaa (1992). "Distributed and Shared Memory Block Algorithms for the Triangular Sylvester Equation with sep^{-1} Estimators," *SIAM J. Matrix Anal. Applic. 13*, 90–101.
J. Gardiner, M.R. Wette, A.J. Laub, J.J. Amato, and C.B. Moler (1992). "Algorithm 705: A FORTRAN-77 Software Package for Solving the Sylvester Matrix Equation $AXB^T + CXD^T = E$," *ACM Trans. Math. Softw. 18*, 232–238.
V. Simoncini (1996). "On the Numerical Solution of AX -XB =C," *BIT 36*, 814–830.
C.H. Bischof, B.N Datta, and A. Purkayastha (1996). "A Parallel Algorithm for the Sylvester Observer Equation," *SIAM J. Sci. Comput. 17*, 686–698.
D. Calvetti, B. Lewis, L. Reichel (2001). "On the Solution of Large Sylvester-Observer Equations," *Num. Lin. Alg. 8*, 435–451.

The constrained Sylvester equation problem is considered in:

J.B. Barlow, M.M. Monahemi, and D.P. O'Leary (1992). "Constrained Matrix Sylvester Equations," *SIAM J. Matrix Anal. Applic. 13*, 1–9.
A.R. Ghavimi and A.J. Laub (1996). "Numerical Methods for Nearly Singular Constrained Matrix Sylvester Equations." *SIAM J. Matrix Anal. Applic. 17*, 212–221.

The Lyapunov problem $FX + XF^T = -C$ where C is non-negative definite has a very important role to play in control theory, see:

G. Hewer and C. Kenney (1988). "The Sensitivity of the Stable Lyapunov Equation," *SIAM J. Control Optim 26*, 321–344.
A.R. Ghavimi and A.J. Laub (1995). "Residual Bounds for Discrete-Time Lyapunov Equations," *IEEE Trans. Autom. Contr. 40*, 1244–1249.
J.-R. Li and J. White (2004). "Low-Rank Solution of Lyapunov Equations," *SIAM Review 46*, 693–713.

Several authors have considered generalizations of the Sylvester equation, i.e., $\Sigma F_i X G_i = C$. These include:

P. Lancaster (1970). "Explicit Solution of Linear Matrix Equations," *SIAM Review 12*, 544–566.
H. Wimmer and A.D. Ziebur (1972). "Solving the Matrix Equations $\Sigma f_p(A)g_p(A) = C$," *SIAM Review 14*, 318–323.
W.J. Vetter (1975). "Vector Structures and Solutions of Linear Matrix Equations," *Lin. Alg. Applic. 10*, 181–188.

7.7 The Generalized Eigenvalue Problem

If $A, B \in \mathbb{C}^{n \times n}$, then the set of all matrices of the form $A - \lambda B$ with $\lambda \in \mathbb{C}$ is a *pencil*. The *generalized eigenvalues* of $A - \lambda B$ are elements of the set $\lambda(A, B)$ defined by

$$\lambda(A, B) = \{z \in \mathbb{C} : \det(A - zB) = 0\}.$$

If $\lambda \in \lambda(A, B)$ and $0 \neq x \in \mathbb{C}^n$ satisfies

$$Ax = \lambda Bx, \qquad (7.7.1)$$

then x is an *eigenvector* of $A - \lambda B$. The problem of finding nontrivial solutions to (7.7.1) is the *generalized eigenvalue problem* and in this section we survey some of its mathematical properties and derive a stable method for its solution. We briefly discuss how a polynomial eigenvalue problem can be converted into an equivalent generalized eigenvalue problem through a linearization process.

7.7.1 Background

The first thing to observe about the generalized eigenvalue problem is that there are n eigenvalues if and only if $\mathsf{rank}(B) = n$. If B is rank deficient then $\lambda(A, B)$ may be finite, empty, or infinite:

$$A = \begin{bmatrix} 1 & 2 \\ 0 & 3 \end{bmatrix}, \quad B = \begin{bmatrix} 1 & 0 \\ 0 & 0 \end{bmatrix} \quad \Rightarrow \quad \lambda(A, B) = \{1\},$$

$$A = \begin{bmatrix} 1 & 2 \\ 0 & 3 \end{bmatrix}, \quad B = \begin{bmatrix} 0 & 1 \\ 0 & 0 \end{bmatrix} \quad \Rightarrow \quad \lambda(A, B) = \emptyset,$$

$$A = \begin{bmatrix} 1 & 2 \\ 0 & 0 \end{bmatrix}, \quad B = \begin{bmatrix} 1 & 0 \\ 0 & 0 \end{bmatrix} \quad \Rightarrow \quad \lambda(A, B) = \mathbb{C}.$$

Note that if $0 \neq \lambda \in \lambda(A, B)$, then $(1/\lambda) \in \lambda(B, A)$. Moreover, if B is nonsingular, then $\lambda(A, B) = \lambda(B^{-1}A, I) = \lambda(B^{-1}A)$. This last observation suggests one method for solving the $A - \lambda B$ problem if B is nonsingular:

Step 1. Solve $BC = A$ for C using (say) Gaussian elimination with pivoting.

Step 2. Use the QR algorithm to compute the eigenvalues of C.

In this framework, C is affected by roundoff errors of order $\mathbf{u}\| A \|_2 \| B^{-1} \|_2$. If B is ill-conditioned, then this precludes the possibility of computing any generalized eigenvalue accurately—even those eigenvalues that may be regarded as well-conditioned. For example, if

$$A = \begin{bmatrix} 1.746 & .940 \\ 1.246 & 1.898 \end{bmatrix} \quad \text{and} \quad B = \begin{bmatrix} .780 & .563 \\ .913 & .659 \end{bmatrix},$$

then $\lambda(A,B) = \{2,\ 1.07 \times 10^6\}$. With 7-digit floating point arithmetic, we find $\lambda(\mathrm{fl}(AB^{-1})) = \{1.562539,\ 1.01 \times 10^6\}$. The poor quality of the small eigenvalue is because $\kappa_2(B) \approx 2 \times 10^6$. On the other hand, we find that

$$\lambda(I, \mathrm{fl}(A^{-1}B)) \approx \{2.000001,\ 1.06 \times 10^6\}.$$

The accuracy of the small eigenvalue is improved because $\kappa_2(A) \approx 4$.

The example suggests that we seek an alternative approach to the generalized eigenvalue problem. One idea is to compute well-conditioned Q and Z such that the matrices

$$A_1 = Q^{-1}AZ, \qquad B_1 = Q^{-1}BZ \qquad\qquad (7.7.2)$$

are each in canonical form. Note that $\lambda(A,B) = \lambda(A_1, B_1)$ since

$$Ax = \lambda Bx \quad\Leftrightarrow\quad A_1 y = \lambda B_1 y, \quad x = Zy.$$

We say that the pencils $A - \lambda B$ and $A_1 - \lambda B_1$ are *equivalent* if (7.7.2) holds with nonsingular Q and Z.

As in the standard eigenproblem $A - \lambda I$ there is a choice between canonical forms. Corresponding to the Jordan form is a decomposition of Kronecker in which both A_1 and B_1 are block diagonal with blocks that are similar in structure to Jordan blocks. The Kronecker canonical form poses the same numerical challenges as the Jordan form, but it provides insight into the mathematical properties of the pencil $A - \lambda B$. See Wilkinson (1978) and Demmel and Kågström (1987) for details.

7.7.2 The Generalized Schur Decomposition

From the numerical point of view, it makes to insist that the transformation matrices Q and Z be unitary. This leads to the following decomposition described in Moler and Stewart (1973).

Theorem 7.7.1 (Generalized Schur Decomposition). *If A and B are in $\mathbb{C}^{n\times n}$, then there exist unitary Q and Z such that $Q^H AZ = T$ and $Q^H BZ = S$ are upper triangular. If for some k, t_{kk} and s_{kk} are both zero, then $\lambda(A,B) = \mathbb{C}$. Otherwise*

$$\lambda(A,B) = \{t_{ii}/s_{ii} : s_{ii} \neq 0\}.$$

Proof. Let $\{B_k\}$ be a sequence of nonsingular matrices that converge to B. For each k, let

$$Q_k^H (AB_k^{-1})Q_k = R_k$$

be a Schur decomposition of AB_k^{-1}. Let Z_k be unitary such that

$$Z_k^H (B_k^{-1}Q_k) = S_k^{-1}$$

is upper triangular. It follows that $Q_k^H AZ_k = R_k S_k$ and $Q_k^H B_k Z_k = S_k$ are also upper triangular. Using the Bolzano-Weierstrass theorem, we know that the bounded sequence $\{(Q_k, Z_k)\}$ has a converging subsequence,

$$\lim_{i \to \infty} (Q_{k_i}, Z_{k_i}) = (Q, Z).$$

It is easy to show that Q and Z are unitary and that $Q^H AZ$ and $Q^H BZ$ are upper triangular. The assertions about $\lambda(A, B)$ follow from the identity

$$\det(A - \lambda B) = \det(QZ^H) \prod_{i=1}^{n}(t_{ii} - \lambda s_{ii})$$

and that completes the proof of the theorem. $\square$

If A and B are real then the following decomposition, which corresponds to the real Schur decomposition (Theorem 7.4.1), is of interest.

Theorem 7.7.2 (Generalized Real Schur Decomposition). *If A and B are in $\mathbb{R}^{n \times n}$ then there exist orthogonal matrices Q and Z such that $Q^T AZ$ is upper quasi-triangular and $Q^T BZ$ is upper triangular.*

Proof. See Stewart (1972). $\square$

In the remainder of this section we are concerned with the computation of this decomposition and the mathematical insight that it provides.

7.7.3 Sensitivity Issues

The generalized Schur decomposition sheds light on the issue of eigenvalue sensitivity for the $A - \lambda B$ problem. Clearly, small changes in A and B can induce large changes in the eigenvalue $\lambda_i = t_{ii}/s_{ii}$ if s_{ii} is small. However, as Stewart (1978) argues, it may not be appropriate to regard such an eigenvalue as "ill-conditioned." The reason is that the reciprocal $\mu_i = s_{ii}/t_{ii}$ might be a very well-behaved eigenvalue for the pencil $\mu A - B$. In the Stewart analysis, A and B are treated symmetrically and the eigenvalues are regarded more as ordered pairs (t_{ii}, s_{ii}) than as quotients. With this point of view it becomes appropriate to measure eigenvalue perturbations in the *chordal metric* chord(a, b) defined by

$$\text{chord}(a, b) = \frac{|a - b|}{\sqrt{1 + a^2}\sqrt{1 + b^2}}.$$

Stewart shows that if λ is a distinct eigenvalue of $A - \lambda B$ and λ_ϵ is the corresponding eigenvalue of the perturbed pencil $\tilde{A} - \lambda \tilde{B}$ with $\| A - \tilde{A} \|_2 \approx \| B - \tilde{B} \|_2 \approx \epsilon$, then

$$\text{chord}(\lambda, \lambda_\epsilon) \leq \frac{\epsilon}{\sqrt{(y^H Ax)^2 + (y^H Bx)^2}} + O(\epsilon^2)$$

where x and y have unit 2-norm and satisfy $Ax = \lambda Bx$ and $y^H A = \lambda y^H B$. Note that the denominator in the upper bound is symmetric in A and B. The "truly" ill-conditioned eigenvalues are those for which this denominator is small.

The extreme case when both t_{kk} and s_{kk} are zero for some k has been studied by Wilkinson (1979). In this case, the remaining quotients t_{ii}/s_{ii} can take on arbitrary values.

7.7.4   Hessenberg-Triangular Form

The first step in computing the generalized real Schur decomposition of the pair (A, B) is to reduce A to upper Hessenberg form and B to upper triangular form via orthogonal transformations. We first determine an orthogonal U such that $U^T B$ is upper triangular. Of course, to preserve eigenvalues, we must also update A in exactly the same way. Let us trace what happens in the $n = 5$ case.

$$A \leftarrow U^T A = \begin{bmatrix} \times & \times & \times & \times & \times \\ \times & \times & \times & \times & \times \\ \times & \times & \times & \times & \times \\ \times & \times & \times & \times & \times \\ \times & \times & \times & \times & \times \end{bmatrix}, \quad B \leftarrow U^T B = \begin{bmatrix} \times & \times & \times & \times & \times \\ 0 & \times & \times & \times & \times \\ 0 & 0 & \times & \times & \times \\ 0 & 0 & 0 & \times & \times \\ 0 & 0 & 0 & 0 & \times \end{bmatrix}.$$

Next, we reduce A to upper Hessenberg form while preserving B's upper triangular form. First, a Givens rotation Q_{45} is determined to zero a_{51}:

$$A \leftarrow Q_{45}^T A = \begin{bmatrix} \times & \times & \times & \times & \times \\ \times & \times & \times & \times & \times \\ \times & \times & \times & \times & \times \\ \times & \times & \times & \times & \times \\ 0 & \times & \times & \times & \times \end{bmatrix}, \quad B \leftarrow Q_{45}^T B = \begin{bmatrix} \times & \times & \times & \times & \times \\ 0 & \times & \times & \times & \times \\ 0 & 0 & \times & \times & \times \\ 0 & 0 & 0 & \times & \times \\ 0 & 0 & 0 & \times & \times \end{bmatrix}.$$

The nonzero entry arising in the (5,4) position in B can be zeroed by postmultiplying with an appropriate Givens rotation Z_{45}:

$$A \leftarrow A Z_{45} = \begin{bmatrix} \times & \times & \times & \times & \times \\ \times & \times & \times & \times & \times \\ \times & \times & \times & \times & \times \\ \times & \times & \times & \times & \times \\ 0 & \times & \times & \times & \times \end{bmatrix}, \quad B \leftarrow B Z_{45} = \begin{bmatrix} \times & \times & \times & \times & \times \\ 0 & \times & \times & \times & \times \\ 0 & 0 & \times & \times & \times \\ 0 & 0 & 0 & \times & \times \\ 0 & 0 & 0 & 0 & \times \end{bmatrix}.$$

Zeros are similarly introduced into the $(4, 1)$ and $(3, 1)$ positions in A:

$$A \leftarrow Q_{34}^T A = \begin{bmatrix} \times & \times & \times & \times & \times \\ \times & \times & \times & \times & \times \\ \times & \times & \times & \times & \times \\ 0 & \times & \times & \times & \times \\ 0 & \times & \times & \times & \times \end{bmatrix}, \quad B \leftarrow Q_{34}^T B = \begin{bmatrix} \times & \times & \times & \times & \times \\ 0 & \times & \times & \times & \times \\ 0 & 0 & \times & \times & \times \\ 0 & 0 & \times & \times & \times \\ 0 & 0 & 0 & 0 & \times \end{bmatrix},$$

$$A \leftarrow A Z_{34} = \begin{bmatrix} \times & \times & \times & \times & \times \\ \times & \times & \times & \times & \times \\ \times & \times & \times & \times & \times \\ 0 & \times & \times & \times & \times \\ 0 & \times & \times & \times & \times \end{bmatrix}, \quad B \leftarrow B Z_{34} = \begin{bmatrix} \times & \times & \times & \times & \times \\ 0 & \times & \times & \times & \times \\ 0 & 0 & \times & \times & \times \\ 0 & 0 & 0 & \times & \times \\ 0 & 0 & 0 & 0 & \times \end{bmatrix},$$

$$A \leftarrow Q_{23}^T A = \begin{bmatrix} \times & \times & \times & \times & \times \\ \times & \times & \times & \times & \times \\ 0 & \times & \times & \times & \times \\ 0 & \times & \times & \times & \times \\ 0 & \times & \times & \times & \times \end{bmatrix}, \quad B \leftarrow Q_{23}^T B = \begin{bmatrix} \times & \times & \times & \times & \times \\ 0 & \times & \times & \times & \times \\ 0 & \times & \times & \times & \times \\ 0 & 0 & 0 & \times & \times \\ 0 & 0 & 0 & 0 & \times \end{bmatrix},$$

$$
A \leftarrow AZ_{23} =
\begin{bmatrix}
\times & \times & \times & \times & \times \\
\times & \times & \times & \times & \times \\
0 & \times & \times & \times & \times \\
0 & \times & \times & \times & \times \\
0 & \times & \times & \times & \times
\end{bmatrix}, \quad
B \leftarrow BZ_{23} =
\begin{bmatrix}
\times & \times & \times & \times & \times \\
0 & \times & \times & \times & \times \\
0 & 0 & \times & \times & \times \\
0 & 0 & 0 & \times & \times \\
0 & 0 & 0 & 0 & \times
\end{bmatrix}.
$$

A is now upper Hessenberg through its first column. The reduction is completed by zeroing a_{52}, a_{42}, and a_{53}. Note that two orthogonal transformations are required for each a_{ij} that is zeroed—one to do the zeroing and the other to restore B's triangularity. Either Givens rotations or 2-by-2 modified Householder transformations can be used. Overall we have:

Algorithm 7.7.1 (Hessenberg-Triangular Reduction) Given A and B in $\mathbb{R}^{n \times n}$, the following algorithm overwrites A with an upper Hessenberg matrix $Q^T A Z$ and B with an upper triangular matrix $Q^T B Z$ where both Q and Z are orthogonal.

> Compute the factorization $B = QR$ using Algorithm 5.2.1 and overwrite
> $\quad A$ with $Q^T A$ and B with $Q^T B$.
> **for** $j = 1{:}n - 2$
>   **for** $i = n{:} - 1{:}j + 2$
> $[c, s] = \mathsf{givens}(A(i - 1, j), A(i, j))$
> $A(i - 1{:}i, j{:}n) = \begin{bmatrix} c & s \\ -s & c \end{bmatrix}^T A(i - 1{:}i, j{:}n)$
> $B(i - 1{:}i, i - 1{:}n) = \begin{bmatrix} c & s \\ -s & c \end{bmatrix}^T B(i - 1{:}i, i - 1{:}n)$
> $[c, s] = \mathsf{givens}(-B(i, i), B(i, i - 1))$
> $B(1{:}i, i - 1{:}i) = B(1{:}i, i - 1{:}i) \begin{bmatrix} c & s \\ -s & c \end{bmatrix}$
> $A(1{:}n, i - 1{:}i) = A(1{:}n, i - 1{:}i) \begin{bmatrix} c & s \\ -s & c \end{bmatrix}$
>   **end**
> **end**

This algorithm requires about $8n^3$ flops. The accumulation of Q and Z requires about $4n^3$ and $3n^3$ flops, respectively.

The reduction of $A - \lambda B$ to Hessenberg-triangular form serves as a "front end" decomposition for a generalized QR iteration known as the QZ iteration which we describe next.

7.7.5   Deflation

In describing the QZ iteration we may assume without loss of generality that A is an unreduced upper Hessenberg matrix and that B is a nonsingular upper triangular matrix. The first of these assertions is obvious, for if $a_{k+1,k} = 0$ then

$$A - \lambda B = \begin{bmatrix} A_{11} - \lambda B_{11} & A_{12} - \lambda B_{12} \\ 0 & A_{22} - \lambda B_{22} \end{bmatrix} \begin{matrix} k \\ n-k \end{matrix} ,$$
$$\qquad\qquad\qquad k \qquad\qquad n-k$$

and we may proceed to solve the two smaller problems $A_{11} - \lambda B_{11}$ and $A_{22} - \lambda B_{22}$. On the other hand, if $b_{kk} = 0$ for some k, then it is possible to introduce a zero in A's $(n, n-1)$ position and thereby deflate. Illustrating by example, suppose $n = 5$ and $k = 3$:

$$A = \begin{bmatrix} \times & \times & \times & \times & \times \\ \times & \times & \times & \times & \times \\ 0 & \times & \times & \times & \times \\ 0 & 0 & \times & \times & \times \\ 0 & 0 & 0 & \times & \times \end{bmatrix}, \qquad B = \begin{bmatrix} \times & \times & \times & \times & \times \\ 0 & \times & \times & \times & \times \\ 0 & 0 & 0 & \times & \times \\ 0 & 0 & 0 & \times & \times \\ 0 & 0 & 0 & 0 & \times \end{bmatrix}.$$

The zero on B's diagonal can be "pushed down" to the $(5,5)$ position as follows using Givens rotations:

$$A \leftarrow Q_{34}^T A = \begin{bmatrix} \times & \times & \times & \times & \times \\ \times & \times & \times & \times & \times \\ 0 & \times & \times & \times & \times \\ 0 & \times & \times & \times & \times \\ 0 & 0 & 0 & \times & \times \end{bmatrix}, \qquad B \leftarrow Q_{34}^T B = \begin{bmatrix} \times & \times & \times & \times & \times \\ 0 & \times & \times & \times & \times \\ 0 & 0 & 0 & \times & \times \\ 0 & 0 & 0 & 0 & \times \\ 0 & 0 & 0 & 0 & \times \end{bmatrix},$$

$$A \leftarrow A Z_{23} = \begin{bmatrix} \times & \times & \times & \times & \times \\ \times & \times & \times & \times & \times \\ 0 & \times & \times & \times & \times \\ 0 & 0 & \times & \times & \times \\ 0 & 0 & 0 & \times & \times \end{bmatrix}, \qquad B \leftarrow B Z_{23} = \begin{bmatrix} \times & \times & \times & \times & \times \\ 0 & \times & \times & \times & \times \\ 0 & 0 & 0 & \times & \times \\ 0 & 0 & 0 & 0 & \times \\ 0 & 0 & 0 & 0 & \times \end{bmatrix},$$

$$A \leftarrow Q_{45}^T A = \begin{bmatrix} \times & \times & \times & \times & \times \\ \times & \times & \times & \times & \times \\ 0 & \times & \times & \times & \times \\ 0 & 0 & \times & \times & \times \\ 0 & 0 & \times & \times & \times \end{bmatrix}, \qquad B \leftarrow Q_{45}^T B = \begin{bmatrix} \times & \times & \times & \times & \times \\ 0 & \times & \times & \times & \times \\ 0 & 0 & 0 & \times & \times \\ 0 & 0 & 0 & 0 & \times \\ 0 & 0 & 0 & 0 & 0 \end{bmatrix},$$

$$A \leftarrow A Z_{34} = \begin{bmatrix} \times & \times & \times & \times & \times \\ \times & \times & \times & \times & \times \\ 0 & \times & \times & \times & \times \\ 0 & 0 & \times & \times & \times \\ 0 & 0 & 0 & \times & \times \end{bmatrix}, \qquad B \leftarrow B Z_{34}^T = \begin{bmatrix} \times & \times & \times & \times & \times \\ 0 & \times & \times & \times & \times \\ 0 & 0 & \times & \times & \times \\ 0 & 0 & 0 & 0 & \times \\ 0 & 0 & 0 & 0 & 0 \end{bmatrix},$$

$$A \leftarrow A Z_{45} = \begin{bmatrix} \times & \times & \times & \times & \times \\ \times & \times & \times & \times & \times \\ 0 & \times & \times & \times & \times \\ 0 & 0 & \times & \times & \times \\ 0 & 0 & 0 & 0 & \times \end{bmatrix}, \qquad B \leftarrow B Z_{45} = \begin{bmatrix} \times & \times & \times & \times & \times \\ 0 & \times & \times & \times & \times \\ 0 & 0 & \times & \times & \times \\ 0 & 0 & 0 & \times & \times \\ 0 & 0 & 0 & 0 & 0 \end{bmatrix}.$$

This zero-chasing technique is perfectly general and can be used to zero $a_{n,n-1}$ regardless of where the zero appears along B's diagonal.

7.7.6   The QZ Step

We are now in a position to describe a QZ step. The basic idea is to update A and B as follows

$$(\bar{A} - \lambda \bar{B}) = \bar{Q}^T (A - \lambda B) \bar{Z},$$

where $\bar{A}$ is upper Hessenberg, $\bar{B}$ is upper triangular, $\bar{Q}$ and $\bar{Z}$ are each orthogonal, and $\bar{A}\bar{B}^{-1}$ is essentially the same matrix that would result if a Francis QR step (Algorithm 7.5.1) were explicitly applied to AB^{-1}. This can be done with some clever zero-chasing and an appeal to the implicit Q theorem.

Let $M = AB^{-1}$ (upper Hessenberg) and let v be the first column of the matrix $(M - aI)(M - bI)$, where a and b are the eigenvalues of M's lower 2-by-2 submatrix. Note that v can be calculated in $O(1)$ flops. If P_0 is a Householder matrix such that $P_0 v$ is a multiple of e_1, then

$$A \leftarrow P_0 A = \begin{bmatrix} \times & \times & \times & \times & \times & \times \\ \times & \times & \times & \times & \times & \times \\ \times & \times & \times & \times & \times & \times \\ 0 & 0 & \times & \times & \times & \times \\ 0 & 0 & 0 & \times & \times & \times \\ 0 & 0 & 0 & 0 & \times & \times \end{bmatrix}, \quad B \leftarrow P_0 B = \begin{bmatrix} \times & \times & \times & \times & \times & \times \\ \times & \times & \times & \times & \times & \times \\ \times & \times & \times & \times & \times & \times \\ 0 & 0 & 0 & \times & \times & \times \\ 0 & 0 & 0 & 0 & \times & \times \\ 0 & 0 & 0 & 0 & 0 & \times \end{bmatrix}.$$

The idea now is to restore these matrices to Hessenberg-triangular form by chasing the unwanted nonzero elements down the diagonal.

To this end, we first determine a pair of Householder matrices Z_1 and Z_2 to zero b_{31}, b_{32}, and b_{21}:

$$A \leftarrow AZ_1 Z_2 = \begin{bmatrix} \times & \times & \times & \times & \times & \times \\ \times & \times & \times & \times & \times & \times \\ \times & \times & \times & \times & \times & \times \\ \times & \times & \times & \times & \times & \times \\ 0 & 0 & 0 & \times & \times & \times \\ 0 & 0 & 0 & 0 & \times & \times \end{bmatrix}, \quad B \leftarrow BZ_1 Z_2 = \begin{bmatrix} \times & \times & \times & \times & \times & \times \\ 0 & \times & \times & \times & \times & \times \\ 0 & 0 & \times & \times & \times & \times \\ 0 & 0 & 0 & \times & \times & \times \\ 0 & 0 & 0 & 0 & \times & \times \\ 0 & 0 & 0 & 0 & 0 & \times \end{bmatrix}.$$

Then a Householder matrix P_1 is used to zero a_{31} and a_{41}:

$$A \leftarrow P_1 A = \begin{bmatrix} \times & \times & \times & \times & \times & \times \\ \times & \times & \times & \times & \times & \times \\ 0 & \times & \times & \times & \times & \times \\ 0 & \times & \times & \times & \times & \times \\ 0 & 0 & 0 & \times & \times & \times \\ 0 & 0 & 0 & 0 & \times & \times \end{bmatrix}, \quad B \leftarrow P_1 B = \begin{bmatrix} \times & \times & \times & \times & \times & \times \\ 0 & \times & \times & \times & \times & \times \\ 0 & \times & \times & \times & \times & \times \\ 0 & \times & \times & \times & \times & \times \\ 0 & 0 & 0 & 0 & \times & \times \\ 0 & 0 & 0 & 0 & 0 & \times \end{bmatrix}.$$

Notice that with this step the unwanted nonzero elements have been shifted down and to the right from their original position. This illustrates a typical step in the QZ iteration. Notice that $Q = Q_0 Q_1 \cdots Q_{n-2}$ has the same first column as Q_0. By the way the initial Householder matrix was determined, we can apply the implicit Q theorem and assert that $AB^{-1} = Q^T (AB^{-1}) Q$ is indeed essentially the same matrix that we would obtain by applying the Francis iteration to $M = AB^{-1}$ directly. Overall we have the following algorithm.

Algorithm 7.7.2 (The QZ Step) Given an unreduced upper Hessenberg matrix $A \in \mathbb{R}^{n \times n}$ and a nonsingular upper triangular matrix $B \in \mathbb{R}^{n \times n}$, the following algorithm overwrites A with the upper Hessenberg matrix $Q^T A Z$ and B with the upper triangular matrix $Q^T B Z$ where Q and Z are orthogonal and Q has the same first column as the orthogonal similarity transformation in Algorithm 7.5.1 when it is applied to AB^{-1}.

> Let $M = AB^{-1}$ and compute $(M - aI)(M - bI)e_1 = [x, y, z, 0, \ldots, 0]^T$
> where a and b are the eigenvalues of M's lower 2-by-2.

for $k = 1{:}n - 2$

$\quad$ Find Householder Q_k so $\quad Q_k \begin{bmatrix} x \\ y \\ z \end{bmatrix} = \begin{bmatrix} * \\ 0 \\ 0 \end{bmatrix}$.

$\quad A = \mathrm{diag}(I_{k-1}, Q_k, I_{n-k-2}) \cdot A$

$\quad B = \mathrm{diag}(I_{k-1}, Q_k, I_{n-k-2}) \cdot B$

$\quad$ Find Householder Z_{k1} so $\begin{bmatrix} b_{k+2,k} \mid b_{k+2,k+1} \mid b_{k+2,k+2} \end{bmatrix} Z_{k1} = \begin{bmatrix} 0 \mid 0 \mid * \end{bmatrix}$.

$\quad A = A \cdot \mathrm{diag}(I_{k-1}, Z_{k1}, I_{n-k-2})$

$\quad B = B \cdot \mathrm{diag}(I_{k-1}, Z_{k1}, I_{n-k-2})$

$\quad$ Find Householder Z_{k2} so $\begin{bmatrix} b_{k+1,k} \mid b_{k+1,k+1} \end{bmatrix} Z_{k2} = \begin{bmatrix} 0 \mid * \end{bmatrix}$.

$\quad A = A \cdot \mathrm{diag}(I_{k-1}, Z_{k2}, I_{n-k-1})$

$\quad B = B \cdot \mathrm{diag}(I_{k-1}, Z_{k2}, I_{n-k-1})$

$\quad x = a_{k+1,k};\ y = a_{k+2,k}$

$\quad$ **if** $k < n - 2$

$\quad\quad z = a_{k+3,k}$

$\quad$ **end**

end

Find Householder Q_{n-1} so $\quad Q_{n-1} \begin{bmatrix} x \\ y \end{bmatrix} = \begin{bmatrix} * \\ 0 \end{bmatrix}$.

$A = \mathrm{diag}(I_{n-2}, Q_{n-1}) \cdot A$

$B = \mathrm{diag}(I_{n-2}, Q_{n-1}) \cdot B$.

Find Householder Z_{n-1} so $\begin{bmatrix} b_{n,n-1} \mid b_{nn} \end{bmatrix} Z_{n-1} = \begin{bmatrix} 0 \mid * \end{bmatrix}$.

$A = A \cdot \mathrm{diag}(I_{n-2}, Z_{n-1})$

$B = B \cdot \mathrm{diag}(I_{n-2}, Z_{n-1})$

This algorithm requires $22n^2$ flops. Q and Z can be accumulated for an additional $8n^2$ flops and $13n^2$ flops, respectively.

7.7.7   The Overall QZ Process

By applying a sequence of QZ steps to the Hessenberg-triangular pencil $A - \lambda B$, it is possible to reduce A to quasi-triangular form. In doing this it is necessary to monitor A's subdiagonal and B's diagonal in order to bring about decoupling whenever possible. The complete process, due to Moler and Stewart (1973), is as follows:

Algorithm 7.7.3 Given $A \in \mathbb{R}^{n \times n}$ and $B \in \mathbb{R}^{n \times n}$, the following algorithm computes orthogonal Q and Z such that $Q^T A Z = T$ is upper quasi-triangular and $Q^T B Z = S$ is upper triangular. A is overwritten by T and B by S.

Using Algorithm 7.7.1, overwrite A with $Q^T A Z$ (upper Hessenberg) and B with $Q^T B Z$ (upper triangular).

until $q = n$

Set to zero subdiagonal entries that satisfy $|a_{i,i-1}| \leq \epsilon(|a_{i-1,i-1}| + |a_{ii}|)$.

Find the largest nonnegative q and the smallest nonnegative p such that if

$$A = \begin{bmatrix} A_{11} & A_{12} & A_{13} \\ 0 & A_{22} & A_{23} \\ 0 & 0 & A_{33} \end{bmatrix} \begin{matrix} p \\ n-p-q \\ q \end{matrix}$$
$$\begin{matrix} p & n-p-q & q \end{matrix}$$

then A_{33} is upper quasi-triangular and A_{22} is upper Hessenberg and unreduced.

Partition B conformably:

$$B = \begin{bmatrix} B_{11} & B_{12} & B_{13} \\ 0 & B_{22} & B_{23} \\ 0 & 0 & B_{33} \end{bmatrix} \begin{matrix} p \\ n-p-q \\ q \end{matrix}$$
$$\begin{matrix} p & n-p-q & q \end{matrix}$$

if $q < n$
    if B_{22} is singular
        Zero $a_{n-q,n-q-1}$
    else
        Apply Algorithm 7.7.2 to A_{22} and B_{22} and update:
            $A = \text{diag}(I_p, Q, I_q)^T A \cdot \text{diag}(I_p, Z, I_q)$
            $B = \text{diag}(I_p, Q, I_q)^T B \cdot \text{diag}(I_p, Z, I_q)$
    end
  end
end

This algorithm requires $30n^3$ flops. If Q is desired, an additional $16n^3$ are necessary. If Z is required, an additional $20n^3$ are needed. These estimates of work are based on the experience that about two QZ iterations per eigenvalue are necessary. Thus, the convergence properties of QZ are the same as for QR. The speed of the QZ algorithm is not affected by rank deficiency in B.

The computed S and T can be shown to satisfy

$$Q_0^T(A + E)Z_0 = T, \qquad Q_0^T(B + F)Z_0 = S,$$

where Q_0 and Z_0 are exactly orthogonal and $\| E \|_2 \approx \mathbf{u} \| A \|_2$ and $\| F \|_2 \approx \mathbf{u} \| B \|_2$.

7.7.8   Generalized Invariant Subspace Computations

Many of the invariant subspace computations discussed in §7.6 carry over to the generalized eigenvalue problem. For example, approximate eigenvectors can be found via inverse iteration:

$q^{(0)} \in \mathbb{C}^{n \times n}$ given.

for $k = 1, 2, \ldots$

        Solve $(A - \mu B)z^{(k)} = Bq^{(k-1)}$.

        Normalize: $q^{(k)} = z^{(k)}/\| z^{(k)} \|_2$.

        $\lambda^{(k)} = [q^{(k)}]^H Aq^{(k)} \,/\, [q^{(k)}]^H Aq^{(k)}$

end

If B is nonsingular, then this is equivalent to applying (7.6.1) with the matrix $B^{-1}A$. Typically, only a single iteration is required if μ is an approximate eigenvalue computed by the QZ algorithm. By inverse iterating with the Hessenberg-triangular pencil, costly accumulation of the Z-transformations during the QZ iteration can be avoided.

    Corresponding to the notion of an invariant subspace for a single matrix, we have the notion of a *deflating* subspace for the pencil $A - \lambda B$. In particular, we say that a k-dimensional subspace $S \subseteq \mathbb{C}^n$ is deflating for the pencil $A - \lambda B$ if the subspace $\{ Ax + By : x, y \in S \}$ has dimension k or less. Note that if

$$Q^H AZ = T, \qquad Q^H BZ = S$$

is a generalized Schur decomposition of $A - \lambda B$, then the columns of Z in the generalized Schur decomposition define a family of deflating subspaces. Indeed, if

$$Q = [\, q_1 \,|\cdots|\, q_n \,], \qquad Z = [\, z_1 \,|\cdots|\, z_n \,]$$

are column partitionings, then

$$\mathsf{span}\{Az_1, \ldots, Az_k\} \subseteq \mathsf{span}\{q_1, \ldots, q_k\},$$

$$\mathsf{span}\{Bz_1, \ldots, Bz_k\} \subseteq \mathsf{span}\{q_1, \ldots, q_k\},$$

for $k = 1{:}n$. Properties of deflating subspaces and their behavior under perturbation are described in Stewart (1972).

7.7.9   A Note on the Polynomial Eigenvalue Problem

More general than the generalized eigenvalue problem is the *polynomial eigenvalue problem*. Here we are given matrices $A_0, \ldots, A_d \in \mathbb{C}^{n \times n}$ and determine $\lambda \in \mathbb{C}$ and $0 \neq x \in \mathbb{C}^n$ so that

$$P(\lambda)x = 0 \tag{7.7.3}$$

where the λ-*matrix* $P(\lambda)$ is defined by

$$P(\lambda) = A_0 + \lambda A_1 + \cdots + \lambda^d A_d. \tag{7.7.4}$$

We assume $A_d \neq 0$ and regard d as the *degree* of $P(\lambda)$. The theory behind the polynomial eigenvalue problem is nicely developed in Lancaster (1966).

It is possible to convert (7.7.3) into an equivalent linear eigenvalue problem with larger dimension. For example, suppose $d = 3$ and

$$
L(\lambda) = \begin{bmatrix} 0 & 0 & A_0 \\ -I & 0 & A_1 \\ 0 & -I & A_2 \end{bmatrix} + \lambda \begin{bmatrix} I & 0 & 0 \\ 0 & I & 0 \\ 0 & 0 & A_3 \end{bmatrix}.
\tag{7.7.5}
$$

If

$$
L(\lambda) \begin{bmatrix} u_1 \\ u_2 \\ x \end{bmatrix} = \begin{bmatrix} 0 \\ 0 \\ 0 \end{bmatrix},
$$

then

$$
0 = A_0 x + \lambda u_1 = A_0 + \lambda(A_1 x + \lambda u_2) = A_0 + \lambda(A_1 x + \lambda(A_2 + \lambda A_3))x = P(\lambda)x.
$$

In general, we say that $L(\lambda)$ is a *linearization* of $P(\lambda)$ if there are dn-by-dn λ-matrices $S(\lambda)$ and $T(\lambda)$, each with constant nonzero determinants, so that

$$
S(\lambda) \begin{bmatrix} P(\lambda) & 0 \\ 0 & I_{(d-1)n} \end{bmatrix} T(\lambda) = L(\lambda)
\tag{7.7.6}
$$

has unit degree. With this conversion, the $A - \lambda B$ methods just discussed can be applied to find the required eigenvalues and eigenvectors.

Recent work has focused on how to choose the λ-transformations $S(\lambda)$ and $T(\lambda)$ so that special structure in $P(\lambda)$ is reflected in $L(\lambda)$. See Mackey, Mackey, Mehl, and Mehrmann (2006). The idea is to think of (7.7.6) as a factorization and to identify the transformations that produce a properly structured $L(\lambda)$. To appreciate this solution framework it is necessary to have a facility with λ-matrix manipulation and to that end we briefly examine the λ-matrix transformations behind the above linearization. If

$$
P_1(\lambda) = A_1 + \lambda A_2 + \cdots + \lambda^{d-1} A_d
$$

then

$$
P(\lambda) = A_0 + \lambda P_1(\lambda)
$$

and it is easy to verify that

$$
\begin{bmatrix} I_n & -\lambda I_n \\ 0 & I_n \end{bmatrix} \begin{bmatrix} A_0 + \lambda P_1(\lambda) & 0 \\ 0 & I_n \end{bmatrix} \begin{bmatrix} 0 & I_n \\ -I_n & P_1(\lambda) \end{bmatrix} = \begin{bmatrix} \lambda I_n & A_0 \\ -I_n & P_1(\lambda) \end{bmatrix}.
$$

Notice that the transformation matrices have unit determinant and that the λ-matrix on the right-hand side has degree $d - 1$. The process can be repeated. If

$$
P_2(\lambda) = A_2 + \lambda A_3 + \cdots + \lambda^{d-2} A_d
$$

then

$$
P_1(\lambda) = A_1 + \lambda P_2(\lambda)
$$

and

$$
\begin{bmatrix}
I_n & 0 & 0 \\
0 & I_n & -\lambda I_n \\
0 & 0 & I_n
\end{bmatrix}
\begin{bmatrix}
\lambda I_n & A_0 & 0 \\
-I_n & P_1(\lambda) & 0 \\
0 & 0 & I_n
\end{bmatrix}
\begin{bmatrix}
I_n & 0 & 0 \\
0 & 0 & I_n \\
0 & -I_n & P_2(\lambda)
\end{bmatrix}
=
$$

$$
\begin{bmatrix}
\lambda I_n & 0 & A_0 \\
-I_n & \lambda I_n & A_1 \\
0 & -I_n & P_2(\lambda)
\end{bmatrix} .
$$

Note that the matrix on the right has degree $d-2$. A straightforward induction argument can be assembled to establish that if the dn-by-dn matrices $S(\lambda)$ and $T(\lambda)$ are defined by

$$
S(\lambda) = \begin{bmatrix}
I_n & -\lambda I_n & 0 & \cdots & 0 \\
0 & I_n & -\lambda I_n & & \vdots \\
0 & & \ddots & \ddots & \\
\vdots & & & I_n & -\lambda I_n \\
0 & 0 & \cdots & 0 & I_n
\end{bmatrix} , \quad
T(\lambda) = \begin{bmatrix}
0 & 0 & 0 & \cdots & I \\
-I_n & 0 & & & P_1(\lambda) \\
0 & -I_n & \ddots & & \vdots \\
\vdots & & \ddots & \ddots & P_{d-2}(\lambda) \\
0 & 0 & \cdots & -I_n & P_{d-1}(\lambda)
\end{bmatrix}
$$

where

$$
P_k(\lambda) = A_k + \lambda A_{k+1} + \cdots + \lambda^{d-k} A_d,
$$

then

$$
S(\lambda)
\begin{bmatrix}
P(\lambda) & 0 \\
0 & I_{(d-1)n}
\end{bmatrix}
T(\lambda) =
\begin{bmatrix}
\lambda I_n & 0 & 0 & \cdots & A_0 \\
-I_n & \lambda I_n & & & A_1 \\
0 & -I_n & \ddots & & \vdots \\
\vdots & & \ddots & \lambda I_n & A_{d-2} \\
0 & 0 & \cdots & -I_n & A_{d-1} + \lambda A_d
\end{bmatrix} .
$$

Note that, if we solve the linearized problem using the QZ algorithm, then $O((dn)^3)$ flops are required.

Problems

P7.7.1 Suppose A and B are in $\mathbb{R}^{n \times n}$ and that

$$
U^T B V = \begin{bmatrix}
D & 0 \\
0 & 0
\end{bmatrix} \begin{matrix} r \\ n-r \end{matrix} , \quad
U = [\, U_1 \mid U_2 \,] , \quad
V = [\, V_1 \mid V_2 \,] ,
$$
$$
\quad\quad\quad\quad r \quad n-r
$$

is the SVD of B, where D is r-by-r and $r = \mathsf{rank}(B)$. Show that if $\lambda(A, B) = \mathbb{C}$ then $U_2^T A V_2$ is singular.

P7.7.2 Suppose A and B are in $\mathbb{R}^{n \times n}$. Give an algorithm for computing orthogonal Q and Z such that $Q^T A Z$ is upper Hessenberg and $Z^T B Q$ is upper triangular.

P7.7.3 Suppose

$$A = \begin{bmatrix} A_{11} & A_{12} \\ 0 & A_{22} \end{bmatrix} \quad \text{and} \quad B = \begin{bmatrix} B_{11} & B_{12} \\ 0 & B_{22} \end{bmatrix}$$

with $A_{11}, B_{11} \in \mathbb{R}^{k \times k}$ and $A_{22}, B_{22} \in \mathbb{R}^{j \times j}$. Under what circumstances do there exist

$$X = \begin{bmatrix} I_k & X_{12} \\ 0 & I_j \end{bmatrix} \quad \text{and} \quad Y = \begin{bmatrix} I_k & Y_{12} \\ 0 & I_j \end{bmatrix}$$

so that $Y^{-1}AX$ and $Y^{-1}BX$ are both block diagonal? This is the *generalized Sylvester equation problem*. Specify an algorithm for the case when A_{11}, A_{22}, B_{11}, and B_{22} are upper triangular. See Kågström (1994).

P7.7.4 Suppose $\mu \notin \lambda(A, B)$. Relate the eigenvalues and eigenvectors of $A_1 = (A - \mu B)^{-1}A$ and $B_1 = (A - \mu B)^{-1}B$ to the generalized eigenvalues and eigenvectors of $A - \lambda B$.

P7.7.5 What does the generalized Schur decomposition say about the pencil $A - \lambda A^T$? Hint: If $T \in \mathbb{R}^{n \times n}$ is upper triangular, then $\mathcal{E}_n T \mathcal{E}_n$ is lower triangular where $\mathcal{E}_n$ is the exchange permutation defined in §1.2.11.

P7.7.6 Prove that

$$L_1(\lambda) = \begin{bmatrix} A_3 + \lambda A_4 & A_2 & A_1 & A_0 \\ -I_n & 0 & 0 & 0 \\ 0 & -I_n & 0 & 0 \\ 0 & 0 & -I_n & 0 \end{bmatrix}, \quad L_2(\lambda) = \begin{bmatrix} A_3 + \lambda A_4 & -I_n & 0 & 0 \\ A_2 & 0 & -I_n & 0 \\ A_1 & 0 & 0 & -I_n \\ A_0 & 0 & 0 & 0 \end{bmatrix}$$

are linearizations of

$$P(\lambda) = A_0 + \lambda A_1 + \lambda^2 A_2 + \lambda^3 A_3 + \lambda^4 A_4.$$

Specify the λ-matrix transformations that relate $\text{diag}(P(\lambda), I_{3n})$ to both $L_1(\lambda)$ and $L_2(\lambda)$.

Notes and References for §7.7

For background to the generalized eigenvalue problem we recommend Stewart(IMC), Stewart and Sun (MPT), and Watkins (MEP) and:

B. Kågström and A. Ruhe (1983). *Matrix Pencils*, Proceedings Pite Havsbad, 1982, Lecture Notes in Mathematics Vol. 973, Springer-Verlag, New York.

QZ-related papers include:

C.B. Moler and G.W. Stewart (1973). "An Algorithm for Generalized Matrix Eigenvalue Problems," *SIAM J. Numer. Anal. 10*, 241–256.
L. Kaufman (1974). "The LZ Algorithm to Solve the Generalized Eigenvalue Problem," *SIAM J. Numer. Anal. 11*, 997–1024.
R.C. Ward (1975). "The Combination Shift QZ Algorithm," *SIAM J. Numer. Anal. 12*, 835–853.
C.F. Van Loan (1975). "A General Matrix Eigenvalue Algorithm," *SIAM J. Numer. Anal. 12*, 819–834.
L. Kaufman (1977). "Some Thoughts on the QZ Algorithm for Solving the Generalized Eigenvalue Problem," *ACM Trans. Math. Softw. 3*, 65–75.
R.C. Ward (1981). "Balancing the Generalized Eigenvalue Problem," *SIAM J. Sci. Stat. Comput. 2*, 141–152.
P. Van Dooren (1982). "Algorithm 590: DSUBSP and EXCHQZ: Fortran Routines for Computing Deflating Subspaces with Specified Spectrum," *ACM Trans. Math. Softw. 8*, 376–382.
K. Dackland and B. Kågström (1999). "Blocked Algorithms and Software for Reduction of a Regular Matrix Pair to Generalized Schur Form," *ACM Trans. Math. Softw. 25*, 425–454.
D.S. Watkins (2000). "Performance of the QZ Algorithm in the Presence of Infinite Eigenvalues," *SIAM J. Matrix Anal. Applic. 22*, 364–375.
B. Kågström, D. Kressner, E.S. Quintana-Orti, and G. Quintana-Orti (2008). "Blocked Algorithms for the Reduction to Hessenberg-Triangular Form Revisited," *BIT 48*, 563–584.

Many algorithmic ideas associated with the $A - \lambda I$ problem extend to the $A - \lambda B$ problem:

A. Jennings and M.R. Osborne (1977). "Generalized Eigenvalue Problems for Certain Unsymmetric Band Matrices," *Lin. Alg. Applic. 29*, 139–150.

V.N. Kublanovskaya (1984). "AB Algorithm and Its Modifications for the Spectral Problem of Linear Pencils of Matrices," *Numer. Math. 43*, 329–342.

Z. Bai, J. Demmel, and M. Gu (1997). "An Inverse Free Parallel Spectral Divide and Conquer Algorithm for Nonsymmetric Eigenproblems," *Numer. Math. 76*, 279–308.

G.H. Golub and Q. Ye (2000). "Inexact Inverse Iteration for Generalized Eigenvalue Problems," *BIT 40*, 671–684.

F. Tisseur (2001). "Newton's Method in Floating Point Arithmetic and Iterative Refinement of Generalized Eigenvalue Problems," *SIAM J. Matrix Anal. Applic. 22*, 1038–1057.

D. Lemonnier and P. Van Dooren (2006). "Balancing Regular Matrix Pencils," *SIAM J. Matrix Anal. Applic. 28*, 253–263.

R. Granat, B. Kågström, and D. Kressner (2007). "Computing Periodic Deflating Subspaces Associated with a Specified Set of Eigenvalues," *BIT 47*, 763–791.

The perturbation theory for the generalized eigenvalue problem is treated in:

G.W. Stewart (1972). "On the Sensitivity of the Eigenvalue Problem $Ax = \lambda Bx$," *SIAM J. Numer. Anal. 9*, 669–686.

G.W. Stewart (1973). "Error and Perturbation Bounds for Subspaces Associated with Certain Eigenvalue Problems," *SIAM Review 15*, 727–764.

G.W. Stewart (1975). "Gershgorin Theory for the Generalized Eigenvalue Problem $Ax = \lambda Bx$," *Math. Comput. 29*, 600–606.

A. Pokrzywa (1986). "On Perturbations and the Equivalence Orbit of a Matrix Pencil," *Lin. Alg. Applic. 82*, 99–121.

J. Sun (1995). "Perturbation Bounds for the Generalized Schur Decomposition," *SIAM J. Matrix Anal. Applic. 16*, 1328–1340.

R. Bhatia and R.-C. Li (1996). "On Perturbations of Matrix Pencils with Real Spectra. II," *Math. Comput. 65*, 637–645.

J.-P. Dedieu (1997). "Condition Operators, Condition Numbers, and Condition Number Theorem for the Generalized Eigenvalue Problem," *Lin. Alg. Applic. 263*, 1–24.

D.J. Higham and N.J. Higham (1998). "Structured Backward Error and Condition of Generalized Eigenvalue Problems," *SIAM J. Matrix Anal. Applic. 20*, 493–512.

R. Byers, C. He, and V. Mehrmann (1998). "Where is the Nearest Non-Regular Pencil?," *Lin. Alg. Applic. 285*, 81–105.

V. Frayss and V. Toumazou (1998). "A Note on the Normwise Perturbation Theory for the Regular Generalized Eigenproblem," *Numer. Lin. Alg. 5*, 1–10.

R.-C. Li (2003). "On Perturbations of Matrix Pencils with Real Spectra, A Revisit," *Math. Comput. 72*, 715–728.

S. Bora and V. Mehrmann (2006). "Linear Perturbation Theory for Structured Matrix Pencils Arising in Control Theory," *SIAM J. Matrix Anal. Applic. 28*, 148–169.

X.S. Chen (2007). "On Perturbation Bounds of Generalized Eigenvalues for Diagonalizable Pairs," *Numer. Math. 107*, 79–86.

The *Kronecker structure* of the pencil $A - \lambda B$ is analogous to Jordan structure of $A - \lambda I$ and it can provide useful information about the underlying application. Papers concerned with this important decomposition include:

J.H. Wilkinson (1978). "Linear Differential Equations and Kronecker's Canonical Form," in *Recent Advances in Numerical Analysis*, C. de Boor and G.H. Golub (eds.), Academic Press, New York, 231–265.

J.H. Wilkinson (1979). "Kronecker's Canonical Form and the QZ Algorithm," *Lin. Alg. Applic. 28*, 285–303.

P. Van Dooren (1979). "The Computation of Kronecker's Canonical Form of a Singular Pencil," *Lin. Alg. Applic. 27*, 103–140.

J.W. Demmel (1983). "The Condition Number of Equivalence Transformations that Block Diagonalize Matrix Pencils," *SIAM J. Numer. Anal. 20*, 599–610.

J.W. Demmel and B. Kågström (1987). "Computing Stable Eigendecompositions of Matrix Pencils," *Linear Alg. Applic. 88/89*, 139–186.

B. Kågström (1985). "The Generalized Singular Value Decomposition and the General $A - \lambda B$ Problem," *BIT 24*, 568–583.

B. Kågström (1986). "RGSVD: An Algorithm for Computing the Kronecker Structure and Reducing Subspaces of Singular $A - \lambda B$ Pencils," *SIAM J. Sci. Stat. Comput. 7*, 185–211.

J. Demmel and B. Kågström (1986). "Stably Computing the Kronecker Structure and Reducing Subspaces of Singular Pencils $A - \lambda B$ for Uncertain Data," in *Large Scale Eigenvalue Problems*, J. Cullum and R.A. Willoughby (eds.), North-Holland, Amsterdam.

T. Beelen and P. Van Dooren (1988). "An Improved Algorithm for the Computation of Kronecker's Canonical Form of a Singular Pencil," *Lin. Alg. Applic. 105*, 9–65.

E. Elmroth and B. Kågström(1996). "The Set of 2-by-3 Matrix Pencils — Kronecker Structures and Their Transitions under Perturbations," *SIAM J. Matrix Anal. Applic. 17*, 1–34.

A. Edelman, E. Elmroth, and B. Kågström (1997). "A Geometric Approach to Perturbation Theory of Matrices and Matrix Pencils Part I: Versal Deformations," *SIAM J. Matrix Anal. Applic. 18*, 653–692.

E. Elmroth, P. Johansson, and B. Kågström (2001). "Computation and Presentation of Graphs Displaying Closure Hierarchies of Jordan and Kronecker Structures," *Num. Lin. Alg. 8*, 381–399.

Just as the Schur decomposition can be used to solve the Sylvester equation problem $A_1 X - X A_2 = B$, the generalized Schur decomposition can be used to solve the generalized Sylvester equation problem where matrices X and Y are sought so that $A_1 X - Y A_2 = B_1$ and $A_3 X - Y A_4 = B_2$, see:

W. Enright and S. Serbin (1978). "A Note on the Efficient Solution of Matrix Pencil Systems," *BIT 18*, 276–81.

B. Kågström and L. Westin (1989). "Generalized Schur Methods with Condition Estimators for Solving the Generalized Sylvester Equation," *IEEE Trans. Autom. Contr. AC-34*, 745–751.

B. Kågström (1994). "A Perturbation Analysis of the Generalized Sylvester Equation $(AR - LB, DR - LE) = (C, F)$," *SIAM J. Matrix Anal. Applic. 15*, 1045–1060.

J.-G. Sun (1996). "Perturbation Analysis of System Hessenberg and Hessenberg-Triangular Forms," *Lin. Alg. Applic. 241-3*, 811–849.

B. Kågström and P. Poromaa (1996). "LAPACK-style Algorithms and Software for Solving the Generalized Sylvester Equation and Estimating the Separation Between Regular Matrix Pairs," *ACM Trans. Math. Softw. 22*, 78–103.

I. Jonsson and B. Kågström (2002). "Recursive Blocked Algorithms for Solving Triangular Systems–Part II: Two-sided and Generalized Sylvester and Lyapunov Matrix Equations," *ACM Trans. Math. Softw. 28*, 416–435.

R. Granat and B. Kågström (2010). "Parallel Solvers for Sylvester-Type Matrix Equations with Applications in Condition Estimation, Part I: Theory and Algorithms," *ACM Trans. Math. Softw. 37*, Article 32.

Rectangular generalized eigenvalue problems also arise. In this setting the goal is to reduce the rank of $A - \lambda B$, see:

G.W. Stewart (1994). "Perturbation Theory for Rectangular Matrix Pencils," *Lin. Alg. Applic. 208/209*, 297–301.

G. Boutry, M. Elad, G.H. Golub, and P. Milanfar (2005). "The Generalized Eigenvalue Problem for Nonsquare Pencils Using a Minimal Perturbation Approach," *SIAM J. Matrix Anal. Applic. 27*, 582–601.

D. Chu and G.H. Golub (2006). "On a Generalized Eigenvalue Problem for Nonsquare Pencils," *SIAM J. Matrix Anal. Applic. 28*, 770–787.

References for the polynomial eigenvalue problem include:

P. Lancaster (1966). *Lambda-Matrices and Vibrating Systems*, Pergamon Press, Oxford, U.K.

I. Gohberg, P. Lancaster, and L. Rodman (1982). *Matrix Polynomials*, Academic Press, New York.

F. Tisseur (2000). "Backward Error and Condition of Polynomial Eigenvalue Problems," *Lin. Alg. Applic. 309*, 339–361.

J.-P. Dedieu and F. Tisseur (2003). "Perturbation Theory for Homogeneous Polynomial Eigenvalue Problems," *Lin. Alg. Applic. 358*, 71–94.

N.J. Higham, D.S. Mackey, and F. Tisseur (2006). "The Conditioning of Linearizations of Matrix Polynomials," *SIAM J. Matrix Anal. Applic. 28*, 1005–1028.

D.S. Mackey, N. Mackey, C. Mehl, V. Mehrmann (2006). "Vector Spaces of Linearizations for Matrix Polynomials," *SIAM J. Matrix Anal. Applic. 28*, 971–1004.

The structured quadratic eigenvalue problem is discussed briefly in §8.7.9.

7.8 Hamiltonian and Product Eigenvalue Problems

Two structured unsymmetric eigenvalue problems are considered. The Hamiltonian matrix eigenvalue problem comes with its own special Schur decomposition. Orthogonal symplectic similarity transformations are used to bring about the required reduction. The product eigenvalue problem involves computing the eigenvalues of a product like $A_1 A_2^{-1} A_3$ without actually forming the product or the designated inverses. For detailed background to these problems, see Kressner (NMGS) and Watkins (MEP).

7.8.1 Hamiltonian Matrix Eigenproblems

Hamiltonian and symplectic matrices are introduced in §1.3.10. Their 2-by-2 block structure provide a nice framework for practicing block matrix manipulation, see P1.3.2 and P2.5.4. We now describe some interesting eigenvalue problems that involve these matrices. For a given n, we define the matrix $J \in \mathbb{R}^{2n \times 2n}$ by

$$ J = \begin{bmatrix} 0 & I_n \\ -I_n & 0 \end{bmatrix} $$

and proceed to work with the families of 2-by-2 block structured matrices that are displayed in Figure 7.8.1. We mention four important facts concerning these matrices.

Family	Definition	What They Look Like	
Hamiltonian	$JM = (JM)^T$	$M = \begin{bmatrix} A & G \\ F & -A^T \end{bmatrix}$	G symmetric F symmetric
Skew Hamiltonian	$JN = -(JN)^T$	$N = \begin{bmatrix} A & G \\ F & A^T \end{bmatrix}$	G skew-symmetric F skew-symmetric
Symplectic	$JS = S^{-T}J$	$S = \begin{bmatrix} S_{11} & S_{12} \\ S_{21} & S_{22} \end{bmatrix}$	$S_{11}^T S_{21}$ symmetric $S_{22}^T S_{12}$ symmetric $S_{11}^T S_{22} = I + S_{21}^T S_{12}$
Orthogonal Symplectic	$JQ = QJ$	$Q = \begin{bmatrix} Q_1 & Q_2 \\ -Q_2 & Q_1 \end{bmatrix}$	$Q_1^T Q_2$ symmetric $I = Q_1^T Q_1 + Q_2^T Q_2$

Figure 7.8.1. *Hamiltonian and symplectic structures*

(1) Symplectic similarity transformations preserve Hamiltonian structure:

$$ J(S^{-1}MS) = (JS^{-1}J^T)(JMJ^T)(JS) = -S^T M^T S^{-T} J = (J(S^{-1}MS))^T. $$

(2) The square of a Hamiltonian matrix is skew-Hamiltonian:

$$JM^2 = (JMJ^T)(JM) = -M^T(JM)^T = -M^{2T}J^T = -(JM^2)^T.$$

(3) If M is a Hamiltonian matrix and $\lambda \in \lambda(M)$, then $-\lambda \in \lambda(M)$:

$$M \begin{bmatrix} u \\ v \end{bmatrix} = \lambda \begin{bmatrix} u \\ v \end{bmatrix} \quad \Rightarrow \quad M^T \begin{bmatrix} v \\ -u \end{bmatrix} = -\lambda \begin{bmatrix} v \\ -u \end{bmatrix}.$$

(4) If S is symplectic and $\lambda \in \lambda(S)$, then $1/\lambda \in \lambda(S)$:

$$S \begin{bmatrix} u \\ v \end{bmatrix} = \lambda \begin{bmatrix} u \\ v \end{bmatrix} \quad \Rightarrow \quad S^T \begin{bmatrix} v \\ -u \end{bmatrix} = \frac{1}{\lambda} \begin{bmatrix} v \\ -u \end{bmatrix}.$$

Symplectic versions of Householder and Givens transformations have a prominanent role to play in Hamiltonian matrix computations. If $P = I_n - 2vv^T$ is a Householder matrix, then $\mathrm{diag}(P, P)$ is a symplectic orthogonal matrix. Likewise, if $G \in \mathbb{R}^{2n \times 2n}$ is a Givens rotation that involves planes i and $i+n$, then G is a symplectic orthogonal matrix. Combinations of these transformations can be used to introduce zeros. For example, a Householder-Givens-Householder sequence can do this:

$$\begin{bmatrix} \times \\ \times \\ \times \\ \times \\ \hline \times \\ \times \\ \times \\ \times \end{bmatrix} \xrightarrow{\mathrm{diag}(P_1, P_1)} \begin{bmatrix} \times \\ \times \\ \times \\ \times \\ \hline \times \\ 0 \\ 0 \\ 0 \end{bmatrix} \xrightarrow{G_{1,5}} \begin{bmatrix} \times \\ \times \\ \times \\ \times \\ \hline 0 \\ 0 \\ 0 \\ 0 \end{bmatrix} \xrightarrow{\mathrm{diag}(P_2, P_2)} \begin{bmatrix} \times \\ 0 \\ 0 \\ 0 \\ \hline 0 \\ 0 \\ 0 \\ 0 \end{bmatrix}.$$

This kind of vector reduction can be sequenced to produce a constructive proof of a structured Schur decomposition for Hamiltonian matrices. Suppose λ is a real eigenvalue of a Hamiltonian matrix M and that $x \in \mathbb{R}^{2n}$ is a unit 2-norm vector with $Mx = \lambda x$. If $Q_1 \in \mathbb{R}^{2n \times 2n}$ is an orthogonal symplectic matrix and $Q_1^T x = e_1$, then it follows from $(Q_1^T M Q_1)(Q_1^T x) = \lambda(Q_1^T x)$ that

$$Q_1^T M Q_1 = \begin{bmatrix} \lambda & \times & \times & \times & \times & \times & \times & \times \\ 0 & \times & \times & \times & \times & \times & \times & \times \\ 0 & \times & \times & \times & \times & \times & \times & \times \\ 0 & \times & \times & \times & \times & \times & \times & \times \\ \hline 0 & 0 & 0 & 0 & -\lambda & 0 & 0 & 0 \\ 0 & \times & \times & \times & \times & \times & \times & \times \\ 0 & \times & \times & \times & \times & \times & \times & \times \\ 0 & \times & \times & \times & \times & \times & \times & \times \end{bmatrix}.$$

The "extra" zeros follow from the Hamiltonian structure of $Q_1^T M Q_1$. The process can be repeated on the 6-by-6 Hamiltonian submatrix defined by rows and columns 2-3-4-6-7-8. Together with the assumption that M has no purely imaginary eigenvalues, it is possible to show that an orthogonal symplectic matrix Q exists so that

$$Q^T M Q = \begin{bmatrix} Q_1 & Q_2 \\ -Q_2 & Q_1 \end{bmatrix}^T \begin{bmatrix} A & F \\ G & -A^T \end{bmatrix} \begin{bmatrix} Q_1 & Q_2 \\ -Q_2 & Q_1 \end{bmatrix} = \begin{bmatrix} T & R \\ 0 & -T^T \end{bmatrix} \quad (7.8.1)$$

where $T \in \mathbb{R}^{n \times n}$ is upper quasi-triangular. This is the *real Hamiltonian-Schur decomposition*. See Paige and Van Loan (1981) and, for a more general version, Lin, Mehrmann, and Xu (1999).

One reason that the Hamiltonian eigenvalue problem is so important is its connection to the *algebraic Ricatti equation*

$$G + XA + A^T X - XFX = 0. \quad (7.8.2)$$

This quadratic matrix problem arises in optimal control and a symmetric solution is sought so that the eigenvalues of $A - FX$ are in the open left half plane. Modest assumptions typically ensure that M has no eigenvalues on the imaginary axis and that the matrix Q_1 in (7.8.1) is nonsingular. If we compare (2,1) blocks in (7.8.1), then

$$Q_2^T A Q_1 - Q_2^T F Q_2 + Q_1^T G Q_1 + Q_1^T A^T Q_2 = 0.$$

It follows from $I_n = Q_1^T Q_1 + Q_2^T Q_2$ that $X = Q_2 Q_1^{-1}$ is symmetric and that it satisfies (7.8.2). From (7.8.1) it is easy to show that $A - FX = Q_1 T Q_1^{-1}$ and so the eigenvalues of $A - FX$ are the eigenvalues of T. It follows that the desired solution to the algebraic Ricatti equation can be obtained by computing the real Hamiltonian-Schur decomposition and ordering the eigenvalues so that $\lambda(T)$ is in the left half plane.

How might the real Hamiltonian-Schur form be computed? One idea is to reduce M to some condensed Hamiltonian form and then devise a structure-preserving QR-iteration. Regarding the former task, it is easy to compute an orthogonal symplectic U_0 so that

$$U_0^T M U_0 = \begin{bmatrix} H & R \\ D & -H^T \end{bmatrix} \quad (7.8.3)$$

where $H \in \mathbb{R}^{n \times n}$ is upper Hessenberg and D is diagonal. Unfortunately, a structure-preserving QR iteration that maintains this condensed form has yet to be devised. This impasse prompts consideration of methods that involve the skew-Hamiltonian matrix $N = M^2$. Because the (2,1) block of a skew-Hamiltonian matrix is skew-symmetric, it has a zero diagonal. Symplectic similarity transforms preserve skew-Hamiltonian structure, and it is straightforward to compute an orthogonal symplectic matrix V_0 such that

$$V_0^T M^2 V_0 = \begin{bmatrix} H & R \\ 0 & H^T \end{bmatrix}, \quad (7.8.4)$$

where H is upper Hessenberg. If $U^T H U = T$ is the real Schur form of H and and $Q = V_0 \cdot \text{diag}(U, U)$, then

$$Q^T M^2 Q = \begin{bmatrix} T & U^T R U \\ 0 & T^T \end{bmatrix}$$

is the *real skew-Hamiltonian Schur form*. See Van Loan (1984). It does *not* follow that $Q^T M Q$ is in Schur-Hamiltonian form. Moreover, the quality of the computed small eigenvalues is not good because of the explicit squaring of M. However, these shortfalls can be overcome in an efficient numerically sound way, see Chu, Lie, and Mehrmann (2007) and the references therein. Kressner (NMSE, p. 175–208) and Watkins (MEP, p. 319–341) have in-depth treatments of the Hamiltonian eigenvalue problem.

7.8.2 Product Eigenvalue Problems

Using SVD and QZ, we can compute the eigenvalues of $A^T A$ and $B^{-1} A$ without forming products or inverses. The intelligent computation of the Hamiltonian-Schur decomposition involves a correspondingly careful handling of the product M-times-M. In this subsection we further develop this theme by discussing various product decompositions. Here is an example that suggests how we might compute the Hessenberg decomposition of

$$A = A_3 A_2 A_1$$

where $A_1, A_2, A_3 \in \mathbb{R}^{n \times n}$. Instead of forming this product explicitly, we compute orthogonal $U_1, U_2, U_3 \in \mathbb{R}^{n \times n}$ such that

$$
\begin{aligned}
U_1^T A_3 U_3 &= H_3 \quad &\text{(upper Hessenberg)}, \\
U_3^T A_2 U_2 &= T_2 \quad &\text{(upper triangular)}, \\
U_2^T A_1 U_1 &= T_1 \quad &\text{(upper triangular)}.
\end{aligned}
\tag{7.8.5}
$$

It follows that

$$U_1^T A U_1 = (U_1^T A_3 U_3)(U_3^T A_2 U_2)(U_2^T A_1 U_1) = H_3 T_2 T_1$$

is upper Hessenberg. A procedure for doing this would start by computing the QR factorizations

$$Q_2^T A_1 = R_1, \qquad Q_3^T (A_2 Q_2) = R_2.$$

If $\tilde{A}_3 = A_3 Q_3$, then $A = \tilde{A}_3 R_2 R_1$. The next phase involves reducing $\tilde{A}_3$ to Hessenberg form with Givens transformations coupled with "bulge chasing" to preserve the triangular structures already obtained. The process is similar to the reduction of $A - \lambda B$ to Hessenberg-triangular form; see §7.7.4.

Now suppose we want to compute the real Schur form of A

$$
\begin{aligned}
Q_1^T A_3 Q_3 &= T_3 \quad &\text{(upper quasi-triangular)}, \\
Q_3^T A_2 Q_2 &= T_2 \quad &\text{(upper triangular)}, \\
Q_2^T A_1 Q_1 &= T_1 \quad &\text{(upper triangular)},
\end{aligned}
\tag{7.8.6}
$$

where $Q_1, Q_2, Q_3 \in \mathbb{R}^{n \times n}$ are orthogonal. Without loss of generality we may assume that $\{A_3, A_2, A_1\}$ is in Hessenberg-triangular-triangular form. Analogous to the QZ iteration, the next phase is to produce a sequence of converging triplets

$$\{A_3^{(k)}, A_2^{(k)}, A_1^{(k)}\} \to \{T_3, T_2, T_1\} \tag{7.8.7}$$

with the property that all the iterates are in Hessenberg-triangular-triangular form.

Product decompositions (7.8.5) and (7.8.6) can be framed as structured decompositions of block-cyclic 3-by-3 matrices. For example, if

$$
U = \begin{bmatrix} U_1 & 0 & 0 \\ 0 & U_2 & 0 \\ 0 & 0 & U_3 \end{bmatrix}
$$

then we have the following restatement of (7.8.5):

$$
U^T \begin{bmatrix} 0 & 0 & A_3 \\ A_1 & 0 & 0 \\ 0 & A_2 & 0 \end{bmatrix} U = \begin{bmatrix} 0 & 0 & H_3 \\ T_1 & 0 & 0 \\ 0 & T_2 & 0 \end{bmatrix} = \tilde{H}.
$$

Consider the zero-nonzero structure of this matrix for the case $n = 4$:

$$
\tilde{H} = \left[\begin{array}{cccc|cccc|cccc}
0 & 0 & 0 & 0 & 0 & 0 & 0 & 0 & \times & \times & \times & \times \\
0 & 0 & 0 & 0 & 0 & 0 & 0 & 0 & \times & \times & \times & \times \\
0 & 0 & 0 & 0 & 0 & 0 & 0 & 0 & 0 & \times & \times & \times \\
0 & 0 & 0 & 0 & 0 & 0 & 0 & 0 & 0 & 0 & \times & \times \\
\hline
\times & \times & \times & \times & 0 & 0 & 0 & 0 & 0 & 0 & 0 & 0 \\
0 & \times & \times & \times & 0 & 0 & 0 & 0 & 0 & 0 & 0 & 0 \\
0 & 0 & \times & \times & 0 & 0 & 0 & 0 & 0 & 0 & 0 & 0 \\
0 & 0 & 0 & \times & 0 & 0 & 0 & 0 & 0 & 0 & 0 & 0 \\
\hline
0 & 0 & 0 & 0 & \times & \times & \times & \times & 0 & 0 & 0 & 0 \\
0 & 0 & 0 & 0 & 0 & \times & \times & \times & 0 & 0 & 0 & 0 \\
0 & 0 & 0 & 0 & 0 & 0 & \times & \times & 0 & 0 & 0 & 0 \\
0 & 0 & 0 & 0 & 0 & 0 & 0 & \times & 0 & 0 & 0 & 0
\end{array}\right].
$$

Using the perfect shuffle $\mathcal{P}_{34}$ (see §1.2.11) we also have

$$
\mathcal{P}_{34} \tilde{H} \mathcal{P}_{34} = \left[\begin{array}{ccc|ccc|ccc|ccc}
0 & 0 & \times & 0 & 0 & \times & 0 & 0 & \times & 0 & 0 & \times \\
\times & 0 & 0 & \times & 0 & 0 & \times & 0 & 0 & \times & 0 & 0 \\
0 & \times & 0 & 0 & \times & 0 & 0 & \times & 0 & 0 & \times & 0 \\
\hline
0 & 0 & \times & 0 & 0 & \times & 0 & 0 & \times & 0 & 0 & \times \\
0 & 0 & 0 & \times & 0 & 0 & \times & 0 & 0 & \times & 0 & 0 \\
0 & 0 & 0 & 0 & \times & 0 & 0 & \times & 0 & 0 & \times & 0 \\
\hline
0 & 0 & 0 & 0 & 0 & \times & 0 & 0 & \times & 0 & 0 & \times \\
0 & 0 & 0 & 0 & 0 & 0 & \times & 0 & 0 & \times & 0 & 0 \\
0 & 0 & 0 & 0 & 0 & 0 & 0 & \times & 0 & 0 & \times & 0 \\
\hline
0 & 0 & 0 & 0 & 0 & 0 & 0 & 0 & \times & 0 & 0 & \times \\
0 & 0 & 0 & 0 & 0 & 0 & 0 & 0 & 0 & \times & 0 & 0 \\
0 & 0 & 0 & 0 & 0 & 0 & 0 & 0 & 0 & 0 & \times & 0
\end{array}\right].
$$

Note that this is a highly structured 12-by-12 upper Hessenberg matrix. This connection makes it possible to regard the product-QR iteration as a structure-preserving

QR iteration. For a detailed discussion about this connection and its implications for both analysis and computation, see Kressner (NMSE, pp. 146–174) and Watkins(MEP, pp. 293–303). We mention that with the "technology" that has been developed, it is possible to solve product eigenvalue problems where the factor matrices that define A are rectangular. Square nonsingular factors can also participate through their inverses, e.g., $A = A_3 A_2^{-1} A_1$.

Problems

P7.8.1 What can you say about the eigenvalues and eigenvectors of a symplectic matrix?

P7.8.2 Suppose $S_1, S_2 \in \mathbb{R}^{n \times n}$ are both skew-symmetric and let $A = S_1 S_2$. Show that the nonzero eigenvalues of A are not simple. How would you compute these eigenvalues?

P7.8.3 Relate the eigenvalues and eigenvectors of

$$
A = \begin{bmatrix} 0 & A_1 & 0 & 0 \\ 0 & 0 & A_2 & 0 \\ 0 & 0 & 0 & A_3 \\ A_4 & 0 & 0 & 0 \end{bmatrix}.
$$

to the eigenvalues and eigenvectors of $\tilde{A} = A_1 A_2 A_3 A_4$. Assume that the diagonal blocks are square.

Notes and References for §7.8

The books by Kressner(NMSE) and Watkins (MEP) have chapters on product eigenvalue problems and Hamiltonian eigenvalue problems. The sometimes bewildering network of interconnections that exist among various structured classes of matrices is clarified in:

A. Bunse-Gerstner, R. Byers, and V. Mehrmann (1992). "A Chart of Numerical Methods for Structured Eigenvalue Problems," *SIAM J. Matrix Anal. Applic. 13*, 419–453.

Papers concerned with the Hamiltonian Schur decomposition include:

A.J. Laub and K. Meyer (1974). "Canonical Forms for Symplectic and Hamiltonian Matrices," *J. Celestial Mechanics 9*, 213–238.

C.C. Paige and C. Van Loan (1981). "A Schur Decomposition for Hamiltonian Matrices," *Lin. Alg. Applic. 41*, 11–32.

V. Mehrmann (1991). *Autonomous Linear Quadratic Control Problems, Theory and Numerical Solution*, Lecture Notes in Control and Information Sciences No. 163, Springer-Verlag, Heidelberg.

W.-W. Lin, V. Mehrmann, and H. Xu (1999). "Canonical Forms for Hamiltonian and Symplectic Matrices and Pencils," *Lin. Alg. Applic. 302/303*, 469–533.

Various methods for Hamiltonian eigenvalue problems have been devised that exploit the rich underlying structure, see:

C. Van Loan (1984). "A Symplectic Method for Approximating All the Eigenvalues of a Hamiltonian Matrix," *Lin. Alg. Applic. 61*, 233–252.

R. Byers (1986) "A Hamiltonian QR Algorithm," *SIAM J. Sci. Stat. Comput. 7*, 212–229.

P. Benner, R. Byers, and E. Barth (2000). "Algorithm 800: Fortran 77 Subroutines for Computing the Eigenvalues of Hamiltonian Matrices. I: the Square-Reduced Method," *ACM Trans. Math. Softw. 26*, 49–77.

H. Fassbender, D.S. Mackey and N. Mackey (2001). "Hamilton and Jacobi Come Full Circle: Jacobi Algorithms for Structured Hamiltonian Eigenproblems," *Lin. Alg. Applic. 332-4*, 37–80.

D.S. Watkins (2006). "On the Reduction of a Hamiltonian Matrix to Hamiltonian Schur Form," *ETNA 23*, 141–157.

D.S. Watkins (2004). "On Hamiltonian and Symplectic Lanczos Processes," *Lin. Alg. Applic. 385*, 23–45.

D. Chu, X. Liu, and V. Mehrmann (2007). "A Numerical Method for Computing the Hamiltonian Schur Form," *Numer. Math. 105*, 375–412.

Generalized eigenvalue problems that involve Hamiltonian matrices also arise:

P. Benner, V. Mehrmann, and H. Xu (1998). "A Numerically Stable, Structure Preserving Method for Computing the Eigenvalues of Real Hamiltonian or Symplectic Pencils," *Numer. Math. 78*, 329–358.

C. Mehl (2000). "Condensed Forms for Skew-Hamiltonian/Hamiltonian Pencils," *SIAM J. Matrix Anal. Applic. 21*, 454–476.

V. Mehrmann and D.S. Watkins (2001). "Structure–Preserving Methods for Computing Eigenpairs of Large Sparse Skew–Hamiltonian/Hamiltonian Pencils," *SIAM J. Sci. Comput. 22*, 1905–1925.

P. Benner and R. Byers, V. Mehrmann, and H. Xu (2002). "Numerical Computation of Deflating Subspaces of Skew-Hamiltonian/Hamiltonian Pencils," *SIAM J. Matrix Anal. Applic. 24*, 165–190.

Methods for symplectic eigenvalue problems are discussed in:

P. Benner, H. Fassbender and D.S. Watkins (1999). "SR and SZ Algorithms for the Symplectic (Butterfly) Eigenproblem," *Lin. Alg. Applic. 287*, 41–76.

The Golub-Kahan SVD algorithm that we discuss in the next chapter does not form $A^T A$ or AA^T despite the rich connection to the Schur decompositions of those matrices. From that point on there has been an appreciation for the numerical dangers associated with explicit products. Here is a sampling of the literature:

C. Van Loan (1975). "A General Matrix Eigenvalue Algorithm," *SIAM J. Numer. Anal. 12*, 819–834.

M.T. Heath, A.J. Laub, C.C. Paige, and R.C. Ward (1986). "Computing the SVD of a Product of Two Matrices," *SIAM J. Sci. Stat. Comput. 7*, 1147–1159.

R. Mathias (1998). "Analysis of Algorithms for Orthogonalizing Products of Unitary Matrices," *Num. Lin. Alg. 3*, 125–145.

G. Golub, K. Solna, and P. Van Dooren (2000). "Computing the SVD of a General Matrix Product/Quotient," *SIAM J. Matrix Anal. Applic. 22*, 1–19.

D.S. Watkins (2005). "Product Eigenvalue Problems," *SIAM Review 47*, 3–40.

R. Granat and B. Kgstrom (2006). "Direct Eigenvalue Reordering in a Product of Matrices in Periodic Schur Form," *SIAM J. Matrix Anal. Applic. 28*, 285–300.

Finally we mention that there is a substantial body of work concerned with structured error analysis and structured perturbation theory for structured matrix problems, see:

F. Tisseur (2003). "A Chart of Backward Errors for Singly and Doubly Structured Eigenvalue Problems," *SIAM J. Matrix Anal. Applic. 24*, 877–897.

R. Byers and D. Kressner (2006). "Structured Condition Numbers for Invariant Subspaces," *SIAM J. Matrix Anal. Applic. 28*, 326–347.

M. Karow, D. Kressner, and F. Tisseur (2006). "Structured Eigenvalue Condition Numbers," *SIAM J. Matrix Anal. Applic. 28*, 1052–1068.

7.9 Pseudospectra

If the purpose of computing is insight, then it is easy to see why the well-conditioned eigenvector basis is such a valued commodity, for in many matrix problems, replacement of A with its diagonalization $X^{-1}AX$ leads to powerful, analytic simplifications. However, the insight-through-eigensystem paradigm has diminished impact in problems where the matrix of eigenvectors is ill-conditioned or nonexistent. Intelligent invariant subspace computation as discussed in §7.6 is one way to address the shortfall; pseudospectra are another. In this brief section we discuss the essential ideas behind the theory and computation of pseudospectra. The central message is simple: if you are working with a nonnormal matrix, then a graphical pseudospectral analysis effectively tells you just how much to trust the eigenvalue/eigenvector "story."

A slightly awkward feature of our presentation has to do with the positioning of this section in the text. As we will see, SVD calculations are an essential part of the pseudospectra scene and we do not detail dense matrix algorithms for that important decomposition until the next chapter. However, it makes sense to introduce the pseudospectra concept here at the end of Chapter 7 while the challenges of the unsymmetric eigenvalue problem are fresh in mind. Moreover, with this "early" foundation we can subsequently present various pseudospectra insights that concern the behavior of the matrix exponential (§9.3), the Arnoldi method for sparse unsymmetric eigenvalue problems (§10.5), and the GMRES method for sparse unsymmetric linear systems (§11.4).

For maximum generality, we investigate the pseudospectra of complex, nonnormal matrices. The definitive pseudospectra reference is Trefethen and Embree (SAP). Virtually everything we discuss is presented in greater detail in that excellent volume.

7.9.1 Motivation

In many settings, the eigenvalues of a matrix "say something" about an underlying phenomenon. For example, if

$$A = \begin{bmatrix} \lambda_1 & M \\ 0 & \lambda_2 \end{bmatrix}, \qquad M > 0,$$

then

$$\lim_{k \to \infty} \| A^k \|_2 = 0$$

if and only if $|\lambda_1| < 1$ and $|\lambda_2| < 1$. This follows from Lemma 7.3.1, a result that we needed to establish the convergence of the QR iteration. Applied to our 2-by-2 example, the lemma can be used to show that

$$\| A^k \|_2 \le \frac{M}{\epsilon} (\rho(A) + \epsilon)^k$$

for any $\epsilon > 0$ where $\rho(A) = \max\{|\lambda_1|, |\lambda_2|\}$ is the spectral radius. By making ϵ small enough in this inequality, we can draw a conclusion about the asymptotic behavior of A^k:

> If $\rho(A) < 1$, then asymptotically A^k converges to zero as $\rho(A)^k$. (7.9.1)

However, while the eigenvalues adequately predict the limiting behavior of $\| A^k \|_2$, they do not (by themselves) tell us much about what is happening if k is small. Indeed, if $\lambda_1 \ne \lambda_2$, then using the diagonalization

$$A = \begin{bmatrix} 1 & M/(\lambda_2 - \lambda_1) \\ 0 & 1 \end{bmatrix} \begin{bmatrix} \lambda_1 & 0 \\ 0 & \lambda_2 \end{bmatrix} \begin{bmatrix} 1 & M/(\lambda_2 - \lambda_1) \\ 0 & 1 \end{bmatrix}^{-1} \qquad (7.9.2)$$

we can show that

$$A^k = \begin{bmatrix} \lambda_1^k & M \sum_{j=0}^{k-1} \lambda_1^{k-1-j} \lambda_2^j \\ 0 & \lambda_2^k \end{bmatrix}. \qquad (7.9.3)$$

Consideration of the (1,2) entry suggests that A^k may grow before decay sets in. This is affirmed in Figure 7.9.1 where the size of $\| A^k \|_2$ is tracked for the example

$$A = \begin{bmatrix} 0.999 & 1000 \\ 0.0 & 0.998 \end{bmatrix}.$$

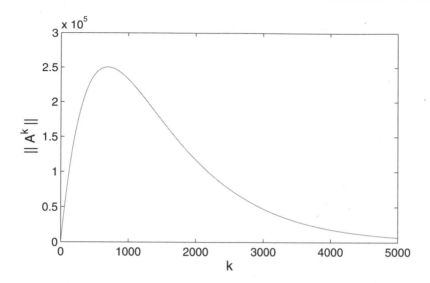

Figure 7.9.1. $\| A^k \|_2$ *can grow even if* $\rho(A) < 1$

Thus, it is perhaps better to augment (7.9.1) as follows:

If $\rho(A) < 1$, then aymptotically A^k converges to zero like $\rho(A)^k$. However, A^k may grow substantially before exponential decay sets in.

(7.9.4)

This example with its ill-conditioned eigenvector matrix displayed in (7.9.2), points to just why classical eigenvalue analysis is not so informative for nonnormal matrices. *Ill-conditioned eigenvector bases create a discrepancy between how A behaves and how its diagonalization XAX^{-1} behaves.* Pseudospectra analysis and computation narrow this gap.

7.9.2   Definitions

The pseudospectra idea is a generalization of the eigenvalue idea. Whereas the spectrum $\Lambda(A)$ is the set of all $z \in \mathbb{C}$ that make $\sigma_{min}(A - \lambda I)$ zero, the ϵ-*pseudospectrum* of a matrix $A \in \mathbb{C}^{n \times n}$ is the subset of the complex plane defined by

$$\Lambda_\epsilon(A) \;=\; \{ z \in \mathbb{C} : \sigma_{\min}(A - \lambda I) \leq \epsilon \}. \tag{7.9.5}$$

If $\lambda \in \Lambda_\epsilon(A)$, then λ is an ϵ-*pseudoeigenvalue* of A. A unit 2-norm vector v that satisfies $\| (A - \lambda I)v \|_2 = \epsilon$ is a corresponding ϵ-*pseudoeigenvector*. Note that if ϵ is zero, then $\Lambda_\epsilon(A)$ is just the set of A's eigenvalues, i.e., $\Lambda_0(A) = \Lambda(A)$.

We mention that because of their interest in what pseudospectra say about general linear operators, Trefethen and Embree (2005) use a strict inequality in the definition (7.9.5). The distinction has no impact in the matrix case.

Equivalent definitions of $\Lambda_\epsilon(\cdot)$ include

$$\Lambda_\epsilon(A) = \left\{ z \in \mathbb{C} : \| (zI - A)^{-1} \|_2 \geq \frac{1}{\epsilon} \right\} \tag{7.9.6}$$

which highlights the *resolvent* $(zI - A)^{-1}$ and

$$\Lambda_\epsilon(A) = \{ z \in \mathbb{C} : z \in \Lambda(A + E), \; \| E \|_2 \leq \epsilon \} \tag{7.9.7}$$

which characterize pseudspectra as (traditional) eigenvalues of nearby matrices. The equivalence of these three definitions is a straightforward verification that makes use of Chapter 2 facts about singular values, 2-norms, and matrix inverses. We mention that greater generality can be achieved in (7.9.6) and (7.9.7) by replacing the 2-norm with an arbitrary matrix norm.

7.9.3 Display

The pseudospectrum of a matrix is a visible subset of the complex plane so graphical display has a critical role to play in pseudospectra analysis. The MATLAB-based *Eigtool* system developed by Wright(2002) can be used to produce pseudospectra plots that are as pleasing to the eye as they are informative. Eigtool's pseudospectra plots are contour plots where each contour displays the z-values associated with a specified value of ϵ. Since

$$\epsilon_1 \leq \epsilon_2 \quad \Rightarrow \quad \Lambda_{\epsilon_1} \subseteq \Lambda_{\epsilon_2}$$

the typical pseudospectral plot is basically a topographical map that depicts the function $f(z) = \sigma_{\min}(zI - A)$ in the vicinity of the eigenvalues.

We present three Eigtool-produced plots that serve as illuminating examples. The first involves the n-by-n Kahan matrix $\mathsf{Kah}_n(s)$, e.g.,

$$\mathsf{Kah}_5(s) = \begin{bmatrix} 1 & -c & -c & -c & -c \\ 0 & s & -sc & -sc & -sc \\ 0 & 0 & s^2 & -s^2c & -s^2c \\ 0 & 0 & 0 & s^3 & -s^3c \\ 0 & 0 & 0 & 0 & s^4 \end{bmatrix}, \qquad c^2 + s^2 = 1.$$

Recall that we used these matrices in §5.4.3 to show that QR with column pivoting can fail to detect rank deficiency. The eigenvalues $\{1, s, s^2, \ldots, s^{n-1}\}$ of $\mathsf{Kah}_n(s)$ are extremely sensitive to perturbation. This is revealed by considering the $\epsilon = 10^{-6}$ contour that is displayed in Figure 7.9.2 together with $\Lambda(\mathsf{Kah}_n(s))$.

The second example is the Demmel matrix $\mathsf{Dem}_n(\beta)$, e.g.,

$$\mathsf{Dem}_5(\beta) = - \begin{bmatrix} 1 & \beta & \beta^2 & \beta^3 & \beta^4 \\ 0 & 1 & \beta & \beta^2 & \beta^3 \\ 0 & 0 & 1 & \beta & \beta^2 \\ 0 & 0 & 0 & 1 & \beta \\ 0 & 0 & 0 & 0 & 1 \end{bmatrix}.$$

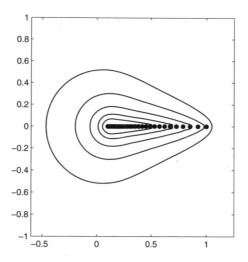

Figure 7.9.2. $\Lambda_\epsilon(\mathsf{Kah}_{30}(s))$ *with* $s^{29} = 0.1$ *and contours for* $\epsilon = 10^{-2}, \ldots, 10^{-6}$

The matrix $\mathrm{Dem}_n(\beta)$ is defective and has the property that very small perturbations can move an original eigenvalue to a position that are relatively far out on the imaginary axis. See Figure 7.9.3. The example is used to illuminate the nearness-to-instability problem presented in P7.9.13.

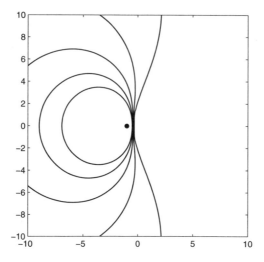

Figure 7.9.3. $\Lambda_\epsilon(\mathsf{Dem}_{50}(\beta))$ *with* $\beta^{49} = 10^8$ *and contours for* $\epsilon = 10^{-2}, \ldots, 10^{-6}$

The last example concerns the pseudospectra of the MATLAB "Gallery(5)" matrix:

$$
G_5 = \begin{bmatrix}
-9 & 11 & -21 & 63 & -252 \\
70 & -69 & 141 & -421 & 1684 \\
-575 & 575 & -1149 & 3451 & -13801 \\
3891 & -3891 & 7782 & -23345 & 93365 \\
1024 & -1024 & 2048 & -6144 & 24572
\end{bmatrix}.
$$

Notice in Figure 7.9.4 that $\Lambda_{10^{-13.5}}(G_5)$ has five components. In general, it can be

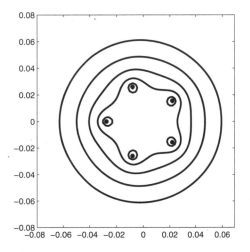

Figure 7.9.4. $\Lambda_\epsilon(G_5)$ *with contours for* $\epsilon = 10^{-11.5}, 10^{-12}, \ldots, 10^{-13.5}, 10^{-14}$

shown that each connected component of $\Lambda_\epsilon(A)$ contains at least one eigenvalue of A.

7.9.4 Some Elementary Properties

Pseudospectra are subsets of the complex plane so we start with a quick summary of notation. If S_1 and S_2 are subsets of the complex plane, then their sum $S_1 + S_2$ is defined by

$$
S_1 + S_2 = \{\, s : s = s_1 + s_2,\ s_1 \in S_1,\ s_2 \in S_2 \,\}.
$$

If S_1 consists of a single complex number α, then we write $\alpha + S_2$. If S is a subset of the complex plane and β is a complex number, then $\beta \cdot S$ is defined by

$$
\beta \cdot S = \{\, \beta z : z \in S \,\}.
$$

The disk of radius ϵ centered at the origin is denoted by

$$
\Delta_\epsilon = \{\, z : |z| \le \epsilon \,\}.
$$

Finally, the distance from a complex number z_0 to a set of complex numbers S is defined by

$$
\mathsf{dist}(z_0, S) = \min\{\, |z_0 - z| : z \in S \,\}.
$$

Our first result is about the effect of translation and scaling. For eigenvalues we have

$$\Lambda(\alpha I + \beta A) \; = \; \alpha + \beta \cdot \Lambda(A).$$

The following theorem establishes an analogous result for pseudospectra.

Theorem 7.9.1. *If* $\alpha, \beta \in \mathbb{C}$ *and* $A \in \mathbb{C}^{n \times n}$, *then* $\Lambda_{\epsilon |\beta|}(\alpha I + \beta A) \; = \; \alpha + \beta \cdot \Lambda_{\epsilon}(A)$.

Proof. Note that

$$
\begin{aligned}
\Lambda_{\epsilon}(\alpha I + A) \; &= \; \{\, z : \| \, (zI - (\alpha I + A))^{-1} \, \| \geq 1/\epsilon \,\} \\
&= \; \{\, z : \| \, ((z - \alpha)I - A)^{-1} \, \| \geq 1/\epsilon \,\} \\
&= \; \alpha + \{\, z - \alpha : \| \, ((z - \alpha)I - A)^{-1} \, \| \geq 1/\epsilon \,\} \\
&= \; \alpha + \{\, z : \| \, (zI - A)^{-1} \, \| \geq 1/\epsilon \,\} \; = \; \Lambda_{\epsilon}(A)
\end{aligned}
$$

and

$$
\begin{aligned}
\Lambda_{\epsilon |\beta|}(\beta \cdot A) \; &= \; \{\, z : \| \, (zI - \beta A)^{-1} \, \| \; \geq \; 1/|\beta|\epsilon \,\} \\
&= \; \{\, z : \| \, (z/\beta)I - A)^{-1} \, \| \; \geq \; 1/\epsilon \,\} \\
&= \; \beta \cdot \{\, z/\beta : \| \, (z/\beta)I - A)^{-1} \, \| \; \geq \; 1/\epsilon \,\} \\
&= \; \beta \cdot \{\, z : \| \, zI - A)^{-1} \, \| \; \geq \; 1/\epsilon \,\} \; = \; \beta \cdot \Lambda_{\epsilon}(A).
\end{aligned}
$$

The theorem readily follows by composing these two results. $\square$

General similarity transforms preserve eigenvalues but not ϵ-pseudoeigenvalues. However, a simple inclusion property holds in the pseudospectra case.

Theorem 7.9.2. *If* $B = X^{-1}AX$, *then* $\Lambda_{\epsilon}(B) \subseteq \Lambda_{\epsilon \kappa_2(X)}(A)$.

Proof. If $z \in \Lambda_{\epsilon}(B)$, then

$$
\frac{1}{\epsilon} \leq \| \, (zI - B)^{-1} \, \| \; = \; \| \, X^{-1}(zI - A)^{-1}X^{-1} \, \| \; \leq \; \kappa_2(X) \| \, (zI - A)^{-1} \, \|,
$$

from which the theorem follows. $\square$

Corollary 7.9.3. *If* $X \in \mathbb{C}^{n \times n}$ *is unitary and* $A \in \mathbb{C}^{n \times n}$, *then* $\Lambda_{\epsilon}(X^{-1}AX) = \Lambda_{\epsilon}(A)$.

Proof. The proof is left as an exercise. $\square$

The ϵ-pseudospectrum of a diagonal matrix is the union of ϵ-disks.

Theorem 7.9.4. *If* $D = \mathrm{diag}(\lambda_1, \ldots, \lambda_n)$, *then* $\Lambda_{\epsilon}(D) \; = \; \{\lambda_1, \ldots, \lambda_n\} + \Delta_{\epsilon}$.

Proof. The proof is left as an exercise. $\square$

Corollary 7.9.5. *If $A \in \mathbb{C}^{n \times n}$ is normal, then $\Lambda_\epsilon(A) = \Lambda(A) + \Delta_\epsilon$.*

Proof. Since A is normal, it has a diagonal Schur form $Q^H A Q = \mathrm{diag}(\lambda_1, \dots, \lambda_n) = D$ with unitary Q. The proof follows from Theorem 7.9.4. $\square$

If $T = (T_{ij})$ is a 2-by-2 block triangular matrix, then $\Lambda(T) = \Lambda(T_{11}) \cup \Lambda(T_{22})$. Here is the pseudospectral analog:

Theorem 7.9.6. *If*

$$
T = \begin{bmatrix} T_{11} & T_{12} \\ 0 & T_{22} \end{bmatrix}
$$

with square diagonal blocks, then $\Lambda_\epsilon(T_{11}) \cup \Lambda_\epsilon(T_{22}) \subseteq \Lambda_\epsilon(T)$.

Proof. The proof is left as an exercise. $\square$

Corollary 7.9.7. *If*

$$
T = \begin{bmatrix} T_{11} & 0 \\ 0 & T_{22} \end{bmatrix}
$$

with square diagonal blocks, then $\Lambda_\epsilon(T) = \Lambda_\epsilon(T_{11}) \cup \Lambda_\epsilon(T_{22})$.

Proof. The proof is left as an exercise. $\square$

The last property in our gallery of facts connects the resolvant $(z_0 I - A)^{-1}$ to the distance that separates z_0 from $\Lambda_\epsilon(A)$.

Theorem 7.9.8. *If $z_0 \in \mathbb{C}$ and $A \in \mathbb{C}^{n \times n}$, then*

$$
\mathrm{dist}(z_0, \Lambda_\epsilon(A)) \geq \frac{1}{\| (z_0 I - A)^{-1} \|_2} - \epsilon.
$$

Proof. For any $z \in \Lambda_\epsilon(A)$ we have from Corollary 2.4.4 and (7.9.6) that

$$
\epsilon \geq \sigma_{\min}(zI - A) = \sigma_{\min}((z_0 I - A) - (z - z_0)I) \geq \sigma_{\min}(z_0 I - A) - |z - z_0|
$$

and thus

$$
|z - z_0| \geq \frac{1}{\|(z_0 I - A)^{-1}\|} - \epsilon.
$$

The proof is completed by minimizing over all $z \in \Lambda_{\epsilon(A)}$. $\square$

7.9.5 Computing Pseudospectra

The production of a pseudospectral contour plot such as those displayed above requires sufficiently accurate approximations of $\sigma_{\min}(zI - A)$ on a grid that consists of (perhaps)

1000's of z-values. As we will see in §8.6, the computation of the complete SVD of an n-by-n dense matrix is an $O(n^3)$ endeavor. Fortunately, steps can be taken to reduce each grid point calculation to $O(n^2)$ or less by exploiting the following ideas:

1. Avoid SVD-type computations in regions where $\sigma_{\min}(zI - A)$ is slowly varying. See Gallestey (1998).

2. Exploit Theorem 7.9.6 by ordering the eigenvalues so that the invariant subspace associated with $\Lambda(T_{11})$ captures the essential behavior of $(zI - A)^{-1}$. See Reddy, Schmid, and Henningson (1993).

3. Precompute the Schur decomposition $Q^H A Q = T$ and apply a $\sigma_{\min}$ algorithm that is efficient for triangular matrices. See Lui (1997).

We offer a few comments on the last strategy since it has much in common with the condition estimation problem that we discussed in §3.5.4. The starting point is to recognize that since Q is unitary,

$$\sigma_{\min}(zI - A) = \sigma_{\min}(zI - T).$$

The triangular structure of the transformed problem makes it possible to obtain a satisfactory estimate of $\sigma_{\min}(zI - A)$ in $O(n^2)$ flops. If d is a unit 2-norm vector and $(zI - T)y = d$, then it follows from the SVD of $zI - T$ that

$$\sigma_{\min}(zI - T) \leq \frac{1}{\| y \|_2}.$$

Let $u_{\min}$ be a left singular vector associated with $\sigma_{\min}(zI - T)$. If d is has a significant component in the direction of $u_{\min}$, then

$$\sigma_{\min}(zI - T) \approx \frac{1}{\| y \|_2}.$$

Recall that Algorithm 3.5.1 is a cheap heuristic procedure that dynamically determines the right hand side vector d so that the solution to a given triangular system is large in norm. This is tantamount to choosing d so that it is rich in the direction of $u_{\min}$. A complex arithmetic, 2-norm variant of Algorithm 3.5.1 is outlined in P7.9.13. It can be applied to $zI - T$. The resulting d-vector can be refined using inverse iteration ideas, see Toh and Trefethen (1996) and §8.2.2. Other approaches are discussed by Wright and Trefethen (2001).

7.9.6   Computing the ϵ-Pseudospectral Abscissa and Radius

The ϵ-pseudospectral abscissa of a matrix $A \in \mathbb{C}^{n \times n}$ is the rightmost point on the boundary of Λ_ϵ:

$$\alpha_\epsilon(A) = \max_{z \in \Lambda_\epsilon(A)} \text{Re}(z). \tag{7.9.8}$$

Likewise, the ϵ-pseudospectral radius is the point of largest magnitude on the boundary of Λ_ϵ:

$$\rho_\epsilon(A) = \max_{z \in \Lambda_\epsilon(A)} |z|. \tag{7.9.9}$$

These quantities arise in the analysis of dynamical systems and effective iterative algorithms for their estimation have been proposed by Burke, Lewis, and Overton (2003) and Mengi and Overton (2005). A complete presentation and analysis of their very clever optimization procedures, which build on the work of Byers (1988), is beyond the scope of the text. However, at their core they involve interesting intersection problems that can be reformulated as structured eigenvalue problems. For example, if $i \cdot r$ is an eigenvalue of the matrix

$$M = \begin{bmatrix} ie^{i\theta} A^H & -\epsilon I \\ \epsilon I & ie^{-i\theta} A \end{bmatrix}, \tag{7.9.10}$$

then ϵ is a singular value of $A - re^{i\theta} I$. To see this, observe that if

$$\begin{bmatrix} ie^{i\theta} A^H & -\epsilon I \\ \epsilon I & ie^{-i\theta} A \end{bmatrix} \begin{bmatrix} f \\ g \end{bmatrix} = i \cdot r \begin{bmatrix} f \\ g \end{bmatrix},$$

then

$$(A - re^{i\theta} I)^H (A - re^{i\theta} I) g = \epsilon^2 g.$$

The complex version of the SVD (§2.4.4) says that ϵ is a singular value of $A - re^{i\theta} I$. It can be shown that if $ir_{\max}$ is the largest pure imaginary eigenvalue of M, then

$$\epsilon = \sigma_{\min}(A - r_{\max} e^{i\theta} I).$$

This result can be used to compute the intersection of the ray $\{ re^{i\theta} : R \geq 0 \}$ and the boundary of $\Lambda_\epsilon(A)$. This computation is at the heart of computing the ϵ-pseudospectral radius. See Mengi and Overton (2005).

7.9.7 Matrix Powers and the ϵ-Pseudospectral Radius

At the start of this section we used the example

$$A = \begin{bmatrix} 0.999 & 1000 \\ 0.000 & 0.998 \end{bmatrix}$$

to show that $\| A^k \|_2$ can grow even though $\rho(A) < 1$. This kind of transient behavior can be anticipated by the pseudospectral radius. Indeed, it can be shown that for any $\epsilon > 0$,

$$\sup_{k \geq 0} \| A^k \|_2 \geq \frac{\rho_\epsilon(A) - 1}{\epsilon}. \tag{7.9.11}$$

See Trefethen and Embree (SAP, pp. 160–161). This says that transient growth will occur if there is a contour $\{z : \| (\| zI - A)^{-1} \| = 1/\epsilon \}$ that extends beyond the unit disk. For the above 2-by-2 example, if $\epsilon = 10^{-8}$, then $\rho_\epsilon(A) \approx 1.0017$ and the inequality (7.9.11) says that for some k, $\| A^k \|_2 \geq 1.7 \times 10^5$. This is consistent with what is displayed in Figure 7.9.1.

Problems

P7.9.1 Show that the definitions (7.9.5), (7.9.6), and (7.9.7) are equivalent.

P7.9.2 Prove Corollary 7.9.3.

P7.9.3 Prove Theorem 7.9.4.

P7.9.4 Prove Theorem 7.9.6.

P7.9.5 Prove Corollary 7.9.7.

P7.9.6 Show that if $A, E \in \mathbb{C}^{n \times n}$, then $\Lambda_\epsilon(A + E) \subseteq \Lambda_{\epsilon + \| E \|_2}(A)$.

P7.9.7 Suppose $\sigma_{\min}(z_1 I - A) = \epsilon_1$ and $\sigma_{\min}(z_2 I - A) = \epsilon_2$. Prove that there exists a real number μ so that if $z_3 = (1 - \mu)z_1 + \mu z_2$, then $\sigma_{\min}(z_3 I - A) = (\epsilon_1 + \epsilon_2)/2$?

P7.9.8 Suppose $A \in \mathbb{C}^{n \times n}$ is normal and $E \in \mathbb{C}^{n \times n}$ is nonnormal. State and prove a theorem about $\Lambda_\epsilon(A + E)$.

P7.9.9 Explain the connection between Theorem 7.9.2 and the Bauer-Fike Theorem (Theorem 7.2.2).

P7.9.10 Define the matrix $J \in \mathbb{R}^{2n \times 2n}$ by

$$J = \begin{bmatrix} 0 & I_n \\ -I_n & 0 \end{bmatrix}.$$

(a) The matrix $H \in \mathbb{R}^{2n \times 2n}$ is a Hamiltonian matrix if $J^T H J = -H^T$. It is easy to show that if H is Hamiltonian and $\lambda \in \Lambda(H)$, then $-\lambda \in \Lambda(H)$. Does it follow that if $\lambda \in \Lambda_\epsilon(H)$, then $-\lambda \in \Lambda_\epsilon(H)$?
(b) The matrix $S \in \mathbb{R}^{2n \times 2n}$ is a symplectic matrix if $J^T S J = S^{-T}$. It is easy to show that if S is symplectic and $\lambda \in \Lambda(S)$, then $1/\lambda \in \Lambda(S)$. Does it follow that if $\lambda \in \Lambda_\epsilon(S)$, then $1/\lambda \in \Lambda_\epsilon(S)$?

P7.9.11 Unsymmetric Toeplitz matrices tend to have very ill-conditioned eigensystems and thus have interesting pseudospectral properties. Suppose

$$A = \begin{bmatrix} 0 & 1 & \cdots & 0 \\ \alpha & 0 & \ddots & \vdots \\ \vdots & \ddots & \ddots & 1 \\ 0 & \cdots & \alpha & 0 \end{bmatrix}.$$

(a) Construct a diagonal matrix S so that $S^{-1}AS = B$ is symmetric and tridiagonal with 1's on its subdiagonal and superdiagonal. (b) What can you say about the condition of A's eigenvector matrix?

P7.9.12 A matrix $A \in \mathbb{C}^{n \times n}$ is *stable* if all of its eigenvalues have negative real parts. Consider the problem of minimizing $\| E \|_2$ subject to the constraint that $A + E$ has an eigenvalue on the imaginary axis. Explain why this optimization problem is equivalent to minimizing $\sigma_{\min}(irI - A)$ over all $r \in \mathbb{R}$. If E_* is a minimizing E, then $\| E \|_2$ can be regarded as measure of A's nearness to instability. What is the connection between A's nearness to instability and $\alpha_\epsilon(A)$?

P7.9.13 This problem is about the cheap estimation of the minimum singular value of a matrix, a critical computation that is performed over an over again during the course of displaying the pseudospectrum of a matrix. In light of the discussion in §7.9.5, the challenge is to estimate the smallest singular value of an upper triangular matrix $U = T - zI$ where T is the Schur form of $A \in \mathbb{R}^{n \times n}$. The condition estimation ideas of §3.5.4 are relevant. We want to determine a unit 2-norm vector $d \in \mathbb{C}^n$ such that the solution to $Uy = d$ has a large 2-norm for then $\sigma_{\min}(U) \approx 1/\| y \|_2$. (a) Suppose

$$U = \begin{bmatrix} u_{11} & u^H \\ 0 & U_1 \end{bmatrix} \qquad y = \begin{bmatrix} \tau \\ z \end{bmatrix} \qquad d = \begin{bmatrix} c \\ sd_1 \end{bmatrix}$$

where $u_{11}, \tau \in \mathbb{C}$, $u, z, d_1 \in \mathbb{C}^{n-1}$, $U_1 \in \mathbb{C}^{(n-1) \times (n-1)}$, $\| d_1 \|_2 = 1$, $U_1 y_1 = d_1$, and $c^2 + s^2 = 1$. Give an algorithm that determines c and s so that if $Uy = d$, then $\| y \|_2$ is as large as possible. Hint: This is a 2-by-2 SVD problem. (b) Using part (a), develop a nonrecursive method for estimating $\sigma_{\min}(U(k{:}n, k{:}n))$ for $k = n{:} - 1{:}1$.

Notes and References for §7.7

Besides Trefethen and Embree (SAP), the following papers provide a nice introduction to the pseudospectra idea:

M. Embree and L.N. Trefethen (2001). "Generalizing Eigenvalue Theorems to Pseudospectra Theorems," *SIAM J. Sci. Comput. 23*, 583–590.

L.N. Trefethen (1997). "Pseudospectra of Linear Operators," *SIAM Review 39*, 383–406.

For more details concerning the computation and display of pseudoeigenvalues, see:

S.C. Reddy, P.J. Schmid, and D.S. Henningson (1993). "Pseudospectra of the Orr-Sommerfeld Operator," *SIAM J. Applic. Math. 53*, 15–47.

S.-H. Lui (1997). "Computation of Pseudospectra by Continuation," *SIAM J. Sci. Comput. 18*, 565–573.

E. Gallestey (1998). "Computing Spectral Value Sets Using the Subharmonicity of the Norm of Rational Matrices," *BIT, 38*, 22–33.

L.N. Trefethen (1999). "Computation of Pseudospectra," *Acta Numerica 8*, 247–295.

T.G. Wright (2002). *Eigtool*, http://www.comlab.ox.ac.uk/pseudospectra/eigtool/.

Interesting extensions/generalizations/applications of the pseudospectra idea include:

L. Reichel and L.N. Trefethen (1992). "Eigenvalues and Pseudo-Eigenvalues of Toeplitz Matrices," *Lin. Alg. Applic. 164–164*, 153–185.

K-C. Toh and L.N. Trefethen (1994). "Pseudozeros of Polynomials and Pseudospectra of Companion Matrices," *Numer. Math. 68*, 403–425.

F. Kittaneh (1995). "Singular Values of Companion Matrices and Bounds on Zeros of Polynomials," *SIAM J. Matrix Anal. Applic. 16*, 333–340.

N.J. Higham and F. Tisseur (2000). "A Block Algorithm for Matrix 1-Norm Estimation, with an Application to 1-Norm Pseudospectra," *SIAM J. Matrix Anal. Applic. 21*, 1185–1201.

T.G. Wright and L.N. Trefethen (2002). "Pseudospectra of Rectangular matrices," *IMA J. Numer. Anal. 22*, 501–519.

R. Alam and S. Bora (2005). "On Stable Eigendecompositions of Matrices," *SIAM J. Matrix Anal. Applic. 26*, 830–848.

Pseudospectra papers that relate to the notions of controllability and stability of linear systems include:

J.V. Burke and A.S. Lewis. and M.L. Overton (2003). "Optimization and Pseudospectra, with Applications to Robust Stability," *SIAM J. Matrix Anal. Applic. 25*, 80–104.

J.V. Burke, A.S. Lewis, and M.L. Overton (2003). "Robust Stability and a Criss–Cross Algorithm for Pseudospectra," *IMA J. Numer. Anal. 23*, 359–375.

J.V. Burke, A.S. Lewis and M.L. Overton (2004). "Pseudospectral Components and the Distance to Uncontrollability," *SIAM J. Matrix Anal. Applic. 26*, 350–361.

The following papers are concerned with the computation of the numerical radius, spectral radius, and field of values:

C. He and G.A. Watson (1997). "An Algorithm for Computing the Numerical Radius," *IMA J. Numer. Anal. 17*, 329–342.

G.A. Watson (1996). "Computing the Numerical Radius" *Lin. Alg. Applic. 234*, 163–172.

T. Braconnier and N.J. Higham (1996). "Computing the Field of Values and Pseudospectra Using the Lanczos Method with Continuation," *BIT 36*, 422–440.

E. Mengi and M.L. Overton (2005). "Algorithms for the Computation of the Pseudospectral Radius and the Numerical Radius of a Matrix," *IMA J. Numer. Anal. 25*, 648–669.

N. Guglielmi and M. Overton (2011). "Fast Algorithms for the Approximation of the Pseudospectral Abscissa and Pseudospectral Radius of a Matrix," *SIAM J. Matrix Anal. Applic. 32*, 1166–1192.

For more insight into the behavior of matrix powers, see:

P. Henrici (1962). "Bounds for Iterates, Inverses, Spectral Variation, and Fields of Values of Nonnormal Matrices," *Numer. Math.4*, 24–40.

J. Descloux (1963). "Bounds for the Spectral Norm of Functions of Matrices," *Numer. Math. 5*, 185–90.

T. Ransford (2007). "On Pseudospectra and Power Growth," *SIAM J. Matrix Anal. Applic. 29*, 699–711.

As an example of what pseudospectra can tell us about highly structured matrices, see:

L. Reichel and L.N. Trefethen (1992). "Eigenvalues and Pseudo-eigenvalues of Toeplitz Matrices," *Lin. Alg. Applic. 162/163/164*, 153–186.

Chapter 8

Symmetric Eigenvalue Problems

8.1 Properties and Decompositions

8.2 Power Iterations

8.3 The Symmetric QR Algorithm

8.4 More Methods for Tridiagonal Problems

8.5 Jacobi Methods

8.6 Computing the SVD

8.7 Generalized Eigenvalue Problems with Symmetry

The symmetric eigenvalue problem with its rich mathematical structure is one of the most aesthetically pleasing problems in numerical linear algebra. We begin with a brief discussion of the mathematical properties that underlie the algorithms that follow. In §8.2 and §8.3 we develop various power iterations and eventually focus on the symmetric QR algorithm. Methods for the important case when the matrix is tridiagonal are covered in §8.4. These include the method of bisection and a divide and conquer technique. In §8.5 we discuss Jacobi's method, one of the earliest matrix algorithms to appear in the literature. This technique is of interest because it is amenable to parallel computation and because of its interesting high-accuracy properties. The computation of the singular value decomposition is detailed in §8.6. The central algorithm is a variant of the symmetric QR iteration that works on bidiagonal matrices.

In §8.7 we discuss the generalized eigenvalue problem $Ax = \lambda Bx$ for the important case when A is symmetric and B is symmetric positive definite. The generalized singular value decomposition $A^T Ax = \mu^2 B^T Bx$ is also covered. The section concludes with a brief examination of the quadratic eigenvalue problem $(\lambda^2 M + \lambda C + K)x = 0$ in the presence of symmetry, skew-symmetry, and definiteness.

Reading Notes

Knowledge of Chapters 1-3 and §5.1–§5.2 are assumed. Within this chapter there are the following dependencies:

$$\S 8.4$$
$$\uparrow$$
$$\S 8.1 \quad \rightarrow \quad \S 8.2 \quad \rightarrow \quad \S 8.3 \quad \rightarrow \quad \S 8.6 \quad \rightarrow \quad \S 8.7$$
$$\downarrow$$
$$\S 8.5$$

Many of the algorithms and theorems in this chapter have unsymmetric counterparts in Chapter 7. However, except for a few concepts and definitions, our treatment of the symmetric eigenproblem can be studied before reading Chapter 7.

Complementary references include Wilkinson (AEP), Stewart (MAE), Parlett (SEP), and Stewart and Sun (MPA).

8.1   Properties and Decompositions

In this section we summarize the mathematics required to develop and analyze algorithms for the symmetric eigenvalue problem.

8.1.1   Eigenvalues and Eigenvectors

Symmetry guarantees that all of A's eigenvalues are real and that there is an orthonormal basis of eigenvectors.

Theorem 8.1.1 (Symmetric Schur Decomposition). *If $A \in \mathbb{R}^{n \times n}$ is symmetric, then there exists a real orthogonal Q such that*

$$Q^T A Q = \Lambda = \text{diag}(\lambda_1, \ldots, \lambda_n).$$

Moreover, for $k = 1{:}n$, $AQ(:, k) = \lambda_k Q(:, k)$. Compare with Theorem 7.1.3.

Proof. Suppose $\lambda_1 \in \lambda(A)$ and that $x \in \mathbb{C}^n$ is a unit 2-norm eigenvector with $Ax = \lambda_1 x$. Since $\lambda_1 = x^H A x = x^H A^H x = \overline{x^H A x} = \overline{\lambda_1}$ it follows that $\lambda_1 \in \mathbb{R}$. Thus, we may assume that $x \in \mathbb{R}^n$. Let $P_1 \in \mathbb{R}^{n \times n}$ be a Householder matrix such that $P_1^T x = e_1 = I_n(:, 1)$. It follows from $Ax = \lambda_1 x$ that $(P_1^T A P_1)e_1 = \lambda e_1$. This says that the first column of $P_1^T A P_1$ is a multiple of e_1. But since $P_1^T A P_1$ is symmetric, it must have the form

$$P_1^T A P_1 = \begin{bmatrix} \lambda_1 & 0 \\ 0 & A_1 \end{bmatrix}$$

where $A_1 \in \mathbb{R}^{(n-1) \times (n-1)}$ is symmetric. By induction we may assume that there is an orthogonal $Q_1 \in \mathbb{R}^{(n-1) \times (n-1)}$ such that $Q_1^T A_1 Q_1 = \Lambda_1$ is diagonal. The theorem follows by setting

$$Q = P_1 \begin{bmatrix} 1 & 0 \\ 0 & Q_1 \end{bmatrix} \qquad \text{and} \qquad \Lambda = \begin{bmatrix} \lambda_1 & 0 \\ 0 & \Lambda_1 \end{bmatrix}$$

and comparing columns in the matrix equation $AQ = Q\Lambda$. $\square$

For a symmetric matrix A we shall use the notation $\lambda_k(A)$ to designate the kth largest eigenvalue, i.e.,

$$\lambda_n(A) \leq \cdots \leq \lambda_2(A) \leq \lambda_1(A).$$

It follows from the orthogonal invariance of the 2-norm that A has singular values $\{|\lambda_1(A)|, \ldots, |\lambda_n(A)|\}$ and

$$\| A \|_2 = \max\{ |\lambda_1(A)| , |\lambda_n(A)| \}.$$

The eigenvalues of a symmetric matrix have a *minimax* characterization that revolves around the quadratic form $x^T A x / x^T x$.

Theorem 8.1.2 (Courant-Fischer Minimax Theorem). *If $A \in \mathbb{R}^{n \times n}$ is symmetric, then*

$$\lambda_k(A) = \max_{\dim(S)=k} \min_{0 \neq y \in S} \frac{y^T A y}{y^T y}$$

for $k = 1:n$.

Proof. Let $Q^T A Q = \text{diag}(\lambda_i)$ be the Schur decomposition with $\lambda_k = \lambda_k(A)$ and $Q = [\, q_1 \mid \cdots \mid q_n \,]$. Define

$$S_k = \text{span}\{q_1, \ldots, q_k\},$$

the invariant subspace associated with $\lambda_1, \ldots, \lambda_k$. It is easy to show that

$$\max_{\dim(S)=k} \min_{0 \neq y \in S} \frac{y^T A y}{y^T y} \geq \min_{0 \neq y \in S_k} \frac{y^T A y}{y^T y} = q_k^T A q_k = \lambda_k(A).$$

To establish the reverse inequality, let S be any k-dimensional subspace and note that it must intersect $\text{span}\{q_k, \ldots, q_n\}$, a subspace that has dimension $n - k + 1$. If $y_* = \alpha_k q_k + \cdots + \alpha_n q_n$ is in this intersection, then

$$\min_{0 \neq y \in S} \frac{y^T A y}{y^T y} \leq \frac{y_*^T A y_*}{y_*^T y_*} \leq \lambda_k(A).$$

Since this inequality holds for all k-dimensional subspaces,

$$\max_{\dim(S)=k} \min_{0 \neq y \in S} \frac{y^T A y}{y^T y} \leq \lambda_k(A)$$

thereby completing the proof of the theorem. $\square$

Note that if $A \in \mathbb{R}^{n \times n}$ is symmetric positive definite, then $\lambda_n(A) > 0$.

8.1.2 Eigenvalue Sensitivity

An important solution framework for the symmetric eigenproblem involves the production of a sequence of orthogonal transformations $\{Q_k\}$ with the property that the matrices $Q_k^T A Q_k$ are progressively "more diagonal." The question naturally arises, how well do the diagonal elements of a matrix approximate its eigenvalues?

Theorem 8.1.3 (Gershgorin). *Suppose $A \in \mathbb{R}^{n \times n}$ is symmetric and that $Q \in \mathbb{R}^{n \times n}$ is orthogonal. If $Q^T A Q = D + F$ where $D = \mathrm{diag}(d_1, \ldots, d_n)$ and F has zero diagonal entries, then*

$$\lambda(A) \subseteq \bigcup_{i=1}^{n} [d_i - r_i, d_i + r_i]$$

where $r_i = \sum_{j=1}^{n} |f_{ij}|$ for $i = 1{:}n$. Compare with Theorem 7.2.1.

Proof. Suppose $\lambda \in \lambda(A)$ and assume without loss of generality that $\lambda \neq d_i$ for $i = 1{:}n$. Since $(D - \lambda I) + F$ is singular, it follows from Lemma 2.3.3 that

$$1 \leq \| (D - \lambda I)^{-1} F \|_\infty = \sum_{j=1}^{n} \frac{|f_{kj}|}{|d_k - \lambda|} = \frac{r_k}{|d_k - \lambda|}$$

for some k, $1 \leq k \leq n$. But this implies that $\lambda \in [d_k - r_k, d_k + r_k]$. $\square$

    The next results show that if A is perturbed by a symmetric matrix E, then its eigenvalues do not move by more than $\| E \|_F$.

Theorem 8.1.4 (Wielandt-Hoffman). *If A and $A + E$ are n-by-n symmetric matrices, then*

$$\sum_{i=1}^{n} \left(\lambda_i(A + E) - \lambda_i(A)\right)^2 \leq \| E \|_F^2 .$$

Proof. See Wilkinson (AEP, pp. 104–108), Stewart and Sun (MPT, pp. 189–191), or Lax (1997, pp. 134–136). $\square$

Theorem 8.1.5. *If A and $A + E$ are n-by-n symmetric matrices, then*

$$\lambda_k(A) + \lambda_n(E) \leq \lambda_k(A + E) \leq \lambda_k(A) + \lambda_1(E), \qquad k = 1{:}n.$$

Proof. This follows from the minimax characterization. For details see Wilkinson (AEP, pp. 101–102) or Stewart and Sun (MPT, p. 203). $\square$

Corollary 8.1.6. *If A and $A + E$ are n-by-n symmetric matrices, then*

$$|\lambda_k(A + E) - \lambda_k(A)| \leq \| E \|_2$$

for $k = 1{:}n$.

Proof. Observe that

$$|\lambda_k(A + E) - \lambda_k(A)| \leq \max\{|\lambda_n(E)|, |\lambda_1(E)\|\} = \| E \|_2$$

for $k = 1{:}n$. $\square$

A pair of additional perturbation results that are important follow from the minimax property.

Theorem 8.1.7 (Interlacing Property). *If $A \in \mathbb{R}^{n \times n}$ is symmetric and $A_r = A(1{:}r, 1{:}r)$, then*

$$\lambda_{r+1}(A_{r+1}) \leq \lambda_r(A_r) \leq \lambda_r(A_{r+1}) \leq \cdots \leq \lambda_2(A_{r+1}) \leq \lambda_1(A_r) \leq \lambda_1(A_{r+1})$$

for $r = 1{:}n-1$.

Proof. Wilkinson (AEP, pp. 103–104). $\square$

Theorem 8.1.8. *Suppose $B = A + \tau cc^T$ where $A \in \mathbb{R}^{n \times n}$ is symmetric, $c \in \mathbb{R}^n$ has unit 2-norm, and $\tau \in \mathbb{R}$. If $\tau \geq 0$, then*

$$\lambda_i(B) \in [\lambda_i(A),\ \lambda_{i-1}(A)], \qquad i = 2{:}n,$$

while if $\tau \leq 0$ then

$$\lambda_i(B) \in [\lambda_{i+1}(A),\ \lambda_i(A)], \qquad i = 1{:}n-1.$$

In either case, there exist nonnegative $m_1, \ldots, m_n$ such that

$$\lambda_i(B) = \lambda_i(A) + m_i\tau, \qquad i = 1{:}n$$

with $m_1 + \cdots + m_n = 1$.

Proof. Wilkinson (AEP, pp. 94–97). See also P8.1.8. $\square$

8.1.3 Invariant Subspaces

If $S \subseteq \mathbb{R}^n$ and $x \in S \Rightarrow Ax \in S$, then S is an *invariant subspace* for $A \in \mathbb{R}^{n \times n}$. Note that if $x \in \mathbb{R}^i$s an eigenvector for A, then $S = \mathsf{span}\{x\}$ is 1-dimensional invariant subspace. Invariant subspaces serve to "take apart" the eigenvalue problem and figure heavily in many solution frameworks. The following theorem explains why.

Theorem 8.1.9. *Suppose $A \in \mathbb{R}^{n \times n}$ is symmetric and that*

$$Q = [\ Q_1 \mid Q_2\ ] \atop \quad\ r \quad\ n-r$$

is orthogonal. If $\mathsf{ran}(Q_1)$ is an invariant subspace, then

$$Q^T A Q = D = \begin{bmatrix} D_1 & 0 \\ 0 & D_2 \end{bmatrix} \begin{matrix} r \\ n-r \end{matrix} \qquad (8.1.1)$$
$$\qquad\qquad\quad r \quad\ n-r$$

and $\lambda(A) = \lambda(D_1) \cup \lambda(D_2)$. Compare with Lemma 7.1.2.

Proof. If

$$Q^T A Q = \begin{bmatrix} D_1 & E_{21}^T \\ E_{21} & D_2 \end{bmatrix},$$

then from $AQ = QD$ we have $AQ_1 - Q_1 D_1 = Q_2 E_{21}$. Since $\mathsf{ran}(Q_1)$ is invariant, the columns of $Q_2 E_{21}$ are also in $\mathsf{ran}(Q_1)$ and therefore perpendicular to the columns of Q_2. Thus,

$$0 = Q_2^T (AQ_1 - Q_1 D_1) = Q_2^T Q_2 E_{21} = E_{21}.$$

and so (8.1.1) holds. It is easy to show

$$\mathsf{det}(A - \lambda I_n) = \mathsf{det}(Q^T A Q - \lambda I_n) = \mathsf{det}(D_1 - \lambda I_r) \cdot \mathsf{det}(D_2 - \lambda I_{n-r})$$

confirming that $\lambda(A) = \lambda(D_1) \cup \lambda(D_2)$. $\square$

The sensitivity to perturbation of an invariant subspace depends upon the separation of the associated eigenvalues from the rest of the spectrum. The appropriate measure of separation between the eigenvalues of two symmetric matrices B and C is given by

$$\mathsf{sep}(B, C) = \min_{\substack{\lambda \in \lambda(B) \\ \mu \in \lambda(C)}} |\lambda - \mu|. \tag{8.1.2}$$

With this definition we have the following result.

Theorem 8.1.10. *Suppose A and $A + E$ are n-by-n symmetric matrices and that*

$$Q = [\, Q_1 \,|\, Q_2 \,] \atop r \quad n-r$$

is an orthogonal matrix such that $\mathsf{ran}(Q_1)$ is an invariant subspace for A. Partition the matrices $Q^T A Q$ and $Q^T E Q$ as follows:

$$Q^T A Q = \begin{bmatrix} D_1 & 0 \\ 0 & D_2 \end{bmatrix} \begin{matrix} r \\ n-r \end{matrix} \quad , \quad Q^T E Q = \begin{bmatrix} E_{11} & E_{21}^T \\ E_{21} & E_{22} \end{bmatrix} \begin{matrix} r \\ n-r \end{matrix} \quad .$$
$$\phantom{Q^T A Q = \begin{bmatrix}} r \quad n-r r \quad n-r$$

If $\mathsf{sep}(D_1, D_2) > 0$ and

$$\| E \|_F \le \frac{\mathsf{sep}(D_1, D_2)}{5},$$

then there exists a matrix $P \in \mathbb{R}^{(n-r) \times r}$ with

$$\| P \|_F \le \frac{4}{\mathsf{sep}(D_1, D_2)} \| E_{21} \|_F$$

such that the columns of $\hat{Q}_1 = (Q_1 + Q_2 P)(I + P^T P)^{-1/2}$ define an orthonormal basis for a subspace that is invariant for $A + E$. Compare with Theorem 7.2.4.

Proof. This result is a slight adaptation of Theorem 4.11 in Stewart (1973). The matrix $(I + P^T P)^{-1/2}$ is the inverse of the square root of $I + P^T P$. See §4.2.4. $\square$

Corollary 8.1.11. *If the conditions of the theorem hold, then*

$$\text{dist}(\,\text{ran}(Q_1), \text{ran}(\hat{Q}_1)) \;\leq\; \frac{4}{\text{sep}(D_1, D_2)} \,\|\, E_{21} \,\|_F.$$

Compare with Corollary 7.2.5.

Proof. It can be shown using the SVD that

$$\|\, P(I + P^T P)^{-1/2} \,\|_2 \;\leq\; \|\, P \,\|_2 \;\leq\; \|\, P \,\|_F. \tag{8.1.3}$$

Since $Q_2^T \hat{Q}_1 = P(I + P^T P)^{-1/2}$ it follows that

$$\text{dist}(\text{ran}(Q_1), \text{ran}(\hat{Q}_1)) = \|\, Q_2^T \hat{Q}_1 \,\|_2 \;=\; \|\, P(I + P^H P)^{-1/2} \,\|_2$$

$$\leq \|\, P \,\|_2 \;\leq\; 4\|\, E_{21} \,\|_F / \text{sep}(D_1, D_2)$$

completing the proof.   □

Thus, the reciprocal of $\text{sep}(D_1, D_2)$ can be thought of as a condition number that measures the sensitivity of $\text{ran}(Q_1)$ as an invariant subspace.

The effect of perturbations on a single eigenvector is sufficiently important that we specialize the above results to this case.

Theorem 8.1.12. *Suppose A and $A + E$ are n-by-n symmetric matrices and that*

$$Q = [\, q_1 \mid Q_2 \,]$$
$$ 1 \quad n-1$$

is an orthogonal matrix such that q_1 is an eigenvector for A. Partition the matrices $Q^T AQ$ and $Q^T EQ$ as follows:

$$Q^T AQ = \begin{bmatrix} \lambda & 0 \\ 0 & D_2 \end{bmatrix} \begin{matrix} 1 \\ n-1 \end{matrix} \,, \qquad Q^T EQ = \begin{bmatrix} \epsilon & e^T \\ e & E_{22} \end{bmatrix} \begin{matrix} 1 \\ n-1 \end{matrix} \,.$$
$$\phantom{Q^T AQ = \begin{bmatrix} \lambda \end{bmatrix}} 1 \quad n-1 1 \quad n-1$$

If

$$d = \min_{\mu \in \lambda(D_2)} \; |\lambda - \mu| \;>\; 0$$

and

$$\|\, E \,\|_F \;\leq\; \frac{d}{5},$$

then there exists $p \in \mathbb{R}^{n-1}$ satisfying

$$\|\, p \,\|_2 \;\leq\; \frac{4}{d} \,\|\, e \,\|_2$$

such that $\hat{q}_1 = (q_1 + Q_2 p)/\sqrt{1 + p^T p}$ is a unit 2-norm eigenvector for $A + E$. Moreover,

$$\text{dist}(\,\text{span}\{\, q_1 \,\}, \text{span}\{\hat{q}_1\}) \;=\; \sqrt{1 - (q_1^T \hat{q}_1)^2} \;\leq\; \frac{4}{d} \,\|\, e \,\|_2.$$

Compare with Corollary 7.2.6.

Proof. Apply Theorem 8.1.10 and Corollary 8.1.11 with $r = 1$ and observe that if $D_1 = (\lambda)$, then $d = \mathsf{sep}(D_1, D_2)$. $\square$

8.1.4 Approximate Invariant Subspaces

If the columns of $Q_1 \in \mathbb{R}^{n \times r}$ are independent and the *residual matrix* $R = AQ_1 - Q_1 S$ is small for some $S \in \mathbb{R}^{r \times r}$, then the columns of Q_1 define an approximate invariant subspace. Let us discover what we can say about the eigensystem of A when in the possession of such a matrix.

Theorem 8.1.13. *Suppose* $A \in \mathbb{R}^{n \times n}$ *and* $S \in \mathbb{R}^{r \times r}$ *are symmetric and that*

$$AQ_1 - Q_1 S = E_1$$

where $Q_1 \in \mathbb{R}^{n \times r}$ *satisfies* $Q_1^T Q_1 = I_r$. *Then there exist* $\mu_1, \ldots, \mu_r \in \lambda(A)$ *such that*

$$|\mu_k - \lambda_k(S)| \leq \sqrt{2} \, \| E_1 \|_2$$

for $k = 1{:}r$.

Proof. Let $Q_2 \in \mathbb{R}^{n \times (n-r)}$ be any matrix such that $Q = [\, Q_1 \mid Q_2 \,]$ is orthogonal. It follows that

$$Q^T A Q = \begin{bmatrix} S & 0 \\ 0 & Q_2^T A Q_2 \end{bmatrix} + \begin{bmatrix} Q_1^T E_1 & E_1^T Q_2 \\ Q_2^T E_1 & 0 \end{bmatrix} \equiv B + E$$

and so by using Corollary 8.1.6 we have $|\lambda_k(A) - \lambda_k(B)| \leq \| E \|_2$ for $k = 1{:}n$. Since $\lambda(S) \subseteq \lambda(B)$, there exist $\mu_1, \ldots, \mu_r \in \lambda(A)$ such that $|\mu_k - \lambda_k(S)| \leq \| E \|_2$ for $k = 1{:}r$. The theorem follows by noting that for any $x \in \mathbb{R}^r$ and $y \in \mathbb{R}^{n-r}$ we have

$$\left\| E \begin{bmatrix} x \\ y \end{bmatrix} \right\|_2 \leq \| E_1 x \|_2 + \| E_1^T Q_2 y \|_2 \leq \| E_1 \|_2 \| x \|_2 + \| E_1 \|_2 \| y \|_2$$

from which we readily conclude that $\| E \|_2 \leq \sqrt{2} \| E_1 \|_2$. $\square$

The eigenvalue bounds in Theorem 8.1.13 depend on $\| AQ_1 - Q_1 S \|_2$. Given A and Q_1, the following theorem indicates how to choose S so that this quantity is minimized in the Frobenius norm.

Theorem 8.1.14. *If* $A \in \mathbb{R}^{n \times n}$ *is symmetric and* $Q_1 \in \mathbb{R}^{n \times r}$ *has orthonormal columns, then*

$$\min_{S \in \mathbb{R}^{r \times r}} \| AQ_1 - Q_1 S \|_F = \| (I - Q_1 Q_1^T) A Q_1 \|_F$$

and $S = Q_1^T A Q_1$ *is the minimizer.*

Proof. Let $Q_2 \in \mathbb{R}^{n \times (n-r)}$ be such that $Q = [\ Q_1, \ Q_2\ ]$ is orthogonal. For any $S \in \mathbb{R}^{r \times r}$ we have

$$\| AQ_1 - Q_1 S \|_F^2 \ = \ \| Q^T AQ_1 - Q^T Q_1 S \|_F^2 \ = \ \| Q_1^T AQ_1 - S \|_F^2 + \| Q_2^T AQ_1 \|_F^2.$$

Clearly, the minimizing S is given by $S = Q_1^T AQ_1$. $\square$

This result enables us to associate any r-dimensional subspace $\mathsf{ran}(Q_1)$, with a set of r "optimal" eigenvalue-eigenvector approximates.

Theorem 8.1.15. *Suppose $A \in \mathbb{R}^{n \times n}$ is symmetric and that $Q_1 \in \mathbb{R}^{n \times r}$ satisfies $Q_1^T Q_1 = I_r$. If*

$$Z^T (Q_1^T AQ_1) Z \ = \ \mathrm{diag}(\theta_1, \ldots, \theta_r) \ = \ D$$

is the Schur decomposition of $Q_1^T AQ_1$ and $Q_1 Z = [\ y_1 \mid \cdots \mid y_r\ ]$, then

$$\| Ay_k - \theta_k y_k \|_2 \ = \ \| (I - Q_1 Q_1^T) AQ_1 Ze_k \|_2 \ \leq \ \| (I - Q_1 Q_1^T) AQ_1 \|_2$$

for $k = 1{:}r$.

Proof. It is easy to show that

$$Ay_k - \theta_k y_k \ = \ AQ_1 Ze_k - Q_1 ZDe_k \ = \ (AQ_1 - Q_1(Q_1^T AQ_1)) Ze_k.$$

The theorem follows by taking norms. $\square$

In Theorem 8.1.15, the θ_k are called *Ritz values*, the y_k are called *Ritz vectors*, and the (θ_k, y_k) are called *Ritz pairs*.

The usefulness of Theorem 8.1.13 is enhanced if we weaken the assumption that the columns of Q_1 are orthonormal. As can be expected, the bounds deteriorate with the loss of orthogonality.

Theorem 8.1.16. *Suppose $A \in \mathbb{R}^{n \times n}$ is symmetric and that*

$$AX_1 - X_1 S \ = \ F_1,$$

where $X_1 \in \mathbb{R}^{n \times r}$ and $S = X_1^T AX_1$. If

$$\| X_1^T X_1 - I_r \|_2 = \tau < 1, \tag{8.1.4}$$

then there exist $\mu_1, \ldots, \mu_r \in \lambda(A)$ such that

$$|\mu_k - \lambda_k(S)| \ \leq \ \sqrt{2}\, (\| F_1 \|_2 \ + \ \tau(2 + \tau) \| A \|_2)$$

for $k = 1{:}r$.

Proof. For any $Q \in \mathbb{R}^{n \times r}$ with orthonormal columns, define $E_1 \in \mathbb{R}^{n \times r}$ by

$$E_1 \ = \ AQ - QS.$$

It follows that

$$E_1 \ = \ A(Q - X_1) - (Q - X_1)S \ + \ F_1$$

and so

$$\| E_1 \|_2 \leq \| F_1 \|_2 + \| Q - X \|_2 \| A \|_2 \left(1 + \| X_1 \|_2^2 \right). \tag{8.1.5}$$

Note that

$$\| X_1 \|_2^2 = \| X_1^T X_1 \|_2 \leq \| X^T X_1 - I_r \|_2 + \| I_r \|_2 = 1 + \tau. \tag{8.1.6}$$

Let $U^T X_1 V = \Sigma = \mathrm{diag}(\sigma_1, \ldots, \sigma_r)$ be the thin SVD of X_1. It follows from (8.1.4) that

$$\| \Sigma^2 - I_r \|_2 = \tau$$

and thus $1 - \sigma_r^2 = \tau$. This implies

$$\| Q - X_1 \|_2 = \| U(I_r - \Sigma)V^T \|_2 = \| I_r - \Sigma \|_2 = 1 - \sigma_r \leq 1 - \sigma_r^2 = \tau. \tag{8.1.7}$$

The theorem is established by substituting (8.1.6) and (8.1.7) into (8.1.5) and using Theorem 8.1.13.   □

8.1.5   The Law of Inertia

The *inertia* of a symmetric matrix A is a triplet of nonnegative integers (m, z, p) where m, z, and p are respectively the numbers of negative, zero, and positive eigenvalues.

Theorem 8.1.17 (Sylvester Law of Inertia).   *If $A \in \mathbb{R}^{n \times n}$ is symmetric and $X \in \mathbb{R}^{n \times n}$ is nonsingular, then A and $X^T A X$ have the same inertia.*

Proof. Suppose for some r that $\lambda_r(A) > 0$ and define the subspace $S_0 \subseteq \mathbb{R}^n$ by

$$S_0 = \mathsf{span}\{X^{-1}q_1, \ldots, X^{-1}q_r\}, \qquad q_i \neq 0,$$

where $Aq_i = \lambda_i(A)q_i$ and $i = 1{:}r$. From the minimax characterization of $\lambda_r(X^T A X)$ we have

$$\lambda_r(X^T A X) = \max_{\dim(S)=r} \min_{y \in S} \frac{y^T (X^T A X)y}{y^T y} \geq \min_{y \in S_0} \frac{y^T (X^T A X)y}{y^T y}.$$

Since

$$y \in \mathbb{R}^n \implies \frac{y^T (X^T X)y}{y^T y} \geq \sigma_n(X)^2 \qquad y \in S_0 \implies \frac{y^T (X^T A X)y}{y^T (X^T X)y} \geq \lambda_r(A),$$

it follows that

$$\lambda_r(X^T A X) \geq \min_{y \in S_0} \left\{ \frac{y^T (X^T A X)y}{y^T (X^T X)y} \frac{y^T (X^T X)y}{y^T y} \right\} \geq \lambda_r(A)\sigma_n(X)^2.$$

An analogous argument with the roles of A and $X^T A X$ reversed shows that

$$\lambda_r(A) \geq \lambda_r(X^T A X)\sigma_n(X^{-1})^2 = \frac{\lambda_r(X^T A X)}{\sigma_1(X)^2}.$$

Thus, $\lambda_r(A)$ and $\lambda_r(X^T A X)$ have the same sign and so we have shown that A and $X^T A X$ have the same number of positive eigenvalues. If we apply this result to $-A$, we conclude that A and $X^T A X$ have the same number of negative eigenvalues. Obviously, the number of zero eigenvalues possessed by each matrix is also the same. □

A transformation of the form $A \to X^T A X$ where X is nonsingular is called a *conguence transformation*. Thus, a congruence transformation of a symmetric matrix preserves inertia.

Problems

P8.1.1 Without using any of the results in this section, show that the eigenvalues of a 2-by-2 symmetric matrix must be real.

P8.1.2 Compute the Schur decomposition of $A = \begin{bmatrix} 1 & 2 \\ 2 & 3 \end{bmatrix}$.

P8.1.3 Show that the eigenvalues of a Hermitian matrix ($A^H = A$) are real. For each theorem and corollary in this section, state and prove the corresponding result for Hermitian matrices. Which results have analogs when A is skew-symmetric? Hint: If $A^T = -A$, then iA is Hermitian.

P8.1.4 Show that if $X \in \mathbb{R}^{n \times r}$, $r \le n$, and $\| X^T X - I \|_2 = \tau < 1$, then $\sigma_{\min}(X) \ge 1 - \tau$.

P8.1.5 Suppose $A, E \in \mathbb{R}^{n \times n}$ are symmetric and consider the Schur decomposition $A + tE = QDQ^T$ where we *assume* that $Q = Q(t)$ and $D = D(t)$ are continuously differentiable functions of $t \in \mathbb{R}$. Show that $\dot{D}(t) = \text{diag}(Q(t)^T E Q(t))$ where the matrix on the right is the diagonal part of $Q(t)^T E Q(t)$. Establish the Wielandt-Hoffman theorem by integrating both sides of this equation from 0 to 1 and taking Frobenius norms to show that

$$\| D(1) - D(0) \|_F \le \int_0^1 \| \text{diag}(Q(t)^T E Q(t) \|_F \, dt \le \| E \|_F.$$

P8.1.6 Prove Theorem 8.1.5.

P8.1.7 Prove Theorem 8.1.7.

P8.1.8 Prove Theorem 8.1.8 using the fact that the trace of a square matrix is the sum of its eigenvalues.

P8.1.9 Show that if $B \in \mathbb{R}^{m \times m}$ and $C \in \mathbb{R}^{n \times n}$ are symmetric, then $\text{sep}(B, C) = \min \| BX - XC \|_F$ where the min is taken over all matrices $X \in \mathbb{R}^{m \times n}$.

P8.1.10 Prove the inequality (8.1.3).

P8.1.11 Suppose $A \in \mathbb{R}^{n \times n}$ is symmetric and $C \in \mathbb{R}^{n \times r}$ has full column rank and assume that $r \ll n$. By using Theorem 8.1.8 relate the eigenvalues of $A + CC^T$ to the eigenvalues of A.

P8.1.12 Give an algorithm for computing the solution to

$$\min_{\substack{\text{rank}(S) = 1 \\ S = S^T}} \| A - S \|_F.$$

Note that if $S \in \mathbb{R}^{n \times n}$ is a symmetric rank-1 matrix then either $S = vv^T$ or $S = -vv^T$ for some $v \in \mathbb{R}^n$.

P8.1.13 Give an algorithm for computing the solution to

$$\min_{\substack{\text{rank}(S) = 2 \\ S = -S^T}} \| A - S \|_F.$$

P8.1.14 Give an example of a real 3-by-3 normal matrix with integer entries that is neither orthogonal, symmetric, nor skew-symmetric.

Notes and References for §8.1

The perturbation theory for the symmetric eigenproblem is surveyed in Wilkinson (AEP, Chap. 2), Parlett (SEP, Chaps. 10 and 11), and Stewart and Sun (MPT, Chaps. 4 and 5). Some representative papers in this well-researched area include:

G.W. Stewart (1973). "Error and Perturbation Bounds for Subspaces Associated with Certain Eigenvalue Problems," *SIAM Review 15*, 727–764.
C.C. Paige (1974). "Eigenvalues of Perturbed Hermitian Matrices," *Lin. Alg. Applic. 8*, 1–10.
W. Kahan (1975). "Spectra of Nearly Hermitian Matrices," *Proc. AMS 48*, 11–17.
A. Schonhage (1979). "Arbitrary Perturbations of Hermitian Matrices," *Lin. Alg. Applic. 24*, 143–49.
D.S. Scott (1985). "On the Accuracy of the Gershgorin Circle Theorem for Bounding the Spread of a Real Symmetric Matrix," *Lin. Alg. Applic. 65*, 147–155
J.-G. Sun (1995). "A Note on Backward Error Perturbations for the Hermitian Eigenvalue Problem," *BIT 35*, 385–393.
Z. Drmač (1996). On Relative Residual Bounds for the Eigenvalues of a Hermitian Matrix," *Lin. Alg. Applic. 244*, 155-163.
Z. Drmač and V. Hari (1997). "Relative Residual Bounds For The Eigenvalues of a Hermitian Semidefinite Matrix," *SIAM J. Matrix Anal. Applic. 18*, 21–29.
R.-C. Li (1998). "Relative Perturbation Theory: I. Eigenvalue and Singular Value Variations," *SIAM J. Matrix Anal. Applic. 19*, 956–982.
R.-C. Li (1998). "Relative Perturbation Theory: II. Eigenspace and Singular Subspace Variations," *SIAM J. Matrix Anal. Applic. 20*, 471–492.
F.M. Dopico, J. Moro and J.M. Molera (2000). "Weyl-Type Relative Perturbation Bounds for Eigensystems of Hermitian Matrices," *Lin. Alg. Applic. 309*, 3–18.
J.L. Barlow and I. Slapničar (2000). "Optimal Perturbation Bounds for the Hermitian Eigenvalue Problem," *Lin. Alg. Applic. 309*, 19–43.
N. Truhar and R.-C. Li (2003). "A sin(2θ) Theorem for Graded Indefinite Hermitian Matrices," *Lin. Alg. Applic. 359*, 263–276.
W. Li and W. Sun (2004). "The Perturbation Bounds for Eigenvalues of Normal Matrices," *Num. Lin. Alg. 12*, 89–94.
C.-K. Li and R.-C. Li (2005). "A Note on Eigenvalues of Perturbed Hermitian Matrices," *Lin. Alg. Applic. 395*, 183–190.
N. Truhar (2006). "Relative Residual Bounds for Eigenvalues of Hermitian Matrices," *SIAM J. Matrix Anal. Applic. 28*, 949–960.

An elementary proof of the Wielandt-Hoffman theorem is given in:

P. Lax (1997). *Linear Algebra*, Wiley-Interscience, New York.

For connections to optimization and differential equations, see:

P. Deift, T. Nanda, and C. Tomei (1983). "Ordinary Differential Equations and the Symmetric Eigenvalue Problem," *SIAM J. Numer. Anal. 20*, 1–22.
M.L. Overton (1988). "Minimizing the Maximum Eigenvalue of a Symmetric Matrix," *SIAM J. Matrix Anal. Applic. 9*, 256-268.
T. Kollo and H. Neudecker (1997). "The Derivative of an Orthogonal Matrix of Eigenvectors of a Symmetric Matrix," *Lin. Alg. Applic. 264*, 489–493.

8.2 Power Iterations

Assume that $A \in \mathbb{R}^{n \times n}$ is symmetric and that $U_0 \in \mathbb{R}^{n \times n}$ is orthogonal. Consider the following *QR iteration*:

$$T_0 = U_0^T A U_0$$

$$\textbf{for } k = 1, 2, \ldots$$

$$T_{k-1} = U_k R_k \quad \text{(QR factorization)} \tag{8.2.1}$$

$$T_k = R_k U_k$$

$$\textbf{end}$$

Since $T_k = R_k U_k = U_k^T (U_k R_k) U_k = U_k^T T_{k-1} U_k$ it follows by induction that

$$T_k = (U_0 U_1 \cdots U_k)^T A (U_0 U_1 \cdots U_k). \qquad (8.2.2)$$

Thus, each T_k is orthogonally similar to A. Moreover, the T_k almost always converge to diagonal form and so it can be said that (8.2.1) almost always converges to a Schur decomposition of A. In order to establish this remarkable result we first consider the power method and the method of orthogonal iteration.

8.2.1 The Power Method

Given a unit 2-norm $q^{(0)} \in \mathbb{R}^n$, the *power method* produces a sequence of vectors $q^{(k)}$ as follows:

> **for** $k = 1, 2, \ldots$
> $\qquad z^{(k)} = A q^{(k-1)}$
> $\qquad q^{(k)} = z^{(k)} / \| z^{(k)} \|_2$ $\qquad\qquad\qquad\qquad$ (8.2.3)
> $\qquad \lambda^{(k)} = [q^{(k)}]^T A q^{(k)}$
> **end**

If $q^{(0)}$ is not "deficient" and A's eigenvalue of maximum modulus is unique, then the $q^{(k)}$ converge to an eigenvector.

Theorem 8.2.1. *Suppose $A \in \mathbb{R}^{n \times n}$ is symmetric and that*

$$Q^T A Q = \text{diag}(\lambda_1, \ldots, \lambda_n)$$

where $Q = [\, q_1 \,|\, \cdots \,|\, q_n \,]$ is orthogonal and $|\lambda_1| > |\lambda_2| \geq \cdots \geq |\lambda_n|$. Let the vectors $q^{(k)}$ be specified by (8.2.3) and define $\theta_k \in [0, \pi/2]$ by

$$\cos(\theta_k) = \left| q_1^T q^{(k)} \right|.$$

If $\cos(\theta_0) \neq 0$, then for $k = 0, 1, \ldots$ we have

$$|\sin(\theta_k)| \leq \tan(\theta_0) \left| \frac{\lambda_2}{\lambda_1} \right|^k, \qquad (8.2.4)$$

$$|\lambda^{(k)} - \lambda_1| \leq \max_{2 \leq i \leq n} |\lambda_1 - \lambda_i| \tan(\theta_0)^2 \left| \frac{\lambda_2}{\lambda_1} \right|^{2k}. \qquad (8.2.5)$$

Proof. From the definition of the iteration, it follows that $q^{(k)}$ is a multiple of $A^k q^{(0)}$ and so

$$|\sin(\theta_k)|^2 = 1 - \left(q_1^T q^{(k)} \right)^2 = 1 - \left(\frac{q_1^T A^k q^{(0)}}{\| A^k q^{(0)} \|_2} \right)^2.$$

If $q^{(0)}$ has the eigenvector expansion $q^{(0)} = a_1 q_1 + \cdots + a_n q_n$, then

$$|a_1| = |q_1^T q^{(0)}| = \cos(\theta_0) \neq 0,$$

$$a_1^2 + \cdots + a_n^2 = 1,$$

and

$$A^k q^{(0)} = a_1 \lambda_1^k q_1 + a_2 \lambda_2^k q_2 + \cdots + a_n \lambda_n^k q_n.$$

Thus,

$$|\sin(\theta_k)|^2 = 1 - \frac{a_1^2 \lambda_1^{2k}}{\displaystyle\sum_{i=1}^{n} a_i^2 \lambda_i^{2k}} = \frac{\displaystyle\sum_{i=2}^{n} a_i^2 \lambda_i^{2k}}{\displaystyle\sum_{i=1}^{n} a_i^2 \lambda_i^{2k}} \leq \frac{\displaystyle\sum_{i=2}^{n} a_i^2 \lambda_i^{2k}}{a_1^2 \lambda_1^{2k}}$$

$$= \frac{1}{a_1^2} \sum_{i=2}^{n} a_i^2 \left(\frac{\lambda_i}{\lambda_1}\right)^{2k} \leq \frac{1}{a_1^2} \left(\sum_{i=2}^{n} a_i^2\right) \left(\frac{\lambda_2}{\lambda_1}\right)^{2k}$$

$$= \frac{1 - a_1^2}{a_1^2} \left(\frac{\lambda_2}{\lambda_1}\right)^{2k} = \tan(\theta_0)^2 \left(\frac{\lambda_2}{\lambda_1}\right)^{2k}.$$

This proves (8.2.4). Likewise,

$$\lambda^{(k)} = \left[q^{(k)}\right]^T A q^{(k)} = \frac{\left[q^{(0)}\right]^T A^{2k+1} q^{(0)}}{\left[q^{(0)}\right]^T A^{2k} q^{(0)}} = \frac{\displaystyle\sum_{i=1}^{n} a_i^2 \lambda_i^{2k+1}}{\displaystyle\sum_{i=1}^{n} a_i^2 \lambda_i^{2k}}$$

and so

$$\left|\lambda^{(k)} - \lambda_1\right| = \left|\frac{\displaystyle\sum_{i=2}^{n} a_i^2 \lambda_i^{2k} (\lambda_i - \lambda_1)}{\displaystyle\sum_{i=1}^{n} a_i^2 \lambda_i^{2k}}\right| \leq \max_{2 \leq i \leq n} |\lambda_1 - \lambda_i| \cdot \frac{1}{a_1^2} \cdot \sum_{i=2}^{n} a_i^2 \left(\frac{\lambda_i}{\lambda_1}\right)^{2k}$$

$$\leq \max_{2 \leq i \leq n} |\lambda_1 - \lambda_n| \cdot \tan(\theta_0)^2 \cdot \left(\frac{\lambda_2}{\lambda_1}\right)^{2k},$$

completing the proof of the theorem.   □

Computable error bounds for the power method can be obtained by using Theorem 8.1.13. If

$$\| A q^{(k)} - \lambda^{(k)} q^{(k)} \|_2 = \delta,$$

then there exists $\lambda \in \lambda(A)$ such that $|\lambda^{(k)} - \lambda| \leq \sqrt{2}\, \delta$.

8.2.2 Inverse Iteration

If the power method (8.2.3) is applied with A replaced by $(A - \lambda I)^{-1}$, then we obtain the method of *inverse iteration*. If λ is very close to a distinct eigenvalue of A, then $q^{(k)}$ will be much richer in the corresponding eigenvector direction than its predecessor $q^{(k-1)}$:

$$\left. \begin{array}{l} x = \displaystyle\sum_{i=1}^{n} a_i q_i \\[2mm] A q_i = \lambda_i q_i, \ i = 1{:}n \end{array} \right\} \quad \Rightarrow \quad (A - \lambda I)^{-1} x = \sum_{i=1}^{n} \frac{a_i}{\lambda_i - \lambda} q_i.$$

Thus, if λ is reasonably close to a well-separated eigenvalue λ_j, then inverse iteration will produce iterates that are increasingly in the direction of q_j. Note that inverse iteration requires at each step the solution of a linear system with matrix of coefficients $A - \lambda I$.

8.2.3 Rayleigh Quotient Iteration

Suppose $A \in \mathbb{R}^{n \times n}$ is symmetric and that x is a given nonzero n-vector. A simple differentiation reveals that

$$\lambda = r(x) \equiv \frac{x^T A x}{x^T x}$$

minimizes $\| (A - \lambda I) x \|_2$. (See also Theorem 8.1.14.) The scalar $r(x)$ is called the *Rayleigh quotient* of x. Clearly, if x is an approximate eigenvector, then $r(x)$ is a reasonable choice for the corresponding eigenvalue. Combining this idea with inverse iteration gives rise to the *Rayleigh quotient iteration* where $x_0 \neq 0$ is given.

for $k = 0, 1, \ldots$

$\qquad \mu_k = r(x_k)$ 　　　　　　　　　　　　　　　　　　　　(8.2.6)

$\qquad$ Solve $(A - \mu_k I) z_{k+1} = x_k$ for z_{k+1}

$\qquad x_{k+1} = z_{k+1} / \| z_{k+1} \|_2$

end

The Rayleigh quotient iteration almost always converges and when it does, the rate of convergence is cubic. We demonstrate this for the case $n = 2$. Without loss of generality, we may assume that $A = \mathrm{diag}(\lambda_1, \lambda_2)$, with $\lambda_1 > \lambda_2$. Denoting x_k by

$$x_k = \begin{bmatrix} c_k \\ s_k \end{bmatrix}, \qquad c_k^2 + s_k^2 = 1,$$

it follows that $\mu_k = \lambda_1 c_k^2 + \lambda_2 s_k^2$ in (8.2.6) and

$$z_{k+1} = \frac{1}{\lambda_1 - \lambda_2} \begin{bmatrix} c_k / s_k^2 \\ -s_k / c_k^2 \end{bmatrix}.$$

A calculation shows that

$$c_{k+1} = \frac{|c_k|^3}{\sqrt{c_k^6 + s_k^6}}, \qquad s_{k+1} = \frac{|s_k|^3}{\sqrt{c_k^6 + s_k^6}}. \tag{8.2.7}$$

From these equations it is clear that the x_k converge cubically to either span$\{e_1\}$ or span$\{e_2\}$ provided $|c_k| \neq |s_k|$. Details associated with the practical implementation of the Rayleigh quotient iteration may be found in Parlett (1974).

8.2.4   Orthogonal Iteration

A straightforward generalization of the power method can be used to compute higher-dimensional invariant subspaces. Let r be a chosen integer that satisfies $1 \leq r \leq n$. Given an n-by-r matrix Q_0 with orthonormal columns, the method of *orthogonal iteration* generates a sequence of matrices $\{Q_k\} \subseteq \mathbb{R}^{n \times r}$ as follows:

> **for** $k = 1, 2, \ldots$
>
> $\qquad Z_k = AQ_{k-1}$ $\hspace{4cm}$ (8.2.8)
>
> $\qquad Q_k R_k = Z_k$ $\qquad$ (QR factorization)
>
> **end**

Note that, if $r = 1$, then this is just the power method. Moreover, the sequence $\{Q_k e_1\}$ is precisely the sequence of vectors produced by the power iteration with starting vector $q^{(0)} = Q_0 e_1$.

In order to analyze the behavior of (8.2.8), assume that

$$Q^T A Q \;=\; D \;=\; \mathrm{diag}(\lambda_i), \qquad |\lambda_1| \geq |\lambda_2| \geq \cdots \geq |\lambda_n| \tag{8.2.9}$$

is a Schur decomposition of $A \in \mathbb{R}^{n \times n}$. Partition Q and D as follows:

$$Q \;=\; [\; Q_\alpha \mid Q_\beta \;]_{\substack{\\ r \quad n-r}}, \qquad\qquad D \;=\; \begin{bmatrix} D_1 & 0 \\ 0 & D_2 \end{bmatrix} \begin{matrix} r \\ n-r \end{matrix} . \tag{8.2.10}$$

$$ \substack{r \quad n-r}$$

If $|\lambda_r| > |\lambda_{r+1}|$, then

$$D_r(A) \;=\; \mathsf{ran}(Q_\alpha)$$

is the *dominant* invariant subspace of dimension r. It is the unique invariant subspace associated with the eigenvalues $\lambda_1, \ldots, \lambda_r$.

The following theorem shows that with reasonable assumptions, the subspaces $\mathsf{ran}(Q_k)$ generated by (8.2.8) converge to $D_r(A)$ at a rate proportional to $|\lambda_{r+1}/\lambda_r|^k$.

Theorem 8.2.2. *Let the Schur decomposition of $A \in \mathbb{R}^{n \times n}$ be given by (8.2.9) and (8.2.10) with $n \geq 2$. Assume $|\lambda_r| > |\lambda_{r+1}|$ and that d_k is defined by*

$$d_k \;=\; \mathsf{dist}(D_r(A), \mathsf{ran}(Q_k)), \qquad k \geq 0.$$

If

$$d_0 < 1, \tag{8.2.11}$$

then the matrices Q_k generated by (8.2.8) satisfy

$$d_k \;\leq\; \left| \frac{\lambda_{r+1}}{\lambda_r} \right|^k \frac{d_0}{\sqrt{1 - d_0^2}}. \tag{8.2.12}$$

Compare with Theorem 7.3.1.

Proof. We mention at the start that the condition (8.2.11) means that no vector in the span of Q_0's columns is perpendicular to $D_r(A)$.

Using induction it can be shown that the matrix Q_k in (8.2.8) satisfies

$$A^k Q_0 = Q_k (R_k \cdots R_1).$$

This is a QR factorization of $A^k Q_0$ and upon substitution of the Schur decomposition (8.2.9)-(8.2.10) we obtain

$$\begin{bmatrix} D_1^k & 0 \\ 0 & D_2^k \end{bmatrix} \begin{bmatrix} Q_\alpha^T Q_0 \\ Q_\beta^T Q_0 \end{bmatrix} = \begin{bmatrix} Q_\alpha^T Q_k \\ Q_\beta^T Q_k \end{bmatrix} (R_k \cdots R_1).$$

If the matrices V_k and W_k are defined by

$$V_k = Q_\alpha^T Q_0,$$
$$W_k = Q_\beta^T Q_0,$$

then

$$D_1^k V_0 = V_k (R_k \cdots R_1), \tag{8.2.13}$$
$$D_2^k W_0 = W_k (R_k \cdots R_1). \tag{8.2.14}$$

Since

$$\begin{bmatrix} V_k \\ W_k \end{bmatrix} = \begin{bmatrix} Q_\alpha^T Q_k \\ Q_\beta^T Q_k \end{bmatrix} = [Q_\alpha \mid Q_\beta]^T Q_k = Q^T Q_k,$$

it follows from the thin CS decomposition (Theorem 2.5.2) that

$$1 = \sigma_{\min}(V_k)^2 + \sigma_{\max}(W_k)^2 = \sigma_{\min}(V_k)^2 + d_k^2.$$

A consequence of this is that

$$\sigma_{\min}(V_0)^2 = 1 - \sigma_{\max}(W_0)^2 = 1 - d_0^2 > 0.$$

It follows from (8.2.13) that the matrices V_k and $(R_k \cdots R_1)$ are nonsingular. Using both that equation and (8.2.14) we obtain

$$W_k = D_2^k W_0 (R_k \cdots R_1)^{-1} = D_2^k W_0 (D_1^k V_0)^{-1} V_k = D_2^k (W_0 V_0^{-1}) D_1^{-k} V_k$$

and so

$$d_k = \| W_k \|_2 \leq \| D_2^k \|_2 \cdot \| W_0 \|_2 \cdot \| V_0^{-1} \|_2 \cdot \| D_1^{-k} \|_2 \cdot \| V_k \|_2$$

$$\leq |\lambda_{r+1}|^k \cdot d_0 \cdot \frac{1}{1 - d_0^2} \cdot \frac{1}{|\lambda_r|^k},$$

from which the theorem follows. $\square$

8.2.5   The QR Iteration

Consider what happens if we apply the method of orthogonal iteration (8.2.8) with $r = n$. Let $Q^T A Q = \text{diag}(\lambda_1, \ldots, \lambda_n)$ be the Schur decomposition and assume

$$|\lambda_1| > |\lambda_2| > \cdots > |\lambda_n|.$$

If $Q = [\, q_1 \mid \cdots \mid q_n \,]$, $Q_k = [\, q_1^{(k)} \mid \cdots \mid q_n^{(k)} \,]$, and

$$\text{dist}(D_i(A), \text{span}\{q_1^{(0)}, \ldots, q_i^{(0)}\}) \; < \; 1 \qquad (8.2.15)$$

for $i = 1{:}n - 1$, then it follows from Theorem 8.2.2 that

$$\text{dist}(\text{span}\{q_1^{(k)}, \ldots, q_i^{(k)}\}, \text{span}\{q_1, \ldots, q_i\}) \; = \; O\left(\left|\frac{\lambda_{i+1}}{\lambda_i}\right|^k\right)$$

for $i = 1{:}n - 1$. This implies that the matrices T_k defined by

$$T_k \; = \; Q_k^T A Q_k$$

are converging to diagonal form. Thus, it can be said that the method of orthogonal iteration computes a Schur decomposition if $r = n$ and the original iterate $Q_0 \in \mathbb{R}^{n \times n}$ is not deficient in the sense of (8.2.11).

The QR iteration arises by considering how to compute the matrix T_k directly from its predecessor T_{k-1}. On the one hand, we have from (8.2.8) and the definition of T_{k-1} that

$$T_{k-1} = Q_{k-1}^T A Q_{k-1} = Q_{k-1}^T (A Q_{k-1}) = (Q_{k-1}^T Q_k) R_k.$$

On the other hand,

$$T_k = Q_k^T A Q_k = (Q_k^T A Q_{k-1})(Q_{k-1}^T Q_k) = R_k (Q_{k-1}^T Q_k).$$

Thus, T_k is determined by computing the QR factorization of T_{k-1} and then multiplying the factors together in reverse order. This is precisely what is done in (8.2.1).

Note that a single QR iteration involves $O(n^3)$ flops. Moreover, since convergence is only linear (when it exists), it is clear that the method is a prohibitively expensive way to compute Schur decompositions. Fortunately, these practical difficulties can be overcome, as we show in the next section.

Problems

P8.2.1   Suppose $A_0 \in \mathbb{R}^{n \times n}$ is symmetric and positive definite and consider the following iteration:

> **for** $k = 1, 2, \ldots$
> $\qquad A_{k-1} = G_k G_k^T \qquad$ (Cholesky factorization)
> $\qquad A_k = G_k^T G_k$
> **end**

(a) Show that this iteration is defined. (b) Show that if

$$A_0 = \begin{bmatrix} a & b \\ b & c \end{bmatrix}$$

with $a \geq c$ has eigenvalues $\lambda_1 \geq \lambda_2 > 0$, then the A_k converge to $\text{diag}(\lambda_1, \lambda_2)$.

P8.2.2 Prove (8.2.7).

P8.2.3 Suppose $A \in \mathbb{R}^{n \times n}$ is symmetric and define the function $f{:}\mathbb{R}^{n+1} \to \mathbb{R}^{n+1}$ by

$$f\left(\left[\begin{array}{c} x \\ \lambda \end{array} \right] \right) = \left[\begin{array}{c} Ax - \lambda x \\ (x^T x - 1)/2 \end{array} \right]$$

where $x \in \mathbb{R}^n$ and $\lambda \in \mathbb{R}$. Suppose x_+ and λ_+ are produced by applying Newton's method to f at the "current point" defined by x_c and λ_c. Give expressions for x_+ and λ_+ assuming that $\| x_c \|_2 = 1$ and $\lambda_c = x_c^T A x_c$.

Notes and References for §8.2

The following references are concerned with the method of orthogonal iteration, which is also known as the method of simultaneous iteration:

G.W. Stewart (1969). "Accelerating The Orthogonal Iteration for the Eigenvalues of a Hermitian Matrix," *Numer. Math. 13*, 362–376.

M. Clint and A. Jennings (1970). "The Evaluation of Eigenvalues and Eigenvectors of Real Symmetric Matrices by Simultaneous Iteration," *Comput. J. 13*, 76–80.

H. Rutishauser (1970). "Simultaneous Iteration Method for Symmetric Matrices," *Numer. Math. 16*, 205–223.

References for the Rayleigh quotient method include:

J. Vandergraft (1971). "Generalized Rayleigh Methods with Applications to Finding Eigenvalues of Large Matrices," *Lin. Alg. Applic. 4*, 353–368.

B.N. Parlett (1974). "The Rayleigh Quotient Iteration and Some Generalizations for Nonnormal Matrices," *Math. Comput. 28*, 679-693.

S. Batterson and J. Smillie (1989). "The Dynamics of Rayleigh Quotient Iteration," *SIAM J. Numer. Anal. 26*, 624–636.

C. Beattie and D.W. Fox (1989). "Localization Criteria and Containment for Rayleigh Quotient Iteration," *SIAM J. Matrix Anal. Applic. 10*, 80–93.

P.T.P. Tang (1994). "Dynamic Condition Estimation and Rayleigh-Ritz Approximation," *SIAM J. Matrix Anal. Applic. 15*, 331–346.

D. P. O'Leary and G. W. Stewart (1998). "On the Convergence of a New Rayleigh Quotient Method with Applications to Large Eigenproblems," *ETNA 7*, 182–189.

J.-L. Fattebert (1998). "A Block Rayleigh Quotient Iteration with Local Quadratic Convergence," *ETNA 7*, 56–74.

Z. Jia and G.W. Stewart (2001). "An Analysis of the Rayleigh-Ritz Method for Approximating Eigenspaces," *Math. Comput. 70*, 637–647.

V. Simoncini and L. Eldén (2002). "Inexact Rayleigh Quotient-Type Methods for Eigenvalue Computations," *BIT 42*, 159–182.

P.A. Absil, R. Mahony, R. Sepulchre, and P. Van Dooren (2002). "A Grassmann-Rayleigh Quotient Iteration for Computing Invariant Subspaces," *SIAM Review 44*, 57–73.

Y. Notay (2003). "Convergence Analysis of Inexact Rayleigh Quotient Iteration," *SIAM J. Matrix Anal. Applic. 24*, 627–644.

A. Dax (2003). "The Orthogonal Rayleigh Quotient Iteration (ORQI) method," *Lin. Alg. Applic. 358*, 23–43.

R.-C. Li (2004). "Accuracy of Computed Eigenvectors Via Optimizing a Rayleigh Quotient," *BIT 44*, 585–593.

Various Newton-type methods have also been derived for the symmetric eigenvalue problem, see:

R.A. Tapia and D.L. Whitley (1988). "The Projected Newton Method Has Order $1 + \sqrt{2}$ for the Symmetric Eigenvalue Problem," *SIAM J. Numer. Anal. 25*, 1376–1382.

P.A. Absil, R. Sepulchre, P. Van Dooren, and R. Mahony (2004). "Cubically Convergent Iterations for Invariant Subspace Computation," *SIAM J. Matrix Anal. Applic. 26*, 70–96.

8.3   The Symmetric QR Algorithm

The symmetric QR iteration (8.2.1) can be made more efficient in two ways. First, we show how to compute an orthogonal U_0 such that $U_0^T A U_0 = T$ is tridiagonal. With this reduction, the iterates produced by (8.2.1) are all tridiagonal and this reduces the work per step to $O(n^2)$. Second, the idea of shifts are introduced and with this change the convergence to diagonal form proceeds at a cubic rate. This is far better than having the off-diagonal entries going to to zero as $|\lambda_{i+1}/\lambda_i|^k$ as discussed in §8.2.5.

8.3.1   Reduction to Tridiagonal Form

If A is symmetric, then it is possible to find an orthogonal Q such that

$$Q^T A Q = T \tag{8.3.1}$$

is tridiagonal. We call this the *tridiagonal decomposition* and as a compression of data, it represents a very big step toward diagonalization.

We show how to compute (8.3.1) with Householder matrices. Suppose that Householder matrices $P_1, \ldots, P_{k-1}$ have been determined such that if

$$A_{k-1} = (P_1 \cdots P_{k-1})^T A (P_1 \cdots P_{k-1}),$$

then

$$A_{k-1} = \begin{bmatrix} B_{11} & B_{12} & 0 \\ B_{21} & B_{22} & B_{23} \\ 0 & B_{32} & B_{33} \end{bmatrix} \begin{matrix} k-1 \\ 1 \\ n-k \end{matrix}$$
$$\quad\quad\quad k-1 \quad\; 1 \quad\; n-k$$

is tridiagonal through its first $k - 1$ columns. If $\tilde{P}_k$ is an order-$(n - k)$ Householder matrix such that $\tilde{P}_k B_{32}$ is a multiple of $I_{n-k}(:, 1)$ and if $P_k = \text{diag}(I_k, \tilde{P}_k)$, then the leading k-by-k principal submatrix of

$$A_k = P_k A_{k-1} P_k = \begin{bmatrix} B_{11} & B_{12} & 0 \\ B_{21} & B_{22} & B_{23}\tilde{P}_k \\ 0 & \tilde{P}_k B_{32} & \tilde{P}_k B_{33}\tilde{P}_k \end{bmatrix} \begin{matrix} k-1 \\ 1 \\ n-k \end{matrix}$$
$$\quad\quad\quad\quad\quad\quad\quad\quad k-1 \quad\quad 1 \quad\quad n-k$$

is tridiagonal. Clearly, if $U_0 = P_1 \cdots P_{n-2}$, then $U_0^T A U_0 = T$ is tridiagonal.

In the calculation of A_k it is important to exploit symmetry during the formation of the matrix $\tilde{P}_k B_{33} \tilde{P}_k$. To be specific, suppose that $\tilde{P}_k$ has the form

$$\tilde{P}_k = I - \beta v v^T, \qquad \beta = 2/v^T v, \quad 0 \neq v \in \mathbb{R}^{n-k}.$$

Note that if $p = \beta B_{33} v$ and $w = p - (\beta p^T v / 2) v$, then

$$\tilde{P}_k B_{33} \tilde{P}_k = B_{33} - v w^T - w v^T.$$

Since only the upper triangular portion of this matrix needs to be calculated, we see that the transition from A_{k-1} to A_k can be accomplished in only $4(n - k)^2$ flops.

Algorithm 8.3.1 (Householder Tridiagonalization) Given a symmetric $A \in \mathbb{R}^{n \times n}$, the following algorithm overwrites A with $T = Q^T A Q$, where T is tridiagonal and $Q = H_1 \cdots H_{n-2}$ is the product of Householder transformations.

> **for** $k = 1{:}n-2$
> $\qquad [v, \beta] = \mathsf{house}(A(k+1{:}n, k))$
> $\qquad p = \beta A(k+1{:}n, k+1{:}n) v$
> $\qquad w = p - (\beta p^T v / 2) v$
> $\qquad A(k+1, k) = \| A(k+1{:}n, k) \|_2; \; A(k, k+1) = A(k+1, k)$
> $\qquad A(k+1{:}n, k+1{:}n) = A(k+1{:}n, k+1{:}n) - vw^T - wv^T$
> **end**

This algorithm requires $4n^3/3$ flops when symmetry is exploited in calculating the rank-2 update. The matrix Q can be stored in factored form in the subdiagonal portion of A. If Q is explicitly required, then it can be formed with an additional $4n^3/3$ flops. Note that if T has a zero subdiagonal, then the eigenproblem splits into a pair of smaller eigenproblems. In particular, if $t_{k+1,k} = 0$, then

$$\lambda(T) = \lambda(T(1{:}k, 1{:}k)) \cup \lambda(T(k+1{:}n, k+1{:}n)).$$

If T has no zero subdiagonal entries, then it is said to be *unreduced*.

Let $\hat{T}$ denote the computed version of T obtained by Algorithm 8.3.1. It can be shown that $\hat{T} = \tilde{Q}^T (A + E) \tilde{Q}$ where $\tilde{Q}$ is exactly orthogonal and E is a symmetric matrix satisfying $\| E \|_F \leq c\mathbf{u}\| A \|_F$ where c is a small constant. See Wilkinson (AEP, p. 297).

8.3.2 Properties of the Tridiagonal Decomposition

We prove two theorems about the tridiagonal decomposition both of which have key roles to play in the following. The first connects (8.3.1) to the QR factorization of a certain *Krylov matrix*. These matrices have the form

$$K(A, v, k) = \left[v \mid Av \mid \cdots \mid A^{k-1} v \right], \qquad A \in \mathbb{R}^{n \times n}, \; v \in \mathbb{R}^n.$$

Theorem 8.3.1. *If $Q^T A Q = T$ is the tridiagonal decomposition of the symmetric matrix $A \in \mathbb{R}^{n \times n}$, then $Q^T K(A, Q(:, 1), n) = R$ is upper triangular. If R is nonsingular, then T is unreduced. If R is singular and k is the smallest index so $r_{kk} = 0$, then k is also the smallest index so $t_{k,k-1}$ is zero. Compare with Theorem 7.4.3.*

Proof. It is clear that if $q_1 = Q(:, 1)$, then

$$Q^T K(A, Q(:, 1), n) = \left[Q^T q_1 \mid (Q^T A Q)(Q^T q_1) \mid \cdots \mid (Q^T A Q)^{n-1}(Q^T q_1) \right]$$

$$= \left[e_1 \mid T e_1 \mid \cdots \mid T^{n-1} e_1 \right] = R$$

is upper triangular with the property that $r_{11} = 1$ and $r_{ii} = t_{21} t_{32} \cdots t_{i,i-1}$ for $i = 2{:}n$. Clearly, if R is nonsingular, then T is unreduced. If R is singular and r_{kk} is its first zero diagonal entry, then $k \geq 2$ and $t_{k,k-1}$ is the first zero subdiagonal entry. $\quad\square$

The next result shows that Q is essentially unique once $Q(:,1)$ is specified.

Theorem 8.3.2 (Implicit Q Theorem). *Suppose* $Q = [\, q_1 \mid \cdots \mid q_n \,]$ *and* $V = [\, v_1 \mid \cdots \mid v_n \,]$ *are orthogonal matrices with the property that both* $Q^T A Q = T$ *and* $V^T A V = S$ *are tridiagonal where* $A \in \mathbb{R}^{n \times n}$ *is symmetric. Let* k *denote the smallest positive integer for which* $t_{k+1,k} = 0$, *with the convention that* $k = n$ *if* T *is unreduced. If* $v_1 = q_1$, *then* $v_i = \pm q_i$ *and* $|t_{i,i-1}| = |s_{i,i-1}|$ *for* $i = 2{:}k$. *Moreover, if* $k < n$, *then* $s_{k+1,k} = 0$. *Compare with Theorem 7.4.2.*

Proof. Define the orthogonal matrix $W = Q^T V$ and observe that $W(:,1) = I_n(:,1) = e_1$ and $W^T T W = S$. By Theorem 8.3.1, $W^T \cdot K(T, e_1, k)$ is upper triangular with full column rank. But $K(T, e_1, k)$ is upper triangular and so by the essential uniqueness of the thin QR factorization, $W(:,1{:}k) = I_n(:,1{:}k) \cdot \text{diag}(\pm 1, \ldots, \pm 1)$. This says that $Q(:,i) = \pm V(:,i)$ for $i = 1{:}k$. The comments about the subdiagonal entries follow since $t_{i+1,i} = Q(:,i+1)^T A Q(:,i)$ and $s_{i+1,i} = V(:,i+1)^T A V(:,i)$ for $i = 1{:}n-1$.   □

8.3.3   The QR Iteration and Tridiagonal Matrices

We quickly state four facts that pertain to the QR iteration and tridiagonal matrices. Complete verifications are straightforward.

- *Preservation of Form.* If $T = QR$ is the QR factorization of a symmetric tridiagonal matrix $T \in \mathbb{R}^{n \times n}$, then Q has lower bandwidth 1 and R has upper bandwidth 2 and it follows that $T_+ = RQ = Q^T(QR)Q = Q^T T Q$ is also symmetric and tridiagonal.

- *Shifts.* If $s \in \mathbb{R}$ and $T - sI = QR$ is the QR factorization, then $T_+ = RQ + sI = Q^T T Q$ is also tridiagonal. This is called a *shifted* QR step.

- *Perfect Shifts.* If T is unreduced, then the first $n-1$ columns of $T - sI$ are independent regardless of s. Thus, if $s \in \lambda(T)$ and $QR = T - sI$ is a QR factorization, then $r_{nn} = 0$ and the last column of $T_+ = RQ + sI$ equals $sI_n(:,n) = se_n$.

- *Cost.* If $T \in \mathbb{R}^{n \times n}$ is tridiagonal, then its QR factorization can be computed by applying a sequence of $n-1$ Givens rotations:

$$
\begin{aligned}
&\textbf{for } k = 1{:}n-1 \\
&\qquad [c, s] = \mathsf{givens}(t_{kk}, t_{k+1,k}) \\
&\qquad m = \min\{k+2, n\} \\
&\qquad T(k{:}k+1, k{:}m) = \begin{bmatrix} c & s \\ -s & c \end{bmatrix}^T T(k{:}k+1, k{:}m) \\
&\textbf{end}
\end{aligned}
$$

This requires $O(n)$ flops. If the rotations are accumulated, then $O(n^2)$ flops are needed.

8.3.4   Explicit Single-Shift QR Iteration

If s is a good approximate eigenvalue, then we suspect that the $(n, n-1)$ will be small after a QR step with shift s. This is the philosophy behind the following iteration:

$$T = U_0^T A U_0 \qquad \text{(tridiagonal)}$$
$$\textbf{for } k = 0, 1, \ldots$$
$$\qquad \text{Determine real shift } \mu. \qquad\qquad\qquad (8.3.2)$$
$$\qquad T - \mu I = UR \qquad \text{(QR factorization)}$$
$$\qquad T = RU + \mu I$$
$$\textbf{end}$$

If

$$T = \begin{bmatrix} a_1 & b_1 & & \cdots & & 0 \\ b_1 & a_2 & \ddots & & & \vdots \\ & \ddots & \ddots & \ddots & & \\ \vdots & & \ddots & \ddots & b_{n-1} \\ 0 & \cdots & & & b_{n-1} & a_n \end{bmatrix},$$

then one reasonable choice for the shift is $\mu = a_n$. However, a more effective choice is to shift by the eigenvalue of

$$T(n-1{:}n, n-1{:}n) = \begin{bmatrix} a_{n-1} & b_{n-1} \\ b_{n-1} & a_n \end{bmatrix}$$

that is closer to a_n. This is known as the *Wilkinson shift* and it is given by

$$\mu = a_n + d - \text{sign}(d)\sqrt{d^2 + b_{n-1}^2} \qquad (8.3.3)$$

where $d = (a_{n-1} - a_n)/2$. Wilkinson (1968) has shown that (8.3.2) is cubically convergent with either shift strategy, but gives heuristic reasons why (8.3.3) is preferred.

8.3.5   Implicit Shift Version

It is possible to execute the transition from T to $T_+ = RU + \mu I = U^T T U$ without explicitly forming the matrix $T - \mu I$. This has advantages when the shift is much larger than some of the a_i. Let $c = \cos(\theta)$ and $s = \sin(\theta)$ be computed such that

$$\begin{bmatrix} c & s \\ -s & c \end{bmatrix}^T \begin{bmatrix} a_1 - \mu \\ b_1 \end{bmatrix} = \begin{bmatrix} \times \\ 0 \end{bmatrix}.$$

If we set $G_1 = G(1, 2, \theta)$, then $G_1 e_1 = U e_1$ and

$$T \leftarrow G_1^T T G_1 = \begin{bmatrix} \times & \times & + & 0 & 0 & 0 \\ \times & \times & \times & 0 & 0 & 0 \\ + & \times & \times & \times & 0 & 0 \\ 0 & 0 & \times & \times & \times & 0 \\ 0 & 0 & 0 & \times & \times & \times \\ 0 & 0 & 0 & 0 & \times & \times \end{bmatrix}.$$

We are thus in a position to apply the implicit Q theorem provided we can compute rotations $G_2, \ldots, G_{n-1}$ with the property that if $Z = G_1 G_2 \cdots G_{n-1}$, then $Ze_1 = G_1 e_1 = Ue_1$ *and* $Z^T T Z$ is tridiagonal. Note that the first column of Z and U are identical provided we take each G_i to be of the form $G_i = G(i, i+1, \theta_i)$, $i = 2{:}n-1$. But G_i of this form can be used to chase the unwanted nonzero element "+" out of the matrix $G_1^T T G_1$ as follows:

$$
\xrightarrow{G_2}
\begin{bmatrix}
\times & \times & 0 & 0 & 0 & 0 \\
\times & \times & \times & + & 0 & 0 \\
0 & \times & \times & \times & 0 & 0 \\
0 & + & \times & \times & \times & 0 \\
0 & 0 & 0 & \times & \times & \times \\
0 & 0 & 0 & 0 & \times & \times
\end{bmatrix}
\xrightarrow{G_3}
\begin{bmatrix}
\times & \times & 0 & 0 & 0 & 0 \\
\times & \times & \times & 0 & 0 & 0 \\
0 & \times & \times & \times & + & 0 \\
0 & 0 & \times & \times & \times & 0 \\
0 & 0 & + & \times & \times & \times \\
0 & 0 & 0 & 0 & \times & \times
\end{bmatrix}
$$

$$
\xrightarrow{G_4}
\begin{bmatrix}
\times & \times & 0 & 0 & 0 & 0 \\
\times & \times & \times & 0 & 0 & 0 \\
0 & \times & \times & \times & 0 & 0 \\
0 & 0 & \times & \times & \times & + \\
0 & 0 & 0 & \times & \times & \times \\
0 & 0 & 0 & + & \times & \times
\end{bmatrix}
\xrightarrow{G_5}
\begin{bmatrix}
\times & \times & 0 & 0 & 0 & 0 \\
\times & \times & \times & 0 & 0 & 0 \\
0 & \times & \times & \times & 0 & 0 \\
0 & 0 & \times & \times & \times & 0 \\
0 & 0 & 0 & \times & \times & \times \\
0 & 0 & 0 & 0 & \times & \times
\end{bmatrix}.
$$

Thus, it follows from the implicit Q theorem that the tridiagonal matrix $Z^T T Z$ produced by this zero-chasing technique is essentially the same as the tridiagonal matrix T obtained by the explicit method. (We may assume that all tridiagonal matrices in question are unreduced for otherwise the problem decouples.)

Note that at any stage of the zero-chasing, there is only one nonzero entry outside the tridiagonal band. How this nonzero entry moves down the matrix during the update $T \leftarrow G_k^T T G_k$ is illustrated in the following:

$$
\begin{bmatrix}
1 & 0 & 0 & 0 \\
0 & c & s & 0 \\
0 & -s & c & 0 \\
0 & 0 & 0 & 1
\end{bmatrix}^T
\begin{bmatrix}
a_k & b_k & z_k & 0 \\
b_k & a_p & b_p & 0 \\
z_k & b_p & a_q & b_q \\
0 & 0 & b_q & a_r
\end{bmatrix}
\begin{bmatrix}
1 & 0 & 0 & 0 \\
0 & c & s & 0 \\
0 & -s & c & 0 \\
0 & 0 & 0 & 1
\end{bmatrix}
=
\begin{bmatrix}
a_k & b_k & 0 & 0 \\
b_k & a_p & b_p & z_p \\
0 & b_p & a_q & b_q \\
0 & z_p & b_q & a_r
\end{bmatrix}.
$$

Here $(p, q, r) = (k+1, k+2, k+3)$. This update can be performed in about 26 flops once c and s have been determined from the equation $b_k s + z_k c = 0$. Overall, we obtain

Algorithm 8.3.2 (Implicit Symmetric QR Step with Wilkinson Shift) Given an unreduced symmetric tridiagonal matrix $T \in \mathbb{R}^{n \times n}$, the following algorithm overwrites T with $Z^T T Z$, where $Z = G_1 \cdots G_{n-1}$ is a product of Givens rotations with the property that $Z^T(T - \mu I)$ is upper triangular and μ is that eigenvalue of T's trailing 2-by-2 principal submatrix closer to t_{nn}.

$$
\begin{aligned}
&d = (t_{n-1,n-1} - t_{nn})/2 \\
&\mu = t_{nn} - t_{n,n-1}^2 \Big/ \left(d + \text{sign}(d)\sqrt{d^2 + t_{n,n-1}^2} \right) \\
&x = t_{11} - \mu \\
&z = t_{21}
\end{aligned}
$$

for $k = 1{:}n - 1$

    $[\, c, s \,] = \mathsf{givens}(x, z)$

    $T = G_k^T T G_k$, where $G_k = G(k, k+1, \theta)$

    if $k < n - 1$

        $x = t_{k+1,k}$

        $z = t_{k+2,k}$

    end

end

This algorithm requires about $30n$ flops and n square roots. If a given orthogonal matrix Q is overwritten with $QG_1 \cdots G_{n-1}$, then an additional $6n^2$ flops are needed. Of course, in any practical implementation the tridiagonal matrix T would be stored in a pair of n-vectors and not in an n-by-n array.

Algorithm 8.3.2 is the basis of the symmetric QR algorithm—the standard means for computing the Schur decomposition of a dense symmetric matrix.

Algorithm 8.3.3 (Symmetric QR Algorithm) Given $A \in \mathbb{R}^{n \times n}$ (symmetric) and a tolerance tol greater than the unit roundoff, this algorithm computes an approximate symmetric Schur decomposition $Q^T A Q = D$. A is overwritten with the tridiagonal decomposition.

    Use Algorithm 8.3.1, compute the tridiagonalization

        $T = (P_1 \cdots P_{n-2})^T A (P_1 \cdots P_{n-2})$

        Set $D = T$ and if Q is desired, form $Q = P_1 \cdots P_{n-2}$. (See §5.1.6.)

until $q = n$

    For $i = 1{:}n - 1$, set $d_{i+1,i}$ and $d_{i,i+1}$ to zero if

        $|d_{i+1,i}| = |d_{i,i+1}| \leq \mathsf{tol}\,(|d_{ii}| + |d_{i+1,i+1}|)$

    Find the largest q and the smallest p such that if

$$
D = \begin{array}{c} \\ \\ \\ \end{array}
\left[
\begin{array}{ccc}
D_{11} & 0 & 0 \\
0 & D_{22} & 0 \\
0 & 0 & D_{33}
\end{array}
\right]
\begin{array}{c} p \\ n-p-q \\ q \end{array}
$$

$$
\quad\;\; \begin{array}{ccc} p & n-p-q & q \end{array}
$$

    then D_{33} is diagonal and D_{22} is unreduced.

    if $q < n$

        Apply Algorithm 8.3.2 to D_{22}:

            $D = \mathrm{diag}(I_p, Z, I_q)^T \cdot D \cdot \mathrm{diag}(I_p, Z, I_q)$

        If Q is desired, then $Q = Q \cdot \mathrm{diag}(I_p, Z, I_q)$.

    end

end

This algorithm requires about $4n^3/3$ flops if Q is not accumulated and about $9n^3$ flops if Q is accumulated.

The computed eigenvalues $\hat{\lambda}_i$ obtained via Algorithm 8.3.3 are the exact eigenvalues of a matrix that is near to A:

$$Q_0^T(A + E)Q_0 = \text{diag}(\hat{\lambda}_i), \qquad Q_0^T Q_0 = I, \qquad \| E \|_2 \approx \mathbf{u}\| A \|_2.$$

Using Corollary 8.1.6 we know that the absolute error in each $\hat{\lambda}_i$ is small in the sense that

$$|\hat{\lambda}_i - \lambda_i| \approx \mathbf{u}\| A \|_2.$$

If $\hat{Q} = [\, \hat{q}_1 \mid \cdots \mid \hat{q}_n \,]$ is the computed matrix of orthonormal eigenvectors, then the accuracy of $\hat{q}_i$ depends on the separation of λ_i from the remainder of the spectrum. See Theorem 8.1.12.

If all of the eigenvalues and a few of the eigenvectors are desired, then it is cheaper not to accumulate Q in Algorithm 8.3.3. Instead, the desired eigenvectors can be found via inverse iteration with T. See §8.2.2. Usually just one step is sufficient to get a good eigenvector, even with a random initial vector.

If just a few eigenvalues and eigenvectors are required, then the special techniques in §8.4 are appropriate.

8.3.6   The Rayleigh Quotient Connection

It is interesting to identify a relationship between the Rayleigh quotient iteration and the symmetric QR algorithm. Suppose we apply the latter to the tridiagonal matrix $T \in \mathbb{R}^{n \times n}$ with shift $\sigma = e_n^T T e_n = t_{nn}$. If $T - \sigma I = QR$, then we obtain $T_+ = RQ + \sigma I$. From the equation $(T - \sigma I)Q = R^T$ it follows that

$$(T - \sigma I)q_n = r_{nn}e_n,$$

where q_n is the last column of the orthogonal matrix Q. Thus, if we apply (8.2.6) with $x_0 = e_n$, then $x_1 = q_n$.

8.3.7   Orthogonal Iteration with Ritz Acceleration

Recall from §8.2.4 that an orthogonal iteration step involves a matrix-matrix product and a QR factorization:

$$Z_k = A\tilde{Q}_{k-1},$$
$$\tilde{Q}_k R_k = Z_k \quad \text{(QR factorization)}$$

Theorem 8.1.14 says that we can minimize $\| A\tilde{Q}_k - \tilde{Q}_k S \|_F$ by setting S equal to

$$S_k = \tilde{Q}_k^T A \tilde{Q}_k.$$

If $U_k^T S_k U_k = D_k$ is the Schur decomposition of $S_k \in \mathbb{R}^{r \times r}$ and $Q_k = \tilde{Q}_k U_k$, then

$$\| A Q_k - Q_k D_k \|_F = \| A\tilde{Q}_k - \tilde{Q}_k S_k \|_F$$

showing that the columns of Q_k are the best possible basis to take after k steps from the standpoint of minimizing the residual. This defines the *Ritz acceleration* idea:

$Q_0 \in \mathbb{R}^{n \times r}$ given with $Q_0^T Q_0 = I_r$

for $k = 1, 2, \ldots$

$\qquad Z_k = A Q_{k-1}$

$\qquad \tilde{Q}_k R_k = Z_k \qquad$ (QR factorization)

$\qquad S_k = \tilde{Q}_k^T A \tilde{Q}_k \qquad\qquad\qquad\qquad\qquad$ (8.3.6)

$\qquad U_k^T S_k U_k = D_k \qquad$ (Schur decomposition)

$\qquad Q_k = \tilde{Q}_k U_k$

end

It can be shown that if

$$D_k = \text{diag}(\theta_1^{(k)}, \ldots, \theta_r^{(k)})], \qquad |\theta_1^{(k)}| \geq \cdots \geq |\theta_r^{(k)}|,$$

then

$$|\theta_i^{(k)} - \lambda_i(A)| = O\left(\left|\frac{\lambda_{r+1}}{\lambda_i}\right|^k\right), \qquad i = 1{:}r.$$

Recall that Theorem 8.2.2 says the eigenvalues of $\tilde{Q}_k^T A \tilde{Q}_k$ converge with rate $|\lambda_{r+1}/\lambda_r|^k$. Thus, the Ritz values converge at a more favorable rate. For details, see Stewart (1969).

Problems

P8.3.1 Suppose λ is an eigenvalue of a symmetric tridiagonal matrix T. Show that if λ has algebraic multiplicity k, then at least $k - 1$ of T's subdiagonal elements are zero.

P8.3.2 Suppose A is symmetric and has bandwidth p. Show that if we perform the shifted QR step $A - \mu I = QR$, $A = RQ + \mu I$, then A has bandwidth p.

P8.3.3 Let

$$A = \begin{bmatrix} w & x \\ x & z \end{bmatrix}$$

be real and suppose we perform the following shifted QR step: $A - zI = UR$, $\tilde{A} = RU + zI$. Show that

$$\tilde{A} = \begin{bmatrix} \tilde{w} & \tilde{x} \\ \tilde{x} & \tilde{z} \end{bmatrix}$$

where

$$\tilde{w} = w + x^2(w - z)/[(w - z)^2 + x^2],$$
$$\tilde{z} = z - x^2(w - z)/[(w - z)^2 + x^2],$$
$$\tilde{x} = -x^3/[(w - z)^2 + x^2].$$

P8.3.4 Suppose $A \in \mathbb{C}^{n \times n}$ is Hermitian. Show how to construct unitary Q such that $Q^H A Q = T$ is real, symmetric, and tridiagonal.

P8.3.5 Show that if $A = B + iC$ is Hermitian, then

$$M = \begin{bmatrix} B & -C \\ C & B \end{bmatrix}$$

is symmetric. Relate the eigenvalues and eigenvectors of A and M.

P8.3.6 Rewrite Algorithm 8.3.2 for the case when A is stored in two n-vectors. Justify the given flop count.

P8.3.7 Suppose $A = S + \sigma u u^T$ where $S \in \mathbb{R}^{n \times n}$ is skew-symmetric ($S^T = -S$), $u \in \mathbb{R}^n$ has unit

2-norm, and $\sigma \in \mathbb{R}$. Show how to compute an orthogonal Q such that $Q^T A Q$ is tridiagonal and $Q^T u = e_1$.

P8.3.8 Suppose

$$
C = \left[\begin{array}{cc} 0 & B^T \\ B & 0 \end{array} \right]
$$

where $B \in \mathbb{R}^{n \times n}$ is upper bidiagonal. Determine a perfect shuffle permutation $P \in \mathbb{R}^{2n \times 2n}$ so that $T = PCP^T$ is tridiagonal with a zero diagonal.

Notes and References for §8.3

Historically important Algol specifications related to the algorithms in this section include:

R.S. Martin and J.H. Wilkinson (1967). "Solution of Symmetric and Unsymmetric Band Equations and the Calculation of Eigenvectors of Band Matrices," *Numer. Math. 9*, 279–301.

H. Bowdler, R.S. Martin, C. Reinsch, and J.H. Wilkinson (1968). "The QR and QL Algorithms for Symmetric Matrices," *Numer. Math. 11*, 293–306.

A. Dubrulle, R.S. Martin, and J.H. Wilkinson (1968). "The Implicit QL Algorithm," *Numer. Math. 12*, 377–383.

R.S. Martin and J.H. Wilkinson (1968). "Householder's Tridiagonalization of a Symmetric Matrix," *Numer. Math. 11*, 181–195.

C. Reinsch and F.L. Bauer (1968). "Rational QR Transformation with Newton's Shift for Symmetric Tridiagonal Matrices," *Numer. Math. 11*, 264–272.

R.S. Martin, C. Reinsch, and J.H. Wilkinson (1970). "The QR Algorithm for Band Symmetric Matrices," *Numer. Math. 16*, 85–92.

The convergence properties of Algorithm 8.3.3 are detailed in Lawson and Hanson (SLE), see:

J.H. Wilkinson (1968). "Global Convergence of Tridiagonal QR Algorithm With Origin Shifts," *Lin. Alg. Applic. 1*, 409–420.

T.J. Dekker and J.F. Traub (1971). "The Shifted QR Algorithm for Hermitian Matrices," *Lin. Alg. Applic. 4*, 137–154.

W. Hoffman and B.N. Parlett (1978). "A New Proof of Global Convergence for the Tridiagonal QL Algorithm," *SIAM J. Numer. Anal. 15*, 929–937.

S. Batterson (1994). "Convergence of the Francis Shifted QR Algorithm on Normal Matrices," *Lin. Alg. Applic. 207*, 181–195.

T.-L. Wang (2001). "Convergence of the Tridiagonal QR Algorithm," *Lin. Alg. Applic. 322*, 1–17.

Shifting and deflation are critical to the effective implementation of the symmetric QR iteration, see:

F.L. Bauer and C. Reinsch (1968). "Rational QR Transformations with Newton Shift for Symmetric Tridiagonal Matrices," *Numer. Math. 11*, 264–272.

G.W. Stewart (1970). "Incorporating Origin Shifts into the QR Algorithm for Symmetric Tridiagonal Matrices," *Commun. ACM 13*, 365–367.

I.S. Dhillon and A.N. Malyshev (2003). "Inner Deflation for Symmetric Tridiagonal Matrices," *Lin. Alg. Applic. 358*, 139–144.

The efficient reduction of a general band symmetric matrix to tridiagonal form is a challenging computation from several standpoints:

H.R. Schwartz (1968). "Tridiagonalization of a Symmetric Band Matrix," *Numer. Math. 12*, 231–241.

C.H. Bischof and X. Sun (1996). "On Tridiagonalizing and Diagonalizing Symmetric Matrices with Repeated Eigenvalues," *SIAM J. Matrix Anal. Applic. 17*, 869–885.

L. Kaufman (2000). "Band Reduction Algorithms Revisited," *ACM Trans. Math. Softw. 26*, 551–567.

C.H. Bischof, B. Lang, and X. Sun (2000). "A Framework for Symmetric Band Reduction," *ACM Trans. Math. Softw. 26*, 581–601.

Finally we mention that comparable techniques exist for skew-symmetric and general normal matrices, see:

R.C. Ward and L.J. Gray (1978). "Eigensystem Computation for Skew-Symmetric and A Class of Symmetric Matrices," *ACM Trans. Math. Softw. 4*, 278–285.

C.P. Huang (1981). "On the Convergence of the QR Algorithm with Origin Shifts for Normal Matrices," *IMA J. Numer. Anal. 1*, 127–133.

S. Iwata (1998). "Block Triangularization of Skew-Symmetric Matrices," *Lin. Alg. Applic. 273*, 215–226.

8.4 More Methods for Tridiagonal Problems

In this section we develop special methods for the symmetric tridiagonal eigenproblem. The tridiagonal form

$$
T = \begin{bmatrix}
\alpha_1 & \beta_1 & & \cdots & & 0 \\
\beta_1 & \alpha_2 & \ddots & & & \vdots \\
& \ddots & \ddots & \ddots & & \\
\vdots & & \ddots & \ddots & & \beta_{n-1} \\
0 & \cdots & & & \beta_{n-1} & \alpha_n
\end{bmatrix}
\tag{8.4.1}
$$

can be obtained by Householder reduction (cf. §8.3.1). However, symmetric tridiagonal eigenproblems arise naturally in many settings.

We first discuss bisection methods that are of interest when selected portions of the eigensystem are required. This is followed by the presentation of a divide-and-conquer algorithm that can be used to acquire the full symmetric Schur decomposition in a way that is amenable to parallel processing.

8.4.1 Eigenvalues by Bisection

Let T_r denote the leading r-by-r principal submatrix of the matrix T in (8.4.1). Define the polynomial $p_r(x)$ by

$$
p_r(x) = \det(T_r - xI)
$$

for $r = 1{:}n$. A simple determinantal expansion shows that

$$
p_r(x) = (\alpha_r - x)p_{r-1}(x) - \beta_{r-1}^2 p_{r-2}(x)
\tag{8.4.2}
$$

for $r = 2{:}n$ if we set $p_0(x) = 1$. Because $p_n(x)$ can be evaluated in $O(n)$ flops, it is feasible to find its roots using the method of bisection. For example, if tol is a small positive constant, $p_n(y){\cdot}p_n(z) < 0$, and $y < z$, then the iteration

> **while** $|y - z| > $ tol${\cdot}(|y| + |z|)$
> > $x = (y + z)/2$
> > **if** $p_n(x){\cdot}p_n(y) < 0$
> > > $z = x$
> > **else**
> > > $y = x$
> > **end**
> **end**

is guaranteed to terminate with $(y+z)/2$ an approximate zero of $p_n(x)$, i.e., an approximate eigenvalue of T. The iteration converges linearly in that the error is approximately halved at each step.

8.4.2   Sturm Sequence Methods

Sometimes it is necessary to compute the kth largest eigenvalue of T for some prescribed value of k. This can be done efficiently by using the bisection idea and the following classical result:

Theorem 8.4.1 (Sturm Sequence Property). *If the tridiagonal matrix in (8.4.1) has no zero subdiagonal entries, then the eigenvalues of T_{r-1} strictly separate the eigenvalues of T_r:*

$$\lambda_r(T_r) < \lambda_{r-1}(T_{r-1}) < \lambda_{r-1}(T_r) < \cdots < \lambda_2(T_r) < \lambda_1(T_{r-1}) < \lambda_1(T_r).$$

Moreover, if $a(\lambda)$ denotes the number of sign changes in the sequence

$$\{\, p_0(\lambda),\, p_1(\lambda), \ldots,\, p_n(\lambda) \,\},$$

then $a(\lambda)$ equals the number of T's eigenvalues that are less than λ. Here, the polynomials $p_r(x)$ are defined by (8.4.2) and we have the convention that $p_r(\lambda)$ has the opposite sign from $p_{r-1}(\lambda)$ if $p_r(\lambda) = 0$.

Proof. It follows from Theorem 8.1.7 that the eigenvalues of T_{r-1} weakly separate those of T_r. To prove strict separation, suppose that $p_r(\mu) = p_{r-1}(\mu) = 0$ for some r and μ. It follows from (8.4.2) and the assumption that the matrix T is unreduced that

$$p_0(\mu) \;=\; p_1(\mu) \;=\; \cdots \;=\; p_r(\mu) \;=\; 0,$$

a contradiction. Thus, we must have strict separation. The assertion about $a(\lambda)$ is established in Wilkinson (AEP, pp. 300–301). □

Suppose we wish to compute $\lambda_k(T)$. From the Gershgorin theorem (Theorem 8.1.3) it follows that $\lambda_k(T) \in [y, z]$ where

$$y \;=\; \min_{1 \le i \le n} a_i - |b_i| - |b_{i-1}|, \qquad\qquad z \;=\; \max_{1 \le i \le n} a_i + |b_i| + |b_{i-1}|$$

and we have set $b_0 = b_n = 0$. Using $[y, z]$ as an initial bracketing interval, it is clear from the Sturm sequence property that the iteration

> **while** $|z - y| > \mathbf{u}(|y| + |z|)$
> $x = (y + z)/2$
>   **if** $a(x) \ge n - k$ (8.4.3)
> $z = x$
>   **else**
> $y = x$
>   **end**
> **end**

produces a sequence of subintervals that are repeatedly halved in length but which always contain $\lambda_k(T)$.

During the execution of (8.4.3), information about the location of other eigenvalues is obtained. By systematically keeping track of this information it is possible to devise an efficient scheme for computing contiguous subsets of $\lambda(T)$, e.g., $\{\lambda_k(T), \lambda_{k+1}(T), \dots, \lambda_{k+j}(T)\}$. See Barth, Martin, and Wilkinson (1967).

If selected eigenvalues of a general symmetric matrix A are desired, then it is necessary first to compute the tridiagonalization $T = U_0^T A U_0$ before the above bisection schemes can be applied. This can be done using Algorithm 8.3.1 or by the Lanczos algorithm discussed in §10.2. In either case, the corresponding eigenvectors can be readily found via inverse iteration since tridiagonal systems can be solved in $O(n)$ flops. See §4.3.6 and §8.2.2.

In those applications where the original matrix A already has tridiagonal form, bisection computes eigenvalues with small relative error, regardless of their magnitude. This is in contrast to the tridiagonal QR iteration, where the computed eigenvalues $\tilde{\lambda}_i$ can be guaranteed only to have small absolute error: $|\tilde{\lambda}_i - \lambda_i(T)| \approx \mathbf{u}\| T \|_2$

Finally, it is possible to compute specific eigenvalues of a symmetric matrix by using the LDL^T factorization (§4.3.6) and exploiting the Sylvester inertia theorem (Theorem 8.1.17). If

$$A - \mu I = LDL^T, \qquad A = A^T \in \mathbb{R}^{n \times n},$$

is the LDL^T factorization of $A - \mu I$ with $D = \text{diag}(d_1, \dots, d_n)$, then the number of negative d_i equals the number of $\lambda_i(A)$ that are less than μ. See Parlett (SEP, p. 46) for details.

8.4.3 Eigensystems of Diagonal Plus Rank-1 Matrices

Our next method for the symmetric tridiagonal eigenproblem requires that we be able to compute efficiently the eigenvalues and eigenvectors of a matrix of the form $D + \rho z z^T$ where $D \in \mathbb{R}^{n \times n}$ is diagonal, $z \in \mathbb{R}^n$, and $\rho \in \mathbb{R}$. This problem is important in its own right and the key computations rest upon the following pair of results.

Lemma 8.4.2. *Suppose* $D = \text{diag}(d_1, \dots, d_n) \in \mathbb{R}^{n \times n}$ *with*

$$d_1 > \dots > d_n.$$

Assume that $\rho \neq 0$ *and that* $z \in \mathbb{R}^n$ *has no zero components. If*

$$(D + \rho z z^T)v = \lambda v, \qquad v \neq 0,$$

then $z^T v \neq 0$ *and* $D - \lambda I$ *is nonsingular.*

Proof. If $\lambda \in \lambda(D)$, then $\lambda = d_i$ for some i and thus

$$0 = e_i^T[(D - \lambda I)v + \rho(z^T v)z] = \rho(z^T v)z_i.$$

Since ρ and z_i are nonzero, it follows that $0 = z^T v$ and so $Dv = \lambda v$. However, D has distinct eigenvalues and therefore $v \in \text{span}\{e_i\}$. This implies $0 = z^T v = z_i$, a contradiction. Thus, D and $D + \rho z z^T$ have no common eigenvalues and $z^T v \neq 0$. $\square$

Theorem 8.4.3. *Suppose* $D = \text{diag}(d_1, \ldots, d_n) \in \mathbb{R}^{n \times n}$ *and that the diagonal entries satisfy* $d_1 > \cdots > d_n$. *Assume that* $\rho \neq 0$ *and that* $z \in \mathbb{R}^n$ *has no zero components. If* $V \in \mathbb{R}^{n \times n}$ *is orthogonal such that*

$$V^T(D + \rho z z^T)V = \text{diag}(\lambda_1, \ldots, \lambda_n)$$

with $\lambda_1 \geq \cdots \geq \lambda_n$ *and* $V = [\, v_1 \,|\, \cdots \,|\, v_n \,]$, *then*

(a) *The* λ_i *are the n zeros of* $f(\lambda) = 1 + \rho z^T(D - \lambda I)^{-1}z$.

(b) *If* $\rho > 0$, *then* $\lambda_1 > d_1 > \lambda_2 > \cdots > \lambda_n > d_n$.
If $\rho < 0$, *then* $d_1 > \lambda_1 > d_2 > \cdots > d_n > \lambda_n$.

(c) *The eigenvector* v_i *is a multiple of* $(D - \lambda_i I)^{-1}z$.

Proof. If $(D + \rho z z^T)v = \lambda v$, then

$$(D - \lambda I)v + \rho(z^T v)z = 0. \tag{8.4.4}$$

We know from Lemma 8.4.2 that $D - \lambda I$ is nonsingular. Thus,

$$v \in \text{span}\{(D - \lambda I)^{-1}z\},$$

thereby establishing (c). Moreover, if we apply $z^T(D - \lambda I)^{-1}$ to both sides of equation (8.4.4) we obtain

$$(z^T v) \cdot (1 + \rho z^T(D - \lambda I)^{-1}z) = 0.$$

By Lemma 8.4.2, $z^T v \neq 0$ and so this shows that if $\lambda \in \lambda(D + \rho z z^T)$, then $f(\lambda) = 0$. We must show that all the zeros of f are eigenvalues of $D + \rho z z^T$ and that the interlacing relations (b) hold.

To do this we look more carefully at the equations

$$f(\lambda) = 1 + \rho\left(\frac{z_1^2}{d_1 - \lambda} + \cdots + \frac{z_n^2}{d_n - \lambda}\right),$$

$$f'(\lambda) = \rho\left(\frac{z_1^2}{(d_1 - \lambda)^2} + \cdots + \frac{z_n^2}{(d_n - \lambda)^2}\right).$$

Note that f is monotone in between its poles. This allows us to conclude that, if $\rho > 0$, then f has precisely n roots, one in each of the intervals

$$(d_n, d_{n-1}), \ldots, (d_2, d_1), \ (d_1, \infty).$$

If $\rho < 0$, then f has exactly n roots, one in each of the intervals

$$(-\infty, d_n), (d_n, d_{n-1}), \ldots, (d_2, d_1).$$

Thus, in either case the zeros of f are exactly the eigenvalues of $D + \rho v v^T$. $\square$

The theorem suggests that in order to compute V we must find the roots $\lambda_1, \ldots, \lambda_n$ of f using a Newton-like procedure and then compute the columns of V by normalizing the vectors $(D - \lambda_i I)^{-1} z$ for $i = 1{:}n$. The same plan of attack can be followed even if there are repeated d_i and zero z_i.

Theorem 8.4.4. *If $D = \mathrm{diag}(d_1, \ldots, d_n)$ and $z \in \mathbb{R}^n$, then there exists an orthogonal matrix V_1 such that if $V_1^T D V_1 = \mathrm{diag}(\mu_1, \ldots, \mu_n)$ and $w = V_1^T z$ then*

$$\mu_1 > \mu_2 > \cdots > \mu_r \geq \mu_{r+1} \geq \cdots \geq \mu_n \, ;$$

$w_i \neq 0$ *for* $i = 1{:}r$, *and* $w_i = 0$ *for* $i = r + 1{:}n$.

Proof. We give a constructive proof based upon two elementary operations. The first deals with repeated diagonal entries while the second handles the situation when the z-vector has a zero component.

Suppose $d_i = d_j$ for some $i < j$. Let $G(i, j, \theta)$ be a Givens rotation in the (i, j) plane with the property that the jth component of $G(i, j, \theta)^T z$ is zero. It is not hard to show that $G(i, j, \theta)^T D \, G(i, j, \theta) = D$. Thus, we can zero a component of z if there is a repeated d_i.

If $z_i = 0$, $z_j \neq 0$, and $i < j$, then let P be the identity with columns i and j interchanged. It follows that $P^T D P$ is diagonal, $(P^T z)_i \neq 0$, and $(P^T z)_j = 0$. Thus, we can permute all the zero z_i to the "bottom."

It is clear that the repetition of these two maneuvers will render the desired canonical structure. The orthogonal matrix V_1 is the product of the rotations that are required by the process. $\square$

See Barlow (1993) and the references therein for a discussion of the solution procedures that we have outlined above.

8.4.4 A Divide-and-Conquer Framework

We now present a divide-and-conquer method for computing the Schur decomposition

$$Q^T T Q = \Lambda = \mathrm{diag}(\lambda_1, \ldots, \lambda_n), \qquad Q^T Q = I, \tag{8.4.5}$$

for tridiagonal T that involves (a) "tearing" T in half, (b) computing the Schur decompositions of the two parts, and (c) combining the two half-sized Schur decompositions into the required full-size Schur decomposition. The overall procedure, developed by Dongarra and Sorensen (1987), is suitable for parallel computation.

We first show how T can be "torn" in half with a rank-1 modification. For simplicity, assume $n = 2m$ and that $T \in \mathbb{R}^{n \times n}$ is given by (8.4.1). Define $v \in \mathbb{R}^n$ as follows

$$v = \begin{bmatrix} e_m^{(m)} \\ \theta \, e_1^{(m)} \end{bmatrix}, \qquad \theta \in \{-1, +1\}. \tag{8.4.6}$$

Note that for all $\rho \in \mathbb{R}$ the matrix $\widetilde{T} = T - \rho v v^T$ is identical to T except in its "middle four" entries:

$$\widetilde{T}(m{:}m+1, m{:}m+1) = \begin{bmatrix} \alpha_m - \rho & \beta_m - \rho\theta \\ \beta_m - \rho\theta & \alpha_{m+1} - \rho\theta^2 \end{bmatrix}.$$

If we set $\rho\theta = \beta_m$, then

$$T = \begin{bmatrix} T_1 & 0 \\ 0 & T_2 \end{bmatrix} + \rho vv^T,$$

where

$$T_1 = \begin{bmatrix} \alpha_1 & \beta_1 & & \cdots & & 0 \\ \beta_1 & \alpha_2 & \ddots & & & \vdots \\ & \ddots & \ddots & \ddots & & \\ \vdots & & \ddots & \ddots & \beta_{m-1} \\ 0 & \cdots & & \beta_{m-1} & \tilde{\alpha}_m \end{bmatrix}, \quad T_2 = \begin{bmatrix} \tilde{\alpha}_{m+1} & \beta_{m+1} & & \cdots & & 0 \\ \beta_{m+1} & \alpha_{m+2} & \ddots & & & \vdots \\ & \ddots & \ddots & \ddots & & \\ \vdots & & \ddots & \ddots & \beta_{n-1} \\ 0 & \cdots & & \beta_{n-1} & \alpha_n \end{bmatrix},$$

and $\tilde{a}_m = a_m - \rho$ and $\tilde{a}_{m+1} = a_{m+1} - \rho\theta^2$.

Now suppose that we have m-by-m orthogonal matrices Q_1 and Q_2 such that $Q_1^T T_1 Q_1 = D_1$ and $Q_2^T T_2 Q_2 = D_2$ are each diagonal. If we set

$$U = \begin{bmatrix} Q_1 & 0 \\ 0 & Q_2 \end{bmatrix},$$

then

$$U^T T U = U^T \left(\begin{bmatrix} T_1 & 0 \\ 0 & T_2 \end{bmatrix} + \rho vv^T \right) U = D + \rho zz^T$$

where

$$D = \begin{bmatrix} D_1 & 0 \\ 0 & D_2 \end{bmatrix}$$

is diagonal and

$$z = U^T v = \begin{bmatrix} Q_1^T e_m \\ \theta Q_2^T e_1 \end{bmatrix}.$$

Comparing these equations we see that the effective synthesis of the two half-sized Schur decompositions requires the quick and stable computation of an orthogonal V such that

$$V^T (D + \rho zz^T) V = \Lambda = \text{diag}(\lambda_1, \dots, \lambda_n)$$

which we discussed in §8.4.3.

8.4.5   A Parallel Implementation

Having stepped through the tearing and synthesis operations, we can now illustrate how the overall process can be implemented in parallel. For clarity, assume that $n = 8N$ for some positive integer N and that three levels of tearing are performed. See Figure 8.4.1. The indices are specified in binary and at each node the Schur decomposition of a tridiagonal matrix $T(b)$ is obtained from the eigensystems of the tridiagonals $T(b0)$ and $T(b1)$. For example, the eigensystems for the N-by-N matrices $T(110)$ and $T(111)$ are combined to produce the eigensystem for the $2N$-by-$2N$ tridiagonal matrix $T(11)$. What makes this framework amenable to parallel computation is the independence of the tearing/synthesis problems that are associated with each level in the tree.

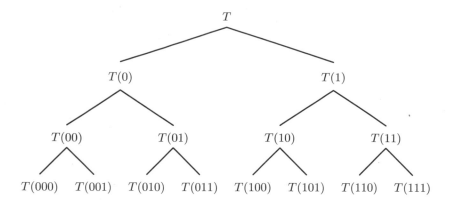

Figure 8.4.1. *The divide-and-conquer framework*

8.4.6   An Inverse Tridiagonal Eigenvalue Problem

For additional perspective on symmetric trididagonal matrices and their rich eigen-structure we consider an *inverse eigenvalue problem*. Assume that $\lambda_1, \ldots, \lambda_n$ and $\tilde{\lambda}_1, \ldots, \tilde{\lambda}_{n-1}$ are given real numbers that satisfy

$$\lambda_1 > \tilde{\lambda}_1 > \lambda_2 > \cdots > \lambda'_{n-1} > \tilde{\lambda}_{n-1} > \lambda_n . \tag{8.4.7}$$

The goal is to compute a symmetric tridiagonal matrix $T \in \mathbb{R}^{n \times n}$ such that

$$\lambda(T) = \{\lambda_1, \ldots, \lambda_n, \} , \tag{8.4.8}$$

$$\lambda(T(2{:}n, 2{:}n)) = \{\tilde{\lambda}_1, \ldots, \tilde{\lambda}_{n-1}\}. \tag{8.4.9}$$

Inverse eigenvalue problems arise in many applications and generally involve computing a matrix that has specified spectral properties. For an overview, see Chu and Golub (2005). Our example is taken from Golub (1973).

The problem we are considering can be framed as a Householder tridiagonalization problem with a constraint on the orthogonal transformation. Define

$$\Lambda = \text{diag}(\lambda_1, \ldots, \lambda_n)$$

and let Q be orthogonal so that $Q^T \Lambda Q = T$ is tridiagonal. There are an infinite number of possible Q-matrices that do this and in each case the matrix T satisfies (8.4.8). The challenge is to choose Q so that (8.4.9) holds as well. Recall that a tridiagonalizing Q is essentially determined by its first column because of the implicit-Q- theorem (Theorem 8.3.2). Thus, the problem is solved if we can figure out a way to compute $Q(:,1)$ so that (8.4.9) holds.

The starting point in the derivation of the method is to realize that the eigenvalues of $T(2{:}n, 2{:}n)$ are the stationary values of $x^T T x$ subject to the constraints $x^T x = 1$ and $e_1^T x = 0$. To characterize these stationary values we use the method of Lagrange multipliers and set to zero the gradient of

$$\phi(x, \lambda, \mu) = x^T T x - \lambda(x^T x - 1) + 2\mu x^T e_1$$

which gives $(T - \lambda I)x = -\mu e_1$. Because λ is an eigenvalue of $T(2{:}n, 2{:}n)$ it is not an eigenvalue of T and so

$$x = -\mu(T - \lambda I)^{-1}e_1.$$

Since $e_1^T x = 0$, it follows that

$$0 = e_1^T(T - \lambda I)^{-1}e_1 = e_1^T(Q^T \Lambda Q - \lambda I)^{-1}e_1 = \sum_{i=1}^{n} \frac{d_i^2}{\lambda_i - \lambda} \tag{8.4.10}$$

where

$$Q(:, 1) = \begin{bmatrix} d_1 \\ \vdots \\ d_n \end{bmatrix}. \tag{8.4.11}$$

By multiplying both sides of equation (8.4.10) by $(\lambda_1 - \lambda) \cdots (\lambda_n - \lambda)$, we can conclude that $\tilde{\lambda}_1, \ldots, \tilde{\lambda}_{n-1}$ are the zeros of the polynomial

$$p(\lambda) = \sum_{i=1}^{n} d_i^2 \prod_{\substack{j=1 \\ j \neq i}}^{n} (\lambda_j - \lambda).$$

It follows that

$$p(\lambda) = \alpha \cdot \prod_{j=1}^{n-1} (\tilde{\lambda}_j - \lambda)$$

for some scalar α. By comparing the coefficient of λ^{n-1} in each of these expressions for $p(\lambda)$ and noting from (8.4.11) that $d_1^2 + \cdots + d_n^2 = 1$, we see that $\alpha = 1$. From the equation

$$\sum_{i=1}^{n} d_i^2 \prod_{\substack{j=1 \\ j \neq i}}^{n} (\lambda_j - \lambda) = \prod_{j=1}^{n-1} (\tilde{\lambda}_j - \lambda)$$

we immediately see that

$$d_k^2 = \prod_{j=1}^{n-1} (\tilde{\lambda}_j - \lambda_k) \Bigg/ \prod_{\substack{j=1 \\ j \neq k}}^{n-1} (\lambda_j - \lambda_k), \qquad k = 1{:}n. \tag{8.4.12}$$

It is easy to show using (8.4.7) that the quantity on the right is positive and thus (8.4.11) can be used to determine the components of $d = Q(:, 1)$ up to with a factor of ± 1. Once this vector is available, then we can determine the required tridiagonal matrix T as follows:

Step 1. Let P be a Householder matrix so that $Pd = \pm 1$ and set $A = P^T \Lambda P$.

Step 2. Compute the tridiagonalization $Q_1^T A Q_1 = T$ via Algorithm 8.3.1 and observe from the implementation that $Q_1(:, 1) = e_1$.

Step 3. Set $Q = P Q_1$.

It follows that $Q(:,1) = P(Q_1 e_1) = P e_1 = \pm d$. The sign does not matter.

Problems

P8.4.1 Suppose λ is an eigenvalue of a symmetric tridiagonal matrix T. Show that if λ has algebraic multiplicity k, then T has at least $k - 1$ subdiagonal entries that are zero.

P8.4.2 Give an algorithm for determining ρ and θ in (8.4.6) with the property that $\theta \in \{-1, 1\}$ and $\min\{\,|a_m - \rho|, |a_{m+1} - \rho|\,\}$ is maximized.

P8.4.3 Let $p_r(\lambda) = \det(T(1{:}r, 1{:}r) - \lambda I_r)$ where T is given by (8.4.1). Derive a recursion for evaluating $p'_n(\lambda)$ and use it to develop a Newton iteration that can compute eigenvalues of T.

P8.4.4 If T is positive definite, does it follow that the matrices T_1 and T_2 in §8.4.4 are positive definite?

P8.4.5 Suppose $A = S + \sigma u u^T$ where $S \in \mathbb{R}^{n \times n}$ is skew-symmetric, $u \in \mathbb{R}^n$, and $\sigma \in \mathbb{R}$. Show how to compute an orthogonal Q such that $Q^T A Q = T + \sigma e_1 e_1^T$ where T is tridiagonal and skew-symmetric.

P8.4.6 Suppose λ is a known eigenvalue of a unreduced symmetric tridiagonal matrix $T \in \mathbb{R}^{n \times n}$. Show how to compute $x(1{:}n-1)$ from the equation $Tx = \lambda x$ given that $x_n = 1$.

P8.4.7 Verify that the quantity on the right-hand side of (8.4.12) is positive.

P8.4.8 Suppose that

$$A = \begin{bmatrix} D & v \\ v^T & d_n \end{bmatrix}$$

where $D = \operatorname{diag}(d_1, \ldots, d_{n-1})$ has distinct diagonal entries and $v \in \mathbb{R}^{n-1}$ has no zero entries. (a) Show that if $\lambda \in \lambda(A)$, then $D - \lambda I_{n-1}$ is nonsingular. (b) Show that if $\lambda \in \lambda(A)$, then λ is a zero of

$$f(\lambda) = \lambda + \sum_{k=1}^{n-1} \frac{v_k^2}{d_k - \lambda} - d_n.$$

Notes and References for §8.4

Bisection/Sturm sequence methods are discussed in:

W. Barth, R.S. Martin, and J.H. Wilkinson (1967). "Calculation of the Eigenvalues of a Symmetric Tridiagonal Matrix by the Method of Bisection," *Numer. Math. 9*, 386–393.

K.K. Gupta (1972). "Solution of Eigenvalue Problems by Sturm Sequence Method," *Int. J. Numer. Meth. Eng. 4*, 379–404.

J.W. Demmel, I.S. Dhillon, and H. Ren (1994) "On the Correctness of Parallel Bisection in Floating Point," *ETNA 3*, 116–149.

Early references concerned with the divide-and-conquer framework that we outlined include:

J.R. Bunch, C.P. Nielsen, and D.C. Sorensen (1978). "Rank-One Modification of the Symmetric Eigenproblem," *Numer. Math. 31*, 31–48.

J.J.M. Cuppen (1981). "A Divide and Conquer Method for the Symmetric Eigenproblem," *Numer. Math. 36*, 177–195.

J.J. Dongarra and D.C. Sorensen (1987). "A Fully Parallel Algorithm for the Symmetric Eigenvalue Problem," *SIAM J. Sci. Stat. Comput. 8*, S139–S154.

Great care must be taken to ensure orthogonality in the computed matrix of eigenvectors, something that is a major challenge when the eigenvalues are close and clustered. The development of reliable implementations is a classic tale that involves a mix of sophisticated theory and clever algorithmic insights, see:

M. Gu and S.C. Eisenstat (1995). "A Divide-and-Conquer Algorithm for the Symmetric Tridiagonal Eigenproblem," *SIAM J. Matrix Anal. Applic. 16*, 172–191.

B.N. Parlett (1996). "Invariant Subspaces for Tightly Clustered Eigenvalues of Tridiagonals," *BIT 36*, 542–562.

B.N. Parlett and I.S. Dhillon (2000). "Relatively Robust Representations of Symmetric Tridiagonals," *Lin. Alg. Applic. 309*, 121–151.

I.S. Dhillon and B.N. Parlett (2003). "Orthogonal Eigenvectors and Relative Gaps," *SIAM J. Matrix Anal. Applic. 25*, 858–899.

I.S. Dhillon and B.N. Parlett (2004). "Multiple Representations to Compute Orthogonal Eigenvectors of Symmetric Tridiagonal Matrices," *Lin. Alg. Applic. 387*, 1–28.

O.A. Marques, B.N. Parlett, and C. Vömel (2005). "Computations of Eigenpair Subsets with the MRRR Algorithm," *Numer. Lin. Alg. Applic. 13*, 643–653.

P. Bientinesi, I.S. Dhillon, and R.A. van de Geijn (2005). "A Parallel Eigensolver for Dense Symmetric Matrices Based on Multiple Relatively Robust Representations," *SIAM J. Sci. Comput. 27*, 43–66.

Various extensions and generalizations of the basic idea have also been proposed:

S. Huss–Lederman, A. Tsao, and T. Turnbull (1997). "A Parallelizable Eigensolver for Real Diagonalizable Matrices with Real Eigenvalues," *SIAM J. Sci. Comput. 18*, 869–885.

B. Hendrickson, E. Jessup, and C. Smith (1998). "Toward an Efficient Parallel Eigensolver for Dense Symmetric Matrices," *SIAM J. Sci. Comput. 20*, 1132–1154.

W.N. Gansterer, J. Schneid, and C.W. Ueberhuber (2001). "A Low-Complexity Divide-and-Conquer Method for Computing Eigenvalues and Eigenvectors of Symmetric Band Matrices," *BIT 41*, 967–976.

W.N. Gansterer, R.C. Ward, and R.P. Muller (2002). "An Extension of the Divide-and-Conquer Method for a Class of Symmetric Block-Tridiagonal Eigenproblems," *ACM Trans. Math. Softw. 28*, 45–58.

W.N. Gansterer, R.C. Ward, R.P. Muller, and W.A. Goddard and III (2003). "Computing Approximate Eigenpairs of Symmetric Block Tridiagonal Matrices," *SIAM J. Sci. Comput. 24*, 65–85.

Y. Bai and R.C. Ward (2007). "A Parallel Symmetric Block-Tridiagonal Divide-and-Conquer Algorithm," *ACM Trans. Math. Softw. 33*, Article 35.

For a detailed treatment of various inverse eigenvalue problems, see:

M.T. Chu and G.H. Golub (2005). *Inverse Eigenvalue Problems*, Oxford University Press, Oxford, U.K.

Selected papers that discuss a range of inverse eigenvalue problems include:

D. Boley and G.H. Golub (1987). "A Survey of Matrix Inverse Eigenvalue Problems," *Inverse Problems 3*, 595–622.

M.T. Chu (1998). "Inverse Eigenvalue Problems," *SIAM Review 40*, 1–39.

C.-K. Li and R. Mathias (2001). "Construction of Matrices with Prescribed Singular Values and Eigenvalues," *BIT 41*, 115–126.

The derivation in §8.4.6 involved the constrained optimization of a quadratic form, an important problem in its own right, see:

G.H. Golub and R. Underwood (1970). "Stationary Values of the Ratio of Quadratic Forms Subject to Linear Constraints," *Z. Angew. Math. Phys. 21*, 318–326.

G.H. Golub (1973). "Some Modified Eigenvalue Problems," *SIAM Review 15*, 318–334.

S. Leon (1994). "Maximizing Bilinear Forms Subject to Linear Constraints," *Lin. Alg. Applic. 210*, 49–58.

8.5 Jacobi Methods

Jacobi methods for the symmetric eigenvalue problem attract current attention because they are inherently parallel. They work by performing a sequence of orthogonal similarity updates $A \leftarrow Q^T A Q$ with the property that each new A, although full, is "more diagonal" than its predecessor. Eventually, the off-diagonal entries are small enough to be declared zero.

After surveying the basic ideas behind the Jacobi approach we develop a parallel Jacobi procedure.

8.5.1 The Jacobi Idea

The idea behind Jacobi's method is to systematically reduce the quantity

$$\text{off}(A) = \sqrt{\sum_{i=1}^{n}\sum_{\substack{j=1\\j\neq i}}^{n} a_{ij}^2} \, ,$$

i.e., the Frobenius norm of the off-diagonal elements. The tools for doing this are rotations of the form

$$J(p,q,\theta) = \begin{bmatrix} 1 & \cdots & 0 & \cdots & 0 & \cdots & 0 \\ \vdots & \ddots & \vdots & & \vdots & & \vdots \\ 0 & \cdots & c & \cdots & s & \cdots & 0 \\ \vdots & & \vdots & \ddots & \vdots & & \vdots \\ 0 & \cdots & -s & \cdots & c & \cdots & 0 \\ \vdots & & \vdots & & \vdots & \ddots & \vdots \\ 0 & \cdots & 0 & \cdots & 0 & \cdots & 1 \end{bmatrix} \begin{matrix} \\ \\ p \\ \\ q \\ \\ \\ \end{matrix}$$

$$\begin{matrix} p & \quad & q \end{matrix}$$

which we call *Jacobi rotations*. Jacobi rotations are no different from Givens rotations; see §5.1.8. We submit to the name change in this section to honor the inventor.

The basic step in a Jacobi eigenvalue procedure involves (i) choosing an index pair (p,q) that satisfies $1 \le p < q \le n$, (ii) computing a cosine-sine pair (c,s) such that

$$\begin{bmatrix} b_{pp} & b_{pq} \\ b_{qp} & b_{qq} \end{bmatrix} = \begin{bmatrix} c & s \\ -s & c \end{bmatrix}^T \begin{bmatrix} a_{pp} & a_{pq} \\ a_{qp} & a_{qq} \end{bmatrix} \begin{bmatrix} c & s \\ -s & c \end{bmatrix} \tag{8.5.1}$$

is diagonal, and (iii) overwriting A with $B = J^T A J$ where $J = J(p,q,\theta)$. Observe that the matrix B agrees with A except in rows and columns p and q. Moreover, since the Frobenius norm is preserved by orthogonal transformations, we find that

$$a_{pp}^2 + a_{qq}^2 + 2a_{pq}^2 = b_{pp}^2 + b_{qq}^2 + 2b_{pq}^2 = b_{pp}^2 + b_{qq}^2.$$

It follows that

$$\text{off}(B)^2 = \| B \|_F^2 - \sum_{i=1}^{n} b_{ii}^2 = \| A \|_F^2 - \sum_{i=1}^{n} a_{ii}^2 + (a_{pp}^2 + a_{qq}^2 - b_{pp}^2 - b_{qq}^2) \tag{8.5.2}$$

$$= \text{off}(A)^2 - 2a_{pq}^2 \, .$$

It is in this sense that A moves closer to diagonal form with each Jacobi step.

Before we discuss how the index pair (p,q) can be chosen, let us look at the actual computations associated with the (p,q) subproblem.

8.5.2   The 2-by-2 Symmetric Schur Decomposition

To say that we diagonalize in (8.5.1) is to say that

$$0 \; = \; b_{pq} \; = \; a_{pq}(c^2 - s^2) + (a_{pp} - a_{qq})cs. \tag{8.5.3}$$

If $a_{pq} = 0$, then we just set $c = 1$ and $s = 0$. Otherwise, define

$$\tau \; = \; \frac{a_{qq} - a_{pp}}{2a_{pq}} \quad \text{and} \quad t \; = \; s/c$$

and conclude from (8.5.3) that $t = \tan(\theta)$ solves the quadratic

$$t^2 + 2\tau t - 1 = 0 \, .$$

It turns out to be important to select the smaller of the two roots:

$$t_{\min} \; = \; \begin{cases} 1/(\tau + \sqrt{1 + \tau^2}) & \text{if } \tau \geq 0, \\ 1/(\tau - \sqrt{1 + \tau^2}) & \text{if } \tau < 0. \end{cases}$$

This is implies that the rotation angle satisfies $|\theta| \leq \pi/4$ and has the effect of maximizing c:

$$c = 1/\sqrt{1 + t_{\min}^2}, \qquad s = t_{\min}\, c \, .$$

This in turn minimizes the difference between A and the update B:

$$\| \, B - A \, \|_F^2 \; = \; 4(1 - c) \sum_{\substack{i=1 \\ i \neq p,q}}^{n} (a_{ip}^2 + a_{iq}^2) \; + \; 2a_{pq}^2/c^2.$$

We summarize the 2-by-2 computations as follows:

Algorithm 8.5.1   Given an n-by-n symmetric A and integers p and q that satisfy $1 \leq p < q \leq n$, this algorithm computes a cosine-sine pair $\{c, s\}$ such that if $B = J(p, q, \theta)^T A J(p, q, \theta)$, then $b_{pq} = b_{qp} = 0$.

> **function** $[c,\, s] = \mathsf{symSchur2}(A, p, q)$
>> **if** $A(p, q) \neq 0$
>>> $\tau = (A(q, q) - A(p, p))/(2A(p, q))$
>>> **if** $\tau \geq 0$
>>>> $t = 1/(\tau + \sqrt{1 + \tau^2})$
>>> **else**
>>>> $t = 1/(\tau - \sqrt{1 + \tau^2})$
>>> **end**
>>> $c = 1/\sqrt{1 + t^2},\ s = tc$
>> **else**
>>> $c = 1,\ s = 0$
>> **end**

8.5.3 The Classical Jacobi Algorithm

As we mentioned above, only rows and columns p and q are altered when the (p, q) subproblem is solved. Once symSchur2 determines the 2-by-2 rotation, then the update $A \leftarrow J(p, q, \theta)^T A J(p, q, \theta)$ can be implemented in $6n$ flops if symmetry is exploited.

How do we choose the indices p and q? From the standpoint of maximizing the reduction of off(A) in (8.5.2), it makes sense to choose (p, q) so that a_{pq}^2 is maximal. This is the basis of the *classical* Jacobi algorithm.

Algorithm 8.5.2 (Classical Jacobi) Given a symmetric $A \in \mathbb{R}^{n \times n}$ and a positive tolerance tol, this algorithm overwrites A with $V^T A V$ where V is orthogonal and off$(V^T A V) \leq tol \cdot \| A \|_F$.

$V = I_n$, $\delta = $ tol $\cdot \| A \|_F$
while off$(A) > \delta$
      Choose (p, q) so $|a_{pq}| = \max_{i \neq j} |a_{ij}|$
      $[c, s] = $ symSchur2(A, p, q)
      $A = J(p, q, \theta)^T A J(p, q, \theta)$
      $V = V J(p, q, \theta)$
end

Since $|a_{pq}|$ is the largest off-diagonal entry,

$$\text{off}(A)^2 \leq N(a_{pq}^2 + a_{qp}^2)$$

where

$$N = \frac{n(n-1)}{2}.$$

From (8.5.2) it follows that

$$\text{off}(B)^2 \leq \left(1 - \frac{1}{N}\right) \text{off}(A)^2.$$

By induction, if $A^{(k)}$ denotes the matrix A after k Jacobi updates, then

$$\text{off}(A^{(k)})^2 \leq \left(1 - \frac{1}{N}\right)^k \text{off}(A^{(0)})^2.$$

This implies that the classical Jacobi procedure converges at a linear rate.

However, the asymptotic convergence rate of the method is considerably better than linear. Schonhage (1964) and van Kempen (1966) show that for k large enough, there is a constant c such that

$$\text{off}(A^{(k+N)}) \leq c \cdot \text{off}(A^{(k)})^2,$$

i.e., quadratic convergence. An earlier paper by Henrici (1958) established the same result for the special case when A has distinct eigenvalues. In the convergence theory for the Jacobi iteration, it is critical that $|\theta| \leq \pi/4$. Among other things this precludes the possibility of interchanging nearly converged diagonal entries. This follows from

the formulae $b_{pp} = a_{pp} - ta_{pq}$ and $b_{qq} = a_{qq} + ta_{pq}$, which can be derived from Equation
(8.5.1) and the definition $t = \sin(\theta)/\cos(\theta)$.

It is customary to refer to N Jacobi updates as a *sweep*. Thus, after a sufficient
number of iterations, quadratic convergence is observed when examining $\text{off}(A)$ after
every sweep.

There is no rigorous theory that enables one to predict the number of sweeps that
are required to achieve a specified reduction in $\text{off}(A)$. However, Brent and Luk (1985)
have argued heuristically that the number of sweeps is proportional to $\log(n)$ and this
seems to be the case in practice.

8.5.4   The Cyclic-by-Row Algorithm

The trouble with the classical Jacobi method is that the updates involve $O(n)$ flops
while the search for the optimal (p, q) is $O(n^2)$. One way to address this imbalance is
to fix the sequence of subproblems to be solved in advance. A reasonable possibility is
to step through all the subproblems in row-by-row fashion. For example, if $n = 4$ we
cycle as follows:

$$(p, q) \;=\; (1,2), (1,3), (1,4), (2,3), (2,4), (3,4), (1,2), \ldots.$$

This ordering scheme is referred to as *cyclic by row* and it results in the following
procedure:

Algorithm 8.5.3 (Cyclic Jacobi) Given a symmetric matrix $A \in \mathbb{R}^{n \times n}$ and a positive
tolerance tol, this algorithm overwrites A with $V^T A V$ where V is orthogonal and
$\text{off}(V^T A V) \leq \text{tol} \cdot \| A \|_F$.

$$V = I_n, \quad \delta = \text{tol} \cdot \| A \|_F$$
$$\textbf{while } \text{off}(A) > \delta$$
$$\qquad \textbf{for } p = 1{:}n - 1$$
$$\qquad\qquad \textbf{for } q = p + 1{:}n$$
$$\qquad\qquad\qquad [c, s] \;=\; \text{symSchur2}(A, p, q)$$
$$\qquad\qquad\qquad A \;=\; J(p, q, \theta)^T A J(p, q, \theta)$$
$$\qquad\qquad\qquad V \;=\; V J(p, q, \theta)$$
$$\qquad\qquad \textbf{end}$$
$$\qquad \textbf{end}$$
$$\textbf{end}$$

The cyclic Jacobi algorithm also converges quadratically. (See Wilkinson (1962) and
van Kempen (1966).) However, since it does not require off-diagonal search, it is
considerably faster than Jacobi's original algorithm.

8.5.5   Error Analysis

Using Wilkinson's error analysis it is possible to show that if r sweeps are required by
Algorithm 8.5.3 and $d_1, \ldots, d_n$ specify the diagonal entries of the final, computed A

matrix, then

$$\sum_{i=1}^{n} (d_i - \lambda_i)^2 \leq (\text{tol} + k_r \mathbf{u}) \| A \|_F$$

for some ordering of A's eigenvalues λ_i. The parameter k_r depends mildly on r.

Although the cyclic Jacobi method converges quadratically, it is not generally competitive with the symmetric QR algorithm. For example, if we just count flops, then two sweeps of Jacobi are roughly equivalent to a complete QR reduction to diagonal form with accumulation of transformations. However, for small n this liability is not very dramatic. Moreover, if an approximate eigenvector matrix V is known, then $V^T A V$ is almost diagonal, a situation that Jacobi can exploit but not QR.

Another interesting feature of the Jacobi method is that it can compute the eigenvalues with small *relative* error if A is positive definite. To appreciate this point, note that the Wilkinson analysis cited above coupled with the §8.1 perturbation theory ensures that the computed eigenvalues $\hat{\lambda}_1 \geq \cdots \geq \hat{\lambda}_n$ satisfy

$$\frac{|\hat{\lambda}_i - \lambda_i(A)|}{\lambda_i(A)} \approx \mathbf{u} \frac{\| A \|_2}{\lambda_i(A)} \leq \mathbf{u} \, \kappa_2(A).$$

However, a refined, componentwise error analysis by Demmel and Veselić (1992) shows that in the positive definite case

$$\frac{|\hat{\lambda}_i - \lambda_i(A)|}{\lambda_i(A)} \approx \mathbf{u} \, \kappa_2(D^{-1} A D^{-1}) \tag{8.5.4}$$

where $D = \text{diag}(\sqrt{a_{11}}, \ldots, \sqrt{a_{nn}})$ and this is generally a much smaller approximating bound. The key to establishing this result is some new perturbation theory and a demonstration that if A_+ is a computed Jacobi update obtained from the current matrix A_c, then the eigenvalues of A_+ are relatively close to the eigenvalues of A_c in the sense of (8.5.4). To make the whole thing work in practice, the termination criterion is not based upon the comparison of $\text{off}(A)$ with $\mathbf{u} \| A \|_F$ but rather on the size of each $|a_{ij}|$ compared to $\mathbf{u} \sqrt{a_{ii} a_{jj}}$.

8.5.6 Block Jacobi Procedures

It is usually the case when solving the symmetric eigenvalue problem on a p-processor machine that $n \gg p$. In this case a block version of the Jacobi algorithm may be appropriate. Block versions of the above procedures are straightforward. Suppose that $n = rN$ and that we partition the n-by-n matrix A as follows:

$$A = \begin{bmatrix} A_{11} & \cdots & A_{1N} \\ \vdots & & \vdots \\ A_{N1} & \cdots & A_{NN} \end{bmatrix}.$$

Here, each A_{ij} is r-by-r. In a block Jacobi procedure the (p, q) subproblem involves computing the $2r$-by-$2r$ Schur decomposition

$$\begin{bmatrix} V_{pp} & V_{pq} \\ V_{qp} & V_{qq} \end{bmatrix}^T \begin{bmatrix} A_{pp} & A_{pq} \\ A_{qp} & A_{qq} \end{bmatrix} \begin{bmatrix} V_{pp} & V_{pq} \\ V_{qp} & V_{qq} \end{bmatrix} = \begin{bmatrix} D_{pp} & 0 \\ 0 & D_{qq} \end{bmatrix}$$

and then applying to A the block Jacobi rotation made up of the V_{ij}. If we call this block rotation V, then it is easy to show that

$$\text{off}(V^T A V)^2 = \text{off}(A)^2 - \left(2\| A_{pq} \|_F^2 + \text{off}(A_{pp})^2 + \text{off}(A_{qq})^2 \right).$$

Block Jacobi procedures have many interesting computational aspects. For example, there are several ways to solve the subproblems, and the choice appears to be critical. See Bischof (1987).

8.5.7   A Note on the Parallel Ordering

The Block Jacobi approach to the symmetric eigenvalue problem has an inherent parallelism that has attracted significant attention. The key observation is that the (i_1, j_1) subproblem is independent of the (i_2, j_2) subproblem *if* the four indices i_1, j_1, i_2, and j_2 are distinct. Moreover, if we regard the A as a $2m$-by-$2m$ block matrix, then it is possible to partition the set of off-diagonal index pairs into a collection of $2m - 1$ *rotation sets*, each of which identifies m, nonconflicting subproblems.

A good way to visualize this is to imagine a chess tournament with $2m$ players in which everybody must play everybody else exactly once. Suppose $m = 4$. In "round 1" we have Player 1 versus Player 2, Player 3 versus Player 4, Player 5 versus Player 6, and Player 7 versus Player 8. Thus, there are four tables of action:

1	3	5	7
2	4	6	8

This corresponds to the first rotation set:

$$rot.set(1) \;=\; \{\, (1,2), (3,4), (5,6), (7,8) \,\}.$$

To set up rounds 2 through 7, Player 1 stays put and Players 2 through 8 move from table to table in merry-go-round fashion:

1	2	3	5
4	6	8	7

$$rot.set(2) \;=\; \{(1,4), (2,6), (3,8), (5,7)\},$$

1	4	2	3
6	8	7	5

$$rot.set(3) \;=\; \{(1,6), (4,8), (2,7), (3,5)\},$$

1	6	4	2
8	7	5	3

$$rot.set(4) \;=\; \{(1,8), (6,7), (4,5), (2,3)\},$$

1	8	6	4
7	5	3	2

$$rot.set(5) \;=\; \{(1,7), (5,8), (3,6), (2,4)\},$$

1	7	8	6
5	3	2	4

$$rot.set(6) \;=\; \{(1,5), (3,7), (2,8), (4,6)\},$$

1	5	7	8
3	2	4	6

$$rot.set(7) \;=\; \{(1,3), (2,5), (4,7), (6,8)\}.$$

Taken in order, the seven rotation sets define the parallel ordering of the 28 possible off-diagonal index pairs.

For general m, a multiprocessor implementation would involve solving the sub-problems within each rotation set in parallel. Although the generation of the subproblem rotations is independent, some synchronization is required to carry out the block similarity transform updates.

Problems

P8.5.1 Let the scalar γ be given along with the matrix

$$A = \left[\begin{array}{cc} w & x \\ x & z \end{array} \right].$$

It is desired to compute an orthogonal matrix

$$J = \left[\begin{array}{cc} c & s \\ -s & c \end{array} \right]$$

such that the (1, 1) entry of $J^T A J$ equals γ. Show that this requirement leads to the equation

$$(w - \gamma)\tau^2 - 2x\tau + (z - \gamma) = 0,$$

where $\tau = c/s$. Verify that this quadratic has real roots if γ satisfies $\lambda_2 \leq \gamma \leq \lambda_1$, where λ_1 and λ_2 are the eigenvalues of A.

P8.5.2 Let $A \in \mathbb{R}^{n \times n}$ be symmetric. Give an algorithm that computes the factorization

$$Q^T A Q = \gamma I + F$$

where Q is a product of Jacobi rotations, $\gamma = \text{tr}(A)/n$, and F has zero diagonal entries. Discuss the uniqueness of Q.

P8.5.3 Formulate Jacobi procedures for (a) skew-symmetric matrices and (b) complex Hermitian matrices.

P8.5.4 Partition the n-by-n real symmetric matrix A as follows:

$$A = \left[\begin{array}{cc} a & v^T \\ v & A_1 \end{array} \right] \begin{array}{c} 1 \\ n-1 \end{array} \quad .$$
$$\quad\quad 1 \quad n-1$$

Let Q be a Householder matrix such that if $B = Q^T A Q$, then $B(3{:}n, 1) = 0$. Let $J = J(1, 2, \theta)$ be determined such that if $C = J^T B J$, then $c_{12} = 0$ and $c_{11} \geq c_{22}$. Show $c_{11} \geq a + \| v \|_2$. La Budde (1964) formulated an algorithm for the symmetric eigenvalue probem based upon repetition of this Householder-Jacobi computation.

P8.5.5 When implementing the cyclic Jacobi algorithm, it is sensible to skip the annihilation of a_{pq} if its modulus is less than some small, sweep-dependent parameter because the net reduction in off(A) is not worth the cost. This leads to what is called the *threshold Jacobi method*. Details concerning this variant of Jacobi's algorithm may be found in Wilkinson (AEP, p. 277). Show that appropriate thresholding can guarantee convergence.

P8.5.6 Given a positive integer m, let $M = (2m - 1)m$. Develop an algorithm for computing integer vectors $i, j \in \mathbb{R}^M$ so that $(i_1, j_1), \ldots, (i_M, j_M)$ defines the parallel ordering.

Notes and References for §8.5

Jacobi's original paper is one of the earliest references found in the numerical analysis literature:

C.G.J. Jacobi (1846). "Uber ein Leichtes Verfahren Die in der Theorie der Sacularstroungen Vorkommendern Gleichungen Numerisch Aufzulosen," *Crelle's J. 30*, 51–94.

Prior to the QR algorithm, the Jacobi technique was the standard method for solving dense symmetric eigenvalue problems. Early references include:

M. Lotkin (1956). "Characteristic Values of Arbitrary Matrices," *Quart. Appl. Math. 14*, 267–275.

D.A. Pope and C. Tompkins (1957). "Maximizing Functions of Rotations: Experiments Concerning Speed of Diagonalization of Symmetric Matrices Using Jacobi's Method," *J. ACM 4*, 459–466.

C.D. La Budde (1964). "Two Classes of Algorithms for Finding the Eigenvalues and Eigenvectors of Real Symmetric Matrices," *J. ACM 11*, 53–58.

H. Rutishauser (1966). "The Jacobi Method for Real Symmetric Matrices," *Numer. Math. 9*, 1–10.

See also Wilkinson (AEP, p. 265) and:

J.H. Wilkinson (1968). "Almost Diagonal Matrices with Multiple or Close Eigenvalues," *Lin. Alg. Applic. 1*, 1–12.

Papers that are concerned with quadratic convergence include:

P. Henrici (1958). "On the Speed of Convergence of Cyclic and Quasicyclic Jacobi Methods for Computing the Eigenvalues of Hermitian Matrices," *SIAM J. Appl. Math. 6*, 144–162.

E.R. Hansen (1962). "On Quasicyclic Jacobi Methods," *J. ACM 9*, 118–135.

J.H. Wilkinson (1962). "Note on the Quadratic Convergence of the Cyclic Jacobi Process," *Numer. Math. 6*, 296–300.

E.R. Hansen (1963). "On Cyclic Jacobi Methods," *SIAM J. Appl. Math. 11*, 448–459.

A. Schonhage (1964). "On the Quadratic Convergence of the Jacobi Process," *Numer. Math. 6*, 410–412.

H.P.M. van Kempen (1966). "On Quadratic Convergence of the Special Cyclic Jacobi Method," *Numer. Math. 9*, 19–22.

P. Henrici and K. Zimmermann (1968). "An Estimate for the Norms of Certain Cyclic Jacobi Operators," *Lin. Alg. Applic. 1*, 489–501.

K.W. Brodlie and M.J.D. Powell (1975). "On the Convergence of Cyclic Jacobi Methods," *J. Inst. Math. Applic. 15*, 279–287.

The ordering of the subproblems within a sweep is important:

W.F. Mascarenhas (1995). "On the Convergence of the Jacobi Method for Arbitrary Orderings," *SIAM J. Matrix Anal. Applic. 16*, 1197–1209.

Z. Dramač (1996). "On the Condition Behaviour in the Jacobi Method," *SIAM J. Matrix Anal. Applic. 17*, 509–514.

V. Hari (2007). "Convergence of a Block-Oriented Quasi-Cyclic Jacobi Method," *SIAM J. Matrix Anal. Applic. 29*, 349–369.

Z. Drmač (2010). "A Global Convergence Proof for Cyclic Jacobi Methods with Block Rotations," *SIAM J. Matrix Anal. Applic. 31*, 1329–1350.

Detailed error analyses that establish the high accuracy of Jacobi's method include:

J. Barlow and J. Demmel (1990). "Computing Accurate Eigensystems of Scaled Diagonally Dominant Matrices," *SIAM J. Numer. Anal. 27*, 762–791.

J.W. Demmel and K. Veselić (1992). "Jacobi's Method is More Accurate than QR," *SIAM J. Matrix Anal. Applic. 13*, 1204–1245.

W.F. Mascarenhas (1994). "A Note on Jacobi Being More Accurate than QR," *SIAM J. Matrix Anal. Applic. 15*, 215–218.

R. Mathias (1995). "Accurate Eigensystem Computations by Jacobi Methods," *SIAM J. Matrix Anal. Applic. 16*, 977–1003.

K. Veselić (1996). "A Note on the Accuracy of Symmetric Eigenreduction Algorithms," *ETNA 4*, 37–45.

F.M. Dopico, J.M. Molera, and J. Moro (2003). "An Orthogonal High Relative Accuracy Algorithm for the Symmetric Eigenproblem," *SIAM J. Matrix Anal. Applic. 25*, 301–351.

F.M. Dopico, P. Koev, and J.M. Molera (2008). "Implicit Standard Jacobi Gives High Relative Accuracy," *Numer. Math. 113*, 519–553.

Attempts have been made to extend the Jacobi iteration to other classes of matrices and to push through corresponding convergence results. The case of normal matrices is discussed in:

H.H. Goldstine and L.P. Horowitz (1959). "A Procedure for the Diagonalization of Normal Matrices," *J. ACM 6*, 176–195.

G. Loizou (1972). "On the Quadratic Convergence of the Jacobi Method for Normal Matrices," *Comput. J. 15*, 274–276.

M.H.C. Paardekooper (1971). "An Eigenvalue Algorithm for Skew Symmetric Matrices," *Numer. Math. 17*, 189–202.

A. Ruhe (1972). "On the Quadratic Convergence of the Jacobi Method for Normal Matrices," *BIT 7*, 305–313.

D. Hacon (1993). "Jacobi's Method for Skew-Symmetric Matrices," *SIAM J. Matrix Anal. Applic. 14*, 619–628.

Essentially, the analysis and algorithmic developments presented in the text carry over to the normal case with minor modification. For non-normal matrices, the situation is considerably more difficult:

J. Greenstadt (1955). "A Method for Finding Roots of Arbitrary Matrices," *Math. Tables and Other Aids to Comp. 9*, 47–52.

C.E. Froberg (1965). "On Triangularization of Complex Matrices by Two Dimensional Unitary Tranformations," *BIT 5*, 230–234.

J. Boothroyd and P.J. Eberlein (1968). "Solution to the Eigenproblem by a Norm-Reducing Jacobi-Type Method (Handbook)," *Numer. Math. 11*, 1–12.

A. Ruhe (1968). "On the Quadratic Convergence of a Generalization of the Jacobi Method to Arbitrary Matrices," *BIT 8*, 210–231.

A. Ruhe (1969). "The Norm of a Matrix After a Similarity Transformation," *BIT 9*, 53–58.

P.J. Eberlein (1970). "Solution to the Complex Eigenproblem by a Norm-Reducing Jacobi-type Method," *Numer. Math. 14*, 232–245.

C.P. Huang (1975). "A Jacobi-Type Method for Triangularizing an Arbitrary Matrix," *SIAM J. Numer. Anal. 12*, 566–570.

V. Hari (1982). "On the Global Convergence of the Eberlein Method for Real Matrices," *Numer. Math. 39*, 361–370.

G.W. Stewart (1985). "A Jacobi-Like Algorithm for Computing the Schur Decomposition of a Non-hermitian Matrix," *SIAM J. Sci. Stat. Comput. 6*, 853–862.

C. Mehl (2008). "On Asymptotic Convergence of Nonsymmetric Jacobi Algorithms," *SIAM J. Matrix Anal. Applic. 30*, 291–311.

Jacobi methods for complex symmetric matrices have also been developed, see:

J.J. Seaton (1969). "Diagonalization of Complex Symmetric Matrices Using a Modified Jacobi Method," *Comput. J. 12*, 156–157.

P.J. Eberlein (1971). "On the Diagonalization of Complex Symmetric Matrices," *J. Inst. Math. Applic. 7*, 377–383.

P. Anderson and G. Loizou (1973). "On the Quadratic Convergence of an Algorithm Which Diagonalizes a Complex Symmetric Matrix," *J. Inst. Math. Applic. 12*, 261–271.

P. Anderson and G. Loizou (1976). "A Jacobi-Type Method for Complex Symmetric Matrices (Handbook)," *Numer. Math. 25*, 347–363.

Other extensions include:

N. Mackey (1995). "Hamilton and Jacobi Meet Again: Quaternions and the Eigenvalue Problem," *SIAM J. Matrix Anal. Applic. 16*, 421–435.

A.W. Bojanczyk (2003). "An Implicit Jacobi-like Method for Computing Generalized Hyperbolic SVD," *Lin. Alg. Applic. 358*, 293–307.

For a sampling of papers concerned with various aspects of parallel Jacobi, see:

A. Sameh (1971). "On Jacobi and Jacobi-like Algorithms for a Parallel Computer," *Math. Comput. 25*, 579–590.

D.S. Scott, M.T. Heath, and R.C. Ward (1986). "Parallel Block Jacobi Eigenvalue Algorithms Using Systolic Arrays," *Lin. Alg. Applic. 77*, 345–356.

P.J. Eberlein (1987). "On Using the Jacobi Method on a Hypercube," in *Hypercube Multiprocessors*, M.T. Heath (ed.), SIAM Publications, Philadelphia.

G. Shroff and R. Schreiber (1989). "On the Convergence of the Cyclic Jacobi Method for Parallel Block Orderings," *SIAM J. Matrix Anal. Applic. 10*, 326–346.

M.H.C. Paardekooper (1991). "A Quadratically Convergent Parallel Jacobi Process for Diagonally Dominant Matrices with Nondistinct Eigenvalues," *Lin. Alg. Applic. 145*, 71–88.

T. Londre and N.H. Rhee (2005). "Numerical Stability of the Parallel Jacobi Method," *SIAM J. Matrix Anal. Applic. 26*, 985–1000.

8.6   Computing the SVD

If $U^T A V = B$ is the bidiagonal decomposition of $A \in \mathbb{R}^{m \times n}$, then $V^T(A^T A)V = B^T B$ is the tridiagonal decomposition of the symmetric matrix $A^T A \in \mathbb{R}^{n \times n}$. Thus, there is an intimate connection between Algorithm 5.4.2 (Householder bidiagonalization) and Algorithm 8.3.1 (Householder tridiagonalization). In this section we carry this a step further and show that there is a bidiagonal SVD procedure that corresponds to the symmetric tridiagonal QR iteration. Before we get into the details, we catalog some important SVD properties that have algorithmic ramifications.

8.6.1   Connections to the Symmetric Eigenvalue Problem

There are important relationships between the singular value decomposition of a matrix A and the Schur decompositions of the symmetric matrices

$$S_1 = A^T A, \qquad S_2 = AA^T \qquad S_3 = \begin{bmatrix} 0 & A^T \\ A & 0 \end{bmatrix}.$$

Indeed, if

$$U^T A V = \text{diag}(\sigma_1, \ldots, \sigma_n)$$

is the SVD of $A \in \mathbb{R}^{m \times n}$ $(m \geq n)$, then

$$V^T(A^T A)V = \text{diag}(\sigma_1^2, \ldots, \sigma_n^2) \in \mathbb{R}^{n \times n} \tag{8.6.1}$$

and

$$U^T(AA^T)U = \text{diag}(\sigma_1^2, \ldots, \sigma_n^2, \underbrace{0, \ldots, 0}_{m-n}) \in \mathbb{R}^{m \times m} \tag{8.6.2}$$

Moreover, if

$$U = \underset{n \quad\;\; m-n}{[\ U_1 \mid U_2\ ]}$$

and we define the orthogonal matrix $Q \in \mathbb{R}^{(m+n) \times (m+n)}$ by

$$Q = \frac{1}{\sqrt{2}} \begin{bmatrix} V & V & 0 \\ U_1 & -U_1 & \sqrt{2}\,U_2 \end{bmatrix},$$

then

$$Q^T \begin{bmatrix} 0 & A^T \\ A & 0 \end{bmatrix} Q = \text{diag}(\sigma_1, \ldots, \sigma_n, -\sigma_1, \ldots, -\sigma_n, \underbrace{0, \ldots, 0}_{m-n}). \tag{8.6.3}$$

These connections to the symmetric eigenproblem allow us to adapt the mathematical and algorithmic developments of the previous sections to the singular value problem. Good references for this section include Lawson and Hanson (SLS) and Stewart and Sun (MPT).

8.6.2 Perturbation Theory and Properties

We first establish perturbation results for the SVD based on the theorems of §8.1. Recall that $\sigma_i(A)$ denotes the ith largest singular value of A.

Theorem 8.6.1. *If $A \in \mathbb{R}^{m \times n}$, then for $k = 1{:}\min\{m, n\}$*

$$
\sigma_k(A) = \min_{\dim(S)=n-k+1} \; \max_{\substack{x \in S \\ y \in \mathbb{R}^m}} \frac{y^T A x}{\| x \|_2 \| y \|_2} = \max_{\dim(S)=k} \; \min_{x \in S} \frac{\| A x \|_2}{\| x \|_2}.
$$

In this expression, S is a subspace of $\mathbb{R}^n$.

Proof. The rightmost characterization follows by applying Theorem 8.1.2 to $A^T A$. For the remainder of the proof see Xiang (2006). $\square$

Corollary 8.6.2. *If A and $A + E$ are in $\mathbb{R}^{m \times n}$ with $m \geq n$, then for $k = 1{:}n$*

$$
|\sigma_k(A + E) - \sigma_k(A)| \leq \sigma_1(E) = \| E \|_2.
$$

Proof. Define $\widetilde{A}$ and $\widetilde{E}$ by

$$
\widetilde{A} = \begin{bmatrix} 0 & A^T \\ A & 0 \end{bmatrix}, \qquad \widetilde{A} + \widetilde{E} = \begin{bmatrix} 0 & (A+E)^T \\ A+E & 0 \end{bmatrix}. \tag{8.6.4}
$$

The corollary follows by applying Corollary 8.1.6 with A replaced by $\widetilde{A}$ and $A + E$ replaced by $\widetilde{A} + \widetilde{E}$. $\square$

Corollary 8.6.3. *Let $A = [\, a_1 \mid \cdots \mid a_n \,] \in \mathbb{R}^{m \times n}$ be a column partitioning with $m \geq n$. If $A_r = [\, a_1 \mid \cdots \mid a_r \,]$, then for $r = 1{:}n-1$*

$$
\sigma_1(A_{r+1}) \geq \sigma_1(A_r) \geq \sigma_2(A_{r+1}) \geq \cdots \geq \sigma_r(A_{r+1}) \geq \sigma_r(A_r) \geq \sigma_{r+1}(A_{r+1}).
$$

Proof. Apply Corollary 8.1.7 to $A^T A$. $\square$

The next result is a Wielandt-Hoffman theorem for singular values:

Theorem 8.6.4. *If A and $A + E$ are in $\mathbb{R}^{m \times n}$ with $m \geq n$, then*

$$
\sum_{k=1}^{n} (\sigma_k(A + E) - \sigma_k(A))^2 \leq \| E \|_F^2.
$$

Proof. Apply Theorem 8.1.4 with A and E replaced by the matrices $\widetilde{A}$ and $\widetilde{E}$ defined by (8.6.4). $\square$

For $A \in \mathbb{R}^{m \times n}$ we say that the k-dimensional subspaces $S \subseteq \mathbb{R}^n$ and $T \subseteq \mathbb{R}^m$ form a *singular subspace pair* if $x \in S$ and $y \in T$ imply $Ax \in T$ and $A^T y \in S$. The following result is concerned with the perturbation of singular subspace pairs.

Theorem 8.6.5. *Let $A, E \in \mathbb{R}^{m \times n}$ with $m \geq n$ be given and suppose that $V \in \mathbb{R}^{n \times n}$ and $U \in \mathbb{R}^{m \times m}$ are orthogonal. Assume that*

$$
V = [\; V_1 \mid V_2 \;] \; , \qquad U = [\; U_1 \mid U_2 \;] \; ,
$$
$$
\quad\; r \quad n-r \qquad\qquad\quad\; r \quad m-r
$$

and that $\mathsf{ran}(V_1)$ *and* $\mathsf{ran}(U_1)$ *form a singular subspace pair for A. Let*

$$
U^T A V = \begin{bmatrix} A_{11} & 0 \\ 0 & A_{22} \end{bmatrix} \begin{matrix} r \\ m-r \end{matrix} \; , \qquad U^T E V = \begin{bmatrix} E_{11} & E_{12} \\ E_{21} & E_{22} \end{bmatrix} \begin{matrix} r \\ m-r \end{matrix} \; ,
$$
$$
\quad\;\; r \quad\;\; n-r \qquad\qquad\qquad\quad r \quad\;\; n-r
$$

and assume that

$$
\delta = \min_{\substack{\sigma \in \sigma(A_{11}) \\ \gamma \in \sigma(A_{22})}} |\sigma - \gamma| > 0.
$$

If

$$
\| \, E \, \|_F \leq \frac{\delta}{5},
$$

then there exist matrices $P \in \mathbb{R}^{(n-r) \times r}$ and $Q \in \mathbb{R}^{(m-r) \times r}$ satisfying

$$
\left\| \begin{bmatrix} Q \\ P \end{bmatrix} \right\|_F \leq 4 \frac{\| \, E \, \|_F}{\delta}
$$

such that $\mathsf{ran}(V_1 + V_2 Q)$ *and* $\mathsf{ran}(U_1 + U_2 P)$ *is a singular subspace pair for $A + E$.*

Proof. See Stewart (1973, Theorem 6.4).   □

Roughly speaking, the theorem says that $O(\epsilon)$ changes in A can alter a singular subspace by an amount ϵ/δ, where δ measures the separation of the associated singular values.

8.6.3   The SVD Algorithm

We now show how a variant of the QR algorithm can be used to compute the SVD of an $A \in \mathbb{R}^{m \times n}$ with $m \geq n$. At first glance, this appears straightforward. Equation (8.6.1) suggests that we proceed as follows:

Step 1. Form $C = A^T A$,

Step 2. Use the symmetric QR algorithm to compute $V_1^T C V_1 = \mathrm{diag}(\sigma_i^2)$.

Step 3. Apply QR with column pivoting to $A V_1$ obtaining $U^T (A V_1) \Pi = R$.

Since R has orthogonal columns, it follows that $U^T A(V_1 \Pi)$ is diagonal. However, as we saw in §5.3.2, the formation of $A^T A$ can lead to a loss of information. The situation is not quite so bad here, since the original A is used to compute U.

A preferable method for computing the SVD is described by Golub and Kahan (1965). Their technique finds U and V simultaneously by *implicitly* applying the symmetric QR algorithm to $A^T A$. The first step is to reduce A to upper bidiagonal form using Algorithm 5.4.2:

$$U_B^T A V_B = \begin{bmatrix} B \\ 0 \end{bmatrix}, \qquad B = \begin{bmatrix} d_1 & f_1 & & \cdots & & 0 \\ 0 & d_2 & \ddots & & & \vdots \\ & & \ddots & \ddots & \ddots & \\ & & & \ddots & \ddots & f_{n-1} \\ \vdots & & & & \ddots & d_n \\ 0 & \cdots & & & 0 & d_n \end{bmatrix} \in \mathbb{R}^{n \times n}.$$

The remaining problem is thus to compute the SVD of B. To this end, consider applying an implicit-shift QR step (Algorithm 8.3.2) to the tridiagonal matrix $T = B^T B$:

Step 1. Compute the eigenvalue λ of

$$T(m{:}n, m{:}n) = \begin{bmatrix} d_m^2 + f_{m-1}^2 & d_m f_m \\ d_m f_m & d_n^2 + f_m^2 \end{bmatrix}, \qquad m = n-1,$$

that is closer to $d_n^2 + f_m^2$.

Step 2. Compute $c_1 = \cos(\theta_1)$ and $s_1 = \sin(\theta_1)$ such that

$$\begin{bmatrix} c_1 & s_1 \\ -s_1 & c_1 \end{bmatrix}^T \begin{bmatrix} d_1^2 - \lambda \\ d_1 f_1 \end{bmatrix} = \begin{bmatrix} \times \\ 0 \end{bmatrix}$$

and set $G_1 = G(1, 2, \theta_1)$.

Step 3. Compute Givens rotations $G_2, \ldots, G_{n-1}$ so that if $Q = G_1 \cdots G_{n-1}$ then $Q^T T Q$ is tridiagonal and $Q e_1 = G_1 e_1$.

Note that these calculations require the explicit formation of $B^T B$, which, as we have seen, is unwise from the numerical standpoint.

Suppose instead that we apply the Givens rotation G_1 above to B directly. Illustrating with the $n = 6$ case we have

$$B \leftarrow BG_1 = \begin{bmatrix} \times & \times & 0 & 0 & 0 & 0 \\ + & \times & \times & 0 & 0 & 0 \\ 0 & 0 & \times & \times & 0 & 0 \\ 0 & 0 & 0 & \times & \times & 0 \\ 0 & 0 & 0 & 0 & \times & \times \\ 0 & 0 & 0 & 0 & 0 & \times \end{bmatrix}.$$

We then can determine Givens rotations $U_1, V_2, U_2, \ldots, V_{n-1}$, and U_{n-1} to chase the unwanted nonzero element down the bidiagonal:

$$B \leftarrow U_1^T B = \begin{bmatrix} \times & \times & + & 0 & 0 & 0 \\ 0 & \times & \times & 0 & 0 & 0 \\ 0 & 0 & \times & \times & 0 & 0 \\ 0 & 0 & 0 & \times & \times & 0 \\ 0 & 0 & 0 & 0 & \times & \times \\ 0 & 0 & 0 & 0 & 0 & \times \end{bmatrix}, \quad B \leftarrow BV_2 = \begin{bmatrix} \times & \times & 0 & 0 & 0 & 0 \\ 0 & \times & \times & 0 & 0 & 0 \\ 0 & + & \times & \times & 0 & 0 \\ 0 & 0 & 0 & \times & \times & 0 \\ 0 & 0 & 0 & 0 & \times & \times \\ 0 & 0 & 0 & 0 & 0 & \times \end{bmatrix},$$

$$B \leftarrow U_2^T B = \begin{bmatrix} \times & \times & 0 & 0 & 0 & 0 \\ 0 & \times & \times & + & 0 & 0 \\ 0 & 0 & \times & \times & 0 & 0 \\ 0 & 0 & 0 & \times & \times & 0 \\ 0 & 0 & 0 & 0 & \times & \times \\ 0 & 0 & 0 & 0 & 0 & \times \end{bmatrix}, \quad B \leftarrow BV_3 = \begin{bmatrix} \times & \times & 0 & 0 & 0 & 0 \\ 0 & \times & \times & 0 & 0 & 0 \\ 0 & 0 & \times & \times & 0 & 0 \\ 0 & 0 & + & \times & \times & 0 \\ 0 & 0 & 0 & 0 & \times & \times \\ 0 & 0 & 0 & 0 & 0 & \times \end{bmatrix},$$

and so on. The process terminates with a new bidiagonal $\tilde{B}$ that is related to B as follows:

$$\tilde{B} = (U_{n-1}^T \cdots U_1^T) B (G_1 V_2 \cdots V_{n-1}) = \tilde{U}^T B \tilde{V}.$$

Since each V_i has the form $V_i = G(i, i+1, \theta_i)$ where $i = 2{:}n-1$, it follows that $\tilde{V} e_1 = Q e_1$. By the Implicit Q theorem we can assert that $\tilde{V}$ and Q are essentially the same. Thus, we can implicitly effect the transition from T to $\tilde{T} = \tilde{B}^T \tilde{B}$ by working directly on the bidiagonal matrix B.

Of course, for these claims to hold it is necessary that the underlying tridiagonal matrices be unreduced. Since the subdiagonal entries of $B^T B$ are of the form $d_i f_i$, it is clear that we must search the bidiagonal band for zeros. If $f_k = 0$ for some k, then

$$B = \begin{bmatrix} B_1 & 0 \\ 0 & B_2 \end{bmatrix} \begin{matrix} k \\ n-k \end{matrix}$$
$$\begin{matrix} k & n-k \end{matrix}$$

and the original SVD problem decouples into two smaller problems involving the matrices B_1 and B_2. If $d_k = 0$ for some $k < n$, then premultiplication by a sequence of Givens transformations can zero f_k. For example, if $n = 6$ and $k = 3$, then by rotating in row planes (3,4), (3,5), and (3,6) we can zero the entire third row:

$$B = \begin{bmatrix} \times & \times & 0 & 0 & 0 & 0 \\ 0 & \times & \times & 0 & 0 & 0 \\ 0 & 0 & 0 & \times & 0 & 0 \\ 0 & 0 & 0 & \times & \times & 0 \\ 0 & 0 & 0 & 0 & \times & \times \\ 0 & 0 & 0 & 0 & 0 & \times \end{bmatrix} \xrightarrow{(3,4)} \begin{bmatrix} \times & \times & 0 & 0 & 0 & 0 \\ 0 & \times & \times & 0 & 0 & 0 \\ 0 & 0 & 0 & 0 & + & 0 \\ 0 & 0 & 0 & \times & \times & 0 \\ 0 & 0 & 0 & 0 & \times & \times \\ 0 & 0 & 0 & 0 & 0 & \times \end{bmatrix}$$

$$\xrightarrow{(3,5)} \begin{bmatrix} \times & \times & 0 & 0 & 0 & 0 \\ 0 & \times & \times & 0 & 0 & 0 \\ 0 & 0 & 0 & 0 & 0 & + \\ 0 & 0 & 0 & \times & \times & 0 \\ 0 & 0 & 0 & 0 & \times & \times \\ 0 & 0 & 0 & 0 & 0 & \times \end{bmatrix} \xrightarrow{(3,6)} \begin{bmatrix} \times & \times & 0 & 0 & 0 & 0 \\ 0 & \times & \times & 0 & 0 & 0 \\ 0 & 0 & 0 & 0 & 0 & 0 \\ 0 & 0 & 0 & \times & \times & 0 \\ 0 & 0 & 0 & 0 & \times & \times \\ 0 & 0 & 0 & 0 & 0 & \times \end{bmatrix}.$$

If $d_n = 0$, then the last column can be zeroed with a series of column rotations in planes $(n-1, n)$, $(n-2, n)$, ..., $(1, n)$. Thus, we can decouple if $f_1 \cdots f_{n-1} = 0$ or $d_1 \cdots d_n = 0$. Putting it all together we obtain the following SVD analogue of Algorithm 8.3.2.

Algorithm 8.6.1 (Golub-Kahan SVD Step)   Given a bidiagonal matrix $B \in \mathbb{R}^{m \times n}$ having no zeros on its diagonal or superdiagonal, the following algorithm overwrites B with the bidiagonal matrix $\bar{B} = \bar{U}^T B \bar{V}$ where $\bar{U}$ and $\bar{V}$ are orthogonal and $\bar{V}$ is essentially the orthogonal matrix that would be obtained by applying Algorithm 8.3.2 to $T = B^T B$.

> Let μ be the eigenvalue of the trailing 2-by-2 submatrix of $T = B^T B$
> > that is closer to t_{nn}.
>
> $y = t_{11} - \mu$
>
> $z = t_{12}$
>
> **for** $k = 1{:}n - 1$
>
> > Determine $c = \cos(\theta)$ and $s = \sin(\theta)$ such that
> >
> > $$\begin{bmatrix} y & z \end{bmatrix} \begin{bmatrix} c & s \\ -s & c \end{bmatrix} = \begin{bmatrix} * & 0 \end{bmatrix}.$$
> >
> > $B = B \cdot G(k, k+1, \theta)$
> >
> > $y = b_{kk}$
> >
> > $z = b_{k+1,k}$
> >
> > Determine $c = \cos(\theta)$ and $s = \sin(\theta)$ such that
> >
> > $$\begin{bmatrix} c & s \\ -s & c \end{bmatrix}^T \begin{bmatrix} y \\ z \end{bmatrix} = \begin{bmatrix} * \\ 0 \end{bmatrix}.$$
> >
> > $B = G(k, k+1, \theta)^T B$
> >
> > **if** $k < n - 1$
> >
> > > $y = b_{k,k+1}$
> > >
> > > $z = b_{k,k+2}$
> >
> > **end**
>
> **end**

An efficient implementation of this algorithm would store B's diagonal and superdiagonal in vectors $d(1{:}n)$ and $f(1{:}n-1)$, respectively, and would require $30n$ flops and $2n$ square roots. Accumulating U requires $6mn$ flops. Accumulating V requires $6n^2$ flops.

   Typically, after a few of the above SVD iterations, the superdiagonal entry f_{n-1} becomes negligible. Criteria for smallness within B's band are usually of the form

$$|f_i| \leq \mathsf{tol} \cdot (\, |d_i| + |d_{i+1}| \,),$$
$$|d_i| \leq \mathsf{tol} \cdot \| \, B \, \|,$$

where tol is a small multiple of the unit roundoff and $\| \cdot \|$ is some computationally convenient norm. Combining Algorithm 5.4.2 (bidiagonalization), Algorithm 8.6.1, and the decoupling calculations mentioned earlier gives the following procedure.

Algorithm 8.6.2 (The SVD Algorithm)   Given $A \in \mathbb{R}^{m \times n}$ $(m \geq n)$ and ϵ, a small multiple of the unit roundoff, the following algorithm overwrites A with $U^T A V = D + E$, where $U \in \mathbb{R}^{m \times m}$ is orthogonal, $V \in \mathbb{R}^{n \times n}$ is orthogonal, $D \in \mathbb{R}^{m \times n}$ is diagonal, and E satisfies $\| E \|_2 \approx \mathbf{u} \| A \|_2$.

Use Algorithm 5.4.2 to compute the bidiagonalization.

$$\left[\begin{array}{c} B \\ 0 \end{array} \right] \leftarrow (U_1 \cdots U_n)^T A (V_1 \cdots V_{n-2}).$$

until $q = n$

For $i = 1{:}n - 1$, set $b_{i,i+1}$ to zero if $|b_{i,i+1}| \leq \epsilon(|b_{ii}| + |b_{i+1,i+1}|)$.

Find the largest q and the smallest p such that if

$$B = \left[\begin{array}{ccc} B_{11} & 0 & 0 \\ 0 & B_{22} & 0 \\ 0 & 0 & B_{33} \end{array} \right] \begin{array}{c} p \\ n-p-q \\ q \end{array}$$
$$\qquad\quad\; \begin{array}{ccc} p & n-p-q & q \end{array}$$

then B_{33} is diagonal and B_{22} has a nonzero superdiagonal.

if $q < n$

if any diagonal entry in B_{22} is zero, then zero the superdiagonal entry in the same row.

else

Apply Algorithm 8.6.1 to B_{22}.

$B = \mathrm{diag}(I_p, U, I_{q+m-n})^T B \, \mathrm{diag}(I_p, V, I_q)$

end

end

end

The amount of work required by this algorithm depends on how much of the SVD is required. For example, when solving the LS problem, U^T need never be explicitly formed but merely applied to b as it is developed. In other applications, only the matrix $U_1 = U(:, 1{:}n)$ is required. Another variable that affects the volume of work in Algorithm 8.6.2 concerns the R-bidiagonalization idea that we discussed in §5.4.9. Recall that unless A is "almost square," it pays to reduce A to triangular form via QR and before bidiagonalizing. If R-bidiagonalization is used in the SVD context, then we refer to the overall process as the R-SVD. Figure 8.6.1 summarizes the work associated with the various possibilities By comparing the entries in this table (which are meant only as approximate estimates of work), we conclude that the R-SVD approach is more efficient unless $m \approx n$.

8.6.4   Jacobi SVD Procedures

It is straightforward to adapt the Jacobi procedures of §8.5 to the SVD problem. Instead of solving a sequence of 2-by-2 symmetric eigenproblems, we solve a sequence

Required	Golub-Reinsch SVD	R-SVD
Σ	$4mn^2 - 4n^3/3$	$2mn^2 + 2n^3$
Σ, V	$4mn^2 + 8n^3$	$2mn^2 + 11n^3$
Σ, U	$4m^2n - 8mn^2$	$4m^2n + 13n^3$
Σ, U_1	$14mn^2 - 2n^3$	$6mn^2 + 11n^3$
Σ, U, V	$4m^2n + 8mn^2 + 9n^3$	$4m^2n + 22n^3$
Σ, U_1, V	$14mn^2 + 8n^3$	$6mn^2 + 20n^3$

Figure 8.6.1. *Work associated with various SVD-related calculations*

of 2-by-2 SVD problems. Thus, for a given index pair (p, q) we compute a pair of rotations such that

$$\begin{bmatrix} c_1 & s_1 \\ -s_1 & c_1 \end{bmatrix}^T \begin{bmatrix} a_{pp} & a_{pq} \\ a_{qp} & a_{qq} \end{bmatrix} \begin{bmatrix} c_2 & s_2 \\ -s_2 & c_2 \end{bmatrix} = \begin{bmatrix} d_p & 0 \\ 0 & d_q \end{bmatrix}.$$

See P8.6.5. The resulting algorithm is referred to as *two-sided* because each update involves a pre- and a post-multiplication.

A *one-sided* Jacobi algorithm involves a sequence of pairwise column orthogonalizations. For a given index pair (p, q) a Jacobi rotation $J(p, q, \theta)$ is determined so that columns p and q of $AJ(p, q, \theta)$ are orthogonal to each other. See P8.6.8. Note that this corresponds to zeroing the (p, q) and (q, p) entries in $A^T A$. Once AV has sufficiently orthogonal columns, the rest of the SVD (U and Σ) follows from column scaling: $AV = U\Sigma$.

Problems

P8.6.1 Give formulae for the eigenvectors of

$$S = \begin{bmatrix} 0 & A^T \\ A & 0 \end{bmatrix}$$

in terms of the singular vectors of $A \in \mathbb{R}^{m \times n}$ where $m \geq n$.

P8.6.2 Relate the singular values and vectors of $A = B + iC$ $(B, C \in \mathbb{R}^{m \times n})$ to those of

$$\tilde{A} = \begin{bmatrix} B & -C \\ C & B \end{bmatrix}.$$

P8.6.3 Suppose $B \in \mathbb{R}^{n \times n}$ is upper bidiagonal with diagonal entries $d(1{:}n)$ and superdiagonal entries $f(1{:}n - 1)$. State and prove a singular value version of Theorem 8.3.1.

P8.6.4 Assume that $n = 2m$ and that $S \in \mathbb{R}^{n \times n}$ is skew-symmetric and tridiagonal. Show that there exists a permutation $P \in \mathbb{R}^{n \times n}$ such that

$$P^T S P = \begin{bmatrix} 0 & -B^T \\ B & 0 \end{bmatrix}$$

where $B \in \mathbb{R}^{m \times m}$. Describe the structure of B and show how to compute the eigenvalues and eigenvectors of S via the SVD of B. Repeat for the case $n = 2m + 1$.

P8.6.5 (a) Let

$$C = \begin{bmatrix} w & x \\ y & z \end{bmatrix}$$

be real. Give a stable algorithm for computing c and s with $c^2 + s^2 = 1$ such that

$$B = \begin{bmatrix} c & s \\ -s & c \end{bmatrix} C$$

is symmetric. (b) Combine (a) with Algorithm 8.5.1 to obtain a stable algorithm for computing the SVD of C. (c) Part (b) can be used to develop a Jacobi-like algorithm for computing the SVD of $A \in \mathbb{R}^{n \times n}$. For a given (p, q) with $p < q$, Jacobi transformations $J(p, q, \theta_1)$ and $J(p, q, \theta_2)$ are determined such that if

$$B = J(p, q, \theta_1)^T A J(p, q, \theta_2),$$

then $b_{pq} = b_{qp} = 0$. Show

$$\mathrm{off}(B)^2 = \mathrm{off}(A)^2 - a_{pq}^2 - a_{qp}^2.$$

(d) Consider one sweep of a cyclic-by-row Jacobi SVD procedure applied to $A \in \mathbb{R}^{n \times n}$:

> **for** $p = 1{:}n - 1$
>   **for** $q = p + 1{:}n$
> $A = J(p, q, \theta_1)^T A J(p, q, \theta_2)$
>   **end**
> **end**

Assume that the Jacobi rotation matrices are chosen so that $a_{pq} = a_{qp} = 0$ after the (p, q) update. Show that if A is upper (lower) triangular at the beginning of the sweep, then it is lower (upper) triangular after the sweep is completed. See Kogbetliantz (1955). (e) How could these Jacobi ideas be used to compute the SVD of a rectangular matrix?

P8.6.6 Let x and y be in $\mathbb{R}^m$ and define the orthogonal matrix Q by

$$Q = \begin{bmatrix} c & s \\ -s & c \end{bmatrix}.$$

Give a stable algorithm for computing c and s such that the columns of $[\,x \mid y\,] Q$ are orthogonal to each other.

Notes and References for §8.6

For a general perspective and overview of the SVD we recommend:

G.W. Stewart (1993). "On the Early History of the Singular Value Decomposition," *SIAM Review*
    35, 551–566.
A.K. Cline and I.S. Dhillon (2006). "Computation of the Singular Value Decomposition," in *Handbook*
    of Linear Algebra, L. Hogben (ed.), Chapman and Hall, London, §45-1.

A perturbation theory for the SVD is developed in Stewart and Sun (MPT). See also:

P.A. Wedin (1972). "Perturbation Bounds in Connection with the Singular Value Decomposition,"
    BIT 12, 99–111.
G.W. Stewart (1973). "Error and Perturbation Bounds for Subspaces Associated with Certain Eigen-
    value Problems," *SIAM Review 15*, 727–764.
A. Ruhe (1975). "On the Closeness of Eigenvalues and Singular Values for Almost Normal Matrices,"
    Lin. Alg. Applic. 11, 87–94.
G.W. Stewart (1979). "A Note on the Perturbation of Singular Values," *Lin. Alg. Applic. 28*,
    213–216.
G.W. Stewart (1984). "A Second Order Perturbation Expansion for Small Singular Values," *Lin. Alg.
    Applic. 56*, 231–236.
S. Chandrasekaren and I.C.F. Ipsen (1994). "Backward Errors for Eigenvalue and Singular Value
    Decompositions," *Numer. Math. 68*, 215–223.
R.J. Vaccaro (1994). "A Second-Order Perturbation Expansion for the SVD," *SIAM J. Matrix Anal.
    Applic. 15*, 661–671.

J. Sun (1996). "Perturbation Analysis of Singular Subspaces and Deflating Subspaces," *Numer. Math. 73*, 235–263.

F.M. Dopico (2000). "A Note on Sin T Theorems for Singular Subspace Variations *BIT 40*, 395–403.

R.-C. Li and G. W. Stewart (2000). "A New Relative Perturbation Theorem for Singular Subspaces," *Lin. Alg. Applic. 313*, 41–51.

C.-K. Li and R. Mathias (2002). "Inequalities on Singular Values of Block Triangular Matrices," *SIAM J. Matrix Anal. Applic. 24*, 126–131.

F.M. Dopico and J. Moro (2002). "Perturbation Theory for Simultaneous Bases of Singular Subspaces," *BIT 42*, 84–109.

K.A. O'Neil (2005). "Critical Points of the Singular Value Decomposition," *SIAM J. Matrix Anal. Applic. 27*, 459–473.

M. Stewart (2006). "Perturbation of the SVD in the Presence of Small Singular Values," *Lin. Alg. Applic. 419*, 53–77.

H. Xiang (2006). "A Note on the Minimax Representation for the Subspace Distance and Singular Values," *Lin. Alg. Applic. 414*, 470–473.

W. Li and W. Sun (2007). "Combined Perturbation Bounds: I. Eigensystems and Singular Value Decompositions," *SIAM J. Matrix Anal. Applic. 29*, 643–655.

J. Matejaš and V. Hari (2008). "Relative Eigenvalues and Singular Value Perturbations of Scaled Diagonally Dominant Matrices," *BIT 48*, 769–781.

Classical papers that lay out the ideas behind the SVD algorithm include:

G.H. Golub and W. Kahan (1965). "Calculating the Singular Values and Pseudo-Inverse of a Matrix," *SIAM J. Numer. Anal. 2*, 205–224.

P.A. Businger and G.H. Golub (1969). "Algorithm 358: Singular Value Decomposition of the Complex Matrix," *Commun. ACM 12*, 564–565.

G.H. Golub and C. Reinsch (1970). "Singular Value Decomposition and Least Squares Solutions," *Numer. Math. 14*, 403–420.

For related algorithmic developments and analysis, see:

T.F. Chan (1982). "An Improved Algorithm for Computing the Singular Value Decomposition," *ACM Trans. Math. Softw. 8*, 72 83.

J.J.M. Cuppen (1983). "The Singular Value Decomposition in Product Form," *SIAM J. Sci. Stat. Comput. 4*, 216–222.

J.J. Dongarra (1983). "Improving the Accuracy of Computed Singular Values," *SIAM J. Sci. Stat. Comput. 4*, 712–719.

S. Van Huffel, J. Vandewalle, and A. Haegemans (1987). "An Efficient and Reliable Algorithm for Computing the Singular Subspace of a Matrix Associated with its Smallest Singular Values," *J. Comp. Appl. Math. 19*, 313–330.

P. Deift, J. Demmel, L.-C. Li, and C. Tomei (1991). "The Bidiagonal Singular Value Decomposition and Hamiltonian Mechanics," *SIAM J. Numer. Anal. 28*, 1463–1516.

R. Mathias and G.W. Stewart (1993). "A Block QR Algorithm and the Singular Value Decomposition," *Lin. Alg. Applic. 182*, 91–100.

V. Mehrmann and W. Rath (1993). "Numerical Methods for the Computation of Analytic Singular Value Decompositions," *ETNA 1*, 72–88.

Å. Björck, E. Grimme, and P. Van Dooren (1994). "An Implicit Shift Bidiagonalization Algorithm for Ill-Posed Problems," *BIT 34*, 510–534.

K.V. Fernando and B.N. Parlett (1994). "Accurate Singular Values and Differential qd Algorithms," *Numer. Math. 67*, 191–230.

S. Chandrasekaran and I.C.F. Ipsen (1995). "Analysis of a QR Algorithm for Computing Singular Values," *SIAM J. Matrix Anal. Applic. 16*, 520–535.

U. von Matt (1997). "The Orthogonal qd–Algorithm," *SIAM J. Sci. Comput. 18*, 1163–1186.

K.V. Fernando (1998). "Accurately Counting Singular Values of Bidiagonal Matrices and Eigenvalues of Skew-Symmetric Tridiagonal Matrices," *SIAM J. Matrix Anal. Applic. 20*, 373–399.

N.J. Higham (2000). "QR factorization with Complete Pivoting and Accurate Computation of the SVD," *Lin. Alg. Applic. 309*, 153–174.

Divide-and-conquer methods for the bidiagonal SVD problem have been developed that are analogous to the tridiagonal eigenvalue strategies outlined in §8.4.4:

J.W. Demmel and W. Kahan (1990). "Accurate Singular Values of Bidiagonal Matrices," *SIAM J. Sci. Stat. Comput. 11*, 873–912.

E.R. Jessup and D.C. Sorensen (1994). "A Parallel Algorithm for Computing the Singular Value Decomposition of a Matrix," *SIAM J. Matrix Anal. Applic. 15*, 530–548.

M. Gu and S.C. Eisenstat (1995). "A Divide-and-Conquer Algorithm for the Bidiagonal SVD," *SIAM J. Matrix Anal. Applic. 16*, 79–92.

P.R. Willems, B. Lang, and C. Vömel (2006). "Computing the Bidiagonal SVD Using Multiple Relatively Robust Representations," *SIAM J. Matrix Anal. Applic. 28*, 907–926.

T. Konda and Y. Nakamura (2009). "A New Algorithm for Singular Value Decomposition and Its Parallelization," *Parallel Comput. 35*, 331–344.

For structured SVD problems, there are interesting, specialized results, see:

S. Van Huffel and H. Park (1994). "Parallel Tri- and Bidiagonalization of Bordered Bidiagonal Matrices," *Parallel Comput. 20*, 1107–1128.

J. Demmel and P. Koev (2004). "Accurate SVDs of Weakly Diagonally Dominant M-matrices," *Num. Math. 98*, 99–104.

N. Mastronardi, M. Van Barel, and R. Vandebril (2008). "A Fast Algorithm for the Recursive Calculation of Dominant Singular Subspaces," *J. Comp. Appl. Math. 218*, 238–246.

Jacobi methods for the SVD fall into two categories. The two-sided Jacobi algorithms repeatedly perform the update $A \leftarrow U^T A V$ producing a sequence of iterates that are increasingly diagonal.

E.G. Kogbetliantz (1955). "Solution of Linear Equations by Diagonalization of Coefficient Matrix," *Quart. Appl. Math. 13*, 123–132.

G.E. Forsythe and P. Henrici (1960). "The Cyclic Jacobi Method for Computing the Principal Values of a Complex Matrix," *Trans. AMS 94*, 1–23.

C.C. Paige and P. Van Dooren (1986). "On the Quadratic Convergence of Kogbetliantz's Algorithm for Computing the Singular Value Decomposition," *Lin. Alg. Applic. 77*, 301–313.

J.P. Charlier and P. Van Dooren (1987). "On Kogbetliantz's SVD Algorithm in the Presence of Clusters," *Lin. Alg. Applic. 95*, 135–160.

Z. Bai (1988). "Note on the Quadratic Convergence of Kogbetliantz's Algorithm for Computing the Singular Value Decomposition," *Lin. Alg. Applic. 104*, 131–140.

J.P. Charlier, M. Vanbegin, and P. Van Dooren (1988). "On Efficient Implementation of Kogbetliantz's Algorithm for Computing the Singular Value Decomposition," *Numer. Math. 52*, 279–300.

K.V. Fernando (1989). "Linear Convergence of the Row-Cyclic Jacobi and Kogbetliantz Methods," *Numer. Math. 56*, 73–92.

Z. Drmač and K. Veselič (2008). "New Fast and Accurate Jacobi SVD Algorithm I," *SIAM J. Matrix Anal. Applic. 29*, 1322–1342.

The one-sided Jacobi SVD procedures repeatedly perform the update $A \leftarrow A V$ producing a sequence of iterates with columns that are increasingly orthogonal, see:

J.C. Nash (1975). "A One-Sided Tranformation Method for the Singular Value Decomposition and Algebraic Eigenproblem," *Comput. J. 18*, 74–76.

P.C. Hansen (1988). "Reducing the Number of Sweeps in Hestenes Method," in *Singular Value Decomposition and Signal Processing*, E.F. Deprettere (ed.) North Holland, Amsterdam.

K. Veselič and V. Hari (1989). "A Note on a One-Sided Jacobi Algorithm," *Numer. Math. 56*, 627–633.

Careful implementation and analysis has shown that Jacobi SVD has remarkably accuracy:

J. Demmel, M. Gu, S. Eisenstat, I. Slapnicar, K. Veselić, and Z. Drmač (1999). "Computing the Singular Value Decomposition with High Relative Accuracy," *Lin. Alg. Applic. 299*, 21–80.

Z Drmač (1999). "A Posteriori Computation of the Singular Vectors in a Preconditioned Jacobi SVD Algorithm," *IMA J. Numer. Anal. 19*, 191–213.

Z. Drmač (1997). "Implementation of Jacobi Rotations for Accurate Singular Value Computation in Floating Point Arithmetic," *SIAM J. Sci. Comput. 18*, 1200–1222.

F.M. Dopico and J. Moro (2004). "A Note on Multiplicative Backward Errors of Accurate SVD Algorithms," *SIAM J. Matrix Anal. Applic. 25*, 1021–1031.

The parallel implementation of the Jacobi SVD has a long and interesting history:

F.T. Luk (1980). "Computing the Singular Value Decomposition on the ILLIAC IV," *ACM Trans. Math. Softw. 6*, 524–539.

R.P. Brent and F.T. Luk (1985). "The Solution of Singular Value and Symmetric Eigenvalue Problems on Multiprocessor Arrays," *SIAM J. Sci. Stat. Comput. 6*, 69–84.

R.P. Brent, F.T. Luk, and C. Van Loan (1985). "Computation of the Singular Value Decomposition Using Mesh Connected Processors," *J. VLSI Computer Systems 1*, 242–270.

F.T. Luk (1986). "A Triangular Processor Array for Computing Singular Values," *Lin. Alg. Applic. 77*, 259–274.

M. Berry and A. Sameh (1986). "Multiprocessor Jacobi Algorithms for Dense Symmetric Eigenvalue and Singular Value Decompositions," in *Proceedings International Conference on Parallel Processing*, 433–440.

R. Schreiber (1986). "Solving Eigenvalue and Singular Value Problems on an Undersized Systolic Array," *SIAM J. Sci. Stat. Comput. 7*, 441–451.

C.H. Bischof and C. Van Loan (1986). "Computing the SVD on a Ring of Array Processors," in *Large Scale Eigenvalue Problems*, J. Cullum and R. Willoughby (eds.), North Holland, Amsterdam, 51–66.

C.H. Bischof (1987). "The Two-Sided Block Jacobi Method on Hypercube Architectures," in *Hypercube Multiprocessors*, M.T. Heath (ed.), SIAM Publications, Philadelphia, PA.

C.H. Bischof (1989). "Computing the Singular Value Decomposition on a Distributed System of Vector Processors," *Parallel Comput. 11*, 171–186.

M. Beča, G. Okša, M. Vajteršic, and L. Grigori (2010). "On Iterative QR Pre-Processing in the Parallel Block-Jacobi SVD Algorithm," *Parallel Comput. 36*, 297–307.

8.7 Generalized Eigenvalue Problems with Symmetry

This section is mostly about a pair of symmetrically structured versions of the generalized eigenvalue problem that we considered in §7.7. In the *symmetric-definite problem* we seek nontrivial solutions to the problem

$$Ax = \lambda Bx \tag{8.7.1}$$

where $A \in \mathbb{R}^{n \times n}$ is symmetric and $B \in \mathbb{R}^{n \times n}$ is symmetric positive definite. The *generalized singular value problem* has the form

$$A^T A x = \mu^2 B^T B x \tag{8.7.2}$$

where $A \in \mathbb{R}^{m_1 \times n}$ and $B \in \mathbb{R}^{m_2 \times n}$. By setting $B = I_n$ we see that these problems are (respectively) generalizations of the symmetric eigenvalue problem and the singular value problem.

8.7.1 The Symmetric-Definite Generalized Eigenproblem

The *generalized eigenvalues* of the symmetric-definite pair $\{A, B\}$ are denoted by $\lambda(A, B)$ where

$$\lambda(A, B) = \{ \lambda \mid \det(A - \lambda B) = 0 \}.$$

If $\lambda \in \lambda(A, B)$ and x is a nonzero vector that satisfies $Ax = \lambda Bx$, then x is a *generalized eigenvector*.

A symmetric-definite problem can be transformed to an equivalent symmetric-definite problem with a congruence transformation:

$$A - \lambda B \text{ is singular} \quad \Leftrightarrow \quad (X^T A X) - \lambda(X^T B X) \text{ is singular}.$$

Thus, if X is nonsingular, then $\lambda(A, B) = \lambda(X^T A X, X^T B X)$.

For a symmetric-definite pair $\{A, B\}$, it is possible to choose a real nonsingular X so that $X^T A X$ and $X^T B X$ are diagonal. This follows from the next result.

Theorem 8.7.1. *Suppose A and B are n-by-n symmetric matrices, and define $C(\mu)$ by*

$$C(\mu) \; = \; \mu A + (1 - \mu)B \qquad \mu \in \mathbb{R}. \tag{8.7.3}$$

If there exists a $\mu \in [0, 1]$ such that $C(\mu)$ is nonnegative definite and

$$\mathsf{null}(C(\mu)) \; = \; \mathsf{null}(A) \cap \mathsf{null}(B)$$

then there exists a nonsingular X such that both $X^T A X$ and $X^T B X$ are diagonal.

Proof. Let $\mu \in [0, 1]$ be chosen so that $C(\mu)$ is nonnegative definite with the property that $\mathsf{null}(C(\mu)) = \mathsf{null}(A) \cap \mathsf{null}(B)$. Let

$$Q_1^T C(\mu) Q_1 \; = \; \begin{bmatrix} D & 0 \\ 0 & 0 \end{bmatrix}, \qquad D = \mathrm{diag}(d_1, \dots, d_k), \; d_i > 0,$$

be the Schur decomposition of $C(\mu)$ and define $X_1 = Q_1 \cdot \mathrm{diag}(D^{-1/2}, I_{n-k})$. If

$$A_1 \; = \; X_1^T A X_1, \qquad B_1 \; = \; X_1^T B X_1, \qquad C_1 \; = \; X_1^T C(\mu) X_1,$$

then

$$C_1 \; = \; \begin{bmatrix} I_k & 0 \\ 0 & 0 \end{bmatrix} \; = \; \mu A_1 + (1 - \mu) B_1.$$

Since

$$\mathsf{span}\{e_{k+1}, \dots, e_n\} \; = \; \mathsf{null}(C_1) \; = \; \mathsf{null}(A_1) \cap \mathsf{null}(B_1)$$

it follows that A_1 and B_1 have the following block structure:

$$A_1 \; = \; \begin{bmatrix} A_{11} & 0 \\ 0 & 0 \end{bmatrix} \begin{matrix} k \\ n-k \end{matrix} \quad , \qquad B_1 \; = \; \begin{bmatrix} B_{11} & 0 \\ 0 & 0 \end{bmatrix} \begin{matrix} k \\ n-k \end{matrix} \quad .$$
$$\quad\; k \quad n-k \qquad\qquad\qquad k \quad n-k$$

Moreover $I_k = \mu A_{11} + (1 - \mu) B_{11}$.

Suppose $\mu \neq 0$. It then follows that if $Z^T B_{11} Z = \mathrm{diag}(b_1, \dots, b_k)$ is the Schur decomposition of B_{11} and we set

$$X \; = \; X_1 \cdot \mathrm{diag}(Z, I_{n-k})$$

then

$$X^T B X \; = \; \mathrm{diag}(b_1, \dots, b_k, 0, \dots, 0) \; \equiv \; D_B$$

and

$$X^T A X \; = \; \frac{1}{\mu} X^T \left(C(\mu) - (1 - \mu)B \right) X \; = \; \frac{1}{\mu} \left(\begin{bmatrix} I_k & 0 \\ 0 & 0 \end{bmatrix} - (1 - \mu) D_B \right) \; \equiv \; D_A.$$

On the other hand, if $\mu = 0$, then let $Z^T A_{11} Z = \text{diag}(a_1, \ldots, a_k)$ be the Schur decomposition of A_{11} and set $X = X_1 \text{diag}(Z, I_{n-k})$. It is easy to verify that in this case as well, both $X^T A X$ and $X^T B X$ are diagonal. $\quad\square$

Frequently, the conditions in Theorem 8.7.1 are satisfied because either A or B is positive definite.

Corollary 8.7.2. *If $A - \lambda B \in \mathbb{R}^{n \times n}$ is symmetric-definite, then there exists a nonsingular*

$$X = [\, x_1 \,|\, \cdots \,|\, x_n \,]$$

such that

$$X^T A X = \text{diag}(a_1, \ldots, a_n)$$

and

$$X^T B X = \text{diag}(b_1, \ldots, b_n).$$

Moreover, $A x_i = \lambda_i B x_i$ for $i = 1{:}n$ where $\lambda_i = a_i / b_i$.

Proof. By setting $\mu = 0$ in Theorem 8.7.1 we see that symmetric-definite pencils can be simultaneously diagonalized. The rest of the corollary is easily verified. $\quad\square$

Stewart (1979) has worked out a perturbation theory for symmetric pencils $A - \lambda B$ that satisfy

$$c(A, B) = \min_{\|x\|_2 = 1} (x^T A x)^2 + (x^T B x)^2 > 0. \tag{8.7.4}$$

The scalar $c(A, B)$ is called the *Crawford number* of the pencil $A - \lambda B$.

Theorem 8.7.3. *Suppose $A - \lambda B$ is an n-by-n symmetric-definite pencil with eigenvalues*

$$\lambda_1 \geq \lambda_2 \geq \cdots \geq \lambda_n.$$

Suppose E_A and E_B are symmetric n-by-n matrices that satisfy

$$\epsilon^2 = \| E_A \|_2^2 + \| E_B \|_2^2 < c(A, B).$$

Then $(A + E_A) - \lambda(B + E_B)$ is symmetric-definite with eigenvalues

$$\mu_1 \geq \cdots \geq \mu_n$$

that satisfy

$$|\arctan(\lambda_i) - \arctan(\mu_i)| \leq \arctan(\epsilon / c(A, B))$$

for $i = 1{:}n$.

Proof. See Stewart (1979). $\quad\square$

8.7.2   Simultaneous Reduction of A and B

Turning to algorithmic matters, we first present a method for solving the symmetric-definite problem that utilizes both the Cholesky factorization and the symmetric QR algorithm.

Algorithm 8.7.1 Given $A = A^T \in \mathbb{R}^{n \times n}$ and $B = B^T \in \mathbb{R}^{n \times n}$ with B positive definite, the following algorithm computes a nonsingular X such that $X^T A X = \mathrm{diag}(a_1, \ldots, a_n)$ and $X^T B X = I_n$.

> Compute the Cholesky factorization $B = GG^T$ using Algorithm 4.2.2.
>
> Compute $C = G^{-1} A G^{-T}$.
>
> Use the symmetric QR algorithm to compute the Schur decomposition
> $$Q^T C Q = \mathrm{diag}(a_1, \ldots, a_n).$$
> Set $X = G^{-T} Q$.

This algorithm requires about $14n^3$ flops. In a practical implementation, A can be overwritten by the matrix C. See Martin and Wilkinson (1968) for details. Note that

$$\lambda(A, B) = \lambda(A, GG^T) = \lambda(G^{-1} A G^{-T}, I) = \lambda(C) = \{a_1, \ldots, a_n\}.$$

If $\hat{a}_i$ is a computed eigenvalue obtained by Algorithm 8.7.1, then it can be shown that

$$\hat{a}_i \in \lambda(G^{-1} A G^{-T} + E_i)$$

where

$$\| E_i \|_2 \approx \mathbf{u} \| A \|_2 \| B^{-1} \|_2.$$

Thus, if B is ill-conditioned, then $\hat{a}_i$ may be severely contaminated with roundoff error even if a_i is a well-conditioned generalized eigenvalue. The problem, of course, is that in this case, the matrix $C = G^{-1} A G^{-T}$ can have some very large entries if B, and hence G, is ill-conditioned. This difficulty can sometimes be overcome by replacing the matrix G in Algorithm 8.7.1 with $V D^{-1/2}$ where $V^T B V = D$ is the Schur decomposition of B. If the diagonal entries of D are ordered from smallest to largest, then the large entries in C are concentrated in the upper left-hand corner. The small eigenvalues of C can then be computed without excessive roundoff error contamination (or so the heuristic goes). For further discussion, consult Wilkinson (AEP, pp. 337–38).

The condition of the matrix X in Algorithm 8.7.1 can sometimes be improved by replacing B with a suitable convex combination of A and B. The connection between the eigenvalues of the modified pencil and those of the original are detailed in the proof of Theorem 8.7.1.

Other difficulties concerning Algorithm 8.7.1 relate to the fact that $G^{-1} A G^{-T}$ is generally full even when A and B are sparse. This is a serious problem, since many of the symmetric-definite problems arising in practice are large and sparse. Crawford (1973) has shown how to implement Algorithm 8.7.1 effectively when A and B are banded. Aside from this case, however, the simultaneous diagonalization approach is impractical for the large, sparse symmetric-definite problem. Alternate strategies are discussed in Chapter 10.

8.7.3 Other Methods

Many of the symmetric eigenvalue methods presented in earlier sections have symmetric-definite generalizations. For example, the Rayleigh quotient iteration (8.2.6) can be extended as follows:

x_0 given with $\| x_0 \|_2 = 1$
for $k = 0, 1, \ldots$
$$\mu_k = x_k^T A x_k / x_k^T B x_k \qquad (8.7.5)$$
Solve $(A - \mu_k B)z_{k+1} = B x_k$ for z_{k+1}.
$x_{k+1} = z_{k+1}/\| z_{k+1} \|_2$
end

The main idea behind this iteration is that

$$\lambda = \frac{x^T A x}{x^T B x} \qquad (8.7.6)$$

minimizes

$$f(\lambda) = \| A x - \lambda B x \|_B \qquad (8.7.7)$$

where $\| \cdot \|_B$ is defined by $\|z\|_B^2 = z^T B^{-1} z$. The mathematical properties of (8.7.5) are similar to those of (8.2.6). Its applicability depends on whether or not systems of the form $(A - \mu B)z = x$ can be readily solved. Likewise, the same comment pertains to the generalized orthogonal iteration:

$Q_0 \in \mathbb{R}^{n \times p}$ given with $Q_0^T Q_0 = I_p$
for $k = 1, 2, \ldots$ $\qquad (8.7.8)$
Solve $BZ_k = AQ_{k-1}$ for Z_k
$Z_k = Q_k R_k$ (QR factorization, $Q_k \in \mathbb{R}^{n \times p}$, $R_k \in \mathbb{R}^{p \times p}$)
end

This is mathematically equivalent to (7.3.6) with A replaced by $B^{-1}A$. Its practicality strongly depends on how easy it is to solve linear systems of the form $Bz = y$.

8.7.4 The Generalized Singular Value Problem

We now turn our attention to the generalized singular value decomposition introduced in §6.1.6. This decomposition is concerned with the simultaneous diagonalization of two rectangular matrices A and B that are assumed to have the same number of columns. We restate the decomposition here with a simplification that both A and B have at least as many rows as columns. This assumption is not necessary, but it serves to unclutter our presentation of the GSVD algorithm.

Theorem 8.7.4 (Tall Rectangular Version). *If $A \in \mathbb{R}^{m_1 \times n}$ and $B \in \mathbb{R}^{m_2 \times n}$ have at least as many rows as columns, then there exists an orthogonal matrix $U_1 \in \mathbb{R}^{m_1 \times m_1}$, an orthogonal matrix $U_2 \in \mathbb{R}^{m_2 \times m_2}$, and a nonsingular matrix $X \in \mathbb{R}^{n \times n}$ such that*

$$U_1^T AX = \operatorname{diag}(\alpha_1, \ldots, \alpha_n),$$
$$U_2^T BX = \operatorname{diag}(\beta_1, \ldots, \beta_n).$$

Proof. See Theorem 6.1.1. $\square$

The *generalized singular values* of the matrix pair $\{A, B\}$ are defined by

$$\sigma(A, B) \;=\; \{\alpha_1/\beta_1, \ldots, \alpha_n/\beta_n\}.$$

We give names to the columns of X, U_1, and U_2. The columns of X are the *right generalized singular vectors*, the columns of U_1 are the *left-A generalized singular vectors*, and the columns of U_2 are the *left-B generalized singular vectors*. Note that

$$AX(:,k) \;=\; \alpha_k U_1(:,k),$$
$$BX(:,k) \;=\; \beta_k U_2(:,k),$$

for $k = 1{:}n$.

There is a connection between the GSVD of the matrix pair $\{A, B\}$ and the "symmetric-definite-definite" pencil $A^T A - \lambda B^T B$. Since

$$X^T(A^T A \,-\, \lambda B^T B)X \;=\; D_A^T D_A \,-\, \lambda D_B^T D_B \;=\; \mathrm{diag}(\alpha_k^2 - \lambda \beta_k^2),$$

it follows that the right generalized singular vectors of $\{A, B\}$ are the generalized eigenvectors for $A^T A - \lambda B^T B$ and the eigenvalues of the pencil $A^T A - \lambda B^T B$ are squares of the generalized singular values of $\{A, B\}$.

All these GSVD facts revert to familiar SVD facts by setting $B = I_n$. For example, if $B = I_n$, then we can set $X = U_2$ and $U_1^T A X = D_A$ is the SVD.

We mention that the generalized singular values of $\{A, B\}$ are the stationary values of

$$\phi_{A,B}(x) \;=\; \frac{\|\, Ax \,\|_2}{\|\, Bx \,\|_2}$$

and the right generalized singular vectors are the associated stationary vectors. The left-A and left-B generalized singular vectors are stationary vectors associated with the quotient $\|\, y \,\|_2 / \|\, x \,\|_2$ subject to the constraints

$$A^T x = B^T y, \qquad x \perp \mathsf{null}(A^T), \qquad y \perp \mathsf{null}(A^T).$$

See Chu, Funderlic, and Golub (1997).

A GSVD perturbation theory has been developed by Sun (1983, 1998, 2000), Paige (1984), and Li (1990).

8.7.5   Computing the GSVD Using the CS Decomposition

Our proof of the GSVD in Theorem 6.1.1 is constructive and makes use of the CS decomposition. In practice, computing the GSVD via the CS decomposition is a viable strategy.

Algorithm 8.7.2 (GSVD (Tall, Full-Rank Version)) Assume that $A \in \mathbb{R}^{m_1 \times n}$ and $B \in \mathbb{R}^{m_2 \times n}$, with $m_1 \geq n$, $m_2 \geq n$, and $\mathsf{null}(A) \cap \mathsf{null}(B) = \emptyset$. The following algorithm computes an orthogonal matrix $U_1 \in \mathbb{R}^{m_1 \times m_1}$, an orthogonal matrix $U_2 \in \mathbb{R}^{m_2 \times m_2}$, a nonsingular matrix $X \in \mathbb{R}^{n \times n}$, and diagonal matrices $D_A \in \mathbb{R}^{m_1 \times n}$ and $D_B \in \mathbb{R}^{m_1 \times n}$ such that $U_1^T A X = D_A$ and $U_2^T B X = D_B$.

Compute the the QR factorization

$$\left[\begin{array}{c} A \\ B \end{array}\right] = \left[\begin{array}{c} Q_1 \\ Q_2 \end{array}\right] R.$$

Compute the CS decomposition

$$U_1^T Q_1 V = D_A = \text{diag}(\alpha_1, \ldots, \alpha_n),$$
$$U_2^T Q_2 V = D_B = \text{diag}(\beta_1, \ldots, \beta_n).$$

Solve $RX = V$ for X.

The assumption that $\text{null}(A) \cap \text{null}(B) = \emptyset$ is not essential. See Van Loan (1985). Regardless, the condition of the matrix X is an issue that affects accuracy. However, we point out that it is possible to compute designated right generalized singular vector subspaces without having to compute explicitly selected columns of the matrix $X = VR^{-1}$. For example, suppose that we wish to compute an orthonormal basis for the subspace $S = \text{span}\{x_1, \ldots x_k\}$ where $x_i = X(:, i)$. If we compute an orthogonal Z and upper triangular T so $TZ^T = V^T R$, then

$$ZT^{-1} = R^{-1}V = X$$

and $S = \text{span}\{z_1, \ldots z_k\}$ where $z_i = Z(:, i)$. See P5.2.2 concerning the computation of Z and T.

8.7.6 Computing the CS Decomposition

At first glance, the computation of the CS decomposition looks easy. After all, it is just a collection of SVDs. However, there are some complicating numerical issues that need to be addressed. To build an appreciation for this, we step through the "thin" version of the algorithm developed by Van Loan (1985) for the case

$$Q = \left[\begin{array}{c} Q_1 \\ \hline Q_2 \end{array}\right] = \left[\begin{array}{ccccc} \times & \times & \times & \times & \times \\ \times & \times & \times & \times & \times \\ \times & \times & \times & \times & \times \\ \times & \times & \times & \times & \times \\ \times & \times & \times & \times & \times \\ \hline \times & \times & \times & \times & \times \\ \times & \times & \times & \times & \times \\ \times & \times & \times & \times & \times \\ \times & \times & \times & \times & \times \\ \times & \times & \times & \times & \times \end{array}\right].$$

In exact arithmetic, the goal is to compute 5-by-5 orthogonal matrices U_1, U_2, and V so that

$$U_1^T Q_1 V = C = \text{diag}(c_1, c_2, c_3, c_4, c_5),$$
$$U_2^T Q_2 V = S = \text{diag}(s_1, s_2, s_3, s_4, s_5).$$

In floating point, we strive to compute matrices $\widehat{U}_2$, $\widehat{U}_2$ and $\widehat{V}$ that are orthogonal to working precision and which transform Q_1 and Q_2 into nearly diagonal form:

$$\text{fl}(\widehat{U}_1^T Q_1 \widehat{V}) = \text{diag}(\hat{c}_k) + E_1, \qquad \| E_1 \| \approx \mathbf{u}, \tag{8.7.9}$$

$$\text{fl}(\widehat{U}_2^T Q_2 \widehat{V}) = \text{diag}(\hat{s}_k) + E_2, \qquad \| E_2 \| \approx \mathbf{u}. \tag{8.7.10}$$

In what follows, it will be obvious that the computed versions of U_1, U_2 and V are orthogonal to working precision, as they will be "put together" from numerically sound QR factorizations and SVDs. The challenge is to affirm (8.7.9) and (8.7.10).

We start by computing the SVD

$$U_2^T Q_1 V = S$$

followed by the QR factorization

$$U_1 R = Q_1 V.$$

Overwriting Q_2 with S and Q_1 with R gives

$$Q = \left[\begin{array}{ccccc} r_{11} & r_{12} & r_{13} & r_{14} & r_{15} \\ \epsilon_{21} & r_{22} & r_{23} & r_{24} & r_{25} \\ \epsilon_{31} & \epsilon_{32} & r_{33} & r_{34} & r_{35} \\ \epsilon_{41} & \epsilon_{42} & \epsilon_{43} & r_{44} & r_{45} \\ \epsilon_{51} & \epsilon_{52} & \epsilon_{53} & \epsilon_{54} & r_{55} \\ \hline s_1 & \delta_{12} & \delta_{13} & \delta_{14} & \delta_{25} \\ \delta_{21} & s_2 & \delta_{23} & \delta_{24} & \delta_{25} \\ \delta_{31} & \delta_{32} & s_3 & \delta_{34} & \delta_{35} \\ \delta_{41} & \delta_{42} & \delta_{43} & s_4 & \delta_{45} \\ \delta_{51} & \delta_{52} & \delta_{53} & \delta_{54} & s_5 \end{array} \right] \quad \begin{array}{c} \epsilon_{ij} = O(\mathbf{u}), \\[4em] \delta_{ij} = O(\mathbf{u}), \end{array}$$

Since the columns of this matrix are orthonormal to machine precision, it follows that

$$|r_{11} r_{1j}| \approx \mathbf{u}, \qquad j = 2{:}5.$$

Note that if $|r_{11}| = O(1)$, then we may conclude that $|r_{1j}| \approx \mathbf{u}$ for $j = 2{:}5$. This will be the case if (for example) $s_1 \leq 1/\sqrt{2}$ for then

$$|r_{11}| \approx \sqrt{1 - s_1^2} \geq \frac{1}{\sqrt{2}}.$$

With this in mind, let us assume that the singular values $s_1, \ldots, s_5$ are ordered from little to big and that

$$0 \leq s_1 \leq s_2 \leq \frac{1}{\sqrt{2}} < s_3 \leq s_4 \leq s_5. \tag{8.7.11}$$

Working with the near-orthonormality of the columns of Q, we conclude that

$$
Q = \left[\begin{array}{ccc|cc}
c_1 & \epsilon_{12} & \epsilon_{13} & \epsilon_{14} & \epsilon_{15} \\
\epsilon_{21} & c_2 & \epsilon_{23} & \epsilon_{24} & \epsilon_{25} \\
\hline
\epsilon_{31} & \epsilon_{32} & r_{33} & r_{34} & r_{35} \\
\epsilon_{41} & \epsilon_{42} & \epsilon_{43} & r_{44} & r_{45} \\
\epsilon_{51} & \epsilon_{52} & \epsilon_{53} & \epsilon_{54} & r_{55} \\
\hline\hline
s_1 & \delta_{12} & \delta_{13} & \delta_{14} & \delta_{25} \\
\delta_{21} & s_2 & \delta_{23} & \delta_{24} & \delta_{25} \\
\hline
\delta_{31} & \delta_{32} & s_3 & \delta_{34} & \delta_{35} \\
\delta_{41} & \delta_{42} & \delta_{43} & s_4 & \delta_{45} \\
\delta_{51} & \delta_{52} & \delta_{53} & \delta_{54} & s_5
\end{array}\right]
\qquad
\begin{array}{l}
\epsilon_{ij} = O(\mathbf{u}), \\[2em]
\delta_{ij} = O(\mathbf{u}).
\end{array}
$$

Note that

$$
|r_{34}| \approx \frac{\mathbf{u}}{|r_{33}|} \approx \frac{\mathbf{u}}{\sqrt{1 - s_3^2}}.
$$

Since s_3 can be close to 1, we cannot guarantee that r_{34} is sufficiently small. Similar comments apply to r_{35} and r_{45}.

To rectify this we compute the SVD of $Q(3{:}5, 3{:}5)$, taking care to apply the U-matrix across rows 3 to 5 and the V matrix across columns 3:5. This gives

$$
Q = \left[\begin{array}{ccc|cc}
c_1 & \epsilon_{12} & \epsilon_{13} & \epsilon_{14} & \epsilon_{15} \\
\epsilon_{21} & c_2 & \epsilon_{23} & \epsilon_{24} & \epsilon_{25} \\
\hline
\epsilon_{31} & \epsilon_{32} & c_3 & \epsilon_{34} & \epsilon_{35} \\
\epsilon_{41} & \epsilon_{42} & \epsilon_{43} & c_4 & \epsilon_{45} \\
\epsilon_{51} & \epsilon_{52} & \epsilon_{53} & \epsilon_{54} & c_5 \\
\hline\hline
s_1 & \delta_{12} & \delta_{13} & \delta_{14} & \delta_{25} \\
\delta_{21} & s_2 & \delta_{23} & \delta_{24} & \delta_{25} \\
\hline
\delta_{31} & \delta_{32} & t_{33} & t_{34} & t_{35} \\
\delta_{41} & \delta_{42} & t_{43} & t_{44} & t_{45} \\
\delta_{51} & \delta_{52} & t_{53} & t_{54} & t_{55}
\end{array}\right]
\qquad
\begin{array}{l}
\epsilon_{ij} = O(\mathbf{u}), \\[2em]
\delta_{ij} = O(\mathbf{u}).
\end{array}
$$

Thus, by diagonalizing the (2,2) block of Q_1 we fill the (2,2) block of Q_2. However, if we compute the QR factorization of $Q(8{:}10, 3{:}5)$ and apply the orthogonal factor across rows 8:10, then we obtain

$$
Q = \left[
\begin{array}{cc|ccc}
c_1 & \epsilon_{12} & \epsilon_{13} & \epsilon_{14} & \epsilon_{15} \\
\epsilon_{21} & c_2 & \epsilon_{23} & \epsilon_{24} & \epsilon_{25} \\
\hline
\epsilon_{31} & \epsilon_{32} & c_3 & \epsilon_{34} & \epsilon_{35} \\
\epsilon_{41} & \epsilon_{42} & \epsilon_{43} & c_4 & \epsilon_{45} \\
\epsilon_{51} & \epsilon_{52} & \epsilon_{53} & \epsilon_{54} & c_5 \\
\hline
s_1 & \delta_{12} & \delta_{13} & \delta_{14} & \delta_{25} \\
\delta_{21} & s_2 & \delta_{23} & \delta_{24} & \delta_{25} \\
\hline
\delta_{31} & \delta_{32} & t_{33} & t_{34} & t_{35} \\
\delta_{41} & \delta_{42} & t_{43} & t_{44} & t_{45} \\
\delta_{51} & \delta_{52} & t_{53} & t_{54} & t_{55}
\end{array}
\right]
\qquad
\begin{array}{l}
\epsilon_{ij} = O(\mathbf{u}), \\[3em]
\delta_{ij} = O(\mathbf{u}).
\end{array}
$$

Using the near-orthonormality of the columns of Q and the fact that c_3, c_4, and c_5 are all less than $1/\sqrt{2}$, we can conclude (for example) that

$$
|t_{34}| \approx O\left(\frac{\mathbf{u}}{|t_{33}|}\right) \approx O\left(\frac{\mathbf{u}}{\sqrt{1 - c_3^2}}\right) = O(\mathbf{u}).
$$

Using similar arguments we may conclude that both t_{35} and t_{45} are $O(\mathbf{u})$. It follows that the updated Q_1 and Q_2 are diagonal to within the required tolerance and that (8.7.9) and (8.7.10) are achieved as a result.

8.7.7   The Kogbetliantz Approach

Paige (1986) developed a method for computing the GSVD based on the Kogbetliantz Jacobi SVD procedure. At each step a 2-by-2 GSVD problem is solved, a calculation that we briefly examine. Suppose F and G are 2-by-2 and that G is nonsingular. If

$$
U_1^T(FG^{-1})U_2 = \Sigma = \left[\begin{array}{cc} \sigma_1 & 0 \\ 0 & \sigma_2 \end{array}\right]
$$

is the SVD of FG^{-1}, then $\sigma(F, G) = \{\sigma_1, \sigma_2\}$ and

$$
U_1^T F = (U_2^T G)\Sigma.
$$

This says that the rows of $U_1^T F$ are parallel to the corresponding rows of $U_2^T G$. Thus, if Z is orthogonal so that $U_2^T G Z = G_1$ is upper triangular, then $U_1^T F Z = F_1$ is also upper triangular. In the Paige algorithm, these 2-by-2 calculations resonate with the preservation of the triangular form that is key to the Kogbetliantz procedure. Moreover, the A and B input matrices are separately updated and the updates only involve orthogonal transformations. Although some of the calculations are very delicate, the overall procedure is tantamount to applying Kogbetliantz implicitly to the matrix AB^{-1}.

8.7.8 Other Generalizations of the SVD

What we have been calling the "generalized singular value decomposition" is sometimes referred to as the *quotient singular value decomposition* or QSVD. A key feature of the decomposition is that it separately transforms the input matrices A and B in such a way that the generalized singular values and vectors are exposed, sometimes implicitly.

It turns out that there are other ways to generalize the SVD. In the *product singular value decomposition* problem we are given $A \in \mathbb{R}^{m \times n_1}$ and $B \in \mathbb{R}^{m \times n_2}$ and require the SVD of $A^T B$. The challenge is to compute $U^T (A^T B) V = \Sigma$ without actually forming $A^T B$ as that operation can result in a significant loss of information. See Drmač (1998, 2000).

The *restricted singular value decomposition* involves three matrices and is best motivated from a a variational point of view. If $A \in \mathbb{R}^{m \times n}$, $B \in \mathbb{R}^{m \times q}$, and $C \in \mathbb{R}^{n \times p}$, then the restricted singular values of the triplet $\{A, B, C\}$ are the stationary values of

$$\psi_{A,B,C}(x,y) = \frac{y^T A x}{\| By \|_2 \| Cx \|_2}.$$

See Zha (1991), De Moor and Golub (1991), and Chu, De Lathauwer, and De Moor (2000). As with the product SVD, the challenge is to compute the required quantities without forming inverses and products.

All these ideas can be extended to chains of matrices, e.g., the computation of the SVD of a matrix product $A = A_1 A_2 \cdots A_k$ without explicitly forming A. See De Moor and Zha (1991) and De Moor and Van Dooren (1992).

8.7.9 A Note on the Quadratic Eigenvalue Problem

We build on our §7.7.9 discussion of the polynomial eigenvalue problem and briefly consider some structured versions of the quadratic case,

$$\left(\lambda^2 M + \lambda C + K \right) x = 0, \qquad M, C, K \in \mathbb{R}^{n \times n}. \tag{8.7.12}$$

We recommend the excellent survey by Tisseur and Meerbergen (2001) for more detail.

Note that the eigenvalue in (8.7.12) solves the quadratic equation

$$(x^H M x)\lambda^2 + (x^H C x)\lambda + (x^H K x) = 0. \tag{8.7.13}$$

and thus

$$\lambda = \frac{-(x^H C x) \pm \sqrt{(x^H C x)^2 - 4(x^H M x)(x^H K x)}}{2(x^H M x)}, \tag{8.7.14}$$

assuming that $x^H M x \neq 0$. Linearized versions of (8.7.12) include

$$\begin{bmatrix} 0 & N \\ K & C \end{bmatrix} \begin{bmatrix} x \\ u \end{bmatrix} = \lambda \begin{bmatrix} N & 0 \\ 0 & -M \end{bmatrix} \begin{bmatrix} x \\ u \end{bmatrix} \tag{8.7.15}$$

and

$$\begin{bmatrix} -K & 0 \\ 0 & N \end{bmatrix} \begin{bmatrix} x \\ u \end{bmatrix} = \lambda \begin{bmatrix} C & M \\ N & 0 \end{bmatrix} \begin{bmatrix} x \\ u \end{bmatrix} \tag{8.7.16}$$

where $N \in \mathbb{R}^{n \times n}$ is nonsingular.

In many applications, the matrices M and C are symmetric and positive definite and K is symmetric and positive semidefinite. It follows from (8.7.14) that in this case the eigenvalues have nonpositive real part. If we set $N = K$ in (8.7.15), then we obtain the following generalized eigenvalue problem:

$$\begin{bmatrix} 0 & K \\ K & C \end{bmatrix} \begin{bmatrix} x \\ u \end{bmatrix} = \lambda \begin{bmatrix} K & 0 \\ 0 & -M \end{bmatrix} \begin{bmatrix} x \\ u \end{bmatrix}.$$

This is *not* a symmetric-definite problem. However, if the *overdamping condition*

$$\min_{x^T x = 1} (x^T C x)^2 - 4(x^T M x)(x^T K x) = \gamma^2 > 0$$

holds, then it can be shown that there is a scalar $\mu > 0$ so that

$$A(\mu) = \begin{bmatrix} \mu K & K \\ K & C - \mu M \end{bmatrix}$$

is positive definite. It follows from Theorem 8.7.1 that (8.7.16) can be diagonalized by congruence. See Vescelić (1993).

A quadratic eigenvalue problem that arises in the analysis of gyroscopic systems has the property that $M = M^T$ (positive definite), $K = K^T$, and $C = -C^T$. It is easy to see from (8.7.14) that the eigenvalues are all purely imaginary. For this problem we have the structured linearization

$$\begin{bmatrix} 0 & -K \\ M & 0 \end{bmatrix} \begin{bmatrix} u \\ x \end{bmatrix} = \lambda \begin{bmatrix} M & C \\ 0 & M \end{bmatrix} \begin{bmatrix} u \\ x \end{bmatrix}.$$

Notice that this is a Hamiltonian/skew-Hamiltonian generalized eigenvalue problem.

In the *quadratic palindomic problem*, $K = M^T$ and $C = C^T$ and the eigenvalues come in reciprocal pairs, i.e., if $Q(\lambda)$ is singular then so is $Q(1/\lambda)$. In addition, we have the linearization

$$\begin{bmatrix} M^T & M^T \\ C - M & M^T \end{bmatrix} \begin{bmatrix} y \\ z \end{bmatrix} = \lambda \begin{bmatrix} -M & M^T - C \\ -M & -M \end{bmatrix} \begin{bmatrix} y \\ z \end{bmatrix}. \qquad (8.7.17)$$

Note that if this equation holds, then

$$(\lambda^2 M + \lambda C + M^T)(y + z) = 0. \qquad (8.7.18)$$

For a systematic treatment of linearizations for structured polynomial eigenvalue problems, see Mackey, Mackey, Mehl, and Mehrmann (2006).

Problems

P8.7.1 Suppose $A \in \mathbb{R}^{n \times n}$ is symmetric and $G \in \mathbb{R}^{n \times n}$ is lower triangular and nonsingular. Give an efficient algorithm for computing $C = G^{-1}AG^{-T}$.

P8.7.2 Suppose $A \in \mathbb{R}^{n \times n}$ is symmetric and $B \in \mathbb{R}^{n \times n}$ is symmetric positive definite. Give an algorithm for computing the eigenvalues of AB that uses the Cholesky factorization and the symmetric

QR algorithm.

P8.7.3 Relate the principal angles and vectors between $\mathsf{ran}(A)$ and $\mathsf{ran}(B)$ to the eigenvalues and eigenvectors of the generalized eigenvalue problem

$$
\begin{bmatrix} 0 & A^T B \\ B^T A & 0 \end{bmatrix} \begin{bmatrix} y \\ z \end{bmatrix} = \sigma \begin{bmatrix} A^T A & 0 \\ 0 & B^T B \end{bmatrix} \begin{bmatrix} y \\ z \end{bmatrix}.
$$

P8.7.4 Show that if C is real and diagonalizable, then there exist symmetric matrices A and B, B nonsingular, such that $C = AB^{-1}$. This shows that symmetric pencils $A - \lambda B$ are essentially general.

P8.7.5 Show how to convert an $Ax = \lambda Bx$ problem into a generalized singular value problem if A and B are both symmetric and nonnegative definite.

P8.7.6 Given $Y \in \mathbb{R}^{n \times n}$ show how to compute Householder matrices $H_2, \ldots, H_n$ so that $Y H_n \cdots H_2 = T$ is upper triangular. Hint: H_k zeros out the kth row.

P8.7.7 Suppose

$$
\begin{bmatrix} 0 & A \\ A^T & 0 \end{bmatrix} \begin{bmatrix} y \\ z \end{bmatrix} = \lambda \begin{bmatrix} B_1 & 0 \\ 0 & B_2 \end{bmatrix} \begin{bmatrix} y \\ z \end{bmatrix}
$$

where $A \in \mathbb{R}^{m \times n}$, $B_1 \in \mathbb{R}^{m \times m}$, and $B_2 \in \mathbb{R}^{n \times n}$. Assume that B_1 and B_2 are positive definite with Cholesky triangles G_1 and G_2 respectively. Relate the generalized eigenvalues of this problem to the singular values of $G_1^{-1} A G_2^{-T}$

P8.7.8 Suppose A and B are both symmetric positive definite. Show how to compute $\lambda(A, B)$ and the corresponding eigenvectors using the Cholesky factorization and CS decomposition.

P8.7.9 Consider the problem

$$
\min_{\substack{x^T Bx=\beta^2 \\ x^T Cx=\gamma^2}} \| Ax - b \|_2, \qquad A \in \mathbb{R}^{m \times n}, \ b \in \mathbb{R}^m, \ B, C \in \mathbb{R}^{n \times n}.
$$

Assume that B and C are positive definite and that $Z \in \mathbb{R}^{n \times n}$ is a nonsingular matrix with the property that $Z^T BZ = \mathrm{diag}(\lambda_1, \ldots, \lambda_n)$ and $Z^T CZ = I_n$. Assume that $\lambda_1 \geq \cdots \geq \lambda_n$. (a) Show that the the the set of feasible x is empty unless $\lambda_n \leq \beta^2/\gamma^2 \leq \lambda_1$. (b) Using Z, show how the two-constraint problem can be converted to a single-constraint problem of the form

$$
\min_{y^T Wy=\beta^2-\lambda_n\gamma^2} \| \tilde{A}x - b \|_2
$$

where $W = \mathrm{diag}(\lambda_1, \ldots, \lambda_n) - \lambda_n I$.

P8.7.10 Show that (8.7.17) implies (8.7.18).

Notes and References for §8.7

Just how far one can simplify a symmetric pencil $A - \lambda B$ via congruence is thoroughly discussed in:

P. Lancaster and L. Rodman (2005). "Canonical Forms for Hermitian Matrix Pairs under Strict Equivalence and Congruence," *SIAM Review 47*, 407–443.

The sensitivity of the symmetric-definite eigenvalue problem is covered in Stewart and Sun (MPT, Chap. 6). See also:

C.R. Crawford (1976). "A Stable Generalized Eigenvalue Problem," *SIAM J. Numer. Anal. 13*, 854–860.

C.-K. Li and R. Mathias (1998). "Generalized Eigenvalues of a Definite Hermitian Matrix Pair," *Lin. Alg. Applic. 271*, 309–321.

S.H. Cheng and N.J. Higham (1999). "The Nearest Definite Pair for the Hermitian Generalized Eigenvalue Problem," *Lin. Alg. Applic. 302-3*, 63–76.

C.-K. Li and R. Mathias (2006). "Distances from a Hermitian Pair to Diagonalizable and Nondiagonalizable Hermitian Pairs," *SIAM J. Matrix Anal. Applic. 28*, 301–305.

Y. Nakatsukasa (2010). "Perturbed Behavior of a Multiple Eigenvalue in Generalized Hermitian Eigenvalue Problems," *BIT 50*, 109–121.

R.-C. Li, Y. Nakatsukasa, N. Truhar, and S. Xu (2011). "Perturbation of Partitioned Hermitian Definite Generalized Eigenvalue Problems," *SIAM J. Matrix Anal. Applic. 32*, 642–663.

Although it is possible to diagonalize a symmetric-definite pencil, serious numerical issues arise if the congruence transformation is ill-conditioned. Various methods for "controlling the damage" have been proposed including:

R.S. Martin and J.H. Wilkinson (1968). "Reduction of a Symmetric Eigenproblem $Ax = \lambda Bx$ and Related Problems to Standard Form," *Numer. Math. 11*, 99–110.

G. Fix and R. Heiberger (1972). "An Algorithm for the Ill-Conditioned Generalized Eigenvalue Problem," *SIAM J. Numer. Anal. 9*, 78–88.

A. Bunse-Gerstner (1984). "An Algorithm for the Symmetric Generalized Eigenvalue Problem," *Lin. Alg. Applic. 58*, 43–68.

S. Chandrasekaran (2000). "An Efficient and Stable Algorithm for the Symmetric-Definite Generalized Eigenvalue Problem," *SIAM J. Matrix Anal. Applic. 21*, 1202–1228.

P.I. Davies, N.J. Higham, and F. Tisseur (2001). "Analysis of the Cholesky Method with Iterative Refinement for Solving the Symmetric Definite Generalized Eigenproblem," *SIAM J. Matrix Anal. Applic. 23*, 472–493.

F. Tisseur (2004). "Tridiagonal-Diagonal Reduction of Symmetric Indefinite Pairs," *SIAM J. Matrix Anal. Applic. 26*, 215–232.

Exploiting bandedness in A and B can be important, see:

G. Peters and J.H. Wilkinson (1969). "Eigenvalues of $Ax = \lambda Bx$ with Band Symmetric A and B," *Comput. J. 12*, 398-404.

C.R. Crawford (1973). "Reduction of a Band Symmetric Generalized Eigenvalue Problem," *Commun. ACM 16*, 41–44.

L. Kaufman (1993). "An Algorithm for the Banded Symmetric Generalized Matrix Eigenvalue Problem," *SIAM J. Matrix Anal. Applic. 14*, 372–389.

K. Li, T-Y. Li, and Z. Zeng (1994). "An Algorithm for the Generalized Symmetric Tridiagonal Eigenvalue Problem," *Numer. Algorithms 8*, 269–291.

The existence of a positive semidefinite linear combination of A and B was central to Theorem 8.7.1. Interestingly, the practical computation of such a combination has been addressed, see:

C.R. Crawford (1986). "Algorithm 646 PDFIND: A Routine to Find a Positive Definite Linear Combination of Two Real Symmetric Matrices," *ACM Trans. Math. Softw. 12*, 278–282.

C.-H. Guo, N.J. Higham, and F. Tisseur (2009). "An Improved Arc Algorithm for Detecting Definite Hermitian Pairs," *SIAM J. Matrix Anal. Applic. 31*, 1131–1151.

As we mentioned, many techniques for the symmetric eigenvalue problem have natural extensions to the symmetric-definite problem. These include methods based on the Rayleigh quotient idea:

E. Jiang(1990). "An Algorithm for Finding Generalized Eigenpairs of a Symmetric Definite Matrix Pencil," *Lin. Alg. Applic. 132*, 65–91.

R-C. Li (1994). "On Eigenvalue Variations of Rayleigh Quotient Matrix Pencils of a Definite Pencil," *Lin. Alg. Applic. 208/209*, 471–483.

There are also generalizations of the Jacobi method:

K. Veselič (1993). "A Jacobi Eigenreduction Algorithm for Definite Matrix Pairs," *Numer. Math. 64*, 241–268.

C. Mehl (2004). "Jacobi-like Algorithms for the Indefinite Generalized Hermitian Eigenvalue Problem," *SIAM J. Matrix Anal. Applic. 25*, 964–985.

Homotopy methods have also found application:

K. Li and T-Y. Li (1993). "A Homotopy Algorithm for a Symmetric Generalized Eigenproblem," *Numer. Algorithms 4*, 167–195.

T. Zhang and K.H. Law, and G.H. Golub (1998). "On the Homotopy Method for Perturbed Symmetric Generalized Eigenvalue Problems," *SIAM J. Sci. Comput. 19*, 1625–1645.

We shall have more to say about symmetric-definite problems with general sparsity in Chapter 10. If the matrices are banded, then it is possible to implement an effective a generalization of simultaneous iteration, see:

H. Zhang and W.F. Moss (1994). "Using Parallel Banded Linear System Solvers in Generalized Eigenvalue Problems," *Parallel Comput. 20*, 1089–1106.

Turning our attention to the GSVD literature, the original references include:

C.F. Van Loan (1976). "Generalizing the Singular Value Decomposition," *SIAM J. Numer. Anal. 13*, 76–83.

C.C. Paige and M. Saunders (1981). "Towards A Generalized Singular Value Decomposition," *SIAM J. Numer. Anal. 18*, 398–405.

The sensitivity of the GSVD is detailed in Stewart and Sun (MPT) as as well in the following papers:

J.-G. Sun (1983). "Perturbation Analysis for the Generalized Singular Value Problem," *SIAM J. Numer. Anal. 20*, 611–625.

C.C. Paige (1984). "A Note on a Result of Sun J.-Guang: Sensitivity of the CS and GSV Decompositions," *SIAM J. Numer. Anal. 21*, 186–191.

R-C. Li (1993). "Bounds on Perturbations of Generalized Singular Values and of Associated Subspaces," *SIAM J. Matrix Anal. Applic. 14*, 195–234.

J.-G. Sun (1998). "Perturbation Analysis of Generalized Singular Subspaces," *Numer. Math. 79*, 615–641.

J.-G. Sun (2000). "Condition Number and Backward Error for the Generalized Singular Value Decomposition," *SIAM J. Matrix Anal. Applic. 22*, 323–341.

X.S. Chen and W. Li (2008). "A Note on Backward Error Analysis of the Generalized Singular Value Decomposition," *SIAM J. Matrix Anal. Applic. 30*, 1358–1370.

The variational characterization of the GSVD is analyzed in:

M.T. Chu, R.F Funderlic, and G.H. Golub (1997). "On a Variational Formulation of the Generalized Singular Value Decomposition," *SIAM J. Matrix Anal. Applic. 18*, 1082–1092.

Connections between GSVD and the pencil $A - \lambda B$ are discussed in:

B. Kågström (1985). "The Generalized Singular Value Decomposition and the General $A - \lambda B$ Problem," *BIT 24*, 568–583.

Stable methods for computing the CS and generalized singular value decompositions are described in:

G.W. Stewart (1982). "Computing the C-S Decomposition of a Partitioned Orthonormal Matrix," *Numer. Math. 40*, 297–306.

G.W. Stewart (1983). "A Method for Computing the Generalized Singular Value Decomposition," in *Matrix Pencils* , B. Kågström and A. Ruhe (eds.), Springer-Verlag, New York, 207–220.

C.F. Van Loan (1985). "Computing the CS and Generalized Singular Value Decomposition," *Numer. Math. 46*, 479–492.

B.D. Sutton (2012). "Stable Computation of the CS Decomposition: Simultaneous Bidiagonalization," *SIAM. J. Matrix Anal. Applic. 33*, 1–21.

The idea of using the Kogbetliantz procedure for the GSVD problem is developed in:

C.C. Paige (1986). "Computing the Generalized Singular Value Decomposition," *SIAM J. Sci. Stat. Comput. 7*, 1126–1146.

Z. Bai and H. Zha (1993). "A New Preprocessing Algorithm for the Computation of the Generalized Singular Value Decomposition," *SIAM J. Sci. Comp. 14*, 1007–1012.

Z. Bai and J.W. Demmel (1993). "Computing the Generalized Singular Value Decomposition," *SIAM J. Sci. Comput. 14*, 1464–1486.

Other methods for computing the GSVD include:

Z. Drmač (1998). "A Tangent Algorithm for Computing the Generalized Singular Value Decomposition," *SIAM J. Numer. Anal. 35*, 1804–1832.

Z. Drmač and E.R. Jessup (2001). "On Accurate Quotient Singular Value Computation in Floating-Point Arithmetic," *SIAM J. Matrix Anal. Applic. 22*, 853–873.

S. Friedland (2005). "A New Approach to Generalized Singular Value Decomposition," *SIAM J. Matrix Anal. Applic. 27*, 434–444.

Stable methods for computing the product and restricted SVDs are discussed in the following papers:

M.T. Heath, A.J. Laub, C.C. Paige, and R.C. Ward (1986). "Computing the Singular Value Decomposition of a Product of Two Matrices," *SIAM J. Sci. Stat. Comput. 7*, 1147–1159.

K.V. Fernando and S. Hammarling (1988). "A Product-Induced Singular Value Decomposition for Two Matrices and Balanced Realization," in *Linear Algebra in Systems and Control*, B.N. Datta et al (eds), SIAM Publications, Philadelphia, PA.

B. De Moor and H. Zha (1991). "A Tree of Generalizations of the Ordinary Singular Value Decomposition," *Lin. Alg. Applic. 147*, 469–500.

H. Zha (1991). "The Restricted Singular Value Decomposition of Matrix Triplets," *SIAM J. Matrix Anal. Applic. 12*, 172–194.

B. De Moor and G.H. Golub (1991). "The Restricted Singular Value Decomposition: Properties and Applications," *SIAM J. Matrix Anal. Applic. 12*, 401–425.

B. De Moor and P. Van Dooren (1992). "Generalizing the Singular Value and QR Decompositions," *SIAM J. Matrix Anal. Applic. 13*, 993–1014.

H. Zha (1992). "A Numerical Algorithm for Computing the Restricted Singular Value Decomposition of Matrix Triplets," *Lin. Alg. Applic. 168*, 1–25.

G.E. Adams, A.W. Bojanczyk, and F.T. Luk (1994). "Computing the PSVD of Two 2×2 Triangular Matrices," *SIAM J. Matrix Anal. Applic. 15*, 366–382.

Z. Drmač (1998). "Accurate Computation of the Product-Induced Singular Value Decomposition with Applications," *SIAM J. Numer. Anal. 35*, 1969–1994.

D. Chu, L. De Lathauwer, and B. De Moor (2000). "On the Computation of the Restricted Singular Value Decomposition via the Cosine-Sine Decomposition," *SIAM J. Matrix Anal. Applic. 22*, 580–601.

D. Chu and B.De Moor (2000). "On a variational formulation of the QSVD and the RSVD," *Lin. Alg. Applic. 311,* 61–78.

For coverage of structured quadratic eigenvalue problems, see:

P. Lancaster (1991). "Quadratic Eigenvalue Problems," *Lin. Alg. Applic. 150*, 499–506.

F. Tisseur and N.J. Higham (2001). "Structured Pseudospectra for Polynomial Eigenvalue Problems, with Applications," *SIAM J. Matrix Anal. Applic. 23*, 187–208.

F. Tisseur and K. Meerbergen (2001). "The Quadratic Eigenvalue Problem," *SIAM Review 43*, 235–286.

V. Mehrmann and D. Watkins (2002). "Polynomial Eigenvalue Problems with Hamiltonian Structure," *Electr. Trans. Numer. Anal. 13*, 106–118.

U.B. Holz, G.H. Golub, and K.H. Law (2004). "A Subspace Approximation Method for the Quadratic Eigenvalue Problem," *SIAM J. Matrix Anal. Applic. 26*, 498–521.

D.S. Mackey, N. Mackey, C. Mehl, and V. Mehrmann (2006). "Structured Polynomial Eigenvalue Problems: Good Vibrations from Good Linearizations," *SIAM. J. Matrix Anal. Applic. 28,* 1029–1051.

B. Plestenjak (2006). "Numerical Methods for the Tridiagonal Hyperbolic Quadratic Eigenvalue Problem," *SIAM J. Matrix Anal. Applic. 28*, 1157–1172.

E.K.-W. Chu, T.-M. Hwang, W.-W. Lin, and C.-T. Wu (2008). "Vibration of Fast Trains, Palindromic Eigenvalue Problems, and Structure-Preserving Doubling Algorithms," *J. Comp. Appl. Math. 219*, 237–252.

Chapter 9

Functions of Matrices

9.1 Eigenvalue Methods

9.2 Approximation Methods

9.3 The Matrix Exponential

9.4 The Sign, Square Root, and Log of a Matrix

Computing a function $f(A)$ of an n-by-n matrix A is a common problem in many application areas. Roughly speaking, if the scalar function $f(z)$ is defined on $\lambda(A)$, then $f(A)$ is defined by substituting "A" for "z" in the "formula" for $f(z)$. For example, if $f(z) = (1+z)/(1-z)$ and $1 \notin \lambda(A)$, then $f(A) = (I + A)(I - A)^{-1}$.

The computations get particularly interesting when the function f is transcendental. One approach in this more complicated situation is to compute an eigenvalue decomposition $A = YBY^{-1}$ and use the formula $f(A) = Yf(B)Y^{-1}$. If B is sufficiently simple, then it is often possible to calculate $f(B)$ directly. This is illustrated in §9.1 for the Jordan and Schur decompositions.

Another class of methods involves the approximation of the desired function $f(A)$ with an easy-to-calculate function $g(A)$. For example, g might be a truncated Taylor series approximation to f. Error bounds associated with the approximation of matrix functions are given in §9.2.

In §9.3 we discuss the special and very important problem of computing the matrix exponential e^A. The matrix sign, square root, and logarithm functions and connections to the polar decomposition are treated in §9.4.

Reading Notes

Knowledge of Chapters 3 and 7 is assumed. Within this chapter there are the following dependencies:

$$\S 9.1 \quad \rightarrow \quad \S 9.2 \quad \rightarrow \quad \S 9.3$$
$$\downarrow$$
$$\S 9.4$$

Complementary references include Horn and Johnson (TMA) and the definitive text by Higham (FOM). We mention that aspects of the $f(A)$-times-a-vector problem are treated in §10.2.

9.1   Eigenvalue Methods

Here are some examples of matrix functions:

$$p(A) = I + A,$$

$$r(A) = \left(I - \frac{A}{2}\right)^{-1}\left(I + \frac{A}{2}\right), \qquad 2 \notin \lambda(A),$$

$$e^A = \sum_{k=0}^{\infty} \frac{A^k}{k!}.$$

Obviously, these are the matrix versions of the scalar-valued functions

$$p(z) = 1 + z,$$

$$r(z) = (1 - (z/2))^{-1}(1 + (z/2)), \qquad 2 \neq z,$$

$$e^z = \sum_{k=0}^{\infty} \frac{z^k}{k!}.$$

Given an n-by-n matrix A, it appears that all we have to do to define $f(A)$ is to substitute A into the formula for f. However, to make subsequent algorithmic developments precise, we need to be a little more formal. It turns out that there are several equivalent ways to define a function of a matrix. See Higham (FOM, §1.2). Because of its prominence in the literature and its simplicity, we take as our "base" definition one that involves the Jordan canonical form (JCF).

9.1.1   A Jordan-Based Definition

Suppose $A \in \mathbb{C}^{n \times n}$ and let

$$A = X \cdot \mathrm{diag}(J_1, \ldots, J_q) \cdot X^{-1} \tag{9.1.1}$$

be its JCF with

$$J_i = \begin{bmatrix} \lambda_i & 1 & \cdots & \cdots & 0 \\ 0 & \lambda_i & 1 & \cdots & \vdots \\ \vdots & \ddots & \ddots & \ddots & \vdots \\ \vdots & \vdots & \ddots & \ddots & 1 \\ 0 & \cdots & \cdots & 0 & \lambda_i \end{bmatrix} \in \mathbb{C}^{n_i \times n_i}, \qquad i = 1{:}q. \tag{9.1.2}$$

The matrix function $f(A)$ is defined by

$$f(A) = X \cdot \text{diag}(F_1, \ldots, F_q) \cdot X^{-1} \tag{9.1.3}$$

where

$$F_i = \begin{bmatrix} f(\lambda_i) & f^{(1)}(\lambda_i) & \cdots & \cdots & \dfrac{f^{(n_i-1)}(\lambda_i)}{(n_i-1)!} \\ 0 & f(\lambda_i) & \ddots & \cdots & \vdots \\ \vdots & \vdots & \ddots & \ddots & \vdots \\ \vdots & \vdots & \vdots & \ddots & f^{(1)}(\lambda_i) \\ 0 & \cdots & \cdots & \cdots & f(\lambda_i) \end{bmatrix}, \qquad i = 1{:}q, \tag{9.1.4}$$

assuming that all the required derivative evaluations exist.

9.1.2 The Taylor Series Representation

If f can be represented by a Taylor series on A's spectrum, then $f(A)$ can be represented by the same Taylor series in A. To fix ideas, assume that f is analytic in a neighborhood of $z_0 \in \mathbb{C}$ and that for some $r > 0$ we have

$$f(z) = \sum_{k=0}^{\infty} \frac{f^{(k)}(z_0)}{k!}(z - z_0)^k, \qquad |z - z_0| < r. \tag{9.1.5}$$

Our first result applies to a single Jordan block.

Lemma 9.1.1. *Suppose $B \in \mathbb{C}^{m \times m}$ is a Jordan block and write $B = \lambda I_m + E$ where E is its strictly upper bidiagonal part. Given (9.1.5), if $|\lambda - z_0| < r$, then*

$$f(B) = \sum_{k=0}^{\infty} \frac{f^{(k)}(z_0)}{k!}(B - z_0 I_m)^k.$$

Proof. Note that powers of E are highly structured, e.g.,

$$E = \begin{bmatrix} 0 & 1 & 0 & 0 \\ 0 & 0 & 1 & 0 \\ 0 & 0 & 0 & 1 \\ 0 & 0 & 0 & 0 \end{bmatrix}, \quad E^2 = \begin{bmatrix} 0 & 0 & 1 & 0 \\ 0 & 0 & 0 & 1 \\ 0 & 0 & 0 & 0 \\ 0 & 0 & 0 & 0 \end{bmatrix}, \quad E^3 = \begin{bmatrix} 0 & 0 & 0 & 1 \\ 0 & 0 & 0 & 0 \\ 0 & 0 & 0 & 0 \\ 0 & 0 & 0 & 0 \end{bmatrix}.$$

In terms of the Kronecker delta, if $0 \leq p \leq m - 1$, then $[E^p]_{ij} = (\delta_{i,j-p})$. It follows from (9.1.4) that

$$f(B) = \sum_{p=0}^{m-1} f^{(p)}(\lambda) \frac{E^p}{p!}. \tag{9.1.6}$$

On the other hand, if $p > m$, then $E^p = 0$. Thus, for any $k \geq 0$ we have

$$(B - z_0 I)^k = ((\lambda - z_0)I + E)^k = \sum_{p=0}^{k} \frac{k(k-1)\cdots(k-p+1)}{p!} \cdot (\lambda - z_0)^{k-p} \cdot E^p$$

$$= \sum_{p=0}^{\min\{k,m-1\}} \left[\frac{d^p}{d\lambda^p}(\lambda - z_0)^k \right] \frac{E^p}{p!}.$$

If N is a nonnegative integer, then

$$\sum_{k=0}^{N} \frac{f^{(k)}(z_0)}{k!}(B - z_0 I)^k = \sum_{p=0}^{\min\{k,m-1\}} \frac{d^p}{d\lambda^p} \left(\sum_{k=0}^{N} \frac{f^{(k)}(z_0)}{k!}(\lambda - z_0)^k \right) \frac{E^p}{p!}.$$

The lemma follows by taking limits with respect to N and using both (9.1.6) and the Taylor series representation of $f(z)$.   □

A similar result holds for general matrices.

Theorem 9.1.2. *If f has the Taylor series representation (9.1.5) and $|\lambda - z_0| < r$ for all $\lambda \in \lambda(A)$ where $A \in \mathbb{C}^{n \times n}$, then*

$$f(A) = \sum_{k=0}^{\infty} \frac{f^{(k)}(z_0)}{k!}(A - z_0 I)^k.$$

Proof. Let the JCF of A be given by (9.1.1) and (9.1.2). From Lemma 9.1.1 we have

$$f(J_i) = \sum_{k=0}^{\infty} \alpha_k (J_i - z_0 I)^k, \qquad \alpha_k = \frac{f^{(k)}(z_0)}{k!},$$

for $i = 1{:}q$. Using the definition (9.1.3) and (9.1.4) we see that

$$f(A) = X \cdot \mathrm{diag}\left(\sum_{k=0}^{\infty} \alpha_k (J_1 - z_0 I_{n_1})^k, \ldots, \sum_{k=0}^{\infty} \alpha_k (J_q - z_0 I_{n_q})^k \right) \cdot X^{-1}$$

$$= X \cdot \left(\sum_{k=0}^{\infty} \alpha_k (J - z_0 I_n)^k \right) \cdot X^{-1}$$

$$= \sum_{k=0}^{\infty} \alpha_k \left(X(J - z_0 I_n)X^{-1} \right)^k = \sum_{k=0}^{\infty} \alpha_k (A - z_0 I_n)^k,$$

completing the proof of the theorem.   □

Important matrix functions that have simple Taylor series definitions include

$$\exp(A) \;=\; \sum_{k=0}^{\infty} \frac{A^k}{k!},$$

$$\log(I - A) \;=\; \sum_{k=1}^{\infty} \frac{A^k}{k}, \qquad |\lambda| < 1, \; \lambda \in \lambda(A),$$

$$\sin(A) \;=\; \sum_{k=0}^{\infty} (-1)^k \frac{A^{2k+1}}{(2k+1)!},$$

$$\cos(A) \;=\; \sum_{k=0}^{\infty} (-1)^k \frac{A^{2k}}{(2k)!}.$$

For clarity in this section and the next, we consider only matrix functions that have a Taylor series representation. In that case it is easy to verify that

$$A \cdot f(A) \;=\; f(A) \cdot A \tag{9.1.7}$$

and

$$f(X^{-1}AX) \;=\; X \cdot f(A) \cdot X^{-1}. \tag{9.1.8}$$

9.1.3   An Eigenvector Approach

If $A \in \mathbb{C}^{n \times n}$ is diagonalizable, then it is particularly easy to specify $f(A)$ in terms of A's eigenvalues and eigenvectors.

Corollary 9.1.3. *If $A \in \mathbb{C}^{n \times n}$, $A = X \cdot \mathrm{diag}(\lambda_1, \ldots, \lambda_n) \cdot X^{-1}$, and $f(A)$ is defined, then*

$$f(A) \;=\; X \cdot \mathrm{diag}(f(\lambda_1), \ldots, f(\lambda_n)) \cdot X^{-1}. \tag{9.1.9}$$

Proof. This result is an easy consequence of Theorem 9.1.2 since all the Jordan blocks are 1-by-1.   □

Unfortunately, if the matrix of eigenvectors is ill-conditioned, then computing $f(A)$ via (9.1.8) is likely introduce errors of order $\mathbf{u}\,\kappa_2(X)$ because of the required solution of a linear system that involves the eigenvector matrix X. For example, if

$$A \;=\; \begin{bmatrix} 1 + 10^{-5} & 1 \\ 0 & 1 - 10^{-5} \end{bmatrix},$$

then any matrix of eigenvectors is a column-scaled version of

$$X \;=\; \begin{bmatrix} 1 & -1 \\ 0 & 2(1 - 10^{-5}) \end{bmatrix}$$

and has a 2-norm condition number of order 10^5. Using a computer with machine precision $\mathbf{u} \approx 10^{-7}$, we find

$$\text{fl}\left(X^{-1}\text{diag}(\exp(1+10^{-5}), \exp(1-10^{-5}))X\right) = \begin{bmatrix} 2.718307 & 2.750000 \\ 0.000000 & 2.718254 \end{bmatrix}$$

while

$$e^A = \begin{bmatrix} 2.718309 & 2.718282 \\ 0.000000 & 2.718255 \end{bmatrix}.$$

The example suggests that ill-conditioned similarity transformations should be avoided when computing a function of a matrix. On the other hand, if A is a normal matrix, then it has a perfectly conditioned matrix of eigenvectors. In this situation, computation of $f(A)$ via diagonalization is a recommended strategy.

9.1.4   A Schur Decomposition Approach

Some of the difficulties associated with the Jordan approach to the matrix function problem can be circumvented by relying upon the Schur decomposition. If $A = QTQ^H$ is the Schur decomposition of A, then by (9.1.8),

$$f(A) = Qf(T)Q^H.$$

For this to be effective, we need an algorithm for computing functions of upper triangular matrices. Unfortunately, an explicit expression for $f(T)$ is very complicated.

Theorem 9.1.4. *Let $T = (t_{ij})$ be an n-by-n upper triangular matrix with $\lambda_i = t_{ii}$ and assume $f(T)$ is defined. If $f(T) = (f_{ij})$, then $f_{ij} = 0$ if $i > j$, $f_{ij} = f(\lambda_i)$ for $i = j$, and for all $i < j$ we have*

$$f_{ij} = \sum_{(s_0, \ldots, s_k) \in S_{ij}} t_{s_0, s_1} t_{s_1, s_2} \cdots t_{s_{k-1}, s_k} f[\lambda_{s_0}, \ldots, \lambda_{s_k}], \tag{9.1.10}$$

where S_{ij} is the set of all strictly increasing sequences of integers that start at i and end at j, and $f[\lambda_{s_0}, \ldots, \lambda_{s_k}]$ is the kth order divided difference of f at $\{\lambda_{s_0}, \ldots, \lambda_{s_k}\}$.

Proof. See Descloux (1963), Davis (1973), or Van Loan (1975). $\square$

To illustrate the theorem, if

$$T = \begin{bmatrix} \lambda_1 & t_{12} & t_{13} \\ 0 & \lambda_2 & t_{23} \\ 0 & 0 & \lambda_3 \end{bmatrix}$$

then

$$f(T) = \begin{bmatrix} f(\lambda_1) & t_{12} \cdot \dfrac{f(\lambda_2) - f(\lambda_1)}{\lambda_2 - \lambda_1} & F_{13} \\ \\ 0 & f(\lambda_2) & t_{23} \cdot \dfrac{f(\lambda_3) - f(\lambda_2)}{\lambda_3 - \lambda_2} \\ \\ 0 & 0 & f(\lambda_3) \end{bmatrix},$$

where

$$
F_{13} = t_{13} \cdot \frac{f(\lambda_3) - f(\lambda_1)}{\lambda_3 - \lambda_1} + t_{12} t_{23} \cdot \frac{\dfrac{f(\lambda_3) - f(\lambda_2)}{\lambda_3 - \lambda_2} - \dfrac{f(\lambda_2) - f(\lambda_1)}{\lambda_2 - \lambda_1}}{\lambda_3 - \lambda_1}.
$$

The recipes for the upper triangular entries get increasing complicated as we move away from the diagonal. Indeed, if we explicitly use (9.1.10) to evaluate $f(T)$, then $O(2^n)$ flops are required. However, Parlett (1974) has derived an elegant recursive method for determining the strictly upper triangular portion of the matrix $F = f(T)$. It requires only $2n^3/3$ flops and can be derived from the commutivity equation $FT = TF$. Indeed, by comparing (i, j) entries in this equation, we find

$$
\sum_{k=i}^{j} f_{ik} t_{kj} = \sum_{k=i}^{j} t_{ik} f_{kj}, \qquad j > i,
$$

and thus, if t_{ii} and t_{jj} are distinct,

$$
f_{ij} = t_{ij} \frac{f_{jj} - f_{ii}}{t_{jj} - t_{ii}} + \sum_{k=i+1}^{j-1} \frac{t_{ik} f_{kj} - f_{ik} t_{kj}}{t_{jj} - t_{ii}}. \tag{9.1.11}
$$

From this we conclude that f_{ij} is a linear combination of its neighbors in the matrix F that are to its left and below. For example, the entry f_{25} depends upon f_{22}, f_{23}, f_{24}, f_{55}, f_{45}, and f_{35}. Because of this, the entire upper triangular portion of F can be computed superdiagonal by superdiagonal beginning with $\mathrm{diag}(f(t_{11}), \ldots, f(t_{nn}))$. The complete procedure is as follows:

Algorithm 9.1.1 (Schur-Parlett) This algorithm computes the matrix function $F = f(T)$ where T is upper triangular with distinct eigenvalues and f is defined on $\lambda(T)$.

> **for** $i = 1{:}n$
> $\qquad f_{ii} = f(t_{ii})$
> **end**
> **for** $p = 1{:}n - 1$
> $\qquad$ **for** $i = 1{:}n - p$
> $\qquad\qquad j = i + p$
> $\qquad\qquad s = t_{ij}(f_{jj} - f_{ii})$
> $\qquad\qquad$ **for** $k = i + 1{:}j - 1$
> $\qquad\qquad\qquad s = s + t_{ik} f_{kj} - f_{ik} t_{kj}$
> $\qquad\qquad$ **end**
> $\qquad\qquad f_{ij} = s/(t_{jj} - t_{ii})$
> $\qquad$ **end**
> **end**

This algorithm requires $2n^3/3$ flops. Assuming that $A = QTQ^H$ is the Schur decomposition of A, $f(A) = QFQ^H$ where $F = f(T)$. Clearly, most of the work in computing $f(A)$ by this approach is in the computation of the Schur decomposition, unless f is extremely expensive to evaluate.

9.1.5   A Block Schur-Parlett Approach

If A has multiple or nearly multiple eigenvalues, then the divided differences associated with Algorithm 9.1.1 become problematic and it is advisable to use a block version of the method. We outline such a procedure due to Parlett (1974). The first step is to choose Q in the Schur decomposition so that we have a partitioning

$$
T = \begin{bmatrix}
T_{11} & T_{12} & \cdots & T_{1p} \\
0 & T_{22} & \cdots & T_{2p} \\
\vdots & \vdots & \ddots & \vdots \\
0 & 0 & \cdots & T_{pp}
\end{bmatrix}
$$

where $\lambda(T_{ii}) \cap \lambda(T_{jj}) = \emptyset$ and each diagonal block is associated with an eigenvalue cluster. The methods of §7.6 are applicable for this stage of the calculation.

Partition $F = f(T)$ conformably

$$
F = \begin{bmatrix}
F_{11} & F_{12} & \cdots & F_{1p} \\
0 & F_{22} & \cdots & F_{2p} \\
\vdots & \vdots & \ddots & \vdots \\
0 & 0 & \cdots & F_{pp}
\end{bmatrix},
$$

and notice that

$$
F_{ii} = f(T_{ii}), \qquad i = 1{:}p.
$$

Since the eigenvalues of T_{ii} are clustered, these calculations require special methods. Some possibilities are discussed in the next section.

Once the diagonal blocks of F are known, the blocks in the strict upper triangle of F can be found recursively, as in the scalar case. To derive the governing equations, we equate (i, j) blocks in $FT = TF$ for $i < j$ and obtain the following generalization of (9.1.11):

$$
F_{ij}T_{jj} - T_{ii}F_{ij} = T_{ij}F_{jj} - F_{ii}T_{ij} + \sum_{k=i+1}^{j-1} (T_{ik}F_{kj} - F_{ik}T_{kj}). \tag{9.1.12}
$$

This is a Sylvester system whose unknowns are the elements of the block F_{ij} and whose right-hand side is "known" if we compute the F_{ij} one block superdiagonal at a time. We can solve (9.1.12) using the Bartels-Stewart algorithm (Algorithm 7.6.2). For more details see Higham (FOM, Chap. 9).

9.1.6   Sensitivity of Matrix Functions

Does the Schur-Parlett algorithm avoid the pitfalls associated with the diagonalization approach when the matrix of eigenvectors is ill-conditioned? The proper comparison of the two solution frameworks requires an appreciation for the notion of condition as applied to the $f(A)$ problem. Toward that end we define the relative condition of f at matrix $A \in \mathbb{C}^{n \times n}$ is given as

$$
\mathrm{cond}_{\mathrm{rel}}(f, A) = \lim_{\epsilon \to 0} \sup_{\|E\| \le \epsilon \|A\|} \frac{\| f(A + E) - f(A) \|}{\epsilon \| f(A) \|}.
$$

This quantity is essentially a normalized *Frechet derivative* of the mapping $A \to f(A)$ and various heuristic methods have been developed for estimating its value.

It turns out that the careful implementation of the block Schur-Parlett algorithm is usually forward stable in the sense that

$$\frac{\| \hat{F} - f(A) \|}{\| f(A) \|} \approx \mathbf{u} \cdot \mathrm{cond}_{\mathrm{rel}}(f, A)$$

where $\hat{F}$ is the computed version of $f(A)$. The same cannot be said of the diagonalization framework when the matrix of eigenvectors is ill-conditioned. For more details, see Higham (FOM, Chap. 3).

Problems

P9.1.1 Suppose

$$A = \begin{bmatrix} \lambda & \mu_1 \\ \mu_2 & \lambda \end{bmatrix}, \qquad \mu_1 \mu_2 < 0.$$

Use the power series definitions to develop closed form expressions for $\exp(A)$, $\sin(A)$, and $\cos(A)$.

P9.1.2 Rewrite Algorithm 9.1.1 so that $f(T)$ is computed column by column.

P9.1.3 Suppose $A = X\mathrm{diag}(\lambda_i)X^{-1}$ where $X = [\, x_1 \mid \cdots \mid x_n \,]$ and $X^{-1} = [\, y_1 \mid \cdots \mid y_n \,]^H$. Show that if $f(A)$ is defined, then

$$f(A) = \sum_{k=1}^{n} f(\lambda_i) x_i y_i^H.$$

P9.1.4 Show that

$$T = \begin{bmatrix} T_{11} & T_{12} \\ 0 & T_{22} \end{bmatrix} \begin{matrix} p \\ q \end{matrix} \quad \Rightarrow \quad f(T) = \begin{bmatrix} F_{11} & F_{12} \\ 0 & F_{22} \end{bmatrix} \begin{matrix} p \\ q \end{matrix}$$
$$\quad\quad p \quad\, q \quad\quad\quad\quad\quad\quad\quad\quad\, p \quad\, q$$

where $F_{11} = f(T_{11})$ and $F_{22} = f(T_{22})$. Assume $f(T)$ is defined.

Notes and References for §9.1

As we discussed, other definitions of $f(A)$ are possible. However, for the matrix functions typically encountered in practice, all these definitions are equivalent, see:

R.F. Rinehart (1955). "The Equivalence of Definitions of a Matric Function," *Amer. Math. Monthly* *62*, 395–414.

The following papers are concerned with the Schur decomposition and its relationship to the $f(A)$ problem:

C. Davis (1973). "Explicit Functional Calculus," *Lin. Alg. Applic. 6*, 193–199.

J. Descloux (1963). "Bounds for the Spectral Norm of Functions of Matrices," *Numer. Math. 5*, 185–190.

C.F. Van Loan (1975). "A Study of the Matrix Exponential," Numerical Analysis Report No. 10, Department of Mathematics, University of Manchester, England. Available as Report 2006.397 from http://eprints.ma.man.ac.uk/.

Algorithm 9.1.1 and the various computational difficulties that arise when it is applied to a matrix having close or repeated eigenvalues are discuss

B.N. Parlett (1976). "A Recurrence among the Elements of Functions of Triangular Matrices," *Lin. Alg. Applic. 14*, 117–121.

P.I. Davies and N.J. Higham (2003). "A Schur-Parlett Algorithm for Computing Matrix Functions," *SIAM J. Matrix Anal. Applic. 25*, 464–485.

A compromise between the Jordan and Schur approaches to the $f(A)$ problem results if A is reduced to block diagonal form as described in §7.6.3, see:

B. Kågström (1977). "Numerical Computation of Matrix Functions," Department of Information Processing Report UMINF-58.77, University of Ümeå, Sweden.

E.B. Davies (2007). "Approximate Diagonalization," *SIAM J. Matrix Anal. Applic. 29*, 1051–1064.

The sensitivity of matrix functions to perturbation is discussed in:

C.S. Kenney and A.J. Laub (1989). "Condition Estimates for Matrix Functions," *SIAM J. Matrix Anal. Applic. 10*, 191–209.

C.S. Kenney and A.J. Laub (1994). "Small-Sample Statistical Condition Estimates for General Matrix Functions," *SIAM J. Sci. Comput. 15*, 36–61.

R. Mathias (1995). "Condition Estimation for Matrix Functions via the Schur Decomposition," *SIAM J. Matrix Anal. Applic. 16*, 565–578.

9.2   Approximation Methods

We now consider a class of methods for computing matrix functions which at first glance do not appear to involve eigenvalues. These techniques are based on the idea that, if $g(z)$ approximates $f(z)$ on $\lambda(A)$, then $f(A)$ approximates $g(A)$, e.g.,

$$ e^A \approx I + A + \frac{A^2}{2!} + \cdots + \frac{A^q}{q!}. $$

We begin by bounding $\| f(A) - g(A) \|$ using the Jordan and Schur matrix function representations. We follow this discussion with some comments on the evaluation of matrix polynomials.

9.2.1   A Jordan Analysis

The Jordan representation of matrix functions (Theorem 9.1.2) can be used to bound the error in an approximant $g(A)$ of $f(A)$.

Theorem 9.2.1. *Assume that*

$$ A = X \cdot \operatorname{diag}(J_1, \ldots, J_q) \cdot X^{-1} $$

is the JCF of $A \in \mathbb{C}^{n \times n}$ with

$$ J_i = \begin{bmatrix} \lambda_i & 1 & \cdots & \cdots & 0 \\ 0 & \lambda_i & 1 & \vdots & \vdots \\ \vdots & \vdots & \ddots & \ddots & \vdots \\ \vdots & \vdots & \vdots & \ddots & 1 \\ 0 & \cdots & \cdots & \cdots & \lambda_i \end{bmatrix}, \qquad n_i\text{-by-}n_i, $$

for $i = 1{:}q$. If $f(z)$ and $g(z)$ are analytic on an open set containing $\lambda(A)$, then

$$ \| f(A) - g(A) \|_2 \leq \kappa_2(X) \max_{\substack{1 \leq i \leq p \\ 0 \leq r \leq n_i - 1}} n_i \frac{\left| f^{(r)}(\lambda_i) - g^{(r)}(\lambda_i) \right|}{r!}. $$

Proof. Defining $h(z) = f(z) - g(z)$ we have

$$\| f(A) - g(A) \|_2 \;=\; \| X \operatorname{diag}(h(J_1), \ldots, h(J_q)) X^{-1} \|_2 \;\leq\; \kappa_2(X) \max_{1 \leq i \leq q} \| h(J_i) \|_2.$$

Using Theorem 9.1.2 and equation (2.3.8) we conclude that

$$\| h(J_i) \|_2 \;\leq\; n_i \max_{0 \leq r \leq n_i - 1} \frac{| h^{(r)}(\lambda_i) |}{r!}$$

thereby proving the theorem. $\square$

9.2.2 A Schur Analysis

If we use the Schur decomposition $A = QTQ^H$ instead of the Jordan decomposition, then the norm of T's strictly upper triangular portion is involved in the discrepancy between $f(A)$ and $g(A)$.

Theorem 9.2.2. *Let* $Q^H A Q = T = \operatorname{diag}(\lambda_i) + N$ *be the Schur decomposition of* $A \in \mathbb{C}^{n \times n}$, *with* N *being the strictly upper triangular portion of* T. *If* $f(z)$ *and* $g(z)$ *are analytic on a closed convex set* Ω *whose interior contains* $\lambda(A)$, *then*

$$\| f(A) - g(A) \|_F \;\leq\; \sum_{r=0}^{n-1} \delta_r \frac{\| \, |N|^r \, \|_F}{r!}$$

where

$$\delta_r \;=\; \sup_{z \in \Omega} \left| f^{(r)}(z) - g^{(r)}(z) \right|.$$

Proof. Let $h(z) = f(z) - g(z)$ and set $H = (h_{ij}) = h(A)$. Let $S_{ij}^{(r)}$ denote the set of strictly increasing integer sequences $(s_0, \ldots, s_r)$ with the property that $s_0 = i$ and $s_r = j$. Notice that

$$S_{ij} \;=\; \bigcup_{r=1}^{j-i} S_{ij}^{(r)}$$

and so from Theorem 9.1.3, we obtain the following for all $i < j$:

$$h_{ij} \;=\; \sum_{r=1}^{j-1} \sum_{s \in S_{ij}^{(r)}} n_{s_0, s_1} n_{s_1, s_2} \cdots n_{s_{r-1}, s_r} h\left[\lambda_{s_0}, \ldots, \lambda_{s_r} \right].$$

Now since Ω is convex and h analytic, we have

$$\left| h\left[\lambda_{s_0}, \ldots, \lambda_{s_r} \right] \right| \;\leq\; \sup_{z \in \Omega} \frac{\left| h^{(r)}(z) \right|}{r!} \;=\; \frac{\delta_r}{r!}. \tag{9.2.1}$$

Furthermore if $|N|^r = (n_{ij}^{(r)})$ for $r \geq 1$, then it can be shown that

$$
n_{ij}^{(r)} = \begin{cases} 0, & j < i + r, \\ \sum_{s \in S_{ij}^{(r)}} \left| n_{s_0,s_1} n_{s_1,s_2} \cdots n_{s_{r-1},s_r} \right|, & j \geq i + r. \end{cases} \tag{9.2.2}
$$

The theorem now follows by taking absolute values in the expression for h_{ij} and then using (9.2.1) and (9.2.2)   ∏

There can be a pronounced discrepancy between the Jordan and Schur error bounds. For example, if

$$
A = \begin{bmatrix} -.01 & 1 & 1 \\ 0 & 0 & 1 \\ 0 & 0 & .01 \end{bmatrix}.
$$

If $f(z) = e^z$ and $g(z) = 1 + z + z^2/2$, then $\| f(A) - g(A) \| \approx 10^{-5}$ in either the Frobenius norm or the 2-norm. Since $\kappa_2(X) \approx 10^7$, the error predicted by Theorem 9.2.1 is $O(1)$, rather pessimistic. On the other hand, the error predicted by the Schur decomposition approach is $O(10^{-2})$.

Theorems 9.2.1 and 9.2.2 remind us that approximating a function of a nonnormal matrix is more complicated than approximating a function of a scalar. In particular, we see that if the eigensystem of A is ill-conditioned and/or A's departure from normality is large, then the discrepancy between $f(A)$ and $g(A)$ may be considerably larger than the maximum of $|f(z) - g(z)|$ on $\lambda(A)$. Thus, even though approximation methods avoid eigenvalue computations, they evidently appear to be influenced by the structure of A's eigensystem. It is a perfect venue for pseudospectral analysis.

9.2.3   Taylor Approximants

A common way to approximate a matrix function such as e^A is by truncating its Taylor series. The following theorem bounds the errors that arise when matrix functions such as these are approximated via truncated Taylor series.

Theorem 9.2.3. *If $f(z)$ has the Taylor series*

$$
f(z) = \sum_{k=0}^{\infty} \alpha_k z^k
$$

on an open disk containing the eigenvalues of $A \in \mathbb{C}^{n \times n}$, then

$$
\left\| f(A) - \sum_{k=0}^{q} \alpha_k A^k \right\|_2 \leq \frac{n}{(q+1)!} \max_{0 \leq s \leq 1} \| A^{q+1} f^{(q+1)}(As) \|_2 .
$$

Proof. Define the matrix $E(s)$ by

$$
f(As) = \sum_{k=0}^{q} \alpha_k (As)^k + E(s), \qquad 0 \leq s \leq 1. \tag{9.2.3}
$$

If $f_{ij}(s)$ is the (i,j) entry of $f(As)$, then it is necessarily analytic and so

$$f_{ij}(s) = \left(\sum_{k=0}^{q} \frac{f_{ij}^{(k)}(0)}{k!} s^k \right) + \frac{f_{ij}^{(q+1)}(\varepsilon_{ij})}{(q+1)!} s^{q+1} \qquad (9.2.4)$$

where ε_{ij} satisfies $0 \le \varepsilon_{ij} \le s \le 1$.

By comparing powers of s in (9.2.3) and (9.2.4) we conclude that $e_{ij}(s)$, the (i,j) entry of $E(s)$, has the form

$$e_{ij}(s) = \frac{f_{ij}^{(q+1)}(\varepsilon_{ij})}{(q+1)!} s^{q+1}.$$

Now $f_{ij}^{(q-1)}(s)$ is the (i,j) entry of $A^{q+1} f^{(q+1)}(As)$ and therefore

$$|e_{ij}(s)| \le \max_{0 \le s \le 1} \frac{f_{ij}^{(q+1)}(s)}{(q+1)!} \le \max_{0 \le s \le 1} \frac{\| A^{q+1} f^{(q+1)}(As) \|_2}{(q+1)!}.$$

The theorem now follows by applying (2.3.8). $\quad\square$

We mention that the factor of n in the upper bound can be removed with more careful analysis. See Mathias (1993).

In practice, it does *not* follow that greater accuracy results by taking a longer Taylor approximation. For example, if

$$A = \begin{bmatrix} -49 & 24 \\ -64 & 31 \end{bmatrix},$$

then it can be shown that

$$e^A = \begin{bmatrix} -0.735759 & .0551819 \\ -1.471518 & 1.103638 \end{bmatrix}.$$

For $q = 59$, Theorem 9.2.3 predicts that

$$\left\| e^A - \sum_{k=0}^{q} \frac{A^k}{k!} \right\|_2 \le \frac{n}{(q+1)!} \max_{0 \le s \le 1} \left\| A^{q+1} e^{As} \right\|_2 \le 10^{-60}.$$

However, if $\mathbf{u} \approx 10^{-7}$, then we find

$$\mathrm{fl}\left(\sum_{k=0}^{59} \frac{A^k}{k!} \right) = \begin{bmatrix} -22.25880 & -1.4322766 \\ -61.49931 & -3.474280 \end{bmatrix}.$$

The problem is that some of the partial sums have large elements. For example, the matrix $I + A + \cdots + A^{17}/17!$ has entries of order 10^7. Since the machine precision is approximately 10^{-7}, rounding errors larger than the norm of the solution are sustained.

The example highlights the a well known shortcoming of truncated Taylor series approximation–it tends to be effcetive only near the origin. The problem can sometimes be circumvented through a change of scale. For example, by repeatedly using the *double angle* formulae

$$\cos(2A) = 2\cos(A)^2 - I, \qquad \sin(2A) = 2\sin(A)\cos(A),$$

the cosine and sine of a matrix can be built up from Taylor approximations to $\cos(A/2^k)$ and $\sin(A/2^k)$:

$S_0 = $ Taylor approximate to $\sin(A/2^k)$

$C_0 = $ Taylor approximate to $\cos(A/2^k)$

for $j = 1{:}k$

$\qquad S_j = 2S_{j-1}C_{j-1}$

$\qquad C_j = 2C_{j-1}^2 - I$

end

Here k is a positive integer chosen so that, say, $\| A \|_\infty \approx 2^k$. See Serbin and Blalock (1979), Higham and Smith (2003), and Hargreaves and Higham (2005).

9.2.4   Evaluating Matrix Polynomials

Since the approximation of transcendental matrix functions usually involves the evaluation of polynomials, it is worthwhile to look at the details of computing

$$p(A) \;=\; b_0 I + b_1 A + \cdots + b_q A^q$$

where the scalars $b_0, \ldots, b_q \in \mathbb{R}$ are given. The most obvious approach is to invoke Horner's scheme:

Algorithm 9.2.1 Given a matrix A and $b(0{:}q)$, the following algorithm computes the polynomial $F = b_q A^q + \cdots + b_1 A + b_0 I$.

$F = b_q A + b_{q-1} I$

for $k = q - 2{:} - 1{:}0$

$\qquad F = AF + b_k I$

end

This requires $q - 1$ matrix multiplications. However, unlike the scalar case, this summation process is not optimal. To see why, suppose $q = 9$ and observe that

$$p(A) \;=\; A^3(A^3(b_9 A^3 + (b_8 A^2 + b_7 A + b_6 I)) + (b_5 A^2 + b_4 A + b_3 I)) + b_2 A^2 + b_1 A + b_0 I.$$

Thus, $F = p(A)$ can be evaluated with only four matrix multiplications:

$$
\begin{aligned}
A_2 &= A^2, \\
A_3 &= AA_2, \\
F_1 &= b_9 A_3 + b_8 A_2 + b_7 A + b_6 I, \\
F_2 &= A_3 F_1 + b_5 A_2 + b_4 A + b_3 I, \\
F &= A_3 F_2 + b_2 A_2 + b_1 A + b_0 I.
\end{aligned}
$$

In general, if s is any integer that satisfies $1 \leq s \leq \sqrt{q}$, then

$$p(A) = \sum_{k=0}^{r} B_k \cdot (A^s)^k, \qquad r = \text{floor}(q/s), \tag{9.2.5}$$

where

$$B_k = \begin{cases} b_{sk+s-1}A^{s-1} + \cdots + b_{sk+1}A + b_{sk}I, & k = 0{:}r-1, \\ \\ b_q A^{q-sr} + \cdots + b_{sr+1}A + b_{sr}I, & k = r. \end{cases}$$

After $A^2, \ldots, A^s$ are computed, then Horner's rule can be applied to (9.2.5) and the net result is that $p(A)$ can be computed with $s + r - 1$ matrix multiplications. By choosing $s = \text{floor}(\sqrt{q})$, the number of matrix multiplications is approximately minimized. This technique is discussed by Paterson and Stockmeyer (1973). Van Loan (1978) shows how the procedure can be implemented without storage arrays for $A^2, \ldots, A^s$.

9.2.5 Computing Powers of a Matrix

The problem of raising a matrix to a given power deserves special mention. Suppose it is required to compute A^{13}. Noting that $A^4 = (A^2)^2$, $A^8 = (A^4)^2$, and $A^{13} = A^8 A^4 A$, we see that this can be accomplished with just five matrix multiplications. In general we have

Algorithm 9.2.2 (Binary Powering) The following algorithm computes $F = A^s$ where s is a positive integer and $A \in \mathbb{R}^{n \times n}$.

Let $s = \displaystyle\sum_{k=0}^{t} \beta_k 2^k$ be the binary expansion of s with $\beta_t \neq 0$

$Z = A; \ q = 0$

while $\beta_q = 0$

    $Z = Z^2; \ q = q+1$

end

$F = Z$

for $k = q + 1{:}t$

    $Z = Z^2$

    if $\beta_k \neq 0$

        $F = FZ$

    end

end

This algorithm requires at most $2\,\text{floor}[\log_2(s)]$ matrix multiplications. If s is a power of 2, then only $\log_2(s)$ matrix multiplications are needed.

9.2.6 Integrating Matrix Functions

We conclude this section with some remarks about the integration of a parameterized matrix function. Suppose $A \in \mathbb{R}^{n \times n}$ and that $f(At)$ is defined for all $t \in [a, b]$. We can approximate

$$F = \int_a^b f(At)dt \qquad \Leftrightarrow \qquad [F]_{ij} = \int_a^b [f(At)]_{ij}\, dt$$

by applying any suitable quadrature rule. For example, with Simpson's rule, we have

$$F \approx \tilde{F} = \frac{h}{3} \sum_{k=0}^{m} w_k f(A(a + kh)) \qquad (9.2.6)$$

where m is even, $h = (b - a)/m$, and

$$w_k = \begin{cases} 1 & k = 0, m, \\ 4 & k \text{ odd}, \\ 2 & k \text{ even}, k \neq 0, m. \end{cases}$$

If $(d^4/dz^4)f(zt) = f^{(4)}(zt)$ is continuous for $t \in [a, b]$ and if $f^{(4)}(At)$ is defined on this same interval, then it can be shown that $\tilde{F} = F + E$ where

$$\| E \|_2 \leq \frac{nh^4(b - a)}{180} \max_{a \leq t \leq b} \| f^{(4)}(At) \|_2. \qquad (9.2.7)$$

Let f_{ij} and e_{ij} denote the (i, j) entries of F and E, respectively. Under the above assumptions we can apply the standard error bounds for Simpson's rule and obtain

$$|e_{ij}| \leq \frac{h^4(b - a)}{180} \max_{a \leq t \leq b} |e_i^T f^{(4)}(At)e_j|.$$

The inequality (9.2.7) now follows since $\| E \|_2 \leq n \max |e_{ij}|$ and

$$\max_{a \leq t \leq b} |e_i^T f^{(4)}(At)e_j| \leq \max_{a \leq t \leq b} \| f^{(4)}(At) \|_2.$$

Of course, in a practical application of (9.2.6), the function evaluations $f(A(a + kh))$ normally have to be approximated. Thus, the overall error involves the error in approximating $f(A(a + kh)$ as well as the Simpson rule error.

9.2.7   A Note on the Cauchy Integral Formulation

Yet another way to define a function of a matrix $C \in \mathbb{C}^{n \times n}$ is through the Cauchy integral theorem. Suppose $f(z)$ is analytic inside and on a closed contour Γ which encloses $\lambda(A)$. We can define $f(A)$ to be the matrix

$$f(A) = \frac{1}{2\pi i} \oint_\Gamma f(z)(zI - A)^{-1}dz. \qquad (9.2.8)$$

The integral is defined on an element-by-element basis:

$$f(A) = (f_{kj}) \quad \Longrightarrow \quad f_{kj} = \frac{1}{2\pi i} \oint_\Gamma f(z)e_k^T(zI - A)^{-1}e_j dz.$$

Notice that the entries of $(zI - A)^{-1}$ are analytic on Γ and that $f(A)$ is defined whenever $f(z)$ is analytic in a neighborhood of $\lambda(A)$. Using quadrature and other tools, Hale, Higham, and Trefethen (2007) have shown how this characterization can be used in practice to compute certain types of matrix functions.

Problems

P9.2.1 Verify (9.2.2).

P9.2.2 Show that if $\| A \|_2 < 1$, then $\log(I + A)$ exists and satisfies the bound

$$\| \log(I + A) \|_2 \leq \| A \|_2 / (1 - \| A \|_2).$$

P9.2.3 Using Theorem 9.2.3, bound the error in the following approximations:

$$\sin(A) \approx \sum_{k=0}^{q} (-1)^k \frac{A^{2k+1}}{(2k+1)!}, \qquad \cos(A) \approx \sum_{k=0}^{q} (-1)^k \frac{A^{2k}}{(2k)!}.$$

P9.2.4 Suppose $A \in \mathbb{R}^{n \times n}$ is nonsingular and $X_0 \in \mathbb{R}^{n \times n}$ is given. The iteration defined by

$$X_{k+1} = X_k(2I - AX_k)$$

is the matrix analogue of Newton's method applied to the function $f(x) = a - (1/x)$. Use the SVD to analyze this iteration. Do the iterates converge to A^{-1}? Discuss the choice of X_0.

P9.2.5 Assume $A \in \mathbb{R}^{2 \times 2}$. (a) Specify real scalars α and β so that $A^4 = \alpha I + \beta A$. (b) Develop recursive recipes for α_k and β_k so that $A^k = \alpha_k I + \beta_k A$ for $k \geq 2$.

Notes and References for §9.2

The optimality of Horner's rule for polynomial evaluation is discussed in:

M.S. Paterson and L.J. Stockmeyer (1973). "On the Number of Nonscalar Multiplications Necessary to Evaluate Polynomials," *SIAM J. Comput. 2*, 60–66.
D.E. Knuth (1981). *The Art of Computer Programming, Vol. 2. Seminumerical Algorithms*, second edition, Addison-Wesley, Reading, MA.

The Horner evaluation of matrix polynomials is analyzed in:

C.F. Van Loan (1978). "A Note on the Evaluation of Matrix Polynomials," *IEEE Trans. Autom. Control AC-24*, 320–321.

Other aspects of matrix function approximation and evaluation are discussed in:

H. Bolz and W. Niethammer (1988). "On the Evaluation of Matrix Functions Given by Power Series," *SIAM J. Matrix Anal. Applic. 9*, 202–209.
R. Mathias (1993). "Approximation of Matrix-Valued Functions," *SIAM J. Matrix Anal. Applic. 14*, 1061–1063.
N.J. Higham and P.A. Knight (1995). "Matrix Powers in Finite Precision Arithmetic," *SIAM J. Matrix Anal. Applic. 16*, 343–358.
P. Sebastiani (1996). "On the Derivatives of Matrix Powers," *SIAM J. Matrix Anal. Applic. 17*, 640–648.
D.S. Bernstein and C.F. Van Loan (2000). "Rational Matrix Functions and Rank-One Updates," *SIAM J. Matrix Anal. Applic. 22*, 145–154.

For a discussion of methods for computing the sine and cosine of a matrix, see:

S. Serbin and S. Blalock (1979). "An Algorithm for Computing the Matrix Cosine," *SIAM J. Sci. Stat. Comput. 1*, 198–204.
N.J. Higham and M.I. Smit (2003). "Computing the Matrix Cosine," *Numer. Algorithms 34*, 13–26.
G. Hargreaves and N.J. Higham (2005). "Efficient Algorithms for the Matrix Cosine and Sine," *Numer. Algorithms 40*, 383–400.

The computation of $f(A)$ using contour integrals is analyzed in:

N. Hale, N.J. Higham, and L.N. Trefethen (2007). "Computing A^α, $\log(A)$, and Related Matrix Functions by Contour Integrals," *SIAM J. Numer. Anal. 46*, 2505–2523.

9.3   The Matrix Exponential

One of the most frequently computed matrix functions is the exponential

$$e^{At} = \sum_{k=0}^{\infty} \frac{(At)^k}{k!}.$$

Numerous algorithms for computing e^{At} have been proposed, but most of them are of dubious numerical quality, as is pointed out in the survey articles by Moler and Van Loan (1978) and its update Moler and Van Loan (2003). In order to illustrate what the computational difficulties are, we present a "scaling and squaring" method based upon Padé approximation. A brief analysis of the method follows that involves some e^{At} perturbation theory and includes comments about the shortcomings of eigenanalysis in settings where nonnormality prevails.

9.3.1   A Padé Approximation Method

Following the discussion in §9.2, if $g(z) \approx e^z$, then $g(A) \approx e^A$. A very useful class of approximants for this purpose are the Padé functions defined by

$$R_{pq}(z) = D_{pq}(z)^{-1} N_{pq}(z),$$

where

$$N_{pq}(z) = \sum_{k=0}^{p} \frac{(p+q-k)!p!}{(p+q)!k!(p-k)!} z^k$$

and

$$D_{pq}(z) = \sum_{k=0}^{q} \frac{(p+q-k)!q!}{(p+q)!k!(q-k)!} (-z)^k.$$

Notice that

$$R_{po}(z) = 1 + z + \cdots + z^p/p!$$

is the order-p Taylor polynomial.

Unfortunately, the Padé approximants are good only near the origin, as the following identity reveals:

$$e^A = R_{pq}(A) + \frac{(-1)^q}{(p+q)!} A^{p+q+1} D_{pq}(A)^{-1} \int_0^1 u^p (1-u)^q e^{A(1-u)} du. \qquad (9.3.1)$$

However, this problem can be overcome by exploiting the fact that

$$e^A = (e^{A/m})^m.$$

In particular, we can scale A by m such that $F_{pq} = R_{pq}(A/m)$ is a suitably accurate approximation to $e^{A/m}$. We then compute F_{pq}^m using Algorithm 9.2.2. If m is a power of two, then this amounts to repeated squaring and so is very efficient. The success of the overall procedure depends on the accuracy of the approximant

$$F_{pq} = \left(R_{pq} \left(\frac{A}{2^j} \right) \right)^{2^j}.$$

In Moler and Van Loan (1978) it is shown that, if

$$\frac{\| A \|_\infty}{2^j} \leq \frac{1}{2},$$

then there exists an $E \in \mathbb{R}^{n \times n}$ such that $F_{pq} = e^{A+E}$, $AE = EA$, and

$$\| E \|_\infty \leq \varepsilon(p, q) \| A \|_\infty,$$

where

$$\varepsilon(p, q) = 2^{3-(p+q)} \frac{p! q!}{(p+q)!(p+q+1)!}.$$

Using these results it is easy to establish the inequality

$$\frac{\| e^A - F_{pq} \|_\infty}{\| e^A \|_\infty} \leq \epsilon(p, q) \| A \|_\infty e^{\epsilon(p,q)\|A\|_\infty}.$$

The parameters p and q can be determined according to some relative error tolerance. Since F_{pq} requires about $j + \max\{p, q\}$ matrix multiplications, it makes sense to set $p = q$ as this choice minimizes $\epsilon(p, q)$ for a given amount of work. Overall we obtain

Algorithm 9.3.1 (Scaling and Squaring) Given $\delta > 0$ and $A \in \mathbb{R}^{n \times n}$, the following algorithm computes $F = e^{A+E}$ where $\| E \|_\infty \leq \delta \| A \|_\infty$.

$j = \max\{\, 0\,,\, 1 + \text{floor}(\log_2(\| A \|_\infty)) \,\}$

$A = A/2^j$

Let q be the smallest nonnegative integer such that $\epsilon(q, q) \leq \delta$

$D = I$, $N = I$, $X = I$, $c = 1$

for $k = 1{:}q$

    $c = c \cdot (q - k + 1)/((2q - k + 1)k)$

    $X = AX$, $N = N + c \cdot X$, $D = D + (-1)^k c \cdot X$

end

Solve $DF = N$ for F using Gaussian elimination

for $k = 1{:}j$

    $F = F^2$

end

This algorithm requires about $2(q+j+1/3)n^3$ flops. Its roundoff error properties of have been analyzed by Ward (1977). For further analysis and algorithmic improvements, see Higham (2005) and Al-Mohy and Higham (2009).

    The special Horner techniques of §9.2.4 can be applied to quicken the computation of $D = D_{qq}(A)$ and $N = N_{qq}(A)$. For example, if $q = 8$ we have $N_{qq}(A) = U + AV$ and $D_{qq}(A) = U - AV$ where

$$U = c_0 I + c_2 A^2 + (c_4 I + c_6 A^2 + c_8 A^4) A^4$$

and

$$V = c_1 I + c_3 A^2 + (c_5 I + c_7 A^2) A^4.$$

Clearly, N and D can be computed with five matrix multiplications instead of seven as required by Algorithm 9.3.1.

9.3.2   Perturbation Theory

Is Algorithm 9.3.1 stable in the presence of roundoff error? To answer this question we need to understand the sensitivity of the matrix exponential to perturbations in A. The rich structure of this particular matrix function enables us to say more about the condition of the e^A problem than is typically the case for a general matrix function. (See §9.1.6.)

The starting point in the discussion is the initial value problem

$$\dot{X}(t) = AX(t), \qquad X(0) = I,$$

where $A, X(t) \in \mathbb{R}^{n \times n}$. This has the unique solution $X(t) = e^{At}$, a characterization of the matrix exponential that can be used to establish the identity

$$e^{(A+E)t} - e^{At} = \int_0^t e^{A(t-s)} E e^{(A+E)s} ds.$$

From this it follows that

$$\frac{\| e^{(A+E)t} - e^{At} \|_2}{\| e^{At} \|_2} \leq \frac{\| E \|_2}{\| e^{At} \|_2} \int_0^t \| e^{A(t-s)} \|_2 \| e^{(A+E)s} \|_2 ds.$$

Further simplifications result if we bound the norms of the exponentials that appear in the integrand. One way of doing this is through the Schur decomposition. If $Q^H A Q = \mathrm{diag}(\lambda_i) + N$ is the Schur decomposition of $A \in \mathbb{C}^{n \times n}$, then it can be shown that

$$\| e^{At} \|_2 \leq e^{\alpha(A)t} M_S(t), \tag{9.3.2}$$

where

$$\alpha(A) = \max \{ \mathrm{Re}(\lambda) : \lambda \in \lambda(A) \} \tag{9.3.3}$$

is the spectral abscissa and

$$M_S(t) = \sum_{k=0}^{n-1} \frac{\| Nt \|_2^k}{k!}.$$

With a little manipulation it can be shown that

$$\frac{\| e^{(A+E)t} - e^{At} \|_2}{\| e^{At} \|_2} \leq t\| E \|_2 M_S(t)^2 \exp(t M_S(t) \| E \|_2).$$

Notice that $M_S(t) \equiv 1$ if and only if A is normal, suggesting that the matrix exponential problem is "well-behaved" if A is normal. This observation is confirmed by the behavior of the *matrix exponential condition number* $\nu(A, t)$, defined by

$$\nu(A, t) = \max_{\| E \| \leq 1} \left\| \int_0^t e^{A(t-s)} E e^{As} ds \right\|_2 \frac{\| A \|_2}{\| e^{At} \|_2}.$$

This quantity, discussed by Van Loan (1977), measures the sensitivity of the map $A \to e^{At}$ in that for a given t, there is a matrix E for which

$$\frac{\| e^{(A+E)t} - e^{At} \|_2}{\| e^{At} \|_2} \approx \nu(A, t) \frac{\| E \|_2}{\| A \|_2}.$$

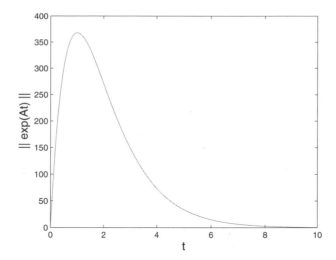

Figure 9.3.1. $\| e^{At} \|_2$ *can grow even if* $\alpha(A) < 0$

Thus, if $\nu(A, t)$ is large, small changes in A can induce relatively large changes in e^{At}. Unfortunately, it is difficult to characterize precisely those A for which $\nu(A, t)$ is large. (This is in contrast to the linear equation problem $Ax = b$, where the ill-conditioned A are neatly described in terms of SVD.) One thing we can say, however, is that $\nu(A, t) \geq t\| A \|_2$, with equality holding for all nonnegative t if and only if the matrix A is normal.

9.3.3   Pseudospectra

Dwelling a little more on the effect of nonnormality, we know from the analysis of §9.2 that approximating e^{At} involves more than just approximating e^{zt} on $\lambda(A)$. Another clue that eigenvalues do not "tell the whole story" in the e^{At} problem has to do with the inability of the spectral abscissa (9.3.3) to predict the size of $\| e^{At} \|_2$ as a function of time. If A is normal, then

$$\| e^{At} \|_2 \; = \; e^{\alpha(A)t}. \tag{9.3.4}$$

Thus, there is uniform decay if the eigenvalues of A are in the open left half plane. But if A is non-normal, then e^{At} can grow before decay sets in. The 2-by-2 example

$$A \; = \; \begin{bmatrix} -1 & 1000 \\ 0 & -1 \end{bmatrix} \quad \Leftrightarrow \quad e^{At} = e^{-t} \begin{bmatrix} 1 & 1000 \cdot t \\ 0 & 1 \end{bmatrix} \tag{9.3.5}$$

plainly illustrates this point in Figure 9.3.1.

Pseudospectra can be used to shed light on the transient growth of $\| e^{At} \|$. For example, it can be shown that for every $\epsilon > 0$,

$$\sup_{t>0} \| e^{At} \|_2 \; \geq \; \frac{\alpha_\epsilon(A)}{\epsilon} \tag{9.3.6}$$

where $\alpha_\epsilon(A)$ is the ϵ-pseudospectral abscissa introduced in (7.8.8):

$$\alpha_\epsilon(A) = \sup_{z \in \Lambda_\epsilon(A)} \text{Re}(z).$$

For the 2-by-2 matrix in (9.3.5), it can be shown that $\alpha_{.01}(A)/.01 \approx 216$, a value that is consistent with the growth curve in Figure 9.3.1. See Trefethen and Embree (SAP, Chap. 15) for more pseudospectral insights into the behavior of $\| e^{At} \|_2$.

9.3.4   Some Stability Issues

With this discussion we are ready to begin thinking about the stability of Algorithm 9.3.1. A potential difficulty arises during the squaring process if A is a matrix whose exponential grows before it decays. If

$$G = R_{qq}\left(\frac{A}{2^j}\right) \approx e^{A/2^j},$$

then it can be shown that rounding errors of order

$$\gamma = \mathbf{u}\| G^2 \|_2 \cdot \| G^4 \|_2 \cdot \| G^8 \|_2 \cdots \| G^{2^{j-1}} \|_2$$

can be expected to contaminate the computed G^{2^j}. If $\| e^{At} \|_2$ has a substantial initial growth, then it may be the case that

$$\gamma \gg \mathbf{u}\| G^{2^j} \|_2 \approx \mathbf{u}\| e^A \|_2,$$

thus ruling out the possibility of small relative errors.

If A is normal, then so is the matrix G and therefore $\| G^m \|_2 = \| G \|_2^m$ for all positive integers m. Thus, $\gamma \approx \mathbf{u}\| G^{2^j} \|_2 \approx \mathbf{u}\| e^A \|_2$ and so the initial growth problems disappear. The algorithm can essentially be guaranteed to produce small relative error when A is normal. On the other hand, it is more difficult to draw conclusions about the method when A is nonnormal because the connection between $\nu(A, t)$ and the initial growth phenomena is unclear. However, numerical experiments suggest that Algorithm 9.3.1 fails to produce a relatively accurate e^A only when $v(A, 1)$ is correspondingly large.

Problems

P9.3.1 Show that $e^{(A+B)t} = e^{At}e^{Bt}$ for all t if and only if $AB = BA$. Hint: Express both sides as a power series in t and compare the coefficient of t.

P9.3.2 Suppose that A is skew-symmetric. Show that both e^A and the (1,1) Padé approximatant $R_{11}(A)$ are orthogonal. Are there any other values of p and q for which $R_{pq}(A)$ is orthogonal?

P9.3.3 Show that if A is nonsingular, then there exists a matrix X such that $A = e^X$. Is X unique?

P9.3.4 Show that if

$$\exp\left(\begin{bmatrix} -A^T & P \\ 0 & A \end{bmatrix} z\right) = \begin{bmatrix} F_{11} & F_{12} \\ 0 & F_{22} \end{bmatrix} \begin{matrix} n \\ n \end{matrix}$$
$$ \begin{matrix} n & n \end{matrix}$$

then

$$F_{11}^T F_{12} = \int_0^z e^{A^T t} P e^{At} dt.$$

P9.3.5 Give an algorithm for computing e^A when $A = uv^T$, $u, v \in \mathbb{R}^n$.

P9.3.6 Suppose $A \in \mathbb{R}^{n \times n}$ and that $v \in \mathbb{R}^n$ has unit 2-norm. Define the function $\phi(t) = \| e^{At} v \|_2^2 / 2$ and show that
$$\dot{\phi}(t) \leq \mu(A) \phi(t)$$
where $\mu(A) = \lambda_1((A + A^T)/2)$. Conclude that
$$\| e^{At} \|_2 \leq e^{\mu(A)t}$$
where $t \geq 0$.

P9.3.7 Suppose $A \in \mathbb{R}^{n \times n}$ has the property that its off-diagonal entries are negative and its column sums are zero. Show that for all t, $F = \exp(At)$ has nonnegative entries and unit column sums.

Notes and References for §9.3

Much of what appears in this section and an extensive bibliography may be found in the following survey articles:

C.B. Moler and C.F. Van Loan (1978). "Nineteen Dubious Ways to Compute the Exponential of a Matrix," *SIAM Review 20,* 801–836.
C.B. Moler and C.F.Van Loan (2003). "Nineteen Dubious Ways to Compute the Exponential of a Matrix, Twenty-Five Years Later," *SIAM Review 45,* 3–49.

Scaling and squaring with Padé approximants (Algorithm 9.3.1) and a careful implementation of the Schur decomposition method (Algorithm 9.1.1) were found to be among the less dubious of the nineteen methods scrutinized. Various aspects of Padé approximation of the matrix exponential are discussed in:

W. Fair and Y. Luke (1970). "Padé Approximations to the Operator Exponential," *Numer. Math. 14,* 379–382.
C.F. Van Loan (1977). "On the Limitation and Application of Padé Approximation to the Matrix Exponential," in *Padé and Rational Approximation,* E.B. Saff and R.S. Varga (eds.), Academic Press, New York.
R.C. Ward (1977). "Numerical Computation of the Matrix Exponential with Accuracy Estimate," *SIAM J. Numer. Anal. 14,* 600–614.
A. Wragg (1973). "Computation of the Exponential of a Matrix I: Theoretical Considerations," *J. Inst. Math. Applic. 11,* 369–375.
A. Wragg (1975). "Computation of the Exponential of a Matrix II: Practical Considerations," *J. Inst. Math. Applic. 15,* 273–278.
L. Dieci and A. Papini (2000). "Padé Approximation for the Exponential of a Block Triangular Matrix," *Lin. Alg. Applic. 308,* 183–202.
M. Arioli, B. Codenotti and C. Fassino (1996). "The Padé Method for Computing the Matrix Exponential," *Lin. Alg. Applic. 240,* 111–130.
N.J. Higham (2005). "The Scaling and Squaring Method for the Matrix Exponential Revisited," *SIAM J. Matrix Anal. Applic. 26,* 1179–1193.
A.H. Al-Mohy and N.J. Higham (2009). "A New Scaling and Squaring Algorithm for the Matrix Exponential," *SIAM J. Matrix Anal. Applic. 31,* 970–989.

A proof of Equation (9.3.1) for the scalar case appears in:

R.S. Varga (1961). "On Higher-Order Stable Implicit Methods for Solving Parabolic Partial Differential Equations," *J. Math. Phys. 40,* 220–231.

There are many applications in control theory calling for the computation of the matrix exponential. In the linear optimal regular problem, for example, various integrals involving the matrix exponential are required, see:

J. Johnson and C.L. Phillips (1971). "An Algorithm for the Computation of the Integral of the State Transition Matrix," *IEEE Trans. Autom. Control AC-16,* 204–205.
C.F. Van Loan (1978). "Computing Integrals Involving the Matrix Exponential," *IEEE Trans. Autom. Control AC-23,* 395–404.

An understanding of the map $A \to \exp(At)$ and its sensitivity is helpful when assessing the performance of algorithms for computing the matrix exponential. Work in this direction includes:

B. Kågström (1977). "Bounds and Perturbation Bounds for the Matrix Exponential," *BIT 17*, 39–57.

C.F. Van Loan (1977). "The Sensitivity of the Matrix Exponential," *SIAM J. Numer. Anal. 14*, 971–981.

R. Mathias (1992). "Evaluating the Fréchet Derivative of the Matrix Exponential," *Numer. Math. 63*, 213–226.

I. Najfeld and T.F. Havel (1995). "Derivatives of the Matrix Exponential and Their Computation," *Adv. Appl. Math. 16*, 321–375.

A.H. Al-Mohy and N.J. Higham (2009). "Computing the Frèchet Derivative of the Matrix Exponential, with an Application to Condition Number Estimation," *SIAM J. Matrix Anal. Applic. 30*, 1639–1657.

A software package for computing small dense and large sparse matrix exponentials in Fortran and MATLAB is presented in the following reference:

R.B. Sidje (1998) "Expokit: a Software Package for Computing Matrix Exponentials," *ACM Trans. Math. Softw. 24*, 130–156.

Consideration of P9.3.2 and P9.3.5 shows that the exponential of a structured matrix can have important properties, see:

J. Xue and Q. Ye (2008). "Entrywise Relative Perturbation Bounds for Exponentials of Essentially Non-negative Matrices," *Numer. Math. 110*, 393–403.

J. Cardoso and F.S. Leite (2010). "Exponentials of Skew-Symmetric Matrices and Logarithms of Orthogonal Matrices," *J. Comput. Appl. Math. 233*, 2867–2875.

9.4   The Sign, Square Root, and Log of a Matrix

The matrix logarithm problem is the inverse of the matrix exponential problem. Not surprisingly, there is an inverse of the scaling and squaring procedure given in §9.3.1 that involves repeated matrix square roots. Thus, before we can discuss $\log(A)$ we need to understand the $\sqrt{A}$ problem. This in turn has connections to the matrix sign function and the polar decomposition.

9.4.1   The Matrix Sign Function

For all $z \in \mathbb{C}$ that are not on the imaginary axis, we define the sign($\cdot$) function by

$$\text{sign}(z) = \begin{cases} -1 & \text{if Re}(z) < 0, \\ +1 & \text{if Re}(z) > 0. \end{cases}$$

The sign of a matrix has a particularly simple form Suppose $A \in \mathbb{C}^{n \times n}$ has no pure imaginary eigenvalues and that the blocks in its JCF $A = XJX^{-1}$ are ordered so that

$$J = \begin{bmatrix} J_1 & 0 \\ 0 & J_2 \end{bmatrix} \begin{matrix} m_1 \\ m_2 \end{matrix}$$
$$\begin{matrix} m_1 & m_2 \end{matrix}$$

where the eigenvalues of $J_1 \in \mathbb{C}^{m_1 \times m_1}$ lie in the open left half plane and the eigenvalues of $J_2 \in \mathbb{C}^{m_2 \times m_2}$ lie in the open right half plane. Noting that all the derivatives of the sign function are zero, it follows from Theorem 9.1.1 that

$$\text{sign}(A) = X \begin{bmatrix} \text{sign}(J_1) & 0 \\ 0 & \text{sign}(J_2) \end{bmatrix} X^{-1} = X \begin{bmatrix} -I_{m_1} & 0 \\ 0 & I_{m_2} \end{bmatrix} X^{-1}.$$

With the partitionings

$$X = [\, X_1 \,|\, X_2 \,] \qquad X^{-H} = [\, Y_1 \,|\, Y_2 \,] \, ,$$
$$ m_1 \quad m_2 \qquad\qquad\qquad m_1 \quad m_2$$

we have

$$\mathrm{sign}(A) = X_2 Y_2^H - X_1 Y_1^H$$

$$I_n = X_1 Y_1^H + X_2 Y_2^H$$

and so

$$X_2 Y_2^H = \frac{1}{2} (I_n + \mathrm{sign}(A)) \,.$$

Suppose apply QR-with-column pivoting to this rank-m_2 matrix:

$$\frac{1}{2} (I_n + \mathrm{sign}(A)) \, \Pi = QR.$$

It follows that $\mathrm{ran}(Q(:, 1{:}m_2)) = \mathrm{ran}(X_2)$, the invariant subspace associated with A's right half-plane eigenvalues. Thus, an approximation of $\mathrm{sign}(A)$ yields approximate invariant subspace information.

A number of iterative methods for computing $\mathrm{sign}(A)$ have been proposed. The fact that $\mathrm{sign}(z)$ is a zero of $g(z) = z^2 - 1$ suggests a matrix analogue of the Newton iteration

$$z_{k+1} = z_k - \frac{g(z_k)}{g'(z_k)} = \frac{1}{2} \left(z_k + \frac{1}{z_k} \right),$$

i.e.,

$$S_0 = A$$
$$\textbf{for } k = 0, 1, \ldots \qquad\qquad\qquad\qquad (9.4.1)$$
$$\qquad S_{k+1} = \left(S_k + S_k^{-1} \right) / 2$$
$$\textbf{end}$$

We proceed to show that this iteration is well-defined and converges to $\mathrm{sign}(A)$, assuming that A has no eigenvalues on the imaginary axis.

Note that if $a + bi$ is an eigenvalue of S_k, then

$$\frac{1}{2} \left(a + bi + \frac{1}{a + bi} \right) = \frac{a}{2} \left(1 + \frac{1}{a^2 + b^2} \right) + \frac{b}{2} \left(1 - \frac{1}{a^2 + b^2} \right) i$$

is an eigenvalue of S_{k+1}. Thus, if S_k is nonsingular, then S_{k+1} is nonsingular. It follows by induction that (9.4.1) is defined. Moreover, $\mathrm{sign}(S_k) = \mathrm{sign}(A)$ because an eigenvalue cannot "jump" across the imaginary axis during the iteration.

To prove that S_k converges to $S = \mathrm{sign}(A)$, we first observe that $SS_k = S_k S$ since both matrices are rational functions of A. Using this commutivity result and the identity $S^2 = S$, it is easy to show that

$$S_{k+1} - S = \frac{1}{2} S_k^{-1} (S_k - S)^2 \qquad\qquad (9.4.2)$$

and

$$S_{k+1} + S = \frac{1}{2} S_k^{-1} (S_k + S)^2 \,. \qquad\qquad (9.4.3)$$

If M is a matrix and sign(M) is defined, then $M + \text{sign}(M)$ is nonsingular because its eigenvalues have the form $\lambda + \text{sign}(\lambda)$ which are clearly nonzero. Thus, the matrix

$$S_k + S \;=\; S_k + \text{sign}(A) \;=\; S_k + \text{sign}(S_k)$$

is nonsingular. By manipulating equations (9.4.2) and (9.4.3) we conclude that if

$$G_k = (S_k - S)(S_k + S)^{-1}, \tag{9.4.4}$$

then $G_{k+1} = G_k^2$. It follows by induction that $G_k = G_0^{2^k}$. If $\lambda \in \lambda(A)$, then

$$\mu \;=\; \frac{\lambda - \text{sign}(\lambda)}{\lambda + \text{sign}(\lambda)}$$

is an eigenvalue of $G_0 = (A - S)(A + S)^{-1}$. Since $|\mu| < 1$ it follows from Lemma 7.3.2 that $G_k \to 0$ and so

$$S_k \;=\; S(I + G_k)(I - G_k)^{-1} \;\to\; S.$$

Taking norms in (9.4.2) we conclude that the rate of convergence is quadratic:

$$\| \, S_{k+1} - S \, \| \;\leq\; \frac{1}{2} \| \, S_k^{-1} \, \| \cdot \| \, S_k - S \, \|^2.$$

The overall efficiency of the method in practice is a concern since $O(n^3)$ flops per iteration are required. To address this issue several enhancements of the basic iteration (9.4.1) have been proposed. One idea is to incorporate the Newton approximation

$$S_k^{-1} \approx S_k(2I - S_k^2).$$

(See P9.4.1.) Using this estimate instead of the actual inverse in (9.4.1) gives update step

$$S_{k+1} = \frac{1}{2}(S_k + S_k(2I - S_k^2)) \;=\; \frac{1}{2}S_k(3I - S_k^2). \tag{9.4.5}$$

This is referred to as the *Newton-Schultz iteration*. Another idea is to introduce a scale factor:

$$S_{k+1} = \frac{1}{2}\left((\mu_k S_k) + (\mu_k S_k)^{-1}\right). \tag{9.4.6}$$

Interesting choices for μ_k include $|\det(S_k)|^{1/n}$, $\sqrt{\rho(S_k^{-1})/\rho(S_k)}$, and $\sqrt{\| \, S_k^{-1} \, \| \| \, S_k \, \|}$ where $\rho(\cdot)$ is the spectral radius. For insights into the effective computation of the matrix sign function and related stability issues, see Kenney and Laub (1991, 1992), Higham (2007), and Higham (FOM, Chap. 5).

9.4.2   The Matrix Square Root

Ambiguity arises in the $f(A)$ problem if the underlying function has branches. For example, if $f(x) = \sqrt{x}$ and

$$A = \begin{bmatrix} 4 & 10 \\ 0 & 9 \end{bmatrix},$$

then

$$A = \begin{bmatrix} 2 & 2 \\ 0 & 3 \end{bmatrix}^2 = \begin{bmatrix} -2 & 10 \\ 0 & 3 \end{bmatrix}^2 = \begin{bmatrix} -2 & -2 \\ 0 & -3 \end{bmatrix}^2 = \begin{bmatrix} 2 & -10 \\ 0 & -3 \end{bmatrix}^2,$$

which shows that there are at least four legitimate choices for $\sqrt{A}$. To clarify the situation we say F is the *principal square root* of A if (a) $F^2 = A$ and (b) the eigenvalues of F have positive real part. We designate this matrix by $A^{1/2}$.

Analogous to the Newton iteration for scalar square roots, $x_{k+1} = (x_k + a/x_k)/2$, we have

$$X_0 = A$$
$$\textbf{for } k = 0, 1, \dots \qquad\qquad (9.4.7)$$
$$X_{k+1} = \left(X_k + X_k^{-1}A \right)/2$$
$$\textbf{end}$$

Notice the similarity between this iteration and the Newton sign iteration (9.4.1). Indeed, by making the substitution $X_k = A^{1/2}S_k$ in (9.4.7) we obtain the Newton sign iteration for $A^{1/2}$. Global convergence and local quadratic convergence follow from what we know about (9.4.1).

Another connection between the matrix sign problem and the matrix square root problem is revealed by applying the Newton sign iteration to the matrix

$$\tilde{A} = \begin{bmatrix} 0 & A \\ I & 0 \end{bmatrix}.$$

Designate the iterates by $\tilde{S}_k$. We show by induction that $\tilde{S}_k$ has the form

$$\tilde{S}_k = \begin{bmatrix} 0 & X_k \\ Y_k & 0 \end{bmatrix}.$$

This is true for $k = 0$ by setting $X_0 = A$ and $Y_0 = I$. To see that the result holds for $k > 0$, observe that

$$\tilde{S}_{k+1} = \frac{1}{2}\left(\tilde{S}_k + \tilde{S}_k^{-1} \right) = \frac{1}{2}\left(\begin{bmatrix} 0 & X_k \\ Y_k & 0 \end{bmatrix} + \begin{bmatrix} 0 & Y_k^{-1} \\ X_k^{-1} & 0 \end{bmatrix} \right)$$

and thus

$$X_{k+1} = \left(X_k + Y_k^{-1} \right)/2, \qquad Y_{k+1} = \left(Y_k + X_k^{-1} \right)/2. \qquad (9.4.8)$$

Another induction argument shows that

$$X_k = AY_k, \qquad k = 0, 1, \dots, \qquad\qquad (9.4.9)$$

and so

$$X_{k+1} = \left(X_k + AX_k^{-1} \right)/2, \qquad Y_{k+1} = \left(Y_k + A^{-1}Y_k^{-1} \right)/2. \qquad (9.4.10)$$

It follows that $X_k \to A^{1/2}$ and $Y_k \to A^{-1/2}$ and we have established the following identity:

$$\text{sign}\left(\begin{bmatrix} 0 & A \\ I & 0 \end{bmatrix}\right) = \begin{bmatrix} 0 & A^{1/2} \\ A^{-1/2} & 0 \end{bmatrix}.$$

Equation (9.4.8) defines the *Denman-Beavers iteration* which turns out to have better numerical properties than (9.4.7). See Meini (2004), Higham (FOM, Chap. 6), and Higham (2008) for an analysis of these and other matrix square root algorithms.

9.4.3 The Polar Decomposition

If $z = a + bi \in \mathbb{C}$ is a nonzero complex number, then its polar representation is a factorization of the form $z = e^{i\theta} r$ where $r = \sqrt{a^2 + b^2}$ and $e^{i\theta} = \cos(\theta) + i\sin(\theta)$ is defined by $(\cos(\theta), \sin(\theta)) = (a/r, b/r)$. The *polar decomposition* of a matrix is similar.

Theorem 9.4.1 (Polar Decomposition). *If $A \in \mathbb{R}^{m \times n}$ and $m \geq n$, then there exists a matrix $U \in \mathbb{R}^{m \times n}$ with orthonormal columns and a symmetric positive semidefinite $P \in \mathbb{R}^{n \times n}$ so that $A = UP$.*

Proof. Suppose $U_A^T A V_A = \Sigma_A$ is the thin SVD of A. It is easy to show that if $U = U_A V_A^T$ and $P = V_A \Sigma_A V_A^T$, then $A = UP$ and U and P have the required properties. $\square$

We refer to U as the *orthogonal polar factor* and P as the *symmetric polar factor*. Note that $P = (A^T A)^{1/2}$ and if $\text{rank}(A) = n$, then $U = A(A^T A)^{-1/2}$. An important application of the polar decomposition is the orthogonal Procrustes problem (see §6.4.1).

Various iterative methods for computing the orthogonal polar factor have been proposed. A quadratically convergent Newton iteration for the square nonsingular case proceeds by repeatedly averaging the current iterate with the inverse of its transpose:

$$X_0 = A \qquad (\text{Assume } A \in \mathbb{R}^{n \times n} \text{ is nonsingular})$$

for $k = 0, 1, \ldots$ $\hspace{4cm}$ (9.4.11)

$$X_{k+1} = \left(X_k + X_k^{-T}\right)/2$$

end

To show that this iteration is well defined we assume that for some k the matrix X_k is nonsingular and that $X_k = U_k P_k$ is its polar decomposition. It follows that

$$X_{k+1} = \frac{1}{2}\left(X_k + X_k^{-T}\right) = \frac{1}{2}\left(U_k P_k + U_k P_k^{-1}\right) = U_k\left(\frac{P_k + P_k^{-1}}{2}\right). \qquad (9.4.12)$$

Since the average of a positive definite matrix and its inverse is also positive definite it follows that X_{k+1} is nonsingular. This shows by induction that (9.4.11) is well-defined and that the P_k satisfy

$$P_{k+1} = (P_k + P_k^{-1})/2, \qquad P_0 = P.$$

This is precisely the Newton sign iteration (9.4.1) with starting matrix $P_0 = P$. Since

$$\| X_k - U \|_2 = \| U(P_k - I) \|_2 = \| P_k - I \|_2$$

and $P_k \to \text{sign}(P) = I$ quadratically, we conclude that X_k matrices in (9.4.11) converge to U quadratically.

Extensions to the rectangular case and various ways to accelerate (9.4.11) are discussed in Higham (1986), Higham and Schreiber (1990), Gander (1990), and Kenney and Laub (1992). In this regard the matrix sign function is (once again) a handy tool for deriving algorithms. Note that if $A = U_A \Sigma_A V_A^T$ is the SVD of $A \in \mathbb{R}^{n \times n}$ and

$$Q = \frac{1}{\sqrt{2}} \begin{bmatrix} U_A & 0 \\ 0 & V_A \end{bmatrix} \begin{bmatrix} I_n & I_n \\ I_n & -I_n \end{bmatrix}$$

then Q is orthogonal and

$$Q^T \begin{bmatrix} 0 & A \\ A^T & 0 \end{bmatrix} Q = \begin{bmatrix} \Sigma_A & 0 \\ 0 & -\Sigma_A \end{bmatrix}.$$

It follows that

$$\text{sign}\left(\begin{bmatrix} 0 & A \\ A^T & 0 \end{bmatrix} \right) = Q \begin{bmatrix} I_n & 0 \\ 0 & -I_n \end{bmatrix} Q^T = \begin{bmatrix} 0 & U \\ U^T & 0 \end{bmatrix}$$

where $U = U_A V_A^T$ is the orthogonal polar factor of A.

There is a well-developed perturbation theory for the polar decomposition. A sample result for square nonsingular matrices due to Li and Sun (2003) says that the orthogonal polar factors U and $\tilde{U}$ for nonsingular $A, \tilde{A} \in \mathbb{R}^{n \times n}$ satisfy the bound

$$\| U - \tilde{U} \|_F \leq \frac{4\| A - \tilde{A} \|_F}{\sigma_{n-1}(A) + \sigma_n(A) + \sigma_{n-1}(\tilde{A}) + \sigma_n(\tilde{A})}.$$

9.4.4 The Matrix Logarithm

Given $A \in \mathbb{R}^{n \times n}$, a solution to the matrix equation $e^X = A$ is a logarithm of A. Note that if $X = \log(A)$, then $X + 2k\pi i$ is also a logarithm. To remove this ambiguity we define the *principal logarithm* as follows. If the real eigenvalues of $A \in \mathbb{R}^{n \times n}$ are all positive then there is a unique real matrix X that satisfies $e^X = A$ with the property that its eigenvalues satisfy $\lambda(X) \subset \{ z \in \mathbb{C} : -\pi < \text{Im}(z) < \pi \}$.

Of course, the eigenvalue-based methods of §9.2 are applicable for the $\log(A)$ problem. We discuss an approximation method that is analogous to Algorithm 9.3.1, the scaling and squaring method for the matrix exponential

As with the exponential, there are a number of different series expansions for the log function that are of computational interest. The simplest is the Maclaurin expansion:

$$\log(A) \approx M_q(A) = \sum_{k=1}^{q} (-1)^{k+1} \frac{(A - I)^k}{k}.$$

To apply this formula we must have $\rho(A - I) < 1$ where $\rho(\cdot)$ is the spectral radius.

The Gregory series expansion for $\log(x)$ yields a rational approximation:

$$\log(A) \approx G_q(A) = -2 \sum_{k=0}^{q} \frac{1}{2k+1} \left((I - A)(I + A)^{-1} \right)^{2k+1}.$$

For this to converge, the real parts of A's eigenvalues must be positive.

Diagonal Padé approximants are also of interest. For example, the (3,3) Padé approximant is given by

$$\log(A) \approx r_{33}(A) = D(A)^{-1} N(A)$$

where

$$D(A) = 60I + 90(A - I) + 36(A - I)^2 + 3(A - I)^3,$$
$$N(A) = 60(A - I) + 60(A - I)^2 + 11(A - I)^3.$$

For an approximation of this type to be effective, the matrix A must be sufficiently close to the identity matrix. Repeated square roots are one way to achieve this:

$k = 0$

$A_0 = A$

while $\| A - I \| > $ tol

$\qquad k = k + 1$

$\qquad A_k = A_{k-1}^{1/2}$

end

The Denman-Beavers iteration (9.4.8) can be invoked to compute the matrix square roots. If we next compute $F \approx \log(A_k)$ by using (say) an appropriately chosen Pade approximant, then $\log(A) = 2^k \log(A_k) \approx 2^k F$. This solution framework is referred to as *inverse scaling and squaring*. There are many details associated with the proper implementation of this procedure and we refer the reader to Cheng, Higham. Kenney, and Laub (2001), Higham (2001), and Higham (FOM, Chap. 11).

Problems

P9.4.1 What does the Newton iteration look like when it is applied to find a root of the function $f(x) = 1/x - a$? Develop an inverse-free Newton iteration for solving the matrix equation $X^{-1} - A$.

P9.4.2 Show that if $\mu_k > 0$ in (9.4.6), then $\text{sign}(S_{k+1}) = \text{sign}(S_k)$.

P9.4.3 Show that $\text{sign}(A) = A(A^2)^{-1/2}$.

P9.4.4 Verify Equation (9.4.9).

P9.4.5 In the Denman-Beavers iteration (9.4.8), define $M_k = X_k Y_k$ and develop a recipe for M_{k+1}.

P9.4.6 Show that if we apply the Newton square root iteration (9.4.9) to a symmetric positive definite matrix A, then $A_k - A_{k+1}$ is positive definite for all k.

P9.4.7 Suppose A is normal. Relate the polar factors of e^A to $S = (A - A^T)/2$ and $T = (A + A^T)/2$.

P9.4.8 Show that the polar decomposition of a nonsingular matrix is unique. Hint: If $A = U_1 P_1$ and $A = U_2 P_2$ are two polar decompositions, then $U_2^T U_1 = P_2 P_1^{-1}$ and $U_1^T U_2 = P_1 P_2^{-1}$ have the same eigenvalues.

P9.4.9 Give a closed-form expression for the polar decomposition $A = UP$ of a real 2-by-2 matrix. Under what conditions is U a rotation?

P9.4.10 Give a closed-form expression for $\log(Q)$ where Q is a 2-by-2 rotation matrix.

P9.4.11 Formulate an $m < n$ version of the polar decomposition for $A \in \mathbb{R}^{m \times n}$.

P9.4.12 Let A by an n-by-n symmetric positive definite matrix. (a) Show that there exists a unique symmetric positive definite X such that $A = X^2$. (b) Show that if $X_0 = I$ and

$$X_{k+1} = (X_k + AX_k^{-1})/2$$

then $X_k \to \sqrt{A}$ quadratically where $\sqrt{A}$ denotes the matrix X in part (a).

P9.4.13 Show that

$$X(t) = C_1 \cos(t\sqrt{A}) + C_2 \sqrt{A^{-1}} \sin(t\sqrt{A})$$

solves the initial value problem $\ddot{X}(t) = -AX(t)$, $X(0) = C_1$, $\dot{X}(0) = C_2$. Assume that A is symmetric positive definite.

Notes and References for §9.4

Everything in this section is covered in greater depth in Higham (FOM). See also:

N.J. Higham (2005). "Functions of Matrices," in *Handbook of Linear Algebra,* L. Hogben (ed.), Chapman and Hall, Boca Raton, FL, §11-1–§11-13.

Papers that discuss the ubiquitous matrix sign function and its applications include:

R. Byers (1987). "Solving the Algebraic Riccati Equation with the Matrix Sign Function," *Linear Alg. Applic. 85,* 267–279.
C.S. Kenney and A.J. Laub (1991). "Rational Iterative Methods for the Matrix Sign Function," *SIAM J. Matrix Anal. Appl. 12,* 273–291.
C.S. Kenney, A.J. Laub, and P.M. Papadopouos (1992). "Matrix Sign Algorithms for Riccati Equations," *IMA J. Math. Control Info. 9,* 331–344.
C.S. Kenney and A.J. Laub (1992). "On Scaling Newton's Method for Polar Decomposition and the Matrix Sign Function," *SIAM J. Matrix Anal. Applic. 13,* 688–706.
R. Byers, C. He, and V. Mehrmann (1997). "The Matrix Sign Function Method and the Computation of Invariant Subspaces," *SIAM J. Matrix Anal. Applic. 18,* 615–632.
Z. Bai and J.W. Demmel (1998). "Using the Matrix Sign Function to Compute Invariant Subspaces," *SIAM J. Matrix Anal. Applic. 19,* 2205–2225.
N.J. Higham (1994). "The Matrix Sign Decomposition and Its Relation to the Polar Decomposition," *Lin. Alg. Applic. 212/213,* 3–20.
N.J. Higham, D.S. Mackey, N. Mackey, and F. Tisseur (2004). "Computing the Polar Decomposition and the Matrix Sign Decomposition in Matrix Groups," *SIAM J. Matrix Anal. Applic. 25,* 1178–1192.

Various aspects of the matrix square root problem are discussed in:

E.D. Denman and A.N. Beavers (1976). "The Matrix Sign Function and Computations in Systems," *Appl. Math. Comput., 2,* 63–94.
Å. Björck and S. Hammarling (1983). "A Schur Method for the Square Root of a Matrix," *Lin. Alg. Applic. 52/53,* 127–140.
N.J. Higham (1986). "Newton's Method for the Matrix Square Root," *Math. Comput. 46,* 537–550.
N.J. Higham (1987). "Computing Real Square Roots of a Real Matrix," *Lin. Alg. Applic. 88/89,* 405–430.
N.J. Higham (1997). "Stable Iterations for the Matrix Square Root," *Numer. Algorithms 15,* 227–242.
Y.Y. Lu (1998). "A Padé Approximation Method for Square Roots of Symmetric Positive Definite Matrices," *SIAM J. Matrix Anal. Applic. 19,* 833–845.
N.J. Higham, D.S. Mackey, N. Mackey, and F. Tisseur (2005). "Functions Preserving Matrix Groups and Iterations for the Matrix Square Root," *SIAM J. Matrix Anal. Applic. 26,* 849–877.
C.-H. Guo and N. J. Higham (2006). "A Schur–Newton Method for the Matrix pth Root and its Inverse," *SIAM J. Matrix Anal. Applic. 28,* 788–804.
B. Meini (2004). "The Matrix Square Root from a New Functional Perspective: Theoretical Results and Computational Issues," *SIAM J. Matrix Anal. Applic. 26,* 362–376.

A. Frommer and B. Hashemi (2009). "Verified Computation of Square Roots of a Matrix," *SIAM J. Matrix Anal. Applic. 31*, 1279–1302.

Computational aspects of the polar decomposition and its generalizations are covered in:

N.J. Higham (1986). "Computing the Polar Decomposition with Applications," *SIAM J. Sci. Statist. Comp. 7*, 1160–1174.
R.S. Schreiber and B.N. Parlett (1988). "Block Reflectors: Theory and Computation," *SIAM J. Numer. Anal. 25*, 189–205.
N.J. Higham and R.S. Schreiber (1990). "Fast Polar Decomposition of an Arbitrary Matrix," *SIAM J. Sci. Statist. Comput. 11*, 648–655.
N.J. Higham and P. Papadimitriou (1994). "A Parallel Algorithm for Computing the Polar Decomposition," *Parallel Comput. 20*, 1161–1173.
A.A. Dubrulle (1999). "An Optimum Iteration for the Matrix Polar Decomposition," *ETNA 8*, 21–25.
A. Zanna and H. Z. Munthe-Kaas (2002). "Generalized Polar Decompositions for the Approximation of the Matrix Exponential," *SIAM J. Matrix Anal. Applic. 23*, 840–862.
B. Laszkiewicz and K. Zietak (2006). "Approximation of Matrices and a Family of Gander Methods for Polar Decomposition," *BIT 46*, 345–366.
R. Byers and H. Xu (2008). "A New Scaling for Newton's Iteration for the Polar Decomposition and Its Backward Stability," *SIAM J. Matrix Anal. Applic. 30*, 822–843.
N.J. Higham, C. Mehl, and F. Tisseur (2010). "The Canonical Generalized Polar Decomposition," *SIAM J. Matrix Anal. Applic. 31*, 2163–2180.

For an analysis as to whether or not the polar decomposition can be computed in a finite number of steps, see:

A. George and Kh. Ikramov (1996). "Is The Polar Decomposition Finitely Computable?," *SIAM J. Matrix Anal. Applic. 17*, 348–354.
A. George and Kh. Ikramov (1997). "Addendum: Is The Polar Decomposition Finitely Computable?," *SIAM J. Matrix Anal. Appl. 18*, 264–264.

There is a considerable literature concerned with how the polar factors change under perturbation:

R. Mathias (1993). "Perturbation Bounds for the Polar Decomposition," *SIAM J. Matrix Anal. Applic. 14*, 588–597.
R.-C. Li (1997). "Relative Perturbation Bounds for the Unitary Polar Factor," *BIT 37*, 67–75.
F. Chaitin-Chatelin, S. Gratton (2000). "On the Condition Numbers Associated with the Polar Factorization of a Matrix," *Numer. Lin. Alg. 7*, 337–354.
W. Li and W. Sun (2003). "New Perturbation Bounds for Unitary Polar Factors," *SIAM J. Matrix Anal. Applic. 25*, 362–372.

Finally, details concerning the matrix logarithm and its computation may be found in:

B.W. Helton (1968). "Logarithms of Matrices," *Proc. AMS 19*, 733–736.
L. Dieci (1996). "Considerations on Computing Real Logarithms of Matrices, Hamiltonian Logarithms, and Skew-Symmetric Logarithms," *Lin. Alg. Applic. 244*, 35–54.
L. Dieci, B. Morini, and A. Papini (1996). "Computational Techniques for Real Logarithms of Matrices," *SIAM J. Matrix Anal. Applic. 17*, 570–593.
C. S. Kenney and A. J. Laub (1998). "A Schur-Fréchet Algorithm for Computing the Logarithm and Exponential of a Matrix," *SIAM J. Matrix Anal. Applic. 19*, 640–663.
L. Dieci (1998). "Real Hamiltonian Logarithm of a Symplectic Matrix," *Lin. Alg. Applic. 281*, 227–246.
L. Dieci and A. Papini (2000). "Conditioning and Padé Approximation of the Logarithm of a Matrix," *SIAM J. Matrix Anal. Applic. 21*, 913–930.
N.J. Higham (2001). "Evaluating Padé Approximants of the Matrix Logarithm," *SIAM J. Matrix Anal. Applic. 22*, 1126–1135.
S.H. Cheng, N.J. Higham, C.S. Kenney, and A.J. Laub (2001). "Approximating the Logarithm of a Matrix to Specified Accuracy," *SIAM J. Matrix Anal. Applic. 22*, 1112–1125.

Chapter 10

Large Sparse Eigenvalue Problems

10.1   **The Symmetric Lanczos Process**

10.2   **Lanczos, Quadrature, and Approximation**

10.3   **Practical Lanczos Procedures**

10.4   **Large Sparse SVD Frameworks**

10.5   **Krylov Methods for Unsymmetric Problems**

10.6   **Jacobi-Davidson and Related Methods**

The Lanczos process computes a sequence of partial tridiagonalizations that are orthogonally related to a given symmetric matrix A. It is of particular interest if A is large and sparse because, instead of updating A along the way as in the Householder method of §8.2, it simply relies on matrix-vector products. Equally important, information about A's extremal eigenvalues tends to emerge fairly early during the iteration, making the method very useful in situations where just a few of A's largest or smallest eigenvalues are desired, together with the corresponding eigenvectors.

The derivation and exact arithmetic attributes of the method are presented in §10.1, including its extraordinary convergence properties. Central to the discussion is the connection to an underlying Krylov subspace that is defined by the starting vector. In §10.2 we point out connections between Gauss quadrature and the Lanczos process that can be used to estimate expressions of the form $u^T f(A)u$ where $f(A)$ is a function of a large, sparse symmetric positive definite matrix A. Unfortunately, a "math book" implementation of the Lanczos method is practically useless because of roundoff error. This makes it necessary to enlist the help of various "workarounds," which we describe in §10.3. A sparse SVD framework based on Golub-Kahan bidiagonalization is detailed in §10.4. We also introduce the idea of a randomized SVD. The last two sections deal with the more difficult unsymmetric problem. The Arnoldi iteration is a Krylov subspace iteration like Lanczos. To make it effective, it is necessary to extract valuable "restart information" from the Hessenberg matrix sequence that it produces. This is discussed in §10.5 together with a brief presentation of the unsymmetric Lanczos framework. In the last section we derive the Jacobi-Davidson method, which combines Newton ideas with Rayleigh-Ritz refinement.

Reading Notes

Familiarity with Chapters 5, 7, and 8 is recommended. Within this chapter there are the following dependencies:

$$\S 10.1 \quad \rightarrow \quad \S 10.3 \quad \rightarrow \quad \S 10.5 \quad \rightarrow \quad \S 10.6$$
$$\downarrow \qquad\qquad \downarrow$$
$$\S 10.2 \qquad\quad \S 10.4$$

General references for this chapter include Parlett (SEP), Stewart (MAE), Watkins (MEP), Chatelin (EOM), Cullum and Willoughby (LALSE), Meurant (LCG), Saad (NMLE), Kressner (NMSE), and EIG_TEMPLATES.

10.1   The Symmetric Lanczos Process

Suppose $A \in \mathbb{R}^{n \times n}$ is large, sparse, and symmetric and assume that a few of its largest and/or smallest eigenvalues are desired. Eigenvalues at either end of the spectrum are referred to as *extremal eigenvalues*. This problem can be addressed by a method attributed to Lanczos (1950). The method generates a sequence of tridiagonal matrices $\{T_k\}$ with the property that the extremal eigenvalues of $T_k \in \mathbb{R}^{k \times k}$ are progressively better estimates of A's extremal eigenvalues. In this section, we derive the technique and investigate some of its exact arithmetic properties.

One way to motivate the Lanczos idea is to be reminded about the shortcomings of the power method that we discussed in §8.2.1. Recall that the power method can be used to find the dominant eigenvalue λ_1 and an associated eigenvector x_1. However, the rate of convergence is dictated by $|\lambda_2/\lambda_1|^k$ where λ_2 is the second largest eigenvalue in absolute value. Unless there is a sufficient magnitude gap between these two eigenvalues, the power method is very slow. Moreover, it does not take advantage of "prior experience." After k steps with initial vector $v^{(0)}$, it has visited the directions defined by the vectors $Av^{(0)}, \ldots, A^k v^{(0)}$. However, instead of searching the span of these vectors for an optimal estimate of x_1, it settles for $A^k v^{(0)}$. The method of orthogonal iteration with Ritz acceleration (§8.3.7) addresses some of these concerns, but it too has a certain disregard for prior iterates. What we need is a method that "learns from experience" and takes advantage of all previously computed matrix-vector products. The Lanczos method fits the bill.

10.1.1   Krylov Subspaces

The derivation of the Lanczos process can proceed in several ways. So that its remarkable convergence properties do not come as a complete surprise, we motivate the method by considering the optimization of the Rayleigh quotient

$$r(x) = \frac{x^T A x}{x^T x}, \qquad x \neq 0.$$

Recall from Theorem 8.1.2 that the maximum and minimum values of $r(x)$ are $\lambda_1(A)$ and $\lambda_n(A)$, respectively. Suppose $\{q_i\} \subseteq \mathbb{R}^n$ is a sequence of orthonormal vectors and define the scalars M_k and m_k by

$$M_k = \lambda_1(Q_k^T A Q_k) = \max_{y \neq 0} \frac{y^T (Q_k^T A Q_k) y}{y^T y} = \max_{\|y\|_2 = 1} r(Q_k y) \leq \lambda_1(A),$$

$$m_k = \lambda_k(Q_k^T A Q_k) = \min_{y \neq 0} \frac{y^T (Q_k^T A Q_k) y}{y^T y} = \min_{\|y\|_2 = 1} r(Q_k y) \geq \lambda_n(A),$$

where $Q_k = [\, q_1 \,|\, \cdots \,|\, q_k \,]$. Since

$$\mathsf{ran}(Q_1) \subset \mathsf{ran}(Q_2) \subset \cdots \subset \mathsf{ran}(Q_n) = \mathbb{R}^n$$

it follows that

$$M_1 \leq M_2 \leq \cdots \leq M_n = \lambda_1(A),$$

$$m_1 \geq m_2 \geq \cdots \geq m_n = \lambda_n(A).$$

Thus, the proposed optimization framework will ultimately converge. However, the challenge is to choose the q-vectors in such a way that M_k and m_k are high-quality estimates well before k equals n.

Searching for a good q_k prompts consideration of the gradient:

$$\nabla r(x) = \frac{2}{x^T x}(Ax - r(x)x). \tag{10.1.1}$$

Suppose $u_k \in \mathsf{span}\{q_1, \ldots, q_k\}$ satisfies $M_k = r(u_k)$. If $\nabla r(u_k) = 0$, then $(r(u_k), u_k)$ is an eigenpair of A. If not, then from the standpoint of making M_{k+1} as large as possible it makes sense to choose the next trial vector q_{k+1} so that

$$\nabla r(u_k) \in \mathsf{span}\{q_1, \ldots, q_{k+1}\}. \tag{10.1.2}$$

This is because $r(x)$ increases most rapidly in the direction of the gradient $\nabla r(x)$. The strategy will guarantee that M_{k+1} is greater than M_k, hopefully by a significant amount. Likewise, if $v_k \in \mathsf{span}\{q_1, \ldots, q_k\}$ satisfies $r(v_k) = m_k$, then it makes sense to require

$$\nabla r(v_k) \in \mathsf{span}\{q_1, \ldots, q_{k+1}\} \tag{10.1.3}$$

since $r(x)$ decreases most rapidly in the direction of $-\nabla r(x)$.

Note that for any $x \in \mathbb{R}^n$ we have

$$\nabla r(x) \in \mathsf{span}\{x, Ax\}.$$

Since the vectors u_k and v_k each belong to $\mathsf{span}\{q_1, \ldots, q_k\}$, it follows that the inclusions (10.1.2) and (10.1.3) are satisfied if

$$\mathsf{span}\{q_1, \ldots, q_k\} = \mathsf{span}\{q_1, Aq_1, \ldots, A^{k-1}q_1\}.$$

This suggests we choose q_{k+1} so that

$$\mathsf{span}\{q_1, \ldots, q_{k+1}\} = \mathsf{span}\{q_1, Aq_1, \ldots, , A^{k-1}q_1, A^k q_1\}$$

and thus we are led to the problem of computing orthonormal bases for the *Krylov subspaces*

$$\mathcal{K}(A, q_1, k) = \mathsf{span}\{q_1, Aq_1, \ldots, A^{k-1}q_1\}.$$

These are just the range spaces of the Krylov matrices

$$K(A, q_1, k) = \left[q_1 \mid Aq_1 \mid A^2q_1 \mid \ldots \mid A^{k-1}q_1 \right]$$

that we introduced in §8.3.2. Note that $\mathcal{K}(A, q_1, k)$ is precisely the subspace that the power method "overlooks" since it merely searches in the direction of $A^{k-1}q_1$.

10.1.2   Tridiagonalization

In order to generate an orthonormal basis for a Krylov subspace we exploit the connection between the tridiagonalization of A and the QR factorization of $K(A, q_1, n)$. Recall from §8.3.2 that if $Q^T AQ = T$ is tridiagonal and $QQ^T = I_n$, then

$$K(A, q_1, n) = QQ^T K(A, q_1, n) = Q\left[e_1 \mid Te_1 \mid T^2e_1 \mid \ldots \mid T^{n-1}e_1 \right]$$

is the QR factorization of $K(A, q_1, n)$ where e_1 and q_1 are respectively the first columns of I_n and Q. Thus, the columns of Q can effectively be generated by tridiagonalizing A with an orthogonal matrix whose first column is q_1.

Householder tridiagonalization, discussed in §8.3.1, can be adapted for this purpose. However, this approach is impractical if A is large and sparse because Householder similarity updates almost always destroy sparsity. As a result, unacceptably large, dense matrices arise during the reduction. This suggests that we try to compute the elements of the tridiagonal matrix $T = Q^T AQ$ directly. Toward that end, designate the columns of Q by

$$Q = \left[q_1 \mid \cdots \mid q_n \right]$$

and the components of T by

$$
T = \begin{bmatrix}
\alpha_1 & \beta_1 & & \cdots & & 0 \\
\beta_1 & \alpha_2 & \ddots & & & \vdots \\
& \ddots & \ddots & \ddots & & \\
\vdots & & \ddots & \ddots & \beta_{n-1} \\
0 & \cdots & & & \beta_{n-1} & \alpha_n
\end{bmatrix}.
$$

Equating columns in $AQ = QT$, we conclude that

$$Aq_k = \beta_{k-1}q_{k-1} + \alpha_k q_k + \beta_k q_{k+1}, \qquad (\beta_0 q_0 \equiv 0),$$

for $k = 1{:}n - 1$. The orthonormality of the q-vectors implies

$$\alpha_k = q_k^T Aq_k.$$

(Another way to see this is that $T_{ij} = q_i^T Aq_j$.) Moreover, if we define the vector r_k by

$$r_k = (A - \alpha_k I)q_k - \beta_{k-1}q_{k-1}$$

and if it is nonzero, then

$$q_{k+1} = r_k/\beta_k$$

where

$$\beta_k = \pm\| r_k \|_2.$$

If $r_k = 0$, then the iteration breaks down but (as we shall see) not without the acquisition of valuable invariant subspace information.

By properly sequencing the above formulae and assuming that $q_1 \in \mathbb{R}^n$ is a given unit vector, we obtain what may be regarded as "Version 0" of the *Lanczos iteration*.

Algorithm 10.1.1 (Lanczos Tridiagonalization) Given a symmetric matrix $A \in \mathbb{R}^{n \times n}$ and a unit 2-norm vector $q_1 \in \mathbb{R}^n$, the following algorithm computes a matrix $Q_k = [q_1 | \ldots | q_k]$ with orthonormal columns and a tridiagonal matrix $T_k \in \mathbb{R}^{k \times k}$ so that $AQ_k = Q_kT_k$. The diagonal and superdiagonal entries of T_k are $\alpha_1, \ldots, \alpha_k$ and $\beta_1, \ldots, \beta_{k-1}$ respectively. The integer k satisfies $1 \le k \le n$.

$$k = 0, \ \beta_0 = 1, \ q_0 = 0, \ r_0 = q_1$$

while $k = 0$ or $\beta_k \neq 0$

$\qquad q_{k+1} = r_k/\beta_k$

$\qquad k = k + 1$

$\qquad \alpha_k = q_k^T A q_k$

$\qquad r_k = (A - \alpha_k I)q_k - \beta_{k-1}q_{k-1}$

$\qquad \beta_k = \| r_k \|_2$

end

There is no loss of generality in choosing β_k to be positive. The q_k vectors are called *Lanczos vectors*. It is important to mention that there are better ways numerically to organize the computation of the Lanczos vectors than Algorithm 10.1.1. See §10.3.1.

10.1.3 Termination and Error Bounds

The Lanczos iteration halts before complete tridiagonalization if q_1 is contained in a proper invariant subspace. This is one of several mathematical properties of the method that we summarize in the following theorem.

Theorem 10.1.1. *The Lanczos iteration (Algorithm 10.1.1) runs until $k = m$, where*

$$m = \mathsf{rank}(K(A, q_1, n)).$$

Moreover, for $k = 1{:}m$ we have

$$AQ_k = Q_kT_k + r_ke_k^T \tag{10.1.4}$$

where $Q_k = [q_1 | \cdots | q_k]$ has orthonormal columns that span $\mathcal{K}(A, q_1, k)$, $e_k = I_n(:, k)$, *and*

$$T_k = \begin{bmatrix} \alpha_1 & \beta_1 & & \cdots & & 0 \\ \beta_1 & \alpha_2 & \ddots & & & \vdots \\ & \ddots & \ddots & \ddots & & \\ \vdots & & \ddots & \ddots & \beta_{k-1} \\ 0 & \cdots & & & \beta_{k-1} & \alpha_k \end{bmatrix}. \tag{10.1.5}$$

Proof. The proof is by induction on k. It clearly holds if $k = 1$. Suppose for some $k > 1$ that the iteration has produced $Q_k = [\, q_1 \mid \cdots \mid q_k \,]$ with orthonormal columns such that

$$\mathsf{ran}(Q_k) = \mathcal{K}(A, q_1, k).$$

It is easy to see from Algorithm 10.1.1 that equation (10.1.4) holds and so

$$Q_k^T A Q_k = T_k + Q_k^T r_k e_k^T. \tag{10.1.6}$$

Suppose i and j are integers that satisfy $1 \leq i \leq j \leq k$. From the equation

$$q_j^T A q_i = q_j^T (\beta_{i-1} q_{i-1} + \alpha_i q_i + \beta_i q_{i+1}) = \beta_{i-1} q_j^T q_{i-1} + \alpha_i q_j^T q_i + \beta_i q_j^T q_{i+1}$$

and the induction assumption $Q_k^T Q_k = I_k$, we see that

$$q_i^T A q_j = q_j^T A q_i = \begin{cases} 0, & \text{if } i < j - 1, \\ \beta_{j-1}, & \text{if } i = j - 1, \\ \alpha_j, & \text{if } i = j. \end{cases}$$

It follows that $Q_k^T A Q_k = T_k$ and so from (10.1.6) we have $Q_k^T r_k = 0$.

If $r_k \neq 0$, then $q_{k+1} = r_k / \| \, r_k \, \|_2$ is orthogonal to $q_1, \ldots, q_k$. It follows that $q_{k+1} \notin \mathcal{K}(A, q_1, k)$ and

$$q_{k+1} \in \mathsf{span}\{A q_k,\, q_k,\, q_{k-1}\} \subseteq \mathcal{K}(A,\, q_1,\, k+1).$$

Thus, $Q_{k+1}^T Q_{k+1} = I_{k+1}$ and

$$\mathsf{ran}(Q_{k+1}) = \mathcal{K}(A, q_1, k+1).$$

On the other hand, if $r_k = 0$, then $A Q_k = Q_k T_k$. This says that $\mathsf{ran}(Q_k) = \mathcal{K}(A, q_1, k)$ is invariant for A and so $k = m = \mathsf{rank}(K(A, q_1, n))$. $\square$

To encounter a zero β_k in the Lanczos iteration is a welcome event in that it signals the computation of an exact invariant subspace. However, valuable approximate invariant subspace information tends to emerge long before the occurrence of a small β. Apparently, more information can be extracted from the tridiagonal matrix T_k and the Krylov subspace spanned by the columns of Q_k.

10.1.4 Ritz Approximations

Recall from §8.1.4 that if S is a subspace of $\mathbb{R}^n$, then with respect to S we say that (θ, y) is a Ritz pair for $A \in \mathbb{R}^{n \times n}$ if $w^T(Ay - \theta y) = 0$ for all $w \in S$. If $S = \mathcal{K}(A, q_1, k)$, then the Lanczos process can be used to compute the associated Ritz values and vectors. Suppose

$$S_k^T T_k S_k = \Theta_k = \mathrm{diag}(\theta_1, \ldots, \theta_k) \tag{10.1.7}$$

is a Schur decomposition of the tridiagonal matrix T_k. If

$$Y_k = [\, y_1 \mid \cdots \mid y_k \,] = Q_k S_k \in \mathbb{R}^{n \times k},$$

then for $i = 1{:}k$ it follows that (θ_i, y_i) is a Ritz pair because

$$Q_k^T(AY_k - Y_k \Theta_k) = (Q_k^T A Q_k)S_k - Q_k^T(Q_k S_k)\Theta_k = T_k S_k - S_k \Theta_k = 0.$$

Two theorems in §8.1 concern Ritz approximation and are of interest to us in the Lanczos setting. Theorem 8.1.14 tells us that the problem of minimizing $\| AQ_k - Q_k B \|_2$ over all k-by-k matrices B is solved by setting $B = T_k = Q_k^T A Q_k$. Thus, the θ_i are the eigenvalues of a "best possible matrix" that happens to be tridiagonal. Theorem 8.1.15 can be used to provide a bound for $\| Ay_i - \theta_i y_i \|_2$. However, we can actually do better. Using (10.1.6) we have

$$Ay_i - \theta_i y_i = (AQ_k - Q_k T_k)S_k e_i = r_k(e_k^T S_k e_i)$$

from which it follows that

$$\| Ay_i - \theta_i y_i \|_2 = |\beta_k| \, |s_{ki}|. \tag{10.1.8}$$

Note that since S_k is orthogonal, $|s_{ki}| \le 1$.

We can use (10.1.8) to obtain a computable error bound. If E is the rank-1 matrix

$$E = -s_{ki} \cdot r_k y_i^T,$$

then

$$(A + E)y_i = \theta_i y_i.$$

It follows from Corollary 8.1.6 that

$$\min_{\mu \in \lambda(A)} |\theta_i - \mu| \le |\beta_k| \, |s_{ki}|$$

for $i = 1{:}k$.

Golub (1974) describes the construction of a more informative rank-1 perturbation E. Use Lanczos tridiagonalization to compute $AQ_k = Q_k T_k + r_k e_k^T$ and then set $E = \tau w w^T$, where $\tau = \pm 1$ and $w = aq_k + br_k$. It follows that

$$(A + E)Q_k = Q_k(T_k + \tau a^2 e_k e_k^T) + (1 + \tau ab)r_k e_k^T.$$

If $0 = 1 + \tau ab$, then

$$\bar{T}_k = T_k + \tau a^2 e_k e_k^T$$

is a tridiagonal matrix whose eigenvalues are also eigenvalues for $A+E$. Using Theorem 8.1.8, it can be shown that the interval $[\lambda_i(\tilde{T}_k), \lambda_{i-1}(\tilde{T}_k)]$ contains an eigenvalue of A for $i = 2{:}k$. These bracketing intervals depend on the choice of τa^2. Suppose we have an approximate eigenvalue λ of A. One possibility is to choose τa^2 so that

$$\det(\tilde{T}_k - \lambda I_k) = (\alpha_k + \tau a^2 - \lambda)p_{k-1}(\lambda) - \beta_{k-1}^2 p_{k-2}(\lambda) = 0$$

where the polynomials $p_i(x) = \det(T_i - xI_i)$ can be evaluated at λ using the three-term recurrence (8.4.2). (This assumes that $p_{k-1}(\lambda) \neq 0$.) The idea of characterizing an approximate eigenvalue λ as an exact eigenvalue of a nearby matrix $A + E$ is discussed in Lehmann (1963) and Householder (1968).

10.1.5   Convergence Theory

The preceding discussion indicates how eigenvalue estimates can be obtained via the Lanczos process, but it reveals nothing about the approximation quality of T_k's eigenvalues as a function of k. Results of this variety have been developed by Kaniel, Paige, Saad, and others and the following theorem is a sample from this body of research.

Theorem 10.1.2. *Let A be an n-by-n symmetric matrix with Schur decomposition*

$$Z^T AZ = \operatorname{diag}(\lambda_1, \dots, \lambda_n), \qquad \lambda_1 \geq \cdots \geq \lambda_n, \qquad Z = \left[\, z_1 \mid \cdots \mid z_n \,\right]. \quad (10.1.9)$$

Suppose k steps of the Lanczos iteration (Algorithm 10.1.1) are performed and that T_k is the tridiagonal matrix (10.1.5). If $\theta_1 = \lambda_1(T_k)$, then

$$\lambda_1 \geq \theta_1 \geq \lambda_1 - (\lambda_1 - \lambda_n)\left(\frac{\tan(\phi_1)}{c_{k-1}(1+2\rho_1)}\right)^2$$

where $\cos(\phi_1) = |q_1^T z_1|$,

$$\rho_1 = \frac{\lambda_1 - \lambda_2}{\lambda_2 - \lambda_n}, \qquad\qquad\qquad (10.1.10)$$

and $c_{k-1}(x)$ is the Chebyshev polynomial of degree $k - 1$.

Proof. From Theorem 8.1.2, we have

$$\theta_1 = \max_{y \neq 0} \frac{y^T T_k y}{y^T y} = \max_{y \neq 0} \frac{(Q_k y)^T A(Q_k y)}{(Q_k y)^T (Q_k y)} = \max_{0 \neq w \in \mathcal{K}(A, q_1, k)} \frac{w^T A w}{w^T w}.$$

Since λ_1 is the maximum of $w^T Aw/w^T w$ over all nonzero w, it follows that $\theta_1 \leq \lambda_1$. To obtain the lower bound for θ_1, note that

$$\theta_1 = \max_{p \in \mathbb{P}_{k-1}} \frac{q_1^T p(A)Ap(A)q_1}{q_1^T p(A)^2 q_1},$$

where $\mathbb{P}_{k-1}$ is the set of degree-$(k{-}1)$ polynomials and $p(x)$ is the *amplifying polynomial*. Given the eigenvector expansion $q_1 = d_1 z_1 + \cdots + d_n z_n$ where $d_i = q_1^T z_i$, it follows that

$$\frac{q_1^T p(A)A\, p(A)q_1}{q_1^T p(A)^2 q_1} = \frac{\displaystyle\sum_{i=1}^{n} d_i^2\, p(\lambda_i)^2 \lambda_i}{\displaystyle\sum_{i=1}^{n} d_i^2\, p(\lambda_i)^2} \geq \frac{\lambda_1 d_1^2\, p(\lambda_1)^2 + \lambda_n \delta^2}{d_1^2\, p(\lambda_1)^2 + \delta^2} = \lambda_1 - \frac{(\lambda_1 - \lambda_n)\delta^2}{d_1^2\, p(\lambda_1)^2 + \delta^2}$$

where

$$\delta^2 = \sum_{i=2}^{n} d_i^2 \, p(\lambda_i)^2.$$

If the polynomial p has the property that it is large at $x = \lambda_1$ compared to its value at $\lambda_2, \dots, \lambda_n$, then we get a better lower bound for the Ritz value θ_1. This is the act of finding an *amplifying polynomial* and a good choice is to set

$$p(x) \;=\; c_{k-1}\left(-1 + 2\frac{x - \lambda_n}{\lambda_2 - \lambda_n}\right)$$

where $c_{k-1}(z)$ is the $(k-1)$st Chebyshev polynomial generated via the recursion

$$c_k(z) \;=\; 2z c_{k-1}(z) - c_{k-2}(z), \qquad c_0 = 1, \; c_1 = z.$$

These polynomials are bounded by unity on $[-1, 1]$, but grow very rapidly outside this interval. By defining $p(x)$ this way, it follows that $|p(\lambda_i)| \le 1$ for $i = 2{:}n$ and $p(\lambda_1) = c_{k-1}(1 + 2\rho_1)$ where ρ_1 is defined by (10.1.10). Thus,

$$\delta^2 \;\le\; \sum_{i=2}^{n} d_i^2 \;=\; 1 - d_1^2$$

and so

$$\theta_1 \;\ge\; \lambda_1 - (\lambda_1 - \lambda_n)\frac{1 - d_1^2}{d_1^2}\frac{1}{\left(c_{k-1}(1 + 2\rho_1)\right)^2}.$$

The desired lower bound is obtained by noting that $\tan(\phi_1)^2 = (1 - d_1^2)/d_1^2$. $\square$

An analogous result pertaining to T_k's smallest eigenvalue is an easy corollary.

Corollary 10.1.3. *Using the same notation as in the theorem, if $\theta_k = \lambda_k(T_k)$, then*

$$\lambda_n \;\le\; \theta_k \;\le\; \lambda_n + (\lambda_1 - \lambda_n)\left(\frac{\tan(\phi_n)}{c_{k-1}(1 + 2\rho_n)}\right)^2$$

where

$$\rho_n \;=\; \frac{\lambda_{n-1} - \lambda_n}{\lambda_1 - \lambda_{n-1}}$$

and $\cos(\phi_n) = q_1^T z_n$.

Proof. Apply Theorem 10.1.2 with A replaced by $-A$. $\square$

The key idea in the proof of Theorem 10.1.2 is to take the amplifying polynomial $p(x)$ to be the translated Chebyshev polynomial, for then $p(A)q_1$ amplifies the component of q_1 in the direction of the eigenvector z_1. A similar idea can be used to obtain bounds for an interior Ritz value θ_i. However, the results are not as satisfactory because the new amplifying polynomial involves the product of the Chebyshev polynomial c_{k-i} and the polynomial $(x - \lambda_1) \cdots (x - \lambda_{i-1})$. For details, see Kaniel (1966) and Paige (1971) and also Saad (1980), who improved the bounds. The main theorem is as follows.

Theorem 10.1.4.   *Using the same notation as Theorem 10.1.2, if $1 \leq i \leq k$ and $\theta_i = \lambda_i(T_k)$, then*

$$\lambda_i \geq \theta_i \geq \lambda_i - (\lambda_1 - \lambda_n)\left(\frac{\kappa_i \tan(\phi_i)}{c_{k-i}(1 + 2\rho_i)}\right)^2$$

where

$$\rho_i = \frac{\lambda_i - \lambda_{i+1}}{\lambda_{i+1} - \lambda_n}, \qquad \kappa_i = \prod_{j=1}^{i-1}\frac{\theta_j - \lambda_n}{\theta_j - \lambda_i}, \qquad \cos(\phi_i) = |q_1^T z_i|.$$

Proof.   See Saad (NMLE, p. 201).   ☐

Because of the κ_i factor and the reduced degree of the amplifying Chebyshev polynomial, it is clear that the bounds deteriorate as i increases.

10.1.6   The Power Method versus the Lanczos Method

It is instructive to compare θ_1 with the corresponding power method estimate of λ_1. (See §8.2.1.) For clarity, assume $\lambda_1 \geq \cdots \geq \lambda_n \geq 0$ in the Schur decomposition (10.1.7). After $k - 1$ power method steps applied to q_1, a vector is obtained in the direction of

$$v = A^{k-1}q_1 = \sum_{i=1}^n d_i\lambda_i^{k-1}z_i$$

along with an eigenvalue estimate

$$\gamma_1 = \frac{v^T A v}{v^T v}.$$

By setting $p(x) = x^{k-1}$ in the proof of Theorem 10.1.2, it is easy to show that

$$\lambda_1 \geq \gamma_1 \geq \lambda_1 - (\lambda_1 - \lambda_n)\tan(\phi_1)^2\left(\frac{\lambda_2}{\lambda_1}\right)^{2(k-1)}. \qquad (10.1.11)$$

Thus, we can compare the quality of the lower bounds for θ_1 and γ_1 by comparing

$$L_{k-1} \equiv \frac{1}{\left[c_{k-1}\left(2\frac{\lambda_1}{\lambda_2} - 1\right)\right]^2} \geq \frac{1}{[c_{k-1}(1 + 2\rho_1)]^2}$$

and

$$R_{k-1} = \left(\frac{\lambda_2}{\lambda_1}\right)^{2(k-1)}.$$

Figure 10.1.1 compares these quantities for various values of k and λ_2/λ_1. The superiority of the Lanczos bound is self-evident. This is not a surprise since θ_1 is the maximum of $r(x) = x^T A x / x^T x$ over *all* of $\mathcal{K}(A, q_1, k)$, while $\gamma_1 = r(v)$ for a particular v in $\mathcal{K}(A, q_1, k)$, namely, $v = A^{k-1}q_1$.

λ_1/λ_2	$k=5$	$k=10$	$k=15$	$k=20$	$k=25$
1.50	$\dfrac{1.1\times10^{-4}}{3.9\times10^{-2}}$	$\dfrac{2.0\times10^{-10}}{6.8\times10^{-4}}$	$\dfrac{3.9\times10^{-16}}{1.2\times10^{-5}}$	$\dfrac{7.4\times10^{-22}}{2.0\times10^{-7}}$	$\dfrac{1.4\times10^{-27}}{3.5\times10^{-9}}$
1.10	$\dfrac{2.7\times10^{-2}}{4.7\times10^{-1}}$	$\dfrac{5.5\times10^{-5}}{1.8\times10^{-1}}$	$\dfrac{1.1\times10^{-7}}{6.9\times10^{-2}}$	$\dfrac{2.1\times10^{-10}}{2.7\times10^{-2}}$	$\dfrac{4.2\times10^{-13}}{1.0\times10^{-2}}$
1.01	$\dfrac{5.6\times10^{-1}}{9.2\times10^{-1}}$	$\dfrac{1.0\times10^{-1}}{8.4\times10^{-1}}$	$\dfrac{1.5\times10^{-2}}{7.6\times10^{-1}}$	$\dfrac{2.0\times10^{-3}}{6.9\times10^{-1}}$	$\dfrac{2.8\times10^{-4}}{6.2\times10^{-1}}$

Figure 10.1.1. L_{k-1}/R_{k-1}

Problems

P10.1.1 Suppose $A \in \mathbb{R}^{n \times n}$ is skew-symmetric. Derive a Lanczos-like algorithm for computing a skew-symmetric tridiagonal matrix T_m such that $AQ_m = Q_m T_m$, where $Q_m^T Q_m = I_m$.

P10.1.2 Let $A \in \mathbb{R}^{n \times n}$ be symmetric and define $r(x) = x^T A x / x^T x$. Suppose $S \subseteq \mathbb{R}^n$ is a subspace with the property that $x \in S$ implies $\nabla r(x) \in S$. Show that S is invariant for A.

P10.1.3 Show that if a symmetric matrix $A \in \mathbb{R}^{n \times n}$ has a multiple eigenvalue, then the Lanczos process terminates prematurely.

P10.1.4 Show that the index m in Theorem 10.1.1 is the dimension of the smallest invariant subspace for A that contains q_1.

P10.1.5 Let $A \in \mathbb{R}^{n \times n}$ be symmetric and consider the problem of determining an orthonormal sequence $q_1, q_2, \ldots$ with the property that once $Q_k = [\, q_1 \mid \cdots \mid q_k \,]$ is known, q_{k+1} is chosen so as to minimize $\mu_k = \| (I - Q_{k+1}Q_{k+1}^T)AQ_k \|_F$. Show that if $\mathrm{span}\{q_1, \ldots, q_k\} = \mathcal{K}(A, q_1, k)$, then it is possible to choose q_{k+1} so $\mu_k = 0$. Explain how this optimization problem leads to the Lanczos iteration.

P10.1.6 Suppose $A \in \mathbb{R}^{n \times n}$ is symmetric and that we wish to compute its largest eigenvalue. Let η be an approximate eigenvector and set $\alpha = \eta^T A\eta / \eta^T \eta$ and $z = A\eta - \alpha\eta$. (a) Show that the interval $[\alpha - \delta, \alpha + \delta]$ must contain an eigenvalue of A where $\delta = \| z \|_2 / \| \eta \|_2$. (b) Consider the new approximation $\bar{\eta} = a\eta + bz$ and determine the scalars a and b so that $\bar{\alpha} = \bar{\eta}^T A\bar{\eta} / \bar{\eta}^T \bar{\eta}$ is maximized. (c) Relate the above computations to the first two steps of the Lanczos process.

P10.1.7 Suppose $T \in \mathbb{R}^{n \times n}$ is tridiagonal and symmetric and that $v \in \mathbb{R}^n$. Show how the Lanczos process can be used (in principle) to compute an orthogonal $Q \in \mathbb{R}^{n \times n}$ in $O(n^2)$ flops such that $Q^T(T + vv^T)Q = \tilde{T}$ is also tridiagonal.

Notes and References for §10.1

Detailed treatments of the symmetric Lanczos algorithm may be found in Parlett (SEP) and Meurant (LCG). The classic reference for the Lanczos method is:

C. Lanczos (1950). "An Iteration Method for the Solution of the Eigenvalue Problem of Linear Differential and Integral Operators," *J. Res. Nat. Bur. Stand. 45*, 255–282.

For details about the convergence of the Ritz values, see:

S. Kaniel (1966). "Estimates for Some Computational Techniques in Linear Algebra," *Math. Comput. 20*, 369–378.

C.C. Paige (1971). "The Computation of Eigenvalues and Eigenvectors of Very Large Sparse Matrices," PhD thesis, University of London.

Y. Saad (1980). "On the Rates of Convergence of the Lanczos and the Block Lanczos Methods," *SIAM J. Numer. Anal. 17*, 687–706.

The connections between Lanczos tridiagonalization, orthogonal polynomials, and the theory of moments are discussed in:

N.J. Lehmann.(1963). "Optimale Eigenwerteinschliessungen," *Numer. Math. 5*, 246–272.
A.S. Householder (1968). "Moments and Characteristic Roots II," *Numer. Math. 11*, 126–128.
G.H. Golub (1974). "Some Uses of the Lanczos Algorithm in Numerical Linear Algebra," in *Topics in Numerical Analysis*, J.J.H. Miller (ed.), Academic Press, New York.
C.C. Paige, B.N. Parlett, and H.A. van der Vorst (1995). "Approximate Solutions and Eigenvalue Bounds from Krylov Subspaces," *Numer. Lin. Alg. Applic. 2*, 115–133.

10.2   Lanczos, Quadrature, and Approximation

To deepen our understanding of the Lanczos process and to build an appreciation for its connections to other areas of applied mathematics, we consider an interesting approximation problem that has broad practical implications. Assume that $A \in \mathbb{R}^{n \times n}$ is a large, sparse, symmetric positive definite matrix whose eigenvalues reside in an interval $[a, b]$. Let $f(\lambda)$ be a given smooth function that is defined on $[a, b]$. Given $u \in \mathbb{R}^n$, our goal is to produce suitably tight lower and upper bounds b and B so that

$$b \leq u^T \cdot f(A) \cdot u \leq B. \tag{10.2.1}$$

In the approach we develop, the bounds are Gauss quadrature rule estimates of a certain integral and the evaluation of the rules requires the eigenvalues and eigenvectors of a Lanczos-produced tridiagonal matrix.

The $u^T f(A) u$ estimation problem has many applications throughout matrix computations. For example, suppose $\hat{x}$ is an approximate solution to the symmetric positive definite system $Ax = b$ and that we have computed the residual $r = b - A\hat{x}$. Note that if $x_* = A^{-1}b$ and $f(\lambda) = 1/\lambda^2$, then

$$\| x_* - \hat{x} \|_2^2 = (x_* - \hat{x})^T (x_* - \hat{x}) = (A^{-1}(b - A\hat{x}))^T (A^{-1}(b - A\hat{x})) = r^T f(A) r.$$

Thus, if we have a $u^T f(A) u$ estimation framework, then we can obtain $Ax = b$ error bounds from residual bounds.

For an in-depth treatment of the material in this section, we refer the reader to the treatise by Golub and Meurant (2010). Our presentation is brief, informal, and stresses the linear algebra highlights.

10.2.1   Reformulation of the Problem

Without an integral in sight, it is mystifying as to why (10.2.1) involves quadrature at all. The key is to regard $u^T f(A) u$ as a *Riemann-Stieltjes integral*. In general, given a suitably nice integrand $f(x)$ and weight function $w(x)$, the Riemann-Stieltjes integral

$$I(f) = \int_a^b f(x) dw(x)$$

is a limit of sums of the form

$$S_N = \sum_{\mu=1}^{N} f(c_\mu)(w(x_\mu) - w(x_{\mu+1}))$$

where $a = x_N < \cdots < x_1 = b$ and $x_{\mu+1} \leq c_\mu \leq x_\mu$. Note that if w is piecewise constant on $[a, b]$, then the only nonzero terms in S_N arise from subintervals that house a "w-jump." For example, suppose $a = \lambda_n < \lambda_2 < \cdots < \lambda_1 = b$ and that

$$w(\lambda) = \begin{cases} w_{n+1} & \text{if } \lambda < a, \\ w_\mu & \text{if } \lambda_\mu \leq \lambda < \lambda_{\mu-1}, \quad \mu = 2{:}n, \\ w_1 & \text{if } b \leq \lambda, \end{cases} \tag{10.2.2}$$

where $0 \leq w_{n+1} \leq \cdots \leq w_1$. By considering the behavior of S_N as $N \to \infty$, we see that

$$\int_a^b f(\lambda) dw(\lambda) = \sum_{\mu=1}^n (w_\mu - w_{\mu+1}) \cdot f(\lambda_\mu). \tag{10.2.3}$$

We are now set to explain why $u^T f(A) u$ is "secretly" a Riemann-Stieltjes integral. Let

$$A = X \Lambda X^T, \qquad \Lambda = \text{diag}(\lambda_1, \ldots, \lambda_n), \tag{10.2.4}$$

be a Schur decomposition of A with $\lambda_n \leq \cdots \leq \lambda_1$. It follows that

$$u^T f(A) u = (X^T u)^T \cdot f(\Lambda) \cdot (X^T u) = \sum_{\mu=1}^n [X^T u]_\mu^2 \cdot f(\lambda_\mu).$$

If we set

$$w_\mu = [X^T u]_\mu^2 + \cdots + [X^T u]_n^2, \qquad \mu = 1{:}n+1, \tag{10.2.5}$$

in (10.2.2), then (10.2.3) becomes

$$\int_a^b f(\lambda) dw(\lambda) = \sum_{\mu=1}^n [X^T u]_\mu^2 \cdot f(\lambda_\mu) = u^T f(A) u. \tag{10.2.6}$$

Our plan is to approximate this integral using Gauss quadrature.

10.2.2 Some Gauss-Type Quadrature Rules and Bounds

Given an accuracy-related parameter k, an interval $[a, b]$, and a weight function $w(\lambda)$, a Gauss-type quadrature rule for the integral

$$I(f) = \int_a^b f(\lambda) \, dw(\lambda)$$

involves a carefully constructed linear combination of f-evaluations across $[a, b]$. The evaluation points (called *nodes*) and the coefficients (called *weights*) that define the linear combination are determined to make the rule correct for polynomials up to a certain degree that is related to k. Here are four examples:

1. Gauss. Compute weights $w_1, \ldots, w_k$ and nodes $t_1, \ldots, t_k$ so if

$$I_G(f) = \sum_{i=1}^k w_i f(t_i) \tag{10.2.7}$$

then $I(f) = I_G(f)$ for all polynomials f that have degree $2k - 1$ or less.

2. Gauss-Radau(a). Compute weights $w_a, w_1, \ldots, w_k$ and nodes $t_1, \ldots, t_k$ so if

$$I_{GR(a)}(f) = w_a f(a) + \sum_{i=1}^{k} w_i f(t_i) \qquad (10.2.8)$$

then $I(f) = I_{GR(a)}(f)$ for all polynomials f that have degree $2k$ or less.

3. Gauss-Radau(b). Compute weights $w_b, w_1, \ldots, w_k$ and nodes $t_1, \ldots, t_k$ so if

$$I_{GR(b)}(f) = w_b f(b) + \sum_{i=1}^{k} w_i f(t_i) \qquad (10.2.9)$$

then $I(f) = I_{GR(b)}(f)$ for all polynomials f that have degree $2k$ or less.

4. Gauss-Lobatto. Compute weights $w_a, w_b, w_1, \ldots, w_k$ and nodes $t_1, \ldots, t_k$ so if

$$I_{GL}(f) = w_a f(a) + w_b f(b) + \sum_{i=1}^{k} w_i f(t_i) \qquad (10.2.10)$$

then $I(f) = I_{GL}(f)$ for all polynomials f that have degree $2k + 1$ or less.

Each of these rules has a neatly specified error. It can be shown that

$$\int_a^b f(\lambda) dw(\lambda) = \begin{cases} I_G(f) & + & R_G(f), \\ I_{GR(a)}(f) & + & R_{GR(a)}(f), \\ I_{GR(b)}(f) & + & R_{GR(b)}(f), \\ I_{GL}(f) & + & R_{GG}(f), \end{cases}$$

where

$$R_G(f) = \frac{f^{(2k)}(\eta)}{(2n)!} \int_a^b \left[\prod_{i=1}^{k} (\lambda - t_i) \right]^2 dw(\lambda), \qquad a < \eta < b,$$

$$R_{GR(a)}(f) = \frac{f^{(2k+1)}(\eta)}{(2k+1)!} \int_a^b (\lambda - a) \left[\prod_{i=1}^{k} (\lambda - t_i) \right]^2 dw(\lambda), \qquad a < \eta < b,$$

$$R_{GR(b)}(f) = \frac{f^{(2k+1)}(\eta)}{(2k+1)!} \int_a^b (\lambda - b) \left[\prod_{i=1}^{k} (\lambda - t_i) \right]^2 dw(\lambda), \qquad a < \eta < b,$$

$$R_{GL}(f) = \frac{f^{(2k+2)}(\eta)}{(2k+2)!} \int_a^b (\lambda - a)(\lambda - b) \left[\prod_{i=1}^{k} (\lambda - t_i) \right]^2 dw(\lambda), \qquad a < \eta < b.$$

If the derivative in the remainder term does not change sign across $[a, b]$, then the rule can be used to produce a bound. For example, if $f(\lambda) = 1/\lambda^2$ and $0 < a < b$, then

$f^{(2k)}$ is positive, $f^{(2k+1)}$ is negative, and we have

$$I_G(f) \leq \int_a^b f(\lambda)dw(\lambda) \leq I_{GR(a)}(f).$$

With this strategy, we can produce lower and upper bounds by selecting and evaluating the right rule. For this to be practical, the behavior of f's higher derivatives must be known and the required rules must be computable.

10.2.3 The Tridiagonal Connection

It turns out that the evaluation of a given Gauss quadrature rule involves a tridiagonal matrix and its eigenvalues and eigenvectors. To develop a strategy that is based upon this connection, we need three facts about orthogonal polynomials and Gauss quadrature.

Fact 1. Given $[a, b]$ and $w(\lambda)$, there is a sequence of polynomials $p_0(\lambda)$, $p_1(\lambda)$, ... that satisfy

$$\int_a^b p_i(\lambda) \cdot p_j(\lambda) \cdot dw(\lambda) = \begin{cases} 1 & \text{if } i = j, \\ 0 & \text{if } i \neq j, \end{cases}$$

with the property that the degree of $p_k(\cdot)$ is k for $k \geq 0$. The polynomials are unique up to a factor of ± 1 and they satisfy a 3-term recurrence

$$\gamma_k p_k(\lambda) = (\lambda - w_k)p_{k-1}(\lambda) - \gamma_{k-1}p_{k-2}(\lambda)$$

where $p_{-1}(\lambda) \equiv 0$ and $p_0(\lambda) \equiv 1$.

Fact 2. The zeros of $p_k(\lambda)$ are the eigenvalues of the tridiagonal matrix

$$T_k = \begin{bmatrix} \omega_1 & \gamma_1 & 0 & \cdots & & 0 \\ \gamma_1 & \omega_2 & \ddots & & & \vdots \\ 0 & \ddots & \ddots & \ddots & & 0 \\ \vdots & & \ddots & \ddots & \omega_{k-1} & \gamma_{k-1} \\ 0 & \cdots & 0 & & \gamma_{k-1} & \omega_k \end{bmatrix}.$$

Since the γ_i are nonzero, it follows from Theorem 8.4.1 that the eigenvalues are distinct.

Fact 3. If

$$S^T T_k S = \text{diag}(\theta_1, \ldots, \theta_k) \tag{10.2.11}$$

is a Schur decomposition of T_k, then the nodes and weights for the Gauss rule (10.2.7) are given by $t_i = \theta_i$ and $w_i = s_{1i}^2$ for $i = 1{:}k$. In other words,

$$I_G(f) = \sum_{i=1}^k s_{1i}^2 \cdot f(\theta_i). \tag{10.2.12}$$

Thus, the only remaining issue is how to construct T_k so that it defines a Gauss rule for (10.2.6).

10.2.4   Gauss Quadrature via Lanczos

We show that if we apply the symmetric Lanczos process (Algorithm 10.1.1) with starting vector $q_1 = u/\| u \|_2$, then the tridiagonal matrices that the method generates are exactly what we need to compute $I_G(f)$.

We first link the Lanczos process to a sequence of orthogonal polynomials. Recall from §10.1.1 that the kth Lanczos vector q_k is in the Krylov subspace $\mathcal{K}(A, q_1, k)$. It follows that $q_k = p_k(A)q_1$ for some degree-k polynomial. From Algorithm 10.1.1 we know that

$$\beta_k q_{k+1} = (A - \alpha_k I)q_k - \beta_{k-1}q_{k-1}$$

where $\beta_0 q_0 \equiv 0$ and so

$$\beta_k p_{k+1}(A)q_1 = (A - \alpha_k I)p_k(A)q_1 - \beta_{k-1}p_{k-1}(A)q_1.$$

From this we conclude that the polynomials satisfy a 3-term recurrence:

$$\beta_k p_{k+1}(\lambda) = (\lambda - \alpha_k)p_k(\lambda) - \beta_{k-1}^2 p_{k-1}(\lambda). \qquad (10.2.13)$$

These polynomials are orthogonal with respect to the $u^T f(A)u$ weight function defined in (10.2.5). To see this, note that

$$
\begin{aligned}
\int_a^b p_i(\lambda)p_j(\lambda)dw(\lambda) &= \sum_{\mu=1}^n [X^T u]_\mu^2 \cdot p_i(\lambda_\mu) \cdot p_j(\lambda_\mu) \\
&= (X^T u)^T \left(p_i(\Lambda) \cdot p_j(\Lambda) \right) \cdot (X^T u) \\
&= u^T \left(X \cdot p_i(\Lambda) \cdot X^T \right) \left(X \cdot p_j(\Lambda) \cdot X^T \right) u \\
&= u^T \left(p_i(A)p_j(A) \right) u \\
&= (p_i(A)u)^T (p_j(A)u) = \| u \|_2^2 \, q_i^T q_j = 0.
\end{aligned}
$$

Coupled with (10.2.13) and Facts 1-3, this result tells us that we can generate an approximation $\sigma = I_G(f)$ to $u^T f(A)u$ as follows:

> *Step 1:* With starting vector $q_1 = u/\| u \|_2$, use the Lanczos process to compute the partial tridiagonalization $AQ_k = Q_k T_k + r_k e_k^T$. (See (10.1.4).)
>
> *Step 2:* Compute the Schur decomposition $S^T T_k S = \text{diag}(\theta_1, \ldots, \theta_k)$.
>
> *Step 3:* Set $\sigma = s_{11}^2 f(\theta_1) + \cdots + s_{1k}^2 f(\theta_k)$.

See Golub and Welsch (1969) for a more rigorous derivation of this procedure.

10.2.5   Computing the Gauss-Radau Rule

Recall from (10.2.1) that we are interested in upper and lower bounds. In light of our remarks at the end of §10.2.2, we need techniques for evaluating other Gauss quadrature rules. By way of illustration, we show how to compute $I_{GR(a)}$ defined in (10.2.8). Guided by Gauss quadrature theory, we run the Lanczos process for k steps as if we were setting out to compute $I_G(f)$. We then must determine $\tilde{\alpha}_{k+1}$ so that if

$$
\tilde{T}_{k+1} = \left[\begin{array}{cccccc|c}
\alpha_1 & \beta_1 & 0 & \cdots & & 0 & 0 \\
\beta_1 & \alpha_2 & \ddots & & & & \vdots \\
0 & \ddots & \ddots & \ddots & & \vdots & \vdots \\
\vdots & & \ddots & \alpha_{k-1} & \beta_{k-1} & 0 & \\
0 & \cdots & & \beta_{k-1} & \alpha_k & \beta_k \\
\hline
0 & \cdots & \cdots & 0 & \beta_k & \tilde{\alpha}_{k+1}
\end{array} \right]
$$

then $a \in \lambda(\tilde{T}_{k+1})$. By considering the top and bottom halves of the equation

$$
\tilde{T}_{k+1} \left[\begin{array}{c} x \\ -1 \end{array} \right] = a \left[\begin{array}{c} x \\ -1 \end{array} \right], \qquad x \in \mathbb{R}^k,
$$

it is easy to verify that $\tilde{\alpha}_{k+1} = a + \beta_{k+1}^2 e_k^T (T_k - aI_k)^{-1} e_k$ works.

10.2.6 The Overall Framework

All the necessary tools are now available to obtain sufficiently accurate upper and bounds in (10.2.1). At the bottom of the loop in Algorithm 10.1.1, we use the current tridiagonal (or an augmented version) to compute the nodes and weights for the lower bound rule. The rule is evaluated to obtain b. Likewise, we use the current tridiagonal (or an augmented version) to compute the nodes and weights for the upper bound rule. The rule is evaluated to obtain B. The **while** loop in Algorithm 10.1.1 can obviously be redesigned to terminate as soon as $B - b$ is sufficiently small.

Problems

P10.2.1 The Chebyschev polynomials are generated by the recursion $p_k(x) = 2xp_{k-1}(x) - p_{k-2}(x)$ and are orthonormal with respect to $w(x) = (1 - x^2)^{-1/2}$ across $[-1, 1]$. What are the zeros of $p_k(x)$?

P10.2.2 Following the strategy used in §10.2.5, show how to compute $I_{GR(b)}$ and $I_{GL}(f)$.

Notes and References for §10.2

For complete coverage of the Gauss quadrature/tridiagonal/Lanczos connection, see:

G.H. Golub and G. Meurant (2010). *Matrices, Moments, and Quadrature with Applications*, Princeton University Press, Princeton, NJ.

Research in this area has a long history:

G.H. Golub (1962). "Bounds for Eigenvalues of Tridiagonal Symmetric Matrices Computed by the LR Method," *Math. Comput. 16*, 438–445.
G.H. Golub and J.H. Welsch (1969). "Calculation of Gauss Quadrature Rules," *Math. Comput. 23*, 221–230.
G.H. Golub (1974). "Bounds for Matrix Moments," *Rocky Mountain J. Math. 4*, 207–211.
C. de Boor and G.H. Golub (1978). "The Numerically Stable Reconstruction of a Jacobi Matrix from Spectral Data," *Lin. Alg. Applic. 21*, 245–260.
J. Kautsky and G.H. Golub (1983). "On the Calculation of Jacobi Matrices," *Lin. Alg. Applic. 52/53*, 439–455.

M. Berry and G.H. Golub (1991). "Estimating the Largest Singular Values of Large Sparse Matrices via Modified Moments," *Numer. Algs. 1*, 353–374.

D.P. Laurie (1996). "Anti-Gaussian Quadrature Rules," *Math. Comput. 65*, 739–747.

Z. Bai and G.H. Golub (1997). "Bounds for the Trace of the Inverse and the Determinant of Symmetric Positive Definite Matrices," *Annals Numer. Math. 4*, 29–38.

M. Benzi and G.H. Golub (1999). "Bounds for the Entries of Matrix Functions with Applications to Preconditioning," *BIT 39*, 417–438.

D. Calvetti, G. H. Golub, W. B. Gragg, and L. Reichel (2000). "Computation of Gauss–Kronrod Quadrature Rules," *Math. Comput. 69*, 1035–1052.

D.P. Laurie (2001). "Computation of Gauss-Type Quadrature Formulas," *J. Comput. Appl. Math. 127*, 201–217.

10.3   Practical Lanczos Procedures

Rounding errors greatly affect the behavior of the Lanczos iteration. The basic difficulty is caused by loss of orthogonality among the Lanczos vectors, a phenomenon that muddies the issue of termination and complicates the relationship between A's eigenvalues and those of the tridiagonal matrices T_k. This troublesome feature, coupled with the advent of Householder's perfectly stable method of tridiagonalization, explains why the Lanczos algorithm was disregarded by numerical analysts during the 1950's and 1960's. However, the pressure to solve large, sparse eigenproblems coupled with the computational insights set forth by Paige (1971) changed all that. With many fewer than n iterations typically required to get good approximate extremal eigenvalues, the Lanczos method became attractive as a sparse matrix technique rather than as a competitor of the Householder approach.

Successful implementation of the Lanczos iteration involves much more than a simple encoding of Algorithm 10.1.1. In this section we present some of the ideas that have been proposed to make the Lanczos procedure viable in practice.

10.3.1   Required Storage and Work

With careful overwriting in Algorithm 10.1.1 and exploitation of the formula

$$\alpha_k \;=\; q_k^T (A q_k - \beta_{k-1} q_{k-1}),$$

the whole Lanczos process can be implemented with just a pair of n-vectors:

> $w = q_1,\ v = Aw,\ \alpha_1 = w^T v,\ v = v - \alpha_1 w,\ \beta_1 = \| v \|_2,\ k = 1$
> **while** $\beta_k \neq 0$
>   **for** $i = 1{:}n$
> $t = w_i,\ w_i = v_i / \beta_k,\ v_i = -\beta_k t$
>   **end**   (10.3.1)
> $v = v + Aw$
> $k = k + 1,\ \alpha_k = w^T v,\ v = v - \alpha_k w,\ \beta_k = \| v \|_2$
> **end**

At the end of the loop body, the array w houses q_k and v houses the residual vector $r_k = A q_k - \alpha_k q_k - \beta_{k-1} q_{k-1}$. See Paige (1972) for a discussion of various Lanczos implementations and their numerical properties. Note that A is not modified during the entire process and that is what makes the procedure so useful for large sparse matrices.

If A has an average of ν nonzeros per row, then approximately $(2\nu + 8)n$ flops are involved in a single Lanczos step. Upon termination the eigenvalues of T_k can be found using the symmetric tridiagonal QR algorithm or any of the special methods of §8.5 such as bisection. The Lanczos vectors are generated in the n-vector w. If eigenvectors are required, then the Lanczos vectors must be saved. Typically, they are stored in secondary memory units.

10.3.2   Roundoff Properties

The development of a practical, easy-to-use Lanczos tridiagonalization process requires an appreciation of the fundamental error analyses of Paige (1971, 1976, 1980). An examination of his results is the best way to motivate the several modified Lanczos procedures of this section.

After j steps of the iteration we obtain the matrix of computed Lanczos vectors $\hat{Q}_k = \begin{bmatrix} \hat{q}_1 & | & \cdots & | & \hat{q}_k \end{bmatrix}$ and the associated tridiagonal matrix

$$
\hat{T}_k \; = \;
\begin{bmatrix}
\hat{\alpha}_1 & \hat{\beta}_1 & & \cdots & & 0 \\
\hat{\beta}_1 & \hat{\alpha}_2 & \ddots & & & \vdots \\
& \ddots & \ddots & \ddots & & \\
\vdots & & \ddots & \ddots & \hat{\beta}_{k-1} \\
0 & \cdots & & & \hat{\beta}_{k-1} & \hat{\alpha}_k
\end{bmatrix} .
$$

Paige (1971, 1976) shows that if $\hat{r}_k$ is the computed analog of r_k, then

$$
A\hat{Q}_k \; = \; \hat{Q}_k \hat{T}_k + \hat{r}_k e_k^T + E_k \tag{10.3.2}
$$

where

$$
\| E_k \|_2 \; \approx \; \mathbf{u} \| A \|_2 . \tag{10.3.3}
$$

This shows that the equation $AQ_k = Q_k T_k + r_k e_k^T$ is satisfied to working precision.

Unfortunately, the picture is much less rosy with respect to the orthogonality among the $\hat{q}_i$. (Normality is not an issue. The computed Lanczos vectors essentially have unit length.) If $\hat{\beta}_k = \mathsf{fl}(\| \hat{r}_k \|_2)$ and we compute $\hat{q}_{k+1} = \mathsf{fl}(\hat{r}_k / \hat{\beta}_k)$, then a simple analysis shows that

$$
\hat{\beta}_k \hat{q}_{k+1} \approx \hat{r}_k + w_k
$$

where

$$
\| w_k \|_2 \approx \mathbf{u} \| \hat{r}_k \|_2 \approx \mathbf{u} \| A \|_2 .
$$

Thus, we may conclude that

$$
|\hat{q}_{k+1}^T \hat{q}_i| \; \approx \; \frac{|\hat{r}_k^T \hat{q}_i| + \mathbf{u} \| A \|_2}{|\hat{\beta}_k|}
$$

for $i = 1{:}k$. In other words, significant departures from orthogonality can be expected when $\hat{\beta}_k$ is small, *even* in the ideal situation where $\hat{r}_k^T \hat{Q}_k$ is zero. A small $\hat{\beta}_k$ implies cancellation in the computation of $\hat{r}_k$. We stress that loss of orthogonality is due to one or several such cancellations and is not the result of the gradual accumulation of roundoff error.

Further details of the Paige analysis are given shortly. Suffice it to say now that loss of orthogonality always occurs in practice and with it, an apparent deterioration in the quality of $\hat{T}_k$'s eigenvalues. This can be quantified by combining (10.3.2) with Theorem 8.1.16. In particular, if we set

$$F_1 = \hat{r}_k e_k^T + E_k, \qquad X_1 = \hat{Q}_k, \qquad S = \hat{T}_k,$$

in that theorem and assume that

$$\tau = \| \hat{Q}_k^T \hat{Q}_k - I_k \|_2$$

satisfies $\tau < 1$, then there exist eigenvalues $\mu_1, \ldots, \mu_k \in \lambda(A)$ such that

$$|\mu_i - \lambda_i(T_k)| \leq \sqrt{2}\,(\| \hat{r}_k \|_2 + \| E_k \|_2 + \tau(2+\tau)\| A \|_2)$$

for $i = 1{:}k$. An obvious way to control the τ factor is to orthogonalize each newly computed Lanczos vector against its predecessors. This leads directly to our first "practical" Lanczos procedure.

10.3.3   Lanczos with Complete Reorthogonalization

Let $r_0, \ldots, r_{k-1} \in \mathbb{R}^n$ be given and suppose that Householder matrices $H_0, \ldots, H_{k-1}$ have been computed such that $(H_0 \cdots H_{k-1})^T [\, r_0 \mid \cdots \mid r_{k-1} \,]$ is upper triangular. Let $[\, q_1 \mid \cdots \mid q_k \,]$ denote the first k columns of the Householder product $(H_0 \cdots H_{k-1})$. Now suppose that we are given a vector $r_k \in \mathbb{R}^n$ and wish to compute a unit vector q_{k+1} in the direction of

$$w = r_k - \sum_{i=1}^{k} (q_i^T r_k) q_i \in \mathsf{span}\{q_1, \ldots, q_k\}^\perp.$$

If a Householder matrix H_k is determined so $(H_0 \cdots H_k)^T [\, r_0 \mid \cdots \mid r_k \,]$ is upper triangular, then it follows that column $(k+1)$ of $H_0 \cdots H_k$ is the desired unit vector.

If we incorporate these Householder computations into the Lanczos process, then we can produce Lanczos vectors that are orthogonal to machine precision:

$r_0 = q_1$ (given unit vector)

Determine Householder H_0 so $H_0 r_0 = e_1$.

for $k = 1{:}n - 1$

$\qquad \alpha_k = q_k^T A q_k$

$\qquad r_k = (A - \alpha_k I)q_k - \beta_{k-1} q_{k-1}, \qquad (\beta_0 q_0 \equiv 0) \hfill (10.3.4)$

$\qquad w = (H_{k-1} \cdots H_0) r_k$

$\qquad$ Determine Householder H_k so $H_k w = [w_1, \ldots, w_k, \beta_k, 0, \ldots, 0]^T$.

$\qquad q_{k+1} = H_0 \cdots H_k e_{k+1}$

end

This is an example of a *complete reorthorgonalization* Lanczos scheme. The idea of using Householder matrices to enforce orthogonality appears in Golub, Underwood, and Wilkinson (1972). That the computed $\hat{q}_i$ in (10.3.4) are orthogonal to working precision follows from the roundoff properties of Householder matrices. Note that by virtue of the definition of q_{k+1}, it makes no difference if $\beta_k = 0$. For this reason, the algorithm may safely run until $k = n - 1$. (However, in practice one would terminate for a much smaller value of k.)

Of course, in any implementation of (10.3.4), one stores the Householder vectors v_k and never explicitly forms the corresponding matrix product. Since we have $H_k(1{:}k, 1{:}k) = I_k$ there is no need to compute the first k components of the vector w in (10.3.4) since we do not use them. (Ideally they are zero.)

Unfortunately, these economies make but a small dent in the computational overhead associated with complete reorthogonalization. The Householder calculations increase the work in the kth Lanczos step by $O(kn)$ flops. Moreover, to compute q_{k+1}, the Householder vectors associated with $H_0, \ldots, H_k$ must be accessed. For large n and k, this usually implies a prohibitive level of memory traffic.

Thus, there is a high price associated with complete reorthogonalization. Fortunately, there are more effective courses of action to take, but these require a greater understanding of just how orthogonality is lost.

10.3.4   Selective Reorthogonalization

A remarkable, ironic consequence of the Paige (1971) error analysis is that loss of orthogonality goes hand in hand with convergence of a Ritz pair. To be precise, suppose the symmetric QR algorithm is applied to $\hat{T}_k$ and renders computed Ritz values $\hat{\theta}_1, \ldots, \hat{\theta}_k$ and a nearly orthogonal matrix of eigenvectors $\hat{S}_k = (\hat{s}_{pq})$. If

$$\hat{Y}_k = \begin{bmatrix} \hat{y}_1 & | & \cdots & | & \hat{y}_k \end{bmatrix} = \mathsf{fl}(\hat{Q}_k \hat{S}_k),$$

then it can be shown that for $i = 1{:}k$ we have

$$|\hat{q}_{k+1}^T \hat{y}_i| \approx \frac{\mathbf{u}\| A \|_2}{|\hat{\beta}_k| \, |\hat{s}_{ki}|} \tag{10.3.5}$$

and

$$\| A\hat{y}_i - \hat{\theta}_i \hat{y}_i \|_2 \approx |\hat{\beta}_k| \, |\hat{s}_{ki}|. \tag{10.3.6}$$

That is, the most recently computed Lanczos vector $\hat{q}_{k+1}$ tends to have a nontrivial and unwanted component in the direction of any converged Ritz vector. Consequently, instead of orthogonalizing $\hat{q}_{k+1}$ against all of the previously computed Lanczos vectors, we can achieve the same effect by orthogonalizing it against the much smaller set of converged Ritz vectors.

The practical aspects of enforcing orthogonality in this way are discussed in Parlett and Scott (1979). In their scheme, known as *selective reorthogonalization*, a computed Ritz pair $\{\hat{\theta}, \hat{y}\}$ is called "good" if it satisfies

$$\| A\hat{y} - \hat{\theta}\hat{y} \|_2 \leq \sqrt{\mathbf{u}}\| A \|_2.$$

As soon as $\hat{q}_{k+1}$ is computed, it is orthogonalized against each good Ritz vector. This is much less costly than complete reorthogonalization, since, at least at first, there are many fewer good Ritz vectors than Lanczos vectors.

One way to implement selective reorthogonalization is to diagonalize $\hat{T}_k$ at each step and then examine the $\hat{s}_{ki}$ in light of (10.3.5) and (10.3.6). A more efficient approach for large k is to estimate the loss-of-orthogonality measure $\| I_k - \hat{Q}_k^T \hat{Q}_k \|_2$ using the following result.

Lemma 10.3.1. *Suppose* $S_+ = [\, S\ d\,]$ *where* $S \in \mathbb{R}^{n \times k}$ *and* $d \in \mathbb{R}^n$. *If*

$$\| I_k - S^T S \|_2 \le \mu \qquad\qquad |1 - d^T d| \ \le\ \delta,$$

then

$$\| I_{k+1} - S_+^T S_+ \|_2 \ \le\ \mu_+$$

where

$$\mu_+ \ =\ \frac{1}{2}\left(\mu + \delta + \sqrt{(\mu - \delta)^2 + 4\| S^T d \|_2^2} \, \right).$$

Proof. See Kahan and Parlett (1974) or Parlett and Scott (1979). □

Thus, if we have a bound for $\| I_k - \hat{Q}_k^T \hat{Q}_k \|_2$, then by applying the lemma with $S = \hat{Q}_k$ and $d = \hat{q}_{k+1}$ we can generate a bound for $\| I_{k+1} - \hat{Q}_{k+1}^T \hat{Q}_{k+1} \|_2$. (In this case $\delta \approx \mathbf{u}$ and we assume that $\hat{q}_{k+1}$ has been orthogonalized against the set of currently good Ritz vectors.) It is possible to estimate the norm of $\hat{Q}_k^T \hat{q}_{k+1}$ from a simple recurrence that spares one the need to access $\hat{q}_1, \ldots, \hat{q}_k$. The overhead is minimal, and when the bounds signal loss of orthogonality, it is time to contemplate the enlargement of the set of good Ritz vectors. Then and only then is $\hat{T}_k$ diagonalized.

10.3.5 The Ghost Eigenvalue Problem

Considerable effort has been spent in trying to develop a workable Lanczos procedure that does not involve any kind of orthogonality enforcement. Research in this direction focuses on the problem of "ghost" eigenvalues. These are multiple eigenvalues of $\hat{T}_k$ that correspond to simple eigenvalues of A. They arise because the iteration essentially restarts itself when orthogonality to a converged Ritz vector is lost. (By way of analogy, consider what would happen during orthogonal iteration (8.2.8) if we "forgot" to orthogonalize.)

The problem of identifying ghost eigenvalues and coping with their presence is discussed by Cullum and Willoughby (1979) and Parlett and Reid (1981). It is a particularly pressing problem in those applications where all of A's eigenvalues are desired, for then the above orthogonalization procedures are expensive to implement.

Difficulties with the Lanczos iteration can be expected even if A has a genuinely multiple eigenvalue. This follows because the $\hat{T}_k$ are unreduced, and unreduced tridiagonal matrices cannot have multiple eigenvalues. The next practical Lanczos procedure that we discuss attempts to circumvent this difficulty.

10.3.6 Block Lanczos Algorithm

Just as the simple power method has a block analogue in simultaneous iteration, so does the Lanczos algorithm have a block version. Suppose $n = rp$ and consider the decomposition

$$
Q^T A Q = \bar{T} = \begin{bmatrix} M_1 & B_1^T & & \cdots & & 0 \\ B_1 & M_2 & \ddots & & & \vdots \\ & \ddots & \ddots & \ddots & & \\ \vdots & & \ddots & \ddots & B_{r-1}^T \\ 0 & \cdots & & & B_{r-1} & M_r \end{bmatrix} \tag{10.3.7}
$$

where

$$
Q = [\, X_1 \mid \cdots \mid X_r \,], \qquad X_i \in \mathbb{R}^{n \times p},
$$

is orthogonal, each $M_i \in \mathbb{R}^{p \times p}$, and each $B_i \in \mathbb{R}^{p \times p}$ is upper triangular. Comparison of blocks in $AQ = Q\bar{T}$ shows that

$$
AX_k = X_{k-1}B_{k-1}^T + X_k M_k + X_{k+1}B_k
$$

for $k = 1{:}r$ assuming $X_0 B_0^T \equiv 0$ and $X_{r+1}B_r \equiv 0$. From the orthogonality of Q we have

$$
M_k = X_k^T A X_k
$$

for $k = 1{:}r$. Moreover, if we define

$$
R_k = AX_k - X_k M_k - X_{k-1}B_{k-1}^T \in \mathbb{R}^{n \times p},
$$

then

$$
X_{k+1}B_k = R_k
$$

is a QR factorization of R_k. These observations suggest that the block tridiagonal matrix $\bar{T}$ in (10.3.7) can be generated as follows:

$X_1 \in \mathbb{R}^{n \times p}$ given with $X_1^T X_1 = I_p$

$M_1 = X_1^T A X_1$

for $k = 1{:}r - 1$ $\hspace{4cm}$ (10.3.8)

$\hspace{1cm} R_k = AX_k - X_k M_k - X_{k-1}B_{k-1}^T \hspace{1cm} (X_0 B_0^T \equiv 0)$

$\hspace{1cm} X_{k+1}B_k = R_k \hspace{1cm}$ (QR factorization of R_k)

$\hspace{1cm} M_{k+1} = X_{k+1}^T A X_{k+1}$

end

At the beginning of the kth pass through the loop we have

$$
A[\, X_1 \mid \cdots \mid X_k \,] = [\, X_1 \mid \cdots \mid X_k \,]\bar{T}_k + R_k [\, 0 \mid \cdots \mid 0 \mid I_p \,], \tag{10.3.9}
$$

where

$$
\bar{T}_k = \begin{bmatrix} M_1 & B_1^T & & \cdots & & 0 \\ B_1 & M_2 & \ddots & & & \vdots \\ & \ddots & \ddots & \ddots & & \\ \vdots & & \ddots & \ddots & B_{k-1}^T \\ 0 & \cdots & & & B_{k-1} & M_k \end{bmatrix}.
$$

Using an argument similar to the one used in the proof of Theorem 10.1.1, we can show that the X_k are mutually orthogonal provided none of the R_k is rank-deficient. However if $\mathsf{rank}(R_k) < p$ for some k, then it is possible to choose the columns of X_{k+1} such that $X_{k+1}^T X_i = 0$, for $i = 1{:}k$. See Golub and Underwood (1977).

Because $\bar{T}_k$ has bandwidth p, it can be efficiently reduced to tridiagonal form using an algorithm of Schwartz (1968). Once tridiagonal form is achieved, the Ritz values can be obtained via the symmetric QR algorithm or any of the special methods of §8.4. In order to decide intelligently when to use block Lanczos, it is necessary to understand how the block dimension affects convergence of the Ritz values. The following generalization of Theorem 10.1.2 sheds light on this issue.

Theorem 10.3.2. *Let A be an n-by-n symmetric matrix with Schur decomposition*

$$Z^T A Z \;=\; \mathsf{diag}(\lambda_1,\dots,\lambda_n), \qquad \lambda_1 \geq \cdots \geq \lambda_n, \qquad Z \;=\; \left[\; z_1 \mid \cdots \mid z_n \;\right].$$

Let $\mu_1 \geq \cdots \geq \mu_p$ be the p largest eigenvalues of the matrix $\bar{T}_k$ obtained after k steps of (10.3.8). Suppose $Z_1 = \left[\; z_1 \mid \cdots \mid z_p \;\right]$ and

$$0 \;<\; \cos(\phi_p) \;=\; \sigma_p(Z_1^T X_1),$$

the smallest singular value of $Z_1^T X_1$. Then for $i = 1{:}p$,

$$\lambda_i \;\geq\; \mu_i \;\geq\; \lambda_i \;-\; (\lambda_1 - \lambda_n)\left(\frac{\tan(\theta_p)}{c_{k-1}\,(1 + 2\rho_i)}\right)^2$$

where

$$\rho_i \;=\; \frac{\lambda_i - \lambda_{p+1}}{\lambda_{p+1} - \lambda_n}$$

and $c_{k-1}(z)$ is the Chebyshev polynomial of degree $k - 1$.

Proof. See Underwood (1975). Compare with Theorem 10.1.2.   □

Analogous inequalities can be obtained for $\bar{T}_k$'s smallest eigenvalues by applying the theorem with A replaced by $-A$. Based on the theorem and scrutiny of (10.3.8), we conclude that

- the error bounds for the Ritz values improve with increased p;

- the amount of work required to compute $\bar{T}_k$'s eigenvalues is proportional to kp^2;

- the block dimension should be at least as large as the largest multiplicity of any sought-after eigenvalue.

Determination of the block dimension in the face of these trade-offs is discussed in detail by Scott (1979). We mention that loss of orthogonality also plagues the block Lanczos algorithm. However, all of the orthogonality enforcement schemes described above can be extended to the block setting.

10.3.7 Block Lanczos Algorithm with Restarting

The block Lanczos algorithm (10.3.8) can be used in an iterative fashion to calculate selected eigenvalues of A. To fix ideas, suppose we wish to calculate the p largest eigenvalues. If $X_1 \in \mathbb{R}^{n \times p}$ is a given matrix having orthonormal columns, then it can be refined as follows:

Step 1. Generate $X_2, \ldots, X_s \in \mathbb{R}^{n \times p}$ via the block Lanczos algorithm.

Step 2. Form $\bar{T}_s = [\, X_1 \mid \cdots \mid X_s \,]^T A \, [\, X_1 \mid \cdots \mid X_s \,]$, an sp-by-sp matrix that has bandwidth p.

Step 3. Compute an orthogonal matrix $U = [\, u_1 \mid \cdots \mid u_{sp} \,]$ such that $U^T \bar{T}_s U = \text{diag}(\theta_1, \ldots, \theta_{sp})$ with $\theta_1 \geq \cdots \geq \theta_{sp}$.

Step 4. Set $X_1^{(\text{new})} = [\, X_1 \mid \cdots \mid X_s \,]\,[\, u_1 \mid \cdots \mid u_p \,]$.

This is the block analog of the *s-step Lanczos algorithm*, which has been extensively analyzed by Cullum and Donath (1974) and Underwood (1975). The same idea can be used to compute several of A's smallest eigenvalues or a mixture of both large and small eigenvalues. See Cullum (1978). The choice of the parameters s and p depends upon storage constraints as well as upon the block-size implications that we discussed above. The value of p can be diminished as the good Ritz vectors emerge. However, this demands that orthogonality to the converged vectors be enforced.

Problems

P10.3.1 Rearrange (10.3.4) and (10.3.8) so that they require one matrix-vector product per iteration.

P10.3.2 If $\text{rank}(R_k) < p$ in (10.3.8), does it follow that $\text{ran}([\, X_1 \mid \cdots \mid X_k \,])$ contains an eigenvector of A?

Notes and References for §10.3

The behavior of the Lanczos method in the presence of roundoff error was originally reported in:

C.C. Paige (1971). "The Computation of Eigenvalues and Eigenvectors of Very Large Sparse Matrices," PhD thesis, University of London.

Important follow-up papers include:

C.C. Paige (1972). "Computational Variants of the Lanczos Method for the Eigenproblem," *J. Inst. Math. Applic. 10*, 373–381.
C.C. Paige (1976). "Error Analysis of the Lanczos Algorithm for Tridiagonalizing a Symmetric Matrix," *J. Inst. Math. Applic. 18*, 341–349.
C.C. Paige (1980). "Accuracy and Effectiveness of the Lanczos Algorithm for the Symmetric Eigenproblem," *Lin. Alg. Applic. 34*, 235–258.

For additional analysis of the method, see Parlett (SEP), Meurant (LCG) as well as:

D.S. Scott (1979). "How to Make the Lanczos Algorithm Converge Slowly," *Math. Comput. 33*, 239–247.
B.N. Parlett, H.D. Simon, and L.M. Stringer (1982). "On Estimating the Largest Eigenvalue with the Lanczos Algorithm," *Math. Comput. 38*, 153–166.
B.N. Parlett and B. Nour-Omid (1985). "The Use of a Refined Error Bound When Updating Eigenvalues of Tridiagonals," *Lin. Alg. Applic. 68*, 179–220.

J. Kuczyński and H. Woźniakowski (1992). "Estimating the Largest Eigenvalue by the Power and Lanczos Algorithms with a Random Start," *SIAM J. Matrix Anal. Applic. 13*, 1094–1122.

G. Meurant and Z. Strakos (2006). "The Lanczos and Conjugate Gradient Algorithms in Finite Precision Arithmetic," *Acta Numerica 15*, 471–542.

A wealth of practical, Lanczos-related information may be found in:

J.K. Cullum and R.A. Willoughby (2002). *Lanczos Algorithms for Large Symmetric Eigenvalue Computations: Vol. I: Theory*, SIAM Publications, Philadelphia, PA.

J. Brown, M. Chu, D. Ellison, and R. Plemmons (1994). *Proceedings of the Cornelius Lanczos International Centenary Conference*, SIAM Publications, Philadelphia, PA.

For a discussion about various reorthogonalization schemes, see:

C.C. Paige (1970). "Practical Use of the Symmetric Lanczos Process with Reorthogonalization," *BIT 10*, 183–195.

G.H. Golub, R. Underwood, and J.H. Wilkinson (1972). "The Lanczos Algorithm for the Symmetric $Ax = \lambda Bx$ Problem," Report STAN-CS-72-270, Department of Computer Science, Stanford University, Stanford, CA.

B.N. Parlett and D.S. Scott (1979). "The Lanczos Algorithm with Selective Orthogonalization," *Math. Comput. 33*, 217–238.

H.D. Simon (1984). "Analysis of the Symmetric Lanczos Algorithm with Reorthogonalization Methods," *Lin. Alg. Applic. 61*, 101–132.

Without any reorthogonalization it is necessary either to monitor the loss of orthogonality and quit at the appropriate instant or else to devise a scheme that will identify unconverged eigenvalues and false multiplicities, see:

W. Kahan and B.N. Parlett (1976). "How Far Should You Go with the Lanczos Process?" in *Sparse Matrix Computations*, J.R. Bunch and D.J. Rose (eds.), Academic Press, New York, 131–144.

J. Cullum and R.A. Willoughby (1979). "Lanczos and the Computation in Specified Intervals of the Spectrum of Large, Sparse Real Symmetric Matrices, in *Sparse Matrix Proc.*, I.S. Duff and G.W. Stewart (eds.), SIAM Publications, Philadelphia, PA.

B.N. Parlett and J.K. Reid (1981). "Tracking the Progress of the Lanczos Algorithm for Large Symmetric Eigenproblems," *IMA J. Num. Anal. 1*, 135–155.

For a restarting framework to be successful, it must exploit the approximate invariant subspace information that has been acquired by the iteration that is about to be shut down, see:

D. Calvetti, L. Reichel, and D.C. Sorensen (1994). "An Implicitly Restarted Lanczos Method for Large Symmetric Eigenvalue Problems," *ETNA 2*, 1–21.

K. Wu and H. Simon (2000). "Thick-Restart Lanczos Method for Large Symmetric Eigenvalue Problems," *SIAM J. Matrix Anal. Applic. 22*, 602–616.

The block Lanczos algorithm is discussed in:

J. Cullum and W.E. Donath (1974). "A Block Lanczos Algorithm for Computing the q Algebraically Largest Eigenvalues and a Corresponding Eigenspace of Large Sparse Real Symmetric Matrices," *Proceedings of the 1974 IEEE Conference on Decision and Control*, Phoenix, AZ, 505–509.

R. Underwood (1975). "An Iterative Block Lanczos Method for the Solution of Large Sparse Symmetric Eigenvalue Problems," Report STAN-CS-75-495, Department of Computer Science, Stanford University, Stanford, CA.

G.H. Golub and R. Underwood (1977). "The Block Lanczos Method for Computing Eigenvalues," in *Mathematical Software III* , J. Rice (ed.), Academic Press, New York, pp. 364–377.

J. Cullum (1978). "The Simultaneous Computation of a Few of the Algebraically Largest and Smallest Eigenvalues of a Large Sparse Symmetric Matrix," *BIT 18*, 265–275.

A. Ruhe (1979). "Implementation Aspects of Band Lanczos Algorithms for Computation of Eigenvalues of Large Sparse Symmetric Matrices," *Math. Comput. 33*, 680–687.

The block Lanczos algorithm generates a symmetric band matrix whose eigenvalues can be computed in any of several ways. One approach is described in:

H.R. Schwartz (1968). "Tridiagonalization of a Symmetric Band Matrix," *Numer. Math. 12*, 231–241.

In some applications it is necessary to obtain estimates of interior eigenvalues. One strategy is to apply Lanczos to the matrix $(A - \mu I)^{-1}$ because the extremal eigenvalues of this matrix are eigenvalues close to μ. However, "shift-and-invert" strategies replace the matrix-vector product in the Lanczos iteration with a large sparse linear equation solve, see:

A.K. Cline, G.H. Golub, and G.W. Platzman (1976). "Calculation of Normal Modes of Oceans Using a Lanczos Method," in *Sparse Matrix Computations*, J.R. Bunch and D.J. Rose (eds), Academic Press, New York, pp. 409–426.

T. Ericsson and A. Ruhe (1980). "The Spectral Transformation Lanczos Method for the Numerical Solution of Large Sparse Generalized Symmetric Eigenvalue Problems," *Math. Comput. 35*, 1251–1268.

R.B. Morgan (1991). "Computing Interior Eigenvalues of Large Matrices," *Lin. Alg. Applic. 154-156*, 289–309.

R.G. Grimes, J.G. Lewis, and H.D. Simon (1994). "A Shifted Block Lanczos Algorithm for Solving Sparse Symmetric Generalized Eigenproblems," *SIAM J. Matrix Anal. Applic. 15*, 228–272.

10.4 Large Sparse SVD Frameworks

The connections between the SVD problem and the symmetric eigenvalue problem are discussed in §8.6.1. In light of that discussion, it is not surprising that there is a Lanczos process for computing selected singular values and vectors of a large, sparse, rectangular matrix A. The basic idea is to generate a bidiagonal matrix B that is orthogonally equivalent to A. We show how to do this in §5.4 using Householder transformations. However, to avoid large dense submatrices along the way, the Lanczos approach generates the bidiagonal entries entries directly.

10.4.1 Golub-Kahan Upper Bidiagonalization

Suppose $A \in \mathbb{R}^{m \times n}$ with $m \geq n$ and recall from §5.4.8 that there exist orthogonal $U \in \mathbb{R}^{m \times m}$ and $V \in \mathbb{R}^{n \times n}$ so that

$$
U^T A V = B = \begin{bmatrix}
\alpha_1 & \beta_1 & \cdots & & \cdots & 0 \\
0 & \alpha_2 & \beta_2 & \cdots & & \vdots \\
\vdots & \ddots & \ddots & \ddots & & \vdots \\
\vdots & & & 0 & \alpha_{n-1} & \beta_{n-1} \\
0 & \cdots & & \cdots & 0 & \alpha_n \\
\hline
& & & 0 & &
\end{bmatrix}. \tag{10.4.1}
$$

Since A and B are orthogonally related, they have the same singular values.

Analogously to our derivation of the symmetric Lanczos procedure in §10.1.1, we proceed to outline a sparse-matrix-friendly method for determining the diagonal and superdiagonal of B. The challenge is to bypass the generally full intermediate matrices associated with the Householder bidiagonalization process (Algorithm 5.4.2). We expect to extract good singular value/vector information long before the full bidiagonalization is complete.

The key is to develop useful recipes for the α's and β's from the matrix equations $AV = UB$ and $A^T U = VB^T$. Given the column partitionings

$$
U = \begin{bmatrix} u_1 \mid \cdots \mid u_m \end{bmatrix}, \qquad V = \begin{bmatrix} v_1 \mid \cdots \mid v_n \end{bmatrix},
$$

we have

$$Av_k = \alpha_k u_k + \beta_{k-1} u_{k-1}, \qquad (10.4.2)$$

$$A^T u_k = \alpha_k v_k + \beta_k v_{k+1} \qquad (10.4.3)$$

for $k = 1{:}n$ with the convention that $\beta_0 u_0 \equiv 0$ and $\beta_n v_{n+1} \equiv 0$. Define the vectors

$$r_k = Av_k - \beta_{k-1} u_{k-1}, \qquad (10.4.4)$$

$$p_k = A^T u_k - \alpha_k v_k. \qquad (10.4.5)$$

Using (10.4.2), (10.4.4), and the orthonormality of the u-vectors, we have

$$\alpha_k = \pm\| r_k \|_2,$$

$$u_k = r_k/\alpha_k, \qquad (\alpha_k \neq 0).$$

Note that if $\alpha_k = 0$, then from (10.4.1) it follows that $A(:, 1{:}k)$ is rank deficient. Similarly we may conclude from (10.4.3) and (10.4.5) that

$$\beta_k = \pm\| p_k \|_2,$$

$$v_{k+1} = p_k/\beta_k, \qquad (\beta_k \neq 0).$$

If $\beta_k = 0$, then it follows from the equations $AV = UB$ and $A^T U = VB^T$ that

$$AU(:, 1{:}k) = V(:, 1{:}k)B(1{:}k, 1{:}k), \qquad (10.4.6)$$

$$A^T V(:, 1{:}k) = U(:, 1{:}k)B(1{:}k, 1{:}k)^T, \qquad (10.4.7)$$

and thus
$$A^T AV(:, 1{:}k) = V(:, 1{:}k)\, B(1{:}k, 1{:}k)^T B(1{:}k, 1{:}k).$$

It follows that $\sigma(B(1{:}k, 1{:}k)) \subseteq \sigma(A)$.

Properly sequenced, the above equations mathematically define the Golub-Kahan process for bidiagonalizing a rectangular matrix.

Algorithm 10.4.1 (Golub-Kahan Bidiagonalization) Given a matrix $A \in \mathbb{R}^{m \times n}$ with full column rank and a unit 2-norm vector $v_c \in \mathbb{R}^n$, the following algorithm computes the factorizations (10.4.6) and (10.4.7) for some k with $1 \leq k \leq n$. The first column of V is v_c.

> $k = 0$, $p_0 = v_c$, $\beta_0 = 1$, $u_0 = 0$
> **while** $\beta_k \neq 0$
>> $v_{k+1} = p_k/\beta_k$
>> $k = k + 1$
>> $r_k = Av_k - \beta_{k-1} u_{k-1}$
>> $\alpha_k = \| r_k \|_2$
>> $u_k = r_k/\alpha_k$
>> $p_k = A^T u_k - \alpha_k v_k$
>> $\beta_k = \| p_k \|_2$
> **end**

This computation was first described by Golub and Kahan (1965). If $V_k = [v_1 | \cdots | v_k]$, $U_k = [u_1 | \cdots | u_k]$, and

$$
B_k = \begin{bmatrix}
\alpha_1 & \beta_1 & \cdots & \cdots & 0 \\
0 & \alpha_2 & \beta_2 & \cdots & \vdots \\
\vdots & \ddots & \ddots & \ddots & 0 \\
\vdots & & 0 & \alpha_{k-1} & \beta_{k-1} \\
0 & \cdots & 0 & 0 & \alpha_k
\end{bmatrix}, \tag{10.4.8}
$$

then after the kth pass through the loop we have

$$
AV_k = U_k B_k, \tag{10.4.9}
$$

$$
A^T U_k = V_k B_k^T + p_k e_k^T, \tag{10.4.10}
$$

assuming that $\alpha_k > 0$. It can be shown that

$$
\text{span}\{v_1, \ldots, v_k\} = \mathcal{K}(A^T A, v_c, k), \tag{10.4.11}
$$

$$
\text{span}\{u_1, \ldots, u_k\} = \mathcal{K}(A A^T, A v_c, k). \tag{10.4.12}
$$

Thus, the symmetric Lanczos convergence theory presented in §10.1.5 can be applied. Good approximations to A's large singular values emerge early, while the small singular values are typically more problematic, especially if there is a cluster near the origin. For further insight, see Luk (1978), Golub, Luk, and Overton (1981), and Björck (NMLS, §7.6).

10.4.2   Ritz Approximations

The Ritz idea can be applied to extract approximate singular values and vectors from the matrices U_k, V_k, and B_k. We simply compute the SVD

$$
F_k^T B_k G_k = \Gamma = \text{diag}(\gamma_1, \ldots, \gamma_k) \tag{10.4.13}
$$

and form the matrices

$$
Y_k = V_k G_k = [y_1 | \cdots | y_k],
$$

$$
Z_k = U_k F_k = [z_1 | \cdots | z_k].
$$

It follows from (10.4.9), (10.4.10), and (10.4.13) that

$$
AY_k = Z_k \Gamma,
$$

$$
A^T Z_k = Y_k \Gamma + p_k e_k^T F_k,
$$

and so for $i = 1{:}k$ we have

$$
Ay_i = \gamma_i z_i, \tag{10.4.14}
$$

$$
A^T z_i = \gamma_i y_i + [F_k]_{ki} \cdot p_k. \tag{10.4.15}
$$

It follows that $A^T A y_i = \gamma_i^2 z_i + [F_k]_{ki} \cdot p_k$ and thus, $\{\gamma_i, y_i\}$ is a Ritz pair for $A^T A$ with respect to $\text{ran}(V_k)$.

10.4.3   The Tridiagonal-Bidiagonal Connection

In §8.6.1 we showed that there is a connection between the SVD of a matrix $A \in \mathbb{R}^{m \times n}$ and the Schur decomposition of the symmetric matrix

$$C = \begin{bmatrix} 0 & A \\ A^T & 0 \end{bmatrix}. \tag{10.4.16}$$

In particular, if σ is a singular value of A, then both σ and $-\sigma$ are eigenvalues of C and the corresponding singular vectors "makeup" the corresponding eigenvectors.

Likewise, a given bidiagonalization of A can be related to a tridiagonalization of C. Assume that $m \geq n$ and that

$$[U_1 \mid U_2]^T A V = \begin{bmatrix} \tilde{B} \\ 0 \end{bmatrix}, \qquad \tilde{B} \in \mathbb{R}^{n \times n},$$

is a bidiagonalization of A with $U_1 \in \mathbb{R}^{m \times n}$, $U_2 \in \mathbb{R}^{m \times (m-n)}$, and $V \in \mathbb{R}^{n \times n}$. Note that

$$Q = \begin{bmatrix} U & 0 \\ 0 & V \end{bmatrix}$$

is orthogonal and

$$\tilde{T} = Q^T C Q = \begin{bmatrix} 0 & \tilde{B} \\ \tilde{B}^T & 0 \end{bmatrix}.$$

This matrix can be symmetrically permuted into tridiagonal form. For example, in the 4-by-3 case, if $P = I_7(:, [5\ 1\ 6\ 2\ 7\ 3\ 4])$, then the reordering $\tilde{T} \rightarrow P\tilde{T}P^T$ has the form

$$\begin{bmatrix} 0 & 0 & 0 & 0 & \alpha_1 & \beta_1 & 0 \\ 0 & 0 & 0 & 0 & 0 & \alpha_2 & \beta_2 \\ 0 & 0 & 0 & 0 & 0 & 0 & \alpha_3 \\ 0 & 0 & 0 & 0 & 0 & 0 & 0 \\ \alpha_1 & 0 & 0 & 0 & 0 & 0 & 0 \\ \beta_1 & \alpha_2 & 0 & 0 & 0 & 0 & 0 \\ 0 & \beta_2 & \alpha_3 & 0 & 0 & 0 & 0 \end{bmatrix} \rightarrow \begin{bmatrix} 0 & \alpha_1 & 0 & 0 & 0 & 0 & 0 \\ \alpha_1 & 0 & \beta_1 & 0 & 0 & 0 & 0 \\ 0 & \beta_1 & 0 & \alpha_2 & 0 & 0 & 0 \\ 0 & 0 & \alpha_2 & 0 & \beta_2 & 0 & 0 \\ 0 & 0 & 0 & \beta_2 & 0 & \alpha_3 & 0 \\ 0 & 0 & 0 & 0 & \alpha_3 & 0 & 0 \\ 0 & 0 & 0 & 0 & 0 & 0 & 0 \end{bmatrix}.$$

This points to an interesting connection between Golub-Kahan bidiagonalization (Algorithm 10.4.1) and Lanczos tridiagonalization (Algorithm 10.1.1). Suppose we apply Algorithm 10.4.1 to $A \in \mathbb{R}^{m \times n}$ with starting vector v_c. Assume that the procedure runs for k steps and produces the bidiagonal matrix B_k displayed in (10.4.8). If we apply Algorithm 10.1.1 to the matrix C defined by (10.4.16) with a starting vector

$$q_1 = \begin{bmatrix} 0 \\ v_c \end{bmatrix} \in \mathbb{R}^{m+n} \tag{10.4.17}$$

then after $2k$ steps the resulting tridiagonal matrix T_{2k} has a zero diagonal and a subdiagonal specified by $[\, \alpha_1,\ \beta_1,\ \alpha_2,\ \beta_2,\ \cdots\ \alpha_{k-1},\ \beta_{k-1},\ \alpha_k \,]$.

10.4.4 Paige-Saunders Lower Bidiagonalization

In §11.4.2 we show how the Golub-Kahan bidiagonalization can be used to solve sparse linear systems and least squares problems. It turns out that in this context, *lower bidiagonalization* is more useful:

$$
U^T A V = B = \begin{bmatrix}
\alpha_1 & 0 & \cdots & & \cdots & 0 \\
\beta_2 & \alpha_2 & 0 & \cdots & & \vdots \\
\vdots & \beta_3 & \ddots & \ddots & & \vdots \\
\vdots & & \ddots & \alpha_{n-1} & 0 \\
0 & \cdots & & \cdots & \beta_{n-1} & \alpha_n \\
0 & \cdots & & \cdots & 0 & \beta_n \\
\hline
& & \multicolumn{2}{c}{\text{\Large 0}} & &
\end{bmatrix}. \tag{10.4.18}
$$

Proceeding as in the derivation of the Golub-Kahan bidiagonalization, we compare columns in the equations $A^T U = V B^T$ and $AV = UB$. If $U = [\, u_1 \mid \cdots \mid u_m \,]$ and $V = [\, v_1 \mid \cdots \mid v_n \,]$ are column partitionings and we define $\beta_1 v_0 \equiv 0$ and $\alpha_{n+1} v_{n+1} \equiv 0$, then for $k = 1{:}n$ we have $A^T u_k = \beta_k v_{k-1} + \alpha_k v_k$ and $A v_k = \alpha_k u_k + \beta_{k+1} u_{k+1}$. Leaving the rest of the derivation to the exercises, we obtain the following.

Algorithm 10.4.2 (Paige-Saunders Bidiagonalization) Given a matrix $A \in \mathbb{R}^{m \times n}$ with the property that $A(1{:}n, 1{:}n)$ is nonsingular and a unit 2-norm vector $u_c \in \mathbb{R}^n$, the following algorithm computes the factorization $AV(:, 1{:}k) = U(:, 1{:}k+1) B(1{:}k+1, 1{:}k)$ where U, V, and B are given by (10.4.18). The first column of U is u_c and the integer k satisfies $1 \le k \le n$.

$$
\begin{aligned}
&k = 1,\ p_0 = u_c,\ \beta_1 = 1,\ v_0 = 0 \\
&\textbf{while } \beta_k > 0 \\
&\qquad u_k = p_{k-1}/\beta_k \\
&\qquad r_k = A^T u_k - \beta_k v_{k-1} \\
&\qquad \alpha_k = \| r_k \|_2 \\
&\qquad v_k = r_k/\alpha_k \\
&\qquad p_k = A v_k - \alpha_k u_k \\
&\qquad \beta_{k+1} = \| p_k \|_2 \\
&\qquad k = k + 1 \\
&\textbf{end}
\end{aligned}
$$

It can be shown that after k passes through the loop we have

$$
AV(:, 1{:}k) = U(:, 1{:}k) B(1{:}k, 1{:}k) + p_k e_k^T \tag{10.4.19}
$$

where $e_k = I_k(:, k)$. See Paige and Saunders (1982) for more details. Their bidiagonalization is equivalent to Golub-Kahan bidiagonalization applied to $[\, b \mid A \,]$.

10.4.5   A Note on Randomized Low-Rank Approximation

The need to extract information from unimaginably large datasets has prompted the development of matrix methods that involve *randomization*. The idea is to develop matrix approximations that are very fast to compute because they rely on limited, random samplings of the given matrix. To give a snapshot of this increasingly important paradigm for large-scale matrix computations, we consider the problem of computing a rank-k approximation to a given matrix $A \in \mathbb{R}^{m \times n}$. For clarity we assume that $k \leq \mathsf{rank}(A)$. Recall that if $A = \tilde{Z} \tilde{\Sigma} \tilde{Y}^T$ is the SVD of A, then

$$\tilde{A}_k = \tilde{Z}_1 \tilde{\Sigma}_1 \tilde{Y}_1^T = \tilde{Z}_1 \tilde{Z}_1^T A \qquad (10.4.20)$$

where $\tilde{Z}_1 = \tilde{Z}(:, 1{:}k)$, $\tilde{\Sigma}_1 = \tilde{\Sigma}(1{:}k, 1{:}k)$, and $\tilde{Y}_1 = \tilde{Y}(:, 1{:}k)$, is the closest rank-$k$ matrix to A as measured in either the 2-norm or Frobenius norm. We assume that A is so large that the Krylov methods just discussed are impractical.

Drineas, Kannan, and Mahoney (2006c) propose a method that approximates the intractable $\tilde{A}_k$ with a rank-k matrix of the form

$$A_k = CUR, \qquad C \in \mathbb{R}^{m \times c}, U \in \mathbb{R}^{c \times r}, R \in \mathbb{R}^{r \times n}, k \leq c, k \leq r \qquad (10.4.21)$$

where the matrices C and R are comprised of randomly chosen values taken from A. The integers c and r are parameters of the method. Discussion of the *CUR decomposition* (10.4.21) nicely illustrates the notion of random sampling in the matrix context and the idea of a probabilistic error bound.

The first step in the CUR framework is to determine C. Each column of this matrix is a scaled, randomly-selected column of A:

Determine column probabilities $q_j = \| A(:, j) \|_2 / \| A \|_F^2$, $j = 1{:}n$.

for $t = 1{:}c$

    Randomly pick $col(t) \in \{1, 2, \ldots, n\}$ with q_α the probability that $col(t) = \alpha$.

    $C(:, t) = A(:, col(t)) / \sqrt{c\, q_{col(t)}}$

end

It follows that $C = A(:, col) D_C$ where $D_C \in \mathbb{R}^{c \times c}$ is a diagonal scaling matrix.

The matrix R is similarly constructed. Each row of this matrix is a scaled, randomly-selected row of A:

Determine row probabilities $p_i = \| A(i, :) \|_2 / \| A \|_F^2$, $i = 1{:}m$.

for $t = 1{:}r$

    Randomly pick $row(t) \in \{1, 2, \ldots, m\}$ with p_α the probability that $row(t) = \alpha$.

    $R(t, :) = A(row(t), :) / \sqrt{r\, p_{row(t)}}$

end

The matrix R has the form $R = D_R A(row, :)$ where $D_R \in \mathbb{R}^{r \times r}$ is a diagonal scaling matrix.

The next step is to choose a rank-k matrix U so that $A_k = CUR$ is close to the best rank-k approximation $\tilde{A}_k$. In the CUR framework, this requires the SVD

$$C = Z\Sigma Y^T = Z_1 \Sigma_1 Y_1^T + Z_2 \Sigma_2 Y_2$$

where $Z_1 = Z(:, 1{:}k)$, $\Sigma_1 = \Sigma(1{:}k, 1{:}k)$, and $Y_1 = Y(:, 1{:}k)$. The matrix U is then given by

$$U = \Phi\Psi^T, \qquad \Phi = Y_1 \Sigma_1^{-2} Y_1^T, \ \Psi = D_R C(row, :).$$

With these definitions, simple manipulations confirm that

$$C\Phi = Z_1 \Sigma_1^{-1} Y_1^T, \tag{10.4.22}$$

$$\Psi^T R = \left(D_R(Z_1(row, :)\Sigma_1 Y_1^T + Z_2(row, :)\Sigma_2 Y_2^T)\right)^T D_R A(row, :), \tag{10.4.23}$$

and

$$CUR = (C\Phi)(\Psi R) = Z_1 \left(D_R Z_1(row, :)\right) \left(D_R A(row, :)\right). \tag{10.4.24}$$

An analysis that critically depends on the selection probabilities $\{q_i\}$ and $\{p_i\}$ shows that $\mathsf{ran}(Z_1) \approx \mathsf{ran}(\tilde{Z}_1)$ and $(D_R Z_1(row, :))^T (D_R A(row, :)) \approx Z_1^T A$. Upon comparison with (10.4.20) we see that $CUR \approx Z_1 Z_1^T A \approx \tilde{Z}_1 \tilde{Z}_1^T A = \tilde{A}_k$. Moreover, given $\epsilon > 0$, $\delta > 0$, and k, it is possible to choose the parameters r and c so that the inequality

$$\| A - CUR \|_F \leq \| A - \tilde{A}_k \|_F + \epsilon \| A \|_F$$

holds with probability $1 - \delta$. Lower bounds for r and c that depend inversely on ϵ and δ are given by Drineas, Kannan, and Mahoney (2006c).

Problems

P10.4.1 Verify Equations (10.4.6), (10.4.7), (10.4.9), and (10.4.10).

P10.4.2 Corresponding to (10.3.1), develop an implementation of Algorithm 10.4.1 that involves a minimum number of vector workspaces.

P10.4.3 Show that if $\mathsf{rank}(A) = n$, then the bidiagonal matrix B in (10.4.18) cannot have a zero on its diagonal.

P10.4.4 Prove (10.4.19). What can you say about $U(:, 1{:}k)$ and $V(:, 1{:}k)$ if $\beta_{k+1} = 0$ in Algorithm 10.4.2?

P10.4.5 Analogous to (10.4.11)-(10.4.12), show that for Algorithm 10.4.2 we have

$$\mathsf{span}\{v_1, \ldots, v_k\} = \mathcal{K}(A^T A, A^T u_c, k), \qquad \mathsf{span}\{u_1, \ldots, u_k\} = \mathcal{K}(A A^T, u_c, k).$$

P10.4.6 Suppose C and q_1 are defined by (10.4.16) and (10.4.17) respectively. (a) Show that

$$\mathcal{K}(C, q_1, 2k) = \mathsf{span}\left\{ \begin{bmatrix} 0 \\ v_c \end{bmatrix}, \begin{bmatrix} Av_c \\ 0 \end{bmatrix}, \begin{bmatrix} 0 \\ A^T Av_c \end{bmatrix}, \ldots, \begin{bmatrix} 0 \\ (A^T A)^{k-1} v_c \end{bmatrix}, \begin{bmatrix} A(A^T A)^{k-1} v_c \\ 0 \end{bmatrix} \right\}.$$

(b) Rigorously prove the claim made in §10.4.3 about the subdiagonal of T_{2k}. (c) State and prove analogous results when the Paige-Saunders bidiagonalization is used.

P10.4.7 Verify Equations 10.4.22–10.4.24.

Notes and References for §10.4

For a more comprehensive treatment of Golub-Kahan bidiagonalization, see Björck (NMLS, §7.6). The relevance of the Lanczos process to the bidiagonalization of a rectangular matrix was first presented in:

G.H. Golub and W. Kahan (1965). "Calculating the Singular Values and Pseudo-Inverse of a Matrix," *SIAM J. Numer. Anal. Ser. B, 2*, 205–224.

The idea of using Golub-Kahan bidiagonalization to solve large sparse linear systems and least squares problems started with the paper:

C.C. Paige (1974). "Bidiagonalization of Matrices and Solution of Linear Equations," *SIAM J. Numer. Anal. 11*, 197–209.

We shall have more to say about this in the next chapter. It is in anticipation of that discussion that we presented the lower bidiagonal scheme, see:

C.C. Paige and M.A. Saunders (1982). "LSQR, An Algorithm for Sparse Linear Equations and Sparse Least Squares," *ACM Trans. Math. Softw. 8*, 43–71.

For practical implementation issues, see:

G.H. Golub, F.T. Luk, and M.L. Overton (1981). "A Block Lanczos Method for Computing the Singular Values and Corresponding Singular Vectors of a Matrix," *ACM Trans. Math. Softw. 7*, 149–169.

J. Cullum, R.A. Willoughby, and M. Lake (1983). "A Lanczos Algorithm for Computing Singular Values and Vectors of Large Matrices," *SIAM J. Sci. Stat. Comput. 4*, 197–215.

M. Berry (1992). "Large-Scale Sparse Singular Value Computations," *International J. Supercomputing Appl. 6*, 13–49.

M. Berry and R.L. Auerbach (1993). "A Block Lanczos SVD Method with Adaptive Reorthogonalization," in *Proceedings of the Cornelius Lanczos International Centenary Conference*, Raleigh, NC, SIAM Publications, Philadelphia, PA.

Z. Jia and D. Niu (2003). "An Implicitly Restarted Refined Bidiagonalization Lanczos Method for Computing a Partial Singular Value Decomposition," *SIAM J. Matrix Anal. Applic. 25*, 246–265.

Interesting applications of the Lanczos bidiagonalization include:

D.P. OLeary and J.A. Simmons (1981). "A Bidiagonalization-Regularization Procedure for Large Scale Discretizations of Ill-Posed Problems," *SIAM J. Sci. Stat. Comput. 2*, 474–489.

D. Calvetti, G.H. Golub, and L. Reichel (1999). "Estimation of the L-curve via Lanczos Bidiagonalization," *BIT 39*, 603–619.

H.D. Simon and H. Zha (2000). "Low-Rank Matrix Approximation Using the Lanczos Bidiagonalization Process with Applications," *SIAM J. Sci. Comput. 21*, 2257–2274.

Our sketch of the CUR decomposition framework is based on:

P. Drineas, R. Kannan, and M.W. Mahoney (2006). "Fast Monte Carlo Algorithms for Matrices III: Computing an Efficient Approximate Decomposition of a Matrix," *SIAM J. Comput. 36*, 184–206.

Additional references concerned with randomization in matrix computations include:

P. Drineas, R. Kannan, and M. W. Mahoney (2006). "Fast Monte Carlo Algorithms for Matrices I: Approximating Matrix Multiplication," *SIAM J. Comput. 36*, 132–157.

P. Drineas, R. Kannan, and M.W. Mahoney (2006). "Fast Monte Carlo Algorithms for Matrices II: Computing Low-Rank Approximations to a Matrix," *SIAM J. Comput. 36*, 158–183.

M.W. Mahoney, M. Maggioni, and P. Drineas (2008). "Tensor-CUR Decompositions For Tensor-Based Data," *SIAM J. Mat. Anal. Applic. 30*, 957–987.

P. Drineas, M.W. Mahoney, and S. Muthukrishnan (2008). "Relative-Error CUR Matrix Decompositions," *SIAM J. Mat. Anal. Applic. 30*, 844–881.

E. Liberty, F. Woolfe, P.-G. Martinsson, V. Rokhlin, and M.Tygert (2008). "Randomized Algorithms for the Low-Rank Approximation of Matrices," *Proc. Natl. Acad. Sci. 104*, 20167–20172.

V. Rokhlin and Mark Tygert (2008). "A Fast Randomized Algorithm for Overdetermined Linear Least-Squares Regression," *Proc. Natl. Acad. Sci. 105*, 13212–13217.

M.W. Mahoney and P. Drineas (2009). "CUR Matrix Decompositions for Improved Data Analysis," *Proc. Natl. Acad. Sci. 106*, 697–702.

D. Achlioptas and F. McSherry (2007). "Fast Computation of Low-Rank Matrix Approximations," *JACM 54(2)*, Article No. 9.

V. Rokhlin, A. Szlam, and M. Tygert (2010). "A Randomized Algorithm for Principal Component Analysis," *SIAM J. Mat. Anal. Applic. 31*, 1100–1124

M.W. Mahoney (2011). "Randomized Algorithms for Matrices and Data," *Foundations and Trends in Machine Learning 3*, 123–224.

N. Halko, P.G. Martinsson, and J.A. Tropp (2011). "Finding Structure with Randomness: Probabilistic Algorithms for Constructing Approximate Matrix Decompositions," *SIAM Review 53*, 217–288

For another perspective on the increasing important role of randomness in matrix computations, see:

A. Edelman and N. Raj Rao (2005). "Random Matrix Theory," *Acta Numerica 14*, 233–297

10.5 Krylov Methods for Unsymmetric Problems

If A is not symmetric, then the orthogonal tridiagonalization $Q^T A Q = T$ does not exist in general. There are two ways to proceed. The Arnoldi approach involves the column-by-column generation of an orthogonal Q such that $Q^T A Q = H$ is the Hessenberg reduction of §7.4. The unsymmetric Lanczos approach computes the columns of matrices Q and P so that $P^T A Q = T$ is tridiagonal and $P^T Q = I$. Methods based on these ideas that are suitable for large, sparse, unsymmetric eigenvalue problems are discussed in this section.

10.5.1 The Basic Arnoldi Process

One way to extend the Lanczos process to unsymmetric matrices is due to Arnoldi (1951) and revolves around the Hessenberg reduction $Q^T A Q = H$. In particular, if $Q = [\, q_1 \mid \cdots \mid q_n \,]$ and we compare columns in $AQ = QH$, then

$$A q_k = \sum_{i=1}^{k+1} h_{ik} q_i, \qquad 1 \le k \le n-1.$$

Isolating the last term in the summation gives

$$h_{k+1,k} q_{k+1} = A q_k - \sum_{i=1}^{k} h_{ik} q_i \equiv r_k$$

where $h_{ik} = q_i^T A q_k$ for $i = 1{:}k$. It follows that if $r_k \ne 0$, then q_{k+1} is specified by

$$q_{k+1} = r_k / h_{k+1,k}$$

where $h_{k+1,k} = \| \, r_k \, \|_2$. These equations define the *Arnoldi process* and in strict analogy to the symmetric Lanczos process (Algorithm 10.1.1) we obtain the following.

Algorithm 10.5.1 (Arnoldi Process) If $A \in \mathbb{R}^{n \times n}$ and $q_1 \in \mathbb{R}^n$ has unit 2-norm, then the following algorithm computes a matrix $Q_t = [q_1, \ldots, q_t] \in \mathbb{R}^{n \times t}$ with orthonormal columns and an upper Hessenberg matrix $H_t = (h_{ij}) \in \mathbb{R}^{t \times t}$ with the property that $AQ_t = Q_t H_t$. The integer t satisfies $1 \le t \le n$.

$$k = 0, \ r_0 = q_1, \ h_{10} = 1$$
$$\textbf{while } (h_{k+1,k} \ne 0)$$
$$\qquad q_{k+1} = r_k / h_{k+1,k}$$
$$\qquad k = k + 1$$
$$\qquad r_k = A q_k$$
$$\qquad \textbf{for } i = 1{:}k$$
$$\qquad \qquad h_{ik} = q_i^T r_k$$
$$\qquad \qquad r_k = r_k - h_{ik} q_i$$
$$\qquad \textbf{end}$$
$$\qquad h_{k+1,k} = \| \, r_k \, \|_2$$
$$\textbf{end}$$
$$t = k$$

The q_k are called *Arnoldi vectors* and they define an orthonormal basis for the Krylov subspace $\mathcal{K}(A, q_1, k)$:

$$\mathsf{span}\{q_1, \ldots, q_k\} = \mathsf{span}\{q_1, Aq_1, \ldots, A^{k-1}q_1\}. \qquad (10.5.1)$$

The situation after k steps is summarized by the equation

$$AQ_k = Q_k H_k + r_k e_k^T \qquad (10.5.2)$$

where $Q_k = [\, q_1 \,|\, \cdots \,|\, q_k \,]$, $e_k = I_k(:, k)$, and

$$
H_k = \begin{bmatrix}
h_{11} & h_{12} & \cdots & & \cdots & h_{1k} \\
h_{21} & h_{22} & \cdots & & \cdots & h_{2k} \\
0 & h_{32} & \ddots & & & \vdots \\
\vdots & & \ddots & \ddots & & \vdots \\
0 & \cdots & & \cdots & h_{k,k-1} & h_{kk}
\end{bmatrix}.
$$

Any decomposition of the form (10.5.2) is a *k-step Arnoldi decomposition* if $Q_k \in \mathbb{R}^{n \times k}$ has orthonormal columns, $H_k \in \mathbb{R}^{k \times k}$ is upper Hessenberg, and $Q_k^T r_k = 0$.

If $y \in \mathbb{R}^k$ is a unit 2-norm eigenvector for H_k and $H_k y = \lambda y$, then from (10.5.2)

$$(A - \lambda I)x = (e_k^T y) r_k$$

where $x = Q_k y$. Since $r_k \in \mathcal{K}(A, q_1, k)^\perp$, it follows that (λ, x) is a Ritz pair for A with respect to $\mathcal{K}(A, q_1, k)$. Note that if $v = (e_k^T y) r_k$, then

$$(A + E)x = \lambda x$$

where $E = -v x^T$ with $\| E \|_2 = |y_k| \|\, r_k \,\|_2$. Recall that in the unsymmetric case, computing an eigenvalue of a nearby matrix does *not* mean that it is close to an exact eigenvalue.

Some numerical properties of the Arnoldi iteration are discussed by Wilkinson (AEP, p. 382). The history of practical Arnoldi-based eigensolvers begins with Saad (1980). Two features of the method distinguish it from the symmetric Lanczos process:

- Arnoldi vectors $q_1, \ldots, q_k$ must all be referenced in step k and the computation of q_{k+1} involves $O(kn)$ flops excluding the matrix-vector product Aq_k Thus, there is a steep penalty associated with the generation of long Arnoldi sequences.

- Extremal eigenvalue information is not as forthcoming as in the symmetric case. There is no unsymmetric Kaniel-Paige-Saad convergence theory.

These realities suggest a framework in which we use the Arnoldi iteration idea with repeated, carefully chosen restarts and a controlled iteration maximum. We described such a framework in conjunction with the block Lanczos procedure in §10.3.7.

10.5.2 Arnoldi with Restarting

Consider running Arnoldi for m steps and then restarting the iteration with a new initial vector q_+ chosen from the span of the Arnoldi vectors $q_1, \ldots, q_m$. Because of the Krylov connection (10.5.1), q_+ has the form

$$q_+ = p(A)q_1$$

for some polynomial of degree $m - 1$. It is instructive to examine the action of $p(A)$ in terms of A's eigenvalues and eigenvectors. Assume for clarity that $A \in \mathbb{R}^{n \times n}$ is diagonalizable and that $Az_i = \lambda_i z_i$ for $i = 1{:}n$. If q_1 has the eigenvector expansion

$$q_1 = a_1 z_1 + \cdots + a_n z_n,$$

then q_+ is a scalar multiple of

$$z = a_1 p(\lambda_1)z_1 + \cdots + a_n p(\lambda_n)z_n.$$

Note that if $p(\lambda_\alpha) \gg p(\lambda_\beta)$, then relatively speaking, q_+ is much richer in the direction of z_α than in the direction of z_β. More generally, by carefully choosing $p(\lambda)$ we can design q_+ so that its component in certain eigenvector directions is emphasized while its component in other eigenvector directions is deemphasized. For example, if

$$p(\lambda) = c \cdot (\lambda - \mu_1)(\lambda - \mu_2) \cdots (\lambda - \mu_p) \tag{10.5.3}$$

where c is a constant, then q_+ is a unit vector in the direction of

$$z = c \cdot \sum_{k=1}^{n} a_k \left(\prod_{i=1}^{p} (\lambda_k - \mu_i) \right) z_k.$$

It follows that z_β is deemphasized relative to z_α if λ_β is near to one of the "filter values" $\mu_1, \ldots, \mu_p$ and λ_α is not. Thus, the act of picking a good restart vector q_+ from $\mathcal{K}(A, q_1, m)$ is the act of picking a filter polynomial that tunes out unwanted portions of the spectrum. Various heuristics for doing this have been developed based on computed Ritz vectors. See Saad (1980, 1984, 1992).

10.5.3 Implicit Restarting

We describe an Arnoldi restarting procedure due to Sorensen (1992) that implicitly determines the filter polynomial (10.5.3) using the QR iteration with shifts. (See §7.5.2.) Suppose $H_c \in \mathbb{R}^{m \times m}$ is upper Hessenberg, $\mu_1, \ldots, \mu_p$ are scalars, and the matrix H_+ is obtained via the shifted QR iteration:

$$H^{(0)} = H_c$$
$$\textbf{for } i = 0{:}p$$
$$\qquad H^{(i-1)} - \mu_i I = V_i R_i \qquad \text{(Givens QR)} \tag{10.5.4}$$
$$\qquad H^{(i)} = R_i V_i + \mu_i I$$
$$\textbf{end}$$
$$H_+ = H^{(p)}$$

Recall from §7.4.2 that each $H^{(i)}$ is upper Hessenberg. Moreover, if

$$V = V_1 \cdots V_p, \tag{10.5.5}$$

then

$$H_+ = V^T H_c V. \tag{10.5.6}$$

The following result shows that the filter polynomial (10.5.3) has a relationship to (10.5.4).

Theorem 10.5.1. *If $V = V_1 \cdots V_p$ and $R = R_p \cdots R_1$ are defined by (10.5.4), then*

$$VR = (H_c - \mu_1 I) \cdots (H_c - \mu_p I). \tag{10.5.7}$$

Proof. We use induction, noting that the theorem is obviously true if $p = 1$. If $\tilde{V} = V_1 \cdots V_{p-1}$ and $\tilde{R} = R_{p-1} \cdots R_1$, then

$$VR = \tilde{V}(V_p R_p)\tilde{R} = \tilde{V}(H^{(p-1)} - \mu_p I)\tilde{R} = \tilde{V}(\tilde{V}^T H_c \tilde{V} - \mu_p I)\tilde{R}$$

$$= (H_c - \mu_p I)\tilde{V}\tilde{R} = (H_c - \mu_p I)(H_c - \mu_1 I) \cdots (H_c - \mu_{p-1}I),$$

where we used the fact that $H^{(p-1)} = \tilde{V}^T H_c \tilde{V}$.   □

Note that the matrix R in (10.5.7) is upper triangular and so it follows that

$$V(:,1) = p(H_c)e_1$$

where $p(\lambda)$ is the filter polynomial (10.5.3) with $c = 1/R(1,1)$.

Now suppose that we have performed m steps of the Arnoldi iteration with starting vector q_1. The Arnoldi factorization (10.5.2) says that we have an upper Hessenberg matrix $H_c \in \mathbb{R}^{m \times m}$ and a matrix $Q_c \in \mathbb{R}^{n \times m}$ with orthonormal columns such that

$$AQ_c = Q_c H_c + r_c e_m^T. \tag{10.5.8}$$

Note that $Q_c(:,1) = q_1$ and $r_c \in \mathbb{R}^n$ has the property that $Q_c^T r_c = 0$. If we apply (10.5.4) to H_c, then by using (10.5.5) and (10.5.6) the preceding Arnoldi factorization transforms to

$$AQ_+ = Q_+ H_+ + r_c e_m^T V \tag{10.5.9}$$

where

$$Q_+ = Q_c V.$$

If q_+ is the first column of this matrix, then

$$q_+ = Q_+(:,1) = Q_c V(:,1) = c \cdot Q_c (H_c - \mu_1 I) \cdots (H_c - \mu_p I)e_1.$$

Equation (10.5.8) implies that

$$(A - \mu I)Q_c e_1 = Q_c(H_c - \mu I)e_1$$

for any $\mu \in \mathbb{R}$ and so

$$q_+ = c(A - \mu_1 I) \cdots (A - \mu_p I)Q_c e_1 = p(A)q_1.$$

This suggests the following framework for repeated restarting:

Repeat:

With starting vector q_1, perform m steps of the Arnoldi iteration

obtaining $Q_c \in \mathbb{R}^{n \times m}$ and $H_c \in \mathbb{R}^{m \times m}$.

Determine filter values $\mu_1, \ldots, \mu_p$.   (10.5.10)

Perform p steps of the shifted QR iteration (10.5.4) obtaining

the Hessenberg matrix H_+ and the orthogonal matrix V .

Replace q_1 with the first column of $Q_c V$.

However, we can do better than this. The orthogonal matrices $V_1, \ldots, V_p$ that arise in (10.5.4) are each upper Hessenberg. (This is easily deduced from the structure of the Givens rotations in Algorithm 5.2.5.) Thus, V has lower bandwidth p and so $V(m, 1{:}m - p - 1) = 0$. It follows from (10.5.9) that if $j = m - p$, then

$$AQ_+(:, 1{:}j) = Q_+(:, 1{:}j)H_+(1{:}j, 1{:}j) + v_{mj}r_c e_j$$

is a j-step Arnoldi decomposition. In other words, we are all set to perform step $j + 1$ of the Arnoldi iteration with starting vector q_+. There is no need to launch the restart from step 1. This leads us to the following modification of (10.5.10):

With starting vector q_1, perform m steps of the Arnoldi iteration obtaining

$Q_c \in \mathbb{R}^{n \times m}$, $H_c \in \mathbb{R}^{m \times m}$, and $r_c \in \mathbb{R}^n$ so $AQ_c = Q_c H_c + r_c e_m^T$.

Repeat:

Determine filter values $\mu_1, \ldots, \mu_p$.

Perform p steps of the shifted QR iteration (10.5.4) applied to H_c

obtaining $H_+ \in \mathbb{R}^{m \times m}$ and $V = (v_{ij}) \in \mathbb{R}^{m \times m}$.

Replace Q_c with the first j columns of $Q_c V$.

Replace H_c with $H_+(1{:}j, 1{:}j)$. .

Replace r_c with $v_{mj}r_c$.

Starting with $AQ_c = Q_c H_c + r_c e_j^T$, perform steps $j + 1, \ldots, j + p = m$ of

the Arnoldi iteration obtaining $AQ_m = Q_m H_m + r_m e_m^T$.

Set $Q_c = Q_m$, $H_c = H_m$, and $r_c = r_m$.

In light of our remarks in §10.5.2, the filter values $\mu_1, \ldots, \mu_p$ should be chosen in the vicinity of A's "unwanted" eigenvalues. In this regard it is possible to formulate useful heuristics that are based on the eigenvalues of the m-by-m Hessenberg matrix H_+. For example, suppose the goal is to find the three smallest eigenvalues of A in absolute value. If $p = m - 3$ and $\lambda(H_+) = \{\tilde{\lambda}_1, \ldots, \tilde{\lambda}_m\}$ with $|\tilde{\lambda}_1| \geq \cdots \geq |\tilde{\lambda}_m|$, then it is reasonable to set $\mu_i = \tilde{\lambda}_i$ for $i = 1{:}p$.

The Arnoldi iteration with implicit restarts has many attractive attributes. For implementation details and further analysis, see Lehoucq and Sorensen (1996), Morgan (1996), and the ARPACK manual by Lehoucq, Sorensen, and Yang (1998).

10.5.4   The Krylov-Schur Algorithm

An alternative restart procedure due to Stewart (2001) relies upon a carefully ordered
Schur decomposition of the Hessenberg matrix H_m that is produced after m steps of
the Arnoldi iteration. Suppose we have computed

$$AQ_m = Q_m H_m + r_m e_m^T$$

and that $m = j+p$, where j is the number of A's eigenvalues that we wish to compute.
Let

$$U^T H_m U = \begin{bmatrix} T_{11} & T_{12} \\ 0 & T_{22} \end{bmatrix}$$

be the Schur decomposition of A and assume that the eigenvalues have been ordered
so that the eigenvalues of $T_{11} \in \mathbb{R}^{j \times j}$ are of interest and the eigenvalues of $T_{22} \in \mathbb{R}^{p \times p}$
are not. (For clarity we ignore the possibility of complex eigenvalues.) The Arnoldi
decomposition above transforms to

$$AQ_+ = Q_+ T + r_c e_m^T U$$

where $Q_+ = Q_m U$. It follows that

$$AQ_+(:,1{:}j) = Q_+(:,1{:}j) T_{11} + r_m u^T$$

where $u^T = U(m, 1{:}j)$. It is possible to determine an orthogonal $Z \in \mathbb{R}^{j \times j}$ so that
$Z^T T_{11} Z$ is upper Hessenberg and $Z^T u = \tau e_j$. (See P10.5.2.) It follows that

$$A(Q_+ Z) = (Q_+ Z)(Z^T T_{11} Z) + r_c (Z^T u)^T$$

is a j-step Arnoldi factorization. We then set Q_j, H_j and r_j to be $Q_+ Z$, $Z^T T_{11} Z$, and
τr_m respectively and perform Arnoldi steps $j+1$ through $j+p = m$. For more detailed
discussion, see Stewart (MAE, Chap. 5) and Watkins (FMC, Chap. 9).

10.5.5   Unsymmetric Lanczos Tridiagonalization

Another way to extend the symmetric Lanczos process is to reduce A to tridiagonal form
using a general similarity transformation. Suppose $A \in \mathbb{R}^{n \times n}$ and that a nonsingular
matrix Q exists such that

$$Q^{-1} A Q = T = \begin{bmatrix} \alpha_1 & \gamma_1 & & \cdots & & 0 \\ \beta_1 & \alpha_2 & \ddots & & & \vdots \\ & \ddots & \ddots & \ddots & & \\ \vdots & & \ddots & \ddots & \gamma_{n-1} \\ 0 & \cdots & & & \beta_{n-1} & \alpha_n \end{bmatrix}.$$

With the column partitionings

$$Q = [\, q_1 \mid \cdots \mid q_n \,],$$
$$Q^{-T} = \tilde{Q} = [\, \tilde{q}_1 \mid \cdots \mid \tilde{q}_n \,],$$

we find upon comparing columns in $AQ = QT$ and $A^T \tilde{Q} = \tilde{Q} T^T$ that

$$Aq_k = \gamma_{k-1} q_{k-1} + \alpha_k q_k + \beta_k q_{k+1}, \qquad \gamma_0 q_0 \equiv 0,$$
$$A^T \tilde{q}_k = \beta_{k-1} \tilde{q}_{k-1} + \alpha_k \tilde{q}_k + \gamma_k \tilde{q}_{k+1}, \qquad \beta_0 \tilde{q}_0 \equiv 0,$$

for $k = 1{:}n - 1$. These equations together with the *biorthogonality condition*

$$\tilde{Q}^T Q = I_n$$

imply

$$\alpha_k = \tilde{q}_k^T A q_k$$

and

$$\beta_k q_{k+1} \equiv r_k = (A - \alpha_k I) q_k - \gamma_{k-1} q_{k-1},$$
$$\gamma_k \tilde{q}_{k+1} \equiv \tilde{r}_k = (A - \alpha_k I)^T \tilde{q}_k - \beta_{k-1} \tilde{q}_{k-1}.$$

There is some flexibility in choosing the scale factors β_k and γ_k. Note that

$$1 = \tilde{q}_{k+1}^T q_{k+1} = (\tilde{r}_k / \gamma_k)^T (r_k / \beta_k).$$

It follows that once β_k is specified, then γ_k is given by

$$\gamma_k = \tilde{r}_k^T r_k / \beta_k.$$

With the "canonical" choice $\beta_k = \| r_k \|_2$ we obtain

> $q_1, \tilde{q}_1$ given unit 2-norm vectors with $\tilde{q}_1^T q_1 \neq 0$
> $k = 0$, $q_0 = 0$, $r_0 = q_1$, $\tilde{q}_0 = 0$, $s_0 = \tilde{q}_1$
> **while** $(r_k \neq 0)$ and $(\tilde{r}_k \neq 0)$ and $(\tilde{r}_k^T r_k \neq 0)$
> $\qquad \beta_k = \| r_k \|_2$
> $\qquad \gamma_k = \tilde{r}_k^T r_k / \beta_k$
> $\qquad q_{k+1} = r_k / \beta_k$
> $\qquad \tilde{q}_{k+1} = \tilde{r}_k / \gamma_k$
> $\qquad k = k + 1$ (10.5.11)
> $\qquad \alpha_k = \tilde{q}_k^T A q_k$
> $\qquad r_k = (A - \alpha_k I) q_k - \gamma_{k-1} q_{k-1}$
> $\qquad \tilde{r}_k = (A - \alpha_k I)^T \tilde{q}_k - \beta_{k-1} \tilde{q}_{k-1}$
> **end**

If

$$T_k = \begin{bmatrix} \alpha_1 & \gamma_1 & & \cdots & & 0 \\ \beta_1 & \alpha_2 & \ddots & & & \vdots \\ & \ddots & \ddots & \ddots & \\ \vdots & & \ddots & \ddots & \gamma_{k-1} \\ 0 & \cdots & & \beta_{k-1} & \alpha_k \end{bmatrix},$$

then the situation at the bottom of the loop is summarized by the equations

$$A\,[\,q_1\,|\cdots|\,q_k\,] \;=\; [\,q_1\,|\cdots|\,q_k\,]\,T_k \;+\; r_k e_k^T, \tag{10.5.12}$$

$$A^T\,[\,\tilde{q}_1\,|\cdots|\,\tilde{q}_k\,] \;=\; [\,\tilde{q}_1\,|\cdots|\,\tilde{q}_k\,]\,T_k^T + \tilde{r}_k e_k^T. \tag{10.5.13}$$

If $r_k = 0$, then the iteration terminates and $\mathsf{span}\{q_1,\ldots,q_k\}$ is an invariant subspace for A. If $\tilde{r}_k = 0$, then the iteration also terminates and $\mathsf{span}\{\tilde{q}_1,\ldots,\tilde{q}_k\}$ is an invariant subspace for A^T. However, if neither of these conditions is true and $\tilde{r}_k^T r_k = 0$, then the tridiagonalization process ends without any invariant subspace information. This is called *serious breakdown*. See Wilkinson (AEP, p. 389) for an early discussion of the matter.

10.5.6   The Look-Ahead Idea

It is interesting to examine the serious breakdown issue in the block version of (10.5.11). For clarity assume that $A \in \mathbb{R}^{n \times n}$ with $n = rp$. Consider the factorization in which we want $\tilde{Q}^T Q = I_n$:

$$\tilde{Q}^T A Q = \begin{bmatrix} M_1 & C_1^T & \cdots & & 0 \\ B_1 & M_2 & \ddots & & \vdots \\ & \ddots & \ddots & \ddots & \\ \vdots & & \ddots & \ddots & C_{r-1}^T \\ 0 & \cdots & & B_{r-1} & M_r \end{bmatrix} \tag{10.5.14}$$

where all the blocks are p-by-p. Let $Q = [\,Q_1\,|\cdots|\,Q_r\,]$ and $\tilde{Q} = [\,\tilde{Q}_1\,|\cdots|\,\tilde{Q}_r\,]$ be conformable partitionings of Q and $\tilde{Q}$. Comparing block columns in the equations $AQ = QT$ and $A^T\tilde{Q} = \tilde{Q}T^T$, we obtain

$$Q_{k+1}B_k \;=\; AQ_k - Q_k M_k - Q_{k-1}C_{k-1}^T \;\;\equiv\; R_k,$$
$$\tilde{Q}_{k+1}C_k \;=\; A^T\tilde{Q}_k - \tilde{Q}_k M_k^T - \tilde{Q}_{k-1}B_{k-1}^T \;\equiv\; S_k.$$

Note that

$$M_k = \tilde{Q}_k^T A Q_k.$$

If $S_k^T R_k = C_k^T \tilde{Q}_{k+1}^T Q_{k+1} B_k \in \mathbb{R}^{p \times p}$ is nonsingular and we compute $B_k, C_k \in \mathbb{R}^{p \times p}$ so that

$$C_k^T B_k = S_k^T R_k,$$

then

$$Q_{k+1} = R_k B_k^{-1}, \tag{10.5.15}$$

$$\tilde{Q}_{k+1} = S_k C_k^{-1} \tag{10.5.16}$$

satisfy $\tilde{Q}_{k+1}^T Q_{k+1} = I_p$. Serious breakdown in this setting is associated with having a singular $S_k^T R_k$.

   One way of solving the serious breakdown problem in (10.5.11) is to go after a factorization of the form (10.5.14) in which the block sizes are dynamically determined. Roughly speaking, in this approach matrices Q_{k+1} and $\tilde{Q}_{k+1}$ are built up column by column with special recursions that culminate in the production of a nonsingular $\tilde{Q}_{k+1}^T Q_{k+1}$. The computations are arranged so that the biorthogonality conditions $\tilde{Q}_i^T Q_{k+1} = 0$ and $Q_i^T \tilde{Q}_{k+1} = 0$ hold for $i = 1{:}k$.

A method of this form belongs to the family of *look-ahead Lanczos* methods. The length of a look-ahead step is the width of the Q_{k+1} and $\tilde{Q}_{k+1}$ that it produces. If that width is one, a conventional block Lanczos step may be taken. Length-2 look-ahead steps are discussed in Parlett, Taylor, and Liu (1985). The notion of *incurable breakdown* is also presented by these authors. Freund, Gutknecht, and Nachtigal (1993) cover the general case along with a host of implementation details. Floating point considerations require the handling of "near" serious breakdown. In practice, each M_k that is 2-by-2 or larger corresponds to an instance of near serious breakdown.

Problems

P10.5.1 Recalling how Theorem 10.1.1 establishes the orthogonality of the Lanczos vectors in Algorithm 10.1.1, state and prove an analogous theorem that does the same thing for the Arnoldi vectors in Algorithm 10.5.1.

P10.5.2 Show that if $C \in \mathbb{R}^{j \times j}$ and $u \in \mathbb{R}^j$, then there exists an orthogonal $Z \in \mathbb{R}^{n \times n}$ so that $Z^T A Z = H$ is upper Hessenberg and the last column of Z is a multiple of u. Hint: Compute a Householder matrix P so that Pu is a multiple of e_j. Then reduce $C = P^T C P$ to upper Hessenberg form by producing a sequence of Householder updates $C = P_i^T C P$ where $C(n-i+1, 1{:}n-i-1)$ is zeroed, $i = 1{:}n-2$.

P10.5.3 Give an example of a starting vector for which the unsymmetric Lanczos iteration (10.5.11) breaks down without rendering any invariant subspace information. Use

$$A = \begin{bmatrix} 1 & 6 & 2 \\ 3 & 0 & 2 \\ 1 & 3 & 5 \end{bmatrix}.$$

P10.5.4 Suppose $H \in \mathbb{R}^{n \times n}$ is upper Hessenberg. Discuss the computation of a unit upper triangular matrix U such that $HU = UT$ where T is tridiagonal.

P10.5.5 Show that the QR algorithm for eigenvalues does not preserve tridiagonal structure in the unsymmetric case.

Notes and References for §10.5

For both analysis and implementation insight, Saad (NMLE) offers the most comprehensive treatment of unsymmetric Krylov methods. Stewart (MAE) and Watkins (MEP) devote entire chapters to the subject and are highly recommended as is the following review article:

D.C. Sorensen (2002). "Numerical Methods for Large Eigenvalue Problems," *Acta Numerica 11*, 519–584.

The original Arnoldi idea first appeared in:

W.E. Arnoldi (1951). "The Principle of Minimized Iterations in the Solution of the Matrix Eigenvalue Problem," *Quarterly of Applied Mathematics 9*, 17–29.

Saad set the stage for the development of practical implementations, see:

Y. Saad (1980). "Variations of Arnoldi's Method for Computing Eigenelements of Large Unsymmetric Matrices.," *Lin. Alg. Applic. 34*, 269–295.

Y. Saad (1984). "Chebyshev Acceleration Techniques for Solving Nonsymmetric Eigenvalue Problems," *Math. Comput. 42*, 567–588.

Y. Saad (1989). "Krylov Subspace Methods on Supercomputers," *SIAM J. Sci. Stat. Comput., 10*, 1200–1232.

References for implicit restarting in the Arnoldi context include:

D.C. Sorensen (1992). "Implicit Application of Polynomial Filters in a k-Step Arnoldi Method," *SIAM J. Matrix Anal. Applic. 13*, 357–385.

R.B. Lehoucq and D.C. Sorensen (1996). "Deflation Techniques for an Implicitly Restarted Iteration," *SIAM J. Matrix Anal. Applic. 17*, 789–821.

R.B. Morgan (1996). "On Restarting the Arnoldi Method for Large Nonsymmetric Eigenvalue Problems," *Math Comput. 65*, 1213–1230.

K. Meerbergen and A. Spence (1997). "Implicitly Restarted Arnoldi with Purification for the Shift-Invert Transformation," *Math. Comput. 218*, 667–689.

R.B. Lehoucq, D. C. Sorensen, and C. Yang (1998). *ARPACK Users' Guide: Solution of Large-Scale Eigenvalue Problems with Implicitly Restarted Arnoldi Methods*, SIAM Publications, Philadelphia, PA.

A. Stathopoulos, Y. Saad, and K. Wu (1998). "Dynamic Thick Restarting of the Davidson and the Implicitly Restarted Arnoldi Methods," *SIAM J. Sci. Comput. 19*, 227–245.

R.B. Lehoucq (2001). "Implicitly Restarted Arnoldi Methods and Subspace Iteration," *SIAM J. Matrix Anal. Applic. 23*, 551–562.

The Krylov-Schur approach to Arnoldi restarting is proposed in:

G.W. Stewart (2001). "A Krylov-Schur Algorithm for Large Eigenproblems," *SIAM J. Matrix Anal. Applic., 23*, 601–614.

The rational Arnoldi process involves the shift-and-invert idea. In this framework Arnoldi is applied to the matrix $(A - \mu I)^{-1}$, see:

A. Ruhe (1984). "Rational Krylov Algorithms for Eigenvalue Computation," *Lin. Alg. Applic. 58*, 391–405.

A. Ruhe (1994). "Rational Krylov Algorithms for Nonsymmetric Eigenvalue Problems II. Matrix Pairs," *Lin. Alg. Applic. 197*, 283–295.

A. Ruhe (1994). "The Rational Krylov Algorithm for Nonsymmetric Eigenvalue Problems III: Complex Shifts for Real Matrices," *BIT 34*, 165–176.

Matrix function problems that involve large sparse matrices can be addressed using Krylov/Lanczos ideas, see:

Y. Saad (1992). "Analysis of Some Krylov Subspace Approximations to the Matrix Exponential," *SIAM J. Numer. Anal. 29*, 209–228.

M. Hochbruck and C. Lubich (1997). "On Krylov Subspace Approximations to the Matrix Exponential Operator," *SIAM J. Numer. Anal. 34*, 1911–1925.

V. Druskin, A. Greenbaum and L. Knizhnerman (1998). "Using Nonorthogonal Lanczos Vectors in the Computation of Matrix Functions," *SIAM J. Sci. Comput. 19*, 38–54.

N. Del Buono, L. Lopez, and R. Peluso (2005). "Computation of the Exponential of Large Sparse Skew–Symmetric Matrices," *SIAM J. Sci. Comp. 27*, 278–293.

M. Eiermann and O.G. Ernst (2006). "A Restarted Krylov Subspace Method for the Evaluation of Matrix Functions," *SIAM J. Numer. Anal. 44*, 2481–2504.

J. van den Eshof and M. Hochbruck (2006). "Preconditioning Lanczos Approximations to the Matrix Exponential," *SIAM J. Sci. Comput. 27*, 1438–1457.

Other Arnoldi-related papers include:

T. Huckle (1994). "The Arnoldi Method for Normal Matrices," *SIAM J. Matrix Anal. Applic. 15*, 479–489.

K.C. Toh and L.N. Trefethen (1996). "Calculation of Pseudospectra by the Arnoldi Iteration," *SIAM J. Sci. Comput. 17*, 1–15.

T.G. Wright and L.N. Trefethen (2001). "Large–Scale Computation of Pseudospectra Using ARPACK and Eigs," *SIAM J. Sci. Comput. 23*, 591–605.

V. Hernandez, J.E. Roman, and A. Tomas (2007). "Parallel Arnoldi Eigensolvers with Enhanced Scalability via Global Communications Rearrangement," *Parallel Computing 33*, 521–540.

The unsymmetric Lanczos process and related look ahead ideas are nicely presented in:

B.N. Parlett, D. Taylor, and Z. Liu (1985). "A Look-Ahead Lanczos Algorithm for Unsymmetric Matrices," *Math. Comput. 44*, 105–124.

R.W. Freund, M. Gutknecht, and N. Nachtigal (1993). "An Implementation of the Look-Ahead Lanczos Algorithm for Non-Hermitian Matrices," *SIAM J. Sci. Stat. Comput. 14*, 137–158.

See also:

Y. Saad (1982). "The Lanczos Biorthogonalization Algorithm and Other Oblique Projection Methods for Solving Large Unsymmetric Eigenproblems," *SIAM J. Numer. Anal. 19*, 485–506.

D.L. Boley, S. Elhay, G.H. Golub and M.H. Gutknecht (1991) "Nonsymmetric Lanczos and Finding Orthogonal Polynomials Associated with Indefinite Weights," *Numer. Algorithms 1*, 21–43.

G.A. Geist (1991). "Reduction of a General Matrix to Tridiagonal Form," *SIAM J. Matrix Anal. Applic. 12*, 362–373.

C. Brezinski, M. Zaglia, and H. Sadok (1991). "Avoiding Breakdown and Near Breakdown in Lanczos Type Algorithms," *Numer. Algorithms 1*, 261–284.

S.K. Kim and A.T. Chronopoulos (1991). "A Class of Lanczos-Like Algorithms Implemented on Parallel Computers," *Parallel Comput. 17*, 763–778.

B.N. Parlett (1992). "Reduction to Tridiagonal Form and Minimal Realizations," *SIAM J. Matrix Anal. Applic. 13*, 567–593.

M. Gutknecht (1992). "A Completed Theory of the Unsymmetric Lanczos Process and Related Algorithms, Part I," *SIAM J. Matrix Anal. Applic. 13*, 594–639.

M. Gutknecht (1994). "A Completed Theory of the Unsymmetric Lanczos Process and Related Algorithms, Part II," *SIAM J. Matrix Anal. Applic. 15*, 15–58.

Z. Bai (1994). "Error Analysis of the Lanczos Algorithm for Nonsymmetric Eigenvalue Problem," *Math. Comput. 62*, 209–226.

T. Huckle (1995). "Low-Rank Modification of the Unsymmetric Lanczos Algorithm," *Math. Comput. 64*, 1577–1588.

Z. Jia (1995). "The Convergence of Generalized Lanczos Methods for Large Unsymmetric Eigenproblems," *SIAM J. Matrix Anal. Applic. 16*, 543–562.

M.T. Chu, R.E. Funderlic, and G.H. Golub (1995). "A Rank-One Reduction Formula and Its Applications to Matrix Factorizations," *SIAM Review 37*, 512–530.

Computing eigenvalues of unsymmetric tridiagonal matrices is discussed in:

D.A. Bini, L. Gemignani, and F. Tisseur (2005). "The Ehrlich-Aberth Method for the Nonsymmetric Tridiagonal Eigenvalue Problem," *SIAM J. Matrix Anal. Applic. 27*, 153–175.

10.6 Jacobi-Davidson and Related Methods

We close the chapter with a brief discussion of the Jacobi-Davidson method, a solution framework that involves a mix of several important ideas. The starting point is a reformulation of the eigenvalue problem as a nonlinear systems problem, a maneuver that enables us to apply Newton-like methods. This leads in a natural way to a method of Jacobi that can be used to compute eigenvalue-eigenvector pairs of symmetric matrices that have a strong diagonal dominance. Eigenproblems of this variety arise in quantum chemistry and it is in that venue where Davidson (1975) developed a very successful generalization of the Jacobi procedure. It builds a (non-Krylov) nested sequence of subspaces and incorporates Ritz approximation. By restricting the Davidson corrections to the orthogonal complement of the current subspace, we arrive at the Jacobi-Davidson method developed by Sleijpen and van der Vorst (1996). Their technique does not require symmetry or diagonal dominance. Thus, in terms of abstraction, exposition in this section starts from the general, descends to the specific, and then climbs back out to the general. All along the way we are driven by practical, algorithmic concerns. Our presentation draws upon the insightful treatments of the Jacobi-Davidson method in Sorensen (2002) and Stewart (MAE, pp. 404–420).

We mention that full appreciation of the Jacobi-Davidson method and its versatility requires an understanding of the next chapter. This is because a critical step in the method requires the approximate solution of a large sparse linear system and preconditioned iterative solvers are typically brought into play. See §11.5.

10.6.1   The Approximate Newton Framework

Consider the n-by-n eigenvalue problem $Ax = \lambda x$ and how we might improve an approximate eigenpair $\{x_c, \lambda_c\}$. Note that if

$$A(x_c + \delta x_c) = (\lambda_c + \delta\lambda_c)(x_c + \delta x_c),$$

then

$$(A - \lambda_c I)\,\delta x_c - \delta\lambda_c x_c = -r_c + \delta\lambda_c \cdot \delta x_c, \tag{10.6.1}$$

where

$$r_c = Ax_c - \lambda_c x_c$$

is the current residual. By ignoring the second-order term $\delta\lambda_c \cdot \delta x_c$ we arrive at the following specification for the corrections δx_c and $\delta\lambda_c$:

$$(A - \lambda_c I)\,\delta x_c - \delta\lambda_c x_c = -r_c. \tag{10.6.2}$$

This is an underdetermined system of nonlinear equations that has a very uninteresting solution obtained by setting $\delta x_c = -x_c$ and $\delta\lambda_c = 0$. To keep away from this situation we add a constraint so that if

$$\begin{bmatrix} x_+ \\ \lambda_+ \end{bmatrix} = \begin{bmatrix} x_c \\ \lambda_c \end{bmatrix} + \begin{bmatrix} \delta x_c \\ \delta\lambda_c \end{bmatrix}, \tag{10.6.3}$$

then the new eigenvector approximation x_+ is nonzero. One way to do this is to require

$$w^T x_+ = 1,$$

where $w \in \mathbb{R}^n$ is an appropriately chosen nonzero vector. Possibilities include $w = x$, which forces x_+ to have unit 2-norm, and $w = e_1$, which forces its first component to be one. Regardless, if x_c is also normalized with respect to w, then

$$w^T \delta x_c = w^T(x_+ - x_c) = 0. \tag{10.6.4}$$

By assembling (10.6.2) and (10.6.4) into a single matrix-vector equation we obtain

$$\begin{bmatrix} A - \lambda_c I & -x_c \\ w^T & 0 \end{bmatrix} \begin{bmatrix} \delta x_c \\ \delta\lambda_c \end{bmatrix} = -\begin{bmatrix} r_c \\ 0 \end{bmatrix}. \tag{10.6.5}$$

This is precisely the Jacobian system that arises if Newton's method is used to find a zero of the function

$$F\left(\begin{bmatrix} x \\ \lambda \end{bmatrix} \right) = \begin{bmatrix} Ax - \lambda x \\ w^T x - 1 \end{bmatrix}.$$

Its solution is easy to specify:

$$\delta\lambda_c = \frac{w^T(A - \lambda_c I)^{-1} r_c}{w^T(A - \lambda_c I)^{-1} x_c}, \tag{10.6.6}$$

$$\delta x_c = -(A - \lambda_c I)^{-1}\left(r_c - \delta\lambda_c x_c \right). \tag{10.6.7}$$

Unfortunately, the required linear equation solving is problematic if A is large and sparse and this prompts us to consider the approximate Newton framework.

The idea behind approximate Newton methods is to replace the Jacobian system with a nearby, look-alike system that is easier to solve. One way to do this in our problem is to approximate A with a matrix M with the proviso that systems of the form $(M - \lambda_c I)z = r$ are "easy" to solve. If $N = M - A$, then (10.6.5) transforms to

$$
\begin{bmatrix} M - \lambda_c I & -x_c \\ w^T & 0 \end{bmatrix} \begin{bmatrix} \delta x_c \\ \delta \lambda_c \end{bmatrix} = - \begin{bmatrix} r_c - N \cdot \delta x_c \\ 0 \end{bmatrix}.
$$

Continuing with the approximate-Newton mentality, let us throw away the inconvenient $N \cdot \delta x_c$ term that is part of the right-hand side. This leaves us with the system

$$
\begin{bmatrix} M - \lambda_c I & -x_c \\ w^T & 0 \end{bmatrix} \begin{bmatrix} \delta x_c \\ \delta \lambda_c \end{bmatrix} = - \begin{bmatrix} r_c \\ 0 \end{bmatrix}, \tag{10.6.8}
$$

and the following compute-friendly recipes for the corrections:

$$
\delta \lambda_c = \frac{w^T (M - \lambda_c I)^{-1} r_c}{w^T (M - \lambda_c I)^{-1} x_c}, \tag{10.6.9}
$$

$$
\delta x_c = -(M - \lambda_c I)^{-1} \left(r_c - \delta \lambda_c x_c \right). \tag{10.6.10}
$$

Of course, by cutting corners in Newton's method we risk losing quadratic convergence. Thus, the design of an approximate Newton strategy must balance the efficiency of the approximate Jacobian solution procedure with a possibly degraded rate of convergence. For an excellent discussion of this tension in the context of the eigenvalue problem, see Stewart (MAE, pp. 396–404).

10.6.2 The Jacobi Orthogonal Component Correction Method

Now suppose

$$
A = \begin{bmatrix} \alpha & c^T \\ c & A_1 \end{bmatrix}, \qquad \alpha \in \mathbb{R},\ c \in \mathbb{R}^{n-1},\ A_1 \in \mathbb{R}^{(n-1) \times (n-1)} \tag{10.6.11}
$$

is symmetric and strongly diagonally dominant. Assume that α is the largest element on the diagonal in absolute value. Our ambition is to compute λ (close to α) and $z \in \mathbb{R}^{n-1}$ so that

$$
\begin{bmatrix} \alpha & c^T \\ c & A_1 \end{bmatrix} \begin{bmatrix} 1 \\ z \end{bmatrix} = \lambda \begin{bmatrix} 1 \\ z \end{bmatrix}. \tag{10.6.12}
$$

Because of the dominance assumption, there is no danger in assuming that the sought-after eigenvector is nicely normalized by setting its first component to 1. Partition δx_c, x_c, and x_+ as follows:

$$
\delta x_c = \begin{bmatrix} \delta \mu_c \\ \delta z_c \end{bmatrix}, \qquad x_c = \begin{bmatrix} 1 \\ z_c \end{bmatrix}, \qquad x_+ = \begin{bmatrix} 1 \\ z_+ \end{bmatrix}.
$$

By substituting (10.6.11) and $w = e_1$ into the Jacobian system (10.6.5), we get

$$
\left[\begin{array}{c|cc}
\alpha - \lambda_c & c^T & -1 \\
\hline
c & A_1 - \lambda_c I & -z_c \\
1 & 0 & 0
\end{array}\right]
\left[\begin{array}{c}
\delta \mu_c \\
\delta z_c \\
\delta \lambda_c
\end{array}\right]
= -
\left[\begin{array}{c}
\alpha + + c^T z_c - \lambda_c \\
\hline
(A_1 - \lambda_c I) z_c + c \\
0
\end{array}\right],
$$

i.e.,

$$
\left[\begin{array}{cc}
A_1 - \lambda_c I & -z_c \\
c^T & -1
\end{array}\right]
\left[\begin{array}{c}
\delta z_c \\
\delta \lambda_c
\end{array}\right]
= -
\left[\begin{array}{c}
(A_1 - \lambda_c I) z_c + c \\
\alpha + c^T z_c - \lambda_c
\end{array}\right].
\tag{10.6.13}
$$

It is easy to verify that this is the Jacobian system that arises if Newton's method is used to compute a zero of

$$
f\left(\left[\begin{array}{c} z \\ \lambda \end{array}\right]\right)
= \left[\begin{array}{cc} \alpha & c^T \\ c & A_1 \end{array}\right]
\left[\begin{array}{c} 1 \\ z \end{array}\right]
- \lambda \left[\begin{array}{c} 1 \\ z \end{array}\right].
$$

If $A_1 = M_1 - N_1$, then (10.6.13) can be rearranged as follows:

$$
(M_1 - \lambda_c I) z_+ = -c + N_1 z_c + \{\delta \lambda_c \cdot z_c + N_1 \cdot \delta z_c\},
$$
$$
\lambda_+ = \alpha + c^T z_+.
$$

The *Jacobi orthogonal component correction (JOCC) method* is defined by ignoring the terms enclosed by the curly brackets and taking M_1 to be the diagonal part of A_1:

$\lambda_1 = \alpha,\ z_1 = 0_{n-1},\ \rho_1 = \| c \|_2,\ k = 1$
while $\rho_k > \mathsf{tol}$
$\quad (M_1 - \lambda_k I) z_{k+1} = -c + N_1 z_k$
$\quad \lambda_{k+1} = \alpha + c^T z_{k+1}$ $\hspace{4cm}$ (10.6.14)
$\quad k = k + 1$
$\quad \rho_k = \| A_1 z_k - \lambda_k z_k + c \|_2$
end

The name of the method stems from the fact that the corrections to the approximate eigenvectors

$$
x_k = \left[\begin{array}{c} 1 \\ z_k \end{array}\right],
$$

are all orthogonal to e_1. Indeed, it is clear from (10.6.14) that each residual

$$
r_k = (A - \lambda_k I) x_k
$$

has a zero first component:

$$
r_k = \left[\begin{array}{cc} \alpha & c^T \\ c & A_1 \end{array}\right]
\left[\begin{array}{c} 1 \\ z_k \end{array}\right]
- \lambda_k \left[\begin{array}{c} 1 \\ z_k \end{array}\right]
= \left[\begin{array}{c} 0 \\ (A_1 - \lambda_k I) z_k + c \end{array}\right].
\tag{10.6.15}
$$

Hence, the termination criterion in (10.6.14) is based on the size of the residual.

Jacobi intended this method to be use in conjunction with his diagonalization procedure for the symmetric eigenvalue problem. As discussed in §8.5, after a sufficient number of sweeps the matrix A is very close to being diagonal. At that point, the JOCC iteration (10.6.14) can be invoked after a possible PAP^T update to maximize the (1,1) entry.

10.6.3 The Davidson Method

As with the JOCC iteration, Davidson's method is applicable to the symmetric diagonally dominant eigenvalue problem (10.6.12). However, it involves a more sophisticated placement of the residual vectors. To motivate the main idea, let M be the diagonal part of A and use (10.6.15) to rewrite the JOCC iteration as follows:

$x_1 = e_1,\ \lambda_1 = x_1^T A x_1,\ r_1 = A x_1 - \lambda_1 x_1,\ V_1 = [\,e_1\,],\ k = 1$

while $\|\,r_k\,\| > $ tol

> Solve the residual correction equation:
> $$(M - \lambda_k I)\delta v_k = -r_k.$$

> Compute an improved eigenpair $\{\lambda_{k+1}, x_{k+1}\}$ so $r_{k+1} \in \mathsf{ran}(V_1)^{\perp}$:
> $$\delta x_k = \delta v_k,\ x_{k+1} = x_k + \delta x_k,\ \lambda_{k+1} = \lambda_k + c^T \delta x_k$$

> $k = k + 1$
>
> $r_k = A x_k - \lambda_k x_k$

end

Davidson's method uses Ritz approximation to ensure that r_k is orthogonal to e_1 and $\delta v_1, \ldots, \delta v_{k-1}$. To acomplish this, the boxed fragment is replaced with the following:

> Expand the current subspace $\mathsf{ran}(V_k)$:
> $$s_{k+1} = (I - V_k V_k^T)\delta v_k$$
> $$v_{k+1} = s_{k+1}/\|\,s_{k+1}\,\|_2,\ V_{k+1} = [\,V_k\,|\,v_{k+1}\,]$$
> Compute an improved eigenpair $\{\lambda_{k+1}, x_{k+1}\}$ so $r_{k+1} \in \mathsf{ran}(V_{k+1})^{\perp}$:
> $$(V_{k+1}^T A V_{k+1})t_{k+1} = \theta_{k+1}t_{k+1} \quad \text{(a suitably chosen Ritz pair)}$$
> $$\lambda_{k+1} = \theta_{k+1},\ x_{k+1} = V_{k+1}t_{k+1}$$

(10.6.16)

There are a number of important issues associated with this method. To begin with, V_k is an n-by-k matrix with orthonormal columns. The transition from V_k to V_{k+1} can be effectively carried out by a modified Gram-Schmidt process. Of course, if k gets too big, then it may be necessary to restart the process using v_k as the initial vector.

Because $r_k = A x_k - \lambda_k x_k = A(V_k t_k) - \theta_k(V_k t_k)$, it follows that

$$V_k^T r_k = (V_k^T A V_k)t_k - \theta_k t_k = 0,$$

i.e., r_k is orthogonal to the range of V_k as required.

We mention that the Davidson algorithm can be generalized by allowing M to be a more involved approximation to A than just its diagonal part. See Crouzeix, Philippe, and Sadkane (1994) for details.

10.6.4 The Jacobi-Davidson Framework

Instead of forcing the correction δx_c to be orthogonal to e_1 as in the Davidson setting, the *Jacobi-Davidson method* insists that δx_c be orthogonal to the current eigenvector approximation x_c. The idea is to expand the current search space in a profitable, unexplored direction.

To see what is involved computationally and to connect with Newton's method, we consider the following modification of (10.6.5):

$$\left[\begin{array}{cc} A - \lambda_c I & -x_c \\ x_c^T & 0 \end{array} \right] \left[\begin{array}{c} \delta x_c \\ \delta \lambda_c \end{array} \right] = - \left[\begin{array}{c} r_c \\ 0 \end{array} \right]. \tag{10.6.17}$$

Note that this is the Jacobian system associated with the function

$$F\left(\left[\begin{array}{c} x \\ \lambda \end{array} \right] \right) = \left[\begin{array}{c} Ax - \lambda x \\ (x^T x - 1)/2 \end{array} \right]$$

given that $x_c^T x_c = 1$. If x_c is so normalized and $\lambda_c = x_c^T A x_c$, then from (10.6.17) we have

$$\begin{aligned}
(I - x_c x_c^T)(A - \lambda_c I)(I - x_c x_c^T)\delta x_c &= -(I - x_c x_c^T)(r_c - \delta \lambda_c x_c) \\
&= -(I - x_c x_c^T)r_c \\
&= -(I - x_c x_c^T)(A x_c - \lambda_c x_c) \\
&= -(I - x_c x_c^T)A x_c \\
&= -(A x_c - \lambda_c x_c) = -r_c.
\end{aligned}$$

Thus, the correction δx_c is obtained by solving the projected system

$$(I - x_c x_c^T)(A - \lambda_c I)(I - x_c x_c^T)\delta x_c = -r_c \tag{10.6.18}$$

subject to the constraint that $x_c^T \delta x_c = 0$.

In Jacobi-Davidson, approximate projected systems are used to expand the current subspace. Compared to the Davidson algorithm, everything remains the same in (10.6.16) except that instead of solving $(M - \lambda_c I)\delta v_k = -r_k$ to determine δv_k, we solve

$$(I - x_k x_k^T)(M - \lambda_k I)(I - x_k x_k^T)\delta v_k = -r_k, \tag{10.6.19}$$

subject to the constraint that $x_k^T \delta v_k = 0$. The resulting framework permits greater flexibility. The initial unit vector x_1 can be arbitrary and various Chapter 11 iterative solvers can be applied to (10.6.19). See Sleijpen and van der Vorst (1996) and Sorensen (2002) for details.

The Jacobi-Davidson framework can be used to solve both symmetric and non-symmetric eigenvalue problems and is important for the way it channels sparse $Ax = b$

technology to the sparse $Ax = \lambda x$ problem. It can be regarded as an approximate Newton iteration that is "steered" to the eigenpair of interest by Ritz calculations. Because an ever-expanding orthonormal basis is maintained, restarting has a key role to play as in the Arnoldi setting (§10.5).

10.6.5 The Trace-Min Algorithm

We briefly discuss the trace-min algorithm that can be used to compute the k smallest eigenvalues and associated eigenvectors for the n-by-n symmetric-definite problem $Ax = \lambda Bx$. It has similarities to the Jacobi-Davidson procedure. The starting point is to realize that if $V_{\text{opt}} \in \mathbb{R}^{n \times k}$ solves

$$\min_{V^T B V = I_k} \text{tr}(V^T A V),$$

then the required eigenvalues/eigenvectors are exposed by $V_{\text{opt}}^T A V_{\text{opt}} = \text{diag}(\mu_1, \ldots, \mu_k)$ and $A V_{\text{opt}}(:,j) = \mu_j B V_{\text{opt}}(:,j)$, for $j = 1{:}k$. The method produces a sequence of V-matrices, each of which satisfies $V^T B V = I_k$. The transition from V_c to V_+ requires the solution of a projected system

$$(I - Q_c Q_c^T) A (I - Q_c Q_c^T) Z_c = A V_c$$

where $Z_c \in \mathbb{R}^{n \times k}$ and $QR = BV_c$ is the thin QR factorization. This system, analogous to the central Jacobi-Davidson update system (10.6.19), can be solved using a suitably preconditioned conjugate gradient iteration. For details, see Sameh and Wisniewski (1982) and Sameh and Tong (2000).

Problems

P10.6.1 How would you solve (10.6.1) assuming that A is upper Hessenberg?

P10.6.2 Assume that

$$A = \begin{bmatrix} \alpha & b \\ b & D + E \end{bmatrix}$$

is an n-by-n symmetric matrix. Assume that D is the diagonal of $A(2{:}n, 2{:}n)$ and that the eigenvalue gap $\delta = \lambda_1(A) - \lambda_2(A)$ is positive. How small must b and E be in order to ensure that $(D + E) - \alpha I$ is diagonally dominant? Use Theorem 8.1.4.

Notes and References for §10.6

For deeper perspectives on the methods of this section, we recommend Stewart (MAE, 404–420) and:

D.C. Sorensen (2002). "Numerical Methods for Large Eigenvalue Problems," *Acta Numerica 11*, 519–584.

Davidson method papers include:

E.R. Davidson (1975). "The Iterative Calculation of a Few of the Lowest Eigenvalues and Corresponding Eigenvectors of Large Real Symmetric Matrices," *J. Comput. Phys. 17*, 87–94.
R.B. Morgan and D.S. Scott (1986). "Generalizations of Davidson's Method for Computing Eigenvalues of Sparse Symmetric Matrices," *SIAM J. Sci. Stat. Comput. 7*, 817–825.
J. Olsen, P. Jorgensen, and J. Simons (1990). "Passing the One-Billion Limit in Full-Configuration (FCI) Interactions," *Chem. Phys. Letters 169*, 463–472.
R.B. Morgan (1992). "Generalizations of Davidson's Method for Computing Eigenvalues of Large Nonsymmetric Matrices," *J. Comput. Phys. 101*, 287–291.

M. Sadkane (1993) "Block-Arnoldi and Davidson Methods for Unsymmetric Large Eigenvalue Problems," *Numer. Math. 64*, 195–211.

M. Crouzeix, B. Philippe, and M. Sadkane (1994). "The Davidson Method," *SIAM J. Sci. Comput. 15*, 62–76.

A. Strathopoulos, Y. Saad, and C.F. Fischer (1995). "Robust Preconditioning for Large, Sparse, Symmetric Eigenvalue Problems," *J. Comput. Appl. Math. 64*, 197–215.

The original Jacobi-Davidson idea appears in:

G.L.G. Sleijpen and H.A. van der Vorst (1996). "A Jacobi-Davidson Iteration Method for Linear Eigenvalue Problems," *SIAM J. Matrix Anal. Applic. 17*, 401–425.

For applications and extensions to other problems, see:

G.L.G. Sleijpen, A.G.L. Booten, D.R. Fokkema, and H.A. van der Vorst (1996). "Jacobi-Davidson Type Methods for Generalized Eigenproblems and Polynomial Eigenproblems," *BIT 36*, 595–633.

G.L.G. Sleijpen, H.A. van der Vorst, and E. Meijerink (1998). "Efficient Expansion of Subspaces in the Jacobi-Davidson Method for Standard and Generalized Eigenproblems," *ETNA 7*, 75–89.

D.R. Fokkema, G.L.G. Sleijpen, and H.A. van der Vorst (1998). "Jacobi-Davidson Style QR and QZ Algorithms for the Reduction of Matrix Pencils," *SIAM J. Sci. Computut. 20*, 94–125.

P. Arbenz and M.E. Hochstenbach (2004). "A Jacobi-Davidson Method for Solving Complex Symmetric Eigenvalue Problems," *SIAM J. Sci. Comput. 25*, 1655–1673.

The trace-min method is detailed in:

A. Sameh and J. Wisniewski (1982). "A Trace Minimization Algorithm for the Generalized Eigenproblem," *SIAM J. Numer. Anal. 19*, 1243–1259.

A. Sameh and Z. Tong (2000). "A Trace Minimization Algorithm for the Symmetric Generalized Eigenproblem," *J. Comput. Appl. Math. 123*, 155–175.

Chapter 11

Large Sparse Linear System Problems

11.1 Direct Methods

11.2 The Classical Iterations

11.3 The Conjugate Gradient Method

11.4 Other Krylov Methods

11.5 Preconditioning

11.6 The Multigrid Framework

This chapter is about solving linear systems and least squares problems when the matrix in question is so large and sparse that we have to rethink our powerful dense factorization strategies. The basic challenge is to live without the standard 2-dimensional array representation where there is a 1:1 correspondence between matrix entries and storage cells.

There is sometimes sufficient structure to actually compute an LU, Cholesky, or QR factorization by using a sparse matrix data structure and by carefully reordering equations and unknowns to control the fill-in of nonzero entries during the factorization process. Methods of this variety are called *direct methods* and they are the subject of §11.1. Our treatment is brief, touching only some of the high points of this well-developed area. A deeper presentation requires much more graph theory and implementation-based insight than we can provide in these few pages.

The rest of the chapter is concerned with the *iterative method* framework. These methods produce a sequence of vectors that typically converge to the solution at a reasonable rate. The matrix A "shows up" only in the context of matrix/vector multiplication. We introduce the strategy in §11.2 through discussion of the "classical" methods of Jacobi, Gauss-Seidel, successive over-relaxation, and Chebyshev. The discrete Poisson problem from §4.8.3 is used to reinforce the major ideas.

Krylov subspace methods are treated in the next two sections. In §11.3 we derive the method of conjugate gradients that is suitable for symmetric positive definite linear systems. The derivation involves the Lanczos process, the method of steepest descent, and the idea of optimizing over a nested sequence of subspaces. Related methods for symmetric indefinite systems, general systems, and least squares problems are covered in §11.4.

It is generally the case that Krylov subspace methods are successful only if there is an effective *preconditioner*. For a given $Ax = b$ problem this essentially requires the design of a matrix M that has two properties. It must capture key features of A and it must be relatively easy to solve systems of the form $Mz = r$. There are several major families of preconditioners and these are surveyed in §11.5 and §11.6, the latter being dedicated to the mesh-coarsening/multigrid framework.

Reading Path

An understanding of the basics about LU, Cholesky, and QR factorizations is essential. Eigenvalue theory and functions of matrices have a prominent role to play in the analysis of iterative $Ax = b$ solvers. The Krylov methods make use of the Lanczos and Arnoldi iterations that we developed in Chapter 10.

Within this chapter, there are the following dependencies:

$$\S11.2 \quad \rightarrow \quad \S11.3 \quad \rightarrow \quad \S11.4 \quad \rightarrow \quad \S11.5$$
$$\downarrow$$
$$\S11.6$$

§11.1 is independent of the others. The books by Axelsson (ISM), Greenbaum (IMSL), Saad (ISPLA), and van der Vorst (IMK) provide excellent background. The software "templates" volume LIN_TEMPLATES (1993) is very useful for its concise presentation of all the major iterative strategies and for the guidance it provides in choosing a suitable method.

11.1   Direct Methods

In this section we examine the direct method framework where the goal is to formulate solution procedures that revolve around careful implementation of the Cholesky, QR, and LU factorizations. Central themes, all of which are detailed more fully by Davis (2006), include the importance of ordering to control fill-in, connections to graph theory, and how to reason about performance in the sparse matrix setting.

It should be noted that the band matrix methods discussed in §4.3 and §4.5 are examples of sparse direct methods.

11.1.1   Representation

Data structures play an important role in sparse matrix computations. Typically, a real vector is used to house the nonzero entries of the matrix and one or two integer vectors are used to specify their "location." The *compressed-column* representation serves as a good illustration. Using a dot-on-grid notation to display sparsity patterns, suppose

$$A = \quad$$ 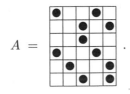 .

The compressed-column representation stores the nonzero entries column by column in a real vector. If A is the matrix, then we denote this vector by $A.val$, e.g.,

$$A.val = \begin{array}{|c|c||c|c|c|c||c|c||c|c|c|} \hline a_{11} & a_{41} & a_{52} & a_{23} & a_{33} & a_{63} & a_{14} & a_{44} & a_{25} & a_{55} & a_{65} \\ \hline \end{array}.$$

An integer vector $A.c$ is used to indicate where each column "begins" in $A.val$:

$$A.c = \begin{array}{|c|c|c|c|c|c|} \hline 1 & 3 & 4 & 7 & 9 & 12 \\ \hline \end{array}.$$

Thus, if $k = A.c(j):A.c(j+1)-1$, then $v = A.val(k)$ is the vector of nonzero components of $A(:,j)$. By convention, the last component of $A.c$ houses $\mathsf{nnz}(A) + 1$ where

$$\mathsf{nnz}(A) = \text{the number of nonzeros in } A.$$

The row indices for the nonzero components in $A(:,1), \ldots, A(:,n)$ are encoded in an integer vector $A.r$, e.g.,

$$A.r = \begin{array}{|c|c||c|c|c|c||c|c||c|c|c|} \hline 1 & 4 & 5 & 2 & 3 & 6 & 1 & 4 & 2 & 5 & 6 \\ \hline \end{array}.$$

In general, if $k = A.c(j):A.c(j+1) - 1$, then $A.val(k) = A(A.r(k),j)$.

Note that the amount of storage required for $A.r$ is comparable to the amount of storage required for the floating-point vector $A.val$. Index vectors represent one of the overheads that distinguish sparse from conventional dense matrix computations.

11.1.2 Operations and Allocations

Consider the gaxpy operation $y = y + Ax$ with A in compressed-column format. If $A \in \mathbb{R}^{m \times n}$ and the dense vectors $y \in \mathbb{R}^m$ and $x \in \mathbb{R}^n$ are conventionally stored, then

$$
\begin{aligned}
&\textbf{for } j = 1{:}n \\
&\quad k = A.c(j){:}A.c(j+1) - 1 \\
&\quad y(A.r(k)) = y(A.r(k)) + A.val(k) \cdot x(j) \\
&\textbf{end}
\end{aligned}
\tag{11.1.1}
$$

overwrites y with $y + Ax$. It is easy to show that $2 \cdot \mathsf{nnz}(A)$ flops are required. Regarding memory access, x is referenced sequentially, y is referenced randomly, and A is referenced through $A.r$ and $A.c$.

A second example highlights the issue of memory allocation. Consider the outer-product update $A = A + uv^T$ where $A \in \mathbb{R}^{m \times n}$, $u \in \mathbb{R}^m$, and $v \in \mathbb{R}^n$ are each stored in compressed-column format. In general, the updated A will have more nonzeros than the original A, e.g.,

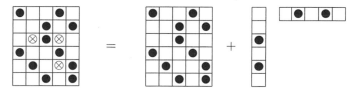

Thus, unlike dense matrix computations where we simply overwrite A with $A + uv^T$ without concern for additional storage, now we must increase the memory allocation

for A in order to house the result. Moreover, the expansion of the vectors $A.val$ and $A.r$ to accommodate the new nonzero entries is a nontrivial overhead. On the other hand, if we can predict the sparsity structure of $A + uv^T$ in advance and allocate space accordingly, then the update can be carried out more efficiently. This amounts to storing zeros in locations that are destined to become nonzero, e.g.,

$$A.val = \boxed{\begin{array}{cc|c|c|cc|c|c|c|c|c|ccc} a_{11} & a_{41} & 0 & a_{52} & a_{23} & a_{33} & a_{63} & a_{14} & 0 & a_{44} & 0 & a_{25} & a_{55} & a_{65} \end{array}},$$

$$A.c = \boxed{\begin{array}{cccccc} 1 & 3 & 5 & 8 & 12 & 15 \end{array}},$$

$$A.r = \boxed{\begin{array}{cc|cc|ccc|c|c|c|cc} 1 & 4 & 3 & 5 & 2 & 3 & 6 & 1 & 3 & 4 & 5 & 2 & 5 & 6 \end{array}}.$$

With this assumption, the outer product update can proceed as follows:

> **for** $\beta = 1:\mathsf{nnz}(v)$
>> $j = v.r(\beta)$
>> $\alpha = 1$
>> **for** $\ell = A.c(j) : A.c(j+1) - 1$
>>> **if** $\alpha \leq \mathsf{nnz}(u)$ `&&` $A.r(\ell) = u.r(\alpha)$ (11.1.2)
>>>> $A.val(\ell) = A.val(\ell) + u.val(\alpha) \cdot v.val(\beta)$
>>>> $\alpha = \alpha + 1$
>>> **end**
>> **end**
> **end**

Note that $A.val(\ell)$ houses a_{ij} and is updated only if $u_i v_j$ is nonzero. The index α is used to reference the nonzero entries of u and is incremented after every access.

The overall success of a sparse matrix procedure typically depends strongly upon how efficiently it predicts and manages the fill-in phenomenon.

11.1.3   Ordering, Fill-In, and the Cholesky Factorization

The first step in the outer-product Cholesky process involves computation of the factorization

$$A = \begin{bmatrix} \alpha & v^T \\ v & B \end{bmatrix} = \begin{bmatrix} \sqrt{\alpha} & 0 \\ v/\sqrt{\alpha} & I \end{bmatrix} \begin{bmatrix} 1 & 0 \\ 0 & A^{(1)} \end{bmatrix} \begin{bmatrix} \sqrt{\alpha} & v^T/\sqrt{\alpha} \\ 0 & I \end{bmatrix} \qquad (11.1.3)$$

where

$$A^{(1)} = B - \frac{vv^T}{\alpha}. \qquad (11.1.4)$$

Recall from §4.2 that this reduction is repeated on the matrix $A^{(1)}$.

Now suppose A is a sparse matrix. From the standpoint of both arithmetic and memory requirements, we have a vested interest in the sparsity of $A^{(1)}$. Since B is sparse, everything hinges on the sparsity of the vector v. Here are two examples that dramatize what is at stake:

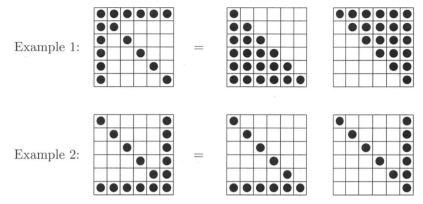

Example 1:   =

Example 2:   =

In Example 1, the vector v associated with the first step is dense and that results in a full $A^{(1)}$. All sparsity is lost and the remaining steps essentially carry out a dense Cholesky factorization. Example 2 tells a happier story. The first v-vector is sparse and the update matrix $A^{(1)}$ has the same "arrow" structure as A. Note that Example 2 can be obtained from Example 1 by a reordering of the form PAP^T where $P = I_n(:, n: -1:1))$. This motivates the *Sparse Cholesky challenge*:

The Sparse Cholesky Challenge

Given a symmetric positive definite matrix $A \in \mathbb{R}^{n \times n}$, efficiently determine a permutation p of $1{:}n$ so that if $P = I_n(:, p)$, then the Cholesky factor in $A(p, p) = PAP^T = GG^T$ is close to being optimally sparse.

Choosing P to actually minimize $\mathsf{nnz}(G)$ is a formidable combinatorial problem and is therefore not a viable option. Fortunately, there are several practical procedures based on heuristics that can be used to determine a good reordering permutation P. These include (1) the Cuthill-McKee ordering, (2) the minimum degree ordering, and (3) the nested dissection ordering. However, before we discuss these strategies, we need to present a few concepts from graph theory.

11.1.4   Graphs and Sparsity

Here is a sparse symmetric matrix A and its *adjacency graph* $\mathcal{G}_A$:

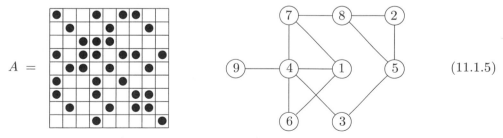

$$A = \qquad\qquad\qquad\qquad\qquad\qquad\qquad\qquad (11.1.5)$$

In an adjacency graph for a symmetric matrix, there is a node for each row, numbered by the row number, and there is an edge between node i and node j if the off-diagonal entry a_{ij} is nonzero. In general, a *graph* $\mathcal{G}(V, E)$ is a set of labeled *nodes* V together with a set of *edges* E, e.g.,

$$V = \{1, 2, 3, 4, 5, 6, 7, 8, 9\},$$

$$E = \{(1, 4), (1, 6), (1, 7), (2, 5), (2, 8), (3, 4), (3, 5), (4, 6), (4, 7), (4, 9), (5, 8), (7, 8)\}.$$

Adjacency graphs for symmetric matrices are *undirected*. This means there is no difference between edge (i, j) and edge (j, i). If P is a permutation matrix, then, except for vertex labeling, the adjacency graphs for A and PAP^T "look the same."

Node i and node j are *neighbors* if there is an edge between them. The *adjacency set* for a node is the set of its neighbors and the cardinality of that set is the *degree* of the node. For the above example we have

Node	1	2	3	4	5	6	7	8	9
Degree	3	2	2	5	3	2	3	3	1

Graph theory is a very powerful language that facilitates reasoning about sparse matrix factorizations. Of particular importance is the use of graphs to predict structure, something that is critical to the design of efficient implementations. For a much deeper appreciation of these issues than what we offer below, see George and Liu (1981), Duff, Erisman, and Reid (1986), and Davis (2006).

11.1.5   The Cuthill-McKee Ordering

Because bandedness is such a tractable form of sparsity, it is natural to approach the Sparse Cholesky challenge by making $\tilde{A} = PAP^T$ as "banded as possible" subject to cost constraints. However, this is too restrictive as Example 2 in §11.1.3 shows. Profile minimization is a better way to induce good sparsity in G. The *profile* of a symmetric $A \in \mathbb{R}^{n \times n}$ is defined by

$$\mathsf{profile}(A) = n + \sum_{i=1}^{n} (i - f_i(A))$$

where the *profile indices* $f_1(A), \ldots, f_n(A)$ are given by

$$f_i(A) = \min\{ j : 1 \le j \le i, \ a_{ij} \ne 0 \}. \tag{11.1.6}$$

For the 9-by-9 example in (11.1.5), $\mathsf{profile}(A) = 37$. We use that matrix to illustrate a heuristic method for approximate profile minimization. The first step is to choose a "starting node" and to relabel it as node 1. For reasons that are given later, we choose node 2 and set $S_0 = \{2\}$:

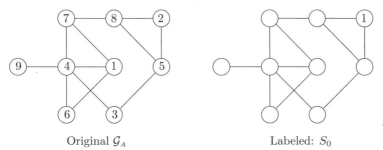

Original $\mathcal{G}_A$ Labeled: S_0

We then proceed to label the remaining nodes as follows:

Label the neighbors of S_0. Those neighbors make up S_1.
Label the unlabeled neighbors of nodes in S_1. Those neighbors make up S_2.
Label the unlabeled neighbors of nodes in S_2. Those neighbors make up S_3.
etc.

If we follow this plan for the example, then $S_1 = \{8, 5\}$, $S_2 = \{7, 3\}$, $S_3 = \{1, 4\}$, and $S_4 = \{6, 9\}$. These are the *level sets* of node 2 and here is how they are determined one after the other:

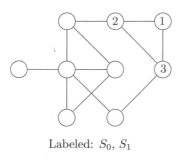

Labeled: S_0, S_1

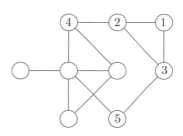

Labeled: S_0, S_1, S_2

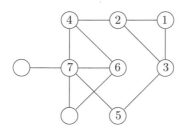

Labeled: S_0, S_1, S_2, S_3

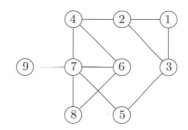

Labeled: S_0, S_1, S_2, S_3, S_4

By "concatenating" the level sets we obtain the *Cuthill-McKee reordering*:

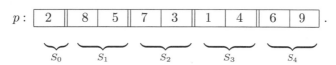

Observe the band structure that is induced by this ordering:

$A(p, p) = $ 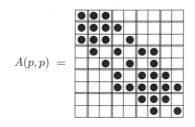 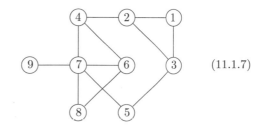   (11.1.7)

Note that $\mathsf{profile}(A(p, p)) = 25$. Moreover, $A(p, p)$ is a 5-by-5 block tridiagonal matrix with square diagonal blocks that have dimension equal to the cardinality of the level sets $S_0, \ldots, S_4$. This suggests why a good choice for S_0 is a node that has "far away" neighbors. Such a node will have a relatively large number of level sets and that means the resulting block tridiagonal matrix $A(p, p)$ will have more diagonal blocks. Heuristically, these blocks will be smaller and that implies a tighter profile. See George and Liu (1981, Chap. 4) for a discussion of this topic and why the *reverse Cuthill-McKee ordering* $p(n{:}{-}1{:}1)$ typically results in less fill-in during the Cholesky process.

11.1.6 The Minimum Degree Ordering

Another effective reordering scheme that is easy to motivate starts with the update recipe (11.1.4) and the observation that the vector v at each step should be as sparse as possible. This version of Cholesky with pivoting for $A = GG^T$ realizes this ambition:

> *Step 0.* $P = I_n$
>
> **for** $k = 1{:}n - 2$
>> *Step 1.* Choose a permutation $P_k \in \mathbb{R}^{(n-k+1)\times(n-k+1)}$ so that if
>>
>> $$P_k \, A(k{:}n, k{:}n) \, P_k^T = \begin{bmatrix} \alpha & v^T \\ v & B \end{bmatrix}$$
>>
>> then v is as sparse as possible
>>
>> *Step 2.* $P = \mathrm{diag}(I_{k-1}, P_k) \cdot P$ $\hspace{3cm}$ (11.1.8)
>>
>> *Step 3.* Reorder $A(k{:}n, k{:}n)$ and each previously computed G-column:
>>
>> $A(k{:}n, k{:}n) = P_k \, A(k{:}n, k{:}n) \, P_k^T$
>>
>> $A(k{:}n, 1{:}k - 1) = P_k \, A(k{:}n, 1{:}k - 1)$
>>
>> *Step 4.* Compute $G(k{:}n, k)$: $A(k{:}n, k) = A(k{:}n, k)/\sqrt{A(k, k)}$
>>
>> *Step 5.* Compute $A^{(k)}$
>>
>> $A(k{+}1{:}n, k{+}1{:}n) = A(k{+}1{:}n, k{+}1{:}n) - A(k{+}1{:}n, k)A(k{+}1{:}n, k)^T$
>
> **end**

The ordering that results from this process is the *minimum degree ordering*. The terminology makes sense because the pivot row in step k is associated with a node in the adjacency graph $\mathcal{G}_{A(k{:}n, k{:}n)}$ whose degree is minimal. Note that this is a greedy heuristic approach to the Sparse Cholesky challenge.

A serious overhead associated with the implementation of (11.1.8) concerns the outer-product update in Step 5. The memory allocation discussion in §11.1.2 suggests that we could make a more efficient procedure if we knew in advance the sparsity structure of the minimum degree Cholesky factor. We could replace Step 0 with

> *Step 0′.* Determine the minimum degree permutation p_{MD} and represent
> $A(p_{MD}, p_{MD})$ with "placeholder" zeros in those locations that fill in.

This would make Steps 1–3 unnecessary and obviate memory requests in Step 5. Moreover, it can happen that a collection of problems need to be solved each with the same sparsity structure. In this case, a single Step 0′ works for the entire collection thereby amortizing the overhead. It turns out that very efficient 0′ procedures have been developed. The basic idea revolves around the intelligent exploitation of two facts that completely characterize the sparsity of the Cholesky factor in $A = GG^T$:

Fact 1: If $j \le i$ and a_{ij} is nonzero, then g_{ij} is nonzero *assuming no numerical cancellation.*

Fact 2: If g_{ik} and g_{jk} are nonzero and $k < j < i$, then g_{ij} is nonzero *assuming no numerical cancellation.* See Parter (1961).

The caveats about no numerical cancellation are required because it is possible for an entry in G to be "luckily zero." For example, Fact 1 follows from the formula

$$g_{ij} \;=\; \left(a_{ij} \;-\; \sum_{k=1}^{j-1} g_{ik} g_{jk} \right) \bigg/ g_{jj},$$

with the assumption that the summation does not equal a_{ij}.

The systematic use of Facts 1 and 2 to determine G's sparsity structure is complicated and involves the construction of an *elimination tree (e-tree)*. Here is an example taken from the detailed presentation by Davis (2006, Chap. 4):

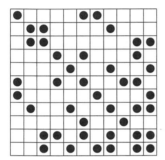
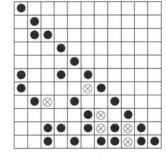
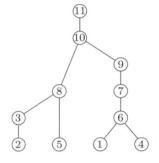

| The matrix A | A's Cholesky factor | A's elimination tree |

The "$\otimes$" entries are nonzero because of Fact 2. For example, g_{76} is nonzero because g_{61} and g_{71} are nonzero. The e-tree captures critical location information. In general, the parent of node i identifies the row of the first subdiagonal nonzero in column i. By encoding this kind of information, the e-tree can be used to answer various path-in-graph questions that relate to fill-in. In addition, the leaf nodes correspond to those columns that can be eliminated independently in a parallel implementation.

11.1.7 Nested Dissection Orderings

Suppose we have a method to determine a permutation P_0 so that $P_0 A P_0^T$ has the following block structure:

$$P_0 A P_0^T \;=\; \begin{bmatrix} A_1 & 0 & C_1 \\ 0 & A_2 & C_2 \\ C_1^T & C_2^T & S \end{bmatrix} \;=\;$$

Through the schematic we are stating "A_1 and A_2 are square and roughly the same size and C_1 and C_2 are relatively thin." Let us refer to this maneuver as a "successful dissection." Suppose $P_{11}A_1P_{11}^T$ and $P_{22}A_2P_{22}^T$ are also successful dissections. If $P = \mathrm{diag}(P_{11}, P_{22}, I) \cdot P_0$, then

The process can obviously be repeated on each of the four big diagonal blocks. Note that the Cholesky factor inherits the recursive block structure

In the end, the ordering produced is an example of a *nested dissection ordering*. These orderings are fill-reducing and work very well on grid-related, elliptic partial differential equation problems; see George and Liu (1981, Chap. 8). In graph terms, the act of finding a successful permutation for a given dissection is equivalent to the problem of finding a good *vertex cut* of $\mathcal{G}(A)$. Davis (2006, pp. 128–130) describes several ways in which this can be done. The payoff is considerable. With standard discretizations, many 2-dimensional problems can be solved with $O(n^{3/2})$ work and $O(n \log n)$ fill-in. For 3-dimensional problems, the typical costs are $O(n^2)$ work and $O(n^{4/3})$ fill-in.

11.1.8   Sparse QR and the Sparse Least Squares Problem

Suppose we want to minimize $\| Ax - b \|_2$ where $A \in \mathbb{R}^{m \times n}$ has full column rank and is sparse. If we are willing and able to form $A^T A$, then we can apply sparse Cholesky technology to the normal equations $A^T A x = A^T b$. In particular, we would compute a permutation P so that $P(A^T A)P^T$ has a sufficiently sparse Cholesky factor. However, aside from the pitfalls of normal equations, the matrix $A^T A$ can be dense even though

A is sparse. (Consider the case when A has a dense row.)

If we prefer to take the QR approach, then it still makes sense to reorder the columns of A, for if $AP^T = QR$ is the thin QR factorization of AP^T, then

$$P(A^T A)P^T = R^T R,$$

i.e., R^T is the Cholesky factor of $P(A^T A)P^T$. However, this poses serious issues that revolve around fill-in and the Q matrix. Suppose Q is determined via Householder QR. Even though P is chosen so that the final matrix R is reasonably sparse, the intermediate Householder updates $A = H_k A$ tend to have high levels of fill-in. A corollary of this is that Q is almost always dense. This can be a show-stopper especially if $m \gg n$ and motivates the *Sparse QR challenge*:

The Sparse QR Challenge

Given a sparse matrix $A \in \mathbb{R}^{m \times n}$, efficiently determine a permutation p of $1{:}n$ so that if $P = I_n(:,p)$, then the R-factor in the thin QR factorization $A(:,p) = AP^T = QR$ is close to being optimally sparse. Use orthogonal transformations to determine R from $A(:,p)$.

Before we show how to address the challenge we establish its relevance to the sparse least squares problem. If $AP^T = QR$ is the thin QR factorization of $A(:,p)$, then the normal equation system $A^T b = A^T A x_{LS}$ transforms to

$$P(A^T b) = (P(A^T A)P^T)P x_{LS} = R^T R P x_{LS}.$$

Solving the normal equations with a QR-produced Cholesky factor constitutes the *seminormal equations* approach to least squares. Observe that it is not necessary to compute Q. If followed by a single step of iterative improvement, then it is possible to show that the computed x_{LS} is just as good as the least squares solution obtained via the QR factorization. Here is the overall solution framework:

Step 1. Determine P so that the Cholesky factor for $P(A^T A)P^T$ is sparse.

Step 2. Carefully compute the matrix R in the thin QR factorization $AP^T = QR$.

Step 3. Solve: $R^T y_0 = P(A^T b)$, $R z_0 = y_0$, $x_0 = P^T z_0$.

Step 4. Improve: $r = b - A x_0$, $R^T y_1 = P(A^T r)$, $R z_1 = y_1$, $e = P^T z_1$, $x_{LS} = x_0 + e$.

To appreciate Steps 3 and 4, think of x_0 as being contaminated by unacceptable levels of error due to the pitfalls of normal equations. Noting that $A^T A x_0 = A^T b - A^T r$ and $A^T A e = A^T r$, we have

$$A^T A(x_0 + e) = A^T b - A^T r + A^T r = A^T b.$$

For a detailed analysis of the seminormal equation approach, see Björck (1987).

Let us return to the Sparse QR challenge and the efficient computaton of R using orthogonal transformations. Recall from §5.2.5 that with the Givens rotation approach there is considerable flexibility with respect to the zeroing order. A strategy for introducing zeros into $A \in \mathbb{R}^{m \times n}$ one row at a time can be organized as follows:

for $i = 2{:}m$
      for $j = 1{:}\min\{i-1, n\}$
          if $a_{ij} \neq 0$

              Compute a Givens rotation G such that $G \begin{bmatrix} a_{jj} \\ a_{ij} \end{bmatrix} = \begin{bmatrix} \times \\ 0 \end{bmatrix}$

$$\text{Update:} \quad \begin{bmatrix} a_{jj} & \cdots & a_{jn} \\ a_{ij} & \cdots & a_{in} \end{bmatrix} = G \begin{bmatrix} a_{jj} & \cdots & a_{jn} \\ a_{ij} & \cdots & a_{in} \end{bmatrix} \qquad (11.1.9)$$

          end
      end
end

The index i names the row that is being "rotated into" the current R matrix. Here is an example that shows how the j-loop oversees that process if $i > n$:

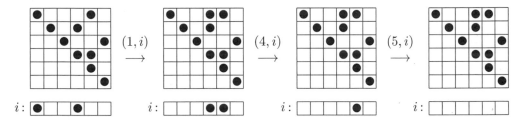

Notice that the rotations can induce fill-in both in R and in the row that is currently being zeroed. Various row-ordering strategies have been proposed to minimize fill-in "along the way" to the final matrix R. See George and Heath (1980) and Björck (NMLS, p. 244). For example, before (11.1.9) is executed, the rows can be arranged so that the first nonzero in each row is never to the left of the first nonzero in the previous row. Rows where the first nonzero element occurs in the same column can be sorted according to the location of the last nonzero element.

11.1.9   Sparse LU

The first step in a pivoted LU procedure applied to $A \in \mathbb{R}^{n \times n}$ computes the factorization

$$PAQ^T = \begin{bmatrix} \alpha & w^T \\ v & B \end{bmatrix} = \begin{bmatrix} 1 & 0 \\ v/\alpha & I_{n-1} \end{bmatrix} \begin{bmatrix} \alpha & w^T \\ 0 & A^{(1)} \end{bmatrix} \qquad (11.1.10)$$

where P and Q are permutation matrices and

$$A^{(1)} = B - \frac{1}{\alpha} v w^T. \qquad (11.1.11)$$

In §3.4 we discussed various choices for P and Q. Stability was the primary issue and everything revolved around making the pivot element α sufficiently large. If A is sparse, then in addition to stability we have to be concerned about the sparsity of $A^{(1)}$. Balancing the tension between stability and sparsity defines the *Sparse LU challenge*:

The Sparse LU Challenge

Given a matrix $A \in \mathbb{R}^{n \times n}$, efficiently determine permutations p and q of 1:n so that if $P = I_n(:, p)$ and $Q = I_n(:, q)$, then the factorization $A(p, q) = PAQ^T = LU$ is reasonably stable and the triangular factors L and U are close to being optimally sparse.

To meet the challenge we must interpolate between a pair of extreme strategies:

- Maximize stability by choosing P and Q so that $|\alpha| = \max |a_{ij}|$.
- Maximize sparsity by choosing P and Q so that $\mathsf{nnz}(A^{(1)})$ is minimized.

Markowitz pivoting provides a framework for doing this. Given a *threshold parameter* τ that satisfies $0 \le \tau \le 1$, choose P and Q in each step of the form (11.1.10) so that $\mathsf{nnz}(A^{(1)})$ is minimized subject to the constraint that $|\alpha| \ge \tau |v_i|$ for $i = 1{:}n - 1$. Small values of τ jeopardize stability but create more opportunities to control fill-in. A typical compromise value is $\tau = 1/10$.

Sometimes there is an advantage to choosing the pivot from the diagonal, i.e., setting $P = Q$. This is the case when the matrix A is *structurally symmetric*. A matrix A is structurally symmetric if a_{ij} and a_{ji} are either both zero or both nonzero. Symmetric matrices whose rows and/or columns are scaled have this property. It is easy so show from (11.1.10) and (11.1.11) that if A is structurally symmetric and $P = Q$, then $A^{(1)}$ is structurally symmetric. The Markowitz strategy can be generalized to express a preference for diagonal pivoting if it is "safe". If a diagonal element is sufficiently large compared to other entries in its column, then P is chosen so that $(PAP^T)_{11}$ is that element and structural symmetry is preserved. Otherwise, a sufficiently large off-diagonal element is brought to the (1,1) position using a PAQ^T update.

Problems

P11.1.1 Give an algorithm that solves an upper triangular system $Tx = b$ given that T is stored in the compressed-column format.

P11.1.2 If both indexing and flops are taken into consideration, is the sparse outer-product update (11.1.2) an $O(\mathsf{nnz}(u) \cdot \mathsf{nnz}(v))$ computation?

P11.1.3 For example (11.1.5), what is the resulting profile if $S_0 = \{9\}$? What if $S_0 = \{4\}$?

P11.1.4 Prove that the Cuthill-McKee ordering permutes A into a block tridiagonal form where the kth diagonal block is r-by-r where r is the cardinality of S_{k-1}.

P11.1.5 (a) What is the resulting profile if the reverse Cuthill-McKee ordering is applied to the example in §11.1.5? (b) What is the elimination tree for the matrix in (11.1.5)?

P11.1.6 Show that if G is the Cholesky factor of A and an element $g_{ij} \ne 0$, then $j \ge f_i$ where f_i is defined by (11.1.6). Conclude that $\mathsf{nnz}(G) \le \mathsf{profile}(A)$.

P11.1.7 Show how the method of seminormal equations can be used efficiently to minimize $\| Mx - d \|_2$ where

$$M = \begin{bmatrix} A_1 & 0 & 0 & C_1 \\ 0 & A_2 & 0 & C_2 \\ 0 & 0 & A_3 & C_3 \end{bmatrix}, \qquad d = \begin{bmatrix} b_1 \\ b_2 \\ b_3 \end{bmatrix},$$

and $A_i \in \mathbb{R}^{m \times n}$, $C_i \in \mathbb{R}^{m \times p}$, and $b_i \in \mathbb{R}^m$ for $i = 1{:}3$. Assume that M has full column rank and that $m > n + p$. Hint: Compute the Q-less QR factorizations of $[A_i\ C_i]$ for $i = 1{:}3$.

Notes and References for §11.1

Early references for direct sparse matrix computations include the following textbooks:

A. George and J.W.-H. Liu (1981). *Computer Solution of Large Sparse Positive Definite Systems*, Prentice-Hall, Englewood Cliffs, NJ.

O. Osterby and Z. Zlatev (1983). *Direct Methods for Sparse Matrices*, Springer-Verlag, New York.

S. Pissanetzky (1984). *Sparse Matrix Technology*, Academic Press, New York.

I.S. Duff, A.M. Erisman, and J.K. Reid (1986). *Direct Methods for Sparse Matrices*, Oxford University Press, London.

A more recent treatment that targets practitioners, provides insight into a range of implementation issues, and has an excellent annotated bibliography is the following:

T.A. Davis (2006). *Direct Methods for Sparse Linear Systems*, SIAM Publications, Philadelphia, PA.

The interplay between graph theory and sparse matrix computations with emphasis on symbolic factorizations that predict fill is nicely set forth in:

J.W.H. Liu (1990). "The Role of Elimination Trees in Sparse Factorizations," *SIAM J. Matrix Anal. Applic. 11*, 134–172.

J.R. Gilbert (1994). "Predicting Structure in Sparse Matrix Computations," *SIAM J. Matrix Anal. Applic. 15*, 62–79.

S.C. Eisenstat and J.W.H. Liu (2008). "Algorithmic Aspects of Elimination Trees for Sparse Unsymmetric Matrices," *SIAM J. Matrix Anal. Applic. 29*, 1363–1381.

Relatively recent papers on profile reduction include:

W.W. Hager (2002). "Minimizing the Profile of a Symmetric Matrix," *SIAM J. Sci. Comput. 23*, 1799–1816.

J.K. Reid and J.A. Scott (2006). "Reducing the Total Bandwidth of a Sparse Unsymmetric Matrix," *SIAM J. Matrix Anal. Applic. 28*, 805–821.

Efficient implementations of the minimum degree idea are discussed in:

P.R. Amestoy, T.A. Davis, and I.S. Duff (1996). "An Approximate Minimum Degree Ordering Algorithm," *SIAM J. Matrix Anal. Applic. 17*, 886–905.

T.A. Davis, J.R. Gilbert, S.I. Larimore, and E.G. Ng (2004). "A Column Approximate Minimum Degree Ordering Algorithm," *ACM Trans. Math. Softw. 30*, 353–376.

For an overview of sparse least squares, see Björck (NMLS, Chap. 6)) and also:

J.A. George and M.T. Heath (1980). "Solution of Sparse Linear Least Squares Problems Using Givens Rotations," *Lin. Alg. Applic. 34*, 69–83.

Å. Björck and I.S. Duff (1980). "A Direct Method for the Solution of Sparse Linear Least Squares Problems," *Lin. Alg. Applic. 34*, 43–67.

A. George and E. Ng (1983). "On Row and Column Orderings for Sparse Least Squares Problems," *SIAM J. Numer. Anal. 20*, 326–344.

M.T. Heath (1984). "Numerical Methods for Large Sparse Linear Least Squares Problems," *SIAM J. Sci. Stat. Comput. 5*, 497–513.

Å. Björck (1987). "Stability Analysis of the Method of Seminormal Equations for Least Squares Problems," *Lin. Alg. Applic. 88/89*, 31–48.

The design of a sparse LU procedure that is also stable is discussed in:

J.W. Demmel, S.C. Eisenstat, J.R. Gilbert, X.S. Li, and J.W.H. Liu (1999). "A Supernodal Approach to Sparse Partial Pivoting," *SIAM J. Matrix Anal. Applic. 20*, 720–755.

L. Grigori, J.W. Demmel, and X.S. Li (2007). "Parallel Symbolic Factorization for Sparse LU with Static Pivoting," *SIAM J. Sci. Comput. 3*, 1289–1314.

L. Grigori, J.R. Gilbert, and M. Cosnard (2008). "Symbolic and Exact Structure Prediction for Sparse Gaussian Elimination with Partial Pivoting," *SIAM J. Matrix Anal. Applic. 30*, 1520–1545.

Frontal methods are a way of organizing outer-product updates so that the resulting implementation is rich in dense matrix operations, a maneuver that is critical from the standpoint of performance, see:

J.W.H. Liu (1992). "The Multifrontal Method for Sparse Matrix Solution: Theory and Practice," *SIAM Review 34*, 82–109.

D.J. Pierce and J.G. Lewis (1997). "Sparse Multifrontal Rank Revealing QR Factorization," *SIAM J. Matrix Anal. Applic. 18*, 159–180.

T.A. Davis and I.S. Duff (1999). "A Combined Unifrontal/Multifrontal Method for Unsymmetric Sparse Matrices," *ACM Trans. Math. Softw. 25*, 1–20.

Another important reordering challenge involves permuting to block triangular form, see:

A. Pothen and C.-J. Fan (1990). "Computing the Block Triangular Form of a Sparse Matrix," *ACM Trans. Math. Softw. 16*, 303–324.
I.S. Duff and B. Uçar (2010). "On the Block Triangular Form of Symmetric Matrices," *SIAM Review 52*, 455–470.

Early papers on parallel sparse matrix computations that are filled with interesting ideas include:

M.T. Heath, E. Ng, and B.W. Peyton (1991). "Parallel Algorithms for Sparse Linear Systems," *SIAM Review 33*, 420–460.
J.R. Gilbert and R. Schreiber (1992). "Highly Parallel Sparse Cholesky Factorization," *SIAM J. Sci. Stat. Comput. 13*, 1151–1172.

For a sparse-matrix discussion of condition estimation, error analysis, and related problems, see:

R.G. Grimes and J.G. Lewis (1981). "Condition Number Estimation for Sparse Matrices," *SIAM J. Sci. Stat. Comput. 2*, 384–388.
M. Arioli, J.W. Demmel, and I.S. Duff (1989). "Solving Sparse Linear Systems with Sparse Backward error," *SIAM J. Matrix Anal. Applic. 10*, 165–190.
C.H. Bischof (1990). "Incremental Condition Estimation for Sparse Matrices," *SIAM J. Matrix Anal. Applic. 11*, 312–322.
M.W. Berry, S.A. Pulatova, and G.W. Stewart (2005). "Algorithm 844: Computing Sparse Reduced-Rank Approximations to Sparse Matrices," *ACM Trans. Math. Softw. 31*, 252–269.

11.2 The Classical Iterations

An iterative method for the $Ax = b$ problem generates a sequence of approximate solutions $\{x^{(k)}\}$ that converges to $x = A^{-1}b$. Typically, the matrix A is involved only in the context of matrix-vector multiplication and that is what makes this framework attractive when A is large and sparse. The critical attributes of an iterative method include the rate of convergence, the amount of computation per step, the volume of required storage, and the pattern of memory access. In this section, we present a collection of classical iterative methods, discuss their practical implementation, and prove a few representative theorems that illuminate their behavior.

11.2.1 The Jacobi and Gauss-Seidel Iterations

The simplest iterative method for the $Ax = b$ problem is the *Jacobi iteration*. The 3-by-3 instance of the method can be motivated by rewriting the equations as follows:

$$x_1 = (b_1 - a_{12}x_2 - a_{13}x_3)/a_{11},$$
$$x_2 = (b_2 - a_{21}x_1 - a_{23}x_3)/a_{22},$$
$$x_3 = (b_3 - a_{31}x_1 - a_{32}x_2)/a_{33}.$$

Suppose $x^{(k-1)}$ is a "current" approximation to $x = A^{-1}b$. A natural way to generate a new approximation $x^{(k)}$ is to compute

$$x_1^{(k)} = (b_1 - a_{12}x_2^{(k-1)} - a_{13}x_3^{(k-1)})/a_{11},$$
$$x_2^{(k)} = (b_2 - a_{21}x_1^{(k-1)} - a_{23}x_3^{(k-1)})/a_{22}, \tag{11.2.1}$$
$$x_3^{(k)} = (b_3 - a_{31}x_1^{(k-1)} - a_{32}x_2^{(k-1)})/a_{33}.$$

Clearly, A must have nonzeros along its diagonal for the method to be defined. For general n we have

for $i = 1{:}n$

$$x_i^{(k)} = \left(b_i - \sum_{j=1}^{i-1} a_{ij} x_j^{(k-1)} - \sum_{j=i+1}^{n} a_{ij} x_j^{(k-1)} \right) \Big/ a_{ii} \qquad (11.2.2)$$

end

Note that the most recent solution estimate is not fully exploited in the updating of a particular component. For example, $x_1^{(k-1)}$ is used in the calculation of $x_2^{(k)}$ even though $x_1^{(k)}$ is available. If we revise the process so that the most current estimates of the solution components are always used, then we obtain the *Gauss-Seidel iteration*:

for $i = 1{:}n$

$$x_i^{(k)} = \left(b_i - \sum_{j=1}^{i-1} a_{ij} x_j^{(k)} - \sum_{j=i+1}^{n} a_{ij} x_j^{(k-1)} \right) \Big/ a_{ii} \qquad (11.2.3)$$

end

As with Jacobi, $a_{11}, \ldots, a_{nn}$ must be nonzero for the iteration to be defined.

For both of these methods, the transition from $x^{(k-1)}$ to $x^{(k)}$ can be succinctly described in terms of the strictly lower triangular, diagonal, and strictly upper triangular parts of the matrix A. Denote these three matrices by L_A, D_A, and U_A respectively, e.g.,

$$L_A = \begin{bmatrix} 0 & 0 & 0 \\ a_{21} & 0 & 0 \\ a_{31} & a_{32} & 0 \end{bmatrix}, \; D_A = \begin{bmatrix} a_{11} & 0 & 0 \\ 0 & a_{22} & 0 \\ 0 & 0 & a_{33} \end{bmatrix}, \; U_A = \begin{bmatrix} 0 & a_{12} & a_{13} \\ 0 & 0 & a_{23} \\ 0 & 0 & 0 \end{bmatrix}.$$

It is easy to show that the Jacobi step (11.2.2) has the form

$$M_{\text{J}} \, x^{(k)} = N_{\text{J}} \, x^{(k-1)} + b \qquad (11.2.4)$$

where $M_{\text{J}} = D_A$ and $N_{\text{J}} = -(L_A + U_A)$. On the other hand, the Gauss-Seidel step (11.2.3) is defined by

$$M_{\text{GS}} \, x^{(k)} = N_{\text{GS}} \, x^{(k-1)} + b \qquad (11.2.5)$$

with $M_{\text{GS}} = (D_A + L_A)$ and $N_{\text{GS}} = -U_A$.

11.2.2   Block Versions

The Jacobi and Gauss-Seidel methods have obvious block analogs. For example, if A is a 3-by-3 block matrix with square, nonsingular diagonal blocks, then the system

$$\begin{bmatrix} A_{11} & A_{12} & A_{13} \\ A_{21} & A_{22} & A_{23} \\ A_{31} & A_{32} & A_{33} \end{bmatrix} \begin{bmatrix} x_1 \\ x_2 \\ x_3 \end{bmatrix} = \begin{bmatrix} b_1 \\ b_2 \\ b_3 \end{bmatrix}$$

can be rewritten as follows:

$$A_{11} x_1 = b_1 - A_{12} \, x_2 - A_{13} \, x_3,$$

$$A_{22} x_2 = b_2 - A_{21} \, x_1 - A_{23} \, x_3,$$

$$A_{33} x_3 = b_3 - A_{31} \, x_1 - A_{32} \, x_2.$$

From this we obtain the *block Jacobi* iteration

$$
A_{11}x_1^{(k)} = b_1 - A_{12}\,x_2^{(k-1)} - A_{13}\,x_3^{(k-1)},
$$
$$
A_{22}x_2^{(k)} = b_2 - A_{21}\,x_1^{(k-1)} - A_{23}\,x_3^{(k-1)},
$$
$$
A_{33}x_3^{(k)} = b_3 - A_{31}\,x_1^{(k-1)} - A_{32}\,x_2^{(k-1)},
$$

and the *block Gauss-Seidel* iteration

$$
A_{11}x_1^{(k)} = b_1 - A_{12}\,x_2^{(k-1)} - A_{13}\,x_3^{(k-1)},
$$
$$
A_{22}x_2^{(k)} = b_2 - A_{21}\,x_1^{(k)} - A_{23}\,x_3^{(k-1)},
$$
$$
A_{33}x_3^{(k)} = b_3 - A_{31}\,x_1^{(k)} - A_{32}\,x_2^{(k)}.
$$

In contrast to the point versions of these iterations, a genuine linear system must be solved for $x_i^{(k)}$. These can be solved directly using LU or Cholesky factorizations or approximately solved via some iterative method. Of course, for this framework to make sense, the diagonal blocks must be nonsingular.

11.2.3 Splittings and Convergence

Many iterative methods for the $Ax = b$ problem can be written in the form

$$
Mx^{(k)} = Nx^{(k-1)} + b \tag{11.2.6}
$$

where $A = M - N$ is a *splitting* and $x^{(0)}$ is a starting vector. For the iteration to be practical, it must be easy to solve linear systems that involve M. This is certainly the case for the Jacobi method where M is diagonal and the Gauss-Seidel method where M is lower triangular.

It turns out that the rate of convergence associated with (11.2.6) depends on the eigenvalues of the *iteration matrix*

$$
G = M^{-1}N.
$$

By subtracting the equation $Mx = Nx + b$ from (11.2.6) we obtain

$$
M(x^{(k)} - x) = N(x^{(k-1)} - x).
$$

Thus, there is a simple connection between the error at a given step and the error at the previous step. Indeed, if

$$
e^{(k)} = x^{(k)} - x,
$$

then

$$
e^{(k)} = M^{-1}Ne^{(k-1)} = Ge^{(k-1)} = G^k e^{(0)}. \tag{11.2.7}
$$

Everything hinges on the behavior of G^k as $k \to \infty$. If $\| G \| < 1$ for some choice of norm, then convergence is assured because

$$
\| e^{(k)} \| = \| G^k e^{(0)} \| \leq \| G^k \| \| e^{(0)} \| \leq \| G \|^k \| e^{(0)} \|.
$$

However, it is the largest eigenvalue of G that determines the asymptotic behavior of G^k. For example, if

$$G = \begin{bmatrix} \lambda & \alpha \\ 0 & \lambda \end{bmatrix},$$

then

$$G^k = \begin{bmatrix} \lambda^k & \alpha\lambda^{k-1} \\ 0 & \lambda^k \end{bmatrix}. \tag{11.2.8}$$

We conclude that for this problem $G^k \to 0$ if and only if the eigenvalue λ satisfies $|\lambda| < 1$. Recall from (7.1.1) the definition of spectral radius:

$$\rho(C) \;=\; \max\{\, |\lambda| : \lambda \in \lambda(C) \,\}.$$

The following theorem links the size of $\rho(M^{-1}N)$ to the convergence of (11.2.6).

Theorem 11.2.1.   *Suppose $A = M - N$ is a splitting of a nonsingular matrix $A \in \mathbb{R}^{n\times n}$. Assuming that M is nonsingular, the iteration (11.2.6) converges to $x = A^{-1}b$ for all starting n-vectors $x^{(0)}$ if and only if $\rho(G) < 1$ where $G = M^{-1}N$.*

Proof.   In light of (11.2.7), it suffices to show that $G^k \to 0$ if and only if $\rho(G) < 1$. If $Gx = \lambda x$, then $G^k x = \lambda^k x$. Thus, if $G^k \to 0$, then we must have $|\lambda| < 1$, i.e., the spectral radius of G must be less than 1.

Now assume $\rho(G) < 1$ and let $G = QTQ^H$ be its Schur decomposition. If $D = \operatorname{diag}(t_{11}, \ldots, t_{nn})$ and $E = A - D$, then it follows from (7.3.15) that

$$\| \, G^k \, \|_2 \;\le\; (1+\mu)^{n-1}\left(\rho(G) + \frac{\| \, E \, \|_F}{1+\mu}\right)^k$$

where μ is any nonnegative real number. It is clear that we can choose this parameter so that the upper bound converges to zero. For example, if G is normal, then $E = 0$ and we can set $\mu = 0$. Otherwise, if

$$\mu \;=\; \frac{2\| \, E \, \|_2}{1 - \rho(G)},$$

then it is easy to verify that

$$\| \, G^k \, \|_2 \;\le\; \left(1 + \frac{2\| \, E \, \|_F}{1 - \rho(G)}\right)^{n-1}\left(\frac{1 + \rho(G)}{2}\right)^k \tag{11.2.9}$$

and this guarantees convergence because $1 + \rho(G) < 2$. $\square$

The 2-by-2 example (11.2.8) and the inequality (11.2.9) serve as a reminder that the spectral radius does not tell us everything about the powers of a nonnormal matrix. Indeed, if G is nonnormal, then is possible for G^k (and the error $\| \, x^{(k)} - x \, \|$) to grow considerably before decay sets in. The ϵ-pseudospectral radius introduced in §7.9.6 provides greater insight into this situation.

To summarize what we have learned so far, two attributes are critical if a method of the form (11.2.6) is to be of interest:

- The underlying splitting $A = M - N$ must have the property that linear systems of the form $Mz = d$ are relatively easy to solve.

- A way must be found to guarantee that $\rho(M^{-1}N) < 1$.

To give a flavor for the kind of analysis that attends the second requirement, we state and prove a pair of convergence results that apply to the Jacobi and Gauss-Seidel iterations.

11.2.4 Diagonal Dominance and Jacobi Iteration

One way to establish that the spectral radius of the iteration matrix G is less than one is to show that $\| G \| < 1$ for some choice of norm. This inequality ensures that all of G's eigenvalues are inside the unit circle. As an example of this type of analysis, consider the situation where the Jacobi iteration is applied to a strictly diagonally dominant linear system. Recall from §4.1.1 that $A \in \mathbb{R}^{n \times n}$ has this property if

$$\sum_{\substack{j=1 \\ j \neq i}}^{n} |a_{ij}| < |a_{ii}|, \qquad i = 1{:}n.$$

Theorem 11.2.2. *If $A \in \mathbb{R}^{n \times n}$ is strictly diagonally dominant, then the Jacobi itreation (11.2.4) converges to $x = A^{-1}b$.*

Proof. Since $G_{\text{J}} = -D_A^{-1}(L_A + U_A)$ it follows that

$$\| G_{\text{J}} \|_\infty = \| D_A^{-1}(L_A + U_A) \|_\infty = \max_{1 \leq i \leq n} \sum_{\substack{j=1 \\ j \neq i}}^{n} \left| \frac{a_{ij}}{a_{ii}} \right| < 1.$$

The theorem follows because no eigenvalue of A can be bigger that $\| A \|_\infty$. □

Usually, the "more dominant" the diagonal the more rapid the convergence, but there are counterexamples. See P11.2.3.

11.2.5 Positive Definiteness and Gauss-Seidel Iteration

A more complicated spectral radius argument is needed to show that Gauss-Seidel converges for matrices that are symmetric positive definite.

Theorem 11.2.3. *If $A \in \mathbb{R}^{n \times n}$ is symmetric and positive definite, then the Gauss-Seidel iteration (11.2.5) converges for any $x^{(0)}$.*

Proof. We must verify that the eigenvalues of $G_{\text{GS}} = -(D_A + L_A)^{-1}L_A^T$ are inside the unit circle. This matrix has the same eigenvalues as the matrix

$$G = D_A^{1/2} G_{\text{GS}} D_A^{-1/2} = -(I + L)^{-1}L^T$$

where $L = D_A^{-1/2} L_A D_A^{-1/2}$. If

$$-(I + L)^{-1} L^T v = \lambda v \qquad v^H v = 1$$

then $-v^H L^H v = \lambda(1 + v^H L v)$. If $v^H L v = a + bi$, then

$$|\lambda|^2 = \left| \frac{-a + bi}{1 + a + bi} \right|^2 = \frac{a^2 + b^2}{1 + 2a + a^2 + b^2}.$$

However, since $D_A^{-1/2} A D_A^{-1/2} = I + L + L^T$ is positive definite, it is not hard to show that $0 < 1 + v^H L v + v^H L^T v = 1 + 2a$ and hence that $|\lambda| < 1$. $\square$

We mention that to bound $\rho(M_{\mathrm{GS}}^{-1} N_{\mathrm{GS}})$ away from 1 requires additional information about A. The required analysis can be quite involved.

11.2.6   Discussion of a Model Problem

It is instructive to consider application of the Jacobi and Gauss-Seidel methods to the symmetric positive definite linear system

$$(I_{n_1} \otimes T_{n_2} + T_{n_1} \otimes I_{n_2}) u = b \qquad (11.2.10)$$

where

$$T_m = \begin{bmatrix} 2 & -1 & \cdots & 0 \\ -1 & 2 & \ddots & \vdots \\ \vdots & \ddots & \ddots & -1 \\ 0 & \cdots & -1 & 2 \end{bmatrix} \in \mathbb{R}^{m \times m}. \qquad (11.2.11)$$

Systems with this structure arise from discretization of the Poisson equation on a rectangular grid; see §4.8.3. Recall that it is convenient to think of the solution vector as doubly subscripted. Associated with grid point (i, j) is the unknown $U(i, j)$. When the system is solved, the value of $U(i, j)$ is the average of the values associated with its north, east, south, and west "grid neighbors." Boundary values are known and fixed and this permits us to reformulate (11.2.10) as a 2-dimensional array averaging problem:

Given $U(0{:}n_1 + 1, 0{:}n_2 + 1)$ with fixed values in its top and bottom row and fixed values in its leftmost and rightmost columns, determine $U(1{:}n_1, 1{:}n_2)$ such that

$$U(i, j) = \frac{U(i, j - 1) + U(i, j + 1) + U(i - 1, j) + U(i + 1, j)}{4}$$

for $i = 1{:}n_1$ and $j = 1{:}n_2$.

It is much easier to reason about Jacobi and Gauss-Seidel from this point of view. For example, the update

$$V = U$$

for $i = 1{:}n_1$

    for $j = 1{:}n_2$

        $U(i,j) = (V(i-1,j) + V(i,j+1) + V(i+1,j) + V(i,j-1))/4$

    end

end

corresponds to one step of Jacobi while

for $i = 1{:}n_1$

    for $j = 1{:}n_2$

        $U(i,j) = (U(i-1,j) + U(i,j+1) + U(i+1,j) + U(i,j-1))/4$

    end

end

is the corresponding update associated with Gauss-Seidel. The organization of both methods reflects the ultimate exploitation of matrix structure: *The matrix A is nowhere in sight!* We simply take advantage of the Kronecker structure at the block level and the 1-2-1 structure of the underlying tridiagonal matrices.

The array-update point of view for the model problem that we are considering makes it easy to appreciate why the Jacobi process is typically easier to vectorize and/or parallelize than Gauss-Seidel. The Jacobi update of $U(1{:}n_1, 1{:}n_2)$ is a matrix averaging:

$$\frac{U(1{:}n_1, 0{:}n_2-1) + U(2{:}n_1+1, 1{:}n_2) + U(1{:}n_1, 2{:}n_2+1) + U(0{:}n_1-1, 1{:}n_2)}{4}.$$

The use-the-most-recent-estimate attribute of the Gauss-Seidel method makes it harder to describe the update at such a high level.

Now let us analyze the spectral radius $\rho(M_{\mathrm{J}}^{-1} N_{\mathrm{J}})$. Closed-form expressions for T_m's eigenvalues permit us to determine this important quantity. Note that

$$\mathcal{T}_m = 2I - E_m$$

where

$$E_m = \begin{bmatrix} 0 & 1 & \cdots & 0 \\ 1 & 0 & \ddots & \vdots \\ \vdots & \ddots & \ddots & 1 \\ 0 & \cdots & 1 & 0 \end{bmatrix}.$$

Since

$$A = I_{n_1} \otimes \mathcal{T}_{n_2} + \mathcal{T}_{n_1} \otimes I_{n_2} = 4I_{n_1 n_2} - (I_{n_1} \otimes E_{n_2}) - (E_{n_1} \otimes I_{n_2}), \quad (11.2.12)$$

the Jacobi splitting $A = M_{\mathrm{J}} - N_{\mathrm{J}}$ is given by

$$M_{\mathrm{J}} = 4I_{n_1 n_2},$$

$$N_{\mathrm{J}} = (I_{n_1} \otimes E_{n_2}) + (E_{n_1} \otimes I_{n_2}).$$

Using results from our fast eigensystem discussion in §4.8.6, it can be shown that

$$S_m^{-1} E_m S_m = D_m = \text{diag}(\mu_1^{(m)}, \ldots, \mu_m^{(m)}) \tag{11.2.13}$$

where S_m is the sine transform matrix $[S_m]_{kj} = \sin(kj\pi/(m+1))$ and

$$\mu_k^{(m)} = 2\cos\left(\frac{k\pi}{m+1}\right), \qquad k = 1{:}m. \tag{11.2.14}$$

It follows that

$$(S_{n_1} \otimes S_{n_2})^{-1} \left(M_J^{-1} N_J\right)(S_{n_1} \otimes S_{n_2}) = (I_{n_1} \otimes D_{n_2} + D_{n_1} \otimes I_{n_2})/4.$$

By using the Kronecker structure of this diagonal matrix and (11.2.14), it is easy to verify that

$$\rho(M_J^{-1} N_J) = \frac{2\cos(\pi/(n_1+1)) + 2\cos(\pi/(n_2+1))}{4}. \tag{11.2.15}$$

Note that this quantity approaches unity as n_1 and n_2 increase.

As a final exercise concerning the model problem, we use its special structure to develop an interesting alternative iteration. From (11.2.12) we can write $A = M_x - N_x$ where

$$M_x = 4I_{n_1 n_2} - (I_{n_1} \otimes E_{n_2}), \qquad N_x = (E_{n_1} \otimes I_{n_2}).$$

Likewise, $A = M_y - N_y$ where

$$M_y = 4I_{n_1 n_2} - (E_{n_1} \otimes I_{n_2}), \qquad N_y = (I_{n_1} \otimes E_{n_2}).$$

These two splittings can be paired to produce the following transition from $u^{(k-1)}$ to $u^{(k)}$:

$$\begin{aligned} M_x v^{(k)} &= N_x u^{(k-1)} + b, \\ M_y u^{(k)} &= N_y v^{(k)} + b. \end{aligned} \tag{11.2.16}$$

Each step has a natural interpretation based on the underlying partial differential equation; see §4.8.4. The first step corresponds to treating the north and south values at each grid point as fixed, while the second step corresponds to treating the east and west values at each grid point as fixed. The resulting iteration is an example of an *alternating direction iteration*. See Varga (1962, Chap. 7). Since

$$u^{(k)} - x = (M_y^{-1} N_y)(v^{(k)} - x) = (M_y^{-1} N_y)(M_x^{-1} N_x)(u^{(k-1)} - x)$$

it follows that $e^{(k)} = G^k e^{(0)}$ where

$$\begin{aligned} G &= (M_y^{-1} N_y)(M_x^{-1} N_x) \\ &= (4I_{n_1 n_2} - E_{n_1} \otimes I_{n_2})^{-1}(I_{n_1} \otimes E_{n_2})(4I_{n_1 n_2} - I_{n_1} \otimes E_{n_2})^{-1}(E_{n_1} \otimes I_{n_2}). \end{aligned}$$

Using (11.2.13) and (11.2.14) it is easy to show that

$$\begin{aligned} (S_{n_1} \otimes S_{n_2})^{-1} G(S_{n_1} \otimes S_{n_2}) = \\ (4I_{n_1 n_2} - D_{n_1} \otimes I_{n_2})^{-1}(I_{n_1} \otimes D_{n_2})(4I_{n_1 n_2} - I_{n_1} \otimes D_{n_2})^{-1}(D_{n_1} \otimes I_{n_2}) \end{aligned}$$

is diagonal and that

$$\rho(G) = = \frac{\cos(\pi/(n_1+1))\cos(\pi/(n_2+1))}{(2 - \cos(\pi/(n_1+1)))(2 - \cos(\pi/(n_2+1)))} < 1. \tag{11.2.17}$$

11.2.7 SOR and Symmetric SOR

The Gauss-Seidel iteration is very attractive because of its simplicity. Unfortunately, if the spectral radius of $M_{\mathrm{GS}}^{-1} N_{\mathrm{GS}}$ is close to unity, then it may be prohibitively slow. To address this concern, we consider the parameterized splitting $A = M_\omega - N_\omega$ where

$$M_\omega = \frac{1}{\omega} D_A + L_A \qquad N_\omega = \left(\frac{1}{\omega} - 1 \right) D_A + U_A. \tag{11.2.18}$$

This defines the method of *successive over-relaxation* (SOR):

$$\left(\frac{1}{\omega} D_A + L_A \right) x^{(k)} = \left(\left(\frac{1}{\omega} - 1 \right) D_A + U_A \right) x^{(k-1)} + b. \tag{11.2.19}$$

At the component level we have

> **for** $i = 1{:}n$
> $$x_i^{(k)} = \omega \left(b_i - \sum_{j=1}^{i-1} a_{ij} x_j^{(k)} - \sum_{j=i+1}^{n} a_{ij} x_j^{(k-1)} \right) \Big/ a_{ii} + (1-\omega) x_i^{(k-1)}$$
> **end**

Note that if $\omega = 1$, then this is just the Gauss-Seidel method. The idea is to choose ω so that $\rho(M_\omega^{-1} N_\omega)$ is minimized. A detailed theory on how to do this is developed by Young (1971). For an excellent synopsis of that theory, see Greenbaum (IMSL, p. 149).

Observe that x is updated top to bottom in the SOR step. We can just as easily update from bottom to top:

> **for** $i = n{:} -1{:}1$
> $$x_i^{(k)} = \omega \left(b_i - \sum_{j=1}^{i-1} a_{ij} x_j^{(k-1)} - \sum_{j=i+1}^{n} a_{ij} x_j^{(k)} \right) \Big/ a_{ii} + (1-\omega) \cdot x_i^{(k-1)}$$
> **end**

This defines the *backward SOR iteration*:

$$\left(\frac{1}{\omega} D_A + U_A \right) x^{(k)} = \left(\left(\frac{1}{\omega} - 1 \right) D_A + L_A \right) x^{(k-1)} + b. \tag{11.2.21}$$

Note that this update can be obtained from (11.2.19) simply by interchanging the roles of L and U.

If A is symmetric ($U_A = L_A^T$), then the *symmetric SOR* (SSOR) method is obtained by combining the forward and backward implementations of the update as follows:

$$\left(\frac{1}{\omega} D_A + L_A \right) y^{(k)} = \left(\left(\frac{1}{\omega} - 1 \right) D_A - L_A^T \right) x^{(k-1)} + b, \tag{11.2.22}$$

$$\left(\frac{1}{\omega} D_A + L_A^T \right) x^{(k)} = \left(\left(\frac{1}{\omega} - 1 \right) D_A - L_A \right) y^{(k)} + b. \tag{11.2.23}$$

It can be shown that if

$$M_{\text{SSOR}} = \frac{\omega}{2-\omega}\left(\frac{1}{\omega}D_A + L_A\right)D_A^{-1}\left(\frac{1}{\omega}D_A + L_A^T\right) \tag{11.2.24}$$

then the transition from $x^{(k-1)}$ to $x^{(k)}$ is given by

$$x^{(k)} = x^{(k-1)} + M_{\text{SSOR}}^{-1}(b - Ax^{(k-1)}). \tag{11.2.25}$$

Note that M_{SSOR} is defined if $0 < \omega < 2$ and that it is symmetric. It is also positive definite if A has positive diagonal entries. Here is a result that shows SSOR converges if A is symmetric and positive definite.

Theorem 11.2.4. *Suppose the SSOR method (11.2.22) and (11.2.23) is applied to a symmetric positive definite $Ax = b$ problem and that $0 < \omega < 2$. If*

$$M_\omega = \frac{1}{\omega}D_A + L_A, \qquad N_\omega = \left(\frac{1}{\omega} - 1\right)D_A - L_A^T, \qquad G = M_\omega^{-T}N_\omega^T M_\omega^{-1}N_\omega,$$

then G has real eigenvalues, $\rho(G) < 1$, and

$$(x^{(k)} - x) = G^k(x^{(0)} - x). \tag{11.2.26}$$

Proof. From (11.2.22) and (11.2.23) it follows that

$$y^{(k)} - x = M_\omega^{-1}N_\omega(x^{(k-1)} - x),$$

$$x^{(k)} - x = M_\omega^{-T}N_\omega^T(y^{(k)} - x),$$

from which it is easy to verify (11.2.26). Since D is a diagonal matrix with positive diagonal entries, there is a diagonal matrix D_1 so $D = D_1^2$. If $L_1 = D_1^{-1}LD_1^{-1}$ and $G_1 = D_1 G D_1^{-1}$, then with a little manipulation we have

$$G_1 = (I + \omega L_1^T)^{-1}(I + \omega L_1)^{-1}((1-\omega)I - \omega L_1)((1-\omega)I - \omega L_1^T).$$

We show that if $\lambda \in \lambda(G_1)$, then $0 \le \lambda < 1$. If $G_1 v = \lambda v$, then

$$((1-\omega)I - \omega L_1)((1-\omega)I - \omega L_1^T)v = \lambda(I + \omega L_1)(I + \omega L_1^T)v.$$

This is a generalized singular value problem; see §8.7.4. It follows that λ is real and nonnegative. Assuming that $v \in \mathbb{R}^n$ has unit 2-norm, it is easy to show that

$$\lambda = \frac{\|(1-\omega)v - \omega L_1^T v\|_2^2}{\|v + \omega L_1^T v\|_2^2} = 1 - \omega(2-\omega)\frac{1 + 2v^T L_1^T v}{\|v + \omega L_1^T v\|_2^2}. \tag{11.2.27}$$

To complete the proof, note that $1 + 2v^T L_1^T v = (D_1^{-1}v)^T A(D_1^{-1}v)$ and that this quantity is positive. By hypothesis, $\omega(2-\omega) > 0$ and so we have $\lambda < 1$. $\quad\square$

The original analysis of the symmetric SOR method is in Young (1970).

11.2.8 The Chebyshev Semi-Iterative Method

Another way to accelerate the convergence of certain iterative methods makes use of Chebyshev polynomials. Suppose the iteration $Mx^{(j+1)} = Nx^{(j)} + b$ has been used to generate $x^{(1)}, \ldots, x^{(k)}$ and that we wish to determine coefficients $\nu_j(k)$, $j = 0{:}k$ such that

$$y^{(k)} = \sum_{j=0}^{k} \nu_j(k)x^{(j)} \tag{11.2.28}$$

represents an improvement over $x^{(k)}$. If $x^{(0)} = \cdots = x^{(k)} = x$, then it is reasonable to insist that $y^{(k)} = x$. If the polynomial

$$p_k(z) = \sum_{j=0}^{k} \nu_j(k)z^j$$

satisfies $p_k(1) = 1$, then this criterion is satisfied and

$$y^{(k)} - x = \sum_{j=0}^{k} \nu_j(k)(x^{(j)} - x) = \sum_{j=0}^{k} \nu_j(k)(M^{-1}N)^j e^{(0)} = p_k(G)e^{(0)}$$

where $G = M^{-1}N$. By taking norms in this equation we obtain

$$\| \, y^{(k)} - x \, \|_2 \; \le \; \| \, p_k(G) \, \|_2 \, \| \, e^{(0)} \, \|_2. \tag{11.2.29}$$

This suggests that we can produce an improved approximate solution if we can find a polynomial $p_k(\,\cdot\,)$ that (a) has degree k, (b) satisfies $p_k(1) = 1$, and (c) does a good job of minimizing the upper bound.

To implement this idea, we assume for simplicity that G is symmetric. (There are ways to proceed if this is not the case; see Manteuffel (1977). Let

$$S^T G S \; = \; \text{diag}(\lambda_1, \ldots, \lambda_n) \; = \; \Lambda$$

be a Schur decomposition of G and assume that

$$-1 < \alpha \le \lambda_n \le \cdots \le \lambda_1 \le \beta < 1 \tag{11.2.30}$$

where α and β are *known* estimates. It follows that

$$\| \, p_k(G) \, \|_2 \; = \; \| \, p_k(\Lambda) \, \|_2 \; = \; \max_{\lambda_i \in \lambda(A)} \; |p_k(\lambda_i)| \; \le \; \max_{\alpha \le \lambda \le \beta} \; |p_k(\lambda)|.$$

The degree-k Chebyshev polynomial $c_k(\cdot)$ can be used to design a good choice for $p_k(\,\cdot\,)$. We want a polynomial whose value on $[\alpha, \beta]$ is small subject to the constraint that $p_k(1) = 1$. Recall from the discussion in §10.1.5 that the Chebyshev polynomials are bounded by unity on $[-1, +1]$, but that their value is very large outside this range. As a consequence, if

$$\mu \; = \; -1 + 2\frac{1 - \alpha}{\beta - \alpha} \; = \; 1 + 2\frac{1 - \beta}{\beta - \alpha},$$

then the polynomial

$$p_k(z) = c_k \left(-1 + 2\frac{z - \alpha}{\beta - \alpha} \right) \Big/ c_k(\mu)$$

satisfies $p_k(1) = 1$ and is bounded by $1/|c_k(\mu)|$ on $[\alpha, \beta]$. From the definition of $p_k(z)$ and inequality (11.2.29) we see

$$\| y^{(k)} - x \|_2 \leq \frac{\| x - x^{(0)} \|_2}{|c_k(\mu)|}.$$

The larger the value of μ the greater the acceleration of convergence.

In order for the whole process to be effective, we need a more efficient method for calculating $y^{(k)}$ than (11.2.28). The retrieval of the vectors $x^{(0)}, \ldots, x^{(k)}$ becomes an unacceptable overhead as k increases. Fortunately, it is possible to derive a three-term recurrence among the $y^{(k)}$ by exploiting the three-term recurrence that exists among the Chebyshev polynomials. Assume (for simplicity) that $\alpha = -\beta$ in (11.2.30) and that we are given $x^{(0)} \in \mathbb{R}^n$. Here is how the process plays out when it is used to accelerate the iteration $Mx^{(j+1)} = Nx^{(j)} + b$:

$$c_0 = 1; c_1 = 1/\beta$$
$$y^{(0)} = x^{(0)}, My^{(1)} = Ny^{(0)} + b, r^{(1)} = b - Ay^{(1)}, k = 1$$
while $\| r^{(k)} \| > tol$
$$\qquad c_{k+1} = (2/\beta)c_k - c_{k-1}$$
$$\qquad \omega_{k+1} = 1 + c_{k-1}/c_{k+1}$$
$$\qquad Mz^{(k)} = r^{(k)}$$
$$\qquad y^{(k+1)} = y^{(k-1)} + \omega_{k+1} \left(y^{(k)} + z^{(k)} - y^{(k-1)} \right)$$
$$\qquad k = k + 1$$
$$\qquad r^{(k)} = b - Ay^{(k)}$$
end

Note that $y^{(0)} = x^{(0)}$ and $y^{(1)} = x^{(1)}$, but that thereafter the $x^{(k)}$ are not involved. For the acceleration to be effective we need good lower and upper bounds in (11.2.30) and that is sometimes difficult to accomplish. The method is extensively analyzed in Golub and Varga (1961) and Varga (1962, Chap. 5).

Problems

P11.2.1 Show that the Jacobi iteration converges for 2-by-2 symmetric positive definite systems.

P11.2.2 Show that if $A = M - N$ is singular, then we can never have $\rho(M^{-1}N) < 1$ even if M is nonsingular.

P11.2.3 (Supplied by R.S. Varga) Suppose that

$$A_1 = \begin{bmatrix} 1 & -1/2 \\ -1/2 & 1 \end{bmatrix}, \qquad A_2 = \begin{bmatrix} 1 & -3/4 \\ -1/12 & 1 \end{bmatrix}.$$

Let J_1 and J_2 be the associated Jacobi iteration matrices. Show that $\rho(J_1) > \rho(J_2)$, thereby refuting the claim that greater diagonal dominance implies more rapid Jacobi convergence.

P11.2.4 Suppose $A = T_{n_1} \otimes I_{n_2} \otimes I_{n_3} + I_{n_1} \otimes T_{n_2} \otimes I_{n_3} + I_{n_1} \otimes I_{n_2} \otimes T_{n_3}$. If Jacobi's method is applied to the problem $Au = b$, then what is the spectral radius of the associated iteration matrix?

P11.2.5 A 5-point "stencil" is associated with the matrix $A = I_{n_1} \otimes T_{n_2} + T_{n_1} \otimes I_{n_2}$ and leads to the requirement that $U(i, j)$ be the average of $U(i-1, j)$, $U(i, j+1)$, $U(i+1, j)$, and $U(i, j-1)$. Formulate a 9-point stencil procedure in which $U(i, j)$ is a suitable average of its eight neighbors. (a) Describe the resulting matrix using Kronecker products. (b) If Jacobi's method is used to solve $Au = b$, then what is the spectral radius of the associated iteration matrix?

P11.2.6 Consider the linear system $(I_{n_1} \otimes T_{n_2} + T_{n_1} \otimes I_{n_2})x = b$. What is the spectral radius of the iteration matrix for the block Jacobi iteration if the diagonal blocks are n_2-by-n_2?

P11.2.7 Prove (11.2.13) and (11.2.14).

P11.2.8 Prove (11.2.15).

P11.2.9 Prove (11.2.17).

P11.2.10 Prove (11.2.24) and (11.2.25).

P11.2.11 Consider the 2-by-2 matrix

$$A = \begin{bmatrix} 1 & \rho \\ -\rho & 1 \end{bmatrix}.$$

(a) Under what conditions do we have $\rho(M_{\mathrm{GS}}^{-1} N_{\mathrm{GS}}) < 1$? (b) For what range of ω do we have $\rho(M_\omega^{-1} N_\omega) < 1$? What value of ω minimizes $\rho(M_\omega^{-1} N_\omega)$? (c) Repeat (a) and (b) for the matrix

$$A = \begin{bmatrix} I_n & S \\ -S^T & I_n \end{bmatrix}$$

where $S \in \mathbb{R}^{n \times n}$. Hint: Use the SVD of S.

P11.2.12 We want to investigate the solution of $Au = f$ where $A \neq A^T$. For a model problem, consider the finite difference approximation to

$$-u'' + \sigma u' = 0, \qquad 0 < x < 1,$$

where $u(0) = 10$ and $u(1) = 10 \exp^\sigma$. This leads to the difference equation

$$-u_{i-1} + 2u_i - u_{i+1} + R(u_{i+1} - u_{i-1}) = 0, \qquad i = 1{:}n,$$

where $R = \sigma h/2$, $u_0 = 10$, and $u_{n+1} = 10 e^\sigma$. The number R should be less than 1. What is the spectral radius of $M^{-1}N$ where $M = (A + A^T)/2$ and $N = (A^T - A)/2$?

P11.2.13 Consider the iteration

$$y^{(k+1)} = \omega(By^{(k)} + d - y^{(k-1)}) + y^{(k-1)}$$

where B has Schur decomposition $Q^T B Q = \mathrm{diag}(\lambda_1, \ldots, \lambda_n)$ with $\lambda_1 \geq \cdots \geq \lambda_n$. Assume that $x = Bx + d$. (a) Derive an equation for $e^{(k)} = y^{(k)} - x$. (b) Assume $y^{(1)} = By^{(0)} + d$. Show that $e^{(k)} = p_k(B)e^{(0)}$ where p_k is an even polynomial if k is even and an odd polynomial if k is odd. (c) Write $f^{(k)} = Q^T e^{(k)}$. Derive a difference equation for $f_j^{(k)}$ for $j = 1{:}n$. Try to specify the exact solution for general $f_j^{(0)}$ and $f_j^{(1)}$. (d) Show how to determine an optimal ω.

P11.2.14 Suppose we want to solve the linear least squares problem $\min \| Ax - b \|_2$ where $A \in \mathbb{R}^{m \times n}$, $\mathrm{rank}(A) = r \leq n$, and $b \in \mathbb{R}^m$. Consider the iterative scheme

$$Mx_{i+1} = Nx_i + A^T b$$

where $M = (A^T A + \lambda W)$, $N = \lambda W$, $\lambda > 0$ and $W \in \mathbb{R}^{n \times n}$ is symmetric positive definite. (a) Show that $M^{-1}N$ is diagonalizable and that $\rho(M^{-1}N) < 1$ if $\mathrm{rank}(A) = n$. (b) Suppose $x_0 = 0$ and that $\| v \|_W = \left(v^T W v \right)^{-1/2}$, the "W-norm." Show that regardless of A's rank, the iterates x_i converge to the minimum W-norm solution to the least squares problem. (c) Show that if $\mathrm{rank}(A) = n$ then $\| x_{LS} - x_{i+1} \|_W \leq \| x_{LS} - x_i \|_W$. (d) Show how to implement the iteration give the QR factorization of

$$M = \begin{bmatrix} A \\ \sqrt{\lambda} F \end{bmatrix}$$

where $W = FF^T$ is the Cholesky factorization of W.

P11.2.15 (a) Suppose $T \in \mathbb{R}^{n \times n}$ is tridiagonal with the property that $t_{i,i+1} t_{i+1,i} > 0$ for $i = 1{:}n-1$.

Show that there is a diagonal matrix $D \in \mathbb{R}^{n \times n}$ so that $S = DTD^{-1}$ is symmetric. (b) Consider the following linear system for unknowns $u_1, \ldots, u_n$:

$$-u_{i-1} + 2u_i - u_{i+1} + \frac{\sigma h}{2}(u_{i+1} - u_i) = f_i, \qquad i = 1{:}n.$$

Assume $u_0 \equiv \alpha$, $u_{n+1} \equiv \beta$, $\sigma > 0$, and $h > 0$. Under what conditions can this tridiagonal system be symmetrized using (a)? (c) Give formulae for the eigenvalues of the Jacobi iteration matrix.

Notes and References for §11.2

For detailed treatment of the material in this section, see Greenbaum (IMSL, Chap. 10) or any of the following volumes:

R.S. Varga (1962). *Matrix Iterative Analysis*, Prentice-Hall, Englewood Cliffs, NJ.
D.M. Young (1971). *Iterative Solution of Large Linear Systems*, Academic Press, New York.
L.A. Hageman and D.M. Young (1981). *Applied Iterative Methods*, Academic Press, New York.
W. Hackbusch (1994). *Iterative Solution of Large Sparse Systems of Equations*, Springer-Verlag, New York.

As we mentioned, Young (1971) has the most comprehensive treatment of the SOR method. The object of SOR theory is to guide the user in choosing the relaxation parameter ω. In this setting, the ordering of equations and unknowns is critical, see:

M.J.M. Bernal and J.H. Verner (1968). "On Generalizing of the Theory of Consistent Orderings for Successive Over-Relaxation Methods," *Numer. Math. 12*, 215–222.
D.M. Young (1970). "Convergence Properties of the Symmetric and Unsymmetric Over-Relaxation Methods," *Math. Comput. 24*, 793–807.
D.M. Young (1972). "Generalization of Property A and Consistent Ordering," *SIAM J. Numer. Anal. 9*, 454–463.
R.A. Nicolaides (1974). "On a Geometrical Aspect of SOR and the Theory of Consistent Ordering for Positive Definite Matrices," *Numer. Math. 12*, 99–104.
A. Ruhe (1974). "SOR Methods for the Eigenvalue Problem with Large Sparse Matrices," *Math. Comput. 28*, 695–710.
L. Adams and H. Jordan (1986). "Is SOR Color-Blind?" *SIAM J. Sci. Stat. Comput. 7*, 490–506.
M. Eiermann and R.S. Varga (1993). "Is the Optimal ω Best for the SOR Iteration Method," *Lin. Alg. Applic. 182*, 257–277.
H. Lu (1999). "Stair Matrices and Their Generalizations with Applications to Iterative Methods I: A Generalization of the Successive Overrelaxation Method," *SIAM J. Numer. Anal. 37*, 1–17.

An analysis of the Chebyshev semi-iterative method appears in:

G.H. Golub and R.S. Varga (1961). "Chebyshev Semi-Iterative Methods, Successive Over-Relaxation Iterative Methods, and Second-Order Richardson Iterative Methods, Parts I and II," *Numer. Math. 3*, 147–156, 157–168.

That work is premised on the assumption that the underlying iteration matrix has real eigenvalues. How to proceed when this is not the case is discussed in:

T.A. Manteuffel (1977). "The Tchebychev Iteration for Nonsymmetric Linear Systems," *Numer. Math. 28*, 307–327.
M. Eiermann and W. Niethammer (1983). "On the Construction of Semi-iterative Methods," *SIAM J. Numer. Anal. 20*, 1153–1160.
W. Niethammer and R.S. Varga (1983). "The Analysis of k-step Iterative Methods for Linear Systems from Summability Theory," *Numer. Math. 41*, 177–206.
G.H. Golub and M. Overton (1988). "The Convergence of Inexact Chebyshev and Richardson Iterative Methods for Solving Linear Systems," *Numer. Math. 53*, 571–594.
D. Calvetti, G.H. Golub, and L. Reichel (1994). "An Adaptive Chebyshev Iterative Method for Nonsymmetric Linear Systems Based on Modified Moments," *Numer. Math. 67*, 21–40.
E. Giladi, G.H. Golub, and J.B. Keller (1998). "Inner and Outer Iterations for the Chebyshev Algorithm," *SIAM J. Numer. Anal. 35*, 300–319.

Other methods for unsymmetric problems are discussed in:

M. Eiermann, W. Niethammer, and R.S. Varga (1992). "Acceleration of Relaxation Methods for Non-Hermitian Linear Systems," *SIAM J. Matrix Anal. Applic. 13*, 979–991.

H. Elman and G.H. Golub (1990). "Iterative Methods for Cyclically Reduced Non-Self-Adjoint Linear Systems I," *Math. Comput. 54*, 671–700.

H. Elman and G.H. Golub (1990). "Iterative Methods for Cyclically Reduced Non-Self-Adjoint Linear Systems II," *Math. Comput. 56*, 215–242.

R. Bramley and A. Sameh (1992). "Row Projection Methods for Large Nonsymmetric Linear Systems," *SIAM J. Sci. Statist. Comput. 13*, 168–193.

Iterative methods for complex symmetric systems are detailed in:

O. Axelsson and A. Kucherov (2000). "Real Valued Iterative Methods for Solving Complex Symmetric Linear Systems," *Numer. Lin. Alg. 7*, 197–218.

V.E. Howle and S.A. Vavasis (2005). "An Iterative Method for Solving Complex-Symmetric Systems Arising in Electrical Power Modeling," *SIAM J. Matrix Anal. Applic. 26*, 1150–1178.

Iterative methods for singular systems are discussed in:

A. Dax (1990). "The Convergence of Linear Stationary Iterative Processes for Solving Singular Unstructured Systems of Linear Equations," *SIAM Review 32*, 611–635.

Z.-H. Cao (2001). "A Note on Properties of Splittings of Singular Symmetric Positive Semidefinite Matrices," *Numer. Math. 88*, 603–606.

Papers that are concerned with parallel implementation include:

D.J. Evans (1984). "Parallel SOR Iterative Methods," *Parallel Comput. 1*, 3–18.

N. Patel and H. Jordan (1984). "A Parallelized Point Rowwise Successive Over-Relaxation Method on a Multiprocessor," *Parallel Comput. 1*, 207–222.

R.J. Plemmons (1986). "A Parallel Block Iterative Scheme Applied to Computations in Structural Analysis," *SIAM J. Alg. Disc. Meth. 7*, 337–347.

C. Kamath and A. Sameh (1989). "A Projection Method for Solving Nonsymmetric Linear Systems on Multiprocessors," *Parallel Computing 9*, 291–312.

P. Amodio and F. Mazzia (1995). "A Parallel Gauss-Seidel Method for Block Tridiagonal Linear Systems," *SIAM J. Sci. Comput. 16*, 1451–1461.

We have seen that the condition $\kappa(A)$ is an important issue when direct methods are applied to $Ax = b$. However, the condition of the system also has a bearing on iterative method performance, see:

M. Arioli and F. Romani (1985). "Relations Between Condition Numbers and the Convergence of the Jacobi Method for Real Positive Definite Matrices," *Numer. Math. 46*, 31–42.

M. Arioli, I.S. Duff, and D. Ruiz (1992). "Stopping Criteria for Iterative Solvers," *SIAM J. Matrix Anal. Applic. 13*, 138–144.

Finally, the effect of rounding errors on the methods of this section is treated in:

H. Wozniakowski (1978). "Roundoff-Error Analysis of Iterations for Large Linear Systems," *Numer. Math. 30*, 301–314.

P.A. Knight (1993). "Error Analysis of Stationary Iteration and Associated Problems," Ph.D. thesis, Department of Mathematics, University of Manchester, England.

11.3 The Conjugate Gradient Method

A difficulty associated with the SOR, Chebyshev semi-iterative, and related methods is that they depend upon parameters that are sometimes hard to choose properly. For example, the Chebyshev acceleration scheme requires good estimates of the largest and smallest eigenvalues of the underlying iteration matrix $M^{-1}N$. This can be a very challenging problem unless this matrix is sufficiently structured. In this section and the next we present various Krylov subspace methods that avoid this difficulty.

We start with the well-known conjugate gradient (CG) method due to Hestenes and Stieffel (1952) and which is applicable to symmetric positive definite systems.

There are several ways to motivate and derive the technique. Our approach involves the method of steepest descent, Krylov subspaces, the Lanczos process, and tridiagonal system solving. After developing the Lanczos implementation of the CG process, we proceed to establish its equivalence with the Hestenes-Stieffel formulation.

A brief comment about notation is in order. Most of the methods in the previous section are developed at the (i, j) level and this necessitated the use of superscripts to designate vector iterates. From now on, the derivations in this chapter can proceed at the vector level. Subscripts will be used to designate vector iterates, so instead of $\{x^{(k)}\}$ we now have $\{x_k\}$.

11.3.1   An Optimization Problem

Suppose $A \in \mathbb{R}^{n \times n}$ is symmetric positive definite, $b \in \mathbb{R}^n$, and that we want to compute the solution x_* to

$$Ax = b. \tag{11.3.1}$$

Note that this problem is equivalent to solving the optimization problem

$$\min_{x \, \in \, \mathbb{R}^n} \; \phi(x) \tag{11.3.2}$$

where

$$\phi(x) \; = \; \frac{1}{2} x^T A x \; - \; x^T b. \tag{11.3.3}$$

This is because ϕ is convex and its gradient is given by

$$\nabla \phi(x) = Ax - b.$$

Thus, if x_c is an approximate minimizer of ϕ, then x_c can be regarded as an approximate solution to $Ax = b$. To make this precise, we define the A-norm by

$$\| \, v \, \|_A \; = \; \sqrt{v^T A v} \, . \tag{11.3.4}$$

Since

$$\phi(x_c) \; = \; \frac{1}{2} x_c^T A x_c - x_c^T b \; = \; \frac{1}{2}(x_c - x_*) A(x_c - x_*) \; - \; \frac{1}{2} b^T A^{-1} b$$

and $\phi(x_*) = -b^T A^{-1} b / 2$, it follows that

$$\phi(x_c) \; = \; \frac{1}{2} \| \, x_c - x_* \, \|_A^2 \; + \; \phi(x_*). \tag{11.3.5}$$

Thus, an iteration that produces a sequence of ever-better approximate minimizers for ϕ is an iteration that produces ever-better approximate solutions to $Ax = b$ as measured in the A-norm.

11.3.2   The Method of Steepest Descent

Let us consider the minimization of ϕ using the method of steepest descent with exact line searches. In this method the current approximate minimizer x_c is improved by searching in the direction of the negative gradient, i.e., the direction of most rapid decrease. In particular, the improved approximate minimizer x_+ is given by

$$x_+ = x_c - \mu_c g_c,$$

where $g_c = Ax_c - b$ is the current gradient and μ_c solves

$$\min_{\mu \in \mathbb{R}} \phi(x_c - \mu g_c). \tag{11.3.6}$$

This is an *exact line search framework*. It is easy to show that

$$\mu_c = \frac{g_c^T g_c}{g_c^T A g_c}$$

and

$$\phi(x_+) = \phi(x_c) - \frac{1}{2} \cdot \frac{(g_c^T g_c)^2}{r_c^T A r_c}. \tag{11.3.7}$$

Thus, the objective function is decreased if $r_c \neq 0$. To establish global convergence of the method, define

$$\kappa_c = \frac{g_c^T A g_c}{g_c^T g_c} \cdot \frac{g_c^T A^{-1} g_c}{g_c^T g_c}$$

and observe that $g_c^T A^{-1} g_c = 2\phi(x_c) + b^T A^{-1} b$ and

$$\phi(x_+) = \phi(x_c) - \frac{1}{2} \frac{1}{\kappa_c} g_c^T A^{-1} g_c = \phi(x_c) - \frac{1}{\kappa_c}\left(\phi(x_c) + \frac{1}{2} b^T A^{-1} b\right). \tag{11.3.8}$$

If $\lambda_{\max}(A)$ and $\lambda_{\min}(A)$ are the largest and smallest eigenvalues of A, then we have

$$\kappa_c = \frac{g_c^T A g_c}{g_c^T g_c} \cdot \frac{g_c^T A^{-1} g_c}{g_c^T g_c} \leq \frac{\lambda_{\max}(A)}{\lambda_{\min}(A)} = \kappa_2(A).$$

If we subtract $\phi(x_*) = -(b^T A^{-1} b)/2$ from both sides of (11.3.8) and use (11.3.5), then we obtain

$$\| x_+ - x_* \|_A^2 \leq \left(1 - \frac{1}{\kappa_2(A)}\right) \| x_c - x_* \|_A^2. \tag{11.3.9}$$

It follows by induction that the method of steepest descent with exact line search is globally convergent.

Algorithm 11.3.1 (Steepest Descent with Exact Line Search) Given a symmetric positive definite $A \in \mathbb{R}^{n \times n}$, $b \in \mathbb{R}^n$, $Ax_0 \approx b$, and a termination tolerance τ, the following algorithm produces $x \in \mathbb{R}^n$ so that $\| Ax - b \|_2 \leq \tau$.

> $x = x_0$, $g = Ax - b$
> **while** $\| g \|_2 > \tau$
> $\quad \mu = (g^T g)/(g^T A g)$, $x = x - \mu g$, $g = Ax - b$
> **end**

Unfortunately, a convergence rate characterized by $(1 - 1/\kappa_2(A))^{k/2}$ is typically not good enough unless A is extremely well-conditioned.

11.3.3 A Subspace Strategy

We can improve upon the steepest descent idea by expanding the dimension of the search space each step. To pursue this idea we introduce the notion of an affine space. Formally, if $v \in \mathbb{R}^n$ and $S \subseteq \mathbb{R}^n$ is a subspace, then

$$v + S = \{\, x \mid x = v + s,\, s \in S \,\}.$$

is an *affine space*. Note that in Algorithm 11.3.1, the step-k optimization is over the affine space $x_k + \mathsf{span}\{\nabla\phi(x_k)\}$.

Given $Ax_0 \approx b$, our plan is to produce a nested sequence of subspaces

$$S_1 \subset S_2 \subset S_3 \subset \cdots$$

that satisfy $\dim(S_k) = k$ and to solve the problem

$$\min_{x \,\in\, x_0 + S_k} \phi(x) \tag{11.3.10}$$

each step along the way. If x_k is the step-k minimizer, then because of the nesting we have $\phi(x_1) \ge \phi(x_2) \ge \cdots \ge \phi(x_n) = \phi(x_*)$. Since $S_n = \mathbb{R}^n$, we ultimately obtain $x_* = A^{-1}b$. Even though this is a finite-step solution framework, it may not be attractive if n is extremely large. The challenge is to find a subspace sequence that promotes rapid decrease in the value of ϕ, for then we may be able to terminate the iteration long before k equals n.

With this goal in mind we note that at x_k the function ϕ decreases most rapidly in the direction of the negative gradient. Thus, it makes sense to choose S_{k+1} so that it includes x_k *and* the gradient $g_k = \nabla\phi(x_k) = Ax_k - b$. This strategy guarantees that x_{k+1} is at least as good as a steepest descent update:

$$\min_{x \in x_0 + S_{k+1}} \phi(x) = \phi(x_{k+1}) \le \min_{\mu \in \mathbb{R}} \phi(x_k - \mu g_k) \tag{11.3.11}$$

If x_0 is an initial guess and we define $g_0 = Ax_0 - b$, then since $\nabla\phi(x_k) \in \mathsf{span}\{x_k, Ax_k\}$ it follows that the only way to satisfy this requirement is to set

$$S_k = \mathcal{K}(A, g_0, k) = \mathsf{span}\{g_0, Ag_0, A^2 g_0, \dots, A^{k-1} g_0\}.$$

We can use the Lanczos process (§10.1) to generate these Krylov subspaces.

11.3.4 The Method of Conjugate Gradients: First Version

Recall that after k steps of the Lanczos iteration (Algorithm 10.1.1) we have generated a matrix

$$Q_k = [\, q_1 \mid \cdots \mid q_k \,] \in \mathbb{R}^{n \times k}$$

with orthonormal columns, a tridiagonal matrix

$$T_k = \begin{bmatrix} \alpha_1 & \beta_1 & & \cdots & & 0 \\ \beta_1 & \alpha_2 & \ddots & & & \vdots \\ & \ddots & \ddots & \ddots & & \\ \vdots & & \ddots & \ddots & & \beta_{k-1} \\ 0 & \cdots & & & \beta_{k-1} & \alpha_k \end{bmatrix}, \tag{11.3.12}$$

and a vector $r_k \in \mathrm{ran}(Q_k)^\perp$ so that

$$AQ_k = Q_kT_k + r_ke_k^T. \tag{11.3.13}$$

Note that the tridiagonal matrix

$$Q_k^T AQ_k = T_k$$

is positive definite. The solution to the optimization problem (11.3.10) via Lanczos is particularly simple if we set $q_1 = r_0/\beta_0$ where $r_0 = b - Ax_0 = -g_0$, and $\beta_0 = \| r_0 \|_2$. Since the columns of Q_k span $S_k = \mathcal{K}(A, g_0, k)$, it follows that the act of minimizing ϕ over $x_0 + S_k$ is equivalent to minimizing $\phi(x_0 + Q_ky)$ over all vectors $y \in \mathbb{R}^k$. Since

$$\phi(x_0 + Q_ky) = \frac{1}{2}(x_0 + Q_ky)^T A(x_0 + Q_ky) - (x_0 + Q_ky)^Tb$$

$$= \frac{1}{2}y^T(Q_k^T AQ_k)y - y^T(Q_k^T r_0) + \phi(x_0)$$

and $\beta_0Q_k(:, 1) = r_0$, it follows that the minimizer y_k satisfies

$$T_ky_k = Q_k^T r_0 = \beta_0e_1$$

and so $x_k = x_0 + Q_ky_k$. Building on Algorithm 10.1.1, this leads to a preliminary version of the *conjugate gradient (CG)* method:

$$k = 0, \ r_0 = b - Ax_0, \ \beta_0 = \| r_0 \|_2, \ q_0 = 0$$
while $\beta_k \neq 0$
$$\qquad q_{k+1} = r_k/\beta_k$$
$$\qquad k = k + 1$$
$$\qquad \alpha_k = q_k^T Aq_k \tag{11.3.14}$$
$$\qquad T_ky_k = \beta_0e_1$$
$$\qquad x_k = Q_ky_k$$
$$\qquad r_k = (A - \alpha_kI)q_k - \beta_{k-1}q_{k-1}$$
$$\qquad \beta_k = \| r_k \|_2$$
end
$$x_* = x_k$$

As it stands, this formulation is not suitable for large problems because x_k is computed as an explicit n-by-k matrix-vector product and this requires access to all previously computed Lanczos vectors. However, before we develop a slick recursion for x_k that circumvents this problem, we establish some important properties that are associated with the iteration.

Theorem 11.3.1. *If k_* is the dimension of the smallest invariant subspace that contains r_0, then the conjugate gradient iteration (11.3.14) terminates with $x_{k_*} = x_*$.*

Proof. From Theorem 10.1.1 we know that the Lanczos iteration terminates after generating q_k if $\mathcal{K}(A, q_1, k)$ is an invariant subspace. If $q_1 = r_0/\| r_0 \|_2$, then q_{k_*}

must be generated for otherwise r_0 would be contained in an invariant subspace with dimension less than k_*. Since we can write r_0 as a linear combination of k_* eigenvectors, it follows that the Krylov matrix $[r_0 \mid Ar_0 \mid A^2 r_0 \mid \cdots \mid A^{k_*} r_0]$ has rank k_*. This implies $\beta_{k_*} = 0$ in (11.3.14) and so the iteration terminates with $x_* = x_{k_*}$. $\square$

An important ramification is that early termination can be expected if the matrix A is a low-rank perturbation of the identity matrix.

Corollary 11.3.2. *Assume that $U \in \mathbb{R}^{n \times r}$, $D \in \mathbb{R}^{r \times r}$ is symmetric, and $r < n$. If $A = I_n + UDU^T$ is positive definite and the conjugate gradient iteration (11.3.14) is applied to the problem $Ax = b$, then at most $r + 1$ iterations are required to compute x_*.*

Proof. If $v \in \mathbb{R}^n$ is in the nullspace of U^T, then $Av = v$ and $\lambda = 1$ is an eigenvalue of A with multiplicity at least $n - r$. It follows that A cannot have more than $r + 1$ distinct eigenvalues. Thus, r_0 is contained in an invariant subspace with dimension $r + 1$. $\square$

Recall that our derivation of (11.3.14) begins with a plan to improve upon the method of steepest descent. Instead of determining x_k from a 1-dimensional search in the direction of the $\nabla \phi(x_{k-1})$, the CG method determines x_k by searching over a Krylov subspace that includes $\nabla \phi(x_{k-1})$. It follows that a CG step is at least as good as a steepest descent step, as the following theorem shows.

Theorem 11.3.3. *If x_* is the solution to the symmetric positive definite system $Ax = b$ and x_k and x_{k+1} are produced by the CG method (11.3.14), then*

$$\| x_{k+1} - x_* \|_A \le \left(1 - \frac{1}{\kappa_2(A)} \right)^{1/2} \cdot \| x_k - x_* \|_A.$$

Proof. Setting $x_c = x_k$ in (11.3.9) gives

$$\| x_+ - x_* \|_A \le \left(1 - \frac{1}{\kappa_2(A)} \right)^{1/2} \| x_k - x_* \|_A,$$

where x_+ is the steepest descent successor to x_c. By using inequality (11.3.11) we have $\| x_{k+1} - x_* \|_A \le \| x_+ - x_* \|_A$. $\square$

Just how these mathematical results color practical matters is detailed in §11.5. For now, we continue with our exact arithmetic derivation of the method.

11.3.5   The Method of Conjugate Gradients: Second Version

Returning to the initial version of the CG method in (11.3.14), we work out the details associated with the tridiagonal solve $T_k y_k = \beta_0 e_1$ and the matrix-vector product $x_k = Q_k y_k$. For the overall implementation to be attractive for large sparse A, we need a way to compute x_k without having to access Lanczos vectors $q_1, \ldots, q_k$. Since the tridiagonal matrix $T_k = Q_k^T A Q_k$ is positive definite, it has an LDL^T factorization. By comparing coefficients in $T_k = L_k D_k L_k^T$ where

$$
L_k = \begin{bmatrix} 1 & 0 & 0 & 0 \\ \ell_1 & 1 & 0 & 0 \\ \vdots & \ddots & \ddots & \vdots \\ 0 & \cdots & \ell_{k-1} & 1 \end{bmatrix}, \qquad D_k = \begin{bmatrix} d_1 & 0 & \cdots & 0 \\ 0 & d_2 & & \vdots \\ \vdots & & \ddots & 0 \\ 0 & \cdots & 0 & d_k \end{bmatrix},
$$

we find

$$
\begin{aligned}
&d_1 = \alpha_1 \\
&\textbf{for } i = 2{:}k \\
&\qquad \ell_{i-1} = \beta_{i-1}/d_{i-1} \\
&\qquad d_i = \alpha_i - \ell_{i-1}\beta_{i-1} \\
&\textbf{end}
\end{aligned} \tag{11.3.15}
$$

Given this factorization, we see that if $v_k \in \mathbb{R}^k$ solves

$$
L_k D_k v_k = \beta_0 e_1 \tag{11.3.16}
$$

then $L_k^T y_k = v_k$. If $C_k \in \mathbb{R}^{n \times k}$ satisfies

$$
C_k L_k^T = Q_k, \tag{11.3.17}
$$

then

$$
x_k = x_0 + Q_k y_k = x_0 + C_k L_k^T y_k = x_0 + C_k v_k. \tag{11.3.18}
$$

This is an impractical recipe because the matrix C_k is full and involves all the Lanczos vectors. However, there are simple connections between C_{k-1} and C_k and between v_{k-1} and v_k that can be used to transform (11.3.18) into a very handy update recipe for x_k. Consider the lower bidiagonal system (11.3.16), e.g.,

$$
\left[\begin{array}{ccc|c} d_1 & 0 & 0 & 0 \\ d_1\ell_1 & d_2 & 0 & 0 \\ 0 & d_2\ell_2 & d_3 & 0 \\ \hline 0 & 0 & d_3\ell_3 & d_4 \end{array}\right] \begin{bmatrix} \nu_1 \\ \nu_2 \\ \nu_3 \\ \nu_4 \end{bmatrix} = \begin{bmatrix} \beta_0 \\ 0 \\ 0 \\ 0 \end{bmatrix}.
$$

We conclude that

$$
v_k = \begin{bmatrix} \nu_1 \\ \vdots \\ \nu_{k-1} \\ \hline \nu_k \end{bmatrix} = \left[\begin{array}{c} v_{k-1} \\ \hline \nu_k \end{array}\right] \tag{11.3.19}
$$

where

$$
\nu_k = \begin{cases} \beta_0/d_1 & \text{if } k = 1 \\ -d_{k-1}\ell_{k-1}\nu_{k-1}/d_k & \text{if } k > 1 \end{cases}. \tag{11.3.20}
$$

Next, we consider a column partitioning of equation (11.3.17), e.g.,

$$\left[\begin{array}{c|c|c|c} c_1 & c_2 & c_3 & c_4 \end{array}\right] \begin{bmatrix} 1 & \ell_1 & 0 & 0 \\ 0 & 1 & \ell_2 & 0 \\ 0 & 0 & 1 & \ell_3 \\ 0 & 0 & 0 & 1 \end{bmatrix} = \left[\begin{array}{c|c|c|c} q_1 & q_2 & q_3 & q_4 \end{array}\right].$$

From this we conclude that

$$C_k = \left[\begin{array}{c|c} C_{k-1} & c_k \end{array}\right] \tag{11.3.21}$$

where

$$c_k = \begin{cases} q_1 & \text{if } k = 1 \\ q_k - \ell_{k-1}c_{k-1} & \text{if } k > 1 \end{cases}. \tag{11.3.22}$$

It follows from (11.3.19) and (11.3.21) that

$$x_k = x_0 + C_k v_k = x_0 + C_{k-1}v_{k-1} + \nu_k c_k = x_{k-1} + \nu_k c_k.$$

This is precisely the kind of recursive formula for x_k that we need to make the recipe (11.3.18) attractive for large sparse problems. Combining this expression with (11.3.20) and (11.3.22), we obtain the following implementation of (11.3.14).

Algorithm 11.3.2 (Conjugate Gradients: Lanczos Version) If $A \in \mathbb{R}^{n \times n}$ is symmetric positive definite, $b \in \mathbb{R}^n$, and $Ax_0 \approx b$, then this algorithm computes $x_* \in \mathbb{R}^n$ so that $Ax_* = b$.

$\qquad k = 0,\ r_0 = b - Ax_0,\ \beta_0 = \| r_0 \|_2,\ q_0 = 0,\ c_0 = 0$
$\qquad$ **while** $\beta_k \neq 0$
$\qquad\qquad q_{k+1} = r_k / \beta_k$
$\qquad\qquad k = k + 1$
$\qquad\qquad \alpha_k = q_k^T A q_k$
$\qquad\qquad$ **if** $k = 1$
$\qquad\qquad\qquad d_1 = \alpha_1,\ \nu_1 = \beta_0 / d_1$
$\qquad\qquad\qquad c_k = q_1$
$\qquad\qquad$ **else**
$\qquad\qquad\qquad \ell_{k-1} = \beta_{k-1}/d_{k-1},\ d_k = \alpha_k - \beta_{k-1}\ell_{k-1},\ \nu_k = -\beta_{k-1}\nu_{k-1}/d_k$
$\qquad\qquad\qquad c_k = q_k - \ell_{k-1}c_{k-1}$
$\qquad\qquad$ **end**
$\qquad\qquad x_k = x_{k-1} + \nu_k c_k$
$\qquad\qquad r_k = Aq_k - \alpha_k q_k - \beta_{k-1}q_{k-1}$
$\qquad\qquad \beta_k = \| r_k \|_2$
$\qquad$ **end**
$\qquad x_* = x_k$

Each iteration involves a single matrix-vector product and about $13n$ flops. It can be implemented with just a handful of length-n storage arrays as we discuss in §11.3.8.

11.3.6 The Gradients Are Conjugate

We make some observations about the gradients and search directions that arise during the CG iteration. First, we show that the gradients

$$g_k = Ax_k - b = \nabla\phi(x_k)$$

are mutually orthogonal, a fact that explains the name of the algorithm.

Theorem 11.3.4. *If $x_1, \ldots, x_k$ are generated by Algorithm 11.3.2, then $g_i^T g_j = 0$ for all i and j that satisfy $1 \le i < j \le k$. Moreover, $g_k = \nu_k r_k$ where ν_k and r_k are defined by the algorithm.*

Proof. The partial tridiagonalization (11.3.13) permits us to write

$$g_k = Ax_k - b = A(x_0 + Q_k y_k) - b = -r_0 + (Q_k T_k + r_k e_k^T)y_k.$$

Since $Q_k T_k y_k = \beta_0 Q_k e_1 = r_0$, it follows that

$$g_k = (e_k^T y_k)r_k.$$

Since each r_i is a multiple of q_{i+1}, it follows that the g_i are mutually orthogonal. To show that $g_k = \nu_k r_k$, we must verify that $e_k^T y_k = \nu_k$. From the equation

$$T_k y_k = (L_k D_k)(L_k^T y_k) = \beta_0 e_1$$

we know that $L_k^T y_k = v_k$ where $(L_k D_k)v_k = \beta_0 e_1$. To complete the proof, recall from (11.3.19) that ν_k is the bottom component of v_k and exploit the fact that L_k^T is unit upper bidiagonal. $\square$

The search directions $c_1, \ldots, c_k$ satisfy a different kind of orthogonality property.

Theorem 11.3.5. *If $c_1, \ldots, c_k$ are generated by Algorithm 11.3.2, then*

$$c_i^T A c_j = \begin{cases} 0 & \text{if } i \ne j, \\ d_j & \text{if } i = j, \end{cases}$$

for all i and j that satisfy $1 \le i < j \le k$.

Proof. Since $Q_k = C_k L_k^T$ and $T_k = Q_k^T A Q_k$, we have

$$T_k = L_k(C_k^T A C_k)L_k^T.$$

But $T_k = L_k D_k L_k^T$ and so from the uniqueness of the LDL^T factorization, we have

$$D_k = C_k^T A C_k.$$

The column partitioning $C_k = [c_1 \mid \ldots \mid c_k]$ implies that $c_i^T A c_j = [D_k]_{ij}$. $\square$

The theorem tells us that the search directions $c_1, \ldots, c_k$ are *A-conjugate*.

11.3.7   The Hestenes-Stiefel Formulation

The preceding results permit us to rewrite Algorithm 11.3.2 in a way that avoids explicit reference to the Lanczos vectors and the entries in the ongoing LDL^T factorization. In addition, we will be able to formulate the termination criterion in terms of the linear system residual $b - Ax_k$ instead of the more obscure "Lanczos residual vector" $(A - \alpha_k I)q_k - \beta_{k-1}q_{k-1}$. The key idea is to think of c_k as a search direction and ρ_k as a step length and to recognize that these quantities can be scaled. Consider the search direction update recipe

$$c_k = q_k - \ell_{k-1}c_{k-1}$$

from Algorithm 11.3.2. Since q_k is a multiple of g_{k-1} we see that

$$\boxed{(\text{search direction } k) \; = \; g_{k-1} \; + \; \text{scalar} \times (\text{search direction } k - 1)}$$

If we write this as

$$p_k \; = \; g_{k-1} + \tau_{k-1}p_{k-1}, \tag{11.3.23}$$

then it follows from

$$Ap_k = Ag_{k-1} + \tau_{k-1}Ap_{k-1}$$

and Theorem 11.3.5 that

$$\tau_{k-1} \; = \; -\frac{p_{k-1}Ag_{k-1}}{p_{k-1}^T Ap_{k-1}} \tag{11.3.24}$$

and

$$p_k^T Ag_{k-1} \; = \; p_k^T Ap_k. \tag{11.3.25}$$

Since p_k is a multiple of c_k, the update formula $x_k = x_{k-1} + \rho_k c_k$ in Algorithm 11.3.2 has the form

$$x_k \; = \; x_{k-1} - \mu_k p_k$$

for some scalar μ_k. By applying A to both sides of this equation and subtracting b we get

$$g_k \; = \; g_{k-1} - \mu_k Ap_k.$$

Using Theorem 11.3.4 and equation (11.3.25) we see that

$$\mu_k \; = \; \frac{g_{k-1}^T g_{k-1}}{g_{k-1}^T Ap_k} \; = \; \frac{g_{k-1}^T g_{k-1}}{p_k^T Ap_k}.$$

From the equations $g_{k-1} \; = \; g_{k-2} - \mu_{k-1}Ap_{k-1}$ and $g_{k-1}^T g_{k-2} = 0$, it follows that

$$g_{k-1}^T g_{k-1} = -\mu_{k-1}g_{k-1}^T Ap_{k-1},$$

$$g_{k-2}^T g_{k-2} = \mu_{k-1}g_{k-2}^T Ap_{k-1} \; = \; \mu_{k-1}p_{k-1}^T Ap_{k-1}.$$

Substituting these equations into (11.3.24) gives

$$\tau_{k-1} \; = \; \frac{g_{k-1}^T g_{k-1}}{g_{k-2}^T g_{k-2}}.$$

By exploiting these recipes for p_k, x_k, g_k, μ_k, and τ_{k-1}, *and redefining r_k to be the residual* $b - Ax_k = -g_k$, we can rewrite Algorithm 11.3.2 as follows.

Algorithm 11.3.3 (Conjugate Gradients: Hestenes-Stiefel Version) If $A \in \mathbb{R}^{n \times n}$ is symmetric positive definite, $b \in \mathbb{R}^n$, and $Ax_0 \approx b$, then this algorithm computes $x_* \in \mathbb{R}^n$ so that $Ax_* = b$.

$$k = 0, \ r_0 = b - Ax_0$$
$$\textbf{while } \| \, r_k \, \|_2 > 0$$
$$\qquad k = k + 1$$
$$\qquad \textbf{if } k = 1$$
$$\qquad\qquad p_k = r_0$$
$$\qquad \textbf{else}$$
$$\qquad\qquad \tau_{k-1} = \left(r_{k-1}^T r_{k-1} \right) / \left(r_{k-2}^T r_{k-2} \right)$$
$$\qquad\qquad p_k = r_{k-1} + \tau_{k-1} p_{k-1}$$
$$\qquad \textbf{end}$$
$$\qquad \mu_k = (r_{k-1}^T r_{k-1}) / (p_k^T A p_k)$$
$$\qquad x_k = x_{k-1} + \mu_k p_k$$
$$\qquad r_k = r_{k-1} - \mu_k A p_k$$
$$\textbf{end}$$
$$x_* = x_k$$

This procedure is essentially the form delineated in Hestenes and Stieffel (1952).

11.3.8   A Few Practical Details

Rounding errors lead to a loss of orthogonality among the residuals and finite termination is not guaranteed in floating point. For an extensive analysis of this fact, see Meurant (LCG). Thus, it makes sense to have a termination criterion based on (say) the size of $\| \, r_k \, \| = \| \, b - Ax_k \, \|$. With that in mind and being careful about required vector workspaces, we obtain the following more practical version of Algorithm 11.3.3.

$$k = 0, \ x = x_0, \ r = b - Ax, \ \rho_c = r^T r, \ \delta = \mathsf{tol} \cdot \| \, b \, \|_2$$
$$\textbf{while } \ \sqrt{\rho_c} > \delta$$
$$\qquad k = k + 1$$
$$\qquad \textbf{if } k = 1$$
$$\qquad\qquad p = r$$
$$\qquad \textbf{else} \hspace{6cm} (11.3.26)$$
$$\qquad\qquad \tau \ = \ \rho_c / \rho_-, \ p = r + \tau p$$
$$\qquad \textbf{end}$$
$$\qquad w \ = \ Ap$$
$$\qquad \mu \ = \ \rho_c / p^T w, \ x \ = \ x + \mu p, \ r \ = \ r - \mu w, \ \rho_- = \rho_c, \ \rho_c = r^T r$$
$$\textbf{end}$$

Thus, a CG step requires one matrix-vector product, three saxpys, and two inner products. Four length-n arrays are required. Note that if x_c is the final iterate and x_* is the exact solution, then

$$\| x_c - x_* \| = \| A^{-1}(b - Ax_c) \|_2 \leq \mathsf{tol} \cdot \| A^{-1} \|_2 \| b \|_2 \leq \mathsf{tol} \cdot \kappa_2(A) \| x_* \|.$$

Thus, a stopping criterion ensures a relative error that is bounded by the product of tol and the condition number.

In practice, it is desirable to terminate the iteration long before k approaches n. Trefethen and Bau (NLA, p. 299) show that

$$\| x - x_k \|_A \leq 2 \| x - x_0 \|_A \left(\frac{\sqrt{\kappa_2(A)} - 1}{\sqrt{\kappa_2(A)} + 1} \right)^k . \tag{11.3.27}$$

Of course, it does not take much of a condition number for the upper bound to be hopelessly close to 1, so, by itself, this result does not provide hope for an early exit. However, as we will see in §11.5, there is a way to induce speedy convergence by applying the method to an equivalent "preconditioned" system that is designed in such a way that (11.3.27) and/or Corollary 11.3.2 predict good things.

11.3.9   Conjugate Gradients Applied to $A^T A$ and $A A^T$

There are two obvious ways to convert an unsymmetric $Ax = b$ problem into an equivalent symmetric positive definite problem:

$$Ax = b \equiv \begin{cases} A^T A x = A^T b, \\[2mm] A A^T y = b, \; x = A^T y. \end{cases}$$

Each of these conversions creates an opportunity to apply the method of conjugate gradients.

If we apply CG to the $A^T A x = A^T b$ problem, then at the kth step a vector x_k is produced that minimizes

$$\phi_{A^T A}(x) = \frac{1}{2} x^T (A^T A) x - x^T (A^T b) = \frac{1}{2} \| Ax - b \|_2^2 - \frac{1}{2} b^T b$$

over the affine space

$$S_k = x_0 + \mathcal{K}(A^T A, A^T r_0, k) \tag{11.3.28}$$

where $r_0 = b - Ax_0$. The resulting algorithm is the *conjugate gradient normal equation residual* (CGNR) method.

If we apply the CG method to the "y-problem" $A A^T y = b$, then at the kth step a vector y_k is produced that minimizes

$$\phi_{A A^T}(y) = \frac{1}{2} y^T A A^T y - y^T b = \frac{1}{2} \| A^T y - A^{-1} b \|_2^2 - \frac{1}{2} b^T (A A^T)^{-1} b$$

over the affine space $y_0 + \mathcal{K}(A A^T, r_0, k)$ where $r_0 = b - Ax_0$. Setting $x_k = A^T y_k$, this says that $x = x_k$ minimizes $\| x - x_* \|_2$ over the affine space defined in (11.3.28).

CG	CGNR	CGNE
$r_c = b - Ax_0$	$r_c = b - Ax_0, \ z_c = A^T r_c$	$r_c = b - Ax_c$
$p_c = r_c$	$p_c = z_c$	$p_c = A^T r_c$
$\mu = \dfrac{r_c^T r_c}{p_c^T A p_c}$	$\mu = \dfrac{z_c^T z_c}{(Ap_c)^T (Ap_c)}$	$\mu = \dfrac{r_c^T r_c}{p_c^T p_c}$
$x_+ = x_c + \mu \, p_c$	$x_+ = x_c + \mu \, p_c$	$x_+ = x_c + \mu \, p_c$
$r_+ = r_c - \mu \, A p_c$	$r_+ = r_c - \mu \, A p_c, \ z_+ = A^T r_+$	$r_+ = r_c - \mu \, A p_c$
$\tau = \dfrac{r_+^T r_+}{r_c^T r_c}$	$\tau = \dfrac{z_+^T z_+}{z_c^T z_c}$	$\tau = \dfrac{r_+^T r_+}{r_c^T r_c}$
$p_+ = r_+ + \tau \, p_c$	$p_+ = z_+ + \tau \, p_c$	$p_+ = A^T r_+ + \tau \, p_c$

Figure 11.3.1. *The initializations and update formulae for the conjugate gradient (CG) method, the conjugate gradient normal equation residual (CGNR) method, and the conjugate gradient normal equation error (CGNE) method. The subscript "c" designates "current" while the subscript "+" designates "next".*

The resulting method is called the *conjugate gradient normal equation error* (CGNE) method. It is also known as *Craig's method.*

Simple modifications of the CG update formulae in Algorithm 11.3.3 are required to implement CGNR and CGNE. We tabulate the initializations and updates of the three methods in Figure 11.3.1. Notice that CGNR and CGNE require procedures for A-times-vector *and* A^T-times-vector. See Saad (IMSLS, pp. 251–254) and Greenbaum (IMSL, Chap. 7) for details and perspective on the squaring of the condition number that is associated with these methods. The CGNR method can be applied if A is rectangular. Thus, it provides a normal equation framework for solving sparse, full rank, least squares problems. See Björck (SLE, pp. 288–293) for discussion and analysis. The CGNE method can also be applied to rectangular problems, but the underlying system must be consistent.

Problems

P11.3.1 How many n-vectors are required to implement each of the algorithms in this section?

P11.3.2 Let α_i and β_i be defined by Algorithm 11.3.2. How could those tridiagonal entries be generated as the iteration in Algorithm 11.3.3 proceeds?

P11.3.3 Derive the update formulae for the CGNR and CGNE methods displayed in Figure 11.3.1.

P11.3.4 Show that if the while-loop condition in Algorithm 11.3.3 is changed to

$$\| r_k \| > \mathsf{tol} \, (\| A \| \| x_k \| + \| b \|),$$

then the algorithm produces the exact solution to a nearby $Ax = b$ problem relative to tol.

Notes and References for §11.3

Background texts for the material in this section include Greenbaum (IMSL), Meurant (LCG), and Saad (ISPLA). The original reference for the conjugate gradient method is:

M.R. Hestenes and E. Stiefel (1952). "Methods of Conjugate Gradients for Solving Linear Systems,"
    J. Res. Nat. Bur. Stand. 49, 409–436.

The idea of regarding conjugate gradients as an iterative method began with the following paper:

J.K. Reid (1971). " On the Method of Conjugate Gradients for the Solution of Large Sparse Systems
    of Linear Equations," *in Large Sparse Sets of Linear Equations*, J.K. Reid (ed.), Academic Press,
    New York, 231–254.

Some historical and unifying perspectives are offered in:

G.H. Golub and D.P. O'Leary (1989). "Some History of the Conjugate Gradient and Lanczos Meth-
    ods," *SIAM Review 31*, 50–102.
M.R. Hestenes (1990). "Conjugacy and Gradients," in *A History of Scientific Computing*, Addison-
    Wesley, Reading, MA.
S. Ashby, T.A. Manteuffel, and P.E. Saylor (1992). "A Taxonomy for Conjugate Gradient Methods,"
    SIAM J. Numer. Anal. 27, 1542–1568.

Over the years, many authors have analyzed the method:

G.W. Stewart (1975). "The Convergence of the Method of Conjugate Gradients at Isolated Extreme
    Points in the Spectrum," *Numer. Math. 24*, 85–93.
A. Jennings (1977). "Influence of the Eigenvalue Spectrum on the Convergence Rate of the Conjugate
    Gradient Method," *J. Inst. Math. Applic. 20*, 61–72.
O. Axelsson (1977). "Solution of Linear Systems of Equations: Iterative Methods," in *Sparse Matrix
    Techniques: Copenhagen*, 1976, V.A. Barker (ed.), Springer-Verlag, Berlin.
M.R. Hestenes (1980). *Conjugate Direction Methods in Optimization*, Springer-Verlag, Berlin.
J. Cullum and R. Willoughby (1980). "The Lanczos Phenomena: An Interpretation Based on Conju-
    gate Gradient Optimization," *Lin. Alg. Applic. 29*, 63–90.
A. van der Sluis and H.A. van der Vorst (1986). "The Rate of Convergence of Conjugate Gradients,"
    Numer. Math. 48, 543–560.
A.E. Naiman, I.M. Babuka, and H.C. Elman (1997). "A Note on Conjugate Gradient Convergence,"
    Numer. Math. 76, 209–230.
A.E. Naiman and S. Engelberg (2000). "A Note on Conjugate Gradient Convergence - Part II,"
    Numer. Math. 85, 665–683.
S. Engelberg and A.E. Naiman (2000). "A Note on Conjugate Gradient Convergence - Part III,"
    Numer. Math. 85, 685–696.

For a floating-point discussion of CG, see Meurant (LCG) as well as:

H. Wozniakowski (1980). "Roundoff Error Analysis of a New Class of Conjugate Gradient Algorithms,"
    Lin. Alg. Applic. 29, 509–529.
A. Greenbaum and Z. Strakos (1992). "Predicting the Behavior of Finite Precision Lanczos and
    Conjugate Gradient Computations," *SIAM J. Matrix Anal. Applic. 13*, 121–137.
Z. Strakoš and P. Tichý (2002). "On Error Estimation in the Conjugate Gradient Method and Why
    it Works in Finite Precision Computations," *ETNA 13*, 56–80.
G. Meurant and Z. Strakoš (2006). "The Lanczos and Conjugate Gradient Algorithms in Finite
    Precision Arithmetic," *Acta Numerica 15*, 471–542.

The family of CG-related methods is very large and the following is a small subset of the literature:

G.W. Stewart (1973). "Conjugate Direction Methods for Solving Systems of Linear Equations,"
    Numer. Math. 21, 284–297.
D.P. O'Leary (1980). "The Block Conjugate Gradient Algorithm and Related Methods," *Lin. Alg.
    Applic. 29*, 293–322.
J.E. Dennis Jr. and K. Turner (1987). "Generalized Conjugate Directions," *Lin. Alg. Applic. 88/89*,
    187–209.
A. Bunse-Gerstner and R. Stover (1999). "On a Conjugate Gradient-Type Method for Solving Complex
    Symmetric Linear Systems," *Lin. Alg. Applic. 287*, 105–123.
T. Barth and T. Manteuffel (2000). "Multiple Recursion Conjugate Gradient Algorithms Part I:
    Sufficient Conditions," *SIAM J. Matrix Anal. Applic. 21*, 768–796.
C. Li (2001). "CGNR Is an Error Reducing Algorithm," *SIAM J. Sci. Comput. 22*, 2109–2112.
A.A. Dubrulle (2001). "Retooling the Method of Block Conjugate Gradients," *ETNA 12*, 216–233.

W.W. Hager and H. Zhang (2006). "Algorithm 851: CG_DESCENT, a Conjugate Gradient Method with Guaranteed Descent," *ACM Trans. Math. Softw. 32*, 113–137.

Y. Saad (2006). "Filtered Conjugate Residual-type Algorithms with Applications," *SIAM J. Matrix Anal. Applic. 28*, 845–870.

The use of the method to solve certain eigenvalue problems is detailed in:

A. Ruhe and T. Wiberg (1972). "The Method of Conjugate Gradients Used in Inverse Iteration," *BIT 12*, 543–554.

A. Edelman and S.T. Smith (1996). "On Conjugate Gradient-Like Methods for Eigen-Like Problems," *BIT 36*, 494–508.

The design of sensible stopping criteria has many subtleties, see:

S.F. Ashby, M.J. Holst, A. Manteuffel, and P.E. Saylor (2001). "The Role of the Inner Product in Stopping Criteria for Conjugate Gradient Iterations," *BIT 41*, 26–52.

M. Arioli (2004). "A Stopping Criterion for the Conjugate Gradient Algorithm in a Finite Element Method Framework," *Numer. Math. 97*, 1–24.

11.4 Other Krylov Methods

The conjugate gradient method can be regarded as a clever pairing of the symmetric Lanczos process and the LDL^T factorization. The "cleverness" is associated with the recursions that support an economical transition from x_{k-1} to x_k. In this section we move beyond symmetric positive definite systems and present instances of the same paradigm for more general problems:

$$\left(\begin{array}{c} \text{Krylov} \\ \text{process} \end{array} \right) + \left(\begin{array}{c} \text{Matrix} \\ \text{factorization} \end{array} \right) + \left(\begin{array}{c} \text{Clever} \\ \text{recursions} \end{array} \right) = \left(\begin{array}{c} \text{Sparse} \\ \text{matrix} \\ \text{method} \end{array} \right).$$

Methods for the symmetric indefinite problem (MINRES, SYMMLQ), the least squares problem (LSQR, LSMR), and the square $Ax = b$ problem (GMRES, QMR, BiCG, CGS, BiCGStab) are briefly discussed. The Lanczos, Arnoldi, and unsymmetric Lanczos iterations are in the mix. Our goal is to communicate the main idea behind these methods. For deeper insight, practical intuition, and analysis, see Saad (ISPLA), Greenbaum (IMSL), van der Vorst (IMK), Freund, Golub, and Nachtigal (1992), and LIN_TEMPLATES.

11.4.1 MINRES and SYMMLQ for Symmetric Systems

Assume that $A \in \mathbb{R}^{n \times n}$ is symmetric indefinite, i.e., $\lambda_{\min}(A) < 0 < \lambda_{\max}(A)$. A consequence of this is that we cannot recast the $Ax = b$ problem as a minimization problem associated with $\phi(x) = x^T A x / 2 - x^T b$. Indeed, this function has no lower bound. If $Ax = \lambda_{\min} x$, then $\phi(\alpha x) = \alpha^2 \lambda_{\min} - \alpha x^T b$ approaches $-\infty$ as α gets big.

This suggests that we switch to a more workable objective function. Instead of adopting the CG strategy of minimizing ϕ over the affine space $x_0 + \mathcal{K}(A, r_0, k)$, we propose to solve

$$\min_{x \in x_0 + \mathcal{K}(A, r_0, k)} \| b - Ax \|_2. \qquad (11.4.1)$$

at each step. As in CG, we use the Lanczos process to generate the Krylov subspaces, setting $q_1 = r_0/\beta_0$ where $r_0 = b - Ax_0$ and $\beta_0 = \| g_0 \|_2$. After k steps we have

$$AQ_k = Q_k T_k + \beta_k q_{k+1} e_k^T.$$

That is,

$$AQ_k = Q_{k+1}H_k, \tag{11.4.2}$$

where $H_k \in \mathbb{R}^{k+1 \times k}$ is the Hessenberg matrix

$$H_k = \begin{bmatrix} \alpha_1 & \beta_2 & \cdots & \cdots & 0 \\ \beta_1 & \alpha_2 & \ddots & & 0 \\ \vdots & \ddots & \ddots & & \vdots \\ \vdots & & \ddots & & \beta_{k-1} \\ 0 & \cdots & \cdots & \beta_{k-1} & \alpha_k \\ \hline 0 & \cdots & \cdots & 0 & \beta_k \end{bmatrix}. \tag{11.4.3}$$

Writing $x = x_0 + Q_k y$ and recalling that $\mathsf{ran}(Q_k) = \mathcal{K}(A, r_0, k)$, we see that the optimization (11.4.1) involves minimizing

$$\| A(x_0 + Q_k y) - b \|_2 = \| Q_{k+1}H_k y - (b - Ax_0) \|_2 = \| H_k y - \beta_0 e_1 \|_2$$

over all $y \in \mathbb{R}^k$. To solve this problem we take a hint from §5.2.6 and use the Givens QR factorization procedure. Suppose $G_1, \ldots, G_k$ are Givens rotations such that

$$G_k^T \cdots G_1^T H_k = \begin{bmatrix} R_k \\ \hline 0 \end{bmatrix}, \qquad R_k \in \mathbb{R}^{k \times k},$$

is upper triangular. If

$$G_k^T \cdots G_1^T (\beta_0 e_1) = \begin{bmatrix} p_k \\ \rho_k \end{bmatrix}, \qquad p_k \in \mathbb{R}^k,$$

and $y_k \in \mathbb{R}^k$ solves $R_k y_k = p_k$, then $x_k = x_0 + Q_k y_k$ solves (11.4.1) and the norm of the residual is given by $\| b - Ax_k \|_2 = |\rho_k|$. The transition

$$\{H_{k-1}, R_{k-1}, p_{k-1}, \rho_{k-1}\} \quad \rightarrow \quad \{H_k, R_k, p_k, \rho_k\}$$

can be realized with $O(1)$ flops after the kth Lanczos step is performed. The Givens rotation G_k can be determined from β_k and $[R_{k-1}]_{k-1,k-1}$. Note that after step $k-1$ we already have the first $k-2$ rows of R_k and the first $k-2$ components of p_k. The matrix R_k has upper bandwidth 2 and so the triangular system that determines y_k can be solved with $O(k)$ flops. Thus, in computing $x_k = x_0 + Q_k y_k$ each step is not essential. On the other hand, it is possible to work out an $O(n)$ transition from x_{k-1} to x_k through recursions that involve Q_k and the QR factorization of H_k. (This corresponds to the LDL^T-plus-Q_k recursions associated with CG developed in §11.3.5.) Either way, there is no need to access all the Lanczos vectors each step. Properly implemented, we have the MINRES method of Paige and Saunders (1975).

   An alternative approach developed by the same authors works with the LQ *factorization* of the tridiagonal matrix T_k. We mimic the §11.3.4 in the CG derivation leading to (11.3.14). However, the solution of the tridiagonal system

$$T_k y_k = \beta_0 e_1 \tag{11.4.4}$$

is problematic because T_k is no longer positive definite. This means that the LDL^T factorization, together with the associated recursions, is no longer safe to use.

A way around this difficulty is to work with the transpose of the matrix equation $AQ_{k-1} = Q_k H_{k-1}$. Suppose $x_k = x_0 + Q_k y_k$ where y_k is the minimum-norm solution to the $(k-1)$-by-k underdetermined system

$$H_{k-1}^T y_k = \beta_0 e_1. \tag{11.4.5}$$

It follows from $r_0 = \beta_0 Q_{k-1} e_1$, $r_k = r_0 - AQ_{k-1} y_k$, and $Q_{k-1}^T A = H_{k-1}^T Q_k^T$ that

$$Q_{k-1}^T r_k = \beta_0 e_1 - H_{k-1}^T y_k = 0.$$

Thus, the residual $r_k = b - Ax_k$ is orthogonal to $q_1, \ldots, q_{k-1}$. Note that the underdetermined system (11.4.5) has full row rank and that y_k can be determined via a Givens rotation lower triangularization, e.g.,

$$
\begin{bmatrix}
\alpha_1 & \beta_1 & 0 & 0 & 0 \\
\beta_1 & \alpha_2 & \beta_2 & 0 & 0 \\
0 & \beta_2 & \alpha_3 & \beta_3 & 0 \\
0 & 0 & \beta_3 & \alpha_4 & \beta_4
\end{bmatrix} G_1 G_2 G_3 G_4 =
\begin{bmatrix}
\times & 0 & 0 & 0 & 0 \\
\times & \times & 0 & 0 & 0 \\
\times & \times & \times & 0 & 0 \\
0 & \times & \times & \times & 0
\end{bmatrix} = \begin{bmatrix} L_4 & | & 0 \end{bmatrix}.
$$

This is an *LQ factorization* and in general we have

$$H_{k-1}^T G_1 \cdots G_{k-1} = \begin{bmatrix} L_{k-1} & | & 0 \end{bmatrix}$$

where L_{k-1} is lower triangular. (This is just the transpose of the Givens QR factorization of H_{k-1}.) If $w_{k-1} \in \mathbb{R}^{k-1}$ solves the necessarily nonsingular system $L_{k-1} w_{k-1} = \beta_0 e_1$, then

$$y_k = G_1 \cdots G_{k-1} \begin{bmatrix} w_{k-1} \\ 0 \end{bmatrix}.$$

The special structure of L_{k-1} (it has lower bandwidth equal to 2) and the Givens rotation sequence make it possible to realize the transition from x_k to x_{k+1} with $O(n)$ work in a way that does not require access to all the Lanczos vectors. Collectively, these ideas define the SYMMLQ method of Paige and Saunders (1975).

11.4.2 LSQR and LSMR for Least Squares Problems

We show how the sparse least squares problem $\min \| Ax - b \|_2$ can be solved using the Paige-Saunders lower bidiagonalization process described in §10.4.4. Indeed, if we apply Algorithm 10.4.2 with $u_1 = r_0/\beta_0$ where $r_0 = b - Ax_0$ and $\beta_0 = \| r_0 \|_2$, then after k steps we have a partial factorization of the form

$$AV_k = U_k B_k + p_k e_k^T$$

where $V = [\, v_1 \,|\, \cdots \,|\, v_k \,] \in \mathbb{R}^{n \times k}$ has orthonormal columns, $U = [\, u_1 \,|\, \cdots \,|\, u_k \,] \in \mathbb{R}^{m \times k}$ has orthonormal columns, and $B_k \in \mathbb{R}^{k \times k}$ is lower bidiagonal. If $p_k \in \mathbb{R}^m$ is nonzero, then we can write

$$AV_k = U_{k+1} \tilde{B}_k$$

where $\tilde{B}_k \in \mathbb{R}^{k+1 \times k}$ is given by

$$
\tilde{B}_k \;=\;
\begin{bmatrix}
\alpha_1 & 0 & \cdots & \cdots & 0 \\
\beta_1 & \alpha_2 & \ddots & & 0 \\
\vdots & \ddots & & & \vdots \\
\vdots & & & \ddots & 0 \\
0 & \cdots & \cdots & \beta_{k-1} & \alpha_k \\
0 & \cdots & \cdots & 0 & \beta_k
\end{bmatrix}.
\tag{11.4.6}
$$

It can be shown that $\mathrm{span}\{v_1, \ldots, v_k\} = \mathcal{K}(A^T A, A^T r_0, k)$. In the LSQR method of Paige and Saunders (1982), the kth approximate minimizer x_k solves the problem

$$
\min_{x \in x_0 + \mathcal{K}(A^T A, A^T r_0, k)} \| Ax - b \|_2.
\tag{11.4.7}
$$

Thus, $x_k = x_0 + V_k y_k$ where $y_k \in \mathbb{R}^k$ is the minimizer of

$$
\| A(x_0 + V_k y) - b \|_2 \;=\; \| U_{k+1} \tilde{B}_k y - (b - Ax_0) \|_2 \;=\; \| \tilde{B}_k y - \beta_0 e_1 \|_2.
$$

Givens QR can be used to solve this problem just as it is used in the MINRES context above. Suppose

$$
G_k^T \cdots G_1^T \tilde{B}_k \;=\; \begin{bmatrix} R_k \\ 0 \end{bmatrix},
\qquad\qquad
G_k^T \cdots G_1^T (\beta_1 e_1) \;=\; \begin{bmatrix} p_k \\ \rho_k \end{bmatrix},
$$

where $G_1, \ldots, G_k$ are Givens rotations, $R_k \in \mathbb{R}^{k \times k}$ is upper triangular, $p_k \in \mathbb{R}^k$, and $\rho_k \in \mathbb{R}$. Then, y_k solves $R_k y = p_k$ and

$$
x_k \;=\; x_0 + V_k y_k \;=\; x_0 + W_k p_k
$$

where $W_k = V_k R_k^{-1}$. It is possible to compute x_k from x_{k-1} via a simple recursion that involves the last column of W_k. Overall, we obtain the LSQR method of Paige and Saunders (1982). It requires only a few vectors of storage to implement.

The *LSMR method* provides an alternative to the LSQR method and is mathematically equivalent to MINRES applied to the normal equations $A^T A x = A^T b$. Like LSQR, the technique can be used to solve least squares problems, regularized least squares problems, undetermined systems, and square unsymmetric systems. The 2-norms of the vectors $r_k = b - Ax_k$ and $A^T r_k$ decrease monotonically, which allows for tractable early-termination. See Fong and Saunders (2011) for more details.

11.4.3   GMRES for General $Ax = b$

The Paige-Saunders MINRES method (§11.4.1) is a Lanczos-based technique that can be used to solve symmetric $Ax = b$ problems. The kth iterate x_k minimizes $\| Ax - b \|_2$ over $x_0 + \mathcal{K}(A, b, k)$. We now present an Arnoldi-based iteration that does the same thing and is applicable to general linear systems. The method is referred to as the *generalized minimum residual (GMRES) method* and is due to Saad and Shultz (1986).

After k steps of the Arnoldi iteration (Algorithm 10.5.1) it is easy to confirm using (10.5.2) that

$$AQ_k = Q_{k+1}\tilde{H}_k \qquad (11.4.8)$$

where the columns of

$$Q_{k+1} = [\, Q_k \mid q_{k+1} \,]$$

are the orthonormal Arnoldi vectors and the upper Hessenberg matrix $\tilde{H}_k$ is given by

$$\tilde{H}_k = \begin{bmatrix} h_{11} & h_{12} & \cdots & \cdots & h_{1k} \\ h_{21} & h_{22} & \cdots & \cdots & h_{2k} \\ 0 & \ddots & \ddots & & \vdots \\ \vdots & & \ddots & \ddots & \vdots \\ 0 & \cdots & \cdots & h_{k,k-1} & h_{kk} \\ 0 & \cdots & \cdots & 0 & h_{k+1,k} \end{bmatrix} \in \mathbb{R}^{k+1 \times k}.$$

Moreover, if $q_1 = r_0/\beta_0$ where $r_0 = b - Ax_0$ and $\beta_0 = \|\, r_0\,\|_2$, then

$$\mathsf{span}\{q_1, \ldots, q_k\} = \mathcal{K}(A, r_0, k).$$

In step k, the GMRES method requires minimization of $\|\, Ax - b\,\|_2$ over the affine space $x_0 + \mathcal{K}(A, r_0, k)$. As with MINRES, we must find a vector $y \in \mathbb{R}^k$ so that

$$\|\, A(x_0 + Q_k y) - b\,\|_2 = \|\, Q_{k+1}\tilde{H}_k y - (b - Ax_0)\,\|_2 = \|\, \tilde{H}_k y - \beta_0 e_1\,\|_2$$

is minimized. If y_k is the solution to this $(k+1)$-by-k least squares problem, then the k-th GMRES iterate is given by $x_k = x_0 + Q_k y_k$. Note that if Givens rotations $G_1, \ldots, G_k$ have been determined so that

$$G_k^T \cdots G_1^T \tilde{H}_k = \begin{bmatrix} R_k \\ \hline 0 \end{bmatrix}, \qquad R_k \in \mathbb{R}^{k \times k}, \qquad (11.4.9)$$

is upper triangular and we set

$$G_k^T \cdots G_1^T (\beta_0 e_1) = \begin{bmatrix} p_k \\ \hline \rho_k \end{bmatrix}, \qquad (11.4.10)$$

where $p_k \in \mathbb{R}^k$ and $\rho_k \in \mathbb{R}$, then $R_k y_k = p_k$ and

$$|\rho_k| = \|\, Ax_k - b\,\|_2.$$

The transition

$$\{R_{k-1}, p_{k-1}, \rho_{k-1}\} \rightarrow \{R_k, p_k, \rho_k\}$$

is a particularly simple update that involves the generation of a single rotation G_k and exploitation of the identities $R_{k-1} = R_k(1{:}k-1, 1{:}k-1)$ and $p_k(1{:}k-1) = p_{k-1}$.

As a procedure for large sparse problems, the GMRES method inherits the usual Arnoldi concern: the computation of $H(1{:}k+1, k)$ requires $O(kn)$ flops and access to all previously computed Arnoldi vectors. For this reason it is necesssary to build a restart strategy around the following, m-step GMRES building block:

Algorithm 11.4.2 (m-step GMRES) If $A \in \mathbb{R}^{n \times n}$ is nonsingular, $b \in \mathbb{R}^n$, $Ax_0 \approx b$, and m is a positive iteration limit, then this algorithm computes $\tilde{x} \in \mathbb{R}^n$ where either $\tilde{x}$ solves $Ax = b$ or minimizes $\| Ax - b \|_2$ over the affine space $x_0 + \mathcal{K}(A, r_0, m)$ where $r_0 = b - Ax_0$.

> $k = 0$, $r_0 = b - Ax_0$, $\beta_0 = \| r_0 \|_2$
> **while** $(\beta_k > 0)$ and $k < m$
> > $q_{k+1} = r_k / \beta_k$
> > $k = k + 1$
> > $r_k = Aq_k$
> > **for** $i = 1{:}k$
> > > $h_{ik} = q_i^T r_k$ (11.4.11)
> > > $r_k = r_k - h_{ik} q_i$
> > **end**
> > $\beta_k = \| r_k \|_2$
> > $h_{k+1,k} = \beta_k$
> > Apply $G_1, \ldots, G_{k-1}$ to $H(1{:}k, k)$ and determine G_k, R_k, p_k, and ρ_k
> **end**
> Solve $R_k y_k = p_k$ and set $\tilde{x} = x_0 + Q_k y_k$

If $\tilde{x}$ is not good enough, then the process can be repeated with the new x_0 set to $\tilde{x}$. There are many important implementation details associated with this framework, see Saad (IMSLA, pp. 164–184) and van der Vorst (IMK, pp. 65–84).

11.4.4 Optimizing from the Polynomial Point of View

Before we present the next group of methods, it is instructive to connect the Krylov framework with polynomial approximation. Suppose the columns of $Q_k \in \mathbb{R}^{n \times k}$ span $\mathcal{K}(A, q_1, k)$. It follows that if $y \in \mathbb{R}^k$, then $Q_k y = \varphi(A) q_1$ for some polynomial φ that has degree $k - 1$ or less. This is because

$$Q_k = [\, q_1 \mid Aq_1 \mid \cdots \mid A^{k-1} q_1 \,] B$$

for some nonsingular $B \in \mathbb{R}^{k \times k}$ and so if $\alpha = By$, then

$$Q_k y = [\, q_1 \mid Aq_1 \mid \cdots \mid A^{k-1} q_1 \,] \alpha = (\alpha_1 I + \alpha_2 A + \cdots + \alpha_k A^{k-1}) q_1.$$

Thus, the GMRES (and MINRES) optimization can be rephrased as a polynomial optimization problem. If $\mathbb{P}_k$ denotes the set of all degree-k polynomials, then we have

$$\min_{x \in x_0 + \mathcal{K}(A, r_0, k)} \| b - Ax \|_2 = \min_{\varphi \in \mathbb{P}_{k-1}} \| b - A(x_0 + \varphi(A)) r_0 \|_2$$

$$= \min_{\varphi \in \mathbb{P}_{k-1}} \| (I - A \cdot \varphi(A)) r_0 \|_2$$

$$= \min_{\psi \in \mathbb{P}_k, \psi(0) = 1} \| \psi(A) r_0 \|_2 .$$

This point of view figures heavily in the analysis of various Krylov subspace methods and can also be used to suggest alternative strategies.

11.4.5 BiCG, CGS, BiCGstab, and QMR for General $Ax = b$

Just as the Arnoldi iteration underwrites GMRES, the unsymmetric Lanczos process (10.5.11) underwrites the next cohort of methods that we present. Suppose we complete k steps of (10.5.11) with $q_1 = r_0/\beta_0$, $r_0 = b - Ax_0$, $\beta_0 = \| r_0 \|_2$, and $r_0^T \tilde{r}_0 \neq 0$. This means we have the partial factorizations

$$AQ_k = Q_kT_k + r_ke_k^T, \qquad \tilde{Q}_k^T r_k = 0, \qquad (11.4.12)$$

$$A^T\tilde{Q}_k = \tilde{Q}_kT_k^T + \tilde{r}_ke_k^T, \qquad Q_k^T\tilde{r}_k = 0, \qquad (11.4.13)$$

where

$$Q_k = [\, q_1 \,|\, \cdots \,|\, q_k \,], \qquad \mathsf{ran}(Q_k) = \mathcal{K}(A, r_0, k),$$
$$\tilde{Q}_k = [\, \tilde{q}_1 \,|\, \cdots \,|\, \tilde{q}_k \,], \qquad \mathsf{ran}(\tilde{Q}_k) = \mathcal{K}(A^T, \tilde{r}_0, k).$$

In addition, $\tilde{Q}_k^T Q_k = I_k$ and $\tilde{Q}_k^T AQ_k = T_k \in \mathbb{R}^{k \times k}$ is tridiagonal. Vectors q_{k+1} and $\tilde{q}_{k+1}$ and scalars β_k and τ_k satisfy

$$\beta_kq_{k+1} = r_k, \qquad \tau_k\tilde{q}_{k+1} = \tilde{r}_k$$

and can be generated with access to just the last two columns of Q_k and $\tilde{Q}_k$.

In step k of the *biconjugate gradient (BiCG) method*, an iterate $x_k = x_0 + Q_ky_k$ is produced where $y_k \in \mathbb{R}^k$ solves the k-by-k tridiagonal system

$$T_ky_k = \tilde{Q}_k^T r_0.$$

It follows that

$$\tilde{Q}_k^T(b - Ax_k) = \tilde{Q}_k^T(b - A(x_0 + Q_ky_k)) = \tilde{Q}_k^T r_0 - T_ky_k = 0.$$

Thus, the residual associated with x_k is orthogonal to the range of $\tilde{Q}_k$.

Assume that T_k has an LU factorization $T_k = L_kU_k$ and note that L_k is unit lower bidiagonal and U_k is upper bidiagonal. It follows that

$$x_k = x_0 + Q_kT_k^{-1}\tilde{Q}_k^T r_0 = (Q_kU_k^{-1})(L_k^{-1}(\tilde{Q}_k^T r_0)).$$

Analogously to how we derived the CG algorithm, it is possible to develop simple connections between the matrix $(Q_kU_k^{-1})$ and its predecessor and between the vector $(L_k^{-1}(\tilde{Q}_k^T r_0))$ and its predecessor. The end result is a procedure that can generate x_k through simple recursions, which we report in Figure 11.4.1. We mention that the BiCG method is subject to serious breakdown because of its dependence on the unsymmetric Lanczos process. However, with the look-ahead idea discussed in §10.5.6, it is possible to overcome some of these difficulties. Notice that BiCG collapses to CG if A is symmetric positive definite and $\tilde{r}_0 = r_0$. Also observe the similarity between the r and $\tilde{r}$ updates and the p and $\tilde{p}$ updates.

A negative aspect of the BiCG method is that it requires procedures for both A-times-vector and A^T-times-vector. (In some applications the latter is a challenge.)

BiCG	CGS	BiCGstab
$r_0 = b - Ax_0$	$r_0 = b - Ax_0$	$r_0 = b - Ax_0$
$\tilde{r}_0^T r_0 \neq 0$	$\tilde{r}_0^T \tilde{r}_0 \neq 0$	$\tilde{r}_0^T \tilde{r}_0 \neq 0$
$x_c = x_0$	$x_c = x_0$	$x_c = x_0$
$p_c = r_c = r_0$	$p_c = r_c = r_0$	$p_c = r_c = r_0$
$\tilde{p}_c = \tilde{r}_c = \tilde{r}_0$	$u_c = r_c$	
$\mu = \dfrac{\tilde{r}_c^T r_c}{\tilde{p}_c^T A p_c}$	$\mu = \dfrac{\tilde{r}_0^T r_c}{\tilde{r}_0^T A p_c}$	$\mu = \dfrac{\tilde{r}_0^T r_c}{\tilde{r}_0^T A p_c}$
$x_+ = x_c + \mu\, p_c$	$q_c = u_c - \mu\, A p_c$	$s_c = r_c - \mu\, A p_c$
$r_+ = r_c - \mu\, A p_c$	$x_+ = x_c + \mu\,(u_c + q_c)$	$\omega = \dfrac{s_c^T A s_c}{(A s_c)^T (A s_c)}$
$\tilde{r}_+ = \tilde{r}_c - \mu\, A^T \tilde{p}_c$	$r_+ = r_c - \mu\, A(u_c + q_c)$	$x_+ = x_c + \mu\, p_c + \omega s_c$
$\tau = \dfrac{\tilde{r}_+^T r_+}{\tilde{r}_c^T r_c}$	$\tau = \dfrac{\tilde{r}_0^T r_+}{\tilde{r}_0^T r_c}$	$r_+ = s_c - \omega A s_c$
$p_+ = r_+ + \tau\, p_c$	$u_+ = r_+ + \tau\, q_c$	$\tau = \dfrac{(\tilde{r}_0^T r_+)\,\mu}{(\tilde{r}_0^T r_c)\,\omega}$
$\tilde{p}_+ = \tilde{r}_+ + \tau\tilde{p}_c$	$p_+ = u_+ + \tau(q_c + \tau p_c)$	$p_+ = r_+ + \tau(p_c - \omega A p_c)$

Figure 11.4.1. *The initializations and update formulae for the biconjugate gradient (BiCG) method, the conjugate gradient squared (CGS) method, and the biconjugate gradient stablilized (BiCGstab) method. The subscript "c" designates "current" while the subscript "+" designates "next".*

The *conjugate gradient squared (CGS) method* circumvents this problem and has some interesting convergence properties as well. The derivation of the method uses the polynomial point of view that we outlined in the previous section. It is easy to conclude from Figure 11.4.1 that after k steps of the procedure we have degree-k polynomials ψ_k and φ_k so that

$$
\begin{aligned}
r_k &= \psi_k(A)r_0, & p_k &= \varphi_k(A)r_0, \\
\tilde{r}_k &= \psi_k(A^T)\tilde{r}_0, & \tilde{p}_k &= \varphi_k(A^T)\tilde{r}_0,
\end{aligned}
\qquad (11.4.14)
$$

and $\psi_k(0) = \varphi_k(0) = 1$. This enables us to characterize expressions like $\tilde{r}_k^T r_k$ and $\tilde{p}_k^T A p_k$ in a way that involves only A-times-vector:

$$
\begin{aligned}
\tilde{r}_k^T r_k &= \left(\psi_k(A^T)\tilde{r}_0\right)^T \left(\psi_k(A)r_0\right) &= \tilde{r}_0^T \left(\psi_k^2(A)r_0\right), \\
\tilde{p}_k^T A p_k &= \left(\varphi_k(A^T)\tilde{r}_0\right)^T A \left(\varphi_k(A)r_0\right) &= \tilde{r}_0^T \left(A\varphi_k^2(A)r_0\right).
\end{aligned}
$$

It is possible to develop simple recursions among the polynomials $\{\psi_k\}$ and $\{\varphi_k\}$ that facilitate the transitions

$$r_{k-1} = \psi_{k-1}^2(A)r_0 \;\rightarrow\; \psi_k^2(A)r_0 \;=\; r_k,$$
$$p_{k-1} = \varphi_{k-1}^2(A)r_0 \;\rightarrow\; \varphi_k^2(A)r_0 \;=\; p_k.$$

This leads to the *conjugate gradient squared (CGS)* method of Sonneveld (1989). It produces iterates x_k whose residuals r_k satisfy $r_k = \psi_k(A)^2 r_0$. Note from Figure 11.4.1 that the updates rely on only matrix-vector products that involve only A. Because of the squaring of the BiCG residual polynomial ψ_k, the method typically outperforms BiCG when it works, i.e., $(\|\,\psi_k(A)^2 r_0\,\|_2 \ll \|\,\psi_k(A)r_0\,\|_2)$. By the same token, it typically underperforms when BiCG struggles.

A third member in this family of $Ax = b$ solvers is the *BiCGstab method* of van der Vorst (1992). It addresses the sometimes erratic behavior of BiCG by producing iterates x_k whose residuals satisfy

$$r_k \;=\; (1 - \omega_k A)\cdots(1 - \omega_1 A)\psi_k(A)r_0$$

where ψ_k is the BiCG residual polynomial defined in (11.4.14). The parameter ω_k is chosen in step k to minimize $\|\,r_k\,\|_2$ given $\omega_1, \ldots, \omega_{k-1}$ and the vector $\psi_k(A)r_0$. The computations associated with this transpose-free method are given in Figure 11.4.1.

Yet another iteration that is built upon the unsymmetric Lanczos process is the *quasi-minimum residual (QMR) method* of Freund and Nachtigal (1991). As in BiCG, the kth iterate has the form $x_k = x_0 + Q_k y_k$ where Q_k is specified by (11.4.12). This equation can be rewritten as $AQ_k = Q_{k+1}\tilde{T}_k$ where $\tilde{T}_k \in \mathbb{R}^{k+1 \times k}$ is tridiagonal. It follows that if $q_1 = r_0/\beta_0$ where $r_0 = b - Ax_0$ and $\beta_0 = \|\,r_0\,\|_2$, then

$$b - A(x_0 + Q_k y) \;=\; r_0 - AQ_k y \;=\; r_0 - Q_{k+1}\tilde{T}_k y \;=\; Q_{k+1}(\beta_0 e_1 - \tilde{T}_k y).$$

In QMR, y is chosen to minimize $\|\,\beta_0 e_1 - \tilde{T}_k y\,\|_2$. Note that GMRES minimizes the same quantity because Q_{k+1} has orthonormal columns in Arnoldi.

Problems

P11.4.1 Assume that the cost of a length-n inner product or saxpy is one unit. Assume that $A \in \mathbb{R}^{n \times n}$ and that the matrix-vector products involving A and A^T cost α and β units, respectively. Compare the per iteration cost associated with the BiCG, CGS, and BiCGstab methods.

P11.4.2 Suppose $A \in \mathbb{R}^{n \times n}$ and $v \in \mathbb{R}^n$ are given. How can we choose ω to minimize $\|\,(I - \omega A)v\,\|_2$?

P11.4.3 Give an algorithm that computes $\psi_k(a)$ where $a \in \mathbb{R}$ and ψ_k is defined by (11.4.14).

Notes and References for §11.4

For general systems, we have avoided the when-to-use-what-method question because there are no clear-cut answers. For guidance we recommend LIN_TEMPLATES, Greenbaum (IMSL), Saad (ISPLA), and van der Vorst (IKM), each of which provides a great deal of insight. See also:

R.W. Freund, G.H. Golub, and N.M. Nachtigal (1992). "Iterative Solution of Linear Systems," *Acta Numerica 1*, 57–100.

The MINRES, SYMMLQ, and LSQR frameworks due to Paige and Saunders initiated one of the most important threads of Krylov method research:

C.C. Paige and M.A. Saunders (1975). "Solution of Sparse Indefinite Systems of Linear Equations," *SIAM J. Numer. Anal. 12*, 617–629.

C.C. Paige and M.A. Saunders (1982). "LSQR: An Algorithm for Sparse Linear Equations and Sparse Least Squares," *ACM Trans. Math. Softw. 8*, 43–71.

M.A. Saunders, H.D. Simon, and E.L. Yip (1988). "Two Conjugate-Gradient Type Methods for Unsymmetric Linear Systems," *SIAM J. Numer. Anal. 25*, 927–940.

C.C. Paige, B.N. Parlett, and H.A. van der Vorst (1995). "Approximate Solutions and Eigenvalue Bounds for Krylov Subspaces," *Numer. Lin. Alg. Applic. 3*, 115–133.

M.A. Saunders (1997). "Computing Projections with LSQR," *BIT 37*, 96–104.

F.A. Dul (1998). "MINRES and MINERR Are Better than SYMMLQ in Eigenpair Computations," *SIAM J. Sci. Comput. 19*, 1767–1782.

S.J. Benbow (1999). "Solving Generalized Least-Squares Problems with LSQR," *SIAM J. Matrix Anal. Applic. 21*, 166–177.

M. Kilmer and G.W. Stewart (2000). "Iterative Regularization and MINRES," *SIAM J. Matrix Anal. Appl. 21*, 613–628.

L. Reichel and Q. Ye (2008). "A Generalized LSQR Algorithm," *Numer. Lin. Alg. Applic. 15*, 643–660.

X.-W. Chang, C.C. Paige, and D. Titley-Peloquin (2009). "Stopping Criteria for the Iterative Solution of Linear Least Squares Problems," *SIAM J. Matrix Anal. Applic. 31*, 831–852.

S.-C. Choi, C.C. Paige, and M.A. Saunders (2011). "MINRES-QLP: A Krylov Subspace Method for Indefinite or Singular Symmetric Systems," *SIAM J. Sci. Comput. 33*, 1810–1836.

D.C.-L. Fong and M.A. Saunders (2011). "LSMR: An Iterative Algorithm for Sparse Least-Squares Problems," *SIAM J. Sci. Comput. 33*, 2950–2971.

The original GMRES paper is set forth in:

Y. Saad and M. Schultz (1986). "GMRES: A Generalized Minimum Residual Algorithm for Solving Unsymmetric Linear Systems," *SIAM J. Sci. Stat. Comput. 7*, 856–869.

and there is a great deal of follow-up analysis:

S.L. Campbell, I.C.F. Ipsen, C.T. Kelley, and C.D. Meyer (1996). "GMRES and the Minimal Polynomial," *BIT 36*, 664–675.

A. Greenbaum, V. Ptak, and Z. Strakoš (1996). "Any Nonincreasing Convergence Curve is Possible for GMRES," *SIAM J. Matrix Anal. Applic. 17*, 465–469.

K.-C. Toh (1997). "GMRES vs. Ideal GMRES," *SIAM J. Matrix Anal. Applic. 18*, 30–36.

M. Arioli, V. Ptak, and Z. Strakoš (1998). "Krylov Sequences of Maximal Length and Convergence of GMRES," *BIT 38*, 636–643.

Y. Saad (2000). "Further Analysis of Minimum Residual Iterations," *Numer. Lin. Alg. 7*, 67–93.

I.C.F. Ipsen (2000). "Expressions and Bounds for the GMRES Residual," *BIT 40*, 524–535.

D. Calvetti, B. Lewis, and L. Reichel (2002). "On the Regularizing Properties of the GMRES Method," *Numer. Math. 91*, 605–625.

J. Liesen, M. Rozloznik, and Z. Strakoš (2002). "Least Squares Residuals and Minimal Residual Methods," *SIAM J. Sci. Comput. 23*, 1503–1525.

J. Liesen and P. Tichý (2004). "The Worst-Case GMRES for Normal Matrices," *BIT 44*, 79–98.

C.C. Paige, M. Rozloznik, and Z. Strakoš (2006). "Modified Gram-Schmidt (MGS), Least Squares, and Backward Stability of MGS-GMRES," *SIAM J. Matrix Anal. Applic 28*, 264–284.

For pseudosprectral analysis of the method, see Trefethen and Embree (SAP, Chap. 26) as well as

M. Embree (1999). "Convergence of Krylov Subspace Methods for Non-Normal Matrices," PhD Thesis, Oxford University.

References concerned with the critical issue of restarting include:

R.B. Morgan (1995). "A Restarted GMRES Method Augmented with Eigenvectors," *SIAM J. Matrix Anal. Applic. 16*, 1154–1171.

A. Frommer and U. Glassner (1998). "Restarted GMRES for Shifted Linear Systems," *SIAM J. Sci. Comput. 19*, 15–26.

V. Simoncini (1999). "A New Variant of Restarted GMRES," *Numer. Lin. Alg. 6*, 61–77.

R.B. Morgan (2000). "Implicitly Restarted GMRES and Arnoldi Methods for Nonsymmetric Systems of Equations," *SIAM J. Matrix Anal. Applic. 21*, 1112–1135.

K. Moriya and T. Nodera (2000). "The DEFLATED-GMRES(m,k) Method with Switching the Restart Frequency Dynamically," *Numer. Lin. Alg. 7*, 569–584.

J. Zitko (2000). "Generalization of Convergence Conditions for a Restarted GMRES," *Numer. Lin. Alg. 7*, 117–131.

R.B. Morgan (2002). "GMRES with Deflated Restarting," *SIAM J. Sci. Comput. 24*, 20–37.

M. Embree (2003). "The Tortoise and the Hare Restart GMRES," *SIAM Review 45*, 259–266.

J. Zitko (2004). "Convergence Conditions for a Restarted GMRES Method Augmented with Eigenspaces," *Numer. Lin. Alg. 12*, 373–390.

Various practical issues concerned with GMRES implementation are covered in:

H.F. Walker (1988). "Implementation of the GMRES Method Using Householder Transformations," *SIAM J. Sci. Stat. Comput. 9*, 152–163.

A. Greenbaum, M. Rozloznik, and Z. Strako (1997). "Numerical Behaviour of the Modified Gram-Schmidt GMRES Implementation," *BIT 37*, 706–719.

P.N. Brown and H.F. Walker (1997). "GMRES On (Nearly) Singular Systems," *SIAM J. Matrix Anal. Applic. 18*, 37–51.

K. Burrage and J. Erhel (1998). "On the Performance of Various Adaptive Preconditioned GMRES Strategies," *Numer. Lin. Alg. 5*, 101–121.

Y. Saad and K. Wu (1998). "DQGMRES: a Direct Quasi-minimal Residual Algorithm Based on Incomplete Orthogonalization," *Numer. Lin. Alg. 3*, 329–343.

M. Sosonkina, L.T. Watson, R.K. Kapania, and H.F. Walker (1999). "A New Adaptive GMRES Algorithm for Achieving High Accuracy," *Numer. Lin. Alg. 5*, 275–297.

J. Liesen (2000). "Computable Convergence Bounds for GMRES," *SIAM J. Matrix Anal. Applic. 21*, 882–903.

V. Frayss, L. Giraud, S. Gratton, and J. Langou (2005). "Algorithm 842: A Set of GMRES Routines for Real and Complex Arithmetics on High Performance Computers," *ACM Trans. Math. Softw. 31*, 228–238.

A.H. Baker, E.R. Jessup and T. Manteuffel (2005). "A Technique for Accelerating the Convergence of Restarted GMRES," *SIAM J. Matrix Anal. Applic. 26*, 962–984.

L. Reichel and Q. Ye (2005). "Breakdown-free GMRES for Singular Systems," *SIAM J. Matrix Anal. Applic. 26*, 1001–1021.

There is a block version of the GMRES method, see:

V. Simoncini and E. Gallopoulos (1996). "Convergence Properties of Block GMRES and Matrix Polynomials," *Lin. Alg. Applic. 247*, 97–119.

A.H. Baker, J.M. Dennis, and E.R. Jessup (2006). "On Improving Linear Solver Performance: A Block Variant of GMRES," *SIAM J. Sci. Comput. 27*, 1608–1626.

M. Robb and M. Sadkane (2006). "Exact and Inexact Breakdowns in the Block GMRES Method," *Lin. Alg. Applic. 419*, 265–285.

Original references associated with the BiCG, CGS, QMR, and BiCGstab methods include:

C. Lanczos (1952). "Solution of Systems of Linear Equations by Minimized Iterations," *J. Res. Nat. Bur. Stand. 49*, 33-53.

R. Fletcher (1975). "Conjugate Gradient Methods for Indefinite Systems," in *Proceedings of the Dundee Biennial Conference on Numerical Analysis, 1974*, G.A. Watson (ed), Springer-Verlag, New York.

P. Sonneveld (1989). "CGS: A Fast Lanczos-Type Solver for Nonsymmetric Linear Systems," *SIAM J. Sci. Stat. Comput. 10*, 36–52.

R. Freund and N. Nachtigal (1991). "QMR: A Quasi-Minimal Residual Method for Non-Hermitian Linear Systems," *Numer. Math. 60*, 315–339.

H.A. van der Vorst (1992). "Bi-CGSTAB: A Fast and Smoothly Converging Variant of Bi-CG for the Solution of Nonsymmetric Linear Systems," *SIAM J. Sci. Stat. Comput. 13*, 631–644.

Subsequent papers that pertain to these methods include:

G.L.G. Sleijpen and D.R. Fokkema (1993). "BiCGstab(l) for Linear Equations Involving Unsymmetric Matrices with Complex Spectrum," *ETNA 1*, 11–32.

R. Freund (1993). "A Transpose Free Quasi-Minimum Residual Algoroithm for Non-Hermitian Linear Systems," *SIAM J. Sci. Comput. 14*, 470–482.

R.W. Freund and N.M. Nachtigal (1996). "QMRPACK: a Package of QMR Algorithms," *ACM Trans. Math. Softw.* *22*, 46–77.

M.-C. Yeung and T.F. Chan (1999). "ML(k)BiCGSTAB: A BiCGSTAB Variant Based on Multiple Lanczos Starting Vectors," *SIAM J. Sci. Comput.* *21*, 1263–1290.

M. Kilmer, E. Miller, and C. Rappaport (2001). "QMR-Based Projection Techniques for the Solution of Non–Hermitian Systems with Multiple Right–Hand Sides," *SIAM J. Sci. Comput.* *23*, 761–780.

A. El Guennouni, K. Jbilou, and H. Sadok (2003). "A Block Version of BiCGSTAB for Linear Systems with Multiple Right-Hand Sides," *ETNA 16*, 129–142.

G.L.G. Sleijpen, P. Sonneveld, and M.B. van Gijzen (2009). "BiCGSTAB as an Induced Dimension Reduction Method," *Appl. Numer. Math. 60*, 1100–1114.

M.H. Gutknecht (2010). "IDR Explained," *ETNA 36*, 126–148.

11.5 Preconditioning

In general, a Krylov method for $Ax = b$ converges more rapidly if $A \in \mathbb{R}^{n \times n}$ "looks like the identity" and preconditioning can be thought of as a way to bring this about. A matrix can look like the identity in several ways. For example, if A is symmetric positive definite such that $A \approx I + \Delta A$, and $\mathsf{rank}(\Delta A) = k_* \ll n$, then Theorem 11.3.1 plus intuition says that the CG method should produce a good approximate solution after about k_* steps. In this section we identify several major preconditioning strategies and briefly discuss some of their key attributes. Our goal is to impart a sense of what it takes to design or invoke a good preconditioner—an absolutely essential skill to have in many problem settings. For a more in-depth treatment, see Saad (IMSLS), Greenbaum (IMSL), van der Vorst (IMK) and LIN_TEMPLATES.

11.5.1 The Basic Idea

Suppose $M = M_1 M_2$ is nonsingular and consider the linear system $\tilde{A}\tilde{x} = \tilde{b}$ where

$$\tilde{A} = M_1^{-1} A M_2^{-1}, \qquad \tilde{b} = M_1^{-1} b.$$

Note that if M looks like A, then $\tilde{A}$ looks like I. The proposal is to solve the "tilde problem" with a suitably chosen Krylov procedure and then determine x by solving $M_2 x = \tilde{x}$. The matrix M is called a *preconditioner* and it must have two attributes for this solution framework to be of interest:

Criterion 1. M must capture the essence of A, for if $M \approx A$, then we have $I \approx M_1^{-1} A M_2^{-1} = \tilde{A}$. (In settings where M is specified through its inverse, it is more appropriate to say that M^{-1} captures the essence of A^{-1}.)

Criterion 2. It must be easy to solve linear systems that involve the matrices M_1 and M_2 because the Krylov process involves the operation $(M_1^{-1} A M_2^{-1})$-times-vector.

Having a good preconditioner means fewer iterations. However, the cost of an iteration is an issue, as is the overhead associated with the construction of M_1 and M_2. Thus, the enthusiasm for a preconditioner depends upon the strength of the inequality

$$\begin{pmatrix} \text{Set up} \\ M \\ \text{cost} \end{pmatrix} + \begin{pmatrix} \text{Single} \\ \tilde{A}\text{-iteration} \\ \text{cost} \end{pmatrix} \cdot \begin{pmatrix} \text{Number} \\ \text{of } \tilde{A} \\ \text{iterations} \end{pmatrix} < \begin{pmatrix} \text{Single} \\ A\text{-iteration} \\ \text{cost} \end{pmatrix} \cdot \begin{pmatrix} \text{Number} \\ \text{of } A \\ \text{iterations} \end{pmatrix}.$$

There are several ways in which a preconditioner M can capture the essence of A. The difference $A - M$ could be small in norm or low in rank. More generally, if

$$A = [\text{ friendly/important part }] + [\text{ troublesome/lesser part }],$$

then the important part is an obvious candidate for a preconditioner subject to the constraint imposed by Criterion 2. For example, if A is symmetric positive definite, then its diagonal qualifies as an important part that is computationally friendly.

11.5.2 The Preconditioned CG and GMRES Methods

Before we step through the various ways that a linear system can be preconditioned, we show how the CG and GMRES iterations transform in the presence of a preconditioner. For details related to other preconditioned Krylov methods, see LIN_TEMPLATES.

Suppose $M \in \mathbb{R}^{n \times n}$ is a symmetric positive definite matrix that we choose to regard as a preconditioner for the symmetric positive definite linear systems $Ax = b$. Recall that there is a unique symmetric positive definite matrix C such that $M = C^2$. See §4.2.4. If

$$\tilde{A} = C^{-1}AC^{-1}, \qquad \tilde{b} = C^{-1}b,$$

then we can solve $Ax = b$ by applying CG to the symmetric positive definite system $\tilde{A}\tilde{x} = \tilde{b}$ and then solving $Cx = \tilde{x}$. For this to be a practical strategy, we must be able execute CG efficiently when it is applied to this "tilde" problem. Referring to Figure 11.3.1, here are the CG update formulae in this case:

$$
\begin{aligned}
\mu &= (\tilde{r}_c^T \tilde{r}_c) / (\tilde{p}_c^T \tilde{A} \tilde{p}_c), \\
\tilde{x}_+ &= \tilde{x}_c - \mu \tilde{p}_c, \\
\tilde{r}_+ &= \tilde{r}_c + \mu \tilde{A} \tilde{p}_c, \\
\tau &= (\tilde{r}_+^T \tilde{r}_+) / (\tilde{r}_c^T \tilde{r}_c), \\
\tilde{p}_+ &= \tilde{r}_c + \tau \tilde{p}_c.
\end{aligned}
\tag{11.5.1}
$$

Typically $\tilde{A}$ is dense and so we must clearly reformulate these five steps if a suitable level of efficiency is to be reached. Note that if $x_c = C^{-1}\tilde{x}_c$ and $r_c = b - Ax_c$, then

$$\tilde{r}_c = \tilde{b} - \tilde{A}\tilde{x}_c = C^{-1}(b - Ax_c) = C^{-1}r_c.$$

By substituting this formula together with $\tilde{r}_+ = C^{-1}r_+$ and the definition of $\tilde{A}$ into (11.5.1) we obtain

$$
\begin{aligned}
\mu &= (r_c^T M^{-1} r_c) / (C^{-1}\tilde{p}_c)^T A (C^{-1}\tilde{p}_c), \\
Cx_+ &= Cx_c - \mu \tilde{p}_c, \\
C^{-1}r_+ &= C^{-1}r_c + \mu C^{-1}AC^{-1}\tilde{p}_c, \\
\tau &= (r_+^T M^{-1} r_+) / (r_c^T M^{-1} r_c), \\
\tilde{p}_+ &= C^{-1}r_c + \tau \tilde{p}_c.
\end{aligned}
$$

If we define $p_c = C^{-1}\tilde{p}_c$ and set $z_c = M^{-1}r_c$, then this transforms to

Solve $M z_c = r_c$,

$\mu = (r_c^T z_c) / (p_c^T A p_c)$,

$x_+ = x_c - \mu p_c$,

$r_+ = r_c + \mu A p_c$,

$\tau = (r_+^T z_+) / (r_c^T z_c)$,

$p_+ = z_c + \tau p_c$,

and we arrive at the method of *preconditioned conjugate gradients* (PCG). Note that although the square root matrix $C = M^{1/2}$ figured heavily in the derivation of PCG, in the end its action is felt only through the preconditioner $M = C^2$.

Algorithm 11.5.1 (Preconditioned Conjugate Gradients) If $A \in \mathbb{R}^{n \times n}$ and $M \in \mathbb{R}^{n \times n}$ are symmetric positive definite, $b \in \mathbb{R}^n$, and $A x_0 \approx b$, then this algorithm computes $x_* \in \mathbb{R}^n$ so that $A x_* = b$.

$k = 0$, $r_0 = b - A x_0$, $\boxed{\text{Solve } M z_0 = r_0}$

 while $\| r_k \|_2 > 0$

  $k = k + 1$

  if $k = 1$

   $p_k = z_0$

  else

   $\tau = (r_{k-1}^T z_{k-1}) / (r_{k-2}^T z_{k-2})$

   $p_k = z_{k-1} + \tau p_{k-1}$

  end

  $\mu = (r_{k-1}^T z_{k-1})/(p_k^T A p_k)$

  $x_k = x_{k-1} - \mu p_k$

  $r_k = r_{k-1} - \mu A p_k$

  $\boxed{\text{Solve } M z_k = r_k}$

 end

 $x_* = x_k$

To highlight the difference between PCG and CG (Algorithm 11.3.2) we have boxed the *preconditioner system* $M z = r$. It follows that the volume of work associated with a PCG iteration is essentially the volume of work associated with an ordinary CG iteration plus the cost of solving the preconditioner system. It can be shown that the residuals and search directions satisfy

$$r_j^T M^{-1} r_i = 0, \qquad p_j^T (C^{-1} A C^{-1}) p_i = 0, \tag{11.5.2}$$

for all $i \neq j$.

We now turn our attention to the preconditioned GMRES method. The idea is to apply the method to the system $(M^{-1} A) x = (M^{-1} b)$. Modifying Algorithm 11.4.2 in this way yields the following procedure:

Algorithm 11.5.2 (Preconditioned m-step GMRES) If $A \in \mathbb{R}^{n \times n}$ and $M \in \mathbb{R}^{n \times n}$ are nonsingular, $b \in \mathbb{R}^n$, $Ax_0 \approx b$, and m is a positive iteration limit, then this algorithm computes $\tilde{x} \in \mathbb{R}^n$ where either $\tilde{x}$ solves $Ax = b$ or minimizes $\| M^{-1}(Ax - b) \|_2$ over the affine space $x_0 + \mathcal{K}(M^{-1}A, M^{-1}r_0, m)$ where $r_0 = b - Ax_0$.

$k = 0$, $r_0 = b - Ax_0$, $\boxed{\text{Solve } Mz_0 = r_0}$, $\beta_0 = \| z_0 \|_2$

while $(\beta_k > 0)$ and $k < m$

   $q_{k+1} = z_k / \beta_k$

   $k = k + 1$

   $\boxed{\text{Solve } Mz_k = Aq_k}$

   for $i = 1{:}k$

      $h_{ik} = q_i^T z_k$

      $z_k = z_k - h_{ik}q_i$

   end

   $\beta_k = \| z_k \|_2$, $h_{k+1,k} = \beta_k$

   Apply $G_1, \ldots, G_{k-1}$ to $H(1{:}k, k)$ and determine G_k, R_k, p_k, and ρ_k.

end

   Solve $R_k y_k = p_k$ and set $\tilde{x} = x_0 + Q_k y_k$.

Note that $\rho_k = \| M^{-1}(b - Ax_k) \|_2$ in this formulation.

11.5.3   Jacobi and SSOR Preconditioners

We now begin a tour of the major preconditioning strategies. Since some strategies help motivate others, the order of presentation is pedagogical. It does not indicate relative importance, nor does it reflect how the research on preconditioning evolved.

Suppose $A \in \mathbb{R}^{n \times n}$ is diagonally dominant or positive definite. For such a matrix, the diagonal tells much of the story and so it makes a certain amount of sense to consider perhaps the simplest preconditioner of all:

$$M = \text{diag}(a_{11}, \ldots, a_{nn}).$$

Diagonal preconditioners are called *Jacobi preconditioners*. Recall from §11.2.2 that Jacobi's method is based on the splitting $A = M - N$ where M is the diagonal of A. Indeed, for any iteration of the form $Mx_+ = Nx_c + b$, we can regard M as a preconditioner. The requirement that

$$\rho(M^{-1}N) = \rho(M^{-1}(M - A)) = \rho(I - M^{-1}A) < 1$$

is a way of saying that M^{-1} must "look like" A^{-1}. In this context, the SSOR preconditioner

$$M = (D - \omega L)D^{-1}(D - \omega L)^T$$

is attractive for certain symmetric positive definite systems. Note that in this case M is also symmetric positive definite and so it can be used with PCG.

If $A = (A_{ij})$ is a p-by-p block matrix that is (block) diagonally dominant or positive definite, then the block Jacobi preconditioner $M = \text{diag}(A_{11}, \ldots, A_{pp})$ is sometimes a natural choice.

11.5.4   Normwise-Near Preconditioners

Sometimes A is near a data-sparse matrix for which there is a fast solution procedure. Circulant preconditioners for symmetric Toeplitz systems are a nice example. For $a \in \mathbb{R}^n$ define the Toeplitz matrix $T(a) \in \mathbb{R}^{n \times n}$ and the circulant matrix $C(a) \in \mathbb{R}^{n \times n}$ by

$$
T(a) \;=\;
\begin{bmatrix}
a_0 & a_1 & a_2 & a_3 \\
a_1 & a_0 & a_1 & a_2 \\
a_2 & a_1 & a_0 & a_1 \\
a_3 & a_2 & a_1 & a_0
\end{bmatrix},
\qquad
C(a) \;=\;
\begin{bmatrix}
a_0 & a_1 & a_2 & a_3 \\
a_3 & a_0 & a_1 & a_2 \\
a_2 & a_3 & a_0 & a_1 \\
a_1 & a_2 & a_3 & a_0
\end{bmatrix},
\qquad (n = 4).
$$

Suppose we determine $\tilde{a}$ so that $\| T(a) - C(\tilde{a}) \|_F$ is minimized. A case can be made that $M = C(\tilde{a})$ captures the essence of $T(a)$ and thus has potential as a preconditioner for the Toeplitz system $T(a)x = b$. Recall from §4.8.2 that circulant linear systems can be solved in $n \log n$ time using the fast Fourier transform. This style of Toeplitz system preconditioning was proposed by Chan (1988).

Because of their importance, there is a large body of work concerned with preconditioners for Toeplitz systems. An idea due to Chan and Strang (1989) is to set $M = \overset{\centerdot}{C}(\tilde{a})$ where

$$
\tilde{a} \;=\;
\begin{bmatrix}
a(0{:}m) \\
a(m-1{:}-1{:}0)
\end{bmatrix}
$$

assuming that $n = 2m$ and $A = T(a)$ is positive definite. Intuition tells us that A's central diagonals carry most of the information and so it makes sense that they define the preconditioner $C(\tilde{a})$.

11.5.5   Sparse Approximate Inverse Preconditioners

Instead of determining M so $\| A - M \|_F$ is small, we can address Criterion 1 above by choosing M^{-1} so that $\| AM^{-1} - I \|_F$ is small. This is the idea behind sparse approximate inverse preconditioners. To be precise about the nature of the approximation, we define the $\mathsf{sp}(\cdot)$ operator. For any $T \in \mathbb{R}^{n \times n}$ define $\mathsf{sp}(T) \in \mathbb{R}^{n \times n}$ by

$$
[\,\mathsf{sp}(T)\,]_{ij} \;=\;
\begin{cases}
1 & \text{if } t_{ij} \neq 0 \\
0 & \text{otherwise}
\end{cases}.
$$

Suppose $Z \in \mathbb{R}^{n \times n}$ is a given n-by-n matrix of zeros and ones with a manageable sparsity pattern and that we solve the constrained least squares problem

$$
\min_{\mathsf{sp}(T) = Z} \; \| AT - I \|_F.
$$

The constraint says that T is to have the same zero-nonzero structure as Z. Thus, the preconditioner M is specified through its inverse: $M^{-1} = T$. A fringe benefit of this type of preconditioner design is that the $Mz = r$ system is solved via matrix-vector multiplication: $z = Tr$. This is what makes this preconditioning approach attractive from the parallel computing point of view. Moreover, the actual columns of T can be computed in parallel because they are independent of each other.

It is important to appreciate that $T(:, k)$ is a constrained minimizer of $\| A\tau - e_k \|_2$. Let *cols* be the subvector of $1:n$ that identifies the nonzero components of $T(:, k)$. (These indices are determined by $Z(:, k)$.) Let *rows* be a subset of $1:n$ that identifies the nonzero rows in $A(:, cols)$. If τ solves the (generally very small) LS problem

$$\min \| A(rows, cols)\tau - e_k(rows) \|_2$$

then $T(:, k)$ is zero except $T(rows, k) = \tau$. We mention that the sparsity pattern Z can be determined dynamically. For example, after completing the above column-k calculation, it is possible to expand *col* cheaply to include more nonzeros in $T(:, k)$. See Grote and Huckle (1997). Updating QR factorizations is part of their method.

11.5.6 Polynomial Preconditioners

Suppose $A = M_1 - N_1$ is a splitting and that $\rho(G) < 1$ where $G = M_1^{-1}N_1$. Since $A = M_1(I - G)$, it follows that

$$A^{-1} = (I - G)^{-1}M_1^{-1} = \left(\sum_{k=0}^{\infty} G^k\right) M_1^{-1}.$$

This suggests another way to generate a preconditioner whose inverse resembles the inverse of A. We simply truncate the infinite series:

$$M^{-1} = \left(\sum_{k=0}^{m} G^k\right) M_1^{-1}.$$

It follows that

$$z = \left(I + G + G^2 + \cdots G^m\right) M_1^{-1}r$$

solves $Mz = r$. Moreover, there is a very simple way to compute this vector:

$z_c = 0$
for $k = 1:m$
    $M_1 z_+ = N_1 z_c + r$
    $z_c = z_+$
end
$z = z_c$

To see why this works, we note that $z_+ = Gz_c + d$ where $M_1 d = r$, and apply induction:

$$z_+ = Gz_c + d = G\left(I + G + \cdots + G^{k-1}\right) d + d = \left(I + G + \cdots G^k\right) d.$$

Thus, the $Mz = r$ calculation requires m steps of the iteration $M_1 z_+ = N_1 z_c + r$.

In the *polynomial preconditioner* paradigm, the given system $Ax = b$ is replaced by $M^{-1}Ax = M^{-1}b$ where the preconditioner M is defined by

$$M^{-1} = p(M_1^{-1}A)M_1^{-1}. \tag{11.5.3}$$

Here, p is a polynomial and M_1 is itself a preconditioner, e.g., the diagonal of A. In the above example, p was determined by the parameter m and the chosen M_1.

We mention that there are more sophisticated ways to design a good polynomial preconditioner. With $M_1 = I$ for clarity in (11.5.3), the goal is for $p(A)$ to look like A^{-1}, i.e., we want $I \approx p(A)A$. Note that $I - p(A)A = q(A)$ where $q(z) = 1 - zp(z)$, so the challenge is to find $q \in \mathbb{P}_{m+1}$ with the property that $q(0) = 1$ and $q(A)$ is small. There are several ways to address this optimization problem in practice, see Ashby, Manteuffel, and Otto (1992) and Saad(1985).

11.5.7   PCG—Again

The polynomial preconditioner discussion points to an important connection between the classical iterations and the preconditioned conjugate gradient algorithm. Many iterative methods have as their basic step

$$x_k = x_{k-2} + \omega_k(\gamma_{k-1}z_{k-1} + x_{k-1} - x_{k-2}) \qquad (11.5.4)$$

where $Mz_{k-1} = r_{k-1} = b - Ax_{k-1}$. For example, if we set $\omega_k = 1$ and $\gamma_k = 1$, then

$$x_k = M^{-1}(b - Ax_{k-1}) + x_{k-1},$$

i.e., $Mx_k = Nx_{k-1} + b$, where $A = M - N$. Following Concus, Golub, and O'Leary (1976), it is also possible to organize the preconditioned CG method with a central step of the form (11.5.4):

$x_{-1} = 0; \ k = 0; \ r_0 = b - Ax_0$

while $r_k \neq 0$

    $k = k + 1$

    Solve $Mz_{k-1} = r_{k-1}$ for z_{k-1}

    $\gamma_{k-1} = z_{k-1}^T Mz_{k-1}/z_{k-1}^T Az_{k-1}$

    if $k = 1$

        $\omega_1 = 1$

    else

$$\omega_k = \left(1 - \frac{\gamma_{k-1}}{\gamma_{k-2}} \frac{z_{k-1}^T Mz_{k-1}}{z_{k-2}^T Mz_{k-2}} \frac{1}{\omega_{k-1}}\right)^{-1}$$

    end

    $x_k = x_{k-2} + \omega_k(\gamma_{k-1}z_{k-1} + x_{k-1} - x_{k-2})$

    $r_k = b - Ax_k$

end

$x = x_k$

Thus, we can think of the scalars ω_k and γ_k in this iteration as acceleration parameters that can be chosen to speed the convergence of the iteration $Mx_k = Nx_{k-1} + b$. Hence, any iterative method based on the splitting $A = M - N$ can be accelerated by the conjugate gradient algorithm as long as M (the preconditioner) is symmetric and positive definite.

11.5.8 Incomplete Cholesky Preconditioners

Assume that $A \in \mathbb{R}^{n \times n}$ is symmetric positive definite and that we are driven to consider the PCG method because A's Cholesky factor G has many more nonzero entries than the lower triangular portion of A. A natural idea for a preconditioner is to set $M = HH^T$ where H is a sufficiently sparse lower triangular matrix so that if

$$R = HH^T - A \tag{11.5.6}$$

then

$$a_{ij} \neq 0 \Rightarrow r_{ij} = 0. \tag{11.5.7}$$

This means that $[HH^T]_{ij} = a_{ij}$ for all nonzero a_{ij}. In this sense, $M = HH^T$ captures the essence of A. To articulate what we mean by a "sufficiently sparse" H matrix, we specify a set P of subdiagonal index pairs and insist that

$$(i, j) \in P \Rightarrow h_{ij} = 0. \tag{11.5.8}$$

Given P, any matrix H that satisfies (11.5.6)–(11.5.8) is an *incomplete Cholesky factor* of A.

It turns out that it is not always possible to compute H given P. To see what the issues are consider the outer-product implementation of the Cholesky factorization. Recall from §4.2 that it involves repeated application of the factorization

$$\begin{bmatrix} \alpha & v^T \\ v & B \end{bmatrix} = \begin{bmatrix} \sqrt{\alpha} & 0 \\ w & I_{n-1} \end{bmatrix} \begin{bmatrix} 1 & 0 \\ 0 & A_1 \end{bmatrix} \begin{bmatrix} \sqrt{\alpha} & w^T \\ 0 & I_{n-1} \end{bmatrix} \tag{11.5.9}$$

where $w = v/\sqrt{\alpha}$ and $A_1 = B - ww^T$. Indeed, if G_1 is the Cholesky factor of A_1, then

$$G = \begin{bmatrix} \sqrt{\alpha} & 0 \\ w & G_1 \end{bmatrix}$$

is the Cholesky factor of A. Now suppose $Z \in \mathbb{R}^{n \times n}$ is a matrix of zeros and ones with $z_{ij} = z_{ji} = 0$ if and only if $(i, j) \in P$. To ensure the existence of an incomplete Cholesky factor with respect to P, we need to guarantee that the following recursive function works:

function $H = \text{incChol}(A, Z, n)$
    if $n = 1$
        $H = \sqrt{A}$
    else
        $\alpha = A(1,1)$, $v = A(2{:}n, 1)$, $B = A(2{:}n, 2{:}n)$
        $w = (v/\sqrt{\alpha}) .* Z(2{:}n, 1)$
        $A_1 = (B - ww^T) .* Z(2{:}n, 2{:}n)$, $H_1 = \text{incChol}(A_1, Z(2{:}n, 2{:}n), n-1)$
        $H = \begin{bmatrix} \sqrt{\alpha} & 0 \\ w & H_1 \end{bmatrix}$
    end

If Z is the matrix of all 1's, then this is just a recursive form of Cholesky factorization. (Set $r = 1$ in Algorithm 4.2.4). As it stands, it is Cholesky with forced zeros in both the w and A_1 calculations. It is easy to show that *if* the algorithm runs to completion, then Equations (11.5.6), (11.5.7), and (11.5.8) are satisfied. One way to guarantee that this happens is to show that A_1 is positive definite. This turns out to be the case if A is a *Stieltjes matrix*. A matrix $A \in \mathbb{R}^{n \times n}$ is a Stieltjes matrix if it is symmetric positive definite and has nonpositive off-diagonal entries. This property holds in many applications. For example, the model problem matrices in §4.8.3 are Stieltjes matrices. Using the notation $C \geq 0$ to mean that matrix C has nonnegative entries, we show that if A is a Stieltjes, then $A^{-1} \geq 0$.

Lemma 11.5.1. *If $A \in \mathbb{R}^{n \times n}$ is a Stieltjes matrix, then $A^{-1} \geq 0$.*

Proof. Write $A = D - E$ where D and $-E$ are the diagonal and off-diagonal parts. Since $A = D^{1/2}(I - F)D^{1/2}$ it follows that the spectral radius of $F = D^{-1/2}ED^{-1/2}$ satisfies $\rho(F) < 1$. Thus, the entries of

$$A^{-1} = D^{-1/2} \left(\sum_{k=0}^{\infty} F^k \right) D^{-1/2}$$

are clearly nonnegative.   □

The following result is what we need to guarantee that the function incChol does not break down.

Theorem 11.5.2. *If*

$$A = \begin{bmatrix} \alpha & v^T \\ v & B \end{bmatrix}, \qquad \alpha \in \mathbb{R}, \ v \in \mathbb{R}^{n-1}, \ B \in \mathbb{R}^{(n-1) \times (n-1)},$$

is a Stieltjes matrix and $\tilde{v} \in \mathbb{R}^{n-1}$ is obtained from v by setting any subset of its components to zero, then

$$\tilde{B} = B - \frac{\tilde{v}\tilde{v}^T}{\alpha}$$

is a Stieltjes matrix.

Proof. It is clear that $\tilde{B} = \left(\tilde{b}_{ij} \right)$ has nonpositive off-diagonal entries since $\tilde{v} \leq 0$ and

$$\tilde{b}_{ij} = b_{ij} - \frac{\tilde{v}_i \tilde{v}_j}{\alpha}.$$

Our task is to show that $\tilde{B}$ is positive definite.
     Since A is positive definite it follows that if

$$x = \frac{1}{\sqrt{\alpha}} \begin{bmatrix} 1 \\ -B^{-1}v \end{bmatrix}$$

then

$$0 < x^T A x = 1 - \frac{v^T B^{-1} v}{\alpha}.$$

Since $B^{-1} \geq 0$ and $v \leq 0$, we have $\tilde{v}^T B^{-1} \tilde{v} \leq v^T B^{-1} v$ and so

$$\gamma \equiv 1 - \frac{\tilde{v}^T B^{-1} \tilde{v}}{\alpha} \geq 1 - \frac{v^T B^{-1} v}{\alpha} > 0.$$

Using the Sherman-Morrison formula

$$\tilde{B}^{-1} = \left(B - \frac{\tilde{v}\tilde{v}^T}{\alpha} \right)^{-1} = B^{-1} + \frac{1}{\gamma} B^{-1} \frac{\tilde{v}\tilde{v}^T}{\alpha} B^{-1}$$

we see that $\tilde{B}$ is positive definite. $\square$

A theorem of this variety can be found in the landmark paper by Meijerink and van der Vorst (1977).

So far we have just discussed *incomplete Cholesky by position*. The sparsity pattern for the incomplete factor is determined in advance through the set P and does *not* depend on the values in A. An alternative approach makes use of a *drop tolerance* $\tau > 0$, which is used to determine whether or not a "potential" h_{ij} is set to zero. As an example of this strategy, suppose we compute the matrix A_1 in incChol as follows:

$$[A_1]_{ij} = \begin{cases} 0 & \text{if } |b_{ij} - w_i w_j| < \tau \sqrt{b_{ii} b_{jj}} \text{ ,} \\ b_{ij} - w_i w_j & \text{if } |b_{ij} - w_i w_j| \geq \tau \sqrt{b_{ii} b_{jj}} \text{ .} \end{cases}$$

The idea is to drop unimportant entries in the update if they are small in a relative sense. Care has to be exercised in the selection of τ so as not to induce an unacceptable level of fill-in. (Larger values of τ reduce fill-in.) The drop tolerance approach is an example of *incomplete Cholesky by value*.

Lin and Moré (1999) describe a strategy that combines the best features of incomplete Cholesky by position and incomplete Cholesky by value. Recall in gaxpy Cholesky (§4.2.5) that the triangular factor G is computed column by column. The idea is to adapt that procedure so that $H(j{:}n, j)$ has at most $N_j + p$ nonzeros, where N_j is the number of nonzeros in $A(j{:}n, j)$ and p is a nonnegative integer:

> **for** $j = 1{:}n$
> 　　$v(j{:}n) = A(j{:}n, j) - H(j{:}n, 1{:}j{-}1) H(j, 1{:}j{-}1)^T$
> 　　$H(j, j) = \sqrt{v(j)}$
> 　　$N_j = $ number of nonzeros in $A(j{:}n, j)$
> 　　Set to zero each component of $v(j+1{:}n)$ that is not one of the $N_j + p$
> 　　　　largest entries in $|v(j{:}n)|$.
> 　　$H(j+1{:}n, j) = v(j+1{:}n)/H(j, j)$
> **end**

It follows that the number of nonzeros in H is bounded by $pn + N_1 + \cdots + N_n$. Thus, the value of p can be set in accordance with available memory. Note that $H(j{:}n, j)$ is defined by the "most important" entries in $v(j{:}n)$. The gaxpy computation of this vector is a sparse gaxpy, and it is critical that this structure be exploited.

The incomplete factorization idea has been highly studied. Research themes include extension to LU, stability, and ways to increase the "mass" of the diagonal to guarantee existence. Particularly important has been the development of ILU(ℓ) preconditioners, which control fill-in by bounding the number of times that an a_{ij} is allowed to be updated. See Benzi (2002).

11.5.9   Incomplete Block Preconditioners

The incomplete factorization idea can be applied at the block level. For example, an incomplete block Cholesky factor $H = (H_{ij})$ of a block symmetric positive definite matrix $A = (A_{ij})$ could be obtained by forcing H_{ij} to be zero if A_{ij} is zero. However, there is another level of opportunity if the individual A_{ij} are themselves sparse, for then it may be necessary to impose constraints on the sparsity structure of the H_{ij}.

To illustrate this in a simple familiar setting, let us build an incomplete Cholesky factorization for a block tridiagonal matrix whose diagonal blocks are tridiagonal and whose subdiagonal and superdiagonal blocks are diagonal. (The §4.8.3 model problem matrices have this structure.) With

$$
A \;=\; \begin{bmatrix} A_1 & E_1^T & 0 \\ E_1 & A_2 & E_2^T \\ 0 & E_2 & A_3 \end{bmatrix} \;=\; \begin{bmatrix} G_1 & 0 & 0 \\ F_1 & G_2 & 0 \\ 0 & F_2 & G_3 \end{bmatrix} \begin{bmatrix} G_1^T & F_1^T & 0 \\ 0 & G_2^T & F_2^T \\ 0 & 0 & G_3^T \end{bmatrix},
$$

here are the recipes for the G_k and F_k if A is p-by-p as a block matrix:

$G_1 G_1^T \;=\; A_1$
for $k = 1{:}p-1$
    $F_k \;=\; E_k G_k^{-T}$
    $G_{k+1} G_{k+1}^T \;=\; A_{k+1} \;-\; E_k (G_k G_k^T)^{-1} E_k^T$
end

Except for G_1, all the Cholesky factor blocks are dense. A way around this difficulty is to replace $(G_k G_k^T)^{-1}$ with a suitably chosen tridiagonal approximation Λ_k:

$\tilde{G}_1 \tilde{G}_1^T \;=\; A_1$
for $k = 1{:}p-1$
    $\tilde{F}_k \;=\; E_k \tilde{G}_k^{-T}$ (11.5.10)
    $\tilde{G}_{k+1} \tilde{G}_{k+1}^T \;=\; A_{k+1} \;-\; E_k \Lambda_k E_k^T$
end

Note that with this strategy, each $\tilde{G}_k$ is lower bidiagonal. The $\tilde{F}_k$ are full, but they do not have to actually be formed in order to solve systems that involve the incomplete factors. For example,

$$
\begin{bmatrix} \tilde{G}_1 & 0 & 0 \\ \tilde{F}_1 & \tilde{G}_2 & 0 \\ 0 & \tilde{F}_2 & \tilde{G}_3 \end{bmatrix} \begin{bmatrix} w_1 \\ w_2 \\ w_3 \end{bmatrix} = \begin{bmatrix} r_1 \\ r_2 \\ r_3 \end{bmatrix}, \qquad
\begin{aligned}
\tilde{G}_1 w_1 &= r_1, \\
\tilde{G}_2 w_2 &= r_2 - E_1 \tilde{G}_1^{-T} w_1, \\
\tilde{G}_3 w_3 &= r_3 - E_2 \tilde{G}_2^{-T} w_2.
\end{aligned}
$$

Each w_k requires a $\tilde{G}_k$-system solution and a $\tilde{G}_k^T$-system solution.

There remains the issue of choosing $\Lambda_1, \ldots, \Lambda_{p-1}$. The central problem is how to determine a symmetric tridiagonal Λ so that if $T \in \mathbb{R}^{m \times m}$ is symmetric positive definite and tridiagonal itself, then $\Lambda \approx T^{-1}$. Possibilities include:

- Let $\Lambda = \text{diag}(1/t_{11}, \ldots, 1/t_{mm})$.

- Let Λ be the tridiagonal part of T^{-1}, an $O(m)$ computation. See P11.5.5.

- Let $\Lambda = U^T U$ where U is the lower bidiagonal portion of K^{-1} where $T = KK^T$ is the Cholesky factorization. This is an $O(m)$ computation. See P11.5.6.

For a discussion of these approximations and what they imply about the associated preconditioners, see Concus, Golub, and Meurant (1985).

11.5.10 Saddle Point System Preconditioners

A nonsingular 2-by-2 block system of the form

$$
K = \begin{bmatrix} A & B_1^T \\ B_2 & -C \end{bmatrix} \begin{bmatrix} x \\ y \end{bmatrix} = \begin{bmatrix} f \\ g \end{bmatrix},
$$

where $A \in \mathbb{R}^{n \times n}$ is positive semidefinite and $C \in \mathbb{R}^{m \times m}$ is symmetric and positive semidefinite is an example of a *saddle point problem*. Equilibrium systems (§4.4.6) are a special case.

Problems with saddle point structure arise in many applications and there is a host of solution frameworks. Various special cases create multiple possibilities for a preconditioner. For example, if A is nonsingular and $C = 0$, then

$$
\begin{bmatrix} A & B_1 \\ B_2^T & 0 \end{bmatrix} = \begin{bmatrix} I & 0 \\ B_2^T A^{-1} & I \end{bmatrix} \begin{bmatrix} A & 0 \\ 0 & S \end{bmatrix} \begin{bmatrix} I & A^{-1}B_1 \\ 0 & I \end{bmatrix}, \qquad S = -B_2^T A^{-1} B_1.
$$

Possible preconditioners include

$$
M = \begin{bmatrix} \tilde{A} & 0 \\ 0 & \tilde{S} \end{bmatrix} \text{ or } \begin{bmatrix} \tilde{A} & B_1 \\ 0 & S \end{bmatrix} \text{ or } \begin{bmatrix} \tilde{A} & 0 \\ B_2^T & \tilde{S} \end{bmatrix} \begin{bmatrix} I & \tilde{A}^{-1}B_1 \\ 0 & I \end{bmatrix}
$$

where $\tilde{A} \approx A$ and $\tilde{S} \approx S$.

If A and C are positive definite, $H_1 = (A + A^T)/2$, $H_2 = (A - A^T)/2$, and $B = B_1 = B_2$, then

$$
\begin{bmatrix} A & B \\ -B^T & C \end{bmatrix} = \begin{bmatrix} H_1 & 0 \\ 0 & C \end{bmatrix} + \begin{bmatrix} H_2 & B \\ -B^T & 0 \end{bmatrix} \equiv K_1 + K_2
$$

is a symmetric positive definite/skew-symmetric splitting. Preconditioners based on

$$
M = (\alpha I + K_2)^{-1}(\alpha I - K_1)(\alpha I + K_1)^{-1}(\alpha I - K_2)
$$

where $\alpha > 0$ have been shown to be effective. See the saddle point problem survey by Benzi, Golub, and Liesen (2005) for more details. Note that the above strategies are specialized ILU strategies.

11.5.11   Domain Decomposition Preconditioners

Domain decomposition is a framework that can be used to design a preconditioner for an $Ax = b$ problem that arises from a discretized boundary value problem (BVP). Here are the main ideas behind the strategy:

Step 1. Express the given "complicated" BVP domain Ω as a union of smaller, "simpler" subdomains $\Omega_1, \ldots, \Omega_s$.

Step 2. Consider what the discretized BVP "looks like" on each subdomain. Presumably, these subproblems are easier to solve because they are smaller and have a computationally friendly geometry.

Step 3. Build the preconditioner M out of the subdomain matrix problems, paying attention to the ordering of the unknowns and how the subdomain solutions relate to one another and the overall solution.

We illustrate this strategy by considering the Poisson problem $\Delta u = f$ on an L-shaped domain Ω with Dirichlet boundary conditions. (For discretization strategies and solution procedures that are applicable to rectangular domains, see §4.8.4.)

Refer to Figure 11.5.1 where we have subdivided Ω into three non-overlapping rectangular subdomains Ω_1, Ω_2, and Ω_3. As a result of this subdivision, there are five

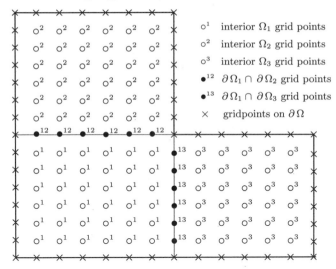

Figure 11.5.1. *The Nonoverlapping subdomain framework*

"types" of gridpoints (and unknowns). With proper ordering, this leads to a block linear system of the form

$$Au = \begin{bmatrix} A_1 & 0 & 0 & B & C \\ 0 & A_2 & 0 & D & 0 \\ 0 & 0 & A_3 & 0 & E \\ F & H & 0 & Q_4 & 0 \\ G & 0 & K & 0 & Q_5 \end{bmatrix} \begin{bmatrix} u_{\circ^1} \\ u_{\circ^2} \\ u_{\circ^3} \\ u_{\bullet^{12}} \\ u_{\bullet^{13}} \end{bmatrix} = \begin{bmatrix} f_{\circ^1} \\ f_{\circ^2} \\ f_{\circ^3} \\ f_{\bullet^{12}} \\ f_{\bullet^{13}} \end{bmatrix} = f \qquad (11.5.11)$$

where A_1, A_2, and A_3 have the discrete Laplacian structure encountered in §4.8.4. Our notation is intuitive: $u_{\bullet 12}$ is the vector of unknowns associated with the $\bullet^{12}$ grid points. Note that A can be factored as

$$A = \begin{bmatrix} I & 0 & 0 & 0 & 0 \\ 0 & I & 0 & 0 & 0 \\ 0 & 0 & I & 0 & 0 \\ FA_1^{-1} & HA_2^{-1} & 0 & I & 0 \\ GA_1^{-1} & 0 & KA_3^{-1} & 0 & I \end{bmatrix} \begin{bmatrix} A_1 & 0 & 0 & B & C \\ 0 & A_2 & 0 & D & 0 \\ 0 & 0 & A_3 & 0 & E \\ 0 & 0 & 0 & S_4 & 0 \\ 0 & 0 & 0 & 0 & S_5 \end{bmatrix} = LU,$$

where S_4 and S_5 are the Schur complements

$$S_4 = Q_4 - FA_1^{-1}B - HA_2^{-1}D,$$
$$S_5 = Q_5 - GA_1^{-1}C - KA_3^{-1}E.$$

If it were not for these typically expensive, dense blocks, the system $Au = f$ could be solved very efficiently via this LU factorization. Fortunately, there are many ways to manage problematic Schur complements. See Saad (IMSLE, pp. 456–465). With appropriate approximations

$$\tilde{S}_4 \approx S_4, \qquad \tilde{S}_5 \approx S_5,$$

we are led to a block ILU preconditioner of the form $M = LU_M$ where

$$U_M = \begin{bmatrix} A_1 & 0 & 0 & B & C \\ 0 & A_2 & 0 & D & 0 \\ 0 & 0 & A_3 & 0 & E \\ 0 & 0 & 0 & \tilde{S}_4 & 0 \\ 0 & 0 & 0 & 0 & \tilde{S}_5 \end{bmatrix}.$$

With sufficient structure, fast Poisson solvers can be used during the L-solves while the efficiency of the U_M solver would depend upon the nature of the Schur complement approximations.

Although the example is simple, it highlights one of the essential ideas behind *nonoverlapping domain decomposition preconditioners* like M. Bordered block diagonal systems must be solved where (a) each diagonal block is associated with a subdomain and (b) the border is relatively "thin" because in the partitioning of the overall domain, the number of domain-coupling unknowns is typically an order of magnitude less than the total number of unknowns. A consequence of (b) is that $A - M$ has low rank and this translates into rapid convergence in a Krylov setting. There are also significant opportunities for parallel computation because of the nearly decoupled subdomain computations. See Bjorstad, Gropp, and Smith (1996).

A similar strategy involves *overlapping subdomains* and we continue with the same example to illustrate the main ideas. Figure 11.5.2 displays a partitioning of the same L-shaped domain into three overlapping subdomains. With proper ordering we obtain

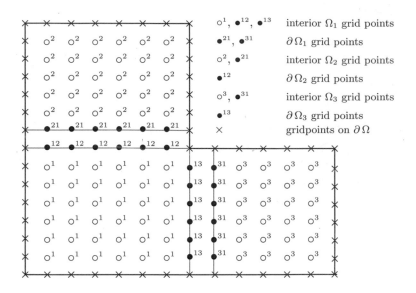

Figure 11.5.2. *The overlapping Schwarz framework*

a block linear system of the form

$$Au = \begin{bmatrix} A_1 & 0 & 0 & B_1 & 0 & C_1 & 0 \\ 0 & A_2 & 0 & 0 & B_2 & 0 & 0 \\ 0 & 0 & A_3 & 0 & 0 & 0 & C_2 \\ F_1 & 0 & 0 & Q_4 & D & 0 & 0 \\ 0 & F_2 & 0 & H & \tilde{Q}_4 & 0 & 0 \\ G_1 & 0 & 0 & 0 & 0 & Q_5 & E \\ 0 & 0 & G_2 & 0 & 0 & K & \tilde{Q}_5 \end{bmatrix} \begin{bmatrix} u_{\circ^1} \\ u_{\circ^2} \\ u_{\circ^3} \\ u_{\bullet^{12}} \\ u_{\bullet^{21}} \\ u_{\bullet^{13}} \\ u_{\bullet^{31}} \end{bmatrix} = \begin{bmatrix} f_{\circ^1} \\ f_{\circ^2} \\ f_{\circ^3} \\ f_{\bullet^{12}} \\ f_{\bullet^{21}} \\ f_{\bullet^{13}} \\ f_{\bullet^{31}} \end{bmatrix} = f.$$

In the *multiplicative Schwarz* approach we cycle through the subdomains improving the interior unknowns along the way. For example, fixing all but the interior Ω_1 unknowns, we solve

$$\begin{bmatrix} A_1 & B_1 & C_1 \\ F_1 & Q_4 & 0 \\ G_1 & 0 & Q_5 \end{bmatrix} \begin{bmatrix} u_{\circ^1} \\ u_{\bullet^{12}} \\ u_{\bullet^{13}} \end{bmatrix} = \begin{bmatrix} f_{\circ^1} \\ f_{\bullet^{12}} \\ f_{\bullet^{13}} \end{bmatrix} - \begin{bmatrix} 0 \\ Du_{\bullet^{21}} \\ Eu_{\bullet^{31}} \end{bmatrix}.$$

After updating $u_{\circ^1}$, $u_{\bullet^{12}}$, and $u_{\bullet^{13}}$ we proceed to fix all but the interior Ω_2 unknowns and solve

$$\begin{bmatrix} A_2 & B_2 \\ F_2 & \tilde{Q}_4 \end{bmatrix} \begin{bmatrix} u_{\circ^2} \\ u_{\bullet^{21}} \end{bmatrix} = \begin{bmatrix} f_{\circ^2} \\ f_{\bullet^{21}} \end{bmatrix} - \begin{bmatrix} 0 \\ Hu_{\bullet^{12}} \end{bmatrix},$$

and update $u_{\circ^2}$ and $u_{\bullet^{21}}$. Finally, we fix all but the interior Ω_3 unknowns and obtain improved versions by solving

$$
\begin{bmatrix} A_3 & C_2 \\ G_2 & \tilde{Q}_5 \end{bmatrix} \begin{bmatrix} u_{\circ 3} \\ u_{\bullet 31} \end{bmatrix} = \begin{bmatrix} f_{\circ 3} \\ f_{\bullet 31} \end{bmatrix} - \begin{bmatrix} 0 \\ K u_{\bullet 13} \end{bmatrix}.
$$

This completes one cycle of multiplicative Schwarz. It is Gauss-Seidel-like in that the most recent values of the current solution are used in each of the three subdomain solves. In the *additive Schwarz* approach, no part if the solution vector u is updated until after the last subdomain solve. This Jacobi-like approach has certain advantages from the standpoint of parallel computing.

For either the multiplicative or additive approach, it is possible to relate $u^{(\text{new})}$ to $u^{(\text{old})}$ via an expression of the form

$$
u^{(\text{new})} = u^{(\text{old})} + M^{-1}(f - Au^{(\text{old})}),
$$

which opens the door to a new family of preconditioning techniques. The geometry of the subdomains and the extent of their overlap critically affects efficiency. Simple geometries can clear a path to fast subdomain solving. Overlap promotes the flow of information between the subdomains but leads to more complicated preconditioners. For an in-depth review of domain decomposition ideas, see Saad (IMSLE, pp. 451–493).

Problems

P11.5.1 Verify (11.5.2).

P11.5.2 Suppose $H \in \mathbb{R}^{n \times n}$ is large sparse upper Hessenberg matrix and that we want to solve $Hx = b$. Note that $H([2{:}n\,1], :)$ has the form $R + e_n v^T$ where R is upper triangular and $v \in \mathbb{R}^n$. Show how GMRES with preconditioner R can (in principle) be used to solve the system in two iterations.

P11.5.3 Show that

$$
A = \begin{bmatrix} 1 & 1 & 3 & 0 \\ 1 & 2 & 0 & 3 \\ 3 & 0 & 19 & -8 \\ 0 & 3 & -8 & 11 \end{bmatrix} = \begin{bmatrix} 1 & 0 & 0 & 0 \\ 1 & 1 & 0 & 0 \\ 3 & -3 & 1 & 0 \\ 0 & 3 & 1 & 1 \end{bmatrix} \begin{bmatrix} 1 & 1 & 3 & 0 \\ 0 & 1 & -3 & 3 \\ 0 & 0 & 1 & 1 \\ 0 & 0 & 0 & 1 \end{bmatrix}
$$

does not have an incomplete Cholesky factorization if $P = \{(4,1), (3,2)\}$.

P11.5.4 Prove that Equations (11.5.6)–(11.5.8) hold if **incChol** executes without breakdown.

P11.5.5 Suppose $T \in \mathbb{R}^{m \times m}$ is symmetric, tridiagonal, and positive definite. There exist $u, v \in \mathbb{R}^m$ so that

$$
[T^{-1}]_{ij} = u_i v_j
$$

for all i and j that satisfy $1 \le j < i < m$. Give an $O(m)$ algorithm for computing u and v.

P11.5.6 Suppose $B \in \mathbb{R}^{m \times m}$ is a nonsingular, lower bidiagonal matrix. Give an $O(m)$ algorithm for computing the lower bidiagonal portion of B^{-1}.

P11.5.7 Consider the computation (11.5.10). Suppose $A_1, \ldots, A_p$ are symmetric with bandwidth q and that $E_1, \ldots, E_{p-1}$ have upper bandwidth 0 and lower bandwidth r. What bandwidth constraints on $\Lambda_1, \ldots, \Lambda_p$ are necessary if $G_1, \ldots, G_p$ are to have lower bandwidth q?

P11.5.8 This problem provides further insight into both the multiplicative Schwarz and additive Schwarz frameworks. Consider the block tridiagonal system

$$
Au = \begin{bmatrix} A_{11} & A_{12} & 0 \\ A_{21} & A_{22} & A_{23} \\ 0 & A_{31} & A_{33} \end{bmatrix} \begin{bmatrix} u_1 \\ u_2 \\ u_3 \end{bmatrix} = \begin{bmatrix} f_1 \\ f_2 \\ f_3 \end{bmatrix} = f
$$

where we assume that A_{22} is much smaller than either A_{11} and A_{33}. Assume that an approximate solution $u^{(k)}$ is improved to $u^{(k+1)}$ via the following multiplicative Schwarz update procedure:

$$
\begin{bmatrix} A_{11} & A_{12} \\ A_{21} & A_{22} \end{bmatrix} \begin{bmatrix} \Delta_1^{(k)} \\ \widetilde{\Delta}_2^{(k)} \end{bmatrix} = \begin{bmatrix} f_1 \\ f_2 \end{bmatrix} - \begin{bmatrix} A_{11} & A_{12} & 0 \\ A_{21} & A_{22} & A_{23} \end{bmatrix} \begin{bmatrix} u_1^{(k)} \\ u_2^{(k)} \\ u_3^{(k)} \end{bmatrix},
$$

$$
\begin{bmatrix} A_{22} & A_{23} \\ A_{32} & A_{33} \end{bmatrix} \begin{bmatrix} \Delta_2^{(k)} \\ \Delta_3^{(k)} \end{bmatrix} = \begin{bmatrix} f_2 \\ f_3 \end{bmatrix} - \begin{bmatrix} A_{21} & A_{22} & A_{23} \\ 0 & A_{32} & A_{33} \end{bmatrix} \begin{bmatrix} u_1^{(k)} + \Delta_1^{(k)} \\ u_2^{(k)} + \widetilde{\Delta}_2^{(k)} \\ u_3^{(k)} \end{bmatrix},
$$

$$
\begin{bmatrix} u_1^{(k+1)} \\ u_2^{(k+1)} \\ u_3^{(k+1)} \end{bmatrix} = \begin{bmatrix} u_1^{(k)} \\ u_2^{(k)} \\ u_3^{(k)} \end{bmatrix} + \begin{bmatrix} \Delta_1^{(k)} \\ \Delta_2^{(k)} \\ \Delta_3^{(k)} \end{bmatrix}.
$$

(a) Determine a matrix M so that $u^{(k+1)} = u^{(k)} + M^{-1}(f - Au^{(k)})$. (b) Repeat for the additive Schwarz update:

$$
\begin{bmatrix} A_{11} & A_{12} \\ A_{21} & A_{22} \end{bmatrix} \begin{bmatrix} \Delta_1^{(k)} \\ \widetilde{\Delta}_2^{(k)} \end{bmatrix} = \begin{bmatrix} f_1 \\ f_2 \end{bmatrix} - \begin{bmatrix} A_{11} & A_{12} & 0 \\ A_{21} & A_{22} & A_{23} \end{bmatrix} \begin{bmatrix} u_1^{(k)} \\ u_2^{(k)} \\ u_3^{(k)} \end{bmatrix},
$$

$$
\begin{bmatrix} A_{22} & A_{23} \\ A_{32} & A_{33} \end{bmatrix} \begin{bmatrix} \Delta_2^{(k)} \\ \Delta_3^{(k)} \end{bmatrix} = \begin{bmatrix} f_2 \\ f_3 \end{bmatrix} - \begin{bmatrix} A_{21} & A_{22} & A_{23} \\ 0 & A_{32} & A_{33} \end{bmatrix} \begin{bmatrix} u_1^{(k)} \\ u_2^{(k)} \\ u_3^{(k)} \end{bmatrix},
$$

$$
\begin{bmatrix} u_1^{(k+1)} \\ u_2^{(k+1)} \\ u_3^{(k+1)} \end{bmatrix} = \begin{bmatrix} u_1^{(k)} \\ u_2^{(k)} \\ u_3^{(k)} \end{bmatrix} + \begin{bmatrix} \Delta_1^{(k)} \\ \widetilde{\Delta}_2 + \Delta_2^{(k)} \\ \Delta_3^{(k)} \end{bmatrix}.
$$

For further discussion, see Greenbaum (IMSL, pp. 198–201).

Notes and References for §11.5

Early papers concerned with preconditioning include:

O. Axelsson (1972). "A Generalized SSOR Method," *BIT 12*, 443–467.

D.J. Evans (1973). "The Analysis and Application of Sparse Matrix Algorithms in the Finite Element Method," in *The Mathematics of Finite Elements and Applications*, J.R. Whiteman (ed), Academic Press, New York, 427–447.

R.H. Bartels and J.W. Daniel (1974). "A Conjugate Gradient Approach to Nonlinear Elliptic Boundary Value Problems," in *Conference on the Numerical Solution of Differential Equations, Dundee, 1973*, G.A. Watson (ed), Springer Verlag, New York.

R.S. Chandra, S.C. Eisenstat, and M.H. Shultz (1975). "Conjugate Gradient Methods for Partial Differential Equations," in *Advances in Computer Methods for Partial Differential Equations*, R. Vichnevetsky (ed), Rutgers University, New Brunswick, NJ.

O. Axelsson (1976). "A Class of Iterative Methods for Finite Element Equations," *Computer Methods in Applied Mechanics and Engineering 9*, 123–137.

P. Concus, G.H. Golub, and D.P. O'Leary (1976). "A Generalized Conjugate Gradient Method for the Numerical Solution of Elliptic Partial Differential Equations," in *Sparse Matrix Computations*, J.R. Bunch and D.J. Rose (eds), Academic Press, New York, 309–332.

J. Douglas Jr. and T. Dupont (1976). "Preconditioned Conjugate Gradient Iteration Applied to Galerkin Methods for a Mildly-Nonlinear Dirichlet Problem," in *Sparse Matrix Computations*, J.R. Bunch and D.J. Rose (eds), Academic Press, New York, 333–348.

For an overview of preconditioning techniques, see Greenbaum (IMSL), Meurant (LCG), Saad (IS-PLA), van der Vorst (IMK), LIN_TEMPLATES as well as the following surveys:

O. Axelsson (1985). "A Survey of Preconditioned Iterative Methods for Linear Systems of Equations," *BIT 25*, 166–187.

M. Benzi (2002). "Preconditioning for Large Linear Systems: A Survey," *J. Comp. Phys. 182*, 418–477.

Papers concerned with sparse approximate inverse preconditioners include:

M. Benzi, C.D. Meyer, and M. Tuma (1996). "A Sparse Approximate Inverse Preconditioner for the Conjugate Gradient Method," *SIAM J. Sci. Comput. 17*, 1135–1149.

E. Chow and Y. Saad (1997). "Approximate Inverse Techniques for Block–Partitioned Matrices," *SIAM J. Sci. Comput. 18*, 1657–1675.

M.J. Grote and T. Huckle (1997). "Parallel Preconditioning with Sparse Approximate Inverses," *SIAM J. Sci. Comput. 18*, 838–853.

N.I.M. Gould and J.A. Scott (1998). "Sparse Approximate-Inverse Preconditioners Using Norm-Minimization Techniques," *SIAM J. Sci. Comput. 19*, 605–625.

M. Benzi and M. Tuma (1998). "A Sparse Approximate-Inverse Preconditioner for Nonsymmetric Linear Systems," *SIAM J. Sci. Comput. 19*, 968–994.

Various aspects of polynomial preconditioners are discussed in:

O.G. Johnson, C.A. Micchelli, and G. Paul (1983). "Polynomial Preconditioners for Conjugate Gradient Calculations," *SIAM J. Numer. Anal. 20*, 362–376.

L. Adams (1985). "m-step Preconditioned Congugate Gradient Methods," *SIAM J. Sci. Stat. Comput. 6*, 452–463.

S. Ashby, T. Manteuffel, and P. Saylor (1989). "Adaptive Polynomial Preconditioning for Hermitian Indefinite Linear Systems," *BIT 29*, 583–609.

R.W. Freund (1990). "On Conjugate Gradient Type Methods and Polynomial Preconditioners for a Class of Complex Non-Hermitian Matrices," *Numer. Math. 57*, 285–312.

S. Ashby, T. Manteuffel, and J. Otto (1992). "A Comparison of Adaptive Chebyshev and Least Squares Polynomial Preconditioning for Hermitian Positive Definite Linear Systems," *SIAM J. Sci. Stat. Comput. 13*, 1–29.

The incomplete Cholesky factorization idea is set forth and analyzed in:

J.A. Meijerink and H.A. van der Vorst (1977). "An Iterative Solution Method for Linear Equation Systems of Which the Coefficient Matrix is a Symmetric M-Matrix," *Math. Comput. 31*, 148–162.

T.A. Manteuffel (1980). "An Incomplete Factorization Technique for Positive Definite Linear Systems," *Math. Comput. 34*, 473–497.

C.-J. Lin and J.J. Moré (1999). "Incomplete Cholesky Factorizations with Limited Memory," *SIAM J. Sci. Comput. 21*, 24–45.

Likewise, for the incomplete LU factorization strategy we have:

M. Bollhofer and Y. Saad (2006). "Multilevel Preconditioners Constructed From Inverse-Based ILUs," *SIAM J. Sci. Comput. 27*, 1627–1650.

H. Elman (1986). "A Stability Analysis of Incomplete LU Factorization," *Math. Comput. 47*, 191–218.

Incomplete QR factorizations have also been devised. See Björck (NMLS, pp. 297–299) as well as:

Z. Jia (1998). "On IOM(q): The Incomplete Orthogonalization Method for Large Unsymmetric Linear Systems," *Numer. Lin. Alg. 3*, 491–512.

Z.-Z. Bai, I.S. Duff, and A.J. Wathen (2001). "A Class of Incomplete Orthogonal Factorization Methods. I: Methods and Theories," *BIT 41*, 53–70.

Incomplete block factorizations are discussed in:

G. Roderigue and D. Wolitzer (1984). "Preconditioning by Incomplete Block Cyclic Reduction," *Math. Comput. 42*, 549–566.

P. Concus, G.H. Golub, and G. Meurant (1985). "Block Preconditioning for the Conjugate Gradient Method," *SIAM J. Sci. Stat. Comput. 6*, 220–252.

O. Axelsson (1985). "Incomplete Block Matrix Factorization Preconditioning Methods. The Ultimate Answer?", *J. Comput. Appl. Math. 12–13*, 3–18.

O. Axelsson (1986). "A General Incomplete Block Matrix Factorization Method," *Lin. Alg. Applic. 74*, 179–190.

The analysis of incomplete factorizations is both difficult and important, see:

Y. Notay (1992). "On the Robustness of Modified Incomplete Factorization Methods," *J. Comput. Math.* *40*, 121–141.

H. Lu and O. Axelsson (1997). "Conditioning Analysis of Block Incomplete Factorizations and Its Application to Elliptic Equations," *Numer. Math.* *78*, 189–209.

M. Bollhofer and Y. Saad (2002). "On the Relations between ILUs and Factored Approximate Inverses," *SIAM J. Matrix Anal. Applic.* *24*, 219–237.

Numerous vector/parallel implementations of the preconditioned CG method have been developed, see:

G. Meurant (1984). "The Block Preconditioned Conjugate Gradient Method on Vector Computers," *BIT 24*, 623–633.

C.C. Ashcraft and R. Grimes (1988). "On Vectorizing Incomplete Factorization and SSOR Preconditioners," *SIAM J. Sci. Stat. Comp. 9,* 122–151.

U. Meier and A. Sameh (1988). "The Behavior of Conjugate Gradient Algorithms on a Multivector Processor with a Hierarchical Memory," *J. Comput. Appl. Math. 24,* 13–32.

H. van der Vorst (1989). "High Performance Preconditioning," *SIAM J. Sci. Stat. Comput. 10,* 1174–1185.

V. Eijkhout (1991). "Analysis of Parallel Incomplete Point Factorizations," *Lin. Alg. Applic. 154–156,* 723–740.

Preconditioners for large Toeplitz systems are discussed in:

T.F. Chan (1988). "An Optimal Circulant Preconditioner for Toeplitz Systems," *SIAM. J. Sci. Stat. Comput. 9,* 766–771.

R.H. Chan and G. Strang (1989). "Toeplitz Equations by Conjugate Gradients with Circulant Preconditioner," *SIAM J. Sci. Stat. Comput. 10,* 104–119.

T. Huckle (1992). "Circulant and Skew-circulant Matrices for Solving Toeplitz Matrix Problems," *SIAM J. Matrix Anal. Applic. 13,* 767–777.

R.H. Chan, J.G. Nagy, and R.J. Plemmons (1994). "Circulant Preconditioned Toeplitz Least Squares Iterations," *SIAM J. Matrix Anal. Applic. 15,* 80–97.

T.F. Chan and J.A. Olkin (1994). "Circulant Preconditioners for Toeplitz Block Matrices," *Numer. alg. 6,* 89–101.

R.H. Chan and M.K. Ng (1996). "Conjugate Gradient Methods for Toeplitz Systems," *SIAM Review 38,* 427–482.

R.H. Chan and X.-Q. Jin (2007). *An Introduction to Iterative Toeplitz Solvers,* SIAM Publications, Philadelphia, PA.

Preconditioners based on the splitting of a matrix into the sum of its symmetric and skew-symmetric parts is covered in the following papers:

Z.-Z. Bai, G.H. Golub, and M.K. Ng (2003). "Hermitian and Skew-Hermitian Splitting Methods for Non-Hermitian Positive Definite Linear Systems," *SIAM J. Matrix Anal. Applic. 24,* 603–626.

Z.-Z. Bai, G.H. Golub, and J.-Y. Pan (2004). "Preconditioned Hermitian and Skew-Hermitian Splitting Methods for Non-Hermitian Positive Semidefinite Linear Systems," *Numer. Math. 98,* 1–32.

Z.-Z. Bai, G.H. Golub, L.-Z. Lu, and J.-F. Yin (2005). "Block Triangular and Skew-Hermitian Splitting Methods for Positive-Definite Linear Systems," *SIAM J. Sci. Comput. 26,* 844–863.

For a discussion of saddle point systems and their preconditioning, see:

M. Benzi, G.H. Golub, and J. Liesen (2005). "Numerical Solution of Saddle Point Problems," *Acta Numerica 14,* 1–137.

G.H. Golub, C. Greif, and J.M. Varah (2005). "An Algebraic Analysis of a Block Diagonal Preconditioner for Saddle Point Systems," *SIAM J. Matrix Anal. Applic. 27,* 779–792.

H.S. Dollar, N.I.M. Gould, W.H.A. Schilders, and A.J. Wathen (2006). "Implicit-Factorization Preconditioning and Iterative Solvers for Regularized Saddle-Point Systems," *SIAM J. Matrix Anal. Applic. 28,* 170–189.

C. Greif and D. Schtzau (2006). "Preconditioners for Saddle Point Linear Systems with Highly Singular (1,1) Blocks," *ETNA 22,* 114–121.

M.A. Botchev and G.H. Golub (2006). "A Class of Nonsymmetric Preconditioners for Saddle Point Problems," *SIAM J. Matrix Anal. Applic. 27,* 1125–1149.

The handling of problematic Schur complements has attracted much attention. For an appreciation of the challenge and what to do about it, see:

H. Elman (1989). "Approximate Schur Complement Preconditioners on Serial and Parallel Computers," *SIAM J. Sci. Stat. Comput. 10*, 581–605.

S.C. Brenner (1999). "The Condition Number of the Schur Complement in Domain Decomposition," *Numer. Math. 83*, 187–203.

F. Zhang (2005). *The Schur Complement and its Applications*, Springer-Verlag, New York.

Z. Li and Y. Saad (2006). "SchurRAS: A Restricted Version of the Overlapping Schur Complement Preconditioner," *SIAM J. Sci. Comput. 27*, 1787–1801.

For an overview of the domain decomposition paradigm, see Demmel (ANLA, pp. 347–356) as well as:

T.F. Chan and T.P. Mathew (1994). "Domain Decomposition Algorithms," *Acta Numerica 3*, 61–143.

W.D. Gropp and D.E. Keyes (1992). "Domain Decomposition with Local Mesh Refinement," *SIAM J. Sci. Statist. Comput. 13*, 967–993.

D.E. Keyes, T.F. Chan, G. Meurant, J.S. Scroggs, and R.G. Voigt (eds) (1992). *Domain Decomposition Methods for Partial Differential Equations*, SIAM Publications, Philadelphia, PA.

T.F. Chan and D. Goovaerts (1992). "On the Relationship Between Overlapping and Nonoverlapping Domain Decomposition Methods," *SIAM J. Matrix Anal. Applic. 13*, 663–670.

B. Smith, P. Bjorstad, and W. Gropp (1996). *Domain Decomposition–Parallel Multilevel Methods for Elliptic Partial Differential Equations*, Cambridge University Press, Cambridge, U.K.

J. Xu and J. Xou (1998) "Some Nonoverlapping Domain Decomposition Methods," *SIAM Review 40*, 857–914.

A. Tosseli and O. Widlund (2010). *Domain Decomposition Methods: Theory and Algorithms*, Springer-Verlag, New York.

For insight into the role of preconditioning for least squares problems and more generally in numerical optimization, see:

P.E. Gill, W. Murray, D.B. Ponceleón, and M.A. Saunders (1992). "Preconditioners for Indefinite Systems Arising in Optimization," *SIAM J. Matrix Anal. Applic. 13*, 292–311.

A. Björck and J. Y. Yuan (1999). "Preconditioners for Least Squares Problems by LU Factorization," *ETNA 8*, 26–35.

M. Benzi and M. Tuma (2003). "A Robust Preconditioner with Low Memory Requirements for Large Sparse Least Squares Problems," *SIAM J. Sci. Comput. 25*, 499–512.

M. Jacobsen, P. C. Hansen, and M. A. Saunders (2003). "Subspace Preconditioned LSQR for Discrete Ill-Posed Problems," *BIT 43*, 975–989.

O. Axelsson and M. Neytcheva (2003). "Preconditioning Methods for Linear Systems Arising in Constrained Optimization Problems," *Numer. Lin. Alg. Applic. 10*, 3–31.

A.R.L. Oliveira and D.C. Sorensen (2004). "A New Class of Preconditioners for Large-Scale Linear Systems from Interior Point Methods for Linear Programming," *Lin. Alg. Applic. 394,* 1–24.

Other ideas associated with preconditioning include inexact solution of the preconditioned system $Mz = r$ and variation of M from iteration to iteration, see:

J. Baglama, D. Calvetti, G. H. Golub, and L. Reichel (1998). "Adaptively Preconditioned GMRES Algorithms," *SIAM J. Sci. Comput. 20*, 243–269.

G.H. Golub and Q.Ye (1999). "Inexact Preconditioned Conjugate Gradient Method with Inner-Outer Iteration," *SIAM J. Sci. Comput. 21*, 1305–1320.

Y. Notay (2000). "Flexible Conjugate Gradients," *SIAM J. Sci. Comput. 22*, 1444–1460.

Error estimation in the preconditioned CG context is discussed in:

O. Axelsson and I. Kaporin (2001). "Error Norm Estimation and Stopping Criteria in Preconditioned Conjugate Gradient Iterations," *Numer. Lin. Alg. 8*, 265–286.

Z. Strakos and P. Tichy (2005). "Error Estimation in Preconditioned Conjugate Gradients," *BIT 45*, 789–817.

11.6   The Multigrid Framework

Let $A^h u^h = b^h$ be a linear system that arises when an elliptic boundary value problem is discretized on a structured grid. The discrete Poisson problems that we discussed in §4.8.3 and §4.8.4 are examples. The superscript "h" is a reminder that the size of the system depends on the fineness of the grid, i.e., the spacing between gridpoints.

The multigrid idea exploits relationships between the "fine grid" solution u^h and its smaller, "coarse grid" analog u^{2h}. Given a current approximate solution u_c^h, the overall framework involves recursive application of the following strategy:

Pre-smooth. With $u_0^h = u_c^h$, perform p_1 steps of a suitable iterative method $u_k^h = G u_{k-1}^h + c$ to produce u_p^h, an error-smoothed version of u_c^h.

Step 1. Compute the current fine-grid residual $r^h = b^h - A^h u_{p_1}^h$. This vector will be rich in certain eigenvector directions and nearly orthogonal to others.

Step 2. Map $r^h \in \mathbb{R}^n$ to $r^{2h} \in \mathbb{R}^m$, a vector that defines what the fine-grid residual looks like on the coarse grid corresponding to $2h$. This will involve an averaging process.

Step 3. Solve the much smaller coarse-grid correction system $A^{2h} z^{2h} = r^{2h}$.

Step 4. Map $z^{2h} \in \mathbb{R}^m$ to $z^h \in \mathbb{R}^n$, a vector that defines what the correction looks like on the fine grid. This will involve interpolation.

Step 5. Update u_c^h to $u_+^h = u_c^h + z^h$.

Post-smooth. With $u_0^h = u_+^h$, perform p_2 steps of a suitable iterative method $u_k^h = G u_{k-1}^h + c$ to produce $u_{++}^h = u_r^h$, an error-smoothed version of u_+^h.

Our plan is to discuss the key issues associated with this paradigm using the 1-dimensional model problem introduced in §4.8.3. The weighted Jacobi method is developed for the pre-smooth and post-smooth steps. Its properties clarify the eigenvector comment in Step 1. After defining the mappings $r^h \to r^{2h}$ and $z^{2h} \to z^h$ associated with Steps 2 and 4, we explain why the Step 5 update results in an improved solution.

Recursion enters the picture through Step 3 as we can apply the same solution strategy to the similar, smaller system $A^{2h} z^{2h} = r^{2h}$. It is through this recursion that we arrive at the overall multigrid framework: the $4h$-grid problem helps solve the $2h$-grid problem, the $8h$-grid problem helps solve the $4h$-grid problem, etc. Depending upon its implementation, the process can be used to either precondition or completely solve the top-level $A^h u^h = b^h$ problem.

The tutorial by Briggs, Henson, and McCormick (2000) provides an excellent introduction to the multigrid framework that was originally proposed in Brandt (1977). For shorter introductions, see Strang (2007, pp. 571–585), Greenbaum (IMSL, pp. 183–197)), Saad (IMSLA, pp. 407–450), and Demmel (ANLA, pp. 331–347).

11.6.1   A Model Problem and the Matrices A^h and Q^h

Consider the problem of finding a function $u(x)$ of $[0, 1]$ that satisfies

$$\frac{d^2 u(x)}{dx^2} = F(x), \qquad u(0) = u(1) = 0. \tag{11.6.1}$$

Our goal is to approximate the solution to (11.6.1) at $x = h,\ 2h, \ldots, nh$ using the discretization strategy set forth in §4.8.3. Here and throughout this section,

$$n = 2^k - 1, \qquad m = 2^{k-1} - 1, \qquad h = 1/2^k.$$

This leads to a linear system

$$A^h u^h = b^h \tag{11.6.2}$$

where $b^h \in \mathbb{R}^n$ and $A^h \in \mathbb{R}^{n \times n}$ is defined by

$$A^h = \frac{1}{h^2}
\begin{bmatrix}
2 & -1 & \cdots & \cdots & 0 \\
-1 & 2 & \ddots & & \vdots \\
\vdots & \ddots & \ddots & \ddots & \vdots \\
\vdots & & \ddots & \ddots & -1 \\
0 & 0 & \cdots & -1 & 2
\end{bmatrix}. \tag{11.6.3}$$

Note that A^h is a multiple of $\mathcal{T}_n^{DD}$, a matrix that we defined in (4.8.7). It has a completely known Schur decomposition

$$(Q^h)^T A^h Q^h = \Lambda^h = \mathrm{diag}(\lambda^h), \tag{11.6.4}$$

where the vector of eigenvalues $\lambda^h \in \mathbb{R}^n$ is given by

$$\lambda_j^h = \frac{4}{h^2} \cdot \sin^2\left(\frac{j\pi}{2(n+1)}\right), \qquad j = 1{:}n, \tag{11.6.5}$$

and the orthogonal eigenvector matrix $Q^h = [\, q_1 \mid \cdots \mid q_n \,]$ is prescribed by

$$q_j = \sqrt{\frac{2}{n+1}}
\begin{bmatrix}
\sin(\theta_j) \\
\vdots \\
\sin(n\theta_j)
\end{bmatrix}, \qquad \theta_j = \frac{j\pi}{n+1}. \tag{11.6.6}$$

The components of this vector involve samplings of the function $\sin(j\pi x)$. As j increases, this function is increasingly oscillatory, prompting us to split the eigenmodes in half. We regard q_j as a *low-frequency* eigenvector if $1 \le j \le m$ and as a *high-frequency* eigenvector if $j > m$.

To facilitate the divide-and-conquer derivations that follow, we identify some critical patterns associated with Q^h and Λ^h. If

$$S^h = \mathrm{diag}(s_1^2, \ldots, s_m^2), \qquad s_j = \sin\left(\frac{j\pi}{2(n+1)}\right), \tag{11.6.7}$$

$$C^h = \mathrm{diag}(c_1^2, \ldots, c_m^2), \qquad c_j = \cos\left(\frac{j\pi}{2(n+1)}\right), \tag{11.6.8}$$

then

$$\Lambda^h = \frac{4}{h^2}
\begin{bmatrix}
S^h & 0 & 0 \\
0 & 1/2 & 0 \\
0 & 0 & \mathcal{E}_m C^h \mathcal{E}_m
\end{bmatrix} \tag{11.6.9}$$

where $\mathcal{E}_m$ is the m-by-m exchange permutation. Regarding Q^h, it houses scaled copies of its m-by-m analog Q^{2h}:

$$Q^h(2{:}2{:}2m, :) \;=\; \left[\; Q^{2h} \mid 0 \mid -Q^{2h}\mathcal{E}_m \;\right]/\sqrt{2}. \qquad (11.6.10)$$

These results follow from the definitions (11.6.5)–(11.6.8) and trigonometric identities.

11.6.2 Damping Error with the Weighted Jacobi Method

Critical to the multigrid framework is the role of the *smoothing iteration*. The term "smoother" is applied to an iterative method that is particularly successful at damping out the high-frequency eigenvector components of the error. To illustrate this part of the process, we introduce the *weighted Jacobi method*. If $L = \mathrm{tril}(A, -1)$, $D = \mathrm{diag}(a_{ii})$, and $U = \mathrm{triu}(A, 1)$, then the iterates for this method are defined by

$$u^{(k)} \;=\; Gu^{(k-1)} + c,$$

where $c = \omega D^{-1}b$, $G = (1 - \omega)I - \omega D^{-1}(L + U)$, and ω is a free parameter that we assume satisfies $0 < \omega \le 1$. Note that if $\omega = 1$, then the method reverts to the simple Jacobi iteration (11.2.2). Other iterations can be used, but the weighted Jacobi method is simple and adequately communicates the role of the smoother in multigrid.

   If we apply the weighted Jacobi method to (11.6.2), then it is easy to verify that the iteration matrix is given by

$$G^{h,\omega} \;=\; I_n - \frac{\omega h^2}{2}A^h. \qquad (11.6.11)$$

By using (11.6.4) and (11.6.5) we see that its Schur decomposition is given by

$$(Q^h)^T G^{h,\omega} Q^h \;=\; \mathrm{diag}(\tau^{h,\omega}), \qquad \tau_j^{h,\omega} \;=\; 1 - 2\omega \sin^2\left(\frac{j\pi}{2(n+1)}\right). \qquad (11.6.12)$$

It follows that $\rho(G^{h,\omega}) < 1$ because we assume $0 < \omega \le 1$ to guarantee convergence. The explicit Schur decomposition enables us to track the error in each eigenvector direction given a starting vector u_0^h:

$$u_0^h - u^h \;=\; \sum_{j=1}^n \alpha_j \cdot q_j \;\Rightarrow\; (u_p^h - u^h) \;=\; (G^{h,\omega})^p\,(u_0^h - u^h) \;=\; \sum_{j=1}^n \alpha_j \cdot (\tau_j^{h,\omega})^p \cdot q_j.$$

Thus, the component of the error in the direction of the eigenvector q_j tends to zero like $|\tau_j^{h,\omega}|^p$. These rates depend on ω and vary with j. We now ask, is there a smart way to choose the value of ω so that the error is rapidly diminished in each eigenvector direction?

   Assume that $n \gg 1$ and consider (11.6.12). For small j we see that $\tau_j^{h,\omega}$ is close to unity regardless of the value of ω. On the other hand, we can move the "large j" eigenvalues toward the origin by choosing a smaller value of ω. These qualitative observations suggest that we choose ω to minimize

$$\mu(\omega) \;=\; \max\{\,|\tau_{m+1}^{h,\omega}|, \ldots, |\tau_n^{h,\omega}|\,\}.$$

In other words, ω should be chosen to promote rapid damping in the direction of the high-frequency eigenvectors. Because the damping rates associated with the low-frequency eigenvectors are much less affected by the choice of ω, they are left out of the optimization. Since

$$-1 < \tau_n^{h,\omega} < \cdots < \tau_{m+1}^{h,\omega} < \cdots < \tau_1^{h,\omega} < 1,$$

it is easy to see that the optimum ω should make $\tau_{m+1}^{h,\omega}$ and $\tau_n^{h,\omega}$ equal in magnitude but opposite in sign, i.e.,

$$-1 + 2\omega \sin^2\left(\frac{n\pi}{2(n+1)}\right) = -\left(-1 + 2\omega \sin^2\left(\frac{(m+1)\pi}{2(n+1)}\right)\right).$$

This is essentially solved by setting $\omega_{opt} = 2/3$. With this choice, $\mu(2/3) = 1/3$ and so

$$\left(\begin{array}{c} p\text{-th iterate error in} \\ \text{high-frequency directions} \end{array}\right) \le \left(\frac{1}{3}\right)^p \left(\begin{array}{c} \text{Starting vector error in} \\ \text{high-frequency directions} \end{array}\right).$$

11.6.3　Interactions Between the Fine and Coarse Grids

Suppose for some modest value of p we use the weighted Jacobi iteration to obtain an approximate solution u_p^h to $A^h u^h = b^h$. We can estimate its error by approximately solving $A^h z = r^h = b^h - A^h u_p^h$. From the discussion in the previous section we know that the residual $r^h = A^h(u^h - u_p^h)$ resides mostly in the span of the low-frequency eigenvectors. Because r^h is smooth, there is not much happening from one gridpoint to the next and it is well-approximated on the coarse grid. This suggests that we might get a good approximation to the error in u_p^h by solving the coarse-grid version of $A^h z = r^h$. To that end, we need to detail how vectors are transformed when we switch grids. Note that on the fine grid, gridpoint $2j$ is coarse gridpoint j:

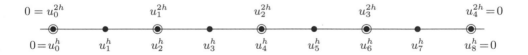

To map values from the fine grid (with $n = 2^k - 1$ gridpoints) to the coarse-grid (with $m = 2^{k-1} - 1$ gridpoints), we use an m-by-n *restriction matrix* R_h^{2h}. Similarly, to generate fine-grid values from coarse-grid values, we use an n-by-m *prolongation matrix* P_{2h}^h. Before these matrices are formally defined, we display the case when $n = 7$ and $m = 3$:

$$R_h^{2h} = \frac{1}{4} \begin{bmatrix} 1 & 2 & 1 & 0 & 0 & 0 & 0 \\ 0 & 0 & 1 & 2 & 1 & 0 & 0 \\ 0 & 0 & 0 & 0 & 1 & 2 & 1 \end{bmatrix}, \qquad P_{2h}^h = \frac{1}{2} \begin{bmatrix} 1 & 0 & 0 \\ 2 & 0 & 0 \\ 1 & 1 & 0 \\ 0 & 2 & 0 \\ 0 & 1 & 1 \\ 0 & 0 & 2 \\ 0 & 0 & 1 \end{bmatrix}. \qquad (11.6.13)$$

The intuition behind these choices is easy to see. The operation $u^{2h} = R_h^{2h} u^h$ takes a fine-grid vector of values and produces a coarse-grid vector of values using a weighted average around each even-indexed component:

$$\begin{bmatrix} u_1^{2h} \\ u_2^{2h} \\ u_3^{2h} \end{bmatrix} = R_h^{2h} \begin{bmatrix} u_1^h \\ u_2^h \\ u_3^h \\ u_4^h \\ u_5^h \\ u_6^h \\ u_7^h \end{bmatrix} = \begin{bmatrix} (u_1^h + 2u_2^h + u_3^h)/4 \\ (u_3^h + 2u_4^h + u_5^h)/4 \\ (u_5^h + 2u_6^h + u_7^h)/4 \end{bmatrix}.$$

The prolongation matrix generates "missing" fine-grid values by averaging adjacent coarse grid values:

$$\begin{bmatrix} u_1^h \\ u_2^h \\ u_3^h \\ u_4^h \\ u_5^h \\ u_6^h \\ u_7^h \end{bmatrix} = P_{2h}^h \begin{bmatrix} u_1^{2h} \\ u_2^{2h} \\ u_3^{2h} \end{bmatrix} = \begin{bmatrix} (u_0^{2h} + u_1^{2h})/2 \\ u_1^{2h} \\ (u_1^{2h} + u_2^{2h})/2 \\ u_2^{2h} \\ (u_2^{2h} + u_3^{2h})/2 \\ u_3^{2h} \\ (u_3^{2h} + u_4^{2h})/2 \end{bmatrix}.$$

The special end-conditions make sense because we are assuming that the solution to the model problem is zero at the endpoints.

For general $n = 2^k - 1$ and $m = 2^{k-1} - 1$, we define the matrices $R_h^{2h} \in \mathbb{R}^{m \times n}$ and $P_{2h}^h \in \mathbb{R}^{n \times m}$ by

$$R_h^{2h} = \frac{1}{4} B^h(2{:}2{:}2m, :), \qquad P_{2h}^h = \frac{1}{2} B^h(:, 2{:}2{:}2m), \tag{11.6.14}$$

where

$$B^h = 4I_n - h^2 A^h. \tag{11.6.15}$$

The connection between the even-indexed columns of this matrix and P_{2h}^h and R_h^{2h} is clear from the example

$$B^h = \begin{bmatrix} 2 & 1 & 0 & 0 & 0 & 0 & 0 \\ 1 & 2 & 1 & 0 & 0 & 0 & 0 \\ 0 & 1 & 2 & 1 & 0 & 0 & 0 \\ 0 & 0 & 1 & 2 & 1 & 0 & 0 \\ 0 & 0 & 0 & 1 & 2 & 1 & 0 \\ 0 & 0 & 0 & 0 & 1 & 2 & 1 \\ 0 & 0 & 0 & 0 & 0 & 1 & 2 \end{bmatrix}, \qquad (n = 7).$$

With the restriction and prolongation operators defined and letting $WJ(k, u_0)$ denote the kth iterate of the weighted Jacobi iteration applied to $A^h u = b^h$ with starting vector u_0, we can make precise the 2-grid multigrid framework:

$$
\begin{aligned}
\textit{Pre-smooth:} \quad & u_{p_1}^h = WJ(p_1, u_c^h), \\
\textit{Fine-grid residual:} \quad & r^h = b^h - A^h u_{p_1}^h, \\
\textit{Restriction:} \quad & r^{2h} = R_h^{2h} r^h, \\
\textit{Coarse-grid correction:} \quad & A^{2h} z^{2h} = r^{2h}, \\
\textit{Prolongation:} \quad & z^h = P_{2h}^h z^{2h}, \\
\textit{Update:} \quad & u_+^h = u_c^h + z^h, \\
\textit{Post-smooth:} \quad & u_{++}^h = WJ(p_2, u_+^h).
\end{aligned}
\tag{11.6.16}
$$

By assembling the middle five equations, we see that

$$
u_+^h = u_p^h + P_{2h}^h (A^{2h})^{-1} R_h^{2h} A^h (u^h - u_{p_1}^h)
$$

and so

$$
(u_+^h - u^h) = E_h(u_{p_1}^h - u^h) \tag{11.6.17}
$$

where

$$
E^h = I_n - P_{2h}^h (A^{2h})^{-1} R_h^{2h} A^h \tag{11.6.18}
$$

can be thought of as a 2-grid error operator. Accounting for the damping in the weighted Jacobi smoothing steps, we have

$$
(u_p^h - u^h) = (G^h)^p (u_c^h - u^h), \qquad p \in \{p_1, p_2\},
$$

where $G^h = G^{h,2/3}$, the optimal-ω iteration matrix. From this we conclude that

$$
(u_{++}^h - u^h) = (G^h)^{p_2} E^h (G^h)^{p_1} (u_c^h - u^h). \tag{11.6.19}
$$

To appreciate how the components of the error diminish, we need to understand what E^h does to the eigenvectors $q_1, \ldots, q_n$. The following lemma is critical to the analysis.

Lemma 11.6.1. *If $n = 2^k - 1$ and $m = 2^{k-1} - 1$, then*

$$
(Q^h)^T P_{2h}^h Q^{2h} = \sqrt{2}
\begin{bmatrix}
C^h \\
0 \\
-\mathcal{E}_m S^h
\end{bmatrix}, \quad
(Q^{2h})^T R_h^{2h} Q^h = \sqrt{\frac{1}{2}}
\begin{bmatrix}
C^h \\
0 \\
-\mathcal{E}_m S^h
\end{bmatrix}^T
\tag{11.6.20}
$$

where the diagonal matrices S^h and C^h are defined by (11.6.7) and (11.6.8).

Proof. From (11.6.4), (11.6.9), and (11.6.15) we have

$$
(Q^h)^T B^h Q^h = 4I_n - h^2 \Lambda^h = 4
\begin{bmatrix}
C^h & 0 & 0 \\
0 & 1/2 & 0 \\
0 & 0 & \mathcal{E}_m S^h \mathcal{E}_m
\end{bmatrix}
\equiv D^h.
$$

Define the index vector $idx = 2{:}2{:}2m$. Since $(Q^h)^T B^h = D^h (Q^h)^T$, it follows from (11.6.10) that

$$(Q^h)^T B^h(:, idx) = D^h Q^h (idx, :)^T = \sqrt{\frac{1}{2}} D^h \begin{bmatrix} I_m \\ 0 \\ -\mathcal{E}_m \end{bmatrix} (Q^{2h})^T.$$

Thus,

$$(Q^h)^T B^h(:, idx) Q^{2h} = \frac{4}{\sqrt{2}} \begin{bmatrix} C^h & 0 & 0 \\ 0 & 1/2 & 0 \\ 0 & 0 & \mathcal{E}_m S^h \mathcal{E}_m \end{bmatrix} \begin{bmatrix} I_m \\ 0 \\ -\mathcal{E}_m \end{bmatrix} = \frac{4}{\sqrt{2}} \begin{bmatrix} C^h \\ 0 \\ -\mathcal{E}_m S^h \end{bmatrix}.$$

The lemma follows since $P_{2h}^h = B^h(:, idx)/2$ and $R_h^{2h} = B^h(:, idx)^T/4$.   □

With these diagonal-like decompositions we can expose the structure of E^h.

Theorem 11.6.2. *If $n = 2^k - 1$ and $m = 2^{k-1} - 1$, then*

$$E^h Q^h = Q^h \begin{bmatrix} S^h & 0 & C^h \mathcal{E}_m \\ 0 & 1 & 0 \\ \mathcal{E}_m S^h & 0 & \mathcal{E}_m C^h \mathcal{E}_m \end{bmatrix}. \qquad (11.6.21)$$

Proof. From (11.6.18) it follows that

$$(Q^h)^T E^h Q^h = I_n - ((Q^h)^T P_{2h}^h Q^{2h})((Q^{2h})^T A^{2h} Q^{2h})^{-1}((Q^{2h})^T R_h^{2h} Q^h)((Q^h)^T A^h Q^h).$$

The proof follows by substituting (11.6.4), (11.6.9), (11.6.20), and

$$(Q^{2h})^T A^{2h} Q^{2h} = \frac{1}{2h^2}(I_m - \sqrt{C^h})$$

into this equation and using trigonometric identities.   □

The block matrix (11.6.21) has the form

$$\begin{bmatrix} S^h & 0 & C^h \mathcal{E}_m \\ 0 & 1 & 0 \\ \mathcal{E}_m S^h & 0 & \mathcal{E}_m C^h \mathcal{E}_m \end{bmatrix} = \left[\begin{array}{ccc|c|ccc} s_1^2 & 0 & 0 & 0 & 0 & 0 & c_1^2 \\ 0 & s_2^2 & 0 & 0 & 0 & c_2^2 & 0 \\ 0 & 0 & s_3^2 & 0 & c_3^2 & 0 & 0 \\ \hline 0 & 0 & 0 & 1 & 0 & 0 & 0 \\ \hline 0 & 0 & s_3^2 & 0 & c_3^2 & 0 & 0 \\ 0 & s_2^2 & 0 & 0 & 0 & c_2^2 & 0 \\ s_1^2 & 0 & 0 & 0 & 0 & 0 & c_1^2 \end{array} \right], \qquad (n = 7),$$

from which it is easy to see that

$$\begin{aligned} E^h q_j &= s_j^2 (q_j + q_{n-j+1}), & j = 1{:}m, \\ E^h q_{m+1} &= q_{m+1}, & (11.6.22) \\ E^h q_{n-j+1} &= c_j^2 (q_j + q_{n-j+1}), & j = 1{:}m. \end{aligned}$$

This enables us to examine the eigenvector components in the error equation (11.6.19) because we also know from §11.6.2 that $G^h q_j = \tau_j q_j$ where $\tau_j = \tau_j^{h,2/3}$. Thus, if the initial error has the eigenvector expansion

$$u_c^h - u^h = \underbrace{\sum_{j=1}^{m} \alpha_j\, q_j}_{\text{low frequency}} + \underbrace{\alpha_{m+1}\, q_{m+1} + \sum_{j=1}^{m} \alpha_{n-j+1}\, q_{n-j+1}}_{\text{high frequency}}$$

and we execute (11.6.16), then the error in u_{++}^h is given by

$$u_{++}^h - u^h = \sum_{j=1}^{m} \tilde{\alpha}_j\, q_j + \tilde{\alpha}_{m+1}\, q_{m+1} + \sum_{j=1}^{m} \tilde{\alpha}_{n-j+1}\, q_{n-j+1},$$

where

$$\tilde{\alpha}_j = \left(\alpha_j\, \tau_j^{p_1}\, s_j^2 + \alpha_{n-j+1}\, \tau_{n-j+1}^{p_1}\, c_j^2 \right) \tau_j^{p_2}, \qquad j = 1{:}m,$$

$$\tilde{\alpha}_{m+1} = \alpha_{m+1}\, \tau_{m+1}^{p_1+p_2},$$

$$\tilde{\alpha}_{n-j+1} = \left(\alpha_j\, \tau_j^{p_1}\, s_j^2 + \alpha_{n-j+1}\, \tau_{n-j+1}^{p_1}\, c_j^2 \right) \tau_{n-j+1}^{p_2}, \quad j = 1{:}m.$$

It is important to appreciate the damping factors in these expressions. By virtue of the weighted Jacobi iteration design, $|\tau_{n-j+1}| \le 1/3$ for $j = 1{:}m$. From the definition of s_j in (11.6.7), we also have $s_j^2 \le 1/2$. It follows from the $\tilde{\alpha}$ recipes that high-frequency error is nicely damped by fine-grid smoothing and that low-frequency error is attenuated by the coarse-grid operations. This interplay together with the fact that the s_j and τ_{n-j+1} bounds are independent of n are what make the multigrid framework so powerful.

11.6.4 V-Cycles and Other Recursive Strategies

If the coarse-grid system in (11.6.16) is solved recursively, then we can encapsulate the overall process as follows given that $A^h u_c^h \approx b^h$:

> **function** $u_{++}^h = \mathsf{mgV}(u_c^h, b^h, h)$
>
> **if** $h \ge h_{\max}$
>
> $\qquad u_{++}^h = WJ(u_c^h, p_0) \qquad$ (for example)
>
> **else**
>
> $\qquad u_{p_1}^h = WJ(u_c^h, p_1)$
>
> $\qquad r^h = b^h - A^h u_{p_1}^h$
>
> $\qquad r^{2h} = R_h^{2h} r^h$
>
> $\qquad z^{2h} = \mathsf{mgV}(0, r^{2h}, 2h)$
>
> $\qquad u_+^h = u_p^h + P_{2h}^h z^{2h}$
>
> $\qquad u_{++}^h = WJ(u_+^c, p_2)$
>
> **end**

Note that the base case ($h \geq h_{\max}$) is defined by a "coarse-enough," gridpoint-spacing parameter $h_{\max}$ and that the solution of the (possibly small) linear system at that level can be obtained in various ways. Figure 11.6.1 depicts the flow of events called a *V-cycle*, if $h_{\max} = 16h$. Five grids are used and the process starts by recurring four

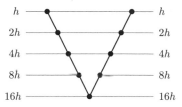

Figure 11.6.1. *A V-cycle*

times before the correction equation is solved. This is done on the $16h$-grid. After that, the corrections are mapped upwards through four levels, eventually generating a solution to the top-level h-grid problem.

Examination of mgV reveals that a V-cycle involves $O(n)$ flops, a hint that the multigrid framework is incredibly efficient. The coefficient of n in the complexity assessment depends on the iteration parameters p_0, p_1 and p_2. *However, the rate of error damping is independent of n, which means that these error-control parameters are not affected by the size of the problem.*

The V-cycle that we illustrated is but one of several strategies for moving in between grids during the course of a multigrid solve. The pattern for *full multigrid* is depicted in Figure 11.6.2. Here, the coarse-grid system is used to obtain a starting value

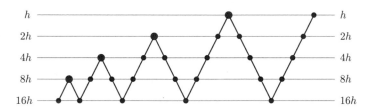

Figure 11.6.2. *Full multigrid*

for its fine-grid neighbor and then a V-cycle is performed to obtain an improvement. The process is repeated.

11.6.5   A Rich Design Space

The multigrid framework is rich with options, some of which are not obvious from our simple, model-problem treatment. For general elliptic boundary value problems on complicated domains, there are several critical decisions that need to be made if the overall procedure is to be effective:

- Determine how to extract the coarse grid from the fine grid, e.g., every other grid-point in each coordinate direction or every other gridpoint in just one coordinate direction.

- Determine the right restriction and prolongation operators.

- Determine the right smoother, e.g., (blocked) weighted Jacobi or Gauss-Seidel.

- Determine the number of pre-smoothing steps and post-smoothing steps.

- Determine the depth and "shape" of the recursion, i.e., the number of participating grids and the order in which they are visited.

- Determine a base-case strategy, i.e., should bottom-level linear systems be solved exactly or approximately?

With so many implementation parameters, it is not surprising that the multigrid framework can be tuned to address a very broad range of problems.

Problems

P11.6.1 Prove (11.6.9) and (11.6.10).

P11.6.2 Fill in the details that are left out of the proof of Theorem 11.6.2.

P11.6.3 Using (11.6.21), determine the SVD of the matrix E^h.

P11.6.4 What are the analogues of P_{2h}^h and R_h^{2h} for the 2-dimensional Poisson problem on a rectangle with Dirichlet boundary conditions? What does the matrix E^h look like in this case? State and prove analogues of Lemma 11.6.1 and Theorem 11.6.2.

Notes and References for §11.6

The multigrid framework was originally set forth in:

A. Brandt (1977). "Multilevel Adaptive Solutions to Boundary Value Problems," *Math. Comput. 31*, 333–390.

For an excellent, highly intuitive introduction, see:

G. Strang (2007). *Computational Science and Engineering*, Wellesley-Cambridge Press, Wellesley, MA.

More in-depth treatments include:

P. Wesseling (1982). *An Introduction to Multigrid Methods,* Wiley, Chichester, U.K.
W. Hackbusch (1985). *Multi-Grid Methods and Applications*, Springer-Verlag, Berlin.
S.F. McCormick (1987). *Multigrid Methods*, SIAM Publications, Philadelphia, PA.
J.H. Bramble (1993). *Multigrid Methods*, Longman Scientific and Technical, Harlow, U.K.
W.L. Briggs, V.E. Henson, and S.F. McCormick (2000). *A Multigrid Tutorial*, second edition, SIAM Publications, Philadelphia, PA.
U. Trottenberg, C. Osterlee, and A. Schuller (2001). *Multigrid*, Academic Press, London.
Y. Shapira (2003). *Matrix-Based Multigrid,* second edition, Springer, New York.

Multigrid can be used as a preconditioning strategy. The coarse-grid problem serves as the easy-to-solve system that "captures the essence" of the fine-grid system, see:

J. Xu (1992). "Iterative Methods by Space Decomposition and Subspace Correction," *SIAM Review 34*, 581–613.
T.F. Chan and B.F. Smith (1994). "Domain Decomposition and Multigrid Algorithms for Elliptic Problems on Unstructured Meshes," *ETNA 2*, 171–182.
B. Lee (2009). "Guidance for Choosing Multigrid Preconditioners for Systems of Elliptic Partial Differential Equations," *SIAM J. Sci. Comput. 31*, 2803–2831.

The multigrid idea can be extended to "gridless" problems. The resulting framework of *algebraic multigrid* methods has met with considerable success in certain application settings, see:

A. Brandt, S.F. McCormick, and J. Ruge (1984). "Algebraic Multigrid (AMG) for Sparse Matrix
   Equations," in *Sparsity and Its Applications*, D.J. Evans (ed.), Cambridge University Press, Cambridge.

J.W. Ruge and K. Stuben (1987). "Algebraic Multigrid," in *Multigrid Methods*, Vol. 3, Frontiers in
   Applied Mathematics, S.F. McCormick (ed.), SIAM Publications, Philadelphia, PA.

Chapter 12

Special Topics

12.1   **Linear Systems with Displacement Structure**

12.2   **Structured-Rank Problems**

12.3   **Kronecker Product Computations**

12.4   **Tensor Unfoldings and Contractions**

12.5   **Tensor Decompositions and Iterations**

Prominent themes in this final chapter include data sparsity, low-rank approximation, exploitation of structure, the importance of representation, and large-scale problems. We revisit (unsymmetric) Toeplitz systems in §12.1 and show how fast stable methods can be developed through a clever data-sparse representation. The ideas extend to other types of structured matrices. Representation is also central to the $O(n)$ methods developed in §12.2 for matrices that have low-rank off-diagonal blocks.

The next three sections form a sequence. The Kronecker product section has general utility, but it is used very heavily in both §12.4 and §12.5 which together provide a brief introduction to the rapidly developing field of tensor computations.

Reading Path

Within this chapter, there are the following dependencies

$$
\begin{array}{ccccccc}
\S 3.1\text{-}\S 3.4,\ \S 4.7 & \rightarrow & \S 12.1 & & & & \S 5.1\text{-}\S 5.3 \\
\S 3.1\text{-}\S 3.4,\ \S 5.1\text{-}\S 5.3 & \rightarrow & \S 12.2 & & & & \downarrow \\
\S 1.4 & \rightarrow & \S 12.3 & \rightarrow & \S 12.4 & \rightarrow & \S 12.5
\end{array}
$$

The schematic also hints at the minimum "prerequisites" for each topic.

12.1 Linear Systems with Displacement Structure

If $A \in \mathbb{R}^{n \times n}$ has rank r, then it has a (non-unique) *product representation* of the form UV^T where $U, V \in \mathbb{R}^{n \times r}$. Note that if $r \ll n$, then the product representation is much more compact than the *explicit representation* that encodes each a_{ij}. In addition to the obvious storage economies, the product representation supports fast computation. If the product representation is fully utilized, then the n-by-n matrix-matrix product $AB = U(V^T B)$ is $O(n^2 r)$ instead of $O(n^3)$. Likewise, by applying the Sherman-Morrison-Woodbury formula, the solution to a linear system of the form $(I+UV^T)x = b$ is $O(nr + r^3)$ instead of $O(n^3)$. The message is simple in both cases: work with U and V and not their explicit product UV^T.

In this section we continue in this direction by discussing "low-rank" way to represent Cauchy, Toeplitz, and Hankel matrices together with some of their generalizations. The data-sparse representation supports fast stable linear equation solving. The key idea is to turn explicit rank-1 updates that are at the heart of Gaussian elimination into equivalent, inexpensive updates of their representation. Our presentation is based on Gohberg, Kailath, and Olshevsky (1995) and Gu (1998).

12.1.1 Displacement Rank

If $F, G \in \mathbb{R}^{n \times n}$ and the *Sylvester map*

$$X \to FX - XG \tag{12.1.1}$$

is nonsingular, then the $\{F, G\}$-*displacement rank* of $A \in \mathbb{R}^{n \times n}$ is defined by

$$\text{rank}_{\{F,G\}}(A) = \text{rank}(FA - AG). \tag{12.1.2}$$

Recall from §7.6.3 that the Sylvester map is nonsingular provided $\lambda(F) \cup \lambda(G) = \emptyset$. Note that if $\text{rank}_{\{F,G\}}(A) = r$, then we can write

$$FA - AG = RS^T, \qquad R, S \in \mathbb{R}^{n \times r}. \tag{12.1.3}$$

The matrices R and S are *generators* for A with respect to F and G, a term that makes sense since we can generate A (or part of A) by working with this equation. If $r \ll n$, then R and S define a data-sparse representation for A. Of course, for this representation to be of interest F and G must be sufficiently simple so that the reconstruction of A via (12.1.3) is cheap.

12.1.2 Cauchy-Like Matrices

If $\omega \in \mathbb{R}^n$ and $\lambda \in \mathbb{R}^n$ and $\omega_k \neq \lambda_j$ for all k and j, then the n-by-n matrix $A = (a_{kj})$ defined by

$$a_{kj} = \frac{1}{\omega_k - \lambda_j}$$

is a *Cauchy matrix*. Note that if

$$\Omega = \text{diag}(\omega_1, \ldots, \omega_n), \qquad \Lambda = \text{diag}(\lambda_1, \ldots, \lambda_n),$$

then

$$[\Omega A - A\Lambda]_{kj} = \frac{\omega_k}{\omega_k - \lambda_j} - \frac{\lambda_j}{\omega_k - \lambda_j} = 1.$$

If $e \in \mathbb{R}^n$ is the vector of all 1's, then

$$\Omega A - A\Lambda = ee^T$$

and thus $\mathsf{rank}_{\{\Omega, \Lambda\}}(A) = 1$.

More generally, if $R \in \mathbb{R}^{n \times r}$ and $S \in \mathbb{R}^{n \times r}$ have rank r, then any matrix A that satisfies

$$\Omega A - A\Lambda = RS^T \qquad (12.1.4)$$

is a *Cauchy-like matrix*. This just means that

$$a_{kj} = \frac{r_k^T s_j}{\omega_k - \lambda_j}$$

where

$$R^T = [\ r_1 \mid \cdots \mid r_n\ ], \qquad S^T = [\ s_1 \mid \cdots \mid s_n\ ]$$

are column partitionings. Note that R and S are generators with respect to Ω and Λ and that $O(r)$ flops are required to reconstruct a matrix entry a_{kj} from (12.1.4).

12.1.3 The Apparent Loss of Structure

Suppose

$$A = \begin{bmatrix} \alpha & g^T \\ f & B \end{bmatrix}, \qquad \alpha \in \mathbb{R},\ f, g \in \mathbb{R}^{n-1},\ B \in \mathbb{R}^{(n-1) \times (n-1)},$$

and assume $\alpha \neq 0$. The first step in Gaussian elimination produces

$$A_1 = B - \frac{1}{\alpha} f g^T$$

and the factorization

$$A = \begin{bmatrix} 1 & 0 \\ f/\alpha & I_{n-1} \end{bmatrix} \begin{bmatrix} \alpha & g^T \\ 0 & A_1 \end{bmatrix}.$$

Let us examine the structure of A_1 given that A is a Cauchy matrix. If $n = 4$ and $a_{kj} = 1/(\omega_k - \lambda_j)$, then

$$A_1 = \begin{bmatrix} \dfrac{1}{\omega_2 - \lambda_2} & \dfrac{1}{\omega_2 - \lambda_3} & \dfrac{1}{\omega_2 - \lambda_4} \\[2mm] \dfrac{1}{\omega_3 - \lambda_2} & \dfrac{1}{\omega_3 - \lambda_3} & \dfrac{1}{\omega_3 - \lambda_4} \\[2mm] \dfrac{1}{\omega_4 - \lambda_2} & \dfrac{1}{\omega_4 - \lambda_3} & \dfrac{1}{\omega_4 - \lambda_4} \end{bmatrix} - \begin{bmatrix} \dfrac{\omega_1 - \lambda_1}{\omega_2 - \lambda_1} \\[2mm] \dfrac{\omega_1 - \lambda_1}{\omega_3 - \lambda_1} \\[2mm] \dfrac{\omega_1 - \lambda_1}{\omega_4 - \lambda_1} \end{bmatrix} \begin{bmatrix} \dfrac{1}{\omega_1 - \lambda_2} \\[2mm] \dfrac{1}{\omega_1 - \lambda_3} \\[2mm] \dfrac{1}{\omega_1 - \lambda_4} \end{bmatrix}^T.$$

If we choose to work with the explicit representation of A, then for general n this update requires $O(n^2)$ work even though it is highly structured and involves $O(n)$ data. And worse, all subsequent steps in the factorization process essentially deal with general matrices rendering an LU computation that is $O(n^3)$.

12.1.4 Displacement Rank and Rank-1 Updates

The situation is much happier if we replace the explicit transition from A to A_1 with a transition that involves updating data sparse representations. The key to developing a fast LU factorization for a Cauchy-like matrix is to recognize that rank-1 updates preserve displacement rank. Here is the result that makes it all possible.

Theorem 12.1.1. *Suppose $A \in \mathbb{R}^{n \times n}$ satisfies*

$$\Omega A - A\Lambda = RS^T \tag{12.1.5}$$

where $R, S \in \mathbb{R}^{n \times r}$ and

$$\Omega = \mathrm{diag}(\omega_1, \ldots, \omega_n), \qquad \Lambda = \mathrm{diag}(\lambda_1, \ldots, \lambda_n)$$

have no common diagonal entries. If

$$A = \begin{bmatrix} \alpha & g^T \\ f & B \end{bmatrix}, \qquad R = \begin{bmatrix} r_1^T \\ R_1 \end{bmatrix}, \qquad S = \begin{bmatrix} s_1^T \\ S_1 \end{bmatrix}$$

are conformably partitioned, $\alpha \neq 0$, and

$$\Omega_1 = \mathrm{diag}(\omega_2, \ldots, \omega_n), \qquad \Lambda_1 = \mathrm{diag}(\lambda_2, \ldots, \lambda_n),$$

then

$$\Omega_1 A_1 - A_1 \Lambda_1 = \tilde{R}_1 \tilde{S}_1^T \tag{12.1.6}$$

where

$$A_1 = B - \frac{fg^T}{\alpha}, \qquad \tilde{R}_1 = R_1 - \frac{1}{\alpha} f r_1^T, \qquad \tilde{S}_1 = S_1 - \frac{1}{\alpha} g s_1^T.$$

Proof. By comparing blocks in (12.1.5) we see that

$$(1,1) : (\omega_1 - \lambda_1)\alpha = r_1^T s_1, \qquad (1,2) : g^T \Lambda_1 = \omega_1 g^T - r_1^T S_1^T,$$

$$(2,1) : \Omega_1 f = R_1 s_1 + \lambda_1 f, \qquad (2,2) : \Omega_1 B - B\Lambda_1 = R_1 S_1^T,$$

and so

$$\begin{aligned}
\Omega_1 A_1 - A_1 \Lambda_1 &= \Omega_1 \left(B - \frac{1}{\alpha} fg^T \right) - \left(B - \frac{1}{\alpha} fg^T \right) \Lambda_1 \\
&= (\Omega_1 B - B\Lambda_1) - \frac{1}{\alpha} \left((\Omega_1 f)g^T - f(g^T \Lambda_1) \right) \\
&= R_1 S_1^T - \frac{1}{\alpha} \left((R_1 s_1 + \lambda_1 f)g^T - f(\omega_1 g^T - r_1^T S_1^T) \right) \\
&= R_1 S_1^T - \frac{1}{\alpha} \left((R_1 s_1)g^T + f(r_1^T S_1^T) - \frac{r_1^T s_1}{\alpha} fg^T \right) \\
&= \left(R_1 - \frac{1}{\alpha} f r_1^T \right) \left(S_1 - \frac{1}{\alpha} g s_1^T \right)^T = \tilde{R}_1 \tilde{S}_1^T.
\end{aligned}$$

This confirms (12.1.6) and completes the proof of the theorem. □

The theorem says that

$$\text{rank}_{\{\Omega,\Lambda\}}(A) \leq r \quad \Rightarrow \quad \text{rank}_{\{\Omega_1,\Lambda_1\}}(A_1) \leq r.$$

This suggests that instead of updating A explicitly to get A_1 at a cost of $O(n^2)$ flops, we should update A's representation $\{\Omega, \Lambda, R, S\}$ at a cost of $O(nr)$ flops to get A_1's representation $\{\Omega_1, \Lambda_1, \tilde{R}_1, \tilde{S}_1\}$.

12.1.5 Fast LU for Cauchy-Like Matrices

Based on Theorem 12.1.1 we can specify a fast LU procedure for Cauchy-like matrices. If A satisfies (12.1.5) and has an LU factorization, then it can be computed using the function **LUdisp** defined as follows:

Algorithm 12.1.1 If $\omega \in \mathbb{R}^n$ and $\lambda \in \mathbb{R}^n$ have no common components, $R, S \in \mathbb{R}^{n \times r}$, and $\Omega A - A\Lambda = RS^T$ where $\Omega = \text{diag}(\omega_1, \ldots, \omega_n)$ and $\Lambda = \text{diag}(\lambda_1, \ldots, \lambda_n)$, then the following function computes the LU factorization $A = LU$.

> **function** $[L, U] = \text{LUdisp}(\omega, \lambda, R, S, n)$
> $r_1^T = R(1, :), \ R_1 = R(2{:}n, :)$
> $s_1^T = S(1, :), \ S_1 = S(2{:}n, :)$
> **if** $n = 1$
> $L = 1$
> $U = r_1^T s_1 / (\omega_1 - \lambda_1)$
> **else**
> $a = (Rs_1) \ ./ \ (\omega - \lambda_1)$
> $\alpha = a_{11}$
> $f = a(2{:}n)$
> $g = (S_1 r_1) \ ./ \ (\omega_1 - \lambda(2{:}n))$
> $\tilde{R}_1 = R_1 - f r_1^T / \alpha$
> $\tilde{S}_1 = S_1 - g s_1^T / \alpha$
> $[L_1, U_1] = \text{LUdisp}(\omega(2{:}n), \lambda(2{:}n), \tilde{R}_1, \tilde{S}_1, n-1)$
> $L = \begin{bmatrix} 1 & 0 \\ f/\alpha & L_1 \end{bmatrix}$
> $U = \begin{bmatrix} \alpha & g^T \\ 0 & U_1 \end{bmatrix}$
> **end**

The nonrecursive version would have the following structure:

> Let $R^{(1)}$ and $S^{(1)}$ be the generators of $A = A^{(1)}$ with respect to diag(ω)
> and diag(λ).

> **for** $k = 1{:}n-1$
>> Use $\omega(k{:}n)$, $\lambda(k{:}n)$, $R^{(k)}$ and $S^{(k)}$ to compute the first row and column of
>> $$A^{(k)} \;=\; \begin{bmatrix} \alpha & g^T \\ f & B \end{bmatrix}.$$
>> $L(k+1{:}n, k) = f/\alpha$, $U(k,k) = \alpha$, $U(k, k+1{:}n) = g^T$
>> Determine the generators $R^{(k+1)}$ and $S^{(k+1)}$ of $A^{(k+1)} = B - fg^T/\alpha$
>> with respect to diag$(\omega(k{:}n))$ and diag$(\lambda(k{:}n))$.

> **end**
> $$U(n,n) = R^{(n)}{\cdot}S^{(n)}/(\omega_n - \lambda_n)$$

A careful accounting reveals that $2n^2 r$ flops are required.

12.1.6 Pivoting

The procedure just developed has numerical difficulties if a small α shows up during the recursion. To guard against this we show how to incorporate a pivoting strategy. Suppose $A \in \mathbb{R}^{n \times n}$ is a Cauchy-like matrix that satisfies the displacement equation

$$\Omega A - A\Lambda = RS^T$$

for diagonal matrices Ω and Λ and n-by-r matrices R and S. If P and Q are n-by-n permutations, then

$$(P\Omega P^T(PAQ^T) - (PAQ^T)(Q\Lambda Q^T) = (PR)(QS)^T.$$

This shows that

$$\tilde{A} = PAQ^T$$

is a Cauchy-like matrix having generators

$$\tilde{R} = PR, \qquad\qquad \tilde{S} = QS$$

with respect to the diagonal matrices

$$\tilde{\Omega} = P\Omega P^T, \qquad\qquad \tilde{\Lambda} = Q\Lambda Q^T.$$

Thus, it is easy to track row and column permutations in the the displacement representation:

$$A \;\to\; PAQ^T, \quad \equiv \quad \{\Omega, \Lambda, R, S\} \;\to\; \{P\Omega P^T, Q\Lambda Q^T, PR, QS\}.$$

By taking advantage of this, it is a simple matter to incorporate partial pivoting in LUdisp and to emerge with the factorization $PA = LU$:

Algorithm 12.1.2 If $\omega \in \mathbb{R}^n$ and $\lambda \in \mathbb{R}^n$ have no common components, $R, S \in \mathbb{R}^{n \times r}$, and $\Omega A - A\Lambda = RS^T$, then the following function computes the LU-with-pivoting factorization $PA = LU$, where $\Omega = \text{diag}(\omega_1, \ldots, \omega_n)$ and $\Lambda = \text{diag}(\lambda_1, \ldots, \lambda_n)$.

function $[L, U, P] = \mathsf{LUdispPiv}(\omega, \lambda, R, S, n)$

Define r_1, R_1, s_1 and S_1 by $R = \begin{bmatrix} r_1^T \\ R_1 \end{bmatrix}$ and $S = \begin{bmatrix} s_1^T \\ S_1 \end{bmatrix}$.

if $n = 1$

    $L = 1$

    $U = r_1^T s_1 / (\omega_1 - \lambda_1)$

else

    $a = (Rs_1) \ ./ \ (\omega - \lambda_1)$

    Determine permutation $P \in \mathbb{R}^{n \times n}$ so that $[Pa]_1$ is maximal and

        update: $a = Pa$, $R = PR$, $\omega = P\omega$.

    $\alpha = a_1$

    $f = a(2{:}n)$

    $g = (S_1 r_1) \ ./ \ (\omega_1 - \lambda(2{:}n))$

    $\tilde{R}_1 = R_1 - f r_1^T / \alpha$

    $\tilde{S}_1 = S_1 - g s_1^T / \alpha$

    $[L_1, U_1, P_1] = \mathsf{LUdispPiv}(\omega(2{:}n), \lambda(2{:}n), \tilde{R}_1, \tilde{S}_1, n - 1)$

    $L = \begin{bmatrix} 1 & 0 \\ P_1 f / \alpha & L_1 \end{bmatrix}$

    $U = \begin{bmatrix} \alpha & g^T \\ 0 & U_1 \end{bmatrix}$

    $P = \begin{bmatrix} 1 & 0 \\ 0 & P_1 \end{bmatrix} P$

end

The processing of the recursive call is based on the fact that if

$$PA = \begin{bmatrix} \alpha & g^T \\ f & B \end{bmatrix} = \begin{bmatrix} 1 & 0 \\ f/\alpha & I_{n-1} \end{bmatrix} \begin{bmatrix} \alpha & g^T \\ 0 & A_1 \end{bmatrix}, \qquad A_1 = B - \frac{1}{\alpha} f g^T,$$

and $P_1 A_1 = L_1 U_1$, then

$$\begin{bmatrix} 1 & 0 \\ 0 & P_1 \end{bmatrix} PA = \begin{bmatrix} 1 & 0 \\ P_1 f/\alpha & L_1 \end{bmatrix} \begin{bmatrix} \alpha & g^T \\ 0 & U_1 \end{bmatrix}.$$

For LUdispPiv implementation details and a proof of its stability, see Gu (1998).

12.1.7   Toeplitz-Like Matrices and Hankel-Like Matrices

Recall from §4.7 that a Toeplitz matrix is constant along each of its diagonals. For example, if $c \in \mathbb{R}^{n-1}$, $\tau \in \mathbb{R}$, and $r \in \mathbb{R}^{n-1}$ are given, then the matrix $T \in \mathbb{R}^{n \times n}$ defined by

$$t_{ij} = \begin{cases} c_{i-j} & \text{if } i > j, \\ \tau & \text{if } i = j, \\ r_{j-i} & \text{if } j > i, \end{cases}$$

is Toeplitz, e.g.,

$$T = \begin{bmatrix} \tau & r_1 & r_2 & r_3 & r_4 \\ c_1 & \tau & r_1 & r_2 & r_3 \\ c_2 & c_1 & \tau & r_1 & r_2 \\ c_3 & c_2 & c_1 & \tau & r_1 \\ c_4 & c_3 & c_2 & c_1 & \tau \end{bmatrix}.$$

To expose the low-displacement-rank structure of a Toeplitz matrix, we define matrices Z_ϕ and $Y_{\gamma,\delta}$ analogously to their $n = 5$ instances:

$$Z_\phi = \begin{bmatrix} 0 & 0 & 0 & 0 & \phi \\ 1 & 0 & 0 & 0 & 0 \\ 0 & 1 & 0 & 0 & 0 \\ 0 & 0 & 1 & 0 & 0 \\ 0 & 0 & 0 & 1 & 0 \end{bmatrix}, \quad Y_{\gamma,\delta} = \begin{bmatrix} \gamma & 1 & 0 & 0 & 0 \\ 1 & 0 & 1 & 0 & 0 \\ 0 & 1 & 0 & 1 & 0 \\ 0 & 0 & 1 & 0 & 1 \\ 0 & 0 & 0 & 1 & \delta \end{bmatrix}. \qquad (12.1.7)$$

It can be shown that

$$Z_1 T - T Z_{-1} = \begin{bmatrix} \times & \times & \times & \times & \times \\ 0 & 0 & 0 & 0 & \times \\ 0 & 0 & 0 & 0 & \times \\ 0 & 0 & 0 & 0 & \times \\ 0 & 0 & 0 & 0 & \times \end{bmatrix}, \qquad \operatorname{rank}_{\{Z_1, Z_{-1}\}}(T) \leq 2, \qquad (12.1.8)$$

$$Y_{00} T - T Y_{11} = \begin{bmatrix} \times & \times & \times & \times & \times \\ \times & 0 & 0 & 0 & \times \\ \times & 0 & 0 & 0 & \times \\ \times & 0 & 0 & 0 & \times \\ \times & \times & \times & \times & \times \end{bmatrix}, \qquad \operatorname{rank}_{\{Y_{00}, Y_{11}\}}(T) \leq 4. \qquad (12.1.9)$$

Furthermore, $\lambda(Z_{-1}) \cup \lambda(Z_1) = \emptyset$ and $\lambda(Y_{00}) \cup \lambda(Y_{11}) = \emptyset$.

A *Hankel* matrix is constant along its antidiagonals, e.g.,

$$H = \begin{bmatrix} c_4 & c_3 & c_2 & c_1 & \tau \\ c_3 & c_2 & c_1 & \tau & r_1 \\ c_2 & c_1 & \tau & r_1 & r_2 \\ c_1 & \tau & r_1 & r_2 & r_3 \\ \tau & r_1 & r_2 & r_3 & r_4 \end{bmatrix}.$$

Note that if $H \in \mathbb{R}^{n \times n}$ is Hankel, then $\mathcal{E}_n H$ is Toeplitz, and so it is not surprising that

Hankel and Toeplitz matrices have similar displacement rank properties:

$$Z_1^T H - H Z_{-1} = \begin{bmatrix} 0 & 0 & 0 & 0 & \times \\ 0 & 0 & 0 & 0 & \times \\ 0 & 0 & 0 & 0 & \times \\ 0 & 0 & 0 & 0 & \times \\ \times & \times & \times & \times & \times \end{bmatrix}, \quad \mathrm{rank}_{\{Z_1^T, Z_{-1}\}}(H) \le 2, \quad (12.1.10)$$

$$Y_{00} H - H Y_{11} = \begin{bmatrix} \times & \times & \times & \times & \times \\ \times & 0 & 0 & 0 & \times \\ \times & 0 & 0 & 0 & \times \\ \times & 0 & 0 & 0 & \times \\ \times & \times & \times & \times & \times \end{bmatrix}, \quad \mathrm{rank}_{\{Y_{00}, Y_{11}\}}(H) \le 4. \quad (12.1.11)$$

It follows from (12.1.9) and (12.1.11) that if $A = T + H$ is the sum of a Toeplitz matrix and a Hankel matrix, then $\mathrm{rank}_{\{Y_{00}, Y_{11}\}}(A) \le 4$.

The classes of Toeplitz, Hankel, and Toeplitz-plus-Hankel matrices can be expanded through the notion of low displacement rank. Analogous to how we defined Cauchy-like matrices in (12.1.4) we have the following, assuming that $R \in \mathbb{R}^{n \times r}$, $S \in \mathbb{R}^{n \times r}$, and $r \ll n$:

$$\left\{ \begin{array}{l} Z_1 A - A Z_{-1} = R S^T \\ Z_1^T A - A Z_{-1} = R S^T \\ Y_{00} A - A Y_{11} = R S^T \end{array} \right\} \text{ means that } A \text{ is } \left\{ \begin{array}{l} \text{Toeplitz-like} \\ \text{Hankel-like} \\ \text{Toeplitz-plus-Hankel-like} \end{array} \right\}.$$

Our next task is to show that a linear system with any of these properties can be efficiently converted to a Cauchy-like system and solved with $O(n^2 r)$ work.

12.1.8 Fast Solvers via Conversion to Cauchy-Like Form

Suppose

$$FA - AG = R S^T, \qquad A, F, G \in \mathbb{R}^{n \times n}, \ R, S \in \mathbb{R}^{n \times r}, \ r \ll n,$$

and that F and G are diagonalizable:

$$X_F^{-1} F X_F = \mathrm{diag}(\omega_1, \ldots, \omega_n) = \Omega,$$

$$X_G^{-1} G X_G = \mathrm{diag}(\lambda_1, \ldots, \lambda_n) = \Lambda.$$

For clarity we assume that F and G have real eigenvalues. It follows from

$$(X_F^{-1} F X_F)(X_F^{-1} A X_G) - (X_F^{-1} A X_G)(X_G^{-1} G X_F) = (X_F^{-1} R)(X_G^T S)^T$$

that

$$\Omega \tilde{A} - \tilde{A} \Lambda = \tilde{R} \tilde{S}^T$$

where $\tilde{A} = X_F^{-1} A X_G$, $\tilde{R} = X_F^{-1} R$, and $\tilde{S} = X_G^T S$ Thus, $\tilde{A}$ is Cauchy-like and we can go about solving the given linear system $Ax = b$ as follows:

Step 1.   Compute $\tilde{R} = X_F^{-1} R$, $\tilde{S} = X_G^T S$, $\tilde{b} = X_F^{-1} b$, and $\tilde{A} = X_F^{-1} A X_G$..

Step 2.   Use Algorithm 12.1.2 to compute $P\tilde{A} = LU$.

Step 3.   Use $P\tilde{A} = LU$ to solve $\tilde{A}\tilde{x} = \tilde{b}$.

Step 4.   Compute $x = X_G \tilde{x}$.

This will not be an attractive framework unless the matrices F and G have fast eigensystems, a concept introduced in §4.8. Fortunately, this is the case for the matrices Z_1, Z_{-1}, Y_{00} and Y_{11}. For example,

$$\mathcal{S}_n^T Y_{00}\, \mathcal{S}_n = 2 \cdot \text{diag}\left(\cos\left(\frac{\pi}{n+1}\right), \ldots, \cos\left(\frac{n\pi}{n+1}\right)\right), \qquad (12.1.12)$$

$$\mathcal{C}_n^T Y_{11}\, \mathcal{C}_n = 2 \cdot \text{diag}\left(1, \cos\left(\frac{\pi}{n}\right), \ldots, \cos\left(\frac{(n-1)\pi}{n}\right)\right), \qquad (12.1.13)$$

where $\mathcal{S}_n$ is the sine transform (DST-I) matrix

$$[\,\mathcal{S}_n\,]_{kj} = \sqrt{\frac{2}{n+1}} \cdot \sin\left(\frac{kj\pi}{n+1}\right),$$

and $\mathcal{C}_n$ is the cosine transform (DCT-II) matrix

$$[\,\mathcal{C}_n\,]_{kj} = \sqrt{\frac{2}{n}} \cdot \cos\left(\frac{(2k-1)(j-1)\pi}{2n}\right) q_j, \qquad q_j = \begin{cases} 1/\sqrt{2} & \text{if } j = 1, \\ 1 & \text{if } j > 1. \end{cases}$$

This allows products like $\mathcal{S}_n R$ and $\mathcal{C}_n^T S$ to be computed with $O(rn\log n)$ flops. In short, Step 3 in the above framework is the most expensive step in the process and it involves $O(n^2 r)$ work. See Gohberg, Kailath, and Olshevsky (1995) and Gu (1998) for details and related references.

Problems

P12.1.1   Refer to (12.1.8) and (12.1.9). (a) Show that if $Z_1 X - X Z_{-1} = 0$, then $X = 0$. (b) Show that if $Y_{00} X - X Y_{11} = 0$, then $X = 0$.

P12.1.2   Develop a nonrecursive version of Algorithm 12.1.2.

P12.1.3   (a) If $T \in \mathbb{R}^{n \times n}$ is Toeplitz, show how to compute $R, S \in \mathbb{R}^{n \times 2}$ so that $Z_1 T - T Z_{-1} = RS^T$. (b) Suppose $R, S \in \mathbb{R}^{n \times r}$ and $T \in \mathbb{R}^{n \times n}$ satisfy $Z_1 T - T Z_{-1} = RS^T$. Give an algorithm that computes $u = T(:, 1)$ and $v = T(1, :)^T$.

P12.1.4   (a) If $T \in \mathbb{R}^{n \times n}$ is Toeplitz, show how to compute $R, S \in \mathbb{R}^{n \times 4}$ so that $Y_{00} T - T Y_{11} = RS^T$. (b) Suppose $R, S \in \mathbb{R}^{n \times r}$ and $T \in \mathbb{R}^{n \times n}$ satisfy $Y_{00} T - T Y_{11} = RS^T$. Give an algorithm that computes $u = T(:, 1)$ and $v = T(1, :)^T$.

P12.1.5   Verify(12.1.13).

P12.1.6   Show that if $A \in \mathbb{R}^{n \times n}$ is defined by

$$a_{ij} = \int_a^b \cos(k\theta)\cos(j\theta)d\theta$$

then A is the sum of a Hankel matrix and Toeplitz matrix. Hint: Make use of the identity $\cos(u+v) = \cos(u)\cos(v) - \sin(u)\sin(v)$.

Notes and References for §12.1

For a general introduction to the area of fast algorithms for structured matrices we recommend:

T. Kailath and A. H. Sayed (eds) (1999). *Fast Reliable Algorithms for Matrices with Structure*, SIAM Publications, Philadelphia, PA.

V. Olshevsky (ed.) (2000). *Structured Matrices in Mathematics, Computer Science, and Engineering I and II*, AMS Contemporary Mathematics Vol. 280/281, AMS, Providence, RI.

D.A. Bini, V. Mehrmann, V. Olshevsky, E.E. Tyrtyshnikov, and M. Van Barel (eds.) (2010). *Structured Matrices and Applications–The Georg Heinig Memorial Volume*, Birkhauser-Springer, Basel, Switzerland.

Papers concerned with the development of fast stable solvers for structured matrices include:

T. Kailath, S. Kung, and M. Morf (1979). "Displacement Ranks of Matrices and Linear Equations," *J. Math. Anal. Applic. 68*, 395–407.

J. Chun and T. Kailath (1991). "Displacement Structure for Hankel, Vandermonde, and Related Matrices," *Lin. Alg. Applic. 151*, 199–227.

T. Kailath and A.H. Sayed (1995). "Displacement Structure: Theory and Applications," *SIAM Review 37*, 297–386.

I. Gohberg, T. Kailath, and V. Olshevsky (1995). "Fast Gaussian Elimination with Partial Pivoting for Matrices with Displacement Structure," *Math. Comput. 212*, 1557–1576.

T. Kailath and V. Olshevsky (1997). "Displacement-Structure Approach to Polynomial Vandermonde and Related Matrices," *Lin. Alg. Applic. 261*, 49–90.

G. Heinig (1997). "Matrices with Higher-Order Displacement Structure," *Lin. Alg. Applic. 278*, 295–301.

M. Gu (1998). "Stable and Efficient Algorithms for Structured Systems of Linear Systems," *SIAM J. Matrix Anal. Applic. 19*, 279–306.

S. Chandrasekaran, M. Gu, X. Sun, J. Xia, and J. Zhu (2007). "A Superfast Algorithm for Toeplitz Systems of Linear Equations," *SIAM J. Matrix Anal. Applic. 29*, 1247–1266.

Displacement rank ideas can be extended to least squares problems:

R.H. Chan, J.G. Nagy, and R.J. Plemmons (1994). "Displacement Preconditioner for Toeplitz Least Squares Iterations," *ETNA 2*, 44–56.

M. Gu (1998). "New Fast Algorithms for Structured Linear Least Squares Problems," *SIAM J. Matrix Anal. Applic. 20*, 244–269.

G. Rodriguez (2006). "Fast Solution of Toeplitz- and Cauchy-Like Least-Squares Problems," *SIAM J. Matrix Anal. Applic. 28*, 724–748.

For insight into the application low-displacement-rank preconditioners, see:

I. Gohberg and V. Olshevsky (1994). "Complexity of Multiplication with Vectors for Structured Matrices," *Linear Alg. Applic. 202*, 163–192.

M.E. Kilmer and D.P. O'Leary (1999). "Pivoted Cauchy-like Preconditioners for Regularized Solution of Ill-Posed Problems," *SIAM J. Sci. Comput. 21*, 88–110.

T. Kailath and V. Olshevsky (2005). "Displacement Structure Approach to Discrete-Trigonometric-Transform Based Preconditioners of G. Strang Type and of T. Chan Type," *SIAM J. Matrix Anal. Applic. 26*, 706–734.

12.2 Structured-Rank Problems

Just as a sparse matrix has lots of zero entries, a *structured rank* matrix has lots of low-rank submatrices. For example, it could be that all off-diagonal blocks have unit rank. In this section we identify some important structured rank matrix problems and point to how they can be solved very quickly with data-sparse representations. To avoid complicated notation, we adopt a small-n, proof-by-example style of exposition. Readers who prefer for more detail and rigor should consult the definitive, two-volume treatise by Vandebril, Van Barel, and Mastronardi (2008).

12.2.1   Semiseparable Matrices

A matrix $A \in \mathbb{R}^{n \times n}$ is *semiseparable* if every block that does not "cross" the diagonal has unit rank or less. This means

$$j_2 \leq i_1 \text{ or } i_2 \leq j_1 \quad \Rightarrow \quad \mathsf{rank}(A(i_1{:}i_2, j_1{:}j_2)) \leq 1. \tag{12.2.1}$$

The rank-1 blocks of interest in a semiseparable matrix are wholly contained in either its upper triangular part or its lower triangular part, e.g.,

$$\begin{bmatrix} \times & \times & a_{13} & a_{14} & \times & \times \\ \times & \times & a_{23} & a_{24} & \times & \times \\ \times & \times & a_{33} & a_{34} & \times & \times \\ \times & \times & \times & \times & \times & \times \\ a_{51} & a_{52} & \times & \times & \times & \times \\ a_{61} & a_{62} & \times & \times & \times & \times \end{bmatrix}, \qquad \begin{array}{l} \mathsf{rank}(A(1{:}3, 3{:}4)) \leq 1, \\[4pt] \mathsf{rank}(A(5{:}6, 1{:}2)) \leq 1. \end{array}$$

Semiseparable matrices are data-sparse and enormous savings can be realized when their structure is exploited. For example, we will show that the factorizations $A = LU$ and $A = QR$ for semiseparable A require just $O(n)$ flops to compute and $O(n)$ flops to represent.

An important example of a semiseparable matrix is the inverse of a unit bidiagonal matrix. Given $r \in \mathbb{R}^{n-1}$ we define $B(r) \in \mathbb{R}^{n \times n}$ by

$$B(r) = \begin{bmatrix} 1 & -r_1 & 0 & 0 & 0 \\ 0 & 1 & -r_2 & 0 & 0 \\ 0 & 0 & 1 & -r_3 & 0 \\ 0 & 0 & 0 & 1 & -r_4 \\ 0 & 0 & 0 & 0 & 1 \end{bmatrix}. \tag{12.2.2}$$

Observe that any submatrix extracted from the upper triangular portion of

$$B(r)^{-1} = \begin{bmatrix} 1 & r_1 & r_1 r_2 & r_1 r_2 r_3 & r_1 r_2 r_3 r_4 \\ 0 & 1 & r_2 & r_2 r_3 & r_2 r_3 r_4 \\ 0 & 0 & 1 & r_3 & r_3 r_4 \\ 0 & 0 & 0 & 1 & r_4 \\ 0 & 0 & 0 & 0 & 1 \end{bmatrix} \tag{12.2.3}$$

has unit rank. If $x \in \mathbb{R}^n$ and $r = x(2{:}n)\,./\,x(1{:}n-1)$ is defined, then

$$B(r)^T x = x_1 e_1.$$

Thus, the matrix $B(r)$ can (in principle) be used to introduce zeros into a vector.

12.2.2 Quasiseparable Matrices

Certain products of Givens rotations exhibit rank structure, but we frame the key fact in more general terms. If $\alpha, \beta, \gamma, \delta \in \mathbb{R}^{n-1}$ and

$$M_k = \text{diag}(I_{k-1}, \tilde{M}_k, I_{n-k-1}), \qquad \tilde{M}_k = \begin{bmatrix} \alpha_k & \beta_k \\ \gamma_k & \delta_k \end{bmatrix},$$

for $k = 1{:}n-1$, then the matrix $M = M_1 \cdots M_{n-1}$ is fully illustrated by

$$M = M_1 M_2 M_3 M_4 = \begin{bmatrix} \alpha_1 & \beta_1\alpha_2 & \beta_1\beta_2\alpha_3 & \beta_1\beta_2\beta_3\alpha_4 & \beta_1\beta_2\beta_3\beta_4 \\ \gamma_1 & \delta_1\alpha_2 & \delta_1\beta_2\alpha_3 & \delta_1\beta_2\beta_3\alpha_4 & \delta_1\beta_2\beta_3\beta_4 \\ 0 & \gamma_2 & \delta_2\alpha_3 & \delta_2\beta_3\alpha_4 & \delta_2\beta_3\beta_4 \\ 0 & 0 & \gamma_3 & \delta_3\alpha_4 & \delta_3\beta_4 \\ 0 & 0 & 0 & \gamma_4 & \delta_4 \end{bmatrix}. \qquad (12.2.4)$$

It has the property that off-diagonal blocks have unit rank or less provided they do not "intersect" the diagonal. *Quasiseparable* matrices have this property and if A is such a matrix, then

$$j_2 < i_1 \ \text{ or } \ i_2 < j_1 \ \Rightarrow \ \text{rank}(A(i_1{:}i_2, j_1{:}j_2)) \le 1. \qquad (12.2.5)$$

By comparing this with (12.2.1), it is clear that the class of semiseparable matrices is a subset of the class of quasiseparable matrices.

12.2.3 Two Representations

The MATLAB tril and triu notation is very handy when formulating a quasiseparable matrix computation. If $A \in \mathbb{R}^{m \times n}$, then a_{ij} is on its *k*th *diagonal* if $j = i + k$. The matrix $B = \text{tril}(A, k)$ is obtained from A by setting to zero all its entries above the *k*th diagonal while $B = \text{triu}(A, k)$ is obtained from A by setting to zero all its entries below the *k*th diagonal. If $k = 0$, then we simply write $\text{tril}(A)$ and $\text{triu}(A)$. We also use the notation $\text{diag}(d)$ to designate the diagonal matrix $\text{diag}(d_1, \ldots, d_n)$ where $d \in \mathbb{R}^n$. Note that if $u, v, d, p, q \in \mathbb{R}^n$, then the matrix

$$A = \text{tril}(uv^T, -1) + \text{diag}(d) + \text{triu}(pq^T, 1) \qquad (12.2.6)$$

is quasiseparable, e.g.,

$$A = \begin{bmatrix} d_1 & p_1q_2 & p_1q_3 & p_1q_4 & p_1q_5 \\ u_2v_1 & d_2 & p_2q_3 & p_2q_4 & p_2q_5 \\ u_3v_1 & u_3v_2 & d_3 & p_3q_4 & p_3q_5 \\ u_4v_1 & u_4v_2 & u_4v_3 & d_4 & p_4q_5 \\ u_5v_1 & u_5v_2 & u_5v_3 & u_5v_4 & d_5 \end{bmatrix}.$$

Should it be the case that $d = u.*v = p.*q$, then this matrix is semiseparable. The representation (12.2.6) is referred to as the *generator representation*.

Not every quasiseparable matrix has a generator representation. For example, if $A = B(r)$ and r has nonzero entries, then it is impossible to find $u, v, d, p, q \in \mathbb{R}^n$ so that (12.2.6) holds. To address this shortcoming, we use the fact that

$$\begin{pmatrix} \text{Quasiseparable} \\ \text{Matrix} \end{pmatrix} .* \begin{pmatrix} \text{Quasiseparable} \\ \text{Matrix} \end{pmatrix} = \begin{pmatrix} \text{Quasiseparable} \\ \text{Matrix} \end{pmatrix}, \quad (12.2.7)$$

and embellish (12.2.6) with a pair of inverse bidiagonal factors. It can be shown that if $A \in \mathbb{R}^{n \times n}$ is quasiseparable, then there exist $u, v, d, p, q \in \mathbb{R}^n$ and $t, r \in \mathbb{R}^{n-1}$ such that

$$A = \mathsf{tril}(uv^T, -1) .* B(t)^{-T} + \mathsf{diag}(d) + \mathsf{triu}(pq^T, 1) .* B(r)^{-1} \quad (12.2.8)$$

$$\equiv \mathbf{S}(u, v, t, d, p, q, r),$$

e.g.,

$$A = \begin{bmatrix} d_1 & p_1 r_1 q_2 & p_1 r_1 r_2 q_3 & p_1 r_1 r_2 r_3 q_4 & p_1 r_1 r_2 r_3 r_4 q_5 \\ u_2 t_1 v_1 & d_2 & p_2 r_2 q_3 & p_2 r_2 r_3 q_4 & p_2 r_2 r_3 r_4 q_5 \\ u_3 t_2 t_1 v_1 & u_3 t_2 v_2 & d_3 & p_3 r_3 q_4 & p_3 r_3 r_4 q_5 \\ u_4 t_3 t_2 t_1 v_1 & u_4 t_3 t_2 v_2 & u_4 t_3 v_3 & d_4 & p_4 r_4 q_5 \\ u_5 t_4 t_3 t_2 t_1 v_1 & u_5 t_4 t_3 t_2 v_2 & u_5 t_4 t_3 v_3 & u_5 t_4 v_4 & d_5 \end{bmatrix}.$$

We refer to (12.2.8) as a *quasiseparable representation* and it has a number of important specializations. If $d = u .* v = p .* q$, then A is semiseparable. If $t = r = \mathbf{1}_{n-1}$, then A is generator representable. If $u = q$, $v = p$, and $t = r$, then A is symmetric. The representation also supports the *semiseparable-plus-diagonal* structure. A matrix $\mathbf{S}(u, v, t, d, p, q, r)$ has this form if d is arbitrary and $u .* v = p .* q$. Here are some inverse-related facts that pertain to semiseparable, quasiseparable, and diagonal-plus-semiseparable matrices:

Fact 1. If A is nonsingular and tridiagonal, then A^{-1} is semiseparable. In addition, if the subdiagonal and superdiagonal entries are nonzero, then A^{-1} is generator-representable.

Fact 2. If A is nonsingular and quasiseparable, then so is A^{-1}.

Fact 3. If $A = D + S$ is nonsingular where D is diagonal and nonsingular and S is semiseparable, then $A^{-1} = D^{-1} + S_1$ where S_1 is semiseparable.

Aspects of the first fact were encountered in §4.3.8.

12.2.4   Computations with Triangular Semiseparable Matrices

Lower and upper triangular matrices that are also semiseparable can be written as follows:

$$L \text{ lower semiseparable} \implies L = \mathbf{S}(u, v, t, u .* v, 0, 0, 0) = \mathsf{tril}(uv^T) .* B(t)^{-T},$$

$$U \text{ upper semiseparable} \implies U = \mathbf{S}(0, 0, 0, p .* q, p, q, r) = \mathsf{triu}(pq^T) .* B(r)^{-1}.$$

Operations with matrices that have this structure can be organized very efficiently. Consider the matrix-vector product

$$y = \left(\text{triu}(pq^T) \cdot * B(r)^{-1} \right) x \qquad (12.2.9)$$

where $x, y, p, q \in \mathbb{R}^n$ and $r \in \mathbb{R}^{n-1}$. This calculation has the form

$$\begin{bmatrix} p_1 q_1 & p_1 r_1 q_2 & p_1 r_1 r_2 q_3 & p_1 r_1 r_2 r_3 q_4 \\ 0 & p_2 q_2 & p_2 r_2 q_3 & p_2 r_2 r_3 q_4 \\ 0 & 0 & p_3 q_3 & p_3 r_3 q_4 \\ 0 & 0 & 0 & p_4 q_4 \end{bmatrix} \begin{bmatrix} x_1 \\ x_2 \\ x_3 \\ x_4 \end{bmatrix} = \begin{bmatrix} y_1 \\ y_2 \\ y_3 \\ y_4 \end{bmatrix}.$$

By grouping the q's with the x's and extracting the p's, we see that

$$\text{diag}(p_1, p_2, p_3, p_4) \begin{bmatrix} 1 & r_1 & r_1 r_2 & r_1 r_2 r_3 \\ 0 & 1 & r_2 & r_2 r_3 \\ 0 & 0 & 1 & r_3 \\ 0 & 0 & 0 & 1 \end{bmatrix} \begin{bmatrix} q_1 x_1 \\ q_2 x_2 \\ q_3 x_3 \\ q_4 x_4 \end{bmatrix} = \begin{bmatrix} y_1 \\ y_2 \\ y_3 \\ y_4 \end{bmatrix}.$$

In other words, (12.2.9) is equivalent to

$$y = p \cdot * \left(B(r)^{-1} (q \cdot * x) \right).$$

Given x, this is clearly an $O(n)$ computation since bidiagonal system solving is $O(n)$. Indeed, y can be computed with just $4n$ flops.

Note that if y is given in (12.2.9) and p and q have nonzero components, then we can solve for x equally fast: $x = (B(r)(y./p))./q$.

12.2.5 The LU Factorization of a Semiseparable Matrix

Suppose $A = \mathbf{S}(u, v, t, u \cdot *v, p, q, r)$ is an n-by-n semiseparable matrix that has an LU factorization. It turns out that both L and U are semiseparable and their respective representations can be computed with $O(n)$ work:

for $k = n-1: -1:1$

    Using A's representation, determine τ_k so that if $\tilde{A} = M_k A$, where

$$M_k = \text{diag}(I_{k-1}, \tilde{M}_k, I_{n-k-1}), \qquad \tilde{M}_k = \begin{bmatrix} 1 & 0 \\ -\tau_k & 1 \end{bmatrix},$$

    then $\tilde{A}(k+1, 1{:}k)$ is zero   (12.2.10)

    Compute the update $A = M_k A$ by updating A's representation

end

$U = A$

Note that if $M = M_1 \cdots M_{n-1}$, then $MA = U$ and $M = B(\tau)$ with $\tau = [\tau_1, \ldots, \tau_{n-1}]^T$. It follows that if $L = M^{-1}$, then L is semiseparable from (12.2.4) and $A = LU$. The challenge is to show that the updates $A = M_k A$ preserve semiseparability.

To see what is involved, suppose $n = 6$ and that we have computed M_5 and M_4 so that

$$M_4 M_5 A = \begin{bmatrix} \times & \times & \times & \times & \times & \times \\ \times & \times & \times & \times & \times & \times \\ \lambda & \lambda & \lambda & \mu & \mu & \mu \\ \lambda & \lambda & \lambda & \mu & \mu & \mu \\ 0 & 0 & 0 & 0 & \times & \times \\ 0 & 0 & 0 & 0 & 0 & \times \end{bmatrix} = \mathbf{S}(u, v, t, u \,.* \, v, p, q, r)$$

is semiseparable. Note that the λ-block and the μ-block are given by

$$\begin{bmatrix} \lambda & \lambda & \lambda \\ \lambda & \lambda & \lambda \end{bmatrix} = \begin{bmatrix} u_3 t_2 t_1 v_1 & u_3 t_2 v_2 & u_3 v_3 \\ u_4 t_3 t_2 t_1 v_1 & u_4 t_3 t_2 v_2 & u_4 t_3 v_3 \end{bmatrix},$$

$$\begin{bmatrix} \mu & \mu & \mu \\ \mu & \mu & \mu \end{bmatrix} = \begin{bmatrix} p_3 r_3 q_4 & p_3 r_3 r_4 q_5 & p_3 r_3 r_4 r_5 q_6 \\ p_4 q_4 & p_4 r_4 q_5 & p_4 r_4 r_5 q_6 \end{bmatrix}.$$

Thus, if

$$\tilde{M}_3 = \begin{bmatrix} 1 & 0 \\ -\tau_3 & 1 \end{bmatrix},$$

then

$$\tilde{M}_3 \begin{bmatrix} \lambda & \lambda & \lambda \\ \lambda & \lambda & \lambda \end{bmatrix} = \begin{bmatrix} u_3 t_2 t_1 v_1 & u_3 t_2 v_2 & u_3 v_3 \\ (u_4 t_3 - \tau_3 u_3) t_2 t_1 v_1 & (u_4 t_3 - \tau_3 u_3) t_2 v_2 & (u_4 t_3 - \tau_3 u_3) v_3 \end{bmatrix},$$

$$\tilde{M}_3 \begin{bmatrix} \mu & \mu & \mu \\ \mu & \mu & \mu \end{bmatrix} = \begin{bmatrix} p_3 r_3 q_4 & p_3 r_3 r_4 q_5 & p_3 r_3 r_4 r_5 q_6 \\ (p_4 - \tau_3 p_3 r_3) q_4 & (p_4 - \tau_3 p_3 r_3) r_4 q_5 & (p_4 - \tau_3 p_3 r_3) r_4 r_5 q_6 \end{bmatrix}.$$

If $u_3 \neq 0$, $\tau_3 = u_4 t_3 / u_3$, and we perform the updates

$$u_4 = 0, \qquad p_4 = p_4 - \tau_3 p_3 r_3,$$

then

$$M_3 M_4 M_5 A = \begin{bmatrix} \times & \times & \times & \times & \times & \times \\ \times & \times & \times & \times & \times & \times \\ \lambda & \lambda & \lambda & \mu & \mu & \mu \\ 0 & 0 & 0 & \tilde{\mu} & \tilde{\mu} & \tilde{\mu} \\ 0 & 0 & 0 & 0 & \times & \times \\ 0 & 0 & 0 & 0 & 0 & \times \end{bmatrix} = \mathbf{S}(u, v, t, u \,.* \, v, p, q, r)$$

is still semiseparable. (The tildes designate updated entries.) Picking up the pattern from this example, we obtain the following $O(n)$ method for computing the LU factorization of a semiseparable matrix.

Algorithm 12.2.1 Assume that $u, v, p, q \in \mathbb{R}^n$ with $u.*v = p.*q$ and that $t, r \in \mathbb{R}^{n-1}$. If $A = \mathbf{S}(u, t, v, u.*v, p, r, q)$ has an LU factorization, then the following algorithm computes $\tilde{p} \in \mathbb{R}^n$ and $\tau \in \mathbb{R}^{n-1}$ so that if $L = B(\tau)^{-T}$ and $U = \mathsf{triu}(\tilde{p}q^T).*B(r)^{-1}$, then $A = LU$.

> **for** $k = n-1: -1:1$
> $\qquad \tau_k = t_k u_{k+1}/u_k$
> $\qquad \tilde{p}_{k+1} = p_{k+1} - p_k \tau_k r_k$
> **end**
> $\tilde{p}_1 = p_1$

This algorithm requires about $5n$ flops. Given our remarks in the previous section about triangular semiseparable matrices, we see that a semiseparable system $Ax = b$ can be solved with $O(n)$ work: $A = LU$, $Ly = b$, $Ux = y$. Note that the vectors τ and $\tilde{p}$ in algorithm 12.2.1 are given by

$$\tau = (u(2{:}n).*t)./u(1{:}n-1)$$

and

$$\tilde{p} = \begin{bmatrix} p_1 \\ p(2{:}n) - p(1{:}n-1).*\tau.*r \end{bmatrix}.$$

Pivoting can be incorporated in Algorithm 12.2.1 to ensure that $|\tau_k| \le 1$ for $k = n-1: -1:1$. At the beginning of step k, if $|u_k| < |u_{k+1}|$, then rows k and $k+1$ are interchanged. The swapping is orchestrated by updating the quasiseparable respresentation of the current A. The end result is an $O(n)$ reduction of the form $M_1 \cdots M_{n-1}A = U$ where U is upper triangular and quasiseparable and $M_k = \mathsf{diag}(I_{k-1}, \tilde{M}_k \tilde{P}_k, I_{n-k-1})$ with

$$\tilde{P}_k = \begin{bmatrix} 1 & 0 \\ 0 & 1 \end{bmatrix} \text{ or } \begin{bmatrix} 0 & 1 \\ 1 & 0 \end{bmatrix}.$$

See Vandebril, Van Barel, and Mastronardi (2008, pp. 165–170) for further details and also how to perform the same tasks when A is quasiseparable.

12.2.6 The Givens-Vector Representation

The QR factorization of a semiseparable matrix is also an $O(n)$ computation. To motivate the algorithm we step through a simple special case that showcases the idea of a structured rank Givens update. Along the way we will discover yet another strategy that can be used to represent a semiseparable matrix.

Assume $A_L \in \mathbb{R}^{n \times n}$ is a lower triangular semiseparable matrix and that $a \in \mathbb{R}^n$ is its first column. We can reduce this column to a multiple of e_1 with a sequence of

$n - 1$ Givens rotations, e.g.,

$$
\begin{bmatrix} c_1 & s_1 & 0 & 0 \\ -s_1 & c_1 & 0 & 0 \\ 0 & 0 & 1 & 0 \\ 0 & 0 & 0 & 1 \end{bmatrix}
\begin{bmatrix} 1 & 0 & 0 & 0 \\ 0 & c_2 & s_2 & 0 \\ 0 & -s_2 & c_2 & 0 \\ 0 & 0 & 0 & 1 \end{bmatrix}
\begin{bmatrix} 1 & 0 & 0 & 0 \\ 0 & 1 & 0 & 0 \\ 0 & 0 & c_3 & s_3 \\ 0 & 0 & -s_3 & c_3 \end{bmatrix}
\begin{bmatrix} a_1 \\ a_2 \\ a_3 \\ a_4 \end{bmatrix}
=
\begin{bmatrix} v_1 \\ 0 \\ 0 \\ 0 \end{bmatrix}.
$$

By moving the rotations to the right-hand side we see that

$$
A_L(:,1) \;=\; \begin{bmatrix} a_1 \\ a_2 \\ a_3 \\ a_4 \end{bmatrix} = v_1 \begin{bmatrix} c_1 \\ c_2 s_1 \\ c_3 s_2 s_1 \\ s_3 s_2 s_1 \end{bmatrix}.
$$

Because this is the first column of a semiseparable matrix, it is not hard to show that there exist "weights" $v_2, \ldots, v_n$ so that

$$
A_L = \begin{bmatrix} c_1 v_1 & 0 & 0 & 0 \\ c_2 s_1 v_1 & c_2 v_2 & 0 & 0 \\ c_3 s_2 s_1 v_1 & c_3 s_2 v_2 & c_3 v_3 & 0 \\ s_3 s_2 s_1 v_1 & s_3 s_2 v_2 & s_3 v_3 & v_4 \end{bmatrix} = B(s)^{-T} .* \mathsf{tril}(cv^T) \qquad (12.2.11)
$$

where

$$
v = \begin{bmatrix} v_1 \\ v_2 \\ v_3 \\ v_4 \end{bmatrix}, \qquad c = \begin{bmatrix} c_1 \\ c_2 \\ c_3 \\ 1 \end{bmatrix}, \qquad s = \begin{bmatrix} s_1 \\ s_2 \\ s_3 \end{bmatrix}.
$$

The encoding (12.2.11) is an example of the *Givens-vector representation* for a triangular semiseparable matrix. It consists of a vector of cosines, a vector of sines, and a vector of weights. By "transposing" this idea, we can similarly represent an upper triangular semiseparable matrix. Thus, for a general semiseparable matrix A we may write

$$
A \;=\; A_L \,+\, A_U,
$$

where

$$
\begin{aligned}
A_L &= \mathsf{tril}(A) = B(s_L)^{-T} .* \mathsf{tril}(c_L v_L^T), \\
A_U &= \mathsf{triu}(A, 1) = B(s_U)^{-1} .* \mathsf{triu}(v_U c_U^T, 1),
\end{aligned}
$$

where c_L, s_L, and v_L (resp. c_U, s_U, and v_U) are the cosine, sine, and weight vectors associated with the lower (resp. upper) triangular part. For more details on the properties and utility of this representation, see Vandebril and Van Barel (2005).

12.2.7   The QR Factorization of a Semiseparable Matrix

The matrix Q in the QR factorization of a semiseparable matrix $A \in \mathbb{R}^{n \times n}$ has a very simple form. Indeed, it is a product of Givens rotations $Q^T = G_1 \cdots G_{n-1}$ where the underlying cosine-sine pairs are precisely those that define Givens representation of A_L. To see this, consider how easy it is to compute the QR factorization of A_L:

$$
\begin{bmatrix} 1 & 0 & 0 & 0 \\ 0 & 1 & 0 & 0 \\ 0 & 0 & c_3 & s_3 \\ 0 & 0 & -s_3 & c_3 \end{bmatrix}
\begin{bmatrix} c_1 v_1 & 0 & 0 & 0 \\ c_2 s_1 v_1 & c_2 v_2 & 0 & 0 \\ c_3 s_2 s_1 v_1 & c_3 s_2 v_2 & c_3 v_3 & 0 \\ s_3 s_2 s_1 v_1 & s_3 s_2 v_2 & s_3 v_3 & v_4 \end{bmatrix}
=
\begin{bmatrix} c_1 v_1 & 0 & 0 & 0 \\ c_2 s_1 v_1 & c_2 v_2 & 0 & 0 \\ s_2 s_1 v_1 & s_2 v_2 & v_3 & s_3 v_4 \\ 0 & 0 & 0 & c_3 v_4 \end{bmatrix},
$$

$$
\begin{bmatrix} 1 & 0 & 0 & 0 \\ 0 & c_2 & s_2 & 0 \\ 0 & -s_2 & c_2 & 0 \\ 0 & 0 & 0 & 1 \end{bmatrix}
\begin{bmatrix} c_1 v_1 & 0 & 0 & 0 \\ c_2 s_1 v_1 & c_2 v_2 & 0 & 0 \\ s_2 s_1 v_1 & s_2 v_2 & v_3 & s_3 v_4 \\ 0 & 0 & 0 & c_3 v_4 \end{bmatrix}
=
\begin{bmatrix} c_1 v_1 & 0 & 0 & 0 \\ s_1 v_1 & v_2 & s_2 v_3 & s_2 s_3 v_4 \\ 0 & 0 & c_2 v_3 & c_2 s_3 v_4 \\ 0 & 0 & 0 & c_3 v_4 \end{bmatrix},
$$

$$
\begin{bmatrix} c_1 & s_1 & 0 & 0 \\ -s_1 & c_1 & 0 & 0 \\ 0 & 0 & 1 & 0 \\ 0 & 0 & 0 & 1 \end{bmatrix}
\begin{bmatrix} c_1 v_1 & 0 & 0 & 0 \\ s_1 v_1 & v_2 & s_2 v_3 & s_2 s_3 v_4 \\ 0 & 0 & c_2 v_3 & c_2 s_3 v_4 \\ 0 & 0 & 0 & c_3 v_4 \end{bmatrix}
=
\begin{bmatrix} v_1 & s_1 v_2 & s_1 s_2 v_3 & s_1 s_2 s_3 v_4 \\ 0 & c_1 v_2 & c_1 s_2 v_3 & c_1 s_2 s_3 v_4 \\ 0 & 0 & c_2 v_3 & c_2 s_3 v_4 \\ 0 & 0 & 0 & c_3 v_4 \end{bmatrix}.
$$

In general, if $\mathsf{tril}(A) = B(s)^{-T} .* \mathsf{tril}(cv^T)$ is a Givens vector representation and

$$
Q^T = G_1 \cdots G_{n-1} \tag{12.2.12}
$$

where

$$
G_k = \mathrm{diag}(I_{k-1}, \tilde{G}_k, I_{n-k-1}), \qquad \tilde{G}_k = \begin{bmatrix} c_k & s_k \\ -s_k & c_k \end{bmatrix}, \tag{12.2.13}
$$

for $k = 1{:}n-1$, then

$$
Q^T \mathsf{tril}(A) = R_L = \mathsf{triu}((\mathcal{D}_n c)v^T) .* B(s)^{-1}. \tag{12.2.14}
$$

(Recall that $\mathcal{D}_n$ is the downshift permutation, see §1.3.x.) Since Q^T is upper Hessenberg, it follows that

$$
Q^T \mathsf{triu}(\mathsf{A}, 1) = R_U
$$

is also upper triangular. Thus,

$$
Q^T A = Q^T (A_L + A_U) = R_L + R_U = R
$$

is the QR factorization of A. Unfortunately, this is not a useful $O(n)$ representation of R from the standpoint of solving $Ax = b$ because the summation gets in the way when we try to solve $(R_L + R_U)x = Q^T b$.

Fortunately, there is a handier way to encode R. Assume for clarity that A has a generator representation

$$
A = \mathsf{tril}(uv^T) + \mathsf{triu}(pq^T), \tag{12.2.15}
$$

where $u, v, p, q \in \mathbb{R}^n$ and $u .* v = p .* q$. We show that R is the upper triangular portion of a rank-2 matrix, i.e.,

$$R = \mathsf{triu}(fg^T + hq^T), \qquad f, g, h \in \mathbb{R}^n. \tag{12.2.16}$$

This means that any submatrix extracted from the upper triangular part of R has rank two or less.

From (12.2.15) we see that the first column of A is a multiple of u. It follows that the Givens rotations that define Q in (12.2.12) can be determined from this vector:

$$G_1 \cdots G_{n-1} u = \begin{bmatrix} \tilde{u}_1 \\ 0 \\ \vdots \\ 0 \end{bmatrix}.$$

Suppose $n = 6$ and that we have computed G_5, G_4 and G_3 so that $A^{(3)} = G_3 G_4 G_5 A$ has the form

$$A^{(3)} = \begin{bmatrix}
u_1 v_1 & p_1 q_2 & p_1 q_3 & p_1 q_4 & p_1 q_5 & p_1 q_6 \\
u_2 v_1 & u_2 v_2 & p_2 q_3 & p_2 q_4 & p_2 q_5 & p_2 q_6 \\
\tilde{u}_3 v_1 & \tilde{u}_3 v_2 & \tilde{f}_3 g_3 + \tilde{h}_3 q_3 & \tilde{f}_3 g_4 + \tilde{h}_3 q_4 & \tilde{f}_3 g_5 + \tilde{h}_3 q_5 & \tilde{f}_3 g_6 + \tilde{h}_3 q_6 \\
0 & 0 & 0 & f_4 g_4 + h_4 q_4 & f_4 g_5 + h_4 q_5 & f_4 g_6 + h_4 q_6 \\
0 & 0 & 0 & 0 & f_5 g_5 + h_5 q_5 & f_5 g_6 + h_5 q_6 \\
0 & 0 & 0 & 0 & 0 & f_6 g_6 + h_6 q_6
\end{bmatrix}.$$

Next, we compute the cosine-sine pair $\{c_2, s_2\}$ so that

$$\tilde{G}_2 \begin{bmatrix} u_2 \\ \tilde{u}_3 \end{bmatrix} = \begin{bmatrix} c_2 & s_2 \\ -s_2 & c_2 \end{bmatrix} \begin{bmatrix} u_2 \\ \tilde{u}_3 \end{bmatrix} = \begin{bmatrix} \tilde{u}_2 \\ 0 \end{bmatrix}.$$

Since

$$\begin{bmatrix} c_2 & s_2 \\ -s_2 & c_2 \end{bmatrix} \begin{bmatrix} p_2 q_j \\ \tilde{f}_3 g_j + \tilde{h}_3 q_j \end{bmatrix} = \begin{bmatrix} c_2 p_2 + s_2 \tilde{h}_3 \\ -s_2 p_2 + c_2 \tilde{h}_3 \end{bmatrix} q_j + \begin{bmatrix} s_2 \tilde{f}_3 \\ c_2 \tilde{f}_3 \end{bmatrix} g_j,$$

for $j = 3{:}6$, it follows that $A^{(2)} = G_2 A^{(3)} = \mathsf{diag}(1, \tilde{G}_2, I_3) A^{(3)}$ has the form

$$A^{(2)} = \begin{bmatrix}
u_1 v_1 & p_1 q_2 & p_1 q_3 & p_1 q_4 & p_1 q_5 & p_1 q_6 \\
\tilde{u}_2 v_1 & \tilde{f}_2 g_2 + \tilde{h}_2 q_2 & \tilde{f}_2 g_3 + \tilde{h}_2 q_3 & \tilde{f}_2 g_4 + \tilde{h}_2 q_4 & \tilde{f}_2 g_5 + \tilde{h}_2 q_5 & \tilde{f}_2 g_6 + \tilde{h}_2 q_6 \\
0 & 0 & f_3 g_3 + h_3 q_3 & f_3 g_4 + h_3 q_4 & f_3 g_5 + h_3 q_5 & f_3 g_6 + h_3 q_6 \\
0 & 0 & 0 & f_4 g_4 + h_4 q_4 & f_4 g_5 + h_4 q_5 & f_4 g_6 + h_4 q_6 \\
0 & 0 & 0 & 0 & f_5 g_5 + h_5 q_5 & f_5 g_6 + h_5 q_6 \\
0 & 0 & 0 & 0 & 0 & f_6 g_6 + h_6 q_6
\end{bmatrix}$$

where

$$\tilde{f}_2 = s_2\tilde{f}_3, \qquad \tilde{f}_3 = c_2\tilde{f}_3, \qquad \tilde{h}_2 = c_2p_2 + s_2\tilde{h}_3, \qquad \tilde{h}_3 = -s_2p_2 + c_2\tilde{h}_3.$$

By considering the transition from $A^{(3)}$ to $A^{(2)}$ via the Givens rotation G_2, we conclude that $\left[A^{(2)}\right]_{22} = \tilde{u}_2v_2$. Since this must equal $\tilde{f}_2g_2 + \tilde{h}_2q_2$ we have

$$g_2 = \frac{\tilde{u}_2v_2 - \tilde{h}_2q_2}{\tilde{f}_2}.$$

By extrapolating from this example and making certain assumptions to guard against divison by zero, we obtain the following QR factorization procedure.

Algorithm 12.2.2 Suppose u, v, p, and q are n-vectors that satisfy $u.*v = p.*q$ and $u_n \neq 0$. If $A = \mathsf{tril}(uv^T) + \mathsf{triu}(pq^T, 1)$, then this algorithm computes cosine-sine pairs $\{c_1, s_1\}, \ldots, \{c_{n-1}, s_{n-1}\}$ and vectors $f, g, h \in \mathbb{R}^n$ so that if Q is defined by (12.2.12) and (12.2.13), then $Q^T A = R = \mathsf{triu}(fg^T + hq^T)$.

$\tilde{u}_n = u_n, \ \tilde{f}_n = u_n, \ g_n = v_n, \ h_n = 0$

for $k = n-1:-1:1$

> Determine c_k and s_k so that $\begin{bmatrix} c_k & s_k \\ -s_k & c_k \end{bmatrix} \begin{bmatrix} u_k \\ \tilde{u}_{k+1} \end{bmatrix} = \begin{bmatrix} \tilde{u}_k \\ 0 \end{bmatrix}.$

> $\tilde{f}_k = s_k\tilde{f}_{k+1}, \ f_{k+1} = c_k\tilde{f}_{k+1}$

> $\begin{bmatrix} h_k \\ h_{k+1} \end{bmatrix} = \begin{bmatrix} c_k & s_k \\ -s_k & c_k \end{bmatrix} \begin{bmatrix} p_k \\ h_{k+1} \end{bmatrix}$

> $g_k = (u_kv_k - h_kq_k)/\tilde{f}_k$

end

$f_1 = \tilde{f}_1$

Regarding the condition that $u_n \neq 0$, it is easy to show by induction that

$$\tilde{f}_k = s_k \cdots s_{n-1}u_n.$$

The s_k are nonzero because $|\tilde{u}_k| = \| u(k{:}n) \|_2 \neq 0$. This algorithm requires $O(n)$ flops and $O(n)$ storage. We stress that there are better ways to implement the QR factorization of a semiseparable matrix than Algorithm 12.2.2. See Van Camp, Mastronardi, and Van Barel (2004). Our goal, as stated above, is to suggest how a structured rank matrix factorization can be organized around Givens rotations. Equally efficient QR factorizations for quasiseparable and semiseparable-plus-diagonal matrices are also possible.

We mention that an n-by-n system of the form $\mathsf{triu}(fg^T + hq^T)x = y$ can be solved in $O(n)$ flops. An induction argument based on the partitioning

$$\begin{bmatrix} f_kg_k + h_kq_k & f_k\tilde{g}^T + h_1\tilde{q}^T \\ 0 & \tilde{f}\tilde{g}^T + \tilde{h}\tilde{q}^T \end{bmatrix} \begin{bmatrix} x_k \\ \tilde{x} \end{bmatrix} = \begin{bmatrix} y_k \\ \tilde{y} \end{bmatrix}$$

where all the "tilde" vectors belong to $\mathbb{R}^{n-k}$ shows why. If $\tilde{x}$, $\alpha = \tilde{g}^T\tilde{x}$, and $\tilde{q}^T\tilde{x}$ are available, then x_k and the updates $\alpha = \alpha + g_kx_k$ and $\beta = \beta + q_kx_k$ require $O(1)$ flops.

12.2.8   Other Rank-Structured Classes

We briefly mention several other rank structures that arise in applications. Fast LU and QR procedures exist in each case.

If p and q are nonnegative integers, then a matrix A is $\{p, q\}$-*semiseparable* if

$$j_2 < i_1 + p \quad \Rightarrow \quad \mathsf{rank}(A(i_1{:}i_2, j_1{:}j_2)) \leq p,$$

$$i_2 > j_1 + q \quad \Rightarrow \quad \mathsf{rank}(A(i_1{:}i_2, j_1{:}j_2)) \leq q.$$

For example, if A is $\{2, 3\}$-semiseparable, then

$$A = \begin{bmatrix} \times & \times & \times & \times & \times & \times & \times \\ a_{21} & a_{22} & a_{23} & \times & \times & \times & \times \\ a_{31} & a_{32} & a_{33} & a_{34} & a_{35} & a_{36} & a_{37} \\ a_{41} & a_{42} & a_{43} & a_{44} & a_{45} & a_{46} & a_{47} \\ \times & \times & \times & a_{54} & a_{55} & a_{56} & a_{57} \\ \times & \times & \times & a_{64} & a_{65} & a_{66} & a_{67} \\ \times & \times & \times & a_{74} & a_{75} & a_{76} & a_{77} \end{bmatrix} \quad \Rightarrow \quad \begin{aligned} & \mathsf{rank}(A(2{:}4, 1{:}3)) \leq 2, \\ & \mathsf{rank}(A(3{:}7, 4{:}7)) \leq 3. \end{aligned}$$

In general, A is $\{p, q\}$-*generator representable* if we have $U, V \in \mathbb{R}^{n \times p}$ and $P, Q \in \mathbb{R}^{n \times q}$ such that

$$\mathsf{tril}(A, p - 1) = \mathsf{tril}(UV^T, p - 1),$$

$$\mathsf{triu}(A, -q + 1) = \mathsf{triu}(PQ^T, -q + 1).$$

If such a matrix is nonsingular, then A^{-1} has lower bandwidth p and upper bandwidth q. If the $\{p, q\}$-semiseparable definition is modified so that the rank-p blocks come from $\mathsf{tril}(A)$ and the rank-q blocks come from $\mathsf{triu}(A)$, then A belongs to the class of *extended* $\{p, q\}$-*separable* matrices. If the $\{p, q\}$-semiseparable definition is modified so that the rank-p blocks come from $\mathsf{tril}(A, -1)$ and the rank-q come from $\mathsf{triu}(A, 1)$, then A belongs to the class of *extended* $\{p, q\}$-*quasiseparable* matrices. A *sequentially semiseparable* matrix is a block matrix that has the following form:

$$A = \begin{bmatrix} D_1 & P_1 Q_2^T & P_1 R_2 Q_3^T & P_1 R_2 R_3 Q_4^T \\ U_2 V_1^T & D_2 & P_2 Q_3^T & P_2 R_3 Q_4^T \\ U_3 T_2 V_1^T & U_3 V_2^T & D_3 & P_3 Q_4^T \\ U_4 T_3 T_2 V_1^T & U_4 T_3 V_2^T & U_4 V_3^T & D_4 \end{bmatrix}. \tag{12.2.17}$$

See Dewilde and van der Veen (1997) and Chandrasekaran et al. (2005). The blocks can be rectangular so least squares problems with this structure can be handled.

Matrices with *hierarchical rank structure* are based on low-rank patterns that emerge through recursive 2-by-2 blockings. (With one level of recursion we would have 2-by-2 block matrix whose diagonal blocks are 2-by-2 block matrices.) Various connections may exist between the low-rank representations of the off-diagonal blocks. The important class of hierarchically semiseparable matrices has a particularly rich and exploitable structure; see Xia (2012).

12.2.9 Semiseparable Eigenvalue Problems and Techniques

Fast versions of various two-sided, eigenvalue-related decompositions also exist. For example, if $A \in \mathbb{R}^{n \times n}$ is symmetric and diagonal-plus-semiseparable, then it is possible to compute the tridiagonalization $Q^T A Q = T$ in $O(n^2)$ flops. The orthogonal matrix Q is a product of Givens rotations each of which participate in a highly-structured update. See Mastronardi, Chandrasekaran, and Van Huffel (2001).

There are also interesting methods for general matrix problems that involve the introduction of semiseparable structures during the solution process. Van Barel, Vanberghen, and van Dooren (2010) approach the product SVD problem through conversion to a semiseparable structure. For example, to compute the SVD of $A = A_1 A_2$ orthogonal matrices U_1, U_2, and U_3 are first computed so that $(U_1^T A_1 U_2)(U_2^T A_2 U_3) = T$ is upper triangular and semiseparable. Vanberghen, Vandebril, and Van Barel (2008) have shown how to compute orthogonal $Q, Z \in \mathbb{R}^{n \times n}$ so that $Q^T B Z = R$ is upper triangular and $Q^T A Z = L$ has the property that $\mathsf{tril}(L)$ is semiseparable. A procedure for reducing the equivalent pencil $L - \lambda R$ to generalized Schur form is also developed.

12.2.10 Eigenvalues of an Orthogonal Upper Hessenberg Matrix

We close with an eigenvalue problem that has quasiseparable structure. Suppose $H \in \mathbb{R}^{n \times n}$ is an upper Hessenberg matrix that is also orthogonal. Our goal is to compute $\lambda(H)$. Note that each eigenvalue is on the unit circle. Without loss of generality we may assume that the subdiagonal entries are nonzero.

If n is odd, then it must have a real eigenvalue because the eigenvalues of a real matrix come in complex conjugate pairs. In this case it is possible to deflate the problem by carefully working with the eigenvector equation $Hx = x$ (or $Hx = -x$). Thus, we may assume that n is even.

For $1 \le k \le n - 1$, define the reflection $G_k \in \mathbb{R}^{n \times n}$ by

$$G_k = G(\phi_k) = \mathsf{diag}\,(I_{k-1}, R(\phi_k), I_{n-k-1})$$

where

$$R(\phi_k) = \left[\begin{array}{cc} -\cos(\phi_k) & \sin(\phi_k) \\ \sin(\phi_k) & \cos(\phi_k) \end{array} \right], \qquad 0 < \phi_k < \pi.$$

These transformations can be used to represent the QR factorization of H. Indeed, as for the Givens process described in §5.2.6, we can compute $G_1, \ldots, G_{n-1}$ so that

$$G_{n-1} \cdots G_1 H = G_n \equiv \mathsf{diag}(1, \ldots, 1, -c_n).$$

The matrix G_n is the "R" matrix. It is diagonal because an orthogonal upper triangular matrix must be diagonal. Since the determinant of a matrix is the product of its eigenvalues, the value of c_n is either $+1$ or -1. If $c_n = -1$, then $\det(H) = -1$, which in turn implies that H has a real eigenvalue and we can deflate the problem. Thus, we may assume that

$$H = G_1 \cdots G_n, \qquad G_n = \mathsf{diag}(1, \ldots, 1, -1), \qquad n = 2m \qquad (12.2.18)$$

and that our goal is to compute

$$\lambda(H) = \{\, \cos(\theta_1) \pm i \cdot \sin(\theta_1), \ldots, \cos(\theta_m) \pm i \cdot \sin(\theta_m) \,\}. \qquad (12.2.19)$$

Note that (12.2.4) and (12.2.18) tell us that H is quasiseparable.

Ammar, Gragg, and Reichel (1986) propose an interesting $O(n^2)$ method that computes the required eigenvalues by setting up a pair of m-by-m bidiagonal SVD problems. Three facts are required:

Fact 1. H is similar to $\tilde{H} = H_o H_e$ where

$$
\begin{aligned}
H_o &= G_1 G_3 \cdots G_{n-1} = \operatorname{diag}(R(\phi_1), R(\phi_3), \ldots, R(\phi_{n-1})), \\
H_e &= G_2 G_4 \cdots G_n \;\;= \operatorname{diag}(1, R(\phi_2), R(\phi_4), \ldots, R(\phi_{n-2}), -1).
\end{aligned}
$$

Fact 2. The matrices

$$
C = \frac{H_o + H_e}{2}, \qquad S = \frac{H_o - H_e}{2}
$$

are symmetric and tridiagonal. Moreover, their eigenvalues are given by

$$
\lambda(C) = \{ \pm\cos(\theta_1/2), \ldots, \pm\cos(\theta_m/2) \},
$$
$$
\lambda(S) = \{ \pm\sin(\theta_1/2), \ldots, \pm\sin(\theta_m/2) \}.
$$

Fact 3. If

$$
Q_o = \operatorname{diag}(R(\phi_1/2), R(\phi_3/2), \ldots, R(\phi_{n-1}/2)),
$$
$$
Q_e = \operatorname{diag}(1, R(\phi_2/2), R(\phi_4/2), \ldots, R(\phi_{n-2}/2), -1),
$$

then perfect shuffle permutations of the matrices

$$
C^{(1)} = Q_o C Q_e, \qquad S^{(1)} = Q_o S Q_e
$$

expose a pair of m-by-m bidiagonal matrices B_C and B_S with the property that

$$
\sigma(B_C) = \{\cos(\theta_1/2), \ldots, \cos(\theta_m/2)\},
$$
$$
\sigma(B_S) = \{\sin(\theta_1/2), \ldots, \sin(\theta_m/2)\}.
$$

Once the bidiagonal matrices B_C and B_S are set up (which involves $O(n)$ work), then their singular values can be computed via Golub-Kahan SVD algorithm. The angle θ_k can be accurately determined from $\sin(\theta_k/2)$ if $0 < \theta_k < \pi/2$ and from $\cos(\theta_k/2)$ otherwise. See Ammar, Gragg, and Reichel (1986) for more details.

Problems

P12.2.1 Rigorously prove that the matrix $B(r)^{-1}$ is semiseparable.

P12.2.2 Prove that A is quasiseparable if and only if $A = \mathbf{S}(u, t, v, d, p, r, q)$ for appropriately chosen vectors u, v, t, d, p, r, and q.

P12.2.3 How many flops are required to execute the n-by-n matrix vector product $y = Ax$ where $A = \mathbf{S}(u, v, t, d, p, q, r)$.

P12.2.4 Refer to (12.2.4). Determine u, v, t, d, p, q, and r so that $M = \mathbf{S}(u, v, t, d, p, q, r)$.

P12.2.5 Suppose $\mathbf{S}(u, v, t, d, v, u, t)$ is symmetric positive definite and semiseparable. Show that its Cholesky factor is semiseparable and give an algorithm for computing its quasiseparable representation.

P12.2.6 Verify the three facts in §12.2.3.

P12.2.7 Develop a fast method for solving the upper triangular system $Tx = y$ where T is the matrix $T = \text{diag}(d) + \text{triu}(pq^T, 1) \cdot * B(r)^{-1}$ with $p, q, d, y \in \mathbb{R}^n$ and $r \in \mathbb{R}^{n-1}$.

P12.2.8 Verify (12.2.7).

P12.2.9 Prove (12.2.14).

P12.2.10 Assume that A is an N-by-N block matrix that has the sequentially separable structure illustrated in (12.2.17). Assume that the blocks are each m-by-m. Give a fast algorithm for computing $y = Ax$ where $x \in \mathbb{R}^{Nm}$.

P12.2.11 It can be shown that

$$
A = \begin{bmatrix} A_1 & B_1^T & 0 & 0 \\ B_1 & A_2 & B_2^T & 0 \\ 0 & B_2 & A_3 & B_3^T \\ 0 & 0 & B_3 & A_4 \end{bmatrix} \Rightarrow A^{-1} = \begin{bmatrix} U_1 V_1^T & V_1 U_2^T & V_1 U_3^T & V_1 U_4^T \\ U_2 V_1^T & U_2 V_2^T & V_2 U_3^T & V_2 U_4^T \\ U_3 V_1^T & U_3 V_2^T & U_3 V_3^T & V_3 U_4^T \\ U_4 V_1^T & U_4 V_2^T & U_4 V_3^T & U_4 V_4^T \end{bmatrix},
$$

assuming that A is symmetric positive definite and that the B_i are nonsingular. Give an algorithm that computes $U_1, \ldots, U_4$ and $V_1, \ldots, V_4$.

P12.2.12 Suppose $a, b, f, g \in \mathbb{R}^n$ and that $A = \text{triu}(ab^T + fg^T)$ is nonsingular. (a) Given $x \in \mathbb{R}^n$, show how to compute efficiently $y = Ax$. (b) Given $y \in \mathbb{R}^n$, show how to compute $x \in \mathbb{R}^n$ so that $Ax = y$. (c) Given $y, d \in \mathbb{R}^n$, show how to compute x so that $y = (A + D)x$ where it is assumed that $D = \text{diag}(d)$ and $A + D$ are nonsingular.

P12.2.13 Verify the three facts in §12.2.10 for the case $n = 8$.

P12.2.14 Show how to compute the eigenvalues of an orthogonal matrix $A \in \mathbb{R}^{n \times n}$ by computing the Schur decompositions of $(A + A^T)/2$ and $(A - A^T)/2$.

Notes and References for §12.2

For all matters concerning structured rank matrix computations, see:

R. Vandebril, M. Van Barel, and N. Mastronardi (2008). *Matrix Computations and Semiseparable Matrices, Vol. I Linear Systems*, Johns Hopkins University Press, Baltimore, MD.

R. Vandebril, M. Van Barel, and N. Mastronardi (2008). *Matrix Computations and Semiseparable Matrices, Vol. II Eigenvalue and Singular Value Methods*, Johns Hopkins University Press, Baltimore, MD.

As we have seen, working with the "right" representation is critically important in order to realize an efficient implementation. For more details, see:

R. Vandebril, M. Van Barel, and N. Mastronardi (2005). "A Note on the Representation and Definition of Semiseparable Matrices," *Num. Lin. Alg. Applic. 12*, 839–858.

References concerned with the fast solution of linear equations and least squares problems with structured rank include:

I. Gohberg, T. Kailath, and I Koltracht (1985) "Linear Complexity Algorithm for Semiseparable Matrices," *Integral Equations Operator Theory 8*, 780–804.

Y. Eidelman and I. Gohberg (1997). "Inversion Formulas and Linear Complexity Algorithm for Diagonal-Plus-Semiseparable Matrices," *Comput. Math. Applic. 33*, 69–79.

P. Dewilde and A.J. van der Veen (1998). *Time-Varying Systems and Computations*, Kluwer Academic, Boston, MA,

S. Chandrasekaran and M. Gu (2003). "Fast and Stable Algorithms for Banded-Plus-Semiseparable Systems of Linear Equations," *SIAM J. Matrix Anal. Applic. 25*, 373–384.

S. Chandrasekaran, P. Dewilde, M. Gu, T. Pals, X. Sun, A.J. Van Der Veen, and D. White (2005). "Some Fast Algorithms for Sequentially Semiseparable Representations," *SIAM J. Matrix Anal. Applic. 27*, 341–364.

E. Van Camp, N. Mastronardi, and M. Van Barel (2004). "Two Fast Algorithms for Solving Diagonal-Plus-Semiseparable Linear Systems," *J. Comput. Appl. Math. 164*, 731–747.

T. Bella, Y. Eidelman, I. Gohberg, V. Koltracht, and V. Olshevsky (2009). "A Fast Bjorck-Pereyra-Type Algorithm for Solving Hessenberg-Quasiseparable-Vandermonde Systems *SIAM. J. Matrix Anal. Applic. 31*, 790–815.

J. Xia and M. Gu (2010). "Robust Approximate Cholesky Factorization of Rank-Structured Symmetric Positive Definite Matrices," *SIAM J. Matrix Anal. Applic. 31*, 2899–2920.

For discussion of methods that exploit hierarchical rank structure, see:

S. Börm, L. Grasedyck, and W. Hackbusch (2003). "Introduction to Hierarchical Matrices with Applications," *Engin. Anal. Boundary Elements 27*, 405–422.

S. Chandrasekaran, M. Gu, and T. Pals (2006). "A Fast ULV Decomposition Solver for Hierarchically Semiseparable Representations," *SIAM J. Matrix Anal. Applic. 28*, 603–622.

S. Chandrasekaran, M. Gu, X. Sun, J. Xia, and J. Zhu (2007). "A Superfast Algorithm for Toeplitz Systems of Linear Equations," *SIAM J. Matrix Anal. Applic. 29*, 1247–1266.

S. Chandrasekaran, M. Gu, J. Xia, and J. Zhu (2007). "A Fast QR Algorithm for Companion Matrices," *Oper. Theory Adv. Applic. 179*, 111–143.

J. Xia, S. Chandrasekaran, M. Gu, and X.S. Li (2010). "Fast algorithms for Hierarchically Semiseparable Matrices," *Numer. Lin. Alg. Applic. 17*, 953–976.

S. Chandrasekaran, P. Dewilde, M. Gu, and N. Somasunderam (2010). "On the Numerical rank of the Off-Diagonal Blocks of Schur Complements of Discretized Elliptic PDEs," *SIAM J. Matrix Anal. Applic. 31*, 2261–2290.

P.G. Martinsson (2011). "A Fast Randomized Algorithm for Computing a Hierarchically Semi-Separable Representation of a Matrix," *SIAM J. Matrix Anal. Applic. 32*, 1251–1274.

J. Xia (2012). "On the Complexity of Some Hierarchical Structured Matrix Algorithms," *SIAM J. Matrix Anal. Applic. 33*, 388–410.

Reductions to tridiagonal, bidiagonal, and Hessenberg form are essential "front ends" for many eigenvalue and singular value procedures. There are ways to proceed when rank structure is present, see:

N. Mastronardi, S. Chandrasekaran, and S. van Huffel (2001). "Fast and Stable Reduction of Diagonal Plus Semi-Separable Matrices to Tridiagonal and Bidiagonal Form," *BIT 41*, 149–157.

M. Van Barel, R. Vandebril, and N. Mastronardi (2005). "An Orthogonal Similarity Reduction of a Matrix into Semiseparable Form," *SIAM J. Matrix Anal. Applic. 27*, 176–197.

M. Van Barel, E. Van Camp, N. Mastronardi (2005). "Orthogonal Similarity Transformation into Block-Semiseparable Matrices of Semiseparability Rank," *Num. Lin. Alg. Applic. 12*, 981–1000.

R. Vandebril, E. Van Camp, M. Van Barel, and N. Mastronardi (2006). "Orthogonal Similarity Transformation of a Symmetric Matrix into a Diagonal-Plus-Semiseparable One with Free Choice of the Diagonal," *Numer. Math. 102*, 709–726.

Y. Eidelman, I. Gohberg, and L. Gemignani (2007). "On the Fast reduction of a Quasiseparable Matrix to Hessenberg and Tridiagonal Forms," *Lin. Alg. Applic. 420*, 86–101.

R. Vandebril, E. Van Camp, M. Van Barel, and N. Mastronardi (2006). "On the Convergence Properties of the Orthogonal Similarity Transformations to Tridiagonal and Semiseparable (Plus Diagonal) Form," *Numer. Math. 104*, 205–239.

Papers concerned with various structured rank eigenvalue iterations include:

R. Vandebril, M. Van Barel, and N. Mastronardi (2004). "A QR Method for Computing the Singular Values via Semiseparable Matrices," *Numer. Math. 99*, 163–195.

R. Vandebril, M. Van Barel, N. Mastronardi (2005). "An Implicit QR algorithm for Symmetric Semiseparable Matrices," *Num. Lin. Alg. 12*, 625–658.

N. Mastronardi, E. Van Camp, and M. Van Barel (2005). "Divide and Conquer Algorithms for Computing the Eigendecomposition of Symmetric Diagonal-plus-Semiseparable Matrices," *Numer. Alg. 39*, 379–398.

Y. Eidelman, I. Gohberg, and V. Olshevsky (2005). "The QR Iteration Method for Hermitian Quasiseparable Matrices of an Arbitrary Order," *Lin. Alg. Applic. 404*, 305–324.

Y. Vanberghen, R. Vandebril, M. Van Barel (2008). "A QZ-Method Based on Semiseparable Matrices," *J. Comput. Appl. Math. 218*, 482–491.

M. Van Barel, Y. Vanberghen, and P. Van Dooren (2010). "Using Semiseparable Matrices to Compute the SVD of a General Matrix Product/Quotient," *J. Comput. Appl. Math. 234*, 3175–3180.

Our discussion of the orthogonal matrix eigenvalue problem is based on:

G.S. Ammar, W.B. Gragg, and L. Reichel (1985). "On the Eigenproblem for Orthogonal Matrices," *Proc. IEEE Conference on Decision and Control*, 1963–1966.

There is an extensive literature concerned with unitary/orthogonal eigenvalue problem including:

P.J. Eberlein and C.P. Huang (1975). "Global Convergence of the QR Algorithm for Unitary Matrices with Some Results for Normal Matrices," *SIAM J. Numer. Anal. 12*, 421–453.
A. Bunse-Gerstner and C. He (1995). "On a Sturm Sequence of Polynomials for Unitary Hessenberg Matrices," *SIAM J. Matrix Anal. Applic. 16*, 1043–1055.
B. Bohnhorst, A. Bunse-Gerstner, and H. Fassbender (2000). "On the Perturbation Theory for Unitary Eigenvalue Problems," *SIAM J. Matrix Anal. Applic. 21*, 809–824.
M. Gu, R. Guzzo, X.-B. Chi, and X.-O. Cao (2003). "A Stable Divide and Conquer Algorithm for the Unitary Eigenproblem," *SIAM J. Matrix Anal. Applic. 25*, 385–404.
M. Stewart (2006). "An Error Analysis of a Unitary Hessenberg QR Algorithm," *SIAM J. Matrix Anal. Applic. 28*, 40–67.
R.J.A. David and D.S. Watkins (2006). "Efficient Implementation of the Multishift QR Algorithm for the Unitary Eigenvalue Problem," *SIAM J. Matrix Anal. Applic. 28*, 623–633.

For a nice introduction to this problem, see Watkins (MEP, pp. 341–346).

12.3 Kronecker Product Computations

The Kronecker product (KP) has a rich algebra that supports a wide range of fast, practical algorithms. It also provides a bridge between matrix computations and tensor computations. This section is a compendium of its most important properties from that point of view. Recall that we introduced the KP in §1.3.6 and identified a few of its properties in §1.3.7 and §1.3.8. Our discussion of fast transforms in §1.4 and the 2-dimensional Poisson problem in §4.8.4 made heavy use of the operation.

12.3.1 Basic Properties

Kronecker product computations are structured block matrix computations. Basic properties are given in §1.3.6–§1.3.8, including

$$
\begin{array}{lll}
\text{Transpose:} & (B \otimes C)^T & = \; B^T \otimes C^T \,, \\
\text{Inverse:} & (B \otimes C)^{-1} & = \; B^{-1} \otimes C^{-1}, \\
\text{Product:} & (B \otimes C)(D \otimes F) & = \; BD \otimes CF \,, \\
\text{Associativity:} & B \otimes (C \otimes D) & = \; (B \otimes C) \otimes D.
\end{array}
$$

Recall that $B \otimes C \neq C \otimes B$, but if $B \in \mathbb{R}^{m_1 \times n_1}$ and $C \in \mathbb{R}^{m_2 \times n_2}$, then

$$
P(B \otimes C)Q^T \; = \; C \otimes B \tag{12.3.1}
$$

where $P = \mathcal{P}_{m_1,m_2}$ and $Q = \mathcal{P}_{n_1,n_2}$ are perfect shuffle permutations, see §1.2.11.

Regarding the Kronecker product of structured matrices, if B is sparse, then $B \otimes C$ has the same sparsity pattern at the block level. If B and C are permutation matrices, then $B \otimes C$ is also a permutation matrix. Indeed, if p and q are permutations of $1{:}m$ and $1{:}n$, then

$$
I_m(p,:) \otimes I_n(q,:) \; = \; I_{mn}(w,:), \qquad w = (1_m \otimes q) \, + \, n{\cdot}(p - 1_m) \otimes \mathbf{1}_n. \tag{12.3.2}
$$

We also have

$$
\begin{aligned}
\text{(orthogonal)} \otimes \text{(orthogonal)} &= \text{(orthogonal)}, \\
\text{(stochastic)} \otimes \text{(stochastic)} &= \text{(stochastic)}, \\
\text{(sym pos def)} \otimes \text{(sym pos def)} &= \text{(sym pos def)}.
\end{aligned}
$$

The inheritance of positive definiteness follows from

$$
\begin{matrix} B = G_B G_B^T \\ C = G_C G_C^T \end{matrix} \quad \Rightarrow \quad B \otimes C = G_B G_B^T \otimes G_C G_C^T = (G_B \otimes G_C)(G_B \otimes G_C)^T.
$$

In other words, the Cholesky factor of $B \otimes C$ is the Kronecker product of the B and C Cholesky factors. Similar results apply to square LU and QR factorizations:

$$
\left. \begin{matrix} P_B B = L_B U_B \\[4pt] P_C C = L_C U_C \end{matrix} \right\} \Rightarrow (P_B \otimes P_C)(B \otimes C) = (L_B \otimes L_C)(U_B \otimes U_C),
$$

$$
\left. \begin{matrix} B = Q_B R_B \\[4pt] C = Q_C R_C \end{matrix} \right\} \Rightarrow B \otimes C = (Q_B \otimes Q_C)(R_B \otimes R_C).
$$

It should be noted that if B and/or C have more rows than columns, then the same can be said about the upper triangular matrices R_B and R_C. In this case, row permutations of $R_B \otimes R_C$ are required to achieve triangular form. On the other hand,

$$
(B \otimes C)(P_B \otimes P_C) = (Q_B \otimes Q_C)(R_B \otimes R_C)
$$

is a thin QR factorization of $B \otimes C$ if $BP_B = Q_B R_B$ and $CP_C = Q_C R_C$ are thin QR factorizations.

The eigenvalues and singular values of $B \otimes C$ have a product connection to the eigenvalues and singular values of B and C:

$$
\lambda(B \otimes C) = \{\, \beta_i \gamma_j : \beta_i \in \lambda(B),\ \gamma_j \in \lambda(C) \,\},
$$

$$
\sigma(B \otimes C) = \{\, \beta_i \gamma_j : \beta_i \in \sigma(B),\ \gamma_j \in \sigma(C) \,\}.
$$

These results are a consequence of the following decompositions:

$$
\left. \begin{matrix} Q_B^H B Q_B = T_B \\[4pt] Q_C^H C Q_C = T_C \end{matrix} \right\} \Rightarrow (Q_B \otimes Q_C)^H (B \otimes C)(Q_B \otimes Q_C) = T_B \otimes T_C, \qquad (12.3.3)
$$

$$
\left. \begin{matrix} U_B^H B V_B = \Sigma_B \\[4pt] U_C^H C V_C = \Sigma_C \end{matrix} \right\} \Rightarrow (U_B \otimes U_C)^H (B \otimes C)(V_B \otimes V_C) = \Sigma_B \otimes \Sigma_C. \qquad (12.3.4)
$$

Note that if $By = \beta y$ and $Cz = \gamma z$, then $(B \otimes C)(y \otimes z) = \beta\gamma\,(y \otimes z)$. Other properties that follow from (12.3.3) and (12.3.4) include

$$
\text{rank}(B \otimes C) = \text{rank}(B) \cdot \text{rank}(C),
$$

$$\det(B \otimes C) = \det(B)^n \cdot \det(C)^m, \qquad B \in \mathbb{R}^{m \times m}, \ C \in \mathbb{R}^{n \times n},$$

$$\mathrm{tr}(B \otimes C) = \mathrm{tr}(B) \cdot \mathrm{tr}(C),$$

$$\| B \otimes C \|_F = \| B \|_F \cdot \| C \|_F,$$

$$\| B \otimes C \|_2 = \| B \|_2 \cdot \| C \|_2.$$

See Horn and Johnson (TMA) for additional KP facts.

12.3.2   The Tracy-Singh Product

We can think of the Kronecker product of two matrices $B = (b_{ij})$ and $C = (c_{ij})$ as the systematic layout of all possible products $b_{ij} c_{k\ell}$, e.g.,

$$
\begin{bmatrix} b_{11} & b_{12} \\ b_{21} & b_{22} \end{bmatrix}
\otimes
\begin{bmatrix} c_{11} & c_{12} \\ c_{21} & c_{22} \end{bmatrix}
=
\left[
\begin{array}{cc|cc}
b_{11}c_{11} & b_{11}c_{12} & b_{12}c_{11} & b_{12}c_{12} \\
b_{11}c_{21} & b_{11}c_{22} & b_{12}c_{21} & b_{12}c_{22} \\
\hline
b_{21}c_{11} & b_{21}c_{12} & b_{22}c_{11} & b_{22}c_{12} \\
b_{21}c_{21} & b_{21}c_{22} & b_{22}c_{21} & b_{22}c_{22}
\end{array}
\right].
$$

However, the Kronecker product of two block matrices $B = (B_{ij})$ and C_{ij}) is *not* the corresponding layout of all possible block-level Kronecker products $B_{ij} \otimes B_{k\ell}$:

$$
\begin{bmatrix} B_{11} & B_{12} \\ B_{21} & B_{22} \end{bmatrix}
\otimes
\begin{bmatrix} C_{11} & C_{12} \\ C_{21} & C_{22} \end{bmatrix}
\neq
\left[
\begin{array}{cc|cc}
B_{11}C_{11} & B_{11}C_{12} & B_{12}C_{11} & B_{12}C_{12} \\
B_{11}C_{21} & B_{11}C_{22} & B_{12}C_{21} & B_{12}C_{22} \\
\hline
B_{21}C_{11} & B_{21}C_{12} & B_{22}C_{11} & B_{22}C_{12} \\
B_{21}C_{21} & B_{21}C_{22} & B_{22}C_{21} & B_{22}C_{22}
\end{array}
\right].
$$

The matrix on the right is an example of the *Tracy-Singh product*. Formally, if we are given the blockings

$$
B =
\begin{bmatrix}
B_{11} & \cdots & B_{1,N_1} \\
\vdots & \ddots & \vdots \\
B_{M_1,1} & \cdots & B_{M_1,N_1}
\end{bmatrix}
\qquad
C =
\begin{bmatrix}
C_{11} & \cdots & C_{1,N_2} \\
\vdots & \ddots & \vdots \\
C_{M_2,1} & \cdots & C_{M_2,N_2}
\end{bmatrix},
\qquad (12.3.5)
$$

with $B_{ij} \in \mathbb{R}^{m_1 \times n_1}$ and $C_{ij} \in \mathbb{R}^{m_2 \times n_2}$, then the Tracy-Singh product is an M_1-by-N_1 block matrix $B \underset{\mathrm{TS}}{\otimes} C$ whose (i,j) block is given by

$$
[\, B \underset{\mathrm{TS}}{\otimes} C \,]_{ij} =
\begin{bmatrix}
B_{ij} \otimes C_{11} & \cdots & B_{ij} \otimes C_{1,N_2} \\
\vdots & \ddots & \vdots \\
B_{ij} \otimes C_{M_2,1} & \cdots & B_{ij} \otimes C_{M_2,N_2}
\end{bmatrix}.
$$

See Tracy and Singh (1972). Given (12.3.5), it can be shown using (12.3.1) that

$$
B \underset{\mathrm{TS}}{\otimes} C = P\,(B \otimes C)\,Q^T
\qquad (12.3.6)
$$

where

$$P = \left(I_{M_1 M_2} \otimes \mathcal{P}_{m_2,m_1}\right)\left(I_{M_1} \otimes \mathcal{P}_{m_1, M_2 m_2}\right), \tag{12.3.7}$$

$$Q = \left(I_{N_1 N_2} \otimes \mathcal{P}_{n_2,n_1}\right)\left(I_{N_1} \otimes \mathcal{P}_{n_1, N_2 n_2}\right). \tag{12.3.8}$$

12.3.3   The Hadamard and Khatri-Rao Products

There are two submatrices of $B \otimes C$ that are particularly important. *The Hadamard Product* is a pointwise product:

$$B \underset{\mathbf{HAD}}{\otimes} C \;=\; B \mathbin{.*} C.$$

Thus, if $B \in \mathbb{R}^{m \times n}$ and $C \in \mathbb{R}^{m \times n}$, then

$$
\begin{bmatrix} b_{11} & b_{12} \\ b_{21} & b_{22} \\ b_{31} & b_{32} \end{bmatrix}
\underset{\mathbf{HAD}}{\otimes}
\begin{bmatrix} c_{11} & c_{12} \\ c_{21} & c_{22} \\ c_{31} & c_{32} \end{bmatrix}
=
\begin{bmatrix} b_{11}c_{11} & b_{12}c_{12} \\ b_{21}c_{21} & b_{22}c_{22} \\ b_{31}c_{31} & b_{32}c_{32} \end{bmatrix}.
$$

The block analog of this is the *Khatri-Rao Product*. If $B = (B_{ij})$ and $C = (C_{ij})$ are each m-by-n block matrices, then

$$B \underset{\mathbf{KR}}{\otimes} C \;=\; (A_{ij}), \qquad A_{ij} = B_{ij} \otimes C_{ij},$$

e.g.,

$$
\begin{bmatrix} B_{11} & B_{12} \\ B_{21} & B_{22} \\ B_{31} & B_{32} \end{bmatrix}
\underset{\mathbf{KR}}{\otimes}
\begin{bmatrix} C_{11} & C_{12} \\ C_{21} & C_{22} \\ C_{31} & C_{32} \end{bmatrix}
=
\begin{bmatrix} B_{11} \otimes C_{11} & B_{12} \otimes C_{12} \\ B_{21} \otimes C_{21} & B_{22} \otimes C_{22} \\ B_{31} \otimes C_{31} & B_{32} \otimes C_{32} \end{bmatrix}.
$$

A particularly important instance of the Khatri-Rao product is based on column partitionings:

$$
\begin{bmatrix} b_1 \mid \cdots \mid b_n \end{bmatrix}
\underset{\mathbf{KR}}{\otimes}
\begin{bmatrix} c_1 \mid \cdots \mid c_n \end{bmatrix}
=
\begin{bmatrix} b_1 \otimes c_1 \mid \cdots \mid b_n \otimes c_n \end{bmatrix}.
$$

For more details on the Khatri-Rao product, see Smilde, Bro, and Geladi (2004).

12.3.4   The Vec and Reshape Operations

In Kronecker product work, matrices are sometimes regarded as vectors and vectors are sometimes turned into matrices. To be precise about these reshapings, we remind the reader about the vec and reshape operations defined in §1.3.7. If $X \in \mathbb{R}^{m \times n}$, then $\mathsf{vec}(X)$ is an nm-by-1 vector obtained by "stacking" X's columns:

$$
\mathsf{vec}(X) \;=\;
\begin{bmatrix} X(:,1) \\ \vdots \\ X(:,n) \end{bmatrix}.
$$

If $B \in \mathbb{R}^{m_1 \times n_1}$, $C \in \mathbb{R}^{m_2 \times n_2}$, and $X \in \mathbb{R}^{n_1 \times m_2}$, then

$$Y = CXB^T \iff \text{vec}(Y) = (B \otimes C) \cdot \text{vec}(X). \tag{12.3.9}$$

Note that the matrix equation

$$F_1 X G_1^T + \cdots + F_p X G_p^T = C \tag{12.3.10}$$

is equivalent to

$$(G_1 \otimes F_1 + \cdots + G_p \otimes F_p) \, \text{vec}(X) = \text{vec}(C). \tag{12.3.11}$$

See Lancaster (1970), Vetter (1975), and also our discussion about block diagonalization in §7.6.3.

The reshape operation takes a vector and turns it into a matrix. If $a \in \mathbb{R}^{mn}$ then

$$A = \text{reshape}(a, m, n) \in \mathbb{R}^{m \times n} \quad \iff \quad \text{vec}(A) = a.$$

Thus, if $u \in \mathbb{R}^m$ and $v \in \mathbb{R}^n$, then $\text{reshape}(v \otimes u, m, n) = uv^T$.

12.3.5   Vec, Perfect Shuffles, and Transposition

There is an important connection between matrix transposition and perfect shuffle permutations. In particular, if $A \in \mathbb{R}^{q \times r}$, then

$$\text{vec}(A^T) = \mathcal{P}_{r,q} \text{vec}(A). \tag{12.3.12}$$

This formulation of matrix transposition provides a handy way to reason about large scale, multipass transposition algorithms that are required when $A \in \mathbb{R}^{q \times r}$ is too large to fit in fast memory. In this situation the transposition must proceed in stages and the overall process corresponds to a factorization of $\mathcal{P}_{r,q}$. For example, if

$$\mathcal{P}_{r,q} = \Gamma_t \cdots \Gamma_1 \tag{12.3.13}$$

where each Γ_k is a "data-motion-friendly" permutation, then $B = A^T$ can be computed with t passes through the data:

$a = \text{vec}(A)$

for $k = 1{:}t$

    $a = \Gamma_k a$

end

$B = \text{reshape}(a, q, r)$

The idea is to choose a factorization (12.3.13) so that the data motion behind the operation kth pass, i.e., $a \leftarrow \Gamma_k a$, is in harmony with the architecture of the underlying memory hierarchy, i.e., blocks that can fit in cache, etc.

As an illustration, suppose we want to assign A^T to B where

$$A = \begin{bmatrix} A_1 \\ \vdots \\ A_r \end{bmatrix}, \qquad A_k \in \mathbb{R}^{q \times q}.$$

We assume that A is stored by column which means that the A_i are not contiguous in memory. To complete the story, suppose each block comfortably fits in cache but that A cannot. Here is a 2-pass factorization of $\mathcal{P}_{rq,q}$:

$$\mathcal{P}_{q,rq} = \Gamma_2\Gamma_1 = (I_r \otimes \mathcal{P}_{q,q}) (\mathcal{P}_{r,q} \otimes I_q).$$

If $\tilde{a} = \Gamma_1 \cdot \mathsf{vec}(A)$, then

$$\mathsf{reshape}(a, q, rq) = \left[\begin{array}{c|c|c} A_1 & \cdots & A_r \end{array}\right]$$

In other words, after the first pass through the data we have computed the block transpose of A. (The A_i are now contiguous in memory.) To complete the overall task, we must transpose each of these blocks. If $b = \Gamma_2\tilde{a}$, then

$$B = \mathsf{reshape}(b, q, rq) = \left[\begin{array}{c|c|c} A_1^T & \cdots & A_r^T \end{array}\right].$$

See Van Loan (FFT) for more details about perfect shuffle factorizations and multipass matrix transposition algorithms.

12.3.6   The Kronecker Product SVD

Suppose $A \in \mathbb{R}^{m \times n}$ is given with $m = m_1 m_2$ and $n = n_1 n_2$. For these integer factorizations the *nearest Kronecker product* (NKP) problem involves minimizing

$$\phi(B, C) = \| A - B \otimes C \|_F \tag{12.3.14}$$

where $B \in \mathbb{R}^{m_1 \times n_1}$ and $C \in \mathbb{R}^{m_2 \times n_2}$. Van Loan and Pitsianis (1992) show how to solve the NKP problem using the singular value decomposition of a permuted version of A. A small example communicates the main idea. Suppose $m_1 = 3$ and $n_1 = m_2 = n_2 = 2$. By carefully thinking about the sum of squares that define ϕ, we see that

$$\phi(B, C) = \left\| \left[\begin{array}{cc|cc} a_{11} & a_{12} & a_{13} & a_{14} \\ a_{21} & a_{22} & a_{23} & a_{24} \\ \hline a_{31} & a_{32} & a_{33} & a_{34} \\ a_{41} & a_{42} & a_{43} & a_{44} \\ \hline a_{51} & a_{52} & a_{53} & a_{54} \\ a_{61} & a_{62} & a_{63} & a_{64} \end{array}\right] - \left[\begin{array}{cc} b_{11} & b_{12} \\ b_{21} & b_{22} \\ b_{31} & b_{32} \end{array}\right] \otimes \left[\begin{array}{cc} c_{11} & c_{12} \\ c_{21} & c_{22} \end{array}\right] \right\|_F$$

$$= \left\| \left[\begin{array}{cccc} a_{11} & a_{21} & a_{12} & a_{22} \\ a_{31} & a_{41} & a_{32} & a_{42} \\ a_{51} & a_{61} & a_{52} & a_{62} \\ a_{13} & a_{23} & a_{14} & a_{24} \\ a_{33} & a_{43} & a_{34} & a_{44} \\ a_{53} & a_{63} & a_{54} & a_{64} \end{array}\right] - \left[\begin{array}{c} b_{11} \\ b_{21} \\ b_{31} \\ b_{12} \\ b_{22} \\ b_{32} \end{array}\right] \left[\begin{array}{cccc} c_{11} & c_{21} & c_{12} & c_{22} \end{array}\right] \right\|_F .$$

Denote the preceding 6-by-4 matrix by $\mathcal{R}(A)$ and observe that

$$\mathcal{R}(A) = \begin{bmatrix} \mathrm{vec}(A_{11})^T \\ \mathrm{vec}(A_{21})^T \\ \mathrm{vec}(A_{31})^T \\ \mathrm{vec}(A_{12})^T \\ \mathrm{vec}(A_{22})^T \\ \mathrm{vec}(A_{32})^T \end{bmatrix}.$$

It follows that

$$\phi(B,C) = \left\| \, \mathcal{R}(A) - \mathrm{vec}(B)\mathrm{vec}(C)^T \, \right\|_F$$

and so the act of minimizing ϕ is equivalent to finding a nearest rank-1 matrix to $\mathcal{R}(A)$. This problem has a simple SVD solution. Referring to Theorem 2.4.8, if

$$U^T \mathcal{R}(A) V = \Sigma \tag{12.3.15}$$

is the SVD of $\mathcal{R}(A)$, then the optimizing B and C are defined by

$$\mathrm{vec}(B_{\mathrm{opt}}) = \sqrt{\sigma_1}\, U(:,1), \qquad \mathrm{vec}(C_{\mathrm{opt}}) = \sqrt{\sigma_1}\, V(:,1).$$

The scalings are arbitrary. Indeed, if B_{opt} and C_{opt} solve the NKP problem and $\alpha \neq 0$, then $\alpha \cdot B_{\mathrm{opt}}$ and $(1/\alpha) \cdot C_{\mathrm{opt}}$ are also optimal.

In general, if

$$A = \begin{bmatrix} A_{11} & \cdots & A_{1,n_1} \\ \vdots & \ddots & \vdots \\ A_{m_1,1} & \cdots & A_{m_1,n_1} \end{bmatrix} \tag{12.3.16}$$

where each $A_{ij} \in \mathbb{R}^{m_2 \times n_2}$, then $\mathcal{R}(A) \in \mathbb{R}^{m_1 n_1 \times m_2 n_2}$ is defined by

$$\mathcal{R}(A) = \begin{bmatrix} \tilde{A}_1 \\ \vdots \\ \tilde{A}_{n_1} \end{bmatrix}, \qquad \tilde{A}_j = \begin{bmatrix} \mathrm{vec}(A_{1j})^T \\ \vdots \\ \mathrm{vec}(A_{m_1,j})^T \end{bmatrix}.$$

The SVD of $\mathcal{R}(A)$ can be "reshaped" into a special SVD-like expansion for A.

Theorem 12.3.1 (Kronecker Product SVD). *If $A \in \mathbb{R}^{m_1 m_2 \times n_1 n_2}$ is blocked according to (12.3.16) and*

$$\mathcal{R}(A) = U\Sigma V^T = \sum_{k=1}^{r} \sigma_k \cdot u_k v_k^T \tag{12.3.17}$$

is the SVD of $\mathcal{R}(A)$ with $u_k = U(:,k)$, $v_k = V(:,k)$, and $\sigma_k = \Sigma(k,k)$, then

$$A = \sum_{k=1}^{r} \sigma_k \cdot U_k \otimes V_k \tag{12.3.18}$$

where $U_k = \mathrm{reshape}(u_k, m_1, n_1)$ and $V_k = \mathrm{reshape}(v_k, m_2, n_2)$.

Proof. In light of (12.3.18), we must show that

$$A_{ij} = \sum_{k=1}^{r} \sigma_k \cdot U_k(i,j) \cdot V_k.$$

But this follows immediately from (12.3.17) which says that

$$\text{vec}(A_{ij})^T = \sum_{k=1}^{r} \sigma_k \cdot U_k(i,j) v_k^T$$

for all i and j.   □

The integer r in the theorem is the *Kronecker product rank* of A given the blocking (12.3.16). Note that if $\tilde{r} \le r$, then

$$A_{\tilde{r}} = \sum_{k=1}^{\tilde{r}} \sigma_k \, U_k \otimes V_k \qquad\qquad (12.3.19)$$

is the closest matrix to A (in the Frobenius norm) that is the sum of $\tilde{r}$ Kronecker products. If A is large and sparse and $\tilde{r}$ is small, then the Lanzcos SVD iteration can effectively be used to compute the required singular values and vectors of $\mathcal{R}(A)$. See §10.4.

12.3.7   Constrained NKP Problems

If A is structured, then it is sometimes the case that the B and C matrices that solve the NKP problem are similarly structured. For example, if A is symmetric and positive definite, then the same can be said of B_{opt} and C_{opt} (if properly normalized). Likewise, if A is nonnegative, then the optimal B and C can be chosen to be nonnegative. These and other structured NKP problems are discussed in Van Loan and Pitsianis (1992).

We mention that a problem like

$$\min_{B,\,C \text{ Toeplitz}} \| A - B \otimes C \|_F, \qquad B \in \mathbb{R}^{m \times m},\, C \in \mathbb{R}^{n \times n},$$

turns into a constrained nearest rank-1 problem of the form

$$\min_{\substack{F^T \text{vec}(B) = 0 \\ G^T \text{vec}(C) = 0}} \| \mathcal{A} - bc^T \|_F$$

where the nullspaces of F^T and G^T define the vector space of m-by-m and n-by-n Toeplitz matrices respectively. This problem can be solved by computing QR factorizations of F and G followed by a reduced-dimension SVD.

12.3.8 Computing the Nearest $X \otimes X$

Suppose $A \in \mathbb{R}^{m^2 \times m^2}$ and that we want to find $X \in \mathbb{R}^{m \times m}$ so that

$$\phi_{\mathrm{sym}}(X) = \| A - X \otimes X \|_F$$

is minimized. Proceeding as we did with the NKP problem, we can reshape this into a nearest symmetric rank-1 problem:

$$\phi_{\mathrm{sym}}(X) = \| \mathcal{R}(A) - \mathrm{vec}(X) \cdot \mathrm{vec}(X)^T \|_F. \tag{12.3.20}$$

It turns out that the solution X_{opt} is a reshaping of an eigenvector associated with the symmetric part of $\mathcal{R}(A)$.

Lemma 12.3.2. *Suppose $M \in \mathbb{R}^{n \times n}$ and that $Q^T T Q = \mathrm{diag}(\alpha_1, \ldots, \alpha_n)$ is a Schur decomposition of $T = (M + M^T)/2$. If*

$$|\alpha_k| = \max\{|\alpha_1|, \ldots, |\alpha_n|\}$$

then the solution to the problem

$$\min_{\substack{Z = Z^T \\ \mathrm{rank}(Z) = 1}} \| M - Z \|_F$$

is given by $Z_{\mathrm{opt}} = \alpha_k q_k q_k^T$ where $q_k = Q(:, k)$.

Proof. See P12.3.11. $\square$

12.3.9 Computing the Nearest $X \otimes Y - Y \otimes X$

Suppose $A \in \mathbb{R}^{n \times n}$, $n = m^2$ and that we wish to find $X, Y \in \mathbb{R}^{m \times m}$ so that

$$\phi_{\mathrm{skew}}(X, Y) = \| A - (X \otimes Y - Y \otimes X) \|_F$$

is minimized. It can be shown that

$$\phi_{\mathrm{skew}}(X) = \| \mathcal{R}(A) - (\mathrm{vec}(X) \cdot \mathrm{vec}(Y)^T - \mathrm{vec}(Y) \cdot \mathrm{vec}(X)^T \|_F. \tag{12.3.21}$$

The optimizing X and Y can be determined by exploiting the following lemma.

Lemma 12.3.3. *Suppose $M \in \mathbb{R}^{n \times n}$ with skew-symmetric part $S = (M - M^T)/2$. If*

$$S[u \mid v] = [u \mid v] \begin{bmatrix} 0 & \mu \\ -\mu & 0 \end{bmatrix}, \qquad u, v \in \mathbb{R}^n,$$

with $\mu = \rho(S)$, $\| u \|_2 = \| v \|_2 = 1$, and $u^T v = 0$, then $Z_{\mathrm{opt}} = \mu \left(uv^T - vu^T \right)$ minimizes $\| M - Z \|_F$ over all rank-2 skew-symmetric matrices $Z \in \mathbb{R}^{n \times n}$.

Proof. See P12.3.12. $\square$

12.3.10   Some Comments About Multiple Kronecker Products

The Kronecker product of three or more matrices results in a matrix that has a recursive block structure. For example,

$$
B \otimes C \otimes D =
\begin{bmatrix} b_{11} & b_{12} \\ b_{21} & b_{22} \end{bmatrix}
\otimes
\begin{bmatrix}
c_{11} & c_{12} & c_{13} & c_{14} \\
c_{21} & c_{22} & c_{23} & c_{24} \\
c_{31} & c_{32} & c_{33} & c_{34} \\
c_{41} & c_{42} & c_{43} & c_{44}
\end{bmatrix}
\otimes
\begin{bmatrix}
d_{11} & d_{12} & d_{13} \\
d_{21} & d_{22} & d_{23} \\
d_{31} & d_{32} & d_{33}
\end{bmatrix}
$$

is a 2-by-2 block matrix whose entries are 4-by-4 block matrices whose entries are 3-by-3 matrices.

A Kronecker product can be regarded as a data-sparse representation. If $A = B_1 \otimes B_2$ and each B-matrix is m-by-m, then $2m^2$ numbers are used to encode a matrix that has m^4 entries. The data sparsity is more dramatic for multiple Kronecker products. If $A = B_1 \otimes \cdots \otimes B_p$ and $B_i \in \mathbb{R}^{m \times m}$, then pm^2 numbers fully describe A, a matrix with m^{2p} entries.

Order of operation can be important when a multiple Kronecker product is involved and the participating matrices vary in dimension. Suppose $B_i \in \mathbb{R}^{m_i \times n_i}$ for $i = 1{:}p$ and that $M_i = m_1 \cdots m_i$ and $N_i = n_1 \cdots n_i$ for $i = 1{:}p$. The matrix-vector product

$$
y = (B_1 \otimes \cdots B_p)x \qquad x \in \mathbb{R}^{N_p}
$$

can be evaluated in many different orders and the associated flop counts can vary tremendously. The search for an optimal ordering is a dynamic programming problem that involves the recursive analysis of calculations like

$$
\mathsf{reshape}(y, M_p/M_i, M_i) = (B_{i+1} \otimes \cdots \otimes B_p) \cdot \mathsf{reshape}(x, N_p/N_i, N_i) \cdot (B_1 \otimes \cdots B_i)^T.
$$

Problems

P12.3.1 Prove (12.3.1) and (12.3.2).

P12.3.2 Assume that the matrices $A_1, \ldots, A_N \in \mathbb{R}^{m \times n}$. Express the summation

$$
f(x, y) = \sum_{k=1}^{N} (y^T A_k x - b_k)^2
$$

in matrix-vector terms given that $y \in \mathbb{R}^m$, $x \in \mathbb{R}^m$, and $b \in \mathbb{R}^N$.

P12.3.3   A total least squares solution to $(B \otimes C)x \approx b$ requires the computation of the smallest singular value and the associated right singular vector of the augmented matrix $M = [\, B \otimes C \,|\, b \,]$. Outline an efficient procedure for doing this that exploits the Kronecker structure of the data matrix.

P12.3.4   Show how to minimize $\| (A_1 \otimes A_2)x - f \|$ subject to the constraint that $(B_1 \otimes B_2)x = g$. Assume that A_1 and A_2 have more rows than columns and that B_1 and B_2 have more columns than rows. Also assume that each of these four matrices has full rank. See Barrlund (1998).

P12.3.5   Suppose $B \in \mathbb{R}^{n \times n}$ and $C \in \mathbb{R}^{m \times m}$ are unsymmetric and positive definite. Does it follow that $B \otimes C$ is positive definite?

P12.3.6   Show how to construct the normalized SVD of $B \otimes C$ from the normalized SVDs of B and C. Assume that $B \in \mathbb{R}^{m_B \times n_B}$ and $C \in \mathbb{R}^{m_C \times n_C}$ with $m_B \geq n_B$ and $m_C \geq n_C$.

P12.3.7   Show how to solve the linear system $(A \otimes B \otimes C)x = d$ assuming that $A, B, C \in \mathbb{R}^{n \times n}$ are symmetric positive definite.

P12.3.8 (a) Given $A \in \mathbb{R}^{mn \times mn}$ and $B \in \mathbb{R}^{m \times m}$, how would you compute $X \in \mathbb{R}^{n \times n}$ so that

$$\phi_B(X) = \| A - B \otimes X \|_F$$

is minimized? (b) Given $A \in \mathbb{R}^{mn \times mn}$ and $C \in \mathbb{R}^{n \times n}$, how would you compute $X \in \mathbb{R}^{m \times m}$ so that

$$\phi_C(X) = \| A - X \otimes C \|_F$$

is minimized?

P12.3.9 What is the nearest Kronecker product to the matrix $A = I_n \otimes \mathcal{T}_m^{DD} + \mathcal{T}_n^{DD} \otimes I_n$ where $\mathcal{T}_k^{DD}$ is defined in (4.8.7).

P12.3.10 If $A \in \mathbb{R}^{mn \times mn}$ is symmetric and tridiagonal, show how to minimize $\| A - B \otimes C \|_F$ subject to the constraint that $B \in \mathbb{R}^{m \times m}$ and $C \in \mathbb{R}^{n \times n}$ are symmetric and tridiagonal.

P12.3.11 Prove Lemma 12.3.2. Hint: Show

$$\| M - \alpha x x^T \|_F^2 = \| M \|_F^2 - 2\alpha x^T T x + \alpha^2$$

where $T = (M + M^T)/2$.

P12.3.12 Prove Lemma 12.3.3. Hint: Show

$$\| M - (xy^T - yx^T) \|_F^2 = \| M \|_F^2 + 2\| x \|_2^2 \| y \|_2^2 - 2(x^T y)^2 - 4x^T S y$$

where $S = (M - M^T)/2$ and use the real Schur form of S.

P12.3.13 For a symmetric matrix $S \in \mathbb{R}^{n \times n}$, the *symmetric vec operation* is fully defined by

$$S = \begin{bmatrix} s_{11} & s_{12} & s_{13} \\ s_{21} & s_{22} & s_{23} \\ s_{31} & s_{32} & s_{33} \end{bmatrix} \Rightarrow \mathsf{svec}(S) = \begin{bmatrix} s_{11} & \sqrt{2}\,s_{21} & \sqrt{2}\,s_{31} & s_{22} & \sqrt{2}\,s_{32} & s_{33} \end{bmatrix}^T.$$

For symmetric $X \in \mathbb{R}^{n \times n}$ and arbitrary $B, C \in \mathbb{R}^{n \times n}$, the *symmetric Kronecker product* is defined by

$$(B \underset{\mathbf{SYM}}{\otimes} C) \cdot \mathsf{svec}(X) = \mathsf{svec}\left(\frac{1}{2} \left(CXB^T + BXC^T \right) \right).$$

For the case $n = 3$, show that there is a matrix $P \in \mathbb{R}^{9 \times 6}$ with orthonormal columns so that $P^T (B \otimes C) P = B \underset{\mathbf{SYM}}{\otimes} C$. See Vandenberge and Boyd (1996).

P12.3.14 The *bi-alternate product* is defined by

$$B \underset{\mathbf{BI}}{\otimes} C = \frac{1}{2} (B \otimes C + C \otimes B).$$

If $B = I$, $C = A$, then solutions to $AX + XA^T = H$ where H is symmetric or skew-symmetric shed light on A's eigenvalue placement. See Govaerts (2000). Given a matrix M, show how to compute the nearest bi-alternate product to M.

P12.3.15 Given $f \in \mathbb{R}^q$ and $g_i \in \mathbb{R}^{p_i}$ for $i = 1{:}m$, determine a permutation P so that

$$P \left(f \otimes \begin{bmatrix} g_1 \\ \vdots \\ g_m \end{bmatrix} \right) = \begin{bmatrix} f \otimes g_1 \\ \vdots \\ f \otimes g_m \end{bmatrix}.$$

Hint: What does (12.3.1) say when B and C are vectors?

Notes and References for §12.3

The history of the Kronecker product (including why it might better be called the "Zehfuss product") is discussed in:

H.V. Henderson, F. Pukelsheim, and S.R. Searle (1983). "On the History of the Kronecker Product," *Lin. Mult. Alg. 14*, 113–120.

For general background on the operation, see:

F. Stenger (1968), "Kronecker Product Extensions of Linear Operators," *SIAM J. Numer. Anal. 5*, 422–435.

J.W. Brewer (1978). "Kronecker Products and Matrix Calculus in System Theory," *IEEE Trans. Circuits Syst. 25*, 772–781.

A. Graham (1981). *Kronecker Products and Matrix Calculus with Applications*, Ellis Horwood, Chichester, England.

M. Davio (1981), "Kronecker Products and Shuffle Algebra," *IEEE Trans. Comput. c-30*, 116–125.

H.V. Henderson and S.R. Searle (1981). "The Vec-Permutation Matrix, The Vec Operator and Kronecker Products: A Review," *Lin. Multilin. Alg. 9*, 271–288.

H.V. Henderson and S.R. Searle(1998). "Vec and Vech Operators for Matrices, with Some uses in Jacobians and Multivariate Statistics," *Canadian J. of Stat. 7*, 65–81.

C. Van Loan (2000). "The Ubiquitous Kronecker Product," *J. Comput. and Appl. Math. 123*, 85–100.

References concerned with various KP-like operations include:

C.R. Rao and S.K. Mitra (1971). *Generalized Inverse of Matrices and Applications*, John Wiley and Sons, New York.

D.S. Tracy and R.P. Singh (1972). "A New Matrix Product and Its Applications in Partitioned Matrices," *Statistica Neerlandica 26*, 143–157.

P.A. Regalia and S. Mitra (1989). "Kronecker Products, Unitary Matrices, and Signal Processing Applications," *SIAM Review 31*, 586–613.

J. Seberry and X-M Zhang (1993). "Some Orthogonal Matrices Constructed by Strong Kronecker Product Multiplication," *Austral. J. Combin. 7*, 213–224.

W. De Launey and J. Seberry (1994), "The Strong Kronecker Product," *J. Combin. Theory, Ser. A 66*, 192–213.

L. Vandenberghe and S. Boyd (1996). "Semidefinite Programming," *SIAM Review 38*, 27–48.

W. Govaerts (2000). *Numerical Methods for Bifurcations of Dynamical Equilibria*, SIAM Publications, Philadelphia, PA.

A. Smilde, R. Bro, and P. Geladi (2004). *Multiway Analysis*, John Wiley, Chichester, England.

For background on the KP connection to Sylvester-type equations, see:

P. Lancaster (1970). "Explicit Solution of Linear Matrix Equations," *SIAM Review 12*, 544–566.

W.J. Vetter (1975). "Vector Structures and Solutions of Linear Matrix Equations," *Lin. Alg. Applic. 10*, 181–188.

Issues associated with the efficient implementation of KP operations are discussed in:

H.C. Andrews and J. Kane (1970). "Kronecker Matrices, Computer Implementation, and Generalized Spectra," *J. Assoc. Comput. Mach. 17*, 260–268.

V. Pereyra and G. Scherer (1973). "Efficient Computer Manipulation of Tensor Products with Applications to Multidimensional Approximation," *Math. Comput. 27*, 595–604.

C. de Boor (1979). "Efficient Computer Manipulation of Tensor Products," *ACM Trans. Math. Softw. 5*, 173–182.

P.E. Buis and W.R. Dyksen (1996). "Efficient Vector and Parallel Manipulation of Tensor Products," *ACM Trans. Math. Softw. 22*, 18–23.

P.E. Buis and W.R. Dyksen (1996). "Algorithm 753: TENPACK: An LAPACK-based Library for the Computer Manipulation of Tensor Products," *ACM Trans. Math. Softw. 22*, 24–29.

W-H. Steeb (1997). *Matrix Calculus and Kronecker Product with Applications and C++ Programs*, World Scientific Publishing, Singapore.

M. Huhtanen (2006). "Real Linear Kronecker Product Operations," *Lin. Alg. Applic. 417*, 347–361.

The KP is associated with the vast majority fast linear transforms. See Van Loan (FFT) as well as:

C-H Huang, J.R. Johnson, and R.W. Johnson (1991). "Multilinear Algebra and Parallel Programming," *J. Supercomput. 5*, 189–217.

J. Granata, M. Conner, and R. Tolimieri (1992). "'Recursive Fast Algorithms and the Role of the Tensor Product," *IEEE Trans. Signal Process. 40*, 2921–2930.

J. Granata, M. Conner, and R. Tolimieri (1992). "The Tensor Product: A Mathematical Programming Language for FFTs and Other Fast DSP Operations," *IEEE SP Magazine, January*, 40–48.

For a discussion of the role of KP approximation in a variety of situations, see:

C. Van Loan and N.P Pitsianis (1992). "Approximation with Kronecker Products", in *Linear Algebra for Large Scale and Real Time Applications,* M.S. Moonen and G.H. Golub (eds.), Kluwer Publications, Dordrecht, 293–314,

T.F. Andre, R.D. Nowak, and B.D. Van Veen (1997). "Low Rank Estimation of Higher Order Statistics," *IEEE Trans. Signal Process. 45,* 673–685.

R.D. Nowak and B. Van Veen (1996). "Tensor Product Basis Approximations for Volterra Filters," *IEEE Trans. Signal Process. 44,* 36–50.

J. Kamm and J.G. Nagy (1998). "Kronecker Product and SVD Approximations in Image Restoration," *Lin. Alg. Applic. 284,* 177–192.

J.G. Nagy and D.P. O'Leary (1998). "Restoring Images Degraded by Spatially Variant Blur," *SIAM J. Sci. Comput. 19,* 1063–1082.

J. Kamm and J.G. Nagy (2000). "Optimal Kronecker Product Approximation of Block Toeplitz Matrices," *SIAM J. Matrix Anal. Applic. 22,* 155–172.

J.G. Nagy, M.K. Ng, and L. Perrone (2003). "Kronecker Product Approximations for Image Restoration with Reflexive Boundary Conditions," *SIAM J. Matrix Anal. Applic. 25,* 829–841.

A.N. Langville and W.J. Stewart (2004). "A Kronecker Product Approximate Preconditioner for SANs," *Num. Lin. Alg. 11,* 723–752.

E. Tyrtyshnikov (2004). "Kronecker-Product Approximations for Some Function-Related Matrices," *Lin. Alg. Applic. 379,* 423–437.

L. Perrone (2005). "Kronecker Product Approximations for Image Restoration with Anti-Reflective Boundary Conditions," *Num. Lin. Alg. 13,* 1–22.

W. Hackbusch, B.N. Khoromskij, and E.E. Tyrtyshnikov (2005). "Hierarchical Kronecker Tensor-Product Approximations," *J. Numer. Math. 13,* 119–156.

V. Olshevsky, I. Oseledets, and E. Tyrtyshnikov (2006). "Tensor Properties of Multilevel Toeplitz and Related matrices," *Lin. Alg. Applic. 412,* 1–21.

J. Leskovec and C. Faloutsos (2007). "Scalable Modeling of Real Graphs Using Kronecker Multiplication," in *Proc. of the 24th International Conference on Machine Learning,* Corvallis, OR.

J. Leskovic (2011). "Kronecker Graphs," in *Graph Algorithms in the Language of Linear Algebra,* J. Kepner and J. Gilbert (eds), SIAM Publications, Philadelphia, PA, 137–204.

For a snapshot of KP algorithms for linear systems and least squares problems, see:

H. Sunwoo (1996). "Simple Algorithms about Kronecker Products in the Linear Model," *Lin. Alg. Applic. 237–8,* 351–358.

D.W. Fausett, C.T. Fulton, and H. Hashish (1997). "Improved Parallel QR Method for Large Least Squares Problems Involving Kronecker Products," *J. Comput. Appl. Math. 78,* 63–78.

A. Barrlund (1998). "Efficient Solution of Constrained Least Squares Problems with Kronecker Product Structure," *SIAM J. Matrix Anal. Applic. 19,* 154–160.

P. Buchholz and T.R. Dayar (2004). "Block SOR for Kronecker Structured Representations," *Lin. Alg. Applic. 386,* 83–109.

A.W. Bojanczyk and A. Lutoborski (2003). "The Procrustes Problem for Orthogonal Kronecker Products," *SIAM J. Sci. Comput. 25,* 148–163.

C.D.M. Martin and C.F. Van Loan (2006). "Shifted Kronecker Product Systems," *SIAM J. Matrix Anal. Applic. 29,* 184–198.

12.4 Tensor Unfoldings and Contractions

An *order-d tensor* $\mathcal{A} \in \mathbb{R}^{n_1 \times \cdots \times n_d}$ is a real d-dimensional array $\mathcal{A}(1{:}n_1, \ldots, 1{:}n_d)$ where the index range in the kth *mode* is from 1 to n_k. Low-order examples include scalars (order-0), vectors (order-1), and matrices (order-2). Order-3 tensors can be visualized as "Rubik cubes of data," although the dimensions do not have to be equal along each mode. For example, $\mathcal{A} \in \mathbb{R}^{m \times n \times 3}$ might house the red, green, and blue pixel data for an m-by-n image, a "stacking" of three matrices. In many applications, a tensor is used to capture what a multivariate function looks like on a lattice of points, e.g., $\mathcal{A}(i, j, k, \ell) \approx f(w_i, x_j, y_k, z_\ell)$. The function f could be the solution to a complicated partial differential equation or a general mapping from some high-dimensional space of input values to a measurement that is acquired experimentally.

Because of their higher dimension, tensors are harder to reason about than matrices. Notation, which is always important, is critically important in tensor computations where vectors of subscripts and deeply nested summations are the rule. In this section we examine some basic tensor operations and develop a handy, matrix type of notation that can be used to describe them. Kronecker products are central.

Excellent background references include De Lathauwer (1997), Smilde, Bro, and Geladi (2004), and Kolda and Bader (2009).

12.4.1   Unfoldings and Contractions: A Preliminary Look

To *unfold* a tensor is to systematically arrange its entries into a matrix.[3] Here is one possible unfolding of a 2-by-2-by-3-by-4 tensor:

$$
A \;=\;
\left[
\begin{array}{cc|cc|cc|cc}
a_{1111}\ a_{1211} & & a_{1112}\ a_{1212} & & a_{1113}\ a_{1213} & & a_{1114}\ a_{1214} \\
a_{2111}\ a_{2211} & & a_{2112}\ a_{2212} & & a_{2113}\ a_{2213} & & a_{2114}\ a_{2214} \\
\hline
a_{1121}\ a_{1221} & & a_{1122}\ a_{1222} & & a_{1123}\ a_{1223} & & a_{1124}\ a_{1224} \\
a_{2121}\ a_{2221} & & a_{2122}\ a_{2222} & & a_{2123}\ a_{2223} & & a_{2124}\ a_{2224} \\
\hline
a_{1131}\ a_{1231} & & a_{1132}\ a_{1232} & & a_{1133}\ a_{1233} & & a_{1134}\ a_{1234} \\
a_{2131}\ a_{2231} & & a_{2132}\ a_{2232} & & a_{2133}\ a_{2233} & & a_{2134}\ a_{2234}
\end{array}
\right].
$$

Order-4 tensors are interesting because of their connection to block matrices. Indeed, a block matrix $A = (A_{k\ell})$ with equally sized blocks can be regarded as an order-4 tensor $\mathcal{A} = (a_{ijk\ell})$ where $[A_{k\ell}]_{ij} = a_{ijk\ell}$.

Unfoldings have an important role to play in tensor computations for three reasons. (1) Operations between tensors can often be reformulated as a matrix computation between unfoldings. (2) Iterative multilinear optimization strategies for tensor problems typically involve one or more unfoldings per step. (3) Hidden structures within a tensor dataset can sometimes be revealed by discovering patterns within its unfoldings. For these reasons, it is important to develop a facility with tensor unfoldings because they serve as a bridge between matrix computations and tensor computations

Operations between tensors typically involve vectors of indices and deeply nested loops. For example, here is a matrix-multiplication-like computation that combines two order-4 tensors to produce a third order-4 tensor:

for $i_1 = 1{:}n$
    for $i_2 = 1{:}n$
        for $i_3 = 1{:}n$
            for $i_4 = 1{:}n$

$$
\mathcal{C}(i_1, i_2, i_3, i_4) \;=\; \sum_{p=1}^{n} \sum_{q=1}^{n} \mathcal{A}(i_1, p, i_3, q) \mathcal{B}(p, i_2, q, i_4) \qquad (12.4.1)
$$

            end
        end
    end
end

---

[3]The process is sometimes referred to as a *tensor flattening* or a *tensor matricization*.

This is an example of a *tensor contraction*. Tensor contractions are essentially reshaped, multi-indexed matrix multiplications and can be very expensive to compute. (The above example involves $O(n^6)$ flops.) It is increasingly common to have $O(n^d)$ contraction bottlenecks in a simulation. In order to successfully tap into the "culture" of of high-performance matrix computations, it is important to have an intuition about tensor contractions and how they can be organized.

12.4.2 Notation and Definitions

If $\mathcal{A} \in \mathbb{R}^{n_1 \times \cdots \times n_d}$ and $\mathbf{i} = (i_1, \ldots, i_d)$ with $1 \leq i_k \leq n_k$ for $k = 1{:}d$, then

$$\mathcal{A}(\mathbf{i}) \equiv \mathcal{A}(i_1, \ldots, i_k).$$

The vector $\mathbf{i}$ is a *subscript vector*. Bold font is used designate subscript vectors while calligraphic font is used for tensors. For low-order tensors we sometimes use matrix-style subscripting, e.g., $\mathcal{A} = (a_{ijk\ell})$. It is sometimes instructive to write $\mathcal{A}(\mathbf{i}, \mathbf{j})$ for $\mathcal{A}([\,\mathbf{i}\,\mathbf{j}\,])$. Thus,

$$\mathcal{A}([\,2\,5\,3\,4\,7\,]) = \mathcal{A}(2,5,3,4,7) = a_{25347} = a_{253,47} = \mathcal{A}([2,5,3],[4,7])$$

shows the several ways that we can refer to a tensor entry.

We extend the MATLAB colon notation in order to identify subtensors. If $\mathbf{L}$ and $\mathbf{R}$ are subscript vectors with the same dimension, then $\mathbf{L} \leq \mathbf{R}$ means that $L_k \leq R_k$ for all k. The length-d subscript vector of all 1's is denoted by $\mathbf{1}_d$. If the dimension is clear from the context, then we just write $\mathbf{1}$. Suppose $\mathcal{A} \in \mathbb{R}^{n_1 \times \cdots \times n_d}$ with $\mathbf{n} = [n_1, \ldots, n_d]$. If $\mathbf{1} \leq \mathbf{L} \leq \mathbf{R} \leq \mathbf{n}$, then $\mathcal{A}(\mathbf{L}{:}\mathbf{R})$ denotes the subtensor

$$B = \mathcal{A}(L_1{:}R_1, \ldots, L_d{:}R_d).$$

Just as we can extract an order-1 tensor from an order-2 tensor, e.g., $A(:,k)$, so can we extract a lower-order tensor from a given tensor. Thus, if $\mathcal{A} \in \mathbb{R}^{2 \times 3 \times 4 \times 5}$, then

(i) $\mathcal{B} = \mathcal{A}(1, :, 2, 4) \in \mathbb{R}^3$ $\Rightarrow$ $\mathcal{B}(i_2) = \mathcal{A}(1, i_2, 2, 4)$,

(ii) $\mathcal{B} = \mathcal{A}(1, :, 2, :) \in \mathbb{R}^{3 \times 5}$ $\Rightarrow$ $\mathcal{B}(i_2, i_4) = \mathcal{A}(1, i_2, 2, i_4)$,

(iii) $\mathcal{B} = \mathcal{A}(:, :, 2, :) \in \mathbb{R}^{2 \times 3 \times 5}$ $\Rightarrow$ $\mathcal{B}(i_1, i_2, i_4) = \mathcal{A}(i_1, i_2, 2, i_4)$.

Order-1 extractions like (i) are called *fibers*. Order-2 extractions like (ii) are called *slices*. More general extractions like (iii) are called *subtensors*.

It is handy to have a multi-index summation notation. If $\mathbf{n}$ is a length-d index vector, then

$$\sum_{\mathbf{i}=\mathbf{1}}^{\mathbf{n}} \equiv \sum_{i_1=1}^{n_1} \cdots \sum_{i_d=1}^{n_d}.$$

Thus, if $\mathcal{A} \in \mathbb{R}^{n_1 \times \cdots \times n_d}$, then its Frobenius norm is given by

$$\| \mathcal{A} \|_F = \sqrt{\sum_{\mathbf{i}=\mathbf{1}}^{\mathbf{n}} \mathcal{A}(\mathbf{i})^2}.$$

12.4.3   The Vec Operation for Tensors

As with matrices, the $\text{vec}(\cdot)$ operator turns tensors into column vectors, e.g.,

$$
\mathcal{A} \in \mathbb{R}^{2 \times 3 \times 2} \quad \Longrightarrow \quad \text{vec}(\mathcal{A}) =
\begin{bmatrix}
\mathcal{A}(:,1,1) \\
\hline
\mathcal{A}(:,2,1) \\
\hline
\mathcal{A}(:,3,1) \\
\hline
\mathcal{A}(:,1,2) \\
\hline
\mathcal{A}(:,2,2) \\
\hline
\mathcal{A}(:,3,2)
\end{bmatrix}
=
\begin{bmatrix}
a_{111} \\
a_{211} \\
a_{121} \\
a_{221} \\
a_{131} \\
a_{231} \\
a_{112} \\
a_{212} \\
a_{122} \\
a_{222} \\
a_{132} \\
a_{232}
\end{bmatrix}.
$$

Formally, if $\mathcal{A} \in \mathbb{R}^{n_1 \times \cdots \times n_d}$, then

$$
\text{vec}(\mathcal{A}) =
\begin{bmatrix}
\text{vec}(\mathcal{A}^{(1)}) \\
\vdots \\
\text{vec}(\mathcal{A}^{(n_d)})
\end{bmatrix}
\tag{12.4.2}
$$

where $\mathcal{A}^{(k)} \in \mathbb{R}^{n_1 \times \cdots \times n_{d-1}}$ is defined by

$$
\mathcal{A}^{(k)}(i_1, \ldots, i_{d-1}) = \mathcal{A}(i_1, \ldots, i_{d-1}, k)
\tag{12.4.3}
$$

for $k = 1{:}n_d$. Alternatively, if we define the integer-valued function col by

$$
\text{col}(\mathbf{i}, \mathbf{n}) = i_1 + (i_2 - 1)n_1 + (i_3 - 1)n_1 n_2 + \cdots + (i_d - 1)n_1 \cdots n_{d-1},
\tag{12.4.4}
$$

then $a = \text{vec}(\mathcal{A})$ is specified by

$$
a(\text{col}(\mathbf{i}, \mathbf{n})) = \mathcal{A}(\mathbf{i}), \qquad 1 \leq \mathbf{i} \leq \mathbf{n}.
\tag{12.4.5}
$$

12.4.4   Tensor Transposition

If $\mathcal{A} \in \mathbb{R}^{n_1 \times n_2 \times n_3}$, then there are $6 = 3!$ possible transpositions identified by the notation $\mathcal{A}^{< [i\,j\,k] >}$ where $[i\,j\,k]$ is a permutation of $[1\,2\,3]$:

$$
\mathcal{B} =
\left\{
\begin{array}{l}
\mathcal{A}^{< [1\,2\,3] >} \\
\mathcal{A}^{< [1\,3\,2] >} \\
\mathcal{A}^{< [2\,1\,3] >} \\
\mathcal{A}^{< [2\,3\,1] >} \\
\mathcal{A}^{< [3\,1\,2] >} \\
\mathcal{A}^{< [3\,2\,1] >}
\end{array}
\right\}
\quad \Longrightarrow \quad
\left\{
\begin{array}{l}
b_{ijk} \\
b_{ikj} \\
b_{jik} \\
b_{jki} \\
b_{kij} \\
b_{kji}
\end{array}
\right\}
= a_{ijk}.
$$

These transpositions can be defined using the perfect shuffle and the **vec** operator. For example, if $\mathcal{B} = \mathcal{A}^{<[3\ 2\ 1]>}$, then $\text{vec}(\mathcal{B}) = (\mathcal{P}_{n_1,n_2} \otimes I_{n_3})\mathcal{P}_{n_1 n_2, n_3} \cdot \text{vec}(\mathcal{A})$.

In general, if $\mathcal{A} \in \mathbb{R}^{n_1 \times \cdots \times n_d}$ and $\mathbf{p} = [p_1, \ldots, p_d]$ is a permutation of the index vector $1{:}d$, then $\mathcal{A}^{<\mathbf{p}>} \in \mathbb{R}^{n_{p_1} \times \cdots \times n_{p_d}}$ is the $\mathbf{p}$-*transpose* of $\mathcal{A}$ defined by

$$\mathcal{A}^{<\mathbf{P}>}(j_{p_1}, \ldots, j_{p_d}) = \mathcal{A}(j_1, \ldots, j_d), \qquad 1 \leq j_k \leq n_k,\ k = 1{:}d,$$

i.e.,

$$\mathcal{A}^{<\mathbf{P}>}(\mathbf{j}(\mathbf{p})) = \mathcal{A}(\mathbf{j}), \qquad 1 \leq \mathbf{j} \leq \mathbf{n}.$$

For additional tensor transposition discussion, see Ragnarsson and Van Loan (2012).

12.4.5 The Modal Unfoldings

Recall that a tensor unfolding is a matrix whose entries come from the tensor. Particularly important are the *modal unfoldings*. If $\mathcal{A} \in \mathbb{R}^{n_1 \times \cdots \times n_d}$ and $N = n_1 \cdots n_d$, then its *mode-k* unfolding is an n_k-by-(N/n_k) matrix whose columns are the mode-k fibers. To illustrate, here are the three modal unfoldings for $\mathcal{A} \in \mathbb{R}^{4 \times 3 \times 2}$:

$$\mathcal{A}_{(1)} = \begin{bmatrix} a_{111} & a_{121} & a_{131} & a_{112} & a_{122} & a_{132} \\ a_{211} & a_{221} & a_{231} & a_{212} & a_{222} & a_{232} \\ a_{311} & a_{321} & a_{331} & a_{312} & a_{322} & a_{332} \\ a_{411} & a_{421} & a_{431} & a_{412} & a_{422} & a_{432} \end{bmatrix},$$

$$\mathcal{A}_{(2)} = \begin{bmatrix} a_{111} & a_{211} & a_{311} & a_{411} & a_{112} & a_{212} & a_{312} & a_{412} \\ a_{121} & a_{221} & a_{321} & a_{421} & a_{122} & a_{222} & a_{322} & a_{422} \\ a_{131} & a_{231} & a_{331} & a_{431} & a_{132} & a_{232} & a_{332} & a_{432} \end{bmatrix},$$

$$\mathcal{A}_{(3)} = \begin{bmatrix} a_{111} & a_{211} & a_{311} & a_{411} & a_{121} & a_{221} & a_{321} & a_{421} & a_{131} & a_{231} & a_{331} & a_{431} \\ a_{112} & a_{212} & a_{312} & a_{412} & a_{122} & a_{222} & a_{322} & a_{422} & a_{132} & a_{232} & a_{332} & a_{432} \end{bmatrix}.$$

We choose to order the fibers left to right according to the "vec" ordering. To be precise, if $\mathcal{A} \in \mathbb{R}^{n_1 \times \cdots \times n_d}$, then its mode-$k$ unfolding $\mathcal{A}_{(k)}$ is completely defined by

$$\mathcal{A}_{(k)}(i_k, \text{col}(\tilde{\mathbf{i}}_\mathbf{k}, \tilde{\mathbf{n}})) = \mathcal{A}(\mathbf{i}) \tag{12.4.6}$$

where $\tilde{\mathbf{i}}_\mathbf{k} = [i_1, \ldots, i_{k-1}, i_{k+1}, \ldots, i_d]$ and $\tilde{\mathbf{n}}_\mathbf{k} = [n_1, \ldots, n_{k-1}, n_{k+1}, \ldots, n_d]$. The rows of $\mathcal{A}_{(k)}$ are associated with subtensors of $\mathcal{A}$. In particular, we can identify $\mathcal{A}_{(k)}(q, :)$ with the order-$(d-1)$ tensor $\mathcal{A}^{(q)}$ defined by $\mathcal{A}^{(q)}(\tilde{\mathbf{i}}_k) = \mathcal{A}_{(k)}(q, \text{col}(\tilde{\mathbf{i}}_k), \tilde{\mathbf{n}}_k)$.

12.4.6 More General Unfoldings

In general, an unfolding for $\mathcal{A} \in \mathbb{R}^{n_1 \times \cdots \times n_d}$ is defined by choosing a set of row modes and a set of column modes. For example, if $\mathcal{A} \in \mathbb{R}^{2 \times 3 \times 2 \times 2 \times 3}$, $\mathbf{r} = 1{:}3$ and $\mathbf{c} = 4{:}5$, then

$$
\mathcal{A}_{\mathbf{r}\times\mathbf{c}} =
\begin{array}{c}
\begin{array}{cccccc}
(1,1) & (2,1) & (1,2) & (2,2) & (1,3) & (2,3)
\end{array} \\
\left[
\begin{array}{cccccc}
a_{111,11} & a_{111,21} & a_{111,12} & a_{111,22} & a_{111,13} & a_{111,23} \\
a_{211,11} & a_{211,21} & a_{211,12} & a_{211,22} & a_{211,13} & a_{211,23} \\
a_{121,11} & a_{121,21} & a_{121,12} & a_{121,22} & a_{121,13} & a_{121,23} \\
a_{221,11} & a_{221,21} & a_{221,12} & a_{221,22} & a_{221,13} & a_{221,23} \\
a_{131,11} & a_{131,21} & a_{131,12} & a_{131,22} & a_{131,13} & a_{131,23} \\
a_{231,11} & a_{231,21} & a_{231,12} & a_{231,22} & a_{231,13} & a_{231,23} \\
a_{112,11} & a_{112,21} & a_{112,12} & a_{112,22} & a_{112,13} & a_{112,23} \\
a_{212,11} & a_{212,21} & a_{212,12} & a_{212,22} & a_{212,13} & a_{212,23} \\
a_{122,11} & a_{122,21} & a_{122,12} & a_{122,22} & a_{122,13} & a_{122,23} \\
a_{222,11} & a_{222,21} & a_{222,12} & a_{222,22} & a_{222,13} & a_{222,23} \\
a_{132,11} & a_{132,21} & a_{132,12} & a_{132,22} & a_{132,13} & a_{132,23} \\
a_{232,11} & a_{232,21} & a_{232,12} & a_{232,22} & a_{232,13} & a_{232,23}
\end{array}
\right]
\begin{array}{c}
(1,1,1) \\
(2,1,1) \\
(1,2,1) \\
(2,2,1) \\
(1,3,1) \\
(2,3,1) \\
(1,1,2) \\
(2,1,2) \\
(1,2,2) \\
(2,2,2) \\
(1,3,2) \\
(2,3,2)
\end{array}
\end{array}
. \quad (12.4.7)
$$

In general, let $\mathbf{p}$ be a permutation of $1{:}d$ and define the row and column modes by

$$
\mathbf{r} = \mathbf{p}(1{:}e), \qquad \mathbf{c} = \mathbf{p}(e+1{:}d),
$$

where $0 \le e \le d$. This partitioning defines a matrix $\mathcal{A}_{\mathbf{r}\times\mathbf{c}}$ that has $n_{p_1}\cdots n_{p_e}$ rows and $n_{p_{e+1}}\cdots n_{p_d}$ columns and whose entries are defined by

$$
\mathcal{A}_{\mathbf{r}\times\mathbf{c}}(\operatorname{col}(\mathbf{i},\mathbf{n}(\mathbf{r})),\operatorname{col}(\mathbf{j},\mathbf{n}(\mathbf{c}))) = \mathcal{A}(\mathbf{i},\mathbf{j}). \quad (12.4.8)
$$

Important special cases include the modal unfoldings

$$
\mathbf{r} = [\,k\,], \; \mathbf{c} = [1,\ldots,k-1,k+1,\ldots,d] \quad \Longrightarrow \quad \mathcal{A}_{\mathbf{r}\times\mathbf{c}} = \mathcal{A}_{(k)}
$$

and the vec operation

$$
\mathbf{r} = 1{:}d, \; \mathbf{c} = [\,\emptyset\,] \quad \Longrightarrow \quad \mathcal{A}_{\mathbf{r}\times\mathbf{c}} = \operatorname{vec}(\mathcal{A}).
$$

12.4.7 Outer Products

The *outer product* of tensor $\mathcal{B} \in \mathbb{R}^{m_1 \times \cdots \times m_f}$ with tensor $\mathcal{C} \in \mathbb{R}^{n_1 \times \cdots \times n_g}$ is the order-$(f+g)$ tensor $\mathcal{A}$ defined by

$$
\mathcal{A}(\mathbf{i},\mathbf{j}) = \mathcal{B}(\mathbf{i}) \circ \mathcal{C}(\mathbf{j}), \qquad 1 \le \mathbf{i} \le \mathbf{m}, \; 1 \le \mathbf{j} \le \mathbf{n}.
$$

Multiple outer products are similarly defined, e.g.,

$$
\mathcal{A} = \mathcal{B} \circ \mathcal{C} \circ \mathcal{D} \quad \Longrightarrow \quad \mathcal{A}(\mathbf{i},\mathbf{j},\mathbf{k}) = \mathcal{B}(\mathbf{i}) \cdot \mathcal{C}(\mathbf{j}) \cdot \mathcal{D}(\mathbf{k}).
$$

Note that if $\mathcal{B}$ and $\mathcal{C}$ are order-2 tensors (matrices), then

$$
\mathcal{A} = \mathcal{B} \circ \mathcal{C} \quad \Rightarrow \quad \mathcal{A}(i_1,i_2,j_1,j_2) = \mathcal{B}(i_1,i_2) \cdot \mathcal{C}(j_1,j_2)
$$

and

$$
\mathcal{A}_{[\,3\,1\,]\times[\,4\,2\,]} = B \otimes C.
$$

Thus, the Kronecker product of two matrices corresponds to their outer product as tensors.

12.4.8 Rank-1 Tensors

Outer products between order-1 tensors (vectors) are particularly important. We say that $\mathcal{A} \in \mathbb{R}^{n_1 \times \cdots \times n_d}$ is a *rank-1 tensor* if there exist vectors $z^{(1)}, \ldots, z^{(d)} \in \mathbb{R}^{n_k}$ such that

$$\mathcal{A}(\mathbf{i}) = z^{(1)}(i_1) \cdots z^{(d)}(i_d), \qquad 1 \leq \mathbf{i} \leq \mathbf{n}.$$

A small example clarifies the definition and reveals a Kronecker product connection:

$$\mathcal{A} = \begin{bmatrix} u_1 \\ u_2 \end{bmatrix} \circ \begin{bmatrix} v_1 \\ v_2 \\ v_3 \end{bmatrix} \circ \begin{bmatrix} w_1 \\ w_2 \end{bmatrix} \quad \Leftrightarrow \quad \begin{bmatrix} a_{111} \\ a_{211} \\ a_{121} \\ a_{221} \\ a_{131} \\ a_{231} \\ a_{112} \\ a_{212} \\ a_{122} \\ a_{222} \\ a_{132} \\ a_{232} \end{bmatrix} = \begin{bmatrix} u_1 v_1 w_1 \\ u_2 v_1 w_1 \\ u_1 v_2 w_1 \\ u_2 v_2 w_1 \\ u_1 v_3 w_1 \\ u_2 v_3 w_1 \\ u_1 v_1 w_2 \\ u_2 v_1 w_2 \\ u_1 v_2 w_2 \\ u_2 v_2 w_2 \\ u_1 v_3 w_2 \\ u_2 v_3 w_2 \end{bmatrix} = w \otimes v \otimes u.$$

The modal unfoldings of a rank-1 tensor are highly structured. For the above example we have

$$\mathcal{A}_{(1)} = \begin{bmatrix} u_1 v_1 w_1 & u_1 v_2 w_1 & u_1 v_3 w_1 & u_1 v_1 w_2 & u_1 v_2 w_2 & u_1 v_3 w_2 \\ u_2 v_1 w_1 & u_2 v_2 w_1 & u_2 v_3 w_1 & u_2 v_1 w_2 & u_2 v_2 w_2 & u_2 v_3 w_2 \end{bmatrix} = u \otimes (w \otimes v)^T,$$

$$\mathcal{A}_{(2)} = \begin{bmatrix} u_1 v_1 w_1 & u_2 v_1 w_1 & u_1 v_1 w_2 & u_2 v_1 w_2 \\ u_1 v_2 w_1 & u_2 v_2 w_1 & u_1 v_2 w_2 & u_2 v_2 w_2 \\ u_1 v_3 w_1 & u_2 v_3 w_1 & u_1 v_3 w_2 & u_2 v_3 w_2 \end{bmatrix} = v \otimes (w \otimes u)^T,$$

$$\mathcal{A}_{(3)} = \begin{bmatrix} u_1 v_1 w_1 & u_2 v_1 w_1 & u_1 v_2 w_1 & u_2 v_2 w_1 & u_1 v_3 w_1 & u_2 v_3 w_1 \\ u_1 v_1 w_2 & u_2 v_1 w_2 & u_1 v_2 w_2 & u_2 v_2 w_2 & u_1 v_3 w_2 & u_2 v_3 w_2 \end{bmatrix} = w \otimes (v \otimes u)^T.$$

In general, if $z^{(k)} \in \mathbb{R}^{n_k}$ for $k = 1{:}d$ and

$$\mathcal{A} = z^{(1)} \circ \cdots \circ z^{(d)} \in \mathbb{R}^{n_1 \times \cdots \times n_d},$$

then its modal unfoldings are rank-1 matrices:

$$\mathcal{A}_{(k)} = z^{(k)} \cdot \left(z^{(d)} \otimes \cdots z^{(k+1)} \otimes z^{(k-1)} \otimes \cdots z^{(1)} \right)^T. \tag{12.4.9}$$

For general unfoldings of a rank-1 tensor, if $\mathbf{p}$ is a permutation of $1{:}d$, $\mathbf{r} = \mathbf{p}(1{:}e)$, and $\mathbf{c} = \mathbf{p}(e + 1{:}d)$, then

$$\mathcal{A}_{\mathbf{r} \times \mathbf{c}} = \left(z^{(p_e)} \circ \cdots \circ z^{(p_1)} \right) \left(z^{(p_d)} \circ \cdots \circ z^{(p_{e+1})} \right)^T. \qquad (12.4.10)$$

Finally, we mention that any tensor can be expressed as a sum of rank-1 tensors

$$\mathcal{A} \in \mathbb{R}^{n_1 \times \cdots \times n_d} \quad \Longrightarrow \quad \mathcal{A} = \sum_{i=1}^{\mathbf{n}} \mathcal{A}(\mathbf{i}) \, I_{n_1}(:, i_1) \circ \cdots \circ I_{n_d}(:, i_d).$$

An important §12.5 theme is to find more informative rank-1 summations than this!

12.4.9   Tensor Contractions and Matrix Multiplication

Let us return to the notion of a tensor contraction introduced in §12.4.1. The first order of business is to show that a contraction between two tensors is essentially a matrix multiplication between a pair of suitably chosen unfoldings. This is a useful connection because it facilitates reasoning about high-performance implementation.

Consider the problem of computing

$$\mathcal{A}(i, j, \alpha_3, \alpha_4, \beta_3, \beta_4, \beta_5) = \sum_{k=1}^{n_2} \mathcal{B}(i, k, \alpha_3, \alpha_4) \cdot \mathcal{C}(k, j, \beta_3, \beta_4, \beta_5) \qquad (12.4.11)$$

where

$$\begin{aligned}
\mathcal{A} &= \mathcal{A}(1{:}n_1, 1{:}m_2, 1{:}n_3, 1{:}n_4, 1{:}m_3, 1{:}m_4, 1{:}m_5), \\
\mathcal{B} &= \mathcal{B}(1{:}n_1, 1{:}n_2, 1{:}n_3, 1{:}n_4), \\
\mathcal{C} &= \mathcal{C}(1{:}m_1, 1{:}m_2, 1{:}m_3, 1{:}m_4, 1{:}m_5),
\end{aligned}$$

and $n_2 = m_1$. The index k is a *contraction index*. The example shows that in a contraction, the order of the output tensor can be (much) larger than the order of either input tensor, a fact that can prompt storage concerns. For example, if $n_1 = \cdots = n_4 = r$ and $m_1 = \cdots = m_5 = r$ in (12.4.11), then $\mathcal{B}$ and $\mathcal{C}$ are $O(r^5)$ while the output tensor $\mathcal{A}$ is $O(r^7)$.

The contraction (12.4.11) is a collection of related matrix-matrix multiplications. Indeed, at the slice level we have

$$\mathcal{A}(:, :, \alpha_3, \alpha_4, \beta_3, \beta_4, \beta_5) = \mathcal{B}(:, :, \alpha_3, \alpha_4) \cdot \mathcal{C}(:, :, \beta_3, \beta_4, \beta_5).$$

Each $\mathcal{A}$-slice is an n_1-by-m_2 matrix obtained as a product of an n_1-by-n_2 $\mathcal{B}$-slice and an m_1-by-m_2 $\mathcal{C}$-slice.

The summation in a contraction can be over more than just a single mode. To illustrate, assume that

$$\begin{aligned}
\mathcal{B} &= \mathcal{B}(1{:}m_1, 1{:}m_2, 1{:}t_1, 1{:}t_2), \\
\mathcal{C} &= \mathcal{C}(1{:}t_1, 1{:}t_2, 1{:}n_1, 1{:}n_2, 1{:}n_3),
\end{aligned}$$

and define $\mathcal{A} = \mathcal{A}(1{:}m_1, 1{:}m_2, 1{:}n_1, 1{:}n_2, 1{:}n_3)$ by

$$\mathcal{A}(i_1, i_2, j_1, j_2, j_3) \;=\; \sum_{k_1=1}^{t_1} \sum_{k_2=1}^{t_2} \mathcal{B}(i_1, i_2, k_1, k_2) \cdot \mathcal{C}(k_1, k_2, j_1, j_2, j_3). \qquad (12.4.12)$$

Note how "matrix like" this computation becomes with multiindex notation:

$$\mathcal{A}(\mathbf{i}, \mathbf{j}) \;=\; \sum_{\mathbf{k}=1}^{\mathbf{t}} \mathcal{B}(\mathbf{i}, \mathbf{k}) \cdot \mathcal{C}(\mathbf{k}, \mathbf{j}), \qquad \mathbf{1 \le i \le m,\ 1 \le j \le n.} \qquad (12.4.13)$$

A fringe benefit of this formulation is how nicely it connects to the following matrix-multiplication specification of $\mathcal{A}$:

$$\mathcal{A}_{[\,1\,2\,] \times [\,3\,4\,5\,]} \;=\; \mathcal{B}_{[\,1\,2\,] \times [\,3\,4\,]} \cdot \mathcal{C}_{[\,1\,2\,] \times [\,3\,4\,5\,]}.$$

The position of the contraction indices in the example (12.4.12) is convenient from the standpoint of framing the overall operation as a product of two unfoldings. However, it is not necessary to have the contraction indices "on the right" in $\mathcal{B}$ and "on the left" in $\mathcal{C}$ to formulate the operation as a matrix multiplication. For example, suppose

$$\mathcal{B} \;=\; \mathcal{B}(1{:}t_2, 1{:}m_1, 1{:}t_1, 1{:}m_2),$$

$$\mathcal{C} \;=\; \mathcal{C}(1{:}n_2, 1{:}t_2, 1{:}n_3, 1{:}t_1, 1{:}n_1),$$

and that we want to compute the tensor $\mathcal{A} = \mathcal{A}(1{:}m_1, 1{:}m_2, 1{:}n_1, 1{:}n_2, 1{:}n_3)$ defined by

$$\mathcal{A}(i_2, j_3, j_1, i_1, j_2) \;=\; \sum_{k_1=1}^{t_1} \sum_{k_2=1}^{t_2} \mathcal{B}(k_2, i_1, k_1, i_2) \cdot \mathcal{C}(j_2, k_2, j_3, k_1, j_1).$$

It can be shown that this calculation is equivalent to

$$\mathcal{A}_{[\,4\,1\,] \times [\,3\,5\,2\,]} \;=\; \mathcal{B}_{[\,2\,4\,] \times [\,3\,1\,]} \cdot \mathcal{C}_{[\,4\,2\,] \times [\,5\,1\,3\,]}.$$

Hidden behind these formulations are important implementation choices that define the overheads associated with memory access. Are the unfoldings explicitly set up? Are there any particularly good data structures that moderate the cost of data transfer? Etc. Because of their higher dimension, there are typically many more ways to organize a tensor contraction than there are to organize a matrix multiplication.

12.4.10   The Modal Product

A very simple but important family of contractions are the modal products. These contractions involve a tensor, a matrix, and a mode. In particular, if $\mathcal{S} \in \mathbb{R}^{n_1 \times \cdots \times n_d}$, $M \in \mathbb{R}^{m_k \times n_k}$, and $1 \le k \le d$, then $\mathcal{A}$ is the mode-k product of $\mathcal{S}$ and M if

$$\mathcal{A}_{(k)} \;=\; M \cdot \mathcal{S}_{(k)}. \qquad (12.4.14)$$

We denote this operation by

$$\mathcal{A} \;=\; \mathcal{S} \times_k M$$

and remark that

$$\mathcal{A}(\alpha_1, \ldots, \alpha_{k-1}, i, \alpha_{k+1}, \ldots, \alpha_d) = \sum_{j=1}^{n_k} M(i,j) \cdot \mathcal{S}(\alpha_1, \ldots, \alpha_{k-1}, j, \alpha_{k+1}, \ldots, \alpha_d)$$

and

$$\text{vec}(\mathcal{A}) = \left(I_{n_{k+1} \cdots n_d} \otimes M \otimes I_{n_1 \cdots n_{k-1}} \right) \cdot \text{vec}(\mathcal{S}) \tag{12.4.15}$$

are equivalent formulations. Every mode-k fiber in $\mathcal{S}$ is multiplied by the matrix M.

Using (12.4.15) and elementary facts about the Kronecker product, it is easy to show that

$$(\mathcal{S} \times_k F) \times_j G = (\mathcal{S} \times_j G) \times_k F, \tag{12.4.16}$$

$$(\mathcal{S} \times_k F) \times_k G = \mathcal{S} \times_k (FG), \tag{12.4.17}$$

assuming that all the dimensions match up.

12.4.11   The Multilinear Product

Suppose we are given an order-4 tensor $\mathcal{S} \in \mathbb{R}^{n_1 \times n_2 \times n_3 \times n_4}$ and four matrices

$$M_1 \in \mathbb{R}^{m_1 \times n_1}, \qquad M_2 \in \mathbb{R}^{m_2 \times n_2}, \qquad M_3 \in \mathbb{R}^{m_3 \times n_3}, \qquad M_4 \in \mathbb{R}^{m_4 \times n_4}.$$

The computation

$$\mathcal{A}(\mathbf{i}) = \sum_{\mathbf{j}=1}^{\mathbf{n}} \mathcal{S}(\mathbf{j}) \cdot M_1(i_1, j_1) \cdot M_2(i_2, j_2) \cdot M_3(i_3, j_3) \cdot M_4(i_4, j_4) \tag{12.4.18}$$

is equivalent to

$$\text{vec}(\mathcal{A}) = (M_4 \otimes M_3 \otimes M_2 \otimes M_1) \, \text{vec}(\mathcal{S}) \tag{12.4.19}$$

and is an order-4 example of a *multilinear product*. As can be seen in the following table, a multilinear product is a sequence of contractions, each being a modal product:

$a^{(0)} = \text{vec}(S)$	$\mathcal{A}^{(0)} = \mathcal{S}$
$a^{(1)} = \left(I_{n_4} \otimes I_{n_3} \otimes I_{n_2} \otimes M_1 \right) a^{(0)}$	$\mathcal{A}^{(1)}_{(1)} = M_1 \, \mathcal{A}^{(0)}_{(1)}$ (Mode-1 product)
$a^{(2)} = \left(I_{n_4} \otimes I_{n_3} \otimes M_2 \otimes I_{n_1} \right) a^{(1)}$	$\mathcal{A}^{(2)}_{(2)} = M_2 \, \mathcal{A}^{(1)}_{(2)}$ (Mode-2 product)
$a^{(3)} = \left(I_{n_4} \otimes M_3 \otimes I_{n_2} \otimes I_{n_1} \right) a^{(2)}$	$\mathcal{A}^{(3)}_{(3)} = M_3 \, \mathcal{A}^{(2)}_{(3)}$ (Mode-3 product)
$a^{(4)} = \left(M_4 \otimes I_{n_3} \otimes I_{n_2} \otimes I_{n_1} \right) a^{(3)}$	$\mathcal{A}^{(4)}_{(4)} = M_4 \, \mathcal{A}^{(3)}_{(4)}$ (Mode-4 product)
$\text{vec}(\mathcal{A}) = a^{(4)}$	$\mathcal{A} = \mathcal{A}^{(4)}$

The left column specifies what is going on in Kronecker product terms while the right column displays the four required modal products. The example shows that mode-k operations can be sequenced,

$$\mathcal{A} = \mathcal{S} \times_1 M_1 \times_2 M_2 \times_3 M_3 \times_4 M_4,$$

and that their order is immaterial, e.g.,

$$\mathcal{A} = \mathcal{S} \times_4 M_4 \times_1 M_1 \times_2 M_2 \times_3 M_3.$$

This follows from (12.4.16).

Because they are used in §12.5, we summarize two key properties of the multilinear product in the following theorem.

Theorem 12.4.1. *Suppose $\mathcal{S} \in \mathbb{R}^{n_1 \times \cdots \times n_d}$ and $M_k \in \mathbb{R}^{m_k \times n_k}$ for $k = 1{:}d$. If the tensor $\mathcal{A} \in \mathbb{R}^{m_1 \times \cdots \times m_d}$ is the multilinear product*

$$\mathcal{A} = \mathcal{S} \times_1 M_1 \times_2 M_2 \cdots \times_d M_d,$$

then

$$\mathcal{A}_{(k)} = M_k \cdot \mathcal{S}_{(k)} \cdot (M_d \otimes \cdots \otimes M_{k+1} \otimes M_{k-1} \otimes \cdots \otimes M_1)^T.$$

If $M_1, \ldots, M_d$ are all nonsingular, then $\mathcal{S} = \mathcal{A} \times_1 M_1^{-1} \times_2 M_2^{-1} \cdots \times_d M_d^{-1}$.

Proof. The proof involves equations (12.4.16) and (12.4.17) and the vec ordering of the mode-k fibers in $A_{(k)}$. $\quad\square$

12.4.12 Space versus Time

We close with an example from Baumgartner et al. (2005) that highlights the importance of order of operations and what the space-time trade-off can look like when a sequence of contractions is involved. Suppose that $\mathcal{A}$, $\mathcal{B}$, $\mathcal{C}$ and $\mathcal{D}$ are N-by-N-by-N-by-N tensors and that $\mathcal{S}$ is defined as follows:

> **for i = 1_4:N**
>> $s = 0$
>> **for k = 1_6:N**
>>> $s = s + \mathcal{A}(i_1, k_1, i_2, k_2) \cdot \mathcal{B}(i_2, k_3, k_4, k_5) \cdot \mathcal{C}(k_6, k_4, i_4, k_2) \cdot \mathcal{D}(k_1, k_6, k_3, k_5)$
>> **end**
>> $\mathcal{S}(\mathbf{i}) = s$
> **end**

Performed "as is," this is an $O(N^{10})$ calculation. On the other hand, if we can afford an additional pair of N-by-N-by-N-by-N arrays then work is reduced to $O(N^6)$. To see this, assume (for clarity) that we have a function $\mathcal{F} = \mathsf{Contract1}(\mathcal{G}, \mathcal{H})$ that computes the contraction

$$\mathcal{F}(\alpha_1, \alpha_2, \alpha_3, \alpha_4) = \sum_{\beta_1=1}^{N} \sum_{\beta_2=1}^{N} \mathcal{G}(\alpha_1, \beta_1, \alpha_2, \beta_2) \cdot \mathcal{H}(\alpha_3, \alpha_4, \beta_1, \beta_2),$$

a function $\mathcal{F} = \text{Contract2}(\mathcal{G}, \mathcal{H})$ that computes the contraction

$$\mathcal{F}(\alpha_1, \alpha_2, \alpha_3, \alpha_4) = \sum_{\beta_1=1}^{N} \sum_{\beta_2=1}^{N} \mathcal{G}(\alpha_1, \beta_1, \alpha_2, \beta_2) \cdot \mathcal{H}(\beta_2, \beta_1, \alpha_3, \alpha_4),$$

and a function $\mathcal{F} = \text{Contract3}(\mathcal{G}, \mathcal{H})$ that computes the contraction

$$\mathcal{F}(\alpha_1, \alpha_2, \alpha_3, \alpha_4) = \sum_{\beta_1=1}^{N} \sum_{\beta_2=1}^{N} \mathcal{G}(\alpha_2, \beta_1, \alpha_4, \beta_2) \cdot \mathcal{H}(\alpha_1\beta_1, \alpha_3, \beta_2).$$

Each of these order-4 contractions requires $O(N^6)$ flops. By exploiting common subexpressions suggested by the parentheses in

$$((\mathcal{B}(i_2, k_3, k_4, k_5) \cdot \mathcal{D}(k_1, k_6, k_3, k_5)) \cdot \mathcal{C}(k_6, k_4, i_4, k_2)) \cdot \mathcal{A}(i_1, k_1, i_2, k_2),$$

we arrive at the following $O(N^6)$ specification of the tensor $\mathcal{S}$:

$$\mathcal{T}_1 = \text{Contract1}(\mathcal{B}, \mathcal{D})$$
$$\mathcal{T}_2 = \text{Contract2}(\mathcal{T}_1, \mathcal{C})$$
$$\mathcal{S} = \text{Contract3}(\mathcal{T}_2, \mathcal{A})$$

Of course, space-time trade-offs frequently arise in matrix computations. However, at the tensor level the stakes are typically higher and the number of options exponential. Systems that are able to chart automatically an optimal course of action subject to constraints that are imposed by the underlying computer system are therefore of interest. See Baumgartner et al. (2005).

Problems

P12.4.1 Explain why (12.4.1) oversees a block matrix multiplication. Hint. Consider each of the three matrices as n-by-n block matrices with n-by-n blocks.

P12.4.2 Prove that the vec definition (12.4.2) and (12.4.3) is equivalent to the vec definition (12.4.4) and (12.4.5).

P12.4.3 How many fibers are there in the tensor $\mathcal{A} \in \mathbb{R}^{n_1 \times \cdots \times n_d}$? How many slices?

P12.4.5 Prove Theorem 12.4.1.

P12.4.6 Suppose $\mathcal{A} \in \mathbb{R}^{n_1 \times \cdots \times n_d}$ and that $\mathcal{B} = \mathcal{A}^{<\mathbf{p}>}$ where $\mathbf{p}$ is a permutation of $1{:}d$. Specify a permutation matrix P so that $\mathcal{B}_{(k)} = \mathcal{A}_{(p(k))} P$.

P12.4.7 Suppose $\mathcal{A} \in \mathbb{R}^{n_1 \times \cdots \times n_d}$, $N = n_1 \cdots n_d$, and that $\mathbf{p}$ is a permutation of $1{:}d$ that involves swapping a single pair of indices, e.g., [1 4 3 2 5]. Determine a permutation matrix $P \in \mathbb{R}^{N \times N}$ so that if $\mathcal{B} = \mathcal{A}^{<\mathbf{p}>}$, then $\text{vec}(\mathcal{B}) = P \cdot \text{vec}(\mathcal{A})$.

P12.4.8 Suppose $\mathcal{A} \in \mathbb{R}^{n_1 \times \cdots \times n_d}$ and that $\mathcal{A}_{(k)}$ has unit rank for some k. Does it follow that $\mathcal{A}$ is a rank-1 tensor?

P12.4.9 Refer to (12.4.18). Specify an unfolding S of $\mathcal{S}$ and an unfolding A of $\mathcal{A}$ so that $A = (M_1 \otimes M_3)S(M_2 \otimes M_4)$.

P12.4.10 Suppose $\mathcal{A} \in \mathbb{R}^{n_1 \times \cdots \times n_d}$ and that both $\mathbf{p}$ and $\mathbf{q}$ are permutations of $1{:}d$. Give a formula for $\mathbf{r}$ so that $(\mathcal{A}^{<\mathbf{p}>})^{<\mathbf{q}>} = \mathcal{A}^{<\mathbf{r}>}$.

Notes and References for §12.4

For an introduction to tensor computations, see:

L. De Lathauwer (1997). "Signal Processing Based on Multilinear Algebra," PhD Thesis, K.U. Leuven.
A. Smilde, R. Bro, and P. Geladi (2004). *Multiway Analysis*, John Wiley, Chichester, England.
T.G. Kolda and B.W. Bader (2009). "Tensor Decompositions and Applications," *SIAM Review* 51, 455–500.

For results that connect unfoldings, the vec operation, Kronecker products, contractions, and transposition, see:

S. Ragnarsson and C. Van Loan (2012). "Block Tensor Unfoldings," *SIAM J. Matrix Anal. Applic. 33*, 149–169.

MATLAB software that supports tensor computations as described in this section includes the Tensor Toolbox:

B.W. Bader and T.G. Kolda (2006). "Algorithm 862: MATLAB Tensor Classes for Fast Algorithm Prototyping," *ACM Trans. Math. Softw.*, 32, 635–653.
B.W. Bader and T.G. Kolda (2007). "Efficient MATLAB Computations with Sparse and Factored Tensors," *SIAM J. Sci. Comput.* 30, 205–231.

The challenges associated with high-performance, large-scale tensor computations are discussed in:

W. Landry (2003). "Implementing a High Performance Tensor Library," *Scientific Programming 11*, 273–290.
C. Lechner, D. Alic, and S. Husa (2004). "From Tensor Equations to Numerical Code," *Computer Algebra Tools for Numerical Relativity*, Vol. 0411063.
G. Baumgartner, A. Auer, D. Bernholdt, A. Bibireata, V. Choppella, D. Cociorva, X. Gao, R. Harrison, S. Hirata, S. Krishnamoorthy, S. Krishnan, C. Lam, Q. Lu, M. Nooijen, R. Pitzer, J. Ramanujam, P. Sadayappan, and A. Sibiryakov (2005). "Synthesis of High-Performance Parallel Programs for a Class of Ab Initio Quantum Chemistry Models," *Proc. IEEE*, 93, 276–292.

The multiway analysis community and the quantum chemistry/electronic structure community each have their own favored style of tensor notation and it is very different! See:

J.L. Synge and A. Schild (1978). *Tensor Calculus*, Dover Publications, New York.
H.A.L. Kiers (2000). "Towards a Standardized Notation and Terminology in Multiway Analysis," *J. Chemometr. 14*, 105–122.

12.5   Tensor Decompositions and Iterations

Decompositions have three roles to play in matrix computations. They can be used to convert a given problem into an equivalent easy-to-solve problem, they can expose hidden relationships among the a_{ij}, and they can open the door to data-sparse approximation. The role of tensor decompositions is similar and in this section we showcase a few important examples. The matrix SVD has a prominent role to play throughout. The goal is to approximate or represent a given tensor with an illuminating (hopefully short) sum of rank-1 tensors. Optimization problems arise that are multilinear in nature and lend themselves to the alternating least squares framework. These methods work by freezing all but one of the unknowns and improving the free-to-range variable with some tractable linear optimization strategy. Interesting matrix computations arise during this process and that is the focus of our discussion. For a much more complete survey of tensor decompositions, properties, and algorithms, see Kolda and Bader (2009). *Our aim in these few pages is simply to give a snapshot of the "inner loop" linear algebra that is associated with a few of these methods and to build intuition for this increasingly important area of high-dimensional scientific computing.*

Heavy use is made of the Kronecker product and tensor unfoldings. Thus, this section builds upon §12.3 and §12.4. We use order-3 tensors to drive the discussion, but periodically summarize what the theorems and algorithms look like for general-order tensors.

12.5.1 The Higher-Order SVD

Let us think about the SVD of $A \in \mathbb{R}^{m \times n}$, not as

$$A = U\Sigma V^T = \sum_{i=1}^{n} \sigma_i u_i v_i T, \tag{12.5.1}$$

but as $U^T A = \Sigma V^T$. The matrix U structures the rows of $U^T A$ so that they are orthogonal to each other and monotone decreasing in norm:

$$U^T A = \begin{bmatrix} \sigma_1 v_1^T \\ \vdots \\ \sigma_n v_n^T \end{bmatrix}. \tag{12.5.2}$$

The optimality of this structure can be seen by considering the following problem:

$$\max_{Q^T Q = I_r} \| Q^T A \|_F, \qquad Q \in \mathbb{R}^{m \times r}. \tag{12.5.3}$$

It is easy to verify that the maximum value is $\sigma_1^2 + \cdots + \sigma_r^2$ and that it can be attained by setting $Q = U(:, 1{:}r)$. The left singular vector matrix does the best job from the standpoint of getting as much "mass" as possible to the top of the transformed A. And that is what SVD does—it concentrates mass and supports an illuminating rank-1 expansion.

Now suppose $\mathcal{A} \in \mathbb{R}^{n_1 \times n_2 \times n_3}$ and consider the following triplet of SVD's, one for each modal unfolding:

$$U_1^T \mathcal{A}_{(1)} = \Sigma_1 V_1^T, \qquad U_2^T \mathcal{A}_{(2)} = \Sigma_2 V_2^T, \qquad U_3^T \mathcal{A}_{(3)} \Sigma_3 V_3^T. \tag{12.5.4}$$

These define three independent modal products:

$$\mathcal{B}^{(1)} = \mathcal{A} \times_1 U_1, \qquad \mathcal{B}^{(2)} = \mathcal{A} \times_2 U_2, \qquad \mathcal{B}^{(3)} = \mathcal{A} \times_3 U_3. \tag{12.5.5}$$

Using Theorem 12.4.1, we have the following unfoldings:

$$\mathcal{B}^{(1)}_{(1)} = \Sigma_1 V_1^T (U_3 \otimes U_2)^T, \qquad \mathcal{B}^{(2)}_{(2)} = \Sigma_2 V_2^T (U_3 \otimes U_1)^T, \qquad \mathcal{B}^{(3)}_{(3)} = \Sigma_1 V_1^T (U_2 \otimes U_1)^T.$$

Note that each of these matrices has the same kind singular value "grading" that is displayed in (12.5.1). Recalling from §12.4.5 that the rows of an unfolding are subtensors, it is easy to show that

$$\| \mathcal{B}^{(1)}(i, :, :) \|_F = \sigma_i(\mathcal{A}_{(1)}), \qquad i = 1{:}n_1,$$

$$\| \mathcal{B}^{(2)}(:, i, :) \|_F = \sigma_i(\mathcal{A}_{(2)}), \qquad i = 1{:}n_2,$$

$$\| \mathcal{B}^{(3)}(:, :, i) \|_F = \sigma_i(\mathcal{A}_{(3)}), \qquad i = 1{:}n_3.$$

If we assemble these three modal products into a single multilinear product, then we get

$$\mathcal{S} = \mathcal{A} \times_1 U_1^T \times_2 U_2^T \times_3 U_3^T.$$

Because the U_i are orthogonal, we can apply Theorem 12.4.1 and get

$$\mathcal{A} = \mathcal{S} \times_1 U_1 \times_2 U_2 \times_3 U_3.$$

This is the *higher-order SVD* (HOSVD) developed by De Lathauwer, De Moor, and Vandewalle (2000). We summarize some of its important properties in the following theorem.

Theorem 12.5.1 (HOSVD). *If $\mathcal{A} \in \mathbb{R}^{n_1 \times \cdots \times n_d}$ and*

$$\mathcal{A}_{(k)} = U_k \Sigma_k V_k^T, \qquad k = 1{:}d,$$

are the SVDs of its modal unfoldings, then its HOSVD is given by

$$\mathcal{A} = \mathcal{S} \times_1 U_1 \times_2 U_2 \cdots \times_d U_d \qquad (12.5.6)$$

where $\mathcal{S} = \mathcal{A} \times_1 U_1^T \times_2 U_2^T \cdots \times_d U_d^T$. The formulation (12.5.6) is equivalent to

$$\mathcal{A} = \sum_{\mathbf{j}=1}^{\mathbf{n}} \mathcal{S}(\mathbf{j}) \cdot U_1(:, j_1) \circ \cdots \circ U_d(:, j_d), \qquad (12.5.7)$$

$$\mathcal{A}(\mathbf{i}) = \sum_{\mathbf{j}=1}^{\mathbf{n}} \mathcal{S}(\mathbf{j}) \cdot U_1(i_1, j_1) \cdots U_d(i_d, j_d), \qquad (12.5.8)$$

$$\mathsf{vec}(\mathcal{A}) = (U_d \otimes \cdots \otimes U_1) \cdot \mathsf{vec}(\mathcal{S}). \qquad (12.5.9)$$

Moreover,

$$\| \mathcal{S}_{(k)}(i, :) \|_F = \sigma_i(A_{(k)}), \qquad i = 1{:}\mathsf{rank}(A_{(k)}) \qquad (12.5.10)$$

for $k = 1{:}d$.

Proof. We leave the verification of (12.5.7)–(12.5.9) to the reader. To establish (12.5.10), note that

$$\mathcal{S}_{(k)} = U_k^T \mathcal{A}_{(k)} \left(U_d \otimes \cdots \otimes U_{k+1} \otimes U_{k-1} \otimes \cdots \otimes U_1 \right)$$
$$= \Sigma_k V_k^T \left(U_d \otimes \cdots \otimes U_{k+1} \otimes U_{k-1} \otimes \cdots \otimes U_1 \right).$$

It follows that the rows of $S_{(k)}$ are mutually orthogonal and that the singular values of $\mathcal{A}_{(k)}$ are the 2-norms of these rows. $\square$

In the HOSVD, the tensor $\mathcal{S}$ is called the *core tensor*. Note that it is *not* diagonal. However, the inequalities (12.5.10) tell us that, the values in $\mathcal{S}$ tend to be smaller as "distance" from the $(1, 1, \ldots, 1)$ entry increases.

12.5.2   The Truncated HOSVD and Multilinear Rank

If $\mathcal{A} \in \mathbb{R}^{n_1 \times \cdots \times n_d}$, then its *multilinear rank* is a the vector of modal unfolding ranks:

$$\mathsf{rank}_*(\mathcal{A}) = \left[\mathsf{rank}(\mathcal{A}_{(1)}), \ldots, \mathsf{rank}(\mathcal{A}_{(d)}) \right].$$

Note that the summation upper bounds in the HOSVD can be replaced by $\mathbf{rank}_*(\mathcal{A})$. For example, (12.5.7) becomes

$$\mathcal{A} = \sum_{\mathbf{j}=1}^{\mathsf{rank}_*(\mathcal{A})} \mathcal{S}(\mathbf{j}) U_1(:, j_1) \circ \cdots \circ U_d(:, j_d).$$

This suggests a path to low-rank approximation. If $\mathbf{r} \leq \mathsf{rank}_*(\mathcal{A})$ with inquality in at least one component, then we can regard

$$\mathcal{A}^{(\mathbf{r})} = \sum_{\mathbf{j}=1}^{\mathbf{r}} \mathcal{S}(\mathbf{j}) U_1(:, j_1) \circ \cdots \circ U_d(:, j_d)$$

as a *truncated HOSVD approximation* to $\mathcal{A}$. It can be shown that

$$\| \mathcal{A} - \mathcal{A}^{(\mathbf{r})} \|_F^2 \leq \min_{1 \leq k \leq d} \sum_{i=r_k+1}^{\mathsf{rank}(\mathcal{A}_{(k)})} \sigma_i(A_{(k)})^2. \tag{12.5.11}$$

12.5.3   The Tucker Approximation Problem

Suppose $\mathcal{A} \in \mathbb{R}^{n_1 \times n_2 \times n_3}$ and assume that $\mathbf{r} \leq \mathbf{rank}_*(\mathcal{A})$ with inequality in at least one component. Prompted by the optimality properties of the matrix SVD, let us consider the following optimization problem:

$$\min_{\mathcal{X}} \| \mathcal{A} - \mathcal{X} \|_F \tag{12.5.12}$$

such that

$$\mathcal{X} = \sum_{\mathbf{j}=1}^{\mathbf{r}} \mathcal{S}(\mathbf{j}) \cdot U_1(:, j_1) \circ U_2(:, j_2) \circ U_3(:, j_3). \tag{12.5.13}$$

We refer to this as the *Tucker approximation problem*. Unfortunately, the truncated HOSVD tensor $\mathcal{A}^{(\mathbf{r})}$ does not solve the Tucker approximation problem, prompting us to develop an appropriate optimization strategy.

To be clear, we are given $\mathcal{A}$ and $\mathbf{r}$ and seek a core tensor $\mathcal{S}$ that is r_1-by-r_2-by-r_3 and matrices $U_1 \in \mathbb{R}^{n_1 \times r_1}$, $U_2 \in \mathbb{R}^{n_2 \times r_2}$, and $U_3 \in \mathbb{R}^{n_3 \times r_3}$ with orthonormal columns so that the tensor $\mathcal{X}$ defined by (12.5.13) solves (12.5.12). Using Theorem 12.4.1 we know that

$$\| \mathcal{A} - \mathcal{X} \|_F = \| \mathsf{vec}(\mathcal{A}) - (U_3 \otimes U_2 \otimes U_1) \cdot \mathsf{vec}(\mathcal{S}) \|_2.$$

Since $U_3 \otimes U_2 \otimes U_1$ has orthonormal columns, it follows that the "best" $\mathcal{S}$ given any triplet $\{U_1, U_2, U_3\}$ is

$$\mathcal{S} = \left(U_3^T \otimes U_2^T \otimes U_1^T \right) \cdot \mathsf{vec}(\mathcal{A}).$$

Thus, we can remove $\mathcal{S}$ from the search space and simply look for $U = U_3 \otimes U_2 \otimes U_1$ so that

$$\| (I - UU^T) \cdot \mathbf{vec}(\mathcal{A}) \|_F^2 = \| \mathbf{vec}(\mathcal{A}) \|_F^2 - \| U^T \cdot \mathbf{vec}(\mathcal{A}) \|_F^2$$

is minimized. In other words, determine U_1, U_2, and U_3 so that

$$\| (U_3^T \otimes U_2^T \otimes U_1^T) \cdot \mathbf{vec}(\mathcal{A}) \|_F = \begin{cases} \| U_1^T \cdot A_{(1)} \cdot (U_3 \otimes U_2) \|_F \\ \| U_2^T \cdot A_{(2)} \cdot (U_3 \otimes U_1) \|_F \\ \| U_3^T \cdot A_{(3)} \cdot (U_2 \otimes U_1) \|_F \end{cases}$$

is maximized. By freezing any two of the three matrices $\{U_1, U_2, U_3\}$ we can improve the third by solving an optimization problem of the form (12.5.3). This suggests the following strategy:

Repeat:

    Maximize $\| U_1^T \cdot A_{(1)} \cdot (U_3 \otimes U_2) \|_F$ with respect to U_1 by computing the

        SVD $\mathcal{A}_{(1)} \cdot (U_3 \otimes U_2) = \tilde{U}_1 \Sigma_1 V_1^T$. Set $U_1 = \tilde{U}_1(:, 1{:}r_1)$.

    Maximize $\| U_2^T \cdot A_{(2)} \cdot (U_3 \otimes U_1) \|_F$ with respect to U_2 by computing the

        SVD $\mathcal{A}_{(2)} \cdot (U_3 \otimes U_1) = \tilde{U}_2 \Sigma_2 V_2^T$. Set $U_2 = \tilde{U}_2(:, 1{:}r_2)$.

    Maximize $\| U_3^T \cdot A_{(3)} \cdot (U_2 \otimes U_1) \|_F$ with respect to U_3: by computing the

        SVD $\mathcal{A}_{(3)} \cdot (U_2 \otimes U_1) = \tilde{U}_3 \Sigma_3 V_3^T$. Set $U_3 = \tilde{U}_3(:, 1{:}r_3)$.

This is an example of the *alternating least squares framework*. For order-d tensors, there are d optimizations to perform each step:

Repeat:
    for $k = 1{:}d$
        Compute the SVD:
$$\mathcal{A}_{(k)} (U_d \otimes \cdots \otimes U_{k+1} \otimes U_{k-1} \otimes \cdots \otimes U_1) = \tilde{U}_k \Sigma_k V_k^T.$$
        $U_k = \tilde{U}_k(:, 1{:}r_k)$
    end

This is essentially the *Tucker framework*. For implementation details concerning this *nonlinear* iteration, see De Lathauwer, De Moor, and Vandewalle (2000b), Smilde, Bro, and Geladi (2004, pp. 119–123), and Kolda and Bader (2009).

12.5.4   The CP Approximation Problem

A nice attribute of the matrix SVD that is that the "core matrix" in the rank-1 expansion is diagonal. This is not true when we graduate to tensors and work with the

Tucker representation. However, there is an alternate way to extrapolate from the matrix SVD if we prefer "diagonalness" to orthogonality. Given $\mathcal{X} \in \mathbb{R}^{n_1 \times n_2 \times n_3}$ and an integer r, we consider the problem

$$\min_{\mathcal{X}} \| \mathcal{A} - \mathcal{X} \|_F \qquad (12.5.14)$$

such that

$$\mathcal{X} = \sum_{j=1}^{r} \lambda_j \cdot F(:,j) \circ G(:,j) \circ H(:,j) \qquad (12.5.15)$$

where $F \in \mathbb{R}^{n_1 \times r}$, $G \in \mathbb{R}^{n_2 \times r}$, and $H \in \mathbb{R}^{n_3 \times r}$. This is an example of the *CP approximation* problem. We assume that the columns of F, G, and H have unit 2-norm.

The modal unfoldings of the tensor (12.5.15) are neatly characterized through the Khatri-Rao product that we defined in §12.3.3. If

$$F = [\, f_1 \mid \cdots \mid f_r \,], \qquad G = [\, g_1 \mid \cdots \mid g_r \,], \qquad H = [\, h_1 \mid \cdots \mid h_r \,],$$

then

$$\mathcal{X}_{(1)} = \sum_{j=1}^{r} \lambda_j \cdot f_j \otimes (h_j \otimes g_j)^T \; = \; F \cdot \mathrm{diag}(\lambda_j) \cdot (H \odot G)^T,$$

$$\mathcal{X}_{(2)} = \sum_{j=1}^{r} \lambda_j \cdot g_j \otimes (h_j \otimes f_j)^T \; = \; G \cdot \mathrm{diag}(\lambda_j) \cdot (H \odot F)^T,$$

$$\mathcal{X}_{(3)} = \sum_{j=1}^{r} \lambda_j \cdot h_j \otimes (g_j \otimes f_j)^T \; = \; H \cdot \mathrm{diag}(\lambda_j) \cdot (G \odot F)^T.$$

These results follow from the previous section. For example,

$$\mathcal{X}_{(1)} = \sum_{j=1}^{r} \lambda_j \, (f_j \circ g_j \circ h_j)_{(1)} \; = \; \sum_{j=1}^{r} \lambda_j f_j (h_j \otimes g_j)^T$$

$$= [\, \lambda_1 f_1 \mid \cdots \mid \lambda_r f_r \,] [\, h_1 \otimes g_1 \mid \cdots \mid h_r \otimes g_r \,]^T \; = \; F \cdot \mathrm{diag}(\lambda_j) \cdot (H \odot G)^T.$$

Noting that

$$\| \mathcal{A} - \mathcal{X} \|_F \; = \; \| \mathcal{A}_{(1)} - \mathcal{X}_{(1)} \|_F \; = \; \| \mathcal{A}_{(2)} - \mathcal{X}_{(2)} \|_F \; = \; \| \mathcal{A}_{(3)} - \mathcal{X}_{(3)} \|_F,$$

we see that the CP approximation problem can be solved by minimizing any one of the following expressions:

$$\| \mathcal{A}_{(1)} - \mathcal{X}_{(1)} \|_F = \| \mathcal{A}_{(1)} - F \cdot \mathrm{diag}(\lambda_j) \cdot (H \odot G)^T \|_F, \qquad (12.5.16)$$

$$\| \mathcal{A}_{(2)} - \mathcal{X}_{(2)} \|_F = \| \mathcal{A}_{(2)} - G \cdot \mathrm{diag}(\lambda_j) \cdot (H \odot F)^T \|_F, \qquad (12.5.17)$$

$$\| \mathcal{A}_{(3)} - \mathcal{X}_{(3)} \|_F = \| \mathcal{A}_{(3)} - H \cdot \mathrm{diag}(\lambda_j) \cdot (G \odot F)^T \|_F. \qquad (12.5.18)$$

This is a multilinear least squares problem. However, observe that if we fix λ, H, and G in (12.5.16), then $\| \, \mathcal{A}_{(1)} - \mathcal{X}_{(1)} \, \|_F$ is linear in F. Similar comments apply to (12.5.17) and (12.5.18) and we are led to the following alternating least squares minimization strategy:

Repeat:

Let $\tilde{F}$ minimize $\| \, \mathcal{A}_{(1)} - \tilde{F} \cdot (H \odot G)^T \, \|_F$ and for $j = 1{:}r$ set

$$\lambda_j = \| \, \tilde{F}(:,j) \, \|_2 \quad \text{and} \quad F(:,j) \, = \, \tilde{F}(:,j)/\lambda_j.$$

Let $\tilde{G}$ minimize $\| \, \mathcal{A}_{(2)} - \tilde{G} \cdot (H \odot F)^T \, \|_F$ and for $j = 1{:}r$ set

$$\lambda_j = \| \, \tilde{G}(:,j) \, \|_2 \quad \text{and} \quad G(:,j) \, = \, \tilde{G}(:,j)/\lambda_j.$$

Let $\tilde{H}$ minimize $\| \, \mathcal{A}_{(3)} - \tilde{H} \cdot (G \odot F)^T \, \|_F$ and for $j = 1{:}r$ set

$$\lambda_j = \| \, \tilde{H}(:,j) \, \|_2 \quad \text{and} \quad H(:,j) \, = \, \tilde{H}(:,j)/\lambda_j.$$

The update calculations for F, G, and H are highly structured linear least squares problems. The central calculations involve linear least square problems of the form

$$\min \| \, (B \odot C)z - d \, \|_2 \tag{12.5.19}$$

where $B \in \mathbb{R}^{p_B \times q}$, $C \in \mathbb{R}^{p_C \times q}$, and $d \in \mathbb{R}^{p_B p_C}$. This is typically a "tall skinny" LS problem. If we form the Khatri-Rao product and use the QR factorization in the usual way, then $O(p_B p_C q^2)$ flops are required to compute z. On the other hand, the normal equation system corresponding to (12.5.19) is

$$\left((B^T B) \, .\!* \, (C^T C) \right) z \, = \, (B \odot C)^T d \tag{12.5.20}$$

which can be formed and solved via the Cholesky factorization in $O((p_B + p_C)q^2)$ flops.

For general tensors $\mathcal{A} \in \mathbb{R}^{n_1 \times \cdots \times n_d}$ there are d least squares problems to solve per pass. In particular, given $\mathcal{A}$ and r, the CP approximation problem involves finding matrices

$$F^{(k)} = [f_1^{(k)} \mid \cdots \mid f_r^{(k)}] \in \mathbb{R}^{n_k \times r}, \qquad k = 1{:}d,$$

with unit 2-norm columns and a vector $\lambda \in \mathbb{R}^r$ so that if

$$\mathcal{X} \, = \, \sum_{j=1}^r \lambda_j f_j^{(1)} \circ \cdots \circ f_j^{(d)}, \tag{12.5.21}$$

then $\| \, \mathcal{A} - \mathcal{X} \, \|_F$ is minimized. Noting that

$$\mathcal{X}_{(k)} \, = \, F^{(k)} \mathrm{diag}(\lambda) \left(F^{(d)} \odot \cdots \odot F^{(k+1)} \odot F^{(k-1)} \odot \cdots \odot F^{(1)} \right)^T,$$

we obtain the following iteration.

Repeat:

    for $k = 1{:}d$

        Minimize $\| \mathcal{A}_{(k)} - \tilde{F}^{(k)} \left(F^{(d)} \odot \cdots \odot F^{(k+1)} \odot F^{(k-1)} \odot \cdots \odot F^{(1)} \right) \|_F$

        with respect to $\tilde{F}^{(k)}$.

        for $j = 1{:}r$

$$\lambda_j = \| \tilde{F}_{(k)}(:,j) \|_2$$

$$F^{(k)}(:,j) = \tilde{F}_k(:,j)/\lambda_j$$

        end

    end

This is the *CANDECOMP/PARAFAC framework.* For implementation details about this *nonlinear* iteration, see Smilde, Bro, and Geladi (2004, pp. 113–119) and Kolda and Bader (2009).

12.5.5 Tensor Rank

The choice of r in the CP approximation problem brings us to the complicated issue of tensor rank. If

$$\mathcal{A} = \sum_{j=1}^{r} \lambda_j f_j^{(1)} \circ \cdots \circ f_j^{(d)}$$

and no shorter sum-of-rank-1's exists, then we say that $\mathcal{A}$ is a rank-r tensor. Thus, we see that in the CP approximation problem is a problem of finding the best rank-r approximation. Using the CP framework to discover the rank of a tensor is problematic because of the following complications.

Complication 1. The tensor rank problem is NP-hard. See and Hillar and Lim (2012).

Complication 2. The largest rank attainable for an n_1-by-$\cdots$-n_d tensor is called the *maximum rank*. There is no simple formula like $\min\{n_1, \ldots, n_d\}$. Indeed, maximum rank is known for only a handful of special cases.

Complication 3. If the set of rank-k tensors in $\mathbb{R}^{n_1 \times \cdots \times n_d}$ has positive measure, then k is a *typical rank*. The space of $n_1 \times \cdots \times n_d$ can have more than one typical rank. For example, the probability that a random 2-by-2-by-2 tensor has rank 2 is .79, while the probability that it has rank 3 is .21, assuming that the a_{ijk} are normally distributed with mean 0 and variance 1. See de Silva and Lim (2008) and Martin (2011) for detailed analysis of the 2-by-2-by 2 case.

Complication 4. The rank of a particular tensor over the real field may be different than its rank over the complex field.

Complication 5. There exist tensors that can be approximated with arbitrary precision by a tensor of lower rank. Such a tensor is said to be *degenerate*.

Complication 6. If

$$\mathcal{X}_r = \sum_{j=1}^{r+1} \lambda_j U_1(:,j) \circ \cdots \circ U_d(:,j)$$

is the best rank-$(r+1)$ approximation of $\mathcal{A}$, then it does *not* follow that

$$\mathcal{X}_{r+1} = \sum_{j=1}^{r} \lambda_j \hat{U}_1(:,j) \circ \cdots \circ \hat{U}_d(:,j)$$

is the best rank-r approximation of $\mathcal{A}$. See Kolda (2003) for an example. Subtracting the best rank-1 approximation can even increase the rank! See Stegeman and Comon (2009).

See Kolda and Bader (2009) for references on tensor rank and its implications for computation. Examples that illuminate the subtleties associated with tensor rank can be found in the the paper by de Silva and Lim (2008).

12.5.6   Tensor Singular Values: A Variational Approach

The singular values of a matrix $A \in \mathbb{R}^{n_1 \times n_2}$ are the stationary values of

$$\psi_A(u,v) = \frac{u^T A v}{\| u \|_2 \| v \|_2} = \frac{\displaystyle\sum_{i_1=1}^{n_1} \sum_{i_2=1}^{n_2} A(i_1,i_2) u(i_1) v(i_2)}{\| u \|_2 \| v \|_2} \tag{12.5.22}$$

and the associated stationary vectors are the corresponding singular vectors. This follows by looking at the gradient equation $\nabla \psi(u,v) = 0$. Indeed, if u and v are unit vectors, then this equation has the form

$$\nabla \psi_A(u,v) = \begin{bmatrix} Av - \psi_A(u,v)u \\ A^T u - \psi_A(u,v)v \end{bmatrix} = 0.$$

This variational characterization of matrix singular values and vectors extends to tensors; see Lim (2005). Suppose $\mathcal{A} \in \mathbb{R}^{n_1 \times n_2 \times n_3}$ and define

$$\psi_\mathcal{A}(u_1,u_2,u_3) = \frac{\displaystyle\sum_{i=1}^{n} \mathcal{A}(\mathbf{i}) \cdot u_1(i_1) u_2(i_2) u_3(i_3)}{\| u_1 \|_2 \| u_2 \|_2 \| u_3 \|_2}$$

where $u_1 \in \mathbb{R}^{n_1}$, $u_2 \in \mathbb{R}^{n_2}$, and $u_3 \in \mathbb{R}^{n_3}$. It is easy to show that

$$\psi_\mathcal{A}(u_1,u_2,u_3) = \begin{cases} u_1^T \mathcal{A}_{(1)}(u_3 \otimes u_2) / (\| u_1 \|_2 \| u_2 \|_2 \| u_3 \|_2), \\ u_2^T \mathcal{A}_{(2)}(u_3 \otimes u_1) / (\| u_1 \|_2 \| u_2 \|_2 \| u_3 \|_2), \\ u_3^T \mathcal{A}_{(3)}(u_2 \otimes u_1) / (\| u_1 \|_2 \| u_2 \|_2 \| u_3 \|_2). \end{cases}$$

If u_1, u_2, and u_3 are unit vectors, then the equation $\nabla \psi_{\mathcal{A}} = 0$ is

$$
\nabla \psi_{\mathcal{A}} = \begin{bmatrix} \mathcal{A}_{(1)}(u_3 \otimes u_2) \\ \mathcal{A}_{(2)}(u_3 \otimes u_1) \\ \mathcal{A}_{(3)}(u_2 \otimes u_1) \end{bmatrix} - \psi_{\mathcal{A}}(u_1, u_2, u_3) \begin{bmatrix} u_1 \\ u_2 \\ u_3 \end{bmatrix} = 0.
$$

If we can satisfy this equation, then we will call $\psi_{\mathcal{A}}(u_1, u_2, u_3)$ a singular value of the tensor $\mathcal{A}$. If we take a componentwise approach to this this nonlinear system we are led to the following iteration

Repeat:

$$
\tilde{u}_1 = \mathcal{A}_{(1)}(u_3 \otimes u_2), \quad u_1 = \tilde{u}_1/\| \tilde{u}_1 \|_2
$$

$$
\tilde{u}_2 = \mathcal{A}_{(2)}(u_3 \otimes u_1), \quad u_2 = \tilde{u}_2/\| \tilde{u}_2 \|_2
$$

$$
\tilde{u}_3 = \mathcal{A}_{(3)}(u_2 \otimes u_1), \quad u_3 = \tilde{u}_3/\| \tilde{u}_3 \|_2
$$

$$
\sigma = \psi(u_1, u_2, u_3)
$$

This can be thought of as a higher-order power iteration. Upon comparison with the Tucker approximation problem with $\mathbf{r} = [1, 1, \ldots, 1]$, we see that it is a strategy for computing a nearest rank-1 tensor.

12.5.7   Symmetric Tensor Eigenvalues: A Variational Approach

If $C \in \mathbb{R}^{N \times N}$ is symmetric, then its eigenvalues are the stationary values of

$$
\phi_C(x) = \frac{x^T C x}{x^T x} = \frac{\displaystyle\sum_{i_1=1}^{N} \sum_{i_2=1}^{N} C(i_1, i_2) x(i_1) x(i_2)}{x^T x} \tag{12.5.23}
$$

and the corresponding stationary vectors are eigenvectors. This follows by setting the gradient of ϕ_C to zero.

If we are to generalize this notion to tensors, then we need to define what we mean by a symmetric tensor. An order-d tensor $\mathcal{C} \in \mathbb{R}^{N \times \cdots \times N}$ is *symmetric* if for any permutation $\mathbf{p}$ of $1{:}d$ we have

$$
\mathcal{C}(\mathbf{i}) = \mathcal{C}(\mathbf{i}(\mathbf{p})), \qquad 1 \le \mathbf{i} \le N.
$$

For the case $d = 3$ this means $c_{ijk} = c_{ikj} = c_{jik} = c_{jki} = c_{kij} = c_{kji}$ for all i, j, and k that satisfy $1 \le i \le N$, $1 \le j \le N$, and $1 \le k \le N$.

It is easy to generalize (12.5.23) to the case of symmetric tensors. If $\mathcal{C} \in \mathbb{R}^{N \times N \times N}$ is symmetric and $x \in \mathbb{R}^N$ then we define $\phi_{\mathcal{C}}$ by

$$
\phi_{\mathcal{C}}(x) = \frac{\displaystyle\sum_{\mathbf{i}=1}^{N} \mathcal{C}(\mathbf{i}) \cdot x(i_1)\, x(i_2)\, x(i_3)}{\| x \|_2^3} = \frac{x^T \mathcal{C}_{(1)}(x \otimes x)}{\| x \|_2^3}. \tag{12.5.24}
$$

Note that if $\mathcal{C}$ is a symmetric tensor, then all its modal unfoldings are the same. The equation $\nabla \phi_{\mathcal{C}}(x) = 0$ with $\| x \|_2 = 1$ has the form

$$\nabla \phi_{\mathcal{C}}(x) = \mathcal{C}_{(1)}(x \otimes x) - \phi_{\mathcal{C}}(x) \cdot x = 0.$$

If this holds then we refer to $\phi_{\mathcal{C}}(x)$ as an eigenvalue of the tensor $\mathcal{C}$, a concept introduced by Lim (2005) and Li (2005). An interesting framework for solving this nonlinear equation has been proposed by Kolda and Mayo (2012). It involves repetition of the operation sequence

$$\tilde{x} = \mathcal{C}_{(1)}(x \otimes x) + \alpha x, \qquad \lambda = \| \tilde{x} \|_2, \qquad x = \tilde{x}/\lambda$$

where the shift parameter α is determined to ensure convexity and eventual convergence of the iteration. For further discussion of the symmetric tensor eigenvalue problem and various power iterations that can be used to solve it, see Zhang and Golub (2001) and Kofidis and Regalia (2002).

12.5.8 Tensor Networks, Tensor Trains, and the Curse

In many applications, tensor decompositions and their approximations are used to discover things about a high-dimensional data set. In other settings, they are used to address the *curse of dimensionality*, i.e., the challenges associated with a computation that requires $O(n^d)$ work or storage. Whereas "big n" is problematic in matrix computations, "big d" is typically the hallmark of a difficult large-scale tensor computation. For example, it is (currently) impossible to store explicitly an $n_1 \times \cdots \times n_{1000}$ tensor if $n_1 = \cdots = n_{1000} = 2$. In general, a solution framework for an order-d tensor problem suffers from the curse of dimensionality if the associated work and storage are exponential in d.

It is in this context that data-sparse tensor approximation is increasingly important. One way to build a high-order, data-sparse tensor is by connecting a set of low-order tensors with a relatively small set of contractions. This is the notion of a *tensor network*. In a tensor network, the nodes are low-order tensors and the edges are contractions. A special case that communicates the main idea is the *tensor train (TT)* representation, which we proceed to illustrate with an order-5 example. Given the low-order tensor "carriages"

$$\begin{aligned}
\mathcal{G}_1: &\quad n_1 \times r_1, \\
\mathcal{G}_2: &\quad r_1 \times n_2 \times r_2, \\
\mathcal{G}_3: &\quad r_2 \times n_3 \times r_3, \\
\mathcal{G}_4: &\quad r_3 \times n_4 \times r_4, \\
\mathcal{G}_5: &\quad r_4 \times n_5,
\end{aligned}$$

we define the order-5 tensor train $\mathcal{T}$ by

$$\mathcal{T}(\mathbf{i}) = \sum_{k=1}^{\mathbf{r}} \mathcal{G}_1(i_1, k_1) \mathcal{G}_2(k_1, i_2, k_2) \mathcal{G}_3(k_2, i_3, k_3) \mathcal{G}_4(k_3, i_4, k_4) \mathcal{G}_5(k_4, i_5). \qquad (12.5.25)$$

The pattern is obvious from the example. The first and last carriages are matrices and all those in between are order-3 tensors. Adjacent carriages are connected by a single contraction. See Figure 12.5.1.

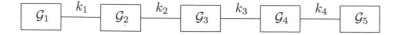

Figure 12.5.1.   *The Order-5 tensor train (12.5.25)*

To appreciate the data-sparsity of an order-d tensor train $\mathcal{T} \in \mathbb{R}^{n_1 \times \cdots \times n_d}$ that is represented through its carriages, assume that $n_1 = \cdots = n_d = n$ and $r_1 = \cdots = r_{d-1} = r \ll n$. It follows that the TT-representation requires $O(dr^2 n)$ memory locations, which is much less than the n^d storage required by the explicit representation.

We present a framework for approximating a given tensor with a data-sparse tensor train. The first order of business is to show that *any* tensor $\mathcal{A}$ as a TT representation. This can be verified by induction. For insight into the proof we consider an order-5 example. Suppose $\mathcal{A} \in \mathbb{R}^{n_1 \times \cdots \times n_5}$ is the result of a contraction between a tensor

$$\mathcal{B}(i_1, i_2, k_2) = \sum_{k_1=1}^{r_1} \mathcal{G}_1(i_1, k_1) \mathcal{G}_2(k_1, i_2, k_2)$$

and a tensor $\mathcal{C}$ as follows

$$\mathcal{A}(i_1, i_2, i_3, i_4, i_5) = \sum_{k_2=1}^{r_2} \mathcal{B}(i_1, i_2, k_2) \mathcal{C}(k_2, i_3, i_4, i_5).$$

If we can express $\mathcal{C}$ as a contraction of the form

$$\mathcal{C}(k_2, i_3, i_4, i_5) = \sum_{k_3=1}^{r_3} \mathcal{G}_3(k_2, i_3, k_3) \tilde{\mathcal{C}}(k_3, i_4, i_5), \tag{12.5.26}$$

then

$$\begin{aligned}
\mathcal{A}(i_1, i_2, i_3, i_4, i_5) &= \sum_{k_2=1}^{r_2} \sum_{k_3=1}^{r_3} \mathcal{B}(i_1, i_2, k_2) \mathcal{G}_3(k_2, i_3, k_3) \tilde{\mathcal{C}}(k_3, i_4, i_5) \\
&= \sum_{k_3=1}^{r_3} \left(\sum_{k_2=1}^{r_2} \mathcal{B}(i_1, i_2, k_2) \mathcal{G}_3(k_2, i_3, k_3) \right) \tilde{\mathcal{C}}(k_3, i_4, i_5) \\
&= \sum_{k_3=1}^{r_3} \tilde{\mathcal{B}}(i_1, i_2, i_3, k_3) \tilde{\mathcal{C}}(k_3, i_4, i_5)
\end{aligned}$$

where

$$\tilde{\mathcal{B}}(i_1, i_2, i_3, k_3) = \sum_{k_1=1}^{r_1} \sum_{k_2=1}^{r_2} \mathcal{G}_1(i_1, k_1) \mathcal{G}_2(k_1, i_2, k_2) \mathcal{G}_3(k_2, i_3, k_3).$$

The transition from writing $\mathcal{A}$ as a contraction of $\mathcal{B}$ and $\mathcal{C}$ to a contraction of $\tilde{\mathcal{B}}$ and $\tilde{\mathcal{C}}$ shows by example how to organize a formal proof that any tensor has a TT-representation. The only remaining issue concerns the "factorization" (12.5.26). It turns out that the tensors $\mathcal{G}_3$ and $\tilde{\mathcal{C}}$ can be determined by computing the SVD of the unfolding

$$C = \mathcal{C}_{[1\ 2]\times[3\ 4]}.$$

Indeed, if $\mathsf{rank}(C) = r_3$ and $C = U_3\Sigma_3 V_3^T$ is the SVD with $\Sigma_3 \in \mathbb{R}^{r_3\times r_3}$, then it can be shown that (12.5.26) holds if we define $\mathcal{G}_3 \in \mathbb{R}^{r_2\times n_3\times r_3}$ and $\tilde{\mathcal{C}} \in \mathbb{R}^{r_3\times n_4\times n_5}$ by

$$\mathsf{vec}(\mathcal{G}_3) = \mathsf{vec}(U_3), \tag{12.5.27}$$

$$\mathsf{vec}(\tilde{\mathcal{C}}) = \mathsf{vec}(\Sigma_3 V_3^T). \tag{12.5.28}$$

By extrapolating from this $d = 5$ discussion we obtain the following procedure due to Oseledets and Tyrtyshnikov (2009) that computes the tensor train representation

$$\mathcal{A}(\mathbf{i}) = \sum_{\mathbf{k}(1:d-1)}^{\mathbf{r}(1:d-1)} \mathcal{G}_1(i_1, k_1)\mathcal{G}_2(k_1, i_2, k_2)\cdots\mathcal{G}_{d-1}(k_{d-2}, i_{d-1}, k_{d-1})\mathcal{G}_d(k_{d-1}, i_d)$$

for any given $\mathcal{A} \in \mathbb{R}^{n_1\times\cdots\times n_d}$:

$M_1 = \mathcal{A}_{(1)}$

SVD: $M_1 = U_1\Sigma_1 V_1^T$ where $\Sigma_1 \in \mathbb{R}^{r_1\times r_1}$ and $r_1 = \mathsf{rank}(M_1)$

for $k = 2{:}d-1$

   $M_k = \mathsf{reshape}(\Sigma_{k-1}V_{k-1}^T, r_{k-1}n_k, n_{k+1}\cdots n_d)$ (12.5.29)

   SVD: $M_k = U_k\Sigma_k V_k^T$ where $\Sigma_k \in \mathbb{R}^{r_k\times r_k}$ and $r_k = \mathsf{rank}(M_k)$

   Define $\mathcal{G}_k \in \mathbb{R}^{r_{k-1}\times n_k\times r_k}$ by $\mathsf{vec}(\mathcal{G}_k) = \mathsf{vec}(U_k)$.

end

$\mathcal{G}_d = \Sigma_{d-1}V_{d-1}^T$

Like the HOSVD, it involves a sequence of SVDs performed on unfoldings.

In its current form, (12.5.29) does *not* in general produce a data-sparse representation. For example, if $d = 5$, $n_1 = \cdots = n_5 = n$, and $M_1, \ldots, M_4$ have full rank, then $r_1 = n$, $r_2 = n^2$, $r_3 = n^2$, and $r_4 = n$. In this case the TT-representation requires the same $O(n^5)$ storage as the explicit representation.

To realize a data-sparse, tensor train approximation, the matrices U_k and $\Sigma_k V_k^T$ are replaced with "thinner" counterparts that are intelligently chosen and cheap to compute. As a result, the r_k's are replaced by (significantly smaller) $\tilde{r}_k$'s. The approximating tensor train involves fewer than $d(n_1 + \cdots + n_d)\cdot(\max \tilde{r}_k)$ numbers. This kind of approximation overcomes the curse of dimensionality assuming that $\max \tilde{r}_k$ does not depend on the modal dimensions. See Oseledets and Tyrtyshnikov (2009) for computational details, successful applications, and discussion about the low-rank approximations of $M_1, \ldots, M_{d-1}$.

Problems

P12.5.1 Suppose $a \in \mathbb{R}^{n_1 n_2 n_3}$. Show how to compute $f \in \mathbb{R}^{n_1}$ and $g \in \mathbb{R}^{n_2}$ so that $\| a - h \otimes g \otimes f \|_2$ is minimized where $h \in \mathbb{R}^{n_3}$ is given. Hint: This is an SVD problem.

P12.5.2 Given $\mathcal{A} \in \mathbb{R}^{n_1\times n_2\times n_3}$ with positive entries, show how to determine $\mathcal{B} = f \circ g \circ h \in \mathbb{R}^{n_1\times n_2\times n_3}$ so that the following function is minimized:

$$\phi(f, g, h) = \sum_{i=1}^{n} |\log(\mathcal{A}(\mathbf{i})) - \log(\mathcal{B}(\mathbf{i}))|^2.$$

P12.5.3 Show that the rank of any unfolding of a tensor $\mathcal{A}$ is never larger than rank($\mathcal{A}$).

P12.5.4 Formulate an HOQRP factorization for a tensor $\mathcal{A} \in \mathbb{R}^{n_1 \times \cdots \times n_d}$ that is based on the QR-with-column-pivoting (QRP) factorizations $\mathcal{A}_{(k)} P_k = Q_k R_k$ for $k = 1{:}d$. Does the core tensor have any special properties?

P12.5.5 Prove (12.5.11).

P12.5.6 Show that (12.5.14) and (12.5.15) are equivalent to minimizing $\| \text{vec}(\mathcal{X}) = (H \odot G \odot F)\lambda \|_2$.

P12.5.7 Justify the flop count that is given for the Cholesky solution of the linear system (12.5.20).

P12.5.8 How many distinct values can there be in a symmetric 3-by-3-by-3 tensor?

P12.5.9 Suppose $\mathcal{A} \in \mathbb{R}^{N \times N \times N \times N}$ has the property that

$$\mathcal{A}(i_1, i_2, i_3, i_4) = \mathcal{A}(i_2, i_1, i_3, i_4) = \mathcal{A}(i_1, i_2, i_4, i_3) = \mathcal{A}(i_3, i_4, i_1, i_2).$$

Note that $\mathcal{A}_{[1\ 3] \times [2\ 4]} = (A_{ij})$ is an N-by-N block matrix with N-by-N blocks. Show that $A_{ij} = A_{ji}$ and $A_{ij}^T = A_{ij}$.

P12.5.10 Develop an order-d version of the iterations presented in §12.5.6. How many flops per iteration are required?

P12.5.11 Show that if $\mathcal{G}_3$ and $\tilde{\mathcal{C}}$ are defined by (12.5.27) and (12.5.28), then (12.5.26) holds.

Notes and References for §12.5

For an in-depth survey of all the major tensor decompositions that are used in multiway analysis together with many pointers to the literature, see:

T.G. Kolda and B.W. Bader (2009). "Tensor Decompositions and Applications," *SIAM Review* 51, 455–500.

Other articles that give perspective on the field of tensor computations include:

L. De Lathauwer and B. De Moor (1998). "From Matrix to Tensor: Multilinear Algebra and Signal Processing," in *Mathematics in Signal Processing IV,* J. McWhirter and I. Proudler (eds.), Clarendon Press, Oxford, 1–15.

P. Comon (2001). "Tensor Decompositions: State of the Art and Applications," in *Mathematics in Signal Processing V*, J. G. McWhirter and I. K. Proudler (eds), Clarendon Press, Oxford, 1–24.

R. Bro (2006). "Review on Multiway Analysis in Chemistry 2000–2005," *Crit. Rev. Analy. Chem. 36*, 279–293.

P. Comon, X. Luciani, A.L.F. de Almeida (2009). "Tensor Decompositions, Alternating Least Squares and Other Tales," *J. Chemometrics 23*, 393-405.

The following two monographs cover both the CP and Tucker models and show how they fit into the larger picture of multiway analysis:

A. Smilde, R. Bro, and P. Geladi (2004). *Multi-Way Analysis: Applications in the Chemical Sciences,* Wiley, Chichester, England.

P.M. Kroonenberg (2008). *Applied Multiway Data Analysis*, Wiley, Hoboken, NJ.

There are several MATLAB toolboxes that are useful for tensor decomposition work, see:

C.A. Anderson and R. Bro (2000). "The N-Way Toolbox for MATLAB," *Chemometrics Intelligent Lab. Syst. 52*, 1–4.

B.W. Bader and T.G. Kolda (2006). "Algorithm 862: MATLAB Tensor Classes for Fast Algorithm Prototyping," *ACM Trans. Math. Softw. 32*, 635–653.

B.W. Bader and T.G. Kolda (2007). "Efficient MATLAB Computations with Sparse and Factored Tensors," *SIAM J. Sci. Comput.* 30, 205–231.

Higher-order SVD-like ideas are presented in:

L.R. Tucker (1966). "Some Mathematical Notes on Three-Mode Factor Analysis," *Psychmetrika 31*, 279–311.

A recasting of Tucker's work in terms of the modern SVD viewpoint with many practical ramifications can be found in the foundational paper:

L. De Lathauwer, B. De Moor and J. Vandewalle (2000). "A Multilinear Singular Value Decomposition," *SIAM J. Matrix Anal. Applic. 21*, 1253–1278.

A sampling of the CANDECOMP/PARAFAC/Tucker literature includes:

R. Bro (1997). "PARAFAC: Tutorial and Applications," *Chemometrics Intelligent Lab. Syst.* 38, 149–171.
T.G. Kolda (2001). "Orthogonal Tensor Decompositions," *SIAM J. Matrix Anal. Applic. 23*, 243–255.
G. Tomasi and R. Bro (2006). "A Comparison of Algorithms for Fitting the PARAFAC Model," *Comput. Stat. Data Analy. 50*, 1700–1734.
L. De Lathauwer (2006). "A Link between the Canonical Decomposition in Multilinear Algebra and Simultaneous Matrix Diagonalization," *SIAM J. Matrix Anal. Applic. 28*, 642–666.
I.V. Oseledets, D.V. Savostianov, and E.E. Tyrtyshnikov (2008). "Tucker Dimensionality Reduction of Three-Dimensional Arrays in Linear Time," *SIAM J. Matrix Anal. Applic. 30*, 939–956.
C.D. Martin and C. Van Loan (2008). "A Jacobi-Type Method for Computing Orthogonal Tensor Decompositions," *SIAM J. Matrix Anal. Applic. 29*, 184–198.

Papers concerned with the tensor rank issue include:

T.G. Kolda (2003). "A Counterexample to the Possibility of an Extension of the Eckart-Young Low-Rank Approximation Theorem for the Orthogonal Rank Tensor Decomposition," *SIAM J. Matrix Anal. Applic. 24*, 762–767.
J.M. Landsberg (2005). "The Border Rank of the Multiplication of 2-by-2 Matrices is Seven," *J. AMS 19*, 447–459.
P. Comon, G.H. Golub, L-H. Lim, and B. Mourrain (2008). "Symmetric Tensors and Symmetric Tensor Rank," *SIAM J. Matrix Anal. Applic. 30*, 1254–1279.
V. de Silva and L.-H. Lim (2008). "Tensor rank and the Ill-Posedness of the Best Low-Rank Approximation Problem," *SIAM J. Matrix Anal. Applic. 30*, 1084–1127.
P. Comon, J.M.F. ten Berg, L. De Lathauwer, and J. Castaing (2008). "Generic Rank and Typical Ranks of Multiway Arrays," *Lin. Alg. Applic. 430*, 2997–3007.
L. Eldin and B. Savas (2011). "Perturbation Theory and Optimality Conditions for the Best Multi-linear Rank Approximation of a Tensor," *SIAM. J. Matrix Anal. Applic. 32*, 1422–1450.
C.D. Martin (2011). "The Rank of a 2-by-2-by-2 Tensor," *Lin. Multil. Alg. 59*, 943–950.
A. Stegeman and P. Comon (2010). "Subtracting a Best Rank-1 Approximation May Increase Tensor Rank," *Lin. Alg. Applic. 433*, 1276-1300.
C.J. Hillar and L.-H. Lim (2012) "Most Tensor Problems Are NP-hard," arXiv:0911.1393.

The idea of defining tensor singular values and eigenvalues through generalized Rayleigh quotients is pursued in the following references:

L.-H. Lim (2005) "Singular Values and Eigenvalues of Tensors: A Variational Approach," Proceedings of the IEEE International Workshop on Computational Advances in Multi-Sensor Adaptive Processing, 129–132.
L. Qi (2005). "Eigenvalues of a Real Supersymmetric Tensor," *J. Symbolic Comput. 40*, 1302–1324.
L. Qi (2006). "Rank and Eigenvalues of a Supersymmetric Tensor, the Multivariate Homogeneous Polynomial and the Algebraic Hypersurface it Defines," *J. Symbolic Comput. 41*, 1309–1327.
L. Qi (2007). Eigenvalues and Invariants of Tensors," *J. Math. Anal. Applic. 325*, 1363–1377.
D. Cartwright and B. Sturmfels (2010). "The Number of Eigenvalues of a Tensor", arXiv:1004.4953v1.

There are a range of rank-1 approximation tensor approximation problems and power methods to solve them, see:

L. De Lathauwer, B. De Moor, and J. Vandewalle (2000). "On the Best Rank-1 and Rank-(r1,r2,...,rN) Approximation of Higher-Order Tensors," *SIAM J. Mat. Anal. Applic.,* 21, 1324–1342.
E. Kofidis and P.A. Regalia (2000). "The Higher-Order Power Method Revisited: Convergence Proofs and Effective Initialization," in Proceedings of the IEEE International Conference on Acoustics, Speech, and Signal Processing, Vol. 5, 2709–2712.
T. Zhang and G. H. Golub (2001). "Rank-one Approximation to High order Tensors," *SIAM J. Mat. Anal. and Applic. 23*, 534–550.

E. Kofidis and P. Regalia (2001). "Tensor Approximation and Signal Processing Applications," in *Structured Matrices in Mathematics, Computer Science, and Engineering I*, V. Olshevsky (ed.), AMS, Providence, RI, 103–133.

E. Kofidis and P.A. Regalia (2002). "On the Best Rank-1 Approximation of Higher-Order Super-Symmetric Tensors," *SIAM J. Matrix Anal. Applic. 23*, 863-884.

L. De Lathauwer and J. Vandewalle (2004). "Dimensionality Reduction in Higher-Order Signal Processing and Rank-(R1;R2;...;RN) Reduction in Multilinear Algebra," *Lin. Alg. Applic. 391*, 31–55.

S. Ragnarsson and C. Van Loan (2012). "Block Tensors and Symmetric Embedding," arXiv:1010.0707v2.

T.G. Kolda and J.R. Mayo (2011). "Shifted Power Method for Computing Tensor Eigenpairs," *SIAM J. Matrix Anal. Applic. 32*, 1095–1124.

Various Newton-like methods have also emerged:

L. Eldén and B. Savas (2009). "A Newton-Grassmann Method for Computing the Best Multi-linear Rank-(R1; R2; R3) Approximation of a Tensor," *SIAM J. Matrix Anal. Applic. 31*, 248–271.

B. Savas and L.-H. Lim (2010) "Quasi-Newton Methods on Grassmannians and Multilinear Approximations of Tensors," *SIAM J. Sci. Comput. 32*, 3352–3393.

M. Ishteva, L. De Lathauwer, P.-A. Absil, and S. Van Huffel (2009). "Differential-Geometric Newton Algorithm for the Best Rank-(R_1, R_2, R_3) Approximation of Tensors", *Numer. Algorithms 51*, 179–194.

Here is a sampling of other tensor decompositions that have recently been proposed:

L. Omberg, G. Golub, and O. Alter (2007). "A Tensor Higher-Order Singular Value Decomposition for Integrative Analysis of Dna Microarray Data from Different Studies," *Proc. Nat. Acad. Sci. 107*, 18371-18376.

L. De Lathauwer (2008). "Decompositions of a Higher-Order Tensor in Block TermsPart II: Definitions and Uniqueness," *SIAM. J. Mat. Anal. Applic. 30*, 1033–1066.

L. De Lathauwer and D. Nion (2008). "Decompositions of a Higher-Order Tensor in Block TermsPart III: Alternating Least Squares Algorithms," *SIAM. J. Mat. Anal. Applic. 30*, 1067–1083.

M.E. Kilmer and C.D. Martin (2010). "Factorization Strategies for Third Order Tensors," *Lin. Alg. Applic. 435*, 641–658.

E. Acar, D.M. Dunlavy, and T.G. Kolda (2011). "A Scalable Optimization Approach for Fitting Canonical Tensor Decompositions," *J. Chemometrics*, 67–86.

E. Acar, D.M. Dunlavy, T.G. Kolda, and M. Mrup (2011). "Scalable Tensor Factorizations for Incomplete Data," *Chemomet. Intell. Lab. Syst. 106*, 41–56.

C. Chi and T. G. Kolda (2012). "On Tensors, Sparsity, and Nonnegative Factorizations," arXiv:1112.2414.

Various tools for managing high-dimensional tensors are discussed in:

S.R. White (1992). "Density Matrix Formulation for Quantum Renormalization Groups," *Phys. Rev. Lett. 69*, 2863–2866.

W. Hackbusch and B.N. Khoromskij (2007). "Tensor-product Approximation to Operators and Functions in High Dimensions," *J. Complexity 23*, 697–714.

I.V. Oseledets and E.E. Tyrtyshnikov (2008). "Breaking the Curse of Dimensionality, or How to Use SVD in Many Dimensions," *SIAM J. Sci. Comput. 31*, 3744–3759.

W. Hackbusch and S. Kuhn (2009). "A New Scheme for the Tensor Representation," *J. Fourier Anal. Applic. 15*, 706–722.

I.V. Oseledets, D.V. Savostyanov, and E.E. Tyrtyshnikov (2009). "Linear Algebra for Tensor Problems," *Computing 85*, 169-188.

I. Oseledets and E. Tyrtyshnikov (2010). "TT-Cross Approximation for Multidimensional Arrays," *Lin. Alg. Applic. 432*, 70–88.

L. Grasedyck (2010). "Hierarchical Singular Value Decomposition of Tensors," *SIAM J. Mat. Anal. Applic. 31*, 2029–2054.

S. Holtz, T. Rohwedder, and R. Schneider (2012). "The Alternating Linear Scheme for Tensor Optimization in the Tensor Train Format," *SIAM J. Sci. Comput. 34*, A683–A713.

For insight into the "curse of dimensionality," see:

G. Beylkin and M.J. Mohlenkamp (2002). "Numerical Operator Calculus in Higher Dimensions," *Proc. Nat. Acad. Sci. 99(16)*, 10246–10251.

G. Beylkin and M.J. Mohlenkamp (2005). "Algorithms for Numerical Analysis in High Dimensions," *SIAM J. Sci. Comput. 26*, 2133–2159.

Index

A-conjugate, 633
A-norm, 629
Aasen's method, 188–90
Absolute value notation, 91
Additive Schwarz, 665
Adjacency set, 602
Affine space, 628
Algebraic multiplicity, 353
Algorithm, 135
Algorithmic detail, xii
Angles between subspaces, 329–31
Antidiagonal, 208
Approximate inverse preconditioner, 654
Approximate Newton method, 590–1
Approximation of a matrix function, 522–3
Arnoldi process, 579–83
   implicit restarting, 581–3
   k-step decomposition, 580
   rational, 588
Arnoldi vectors, 580
Augmented system method, 316

Back substitution, 107
Backward error analysis, 100–1
Backward stable, 136
Backward successive over-relaxation, 619
Balancing, 392
Band algorithms, 176ff
   Cholesky, 180
   Gaussian elimination, 178–9
   Hessenberg LU, 179
   triangular systems, 177–8
Band matrix, 15
   data structures and, 17
   inverse, 182–3
   LU factorization and, 176–7
   pivoting and, 178–9
   profile Cholesky, 184
Bandwidth, 176
barrier, 57
Bartels-Stewart algorithm, 398–400
Basic solution for least squares, 292
Basis, 64
   eigenvector, 400
Bauer-Fike theorem, 357–8
BiCGstab method, 647
Biconjugate gradient method, 645
Bidiagonalization
   Golub-Kahan, 572
   Householder, 284–5
   Paige-Saunders, 575
   upper triangularizing first, 285
Bidiagonal matrix, 15
Big-Oh notation, 12
Binary powering, 527
Bisection methods
   for Toeplitz eigenproblem, 216
   for tridiagonal eigenproblem, 467

BLAS, 12ff
Block algorithms, 196ff
   Cholesky, 168–70
   cyclic reduction, 197–8
   data reuse and, 47ff
   diagonalization, 352–3, 397–400
   Gaussian elimination, 144–6
   Gauss-Seidel, 613
   Jacobi method for eigenvalues, 481–2
   Jacobi method for linear systems, 613
   Lanczos, 566–9
   LU, 118–20, 196–7
   LU with pivoting, 144–6
   multiple right-hand-side triangular, 109–10
   QR factorization, 250–1
   recursive QR factorization, 251
   SPIKE, 197–8
   Tridiagonal, 196ff
   unsymmetric Lanczos, 586
Block-cyclic distribution layout, 50
   and parallel LU, 146
Block diagonal dominance, 197
Block distribution layout, 50
Block Householder, 238–9
Block matrices, 22ff
   data reuse and, 55
   diagonal dominance of, 197
Block tridiagonal systems, 196ff
Bordered linear systems, 202
Bunch-Kaufman algorithm, 192
Bunch-Parlett pivoting, 191

Cache, 46
Cancellation, 97
Cannon's algorithm, 60–1
Canonical correlation problem, 330
Cauchy-like matrix, 682
   conversion to, 689–90
Cauchy-Schwarz inequality, 69
Cayley transform, 68, 245
CGNE, 636–7
CGNR, 636
Characteristic polynomial, 66, 348
   generalized eigenproblem and, 405
Chebyshev polynomials, 621, 653
Chebyshev semi-iterative method, 621–2
Cholesky factorization, 163
   band, 180
   block, 168–70
   downdating and, 338–41
   gaxpy version, 164
   matrix square root and, 163
   profile, 184
   recursive block, 172–3
   stability of, 164–5
Cholesky reduction of $A - \lambda B$, 500
Chordal metric, 407
Circulant systems, 220–2

Classical Gram-Schmidt, 254
Coarse grid role in multigrid, 673
Collatz-Wielandt formula, 373
Colon notation, 6, 16
Column
    deletion or addition in QR, 235–8
    major order, 45
    ordering in QR factorization, 279–80
    orientation, 5, 107–8
    partitioning, 6
    pivoting, 276–7
    weighting in least squares, 306–7
Communication costs, 52ff
Compact WY transformation, 244
Companion matrix, 382–3
Complete orthogonal decomposition, 283
Complete pivoting, 131–3
Complex
    Givens transformation, 243–4
    Householder transformation, 243
    matrices, 13
    matrix multiplication, 29
    QR factorization, 256
    SVD, 80
Complexity of matrix inversion, 174
Componentwise bounds, 92
Compressed column representation, 598–9
Computation/communication ratio, 53ff
Condition estimation, 140, 142–3, 436
Condition of
    eigenvalues, 359–60
    invariant subspaces, 360–1
    least squares problem, 265–7
    linear systems, 87–8
    multiple eigenvalues, 360
    rectangular matrix, 248
    similarity transformation, 354
Confluent Vandermonde matrix, 206
Conformal partition, 23
Congruence transformation, 449
Conjugate
    directions, 633
    transpose, 13
Conjugate gradient method, 625ff
    derivation and properties, 629–30, 633
    Hestenes-Stiefel version, 634–5
    Lanczos version, 632
    practical, 635–6
    pre-conditioned, 651–2
Conjugate gradient squared method, 646
Consistent norms, 71
Constrained least squares, 313–4
Contour integral and $f(A)$, 528–9
Convergence. See under particular algorithm
Courant-Fischer minimax theorem, 441
CP approximation, 735–8
Craig's method, 637
Crawford number, 499
Cross product, 70
Cross-validation, 308
CS decomposition, 84–5, 503–6
    hyperbolic, 344
    subset selection and, 294
    thin version, 84
CUR decomposition, 576
Curse of dimensionality, 741
Cuthill-McKee ordering, 602–4
Cyclic Jacobi method, 480–1
Cyclic reduction, 197–8

Data least squares, 325
Data motion overhead, 53
Data reuse, 46–8
Data sparse, 154
Davidson method, 593–4
Decompositions and factorizations
    Arnoldi, 580
    bidiagonal, 5
    block diagonal, 397–9
    Cholesky, 163
    companion matrix, 382
    complete orthogonal, 283
    CS (general), 85
    CS (thin), 84
    generalized real Schur, 407
    generalized Schur, 406–7
    Hessenberg, 378ff
    Hessenberg-triangular, 408–9
    Jordan, 354
    LDL^T, 165–6
    LU, 114, 128
    QR, 247
    QR (thin version), 248
    real Schur, 376
    Schur, 351
    singular value, 76
    singular value (thin), 80
    symmetric Schur, 440
    tridiagonal, 458–9
Decoupling in eigenproblem, 349–50
Defective eigenvalue, 66, 353
Deflating subspace, 404
Deflation and
    bidiagonal form, 490
    Hessenberg-triangular form, 409–10
    QR algorithm, 385
Denman-Beavers iteration, 539–40
Departure from normality, 351
Derogatory matrix, 383
Determinant, 66, 348
    Gaussian elimination and, 114
    and singularity, 89
    Vandermonde matrix, 206
Diagonal dominance, 154–6, 615
    block, 197
    LU and, 156
Diagonal matrix, 18
Diagonal pivoting method, 191–2
Diagonal plus rank-1, 469–71
Diagonalizable, 67, 353
Differentiation of matrices, 67
Dimension, 64
Direct methods, 598ff
Dirichlet end condition, 222
Discrete cosine transform (DCT), 39
Discrete Fourier transform (DFT), 33–6
    circulant matrices and, 221-2
    factorizations and, 41
    matrix, 34
Discrete Poisson problem
    1-dimensional, 222–4
    2-dimensional, 224-31
Discrete sine transform (DST), 39
Displacement rank, 682
Distance between subspaces, 82
Distributed memory model, 57
Divide-and-conquer algorithms
    cyclic reduction, 197–8
    Strassen, 30–1
    tridiagonal eigenvalue, 471–3

Domain decomposition, 662–5
Dominant
    eigenvalue, 366
    invariant subspace, 368
Dot product, 4, 10
Dot product roundoff, 98
Double implicit shift, 388
Doubling formulae, 526
Downdating Cholesky, 338–41
Drazin inverse, 356
Durbin's algorithm, 210

Eckhart-Young theorem, 79
Eigenproblem
    diagonal plus rank-1, 469–71
    generalized, 405ff, 497ff
    inverse, 473–4
    orthogonal Hessenberg matrix, 703–4
    symmetric, 439ff
    Toeplitz, 214–6
    unsymmetric, 347ff
Eigensystem
    fast, 219
Eigenvalue decompositions
    Jordan, 354
    Schur, 351
Eigenvalues
    algebraic multiplicity, 353
    characteristic polynomial and, 348
    computing selected, 453
    defective, 66
    determinant and, 348
    dominant, 366
    generalized, 405
    geometric multiplicity, 353
    ordering in Schur form, 351, 396–7
    orthogonal Hessenberg, 703–4
    relative perturbation, 365
    repeated, 360
    sensitivity (symmetric case), 441–3
    sensitivity (unsymmetric case), 359–60
    singular values and, 355
    Sturm sequence and, 468
    symmetric tridiagonal, 467ff
    trace, 348
    unstable, 363
Eigenvector, 67
    basis, 400
    dominant, 366
    left, 349
    matrix and condition, 354
    perturbation, 361–2
    right, 349
Elementary Hermitian matrices.
    See Householder matrix
Elementary transformations. See
    Gauss transformations
Equality constained least squares, 315–7
Equilibration, 139
Equilibrium systems, 192–3
Equivalence of vector norms, 69
Error
    absolute, 69
    damping in multigrid, 622-3
    relative, 70
    roundoff, 96–102
Error analysis
    backward, 100
    forward, 100
Euclidean matrix norm. See

Frobenius matrix norm
Exchange permutation matrix, 20
Explicit shift in QR algorithm
    symmetric case, 461
    unsymmetric case, 385–8
Exponential of matrix, 530–6

Factored form representation, 237–8
Factorization. See Decompositions and
    factorizations
Fast methods
    cosine transform, 36ff
    eigensystem, 219, 228–31
    Fourier transform, 33ff
    Givens QR, 245
    Poisson solver, 226–7
    sine transform, 36
Field of values, 349
Fine grid role in multigrid, 673
Floating point
    fl, 96
    fundamental axiom, 96
    maxims, 96–7
    normalized, 94
    numbers, 93
    storage of matrix, 97–8
Flop, 12
Flopcounts, 12, 16
    for square system methods, 298
F-norm, 71
Forward error analysis, 100
Forward substitution, 106
Francis QR step, 390
Frechet derivative, 521
Frobenius matrix norm, 71
Frontal methods, 610
Full multigrid, 678
Function of matrix, 513ff
    eigenvectors and, 517–8
    Schur decomposition and, 518–20
    Taylor series and, 524–6

Gauss-Jordan transformations, 121
Gauss-Radau rule, 560–1
Gauss rules, 557–9
Gauss-Seidel iteration, 611-2
    block, 613
    Poisson equation and, 617
    positive definite systems and, 615
Gauss transformations, 112–3
Gaussian elimination, 111ff
    banded version, 176–9
    block version, 144–5
    complete pivoting and, 131–2
    gaxpy version, 117
    outer product version, 116
    partial pivoting and, 127
    rook pivoting and, 133
    roundoff error and, 122–3
    tournament pivoting and, 150
Gaxpy, 5
    blocked, 25
Gaxpy-rich algorithms
    Cholesky, 164
    Gaussian elimination, 129–30
    LDL^T, 157–8
Gaxpy vs. outer product, 45
Generalized eigenproblem, 405ff
Generalized eigenvalues, 405
    sensitivity, 407

Generalized least squares, 305–6
Generalized Schur decomposition, 406–7
    computation of, 502–3
Generalized singular vectors, 502
Generalized SVD, 309–10, 501–2
    constrained least squares and, 316–7
Generalized Sylvester equation, 417
Generator representation, 693
Geometric multiplicity, 353
Gershgorin theorem, 357, 442
Ghost eigenvalues, 566
givens, 240
Givens QR, 252–3
    parallel, 257
Givens rotations, 239–42
    complex, 243–4
    fast, 245
    rank-revealing decompositions and, 280–2
    square-root free, 246
Global memory, 55
GMRES, 642–4
    m-step, 644
    preconditioned, 652–3
Golub-Kahan
    bidiagonalization, 571–3
    SVD step, 491
Gram-Schmidt
    classical, 254
    modified, 254–5
Graph, 602
Graphs and sparsity, 601–2
Growth in Gaussian elimination, 130–2

Haar wavelet transform, 40ff
    factorization, 41
Hadamard product, 710
Hamiltonian matrix, 29, 420
    eigenvalue problem, 420–1
Hankel-like, 688–9
Hermitian matrix, 18
Hessenberg form, 15
    Arnoldi process and, 579–80
    Householder reduction to, 378–9
    inverse iteration and, 395
    properties, 381–2
    QR factorization and, 253–4
    QR iteration and, 385–6
    unreduced, 381
Hessenberg QR step, 377-8
Hessenberg systems, 179
    LU and, 179
Hessenberg-triangular form, 408–9
Hierarchical memory, 46
Hierarchical rank structure, 702
Higher-order SVD, 732–3
    truncated, 734
Holder inequality, 69
Horner algorithm, 526–7
house, 236
Householder
    bidiagonalization, 284–5
    tridiagonalization, 458–9
Householder matrix, 234–8
    complex, 243
    operations with, 235–7
Hyperbolic
    CS decomposition, 344
    rotations, 339
    transformations, 339

Identity matrix, 19
Ill-conditioned matrix, 88
IEEE arithmetic, 94
Im, 13
Implicit Q theorem
    symmetric matrix version, 460
    unsymmetric matrix version, 381
Implicit symmetric QR step with
    Wilkinson Shift, 461–2
Implicitly restarted Arnoldi
    method, 581–3
Incomplete block preconditioners, 657–60
Incomplete Cholesky, 357–60
Indefinite least squares, 344
Indefinite symmetric matrix, 159
Indefinite systems, 639–41
Independence, 64
Inertia of symmetric matrix, 448
inf, 95
Integrating $f(A)$, 527–8
Interchange permutation, 126
Interlacing property
    singular values, 487
    symmetric eigenvalues, 443
Intersection
    nullspaces, 328–9
    subspaces, 331
Invariant subspace
    approximate, 446–8
    dominant, 378
    perturbation of (symmetric case), 443–5
    perturbation of (unsymmetric case), 361
    Schur vectors and, 351
Inverse, 19
    band matrices and, 182–3
Inverse eigenvalue problems, 473–4
Inverse error analysis. See
    Backward error analysis
Inverse fast transforms
    cosine, 227–8
    Fourier, 220
    sine, 227–8
Inverse iteration
    generalized eigenproblem, 414
    symmetric case, 453
    unsymmetric case, 394–5
Inverse low-rank perturbation, 65
Inverse of matrix,
    perturbation of, 74
    Toeplitz case, 212–3
Inverse orthogonal iteration, 374
Inverse power method, 374
Inverse scaling and squaring, 542
Irreducible, 373
Iteration matrix, 613
Iterative improvement
    fixed precision and, 140
    least squares, 268–9, 272
    linear systems, 139–40
Iterative methods, 611–50

Jacobi iteration for the SVD, 492–3
Jacobi iteration for symmetric
    eigenproblem, 476ff
    classical, 479–80
    cyclic, 480
    error, 480–1
    parallel version, 482–3
Jacobi method for linear systems,
    block version, 613

diagonal dominance and, 615
preconditioning with, 653
Jacobi orthogonal correction method, 591–3
Jacobi rotations, 477
Jacobi-Davidson method, 594–5
Jordan blocks, 400-2
Jordan decomposition, 354
  computation of, 400-2
  matrix functions and, 514, 522-3

Kaniel-Paige-Saad theory, 552–4
Khatri-Rao product, 710
Kogbetiantz algorithm, 506
Kronecker product, 27
  basic properties, 27, 707–8
  multiple, 28, 716
  SVD 712-4
Kronecker structure, 418
Krylov
  matrix, 459
  subspaces, 548
Krylov-Schur algorithm, 584
Krylov subspace methods
  biconjugate gradients, 645
  CG (conjugate gradients), 625ff
  CGNE (conjugate gradient normal equation error), 637–8
  CGNR (conjugate gradient normal equation residual), 637–8
  CGS (conjugate gradient squared), 646
  general linear systems and, 579ff
  GMRES (general minimum residual), 642–5
  MINRES (minimum residual), 639–40
  QMR (quasi-minimum residual), 647
  SYMMLQ, 640–1
Krylov subspace methods for
  general linear systems, 636–7, 642–7
  least squares, 641–2
  singular values, 571–8
  symmetric eigenproblem, 546–56, 562–71
  symmetric indefinite systems, 639–41
  symmetric positive definite systems, 625–39
  unsymmetric eigenproblem, 579–89

Lagrange multipliers, 313
Lanczos tridiagonalization, 546ff
  block version, 566–9
  complete reorthogonalization and, 564–5
  conjugate gradients and, 628–32
  convergence of, 552–4
  Gauss quadrature and, 560–1
  interior eigenvalues and, 553–4
  orthogonality loss, 564
  power method and, 554–5
  practical, 562ff
  Ritz approximation and, 551–2
  roundoff and, 563–4
  selective orthogonalization and, 565–6
  s-step, 569
  termination of, 549
  unsymmetric, 584–7
Lanczos vectors, 549
LDL^T, 156–8
  conjugate gradients and, 631
  with pivoting, 165–6
Leading principal submatrix, 24
Least squares methods, flopcounts for, 293
Least squares problem
  basic solution to, 292
  cross-validation and, 308

equality constraints and, 315–7
full rank, 260ff
generalized, 305–6
indefinite, 344
iterative improvement, 268–9
Khatri-Rao product and, 737
minimum norm solution to, 288–9
quadratic inequality constraint, 313–5
rank deficient, 288ff
residual vs. column independence, 295–6
sensitivity of, 265–7
solution set of, 288
solution via Householder QR, 263–4
sparse, 607–8, 641–2
SVD and, 289
Least squares solution using
  LSQR, 641–2
  modified Gram-Schmidt, 264–5
  normal equations, 262–3
  QR factorization, 263–4
  seminormal equations, 607
  SVD, 289
Left eigenvector, 349
Left-looking, 117
Levels of linear algebra, 12
Level-3 fraction, 109
  block Cholesky, 170
  block LU, 120
  Hessenberg reduction, 380
Levinson algorithm, 211
Linear equation sensitivity, 102, 137ff
Linear independence, 64
Linearization, 415–6
Load balancing, 50ff
Local memory, 50
Local program, 50
Log of a matrix, 541–2
Look-ahead, 217, 586–7
Loop reordering, 9
Loss of orthogonality
  Gram-Schmidt, 254
  Lanczos, 564
Low-rank approximation
  randomized, 576–7
  SVD, 79
LR iteration, 370
LSMR, 642
LSQR, 641–2
LU factorization, 111ff
  band, 177
  block, 196–7
  Cauchy-like, 685–6
  determinant and, 114
  diagonal dominance and, 155
  differentiation of, 120
  existence of, 114
  gaxpy version, 117
  growth factor and, 130–1
  Hessenberg, 179
  mentality, 134
  outer product version, 116
  partial pivoting and, 128
  rectangular matrices and, 118
  roundoff and, 122–3
  semiseparable, 695–7
  sparse, 608–9

Machine precision, 95
Markov chain, 374
Markowitz pivoting, 609

MATLAB, xix
Matrix functions, 513ff
    integrating, 527–8
    Jordan decomposition and, 514–5
    polynomial evaluation and, 526–7
    sensitivity of, 520–1
Matrix multiplication, 2, 8ff
    blocked, 26
    Cannon's algorithm, 60–1
    distributed memory, 50ff
    dot product version, 10
    memory hierarchy and, 47
    outer product version, 11
    parallel, 49ff
    saxpy version, 11
    Strassen, 30–1
    tensor contractions and, 726–7
Matrix norms, 71–3
    consistency, 71
    Frobenius, 71
    relations between, 72–3
    subordinate, 72
Matrix-vector products, 33ff
    blocked, 25
Memory hierarchy, 46
Minimax theorem for
    singular values, 487
    symmetric eigenvalues, 441
Minimum degree ordering, 604–5
Minimum singular value, 78
MINRES, 639–41
Mixed packed format, 171
Mixed precision, 140
Modal product, 727–8
Modal unfoldings, 723
Modified Gram-Schmidt, 254–5
    and least squares, 264–5
Modified LR algorithm, 392
Moore-Penrose conditions, 290
Multigrid, 670ff
Multilinear product, 728–9
Multiple eigenvalues,
    matrix functions and, 520
    unreduced Hessenberg matrices and, 382
Multiple-right-hand-side problem, 108
Multiplicative Schwarz, 664
Multipliers in Gauss transformations, 112

NaN, 95
Nearness to
    Kronecker product, 714–5
    singularity, 88
    skew-hermitian, 449
Nested-dissection ordering, 605–6
Netlib, xix
Neumann end condition, 222
Newton method for Toeplitz eigenvalue, 215
Newton-Schultz iteration, 538
nnz, 599
Node degree, 602
Nonderogatory matrices, 383
Nongeneric total least squares, 324
Nonsingular matrix, 65
Norm
    matrix, 71–3
    vector, 68
Normal equations, 262–3, 268
Normal matrix, 351
    departure from, 351
Normwise-near preconditioners, 654

null, 64
Nullity theorem, 185
Nullspace, 64
    intersection of, 328–9
Numerical radius, 349
Numerical range, 349
Numerical rank
    least squares and, 291
    QR with column pivoting and, 278–9
    SVD and, 275–6

off, 477
Ordering eigenvalues, 396–7
Ordering for sparse matrices
    Cuthill-McKee, 602–4
    minimum degree, 604–6
    nested dissection, 605–7
Orthogonal
    complement, 65
    invariance, 75
    matrix, 66, 234
    Procrustes problem, 327–8
    projection, 82
    symplectic matrix, 420
    vectors, 65
Orthogonal iteration
    symmetric, 454–5, 464–5
    unsymmetric, 367–8, 370–3
Orthogonal matrix representations
    factored form, 237–8
    Givens rotations, 242
    WY block form, 238–9
Orthogonality between subspaces, 65
Orthonormal basis computation, 247
Outer product, 7
    Gaussian elimination and, 115
    LDL^T and, 166
    sparse, 599–600
    between tensors, 724
    versus gaxpy, 45
Overdetermined system, 260

Packed format, 171
Padé approximation, 530–1
PageRank, 374
Parallel computation
    divide and conquer eigensolver, 472–3
    Givens QR, 257
    Jacobi's eigenvalue method, 482–3
    LU, 144ff
    matrix multiplication, 49ff
Parlett-Reid method, 187–8
Parlett-Schur method, 519
    block version, 520
Partitioning
    conformable, 23
    matrix, 5–6
Pencils, equivalence of, 406
Perfect shuffle permutation, 20, 460, 711–2
Periodic end conditions, 222
Permutation matrices, 19ff
Perron-Frobenius theorem, 373
Perron's theorem, 373
Persymmetric matrix, 208
Perturbation results
    eigenvalues (symmetric case), 441–3
    eigenvalues (unsymmetric case), 357–60
    eigenvectors (symmetric case), 445–6
    eigenvectors (unsymmetric case), 361–2
    generalized eigenvalue, 407

invariant subspaces (symmetric case), 444–5
invariant subspaces (unsymmetric case), 361
least squares problem, 265–7
linear equation problem, 82–92
singular subspace pair, 488
singular values, 487
underdetermined systems, 301
Pipelining, 43
Pivoting
    Aasen's method and, 190
    Bunch-Kaufman, 192
    Bunch-Parlett, 191
    Cauchy-like and, 686–7
    column, 276–8
    complete, 131–2
    LU and, 125ff
    Markowitz, 609
    partial, 127
    QR and, 279–80
    rook, 133
    symmetric matrices and, 165–6
    tournament, 150
Plane rotations. *See* Givens rotations
p-norms, 71
    minimization in, 260
Point, line, plane problems, 269–271
Pointwise operations, 3
Polar decomposition, 328, 540–1
Polynomial approximation and GMRES, 644
Polynomial eigenvalue problem, 414–7
Polynomial interpolation, Vandermonde
        systems and, 203–4
Polynomial preconditioner, 655–6
Positive definite systems, 159ff
    Gauss-Seidel and, 615–6
    LDL^T and, 165ff
    properties of, 159–61
    unsymmetric, 161–3
Positive matrix, 373
Positive semidefinite matrix, 159
Post-smoothing in multigrid, 675
Power method, 365ff
    error estimation in, 367
    symmetric case, 451–2
Power series of matrix, 524
Powers of a matrix, 527
Preconditioned
    conjugate gradient method, 651–2, 656ff
    GMRES, 652–3
Preconditioners, 598
    approximate inverse, 654–5
    domain decomposition, 662–5
    incomplete block, 660–1
    incomplete Cholesky, 657–60
    Jacobi and SSOR, 653
    normwise-near, 654
    polynomial, 655
    saddle point, 661
Pre-smoothing role in multigrid, 675
Principal angles and vectors, 329–31
Principal square root, 539
Principal submatrix, 24
Probability vector, 373
Procrustes problem, 327–8
Product eigenvalue problem, 423–5
Product SVD problem, 507
Profile, 602
    Cholesky, 184
    indices, 184, 602
Projections, 82

Prolongation matrix, 673
Pseudo-eigenvalue, 428
Pseudoinverse, 290, 296
Pseudospectra, 426ff
    computing plots, 433–4
    matrix exponential and, 533–4
    properties, 431–3
Pseudospectral abscissa, 434–5
Pseudospectral radius, 434–5

QMR, 647
QR algorithm for eigenvalues
    Hessenberg form and 377–8
    shifts and, 385ff
    symmetric version, 456ff
    tridiagonal form and, 460
    unsymmetric version, 391ff
    Wilkinson shift, 462–3
QR factorization, 246ff
    block Householder, 250–1
    block recursive, 251
    classical Gram-Schmidt and, 254
    column pivoting and, 276–8
    complex, 256
    Givens computation of, 252–3
    Hessenberg matrices and, 253–4
    Householder computation of, 248–9
    least square problem and, 263–4
    modified Gram-Schmidt and, 254–5
    properties of, 246–7
    range space and, 247
    rank of matrix and, 274
    sparse, 606–8
    square systems and, 298–9
    thin version, 248
    tridiagonal matrix and, 460
    underdetermined systems and, 300
    updating, 335–8
Quadratic eigenvalue problem, 507–8
Quadratically constrained least squares, 314–5
Quasidefinite matrix, 194
Quasiseparable matrix, 693
Quotient SVD, 507
QZ algorithm, 412–3
    step, 411-2

ran, 64
Randomization, 576–7
Range of a matrix,
    orthonormal basis for, 247
Rank of matrix, 64
    QR factorization and, 278–9
    SVD and, 275–6
Rank-deficient LS problem, 288ff
    breakdown of QR method, 264
Rank-revealing decomposition, 280–3
Rank-structured matrices, 691ff
Rayleigh quotient iteration, 453–4
    QR algorithm and, 464
    symmetric-definite pencils and, 501
R-bidiagonalization, 285
Re, 13
Real Schur decomposition, 376–7
    generalized, 407
    ordering in, 396–7
Rectangular LU, 118
Recursive algorithms
    block Cholesky, 169
    Strassen, 30–1
Reducible, 373

Regularized least squares, 307ff
Regularized total least squares, 324
Relative error, 69
Relaxation parameter, 619–20
Reorthogonalization
    complete, 564
    selective, 565
Representation, 681–2
    generator, 693
    Givens, 697–8
    quasiseparable, 694
reshape, 28, 711
    and Kronecker product, 28
Residuals vs. accuracy, 138
Restarting
    Arnoldi method and, 581–2
    block Lanczos and, 569
    GMRES and, 644
Restricted generalized SVD, 507
Restricted total least squares, 324
Restriction matrix, 673
Ricatti equation, 422–3
Ridge regression, 307–8
Riemann-Stieltjes integral, 556–7
Right eigenvector, 349
Right-looking, 117
Ritz acceleration, orthogonal
        iteration and, 464–5
Ritz approximation
    eigenvalues, 551-2
    singular values, 573
Rook pivoting, 133
Rotation of subspaces, 327–8
Rotation plus rank-1 (ROPR), 332
Rounding errors. *See under particular*
        algorithm
Roundoff error analysis, 100
    dot product, 98–9
    Wilkinson quote, 99
Row orientation, 5
Row partition, 6
Row scaling, 139
Row weighting in LS problem, 304–5

Saddle point preconditioners, 661
Saxpy, 4, 11
Scaling, linear systems and, 138–9
Scaling and squaring for exp(A), 531
Schur complement, 118–9, 663
Schur decomposition, 67, 350–1
    generalized, 406–7
    matrix functions and, 523–4
    normal matrices and, 351
    real matrices and, 376–7
    symmetric matrices and, 440
    2-by-2 symmetric, 478
Schur vectors, 351
Secular equations, 313–4
Selective reorthogonalizaton, 565–6
Semidefinite systems, 167–8
Semiseparable
    eigenvalue problem, 703–4
    LU factorization, 695–8
    matrix, 682
    plus diagonal, 694
    QR factorization, 698–701
Sensitivity. *See* Perturbation results
sep
    symmetric matrices and, 444
    unsymmetric matrices and, 360

Shared-memory systems, 54–6
Shared-memory traffic, 55–6
Sherman-Morrison formula, 65
Sherman-Morrison-Woodbury formula, 65
Shifts in
    QZ algorithm, 411
    SVD algorithm, 489
    symmetric QR algorithm, 461–2
    unsymmetric QR algorithm, 385–90
Sign function, 536–8
Similar matrices, 67, 349
Similarity transformation, 349
    condition of, 354
    nonunitary, 352–4
Simpson's rule, 528
Simultaneous diagonalization, 499
Simultaneous iteration. *See* orthogonal iteration
Sine of matrix, 526
Singular matrix, 65
Singular subspace pair, 488
Singular value decomposition (SVD), 76–80
    algorithm for, 488–92
    constrained least squares and, 313–4
    generalized, 309–10
    geometry of, 77
    higher-order, 732–3
    Jacobi algorithm for, 492–3
    Lanczos method for, 571ff
    linear systems and, 87–8
    minimum-norm least squares solution, 288–9
    nullspace and, 78
    numerical rank and, 275–6
    perturbation of, 487–8
    principal angles and, 329–31
    projections and, 82
    pseudo-inverse and, 290
    rank of matrix and, 78
    ridge regression and, 307–8
    subset selection and, 293–6
    subspace intersection and, 331
    subspace rotation and, 327–8
    symmetric eigenproblem and, 486
    total least squares and, 321–2
    truncated, 291
Singular values, 76
    condition and, 88
    eigenvalues and, 355
    interlacing property, 48
    minimax characterization, 487
    perturbation of, 487–8
    range and nullspace, 78
    rank and, 78
    smallest, 279–80
Singular vectors, 76
Skeel condition number, 91
    and iterative improvement, 140
Skew-Hamiltonian matrix 420
Skew-Hermitian matrix, 18
Skew-symmetric matrix, 18
span, 64
Sparse factorization challenges
    Cholesky, 601
    LU, 609
    QR, 607
Sparsity, 154
    graphs and, 601–2
Spectral abscissa, 349
Spectral radius, 349, 427, 614
Spectrum of matrix, 348
Speed-up, 53–4

SPIKE framework, 199–201
Splitting, 613
Square root of a matrix, 163
s-step Lanczos, 569
Stable algorithm, 136
Stable matrix, 436
Steepest descent method, 626-7
Stieltjes matrix, 658
Strassen method, 30–1
    error analysis and, 101-2
Strictly diagonally dominant, 155
Stride, 45
Structured rank, 691ff
    types of, 702
Sturm sequence property, 468–9
Submatrix, 24
Subnormal floating point number, 95
Subordinate norm, 72
Subset selection, 293–5
    using QR with column pivoting, 293
Subspace, 64
    angles between, 329–31
    deflating, 414
    distance between, 82-3, 331
    dominant, 368
    intersection, 331
    invariant, 349
    nullspace intersection, 328–9
    orthogonal projections onto, 82
    rotation of, 327–8
Successive over-relaxation (SOR), 619
Sweep, 480
Sylvester equation, 398
    generalized, 417
Sylvester law of inertia, 448
Sylvester map, 682
Symmetric-definite eigenproblem, 497–501
Symmetric eigenproblem, 439ff
    sparse methods, 546ff
Symmetric indefinite methods
    Aasen, 188–90
    Diagonal pivoting, 191–2
    Parlett-Reid, 187–8
Symmetric matrix, 18
Symmetric pivoting, 165
Symmetric positive definite systems, 163ff
Symmetric semidefinite properties, 167–8
Symmetric successive over-relaxation,
    (SSOR), 620
SYMMLQ, 641
Symplectic matrix, 29, 420
symSchur, 478

Taylor approximation of e^A, 530
Taylor series, matrix functions and, 515–7
Tensor
    contractions, 726ff
    eigenvalues, 740–1
    networks, 741
    notation, 721
    rank, 738–9
    rank-1, 725
    singular values, 739–40
    train, 741–3
    transpose, 722-3
    unfoldings, 720
Thin CS decomposition, 84
Thin QR factorization, 248
Thin SVD, 80
Threshold Jacobi, 483

Tikhonov regularization, 309
Toeplitz-like matrix, 688
Toeplitz matrix methods, classical, 208ff
Toroidal network, 58
Total least squares, 320ff
    geometry, 323–4
Tournament pivoting, 150
Trace, 348–9
tr, 348
Trace-min method, 595
Tracy-Singh product, 709
Transition probability matrix, 374
Transpose, 2, 711-2
Trench algorithm, 213
Treppeniteration, 369
Triangular matrices,
    multiplication between, 15
    unit, 110
Triangular systems, 106–11
    band, 177-8
    nonsquare, 109–10
    roundoff and, 124–5
    semiseparable, 694–5
Tridiagonalization,
    connection to bidiagonalization, 574
    Householder, 458–60
    Krylov subspaces and, 459–60
    Lanczos, 548–9
Tridiagonal matrices, 15, 223–4
    QR algorithm and, 460–4
Tridiagonal systems, 180–1
Truncated
    higher-order SVD, 734
    SVD, 291
    total least squares, 324
Tucker approximation problem, 734–5

ULV decomposition, 282–3
ULV updating, 341–3
Underdetermined systems, 134, 299-301
Undirected graph, 602
Unfolding, 723–4
Unit roundoff, 96
Unit stride, 45
Unit vector, 69
Unitary matrix, 80
Unreduced Hessenberg matrices, 381
Unreduced tridiagonal matrices, 459
Unstable eigenvalue, 363
Unsymmetric
    eigenproblem, 347ff
    Lanczos method, 584–7
    positive definite systems, 161–3
    Toeplitz systems, 216–7
Updating
    Cholesky, 338–41
    QR factorization, 334–8
    ULV, 341–3
UTV, 282

Vandermonde systems, 203ff
    confluent, 206
V-cycle, 677–8
vec, 28, 710–11
    for tensors, 722
Vector
    computing, 43ff
    loads and stores, 43
    norms, 68
    operations, 3–4, 44

processing, 43
Vectorization, tridiagonal system solving
    and, 181

Weighted Jacobi iteration, 672–3
Weighting least squares problems
    column, 306–7
    row, 304–5
    See also Scaling
Well-conditioned matrix, 88
Wielandt-Hoffman theorem
    eigenvalues, 442
    singular values, 487

Wilkinson shift, 462–3
Work
    least squares methods and, 293
    linear system methods and, 298
    SVD and, 493
WY representation, 238–9
    compact version, 244

Yule-Walker problem, 201–10